GT	Group Technology	OCR	Optical Character Recognition
GTAW	Gas Tungsten Arc Welding	OS	Operating System
HAZ	Heat Affected Zone	PAW	Plasma Arc Welding (PAC = Cutting) (PAM = Machining)
HERF	High Energy Rate Forming		
HGVS	Human-Guided Vehicle System (fork-lift with driver)	PC	Personal Computer
		PCB	Printed Circuit Board
HIP	Hot Isostatic Pressing	PD	Pitch Diameter
IGES	Initial Graphics Exchange System	PDES	Product Design Exchange Specification
IMPSs	Integrated Manufacturing Production Systems	PLC	Programmable Logic Controller
		PROM	Programmable Read-Only Memory
I/O	Input/Output	PS	Production System
IOCS	Input/Output Control System	P/M	Powder Metallurgy
JIT	Just-In-Time	PVD	Physical Vapor Deposition
LAN	Local Area Network	QC	Quality Control
LASER	Light Amplification by Stimulated Emission of Radiation	QMS	Quality Management System
		RAM	Random Access Memory
LBM	Laser Beam Machining (LBW = Welding) (LBC = Cutting)	RIM	Reaction Injection Molding
		ROM	Read-Only Memory
LED	Light Emitting Diode	SAW	Submerged Arc Welding
LSI	Large Scale Integration	SMAW	Shielded Metal Arc Welding
M	Module	SPC	Statistical Process Control
MAP	Manufacturing Automation Protocol	SQC	Statistical Quality Control
MAPIM	Materials and Processes in Manufacturing	TCM	Thermochemical Machining
MCU	Machine Control Unit	TIG	Tungsten–Inert Gas
MDI	Manual Data Input	TOP	Technical Office Protocol
MIG	Metal–Inert Gas	TPA	Target Point Align
MPS	Manufacturing Production System	TQC	Total Quality Control
mrp	Material Requirements Planning	USM	Ultrasonic Machining (USW = Welding)
MRPII	Material Resources Planning	VA	Value Analysis
MS	Manufacturing System	WAN	Wide Area Network
N	Number	WIMS	Work-in-Manufacturing System
NC	Numerically Controlled	WIP	Work-In-Progress
NDT	NonDestructive Testing (NDE = Evaluation) NDI = Inspection)	YAG	Yttrium-Aluminum Garnet

Materials and Processes in Manufacturing

TH EDITION

Materials and Processes in Manufacturing

E. Paul DeGarmo

Registered Professional Engineer. Professor of Industrial Engineering and Mechanical Engineering Emeritus. University of California-Berkeley

J Temple Black

Registered Professional Engineer. Professor of Industrial Engineering. Auburn University, Auburn, Alabama

Ronald A. Kohser

Professor of Metallurgical Engineering. Univeristy of Missouri-Rolla

Macmillan Publishing Company
NEW YORK
Collier Macmillan Publishers
LONDON

Copyright © 1988, Macmillan Publishing Company, a division of Macmillan, Inc.

PRINTED IN THE UNITED STATES OF AMERICA

Earlier editions copyright © 1957 and 1962 by Macmillan
Publishing Company. Copyright © 1969 by E. Paul DeGarmo,
and copyright © 1974 and 1979 by Darvic Associates, Inc.
Copyright © 1984 by Macmillan Publishing Company

Macmillan Publishing Company
866 Third Avenue, New York, New York 10022

Collier Macmillan Canada, Inc.

LIBRARY OF CONGRESS CATALOGING IN PUBLICATION DATA

DeGarmo, E. Paul
  Materials and processes in manufacturing.

  Includes index.
  1. Manufacturing processes.   2. Materials.
I. Black, J Temple.   II. Kohser, Ronald A.
III. Title.
TS183.D4   1988   671   88-5155
ISBN 0-02-328631-8 (Hardcover Edition)
ISBN 0-02-946140-5 (International Edition)
Printing: 4   5   6   7   8   Year: 0   1   2   3   4   5   6   7

Preface

Manufacturing has been defined as "that which is done to convert stuff into things." While this definition is simplistic in nature, it is really quite correct. Manufacturing begins with "stuff," engineering materials if you will. To produce quality products in an economical fashion, these materials must be properly selected, of high quality, and in the most appropriate shape and condition for the subsequent processing. Then comes the conversion into things—the actual manufacturing processes which convert engineering materials into functionally useable goods. These processes must be selected for compatibility with the selected material and suitability for producing the desired product, in the desired quantity, at the desired rate.

Manufacturing is a "value adding" process, where the conversion of stuff into things adds value to the material. Thus, the objective of a company engaged in manufacturing is to *add value* and do so in the most efficient manner, using the least amount of time, material, money, space, and manpower. To minimize waste and maintain a competitive stature, the processes and operations need to be properly selected and arranged to permit smooth and controlled flow of material through the factory and also provide an optimal degree of product flexibility. Meeting these goals requires a well-designed and efficient manufacturing system.

Philosophy of the Book

Materials and Processes in Manufacturing was written to provide a solid introduction into the fundamentals of manufacturing. The book begins with a survey of engineering materials, the "stuff" that manufacturing begins with, and seeks to provide the basic information that can be used to match the properties of a material to the service requirements of a component. A variety of engineering materials is presented, along with their properties and means of modifying them, such as heat treatment. This section can be used in curriculums that lack preparatory courses in metallurgy, materials science, or strength of materials, or where the student has not yet been exposed to such material. In addition, various chapters in this section can be used as supplements to a basic materials course, providing additional information on topics such as heat treatment, plastics, composites, and material selection.

The remaining two-thirds of the text is devoted to manufacturing processes and the concepts that enable these processes to produce quality products in a competitive manner. Casting, forming, powder metallurgy, material removal, and joining are all developed as families of manufacturing processes. Each section begins with a presentation of the fundamentals on which those processes are based. This is followed by a discussion of the various process alternatives, which can be selected to operate individually or be combined into an integrated system.

Although considerable effort has been made to include many of the new developments in both materials and processes, the major emphasis remains on fundamentals, for these provide an enduring basis for understanding both existing phenomena and that still to come. Another objective of the book is to introduce the broad spectrum of manufacturing processes to individuals who will be involved in the design and manufacture of finished products. The material is presented in a descriptive fashion where the emphasis is on the fundamental workings of the process, its capabilities, typical applications, advantages, and limitations. Furthermore, the system-type interactions of the material and the process are emphasized throughout the text.

History of the Book

Materials and Processes in Manufacturing, by E. Paul DeGarmo, was first published in 1957 and soon became the emulated standard for introductory texts in manufacturing. Through revised editions in 1962, 1969, 1974, and 1979 the book has kept current with the rapidly changing developments in the field and has maintained its stature as the best-received text in manufacturing processes. For the sixth edition, published in 1984 and the present seventh edition, Ronald A. Kohser and J Temple Black have assumed the primary responsibility for the text. The chapters on engineering materials, casting, forming, powder metallurgy, joining, and nondestructive testing were written or revised by Ron Kohser. J Black assumed the responsibility for the introduction, and the chapters on material removal, quality control, and manufacturing and production systems.

Originally, the seventh edition was slated for publication in 1989, but the many pressures and trends in manufacturing prompted an acceleration of that schedule. America's manufacturers are currently battling world-wide competition to produce superior-quality products and deliver them on-time at competitive prices. While there have been some great success stories in recent years, these gains may only be temporary, because many firms have done only what was needed to survive in the short term. While we may not need to replace the current systems with "factories of the future," we do need to work toward developing *"factories with a future."* Strategies, such as integrated manufacturing production systems (IMPSs), Just-In-Time production (JIT), and Total Quality Control (TQC) have all been recognized as appropriate methods for long-term success in manufacturing. Automation and the use of computers in design, process control and inspection have become wide spread. Industrial robots now perform manufacturing operations with repeatability and reliability. These changes affect everyone in the factory, from the designers and manufacturing engineers, to whom this book is directed, to the individual workers and machine operators. In addition, it is important to have a well-informed manufacturing management. The best factories in the USA today are managed by individuals who understand the materials, manufacturing processes, and systems being used in their plant.

New to this Edition

Approximately 30% of the present text is new to the seventh edition. Polymers and composite materials now occupy a far more prominent position in the materials section and a separate chapter was included to present the manufacturing processes that are unique to polymers, composites and ceramics. The section on measurements and quality control has been expanded and now includes non-destructive testing and inspection, statistical process control, and new concepts in total quality assurance. A new chapter presents processes that are unique to the manufacture of electronic components. Material on manufacturing systems has been updated and restructured to reflect the many methods and strategies that are impacting on the manufacturing community. The case studies have been reduced in number, expanded in scope, and regrouped at the back of the book (thereby removing the notion that the information to solve the case is contained in the previous chapter). The case studies can help make students aware of the great importance of properly coordinating design, materials selection and manufacturing in order to produce a satisfactory and failure-free product. The objective of the revision and restructuring has been to produce a state-of-the-art introduction to the materials and processes used in manufacturing.

Intended Audience

The text is designed to be used by engineering (mechanical, manufacturing and industrial) and mechanical technology students, in both 2- and 4-year undergraduate degree programs. In addition, the book has also proved to be useful to engineers and technologists in other disciplines concerned with design and manufacturing (such as aerospace and electronics). It is the intent of the authors to provide a broad-based introduction to a wide-variety of manufacturing processes, rather than an in-depth presentation of any one process or group of processes. For this reason, the text has also been helpful to a variety of factory personnel who desire a reference book that concisely presents the various production alternatives and the advantages and limitations of each. Individuals desiring additional, or in-depth, information on specific materials or processes are referred to the various references listed at the back of the text.

Acknowledgements

In preparing a text of this magnitude, the authors have a great many people to thank. First, Drs. Black and Kohser are indebted to E. Paul DeGarmo for his selecting them to try to "fill his shoes" and continue the fine tradition that he has established with the text. Paul has continued his involvement into the present edition by being the first reader of all manuscript copy. The authors also wish to acknowledge the multitude of assistance, information, and illustrations that has been provided by a variety of industries, professional organizations, and trade associations. The text has become known for the large number of clear and helpful photos and illustrations, many of which were graciously provided by companies and professional organizations. In some cases, equipment is photographed or depicted without safety guards so as to show important details and personnel are not wearing certain items of safety apparel that would be worn during normal operation.

The authors wish to express their thanks to Professors Jim Hool, Marv DeVries, Bernard Jiang, Kofi Nyamekyi, Paul Cohen, Richard Jaeger, Amit Bagchi, Donald Askeland and Robert Wolf for their help and suggestions in improving the text. Angeline Honnell assisted greatly in the writing of the chapters on manufacturing and production systems. William Riffe, William Schoech, Roman Dubrozsky, Glen Gehrig, Dale Wilson, Alphonso Medina, Guna Salvaduray, Chong Chu, Richard Hultin, Roger McNichols, George Harhalakis, Fred Swift, Joe ElGomayel, Gerel Davis and R.A. Wysock provided helpful reviews and suggestions.

Heartfelt thanks are due to our wives (Carol Black and Barb Kohser) and children who endured being "textbook widows and orphans" during the long, hot summer of 1987 when the bulk of this seventh edition was written. Not only did they provide loving support, but Carol and Barb also provided hours of expert proofreading, typing, and editing as the manuscript was prepared.

Further recognition is due to the excellent staff provided by Macmillan Publishing Co. We could not have asked for a better group of people to work with. A special thanks to our editor, Beth Anderson, for putting up with two crazy, procrastinating professors who tried both her patience and her abilities as she sought to coordinate all of the various activities required to produce such a text on a reasonable schedule.

Finally, J Black would like to thank Ron Garrett for being so slow in building his tennis court that he was able to complete his sections of the book without a major distraction and remain somewhat close to schedule.

Contents

1

Introduction

Materials, Manufacturing, and the Standard of Living. The standard of living in any society is determined, primarily, by the *goods* and *services* that are available to its people. In most cases, materials are utilized in the form of manufactured goods. Manufactured goods are typically divided into two classes; *consumer goods* and *producer goods*. Producer goods are those goods manufactured for other companies to use to manufacture either producer or consumer goods. Consumer goods are those purchased directly by the consumer or the general public. For example, someone has to build the machine tool, a lathe, that turns the large rolls that are sold to the rolling mill to be used to roll the sheets of steel which are then formed and become the fenders of your car. Similarly, many service industries depend heavily on the use of manufactured products, just as the agricultural industry is heavily dependent on the use of large farming machines for efficient production.

Converting materials from one form to another adds value to them. The more efficiently materials can be produced and converted into the desired products which function with the prescribed quality, the greater will be the companies' productivity and the better will be the standard of living of the employees.

The history of man has been linked with his ability to work with materials, beginning with the Stone Age and ranging through the eras of copper and bronze, the Iron Age, and recently the age of steel, with our sophisticated ferrous and nonferrous materials. We are now entering the age of tailor-made materials such as composites, as indicated in Figure 1-1. The alloys used in the manufacture of compressor airfoils since 1945 have continuously changed. As materials become more sophisticated, having greater strength and lighter weight, they also become more difficult to manufacture with existing manufacturing methods.

Although materials are no longer used only in their natural state or in modified forms, there is obviously an absolute limit to the amounts of many materials available here on earth. Therefore, although the variety of man-made materials

1

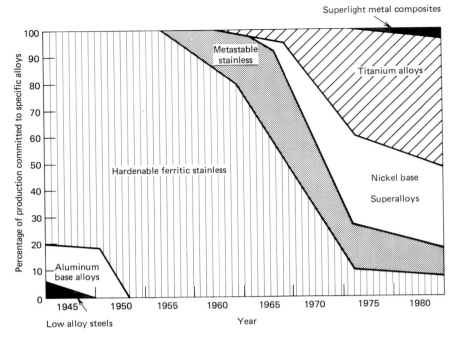

FIGURE 1-1 Changing trends in the mix of alloys used for compressor airfoils in jet engines (*From E. E. Weismantel, AEG-265-8/68[700], Aircraft Engine Group, General Electric Company.*)

continues to increase, resources must be used efficiently and recycled whenever possible. Of course, recycling only postpones the exhaustion date.

Like materials, processes have also proliferated greatly in the last 30 years, with new processes being developed to handle the new materials more efficiently and with less waste. Advances in manufacturing technology often account for improvements in productivity. Even when the technology is proprietary, the competition often gains access to it, usually quite quickly.

Materials, men, and equipment are interrelated factors in manufacturing that must be combined properly in order to achieve low cost, superior quality, and on-time delivery. Typically, as shown in Figure 1-2, *40% of the selling price of a product is manufacturing cost*. Since the selling price is determined by the customer, maintaining the profit often depends on reducing manufacturing cost. Direct labor, usually the target of automation, accounts for only about 12% of manufacturing cost even though many view it as the main factor in increasing productivity. In Chapter 44, a manufacturing strategy is presented that attacks the materials cost, indirect costs, and general administration costs, in addition to labor costs. The materials costs include the cost of storing and handling the materials within the plant. The strategy is called Integrated Manufacturing Production Systems.

Referring again to Figure 1-2, of the total expenses, (selling price less profit), about 68% of dollars are spent on people: about 15% for engineers; 25% for marketing, sales, and general management people; 5% for direct labor, and 10% for indirect labor (55/80 = 68.75%). The average labor cost in manufacturing

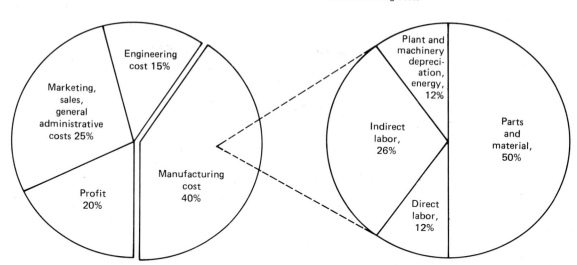

Selling price

Manufacturing costs

FIGURE 1-2 Manufacturing cost is the largest cost in the selling price. The largest manufacturing cost is material costs, not direct labor.

in the United States was around $10 to $11 per hour for the hourly workers in 1986. Reductions in direct labor will have only marginal effects on the people costs. The optimal combination of factors for producing a small quantity of a given product may be very inefficient for a larger quantity of the same product. Consequently, a systems approach, taking all the factors into account, must be used. *This requires a sound and broad understanding of materials, processes, and equipment on the part of the decision makers accompanied by an understanding of the manufacturing systems.* Materials and processes in manufacturing are what this book is all about.

The Roles of Engineers in Manufacturing.   Many engineers have as their function the designing of products. The products are brought into reality through the processing or fabrication of materials. In this capacity designers are a key factor in the material selection and manufacturing procedure. *A design engineer,* better than any other person, should know what the design is to accomplish, what assumptions can be made about service loads and requirements, what service environment the product must withstand, and what appearance the final product is to have. In order to meet these requirements, the material(s) to be used must be selected and specified. In most cases, in order to utilize the material and to enable the product to have the desired form, the designer knows that certain *manufacturing processes* will have to be employed. In many instances, the selection of a specific material may dictate what processing must be used. On the other hand, when certain processes must be used, the design may have to be modified in order for the process to be utilized effectively and economically. Certain dimensional sizes can dictate the processing, and some processes require certain sizes on the parts going into them. In converting the design into reality, many decisions must be made. In most instances, they can be made most effectively at the design stage. It is thus apparent that design engineers are a vital factor in the manufacturing process, and it is indeed a blessing to the company if they can *design for manufacturing*.

Manufacturing engineers select and coordinate specific processes and equipment to be used, or supervise and manage their use. Some design special tooling that is used so that standard machines can be utilized in producing specific products. These engineers must have a broad knowledge of manufacturing processes and of material behavior so that desired operations can be done effectively and efficiently without overloading or damaging machines and without adversely affecting the materials being processed. Although it is not obvious, the most hostile environment the material may ever encounter in its lifetime is the processing environment.

The machines and equipment used in manufacturing and their arrangement in the factory also comprise a design task. *Industrial* or *manufacturing engineers* who design (or lay out) factories have the same concerns of the interrelationship of design, the properties of the materials that the machines are going to process, and the interreaction of the materials and the machines.

Materials engineers devote their major efforts to developing new and better materials. They, too, must be concerned with how these materials can be processed and with the effects that the processing will have on the properties of the materials.

Although their roles may be quite different, it is apparent that a large proportion of engineers must concern themselves with the interrelationships of materials and manufacturing processes.

As an example of the close interrelationship of design, materials selection, and the selection and use of manufacturing processes, consider the standard electrical appliance plug shown in Figure 1-3. Suppose that this plug is sold at the retail store for $4.49. The wholesale outlet sold the plug for $4.00 and the manufacturer probably received about $3.50 for it. As shown in Figure 1-3, it consists of 10 parts. Thus the manufacturer had to produce and assemble the 10 parts for about $1.50—an average of 15 cents per part. Only by giving a great deal of attention to design, selection of materials, selection of processes, selection of equipment used for manufacturing (tooling), and utilization of personnel could such a result be achieved.

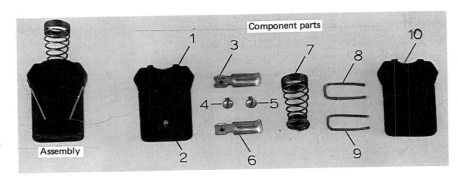

FIGURE 1-3 The component parts that make up the assembly of a low-cost appliance plug.

The appliance plug is a relatively simple product, yet the problems involved in its manufacture are typical of those with which manufacturing industries must deal. The elements of design, materials, and processes are all closely related, each having its effect on the others. For example, if the two plastic shell components were to be fastened together by a rivet instead of by the two U-shaped clips, entirely different machines, processes, and assembly procedures would be required. Such a design change would have a significant impact on the entire manufacturing process and on the cost.

The performance of the clips depends on the selection of the proper material. The material had to be ductile to permit it to be bent without breaking, yet it had to be sufficiently strong and stiff to act as a spring for holding the shells together firmly. It is apparent that both the material and the processing had to be considered when the plug and the clips were designed in order to assure a satisfactory product that could be manufactured economically.

Changing World Competition.   In recent years, major changes in the world of goods manufacturing have taken place. Three of these are

1. World wide competition.
2. Advanced technology.
3. New manufacturing systems structure, strategies, and management.

Worldwide competition is a fact of manufacturing life and it will get stronger in the future. On the other hand, everyone has access to advanced technology provided they have the capital. Technology can always be purchased. Thus, the secret to success in manufacturing is to build a company that can deliver on time to the customers, superior quality products at the lowest possible cost.

Manufacturing and Production Systems.   *Manufacturing* is the economic term for making goods and services available to satisfy human wants. Manufacturing implies creating value by applying useful mental or physical labor.

The *manufacturing processes* are collected together to form a *manufacturing system (MS)*. The manufacturing system takes inputs (see Figure 1-4) and produces products for the customer. The production system includes the manufacturing system and services it. In this text a *production system* will refer to the total company and will include with it the *manufacturing systems*. The following football analogy distinguishes between manufacturing systems and production systems (see Figure 1-5).

Football Analogy. College football is an example of a service industry. Football players are equivalent to manufacturing machine tools. The things that they do, such as punt, pass, run, tackle, and block, are equivalent to operations. Different machines do different operations, and some machines do operations better than others.

The arrangement of machines (often called the plant layout) defines the design of the manufacturing system. In football, this arrangement is called an offensive alignment, or defensive formation. Modern teams use pro "T" sets and "I"

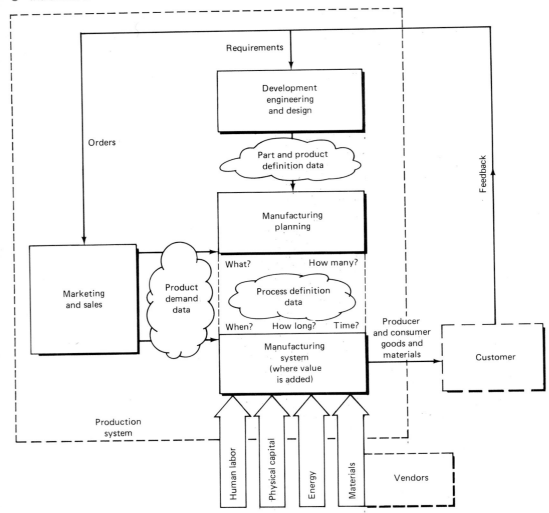

FIGURE 1-4 The production system (PS) takes product demand information and product definition data and uses it to plan the manufacturing system (MS).

formations, but many American factories are still using the "single wing" version of MS design, called the job shop.

The job shop, with some elements of flow shop when the volume is large enough to justify some special-purpose equipment, is the most common manufacturing system. The job shop is functionally arranged; that is, similar processes are put together. The same thing is done in football, with all the linemen segregated from the backs. Coaches are equivalent to foremen, and the head coach is the supervisor.

In the football analogy, the production system would be the athletic department, which sells tickets, runs the training room (machine maintenance and repair), raises operating capital, arranges the materials handling (travel), and

Football players	=	Manufacturing processes or machine tools
Things football players do: Run, punt, pass, block, tackle, catch.	=	Operations on the machine tools like turning, drilling, boring, tapping
Offensive and defensive plays Pro Tee I Back Single Wing	=	Manufacturing systems Job shop Flow job Cellular
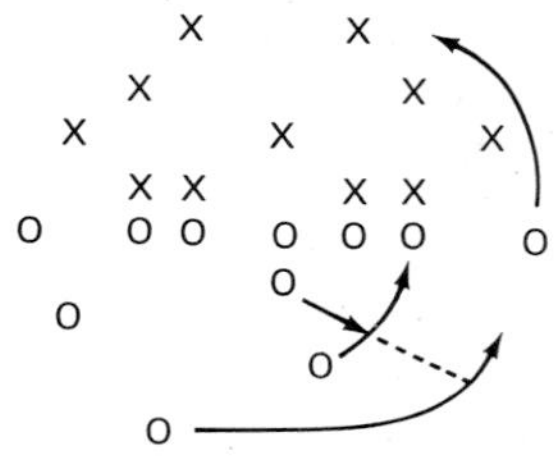 (football play diagram)	=	Layouts of manufacturing systems Functional layout Product layout
Athletic department Maintain field Sell tickets Print programs	=	Production system Design Personnel Accounting Quality control

FIGURE 1-5 Football analogy to manufacturing production systems (MPS).

does whatever is needed to help keep the manufacturing system operating but does not really do any manufacturing. Members of the athletic department never play during a game. They are all indirect labor, managerial and staff employees. In the plant, the production system does not build parts but rather *services* the manufacturing system. This division is also known as "staff," while people who work in manufacturing are called "line."

As shown in Figure 1-4, the production system therefore includes the manufacturing system plus all the other functional areas of the plant for information, design, analysis, and control. These subsystems are somehow connected to each other to produce either goods or services or both. Goods refer to material things. Services are nonmaterial things that we buy to satisfy our wants, our needs, our desires. *Service production systems (SPS)* include transportation, banking, finance, savings and loan, insurance, utilities, health care, education, communication, entertainment, sporting events, and so forth. They are useful labors that do not directly produce a product.

Production terms have a definite rank of importance somewhat like grades in the army. Confusing *system* with *section* is similar to mistaking a colonel for a corporal. In either case, knowledge of rank is necessary (see Table 1.1). The terms tend to overlap because of the inconsistencies of popular usage.

An obvious problem exists here in the terminology of manufacturing and production. The same term can refer to different things. For example, *drill* can refer to the machine tool that does these kinds of operations; the operation itself,

TABLE 1.1 Production Terms for Manufacturing Production Systems (MPSs)

Term	Meaning	Examples
Production system	All aspects of men, machines, materials, and information, considered collectively, needed to manufacture parts or products; integration of all units of the system is critical (see Chapter 43)	Company that makes engines, assembly plant, glassmaking factory, foundry
Manufacturing system (Sequence of operations) (Collection of processes)	A series of manufacturing operations resulting in specific end products; an arrangement or layout of many processes (see Chapter 42)	Rolling steel plates, manufacturing of auto bodies, series of connected operations or processes, a job shop, a flow shop, a continuous process
Machine or machine tool or manufacturing process	A specific piece of equipment designed to accomplish specific processes, often called a *machine tool;* machine tools link together to make a manufacturing system	Spot welder, milling machine, lathe, drill press, forge, drop hammer, die caster
Job (sometimes called a station)	A collection of operations done on machines or a collection of tasks performed by one man at one location on an assembly line	Operation of machines, inspection, final assembly. Forklift driver has the job of moving materials
Operation (sometimes called a process)	A specific action or treatment, the collection of which makes up the job of a worker	Drill, ream, bend, solder, turn, face, mill, extrude, inspect, load
Tools or tooling	Refers to the implements used to hold, cut, shape, or deform the work materials; called *cutting tools* if referring to machining; can refer to *jigs and fixtures* in workholding and *punches and dies* in metal forming	Grinding wheel, drill bit, tap, end milling cutter, die, mold, clamp, three-jaw vise

which can be done on many different kinds of machines; or the cutting tool, which exists in many different forms. It is therefore important to use modifiers whenever possible: "Use the *radial* drill *press* to drill a hole with a 1-inch-diameter spade drill." The emphasis of this book will be directed toward the understanding of the processes, machines, and tools required for manufacturing and how they interact with the materials being processed. In the last three chapters, a brief introduction to the systems aspects is presented.

Production System. The highest ranking term in the hierarchy is *production system.* A production system includes people, money, equipment, materials and supplies, markets, management, and the manufacturing system. In fact, all aspects of commerce (manufacturing, sales, advertising, profit, and distribution) are involved.

Much of the information given for manufacturing production systems (MPS) is relevant to the service production system (SPS). Many *manufacturing production systems* require an SPS for proper product sales. This is particularly true in industries such as the food (restaurant) industry in which customer service is as important as quality and on-time delivery.

Manufacturing Systems. A collection of operations and processes used to obtain a desired product(s) or component(s) is called a *manufacturing system.* The manufacturing system is therefore the design or *arrangement of the manufacturing processes.* Control of a system applies to overall control of the whole, not merely of the individual processes or equipment. The entire manufacturing system must be controlled in order to control inventory levels, product quality, output rates, and so forth.

Five manufacturing system designs can be identified: the job shop, the *flow shop,* the *"linked-cell" shop,* the *project shop,* and the *continuous process.* The latter system primarily deals with liquids, gases (such as an oil refinery) rather than solids or discrete parts.

The most common of these layouts is the *job shop,* characterized by large varieties of components, general-purpose machines, and a functional layout (see Chapter 42). This means that machines are collected by function (all lathes together, all milling machines together) and the parts are routed around the shop in small lots to the various machines.

Flow shops are characterized by larger lots, special-purpose machines, less variety, and more mechanization. Flow shop layouts are typically either continuous or interrupted. If *continuous,* they basically run one large-volume complex item in great quantity and nothing else. The appliance plug was made this way. A transfer line producing an engine block is another typical example. If *interrupted,* the line manufactures large lots but is periodically "changed over" to run a similar but different component.

The *"linked-cell"* manufacturing system is composed of manufacturing cells connected together (linked) using a unique form of inventory and information control (Kanban). See Chapters 42 and 44.

The *project shop* is characterized by the immobility of the item being manufactured. In the construction industry, bridges and roads are good examples. In the manufacture of goods, large airplanes or locomotives are manufactured in project shops. It is necessary that the men, machines, and materials come to the site. The number of end items is not very large, and therefore the lot sizes of the components going into the end item are not large. Thus the job shop usually supplies parts and subassemblies to the project shop in small lots. Naturally, there are many hybrid forms of these manufacturing systems, but the job shop is the most common system.

Because of its design, the job shop has been shown to be the least cost-efficient of all the systems. Component parts in a typical job shop spend only 5% of their time in machines and the rest of the time waiting or being moved from one functional area to the next. Once the part is on the machine, it is actually being processed (that is, having value added to it by the changing of its shape) only about 30%–40% of the time (see Figure 1-6). The rest of the time it is being loaded, unloaded, inspected, and so on. The advent of numerical control machines (see Figure 1-7 and Chapter 29) increased the percentage of time that the machine is making chips because tool movements are programmed and the machines can automatically change tools or load or unload parts. However, there are a number of trends that are forcing manufacturing management

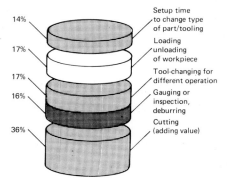

14% Setup time to change type of part/tooling

17% Loading unloading of workpiece

17% Tool-changing for different operation

16% Gauging or inspection, deburring

36% Cutting (adding value)

FIGURE 1-6 The typical utilization of the production time in metal-turning operations with conventional tool-handling, workpiece loading/unloading, setups, and inspection.

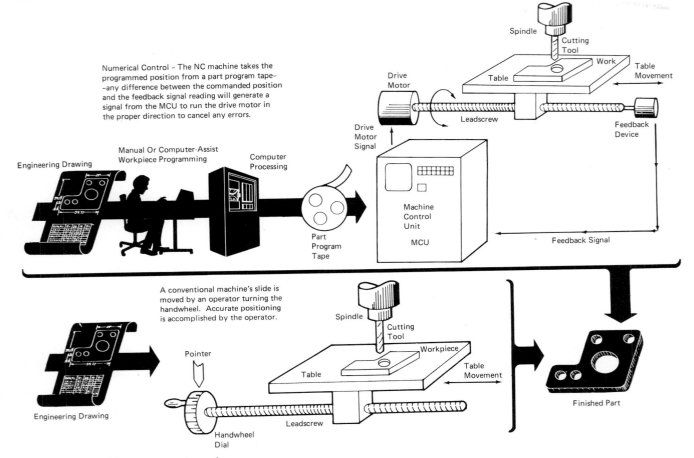

Numerical Control – The NC machine takes the programmed position from a part program tape—any difference between the commanded position and the feedback signal reading will generate a signal from the MCU to run the drive motor in the proper direction to cancel any errors.

Spindle
Cutting Tool
Work
Table Movement
Drive Motor
Table
Leadscrew
Feedback Device
Drive Motor Signal
Machine Control Unit
MCU
Feedback Signal

Engineering Drawing
Manual Or Computer-Assist Workpiece Programming
Computer Processing
Part Program Tape

A conventional machine's slide is moved by an operator turning the handwheel. Accurate positioning is accomplished by the operator.

Spindle
Cutting Tool
Workpiece
Table Movement
Pointer
Table
Leadscrew
Engineering Drawing
Handwheel Dial
Finished Part

FIGURE 1-7 The same part can be made by NC or manual machining. The increased cost of NC can be offset by the decreased manufacturing time and improved quality.

to consider means by which the job shop system itself can be redesigned to improve its overall efficiency. These trends have forced manufacturing companies to convert their batch-oriented job shops into "linked-cells." One of the ways to form a cell is through the use of *group technology*.

Group technology (GT) is a concept whereby similar parts are grouped together into part families. Parts of similar size and shape can often be processed through a *similar set of processes*. A part family based on manufacturing would have the same set or sequences of manufacturing processes. The set of processes is called a cell. Thus, with GT, job shops can be restructured into cells, each cell specializing in a particular family of parts (see Figure 1-8). The parts are handled less, machine setup time is shorter, in-process inventory is lower, and the time needed for parts to get through the manufacturing system is greatly reduced.

Manufacturing Processes. A *manufacturing process* converts unfinished materials to finished products. For example, injection molding, die casting,

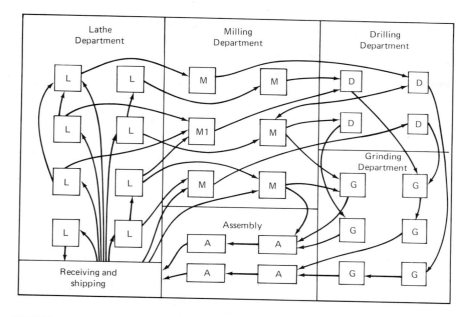

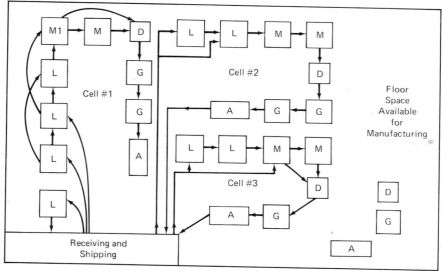

FIGURE 1-8 Group technology restructures the factory floor by grouping processes into cells to process families of parts.

progressive stamping, milling, arc welding, painting, assembling, testing, pasteurizing, homogenizing, and annealing are commonly called *processes* or *manufacturing processes*. The term *processing* often implies continuous products or actions with integrated operations for continuous automatic production.

A *machine tool* is an assembly of related mechanisms on a frame which together produce a desired result. Generally, the motors, controls, and auxiliary devices are included. The cutting tools and workholding devices are considered separately.

A machine tool may do a single process (cutoff saw) or multiple processes, or it may manufacture an entire component. Machine sizes vary from a tabletop drill press to a 1,000-ton forging press.

Job/Station. In the classical system, a *job* is the total of the work or duties a worker performs. A *station* is the work area of a production line worker. (The multifunctional worker concept has changed these traditional definitions.)

A job is a group of related operations and tasks performed at one station or series of stations in cells. For example, the job at a final assembly station may consist of four tasks:

1. Attach carburetor.
2. Connect gas line.
3. Connect vacuum line.
4. Connect accelerator rod.

The job of a turret lathe (a semiautomatic machine) operator may include the following operations and tasks: load, start, index and stop, unload, inspect. The operator's job may also include machine setup. Other machine operations include drilling, reaming, facing, turning, chamfering, and knurling. The operator can run more than one machine or service more that one station.

The terms *job* and *station* have been carried over to unmanned machines. A job, as before, is a group of related operations generally performed at one station, and a station is a position or location in a machine (or process) where specific operations are performed. A simple machine may have only one station. Complex machines can be composed of many stations. The job at a station often includes many simultaneous operations, such as "drill all face holes" by multiple spindle drills.

Operation. An operation is a distinct action performed to produce a desired result or effect. Typical machine operations are loading and unloading. Operations can be divided into suboperational elements. For example, loading is made up of picking up part, placing part in jig, closing jig. However, suboperational elements will not be discussed here.

Operations categorized by function are

1. Materials handling/transporting—change in position.
2. Processing—change in volume and quality, including assembly and disassembly. Processing can include packaging.
3. Packaging—special processing; may be temporary or permanent for shipping. (See also Chapter 41, wherein the term *packaging* is used in a different sense.)
4. Inspecting/testing—compare to the standard or check process behavior.
5. Storing—time lapses without further operations.

These basic operations may occur more than once in some processes, or they may sometimes be omitted. *Remember, it is the manufacturing processes that change the value and quality of the materials.* Defective processes produce poor quality or scrap. Other operations may be necessary but do not, in general, add value, whereas operations performed by machines that do material processing usually do add value.

Treatments.   Treatments operate continuously on the workpiece. They usually alter or modify the product-in-process without tool contact. Heat treating, curing, galvanizing, plating, finishing, (chemical) cleaning, and painting are examples of treatments. Treatments usually do add value to the part.

These processes are often difficult to include in cells because they often have long cycle times, are hazardous to the workers' health, or are unpleasant to be around because of high heat or chemicals. They are often done in large tanks. The cycle time for these processes may *dictate* the cycle times for the entire system. These operations also tend to be material-specific. Many manufactured products are given decorative and protective surface treatments that control the finished appearance. A customer may not buy a new truck with a visible defect in the chrome bumper although this defect does not alter the operation of the car.

Tools, Tooling, Workholders.   The lowest mechanism in the production term rank is the tool. This implement is used to hold, cut, shape, or form the unfinished product. Common hand tools include the saw, hammer, screwdriver, chisel, punch, sandpaper, drill, clamp, file, torch, and grindstone.

Basically, machines are mechanized versions of such hand tools. Most tools are for cutting (drill bits, reamers, single-point turning tools, milling cutters, saw blades, broaches, and grinding wheels). Noncutting tools for forming include extrusion dies, punches, and molds.

Tools also include work holders, jigs, and fixtures. These tools and cutting tools are generally referred to as the *tooling,* which is usually considered separate from machine tools. Cutting tools wear and fail and must be periodically replaced before parts are ruined. The workholding devices must be able to locate and secure the workpieces during processing in a repeatable, mistake-proof way.

Tooling for Measurement and Inspection.   Measuring tools and instruments are also important for manufacturing. Common examples of measuring tools are rulers, calipers, micrometers, and gages. Precision devices that use laser optics or vision systems coupled with sophisticated electronics are becoming commonplace. Vision systems and coordinate measuring machines are becoming critical elements for achieving superior quality.

Integrating Inspection into the Process. The integration of the inspection process into the manufacturing process or the manufacturing system is a critical step toward building products of superior quality. An example will help. Compare an electric typewriter with a computer that does word processing. The electric typewriter is flexible. It types whatever words are wanted in whatever order. It can type in Pica, Elite, or Orator but the font (disk or ball that has the appropriate type size on it) has to be changed according to the size of type wanted. The computer can do all of this but can also, through its software, do italics, darken the words, vary the spacing to justify the right margin, plus many other functions. It immediately checks for incorrect spelling and other defects like repeated words. The software system provides a signal to the hardware to flash the word so that the operator knows something is wrong and can make an immediate correction. If the system were designed to prevent the typist from typing repeated words, then this would be a *Pokayoke*–a defect prevention. Defect prevention is better than immediate defect detection and correction. Ultimately, the system should be able to forecast the probability of a defect, correcting the problem at the source. This means that the typist would have to be removed from the process loop, perhaps by having the system type out what it is told (convert oral to written directly). Pokayoke and source inspection are keys to designing manufacturing systems which produce superior-quality products at low cost.

The Language of Manufacturing or Process Technology.   In manufacturing, material things (goods) are made to satisfy human wants. *Products* result from manufacture. Manufacture also includes conversion processes such as refining, smelting, canning, and mining.

Products can be manufactured by *fabricating* or by processing. *Fabricating* is the manufacture of a product from pieces such as parts, components, or assemblies. Individual products or parts can also be fabricated. Separable discrete items such as tires, nails, spoons, screws, refrigerators, or hinges are fabricated.

Processing is also used to refer to the manufacture of a product by continuous means, or by a continuous series of operations, for a specific purpose. Continous items such as steel strip, beverages, breakfast foods, tubing, chemicals, and petroleum are "processed." Many processed products are marketed as discrete items, such as bottles of beer, bolts of cloth, spools of wire, and sacks of flour.

Separable discrete products, both piece parts and assemblies, are fabricated in a *plant,* a *factory,* or a *mill,* for instance a textile or rolling mill. Products that *flow* (liquids, gases, grains, or powders) are processed in a *plant* or *refinery.* Then *continuous-process industries*—such as petroleum and chemical plants— are sometimes called processing industries or flow industries.

To a lesser extent, the terms *fabricating industries* and *manufacturing industries* are used when referring to fabricators or manufacturers of large products composed of many parts, such as a car, a plane, or a tractor.

Manufacturing often includes continuous-process treatments such as electroplating, heating, demagnetizing, and extrusion forming.

Construction and *agriculture* make goods by means other than manufacturing or processing in factories. Construction is a form of project manufacturing of useful goods. The public may not consider construction as manufacturing because the work is not usually done in a plant or factory, but it can be. There is a company in Delaware that can build a custom house of any design in their factory, truck it to the building site, and assemble it on a foundation in two or three weeks.

Agriculture, fisheries, and commercial fishing produce real goods from useful labor. Lumbering is similar to both agriculture and mining in some respects, and mining should be considered processing. Processes that convert the raw materials from agriculture, fishing, lumbering, and mining into other usable and consumable products are also forms of manufacturing.

Configuration Analysis. In the manufacturing of goods, the primary objective is to produce a component having a desired geometry, size, and finish. Every component has a shape that is bounded by various types of surfaces of certain sizes that are spaced and arranged relative to each other. Consequently, a component is manufactured by producing the surfaces that bound the shape. Surfaces may be

1. Plane or flat.
2. Cylindrical: external or internal.
3. Conical: external or internal.
4. Irregular: curved or warped.

Figure 1-9 illustrates how a shape can be analyzed and broken up into these basic bounding surfaces. Parts are manufactured by using processes that will either (1) remove portions of a rough block of material so as to produce and leave the desired bounding surface, or (2) cause material to form into a stable configuration that has the required bounding surfaces. Consequently, in designing an object, the designer specifies the shape, size, and arrangement of the bounding surface. The part design must be analyzed to determine what materials will provide the desired properties including mating to other components, and what processes can best be employed to obtain the end product at the most reasonable cost.

Understand the Business of the Company. Understanding the process technology of the company is very important for everyone in the company. Manufacturing technology affects the design of the product and the manufacturing system, the way in which the MS can be controlled, the types of people employed, and the materials that can be processed. Table 1.2 outlines the process technology characteristics. One valid criticism of American companies is that

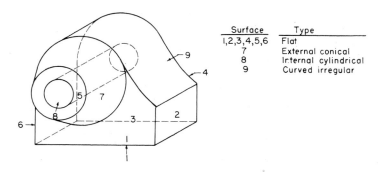

Surface	Type
1,2,3,4,5,6	Flat
7	External conical
8	Internal cylindrical
9	Curved irregular

FIGURE 1-9 Object composed of nine geometric surfaces. Dashed lines are hidden surfaces.

TABLE 1.2   Characterizing a Process Technology

- Mechanics (statics and dynamics of the process).
    How does the process work?
    What are the process mechanics?
    What physically happens, and what makes it happen? (Understand the physics)
- Economics/Costs
    What are the tooling costs, the engineering costs?
    Which costs are short term, which long term?
    What are the setup costs?
- Time spans
    How long does it take to set up?
    How can this time be shortened?
    How long does it take to run a part, once set up?
    What process parameters affect the run time?
- Constraints
    What are the process limits?
    What cannot be done?
    What constrains this process (sizes, speeds, forces, volumes, power, cost)?
    What is very hard to do within an acceptable time/cost frame?
- Uncertainties/Process reliability
    What can go wrong?
    How can this machine fail?
    What do people worry about with this process?
    Is this a reliable, stable process?
- Skills
    What operator skills are critical?
    What is not done automatically?
    How long does it take to learn to do this process?
- Flexibility
    Can this process easily do new parts of a new design or material?
    How does the process react to changes in part design and demand?
    Which changes are easy to do?
- Process capability
    What are the accuracy and precision of the process?
    What tolerances does the process meet? (What is the process capability?)
    How repeatable are those tolerances?

their managers seem to have an aversion to understanding their companies' manufacturing technologies. Failure to understand the company business (that is, its fundamental process technology) can lead to the failure of the company.

The way to overcome technological aversion is to run the process and study the technology. Only someone who has run a drill press can understand the sensitive relationship between feed rate and drill torque and thrust. All processes have these 'know-how' features. Those who run the processes must be part of the decision making for the factory. The CEO who takes a vacation working on the plant floor learning the processes will be well on the way to being the head of a successful company.

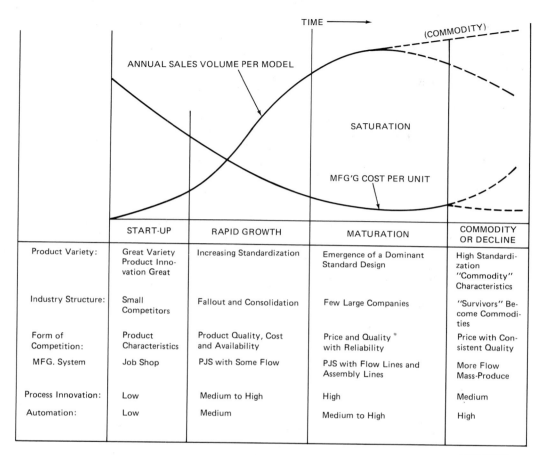

	START-UP	RAPID GROWTH	MATURATION	COMMODITY OR DECLINE
Product Variety:	Great Variety Product Inno- vation Great	Increasing Standardization	Emergence of a Dominant Standard Design	High Standardi- zation "Commodity" Characteristics
Industry Structure:	Small Competitors	Fallout and Consolidation	Few Large Companies	"Survivors" Be- come Commodi- ties
Form of Competition:	Product Characteristics	Product Quality, Cost and Availability	Price and Quality with Reliability	Price with Con- sistent Quality
MFG. System	Job Shop	PJS with Some Flow	PJS with Flow Lines and Assembly Lines	More Flow Mass-Produce
Process Innovation:	Low	Medium to High	High	Medium
Automation:	Low	Medium	Medium to High	High

FIGURE 1-10 Relationship between product life cycle, manufacturing system development/evolution, and manufacturing cost per unit.

Product Life Cycle.

Manufacturing systems are dynamic and change with time. There is a general, traditional relationship between a product's life cycle and the kind of manufacturing system it has. Figure 1-10 simplifies the life cycle into these steps:

Startup	New product or new company, low volume, small company.
Rapid growth	Products become standardized and volume increases rapidly. Company's ability to meet demand stresses its capacity.
Maturation	Standard designs emerge. Process development is very important.
Commodity	Long-life, standard-of-the-industry type of product, *or*
Decline	Product is slowly replaced by improved products.

The maturation of a product in the marketplace generally leads to fewer competitors, with competition based more on price and on-time delivery than on unique product features. As the competitive focus shifts during the different stages of the product life cycle, the requirements placed on manufacturing—

cost, quality, flexibility, and delivery dependability—also change. The stage of the product life cycle affects the product design stability, the length of the product development cycle, the frequency of engineering change order, and the commonality of components, all of which have implications for manufacturing process technology.

In short, the product life cycle concept provides a framework for thinking about the product's evolution through time and the kind of market segments that are likely to develop at various times. The design of the manufacturing system determines the cost per unit, which generally decreases over time with process improvements and increased volumes. The linked-cell approach enables companies to decrease cost per unit significantly while maintaining flexibility and making smooth transitions from low-volume to high-volume manufacturing.

The Basic Manufacturing Processes.   Manufacturing processes can be classified as

- Casting, foundry, or molding processes.
- Forming or metalworking processes.
- Machining (material removal) processes.
- Joining and assembly.
- Surface treatments (finishing).
- Heat treating.
- Other.

These classifications are not mutually exclusive. For example, some finishing processes involve a small amount of metal removal or metal forming. A laser can be used either for joining or for metal removal or heat treating. Occasionally, we have a process such as shearing, which is really metal cutting but is viewed as a (sheet) metal-forming process. Assembly may involve processes other than joining. The categories of process types are far from perfect.

Casting and *molding* processes are covered in Chapters 13–16. Casting uses molten metal and a cavity. The metal retains the desired shape of the mold cavity after solidification. An important advantage of casting and molding is that, in a single step, materials can be converted from a crude form into a desired shape. In most cases, a secondary advantage is that excess or scrap material can easily be recycled. Figure 1-11 illustrates schematically the basic concepts of these processes.

Casting processes commonly are classified into two types: permanent mold (a mold that can be used repeatedly) or nonpermanent mold (a new mold must be prepared for each casting made). Molding processes for plastics and composites are included in Part IV, "Forming Processes."

Forming and *shearing* operations typically utilize material (metal or plastics) that previously has been cast or molded. In many cases the materials pass through a series of forming or shearing operations, so the form of the material for a specific operation may be the result of all the prior operations. The basic purpose of forming and shearing is to modify the shape and size and/or physical properties of the material.

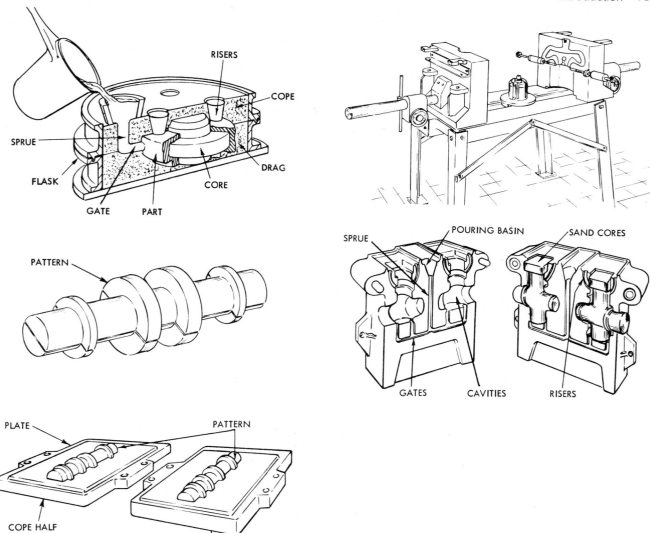

Labels in figure:
RISERS
COPE
SPRUE
DRAG
FLASK
CORE
GATE
PART
PATTERN
PATTERN
PLATE
COPE HALF
DRAG HALF
SPRUE
POURING BASIN
SAND CORES
GATES
CAVITIES
RISERS

FIGURE 1-11 Casting processes. Sand casting on left. Permanent mold casting above.

Some of the forming and shearing processes are shown in Figure 1-12. *Metalforming* and *shearing operations* are done both "hot" and "cold," a reference to the temperature of the material at the time it is being processed with respect to the temperature at which this material can recrystallize (that is, grow new grain structure).

Machining or *metal removal processes* refer to the removal of certain selected areas from a part in order to obtain a desired shape or finish. Chip-making processes are covered in Chapters 21–31. In recent years many new machining processes have been developed. Chips are formed by interaction of a cutting

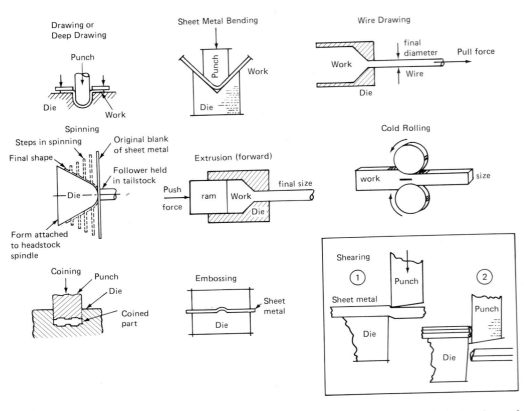

FIGURE 1-12 Common forming and shearing processes.

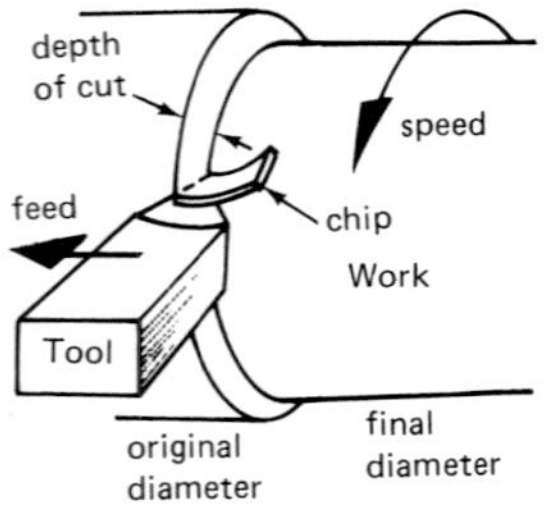

FIGURE 1-13 Single-point metal-cutting process (turning) produces a chip.

tool with the material being machined. Figure 1-13 shows a chip being formed by a single-point cutting tool.

Cutting tools are used to perform the basic and related machining processes that are shown schematically in Figure 1-14. The cutting tools are mounted in machine tools, which provide the required movements of the tool with respect to the work (or vice versa) to accomplish the process desired.

The seven basic machining processes are *shaping, drilling, turning, milling, sawing, broaching,* and *abrasive machining.* Eight basic types of machine tools have been developed to accomplish the basic processes. These are shapers (and planers), drill presses, lathes, boring machines, milling machines, saws, broaches, and grinders. Most of these machine tools are capable of performing more than one of the basic machining processes. This obvious advantage has led to the development of *machining centers* specifically designed to permit several of the basic processes, plus other related processes, to be done on a single machine tool with a single workpiece setup (see Chapter 29).

Included with the machining processes are processes wherein metal is removed by chemical, electrical, electrochemical, or thermal sources. Generally speaking, these nontraditional processes have evolved to fill a specific need when conventional processes were too expensive or too slow when machining very hard materials. One of the first uses of a laser was to machine holes in ultra-

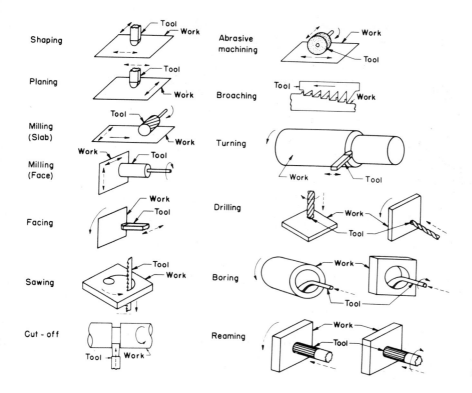

FIGURE 1-14 Schematic representation of basic machining processes.

high-strength metals. It is being used today to drill tiny holes in turbine blades for jet engines. Table 1.3 shows the more common chipless machining processes. Chapter 32 covers these processes.

Perhaps the largest collection of processes, in terms of both diversity and quantity, are the *joining processes,* which include the following:

1. Mechanical fastening.
2. Soldering and brazing.
3. Welding.
4. Press fitting.
5. Shrink fitting.
6. Adhesive bonding.

These processes, covered in Chapters 33–39, are often used in assembly.

Finishing processes are yet another class of processes typically employed for cleaning, removing of burrs left by machining, or providing protective and/or decorative surfaces on workpieces. Surface treatments include chemical and mechanical cleaning, deburring, painting, plating, buffing, galvanizing, and anodizing.

Heat treatment is the heating and cooling of a metal for specific purpose of altering its metallurgical and mechanical properties. Because the changing and controlling of these properties is so important in the processing and performance

TABLE 1.3   Nontraditional Chipless Machining Processes

Process	Metal-removal Mechanism	Examples
Chemical machining, milling, or blanking	Chemical etching	Photoengraving
Electrochemical machining (ECM) or drilling or grinding	High-intensity "reverse electroplating" using high current densities	Machine cavities in dies
Ultrasonic machining (drilling or welding)	Abrasive slurry with grits vibrated into work by ultrasonic means; actually forms chips	Tool and die work (in nonconductors)
Electrodischarge machining (EDM)	Spark erosion of metals by local heating and melting	Drill holes in very hard tool and die materials
Laser beam machining (LBM) or drilling or heat treating	High-energy laser melts and vaporizes metal	Drill holes in turbine blades
Electron beam machining (EBM) or welding or cutting	High-energy electron beam melts and vaporizes metal	Microhole drilling in integrated-circuit boards
Plasma jet machining or cutting or welding	Ionic plasma, very high temperature jets to melt materials	Rapid cutting of plates

of metals, heat treatment is a very important manufacturing process. Each type of metal reacts differently to heat treatment. Consequently, a designer should know not only how a selected metal can be altered by heat treatment but, equally important, *how a selected metal will react, favorably or unfavorably, to any heating or cooling that may be incidental to the manufacturing process.* Heat treating is covered in Chapter 7.

In addition to the processes, there are some other fundamental manufacturing operations that we must consider.

Inspection determines whether the desired objectives stated by the designer in the specifications have been achieved. This activity provides feedback to design and manufacturing with regard to the process behavior. Essential to this inspection function are measurement activities. This material is covered in Chapters 10 and 12.

In *testing,* a product is tried by actual function or operation or by subjection to external effects. Although a test is a form of inspection, it is often not viewed that way. In manufacturing, parts and materials are inspected for conformance to the dimensional and physical specifications.

Testing may simulate the environmental or usage demands to be made on a product after it is placed in service. Complex processes may require many tests and inspections. Testing includes life-cycle tests, destructive tests, nondestructive testing to check for processing defects, (see Chapter 11), wind-tunnel tests, road tests, and overload tests.

Transportation is often referred to as *materials handling* or *conveyance* of the goods and refers to the transporting of unfinished goods, workpieces, and supplies to and from, between, and during manufacturing operations. Loading, positioning, and unloading are also materials-handling operations. Transportation, by truck or train, is materials handling between factories.

Proper manufacturing system design and mechanization can reduce materials handling in countless ways.

Automatic materials handling is a critical part of continuous automatic manufacturing. Until machines can be loaded and unloaded automatically (robotically), continuous automatic production will be very limited.

Material handling, a fundamental operation done by people and by conveyors and loaders, often includes positioning the workpiece within the machine by indexing, shuttle bars, slides, and clamps.

In recent years, wire-guided automated guided vehicles (AGVs) and automatic storage and retrieval systems (AS/RS) have been developed to replace forklift trucks on the factory floor.

Mechanized waste removal (of chips, trimmings, and cutoffs) can be more difficult than handling the product. Chip removal must be done before a tangle of scrap chips damages tooling or creates defective workpieces.

Most texts on manufacturing processes do not mention *packaging,* yet the packaging is often the first thing the customer sees. Also, packaging often maintains the product's quality between completion and use. (Packaging is also used in electronics manufacturing to refer to placing microelectronic chips in containers for mounting on circuit boards.)

Packaging can also prepare the product for delivery to the user. It varies from filling ampules with antibiotics to steel-strapping aluminum ingots into palletized loads. A product may require several packaging operations. For example, Hershey Kisses are (1) individually wrapped in foil, (2) placed in bags, (3) put into boxes, and (4) placed in shipping cartons.

Weighing, filling, sealing, and labeling are packaging operations that are highly automated in many industries. When possible, the cartons or wrapping is formed from material on rolls in the packaging machine.

Packaging is a specialty combining elements of product design (styling), materials handling, and quality control. Some packages cost more than their contents, for example, cosmetics and razor blades.

During *storage,* nothing happens intentionally to the product or part except the passage of time. Part or product deterioration on the shelf is called *shelf life,* meaning that items can rust, age, rot, spoil, embrittle, corrode, creep, and otherwise change in state or structure while, supposedly, nothing is happening to them. Storage is detrimental, wasting the company's time and money. The best strategy is to keep the product moving with as little storage as possible. Storage during processing must be *eliminated,* not automated or computerized. Companies should avoid investing heavily in large automated systems that do not alter the bottom line. Have the outputs improved with respect to the inputs, or has storage increased the costs (indirectly) without improving either the quality or the throughput time?

By not storing a product, the company avoids having to (1) retrieve it, (2) remember where it is, (3) worry about its deteriorating, or (4) pay storage costs. Storage is the biggest waste of all and must be eliminated at every opportunity.

Yardstick for Automation. In 1962, Amber & Amber presented their *"Yardstick for Automation."* The chart that they developed has been updated and is included here as Table 1.4 in a somewhat abbreviated form. The key to the chart is that each level of automation is tied to the human attribute which is being replaced (mechanized or automated) by the machine. Therefore, the A(O) level of automation, in which no human attribute was mechanized, covers the Stone Age through the Iron Age. Two of the earliest machine tools were the crude lathes the Etruscans used for making wooden bowls around 700 B.C. and the windlass-powered broach for machining of grooves into rifle barrels used over 300 years ago.

The first Industrial Revolution can be tied to the development of powered machine tools, dating from 1775, when the energetic "Iron-Mad" John Wilk-

TABLE 1.4 Yardstick for Automation

Orders of Automation	Human Attribute Replaced	Examples
A(0)	*None:* Lever, screw, pulley, wedge	Hand tools, manual machine
A(1)	*Energy:* Muscles replaced	Powered machines and tools, Whitney's milling machine
A(2)	*Dexterity:* Self-feeding	Single-cycle automatics
A(3)	*Diligence:* No feedback	Repeats cycle; open-loop numerical control or automatic screw machine; transfer lines
A(4)	*Judgment:* Positional feedback	Closed loop; numerical control; self-measuring and adjusting
A(5)	*Evaluation:* Adaptive control; deductive analysis; feedback from the process	Computer control; model of process required for analysis and optimization
A(6)	*Learning:* By experience	Limited self-programming; some artificial intelligence (AI); expert systems
A(7)	*Reasoning:* Exhibits intuition; relates causes and effects	Inductive reasoning; Advanced AI in control software
A(8)	*Creativeness:* Performs design unaided	Originality
A(9)	*Dominance:* Supermachine, commands others	Machine is master (Hal in *2001, A Space Odyssey*)

Source: Amber & Amber, *Anatomy of Automation*, Prentice-Hall, Inc., Englewood Cliffs, N.J., 1962. Used by permission, modified by Black.

Outboard bearing

FIGURE 1-15 Model of Wilkinson's horizontal boring machine. (*British Crown Copyright, Science Museum, London.*)

inson, in England, constructed a horizontal boring machine for machining internal cylindrical surfaces, such as in piston-type pumps. In Wilkinson's machine, a model of which is shown in Figure 1-15, the boring bar extended through the casting to be machined and was supported at its outer end by a bearing. Modern boring machines still employ this basic design. Wilkinson reported that his machine could bore a 57-inch-diameter cylinder to such accuracy that nothing greater than an English shilling (about 1/16 in. or 1.59 mm) could be inserted between the piston and the cylinder. This machine tool made Watt's steam engine a reality. At the time of his death, Wilkinson's industrial complex was the largest in the world.

The next A(1) machine tool was developed in 1794 by Henry Maudsley. It was an engine lathe with a practical slide tool rest. This machine tool, shown in Figure 1-16, was the forerunner of the modern engine lathe. The lead screw and change gear mechanism, which enabled threads to be cut, were added about 1800. The first planer was developed in 1817 by Richard Roberts in Manchester,

FIGURE 1-16 Maudsley's screw-cutting lathe. (*British Crown Copyright, Science Museum, London*).

England. Roberts was a student of Maudsley, who also had a hand in the career of Joseph Whitworth, the designer of screw threads. Roberts also added back gears and other improvements to the lathe. The first horizontal milling machine is credited to Eli Whitney in 1818 in New Haven, Connecticut (see Figure 1-17). The development of machine tools that not only could make specific products but could also produce other machines to make other products was fundamental to the first industrial revolution.

While early work in machine tools and precision measurement was done in England, the earliest attempts at interchangeable manufacturing apparently occurred almost simultaneously in Europe and the United States with the development of *filing jigs*, with which duplicate parts could be hand-filed to substantially identical dimensions. In 1798, Eli Whitney, using this technique, was able to obtain and eventually fulfill a contract from the United States government to produce 10,000 army muskets, the parts of each being interchangeable. However, this truly remarkable achievement was accomplished primarily by painstaking handwork and not by specified machines.

Joseph Whitworth, starting about 1830, accelerated the use of Wilkinson's and Maudsley's machine tools by developing precision measuring methods. Later he developed a measuring machine using a large micrometer screw. Still later, he worked toward establishing thread standards and made plug and ring gauges. His work was valuable because precise methods of measurement were the prerequisite for developing interchangeable parts, a requirement for later mass production.

The next significant machine tool was the drill press with automatic feed, developed by John Nasmyth, another student of Maudsley, in 1840 in Manchester. Surface grinding machines came along about 1880 and the era was completed with the development of the bandsaw blades, which could cut metal. In total, there were eight basic machine tools in the first industrial revolution for machining: lathe, milling machine, drill press, broach, boring mill, planer (shaper), grinder, and saw.

The A(2) level of automation was clearly delineated when machine tools became single-cycle, self-feeding machines displaying *dexterity*. Many examples of this level of machine are given in Chapters 21 through 28; they exist in great numbers in many factories. The A(2) level of machine can be loaded with a part and the cycle initiated by the worker. The machine completes the cycle and stops automatically. The A(3) level requires the machine to be *diligent* or repeat cycles automatically. These machines are open loop, meaning that they do not have *feedback*, and are controlled by either an internal fixed program, such as a cam, or are externally programmed with a tape or, more recently, a computer. A(3), A(4), and A(5) levels are basically superimposed on A(2)-level machines, which must be A(1) by definition. The A(3) level includes robots, numerical control (NC) machines which have no feedback, and many special-purpose machine tools.

The A(4) level of automation required that *human judgment* be replaced by a capability in the machine to measure and compare results with desired position or size and make adjustments to minimize errors. This is *feedback* or *closed-*

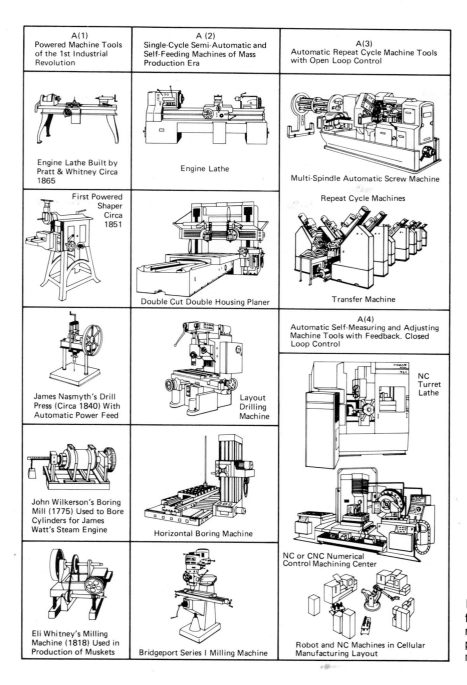

A(1) Powered Machine Tools of the 1st Industrial Revolution	A (2) Single-Cycle Semi-Automatic and Self-Feeding Machines of Mass Production Era	A(3) Automatic Repeat Cycle Machine Tools with Open Loop Control
Engine Lathe Built by Pratt & Whitney Circa 1865	Engine Lathe	Multi-Spindle Automatic Screw Machine
First Powered Shaper Circa 1851	Double Cut Double Housing Planer	Repeat Cycle Machines Transfer Machine
James Nasmyth's Drill Press (Circa 1840) With Automatic Power Feed	Layout Drilling Machine	A(4) Automatic Self-Measuring and Adjusting Machine Tools with Feedback. Closed Loop Control
John Wilkerson's Boring Mill (1775) Used to Bore Cylinders for James Watt's Steam Engine	Horizontal Boring Machine	NC Turret Lathe NC or CNC Numerical Control Machining Center
Eli Whitney's Milling Machine (1818) Used in Production of Muskets	Bridgeport Series I Milling Machine	Robot and NC Machines in Cellular Manufacturing Layout

FIGURE 1-17 Machine tools of the first Industrial Revolution A[1], the mass production era A[2], and examples of A[3] and A[4] levels of automation.

loop control (see Figure 1-7). The first numerical control tool was developed in 1952. It had positional feedback control and is generally recognized as the first A(4) machine tool. By 1958, the first NC *machining center* (see Chapter 29) was being marketed by Kearney and Trecker. This machining center was a compilation of many machine tools capable of performing many processes; in this case, milling, drilling (tapping), and boring (see Figure 1-17). It could automatically change tools to give it greater flexibility. Almost from the start, computers were needed to help program these machines. Within 10 years, NC machine tools became computer numerical control (CNC) machine tools. Thus these machines had their own microprocessor and could be programmed directly. In any event, CNC machines are still A(4) machines.

With the advent of the NC type of machine and more recently the programmable robot, two types of automation were defined. *Hard* or *fixed automation* is exemplified by transfer machines or automatic screw machines, and *flexible* or *programmable automation* is typified by NC machines or robots that can be taught or programmed externally by means of computers. The control was in computer software rather than mechanical hardware.

The A(5) level requires that machines perform *evaluation* of the process itself. Thus the machine must be cognizant of the multiple factors on which the process performance is predicated, evaluate the current setting of the input parameters versus the outputs from the process, and then determine how to alter the inputs to optimize the process. This is called *adaptive control* (AC), see Chapters 29 and 42 for discussions.

There are very few examples of A(5) machines on the factory floor and fewer at the A(6) level wherein the machine control has expert systems capability. *Expert systems* try to infuse the software with the deductive decision-making capability of the human brain by having the system get smarter *through experience*. *Artificial Intelligence* (AI) carries this step higher by *infecting* the control software with programs which exhibit the ability to reason inductively. Levels A(3) through A(4) will be discussed in Chapter 29. Levels A(5) and A(6) are discussed in Chapter 42. Levels A(6) and A(7) are the subjects of intensive worldwide research efforts. Levels A(8) through A(9) will be left to the whims of the science fiction writer.

Automation involves machines, or integrated groups of machines, that automatically perform required machining, forming, assembly, handling, and inspection operations. Through sensing and feedback devices, these systems automatically make necessary corrective adjustments. That is, *man's thinking must be automated*. There are relatively few completely automated systems, but there are numerous examples of highly mechanized machines. While the potential advantage of a completely automated plant are tremendous, in practice, step-by-step automation of individual operations is required. Therefore it is important to have a piecewise plan to convert from the classical job shops to the automated integrated factory of the future. This will be discussed in more detail in Chapter 44. However, the most serious limitation in automation will be available capital, as the initial investments for automation equipment and installations will be large. Because proper engineering economics analysis must be employed to evaluate these investments, students who anticipate a career in manufacturing should consider a course in this area a firm requirement.

 Continually reducing production costs is and always will be the primary objective. New materials are constantly being sought. We are truly entering the age of composites. New processes are needed to deal with new materials which do not pollute the environment. Despite the great advances of recent years, even greater progress may be expected in the future. More attention will be given to eliminating the waste of materials, reducing manufacturing time, and improving quality. The consumer wants products to have high reliability and superior quality.

A New Manufacturing System.   Manufacturing systems and productions systems are introduced in Chapters 42 and 43. The manufacturing process technology described herein is available worldwide. Many countries have about the same level of process development when it comes to manufacturing technology. Much of the technology existing in the world today was developed in the United States. Japan and more recently Taiwan and Korea are making great inroads into American markets, particularly in the automotive and electronics industries. What many people have failed to recognize is that many Japanese companies have developed and promoted a totally different kind of manufacturing production system. In future years, this new system, based on linked-cells, will take its place with the Taylor system of scientific management and the Ford system for mass production. The original working model for this new system is the Toyota Motor Company. The system is known as the Just-in-Time Production System or the Ohno system, after its chief architect, Tiiachi Ohno.

 Many American companies have successfully adopted some version of the Toyota system. The experience of dozens of these companies is amalgamated into ten steps, which, if followed, can make any company a factory with a future. The key steps to developing the Integrated Manufacturing Production System (IMPS) are presented in Chapter 44.

 For the IMPS to work, 100% good units flow rhythmically to subsequent processes without interruption. In order to accomplish this, an integrated quality control (IQC) program has to be developed. The responsibility for quality has been given to production and there is a company-wide attitude toward constant quality improvement. The goal is to make it right (perfect) the first time. Make quality easy to see, stop the line when something goes wrong, and inspect things 100% if necessary to prevent defects from occurring. The results of this system are astonishing, as shown in the 1985 new-car initial quality survey in Table 1.5. Six of the ten problem-free models are Toyotas. In overall average number of problems per vehicle, the Japanese, led by Toyota and Honda, had an average of 169 problems per car versus American cars with 268 and European cars with 267.

 The Japanese became world-class competitors by developing superior design and process technology. Now, Japan is concentrating on new product innovation with the emphasis on high-value-added products. They are spending as much as 14% of sales' dollars for research and development (R&D). The typical

TABLE 1.5 Evidence of the Quality Built into the Toyota Cars Through the Toyota Production System

Total Problems Model	Per 100 Vehicles[a]
Toyota Camry	109
Toyota Cressida	122
Toyota Tercel	127
Toyota Corolla	133
Honda Civic	137
Honda Accord	143
Mercedes-Benz 380	144
Toyota Celica	150
Mercedes-Benz 500	151
Toyota Supra	156
Nissan Sentra	157
Mercedes-Benz 300	159
Honda Civic CRX	165
Mazda RX-7	165
Nissan 200 SX	170
Ford Crown Victoria	171
Chevrolet Sprint	172
Chevrolet Citation II	173
Chevrolet Spectrum	184
Dodge Colt	184
All Japanese models	169
All European models	267
All American models	268

[a]Based on first three months of ownership in 1985.

American company spends less than 5% for research and development. Furthermore, many Japanese firms and firms in other countries as well are now forming joint R&D ventures with United States companies. If the Japanese are able to utilize American research and development as well as they were able to commercialize the design and process technology bought from the United States, many more American companies will be in serious trouble.

Planning for Manufacturing. Low-cost manufacturing does not just happen. There is a close interdependent relationship between the design of a product, selection of materials, selection of processes and equipment, the design of the processes, and tooling selection and design. Each of these steps must be carefully considered, planned, and coordinated before manufacturing starts. This lead time, particularly for complicated products, may take months, even years, and the expenditure of large amounts of money may be involved. Typically, the lead time for a completely new model of an automobile or a modern aircraft may be four to five years.

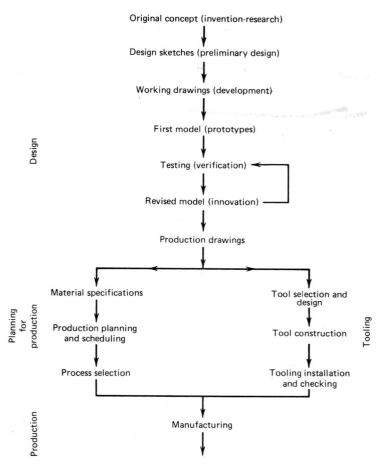

FIGURE 1-18 Traditional steps required to convert an idea into a finished product.

Figure 1-18 shows some of the steps involved in getting one product from the original idea stage to manufacturing. The steps are closely related to each other. For example, the design of the tooling was conditioned by the design of the parts to be produced. It is often possible to simplify the tooling if certain changes are made in the design of the parts or the design of the manufacturing systems. Similarly, the material selection will affect the design of the tooling or the processes selected. Can the design be altered so that it can be produced with tooling already on hand and thus avoid the purchase of new equipment? Close coordination of all the various phases of manufacture is essential if economy is to result. All mistakes and "bugs" should be eliminated during the preliminary phases because changes become more and more costly as work progresses.

With the advent of computers and computer-controlled machines, the integration of the design function and the manufacturing function through the computer is a reality. This is usually called CAD/CAM (computer-aided design/computer-aided manufacturing). The key is a common data base from which detailed drawings can be made for use by both the designer and the manufacturer and from which programs can be generated to make all the tooling. In addition, extensive computer-aided testing and inspection (CATI) of the manufactured parts is taking place. There is no doubt that this trend will continue at ever-accelerating rates as computers become cheaper and smarter.

High Technology Needs a Human Touch.   The most important factor in economical, and successful, manufacturing is the manner in which the resources—men, materials, and capital—are organized and managed so as to provide effective coordination, responsibility, and control. Part of the success of the Toyota system can be attributed to a different management approach. This approach is characterized by a wholistic attitude to people and includes

1. Consensus decision making by management.
2. Decision making at the lowest possible level.
3. Mutual trust, integrity, and loyalty between workers and management.
4. Working in teams or groups.
5. Incentive pay in the form of bonuses for company performance.
6. Stable (even lifetime) employment.
7. Large pool of part-time temporary workers.

There are many companies in the United States that employ some or all of these elements and, obviously, there are many different ways a company can be organized and managed.

The real secret of successful manufacturing lies in designing a simplified system that everyone understands how it works and how it is controlled, with the decision making placed at the correct level. The engineers also must possess a broad fundamental knowledge of design, metallurgy, processing, economics, accounting, and human relations. In the manufacturing game, low-cost mass production is the result of teamwork within an integrated manufacturing production system. This is the key to producing superior quality at less cost with on-time delivery.

Review Questions

1. What role does manufacturing play relative to the standard of living of a country?
2. Are not all goods really consumer goods, depending on how you define the customer? Discuss.
3. Explain the differences between job shop, flow shop, project shop, and cellular shop manufacturing systems.
4. How does a system differ from a process? From a machine tool? From a job? From an operation?
5. Is a cutting tool the same thing as a machine tool? As tooling?
6. What are the major classifications of basic manufacturing processes?

7. Why would it be advantageous if casting could be used to produce a complex-shaped part to be made from a hard-to-machine metal?
8. How is a railroad station like a job station?
9. Since no work is being done on a part when it is in storage, it does not cost you anything. True or false? Explain.
10. List the forming processes used to make a wire coat hanger.
11. Which manufacturing system best describes your college or university? Analogize the elements in this service system to those in the manufacturing system.
12. It is acknowledged that chip-type machining, basically, is an

inefficient process. Yet, it probably is used more than any other to produce desired shapes. Why?

13. What is the level of automation as practiced by the surgeon during an operation in an operating room? (The surgeon is analogous to a plumber who works on body pipes and pieces.)

14. List three purposes of packaging operations.

15. Who invented the first workholders called jigs?

16. Assembly is defined as "the putting together of all the different parts to make a complete machine." Think of (and describe) an assembly process. Is making a club sandwich an assembly process? What about building a house?

17. What kind of assembly process does a slaughter house have?

18. Characterize the process of squeezing toothpaste from a tube—extrusion of toothpaste—using Table 1.2 as a guide line. See Figure 1-12 for assistance.

19. What is a basic difference between mechanization and automation?

20. Give an example of a machine found in the home for each level of automation (See Table 1.4).

21. What difficulties might result if the step "Production Planning and Scheduling" were omitted from the procedure shown in Figure 1-18, assuming a job-shop MS?

22. A company is considering making automobile bumpers from aluminum instead of from steel. List some of the factors it would have to consider in arriving at its decision.

23. Discuss briefly the relationship of design to production.

24. It has been said that low-cost products are more likely to be more carefully designed than high-priced items. Do you think this is true, and why or why not?

25. In a typical metal-cutting job using a conventional machine, which operations add no value to the part?

26. What is a proprietary process?

27. Classify the electric typewriter using the Yardstick for Automation.

28. If the rolls for the cold-rolling mill that produce the sheet metal used in your car cost $300,000, how is it that your car can still cost less than $10,000?

29. Make a list of service production systems, giving an example of each and then explaining the fundamental difference between a SPS and an MPS.

30. In the process of buying a calf, raising it to a cow, and disassembling it into "cuts" of meat for sale, where is "the value added"?

31. What kind of process is chrome plating?

32. Comment on this equation:

$$\frac{\text{Selling price}}{\text{Unit}} = \frac{\text{Manufacturing cost}}{\text{Unit}} + \frac{\text{Profit}}{\text{Unit}}$$

Problems

1. In the manufacture of an $8,000 car, how much do you estimate is spent on direct labor?

2. The average Japanese autoworker makes around $8 to $10 per hour and each car at Toyota has about 30 direct labor hours in it. If the average autoworker in the United States makes $20 per hour, how many man-hours of direct labor are in the typical American car?

3. Using the data in Table A, estimate the percentage of costs of direct labor in the typical medium-sized manufacturing plant. Compare to Figure 1-2.

4. What percentage of costs does power represent in the factories specified in Table A?

5. Suppose the appliance plug in Figure 1-3 were redesigned to be joined by a rivet or by a screw and nut assembly. (You

should be able to find such a redesigned appliance plug at a local discount store.) How much did it cost? How many parts did it have? Make up a "new parts" list and indicate which parts would have to be redesigned and which parts would be eliminated to accommodate the new method of assembly. Es- timate the manufacturing cost of the new plug assuming that manufacturing costs are 40% of the selling price. What is the disadvantage of a riveted plug as opposed to a plug held to- gether by a screw and nut assembly or clips? (Assume the male plug and wire are the same for either design).

TABLE A What It Costs to Operate a Factory

The yearly costs of operating a durable-goods manufacturing plant employing 750 hourly workers, occupying 300,000 square feet and making annual shipments of 33 million pounds to a national market:

Metropolitan Area	Labor Cost (per hour)	Power Cost	Occupancy Cost (millions)	Total Cost (millions)
San Francisco	$13.34	$705,000	$3.3	$32.2
Peoria, Ill.	$13.45	$654,000	$3.4	$31.9
Cleveland	$11.99	$554,000	$3.4	$29.1
Baltimore	$11.43	$639,000	$3.2	$28.1
Minneapolis	$10.96	$472,000	$3.5	$26.9
Los Angeles	$10.47	$733,000	$3.2	$26.3
Chicago	$10.45	$645,000	$3.7	$26.2
Boston	$9.93	$861,000	$3.6	$26.1
Mobile, Ala.	$10.49	$501,000	$2.8	$25.6
Phoenix	$10.18	$590,000	$3.1	$25.4
Atlanta	$10.16	$500,000	$2.9	$25.0
Houston	$10.00	$590,000	$3.0	$24.8
Denver	$10.05	$531,000	$3.3	$24.8
Burlington, Iowa	$8.78	$478,000	$3.2	$22.2

Source: Boyd Co., Princeton, N.J.

Materials

Properties of Materials

In selecting a material, the primary concern of engineers is to match the material properties to the service requirements of the component. Knowing the conditions of load and environment under which the component must operate, engineers must then select an appropriate material, using tabulated test data as the primary guide. They must know what properties they want to consider, how these are determined, and what restrictions or limitations should be placed on their application. Only by having a familiarity with test procedures, capabilities, and limitations can engineers determine whether the listed values of specific properties are, or are not, directly applicable to the problem at hand and then use them intelligently to select a material.

Metallic and Nonmetallic Materials.   Perhaps the most common classification that is encountered in engineering materials is whether the material is *metallic* or *nonmetallic*. The common metallic materials are such metals as iron, copper, aluminum, magnesium, nickel, titanium, lead, tin, and zinc and the alloys of these metals, such as steel, brass, and bronze. They possess the *metallic properties* of luster, high thermal conductivity, and high electrical conductivity; they are relatively ductile; and some have good magnetic properties. Some common nonmetals are wood, brick, concrete, glass, rubber, and plastics. Their properties vary widely, but they generally tend to be less ductile, weaker, and less dense than the metals and have poor electrical and thermal conductivities.

Although metals have traditionally been the more important of the two groups, the nonmetallic group has made great strides, and new nonmetallic materials are continuously being developed. In many cases, metals and nonmetals are now competing materials for the same component. Selection is based on how well each is capable of providing the required properties. Where both can perform adequately, total cost often becomes the deciding factor, where total cost includes both the cost of the material and the cost of fabricating the desired component.

Physical and Mechanical Properties.   One means of distinguishing one material from another is by comparison of *physical properties*. These include such characteristics as density (weight); melting point; optical properties (such as transparency, opaqueness, or color); the thermal properties of specific heat, coefficient of thermal expansion, and thermal conductivity; electrical conductivity; and magnetic properties. In some cases, the physical properties may be of prime importance when one is selecting a material, and some of the more important physical properties will be discussed near the end of this chapter.

More often, however, those properties that describe how a material will respond to applied loads (or forces) will assume the dominant position in material selection. These *mechanical properties* are determined by subjecting prepared specimens to standard laboratory tests designed to evaluate the material's reaction to applied forces. In using the results of such tests, the engineer should remember that they apply only to the specific test conditions that were employed. Actual service conditions of engineered products rarely duplicate the conditions of laboratory testing, so caution should be exercised.

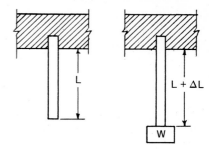

FIGURE 2-1   Tension loading and resultant elongation.

Stress and Strain.   When a load is applied to a component or structure, the material is deformed (*strained*), and internal reactive forces (*stresses*) are produced to resist the applied force. For example, if a weight, W, is suspended from a bar of uniform cross section, as in Figure 2-1, the bar will elongate by an amount ΔL. For a given weight, W, the magnitude of the *elongation*, ΔL, will depend upon the original length of the bar. The amount of deformation of each unit length of the bar, expressed as $e = \Delta L/L$, is called the *unit strain*. Although it is a ratio of a length to another length and is a dimensionless number, it is usually expressed in terms of millimeter per meter or inch per inch, or as a percentage.

Application of the load W also produces reactive stresses within the bar, through which the load is transmitted to the supports. *Stress* is defined as the force or load being transmitted divided by the cross-sectional area transmitting the load. Thus, in Figure 2-1, the stress is $S = W/A$, where A is the cross-sectional area of the supporting bar. It is ordinarily expressed in terms of pounds per square inch (in the English system) or megapascals (in SI units).

In Figure 2-1, the weight tends to stretch or lengthen the bar, so the strain is known as a *tensile strain* and the stress as a *tensile stress*. Other types of loadings produce other types of stress and strain, as illustrated in Figure 2-2. *Compressive forces* tend to shorten the material and produce *compressive stresses and strains*. *Shearing stresses and strains* result from two forces acting on an element of material offset with respect to each other.

Static Properties.   When the loads applied to a material are constant and stationary, or nearly so, they are said to be *static*. In many applications, the load conditions are essentially static, and it becomes important to characterize the behavior of materials under such conditions. Consequently, a number of standardized tests have been developed to determine the *static properties* of materials. The documented test results can then be used to select materials,

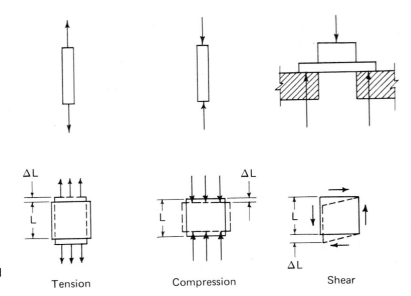

FIGURE 2-2 Examples of tension, compression, and shear loading, and their strain response.

provided that the service conditions are sufficiently similar to those of testing. Even when the service conditions differ from those of testing, the results can be used to qualitatively rate and compare various materials.

The Tensile Test. Considerable information about the properties of a material can be obtained from a uniaxial *tensile test*. A standard specimen is loaded in tension in a testing machine such as the one shown in Figure 2-3. Standard test conditions are used to ensure meaningful and reproducible test results. Standard specimens, the two most common of which are shown in Figure 2-4, are designed to produce uniform uniaxial tension in the central portion and to ensure reduced stresses in the sections that are gripped.

A load, W, is applied and measured by the testing machine, while the elongation (ΔL) or strain over the specified gage length is determined by an external measuring device attached to the specimen. The result is a plot of coordinated load–elongation points, producing a curve of the form in Figure 2-5. Since characteristic loads differ with different-size specimens and elongations vary with different gage lengths, it becomes desirable to remove the size effects and establish a plot that is characteristic of the material's response to the test conditions. If the load is divided by the *original* cross-sectional area and the elongation is divided by the *original* gage length, the size effects are eliminated and the plot becomes known as an *engineering stress-strain curve,* as shown in Figure 2-5. This curve is simply a load–elongation curve with the scales of both axes modified to remove size effects.

In Figure 2-5, it will be noted that, up to a certain stress, the strain is directly proportional to the stress. The stress at which this proportionality ceases to exist is known as the *proportional limit*. Up to this stress, the material obeys *Hooke's law*, which states that stress is directly proportional to strain, and the proportionality constant, or ratio of stress to strain in this region, is known as *Young's*

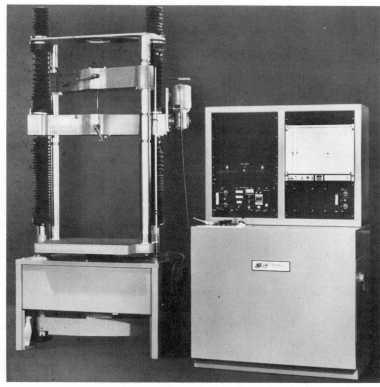

(a)

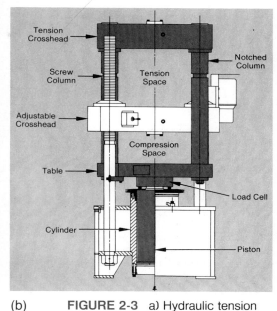

(b)

FIGURE 2-3 a) Hydraulic tension and compression testing machine. b) Schematic of the load frame showing how motion of the darkened yoke can produce tension or compression with respect to the stationary (white) crosspiece. (Courtesy Satec Systems, Inc., Grove City, Pa.)

modulus or the *modulus of elasticity*. This is an inherent and constant property of a given material and is of considerable engineering importance. As a measure of stiffness, it indicates the ability of a given material, for a given cross section, to resist deflection when loaded. It is commonly designated by the symbol E.

Up to a certain stress, if the applied load is removed, the specimen will return to its original length. Thus, from zero stress up to this point, the behavior is elastic and this region of the curve is known as the *elastic region*. The maximum stress for which truly elastic behavior exists is called the *elastic limit*. For some materials, the elastic limit and proportional limit are almost identical. In most cases, however, the elastic limit is slightly higher. Neither quantity, however, should be assigned great engineering significance, for the values are quite dependent upon the sensitivity of the test equipment.

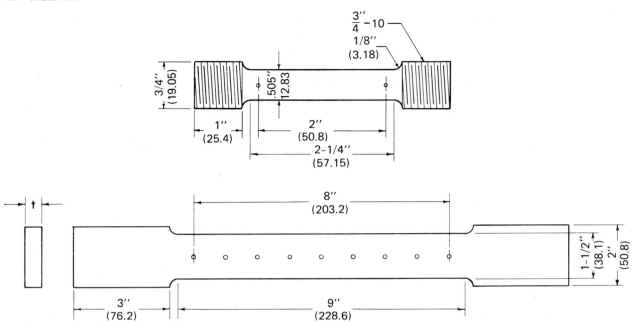

FIGURE 2-4 Two common types of standard tensile test specimens. (Upper) round; (Lower) flat. Dimensions are in inches, with millimeters in parentheses.

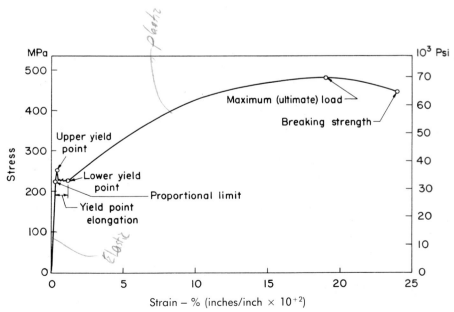

FIGURE 2-5 Engineering stress-strain diagram for a low-carbon steel.

The amount of energy that a unit volume of material can absorb while in the elastic range is called the *resilience* or, in quantitative terms, the *modulus of resilience*. Because energy is the product of force times distance, the area under the load–elongation curve up to the elastic limit is equal to the energy absorbed by the specimen. In dividing load by original area to produce engineering stress, and elongation by gage length to produce engineering strain, the area under the stress–strain curve becomes the energy per unit volume or the modulus of resilience. This energy is potential energy and is therefore released whenever a member is unloaded.

Elongation beyond the elastic limit becomes unrecoverable and is known as *plastic deformation*. Upon removal of all loads, the specimen retains a permanent change in shape. An engineer is usually interested in either the elastic or the plastic response, but rarely both. For most components, plastic flow, except for a slight amount to permit the redistribution of stresses, represents failure of the component through loss of dimensional and tolerance control. However, in manufacturing where plastic deformation is used to shape a product, design stresses must be sufficient to intentionally put the workpiece into the plastic region. Some means is therefore desired to determine the transition from elastic behavior to plastic flow.

Beyond the elastic limit, increases in strain do not require proportionate increases in stress. In some materials, a point may be reached where additional strain occurs without any increase in stress. This point is known as the *yield point* or *yield-point stress*. For low-carbon steels, as in Figure 2-5, two distinct points are significant: the highest stress preceding extensive strain, known as the *upper yield point,* and the lower, relatively constant, "run-out" value, known as the *lower yield point*. The lower value is the one that would appear in tabulated data.

Most materials, however, do not have a well-defined yield point, but have a stress–strain curve of the form shown in Figure 2-6. For such materials, the elastic-to-plastic transition is *defined* by the *offset yield strength*. This is the value of stress that will produce a given and tolerable amount of permanent strain. Deformations used are usually 0.2% or 0.1%, although 0.02% may be used when minute amounts of plastic deformation may lead to component failure. Offset yield strength is then determined by drawing a line parallel to the elastic line, displaced by the offset strain, and reporting the point where it intersects the stress–strain curve, as illustrated in Figure 2-6. The value is reproducible and is independent of equipment sensitivity. However, it is meaningless unless reported with the amount of offset used.

As the straining of the material continues into the plastic range, the material acquires increased load-bearing ability. Since load-bearing ability is equal to strength times cross-sectional area, and the cross-sectional area is decreasing with tensile stretching of the specimen, the material must be increasing in strength. When the mechanism for this strengthening is discussed, in Chapter 3, it will be seen that the strength always continues to increase with deformation. In the tensile test, however, a point is reached where the drop in area with increased strain dominates the increase in strength and the overall load-bearing

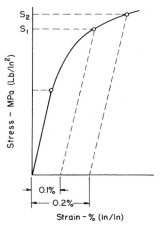

FIGURE 2-6 Stress-strain diagram for a material not having a well-defined yield point, showing the offset method for determining yield strength.

FIGURE 2-7   Standard 0.505-inch-diameter tensile specimen showing a necked region developed prior to failure.

capacity peaks and begins to diminish, as in Figure 2-5. The value of this point on the stress–strain curve is known as the *ultimate strength* or *tensile strength* of the material. The weakest point of the tensile bar at that time continues to be the weakest point by virtue of the decrease in area, and deformation becomes localized. This localized reduction of the cross-sectional area, known as *necking*, is shown in Figure 2-7. It is accompanied by a reduction in the amount of load required to produce additional straining, and the stress–strain curve drops.

If straining is continued far enough, the tensile specimen will ultimately fracture. The stress at which this occurs is called the *breaking strength* or *fracture strength*. For relatively *ductile* materials, the breaking strength is less than the ultimate tensile strength, and necking precedes fracture. For a *brittle* material, fracture usually terminates the stress–strain curve before necking and possibly before the onset of plastic flow.

Ductility and Brittleness.   The extent to which a material exhibits *plasticity* is significant in evaluating its suitability for certain manufacturing processes. Metal deformation processes, for example, require plasticity; the more plastic a material is, the more it can be deformed without rupture. This ability of a material to be deformed plastically without fracture is known as *ductility*.

One of the major ways of evaluating ductility is to consider the *percent elongation* of a tensile-test specimen. As shown in Figure 2-8, however, materials do not elongate uniformly along their entire length when loaded beyond necking. Thus, it has become common practice to report ductility in terms of the percent elongation of a specified gage length. The actual gage length used in the evaluation is of great significance. For the entire 8-inch gage length of Figure 2-8, the elongation becomes 31%. If the center 2-inch segment is considered, elongation becomes 60%. Thus quantitative comparison of material ductility through elongation requires testing of specimens with identical gage lengths.

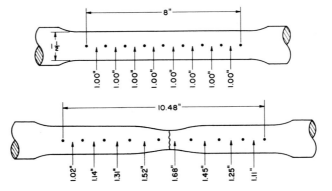

FIGURE 2-8   Final elongation in various segments of a tensile test specimen. The top is the original geometry and the bottom is the shape after fracture.

In many cases, material "failure" is defined as the onset of localized deformation or necking. For example, consider sheet metal being formed into an automobile body component. The operation must be performed in such a way as to maintain uniform thickness, and the material should be evaluated with this in mind. Here, a more meaningful measure of ductility would be *uniform elongation* or *percent elongation prior to necking*. This is determined by constructing a line parallel to the elastic portion of the diagram, passing through the point of highest force or stress. The intercept where the line crosses the strain axis denotes the available uniform elongation. Since the additional elongation that occurs after necking is not considered, uniform elongation is always less than the total elongation at fracture (the generally reported elongation value).

Another indication of ductility is the *percent reduction in area* that occurs in the necked region of the specimen. This is computed as

$$\frac{A_o - A_f}{A_o} \times 100\%$$

where A_o is the original cross-sectional area, and A_f is the smallest area in the necked region, independent of gage length.

Other terms that are often related to the ductility of materials include *malleability, workability,* and *formability*. These terms relate to the ability of a metal to undergo mechanical working processes without rupture. Although plasticity is the controlling property, the terms relate to the material response to specific processes and thus do not describe a material property.

If the material fails with little or no ductility, it is said to be *brittle*. Thus brittleness can be viewed as the opposite of ductility. Brittleness should not be considered as the lack of strength, however, but simply the lack of significant plasticity.

Toughness. *Toughness* is defined as the work per unit volume required to fracture a material and is commonly expressed as a *modulus of toughness*. One means of measuring toughness is through the tensile test, for the total area under the stress–strain curve represents the energy required to produce a fracture in a unit volume of material.

Caution should be exercised in the use of toughness data, however, for the values can vary markedly with different conditions of testing. As will be seen later, variation in temperature and load-application rate can change the nature of a material's stress–strain curve and, hence, the toughness. Toughness is commonly associated with impact or shock loadings. The values obtained from impact tests, however, often fail to correlate with those from static-type tests.

True Stress–True Strain Curves.   The stress strain curve of Figure 2-5 is a plot of *engineering stress, S,* versus *engineering strain, e,* where S is computed as load (W) divided by the *original* cross-sectional area (A_o) and e is the elongation (ΔL) divided by the *original* gage length (L_o). As noted previously and illustrated in Figure 2-7 and 2-8, the cross section of the test bar changes as the test proceeds, first uniformly and then nonuniformly after necking begins.

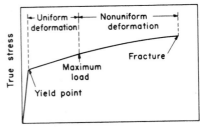

FIGURE 2-9 True stress-True strain curve for an engineering metal.

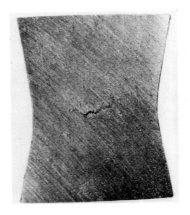

FIGURE 2-10 Section of a tensile test specimen stopped just prior to failure, showing a crack already started in the necked region. (Courtesy E.R. Parker)

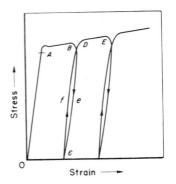

FIGURE 2-11 Stress-strain diagram obtained by unloading and reloading a specimen.

The actual stress within the specimen, therefore, should be based on the instantaneous cross-sectional area and not the original and is greater than the engineering stress shown in Figure 2-5. True stress, σ, can be computed by taking simultaneous readings of load and minimum specimen diameter. The true area can be computed and true stress determined as

$$\sigma = \frac{W}{A}$$

The determination of true strain is somewhat more complex. In place of the change in length divided by the original length that was used to compute engineering strain, true strain is defined as the summation of the incremental strains that occur throughout the test. Thus, for a specimen that has been stretched from length l_o to length l, the *true, natural, or logarithmic strain*, ϵ, would be:

$$\epsilon = \int_{l_o}^{l} \frac{dL}{L} = \ln \frac{l}{l_o} = 2 \ln \frac{D_o}{D}$$

The last equality makes use of the relationship for cylindrical specimens that:

$$\frac{l}{l_o} = \frac{A_o}{A} = \frac{D_o^2}{D^2}$$

Figure 2-9 shows the type of curve that results when the uniaxial tensile test data are plotted in the form of true stress versus true strain. It should be noted that the true stress of the material, a measure of the material strength at that point, continues to rise throughout the test, even after necking. Data beyond the point of necking should be used with extreme caution, however, for once the geometry of the neck begins to form, the stress state in that region becomes a triaxial tension instead of the uniaxial tension assumed for the test. Voids or cracks, such as in Figure 2-10, tend to open in the necked region as a preface to failure. Diameter measurements no longer reflect the true load-bearing area, and the data are further distorted.

Repeated Loading and Unloading.   Let us now use the true stress–true strain diagram of Figure 2-11 to understand how a ductile material, such as steel, behaves when subjected to slow loading and unloading. Unloading and reloading within the elastic region result in simply cycling up and down the linear portion of the diagram between O and A. However, if the initial loading is through point B, unloading from this point follows the path BeC, which is approximately parallel to OA. The specimen now exhibits a permanent strain of OC. Reloading from this point would follow the curve CfD, a slightly different path from that of unloading. The area within the loop formed by the two paths is called the *hysteresis loop* and represents the energy per unit volume that was converted into heat during the unloading and reloading cycle.

Strain Hardening and the Strain-Hardening Exponent.   Upon reloading from point C, elastic behavior is observed through point D. Only when the stress surpasses that of point D does further plastic deformation take place. Thus, point D becomes the new yield point or yield stress, and the material has become stronger than it was in its initial condition. If the test were interrupted at point E, the material would have a new and even higher yield stress. Beyond the elastic region, the true stress–true strain curve actually represents the locus of the yield stress for the various amounts of strain.

When metals are plastically deformed, they *strain-harden* (or *work-harden*); that is, they become harder and the yield stress is raised. This is a progressive phenomenon. As the load is increased to produce plastic deformation, an even greater load will be required to produce further flow. Various materials strain-harden at different rates. That is, for a given deformation, different materials give different increases in strength.

One method of describing this behavior uses the equation

$$\sigma = K\epsilon^n$$

to describe the plastic portion of the true stress–true strain curve. By either a curve fitting operation or by use of a log–log plot of the true stress–true strain data,[1] one can determine the value of n, the *strain hardening coefficient*. As shown in Figure 2-12, a material with a high value of n strain hardens rapidly. Small amounts of deformation induce significant increases in material strength. Materials with a small n value show little change in strength with plastic deformation.

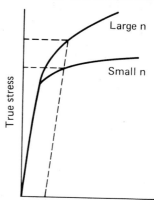

FIGURE 2-12   True stress-true strain curves for metals with large and small strain-hardening. Metals with larger n-values experience larger amounts of strengthening for a given strain.

Damping Capacity.

The hysteresis loop of Figure 2-11 was caused when some of the mechanical energy that was put into the material during the loading and unloading cycle was converted into heat energy. This process produces *mechanical damping*, and materials that possess this property to a high degree are able to absorb mechanical vibrations or damp them out rapidly. This is an important property of materials for certain uses, such as crankshafts and machinery bases. Gray cast iron is used in many applications because of its high damping capacity. Materials with lower damping capacities, such as brass and steel, tend to transmit sound and vibrations.

Hardness.

Hardness is a very important, yet difficult-to-define, property of materials. Numerous tests have been developed around the definition of resistance to permanent indentation under static or dynamic loading. Other tests evaluate resistance to scratching, energy absorption under impact loading, wear resistance, or even resistance to cutting or drilling. Clearly, these phenomena are not the same. Thus, although hardness can be measured by a variety of well-standardized methods, there may be no correlation between the results obtained

---

[1]Taking the logarithms of the equation above results in the equation:

$$\log\sigma = \log K + n\log\epsilon$$

Thus, if true stress–true strain data are plotted on a log–log scale, the slope of the plastic portion will be the value of n.

from the various tests. Caution should be exercised so that the test selected clearly evaluates the phenomenon of interest.

Brinell Hardness Test.   One of the earliest standardized methods of measuring hardness is the *Brinell test*. A hardened steel ball 1 centimeter in diameter is pressed into a smooth surface of material by a standard load of 500, 1500, or 3000 kilograms. The load and ball are removed, and the diameter of the spherical indentation is measured, usually by means of a special grid or a traveling microscope. The Brinell hardness number is equal to the load divided by the spherical surface area of the indentation expressed in kilograms per square millimeter:

$$\text{Brinell hardness number (BHN)} = \frac{\text{load}}{\text{surface area of indentation}}$$

In actual practice, the Brinell hardness number is determined from tables that tabulate the number versus the diameter of the indentation.

The Brinell test is subject to several limitations:

1. It cannot be used on very hard or very soft materials.
2. The test may not be valid for thin specimens. Preferably, the thickness of the material should be at least 10 times the depth of the indentation. Standards specify minimum hardnesses for which tests on thin specimens are valid.
3. The test is not valid for case-hardened surfaces.
4. The test should be conducted on a location far enough removed from the edge of the material so that no edge bulging results.
5. The noticeable indentation may be objectionable on finished parts.
6. The edge of the indentation may not always be clearly defined or may be rather difficult to see on material of certain colors.

Nevertheless, the test is not difficult to conduct and has the advantage that it measures the hardness over a relatively large area and, therefore, does not reflect small-scale variations. It is used to a large extent on iron and steel castings. Figure 2-13 shows a standard Brinell tester. Relatively small, portable testers are also available.

Rockwell Hardness Test.   The widely used *Rockwell hardness test* is similar to the Brinell test in that the hardness value is determined through an indentation produced under a static load. The exact nature of the Rockwell test can be explained in connection with Figure 2-14. A small indenter, either a $1/16$-inch ball or a diamond cone called a *brale,* is first seated firmly against the material by the application of a "minor" load of 10 kilograms. This causes a very slight penetration into the surface. The indicator on the dial of the tester, such as the one shown in Figure 2-15, is then set to zero, and a "major" load is then applied to the indenter to produce a deeper penetration. After the indicating pointer has come to rest, the major load is removed. With the minor load still applied, the tester now indicates the appropriate Rockwell hardness number.

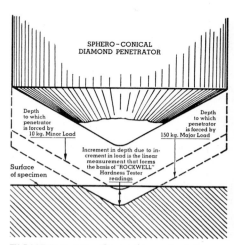

FIGURE 2-13 Brinell hardness tester. (Courtesy Tinius Olsen Testing Machine Co., Inc.)

This number is an indication of the *depth of plastic or permanent penetration* produced by the major load.

Different combinations of major loads and indenters are available and are used for materials of various degrees of hardness. Table 2.1 provides a partial listing of available Rockwell hardness scales. Since multiple scales exist, Rockwell hardness numbers must always be accompanied by a letter indicating the particular combination of load and indenter used in the test. The notation R_C60 (Rockwell C 60) indicates that the brale indenter was used with a major load of 150 kilograms and a reading of 60 was obtained. The B and C scales are used more extensively than the others.

The standard Rockwell tester should not be used on materials less than $\frac{1}{16}$ inch thick, on rough surfaces, or on materials that are not homogeneous, such as gray cast iron. Because of the small size of the indentation, localized variations of roughness, composition, or structure can greatly influence the results. For thinner materials or purposes where a very shallow indentation is desired, the *Rockwell superficial-hardness test* is used. Operating on the same Rockwell principle, the test employs smaller major and minor loads and uses a more sensitive depth-measuring device. This test was designed primarily for determining the hardness of thin sheet metal and the surface hardness of materials that have received such surface treatments as nitriding or carburizing.

In comparison with the Brinell test, the Rockwell test offers the benefit of direct readings in a single step. Because it can be conducted rapidly, it is suitable for routine tests of hardness in mass production. Furthermore, it has the additional advantage that the smaller indentation is often not objectionable to the appearance of the component or is more easily removed in a later operation.

FIGURE 2-14 Operating principle of the Rockwell hardness tester. (Courtesy Wilson Instrument Division, ACCO, Inc.)

FIGURE 2-15 Rockwell hardness tester with digital readout. (Courtesy Page-Wilson Corporation.)

Vickers Hardness Test. The *Vickers hardness test* is similar to the Brinell test, but a square-based pyramid is used as the indenter. As in the Brinell test, the Vickers hardness number is defined as load divided by the surface area of the indentation expressed in kilograms per square millimeter. The advantage of the Vickers approach is the increased accuracy in determining the diagonal of a square as opposed to the diameter of a circle and the assurance that even light loads produce plastic deformation.

Like the other *indentation hardness methods,* the Vickers test offers popular advantages: (1) it is simple to conduct; (2) little time is involved; (3) little surface preparation is required; (4) the test can be done on location; (5) it is relatively inexpensive; and (6) it often provides results that can be used to evaluate material strength, to assess product quality, and so on.

Microhardness Tests. Various microhardness tests have been developed for applications where it is necessary to determine hardness over a very small area of a material. The *Tukon tester,* shown in Figure 2-16, is a machine developed for this purpose. The position for the test is selected under high magnification. A small diamond penetrator is then loaded with a predetermined load of from 25 to 3600 grams. The hardness number, known as a *Knoop hardness number,* is then obtained by dividing the load (in kilograms) by the projected area of the now diamond-shaped indentation (in square millimeters). The length of the indentation is determined by means of a microscope, because the mark is very small. The Vickers test can also be used to determine microhardness.

Durometer Hardness Test. For testing very soft, elastic materials, such as rubbers and nonrigid plastics, a *Durometer* is often used. This instrument, shown in Figure 2-17, measures the resistance of the material to elastic penetration by a spring-loaded conical steel indenter. No permanent deformation occurs. A similar test is used to evaluate molding sands in the casting industry.

Other Hardness Determinations. In the *scleroscope test,* hardness is measured by the rebound of a small, diamond-tipped "hammer" that is dropped from a fixed height onto the surface of the material to be tested. Obviously, this test measures the resilience of a material, and the surface on which the test is made must have a fairly high polish to yield good results. In addition, scleroscope hardness numbers are comparable only among similar materials. A comparison between steel and rubber, therefore, would not be valid.

Hardness has also been defined as the ability of a material to resist being scratched. A crude, but useful, test that employs this principle is the *file test,* wherein one determines whether a material can be cut by a simple metalworking file. The test can be either a pass–fail test using a single file or a semiquantitative evaluation using a series of files that have been pretreated to various levels of known hardness.

Relationships Among the Various Hardness Tests. Since the various tests tend to evaluate somewhat different material phenomena, there is no simple

TABLE 2.1 Loads and Indenters for Rockwell Hardness Tests

Test	Load (kg)	Indenter
A	60	Brale
B	100	$\frac{1}{16}$-in. ball
C	150	Brale
D	100	Brale
F	60	$\frac{1}{16}$-in. ball
G	150	$\frac{1}{16}$-in. ball

relationship between the different types of hardness numbers that can be determined. Approximate relationships have been developed, however, by testing the same material on the various devices. Table 2.2 gives a comparison of hardness values for plain carbon and low-alloy steels. It may be noted that for Rockwell C numbers above 20, the Brinell numbers are approximately 10 times

TABLE 2.2 Hardness Conversion Table

Brinell Number	Vickers Number	Rockwell Number		Scleroscope Number	Tensile Strength 1000	
		C	B		psi	MPa
	940	68		97	368	2537
757[a]	860	66		92	352	2427
722[a]	800	64		88	337	2324
686[a]	745	62		84	324	2234
660[a]	700	60		81	311	2144
615[a]	655	58		78	298	2055
559[a]	595	55		73	276	1903
500	545	52		69	256	1765
475	510	50		67	247	1703
452	485	48		65	238	1641
431	459	46		62	212	1462
410	435	44		58	204	1407
390	412	42		56	196	1351
370	392	40		53	189	1303
350	370	38	110	51	176	1213
341	350	36	109	48	165	1138
321	327	34	108	45	155	1069
302	305	32	107	43	146	1007
285	287	30	105	40	138	951
277	279	28	104	39	134	924
262	263	26	103	37	128	883
248	248	24	102	36	122	841
228	240	20	98	34	116	800
210	222	17	96	32	107	738
202	213	14	94	30	99	683
192	202	12	92	29	95	655
183	192	9	90	28	91	627
174	182	7	88	26	87	600
166	175	4	86	25	83	572
159	167	2	84	24	80	552
153	162		82	23	76	524
148	156		80	22	74	510
140	148		78	22	71	490
135	142		76	21	68	469
131	137		74	20	66	455
126	132		72	20	64	441
121	121		70		62	427
112	114		66		58	

[a]Tungsten carbide ball; others standard ball.

FIGURE 2-16   Microhardness tester. (Courtesy LECO Corporation.)

FIGURE 2-17   Durometer hardness tester. (Courtesy Shore Instrument & Mfg. Company, Inc.)

the Rockwell numbers. Also, for hardnesses below 320 Brinell, the Vickers and Brinell numbers agree closely. Since the relationships will vary with material, mechanical processing, and heat treatment, tables such as Table 2.2 should be used with caution.

Relationship of Hardness to Tensile Strength. Table 2.2 and Figure 2-18 show a comparison of hardness and tensile strength for steel. For plain carbon and very low-alloy steels, the tensile strength (in psi) can be determined fairly well by multiplying the Brinell hardness number by 500. This provides a simple, and very useful, method of determining the approximate tensile strength of a steel by means of a hardness test. For other materials, the relationship will be different and may be too variable to be dependable. For example, the multiplying factor for duraluminum is about 600, whereas for soft brass it is around 800.

Compression Tests. When a material is subjected to compressive forces, the relationships between stress and strain are similar to those for a tension test. Up to a certain value of stress, the material behaves elastically; beyond it, plastic

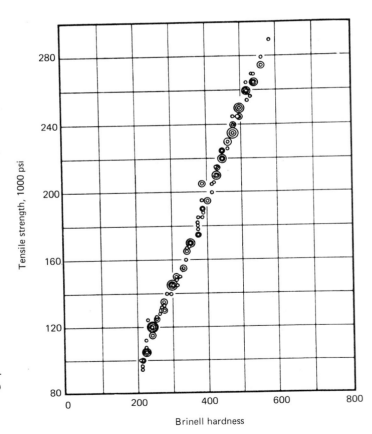

FIGURE 2-18 Relationship of hardness and tensile strength for a group of standard alloy steels. (Courtesy ASM International, Metals Park, Ohio.)

FIGURE 2-19   Failure of wood under compressive loading.

flow occurs. In general, however, the compression test is more difficult and more complex than the standard tensile test. Test specimens must have larger cross-sectional areas to resist bending or buckling. As deformation proceeds, the cross section of the specimen increases, producing a substantial increase in required load (true stress versus true strain is substantially the same for both cases). Frictional effects between the testing machine surfaces and the end surfaces of the specimen tend to alter the results if not properly considered. The selection of the tension or compression mode of testing, however, is largely determined by the type of service to which the material will be subjected.

Failure under compressive loading is generally by buckling or by shear along a plane at 45° to the axis of loading. Figure 2-19 shows a compression failure of a wood specimen.

Dynamic Properties.   In many engineering applications, materials are subjected to dynamic loadings. These may include (1) sudden loads (impacts) or loads that vary rapidly in magnitude; (2) repeated loading and unloading; or (3) frequent changes in the mode of loading, such as from tension to compression. For such operating conditions, the engineer must be concerned with properties other than those determined by the static tests.

Unfortunately, the dynamic tests are not as well standardized or as well controlled as the static tests. In addition, many of the dynamic tests do not give results that can be used directly in design. In these cases, the tests merely permit comparison of materials with respect to the way they respond to certain applied loading conditions. Nevertheless, such tests can serve a very useful purpose, provided one is aware of their limitations.

The Impact Test.   Several tests have been developed to evaluate the fracture resistance of a material under rapidly applied dynamic loads, or *impacts*. Of those tests that have become common, two basic types have emerged: (1) bending impact tests, which include the standard Charpy and Izod tests, and (2) tension impact tests.

The bending impact tests utilize specimens that are supported as beams. In the *Charpy test,* the specimen contains either a V, keyhole, or U notch—the keyhole and V being most common. As show in Figure 2-20, the Charpy test specimen is supported as a simple beam, and the impact is applied to the center, behind the notch, to complete a three-point bending. The *Izod test* specimen is supported as a cantilever beam and is impacted on the end as in Figure 2-21.

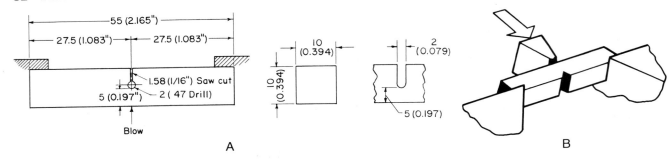

FIGURE 2-20   a) Standard Charpy impact specimens and mode of loading. Illustrated are keyhole and U-notches; dimensions are in millimeters with inches in parentheses.
b) A standard V-notch specimen showing the three-point bending type of impact.

Standard testing machines, such as the one shown in Figure 2-22, apply a predetermined impact energy by means of a swinging pendulum. After breaking or deforming the specimen, the pendulum continues to swing with an energy equal to its original minus that absorbed by the broken specimen. This loss is measured by the angle attained by the pendulum on its upward swing.

Test specimens for bending impacts must be prepared with careful precision to ensure consistent and reproducible results. Notch profile, particularly the radius at the root of V-notch specimens, is extremely critical, for the test measures the energy required to both initiate and propagate a fracture. The effect of notch profile is shown dramatically in Figure 2-23. Here, two specimens have been made from the same piece of steel with the same reduced cross-sectional area. The one with the keyhole notch fractures and absorbs only 43 ft-lb of energy, whereas the other specimen resists fracture and absorbs 65 ft-lb during impact.

Caution should also be used in the application of impact test data for design purposes. Test results apply only to standard specimens containing a standard notch. Changes in the form of the notch or minor variations from the standard specimen geometry can produce significant changes in the results. The test also evaluates material behavior under only one condition of impact rate (speed of the pendulum). Under modified test conditions with faster rates of loading, wider specimens, or sharper notches, many materials lose their energy-absorbing capability and behave in a brittle manner.

The results of bending impact tests, however, are quite valuable in assessing a material's sensitivity to notches and the multiaxial stresses that exist around a notch. In addition, testing can be performed at various temperatures. As will be seen later in this chapter, evaluation of how fracture resistance varies with

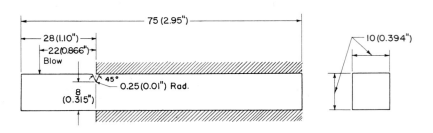

FIGURE 2-21   Izod impact specimen and cantilever mode of loading.

temperature can be valuable input for an individual selecting engineering materials.

The *tensile impact test*, illustrated schematically in Figure 2-24, overcomes many of the objections inherent in the Charpy and Izod tests but is more difficult to perform. With its use, however, the behavior of ductile materials under uniaxial impact loading can be studied without the complications introduced by a notched specimen. The tensile impacts have been supplied by such means as drop weights, modified pendulums, and variable-speed flywheels.

Metal Fatigue and Endurance Limit.

Metals may also fracture when subjected to repeated applications of stress, even though all stresses lie below the ultimate tensile strength and usually below the yield strength as determined by a tensile specimen. This phenomenon, known as *metal fatigue*, may result from the repetition of a particular loading cycle or from an entirely random variation of stress. As such failures probably account for more than 90% of all mechanical fractures, it is important for the engineer to know how materials will react to fatigue conditions.

Although there are an infinite number of possible repeated loadings, the periodic, sinusoidal mode is most suitable for experimental reproduction and subsequent analysis. If one restricts conditions even further by considering only equal-magnitude tension-compression reversals, curves such as that of Figure 2-25 can be developed. If this material were subjected to a normal static tensile test, it would break at about 70,000 psi (480 MPa). However, if it were subjected to a repeated reversing stress of 55,000 psi (380 MPa)—considerably below its breaking stress—it would fail when the loading was repeated about 100,000 times. Similarly, if a reversing stress of 51,000 psi (350 MPa) were applied, 1,000,000 cycles could be sustained before failure. If the applied stress is reduced below 49,000 psi (340 MPa), this steel would not fail regardless of the number of stress applications. Such a curve is known as a *stress-versus-number-of-cycles*, or *S-N*, *curve*. Any point on the curve is the *fatigue strength* corresponding to the given number of cycles of loading. The limiting stress level, below which the material will not fail regardless of the number of cycles of

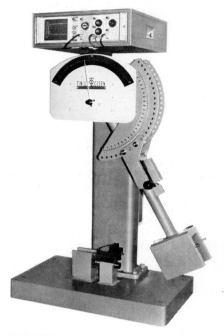

FIGURE 2-22   Impact testing machine. (Courtesy Tinius Olsen Testing Machine Co., Inc.)

FIGURE 2-23   Notched and unnotched impact specimens before and after testing. Both specimens had the same cross-sectional area.

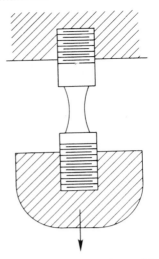

FIGURE 2-24   Tensile impact test schematic.

loading, is known as the *endurance limit* or *endurance strength* and is an important criterion in many design applications.

A different number of cycles of loading is required to reveal the endurance limit of different materials. For steels, about 10 million (10^7) cycles are usually sufficient. Several of the nonferrous metals require 500 million (5×10^8) cycles. Some aluminum alloys require an even greater number, such that no endurance limit is apparent under typical test conditions.

The apparent fatigue strength of materials may be affected by several factors. One of the most important of these is the presence of stress raisers such as small surface cracks, machining marks, or gouges. Data for *S-N* curves are obtained from polished specimens, and the observed lifetime is the cumulative number of cycles required to initiate a fatigue crack and propagate it to failure. If a part contains a surface crack or flaw, the cycles required for crack initiation can be markedly reduced. In addition, stresses concentrate at the tip of the crack producing an accelerated rate of crack growth. Great care should be taken to eliminate stress raisers and surface flaws on parts subjected to cyclic loadings. Proper design and manufacturing practices are often more critical than material selection and heat treatment for fatigue applications.

Another factor worthy of consideration is the temperature of testing. Figure 2-26 shows the shifts in the *S-N* curve for Inconel alloy 625 (Ni-Cr-Fe alloy) as temperature is varied. Since most test data are generated at room temperature, caution should be exercised when the application involves elevated service temperatures.

Fatigue lifetime can also be altered by changes in the environment. When metals are subjected to corrosion during the repeated loadings (a condition known as *corrosion fatigue*), specimen lifetime and the endurance limit are significantly reduced. Special corrosion-resistant coatings, such as zinc or cadmium, may be required for these applications. In addition, tests conducted in a vacuum produce significantly different results from those conducted in air, and further variations have been observed with different levels of humidity. Direct application of test data, therefore, should be done with caution.

Residual stresses in products can also alter fatigue response. If the surface of

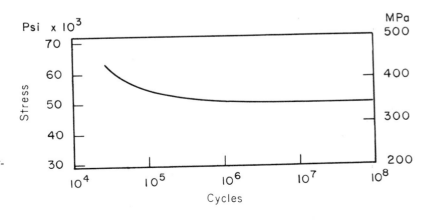

FIGURE 2-25   Typical S-N or endurance limit curve for steel. Specific numbers will vary with the type of steel and treatment.

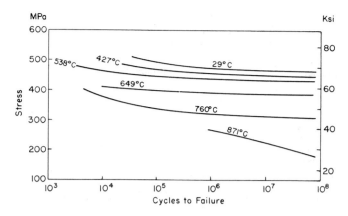

FIGURE 2-26 Fatigue strength of Inconel alloy 625 at various temperatures. (Courtesy Huntington Alloy Products Division, The International Nickel Co., Inc.)

a piece is in a state of compression, as results from shot peening, carburizing, burnishing, or other processing, it becomes more difficult to initiate a fatigue crack, and lifetime is extended. Conversely, processes resulting in residual surface tension, such as welding, can significantly reduce the fatigue lifetime.

If the magnitude of the applied load changes during service (a condition common to many components), the fatigue response of the metal becomes even more complex. Many low-stress cycles may be less damaging to a material than a few high-stress applications. In a contrary manner, if a heavy load stretches and blunts a growing crack, many small loading cycles may be required to "reinitiate" it. Evaluating how materials respond to variations in loading is an area of great importance to design engineers.

Table 2.3 shows the approximate ratio of the endurance limit to ultimate tensile strength for several engineering metals.

TABLE 2.3 Ratio of Endurance Limit to Tensile Strength for Various Materials

Material	Ratio
Steel, AISI 1035	0.46
Steel, screw stock	0.44
Steel, AISI 4140 normalized	0.54
Wrought iron	0.63
Copper, hard	0.33
Beryllium copper (heat-treated)	0.29
Aluminum	0.38
Magnesium	0.38

Fatigue Failures.

Metal components that fail as the result of repeated applications of load and the fatigue phenomenon are commonly called *fatigue failures*. These fractures form a major part of a larger classification known as *progressive fractures*. If the fracture surface of Figure 2-27 is examined closely, two points of fracture initiation can be located. These points usually correspond to discontinuities in the form of a fine surface crack, a sharp corner, machining marks, or even "metallurgical notches" such as an abrupt change in metal structure. Once started, a crack propagates through the metal upon repeated application of load, crack growth being due to the stress at the tip of the crack exceeding the strength of the material. Crack propagation continues until the remaining section of metal no longer has sufficient area to withstand the applied load, at which time complete failure of the remaining section occurs. The section of metal involved in this final failure will have a relatively coarse, granular appearance, whereas the fatigued section between the origin of the crack and the coarse-appearing area will be relatively smooth.

The smooth areas of the fracture will often contain a series of crescent-shaped ridges radiating outward from the origin of the crack. These markings, however, may not be observable under ordinary visual examination. They may be very

FIGURE 2-27 Progressive fracture of an axle within a ball-bearing ring, starting at two points (arrows).

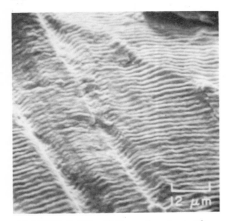

FIGURE 2-28 Fatigue fracture of AISI type 304 stainless steel viewed in a scanning electron microscope at 810X. Well-defined striations are visible. (From "Interpretation of Scanning-Electron-Microscope Fractographs", *Metals Handbook,* Vol. 9, 8th Ed., ASM International, Metals Park, Ohio, 1970, p. 70.)

fine; they may have been obliterated by a rubbing action during repeated cyclic loading; or there may be only a few such marks if failure occurred after only a few cycles of loading ("low-cycle fatigue"). Electron microscope studies of the fracture surface can often reveal these small parallel ridges, or *striations*, which are characteristic of progressive failure. Figure 2-28 presents a typical fatigue fracture at high magnification.

Because the final area of fracture has a crystalline appearance, it has often been said that such failures are due to the metal having "crystallized." Because solid metals are almost always crystalline, such a conclusion is obviously erroneous, and the term should not be applied.

Another common misnomer is to apply the term *fatigue failure* to all fractures having the characteristic progressive-failure appearance. The general appearance of fractures where fatigue is a major factor is often the same as for other fractures where fatigue may be only a minor contributor. Also, the same fracture phenomena may lead to different general appearances depending on the specific conditions of load magnitude, load type (torsional, bending, tension), temperature, and environment. Correct failure analysis requires far more information than can be provided by the examination of a fracture surface.

A final fact regarding failure by fatigue relates to the misconception that failure is time-dependent. The failure of materials under repeated loads below their static strengths is not a function of time but is dependent on the history of loading. High cyclic frequencies can produce failure in relatively short time intervals.

Temperature Effects. It cannot be overemphasized that test data used in design and engineering decisions should be obtained under the conditions that best simulate the conditions of service. Engineers are frequently being confronted with the design of structures, such as aircraft, space vehicles, gas turbines, and nuclear power plants, that require operation under temperatures as low as $-200°F$ ($-130°C$) or as high as $2300°F$ ($1250°C$). Consequently, it is imperative for the designer to know both the short-range and the long-range effects of temperature on the mechanical and physical properties of a material being considered for such applications. From a manufacturing viewpoint, the effects of temperature variations are equally important. Since numerous manufacturing processes involve the use of heat, the processing may tend to alter the properties in a favorable or unfavorable manner. Often a material can be processed successfully, or economically, only because its properties can be changed by heating or cooling.

To a manufacturing engineer, the most important effects of temperature on materials are those relating to the tensile and hardness properties. Figure 2-29 illustrates changes in key data for a medium-carbon steel. Similar effects are shown for magnesium in Figure 2-30. In general terms, an increase in temperature tends to promote a drop in strength and hardness properties and an increase in elongation. For forming operations, these trends are of considerable importance because they permit forming to be done more readily at elevated temperatures where the material is weaker and more ductile.

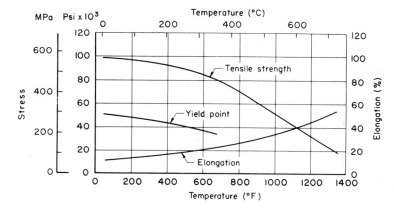

FIGURE 2-29 Some effects of temperature on the tensile properties of a medium-carbon steel.

Figure 2-31 adds another dimension by showing the effects of both temperature and strain rate on the ultimate tensile stress. From this graph, it can be clearly seen that the rate of deformation can strongly influence mechanical properties. Room-temperature, standard-rate tensile test data will be of little use to the engineer concerned with the behavior of a material being hot-rolled at speeds of 5000 feet per minute (1300 meters per minute). The effects of strain rate on the more important yield-strength value are more difficult to evaluate but follow the same trends as tensile strength.

The effect of temperature on impact properties became the subject of intense study when ships, structures, and components fractured unexpectedly in cold environments. Figure 2-32 shows the effect of decreasing temperature on the impact properties of two low-carbon steels. Although of similar compositions, the steels show distinctly different response. The steel indicated by the solid line becomes brittle at temperatures below 25°F (−4°C), whereas the other steel retains its fracture resistance down to −15°F (−26°C). The temperature at which the response goes from high to low energy absorption is known as the *transition temperature* and is useful in evaluating the suitability of materials for certain applications. All steels tend to exhibit the rapid transition in impact

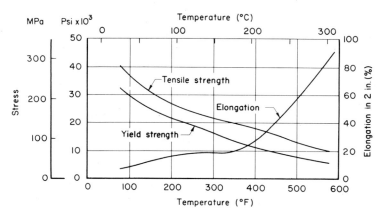

FIGURE 2-30 Effects of temperature on the tensile properties of magnesium.

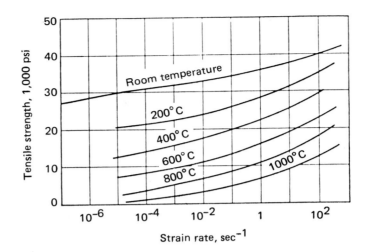

FIGURE 2-31 Effects of temperature and strain rate on the tensile strength of copper. (From A. Nadai and M.J. Manjoine, *J. Appl. Mech.*, Vol. 8, 1941, p. A82, courtesy ASME.)

strength when temperature is decreased, but the temperature at which it occurs varies with the material. Special steels with high nickel contents and several other alloys have been developed for cryogenic applications requiring retention of impact resistance to $-320°F$ ($-195°C$).

Creep.   The long-term effect of temperature is manifest in a phenomenon known as *creep*. If a tensile-type specimen is subjected to a fixed load at an elevated temperature, it will elongate continuously until rupture occurs, even though the applied stress is below the yield strength of the material at the temperature of testing. Although the rate of elongation is small, it is sufficient to be of great importance in the design of equipment such as steam or gas turbines, power plants, and high-temperature pressure vessels that operate at high temperatures for long periods of time.

If a single specimen is tested under fixed load and fixed temperature, a curve such as that of Figure 2-33 is generated. The curve contains three distinct stages: a short-lived initial stage, a rather long second stage of rather linear elongation

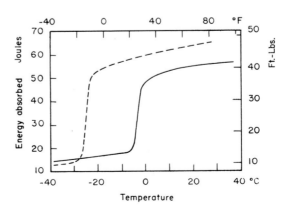

FIGURE 2-32 Effect of temperature on the impact properties of two low-carbon steels.

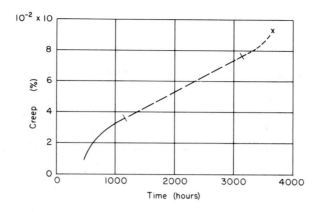

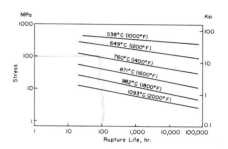

FIGURE 2-33   Creep curve for a single specimen at a fixed temperature, showing the three stages of creep.

rate, and a short-lived third stage leading to fracture. From each test, two significant pieces of engineering data are obtained: the rate of elongation in the second stage, or *creep rate,* and the total elapsed time to rupture. Tests conducted at higher temperatures or with higher applied loads would produce higher creep rates and shorter rupture times.

A very useful engineering tool where creep is a significant factor is a *stress–rupture diagram,* such as the one in Figure 2-34. Rupture-time data from a number of tests at various temperatures and stresses are plotted on a single diagram. Creep-rate data can also be plotted to show the effects of temperature and stress as in Figure 2-35.

In general, the alloying elements of nickel, manganese, molybdenum, tungsten, vanadium, and chromium are helpful in lowering the creep rate of steel. At high temperatures, coarse-grained steels seem to be more creep-resistant than fine-grained steels. Below the lowest recrystallization temperature (a property discussed in Chapter 3), however, the reverse is true, with a fine-grained structure being preferred. Killed steels show superior creep-resisting properties when compared to rimmed steels.

FIGURE 2-34   Stress-rupture diagram of a solution-annealed Incoloy alloy 800 (Fe-Ni-Cr alloy). (Courtesy Huntington Alloy Products Division, The International Nickel Company, Inc.)

Machinability, Formability, and Weldability.

Although many individuals assume that the terms *machinability, formability,* and *weldability* refer to specific material properties, they actually refer to the way a material responds to specific processing techniques and are quite difficult to define. *Machinability,* for example, depends not only on the material being machined, but also on the specific process and the aspects of the process that are of greatest interest. In some cases, one is interested in how easy or fast a metal can be cut, irrespective of the resulting surface finish. In another application, surface finish may be of prime importance. For some processes, the formation of fine chips may be a desirable feature. In others, tool life may be very important. Thus, the term *machinability* may involve several properties of a material, each of varying importance to different individuals.

In a similar manner, *malleability, workability* and *formability* all refer to a material's suitability for plastic deformation processes. However, materials be-

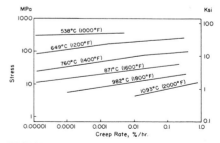

FIGURE 2-35   Creep-rate properties of a solution-annealed Incoloy alloy 800. (Courtesy Huntington Alloy Products Division, The International Nickel Company, Inc.)

have differently at different temperatures. A material with good "hot formability" may behave poorly when deformed at room temperature. Moreover, some materials that flow nicely at low deformation speeds behave in a brittle manner when loaded at rapid rates. Thus, formability must be evaluated for a particular combination of material, process, and process conditions.

Likewise, the "weldability" of a material depends upon the specific welding or joining process being considered.

Fracture Toughness and the Fracture Mechanics Approach.

A discussion of the mechanical properties of material would not be complete without mention of the many tests and design concepts based on the fracture mechanics approach. Instead of treating test specimens as flaw-free materials, fracture mechanics begins with the premise that *all materials contain flaws* or defects of some given size. These may be *material defects,* such as pores, cracks, or inclusions; *manufacturing defects,* in the form of machining-tool marks, arc strikes, or contact damage to external surfaces; or *design defects,* such as abrupt section changes, excessively small fillet radii, and holes. When the specimen is then loaded, the applied stresses are amplified or intensified in the vicinity of these defects, and the result is accelerated failure or failure under unexpected conditions.

Fracture mechanics seeks to identify the conditions under which a defect will grow or propagate to failure and, if possible, the rate of crack or defect growth. In application, the methods concentrate on three principal quantities: (1) the size of the largest or most critical flaw; (2) the applied stress; and (3) the *fracture toughness,* a quantity that describes the fracture resistance of the material or the conditions necessary to induce crack growth. Equations have been developed to relate these three quantities at the onset of crack growth or propagation. Various specimen geometries and flaw locations and orientations have been considered. If nondestructive testing or quality control checks have been applied, the size of the largest flaw that could go undetected is often known. Assuming this flaw to be in the worst possible location and orientation, and knowing the applied stress at that location, the designer can then determine the necessary fracture toughness to prevent this flaw from propagating during service. Conversely, if the material and stress conditions were defined, the size of the maximum permissible flaw could be determined. Inspection conditions could be selected to ensure that flaws greater than this magnitude would not exist. Finally, if a component is found to have a significant flaw and the material is known, then the maximum operating stress can be determined that will ensure no further growth of that flaw.

Thus, according to the philosophy of fracture mechanics, all materials contain flaws or defects, but these defects can be either dormant or growing. Dormant defects are permissible, and the goal of fracture mechanics is to define the conditions under which a given defect will remain dormant. In one 1983 study, it was determined that fracture of materials costs over $119 billion per year in the United States alone. Over 80% of this cost was associated with current methods of fracture prevention, such as overdesign, excessive inspection, and the use of premium quality materials. In addition, a previously mentioned study concluded that over 90% of all dynamic failures of materials were in some way attributable to fatigue. In contrast to the previous method of fatigue testing, where polished specimens were cycled and the lifetime consisted of both crack initiation and crack propagation, fracture mechanics focuses on the growth rate of an already defective material (containing a crack or flaw of known size and shape). The results are far more realistic.

Fracture mechanics is a truly integrated blend of design (applied stresses), inspection (flaw size determination), and materials (fracture toughness). The approach has proved valuable in many areas where fracture could be catastrophic and has shown great refinement and increased acceptance in recent years. The 1983 fracture study concluded that over 29% of the $119 billion cost of fracture ($34.5 billion per year) could be saved with the application of existing technology, such as fracture mechanics.

Physical Properties.   In some engineering applications, the physical properties of an engineering material may be more important than the mechanical properties. For this reason, it is important to become familiar with the types of physical properties that are often evaluated.

In addition to the previously discussed responses of material to variations in temperature, there are three more *thermal properties* that are worthy of consideration. The *heat capacity,* or *specific heat,* of a material is a measure of the amount of energy that must be imparted or extracted to produce a 1-degree change in temperature. This would be quite significant in processes such as casting, where heat must be extracted rapidly to promote solidification, and heat treatment, where quantities of material are heated and cooled. *Thermal conductivity* measures the rate at which heat can be transported or conducted through a material. Although this is often tabulated in reference texts, it is helpful to remember that for metals, thermal conductivity is directly proportional to electrical conductivity. Thus, materials that have good electrical conductivity, such as copper, gold, and aluminum, will also be good transporters of thermal energy. *Thermal expansion* is the final property of significance. Most materials expand on heating and contract on cooling, but the degree of expansion or contraction varies with the material. For components that are machined at room temperature but put in service at elevated temperatures, or castings that solidify at elevated temperatures and then cool, corrections must be incorporated in the manufacture to compensate for the dimensional changes.

Electrical conductivity or resistivity is often an important design consideration and should be evaluated in terms of the material and the variations in the property with levels of purity or changes in temperature.

From the standpoint of *magnetic response,* materials are often classified as diamagnetic, paramagnetic, ferromagnetic, antiferromagnetic, or ferrimagnetic. These terms refer to the way in which the material responds to an applied magnetic field. Terms such as *saturation strength, remanence,* and *magnetic hardness* or *softness* refer to the strength, duration, and nature of this response and can be used to compare and evaluate materials.

Other physical properties of possible importance include *weight or density, melting and boiling points,* and the various *optical properties,* such as the ability to transmit, absorb, or reflect light or other electromagnetic radiation.

Review Questions

1. What are some of the properties commonly associated with metallic materials?
2. What are some of the common physical properties of materials?
3. How are "standard" mechanical properties determined?
4. What are the standard units of reporting or measuring stress and strain in the English system? In the metric system?
5. Stresses and strains are often characterized by the type of forces that produced them. What are some of these categories?
6. What are some of the important properties that relate to the elastic response of materials?
7. What are some of the tensile test properties that relate to the plastic response of materials?
8. What is the physical or mechanical significance of the ultimate strength or tensile strength as measured in a tensile test?
9. Describe two methods of using the tensile test as a means of measuring the ductility of a material.
10. Why might the uniform elongation or percent elongation before necking be a more meaningful measure of useful ductility?
11. Is a brittle material necessarily weak?
12. What is the "toughness" of a material?
13. What is the difference between true stress and engineering stress? True strain and engineering strain?
14. What is strain hardening or work hardening? How might this phenomenon be measured or reported?
15. What are some of the various material characteristics or responses that have been associated with the term *hardness?*
16. What are some of the limitations of the Brinell hardness test?
17. Both the Brinell and Rockwell hardness tests are penetration tests. What would you cite as the major difference between the two methods?
18. What is the purpose of the various microhardness tests?

19. Why might there be a lack of agreement when materials are compared by means of various types of hardness tests?
20. Why is an accurate compression test more difficult to conduct than an accurate tensile test?
21. What are the two most common bending impact tests? How are the specimens supported and loaded during the impact?
22. Why should a designer use extreme caution when applying impact test data for design purposes?
23. By what phenomena can metals fracture when exposed to stresses that lie below their yield strength?
24. *Fatigue strength* and *endurance limit* are both terms that are derived from *S-N* diagrams. Define these terms and describe how they can be determined from a diagram.
25. What are some of the factors that can alter the fatigue lifetime or fatigue behavior of a material?
26. Fatigue markings are caused by the propagation of a crack through a material. What are some of the features that may be responsible for the initiation of a fatigue crack?
27. Why is it important for a designer or engineer to know how a material's properties will change with temperature?
28. Why should one use caution when employing a steel at low (below 0° Fahrenheit) temperatures?
29. What are some of the evaluation tools and quantities that can be used to assess the long-term effect of elevated temperature?
30. Under what conditions of temperature might one prefer a coarse-grained material over one that is fine-grained?
31. Why is there not a single standard means of assessing characteristics like machinability, formability, or weldability?
32. What is the basic premise of the fracture mechanics approach to testing and design?
33. What are the three principal quantities that fracture mechanics tries to relate?
34. What are the three primary thermal properties of a material and what do they measure?

The Nature of Metals and Alloys

The Structure–Property Relationship. The fundamental engineering *properties* of materials presented in Chapter 2 are the direct result of the *structure* of the particular material. Moreover, such properties as strength and ductility are often sensitive to minute variations of structure, some of which are macroscopic, others microscopic, and still others on the atomic scale. To control the properties of materials and intelligently use them to their optimum, the engineer must first have a working knowledge of material structure.

The Basic Structure of Materials. As all materials are composed of the same basic components—*protons, neutrons,* and *electrons*—it is amazing that so many different materials exist with such widely varying properties. This variation is explained by the many possible combinations of these units in a macroscopic assembly. The subatomic components, listed above, combine in different arrangements to form the various elemental *atoms,* each having a *nucleus* of protons and neutrons surrounded by the proper number of electrons to maintain charge neutrality. Atoms then combine in distinctive arrangements to form *molecules* or *crystals*. These units can then be assembled in differing amounts and configurations to form a microscopic-scale structure, or *microstructure,* and ultimately an engineering component. The engineer, therefore, has at his or her disposal a wide variety of metals and nonmetals that possess an almost unlimited range of properties. Because the properties of materials depend upon all levels of structure, as shown schematically in Figure 3-1, it is important for the engineer to understand the entire structure spectrum, from atomic to macroscopic.

Atomic Structure. Experiments have revealed that atoms consist of a relatively dense nucleus composed of positively charged protons and neutral particles of nearly identical mass, known as *neutrons*. Surrounding the nucleus are

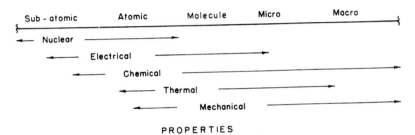

FIGURE 3-1   General relationship of the structural level to various engineering properties.

the negatively charged electrons, which have only $\frac{1}{1839}$ the mass of a neutron and appear in numbers equal to the protons to maintain a net charge balance. Distinct particle groupings produce the known elements, ranging from the relatively simple hydrogen atom to unstable transuranium atoms over 250 times as heavy. Except for density and specific heat, however, the weight of atoms has relatively little influence on engineering properties.

The light electrons that surround the nucleus play a far more significant role in determining properties. Again, experiments reveal that the electrons are arranged in a characteristic structure consisting of shells and subshells, each possessing a distinctive energy. Upon absorbing a small amount of energy, an electron can jump from a low-energy shell near the nucleus to a higher-energy shell farther out. The reverse jump can occur with the release of a distinct amount, or *quantum*, of energy.

Each of the various shells and subshells can contain only a limited number of electrons. The first shell, nearest the nucleus, can contain only 2. The second shell can contain 8; the third, 32. Each shell and subshell is most stable when it is completely filled. For atoms containing electrons in the third shell and beyond, however, relative stability is associated with 8 electrons in the outermost layer.

If, in its outer shell, a normal atom has slightly less than the number of electrons required for stability (for example, seven in its third shell), it will readily accept an electron from another source. It will then have one electron more than the number of protons and becomes a negatively charged atom, or *negative ion*. Extra electrons may cause the formation of ions having negative charges of 1, 2, 3, and more. If an atom has a slight excess of electrons beyond the number required for stability (such as sodium, with one electron in the third shell), it will readily give up the excess electron and become a *positive ion*. The remaining electrons become more strongly bound, so that the removal of electrons is progressively more difficult.

The number of electrons surrounding the nucleus of a neutral atom is called the *atomic number*. More important, however, are those electrons in the outermost shell (or subshell) known as *valence electrons*. These are influential in determining chemical properties, electrical conductivity, some mechanical properties, the nature of interatomic bonding, atom size, and optical characteristics.

Elements with similar electron configurations in their outer shells will tend to have similar properties.

Atomic Bonds. Atoms are rarely found as free and independent units; they are usually linked or bonded to other atoms in some manner as a result of interatomic forces. The electronic structure of the atoms influences the nature of the bond, which may be classed as *primary* (strong) or *secondary* (weak).

The simplest type of primary bond is the *ionic bond*. Electrons break free from atoms with excesses in their valence shell, producing positive ions, and unite with atoms having an incomplete outer shell to form negative ions. The positive and negative ions have a natural attraction for each other, producing a strong bonding force. Figure 3-2 illustrates the process for a bond between sodium and chlorine. In the ionic type of bonding, however, the atoms do not unite in simple pairs. All positively charged atoms attract all negatively charged atoms. Thus, for example, sodium ions surround themselves with negative chlorine ions, and chlorine ions surround themselves with positive sodium ions. The attraction is equal in all directions and results in a three-dimensional structure, such as in Figure 3-3, rather than the simple link of a single bond. For stability in the structure, total charge neutrality must be maintained, so that equal numbers of positive and negative charges are required. General characteristics of materials joined by ionic bonds include moderate to high strength, high hardness, brittleness, high melting point, and electrical insulating properties (all charge transport must be through ion movement).

A second type of primary bond is the *covalent* type. Here the atoms being linked find it impossible to produce completed shells by electron transfer but achieve the same goal by electron sharing. Adjacent atoms share outer-shell electrons so that each achieves a stable electron structure. The shared negative electrons locate between the positive nuclei to form the bonding link. Figure 3-4 illustrates this type of bond for chlorine, where two atoms, each containing seven valence electrons, share a pair to form a stable *molecule*. Stable molecules can also form from the sharing of more than one electron from each atom, as in the case of nitrogen in Figure 3-5a. Atoms need not be identical (as in HF in Figure 3-5b), the sharing need not be equal, and an atom may share electrons with more than one other atom. For elements such as carbon (four valence electrons), one atom may share valence electrons with each of four neighboring carbon atoms. The resulting structure becomes a network of bonded atoms (Figure 3-5c) instead of a finite, well-defined molecule. Like the ionic bond, the covalent bond tends to produce materials with high strength and high melting points. Atom movement within the material (deformation) requires the breaking of distinct bonds, thereby making the material characteristically brittle. Electrical conductivity depends upon bond strength, ranging from conductive tin (weak covalent bond) through semiconductive silicon and germanium to insulating diamond. Engineering materials possessing ionic and covalent bonds tend to be ceramic (refractories or abrasives) or polymeric in nature.

A third type of primary bond can result when a complete outer shell cannot be formed by either electron transfer or electron sharing, and is known as the

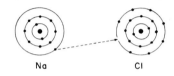
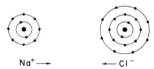

FIGURE 3-2 Ionization of sodium and chlorine, producing stable outer shells by electron transfer.

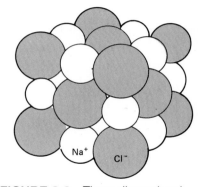

FIGURE 3-3 Three-dimensional structure of the sodium chloride crystal. Note how the various ions are surrounded by ions of the opposite charge.

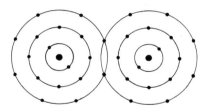

FIGURE 3-4 Formation of a chlorine molecule by a covalent bond.

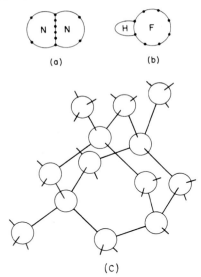

FIGURE 3-5   Examples of covalent bonding in N_2, HF, and diamond.

metallic bond. If there are only a few valence electrons (one, two, or three) in each of a grouping of atoms, these electrons can be removed easily while the remainder are held firmly to the nucleus. The result is a structure of positive ions (nucleus and nonvalence electrons) surrounded by a wandering assortment of universally shared valence electrons (electron cloud or gas), as in Figure 3-6. These highly mobile "free" electrons account for the high electrical and thermal conductivity as well as the opaque optical properties (free electrons can absorb light radiation energies) observed in metals. Moreover, they provide the "cement" necessary to produce the positive–negative–positive attractions necessary for bonding. Bond strength, and therefore material strength, varies over a wide range. More significant, however, is the observation that the positive ions can move within the structure without the breaking of distinct bonds. Materials bonded by metallic bonds can therefore be deformed by atom movement mechanisms and produce a deformed material every bit as strong as the original. This is the basis of metal plasticity, ductility, and many of the shaping processes used in metal fabrication.

Secondary Bonds.   Weak or secondary bonds, known as *van der Waals forces*, can link molecules that possess a nonsymmetrical distribution of charge. Some molecules, such as hydrogen fluoride and water, can be viewed as electric dipoles, in that certain portions of the molecule tend to be more positive or negative than others (an effect referred to as *polarization*). The negative part of one molecule tends to attract the positive part of another to form a weak bond.

Another weak bond can result from momentary polarization caused by random movements of the electrons and the resulting momentary electrical unbalance. This random and momentary polarization leading to attractive forces is called the *dispersion effect.*

A third type of weak bond is the *hydrogen bridge,* where a small hydrogen nucleus, simultaneously attracted to the negative electrons of two different atoms, forms a three-atom link. Such bonds play a significant role in biological systems but, like all secondary bonds, are rarely of engineering significance.

Interatomic Distances.   Since the space occupied by the electron shells is very large compared to the size of the actual electrons, most of the total volume of an atom is vacant space, and the "size of an atom" becomes somewhat

FIGURE 3-6   Schematic of the metallic bond showing the positive ions and associated electron "cloud" for the case of copper.

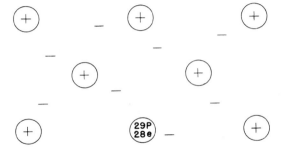

undefined. The interatomic or bonding forces tend to pull the atoms together; at the same time, there are repelling forces between the positive nuclei. Some equilibrium distance exists where the forces of attraction and repulsion are equal.

Thus, atoms can be assigned a distinct size, the equilibrium distance between the centers of two neighboring atoms being considered the sum of the atomic radii. The atomic radius is not a constant, however. Added thermal energy causes the atoms to vibrate about their equilibrium positions, transmitting the energy through the material (thermal conductivity). The repulsive force as atoms approach is much greater than the attractive force when they are separated. Thus, the vibrations, or oscillations, are nonsymmetrical, and the atoms are at larger than normal separations for a greater fraction of time. The result is a phenomenon that we observe macroscopically as thermal expansion. Removal of electrons from the outer shell will decrease the radius, and the addition of electrons will increase it. Consequently, a negative atomic ion is larger than its base atom, and a positive ion is smaller. Atomic radius also changes with the number of adjacent or nearest-neighbor atoms. With more neighbors, there is less attraction to any single neighbor atom, and the interatomic distance is increased. Thus, iron has a slightly different atomic size in its two different crystal forms, as will be discussed in a later chapter.

Atom Arrangements in Materials.   At the next level of material structure, we find that the arrangement of atoms in a material has a significant effect on its properties. Depending on the manner of atomic grouping, materials are classified as having *molecular structures, crystal structures,* or *amorphous structures.*

Molecular structures have a distinct number of atoms that are held together by primary bonds, but they have only relatively weak bonds with other similar groups of atoms. Typical examples of molecules include O_2, H_2O, and C_2H_4 (ethylene). Each molecule is free to act more-or-less independently, so that these materials exhibit relatively low melting and boiling points. Molecular materials tend to be weak, as the molecules can easily move past one another. Upon changes of state from solid to liquid or gas, the molecules remain intact as distinct entities.

Crystal structures are assumed by solid metals and most minerals. Here atoms are arranged in a regular geometric array known as a *space lattice*. These lattices are describable through a unit building block that is essentially repeated throughout space in a periodic manner. Such blocks are known as *unit cells*.

In amorphous structures, such as glass, the atoms have a certain degree of local order but, when viewed as an aggregate, have a more disorganized atom arrangement than the crystalline solids.

Crystal Structure of Metals.   From a manufacturing viewpoint, metals are the most important class of materials. Most often, they are the materials being processed and are used in the machines performing the processing. Consequently, in order to perform manufacturing operations intelligently, it is essential to have a basic knowledge of the fundamental nature of metals and their behavior when subjected to mechanical or thermal treatment.

TABLE 3.1 Types of Lattices of Common Metals at Room Temperature

Metal	Lattice Type
Aluminum	Face-centered cubic
Copper	Face-centered cubic
Gold	Face-centered cubic
Iron	Body-centered cubic
Lead	Face-centered cubic
Magnesium	Hexagonal
Silver	Face-centered cubic
Tin	Body-centered tetragonal
Titanium	Hexagonal

More than 50 of the known chemical elements are classed as metals, and about 40 have commercial importance. These materials are characterized by the metallic bond and possess certain distinguishing characteristics: strength, good electrical and thermal conductivity, luster, the ability to be deformed permanently to a fair degree without fracturing, and a relatively high specific gravity compared with nonmetals. The fact that some metals possess properties different from the general characteristics simply expands their engineering utility.

When metals solidify, they assume a crystalline structure; that is, the atoms arrange themselves in a space lattice. Most metals exist in only one lattice form. A few, however, can exist in the solid state in two or more lattice forms, the particular form depending on the conditions of temperature and pressure. These metals are said to be *allotropic,* and the change from one lattice form to another is called an *allotropic change.* The most notable example of such a metal is iron, where the property makes possible the use of heat-treating procedures to produce a wide range of characteristics. It is largely because of its allotropy that iron is the base of our most important alloys.

Metals are known to solidify into 14 different crystal structures. However, nearly all of the important commercial metals solidify into one of three types of lattices, these being body-centered cubic, face-centered cubic, and hexagonal close-packed. Table 3.1 shows the lattice structure of a number of common metals at room temperature. Figure 3-7, compares these structures to each other and to the easily visualized, but rarely observed, simple cubic structure.

The simple cubic structure of Figure 3-7a can be constructed by placing single atoms at all corners of a cube and subsequently linking identical cubes together. Assuming that the atoms are spheres with atomic radii touching each other, computation reveals that only 52% of available space is occupied. Each atom has only six nearest neighbors. Both of these observations are unfavorable to the metallic bond, where atoms desire the greatest number of nearest neighbors and high-efficiency packing.

If the cube is expanded somewhat to allow the insertion of an additional atom in the center, the *body-centered-cubic* (bcc) structure results, as in Figure 3-7b. Each atom now has eight nearest neighbors, and 68% of the space is occupied. Such a structure is more favorable to metals and is observed in the elements Fe, Cr, Mn, and the others listed in Figure 3-7b.

If Ping-Pong balls, used to simulate atoms, were placed in a box and agitated until a stable arrangement was produced, we would find the structure to consist of layered *close-packed planes,* where each plane looks like Figure 3-8. Two different structures can result, depending upon the sequence in which the various planes are stacked, but both are identical in nearest neighbors (12) and efficiency of occupying space (74%).

One of these sequences produces a structure that can be viewed as an expanded cube with an atom inserted in the center of each face, the *face-centered-cubic* (fcc) structure of Figure 3-7c. Such a structure occurs in many of the most important engineering metals and tends to produce high formability (the ability to be permanently deformed without fracture).

The second of the structures is known as *hexagonal-close-packed* (hcp), where

	Lattice Structure	Unit Cell Schematic	Ping-Pong Ball Model	Number of Nearest Neighbors	Packing Efficiency	Typical Metals
a	Simple cubic	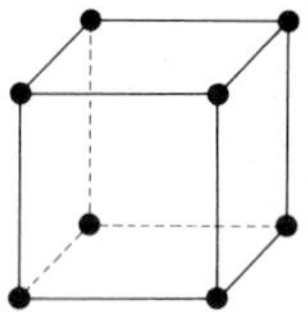		6	52%	None
b	Body-centered cubic	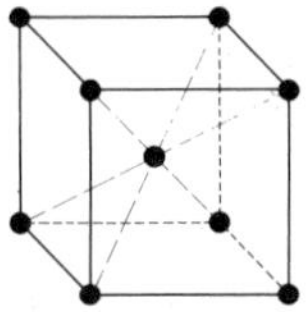		8	68%	Fe, Cr, Mn, Cb, W, Ta, Ti, V, Na, K
c	Face-centered cubic	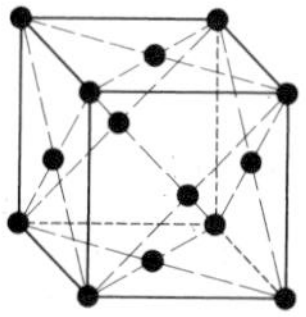		12	74%	Fe, Al, Cu, Ni, Ca, Au, Ag, Pb, Pt
d	Hexagonal close-packed	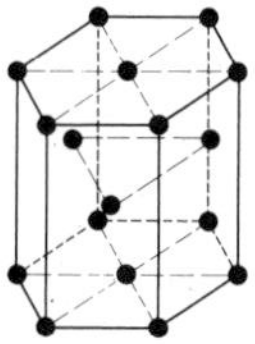		12	74%	Be, Cd, Mg, Zn, Zr

FIGURE 3-7 Comparison of crystal structures: simple cubic, body-centered cubic, face-centered cubic, and hexagonal close-packed.

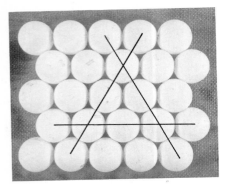

FIGURE 3-8 Close-packed atomic plane showing three directions of closest packing.

the close-packed planes can be clearly identified (see Figure 3-7d). Metals having this structure tend to have poor formability and often require special processing procedures.

Development of Metallic Grains.   As metals solidify, a small particle of solid forms in the liquid, having the lattice structure characteristic of the given material. This particle then acts as a *seed* or *nucleus* onto which other atoms in the vicinity tend to attach themselves. The resulting arrangement is a crystal composed of repetitions of the same basic pattern throughout space, as illustrated in Figure 3-9.

In actual solidification it is expected that many seed or nuclei particles would form independently at various locations in the liquid mass and have random orientations. Each then grows until it begins to interfere with its neighbors, as illustrated in two dimensions in Figure 3-10. Since adjacent lattice structures have different alignments, growth cannot produce a single continuous structure. The small continuous segments of solid are known as *crystals* or *grains,* and the surfaces that divide them (i.e., the surfaces of crystalline discontinuity) are known as *grain boundaries*. The process through which the grain structure is produced is one of *nucleation and growth*.

Grains are the smallest structural units of metal that are observable with ordinary light microscopy. If a piece of metal is polished to a mirror finish with a series of abrasives and then exposed to an attacking chemical for a short time (etched), the grain structure can be seen. The atoms on the grain boundaries are more loosely bonded and tend to react with the chemical more readily than those that are part of the grain interior. When subsequently viewed under reflected light, the attacked boundaries appear dark compared to the relatively unaffected (still flat) grains, as in Figure 3-11. Occasionally, grains are large enough to be seen by the naked eye, as on some galvanized steels, but usually magnification is required.

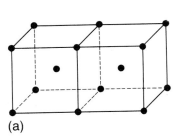

(a)

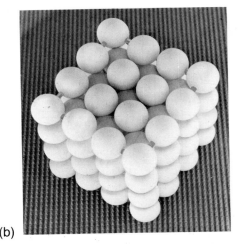

(b)

FIGURE 3-9 Growth of crystals to produce an extended lattice. (a) Line schematic; (b) ping-pong ball model.

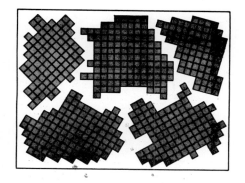 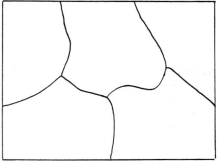

FIGURE 3-10   Schematic representation of the growth of crystals to produce a polycrystalline material.

The number and size of the grains in a metal are a function of two factors: the rate of nucleation and the rate of growth. The greater the rate of nucleation, the smaller the resulting grains. Similarly, the greater the rate of growth, the larger the grain size. Because the overall *grain structure* will influence certain mechanical and physical properties, it is an important property for the engineer to be able to both control and specify. One such specification scheme is the ASTM *grain-size number,* defined by

$$N = 2^{n-1}$$

where $N =$ number of grains per square inch visible in a
          prepared specimen at $100 \times$

    $n =$ ASTM grain-size number

High numbers correspond to smaller grain sizes. Materials with ASTM[1] grain size 7–9 are often desired when good formability is required.

Elastic Deformation of Single Crystals.   To a great extent, the mechanical behavior of materials depends on their crystal structure. Therefore, to understand mechanical behavior, it is important for the engineer to have some understanding of the way crystals react when subjected to mechanical loading. Much of what is known about this subject has been obtained from the study of carefully prepared single crystals that may be several inches long. In general, observation reveals that the behavior of metal crystals depends on (1) the lattice type; (2) the interatomic forces; (3) the spacing between planes of atoms; and (4) the density of atoms on various planes.

If the applied loads are relatively low, the crystal responds by simply stretching or compressing the distance between atoms, as in Figure 3-12. The lattice unit does not change, and the atoms retain their basic positions. The applied load serves only to disrupt the force balance of the atomic bonds so as to transmit the applied load through the body. When the load is removed, balance is restored and the lattice resumes its original size and shape. The response to such loads

FIGURE 3-11   Photomicrograph of alpha ferrite (essentially pure iron) showing grain boundaries; 1000X. (Courtesy USX Corporation.)

[1]ASTM = American Society for Testing and Materials.

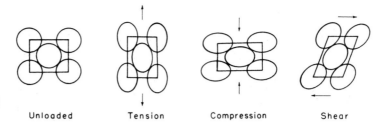

FIGURE 3-12 Distortion of a crystal lattice in response to elastic loadings.

Unloaded   Tension   Compression   Shear

is *elastic* in nature. The amount of stretch or compression (strain) is proportional to the applied load or stress.

Elongation or compression in one direction in response to an applied force also produces an opposite change in dimensions at right angles to that force. The ratio of lateral contraction to axial tensile strain under uniaxial tensile loading is known as *Poisson's ratio*. This ratio is always less than 0.5 and is usually about 0.3.

Plastic Deformation in a Single Crystal.   As the magnitude of applied load is increased, the distortion increases to a point where the atoms must either (1) break bonds to produce a fracture or (2) slide over one another to produce a permanent shift of atom positions. For metallic materials, the second phenomenon generally requires lower loads and thus occurs preferentially in nature. The result is a plastic deformation, where a permanent change in shape occurs without a concurrent deterioration in properties.

Investigation reveals that the mechanism of plastic deformation is the shearing of atomic planes over one another to produce a net displacement. Conceptually, this is similar to the distortion of a deck of playing cards when one card slides over another. As we shall see, the actual mechanism, however, is a progressive one rather than all atoms in a plane shifting simultaneously.

Recalling that a crystal structure is a regular and periodic arrangement of atoms in space, it becomes possible to link atoms into flat planes in a nearly infinite number of ways. Planes having different orientations with respect to the basic lattice will have different atomic densities and different spacing between adjacent parallel planes, as illustrated in Figure 3-13. Given the choice of all possibilities, plastic deformation tends to take place along planes having the highest atomic density and greatest parallel separation. The reason may be seen

FIGURE 3-13 Schematic diagram showing crystalline planes with different atomic densities and interplanar spacings.

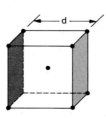
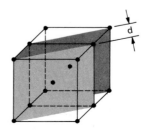

in the simplified Figure 3-14. Planes *A* and *A'* have higher density and greater separation than planes *B* and *B'*. In visualizing relative motion, the atoms of *B* and *B'* would interfere significantly with one another, whereas planes *A* and *A'* do not experience this difficulty.

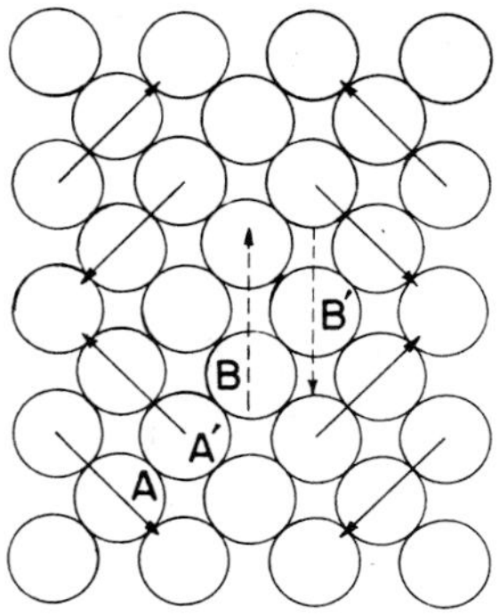

Within the preferred planes are also preferred directions. If sliding occurs in a direction corresponding to close packing of atoms in the plane (as in Figure 3-8), atoms can simply follow one another rather than each having to negotiate its own path. Thus, plastic deformation occurs by the preferential sliding of maximum-density planes (close-packed planes if present) in directions of closest packing. The specific combination of a plane and a direction is called a *slip system,* and the shear deformation is known as *slip*.

The ease with which a given metal may be deformed depends on the ease of shearing one atomic plane over an adjacent one and the favorability with which the plane is oriented with respect to the load. For example, a deck of playing cards does not ''deform'' when laid flat on a table and pressed from the top, or when stacked on edge and pressed uniformly. Only if the deck is skewed with respect to the applied load is sliding produced.

With this understanding, let us now consider the properties of the various crystal structures:

Body-centered-cubic. In the body-centered-cubic structure, there are no close-packed planes. Slip therefore occurs on planes with large interplanar spacings (six of which are illustrated in Figure 3-15) in directions of closest packing (that is, the cube diagonals). If each combination of plane and direction is considered as a slip system, we find that 48 such systems exist. The probability that one of these systems will be oriented for easy shear is great, but the force necessary to effect deformation is rather large. Materials with this structure generally possess high strength with moderate deformation capabilities. (See the typical metals in Figure 3-7.)

Face-centered-cubic. In the face-centered-cubic structure, each unit cube possesses four close-packed planes, as illustrated in Figure 3-15. Each plane contains three close-packed directions (the face diagonals), giving 12 possible slip systems. Again the probability of one system being favorably oriented for shear is great, and the required force is now relatively low. Face-centered metals are relatively weak and possess excellent ductilities, as an inspection of Figure 3-7 will reveal.

FIGURE 3-14   Planar schematic representing the greater deformation resistance of planes of lower atomic density and closer interplanar spacing.

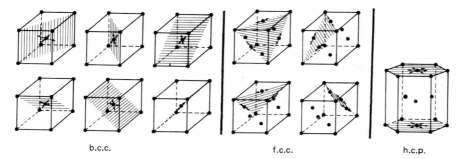

b.c.c.   f.c.c.   h.c.p.

FIGURE 3-15   Slip planes of the various lattice types.

Hexagonal-close-packed. The hexagonal lattice also contains close-packed planes, but only one such plane exists for the lattice. Although this plane contains three close-packed directions and the force required to produce deformation is rather low, the probability of favorable orientation to the applied load is rather small. Metals with the hcp structure tend to have low ductilities and often appear to be brittle.

Dislocation Theory of Slippage.   A theoretical calculation of the strength of metals based on the sliding of atomic planes over one another predicts yield strengths on the order of 3 million pounds per square inch (20,000 MPa). Observed strengths are typically 100 to 150 times less than this value, indicating a significant discrepancy between theory and reality.

Explanation is provided by the fact that plastic deformation does not occur by all the atoms in one plane slipping simultaneously over all atoms in an adjacent plane. Instead, the motion takes place by the progressive slippage of a localized disruption, the disruption being known as a *dislocation.* Consider an analogy. An individual wants to move a carpet a short distance in a given direction. One approach would be to pull on one end and try to "shear the carpet over the floor." This would require a large force acting over a small distance. An alternative approach would be to form and work a wrinkle across the floor to produce a net shift of the whole carpet—a low-force-over-large-distance approach to the same task. In the region of the wrinkle, there is excess carpet with respect to the floor below, and motion of this excess is relatively easy.

It has been shown that metal crystals do not have all their atoms in perfect arrangement but contain various localized imperfections. Two such imperfections are the *edge dislocation* and *screw dislocation,* as illustrated in Figure 3-16. Edge dislocations are the ends of extra half-planes of atoms. Screw dislocations correspond to a partial tearing of crystal planes. In each case, the dislocation is a disruption of the regular, symmetrical arrangement of atoms that can be moved about with rather low applied forces. It is the motion of these microscopic dislocations under applied loads that produces the observed macroscopic plastic deformation.

All engineering metals contain dislocations, usually in abundant quantities. The ease of deformation, therefore, depends on the ease of producing dislocation movement. Barriers to dislocation motion tend to increase the overall strength of the metal. These barriers take the form of other crystal imperfections and may be of the point type (missing atoms or *vacancies,* extra atoms or *interstitials,* or substituting atoms of a different variety as may occur in an alloy), line type (another dislocation), or surface type (crystal grain boundary).

Strain Hardening or Work Hardening.   Many metals possess a unique property in that, after undergoing some deformation, the metal possesses greater resistance to further plastic flow. In essence, metals become stronger when plastically deformed, a phenomenon known as *strain hardening,* or *work hardening.*

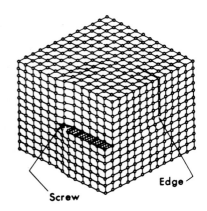

FIGURE 3-16   Schematic representation of screw and edge dislocations.

Understanding of this phenomenon can come from further consideration of the carpet analogy. Suppose this time that the goal is to move the carpet diagonally. The best way would be to move a wrinkle in one direction and then a second one 90° to the first. But suppose that both wrinkles were started simultaneously. We would find that wrinkle 1 would impede the motion of wrinkle 2, and vice versa. In essence, the device that makes deformation easy can also serve to impede the motion of other, similar devices. Returning to metals, we find that plastic deformation is accomplished by the motion of dislocations. As dislocations move, they are more likely to encounter and interact with other such dislocations, thereby producing resistance to further motion. Moreover, mechanisms exist to markedly increase the number of dislocations in a metal during deformation, the effect being an increased probability of interaction.

This phenomenon becomes significant when one considers mechanical working processes operating in the cold-working range. Strength properties of metals can be increased markedly by deformation, so that a deformed inexpensive metal can often substitute for an undeformed more costly one.

Experimental evidence strongly supports the dislocation and slippage theory of deformation. When a load is applied to a metal crystal, deformation begins on the slip system most favorably oriented. The net result is often an observable slip and rotation as illustrated in Figure 3-17 and observed in Figure 3-18. Dislocation motion on one slip system may become blocked as strain hardening produces increased resistance. As the load then increases, further deformation occurs, on alternative systems, which now offer less resistance. This phenomenon, known as *cross slip*, has also been observed.

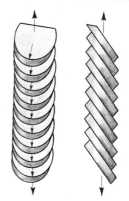

FIGURE 3-17 Schematic representation of slip and rotation resulting from deformation.

Plastic Deformation in Polycrystalline Metals.

Thus far, only the deformation of single crystals has been considered. Metals, as normally encountered, are polycrystalline. Within each crystal of a polycrystalline metal, deformation proceeds in the manner just described. However, since adjacent grains do not have their lattice structures aligned in the same orientation, an applied loading will cause different deformations within the various grains. This type of response is shown in Figure 3-19, where the slip lines in the various grains can be seen. It may be noted that the slip lines do not cross over from one grain to another. The grain boundaries act as barriers to dislocation motion. Finer grain structure—that is, more grains per unit area—generally tends to produce greater strength and hardness coupled with increased impact resistance. This "universal improvement of properties" is a strong motivation for controlling grain size during processing.

Grain Deformation and Fiber Structure.

When a metal is deformed a considerable amount, the grains become elongated in the direction of metal flow, as shown in Figure 3-20. At moderate magnification, a cross section of such a deformed metal may appear fibrous. Concurrent with nonuniformity in structure is a nonuniformity of properties. Because of strain hardening and the fact that the intergranular boundaries are no longer randomly oriented, the strength and

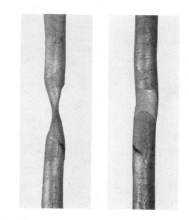

FIGURE 3-18 Front and side views of a single zinc crystal that has been elongated in uniaxial tension. (Courtesy of E.R. Parker.)

FIGURE 3-19 Slip lines in a poly-crystalline material. (From Richard Hertzberg, *Deformation and Fracture Mechanics of Engineering Materials*; courtesy John Wiley & Sons, Inc.)

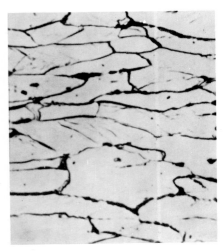

FIGURE 3-20 Deformed grains in cold-worked 1008 steel after 50% reduction by rolling; 1000X. (From *Metals Handbook,* 8th Ed. 1972, ASM International, Metals Park, Ohio.)

other mechanical properties are not the same in all directions. Electrical and magnetic properties may also show directional variation.

The possibility of employing this increase in strength in certain directions is important to both the designer and the manufacturing engineer. Certain processes, such as forging, can be designed to utilize directional properties. Caution should be used, however, for improvement of properties in some directions usually is accompanied by a decline in properties in other directions. Moreover, directional structures may impose serious difficulties in some operations, such as sheet metal drawing.

Fracture of Metals.   Under certain conditions of load, temperature, impact, and deformation, metals may respond by fracture. *Brittle fracture* is the most catastrophic, for it occurs without the prior warning of plastic deformation and propagates rapidly through the metal. Such fractures are usually associated with metals having the bcc or hcp crystal structure. *Ductile fracture* generally occurs when plastic deformation is extended too far. The actual mechanism and type of fracture, however, varies depending on the material, the temperature, the state of stress, and the rate of loading.

Recrystallization.   Plastic deformation increases the energy of a metal by means of creating many additional dislocations and increasing the surface area of the grain boundaries through distortion. If a polycrystalline metal is heated to a high enough temperature after being plastically deformed, new, equiaxed (spherical-shaped), unstrained crystals form from the original distorted structure, as in Figure 3-21. This process of reducing the internal energy is known as *recrystallization*. The temperature at which recrystallization takes place is dif-

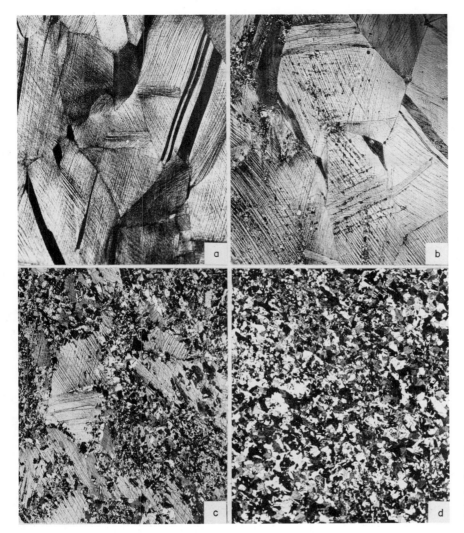

FIGURE 3-21 Recrystallization of 70-30 brass. (a) Cold worked 33%; (b) heated at 580°C (1075°F) for 3 seconds, (c) 4 seconds, and (d) 8 seconds; 45X. (Courtesy J. E. Burke, General Electric Company.)

ferent for each metal and varies with the amount of prior deformation. In general, the greater the amount of deformation, the lower the recrystallization temperature. However, there is a practical lower limit below which recrystallization will not take place in a reasonable length of time. Table 3.2 gives the short-time recrystallization temperatures of several metals.

Noting that metals may fracture if deformed too much, we find it common practice to recrystallize material after certain amounts of cold work. Ductility is restored and the material is ready for further deformation. This process, known as *recrystallization annealing,* enables deformation to be carried out to great lengths without danger of fracture and is important to many manufacturing processes. If metals are deformed at temperatures sufficiently above the recrys-

TABLE 3.2 Lowest Recrystallization Temperature of Common Metals

Metal	Temperature [°F (°C)]
Aluminum	300 (150)
Copper	390 (200)
Gold	390 (200)
Iron	840 (450)
Lead	Below room temperature
Magnesium	300 (150)
Nickel	1100 (590)
Silver	390 (200)
Tin	Below room temperature
Zinc	Room temperature

tallization temperature, working and recrystallization can take place simultaneously, and large deformations are possible.

Having already noted the desirability of fine grain size in improving properties, we find recrystallization to be a means of grain-size control. In metals that do not undergo allotropic changes, a coarse grain structure can be converted to a fine grain structure through recrystallization. The material must first be plastically deformed to provide the driving force. Control of the recrystallization process then establishes the final grain size.

Grain Growth. The recrystallization process tends to produce uniform grains of comparatively small size. If the metal is held at or above the recrystallization temperature for any appreciable time, however, the new grains further increase in size. In effect, some of the grains will become larger at the expense of their neighbors as the material seeks to further decrease the amount of grain-boundary surface area. Since properties tend to diminish with increased grain size, the control of recrystallization is of prime importance.

Hot and Cold Working. When metals are plastically deformed below their recrystallization temperature, the process is called *cold working*. The metal strain hardens and the structure consists of distorted grains. When deformation takes place above the recrystallization temperature, the process is called *hot working*. A recrystallized grain structure continually forms, and no strain hardening is apparent. As shown in Table 3.2, the temperature above which hot working can be performed depends on the material being worked.

Alloys. Up to this point in the chapter, the discussion has been confined to the nature and behavior of pure metals. For most manufacturing applications, however, metals are not used in their pure form, but in the form of *alloys*—materials composed of two or more elements, at least one of which is a metal. The addition of a second element to form an alloy usually results in a change of properties. Knowledge of alloys and their properties is important to the intelligent selection of materials for given applications.

Alloy Types. The response of a metal to an alloy addition may follow any of three mechanisms. The first, and probably the simplest, response occurs when *the two components are insoluble in each other in the solid state*. In this case, the base metal and the alloying element each maintain their individual identities, structures, and properties. The alloy, in effect, assumes a composite structure consisting of two types of building blocks in an intimate mechanical mixture.

The second possibility occurs when *the two elements are soluble in each other in the solid state*. They thus form *a solid solution,* with the alloying element being dissolved in the base metal. These solid solutions may be of two types: (1) *substitutional* and (2) *interstitial*. In the substitutional type, some atoms of the alloy element occupy sites normally occupied by atoms of the host or base metal. Replacement is random in nature, with the alloy atom being free to occupy any atom site in the base lattice. In the interstitial type, the alloy element atoms squeeze into the open spaces between the atoms of the base metal lattice.

The third possibility occurs where *the elements combine to form intermetallic compounds*. In this case, atoms of the alloying element combine with atoms of the base metal *in definite proportions and in definite geometric relationships*. Bonding is primarily of the nonmetallic variety (i.e., ionic or covalent), and the lattice structures are more complex than for metallic materials. Such compounds tend to be hard, brittle, high-strength materials.

Even though alloys are composed of more than one type of atom, their structure is one of lattices and grains just as in pure metals. Their behavior when subjected to loading, therefore, should be similar to that of pure metals, with due provision for the structural modifications. Dislocation movement may now be impeded by the presence of unlike atoms. Grains in composite-type mixtures may show different responses to the same loading, reflecting the different properties of the different chemistry segments.

Atomic Structure and Electrical Properties.

As with mechanical properties, the structure of materials strongly influences their electrical properties. Electrical conductivity involves the movement of valence electrons through the crystalline lattice to produce a net transport of charge. The more perfect the atomic arrangement is in a metal, the higher the conductivity. Conversely, the more lattice imperfections or irregularities, the higher the resistance to electrical conduction.

Electrical resistance depends largely on two factors: (1) lattice imperfections and (2) temperature. Vacant atomic sites, interstitial atoms, substitutional atoms, dislocations, and grain boundaries all act as disruptions to the regularity of a crystalline lattice. Temperature becomes important when one considers the associated atomic vibrations. We have already seen that mechanical energy can displace atoms from their equilibrium positions and stretch bonds. Thermal energy causes atoms to vibrate about their equilibrium positions. This vibration interferes with electron transport, reducing conductivity at higher temperatures. At low temperatures, resistivity becomes primarily a function of crystal imperfections. Thus, the best conductors are pure (defect-free) crystalline solids at low temperatures.

The conductivity of metals is primarily a result of the "free" electrons in the metallic bond. Materials with covalent bonds require bonds to be broken to provide electrons for conduction. Thus, the electrical properties of these materials depend on bond strength. Diamond, for instance, is a strong insulator. Silicon and germanium, however, have weaker bonds that can easily be broken by thermal energy. These materials, when pure, are known as *intrinsic semiconductors*, since moderate amounts of applied energy enable the materials to conduct small amounts of electricity.

The conductivity of the nonmetallic semiconductors can be substantially improved by a process known as *doping*. Both silicon and germanium have four valence electrons and form four covalent bonds. If one of these elements is

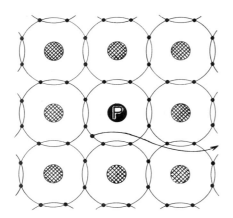

FIGURE 3-22 Schematic representation of an n-type semiconductor, with the excess electron of the phosphorus atom being free for conduction.

replaced with an atom containing five valence electrons, such as phosphorus, the four bonds form, leaving one excess electron, as in Figure 3-22. The extra electron is then free to move about and provide conductivity. Such materials are known as *n-type extrinsic semiconductors.*

A similar effect can be created by the substitution of an atom with three valence electrons, such as aluminum. An electron is now missing from a bond, so that an electron "hole" is created. As a nearby electron jumps into this hole, it creates a hole in the spot it vacated. Movement of electron holes is equivalent to a countermovement of electrons and thus provides conductivity. These materials are known as *p-type semiconductors.*

The control of conductivity through semiconductor devices is the basis of all solid-state electronics and circuitry.

Review Questions

1. Why might an engineer be concerned with the microscopic structure of a material?
2. What is meant by the term *microstructure?*
3. What properties or characteristics of a material are influenced by the valence electrons?
4. What are the three types of primary bonds, and what types of atoms do they unite?
5. What are some general characteristics of ionically bonded materials?
6. What are some general properties and characteristics of covalently bonded materials?
7. What are some unique property features of materials bonded by metallic bonds?
8. What is the role of the valence electrons in the van der Waals type of secondary bond?
9. Why does the interatomic distance increase when a material is heated (thermal expansion)?
10. What is the difference between a crystalline material and one with an amorphous structure?
11. What are some of the general characteristics of metallic materials?
12. What are the three most common crystal structures found in metals?

13. What is the nature of the geometric symmetry observed in a close-packed plane of atoms?
14. What is the efficiency of filling space with spheres in the body-centered-cubic structure? Face-centered-cubic? Hexagonal close-packed?
15. Describe the difference in ductility observed in the two structures formed from close-packed planes.
16. What is one of the most common means of quantifying the grain size of a solid metal?
17. In what way does a metallic crystal respond to low applied loads?
18. What is a slip system in a material? What types of planes and directions tend to be preferred?
19. Based on the ease or difficulty of deformation, what is the dominant mechanical property or characteristic of each of the three most common metal crystal structures?
20. What are some of the common barriers to dislocation movement that can be used to strengthen a metal?
21. What is the mechanism (or mechanisms) responsible for the observed strain hardening of a metal?
22. Why is a fine grain size often desired in an engineering metal?
23. What is the difference between brittle fracture and ductile fracture?

24. In what ways does plastic deformation increase the internal energy of a metal?
25. In what ways can recrystallization be used to enable large amounts of deformation to be performed without fear of fracture?
26. What is the motivation or driving force that causes grain growth to occur?
27. What is the major distinguishing factor between hot and cold working?
28. What are some of the basic structural possibilities that can occur when an alloy addition is made to a base metal?
29. What are the typical mechanical properties of intermetallic compounds?
30. What features in a metal structure tend to impede or reduce electrical conductivity?
31. What types of atoms should be added to silicon or germanium to produce an *n*-type extrinsic semiconductor? A *p*-type extrinsic semiconductor?

Equilibrium Diagrams

As our study of engineering materials becomes more focused on specific metals and alloys, it is increasingly important that the natural characteristics and properties of the material be known for various environmental conditions. What is the basic structure of the material? Is the material uniform throughout, or is it a mixture of two or more distinct components? If there are multiple components, how much of each is present? Is there a component that may impart undesired properties or characteristics? What will happen if temperature is increased or decreased, pressure is changed, or chemistry is varied? The answers to these and other important questions can be obtained through the use of equilibrium phase diagrams.

Phases. Before we move to a discussion of these diagrams, it is imperative that a working definition of the term *phase* first be developed. As a starting definition, a phase is simply a form of material possessing a single characteristic structure and associated characteristic properties. Uniformity of chemistry, structure, and properties is assumed throughout the phase. More rigorously, a phase is *any physically distinct, chemically homogeneous, and mechanically separable portion of a substance*. In lay terms, a phase has a unique structure, uniform composition, and well-defined boundaries or interfaces.

A phase can be continuous (like the air in a room) or discontinuous (like grains of salt in a shaker). A phase can be solid, liquid, or gas. In addition, a phase can be a pure substance or a solution, provided that the structure and composition are uniform throughout. Alcohol and water mix in all proportions and will therefore form a single phase when combined. Oil and water tend to form isolated regions with distinct boundaries and must be regarded as two distinct phases.

Equilibrium Phase Diagram.

The *equilibrium phase diagram* is a *graphic mapping* of the natural tendencies of a material or a material system, assuming infinite time at the conditions specified. Areas of the diagram are assigned to the various phases, with the boundaries indicating the equilibrium conditions of transition.

With the background just developed, let us now consider the types of phase mappings that may be useful. Three *primary variables* are at our disposal: *temperature, pressure,* and *composition.* The simplest type of diagram is a pressure–temperature (P-T) diagram for a fixed composition material.

For simplicity, consider the P-T diagram for water presented in Figure 4-1. With composition fixed, the diagram enables the determination of the stable form of water at any condition of temperature and pressure. When pressure is held constant and temperature is varied, the transition boundaries locate the melting and boiling points. Still other uses can be presented. Locate a temperature where the stable phase at atmospheric pressure is the solid (ice). Now maintain that temperature and begin to decrease the pressure. A transition is encountered wherein solid goes directly to gas with no liquid intermediate. The process is that of freeze drying and is employed in the manufacture of numerous dehydrated products. Through use of the phase diagram, process conditions could be selected that would minimize the amount of cooling and pressure drop that would be required.

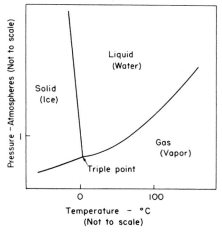

FIGURE 4-1 Schematic pressure-temperature diagram for water.

Temperature–Composition Diagrams.

The water diagram serves as an excellent example of the use of phase diagrams. In engineering applications, however, the P-T phase diagram is rarely used. Most processes are conducted at atmospheric pressure, and variations come primarily in temperature and composition. The most useful mapping, therefore, would be a *temperature–composition phase diagram* at atmospheric pressure. For the remainder of the chapter, it is this second form of diagram that will be considered.

For mapping purposes, temperature is placed on the vertical axis and composition on the horizontal. Figure 4-2 shows the form of such a mapping for the A-B system where the left-hand vertical corresponds to pure material A and the percentage of B increases as we move toward pure B at the right of the diagram. Experimental investigations for filling in the details of the diagram take the basic form of vertical or horizontal scans designed to locate transition points.

Cooling Curves.

Considerable information can be obtained from vertical scans through the diagram in which a fixed composition material is heated and subsequently slow-cooled by removing heat at a uniformly slow rate. Transitions in structure appear as characteristic points in a temperature-versus-time plot of the cooling cycle, known as a *cooling curve.*

For the system composed of sodium chloride (common table salt) and water, five different cooling curves are presented in Figure 4-3. Curve (a) is for pure water being cooled from the liquid state. A smooth continuous line is observed

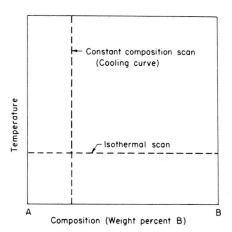

FIGURE 4-2 Mapping for a temperature-composition equilibrium phase diagram.

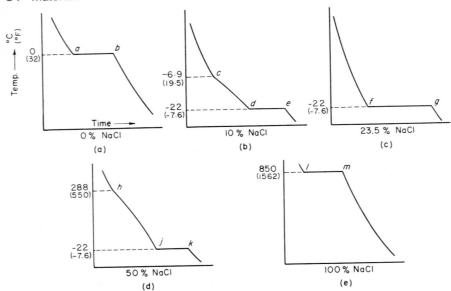

FIGURE 4-3 Cooling curves for various compositions of NaCl-H$_2$O solutions.

for the liquid, the extraction of heat producing a concurrent drop in temperature. When the freezing point is reached (point *a*), the material changes state and releases heat energy during the liquid-to-solid transition. Heat is still being extracted from the system, but its source is the change in state and not a decrease in temperature. Thus, an isothermal, or constant temperature, hold (*a-b*) is observed until solidification is completed. From this point, the newly formed solid experiences a smooth drop in temperature as heat extraction continues. Such a curve is characteristic of pure metals or substances with a distinct melting point.

Curve (b) of Figure 4-3 is the cooling curve for a 10% solution of salt in water. The liquid region undergoes continuous cooling down to point *c*, where the slope abruptly decreases. At this temperature, small particles of ice begin to form, and the slope change is due to the energy released in this transition. The formation of these ice particles leaves the remaining solution richer in salt and imparts a lower freezing temperature to it. Further cooling must take place for additional solid to form. The formation of more solid further lowers the freezing point of the remaining liquid. Instead of possessing a distinct melting point or freezing point, the material is said to have a *freezing range*. When the temperature of point *d* is reached, the remaining liquid solidifies into an intimate mixture of solid salt and solid water (to be discussed later), and an isothermal hold is observed. Further heat extraction from the solid material produces a continuous drop in temperature.

For a solution of 23.5% salt in water, a distinct freezing point is again observed, as shown by curve (*c*) of Figure 4-3. Compositions with richer salt

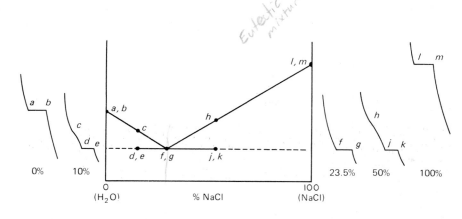

FIGURE 4-4   Partial equilibrium diagram for NaCl and H_2O derived from cooling curve information.

concentrations show phenomena similar to those presented earlier, but with salt being the first substance to solidify from the liquid.

The observed transition points can now be transferred to a temperature–composition diagram with the designating letters corresponding to those on the cooling curves. Figure 4-4 presents such a map, on which several key lines are drawn. Line *a-f-l* denotes the lowest temperature at which the material is totally liquid and is known as the *liquidus line*. Line *d-f-j* is the highest temperature at which a material is completely solid and is called the *solidus line*. Between the liquidus and solidus, two phases coexist, one being liquid and the other being solid. Cooling-curve studies have thus enabled the determination of some key information regarding the system being studied. An equilibrium phase diagram, therefore, can be viewed as a collection of cooling-curve data taken over an entire range of alloy compositions.

Solubility Studies.   The observant reader will note that the ends of the diagram still remain undetermined. Both pure materials have only one transition point below which they appear as a single-phase solid. Can ice retain some salt in solid solution? If so, how much? Can solid salt hold water and still remain a single phase? Completion of the diagram, therefore, requires several horizontal scans to determine any *solubility limits,* the conditions of saturation at various temperatures.

These isothermal scans with variable composition usually require analysis of the specimens by X-ray techniques, microscopy, or other investigative approaches to determine the composition at which transitions occur. As the scan moves away from the pure metal, the first line encountered (provided that the temperature is in the all-solid region) denotes the solubility limit and is known as the *solvus line*. Figure 4-5 presents the equilibrium phase diagram for the lead–tin system, using the conventional notation in which Greek letters are used to label the various single-phase solids. The upper portion closely resembles the salt–water diagram, but the solubility of one metal in the other can be seen at both ends of the diagram.

Having now been exposed to the concepts of equilibrium diagrams, let us move on to consider several specific examples. Presentation will move from the simple to the more complex.

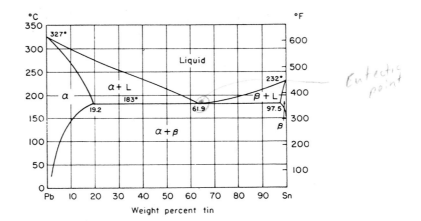

FIGURE 4-5   Lead-tin equilibrium diagram.

Complete Solubility in Both Liquid and Solid States.   If two metals are each completely soluble in the other in both the liquid and solid states, a rather simple diagram results, as illustrated in Figure 4-6 for the copper–nickel system. At temperatures above the liquidus line, the two materials are in liquid solution. Similarly, below the solidus, the materials form a solid solution at all compositions. Between the liquidus and solidus is a two-phase region where liquid and solid solutions coexist.

Partial Solid Solubility.   As might be expected, many materials exhibit neither complete solubility nor complete insolubility in the solid state. Each is soluble in the other up to a certain limit or saturation point, the value of this limit being a function of temperature. Such a diagram has already been observed in the lead–tin system of Figure 4-5.

Consideration of Figure 4-5 shows that the maximum solubility of tin in lead in the solid state is 19.2% by weight. Similarly, tin dissolves up to 2.5wt % lead in solid solution. If the temperature is decreased from this point of maximum solubility, the amount of substance capable of being held in solution generally decreases. If a saturated solution of tin in lead is cooled from 183°C,

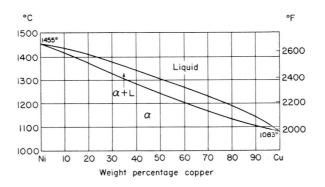

FIGURE 4-6   Copper-nickel equilibrium diagram, showing complete solubility.

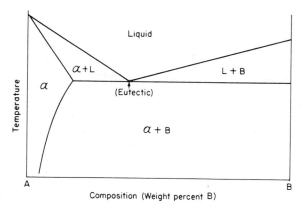

FIGURE 4-7 Equilibrium diagram of two materials, one of which is partially soluble in the other in the solid state.

the material moves from a single-phase region into a two-phase region. Some tin-rich second phase must precipitate from solution. This fact is used to control the properties of many engineering alloys.

Insolubility. If one or both of the components is insoluble in the other, the diagrams also reflect this phenomenon. Figure 4-7 illustrates the case where component A is completely insoluble in component B.

Utilization of Diagrams. Before considering some of the more complex aspects of phase diagrams, let us first return to the complete solubility diagram (Figure 4-8) and develop tools to extract some useful information. For each point of temperature and composition, three pieces of information can be obtained:

1. *The phases present*. The stable phases can be determined by simply locating the point of consideration on the temperature–composition mapping and identifying the region of the diagram in which the point appears.

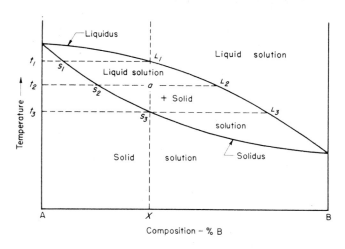

FIGURE 4-8 Equilibrium diagram of two materials that are completely soluble in each other in both the liquid and solid states.

2. *The composition of each phase.* If the point lies in a single-phase region, the composition of the phase is the composition of the material being considered. If the point lies in a two-phase region, a *tie-line* is drawn. A tie-line is simply an isothermal (constant temperature) line drawn through the point of consideration, terminating at the boundaries of the single-phase regions on either side. The composition at which the tie-line intersects the neighboring single-phase regions determines the compositions of these phases in the two-phase mixture. For example, consider point a in Figure 4-8. The tie-line for this point runs from S_2 to L_2. The point S_2 is the intersection of the tie-line with the solid phase, and thus, the solid in the two-phase mixture at a has the composition of point S_2. Similarly, the liquid has the composition of point L_2.

3. *The amount of each phase present.* If the point lies in a single-phase region, there must be 100% of that phase. If the point lies in a two-phase region, the relative amounts can be determined by a *lever-law* calculation using the previously drawn tie-line.

Consider the cooling of alloy X in Figure 4-8 in a manner sufficiently slow so as to approximate equilibrium. At temperatures above t_1, the material is in a single-phase liquid state. Now go to temperature t_1 and draw a tie-line from S_1 to L_1. Any solid forming will be of composition S_1, but common sense tells us that almost all of the material will still be liquid of composition L_1. The entire tie-line lies to the left of composition X. Now drop to temperature t_3, where the material is entirely solid of composition S_3. The tie-line lies to the right of composition X. Extrapolating these observations to intermediate temperatures, such as t_2, we predict that the amount of liquid will be proportional to the relative length of the tie-line to the left of point a. Namely, the amount of liquid at t_2 is

$$\frac{a - S_2}{L_2 - S_2} \times 100\%$$

Similarly, the amount of solid corresponds to the proportion of the tie-line to the right of point a. Since the calculations consider the tie-line as a lever with the phases at each end and a fulcrum at the composition line, they are called *lever-law calculations.*

Other applications of phase diagrams relate to an overall view of the system or the location of the transition points for a specific alloy. For instance, the temperature necessary to put an alloy into a given phase field can easily be determined. Changes that may occur upon the slow heating or slow cooling of a given material can be predicted. In fact, most of the questions posed at the beginning of this chapter can now be answered.

Solidification of Alloy X.   If we use the tools developed above, it now becomes relatively simple to follow the solidification of alloy X in Figure 4-8. At temperature t_1, the first minute amount of solid forms with the chemistry of point S_1. As the temperature drops, more solid forms, but the chemistry of both the solid and the liquid phases is shifting in accordance with the tie-line end points. The chemistry of the liquid follows the liquidus line, and the chemistry of the solid follows the solidus. Finally, at t_3, solidification is complete, and the composition of the single-phase solid is that of alloy X, as required.

The composition of the final solid is different from that of the first solid that was formed. If cooling is sufficiently slow (equilibrium conditions are approached), the composition of the entire mass of solid tends to become uniform at the value predicted by the tie-line. Compositional differences are removed by the phenomenon of diffusion, in which atoms migrate from point to point in the crystal lattice under the energy impetus of elevated temperature. If the cooling rate is rapid, a nonuniform material may result, with the initial solid that formed retaining a composition different from the latter portions of the solid. This structure is referred to as being *cored*.

Three-Phase Reactions.   Several of the diagrams previously presented contain a distinct feature in which phase regions are separated by a horizontal line. These lines are further characterized by either a V intersecting from above or an inverted V intersecting from below. The intersection of the V and the line denotes the location of a *three-phase equilibrium reaction*.

One common type of three-phase reaction, known as a *eutectic*, has already been observed in Figures 4-4, 4-5, and 4-7. Understanding these reactions is possible through the use of the tie-line and lever-law concepts and will be developed by use of the lead–tin diagram of Figure 4-5. Consider any alloy containing between 19.2 and 97.5 wt % tin at a temperature just above the 183°C horizontal line. Tie-line and lever-law computations show that the material contains either a lead-rich or a tin-rich solid and remaining liquid, the liquid having a composition of 61.9 wt % tin. [Note that any liquid will always have a composition of 61.9 wt % at 361°F (183°C), regardless of the overall composition of the alloy.] If we now focus on the liquid and allow it to cool to just below 361°F (183°C), a transition occurs in which liquid of composition 61.9% tin goes to a mixture of lead-rich solid with 19.2% tin and a tin-rich solid containing 97.5% tin. The relative amounts of the two components maintain the overall chemistry. The form for this eutectic transition is similar to a chemical reaction:

$$\text{liquid} \rightarrow \text{solid}_1 + \text{solid}_2$$

Since the two solids have chemistries on either side of the intermediate liquid, a separation must have occurred within the system. Such a separation in a solidifying melt results from two metals that are soluble in the liquid state but only partially soluble in the solid state. Separation requires atom movement, but the distances between the two solids cannot be great. The resulting eutectic structure is an intimate mixture of two single-phase solids and assumes its own characteristic set of physical and mechanical properties. Alloys of the eutectic composition have the lowest melting point of all alloys in a given system and are often used as casting alloys or as filler metal in soldering or brazing applications.

Figure 4-9 summarizes some other types of three-phase reactions that may occur in engineering systems. These include the *peritectic, monotetic,* and *syntetic,* the suffix *-ic* denoting that at least one of the three phases in the reaction is a liquid. If the suffix *-oid* is used, it indicates that all phases involved are solids; the form of the reaction remains the same. Two all-solid reactions can occur: the *eutectoid* and the *peritectoid.* These reactions tend to be a bit more sluggish since all changes must occur in the solid state.

Intermetallic Compounds.   A final feature occurs in systems where the bonding attraction of the component materials is sufficiently strong so that compounds tend to form. These compounds are single-phase solids and tend to break the diagram into recognizable subareas. If components A and B form a compound A_xB_y and the compound cannot tolerate any deviation from that fixed ratio, the intermetallic is known as a *stoichiometric intermetallic* and appears as a single vertical line in the diagram. If some degree of deviation is tolerable, the vertical line expands into a region, and the compound is a *nonstoichiometric intermetallic.* Figure 4-9 shows schematic representations for both types of intermetallic compounds.

In general, intermetallics tend to be hard, brittle materials, these properties being related to their ionic or covalent bonding. If they are present in large quantities or lie along grain boundaries, the overall alloy can be extremely

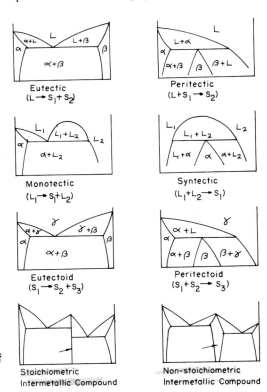

FIGURE 4-9 Summary schematic of three-phase reactions and intermetallic compounds.

brittle. If the same intermetallic can be uniformly distributed throughout the structure in small particles, the effect can be considerable strengthening of the alloy.

Complex Diagrams.   The equilibrium diagrams of actual alloy systems generally belong to one of the basic types just discussed or combinations of them. Often, these diagrams appear to be quite complex and formidable to the manufacturing engineer. By focusing on the particular alloy in question and analyzing specific points using the tie-line and lever-law concepts, even the most complex diagram can be simply dissected. Knowledge of the properties of the various components then enables predictions about the overall product.

The Iron–Carbon Equilibrium Diagram.

Because steel, composed essentially of iron and carbon, is such an important engineering metal, the iron–carbon equilibrium diagram is certainly the most important of those that the average engineer encounters. The diagram most frequently used, however, is not the full iron–carbon diagram, but the iron–iron-carbide diagram of Figure 4-10, where a stoichiometric intermetallic compound, Fe_3C, is used to terminate the carbon range at 6.67 wt % carbon. The names of phases and structures and notations used on the diagram have evolved historically and are used here in their generally accepted form.

Three distinct three-phase reactions can be identified in the diagram. At 2723°F (1495°C), a *peritectic* occurs for alloys with a low weight percentage of

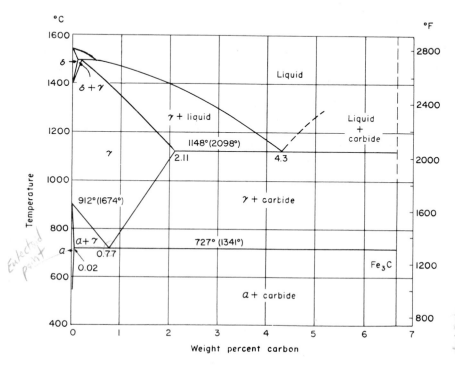

FIGURE 4-10 Iron carbon equilibrium diagram: α = ferrite; γ = austenite; δ = δ-ferrite; Fe_3C = cementite.

carbon. Because of its high temperature and the extensive single-phase gamma region immediately below it, the peritectic reaction rarely assumes any engineering significance. A *eutectic* is observed at 2098°F (1148°C), with the eutectic point at 4.3% carbon. Alloys containing more than 2.11% carbon will experience this eutectic and are classified by the general term *cast irons*. The final three-phase reaction is a *eutectoid* at 1341°F (727°C), with the eutectoid point at 0.77 wt % carbon. Alloys with less than 2.11% carbon can undergo a transition from a single-phase solid solution (γ) through the eutectoid to a two-phase mixture and are known as *steels*. Thus, the point of maximum carbon solubility in iron, 2.11 wt %, forms an arbitrary division between steels and cast irons.

To further our understanding of the diagram, let us now consider the four single-phase components. Three of these occur in pure iron, and the fourth is the carbide at 6.67% carbon. Upon solidification, pure iron forms a body-centered-cubic solid that is stable down to 2541°F (1394°C). Known as *delta ferrite*, this phase is stable only at extremely elevated temperatures and has no significant engineering importance. From 2541°F (1394°C) to 1674°F (912°C), pure iron assumes a face-centered-cubic structure known as austenite (γ) (in honor of the famed metallurgist Sir Robert Austen of England). Key features of austenite are the high formability characteristic of the fcc structure and its high solubility of carbon. Hot forming of steel benefits from these features of formability and compositional uniformity. Moreover, most heat treatment of steel begins with the single-phase austenite structure. *Alpha ferrite*, or more commonly just *ferrite*, is the stable form of iron at temperatures below 1674°F (912°C). This body-centered-cubic structure can hold only 0.02 wt % carbon in solid solution and forces a two-phase mixture in most steels. The only other change upon the further cooling of iron is the nonmagnetic-to-magnetic transition at the Curie point 1418°F (770°C). Because this transition is not associated with any change in phase, it does not appear on the equilibrium phase diagram.

The fourth single-phase region is the brittle intermetallic, Fe_3C, which also goes by the name *cementite*, or iron carbide. Like most intermetallics, it is quite hard and brittle, and care should be exercised in controlling the structures in which it occurs. Alloys with excessive amounts of cementite or cementite in undesirable form tend to have brittle characteristics. Because cementite dissociates prior to melting, its exact melting point is unknown, and the liquidus line remains undetermined in the high-carbon region of the diagram.

A Simplified Iron–Carbon Diagram.   If we focus on the materials normally known as steels, the diagram can be simplified considerably. Those portions of the iron–carbon diagram near the delta (or peritectic) region and those with greater than 2% carbon are of little significance and are deleted. The resulting diagram, such as the one in Figure 4-11, focuses on the eutectoid region and is quite useful in understanding the properties and processing of steel.

The key transition in this diagram is the decomposition of single-phase austenite (γ) to the two-phase ferrite plus carbide structure as temperature drops. Control of this reaction, which arises because of the drastically different carbon

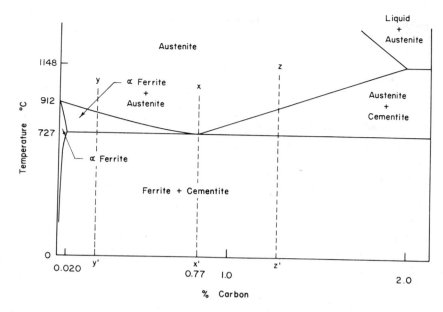

FIGURE 4-11 Simplified iron-carbon phase diagram.

solubilities of austenite and ferrite, enables a wide range of properties to be achieved through heat treatment.

To begin to understand these processes, consider a steel of the eutectoid composition, 0.77% carbon, being slow cooled along line x-x' in Figure 4-11. At the upper temperatures, only austenite is present, the 0.77% carbon being dissolved in solid solution with the iron. When the steel cools to 1341°F (727°C), several changes occur simultaneously. The iron wants to change from the fcc austenite structure to the bcc ferrite structure, but the ferrite can contain only 0.02% carbon in solid solution. The rejected carbon forms the carbon-rich cementite intermetallic with composition Fe_3C. In essence, the net reaction at the eutectoid is

$$\text{austenite} \rightarrow \text{ferrite} + \text{cementite}$$
$$0.77\% \; C \qquad 0.02\% \; C \qquad 6.67\% \; C$$

Since this chemical separation of the carbon component occurs entirely in the solid state, the resulting structure is a fine mechanical mixture of ferrite and cementite. Specimens prepared by polishing and etching in a weak solution of nitric acid and alcohol reveal a lamellar structure of alternating layers or plates that forms on slow cooling. This structure, seen in Figure 4-12, is composed of two distinct phases but has its own set of characteristic properties and goes by the name *pearlite,* because of its resemblance to mother-of-pearl when viewed at low magnification.

Steels having less than the eutectoid amount of carbon (less than 0.77%) are known as *hypoeutectoid steels.* Consider now the transformation of such a material, represented by cooling along line y-y' in Figure 4-11. At high temperatures, the material is entirely austenite, but upon cooling, it enters a region

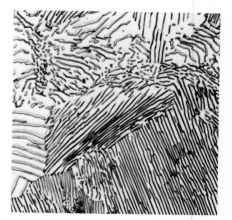

FIGURE 4-12 Pearlite; 1000X. (Courtesy USX Corporation.)

FIGURE 4-13 Photomicrograph of a hypoeutectoid steel showing ferrite (white) and pearlite; 500X. (Courtesy USX Corporation.)

where the stable phases are ferrite and austenite. Tie-line and lever-law calculations show that low-carbon ferrite nucleates and grows, leaving the remaining austenite richer in carbon. At 1341°F (727°C), the austenite is of eutectoid composition (0.77% carbon), and further cooling transforms the remaining austenite to pearlite. The resulting structure is a mixture of primary or proeutectoid ferrite (ferrite that formed before the eutectoid reaction) and regions of pearlite. An example of this structure is shown in Figure 4-13.

Hypereutectoid steels are steels that contain more than the eutectoid amount of carbon. When such a steel cools, as in *z-z′* of Figure 4-11, the process is similar to the hypoeutectoid case, except that the primary or proeutectoid phase is now cementite instead of ferrite. As the carbon-rich phase forms, the remaining austenite decreases in carbon content, reaching the eutectoid composition at 1341°F (727°C). As before, any remaining austenite transforms to pearlite upon slow cooling through this temperature. Figure 4-14 is a photomicrograph of the resulting structure. The continuous network of primary cementite causes this material to be extremely brittle.

It should be remembered, however, that the transitions just described are for equilibrium conditions, which can be approximated by slow cooling. Upon slow heating, these transitions occur in the reverse manner. However, when the alloys are cooled rapidly, entirely different results may be obtained, since sufficient time may not be provided for the normal phase reactions to occur. In these cases, the equilibrium phase diagram is no longer a useful tool for engineering analysis. In view of the importance of rapid cooling processes to the heat treatment of steels and other metals, their characteristics will be discussed in Chapter 5, and new tools will be introduced to aid our understanding.

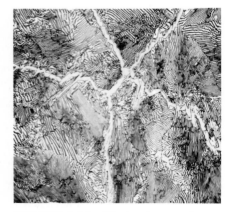

FIGURE 4-14 Photomicrograph of a hypereutectoid steel showing primary cementite along grain boundaries; 500X. (Courtesy USX Corporation.)

Cast Irons. Those alloys of iron and carbon having more than 2.11% carbon are called *cast irons*. More specifically, these are alloys with sufficient carbon so that they experience the eutectic transformation during cooling. Being relatively inexpensive, with good fluidity and rather low liquidus temperatures, they are readily cast and occupy an important place in engineering applications.

Most commercial cast irons also contain a significant amount of silicon, the general composition being 2.0% to 4.0% carbon, 0.5% to 3.0% silicon, less than 1.0% manganese, and less than 0.2% sulfur. Silicon produces two major effects. First, it partially substitutes for carbon, so that use of the phase diagram requires replacing the weight percentage of carbon with a carbon equivalent. Several formulations of carbon equivalent exist, the simplest being the percentage of carbon plus one third of the percentage of silicon. As a second effect, silicon tends to promote the formation of graphite as the carbon-rich single phase instead of the Fe_3C intermetallic. Thus, the eutectic reaction now has two distinct possibilities, as seen in the modified phase diagram of Figure 4-15:

$$liquid \rightarrow austenite + Fe_3C$$

or

$$liquid \rightarrow austenite + graphite$$

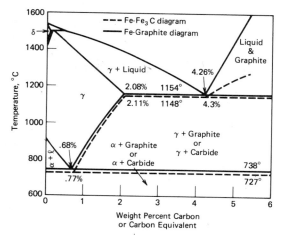

FIGURE 4-15 Iron-carbon diagram showing two possible eutectic reactions. Graphite is the solid line and cementite is the dashed.

The final microstructure, therefore, has two possible extremes: (1) all of the carbon-rich phase in the form of Fe_3C and (2) all of the carbon-rich phase in the form of graphite. In practice, these extremes can be approached in the various types of cast irons by control of the process variables. Graphite formation is promoted by slow cooling, high carbon and silicon contents, heavy section sizes, inoculation practices, and alloy additions of Ni and Cu. Cementite (Fe_3C) is favored by fast cooling, low carbon and silicon levels, thin sections, and alloy additions of Mn, Cr, and Mo.

Four basic types of cast irons are produced, the most common being *gray cast iron*. In this type, most of the carbon is in the form of graphite flakes formed during the eutectic reaction (some carbide may form at the lower eutectoid reaction). When fractured, the freshly exposed surface has a gray appearance (see Figure 4-16), and a graphite smudge can usually be obtained if one rubs a finger across a freshly fractured or machined surface.

FIGURE 4-16 (Left to right) Fractures of gray, white, and malleable iron. (Courtesy Iron Castings Society, Rocky River, Ohio.)

Gray cast iron is the least expensive of the four types and is characterized by those features that promote the formation of graphite. Typical compositions range from 2.5% to 4.0% carbon, 1.0% to 3.0% silicon, and 0.4% to 1.0% manganese. The microstructure consists of three-dimensional graphite flakes dispersed in a matrix of ferrite, pearlite, or other iron-based structure (which forms from austenite during the eutectoid reaction). Several possibilities for the matrix structure will be presented in Chapter 5. Figure 4-17 shows a typical section through gray cast iron. Because the graphite flakes have no appreciable strength, they act essentially as voids in the structure. Moreover, the pointed edges of the flakes act as preexisting notches or crack initiation sites, giving the material its characteristic brittle nature. The size and shape of the graphite flakes have considerable effect on the overall properties of gray cast iron. When maximum strength is desired, small, uniformly distributed flakes are desired with a minimum amount of mutual intersection.

A more effective means of controlling strength is through control of the matrix structure. Several distinct classes of gray iron can be identified on the basis of tensile strength, the class number corresponding to the minimum tensile strength in thousands of pounds per square inch. Class 20 iron (minimum tensile strength of 20,000 psi) consists of high carbon equivalent metal with a ferrite matrix. Higher strengths, up to class 40, are attainable with lower carbon equivalents and a pearlite matrix. Above class 40, alloying is required to provide solid solution strengthening, and heat treatment practices must often be employed. Gray irons can be obtained up through class 80, but in all cases, the presence of graphite flake results in extremely low ductility.

Gray cast irons possess excellent compressive strengths (compressive forces do not promote crack propagation), excellent machinability (graphite acts to break up the chips and to lubricate contact surfaces), good wear resistance

FIGURE 4-17   Photomicrographs of typical gray cast iron; 1000X. (Left) unetched; (right) etched. (Courtesy Bethlehem Steel Corporation.)

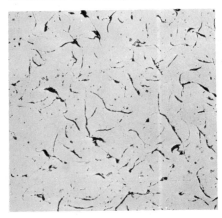

(graphite flakes self-lubricate), and outstanding vibration-damping characteristics (graphite flakes absorb vibration energy). High silicon contents promote good corrosion resistance and the fluidity desired for casting applications. For these reasons, together with its low cost, gray cast iron is an excellent material for large machinery parts that are subjected to high compressive loads and vibration.

White cast iron has essentially all of its carbon in the form of iron carbide and receives its name from the white surface that appears when the material is fractured (see Figure 4-16). Features promoting its formation are those that favor cementite over graphite: a low-carbon equivalent (1.8% to 3.6% carbon, 0.5% to 1.9% silicon, and 0.25% to 0.8% manganese) and rapid cooling.

By virtue of its large amounts of iron carbide, white iron is very hard and brittle and finds application where high abrasion resistance is required. For these uses, it is common to alloy the material so as to produce the hard, wear-resistant *martensite* structure as the iron-rich phase either upon solidification or by subsequent heat treatment. (This structure will be discussed in Chapter 5.) In this manner, both phases contribute to the wear-resistant characteristics.

Other applications involve a white-iron surface layer over a substrate of another material. Mill rolls that require extreme wear resistance may have a white cast iron surface and a steel interior. Variable cooling rates produced by tapered sections of metal chill bars placed in the molding sand can be used to produce white iron surfaces or sections on a gray iron casting. Where different cooling rates are used to produce white and gray cast irons in the same component, there is generally an intervening region of mixed white and gray irons, known as the *mottled zone*.

White iron can also be put through a controlled heat treatment in which the cementite dissociates into its component elements, and some or all of the carbon is converted into irregular graphite spheroids (also referred to as *clump* or *popcorn graphite*). The product, known as *malleable cast iron,* has significantly greater ductility than that exhibited by gray cast iron because the more favorable graphite shape removes the internal notches. The rapid cooling required to produce the starting white-iron structure restricts the size and thickness of malleable iron products. Most weigh less than 10 pounds (4.5 kg).

Two types of malleable iron can be produced, depending on the nature of the thermal cycle. If white iron is heated and held for a prolonged time just below the melting point, the carbon in cementite reverts to graphite. Subsequent slow cooling through the eutectoid reaction causes the carbon-containing austenite to go to ferrite and more graphite. The resulting product is known as *ferritic malleable iron,* which has properties consistent with its structure of irregular graphite spheroids in a ferrite matrix: 10% elongation, 35 ksi (240 MPa) yield strength, 50 ksi (345 MPa) tensile strength, and excellent impact strength, corrosion resistance, and machinability. The heat treatment times, however, are quite lengthy, often involving over 100 hours at elevated temperature. Figure 4-18 shows a typical resulting structure.

If the material is rapidly cooled through the eutectoid transformation after the first thermal hold, the carbon in the austenite does not form additional graphite

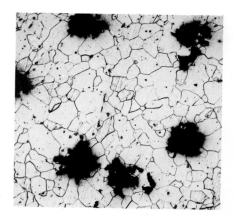

FIGURE 4-18 Photomicrograph of malleable iron. (Courtesy Iron Castings Society.)

but is retained in a pearlite or martensite matrix. These structures are stronger than ferrite, and the resulting product, known as *pearlitic malleable iron*, is characterized by higher strength and lower ductility than its ferritic counterpart. Typical properties range from 1% to 4% elongation, 45 to 85 ksi (310 to 590 MPa) yield strength, and 65 to 105 ksi (450 to 725 MPa) tensile strength, with reduced machinability compared to the ferritic material.

The modified graphite structure of malleable iron provides quite an improvement in properties, but it would be even better if it could be obtained during solidification rather than by prolonged heat treatment at highly elevated temperatures. Certain materials have been found that promote graphite formation and change the morphology of the graphite product. If sufficient magnesium (added in the form of MgFeSi or MgNi alloy) or cesium is added to the liquid iron just before solidification, graphite tends to form as regular spheroids during solidification. The added material is known as a nodulizing agent and the product as *ductile* or *nodular cast iron*. Subsequent control of cooling can produce a wide range of matrix structures, with ferrite or pearlite being the most common (see Figure 4-19). Properties span a wide range from 2% to 18% elongation, 40 to 90 ksi (275 to 620 MPa) yield strength, and 60 to 120 ksi (415 to 825 MPa) tensile strength.

The combination of good ductility, high strength, and castability makes ductile iron a rather attractive engineering material. Unfortunately, the cost of the nodulizer, the higher-grade melting stock, better furnaces, and improved process control required for its manufacture all contribute to the increased cost of ductile iron. Recently, austempered ductile iron has emerged as a significant casting material. It is made by subjecting ductile iron to the austempering treatment described in Chapter 5. Ductile iron with an austenite matrix (known as *austenitic ductile iron*) offers good corrosion resistance, nonmagnetic properties, and dimensional stability at elevated temperatures.

A material known as *compacted graphite iron* (or *vermicular graphite iron*) has also begun to attract considerable attention. Produced by a method similar to that used to produce ductile iron (a Mg-Ce-Ti addition is made), compacted

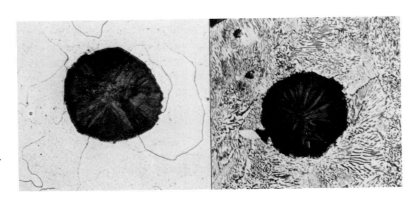

FIGURE 4-19 (Left) Ductile iron with ferrite matrix. (Right) Ductile iron with pearlite matrix; 500X. Note spheroidal graphite nodule.

graphite iron is characterized by a graphite structure intermediate between the flake graphite of gray iron and the nodular graphite of ductile iron and tends to possess the desirable properties and characteristics of each. Depending on composition and section size, tensile strengths of 45 to 75 ksi (310 to 520 MPa) and yield strengths of 35 to 60 ksi (240 to 415 MPa) have been reported, with elongation values averaging 4%. Machinability, castability, damping capacity, and thermal conductivity all approach those of gray cast iron. Impact and fatigue properties approach those of ductile iron. Areas of application tend to be those requiring high strength with good machinability, thermal conductivity, and thermal shock resistance.

Review Questions

1. What are some of the features that are useful in defining a phase?

2. What is an equilibrium phase diagram?

3. What three primary variables are generally considered in equilibrium phase diagrams?

4. Why is a pressure–temperature phase diagram not very useful for many engineering applications?

5. What is a cooling curve?

6. What features in a cooling curve indicate changes in a material's structure?

7. What are the names of the phase diagram lines that denote the upper and lower limits of the freezing range?

8. By what techniques can solubility limits be determined?

9. In general, how does the solubility of one material in another change as temperature is increased?

10. What three pieces of information can be obtained for each point in an equilibrium phase diagram?

11. What feature or point(s) on a tie-line can be used to determine the chemistry (or composition) of the component phases?

12. What tool can be used to compute the relative amounts of the component phases in a two-phase mixture? How does this tool work?

13. What is a cored structure and how is it produced?

14. What is the general form of a eutectic reaction?

15. Why are alloys of eutectic compositions attractive for casting and as filler metals in soldering and brazing?

16. What is a stoichiometric intermetallic compound and how would it appear in a temperature–composition phase diagram?

17. What are the typical mechanical property characteristics of an intermetallic compound?

18. In what form(s) might intermetallic compounds be undesirable in an engineering material? In what form(s) might they be attractive?

19. What feature(s) in the iron–carbon phase diagram are used to denote the difference between cast irons and steels?

20. What features of austenite make it attractive for forming operations and as a starting structure for many heat treatments?

21. Which of the three-phase reactions in the iron–carbon diagram is most important to understanding the behavior of steels? Write this reaction in terms of the interacting phases and their composition.

22. Describe the ability of iron to dissolve carbon in solution when in the form of austenite (the elevated temperature phase) and when in the form of ferrite at room temperature.

23. What is pearlite? What is its structure?

24. What is a hypoeutectoid steel? A hypereutectoid steel?

25. What is the general composition of cast iron? In addition to iron and carbon, what other elements are generally present?

26. What are the two possible high-carbon phases in cast irons? What features tend to favor the formation of each?

27. Describe the microstructure of gray cast iron.

28. Which of the structural units is generally altered to increase the strength of a gray cast iron.

29. What are some of the attractive engineering properties of gray cast iron?

30. What is the dominant mechanical property of white cast iron?

31. What structural feature is primarily responsible for the increased ductility and fracture resistance of malleable cast iron?

32. What requirements of ductile iron manufacture are responsible for its increased cost over materials such as gray cast iron?

33. Compacted graphite iron has a structure and properties intermediate between what two types of already-discussed cast irons?

Heat Treatment

Theory and Processes of Heat Treatment

The presentation on cast irons in Chapter 4 has already shown that variations in the heating or cooling of a material can be used to produce different structures and different properties. Many engineering materials can possess not just one set of properties, but an entire spectrum that can be selected and varied at will. *Heat treatment* is the mechanism by which these properties can be altered without a concurrent change in product shape.

By definition, *heat treatment is the controlled heating and cooling of metals for the purpose of altering their properties*. Because both physical and mechanical properties can be altered by heat treatment, it is one of the most important and widely used manufacturing processes. However, if heat treatment is performed improperly, more harm than good can result.

Although the term *heat treatment* applies only to processes in which the heating and cooling are done for the specific purpose of altering properties, heating and cooling often occur as incidental phases of other manufacturing processes. Obvious examples are processes such as hot forming and welding. Properties are altered in a beneficial or harmful manner, as though an intentional heat treatment cycle had been performed. Thus, the designer who selects material and the engineer who determines its processing must be aware of the possible changes in properties during heating and cooling. Heat treatment must be understood and correlated with the other manufacturing processes if effective results are to be obtained. Both the theory of heat treatment and a survey of the various processes will be presented in this chapter.

Processing Heat Treatments

Although the term *heat treatment* is often associated only with those thermal processes designed to increase strength, the definition permits inclusion of a set of processes that we will call *processing heat treatments*. These are performed with the major goal of preparing the material for fabrication. Specific objectives may be the improvement of machining characteristics, the reduction of forming forces, or the restoration of ductility to enable further processing. Thus, heat treatment becomes even more attractive. The same metal can now be softened for ease of fabrication and then, by another heat treatment, given a totally different set of properties for actual service.

Equilibrium Diagrams as Aids to Heat Treatment.   Most processing heat treatments involve rather slow cooling or extended times at elevated temperatures, thus tending to approximate equilibrium conditions. The resulting structures, therefore, can be reasonably predicted by the use of equilibrium phase diagrams. The diagram indicates the temperatures that must be attained to achieve a desired structure and the change that will occur on subsequent cooling. It should be remembered, however, that the diagram is for true equilibrium conditions, and departures from equilibrium may lead to substantially different results.

Processing Heat Treatments for Steel.   Because most processing heat treatments are applied to plain carbon and low-alloy steels, they will be presented here with the simplified iron–carbon equilibrium diagram of Figure 4-11 serving as a reference guide. Figure 5-1 shows this diagram with the key transition lines labeled in standard notation. The eutectoid line is designated by the symbol A_1, and A_3 designates the boundary between austenite and ferrite + austenite. The transition from austenite to austenite + cementite is designated as the A_{cm} line.

A number of process heat-treating operations are classified under the general term *annealing*. These may be employed to reduce hardness, to remove residual stresses, to improve toughness, to restore ductility, to refine grain size, to reduce segregation, or to alter the mechanical, electrical, or magnetic properties of the material. By producing a certain desired structure, characteristics can be imparted that are favorable to the subsequent operations or application. The temperature, the cooling rate, and the specific details of the process are determined by the material being treated and the objectives of the treatment.

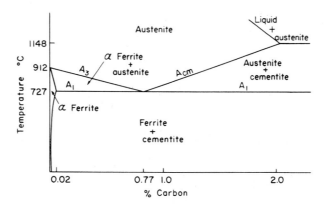

FIGURE 5-1   Simplified iron-carbon phase diagram for steels with transition lines labeled in standard notation.

In *full annealing, hypoeutectoid steels* (less than 0.77% carbon) are heated to 50–100°F (30–60°C) above the A_3 temperature to convert the structure to homogeneous single-phase austenite of uniform composition and temperature, are held at this temperature for a period of time, and are then slowly cooled at a controlled rate to below the A_1 temperature. A general rule is to provide 1 hour at elevated temperature per inch of thickness of the largest section, but energy savings have motivated a reduction of this time. Cooling from the elevated temperature is usually done in the furnace, with the temperature decreasing at a rate of 20–50°F (10–30°C) per hour to at least 50°F (30°C) below the A_1 temperature, followed by air cooling to room temperature. The resulting structure is coarse pearlite (widely spaced lamellae) with excess ferrite in amounts predicted by the phase diagram. The material is quite soft and ductile.

The procedure for hypereutectoid alloys (more than 0.77% carbon) is basically the same, except the original heating is only into the austenite-plus-cementite region (50–100°F above the A_1). If the material is slow-cooled from the pure austenite region, a continuous network of cementite may form on the grain boundaries and make the material brittle. When the material is properly annealed, the structure of a hypereutectoid steel is coarse pearlite plus excess cementite in dispersed spheroidal form.

Full anneals are time-consuming, and considerable energy is needed to maintain the elevated temperatures required during soaking and furnace cooling. When extreme softness is not required and cost savings are desired, *normalizing* may be employed. Here the steel is heated to 100°F (60°C) above the A_3 (hypoeutectoid) or A_{cm} (hypereutectoid), is soaked to produce uniform austenite, and is then removed from the furnace and allowed to cool in still air. The resultant structures and properties depend on the subsequent cooling rate. Although wide variations are possible, depending on the size and geometry of the metal, fine pearlite with excess ferrite or cementite is generally produced.

One should note a key difference between full annealing and normalizing. In the full anneal, the furnace imposes identical cooling conditions at all locations within the metal, which produce identical properties. With normalizing, the cooling is different at different locations. Properties will vary between surface and interior, and different thickness regions will have different properties. When subsequent processing involves automated machining, the added cost of a full anneal may be justified since it produces a product with uniform machinability at all locations.

Where cold working has severely strain-hardened a metal, it is often desirable to restore the ductility, either for service or to permit further processing without danger of fracture. A *process anneal* can be used for this purpose. The metal is heated to a temperature slightly below the A_1, is held long enough at that temperature to achieve softening, and is then cooled at any desired rate (usually in air). Since austenite is not formed, the existing phases simply change their morphology, that is, their size, shape, and distribution. Process anneals are often used to recrystallize low-carbon steel sheets. Because the material is not heated to as high a temperature as in the other processes, a process anneal is somewhat cheaper and more rapid and tends to produce less scaling.

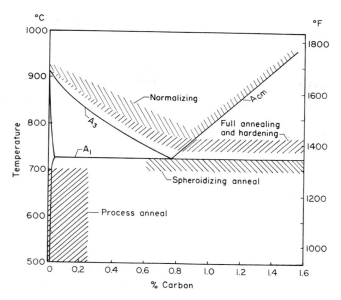

FIGURE 5-2 Graphical summary of process heat treatments for steels on an equilibrium diagram.

A *stress-relief anneal* may be employed to remove residual stresses in large steel castings and welded structures. Parts are heated to temperatures below the A_1 (1000–1200°F or 550–650°C), are held for a period of time, and are then slow-cooled. Times and temperatures vary with the condition of the component.

When high-carbon steels must be prepared for machining or forming, a process known as *spheroidization* is employed. The goal is to produce a structure in which all cementite is in the form of small, well-dispersed spheroids or globules in a ferrite matrix. This can be accomplished by a variety of techniques, including (1) prolonged heating at a temperature just below the A_1 followed by relatively slow cooling; (2) prolonged cycling between temperatures slightly above and slightly below the A_1; or (3) in the case of tool steel or high-alloy steel, heating to 1400–1500°F (750–800°C) or higher and holding at this temperature for several hours, followed by slow cooling.

Although the selection of a processing heat treatment often depends on the desired objectives, steel composition strongly influences the choice. Low-carbon steels (less than 0.3% carbon) are most often normalized or process-annealed. Steels of the medium (0.4% to 0.6%) carbon range are usually full-annealed. Above 0.6% carbon, a spheroidization treatment is generally required. Figure 5-2 provides a graphical summary of the process heat treatments.

Heat Treatments for Nonferrous Metals. Most nonferrous metals do not have the significant phase transitions observed in the iron–carbon system, and for them, process heat treatments do not play such a significant role. Aside from precipitation hardening, which will be discussed later, nonferrous metals are ordinarily heat-treated for three purposes: (1) to obtain a uniform structure (such as to eliminate coring, the segregated solidification structure discussed in Chapter 4); (2) to provide stress relief; or (3) to bring about recrystallization. Coring, which may be present in castings that have cooled too rapidly, can be removed by heating to moderate temperatures and holding for a sufficient period to allow thorough diffusion to take place. Similarly, stresses that result from forming, welding, or brazing can be removed by heating for several hours at relatively low temperatures. Recrystallization, as discussed in Chapter 3, is a function of the particular metal, the degree of prior straining, and the time provided for completion. In general, the more a metal has been strained, the lower the recrystallization temperature or the shorter the time. Without prior straining, recrystallization will not occur and heating will produce only undesirable grain growth.

Heat Treatments Used to Increase Strength

Six major mechanisms are available to increase the strength of metals: (1) solid solution hardening; (2) strain hardening; (3) grain size refinement; (4) precipitation hardening; (5) dispersion hardening; and (6) phase transformations. All can be induced or altered by heat treatment, but not all can be applied to any given metal.

In *solid solution hardening*, a base metal dissolves other atoms in solid solution, either as *substitutional solutions*, where the new atoms occupy sites on the regular crystal lattice, or as *interstitial solutions*, where the new atoms squeeze into "holes" in the base lattice. The amount of strengthening depends on the amount of dissolved solute and the size difference of the atoms involved. Distortion of the host structure makes dislocation motion more difficult.

Strain hardening, as discussed in Chapter 3, produces increased strength by plastic deformation under cold-working conditions.

Because grain boundaries act as barriers to dislocation motion, a metal with small grains tends to be stronger than the same metal with larger grains. Thus, *grain-size refinement* can be used to increase strength, except at elevated temperatures, where failure is by a grain-boundary diffusion-controlled creep mechanism. Grain-size refinement is one of the few processes capable of improving both strength and ductility.

Precipitation hardening, or *age hardening*, is a method whereby strength is obtained from a nonequilibrium structure produced by a three-step (solution treat, quench, and age) heat treatment.

Strength obtained from distinct second-phase particles in a base matrix is called *dispersion hardening*. To be effective, these second phases should be stronger than the matrix, adding strength through both their reinforcing action and by the additional barriers presented to dislocation motion.

Phase transformation strengthening involves alloys that can be heated to form a single high-temperature phase and subsequently transformed to one or more low-temperature phases upon cooling. Where phase transformation is used to increase strength, the cooling is usually rapid and the phases produced are nonequilibrium in nature.

All of the mechanisms just described can be used to increase the strength of nonferrous metals. Solid solution strengthens single-phase metals. Strain hardening is applicable if sufficient ductility is present. Eutectic-forming alloys possess considerable dispersion hardening. The most effective strengthening mechanism for nonferrous metals, however, is precipitation hardening.

Precipitation or Age Hardening.   Some alloy systems, mostly nonferrous, possess a sloping solvus line so that certain alloys can be heated into a single-phase solid solution and, owing to decreasing solubility, will form two distinct phases at lower temperatures. If the heated single phase is rapidly cooled, however, a supersaturated solid solution can be formed in which the material required to form the second phase is trapped in the base lattice. A subsequent *aging* process permits the excess solute atoms to precipitate out of the supersaturated matrix and to produce a controllable nonequilibrium structure.

As an example, consider the aluminum-rich portion of the aluminum–copper phase diagram presented in Figures 5-3 and 5-4. Follow an alloy composed of 96% aluminum and 4% copper, being slowly cooled from 1000°F. On crossing the solvus line (A-B) at 930°F, θ phase is precipitated out of the alpha solid

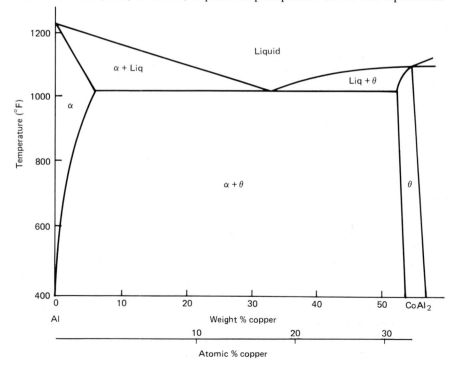

FIGURE 5-3   High-aluminum section of the aluminum-copper equilibrium phase diagram.

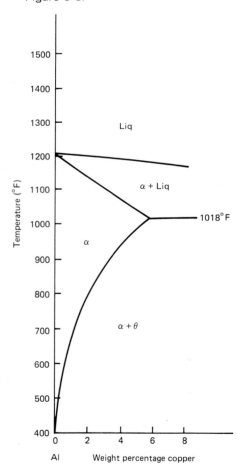

FIGURE 5-4 Enlargement of the solvus-line region of the aluminum-copper equilibrium diagram of Figure 5-3.

solution because the solubility of copper in aluminum decreases from 5.65% at 1018°F to less than 0.2% at room temperature. If this same alloy were rapidly cooled from 1430°F, there would not be sufficient time for the normal two-phase structure to form, and the α phase would be retained in a highly super-saturated state. Because this supersaturated condition is not normal, the excess copper would like to precipitate out of solution and coalesce into θ phase particles, θ phase consisting of an intermetallic compound of the form $CuAl_2$. Atom movement or diffusion is required, and therefore, time at elevated temperature may have to be provided. If the temperature is now raised to around 350°F, precipitation of high-copper clusters will begin to occur. If carefully controlled, this precipitation can produce a quadrupling of the material's yield strength.

Precipitation hardening is a three-step controlled heat treatment. The first step is a heating (*solution treatment*) to put the material into a single-phase solid solution. This heating must not exceed the eutectic temperature or there will be melting in a cored structure. After soaking to achieve a uniform single phase, the material is *quenched* (rapidly cooled), usually in water, to prevent diffusion and to produce the supersaturated solid solution. In this state, the material is rather soft, possibly softer than in the annealed condition. It can be straightened, formed, or machined.

At this point, precipitation-hardening materials can be divided into two types: (1) *naturally aging,* for which room temperature is capable of providing the diffusion necessary to convert the unstable supersaturated structure into the stable two-phase structure, and (2) *artificially aging,* for which elevated temperatures are necessary to promote diffusion. With the first type, refrigeration may be required to retain the after-quench condition of softness, as with aluminum alloy rivets. Removed from the refrigeration, they are easily headed and then proceed to full strength after several days at room temperature. The properties of the second type can be readily controlled by controlling the time and temperature of elevated temperature aging.

Aging is a continuous process that begins with the clustering of solute atoms on distinct planes of the parent lattice. Various transitions may then occur, leading ultimately to the formation of a distinct second phase with its own characteristic crystal structure.

A key concept in this sequence is *coherency.* If the crystallographic planes are continuous in all directions, the solute clusters tend to distort (strain) the adjacent lattice for a sizable distance, and a small cluster appears to be a much larger barrier to dislocation motion. When the clusters reach a certain size, they tend to break free of the parent structure, forming distinct second-phase particles with well-defined interphase boundaries. Coherency is lost, and the mechanism of strengthening becomes *dispersion hardening,* in which the particles present only their physical dimensions as effective dislocation blocks. The material is said to be *overaged.* A benefit of artificial aging, therefore, is the ability to stop the process at any stage by simple quenching. The structure and properties of that stage are retained, provided that the material does not subsequently experience elevated temperature conditions. Artificially aged alloys are quite popular because they allow the tailoring of properties to specific needs.

Precipitation hardening is responsible for the engineering strengths of many aluminum, copper, and magnesium alloys. In many cases, the strength is more than double that observed upon conventional slow cooling. By special processing, some age-hardenable ferrous alloys have also been produced.

Iron-base metals have been heat-treated for centuries, and today, over 90% of all heat treating is performed on steel. The striking changes that resulted from plunging red-hot steel into cold water or some other quenching medium were awe-inspiring to the ancients. Those who did such heat treatment in the making of swords or armor were looked on as possessing unusual powers, and much superstition arose regarding the process. Because quality was directly related to the act of quenching, great importance was placed on the quenching medium that was used. For example, urine was thought to be a superior quenching medium, and that from a red-haired boy was deemed particularly effective.

Strengthening Heat Treatments for Steel

The Isothermal Transformation Diagram. It has only been within the last 100 years that the art of heat treating has begun to turn into a science. One of the major barriers to understanding was the fact that the strengthening treatments were nonequilibrium in nature. Variations in cooling produced variations in structure and properties.

One of the aids to understanding the nonequilibrium processes was the *isothermal transformation* (IT) or *time–temperature–transformation* (T-T-T) diagram, obtained by heating thin specimens of a given metal to form uniform single-phase austenite, "instantaneously" quenching to a temperature where austenite was not the stable phase, holding for variable periods of time, and observing the resultant product via photomicrographs.

For simplicity, consider a carbon steel of eutectoid composition (0.77% carbon) and the resulting T-T-T diagram of Figure 5-5. Above 1341°F (727°C), austenite is the stable phase. Below this temperature, the face-centered austenite would like to transform to body-centered ferrite and carbon-rich cementite. Two factors control the rate of transition: (1) the motivation or driving force for the change and (2) the ability to form the desired products (i.e., the ability to move atoms through diffusion). Figure 5-5 can be interpreted as follows. For any given temperature below 1341°F, zero time corresponds to a sample quenched "instantaneously." The structure is usually unstable austenite. As time passes (moving horizontally across the diagram), a line is encountered representing the start of transformation, and a second line indicates completion of the phase change. At elevated temperatures (just below 1341°F), diffusion is rapid, but the rather sluggish driving force dominates the kinetics. At a low temperature, the driving force is high, but diffusion is quite limited. The phase-transformation kinetics are more rapid at a compromise intermediate temperature than at either

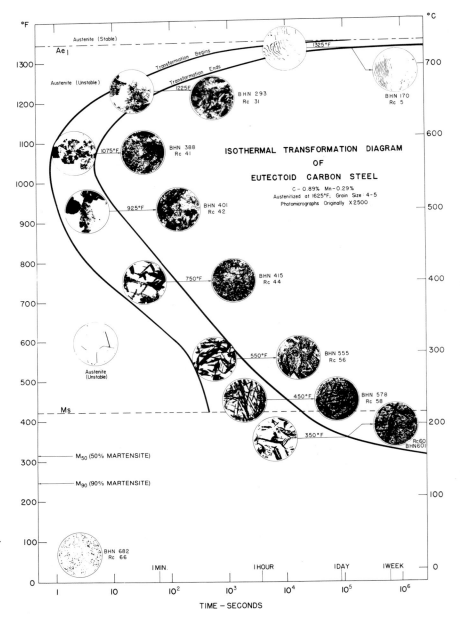

FIGURE 5-5 Isothermal transformation diagram (T-T-T diagram) for eutectoid composition steel. Structures resulting from transformation at various temperatures are shown as insets. (Courtesy USX Corporation.)

extreme, and the result is the often cited *C-curve* terminology. The portion of the C that extends furthest to the left is known as the *nose* of the T-T-T diagram.

Let us now consider the products of the various phase transformations. If the transformation occurs between the A_1 temperature and the nose of the curve, the departure from equilibrium conditions is not very great. Austenite again transforms into ferrite and cementite in the combined structure called *pearlite* (as discussed in Chapter 4). Because diffusion capabilities are greater at higher temperatures, the lamellae spacing (the separation distance between similar layers) is larger for pearlite produced at higher temperatures. The pearlite formed near the A_1 temperature is known as *coarse pearlite*, and the structure formed near the nose is called *fine pearlite*.

If the austenite is quenched to a temperature between the nose and the temperature designated as M_s, another product forms. The transformation is now a significant departure from equilibrium, and the diffusion required to form lamellar pearlite is no longer available. The metal still has the same goal: to change crystal structure from the face-centered austenite to the body-centered ferrite and accommodate the excess carbon in the form of cementite. The resulting structure, however, is not one of alternating plates, but a dispersion of discrete cementite particles in a lathlike or needlelike matrix of ferrite. Electron microscopy may be required to resolve the carbides in this structure, which is known as *bainite*. Because of the fine dispersion of carbide, its strength exceeds that of fine pearlite, and ductility is retained because soft ferrite is the continuous matrix phase.

If the austenite is now quenched to below the M_s temperature, a different type of transformation occurs. The metal still desires to go from the face-centered to the body-centered structure, but it cannot expel the required amount of carbon to form ferrite. Responding to the severe nonequilibrium conditions, it simply undergoes an instantaneous change in crystal structure with no diffusion. The trapped carbon distorts the structure to produce a body-centered tetragonal lattice (distorted body-centered cubic), with the degree of distortion being proportional to the amount of trapped carbon. The new structure, shown in Figure 5-6, is known as *martensite*, and with sufficient carbon, it is exceptionally strong, hard, and brittle. The dislocation motion necessary for metal flow is effectively blocked by the highly distorted lattice.

As shown in Figure 5-7, the hardness and strength of steel in the martensitic condition clearly depend on the carbon content. Below 0.10% carbon, the martensite is not very strong. Hardness is typically 30 to 35 R_c at 0.1% carbon and decreases rapidly with lower amounts of carbon. Since no diffusion occurs in the transformation, higher-carbon-content material forms higher-carbon-containing martensite, with a concurrent increase in strength and hardness and decrease in toughness and ductility. From 0.3% to 0.7% carbon, the hardness increases rapidly. Above 0.7% carbon, however, hardness rises only slightly with increased carbon, a feature related to retained austenite.

The amount of martensite formed on cooling is a function of the lowest temperature encountered and not of the time at that temperature, as shown in Figure 5-8. Returning to Figure 5-5, we see, below the M_s, a temperature designated as M_{50}, at which the structure is 50% martensite and 50% untransformed austenite. At the lower M_{90} temperature, the structure is 90% martensite. If no further cooling is undertaken, the untransformed austenite can remain within the structure indefinitely. This *retained austenite* can cause loss of strength or hardness, dimensional instability or cracking, or brittleness. Since most quenches are to room temperature, retained austenite problems become

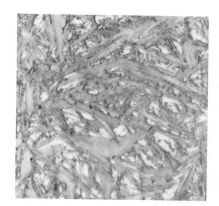

FIGURE 5-6 Photomicrograph of martensite; 1000X. (Courtesy USX Corporation.)

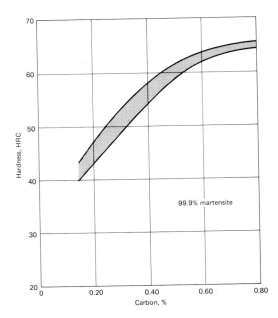

FIGURE 5-7 Effect of carbon on the hardness of martensite.

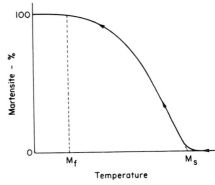

FIGURE 5-8 Schematic representation depicting the amount of martensite formed upon quenching to various temperatures from M_s through M_f.

significant when the martensite finish, or 100% martensite, temperature lies below room temperature. Higher carbon contents and alloy additions both decrease all martensite-related temperatures, and these materials may require refrigeration or a quench in liquid nitrogen to yield full hardness.

Note that all of the transformations that occur below the A_1 are one-way transitions, austenite-to-something. These are the only reactions possible, and it is impossible to convert one product to another without first reheating to above A_1 to again form some stable austenite.

The T-T-T diagram is quite useful in determining the kinetics of transformation and the nature of the product. The left-hand curve shows the elapsed time at constant temperature before transformation begins, and the right-hand curve shows the time required for complete transformation at that temperature. If a hypo- or hypereutectoid steel were considered, additional regions would be added to correspond to the primary equilibrium phases. These regions would not extend below the nose, however, since the nonequilibrium bainite and martensite phases do not have to maintain fixed compositions, as do the phases in the equilibrium diagram. Figure 5-9 is a T-T-T curve for a hypoeutectoid steel.

Tempering of Martensite. Because it lacks good toughness and ductility, medium- or high-carbon martensite is not a useful engineering microstructure, despite its great strength. A subsequent heating, known as *tempering*, is usually required to restore some desired degree of toughness at the expense of a decrease in strength and hardness.

Martensite is a supersaturated solid solution of carbon in alpha ferrite and therefore is a metastable structure. When reheated into the range of 200–1300°F (100–700°C), carbon atoms are rejected from solution, and the structure moves

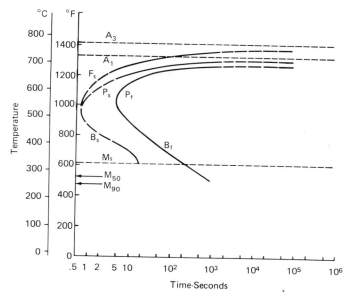

FIGURE 5-9   Isothermal transformation diagram for a hypoeutectoid steel (1050) showing additional region for primary ferrite.

toward a mixture of the stable ferrite and cementite phases. This decomposition of martensite into ferrite and cementite is a time- and temperature-dependent, diffusion-controlled phenomenon with a spectrum of intermediate and transitory conditions. The initial stage, which occurs at 200–400°F (100–200°C), is the precipitation of an intermediate carbide with the composition of $Fe_{2.4}C$, known as *epsilon* (ε) *carbide*. This precipitation allows the matrix to revert to the body-centered-cubic configuration. From 400 to 750°F (200 to 400°C), the structure becomes one of ferrite and cementite. Little change is observable in the microscope, however, for the cementite particles are submicroscopic and the original martensite boundaries are retained. Figure 5-10 shows such a material, which appears as a rather mottled mass with little well-defined structure. Electron-microscopic studies reveal a fine carbide dispersion, which is responsible for the observed softening and improved ductility.

If tempering progresses into the 750–1000°F (400–550°C) range, the martensite boundaries disappear, and a new ferrite structure nucleates and grows. As the precipitated carbides increase in size, the properties move further in the direction of a weaker, but more ductile, material. Figure 5-11 shows some of the newly formed ferrite (white) in the tempered steel.

Above 1000°F (550°C), the new ferrite grains totally consume the original structure, and the cementite particles have become larger and more spherical. Figure 5-12 shows this structure, which has the highest toughness and ductility and the lowest strength of all tempered martensites. Heating to too high a temperature (above the A_1) causes the structure to revert back to stable austenite. By quenching steel to form 100% martensite and then tempering it at various temperatures, an infinite range of structures and corresponding properties can be produced. This is known as the *quench-and-temper process*, and the product is called *tempered martensite*.

FIGURE 5-10   Eutectoid steel, hardened and tempered at 600°F (315°C); 1000X. (Courtesy USX Corporation.)

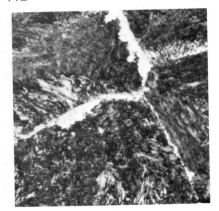

FIGURE 5-11. Eutectoid steel, hardened and tempered at 1000°F (540°C); 1000X. (Courtesy USX Corporation.)

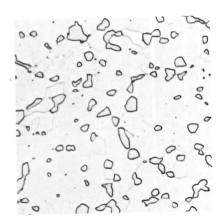

FIGURE 5-12. Photomicrograph of the structure obtained by prolonged tempering above 1000°F (540°C), sometimes called spheroidite; 1000X. (Courtesy USX Corporation.)

FIGURE 5-13 Schematic C-C-T diagram for a eutectoid composition steel, showing several superimposed cooling curves and the resultant structures. (Courtesy USX Corporation.)

Continuous-Cooling Transformations.   Although the T-T-T diagrams have provided significant information about the structures obtained through non-equilibrium thermal processing, they are usually not applicable to direct engineering use because the assumptions of instantaneous cooling followed by constant temperature transformation rarely match reality. A diagram showing the results of continuous cooling at various rates would be far more useful. What will be the result if the temperature is decreased at a rate of 500°F per second, 50°F per second, or 5°F per second?

A *continuous-cooling-transformation* (C-C-T) *diagram*, such as that shown schematically in Figure 5-13, can provide answers to these questions and several others. Critical cooling rates required to obtain products can easily be determined. If cooled fast enough, the structure will be all martensite. A slow cool may produce coarse pearlite and some primary phase. Intermediate rates usually produce mixed structures, the time at any one temperature usually being insufficient for complete transformation. If each structure is regarded as providing a companion set of properties, the wide range of possibilities obtainable through controlled heating and cooling of steel becomes evident.

The Jominy Test for Hardenability.   A tool that is commonly used to assist in the understanding of nonequilibrium heat treatment is the *Jominy end-quench hardenability test* and its associated diagrams. In this test, depicted in

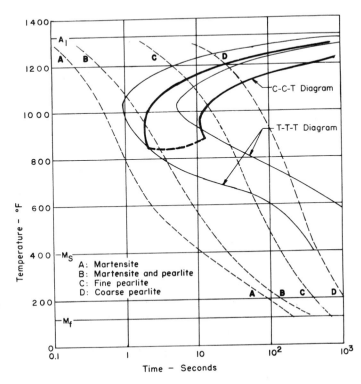

Figure 5-14, an effort is made to reproduce the entire spectrum of cooling rates on a single specimen by quenching a heated bar from one end. The quench is standardized by specifying the specimen geometry, the quench medium (water at 75°F), the internal nozzle diameter (½ inch), the water pressure (that producing a 2½-inch vertical fountain), and the gap between the nozzle and the specimen (½ inch).

After the specimen has been cooled by the quench, a flat region is ground along one side and R_c hardness readings are taken every ¹⁄₁₆th inch along the bar. The resulting data are then plotted as shown in Figure 5-15. As the cooling rate is known for the various locations within the bar, the resulting hardness values can be correlated to the cooling rate that produced them.

Application of the test assumes that identical results will be obtained if the same material undergoes identical cooling histories. If the cooling rate is known for a given location within a part (from experimentation or theory), the properties at that location can be predicted as those at the equivalent cooling-rate location on the Jominy test bar. Likewise, if specific properties are required, the necessary cooling rate can be determined for a selected material. If the cooling rates are restricted, various materials can be compared and a satisfactory alloy selected. Figure 5-16 shows the Jominy curves for several common engineering steels.

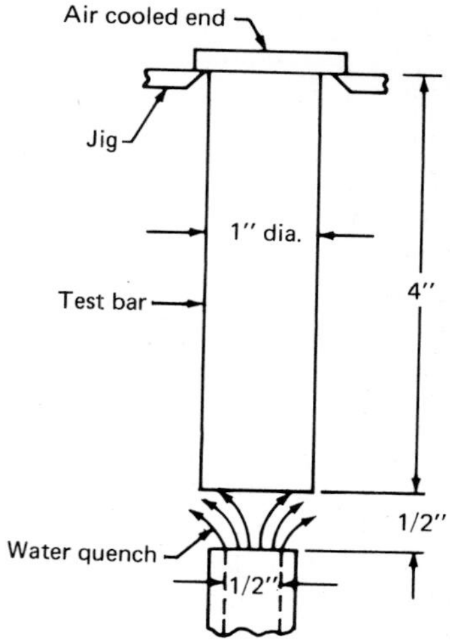

FIGURE 5-14 Schematic diagram of the Jominy hardenability test.

Hardenability.   When one is attempting to understand the heat treatment of steel, several key effects must be understood: the effect of carbon content, the effect of alloy additions, and the effect of various quenching conditions. The first two relate to the material and the third to the process.

Hardness is a mechanical property related to strength and is a strong function of the carbon content of a metal. *Hardenability,* on the other hand, is a measure of the depth to which full hardness can be attained under a normal hardening cycle and is related primarily to the amounts and types of alloying elements. In Figure 5-16, all of the steels have the same carbon content, but they differ in type and amounts of alloy elements. Maximum hardness is the same in all cases, but the depth of hardening varies considerably. Figure 5-17 shows the results for steels containing the same alloying elements but variable amounts of carbon. Note the change in peak hardness.

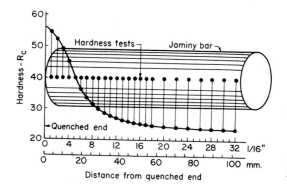

FIGURE 5-15 Typical hardness distribution in Jominy bars.

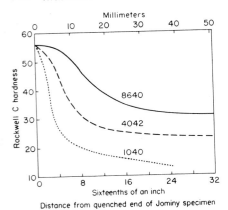

FIGURE 5-16 Jominy hardness curves for engineering steels with the same carbon content and varying types and amounts of alloy elements.

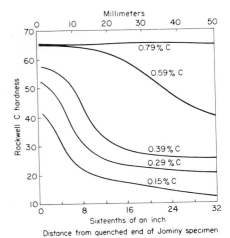

FIGURE 5-17 Jominy hardness curves for engineering steels with identical alloy conditions but variable carbon content.

The primary reason for adding alloy elements to commercial steels is to increase the hardenability, not to improve the strength properties. Steels with greater hardenability need not be cooled as rapidly to achieve a desired level of strength or hardness, and they can be completely hardened in thicker sections.

Materials selection for steels requires an accurate determination of need. Strength and structure tend to be determined by carbon content, the general rule being to stay as low as possible and still meet specifications. Because heat can be extracted only from the surface of a metal, the depth of required hardening sets the conditions for hardenability and quench. For a given quenching condition, different alloys produce different results. Because alloy elements increase the cost of a material, a general rule is to select only what is required to ensure compliance with specifications. Money is often wasted by specifying an alloy steel for an application where a plain carbon steel, or a steel that has lower alloy content (and is thus less costly), would be satisfactory. Another alternative when greater depth of hardness is required is to modify the quench conditions so that a faster cooling rate is achieved. Quench changes may be limited, however, by cracking or warping problems, depending on the shape, size, complexity, and precision of the part being treated.

Quench Media.   Quench media vary in their effectiveness, and one can best understand the variation by considering the three stages of quenching. When a piece of hot metal is first inserted into a tank of liquid quenchant, that adjacent to the metal vaporizes and forms a gaseous layer separating the metal and the liquid. Cooling is slow through this *vapor jacket (first stage)* since all heat transport must now be through a gas. This stage occurs when the metal is above the boiling point of the quenchant. Soon bubbles nucleate and break the jacket; liquid contacts the metal, vaporizes (removing its heat of vaporization from the metal), and forms a bubble; and the process continues. This *second stage of quenching* provides rapid cooling as a result of the large quantities of heat removed by the mechanism. When the metal cools below the boiling point of the quenchant, vaporization can no longer occur. Heat transfer now takes place by conduction across the solid–liquid interface, aided by convection or stirring within the quenchant. This is the *third stage* of quenching.

Water is a fairly good quenching medium because of its high heat of vaporization and the second stage of quenching extending down to 212°F (100°C), usually well into the martensite range or below. Water is also cheap, readily available, easily stored, nontoxic, nonflammable, smokeless, and easy to filter and pump. However, with a water quench, the clinging tendency of the bubbles may cause soft spots in the metal. Agitation is recommended with the use of a water quench. Still other problems with the water quench include its oxidizing nature, its corrosivity, and the tendency to excessive distortion and possible cracking.

Brine (salt water) is a more severe quench medium than water because the salt nucleates bubbles, forcing a more rapid transition through the vapor-jacket stage. Unfortunately, brine tends to accelerate corrosion problems unless completely removed. Sodium or potassium hydroxide can be used when very severe quenching is desired and one wishes to obtain good hardness in low-carbon steels. Various degrees of agitation or spraying of the quenchant can also be used to increase the effectiveness of a given medium.

When a slower cooling rate is desired, oil quenches can be employed. Various oils are available that have high flash points and different degrees of quenching effectiveness. Since the boiling points are often quite high, the transition to third-stage cooling generally precedes the martensite start. The slower cooling through the M_s to M_f temperature range leads to a milder temperature gradient and a reduced likelihood of cracking. Problems associated with oil quenchants include water contamination, smoke, fumes, spill problems, and fire hazard. In addition, quench oils tend to be somewhat expensive.

Quite often, the need is for a quenchant that will cool more rapidly than oils but slower than water or brine. To fill this gap, numerous *polymer quench* solutions (also called *synthetic quenchants*) have been developed. Tailored quenches can be produced by varying the concentration of the quench components (such as a glycol polymer and water). Polymer quenchants provide extremely uniform and reproducible results, are less corrosive than water or brine, are cheaper and less of a fire hazard than oils (no fires, fumes, smoke, or need for air-pollution-control apparatus), and tend to minimize distortion. Many polymer quench mediums are extremely sensitive to concentraton changes, however, and require constant monitoring and control during use.

When slow cooling is desired, molten salt baths can be employed to provide a medium where the quench goes directly to third-stage cooling. Still slower rates can be obtained by cooling in still air, burying the metal in sand, and a variety of other methods.

The Role of Design in the Heat Treatment of Steel.

Design details and material selection play important roles in the satisfactory and economical heat treatment of parts. Proper consideration of these factors usually leads to more simple, more economical, and more reliable products. Failure to relate design and materials to heat-treatment procedure usually produces disappointing or variable results and often results in service failure.

Undesirable design features are (1) nonuniform sections or thicknesses; (2) sharp interior corners; and (3) sharp exterior corners. Because these may easily find their way into the design of parts, the designer should be aware of their effect in heat-treating operations. Undesirable results may include nonuniform structure and properties, undesirable residual stresses, cracking, warping, and dimensional changes.

Heat can only be extracted from a piece of metal through its exposed surface. Thus, if a piece to be hardened has a nonuniform cross section, the thin portion will cool rapidly and fully harden, whereas the thick region may not harden, except on the surface. This surface may even be tempered somewhat by the heat retained in the center of the heavy section. The shape that might be closest to ideal from the viewpoint of quenching would be a doughnut, having a uniform cross section with maximum exposed surface and no sharp corners.

Residual stresses are the often-complex results of the various dimensional

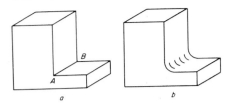

FIGURE 5-18   (Left) Shape containing nonuniform sections joined by a sharp interior corner that may crack during quenching. (Right) Improved design using a large radius to join sections and to avoid cracking in heat treatment.

changes that occur during heat treatment. Thermal contraction during cooling is a well-understood phenomenon. In addition, the various phases and structures that may form are often characterized by different densities, and therefore, a volume expansion or contraction accompanies the phase transformations. When austenite transforms to martensite, there is a volume expansion of up to 4%. Austenite transforming to pearlite also experiences a volume expansion, but it is of smaller magnitude.

If all temperature changes occurred uniformly throughout the part, all changes in dimension would occur simultaneously and the result would be a component free of residual stresses. However, most parts being heat-treated experience nonuniform temperatures during the cooling or quenching process. Cross sections should be designed so that temperature differences are as low as possible and are not concentrated. If this is not possible, slower cooling and a material that will harden in an oil or air quench may be required. Because materials having greater hardenability are invariably more expensive, the design alternative clearly has advantages.

When temperature differences and resultant residual stresses become severe or localized, additional problems, such as cracking or distortion, can result. Figure 5-18a shows an example in which a sharp interior corner was placed at a change of cross section. Upon quenching, stresses concentrate along line *A-B*, and a crack is almost certain to result. When changes in cross section or other transition must be made, they should be gradual, as in the redesigned Figure 5-18b. Generous fillets at interior corners, radiused external corners, and smooth transitions all reduce problems. Use of a more hardenable material or a less severe quench will also aid in preventing unnecessary problems.

Figure 5-19a shows the cross section of a die that consistently cracked during hardening. Eliminating the sharp corners and adding holes to provide a more uniform cross section during quenching, as in Figure 5-19b, eliminated the difficulty.

One of the ominous features about improperly designed heat-treated parts is the fact that the residual stresses may not produce immediate failure but may contribute to a failure at a later time. Applied stresses well within the "safe" designed limit may couple with residual stresses to produce loads sufficient to cause failure. Dimensions may change or warping may result during subsequent machining or grinding operations. Corrosion reactions may be significantly accelerated in the presence of residual stresses. Time and money are lost that could be saved if proper design practices were employed. *If used properly,* heat treatment is an important manufacturing process that enables better results to be obtained with less costly materials. The designer, however, must take into account all the facts and conditions when he or she designs a part, selects a material, and specifies, directly or indirectly, the heat treatment for it.

Techniques Used to Reduce Cracking.   Two variations of rapid quenching have been designed to minimize the high temperature gradients that often result in cracking. Rapid quenching is still required to prevent transformation to the weaker pearlite structure, but instead of quenching through the martensite

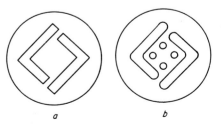

FIGURE 5-19   (Left) Original design. (Right) Improved design to produce more uniform sections.

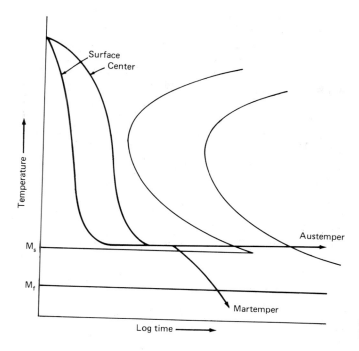

FIGURE 5-20 Schematic representation of the austempering and martempering processes.

transformation, the component is rapidly quenched to a temperature several degrees above the M_s, usually in a bath of molten salt. Holding for a brief time at this temperature enables the piece to come to a nearly uniform temperature. If the material is held at this temperature long enough, the austenite will transform to bainite, a process known as *austempering*. If the material is stabilized and then slow cooled through the martensite transformation, the process is known as *martempering*, or *marquenching*. Here the product is martensite, which must be tempered the same as the martensite formed by rapid quenching. Figure 5-20 depicts these processes schematically on a T-T-T diagram (a misuse of the diagram but good for visualization).

Parts with complicated shapes, undesirable design features, or high precision can benefit from these modified hardening techniques. Straightening can also be performed after stabilization in the quench bath, before final hardening occurs.

Ausforming.   A process often confused with austempering is *ausforming*. Certain alloys tend to retard the pearlite transformation far more than the bainite reaction and produce a T-T-T curve such as that shown in Figure 5-21. If a metal is heated to form austenite and then quenched to the temperature of the ''bay'' between the pearlite and bainite regions, it can retain its austenitic structure for a useful period of time. Deformation can be performed here on an austenitic material, a structure that technically is not stable at the temperatures involved. Benefits include the increased ductility of the face-centered crystal structure, the finer grain size characteristic of recrystallization at a lower tem-

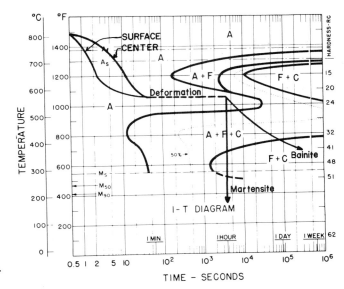

FIGURE 5-21 T-T-T diagram for 4340 steel, showing the "bay" and a schematic of the ausforming process. (Courtesy USX Corporation.)

perature, and some degree of possible strain hardening. Following deformation, the material can be either slowly cooled to produce a bainitic structure or quenched to form martensite, which then is tempered. The material that results shows extremely high strength and ductility, toughness, creep resistance, and fatigue life, far greater than if the processes of deformation and transformation were conducted in their normal sequence. Ausforming is an example of combined heat treatment and deformation, a technique known as *thermomechanical processing*.

Surface Hardening of Steel

Many products require different properties at different locations. Quite frequently, this variation takes the form of a hard, wear-resistant surface coupled with a tough, fracture-resistant core. The methods developed to produce such properties can generally be classified into three basic groups: selective heating of the surface, altered surface chemistry, and deposition of an additional surface layer.

Selective Heating Techniques. If a steel has sufficient carbon content to attain the desired surface hardness, the different properties can be obtained by simply varying the thermal histories of the various regions. *Selective heating* for surface hardness generally requires steels of at least 0.3% carbon. Maximum hardness depends on the carbon content, whereas the depth of hardness depends on the material's hardenability.

Flame hardening employs a high-intensity oxyacetylene flame to raise the surface temperature high enough to reform austenite, which is then quenched and tempered to the desired surface properties. Heat input is quite rapid and is concentrated on the surface. Slow heat transfer and short times leave the interior at low temperature and therefore free from any significant changes.

Considerable flexibility is provided since the rate and depth of heating can be easily varied. Depth of hardening can range from thin skins to over ¼ inch. Flame hardening is often used on large objects, as alternative methods tend to be limited by both size and shape. Equipment varies from crude, hand-held torches to fully automated and computerized units.

Induction heating is particularly well suited to surface hardening, the rate and depth of heating being controlled by the amperage and frequency of the generator. The steel part is placed inside a wound coil, which is then subjected to alternating current. The changing magnetic field induces a current in the steel that flows through the surface layers and heats by electrical resistance. Heating rates are extremely rapid and efficiency is high. The process can be adapted to special shapes and offers the benefits of excellent reproducibility, good quality control, and the possibility of automation. Figure 5-22 shows an induction-hardened gear; hardening has been applied to those areas subject to wear. Distortion during hardening is negligible since rigidity is provided throughout the process by the cool interior of the product.

Laser beam hardening has been used to produce hardened surfaces on uneven geometries. An absorptive coating, such as zinc or manganese phosphate, is first applied to the steel to improve the efficiency of converting light energy into heat. The surface is then scanned with the laser, where beam size, beam intensity, and scanning speed (often as high as 100 inches per minute) have been selected to obtain the desired results. While some heat is removed through transfer into the cool, underlying metal, a water or oil quench is often used on the surface. A 0.4% carbon steel can attain surface hardnesses as high as Rockwell C 65. In addition, the process operates at high speeds, produces little distortion, induces residual compressive stresses on the surface, and can be used to harden selected surface areas while leaving the remaining surfaces unaffected. Computer control and automation can be used to control the process parameters, and conventional mirrors and optics can be used to shape and manipulate the laser beam.

Electron beam hardening is very similar to laser beam hardening. Here, the heat source is a beam of high-energy electrons, rather than a beam of light. The charged particles can be focused and directed by means of electromagnetic controls. Like laser beam treating, the process can be readily automated, and production equipment can perform multiple operations with efficiencies often greater than 90%. Electrons cannot travel in air, however, so the entire operation must be performed in a hard vacuum, and this requirement provides the major limitation. More information on laser and electron beam techniques can be found in the chapters on welding.

Still other selective heating techniques employ immersion in a *lead pot* or *salt bath* as the means of heating the surface.

Techniques Using Altered Surface Chemistry.
When steels contain insufficient carbon to attain the desired surface properties by selective heating,

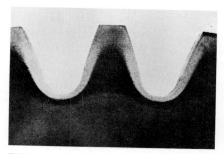

FIGURE 5-22 Section of gear teeth showing induction hardened surfaces. (Courtesy TOCCO Division, Park-Ohio Industries, Inc.)

an alternative approach is to alter the surface chemistry. *Carburizing* is the most common technique within this category. In the *pack-carburizing process*, components are packed in a high-carbon solid medium, enclosed in a gas-tight box, and heated in a furnace for 6 to 72 hours at roughly 1650°F (900°C). The hot carburizing compound produces CO gas, which reacts with the metal, releasing carbon. This, in turn, is readily absorbed into the austenite, which is the stable steel structure at these temperatures. The boxes are then taken from the furnace, and the parts are removed and thermally processed. Direct quenching from the carburization treatment can produce the different surface and core properties due to the different carbon contents of the steel at these locations. A slow cool from carburizing, reheat, and quench or a duplex core-case heat treatment are alternative thermal processes that can produce superior properties. The carbon content of the surface varies from 0.7% to 1.2% depending on the details of the process. Case depth may range from a few thousandths of an inch to over ⅜ inch, but cases over 0.06 inch (1.5 mm) are seldom employed.

Several problems are encountered in pack carburizing. Heating is inefficient; temperature uniformity is questionable; handling is often difficult; and the process is not readily adaptable to continuous operation. *Gas carburizing* overcomes these difficulties by replacing the solid carburizing compound with a carbon-providing gas, usually containing an excess of CO. Although the mechanisms and processing are the same, the operation is faster and more easily controlled. Accuracy and uniformity are increased, and continuous operation is possible. Special types of furnaces are required, however.

In *liquid carburizing* or *cyaniding*, the carbon is supplied by immersing the part in a molten carbon-containing bath. Safety and environmental concerns associated with toxic cyanide fumes have severely limited the use of molten cyanide salts, but nontoxic liquid carburizing compounds have been developed to preserve the technique. Most applications involve the production of thin cases on small parts.

Nitriding produces hardening by the production of alloy nitrides in the surface layers of special steels that contain nitride-forming elements such as aluminum, chromium, molybdenum, or vanadium. The parts are heat-treated and tempered at 1000–1250°F (525–675°C) before nitriding. After cleaning and removal of any decarburized surface material, they are heated in a dissociated ammonia atmosphere (containing nitrogen and hydrogen) for 10 to 40 hours at 950–1150°F (500–625°C). Nitrogen diffusing into the steel then forms alloy nitrides, hardening the metal to a depth of about 0.025 inch (0.65 mm). Very hard cases are formed and distortion is low. No subsequent thermal processing is required. In fact, it should be avoided because the differential thermal expansions and contractions will crack the nitrided case. Finish grinding should also be avoided, if possible, because of the exceptionally thin case. Thus, although the surface hardness is higher than for other hardening methods, the long times at elevated temperatures, coupled with the thin case, restrict the application of nitriding to the production of high-quality cases.

Ionitriding is a plasma process that has emerged as an attractive alternative to the conventional methods. Parts to be treated are placed in an evacuated "furnace," and a direct current potential of 500 to 1000 volts is applied between the parts and the furnace walls. Low-pressure nitrogen gas is introduced into the chamber and becomes ionized. The ions are accelerated toward the product surface, impact, and generate sufficient heat to promote inward diffusion. This is the only heat associated with the process; the "furnace" acts only as a vacuum container and electrode. Figure 5-23 shows the ionized plasma surrounding a stack of gears. The advantages of the process include shorter cycle times, reduced consumption of gases, significantly reduced energy costs, reduced space requirements, and the possibility of total automation. Product quality is improved over that of conventional nitriding and the process is applicable to a wider range of materials.

In *ion carburizing*, a low-pressure methane plasma is created, producing atomic carbon and transferring the carbon to the surface. Production time is generally 30% less than with gas carburizing, and the product is oxide-free because of the near-vacuum conditions. *Ion plating* and *ion implantation* are other new technologies in surface modification.

FIGURE 5-23 A stack of gears undergoing ionitriding. (Courtesy Abar Ipsen Industries.)

Furnaces.   To facilitate production heat treatment, many types of heating equipment have been developed in a wide range of sizes. Furnaces can be either *batch-type* or *continuous-type*. Batch furnaces are those in which the workpiece remains stationary throughout its time in the furnace and may be of either horizontal or vertical design. Continuous furnaces move the components through the heat treatment operation at rates selected to be compatible with other manufacturing operations.

Horizontal batch-type furnaces are often called *box furnaces* because they resemble a rectangular box. As shown in Figure 5-24, a door is provided on one end to permit work to be inserted and removed. Gas or electricity provides the source of heat. For large or long workpieces, *car-bottom furnaces* are used, such as shown in Figure 5-25. The work is loaded onto a refractory-topped flatcar, which is moved on rails into the furnace.

Horizontal furnaces are relatively easy to construct in any size, are easily insulated, and are thermally efficient. However, it is difficult to heat-treat long, slender work in them because of the sagging or warping that is likely to occur. Vertical *pit furnaces*, as shown in Figure 5-26, have been designed for such work. These are cylindrical chambers sunk into the floor with a door on top that can be swung aside to permit the work to be lowered into the furnace. Suspended in this manner, long workpieces are less likely to warp. This type of furnace is also used to heat quantities of small parts, which can be loaded into baskets and lowered into the furnace.

Another furnace type is the *bell furnace,* shown in Figure 5-27. The heating elements are contained within a bottomless bell that is lowered over the work. An airtight inner shell is often employed to contain a protective atmosphere during the heating and cooling cycles and thus to reduce tarnish or oxidation. After the work is heated, the furnace unit can be lifted off and transferred to another batch, the inner shell retaining the controlled atmosphere during cooling.

Heat Treatment Equipment

FIGURE 5-24 Box-type electric heat-treating furnace. (Courtesy Lindberg, A Unit of General Signal.)

FIGURE 5-25   Car-bottom box-type furnace. (Courtesy Hevi Duty Electric Company.)

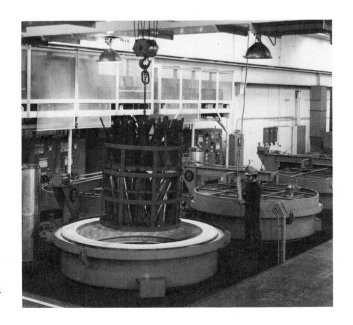

FIGURE 5-26   Vertical pit furnace loading a batch of turbine blade preforms. (Courtesy Flinn & Dreffein Engineering Company.)

FIGURE 5-27   Bell-type heat-treating furnace. (Courtesy Surface Combustion Corporation.)

An insulated cover can be placed over the heated work if slower cooling is desired, as when annealing large batches of steel or other metal.

The *elevator-type furnace* is a modification of the bell furnace, in which the bell is stationary and the work is raised up into it by means of a movable platform that forms the bottom of the furnace. An interesting variation of this furance is one for which there are three vertical positions. In the middle position, the work is loaded onto the platform elevator. In the upper position, the work is in the bell furnace, and in the lower position, it is in a quench tank. Such furnaces are used where work must be quenched as soon as possible after being removed from the furnace.

Continuous furnaces are used for large production runs of the same or similar parts, where a steady flow of workpieces is moved through the furnace by some type of conveyor or push mechanism, the furnace often ending in such a way that the workpiece falls into a quench tank to complete the treatment. Complex cycles of heating, holding, and quenching or cooling can be conducted in an exact and repeatable manner with low labor cost. Circular, continuous furnaces, where the workpieces move on a rotating hearth, are convenient when a single workstation is used to load and unload the furnace.

Nearly all of the above furnaces, if properly designed, can employ artificial gas atmospheres, which can serve to prevent scaling or tarnishing, to prevent decarburization, or to supply carbon or nitrogen for surface modification. While most furnace atmospheres are generated from natural gas, many users have found nitrogen-based atmospheres to offer cost savings, energy savings, increased safety, and environmental attractiveness.

When a liquid heating medium is preferred over gas, *salt bath furnaces* may be employed. Electrically conductive salt can be heated by passing a current between two electrodes suspended in the bath. The electrical currents also cause the bath to circulate and thus to maintain uniform temperature. Nonconductive salt baths can be heated by some form of immersion heater, or the containment vessels can be externally fired. The molten salt serves as a uniform source of heat and can also be selected to prevent scaling or decarburization. The *lead pot* is a similar furnace, where molten lead replaces salt as the heat transfer medium.

The heating rates of gas atmosphere furnaces can be made comparable to liquid baths by incorporating the *fluidized bed* concept. These furnaces consist of a bed of dry, inert particles, such as aluminum oxide, which are heated (by elements in the furnace walls) and fluidized (suspended) in a stream of flowing gas. Products introduced into the bed become engulfed in the particles, which then radiate uniform heat. Temperature and atmosphere can be altered quickly, and high heat transfer rates, high thermal efficiency, and low fuel consumption have been observed. Since atmosphere changes can be performed in minutes, one furnace can be used for nitriding, stress relieving, carburizing, carbonitriding, annealing, and hardening.

Electrical *induction heating* has been applied to many heat-treating operations. Small parts can be through-heated and hardened, as in the other methods. Long products can be heated and quenched continuously as they pass through a coil unit (or as such a unit passes over them). Localized or selective heating can be performed at rapid production rates. Furthermore, a standard induction unit can be adapted to a wide variety of products by simply changing the induction coil and adjusting the equipment settings.

Furnace Controls.   All heat-treating operations should be carried out under rigid controls if uniformity is desired. For this reason, most heat-treating furnaces are equipped with indicating and controlling pyrometers. Some furnaces are also equipped with controllers that regulate the rate of heating or cooling. It should be remembered that it is the temperature of the workpiece and not that of the furnace that controls the result, and it is this temperature that should be monitored. Ample time should be provided to allow the workpiece to attain the uniform temperature of the furnace.

Heat Treatment and Energy.   From a consideration of temperatures and times, it is apparent that heat treatment consumes considerable amounts of energy. However, if one considers the broad picture, heat treatment may actually prove to be an energy conservation measure. Through its use, one can manufacture higher-quality, more durable products and eliminate the need for frequent replacements. Higher strengths may also permit the use of less material in the manufacture of a product and thus may save additional energy.

Further energy savings can often be obtained by integrating the manufacturing operations. For example, a direct quench and temper from hot forging may be used to replace the conventional air-cool, reheat, quench, and temper sequence.

Note, however, that there is some loss of control of the starting austenite structure (temperature, temperature uniformity, and grain size), and this control may be sufficient to justify the additional energy and expense of the conventional method.

Review Questions

1. What is heat treatment?

2. Why should individuals performing hot forming or welding be aware of the effects of heat treatment?

3. What is the major goal of the processing heat treatments? Cite some of the possible specific objectives.

4. Why might equilibrium phase diagrams be useful aids in designing the processing heat treatments?

5. What are some of the possible objectives of annealing operations?

6. Although full annealing treatments can produce extremely soft and ductile structures, what may be some of the objections or undesirable features of these treatments?

7. What might be cited as the major process difference between full annealing and normalizing?

8. What are some of the process heat treatments that can be performed without reaustenizing the material (heating above the A_1 temperature).

9. What types of materials would be candidates for spheroidization? Why?

10. Why is the recrystallization temperature of a metal not a well-defined temperature?

11. What are the six major mechanisms available to increase the strength of a metal?

12. What are the three distinct steps in an age-hardening treatment?

13. What is the difference between natural and artificial aging? Which offers more flexibility? Over which does the engineer have more control?

14. What is the difference between a coherent precipitate and a distinct second-phase particle?

15. What is the time–temperature heating and cooling sequence proposed in the IT or T-T-T diagram? Is this sequence realistic for commercial manufactured items?

16. For steels below the A_1 temperature, what are the stable equilibrium phases as predicted by the equilibrium phase diagram?

17. Based on your understanding of the T-T-T diagram, what are some of the nonequilibrium structures that may be present in heat-treated steels?

18. Which of the low-temperature structures in heat-treated steel forms from a diffusionless phase change?

19. What is the major factor influencing the strength and hardness of steel in the martensitic structure?

20. Why is retained austenite an undesirable structure in heat-treated steels?

21. Why must martensite be tempered before being put into use? What occurs during the tempering operation?

22. What is a C-C-T diagram?

23. Why can the various locations on a Jominy test specimen be correlated with different cooling rates?

24. Explain the concept of equivalent cooling rates (i.e., how Jominy test data can be used to predict the properties of manufactured items).

25. What is hardenability?

26. What alternatives might be possible if it were necessary to harden a product to greater depth?

27. What are the three stages of liquid quenching?

28. Why is an oil quench less likely to produce quench cracks than water or brine?

29. What are some of the attractive qualities of a polymer or synthetic quench?

30. What are some undesirable design features in parts that are to be heat-treated?

31. In what ways might the thermally induced residual stresses (caused by heat treatment) be minimized?

32. Why might residual stresses be undesirable in a heat-treated product?

33. What is the difference between austempering and ausforming?

34. What are some of the methods that can be used to selectively heat-treat the surface of metal parts?

35. What are some of the attractive features of laser-beam surface hardening?

36. Why might gas carburizing be more attractive than pack carburizing in a high-volume production operation?

37. In what ways is ionitriding more attractive than conventional nitriding or carburizing as a means of surface hardening?

38. What are some of the possible functions of the artificial atmospheres used during heat treating?

39. In what ways might the heat treatment of metals actually be an energy conservation measure?

Ferrous Metals and Alloys

Engineering materials are available with a wide range of usable properties and characteristics. Some of these are inherent in the particular material, but many others can be varied by control of the manner of production and processing.

Metals are classic examples of such "history-dependent" materials: their final properties are affected and altered by their past processing history. The particular details of the smelting and refining process control the resulting purity and the type and nature of any influential contaminants. The initial solidification process imparts structural features that may transmit to the final product. Preliminary operations, such as the rolling of sheet or plate, often impart directional variations to properties that should be considered during subsequent processing. Thus, although many "metal users" tend to take the attitude that "Metals come from warehouses," it is important to recognize those aspects of prior processing that can significantly influence further processing and the final properties of the product. Although this text makes no attempt to cover fully the processes and methods involved in the production of engineering metals, some aspects are presented because of their subsequent significance.

This chapter introduces the major ferrous (iron-based) metals and alloys. These materials have been the backbone of civilization, and many varieties have been developed to meet the specific needs of advancing technology. The availability of so many alternatives, however, has often led to poor materials selection. Money can be wasted in the unnecessary selection of an expensive alloy or one that is difficult to fabricate. At other times, such materials may be absolutely necessary. Thus, it is the responsibility of the design and manufacturing engineer to be knowledgeable in the area of engineering materials and to be able to make the best selection from among the available alternatives.

Iron

Iron is the fourth most plentiful element in the earth's crust and, for centuries, has been the most important of the basic engineering metals. The variety of metals and alloys derived from iron has played a central role in the development of civilization and is likely to continue this role in the foreseeable future. Even now, new advances in the technology of iron and iron alloys continue to expand their utility in a myriad of engineering applications.

Iron is rarely found in the metallic state; it occurs in a variety of mineral compounds, knowns as *ores*. The most attractive of these ores are the iron oxides with their companion impurities. These ores are first processed in a manner (chemical reducing reactions) that breaks the iron–oxygen bonds to produce metallic iron. Ore, limestone, coke, and air are continuously introduced into specifically designed furnaces and molten metal is periodically withdrawn.

In the environment of the furnace, other impurity oxides are also reduced. All of the phosphorus and most of the manganese enter the iron. Oxides of silicon and sulfur are partially reduced, and these elements also become part of the resulting metal. Other contaminant elements, such as calcium, magnesium, and aluminum, are collected in the limestone-base slag and are removed from the system. Thus, the resulting pig iron tends to have roughly the following composition:

Carbon	3.0–4.5%
Manganese	0.15–2.5%
Silicon	1.0–3.0%
Sulfur	0.05–0.1%
Phosphorus	0.1–2.0%

A small portion of the pig iron is cast into final shape in this condition. Although the resulting material is referred to as *cast iron*, most commercial cast iron is the result of remelting scrap iron and steel, with the possible addition of some pig iron. The metallurgical properties of cast iron have been presented in Chapter 4, and the melting and utilization in the casting process will be developed in Chapters 13 through 15.

Most pig iron is simply transferred in the molten state to be further processed into steel.

Steel

The manufacture of *steel* is essentially an oxidation process that decreases the amount of carbon, silicon, manganese, phosphorus, and sulfur in a mixture of molten pig iron and steel scrap. In 1856, the *Kelly-Bessemer process* opened up the industry by enabling the manufacture of commercial quantities of steel. The *open-hearth process* surpassed the Bessemer process in tonnage produced in 1908 and was producing over 90% of all steel in 1960. Currently, *oxygen furnaces* of a variety of types and *electric furnaces* produce most of our commercial steels.

In many of these processes, air or oxygen passes over or through the molten metal to drive a variety of exothermic refining reactions. Carbon oxidizes to form gaseous CO or CO_2, which then leaves the melt. Other elements are similarly oxidized and rise to be collected in a removable slag. At the same time, oxygen and other elements from the reaction gases are dissolving in the molten metal and may later become a cause for concern.

Solidification of the Steel.

Regardless of the method by which the steel is made, it must undergo a change of state from liquid to solid before it becomes a usable product. The liquid can be converted directly into finish-shape steel castings or can be solidified into a form suitable for further processing. In most cases, the latter option is exercised through either *continuous casting* or the forming of *ingots*, these shapes being the feedstock material for subsequent forging and rolling operations.

In solidifying the steel, the desire is to obtain metal that is as free of flaws as possible. Most steelmaking furnaces first pour the metal into containment vessels, known as *ladles*. Although these were previously used only as transfer and pouring containers, they have recently emerged as the site for additional processing. *Ladle metallurgy* refers to a variety of processes designed to provide final purification and to fine-tune both the chemistry and the temperature of the melt. Alloy additions can be made, carbon can be further reduced, dissolved gases can be removed, and steps can be taken to control subsequent grain size, to limit inclusion content, to reduce sulfur, and to control the shape of any included sulfides. Stirring, degassing, reheating, and various injection procedures can be practiced to increase the cleanliness of the steel and provide for tighter control of its chemistry and properties.

The processed metal is then poured from these ladles into ingot molds or continuous casters by means of a bottom-pouring process, shown schematically in Figure 6-1. When the metal is extracted from the bottom of the ladle, slag and floating matter are not transferred into the subsequent castings, and a cleaner product results.

When we extend the observation that most contaminants have lower density than the molten metal and will rise to the top, we find that the highest-quality ingots are poured by a method that fills the ingot mold from the bottom. As illustrated in Figure 6-2, the bottom of several ingot molds is connected to a central pouring ingot by ceramic tile tunnels. Hot metal is poured into the center ingot, and as the level rises, hot metal is conveyed through the tunnels to fill the outer molds. Contaminants rise to the top and remain trapped in the central pouring ingot. Thus, the bottom-poured ingots around the outside tend to be of rather high quality.

Figure 6-3 schematically shows the contraction of a metal undergoing cooling and solidification. The major discontinuity at the melting point is known as *solidification shrinkage* and reflects the difference in density of the liquid and solid states of the metal. Thus, when metals solidify, it is expected that a shrinkage void will be observed in the region of the last material to be liquid. In ingots, solidification proceeds inward from the mold walls and upward from the bottom. Shrinkage takes the form of a *pipe* coming in from the top, as

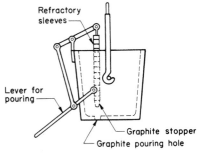

Refractory sleeves

Lever for pouring

Graphite stopper

Graphite pouring hole

FIGURE 6-1 Schematic diagram of a bottom-pouring ladle.

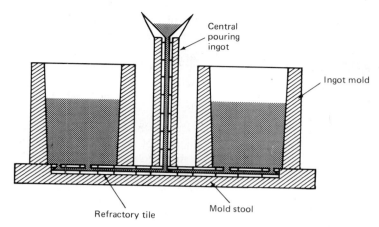

Central pouring ingot

Ingot mold

Refractory tile

Mold stool

FIGURE 6-2 Pouring of ingots by the bottom-pouring process. The bottom of the center mold being poured is connected to the bottom of the remaining molds in the cluster by means of ceramic channels.

illustrated in Figure 6-4. Since the pipe surface has been exposed to the atmosphere at elevated temperature, oxides and surface contaminants form that would prevent the metal from welding back together during subsequent processing. That portion of the ingot containing the pipe must be recycled as scrap; it may well be a substantial fraction of the ingot.

Although the amount of shrinkage cannot be changed, the shape and location can be greatly controlled. One procedure for reducing the amount of pipe is to use a ceramic *hot top* or *exothermic topping* on the top of the ingot mold. If heat is retained at the top, the liquid reservoir at the end of solidification is more of a uniform layer, and the depth of shrinkage is minimized. Variation of the shape of the mold controls not only the solidification shape, but also details of the ingot structure. Tapered ingots having the big end up are commonly used for greatest soundness.

Continuous casting processes have been developed to overcome a number of ingot-related difficulties, such as piping, entrapped slag, and structure variation throughout the length of the solidified product. Figure 6-5 illustrates one of the most common continuous casting procedures, in which molten metal flows from a ladle, through a tundish, into a bottomless, water-cooled mold, usually made of copper. Cooling is controlled so that the outside has solidified before the metal exits the mold. The material is then further cooled by direct water sprays to ensure complete solidification. The cast solid is then either bent and fed horizontally through a short reheat furnace before rolling or is cut to the desired lengths. Mold shape, and thus the shape of the cast product, may vary so that products may be cast with cross sections closer to the desired final shape.

Continuous casting virtually eliminates the problems of piping and mold spatter. From a production viewpoint, it eliminates pouring into the molds, stripping the molds from the solidified metal, handling and reheating the ingots, and then rolling. Instead, a continuously cast shape is produced, ready for reduction by rolling. Cost, energy, and scrap are all significantly reduced. In addition, the products have improved surfaces, more uniform chemical composition, and fewer oxide inclusions.

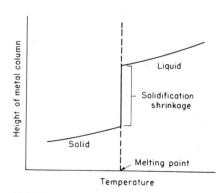

Liquid

Solidification shrinkage

Solid

Melting point

Height of metal column

Temperature

FIGURE 6-3 Height of a column of metal as a function of temperature, showing the significant solidification shrinkage.

FIGURE 6-4 Section of an ingot showing a "pipe" (top) and "segregation" (dark areas). (Courtesy Bethlehem Steel Corporation.)

Figure 6-6 shows a method by which molten steel can be cast into discrete slabs. In this "pressure-pouring" process, pressurized air forces molten metal up into a graphite mold, where it then solidifies. Because the metal is again taken from the bottom of the ladle and is introduced through the bottom of the slab mold, excellent quality is obtained.

Degassification of Ingots. During the oxidation that takes place in the making of steel, considerable amounts of oxygen can dissolve in the molten metal. When this molten metal is then cooled to produce solidification, oxygen and other gases are rejected from the solid as the saturation levels decrease (see the schematic of Figure 6-7). The rejected oxygen links with atomic carbon to produce carbon monoxide gas. A porous structure then results, where the gas-bubble-induced *porosity* has the form of either small, dispersed pores or large blowholes. Pores that are totally internal can be welded shut during subsequent hot forming, but if they are exposed to the air at elevated temperatures, the pore surfaces oxidize and will not weld. Cracks and internal defects may then appear in the finished product.

In many cases, it is desirable to avoid porosity difficulties by removing the oxygen or rendering it nongaseous before solidification. Where high-quality steel is desired, or where subsequent deformation may be inadequate to produce

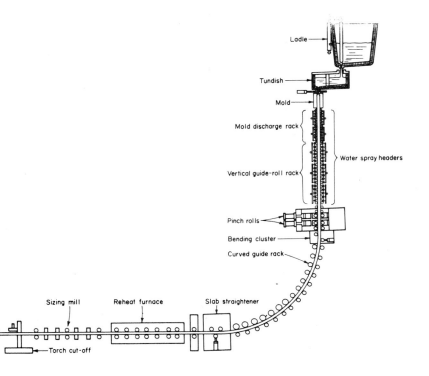

FIGURE 6-5 Schematic representation of the continuous casting process for producing billets, slabs and bars. (Courtesy Materials Engineering.)

welding of the pores, the metal is usually fully *deoxidized* (or *killed*), to produce a killed steel. Aluminum, ferromanganese, or ferrosilicon is added to the molten steel while it is in the ladle to provide material with a higher affinity for oxygen than the carbon. The rejected oxygen simply reacts to produce solid metallic oxides dispersed throughout the structure. High-carbon steels are often fully killed.

For steels with lower carbon contents, a partial deoxidation may be employed to produce a *semikilled steel.* Enough deoxidant is added to partially suppress bubble evolution, but not enough to completely eliminate the effect of oxygen. Some pores still form in the center of the ingot, their volume serving to cancel some of the solidification shrinkage, thus reducing the extent of piping and scrap generation.

For steels of sufficiently low carbon (usually less than 0.2% carbon), a process known as *rimming* may be employed. The steel is only partially deoxidized before solidification. When the material is poured into the ingot mold, the first metal to solidify is almost pure iron, being very low in carbon, oxygen, sulfur, and phosphorus. These elements are rejected from the solid into the remaining liquid. When the concentration of dissolved gases exceeds the saturation point, a layer of CO bubbles evolves. Further solidification produces additional porosity in the inner portions of the ingot, and if properly controlled, the rimming action can produce a porosity level that will approximately compensate for solidification shrinkage.

The outside of the ingot is clean, defect-free metal that provides an excellent blemish-free surface when the ingot is rolled into flat strips or a similar product. The holes on the inside of the rimmed ingots have bright, clean surfaces because they have not been exposed to the air. When hot working is performed, the surfaces weld together to produce a sound product. Figure 6-8 presents a schematic comparison of killed, semikilled, and rimmed ingot structures.

In addition to oxygen, small amounts of other dissolved gases, particularly hydrogen and nitrogen, can have deleterious effects on the performance of steels. This is particularly important in the case of alloy steels because several of the major alloying elements, such as vanadium, columbium, and chromium, tend to increase the solubility of these gases. Consequently, several methods have been devised for *degassing* steels, and considerable quantities of degassed steel are now used for critical applications. Figure 6-9 illustrates a method that is widely used to produce degassed ingots, known as *vacuum degassing.* An ingot mold is enclosed in an evacuated chamber, and the metal stream passes through the vacuum during pouring, the vacuum serving to remove the dissolved gases.

When exceptional purity is required, a *consumable-electrode remelting* process may be employed, as illustrated in Figure 6-10. A solidified metal electrode is remelted by an electric arc in a evacuated chamber and resolidifies into a new form. This process, known as *vacuum arc remelting* (VAR) [or *vacuum induction melting* (VIM) if induction heating replaces the electric arc], is highly effective in removing all dissolved gases, but it does not remove nonmetallic impurities from the metal.

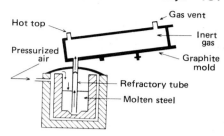

FIGURE 6-6   Schematic representation of the pressure-pouring process for casting ingots and slabs.

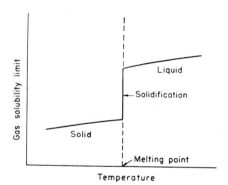

FIGURE 6-7   Gas solubility in a metal as a function of temperature.

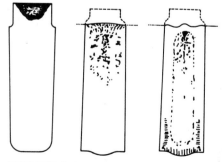

FIGURE 6-8   Schematic comparison of killed, semikilled, and rimmed ingot structures. (Courtesy American Iron and Steel Institute, Washington, D.C.)

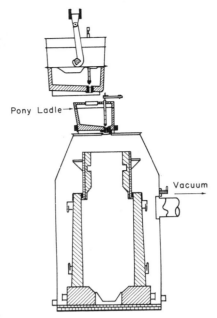

FIGURE 6-9 Method of vacuum degassing steel while pouring ingots.

Extremely clean, gas-free metal can be obtained through the *electroslag remelting process* (ESR), illustrated in Figure 6-11. Here, the electrode is melted and recast by means of an electric current, with the surface of the liquid metal now being covered by a blanket of molten flux. Nonmetallic impurities are collected in the flux blanket, leaving beneath a newly solidified structure with improved quality. (NOTE: This process is simply a large-scale version of the electroslag welding process discussed in Chapter 36.)

Plain-Carbon Steel. Commercial steel, although theoretically an alloy of iron and carbon, actually contains manganese, phosphorus, sulfur, and silicon in significant and detectable quantities. When these four additional elements are present in their normal percentages, the product is referred to as *plain-carbon steel*. Its strength is primarily a function of its carbon content, increasing with increasing carbon. Unfortunately, the ductility of plain-carbon steels decreases as the carbon content is increased, and its hardenability is quite low. In addition, the properties of ordinary carbon steels are impaired by both high and low temperatures, and they are subject to corrosion in most environments.

Plain-carbon steels are generally classed into three subgroups based on their carbon content. *Low-carbon steels* have less than 0.30% carbon, possess good formability and weldability, but lack sufficient hardenability to be hardened to any significant depth. Their structures are usually ferrite and pearlite, and the material is generally used as it comes from the hot-forming or cold-forming processes, or in the as-welded condition. *Medium-carbon steels* have between 0.30% and 0.80% carbon, and they can be quenched to form martensite or bainite if the section size is small and a severe water or brine quench is used. The best balance of properties is obtained at these carbon levels, the high fatigue

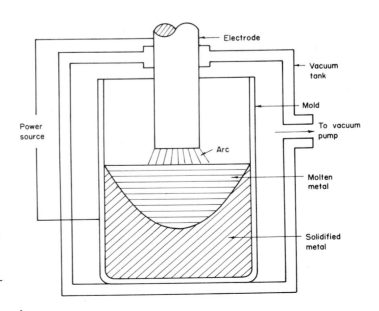

FIGURE 6-10 Method of producing degassed ingots by consumable-electrode vacuum remelting.

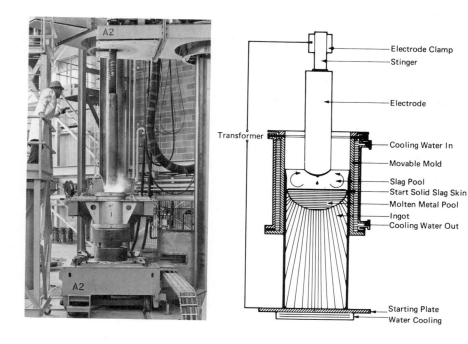

FIGURE 6-11   (Left) Production of an ingot by the electroslag remelting process. (Right) Schematic representation of this process. (Courtesy Carpenter Technology Corporation.)

resistance and toughness of the low-carbon material being in good compromise with the strength and hardness that comes with higher carbon contents. These steels are extremely popular and find numerous mechanical applications. *High-carbon steels* have more than 0.80% carbon. Toughness and formability are quite low, but hardness and wear resistance are high. Severe quenches can form martensite, but hardenability is still poor. Quench cracking is often a problem when the material is pushed to its limit.

Plain-carbon steels are the lowest-cost steel material and should be considered for many applications. Often, however, their limitations become restrictive. When improved material is required, steels can be upgraded by the addition of one or more alloying elements.

Alloy Steels.   The differentiation between "plain-carbon" and "alloy" steel is often somewhat arbitrary. Both contain carbon, manganese, and usually silicon. Copper and boron are also possible additions to both classes. Steels containing more than 1.65% manganese, 0.60% silicon, or 0.60% copper are designated as alloy steels. Also, a steel is considered an alloy steel if a definite amount or minimum of other alloying element is specified or required. The most common alloy elements are chromium, nickel, molybdenum, vanadium, tungsten, cobalt, boron, and copper, as well as manganese, silicon, phosphorus, and sulfur in amounts greater than are normally present.

Effects of the Alloying Elements.   In general, alloying elements are added to steel in small percentages—usually less than 5%—to improve strength or hardenability, or in much larger amounts, often up to 20%, to produce special properties such as corrosion resistance or stability at high or low temperatures. Certain additions may be made during the steelmaking process to remove dissolved oxygen from the melt. Manganese, silicon, and aluminum are frequently used for this deoxidation. Aluminum and, to a lesser extent, vanadium, columbium, and titanium are used to control austenitic grain size. Machinability can be enhanced by additions of sulfur, lead, selenium, and tellurium. Other elements may be added to improve the strength or toughness properties of the product metal. Manganese, silicon, nickel, and copper add strength by forming solid solutions in ferrite. Chromium, vanadium, molybdenum, tungsten, and other elements increase strength by forming dispersed second-phase carbides. Columbium, vanadium, and zirconium can be used for ferrite grain-size control. Nickel and copper are added to low-alloy steels to provide improved corrosion resistance.

For constructional alloy steels, the principal effect of alloying elements is to increase hardenability. The commonly used elements, in order of decreasing effectiveness, are manganese, molybdenum, chromium, silicon, and nickel. Small quantities of vanadium are quite effective, but the response drops off as quantity is increased. Boron is also extremely significant in steels with less than 0.65% carbon.

Each of the various alloy elements can be used to impart distinct properties and characteristics:

Manganese is added to most plain-carbon steels to combine with sulfur and produce soft manganese sulfides. This addition prevents the formation of iron sulfide, which coats grain boundaries and imparts a brittleness to the metal. In alloy steels, manganese increases hardenability, slightly strengthens ferrite, and lowers the martensite transformation temperatures. It is often added in amounts greater than 1.0%. When manganese is added in large percentages (11% to 14%), an austenitic alloy is produced. Its high hardness with good ductility, high strain-hardening capacity, and excellent wear resistance makes it ideal for impact-resisting tools and similar applications.

As previously mentioned, *sulfur* is usually not desired in steel because of the embrittling effect of iron sulfide. In the form of manganese sulfide, however, sulfur is not harmful, provided that the sulfides are not large in quantity and are well dispersed. If manganese sulfide is present in large quantities and in proper form, it can impart desirable machining properties. Some *free-machining steels* have 0.08% to 0.15% sulfur in combination with an increased manganese content.

Nickel is added primarily to increase toughness and impact resistance, primarily at low temperature. It is generally used in amounts from 2% to 5%, often in combination with other alloying elements. Alloys with 12% to 20% nickel and low amounts of carbon possess outstanding corrosion resistance. An iron and 36% nickel alloy, commonly known as *Invar*, has a near-zero thermal expansion coefficient and is used for sensitive measuring devices.

Although large percentages of *chromium* can impart corrosion resistance and heat resistance, in the amounts used in low-alloy steels these effects are minor. In these alloys, chromium serves primarily to increase hardenability and to increase strength. Less than 2% chromium is generally employed. In many alloys, chromium and nickel are used together in a ratio of about one part chromium to two parts nickel. Combined with carbon, chromium carbides provide superior wear resistance.

Molybdenum is used in alloy steels in amounts less than 0.3% to improve hardenability and to increase strength properties, particularly under dynamic and high-temperature conditions. Resistance to temper embrittlement is also attributed to the presence of molybdenum. Molybdenum carbides are extremely stable at elevated temperatures and are used in alloys to retain fine grain size and to provide strength and creep resistance at elevated temperature. Molybdenum carbides are also used in hot-work tool steels, such as those used in forging dies, to impart hardness that will persist at red heat.

Vanadium is another alloying element that forms strong carbides that persist at elevated temperature. Thus, 0.03% to 0.25% vanadium in the form of carbides can effectively inhibit grain growth and increase strength properties (most notably the elastic limit, the yield point, and the impact strength) with almost no loss of ductility.

Another element that forms stable carbides, *tungsten*, is used as a primary alloying element in tool steels that must maintain their hardness at elevated temperatures.

Copper has been known to resist atmospheric corrosion for centuries, but only recently has it been used as an addition to steel (in amounts from 0.10% to 0.50%) to provide this property. Low-carbon sheet steel and structural steels often contain a copper addition to enhance corrosion resistance, but surface quality and hot-working behavior tend to deteriorate somewhat.

In small percentages, *silicon* has an effect on steel similar to that of nickel, increasing the strength properties with little companion loss of ductility. It is an important alloying element in certain high-yield-strength structural steels. It is also used in spring steels (in amounts of about 2%) and promotes the desirable large grain size in steels used for magnetic applications.

Boron is a very powerful hardenability agent, being from 250 to 750 times as effective as nickel, 75 to 125 times as effective as molybdenum, and about 100 times as powerful as chromium. Only a few thousandths of a percent are sufficient to produce the desired effect in low-carbon steels, but the results diminish rapidly with increasing carbon content. As no carbide formation or ferrite strengthening is produced, improved machinability and cold-forming capability often result from the use of boron in place of other hardenability additions.

In addition to its use as a deoxidizer, *aluminum* may be added to steels in amounts of 0.95% to 1.30% to produce a nitriding steel. *Titanium* and *niobium* (columbium) are additional carbide formers. Steels with 0.15% to 0.35% *lead* show substantially improved machinability. *Bismuth* is another alloy that promotes machinability. *Zirconium, cesium,* and *calcium* control the shape of inclusions and thus promote toughness.

TABLE 6.1   Principal Effects of Major Alloying Elements in Steel

Element	Percentage	Primary Function
Manganese	0.25–0.40	Combines with sulfur to prevent brittleness
	>1	Increases hardenability by lowering transformation points and causing transformations to be sluggish
Sulfur	0.08–0.15	Free-machining properties
Nickel	2–5	Toughener
	12–20	Corrosion resistance
Chromium	0.5–2	Increase of hardenability
	4–18	Corrosion resistance
Molybdenum	0.2–5	Stable carbides; inhibits grain growth
Vanadium	0.15	Stable carbides; increases strength while retaining ductility; promotes fine grain structure
Boron	0.001–0.003	Powerful hardenability agent
Tungsten		Hardness at high temperatures
Silicon	0.2–0.7	Increases strength
	2	Spring steels
	Higher percentages	Improves magnetic properties
Copper	0.1–0.4	Corrosion resistance
Aluminum	0.95–1.30	Alloying element in nitriding steels
Titanium	—	Fixes carbon in inert particles
		Reduces martensitic hardness in chromium steels
Lead	—	Improves machinability
Bismuth	—	Improves machinability

Table 6.1 summarizes the basic effects of the common alloying elements in steel. A working knowledge of the information contained in this table is useful to the design engineer in selecting an alloy steel to meet a given set of requirements. Of course, alloying elements are often used in combination; as a result, an immense variety of alloy steels is available. To simplify the situation, a classification system has been developed and has achieved general acceptance in a variety of industries.

The AISI-SAE Classification System.   Undoubtedly the most important group of alloy steels, from the manufacturing viewpoint, is that designated by the AISI identification system. This system, which classifies alloys by chemistry, was started by the Society of Automotive Engineers (SAE) to provide some standardization of the steels used in the automotive industry. It was later adopted and expanded by the American Iron and Steel Institute (AISI) and has become the most universal system in the United States. Both plain-carbon and low-alloy steels are identified by a four-digit number, the first number indicating the major alloying elements and the second number indicating a subgrouping of the major

TABLE 6.2   Some AISI-SAE Standard Steel Designations

AISI Number	Type	Alloying Elements (%)					
		Mn	Ni	Cr	Mo	V	Other
1xxx	Carbon steels						
10xx	Plain carbon						
11xx	Free cutting (S)						
12xx	Free cutting (S) and (P)						
15xx	High manganese						
13xx	High manganese	1.60–1.90					
2xxx	Nickel steels		3.5–5.0				
3xxx	Nickel-chromium		1.0–3.5	0.5–1.75			
4xxx	Molybdenum						
40xx	Mo				0.15–0.30		
41xx	Mo, Cr			0.40–1.10	0.08–0.35		
43xx	Mo, Cr, Ni		1.65–2.00	0.40–0.90	0.20–0.30		
44xx	Mo				0.35–0.60		
46xx	Mo, Ni (low)		0.70–2.00		0.15–0.30		
47xx	Mo, Cr, Ni		0.90–1.20	0.35–0.55	0.15–0.40		
48xx	Mo, Ni (high)		3.25–3.75		0.20–0.30		
5xxx	Chromium						
50xx				0.20–0.60			
51xx				0.70–1.15			
6xxx	Chromium-vanadium						
61xx				0.50–1.10		0.10–0.15	
8xxx	Ni, Cr, Mo						
81xx			0.20–0.40	0.30–0.55	0.08–0.15		
86xx			0.40–0.70	0.40–0.60	0.15–0.25		
87xx			0.40–0.70	0.40–0.60	0.20–0.30		
88xx			0.40–0.70	0.40–0.60	0.30–0.40		
9xxx	Other						
92xx	High silicon						1.20–2.20 Si
93xx	Ni, Cr, Mo		3.00–3.50	1.00–1.40	0.08–0.15		
94xx	Ni, Cr, Mo		0.30–0.60	0.30–0.50	0.08–0.15		

alloy system. Grouping by the first two digits is according to an arbitrary table. The last two digits indicate the approximate carbon content of the metal in "points" of carbon, where 1 point is equivalent to 0.01% carbon. Table 6.2 presents the basic classification. As examples, a 1080 steel would be a plain-carbon steel with 0.80% carbon. Similarly, a 4340 steel would be a Mo-Cr-Ni alloy with 0.40% carbon.

A letter prefix may be used to indicate the process employed to produce the steel, such as electric furnace (E). An X prefix is used to indicate permissible variations in the range of manganese, sulfur, or chromium. The letter B between the second and third digits indicates that the base metal has been supplemented by the addition of boron. Similarly, a letter L in this position indicates a lead addition for enhanced machinability.

The *H-grade AISI steels* are designed by the letter *H* as a suffix to the standard designation. These steels are for use where hardenability is a major requirement and slightly broader variations in steel chemistry are permitted. The steel is supplied to meet hardenability standards as specified by the customer in terms of hardness values at specific locations from the quenched end of a Jominy hardenability specimen.

Other systems of designation, such as the American Society for Testing and Materials (ASTM) and U.S. government (MIL and federal) specification systems focus more on specific applications. Acceptance for a given grade may be based more on physical or mechanical properties than on the chemistry of the metal. Many low-carbon structural and alloy steels are referred to by their ASTM designation.

Selecting Balanced Alloy Steels.   From the previous discussion, it is apparent that two or more alloying elements can produce similar effects. Thus, steels with substantially different chemical compositions can possess almost identical mechanical properties. Figure 6-12 shows quite clearly that steels of quite different composition (chemistry) can have almost identical property ratios *when heat-treated properly*. This fact should be kept in mind by all who select

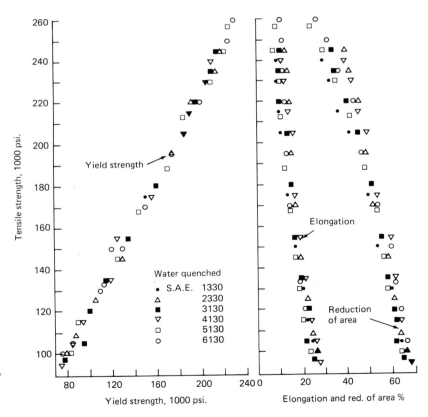

FIGURE 6-12   Straight-line representation of the mechanical properties of properly heat-treated SAE alloy steels. (Courtesy ASM International, Metals Park, Ohio.)

and specify the use of alloy steels. It is particularly important when one realizes that some alloying elements are much more costly than others, and that others may be in short supply because of emergencies or political constraints. Over-specification has often been employed to guarantee success in spite of sloppy manufacturing and heat-treatment practice. The correct steel to use, however, is usually the least expensive one that can be consistently processed to achieve the desired properties. Attaining the desired properties usually involves taking advantage of the effects provided by *all* of the alloy elements. Steels with "balanced" compositions can avoid needlessly large amounts of expensive alloy elements.

An excellent example of what can be achieved can be seen in the series of *EX steels,* originated by the Society of Automotive Engineers in 1963. This series designated new types of alloy steels on a temporary basis. Alloys are subsequently removed from the list when they are promoted to full AISI-SAE status or are dropped for lack of interest. A survey of these new alloys reveals many that have been developed to reduce the need for expensive alloying elements or those that are in short supply, notably nickel and chromium. Others seek to improve a particular attribute of one of the existing standard grades. Table 6.3 lists the compositions and equivalent standard grades (on the basis of hardenability) of several of the EX steels. When the compositions of the EX steel and the composition of the equivalent (from Table 6.2) are compared, the alloy savings become readily apparent.

In selecting alloy steels, it is important to keep use and fabrication in mind. For one product, it might be permissible to increase the carbon content in order to obtain greater strength. For a different application, involving assembly by welding, it would be best to keep the carbon content low and to use a balanced amount of alloy elements to obtain the desired strength without the possible cracking problems associated with a higher-carbon martensite. Steel selection often involves the defining of the required properties, the determination of the best microstructure to provide those properties, and finally the selection of the steel with the best carbon content and hardenability characteristics to achieve that goal. One option is to purchase material on the basis of properties, rather than exact chemical composition. The supplier or producer is then free to provide any material that will possess the desired properties, and substantial cost savings can result. To ensure success, however, it is important that *all* necessary properties be specified.

TABLE 6.3   Compositions and Equivalents of Several EX Steels

EX Number	C	Mn	Cr	Mo	Other	Equivalent AISI Grade
15	0.18–0.23	0.90–1.20	0.40–0.60	0.13–0.20	—	8620
24	0.18–0.23	0.75–1.00	0.45–0.65	0.20–0.30	—	8620
31	0.15–0.20	0.70–0.90	0.45–0.65	0.45–0.60	0.70–1.00Ni	4817

High-Strength–Low Alloy Structural Steels.

There are two general categories of alloy steels: the high-strength–low-alloy (HSLA) types, which rely largely on chemical composition to develop the desired mechanical properties in the as-rolled or normalized condition, and the constructional alloy steels, in which the desired properties are developed by thermal treatment. Many manufactured products require steels with good hardenability, ductility, and fatigue strength—the constructional alloy steels. For structural applications, however, high yield strength, good weldability, and corrosion resistance are most desired, with only limited ductility and virtually no hardenability. Development of steels possessing these properties in the as-rolled condition have made possible substantial cost and weight savings in automobiles, trains, bridges, and buildings.

The low-alloy structural steels have about twice the yield strength of the plain-carbon structural steels. This increase in strength, coupled with resistance to martensite formation in a weld zone, is obtained by adding low percentages of several elements, notably manganese, silicon, niobium (columbium), and vanadium, as well as several others. About 0.2% copper is often added to improve corrosion resistance. Numerous types of these steels have been obtained by the addition of alloying elements in various combinations and quantities. Four of the more common ones are listed in Table 6.4.

HSLA steels represent a significant contribution to the field of structural materials. Through their use, weight savings of 20% to 30% have been achieved without any sacrifice in strength or safety. They are currently produced in sufficient tonnages so that their cost is little more than that of ordinary grades of structural steel.

Quenched-and-Tempered Structural Steel.

The need for even stronger structural steels that can be used as-welded (for products such as submarines and pressure vessels) has led to the development of several alloys that are used

TABLE 6.4 Typical Compositions and Strength Properties of Several Groups of High-Strength Low-Alloy Structural Steels

| Group | Chemical Composition[a] (%) | | | | | Yield | | Tensile | | Elongation in 2 Inches (%) |
	C	Mn	Si	Cb	V	ksi	MPa	ksi	MPa	
Columbium or vanadium	0.20	1.25	0.30	0.01	0.01	55	379	70	483	20
Low manganese-vanadium	0.10	0.50	0.10		0.02	40	276	60	414	35
Manganese-copper	0.25	1.20	0.30			50	345	75	517	20
Manganese-vanadium-copper	0.22	1.25	0.30		0.02	50	345	70	483	22

[a]All have 0.04% P, 0.05% S, and 0.20% Cu.

in the quenched-and-tempered condition. When water-quenched from about 1650°F and tempered at 1150–1200°F to produce a tempered martensite structure, these alloys have a yield strength in the range of 80 to 150 ksi, tensile strengths of 95 to 200 ksi, and elongations between 13% and 20%. When subsequently welded, the material is still tough, because the martensite that forms during welding is of a rather low carbon content (0.15% to 0.20% carbon). These materials have excellent impact resistance at low temperatures and good atmospheric corrosion resistance. Because of their superior strength properties, they can often permit sufficient weight savings to offset the additional cost of the material.

Microalloyed Steels.   Steels containing small amounts of alloying elements like niobium, vanadium, titanium, zirconium, boron, rare earth elements, or combinations of these are being used increasingly as substitutes for heat-treated steels. Known as *microalloyed steels,* they derive their name from the fact that the alloying elements are added only in amounts ranging from 0.05% to 0.15%. Strength and hardness can be increased up to 50% without interfering with the material processing. Weldability can be retained or even improved if the carbon content is simultaneously decreased. In essence, the goal of these materials is to attain maximum strength with minimum carbon, while optimizing weldability, machinability and formability.

Cold-formed microalloyed steels require less cold work to attain the desired level of strength, so they tend to have greater residual ductility. Hot-formed products, such as forgings, can often be used in the air-cooled condition. Many properties match those of quenched-and-tempered materials, but toughness cannot be made comparable. Machinability is often enhanced because of the more uniform hardness, and the ferrite–pearlite structure is often more machinable than the ferrite–carbide structure of quenched-and-tempered steels. Fatigue life and wear resistance can be superior to those of the heat-treated counterparts. Thus, in applications where the properties are adequate, microalloyed steels can provide attractive cost savings. Energy savings can be substantial, straightening or stress relieving after heat treatment is no longer necessary, and quench cracking is not a problem. Because of the increase in material strength, the size and weight of finished products can often be reduced. As a result, the cost of finished forgings could be reduced by 15% to 25%.

Certain precautions must be observed, however, if these materials are to attain their optimum properties. In the high-heat segments of processing, the material must be heated high enough to place all of the alloys in solution. After forming, the products should be rapidly air cooled to 1000–1100°F before they are dropped into collector boxes. Microalloyed steels tend to through-harden on air cooling, so products fail to possess the drop in interior properties generally observed in quenched-and-tempered materials.

Free-Machining Steels.   The increased use of high-speed machining, particularly on automated machine tools, has spurred the use and development of several varieties of *free-machining steels.* These steels machine readily and form small chips when cut. These reduce the rubbing against the cutting tool and the associated friction and heat. The formation of small chips also reduces the likelihood of chip entanglement in the machine and makes chip removal much easier.

Free-machining steels are basically carbon steels that have been modified by an alloy addition to enhance machinability. Sulfur, lead, bismuth, selenium, tellurium, and phosphorus have all been added to enhance machinability. Sulfur (0.08% to 0.33%) combines with manganese (0.7% to 1.6%) to form soft manganese sulfide inclusions. These act as discontinuities in the structure, which serve as sites where broken chips are formed. The inclusions also provide a built-in lubricant that prevents formation of a built-up edge on the cutting tool and imparts an altered cutting geometry (see Chapter 18). Insoluble lead particles serve the same purpose.

The bismuth free-machining steels have recently become an attractive alternative to the previous varieties. Compared with lead, bismuth is a better free-machining agent. It is more environmentally acceptable, has less tendency to form stringers, and can be more uniformly dispersed as its density is closer to that of iron. Machinability is improved because the cutting action generates enough heat to form a thin film of liquid bismuth that lasts for only fractions of a microsecond. Tool life is noticeably extended, and the material is still weldable.

The use of free-machining steels is not without compromise, however. Ductility and impact properties are somewhat reduced over those of the unmodified steels. If this reduction is objectionable, other methods may be employed to enhance machinability. For example, the machinability of steels can be improved by cold working. As the strength and hardness of the metal increase, the chips formed by the cutting process tear away more readily and break up more easily.

Steels for Electrical and Magnetic Applications. Several types of steels are widely used in the electrical industry. *Silicon steels,* which contain 0.5% to 5.0% silicon, have noticeably increased resistivity and permeability. Increased resistivity decreases eddy-current losses, and increased permeability decreases hysteresis losses. When such steels are used in electrical motors, generators, and transformers, power losses and the associated heat problems are reduced. For this reason, silicon steel is often used for the magnetic circuits of electrical equipment. Since silicon causes the steel to become brittle, the amount used should be kept low in applications where the component is subjected to dynamic conditions.

Cobalt increases the magnetic saturation point in a steel when it is added in amounts up to 36%. For this reason, *cobalt alloy steels* are used in electrical equipment where high magnetic densities must exist. Many permanent magnets are made from high-cobalt alloys.

Most recently, amorphous metals have shown considerable promise in this area. Since the material has no grains or grain boundaries, the magnetic domains can move freely in response to magnetic fields, the properties are the same in all directions, and corrosion resistance is improved. The high magnetic strength and low hysteresis losses offer the possibility of smaller, light-weight magnets. When used to replace silicon steel in power transformer cores, this material has the potential of reducing core losses by 50%. (One estimate cites a potential savings in the United States alone of 20 billion kilowatt-hours per year, or about $1 billion.) As the cost of this material has dropped over from $25.00 per pound to approximately $1.50 per pound over the past few years, the potential for application appears bright.

Precoated Steel Sheet.   Traditional sheet metal fabrication has involved the fabrication of components from bare steel, followed by the finishing of these products piece-by-piece. In this system, it is not uncommon for the finishing processes to be the most expensive and time-consuming stages of manufacture.

An alternative to this procedure is to purchase mill-coated sheet material to which the coating was applied by the steel supplier when the material was in the form of a long, continuous strip. Numerous coatings can be specified, including the entire spectrum of dipped and plated metals (including aluminum, zinc, and chromium), vinyls, paints, and other materials. Extra caution must be exercised during fabrication to prevent damage to the coating, but this additional effort and expense are often far less than the cost of finishing individual pieces.

Corrosion-Resistant or Stainless Steels.   *Corrosion-resistant* or *stainless steels* contain sufficient amounts of chromium so that they can no longer be considered low-alloy steels. The corrosion resistance is imparted by the formation of a strongly adherent chromium oxide on the surface of the metal. Good resistance to many of the corrosive media encountered in the chemical industry can be obtained by the addition of 4% to 6% chromium to low-carbon steel. Where improved corrosion resistance and outstanding appearance are required, materials are designed to utilize a superior chromium oxide that forms when the amount of atomic chromium in solution (excluding chromium carbides and other forms where the chromium is unavailable to react with oxygen) exceeds 12%. This category forms what has been commonly called the *true stainless steels*.

Several classification schemes have been devised to categorize these alloys. The American Iron and Steel Institute groups the metals by chemistry and assigns a three-digit number that identifies the basic family and the particular alloy within that family. The material in this text, however, groups these alloys by microstructural families, for it is the basic structure that controls the engineering properties of the metal. Table 6.5 presents the AISI designation scheme for stainless steels and correlates it with the microstructural families.

TABLE 6.5   AISI Designation Scheme for Stainless Steels

Series	Alloys	Structure
200	Chromium, nickel, manganese, or nitrogen	Austenitic
300	Chromium and nickel	Austenitic
400	Chromium only	Ferritic or martensitic
500	Low chromium (<12%)	Martensitic

Chromium is a ferrite stabilizer, the addition of chromium tending to increase the temperature range over which ferrite is the stable structure. If sufficient chromium is added to iron and carbon is kept low, an alloy can be produced that is ferrite at all temperatures below solidification. These alloys are known as *ferritic stainless steels*. They possess rather poor ductility or formability because of the bcc crystal structure, but they are readily weldable. No martensite can form in the welds because there is no possibility of forming austenite that can then transform during cooling.

If insufficient chromium is added to fully eliminate the austenite region, stainless metals can be produced that are austenite at high temperature and ferrite at low. *Martensitic stainless steels* is the name given to these alloys, in which carbon can be dissolved in the austenite, which is then quenched to form a martensitic structure. Variable carbon contents are possible, which produce the various strength levels, with caution being taken to maintain more than 12% chromium in solution.[1] Martensitic stainless steels cost about 1½ times as much as the ferritic alloys, part of this being due to the additional heat treatment, generally consisting of an austenization, quench, stress relief, and temper.

Nickel is an austenite stabilizer, and with sufficient amounts of both chromium and nickel, it is possible to produce a stainless steel in which austenite is the stable structure at room temperature. Known as *austenitic stainless steels*, these alloys may be twice as costly as the ferritic variety, the added expense being attributed to the cost of the alloying nickel and chromium. Manganese and nitrogen can be used as austenite stabilizers and may be substituted for some of the nickel to produce a lower-cost, somewhat lower-quality austenitic stainless steel. Additional cost savings can be achieved by the use of aluminum to replace chromium in its role of imparting corrosion resistance.

Austenitic stainless steels are nonmagnetic and are highly corrosion resistant in almost all media except hydrochloric acid and other halide acids and salts. In addition, they may be polished to a mirror finish and thus combine attractive appearance and good corrosion resistance. Formability is outstanding (a characteristic of the fcc crystal structure), and these steels respond quite well to strengthening by cold work. The response of the popular 18-8 alloy (18% chromium and 8% nickel) to a small amount of cold work is as follows:

	Water Quench	Cold-Rolled 15%
Yield strength	38 ksi (260 MPa)	117 ksi (805 MPa)
Tensile strength	90 ksi (620 MPa)	140 ksi (965 MPa)
Elongation in 2 in.	68%	11%

[1]Slow cools may allow the carbon and chromium to react and form chromium carbides. When this reaction occurs, the chromium is not available to react with oxygen and to form the protective oxide. Martensitic stainless steels may be stainless only in the martensitic condition where the chromium is trapped in atomic solution.

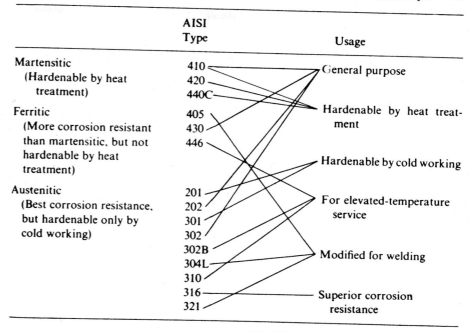

	AISI Type	Usage
Martensitic (Hardenable by heat treatment)	410 420 440C	General purpose
Ferritic (More corrosion resistant than martensitic, but not hardenable by heat treatment)	405 430 446	Hardenable by heat treatment
		Hardenable by cold working
Austenitic (Best corrosion resistance, but hardenable only by cold working)	201 202 301 302 302B 304L 310 316 321	For elevated-temperature service Modified for welding Superior corrosion resistance

FIGURE 6-13 Classification and uses of stainless steels.

These materials are often used in the water-quenched condition, the water quench serving to retain the alloys in solid solution. No phase transformations occur during the quench, since austenite is the stable phase for all temperatures involved.

Austenitic stainless steels are costly materials and should not be specified where the less expensive ferritic or martensitic alloys would be adequate or where a true stainless steel is not required. Figure 6-13 lists several popular alloys from each of the three major classifications and schematically denotes their key properties. Table 6-6 presents the typical compositions of the three basic families of stainless steels.

TABLE 6.6 Typical Compositions of the Ferritic, Martensitic, and Austenitic Stainless Steels

Element	Ferritic	Martensitic	Austenitic[a]
Carbon	0.08–0.20%	0.15–1.2%	0.03–0.25%
Manganese	1–1.5%	1%	2% (5.5–10%)
Silicon	1%	1%	1–2% (0%)
Chromium	11–27%	11.5–18%	16–26%
Nickel			3.5–22%
Phosphorus and sulfur			Normal (0%)
Molybdenum			Some cases
Titanium			Some cases

[a]Values in parentheses are for one type.

A fourth and special class of stainless steels is the *precipitation-hardening* variety. These alloys are basically martensitic or austenitic types, modified by the addition of alloying elements that permit age hardening at relatively low temperatures. By adding age hardening to the existing strengthening mechanisms, these alloys are capable of attaining properties such as a 260 ksi (1790 MPa) yield strength and 265 ksi (1825 MPa) tensile strength with a 2% elongation. However, the precipitation-hardening alloys are among the most expensive of the stainless steels and should be used only when absolutely required.

In addition to the basic types of stainless steels, certain alloys have been developed for specific uses. Ordinary stainless steels are difficult to machine because of their work-hardening properties and their tendency to seize during cutting. Special *free-machining alloys* have been produced within each family, the addition of sulfur or selenium raising the machinability to approximately that of a medium-carbon steel.

Problems with stainless steels generally relate to the loss of corrosion resistance (*sensitization*) when the amount of chromium in solution drops below 12%. Since chromium depletion is usually caused by the formation of chromium carbides along grain boundaries and these carbides form at elevated temperatures, one method of prevention is to keep the carbon content of stainless steels as low as possible, usually less than 0.10%. Another approach is to tie up the carbon with small amounts of "stabilizing" elements, such as titanium or niobium, that have a stronger affinity for carbon than chromium. Rapidly cooling these metals through the range of 900–1500°F (480–820°C) also retards carbide formation.

Another problem with high-chromium stainless steels is an embrittling after long times at elevated temperatures. This is attributed to the formation of *sigma phase*, a brittle structure that forms and coats grain boundaries, producing a brittle crack path through the metal. Stainless steels used in high-temperature service should be checked periodically to detect the formation of sigma phase.

Maraging Steels.   When 15% to 25% nickel and significant amounts of cobalt, molydenum, and titanium are added to a very-low-carbon steel, a material results that can be air quenched from about 1500°F (815°C) to form martensite. This, in turn, can be subsequently age-hardened at 900°F (480°C) to produce yield strengths in excess of 250 ksi (1725 MPa) with elongations in excess of 11%. Such alloys are very useful in applications where ultrahigh strength and good toughness are important.

A typical maraging steel in the 250-ksi (1725-MPa) strength class has a composition of:

0.03%	C	0.10%	Al
18.5%	Ni	0.003%	B
7.5%	Co	0.10%	Si maximum
4.8%	Mo	0.10%	Mn maximum
0.40%	Ti	0.01%	S maximum
0.01%	Zr	0.01%	P maximum

This steel can be hot-worked at temperatures between 1400°F and 2300°F (760–1260°C). When air-cooled from 1500°F (815°C), it has a hardness of about 30R_c and a structure of soft, tough martensite. It is easily machined and, because of its low work-hardening rate, can be cold-worked to a high degree. Aging at 900°F (480°C) for 3 to 6 hours, followed by air cooling, raises the hardness to about 52R_c and produces full strength. Maraging steels can be welded, if welding is followed by the full solution and aging treatment. As would be expected, maraging steels are quite expensive.

Steels for High-Temperature Service.
Continued developments in missiles, jet aircraft, and nuclear power have increased the need for metals that have good strength characteristics, corrosion resistance, and, particularly, creep resistance at high temperatures. Much work has been done to produce both ferrous and nonferrous alloys that have the desired properties at temperatures in excess of 1000°F (550°C).

The ferrous alloys, which are normally restricted to use below 1400°F (760°C), tend to be low-carbon materials with less than 0.1% carbon. One alloy is a modified 18-8 stainless steel, stabilized with either niobium or titanium. A 1000-hour rupture stress of 6 to 7 ksi (40 to 50 MPa) is observed at 1400°F (760°C), with considerably higher strengths at lower temperatures. Iron is also a major component of other high-temperature alloys, but when amounts become less than 50%, the metal can hardly be classified as ferrous in nature. High strengths at high temperature usually require the more expensive nonferrous materials, which will be discussed in a later chapter.

Tool Steels.
Tool steels are metals designed to provide wear resistance and toughness combined with high strength. They are basically high-carbon steels in which the alloy chemistry provides the desired balance of toughness and wear.

Several classification systems have been applied to tool steels, some using chemistry as a basis and others employing hardening method and major mechanical property. The AISI-SAE designation system identifies letter grades by basic features such as the quenching method, the primary applications, a special characteristic, or the specific industry involved. Table 6.7 lists the seven basic families of tool steels and the corresponding AISI-SAE grades. Individual alloys are then listed numerically within the grade to produce a letter–number identification system.

Water-hardening carbon tool steels (W grade) account for a large percentage of all the tool steels used. They are the least expensive and are used for a wide variety of parts that are usually quite small and not subjected to severe usage or elevated temperatures. Because strength and hardness are functions of the carbon content, a wide range of these properties can be obtained through composition variation. These steels must be quenched in water to attain high hardness, and because their hardenability is low, they can be used only for relatively thin sections if full depth of hardness must be obtained. They are also rather brittle, particularly at higher hardness.

TABLE 6.7   **Basic Types of Tool Steel and Corresponding AISI-SAE Grades**

Type	AISI-SAE Grade	Significant Characteristic
1. Water-hardening	W	
2. Cold-work	O	Oil-hardening
	A	Air-hardening medium alloy
	D	High-carbon–high-chromium
3. Shock-resisting	S	
4. High-speed	T	Tungsten base
	M	Molybdenum base
5. Hot-work	H	H1–H19: chromium base
		H20–H39: tungsten base
		H40–H59: molybdenum base
6. Plastic-mold	P	
7. Special-purpose	L	Low alloy
	F	Carbon-tungsten

Typical uses of plain carbon steels, according to carbon content, are as follows:

0.60%–0.75% carbon: Machine parts, chisels, set screws, and similar products in which medium hardness with considerable toughness and shock resistance is required.

0.75%–0.90% carbon: Forging dies, hammers, and sledges.

0.90%–1.10% carbon: General-purpose tooling applications that require good balance of wear resistance and toughness, such as drills, cutters, shear blades, and other heavy-duty cutting edges.

1.10%–1.30% carbon: Small drills, lathe tools, razors, and similar light-duty applications in which extreme hardness is necessary without great toughness.

In applications where improved toughness is desired, small amounts of manganese, silicon, and molybdenum are often added. Vanadium additions of about 0.20% are used to form strong, stable carbides that retain fine grain size during heat treating. One of the main weaknesses of carbon tool steels is the fact that they do not hold their hardness at elevated temperatures. Prolonged exposure to temperatures over 300°F (150°C) results in undesired softening.

When larger parts must be hardened or distortion must be minimized, *oil-* or *air-hardening grades* (O and A designations, respectively) are often employed. Because of the higher hardenability, they can be hardened by less severe quenches, while simultaneously maintaining tighter dimensional tolerances and minimizing the cracking tendency. Manganese tool and die steels form one segment of this group. These contain 0.75% to 1.0% carbon and 1.0% to 2.0% manganese and are moderate in cost. In some cases, manganese is reduced and chromium, silicon, and nickel are added to give greater toughness. These steels are not as hard as the plain manganese types, but they have less tendency to crack because of their greater hardenability and less severe quench requirement.

Chromium tool and die steels are also in this class. The low-chrome tool steels are much the same as plain carbon tool steels, with chromium added to produce increased hardenability and toughness. Here, the chromium levels are usually between 0.5% and 5.0%. High-chromium tool steels are designated by a separate letter (D) and contain between 10% and 18% chromium. Ordinarily, these steels must be annealed before they can be machined. After machining, they are hardened and tempered and usually retain the final hardness at temperatures up to 800°F (425°C). They are used for tools and dies that must withstand hard usage over long periods, often at elevated temperatures. Forging dies, die-casting die blocks, and drawing dies are often made from such steels.

Shock-resisting tool steels (S designation) have been developed for both hot and cold impact applications. Low carbon content (approximately 0.5%) is used to provide toughness, with carbide-forming alloying elements being added to supply the desired abrasion resistance, hardenability, and hot-work characteristics.

High-speed tool steels are used primarily for cutting tools that machine metal and other special applications where retention of hardness at red heat is required. A common type of high-speed steel is the tungsten-based T1 alloy, also known as 18-4-1 because the analysis contains 0.7% carbon, 18% tungsten, 4% chromium, and 1% vanadium. This alloy has a balanced combination of shock resistance and abrasion resistance and is used for a wide variety of cutting applications. Other high-speed tool steels have cobalt added to improve hardness at elevated temperatures. Toughness diminishes and forming problems increase.

The *molybdenum high-speed steels* (M designation) were developed to reduce the amount of tungsten and chromium required to produce the desired properties. The M2 variety is now the most widely used high-speed steel, its higher carbon content and balanced analysis producing properties applicable to a number of general-purpose high-speed uses.

Hot-work tool steels (H designation) have been designed to perform adequately under environments of prolonged high temperature. All employ additions of carbide-forming alloying elements: H1 to H19 are chromium-base types with about 5.0% chromium; H20 to H39 are tungsten-base types with 9% to 18% tungsten coupled with 3% to 4% chromium; and H40 to H59 are molybdenum-based types.

Other types of tool steels include (1) the *plastic mold steels* (P designation) made specifically for the requirements of zinc die casting and plastic molding dies; (2) the *low-alloy special-purpose tool steels* (L designation), such as the L6 extreme toughness variety; and (3) the *carbon–tungsten type* of special-purpose tool steels (F designation), which are water hardening but substantially more wear-resistant than the plain-carbon tool steels.

Alloy Cast Steels and Irons.   The effects of alloying elements are the same regardless of the method used to produce the final shape. To couple the attributes of the casting process and the benefits of alloy material, many alloy cast steels have been developed. In the case of cast iron, however, alloys often perform functions in addition to those cited for steels.

Whereas alloy steels are usually heat-treated, many alloy cast irons are not, except for stress relieving or annealing. If a cast iron is to be quenched, however, chromium, molybdenum, and nickel are frequently added to improve hardenability. In addition, chromium tends to offset the undesirable quenching effects of high carbon, and molybdenum and nickel offsets the high silicon.

If alloy cast irons are not to be heat-treated, the alloy elements are often selected for their ability to alter properties through affecting the formation of graphite or cementite, modifying the morphology of the carbon-rich phase, or simply strengthening the matrix material. Alloys are often added in small amounts to improve strength properties or wear resistance. High-alloy cast irons are often designed to provide corrosion resistance, particularly at high temperatures such as are encountered in the chemical industry.

Nickel promotes graphite formation and tends to produce a finer graphite structure. Chromium, on the other hand, retards graphitization and stabilizes cementite. They are frequently used together in the ratio of 2–3 parts of nickel to 1 part of chromium.

If the silicon content of gray cast iron is lowered, the strength is increased considerably. However, if this decrease in silicon content is carried very far, hard, white cast iron results. The addition of about 2% nickel will minimize the formation of white cast iron. By adjusting the silicon content and adding a small amount of nickel, a cast iron of good strength is obtained without sacrificing machinability.

Molybdenum strengthens gray cast iron to some extent and often forms carbides. In addition, it is used to control the size of the graphite flakes. From 0.5% to 1.0% is ordinarily added.

Among the high-alloy cast irons, *austenitic gray cast iron* is quite common. This contains about 14% nickel, 5% copper, and 2.5% chromium. It possesses good corrosion resistance to many acids and alkalis at temperatures up to about 1500°F (800°C).

Alloy cast irons and steels are generally identified by their ASTM specification number.

Review Questions

1. What is meant by the term *history-dependent material?*
2. What is a ferrous material?
3. When iron ore is reduced to metallic iron, what other elements are generally present in the metal?
4. How does steel differ from pig iron?
5. What are the various forms or means by which molten steel can be solidified?
6. What is ladle metallurgy?

7. What is the advantage of pouring molten metal from the bottom of a ladle?
8. How can the shape and location of solidification shrinkage be controlled in the casting of an ingot?
9. What are some of the attractive economic and processing advantages of continuous casting?
10. What are some of the alternative methods of dealing with dissolved gases in molten steel?

11. What are the attractive metal quality features that are offered by electroslag remelting?
12. What is a plain-carbon steel?
13. What features account for the high-volume use of medium-carbon steels?
14. What are the most common alloy elements added to steel?
15. What are some of the reasons that alloying elements are added to steel?
16. What alloys are particularly effective in increasing the hardenability of steel?
17. Many people think that chromium is added to steel to improve the corrosion resistance. In typical alloy steels, what is the primary contribution of chromium?
18. What are some of the alloy elements that tend to form stable carbides within a steel?
19. What is the significance of the last two digits in a typical four-digit AISI-SAE steel designation?
20. How might fabrication considerations influence the selection of a steel?
21. What is the major difference in the way strength is obtained in the HSLA steels and the constructional alloy steels?
22. What are microalloyed steels?
23. What are some of the potential benefits that may be obtained through the use of microalloyed steels?
24. What are some of the various alloy additions that have been used to improve the machinability of steels?

25. What are some of the compromises associated with the use of free-machining steels?
26. Why have the amorphous metals attracted attention as potential materials for magnetic applications?
27. What is the economic motivation for the use of precoated steels?
28. What structural feature is responsible for the observed corrosion resistance of stainless steels?
29. Which of the three major types of stainless steel is likely to contain significant amounts of carbon? Why?
30. What are some of the unique properties of austenitic stainless steels?
31. What is sensitization of a stainless steel and how can it be prevented?
32. Under what types of service conditions might maraging steels be required?
33. What is a tool steel?
34. How does the AISI-SAE designation system for tool steels differ from that for plain-carbon and alloy steels?
35. What alloys can be used to improve the toughness of tool steels?
36. What types of alloying elements form the primary additions to hot-work tool steels?
37. What are some of the reasons why alloy additions are made to cast irons that are not to undergo heat treatment?

Nonferrous Metals and Alloys

Nonferrous metals and alloys are playing increasingly important roles in modern technology. Because of their number and the fact that their properties vary widely, they provide an almost limitless range of properties for the design engineer. Even though they are not produced in as great tonnages and may be more costly than iron or steel, these metals often possess certain properties or combinations of properties that are not available in ferrous metals:

1. Resistance to corrosion.
2. Ease of fabrication.
3. High electrical and thermal conductivity.
4. Light weight.
5. Color.

Although it is true that corrosion resistance can be obtained in certain ferrous alloys, several of the nonferrous metals possess this property without requiring special and expensive alloying elements. Nearly all of the nonferrous alloys possess at least two of the qualities listed, and some possess all five. For many applications, specific combinations of these properties are highly desirable, and the availability of materials that provide them directly is a strong motivation for the use of nonferrous alloys.

In most cases, the nonferrous alloys are inferior to steel in strength. Also, the modulus of elasticity may be considerably lower, a fact that places them at a distinct disadvantage where stiffness is a necessary characteristic. Fabrication, however, is usually easier than for steel. Those alloys with low melting points are often easy to cast, either in sand molds, permanent molds, or dies. Many alloys have high ductility coupled with low yield points, the ideal combination for easy cold work and high formability. Good machinability is characteristic of many nonferrous alloys. Fabrication savings can often overcome the higher cost of the nonferrous material and favor its use in place of steel. The one fabrication area in which the nonferrous alloys are somewhat inferior to steel is weldability. Because of a number of developments and improvements in welding, however, it is generally possible to produce satisfactory weldments in all of the nonferrous metals.

Copper and Copper Alloys

General Properties and Characteristics. *Copper* is an important engineering metal that has been in use for over 6000 years. Used in its pure state, copper is the backbone of the electrical industry. It is also the major metal in a number of highly important engineering alloys, namely, the brasses and bronzes.

The wide use of copper is based, primarily, on three important properties: its high electrical and thermal *conductivity,* its high *ductility,* and its *corrosion resistance.* Obviously, its excellent conductivity accounts for its importance to the electrical industry. The better grades of conductor copper now have a conductivity rating of about 102%, reflecting metallurgical improvements made since 1913, when the standard conductivity scale was established and pure copper was set at 100%.

Pure copper in its annealed state has a tensile strength of only about 30,000 psi (200 MPa) and an elongation of nearly 60%. By cold working, however, the metal can be hardened and the tensile strength raised to over 65,000 psi (450 MPa) with a decrease in elongation to about 5%. Its relatively low strength and high ductility make copper a very desirable metal where forming operations are necessary. Furthermore, the hardening effects of cold working can be easily removed. The recrystallization temperature of copper is less than 500°F (260°C).

If copper is stressed at high temperatures for a long period of time, it is subject to intercrystalline failure at about half its normal room-temperature strength. Material containing more than 0.3% oxygen is also subject to hydrogen embrittlement when it is exposed to reducing gases above 750°F (400°C).

Commercially Pure Copper. Refined copper containing between 0.2% and 0.05% oxygen (the principal impurity in copper) is called *electrolytic toughpitch (ETP) copper.* It is often used as a base for copper alloys and may be used for electrical applications where the highest conductivity is not required. For superior conductivity, additional refining is required to reduce the oxygen content and to produce *oxygen-free high-conductivity (OFHC) copper.*

Copper-Base Alloys. Copper, as a pure metal, is not used extensively in manufactured products, except in electrical applications. More often, it is the base metal for some alloy to which it imparts its good ductility and corrosion resistance.

Copper-base alloys are commonly identified through a system of numbers standardized by the Copper Development Association (CDA). Table 7.1 presents a breakdown of this system, which has been adopted by the American Society for Testing of Materials (ASTM), the Society of Automotive Engineers (SAE),

TABLE 7.1 **Standard Designations for Copper and Copper Alloys (CDA System)**

Wrought Alloys		Cast Alloys	
100–155	Commercial coppers	833–838	Red brasses and leaded red brasses
162–199	High-copper alloys		
200–299	Copper–zinc alloys (brasses)	842–848	Semired brasses and leaded semired brasses
300–399	Copper–zinc–lead alloys (leaded brasses)	852–858	Yellow brasses and leaded yellow brasses
400–499	Copper–zinc–tin alloys (tin brasses)	861–868	Manganese and leaded manganese bronzes
500–529	Copper–tin alloys (phosphor bronzes)	872–879	Silicon bronzes and silicon brasses
532–548	Copper–tin–lead alloys (leaded phosphor bronzes)	902–917	Tin bronzes
600–642	Copper–aluminum alloys (aluminum bronzes)	922–929	Leaded tin bronzes
		932–945	High-leaded tin bronzes
647–661	Copper–silicon alloys (silicon bronzes)	947–949	Nickel–tin bronzes
		952–958	Aluminum bronzes
667–699	Miscellaneous copper–zinc alloys	962–966	Copper nickels
		973–978	Leaded nickel bronzes
700–725	Copper–nickel alloys		
732–799	Copper–nickel–zinc alloys (nickel silvers)		

and the U.S. government. Alloys numbered from 100 to 190 are mostly copper with less than 2% alloy addition. Numbers 200 to 799 are other wrought alloys. The 800 and 900 series are all casting alloys.

Copper–Zinc Alloys. Zinc is by far the most popular alloying addition to copper, the resulting alloys being known as *brasses*. If the copper content is not over 36%, brass is a single-phase solid solution. Since this structure is identified as the alpha phase, these alloys are often called *alpha brasses*. They are quite ductile and formable, these characteristics increasing with the zinc content up to about 36%. Above 36% zinc, the alloys enter a two-phase region involving the brittle beta phase. Cold-working properties are rather poor for these high-zinc brasses, but deformation is rather easy when performed hot. Like copper, brass is hardenable by cold working and is commercially available in various degrees of hardness.

Table 7.2 lists some of the most common copper–zinc alloys and their composition, properties, and typical uses. Brasses range in color from copper to nearly white, the lower-zinc brasses being more coppery than those with more zinc. The addition of a third element, however, can change the color considerably.

Most brasses have good corrosion resistance. In the 0% to 40% zinc region, the addition of a small amount of tin produces excellent resistance to seawater

TABLE 7.2   Composition, Properties, and Uses of Some Common Copper–Zinc Alloys

CDA Number	Common Name	Composition (%)					Condition	Tensile Strength		Elongation in 2 inches (%)	Typical Uses
		Cu	Zn	Sn	Pb	Mn		ksi	MPa		
220	Commercial bronze	90	10				Soft sheet	38	262	45	Screen wire, hardware, screws, jewelry
							Hard sheet	64	441	4	
							Spring	73	503	3	
240	Low brass	80	20				Annealed sheet	47	324	47	Drawing, architectural work, ornamental
							Hard	75	517	7	
							Spring	91	627	3	
260	Cartridge brass	70	30				Annealed sheet	53	365	54	Munitions, hardware, musical instruments, tubing
							Hard	76	524	7	
							Spring	92	634	3	
270	Yellow brass	65	35				Annealed sheet	46	317	64	Cold forming, radiator cores, springs, screws
							Hard	76	524	7	
280	Muntz metal	60	40				Hot-rolled	54	372	45	Architectural work, condenser tube
							Cold-rolled	80	551	5	
443–445	Admiralty metal	71	28	1			Soft	45	310	60	Condenser tube (salt water), heat exchangers
							Hard	95	655	5	
360	Free-cutting brass	61.5	35.5		3		Soft	47	324	60	Screw-machine parts
							Hard	62	427	20	
675	Manganese bronze	58.5	39	1		0.1	Soft	65	448	33	Clutch disks, pump rods, valve stems, high-strength propellers
							Bars, half hard	84	579	19	

corrosion. Cartridge brass with tin becomes admiralty brass; muntz metal with tin is known as naval brass. Brasses with 20% to 36% zinc are subject to a selective corrosion, known as *dezincification,* when in acid or salt solutions. Another corrosion problem in brasses with more than 15% zinc is season cracking or stress-corrosion cracking. Both stress and exposure to corrosive media are required for this failure to occur. Brasses must often undergo a stress relief to remove the residual stresses induced by cold working prior to being put into service.

Many uses of brass relate to the high electrical and thermal conductivities coupled with adequate strength. Plating characteristics are outstanding and make it an excellent base for decorative chrome or similar coatings. A unique property of alpha brass is its ability to have rubber vulcanized to it without any special treatment except thorough cleaning. It is widely used in mechanical rubber goods because of this property.

Where high machinability is needed, as in automatic screw-machine stock, 2% to 3% lead is added to brass to ensure the production of free-breaking chips.

An alloy containing from 50% to 55% copper and the remainder zinc is often used as a filler metal in brazing. It is an effective agent for joining steel, cast iron, brasses, and copper, producing joints that are nearly as strong as those obtained by welding.

Copper–Tin Alloys.   Alloys of copper and tin, commonly called tin *bronzes,* are considerably more expensive than the brasses because of the high price of tin. Consequently, they have now been replaced to a considerable degree by less expensive nonferrous alloys but are used in certain applications because of special properties.

The term *bronze* is somewhat confusing, since some alloys that contain no tin are called bronzes because of their color. The true bronzes usually contain less than 12% tin. Strength increases with tin content up to about 20%, beyond which the alloys become brittle. Copper–tin alloys are characterized by good strength, toughness, wear resistance, and corrosion resistance. They are often used for bearings, gears, and fittings that are subjected to heavy compressive loads. When copper–tin alloys are used as a bearing material, up to 10% lead is often added.

The most popular wrought bronze is phosphor bronze, which usually contains from 1% to 11% tin. Alloy 521 (CDA) is typical of this class and contains 92% copper, 8% tin, and 0.15% phosphorus. Hard sheet of this material has a tensile strength of 110 ksi (760 MPa) and an elongation of 3%. Soft sheet has a 55 ksi (380 MPa) tensile strength and 65% elongation. It is used for pump parts, gears, springs, and bearings.

Alloy 905 is a commonly used bronze casting alloy containing 88% copper, 10% tin, and 2% zinc. In the cast condition, the tensile strength is about 45 ksi (310 MPa) with an elongation of 45%. It has very good resistance to sea water corrosion and is used on ships for pipe fittings, gears, pump parts, bushings, and bearings.

Copper–Nickel Alloys.   Copper and nickel have complete solid solubility, as seen in Figure 4.6, and a wide range of alloys have been developed. High thermal conductivity coupled with corrosion resistance make these materials a good choice for heat exchangers, cookware, and other heat-transfer applications. *Cupronickels* contain 2% to 30% nickel. *Nickel silvers* have 10% to 30% nickel and at least 5% zinc. An alloy with 45% nickel is known as a *constantan*, and the 67% nickel alloy is called *Monel*.

Other Copper-base Alloys.   The copper alloys discussed previously acquire their strength primarily through solid-solution strengthening and cold work. In the copper alloy family, three alloying elements can produce materials that are precipitation hardenable: aluminum, silicon, and beryllium.

Aluminum bronze alloys usually contain between 6% and 12% aluminum and often 2% to 5% iron. With aluminum contents below 8%, the alloys are very ductile. When aluminum exceeds 9%, hardness approaches that of steel. Still higher aluminum contents result in brittle, but very wear-resistant, materials. A cast aluminum bronze having 86.2% copper, 10.2% aluminum, and 3.3% iron has a tensile strength of 70 to 80 ksi (480 to 550 MPa) and 18% to 22% elongation. By varying the aluminum content and heat treatment, the tensile strength can be varied from about 60 to 125 ksi (415 to 860 MPa). Parts cast in aluminum bronze exhibit large amounts of shrinkage, and castings to be made of this material should be designed with this fact in mind.

Silicon bronzes contain up to 4% silicon and 1.5% zinc (higher zinc contents may be used when the material is to be cast). Strength, formability, machinability, and corrosion resistance are quite good. Tensile strengths can approach 130 ksi (900 MPa) with cold work, whereas the soft metal has a tensile strength of about 55 ksi (380 MPa) and an elongation of 65%. Uses include boiler, tank, and stove applications that require a combination of high strength and corrosion resistance.

Copper–beryllium alloys can be age-hardened to produce the highest strengths of the copper-based metals. They ordinarily contain less than 2% beryllium but are quite expensive to use. When annealed, the material has a yield strength of 25 ksi (170 MPa), a tensile strength of 70 ksi (480 MPa), and an elongation of 50%. After heat treatment, these properties can rise to 160 ksi (1100 MPa), 180 ksi (1250 MPa), and 5%, respectively. Cold work coupled with age hardening can produce even stronger material. The modulus of elasticity is about 18,000 ksi (125,000 MPa), and the endurance limit is around 40 ksi (275 MPa). These properties make the material an excellent choice for electrical contact springs, but cost limits application to small components requiring long life and high reliability. Other applications relate to the unique combination of properties: the material has the strength of steel and is nonsparking, nonmagnetic, and conductive. Recent concerns about the toxicity of beryllium have somewhat reduced the popularity of this material and have created a demand for nonberyllium high-strength alloys.

Aluminum and Aluminum Alloys

General Properties and Characteristics.   Although aluminum has been a commercial metal for just 100 years, it now ranks second to steel in both worldwide quantity and expenditure and is clearly the most important of the nonferrous metals. It has achieved importance in virtually all segments of the world economy, its principal uses being in transportation, construction, electrical applications, containers, consumer durables, and mechanical equipment.

A number of unique and attractive properties account for the engineering significance of aluminum: its *workability*, its *light weight*, its *corrosion resistance*, and its good *electrical and thermal conductivity*. Aluminum has a specific gravity of 2.7, compared to 7.85 for steel, making aluminum about one third the weight of an equal volume of steel. Thus, although aluminum costs more per pound than steel, it is actually cheaper than steel of an equal volume.

Probably the most serious weakness of aluminum from an engineering viewpoint is its relatively *low modulus of elasticity*, roughly one third that of steel. Under identical loadings, an aluminum component deflects three times as much as a steel component of the same design. Since the modulus of elasticity can be modified only slightly by alloying and not at all by heat treatment, it is necessary to design and manufacture aluminum products that place the metal in such a way as to maximize stiffness while using minimal material. Fortunately, this placement can often be done with relative ease because of aluminum's good workability.

Commercially Pure Aluminum.   In its pure state, aluminum is soft, ductile, and not very strong. In the annealed condition, pure aluminum has only about one fifth the strength of hot-rolled structural steel. Thus, commercially pure aluminum is used primarily for its physical, rather than its mechanical, properties.

Electrical-conductor-grade aluminum is used in large quantities and has replaced copper in many applications, such as electrical transmission lines. Commonly designated by the letters *EC*, this grade contains a minimum of 99.45% aluminum and has an electrical conductivity 62% that of copper for the same size wire and 200% that of a copper wire of equal weight.

Aluminums for Mechanical Applications.   For nonelectrical applications, most aluminum is used in the form of alloys. These have much greater strength than pure aluminum yet retain the advantages of light weight, good conductivity, and corrosion resistance. Some alloys are now available that have tensile properties (except for ductility) that are superior to those of low-alloy–high-yield-strength structural steel. On a strength-to-weight basis, most of the aluminum alloys are superior to steel, but wear, creep and fatigue properties are generally rather poor. In addition, aluminum alloys rapidly lose their strength as temperature is increased and should not be considered for applications involving service temperatures much above 500°F (260°C).

The choice of steel or aluminum for any given component is largely a matter of cost, although in many cases the advantages of reduced weight or corrosion resistance may justify an additional expense. Aluminum generally replaces steel or cast iron where there is a strong need for lightness, corrosion resistance, low maintenance expense, or high thermal or electrical conductivity. In modern motor vehicles, aluminum is seeing increased use in body panels, engine blocks, manifolds, and transmission cases, the reduced weight bringing about a significant increase in fuel economy.

Corrosion Resistance of Aluminum and Its Alloys.   Pure aluminum is very reactive and forms a tight, adherent oxide coating on the surface as soon as it is exposed to air. This oxide is resistant to many corrosive media and serves as a corrosion-resistant barrier to protect the underlying metal. Thus, as in stainless steels, the corrosion resistance of the metal is actually a property of the surface oxide. As alloys are added to the aluminum, oxide formation is somewhat retarded. Aluminum alloys, in general, do not have quite the superior corrosion resistance of pure aluminum.

The oxide coating on aluminum alloys also causes some difficulty in relation to its weldability. For consistent-quality resistance welding, it is usually necessary to remove the oxide immediately before welding. During fusion welding processes, the aluminum would oxidize so readily that special fluxes or protective inert-gas atmospheres must be employed. Although welding aluminum may be more difficult than welding steel, suitable techniques have been developed so that high-quality, cost-effective welds can be made with most welding processes.

Wrought-Aluminum Alloys.   Aluminum-base alloys designed for fabrication as solids can be divided into two basic types: those that achieve *strength by solid-solution strengthening and cold working* and those that are *strengthened by heat treatment* (age hardening). Table 7.3 lists some common wrought-aluminum alloys; the standard four-digit designation system for aluminum is used. The first digit indicates the major alloy grouping as follows:

Major Alloying Element	
Aluminum, 99.00% and greater	1xxx
Copper	2xxx
Manganese	3xxx
Silicon	4xxx
Magnesium	5xxx
Magnesium and silicon	6xxx
Zinc	7xxx
Other element	8xxx

The second digit generally indicates a modification of the original alloy, and the last two digits indicate the particular alloy within the family. In the 1xxx series, however, the last three digits are used to denote the purity of the aluminum.

Once the alloy chemistry is specified through this four-digit code, clarification of the alloy condition is made through a *temper designation,* a letter or letter–number suffix:

-F: as fabricated.

-H: strain-hardened.

    -H1: strain-hardened by working to desired dimensions; a second digit, 1 through 9, indicates the degree of hardness, 8 being commercially full-hard and 9 extra-hard.

    -H2: strain-hardened by cold working, followed by partial annealing, second-digit numbers 2 through 8, as above.

    -H3: strain-hardened and stabilized.

-O: annealed.

-T: thermally treated (heat treated).

    -T1: cooled from hot working and naturally aged.

    -T2: cooled from hot working, cold-worked, and naturally aged.

    -T3: solution-heat-treated, cold-worked, and naturally aged.

    -T4: solution-heat-treated and naturally aged.

    -T5: cooled from hot working and artificially aged.

    -T6: solution-heat-treated and artificially aged.

    -T7: solution-heat-treated and stabilized.

    -T8: solution-heat-treated, cold-worked, and artificially aged.

    -T9: solution-heat-treated, artificially aged, and cold-worked.

    -T10: cooled from hot working, cold-worked, and artificially aged.

-W: solution-heat-treated only.

TABLE 7.3 Composition, Typical Properties, and Designations of Some Wrought Aluminum Alloys

| Designation[a] | Composition (%) Aluminum = Balance | | | | | Form Tested | Tensile Strength | | Yield Strength[b] | | Elongation in 2 inches (%) | Brinell Hardness | Uses and Characteristics |
	Cu	Si	Mn	Mg	Others		ksi	MPa	ksi	MPa			
						Work-hardening alloys—not heat-treatable							
1100–0	0.12				99 Al	$\frac{1}{16}$-in. sheet	13	90	5	34	35	23	Commercial Al: good forming properties
1100–H14						$\frac{1}{16}$-in. sheet	16	110	14	97	9	32	Good corrosion resistance, low yield strength
1100–H18						$\frac{1}{16}$-in. sheet	24	165	21	145	5	44	Cooking utensils; sheet and tubing
3003–0	0.12		1.2			$\frac{1}{16}$-in. sheet	16	110	6	41	30	28	Similar to 1100
3003–H14			1.2			$\frac{1}{16}$-in. sheet	22	152	21	145	8	40	Slightly stronger and less ductile
3003–H18						$\frac{1}{16}$-in. sheet	29	200	27	186	4	55	Cooking utensils; sheetmetal work
5052–0				2.5	0.25 Cr	$\frac{1}{16}$-in. sheet	28	193	13	90	25	45	Strongest work-hardening alloy
5052–H32						$\frac{1}{16}$-in. sheet	33	228	28	193	12	60	High yield strength and fatigue limit
5052–H36						$\frac{1}{16}$-in. sheet	40	276	35	241	8	73	Highly stressed sheetmetal products
						Precipitation-hardening alloys—heat-treatable							
2017–0	4.0	0.5	0.7	0.6		$\frac{1}{16}$-in. sheet	26	179	10	69	20	45	Duralumin, original strong alloy
2017–T4						$\frac{1}{16}$-in. sheet	62	428	40	276	20	105	Hardened by quenching and aging

Aluminum alloy[a]	Cu	Si	Mn	Mg	Other	Form	Tensile strength, ksi	MPa	Yield strength[b], ksi	MPa	Elongation, %	Brinell hardness	Uses and characteristics
2024–0	4.4		0.6	1.5		$\frac{1}{16}$-in. sheet	27	186	11	76	20	42	Stronger than 2017
2024–T4						$\frac{1}{16}$-in. sheet	64	441	42	290	19	120	Used widely in aircraft construction
2014–0	4.4	0.8	0.8	0.5		$\frac{1}{2}$-in. extruded shapes	27	186	14	97	12	45	Strong alloy for extruded shapes
2014–T6						Forgings	65	448	55	379	10		Strong forging alloy
2014–T6						$\frac{1}{16}$-in. sheet	70	483	60	413	8	125	Higher yield strength than Alclad 2024
Alclad 2014–T6	4.5	1.0	0.8	0.4		$\frac{1}{16}$-in. sheet	63	434	56	386	7		Clad with heat-treatable alloy[c]
7075–0	1.6		0.2	2.5	{ 0.3 Cr, 5.6 Zn	$\frac{1}{16}$-in. sheet	33	228	15	103	17	60	Alloy of highest strength
7075–T6						$\frac{1}{16}$-in. sheet	76	524	67	462	11	150	Lower ductility than 2024
Alclad 7075–T6						$\frac{1}{16}$-in. sheet	76	524	67	462	11		Strongest Alclad product
7075–T6						$\frac{1}{2}$-in. extruded shapes	80	552	70	483	6		Strongest alloy for extrusions
6061–T6	0.28	0.6		1.0	0.20 Cr	$\frac{1}{2}$-in. extruded shapes	42	290	40	276	12	95	Strong, corrosion resistant
6063–T6		0.4		0.7		$\frac{1}{2}$-in. rod extruded	35	241	31	214	12	80	Good forming properties and corrosion resistance
6151–T6		0.9		0.6	0.25 Cr	Forgings	48	331	43	297	17	90	For intricate forgings
2025–T6	4.5	0.8	0.8			Forgings	55	379	30	207	18	100	Good forgeability, lower cost
2018–T6	4			0.7	2 Ni	Forgings	55	379	40	276	10	100	Strong at elevated temperatures; forged pistons
4032–T6	0.9	12.2		1.1	0.9 Ni	Forgings	55	379	46	317	9	115	Forged aircraft pistons
2011–T3	5.5				(0.5 Bi) 0.5 Pb	$\frac{1}{2}$-in. rod	55	379	43	297	15	95	Free cutting, screw-machine products

[a] O, annealed; T, quenched and aged; H, cold-rolled to hard temper.

[b] Yield strength taken at 0.2% permanent set.

[c] Cladding alloy: 1.0 Mg, 0.7 Si, 0.5 Mn.

Additional digits appended to those listed indicate variations in the basic temper.

It can be noted from Table 7.3 that the work-hardenable alloys (not hardenable by heat treatment) are primarily those in the 1000 (pure aluminum), 3000 (aluminum–manganese), and 500 (aluminum–magnesium) series. Within these series, the 1100, 3003, and 5052 alloys tend to be the most popular.

Because of their higher strengths, the precipitation-hardenable alloys are more numerous and are found primarily in the 2000, 6000, and 7000 series. Alloy 2017, the original *duralumin,* is probably the oldest hardenable aluminum. The 2024 alloy is stronger and has seen considerable use in aircraft applications. Within the 2000 series, ductility does not significantly decrease during the strengthening heat treatment. Some of the newer alloys within the 7000 series have strengths that approach or exceed those of the high-yield-strength structural steels. Ductility, however, is less than that of steel, and fabrication is more difficult than for the 2024-type alloys. Nevertheless, these alloys have also found wide use in aircraft.

Because of their two-phase structure, the heat-treatable alloys tend to have poorer corrosion resistance than either pure aluminum or the single-phase, work-hardenable alloys. Thus, where both high strength and superior corrosion resistance are desired, the wrought aluminum is often produced as *Alclad* material. A thin layer of corrosion-resistant aluminum is bonded to one or both surfaces of the high-strength alloy during rolling, and the material is further processed as a composite.

Because only moderate temperatures are required to lower the strength of these materials, aluminum alloy extrusions and forgings are relatively easy to produce and are manufactured in large quantities. Deep drawing and other sheet-metal-forming operations can be carried out quite easily. In general, the high ductility and low yield strength of the aluminum alloys make them appropriate for almost all forming operations. Good dimensional tolerances and fairly intricate shapes can be produced with relative ease.

The machinability of aluminum-base alloys varies greatly. Most cast alloys can be machined easily. For the wrought alloys, however, special tools and techniques may be desirable if large-scale machining is to be performed. Free-machining alloys, such as 2011, have been developed for screw-machine work. These can be machined at very high speeds and have replaced brass screw-machine stock in many cases.

Aluminum Casting Alloys.   Although its low melting temperature tends to make it suitable for casting, pure aluminum is seldom cast. Its high shrinkage and susceptibility to hot cracking cause considerable difficulty, and scrap is high. By adding small amounts of alloying elements, however, very suitable casting characteristics are obtained and strength is increased. Aluminum alloys are cast in considerable quantity, the principal alloying elements being copper, silicon, and zinc. Table 7.4 lists some common commercial aluminum casting

TABLE 7.4 Composition, Properties, and Designations of Some Aluminum Casting Alloys

Alloy Designation[a]	Process[b]	Composition (%) (Major Alloys > 1%)						Temper	Tensile Strength		Elongation in 2 inches (%)	Uses and Characteristics
		Cu	Si	Mg	Zn	Fe	Other		ksi[c]	MPa[c]		
208	S	4.0	3.0		1.0	1.2		F	19	131	1.5	General-purpose sand castings, can be heat treated
242	S & P	4.0		1.6		1.0	2.0 Ni	T61	40	276	—	Withstands elevated temperatures
295	S	4.5	1.0			1.0		T6	32	221	3.0	Structural castings, heat-treatable
296	P	4.5	2.5			1.2		T6	24	165	2.0	Permanent-mold version of 295
308	P	4.5	5.5			1.0		F	24	166	—	General-purpose permanent mold
319	S & P	3.5	6.0		1.0	1.0		T6	31	214	1.5	Superior casting characteristics
354	P	1.8	9.0					—	—	—	—	High-strength, aircraft
355	S & P	1.3	5.0			1.0		T6	32	221	2.0	High strength and pressure tightness
C355	S & P	1.3	5.0					T61	40	276	3.0	Stronger and more ductile than 355
356	S & P		7.0					T6	30	207	3.0	Excellent castability and impact strength
A356	S & P		7.0					T61	37	255	5.0	Stronger and more ductile than 356
357	S & P		7.0					T6	45	310	3.0	High strength-to-weight castings
359	S & P		9.0					—	—	—	—	High-strength aircraft usage
360	D		9.5			2.0		F	44[d]	303	2.5[d]	Good corrosion resistance and strength
A360	D		9.5			2.0		F	46[d]	317	3.5[d]	Similar to 360
380	D	3.5	8.5		3.0	2.0		F	46[d]	317	2.5[d]	High strength and hardness
A380	D	3.5	8.5		3.0	1.3		F	47[d]	324	3.5[d]	Similar to 380
383	D	1.5	10.5		3.0	1.3		F	45[d]	310	3.5[d]	High strength and hardness
384	D	3.75	11.3		1.0	1.3		F	48[d]	331	2.5	High strength and hardness
413	D	1.0	12.0			2.0		F	43[d]	297	2.5[d]	General purpose, good castability
A413	D	1.0	12.0			1.3		F	42[d]	290	3.5[d]	Similar to 413
443	D	5.25				2.0		F	33[d]	228	9.0[d]	General purpose, good castability
B443	S & P	5.25				2.0		F	17	117	3.0	General-purpose casting alloy
514	S			4.0				F	22	152	6.0	High corrosion resistance
518	D			8.0		1.8		F	45[d]	310	5.0[d]	Good corrosion resistance, strength, and toughness
520	S			10.0				T4	42	290	12.0	High strength with good ductility
535	S			6.9				F	35	241	9.0	Good corrosion resistance and machinability
712	S				5.8			F	34	234	4.0	Good properties without heat treatment
713	S & P				7.5	1.1		F	32	221	3.0	Similar to 712
771	S				7.0			T6	42	290	5.0	Aircraft and computer components
830	S & P	1.0					6.3 Sn 1.0 Ni	T5	16	110	5.0	Bearing alloy

[a] Aluminum Association.
[b] S, sand-cast; P, permanent-mold-cast; D, die-cast.
[c] Minimum figures unless noted.
[d] Typical values.

alloys and employs the designation system of the Aluminum Association. The first digit indicates the alloy group as follows:

Major Alloying Element	
Aluminum, 99.00% and greater	1xx.x
Copper	2xx.x
Silicon with Cu and/or Mg	3xx.x
Silicon	4xx.x
Magnesium	5xx.x
Zinc	7xx.x
Tin	8xx.x
Other elements	9xx.x

The second and third digits identify the particular alloy or aluminum purity, and the last digit, separated by a decimal point, indicates the product form (e.g., casting or ingot). A modification of the original alloy is indicated by a letter before the numerical designation.

Alloys are designed for both properties and process. Where strength requirements are low, as-cast properties are employed. High-strength castings usually require the use of an alloy that can be subsequently heat-treated. Sand casting has the fewest process restrictions. The aluminum alloys used for permanent-mold castings must be designed to have lower coefficients of thermal expansion (or contraction) because the molds offer restraint to the dimensional changes that occur upon cooling. Die-casting alloys require high degrees of fluidity and "castability" because they are often cast into thin sections. Moreover, as die castings are ordinarily not heat-treated, the alloys used are designed to produce rather high "as-cast" strength under rapid cooling conditions. Several of the permanent-mold and die-casting alloys have tensile strengths above 40 ksi (275 MPa).

Aluminum–Lithium Alloys.   In the search for aluminum alloys with higher strength, greater stiffness, and lighter weight, *aluminum–lithium alloys* are emerging as an aerospace "material of the future." Each percentage of lithium (up to 4%) reduces the weight by 3% and increases stiffness by 6%. Alloys have already been developed that have 8% to 10% lower density, 15% to 20% greater stiffness, strengths comparable to those of existing alloys, and good resistance to fatigue crack propagation. Moreover, these alloys can be fabricated just as other aluminum alloys. They are highly machinable, can be welded, and are readily adaptable to forming by forging or extrusion. Some sheet material can even be fabricated by superplastic forming.

Through the use of rapid solidification processing and powder metallurgy practices, aluminum–beryllium–lithium alloys are also emerging. Although still in the developmental stages, they offer the possibility of 20% lower density, 40% greater stiffness and strengths equivalent to those of current alloys (based on a comparison with alloy 7075).

Through material substitution, the redesign of components to take advantage of greater stiffness, and a reduction in the size of related parts (as the entire structure will be lighter), there is a possibility of a 10% to 15% weight savings in aircraft structural applications. On a more graphic basis, the weight of a Boeing 747 could be reduced by about 14,000 pounds. Fuel savings over the life of the airplane would more than compensate for the added manufacturing expense.

Magnesium and Magnesium Alloys

General Properties and Characteristics. *Magnesium* is the lightest of the commercially important metals, having a specific gravity of about 1.75 (compared to 2.7 for aluminum and 7.8 for iron or steel). Like aluminum, magnesium is relatively weak in the pure state and for engineering purposes is almost always used as an alloy. Even in alloy form, however, the metal can be characterized by poor wear, creep, and fatigue properties as well as by poor corrosion resistance. Its modulus of elasticity is even less than that of aluminum, being only about one fifth that of steel. Thick sections are required to provide adequate stiffness, but the alloy is so light that it is often possible to use thicker sections for the required rigidity and still to have a lighter structure than can be obtained with any other metal. Cost per unit volume is low, so the use of thick sections is generally not prohibitive. Moreover, as a great many magnesium components are cast, the thick sections actually become a desirable feature. Corrosion resistance is moderate, unless the metal is exposed to salt water, salt air, or an unfavorable galvanic couple. Enamel or lacquer finishes applied to finished parts can impart adequate corrosion resistance for most applications.

Magnesium alloys also possess limited ductility, a characteristic of the hcp crystal structure. Designers should be aware that brittle fracture is a probable mode of failure when components are loaded beyond the designed conditions. Magnesium automobile wheels are a notable example, and most manufacturers now favor aluminum for lightweight wheels.

On the more positive side, magnesium alloys have a relatively high strength-to-weight ratio, some commercial alloys attaining strengths up to 55 ksi (380 MPa). Nevertheless, the limitations are such that the use of magnesium is generally restricted to applications where light weight is very important. Where aluminum alloys are often used for the load-bearing members of mechanical structures, magnesium alloys are best suited to those applications where lightness is the primary consideration and strength is a secondary requirement.

Magnesium Alloys and Their Fabrication. The designation system for magnesium alloys is not as well standardized as in the case of steels or aluminums, but most producers follow a system using one or two prefix letters, two or three numerals, and a suffix letter. The prefix letters designate the two principal alloying metals according to the following format developed in ASTM specification B275:

A	aluminum	H	thorium	Q	silver
B	bismuth	K	zirconium	R	chromium
C	copper	L	beryllium	S	silicon
D	cadmium	M	manganese	T	tin
E	rare earth	N	nickel	Z	zinc
F	iron	P	lead		

Aluminum, zinc, zirconium, and thorium promote precipitation hardening; manganese improves corrosion resistance; and tin improves castability. Aluminum is the most common alloying element. The numerals correspond to the rounded-off percentages of the two main alloy elements and are arranged in the same order as the letters. The suffix letter distinguishes between different alloys with the same percentage of the principal alloying elements, proceeding alphabetically as compositions become standard. Temper designation is much the same as in the case of aluminum, using -F, -O, -H1, -H2, -T4, -T5, and -T6. Some of the more common magnesium alloys are listed in Table 7.5 together with their properties and uses.

Sand, permanent-mold, and die casting are all well developed for magnesium alloys, die casting being the most popular. Although magnesium is about twice as expensive as aluminum, its hot-chamber die-casting process is easier, more economical, and 40% to 50% faster than the cold-chamber process required for aluminum.

Forming behavior is poor at room temperature, but most conventional processes can be performed when the material is heated to temperatures of 450–700°F (230–370°C). As these temperatures are easily attained and generally do not require a protective atmosphere, many formed and drawn magnesium products are manufactured.

The machinability of magnesium alloys is the best of any commercial metal, and in many applications, the savings in machining costs more than compensate for the increased cost of the material. It is necessary, however, to keep the tools sharp and to provide ample space for the chips.

Magnesium alloys can be spot-welded nearly as easily as aluminum, but scratch brushing or chemical cleaning is necessary before the weld is formed. Fusion welding is carried out most easily by processes using an inert shielding atmosphere of argon or helium gas.

Considerable misinformation exists regarding the fire hazard in processing magnesium alloys. It is true that magnesium alloys are highly combustible when in a finely divided form, such as powder or fine chips, and this hazard should never be ignored. Above 800°F (425°C), a noncombustible, oxygen-free atmosphere is required to suppress burning. Casting operations often require additional precautions because of the reactivity of magnesium with sand and water. In sheet, bar, extruded, or cast form, however, magnesium alloys present no real fire hazard.

TABLE 7.5 **Composition, Properties, and Uses of Common Magnesium Alloys**

| Alloy | Temper | Composition (%) | | | | | | Tensile Strength[a] | | Yield Strength[a] | | Elongation in 2 inches (%) | Uses and Characteristics |
		Al	Rare Earths	Mn	Th	Zn	Zr	ksi	MPa	ksi	MPa		
AM60A	F	6.0		0.13				30	207	17	117	6	Die castings
AM100A	T4	10.0		0.1				34	234	10	69	6	Sand and permanent-mold castings
AZ31B	F	3.0				1.0		32	221	15	103	6	Sheet, plate, extrusions, forgings
AZ61A	F	6.5				1.0		36	248	16	110	7	Sheet, plate, extrusions, forgings
AZ63A	t$	6.0				3.0		34	234	11	76	7	Sand and permanent-mold castings
AZ80A	T5	8.5				0.5		34	234	22	152	2	High-strength forgings, extrusions
AZ81A	T4	7.6				0.7		34	234	11	76	7	Sand and permanent-mold castings
AZ91A	F	9.0				0.7		34	234	23	159	3	Die castings
AZ92A	T4	9.0				2.0		34	234	11	76	6	High-strength sand and permanent-mold castings
EZ33A	T5		3.2			2.6	0.7	30	138	14	97	2	Sand and permanent-mold castings
HK31A	H24				3.2		0.7	33	228	24	166	4	Sheets and plates; castings in T6 temper
HM21A	T5			0.8	2.0			33	228	25	172	3	High-temperature (800°F) sheets, plates, forgings
HZ32A	T5				3.2	2.1		27	186	13	90	4	Sand and permanent-mold castings
ZH62A	T5				1.8	5.7	0.7	35	241	22	152	5	Sand and permanent-mold castings
ZK51A	T5					4.6	0.7	34	234	20	138	5	Sand and permanent-mold castings
ZK60A	T5					5.5	0.45	38	262	20	138	7	Extrusions, forgings

[a]Properties are minimums for the designated temper.

Properties and Applications. As a pure metal, zinc has only one major use: *galvanizing* iron and steel. In this process, steel is acid-cleaned and then coated with a layer of zinc, either by being dipped in a bath of molten metal or by electrolytic plating. The resultant coating provides excellent corrosion resistance, even when it is badly scratched or marred. Moreover, the corrosion resistance will persist until all of the sacrificial zinc has been depleted. Galvanizing accounts for about 35% of all zinc used.

Zinc-Base Alloys

Zinc-base alloys are used primarily in the form of die castings. As a base metal, zinc offers low cost, a melting point of only 715°F (380°C), and the property that it does not adversely affect steel dies when in molten metal contact. Unfortunately, pure zinc is also rather weak and brittle. Thus, when alloys are designed for die casting, the alloy elements are selected for their ability to retain the low melting point while substantially increasing strength and toughness in the as-cast condition. High fluidity and good dimensional stability are additional considerations.

Two of the most popular zinc die-casting alloys are characterized in Table 7.6. Alloy AG40A (also known as *Alloy 903* or *Zamak 3*) is widely used because of its excellent dimensional stability. Alloy AG41A offers higher strength and better corrosion resistance. Zamak 7, a newer alloy, can provide improved castability as a result of a reduced magnesium content. All three alloys offer good tensile strength coupled with exceptional impact resistance. Their development has been responsible for the extensive use of zinc die castings.

The zinc die-casting alloys offer a strength greater than that of all other die-cast metals, with the exception of the copper-base alloys. They can be cast to close dimensional limits with extremely thin sections. Moreover, these alloys are then machinable at a minimum of cost. Resistance to surface corrosion is adequate for a number of applications. Prolonged contact with moisture results in the formation of a white corrosion product, but this can be easily suppressed by some inexpensive surface treatments.

The attractiveness of zinc die casting has been further enhanced by the development of several zinc–aluminum casting alloys with rather large amounts of aluminum (ZA8, ZA12, and ZA27). Initially developed for sand, permanent-

TABLE 7.6 Characteristics of Two Zinc Die-Casting Alloys

	ASTM AG40A (SAE 903) (Zamak 3)	ASTM AG41A (SAE 925) (Zamak 5)
Composition (%)		
Copper	0.25	0.75–1.25
Aluminum	3.5–4.3	3.5–4.3
Magnesium	0.02–0.05	0.03–0.08
Iron, maximum	0.1	0.1
Lead, maximum	0.005	0.005
Cadmium, maximum	0.004	0.004
Tin, maximum	0.003	0.003
Zinc	Remainder	Remainder
Properties		
Tensile strength, as cast (ksi)	41	47.6
Tensile strength, 10 years of aging	35	39.3
Elongation in 2 inches, as cast (%)	10	7
Charpy impact strength, as cast (ft-lb)	43	48
Melting point (°F)	717	717

mold, and graphite-mold casting, these alloys can also be die-cast to achieve higher performance characteristics than those obtained with any of the conventional alloys. Strength, hardness, and wear resistance are all improved, and several of the alloys have demonstrated excellent bearing properties. In view of the lower melting and casting costs, these alloys are becoming attractive alternatives to the conventional aluminum, brass, and bronze casting alloys, as well as cast iron.

Titanium

Titanium is a strong, lightweight, corrosion-resistant metal that has been of commercial importance only since 1948. Because its properties are between those of steel and aluminum, its importance has been increasing rapidly. Its yield strength is about 60 ksi (415 MPa) and can be raised to 190 ksi (1300 MPa) by alloying and heat treatment—a strength comparable to that of many alloy steels. Its density, on the other hand, is only 56% that of steel, and its modulus of elasticity is also about one half that of steel. Mechanical properties are retained up to temperatures of 900°F (480°C), so the metal is often considered a high-temperature engineering material. From the negative side, titanium and its alloys suffer from high cost, fabrication difficulties, a high energy content (they require about 10 times as much energy to produce as steel), and a high reactivity at elevated temperatures (above 900°F).

Titanium alloys are generally grouped into three classes based on their microstructural features. These classes are known as alpha-, beta-, and alpha-beta-titanium alloys, the terms denoting the stable phase or phases at room temperature. Fabrication can be by casting, forging, rolling, or extruding, provided special process modifications and controls are implemented.

The uses of titanium relate primarily to its *high strength-to-weight ratio*, its *good stiffness*, its *corrosion resistance*, and its *retention of mechanical properties at elevated temperatures*. Aerospace applications tend to dominate, but titanium alloys are also used in chemical and electrochemical processing equipment, marine implements, and ordinance equipment. They are often used in place of steel where weight savings are desired and to replace aluminums where high temperature performance is necessary. Some bonding applications utilize the unique property that titanium wets glass and some ceramics.

Nickel-Base Alloys

Nickel-base alloys are noted for their outstanding strength and corrosion resistance, particularly at high temperatures. *Monel* metal, containing about 67% nickel and 30% copper, has been used for years in the chemical and food-processing industries because of its outstanding corrosion characteristics. In fact, Monel probably has better corrosion resistance to more media than any other commercial alloy. It is particularly resistant to salt water, sulfuric acid, and even high-velocity, high-temperature steam. For the latter reason, Monel has been used for steam turbine blades. It can be polished to have an excellent appearance, similar to that of stainless steel, and is often used in ornamental trim and household ware. In its most common form, Monel has a tensile strength ranging from

70 to 170 ksi (480 to 1170 MPa), with a companion elongation of 50% to 2%, respectively.

There are three special grades of Monel that contain small amounts of added alloying elements. K-Monel contains about 3% aluminum and can be precipitation-hardened to a tensile strength of 160 to 180 ksi (1100 to 1240 MPa). H-Monel has 3% silicon added, and S-Monel has 4% silicon. These are used for casting applications and can be precipitation-hardened. To improve the machining characteristics of Monel metal, a special free-machining variety, known as R-Monel, has been produced with about 0.35% sulfur.

Nickel-base alloys are also used for electrical resistors. These materials are primarily nickel–chromium alloys and are known by the trade name of *Nichromes*. One popular alloy contains 80% nickel and 20% chromium. Another has 60% nickel, 16% chromium, and 24% iron. They have excellent resistance to oxidation while retaining strength at red heats.

Still other nickel-base alloys have been developed for extreme high-temperature service. These will be discussed along with other, similar materials in the following section.

Nickel-base materials are generally difficult to cast, but they can be forged and hot-worked. Heating operations, however, are generally performed in controlled atmospheres to prevent intercrystalline embrittlement. Welding operations can be performed with little difficulty.

Nonferrous Alloys for High-Temperature Service

Titanium and titanium alloys have already been cited as useful in providing strength at elevated temperatures, but the maximum temperature for these materials is approximately 900°F (480°C).

Jet engine, gas turbine, rocket, and nuclear applications often require materials that possess high strength, creep resistance, and corrosion resistance at temperatures up to and in excess of 2000°F (1100°C). One class of materials offering these properties is the *superalloys,* some of the more common of which are listed in Table 7.7. It should be noted that nickel, iron and nickel, or cobalt forms the base metal in these alloys. Most are precipitation-hardenable, and yield strengths above 100 ksi (690 MPa) are readily attained. The nickel-base alloys tend to have higher strengths at room temperature, with yield strengths up to 175 ksi (1200 MPa) and ultimate tensile strengths as high as 210 ksi (1450 MPa). These are in comparison with 115 ksi (790 MPa) and 170 ksi (1170 MPa), respectively, for the cobalt-base alloys. The 1000-hour rupture strengths of the nickel-base alloys at 1500°F (815°C) are also higher than those of the cobalt-base material, up to 65 ksi (450 MPa) versus 33 ksi (228 MPa).

Most of the superalloys are very difficult to machine, so methods such as electrodischarge, electrochemical, or ultrasonic machining are often utilized, or they are produced to final shape as investment castings. Powder metallurgy techniques are also being used extensively in the manufacture of superalloy components. Because of their ingredients, all of these alloys are quite expensive, and this expense limits their use to small or critical parts where the cost is not the determining factor.

TABLE 7.7 Some Nonferrous Alloys for High-Temperature Service

Alloy	Composition (%)														
	C	Mn	Si	Cr	Ni	Co	Mo	W	Cb	Ti	Al	B	Zr	Fe	Other
Nickel base															
Hastelloy X	0.1	1.0	1.0	21.8	Balance	2.5	9.0	0.6	—	—	—	—	—	18.5	—
IN-100	0.18	—	—	10.0	Balance	15.0	3.0	—	—	4.7	5.5	0.014	0.06	—	1.0 V
Inconel 601	0.05	0.5	0.25	23.0	Balance	—	—	—	—	—	1.4	—	—	14.1	0.2 Cu
Inconel 718	0.04	0.2	0.2	19.0	Balance	—	3.0	—	5.0	0.9	0.5	—	—	18.5	0.2 Cu
M-252	0.15	0.5	0.5	19.0	Balance	10.0	10.0	—	—	2.6	1.0	0.005	—	—	—
Rene 41	0.09	—	—	19.0	Balance	11.0	10.0	—	—	3.1	1.5	0.01	—	—	—
Rene 80	0.17	—	—	14.0	Balance	9.5	4.0	4.0	—	5.0	3.0	0.015	0.03	—	—
Rene 95	0.15	—	—	14.0	Balance	8.0	3.5	3.5	3.5	2.5	3.5	0.01	0.05	—	—
Udimet 500	0.08	—	—	19.0	Balance	18.0	4	—	—	3.0	3.0	0.005	—	0.5	—
Udimet 700	0.07	—	—	15.0	Balance	18.5	5.0	—	—	3.5	4.4	0.025	—	0.5	—
Waspaloy B	0.07	0.75	0.75	19.5	Balance	13.5	4.3	—	—	3.0	1.4	0.006	0.07	2.0	0.1 Cu
Iron–nickel base															
Illium P	0.20	—	—	28.0	8.0	—	2.0	—	—	—	—	—	—	Balance	3.0 Cu
Incoloy 825	0.03	0.5	0.2	21.5	42.0	—	3.0	—	—	0.9	0.1	—	—	30	2.2 Cu
Incoloy 901	0.05	0.4	0.4	13.5	42.7	—	6.2	—	—	2.5	0.2	—	—	34	—
16-25-6	0.08	1.35	0.7	16.0	25.0	—	6.0	—	—	—	—	—	—	Balance	0.15 N
Cobalt base															
Haynes 150	0.08	0.65	0.75	28.0	—	Balance	—	—	—	—	—	—	—	20.0	—
MAR-M322	1.00	0.10	0.1	21.5	—	Balance	—	9.0	—	0.75	—	—	2.25	—	4.5 Ta
S-816	0.38	1.20	0.4	20.0	20.0	Balance	4.0	4.0	4.0	—	—	—	—	4.0	—
WI-52	0.45	0.5	0.5	21.0	1.0	Balance	—	11.0	2.0	—	—	—	—	2.0	—

Some engineering applications have already exceeded the limits of these alloys, and still others await the development of materials that would make them feasible. For example, one source estimates the temperature of future jet engines to be in excess of 2600°F (1425°C). Materials such as TD-nickel (a nickel alloy containing 2% dispersed thoria) give promise of operating at service temperatures above 2000°F (1100°C). The refractory metals can be used to temperatures as high as 3000°F (1650°C), provided they are protected by a coating that isolates them from gases in their operating environment. These metals, along with their melting points, are *niobium* (4480°F/2470°C); *molybdenum* (4730°F/2610°C); *tantalum* (5430°F/3000°C); *rhenium* (5740°F/3170°C); and *tungsten* (6170°F/3410°C). Figure 7-1 illustrates one such application. Other technologies offering promise for high-temperature service include directionally solidified eutectics, single crystals, engineered ceramics, and advanced coating systems.

The principal uses of *lead* as a pure metal include storage batteries, cable cladding, and sound-dampening shields. These uses consume about 60% of the annual production of the metal. *Tin* is used primarily as a coating on steel to impart corrosion resistance.

Lead, Tin, and Their Alloys

FIGURE 7-1 Niobium coated with fused silicide is used in the afterburner components of the Pratt & Whitney F100 jet pictured above. (Courtesy Pratt & Whitney.)

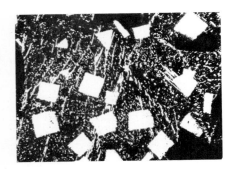

FIGURE 7-2 Photomicrograph of lead babbitt metal, 75X. Square or triangular white masses are SbSn; small white particles are CuSn.

When in the form of lead- or tin-based alloys, the two metals are almost always used together. Bearing material and solder are the two most important uses. An alloy of 84% tin, 8% copper, and 8% antimony is known as *genuine* or *tin babbitt* and is one of the oldest and best bearing materials. Because of the high cost of tin, lead babbitt, composed of 85% lead, 5% tin, 10% antimony, and 0.5% copper, is a more widely used bearing material. For slow speeds and moderate loads, the lead-base babbitts have proved to be quite adequate.

Figure 7-2 shows a photomicrograph of a lead babbitt. The antimony combines with the tin to form hard particles within the softer lead matrix, a structure typical of many bearing metals. The shaft rides on the harder particles with little friction, and the softer matrix acts as a cushion that can distort sufficiently to compensate for misalignment and to ensure a proper fit between the two surfaces.

Soft solders are basically lead–tin alloys with a chemistry near the eutectic composition of 61.9% tin (see Figure 4–5). Although the eutectic alloy has the lowest melting temperature, the high cost of tin has forced many users to specify solders with a lower-than-optimum tin content. A variety of compositions are available, each with its own characteristic melting range.

Some Less-Known Metals and Alloys

Several of the less-known metals and alloys have achieved importance in modern technology as a result of their unique physical and mechanical properties. *Hafnium, thorium,* and *beryllium* are used in nuclear reactors because of their low neutron-absorption characteristics. Depleted *uranium,* because of its very high density (19.1 grams per cubic centimeter), is useful in special applications where maximum weight must be put into a limited space, as in counterweights and flywheels. In addition to its use as a base metal for superalloys, *cobalt* is used as a binder in various powder-based components and sintered carbides.

Zirconium is used for its outstanding corrosion resistance to most acids, chlorides, and organic acids. It offers high strength, good weldability and fatigue resistance, and attractive neutron absorption characteristics. High electrical conductivity can be obtained by alloying with copper. An alloy containing a small percentage of hafnium offers a yield strength of 84 ksi (580 MPa) and a tensile strength of 90 ksi (620 MPa). Some applications of zirconium alloys include heat exchangers, protective claddings, heat shields, and engine components.

Current efforts in aerospace, nuclear, high-temperature, and electronic applications are likely to bring about increased research and development of these and other uncommon metals.

Graphite

Although technically not a metal, *graphite* has considerable potential as an engineering material. It possesses the unique property of having increased strength at elevated temperatures. Recrystallized, polycrystalline graphites have mechanical strengths up to 10 ksi (70 MPa) at room temperature, which double at 4500°F (2500°C).

Large quantities of graphite are used as electrodes in arc furnaces, and other uses are developing rapidly. The addition of small amounts of borides, carbides, nitrides and silicides greatly lowers graphite's oxidation rate at elevated temperatures and improves its mechanical strength. Thus, the material is highly suitable for use as rocket-nozzle inserts and permanent molds for casting various metals. It can be machined quite readily to excellent surface finishes.

Graphite fibers have also found extensive use in composite materials. This application will be discussed in Chapter 8.

Review Questions

1. What types of properties do nonferrous metals possess that are not available in the ferrous metals?
2. In what respects are the nonferrous metals usually inferior to steel?
3. What are the three primary properties of copper and copper alloys that account for their wide use?
4. What are the two primary forms of commercially pure copper, and how do they differ?
5. What are some of the attractive engineering properties that account for the wide use of the copper–zinc alpha brasses?
6. Why may the term *bronze* be potentially confusing when used to designate a metal or alloy?
7. What are some attractive engineering properties of copper–nickel alloys?
8. What are some of the common mechanisms by which strength can be imparted to copper-base alloys?
9. Describe the unique property combination of heat-treated copper–beryllium alloys.
10. What are some of the unique and attractive engineering properties of aluminum and aluminum alloys?
11. How does aluminum compare to copper in terms of electrical conductivity?
12. What are some of the engineering properties that tend to be poor for aluminum alloys?
13. How is the corrosion-resistance mechanism of aluminum and aluminum alloys similar to that observed in stainless steels?
14. What feature in the wrought-aluminum designation scheme is used to denote the condition or structure of a given alloy?
15. What is the primary mechanism used to provide strength to the high-strength, "aircraft-quality" aluminum alloys?
16. For what reason or reasons is pure aluminum seldom cast?
17. What specific material qualities or features might make an

aluminum casting alloy particularly attractive for permanent-mold casting? Die casting?

18. What are some of the features of aluminum–lithium alloys that make them particularly attractive for aerospace applications?

19. What are some of the poor or limiting properties of magnesium and magnesium alloys?

20. What particular design feature or requirement is generally necessary to justify the use of magnesium alloys?

21. In what way can ductility be imparted to magnesium alloys so that they can be formed by conventional processes?

22. What is the primary application of the zinc-base engineering alloys?

23. What are some of the attractive engineering properties of titanium and titanium alloys?

24. What is generally regarded as the upper limit of the temperatures at which titanium alloys retain useful engineering strength?

25. What is the unique feature of Monel alloys compared to other engineering metals?

26. What metals or combinations of metals form the bases of the alloys known as *superalloys?*

27. What class of metals or alloys must be used when the operating temperatures exceed the limits of the superalloys?

28. In the area of lead–tin solders, why have the higher lead materials become more common than the highly desirable eutectic composition?

29. What is one of the unique properties of graphite that makes it attractive for elevated temperature applications?

Nonmetallic Materials: Plastics, Elastomers, Ceramics, and Composites

As a result of their nature, properties, and unique advantages and limitations, nonmetallic materials have always played a significant role in manufacturing. Wood has been a key engineering material down through the centuries, and craftspeople have learned to select and utilize the various types and grades to manufacture quality products. More recently, however, wood has become less significant, and the term *nonmetallic materials* now brings forth an extensive listing of plastics, elastomers, ceramics and composites. Most of these are processed materials and a wide variety of properties and characteristics can be obtained. New materials and variations are being created continuously, and their uses and applications are expanding rapidly. Some observers expect the use of these materials to "explode" in the near future as research and development bring their costs within the realm of the traditional users of steel and other engineering metals. One predicted that high-performance composites, plastics, and advanced ceramics may replace as much as 50% of traditional metals. Most agree that there will be an ever-increasing use of the engineered nonmetallic materials at the expense of the traditional materials currently being used.

Because of the breadth and number of the individual materials that are available, this chapter will not attempt to provide information about all of them. Instead, the emphasis will be on the basic nature and properties of these materials, so that the reader will be able to determine if they are potential candidates for specific products and applications. For detailed information about specific materials within these families, more extensive and dedicated texts, handbooks, and compilations should be consulted.

Plastics

It is difficult to give a precise definition of the term *plastics*. From a technical viewpoint, the term is applied to a group of engineered materials characterized by large molecules that are built up by the joining of small molecules. On a more practical level, these materials are natural or synthetic resins, or their compounds, that can be molded, extruded, cast, or used as thin films or coatings. They offer low density, low tooling costs, good corrosion resistance, cost reduction, and design versatility. From a chemical viewpoint, most are organic substances containing hydrogen, oxygen, carbon, and nitrogen.

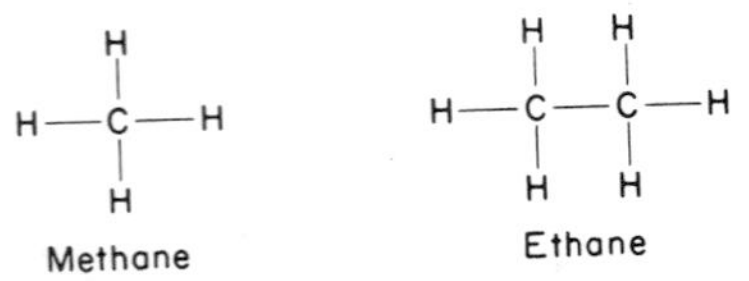

The Molecular Structure of Plastics.   It is helpful to have an understanding of the basic molecular structure of plastics. Most are based on hydrocarbons in which carbon and hydrogen combine in the relationship C_nH_{2n+2}, known as *paraffins*. Theoretically, these hydrocarbons can be linked together indefinitely to form very large molecules, as illustrated in Figure 8-1, the bonds between the atoms being single pairs of shared (covalent) electrons. Because there is no provision for additional atoms to be added to the chain, such molecules are said to be *saturated*. Bonding within the molecule is quite strong, but the attractive forces between adjacent molecules are much weaker.

Carbon and hydrogen can also form molecules in which the carbon atoms are held together by double or triple covalent bonds. Ethylene and acetylene are examples, as illustrated in Figure 8-2. Because such molecules do not have the maximum possible number of hydrogen atoms, they are said to be *unsaturated* and are important in the *polymerization* process, where small molecules join to form large ones with the same constituent atoms.

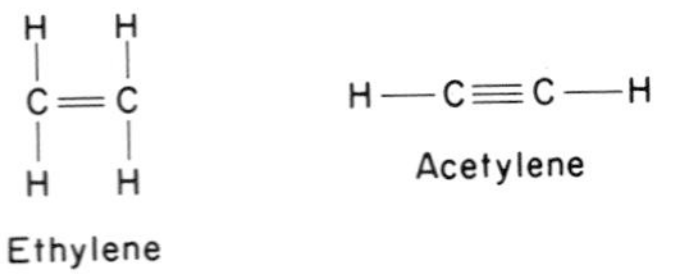

In these organic compounds, four electron pairs surround each carbon atom, and one electron pair is shared with each hydrogen atom. Other atoms or structures can be substituted for carbon and hydrogen, however. Chlorine or even a benzene ring can often take the place of hydrogen. Oxygen, sulfur, or nitrogen can take the place of carbon. Thus, a wide range of organic compounds can be created.

Isomers.   The same kind and number of atoms can also unite in different structural arrangements and thus form different compounds with different engineering properties. Figure 8-3 shows an example of this feature; the two components are known as *isomers*. Isomers can be considered analogous to allotropism or polymorphism in crystalline materials, where the same material can exist in more than one crystal structure.

FIGURE 8-1   Linking of hydrogen and carbon in methane and ethane molecules.

FIGURE 8-2   Covalent bonds in unsaturated ethylene and acetylene molecules.

FIGURE 8-3   Linking of eight hydrogen, one oxygen, and three carbon atoms to form two isomers, propyl and isopropyl alcohol.

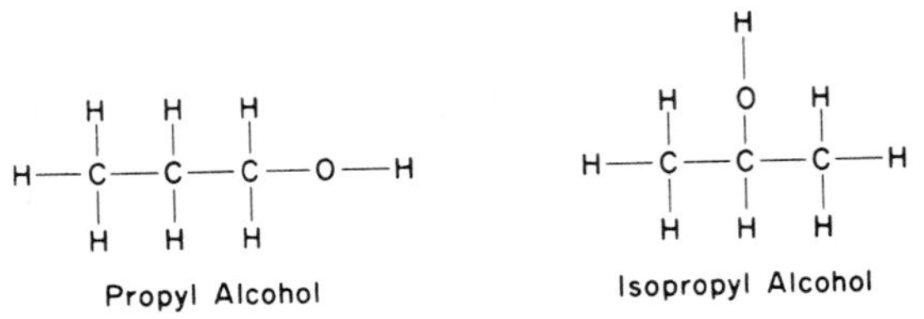

Forming Molecules by Polymerization.

The polymerization process takes place by either an *addition* or a *condensation* mechanism. Figure 8-4 illustrates polymerization by addition, where a number of basic units (*monomers*) are added together to form a large molecule (*polymer*) in which there is a repeated unit (*mer*). Activators, like $H_2O_2 \rightarrow 2\ OH$, initiate and terminate the chain. Thus, the amount of activator relative to the amount of monomer determines the average molecular weight or length of the chain. The average number of mers in the polymer is known as the *degree of polymerization* and ranges from 75 to 750 in most commercial plastics. *Copolymers* are a special category of polymer in which two or more types of mers are combined into the same addition chain. This process, illustrated in Figure 8-5, greatly expands the possibilities of creating new types of plastics with improved physical and mechanical properties.

In contrast to polymerization by addition, in which all of the component atoms appear in the product molecule, *condensation polymerization* occurs as reactive molecules combine with one another in a stepwise fashion, while eliminating small by-product molecules, such as water. Figure 8-6 illustrates the reaction between phenol and formaldehyde to form Bakelite, first discovered in 1910. The structure of condensation polymers is often a three-dimensional framework in which all atoms are linked by strong primary bonds.

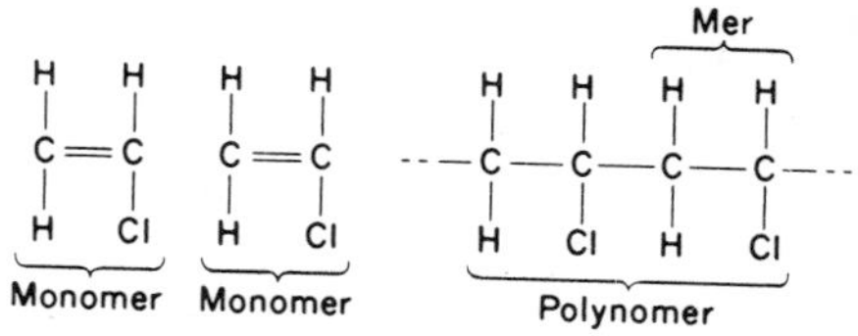

Thermosetting and Thermoplastic Materials.

The terms *thermosetting* and *thermoplastic* refer to the material's response to elevated temperature. Addition polymers can be viewed as long chains of tightly bonded carbon atoms with strongly attached pendants of hydrogen, fluorine, chlorine, or benzene rings. All bonds within the molecules are strong primary bonds. The attraction between neighboring molecules, however, is only by the much weaker van der Waals forces. For these materials, the mechanical and physical properties are largely determined by the intermolecular forces. Because these "secondary bonds" are weakened by elevated temperature, plastics of this type soften with increasing temperature and become harder and stronger when cooled. The softening and hardening of these *thermoplastic* materials can be repeated as often as desired, and no chemical change is involved. Because they contain molecules of different sizes, thermoplastic materials do not have a definite melting temperature; instead, they soften over a range of temperatures.

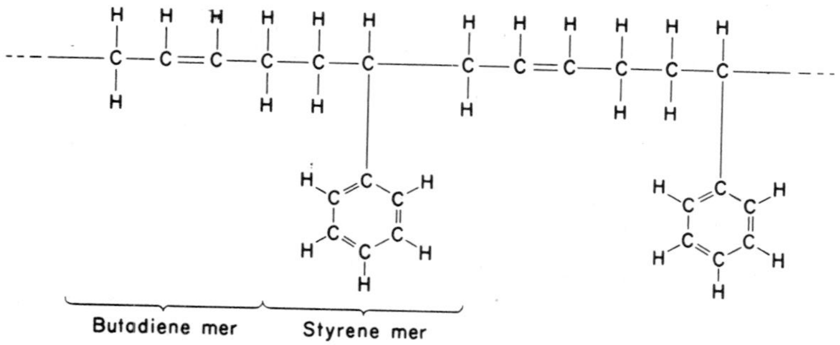

FIGURE 8-4   Polymerization by addition: the uniting of monomers.

FIGURE 8-5   Polymerization by the addition of two kinds of mers: copolymerization.

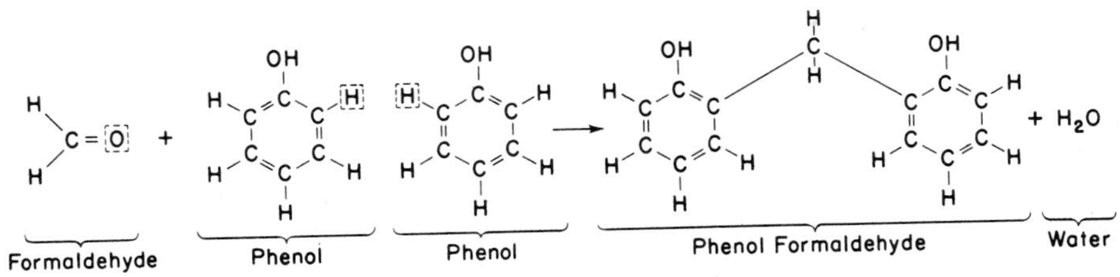

FIGURE 8-6 Formation of phenol-formaldehyde (bakelite) by condensation polymerization.

Since the bonding forces between molecules are much weaker than those within the molecule, deformation occurs by slippage between adjacent molecular chains. Methods for increasing the strength of thermoplastics therefore focus on restricting intermolecular slippage. Longer chains have less freedom of movement and are therefore stronger. Connecting adjacent chains with primary-bond cross-links, as with the sulfur links when vulcanizing rubber, can significantly impede deformation. Finally, because the secondary-bond strength is inversely related to the separation distance of the molecules, processes such as deformation or "crystallization" that produce a tight parallel alignment of adjacent molecules certainly increase strength. Polymers with large side structures, such as chlorine atoms or benzene rings, may be stronger or weaker than those with just hydrogen, depending on whether the dominant effect is the impeded motion or the increased separation distance. Branched polymers, in which the chains divide in a "Y" with primary bonds linking all segments of the chain, can exhibit the same phenomenon, as branching can reduce the density and close packing of the chains.

Thermosetting plastics, on the other hand, are those with a three-dimensional framework structure in which all atoms are connected by strong covalent bonds. These materials generally result from condensation polymerization, in which elevated temperature tends to promote an irreversible reaction, hence the term *thermosetting*. Once these materials are set, additional heatings do not produce softening. Instead, the materials maintain their mechanical properties up to the temperature at which they char or burn. Deformation requires the breaking of primary bonds, so that these plastics tend to be strong, but brittle. As a class, the thermosetting polymers are significantly stronger than the thermoplastics and have a lower ductility, a higher modulus of elasticity, and poorer impact properties.

Whether a polymer is thermosetting or thermoplastic is of great importance to a person who is selecting it for use. However, consideration should be given not only to its behavior in service, but also to how it must be processed. For example, thermoplastics are easily molded. After the material has been formed to the desired shape under conditions of elevated temperature and pressure, the mold must be cooled to cause the plastic to harden and to be able to retain its shape when removed. When products are produced from the thermosetting materials, the mold can remain at an elevated temperature throughout the entire process. The material hardens as a result of the temperature and pressure and can be removed without required cooling of the mold.

Types of Plastics and Their Properties.

Because there are so many varieties of plastics and new ones are being developed almost continuously, it is helpful to have a knowledge of their general properties as well as the specific properties of the basic families. General properties of plastics include:

1. *Light weight.* Most plastics have specific gravities between 1.1 and 1.6, compared with about 1.75 for magnesium. Thus, they are among the lighest of the engineering materials.
2. *Corrosion resistance.* Many plastics perform well in hostile, corrosive environments. A number are notably resistant to acid corrosion.
3. *Electrical resistance.* They are widely used as insulating materials.
4. *Low thermal conductivity.* They are relatively good heat insulators.
5. *Variety of optical properties.* Many plastics have an almost unlimited color range and the color goes throughout, not just on the surface. Both transparent and opaque materials exist.
6. *Formability.* Objects can frequently be produced from plastics in only one operation. Raw material can be converted to the final shape through such processes as casting, extrusion, and molding.
7. *Surface finish.* Excellent surface finishes are often obtained by the same processes that produce the shape. Additional surface-finishing operations may not be required.
8. *Comparatively low cost.* This includes both the material itself and the manufacturing process. Plastics frequently offer reduced tool costs and high rates of production.

Although the attractive features of plastics tend to be in the area of physical properties, the inferior properties often have to do with mechanical strength. None of the plastics possess strength properties that approach those of the engineering metals, but their low density allows them to compete effectively on a strength-to-weight (or specific strength) basis. Many have low impact strength, although several (such as ABS, high-density polyethylene, polycarbonate, and cellulose propionate) have good impact properties. The dimensional stability tends to be greatly inferior to that of metals, and the coefficient of thermal expansion is rather high. Thermoplastics are quite sensitive to heat, and the strength of many drops rapidly as temperatures increase above normal environmental conditions. The thermosetting materials offer good strength retention with increased temperature but have an upper limit of about 500°F (250°C). Although the corrosion resistance of plastics is generally good, they often absorb moisture, and this absorption, in turn, decreases strength. Some thermoplastics can exhibit a 50% drop in tensile strength as the humidity increases from 0% to 100%. Finally, radiation, both ultraviolet and particulate, can markedly alter the properties of plastics. Many plastics used in an outdoor environment have ultimately failed because of the cumulative effect of ultraviolet radiation.

Table 8.1 summarizes the properties of a number of common plastics. From consideration of this table and the above discussion of general properties, it is apparent that plastics are best used in applications that require materials with low to moderate strength, light weight, low electrical and/or thermal conductiv-

TABLE 8.1 Properties and Major Characteristics of Common Types of Plastics

Material	Specific Gravity	Tensile Strength (1000 lb/in²)	Impact Strength Izod ft-lb/in. of Notch	Top Working Temperature [°F (°C)]	Dielectric Strength[b] (volts/mil)	24-Hour Water Absorption (%)	Weatherability	Colorability	Optical Clarity	Chemical Resistance	Injection Molding	Extrusions	Formable Sheet	Film	Fiber	Compressions or Transfer Moldings	Castings	Reinforced Plastics	Moldings	Industrial Thermosetting Laminates	Foam
Thermoplastics																					
ABS material	1.02–1.06	4–8	1.3–10.0		300–400	0.2–0.3	0	×		0	√	√	√								
Acetal	1.4	10	1.5	250 (121)	1200	0.22	0	×		0	√	√	√								
Acrylics	1.12–1.19	5.5–10	0.2–2.3	200 (93)	400–530	0.2–0.4	×	×	×	0	√	√	√	√	√		√				
Cellulose acetate	1.25–1.50	3–8	0.75–4.0	260 (127)	300–600	2.0–6.0	×	×	×		√	√	√	√							
Cellulose acetate butyrate	1.18–1.24	2–6	0.6–3.2	130 (54)	250–350	1.8–2.1	×	×	×		√	√	√	√							
Cellulose propionate	1.19–1.24	1–5	0.8–9	140 (60)	300	1.8–2.1		×	×		√	√	√	√							
Chlorinated polyether	1.4	6	3.3	300 (149)	400	0.01	×			×	√	√	√								
Ethyl cellulose	1.16	3–6	1.8–4.0	150 (66)	350	1.6–2.2		×	×		√	√	√	√							
TFE-fluorocarbon	2.1–2.3	1.5–3	2.5–4.0	500 (260)	450	0	×			×	√	√	√	√		√					
CFE-fluorocarbon	2.1–2.15	4.5–6	3.5–3.6	390 (199)	550	0	×			×	√	√	√	√							
Nylon	1.1–1.2	8–10	2	250 (121)	385–470	0.4–5.5	0	0	0	0	√	√	√	√	√						
Polycarbonate	1.2	9.5	14	250 (121)	400	0.15	×	0	0		√	√	√	√							
Polyethylene	0.96	4	10	200 (93)	440	0.003	0	0	0	×	√	√	√	√							
Polypropylene	0.9–1.27	3.4–5.3	1.02	230 (110)	520–800	0.03		0	×	×	√	√	√	√	√						
Polystyrene	1.05–1.15	5–9	0.3–0.6	190 (88)	400–600	<0.2		×	×	0	√	√	√	√							
Modified polystyrene	1.0–1.1	2.5–6	0.25–11.0	212 (100)	300–600	0.03–0.2		×	×	×	√	√	√	√							√
Vinyl	1.16–1.55	1–5.9	0.25–2.0	220 (104)	25–500	0.2–1		×	×	×	√	√	√	√	√						√
Thermosetting plastics																					
Epoxy	1.1–1.7	4–13	0.4–1.5	325 (163)	500	0.1–0.5	×	×		×						√	√	√	√	√	√
Melamine	1.76–1.98	5–8	0.25–5	350 (177)	460	0.1		×		0						√			√	√	
Phenolic	1.2–1.45	5–9		300 (149)	100–500	0.2–0.6				0						√	√	√	√	√	√
Polyester (other than molding compounds)	1.06–1.46	4–10	0.18–0.4	300 (149)	340–570	0.5		×	0					√	√		√	√			
Polyester (alkyd, DAP)	1.6–1.75	3.2–8	3.6–8		250–350	0.16–0.67										√		√	√	√	
Silicone	2.0	3–5	0.2–3.0	550 (288)		0.4–0.5	×									√	√	√	√	√	√
Urea	1.41–1.80	4–8.5	0.2–0.5	185 (85)	300–600	1–3		×		0						√		√	√	√	√

[a] × denotes a principal reason for its use; 0 indicates a secondary reason.

[b] Short-time ASTM Test.

ity, a wide range of available colors, and ease of fabrication into finished products. No other family of materials can offer this combination of properties. Thus, many plastics are selected for "packaging" or container applications. This classification includes such items as stereo cabinets, clock cases, and household appliance housings, which serve primarily as containers for the interior mechanisms. Applications such as insulation on electrical wires and handles for hot articles capitalize on the low electrical and thermal conductivities. Soft, pliable foamed plastics are used extensively as cushioning material. Rigid foams are used inside sheet metal structures to provide compressive strength. Some plastics are applied as adhesive or bonding agents in the assembly of products, an application to be discussed in Chapter 37. Others can provide inexpensive tooling for applications where pressures, temperatures, and wear requirements are not extreme.

There are many applications in which only one or two of the properties of plastics are sufficient to dictate their use. In other cases, characteristics are desired that are not normally found in plastics, such as high directional strength. To meet these demands, *composite materials* can be designed to incorporate a fabric or fiber reinforcement within a plastic resin. These will be discussed in some detail later in this chapter. For the most part, however, plastics are used where they provide a *combination* of properties not found in other engineering materials. When assessing the suitability of plastics for a given application and then selecting a specific type, one should seek the best overall combination of engineering properties.

The following are some comments about several types of plastics listed in Table 8.1:

ABS: contains acrylonitrile, butadiene, and styrene; low weight, good strength, and very tough; resists heat, weather, and chemicals quite well; dimensionally stable, but flammable.

Acrylics: highest optical clarity, transmitting over 90% of light; common trade names include *Lucite* and *Plexiglas;* high-impact, flexural, tensile, and dielectric strengths; available in a wide range of colors; resist weathering; stretch rather easily.

Cellulose acetate: wide range of colors; good insulating qualities; easily molded; high moisture absorption in most grades and affected by alcohols and alkalies.

Cellulose acetate butyrate: higher impact strength and moisture resistance than cellulose acetate; will withstand rougher usage.

Ethyl cellulose: high electrical resistance and impact strength; retains toughness at low temperatures.

Fluorocarbons: inert to most chemicals; high temperature resistance; very low coefficients of friction (Teflon), used for nonlubricated bearings and nonstick coatings for cooking utensils and electric irons.

Nylon: low coefficient of friction; good strength, abrasion resistance and toughness; excellent dimensional stability; used for bearings and as monofilaments for textiles, fishing line, and ropes.

Polycarbonates: high strength and outstanding toughness.

Polyethylenes: tough, good chemical resistance and high electrical resistance; subject to weathering via ultraviolet light; flammable; used for tubes, pipes, sheeting, and electrical wire insulation.

Polystyrenes: high dimensional stability and low water absorption; best all-around dielectric; burns readily and is adversely affected by citrus juices and cleaning fluids.

Vinyls: wide range of types, from thin, rubbery films to rigid forms; tear resistant; good aging properties; good dimensional stability and water resistance in rigid forms; used for floor and wall coverings, upholstery fabrics, and light-weight water hose.

Epoxies: good strength, toughness, elasticity, chemical resistance, moisture resistance, and dimensional stability; easily compounded to cure at room temperature; used as adhesives, bonding agents, and coatings, and in fiber laminates.

Melamines: excellent resistance to heat, water, and many chemicals; full range of translucent or opaque colors; excellent electrical-arc resistance; tableware, but stained by coffee; used extensively in treating paper and cloth to impart water-repellent properties.

Phenolics: oldest of the plastics, but still widely used; hard, strong, low-cost, and easily molded; resistant to heat and moisture; dimensionally stable; opaque, but with a wide color range; wide variety of forms—sheet, rod, tube, and laminate.

Polyesters: strong and resists environmental influences well; uses include boat and car bodies, pipes, vents and ducts, textiles, adhesives, coatings, and laminates.

Silicones: semiorganic spine molecules with alternating silicon and oxygen atoms; heat- and weather-resistant; low moisture absorption; chemically inert; high dielectric properties; excellent sealants.

Urea formaldehyde: similar properties to phenolics, but available in lighter colors; useful for containers and housings, but not outdoors; used in lighting fixtures because of translucence in thin sections.

Additive Agents in Plastics.   For most uses, other materials are added to plastics to (1) improve their properties; (2) reduce their cost; (3) improve their moldability; and/or (4) impart color. These additive constituents are usually classified as *fillers, plasticizers, lubricants, coloring agents, stabilizers, antioxidants,* and *flame retardants.*

Ordinarily, fillers comprise a large percentage of the total volume of a molded plastic product, being added to improve strength, to reduce shrinkage, to reduce weight, or to provide cost-saving bulk (while often reducing moldability). To a large degree, they determine the general properties of a molded plastic. Selection tends to favor materials that are much less expensive than the plastic resin, but various fillers impart different properties and characteristics. Some of the most common fillers and their properties are

1. *Wood flour:* a general-purpose filler; low cost with fair strength; good moldability.

2. *Cloth fibers:* improved impact strength; fair moldability.
3. *Macerated cloth:* high impact strength; limited moldability.
4. *Glass fibers:* high strength; dimensional stability; translucence.
5. *Asbestos fiber:* heat resistance; dimensional stability.
6. *Mica:* excellent electrical properties and low moisture absorption.

Other fillers are used to impart high strength at elevated temperatures. "Whiskers" of various metals and nonmetals, such as boron, stainless steel, niobium, tantalum, titanium, zirconium, and silicon carbide, have been used. These are from 40 to 200 microinches (1 to 5 microns) in diameter and 0.001 to 0.04 inches (30 to 1000 microns) in length and have high moduli of elasticity and tensile strengths up to 3,000,000 psi (21,000 MPa). More commonly used are filaments of glass, aramid, graphite, or boron, less than 0.1 mm in diameter but of unlimited length. These can provide tensile strengths up to 350 ksi (2450 MPa) with moduli of elasticity up to 60 million psi (420,000 MPa). Glass fiber cloth is used in many composite materials. Carbon black and metal oxide particles are other possible filler materials.

When fillers are used with a plastic resin, the resin acts as the binder, surrounding the filler material and holding the mass together. The surface of a molded part, therefore, is almost pure resin with no exposed filler.

Coloring agents may be either *dyes,* which are soluble in the resins, or insoluble *pigments,* which impart color simply by their presence. In general, dyes are used for transparent plastics and pigments for the opaque ones. As most fillers do not produce attractive colors by themselves, coloring agents are usually needed.

Plasticizers can be added in small amounts to improve the flow of the plastic during molding or to increase the flexibility of thermoplastics by reducing the intermolecular contact and attraction of the polymer chains. When plasticizers are used for molding purposes, the amount used is governed by the intricacy of the mold. In general, the amount should be kept to a minimum because the plasticizer is likely to affect the stability of the finished product through a gradual aging loss. When used for flexibility, plasticizers should be selected with minimum volatility, so as to impart the desired property for as long as possible.

Lubricants can be added in small amounts to improve moldability and to facilitate the removal of parts from the mold. Waxes, stearates, and occasionally soaps are used for this purpose. These are also held to a minimum because of their adverse effect on engineering properties.

Heat, light (especially ultraviolet), and oxidation tend to degrade polymers. Stabilizers and antioxidants can be added to retard these effects. Flame retardants can be added when nonflammability is important.

Plastics as Adhesives.   Structural adhesives capable of withstanding high stresses are common in many industrial applications, and their use has expanded rapidly. They are quite attractive for the bonding of dissimilar materials, such as metals to nonmetals, and have even been used to replace welding or riveting. A wide range of mechanical properties is available through variations in com-

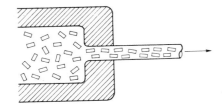

FIGURE 8-7 Schematic representation of the aligning of plastic molecules in the orienting process.

position and additives, and a variety of curing mechanisms can be used. The various features of adhesive bonding will be developed further in Chapter 37.

Oriented Plastics. Because the intermolecular bonds in thermoplastics are much weaker than the atomic bonds within the chains, these plastics can be processed to provide high strength in a given direction by alignment of the molecules parallel to the applied load. This *orienting* process can be accomplished by either stretching, rolling, or extrusion, as illustrated in Figure 8-7. The material is usually heated before the orienting process to aid in overcoming the intermolecular forces and is cooled immediately afterward to "freeze" the molecules in the desired orientation.

Orienting may increase the tensile strength by more than 50%, but a 25% increase is more typical. The elongation may be increased by several hundred percent. If oriented plastics are reheated, however, they tend to deform toward their original shape because of the phenomenon of *viscoelastic memory*.

Plastics for Tooling. Because of their wide range of properties, their ease of conversion into desired shapes, and their excellent properties when loaded in compression, plastics are widely used in tooling to make jigs, fixtures, and forming-die components. Both thermoplastics and thermosets (particularly the cold-setting types) are used for drill and trim jigs, routing and assembly jigs and fixtures, and form blocks for forming. Thermoplastic materials can be used for punch-press punches, these pieces frequently being formed in the female form block. Thermosetting materials have been used for stretch-press dies and tube-bending mandrels. The use of plastics in these tooling applications usually results in less costly tooling, permitting small quantities of products to be produced more economically. In addition, the tooling can often be produced in a much shorter time so that production can begin at an earlier date.

Ablative Coatings. Plastics form the base of many ablative coating materials used on rockets and missiles to provide short-time protection from intense heating conditions, such as those encountered by the early space capsules when reentering the earth's atmosphere. Ablation involves the thermal decomposition of high polymers into low-molecular-weight gaseous products and porous carbon char. Immense amounts of heat are consumed and dissipated in these reactions.

Engineering Plastics. A series of true engineering plastics is emerging with improved thermal properties, first-rate impact and stress resistance, superior electrical characteristics, and excellent processing properties. These materials consist of the following groups: polyamides, polyacetals, polycarbonates, modified polyphenylene oxides, polybutylene terephthalates, polysulfones, and polyetherimides. Although the conventional plastics can be upgraded by stabilizers, fibrous reinforcements, and particulate fillers, there is a companion reduction in other properties. The engineering plastics, on the other hand, offer a more balanced set of properties.

Using advanced technology, materials producers are currently seeking to develop electroconductive polymers with tailored electrical and electronic properties and high-crystalline polymers with properties comparable to those of metals.

Plastics Versus Metals. In many applications, plastics and metals are viewed as competing materials, although their engineering properties are often considerably different. Many of the attractive features of plastics have already been discussed. In addition to these, one may wish to include (1) the ability to be fabricated with lower tooling costs; (2) the ability to be molded at the same rate as product assembly, which reduces inventory; (3) a possible reduction in assembly operations and easier assembly through snap fits, friction welds, or the use of self-tapping fasteners; (4) the ability to reuse manufacturing scrap; and (5) reduced finishing costs.

Metals, on the other hand, are often cheaper, offer faster fabrication speeds and greater impact resistance, are considerably more rigid, and can withstand traditional paint cure temperatures. In addition, resistance to flames, acids, and various solvents is significantly better.

In the automotive industry, plastics now comprise more than 200 pounds (90 kg) of the average 2800-pound (1270-kg) weight of a car. More than 1 billion pounds of plastics were used in United States–manufactured automobiles in 1983. In 1984, this use increased by nearly 28%. Looking to the future, one survey predicts that over 2.2 billion pounds will be used in these applications in 1995.

Elastomers

Elastomers are a class of linear polymers that display an exceptionally large amount of elastic deformation when a force is applied. Many can be stretched to several times their original length. On release of the force, the deformation can be completely recovered, however, as the material quickly returns to approximately its original shape. Moreover, this cycle can be repeated numerous times, as with a common rubber band.

The elastic properties of most engineering materials are the result of a change in distance between adjacent atoms when loads are applied. Hooke's law is commonly obeyed: twice the force produces twice the stretch. When the load is removed, the interatomic forces return all of the atoms to their original position, and the observed elastic deformation is completely recovered.

The elastomers, on the other hand, are generally linear chains, produced by addition polymerization. In these materials, the long polymer chain is in the form of a coil, much like a coil spring. When a force is applied, the polymer stretches by uncoiling in a manner that does not follow Hooke's Law. When the load is removed, the molecules recoil, and the material returns to its original size and shape.

The actual behavior of elastomers, however, is a bit more complex. Although the chains indeed uncoil when placed under load, they also tend to slide over one another to produce a small degree of viscous deformation. When the load is removed, the molecules recoil, but the viscous deformation does not recover and remains as a permanent change in shape.

By cross-linking the coiled molecules, however, it is possible to prevent viscous deformation while retaining the large elastic response. The elasticity or rigidity of the product can be determined by control of the number of cross-links within the material. Small amounts of cross-linking leave the elastomer soft and flexible. Additional cross-linking restricts the uncoiling of the chains, and the material becomes harder, stiffer, and more brittle. Thus, engineering elastomers can be tailored to possess a wide range of properties and stress–strain characteristics.

If elastomers are placed under constant strain, however, some viscous flow occurs over time, even in highly cross-linked material. The dimensions remain fixed, but the force or stress being transmitted through the material tends to decrease. This phenomenon is known as *stress relaxation*. The rate of this relaxation depends on the material, the force, and the temperature.

Rubber. Natural rubber is the oldest commercial elastomer; it is made from the processed sap of a tropical tree. In its crude form, it is an excellent adhesive, and many cements can be made by dissolving it in suitable solvents. Its use as an engineering material dates from 1839, when Charles Goodyear discovered that it could be vulcanized (cross-linked) by the addition of about 30% sulfur and heating to a suitable temperature. The cross-linking restricts the movement of the molecular chains and thus imparts strength. Subsequent research showed that the properties of vulcanized rubber could be further improved by certain additives, such as carbon black, which act as stiffeners, tougheners, and antioxidants. Accelerators have been found that speed up the vulcanization process and enable a reduction in the amount of sulfur, so that most modern rubber compounds contain less than 3%. As in the processing of plastics, softeners and fillers can be added to facilitate processing and to add bulk.

Rubber can be compounded to provide a wide range of characteristics, ranging from soft and gummy to hard, as in ebonite. Where strength is required, textile cords or fabrics can be coated with rubber. The fibers carry the load, and the rubber serves to bond the cords and to isolate them from one another, thus preventing chafing and reducing friction. For severe service, steel wires may be coated to serve as the load-bearing medium. Vehicle tires and heavy-duty conveyor belts are examples of this technology.

Natural rubber compounds are outstanding for their flexibility, their good electrical insulation, their low internal friction, and their resistance to most inorganic acids, salts, and alkalies. However, their resistance to petroleum products, such as oil, gasoline, and naphtha, is poor. In addition, they lose their strength at elevated temperatures. It is not advisable to operate them at temperatures above 175°F (80°C). They also deteriorate fairly rapidly in direct sunlight unless specially compounded.

Artificial Elastomers. An uncertainty in the supply and price of natural rubber led to the development of a number of synthetic or artificial elastomers that have come to assume great commercial importance. Polyisoprene appears to have the same molecular structure as natural rubber, with equal or superior properties. Some artificial elastomers are inferior to natural rubber, and others have distinctly different and frequently superior properties, expanding their usefulness for specific applications. Silicone rubbers, for example, are based on a linear chain of silicon and oxygen atoms. This structure offers high temperature resistance, permitting their use in environments up to 450°F (230°C).

Table 8.2 lists some of the most widely used artificial elastomers (natural rubber is shown for comparison) and gives their typical properties and uses. It should be remembered, however, that the properties of these materials can vary considerably, depending on the details of compounding and processing.

Elastomers for Tooling Applications.   When an elastomer is confined, it acts as a fluid, transmitting force quite uniformly in all directions. This phenomenon often makes it possible to use an elastomer for one half of a die set in metal-forming operations. Elastomers also make it possible to perform bulging and forming of reentrant sections, which would be impossible with rigid dies except through the use of costly multipiece tooling. Because they can be compounded to range from very soft to very hard, hold up well when subjected to compression loading, and can be quickly and economically made into a desired shape, elastomers have become increasingly popular as tool materials. Urethanes are currently the most used for these applications.

Ceramics

Long used in the electrical industry for their high electrical resistance, ceramic materials have also assumed considerable importance in a variety of other engineering applications. Most of these relate to their outstanding physical properties, including their ability to withstand high temperatures (refractories and refractory coatings), to provide a variety of electrical properties (solid-state electronics), and to resist wear (coated cutting tools). In general, ceramics are hard, brittle, high-melting-point materials with low electrical and thermal conductivity, good chemical and thermal stability, good creep resistance, and high compressive strengths. More recently, a family of structural ceramics has emerged, and these materials now provide sufficiently enhanced mechanical properties so that they are attractive for a number of engineering applications.

The Nature and Structure of Ceramics.   Ceramic materials are compounds of metallic and nonmetallic elements (often in the form of oxides, carbides, and nitrides) and exist in a wide variety of compositions and forms. Most have crystal structures, but in contrast to metals, the bonding electrons are generally captive in strong ionic or covalent bonds. The absence of free electrons makes the ceramic materials poor electrical conductors and results in many being transparent in thin sections. Because of the strength of the primary bonds, most ceramics have high melting temperatures.

The crystal structures of ceramic materials can be quite different from those observed in metals. In many ceramics, atoms that differ greatly in size must be accommodated within the structure (as in the sodium chloride crystal of Figure

TABLE 8.2 Properties and Uses of Common Elastomers

Elastomer	Specific Gravity	Durometer Hardness	Tensile Strength (psi)		Elongation (%)		Service Temp. [°F (°C)]		Resistance to[a]			Typical Application
			Pure Gum	Black	Pure Gum	Black	Min.	Max.	Oil	Water Swell	Tear	
Natural rubber	0.93	20–100	2500	4000	750	650	−65 (−54)	180 (82)	P	G	G	Tires, gaskets, hose
Polyacrylate	1.10	40–100	350	2500	600	400	0 (−18)	300 (149)	G	P	F	Oil hose, O-rings
EDPM (ethylene propylene)	0.85	30–100	1	3		500	−40 (−40)	300 (149)	P	G	G	Electric insulation, footwear, hose, belts
Chlorosulfonated polyethylene	1.10	50–90	4	2		400	−65 (−54)	250 (121)	G	E	G	Tank linings, chemical hose, shoe soles and heels
Polychloroprene (neoprene)	1.23	20–90	3500	4000	800	550	−50 (−46)	225 (107)	G	G	G	Wire insulation, belts, hose, gaskets, seals, linings
Polybutadiene	1.93	30–100	1000	3000	800	550	−80 (−62)	212 (100)	P	P	G	Tires, soles and heels, gaskets, seals
Polyisoprene	0.94	20–100	3000	4000	600	600	−65 (−54)	180 (82)	P	G	G	Same as natural rubber
Polysulfide	1.34	20–80	350	1000	600	400	−65 (−54)	180 (82)	E	G	G	Seals, gaskets, diaphragms, valve disks
SBR (styrene budatiene)	0.94	40–100	2			1200	−65 (−54)	225 (107)	P	G	G	Molded mechanical goods, disposable pharmaceutical items
Silicone	1.1	25–90		1200		450	−120 (−84)	450 (232)	F	E	P	Electric insulation, seals, gaskets, O-rings
Epichlorohydrin	1.27	40–90		2		325	−50 (−46)	250 (121)	G	G	G	Diaphragms, seals, molded goods, low-temperature parts
Urethane	0.85	62–95	5000		700		−54 (−65)	212 (100)	E	F	E	Caster wheels, heels, foam padding
Fluoroelastomers	1.65	60–90	1	3		400	−40 (−40)	450 (232)	E	E	F	O-rings, seals, gaskets, roll coverings

[a]P, poor: F, fair: G, good: E, excellent.

3-3), and interstitial sites become extremely important. Charge neutrality must be maintained throughout the structure of ionic materials. Covalent materials can accommodate only a limited number of nearest neighbors, set by the number of shared-electron bonds. These features often force less efficient packing—and hence lower densities—than those observed for metallic materials. As in metals, the same chemistry material can often exist in more than one structural arrangement (polymorphism). Silica (SiO_2), for example, can exist in three forms—quartz, tridymite, and crystobalite—depending on the conditions of temperature and pressure.

Ceramic materials can also exist in the form of chains, similar to the linear molecules in plastics. As in the polymeric materials having this structure, the bonds between the chains are not as strong as those within the chains. Consequently, when forces are applied, cleavage occurs between the chains.

In other ceramics, the atoms bond in the form of sheets and produce layered structures. Relatively weak bonds exist between the sheets, and these surfaces become the preferred sites for fracture. Mica is an example of such a material.

A noncrystalline structure is also possible in solid ceramics. This condition is referred to as the *glassy state,* and the materials are known as *glasses.*

Clay Products. Many ceramic products are based on clay, to which various amounts of quartz and feldspar have been added. Selected proportions are mixed with water and the material is shaped, dried, and fired to produce such products as brick and tile, earthenware, drainage and sewer pipe, china, and porcelain.

Refractory Materials. Refractory materials are ceramics that have been designed to provide acceptable mechanical or chemical properties while at high temperatures. Most are based on stable oxide compounds, the coarse oxide particles being bonded by finer refractory material. Various carbides, nitrides, and borides can also be used in refractory applications.

Three distinct classes of refractories can be identified: acidic, basic, and neutral. Common acidic refractories are based on silica and alumina and can be compounded to provide high temperature resistance along with high hardness and good mechanical properties. (Machinable all-silica ceramics have been used as insulating tiles on the U.S. space shuttle.) Magnesium oxide is the core material for most basic refractories. These are generally more expensive than the acidic materials but are often required in metal-processing applications to provide compatibility with the metal. Neutral refractories are often used to separate the acidic and basic materials because they tend to attack one another. The combination may be desirable, however, when a basic refractory is necessary on the surface for chemical reasons, and the cheaper, acidic material can be used beneath to provide strength and insulation. Table 8.3 presents the compositions of some typical refractories. Figure 8-8 shows a variety of high-strength alumina components.

Ceramics for Electrical and Magnetic Applications. Ceramic materials can also have a variety of useful electrical and magnetic properties. Some ce-

TABLE 8.3   Composition of Some Typical Refractories

Refractory	SiO$_2$	Al$_2$O$_3$	MgO	Fe$_2$O$_3$	Cr$_2$O$_3$
Acidic					
Silica	95–97				
Superduty fire brick	51–53	43–44			
High-alumina fire brick	10–45	50–80			
Basic					
Magnesite			83–93	2–7	
Olivine	43		57		
Neutral					
Chromite	3–13	12–30	10–20	12–25	30–50
Chromite–magnesite	2–8	20–24	30–39	9–12	30–50

Source: Ceramic Data Book, Cahners Publishing Co., 1982.

ramics, such as silicon carbide, are used as resistors and heating elements for furnaces. Others have semiconducting properties and are used for thermistors and rectifiers. Dielectric, piezoelectric, and ferroelectric behavior can be utilized in a number of applications. Barium titanate, for example, is used in capacitors and transducers. High-density clay-based ceramics and aluminum oxide make excellent high-voltage insulators.

Glasses.   Most of the commercial glasses are based on silica (SiO$_2$) with additives that alter the structure or reduce the melting point. Various chemistries can be used to optimize optical properties, thermal stability, and resistance to thermal shock.

Cermets.   Cermets are combinations of metals and ceramics (usually oxides, carbides, or nitrides) bonded together in the same manner in which powder

FIGURE 8-8   A variety of high strength alumina (acid refractory) components, including a filter for molten metal. (Courtesy GTE, Wesgo Division.)

metallurgy parts are produced. This usually involves pressing the powders in molds at pressures ranging from 10 to 40 ksi (70 to 280 MPa) followed by sintering in a controlled atmosphere furnace at about 3000°F (1650°C). Cermets combine the high refractory characteristics of ceramics and the toughness and thermal shock resistance of metals. They are used as crucibles and jet engine nozzles, and in other applications requiring strength and toughness at elevated temperature.

Ceramics for Mechanical Applications—The Structural Ceramics.
Because of their strong interatomic bonding and high shear resistance, ceramic materials tend to have low ductility and high compressive strength. Theoretically, ceramics could also have high tensile strengths, but they generally do not because small cracks, pores, and other defects act as stress concentrators, and their effect cannot be reduced through ductility and plastic flow. Ceramics are sensitive to even small flaws, and as a result, failure typically occurs at tensile stress values between 3 and 30 ksi (20 to 210 MPa).

By elimination or restriction of the various flaws and defects, high tensile strengths can be obtained. Hardness, wear resistance, and strength at elevated temperatures are all attractive properties, as are light weight (specific gravities of 2.3 to 3.85), dimensional stability, corrosion resistance, and chemical inertness. On the negative side, reliability is still rather low. Failure occurs as a result of a brittle fracture mechanism with little, if any, prior warning. The cost is rather high, and it is extremely difficult to form reliable bonds to the various engineering metals. Machining is difficult, so products must be fabricated through the use of net-shape processing.

Structural ceramic materials include silicon nitride, silicon carbide, partially stabilized zirconia (limited to temperatures below 1300°F because of a rapid drop in strength), aluminum oxide, and ceramic composites (generally ceramic fibers, such as silicon carbide, in a glass or glass-ceramic matrix). Silicon carbide and silicon nitride offer excellent strength with moderate toughness. They work well in high-stress, high-temperature applications, such as turbine blades, and may well replace nickel- or cobalt-based superalloys. Figure 8-9 shows gas-turbine rotors made from injection-molded silicon nitride. They are designed to operate at 2300°F (1250°C) without a need for cooling. Figure 8-10 shows a variety of additional silicon nitride products.

Sialon (a *si*licon–*al*uminum–*o*xygen–*n*itrogen ceramic) is stronger than steel, extremely hard, and as light as aluminum. It has good resistance to wear and thermal shock, is an electrical insulator, and retains good tensile and compression strength up to 2550°F (1400°C). However, when overloaded, it will fail by brittle fracture. Its coefficient of thermal expansion is only one third that of steel and one tenth that of plastic, a feature that provides excellent dimensional stability. Areas of application include engine combustion chambers, friction wear pads, bearings, seals, turbocharger rotors, cutting tools, and dies. Manufacture requires powder metallurgy techniques.

A ceramic engine block is an area of considerable interest and research. If perfected, it would allow higher operating temperatures with a companion in-

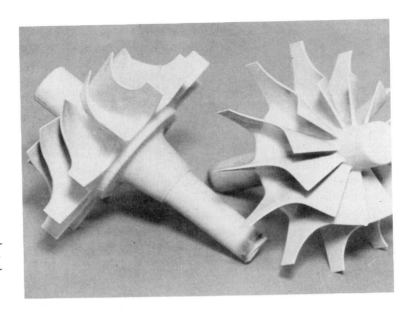

FIGURE 8-9 Gas turbine rotors made of silicon nitride. This light-weight material (1/2 the weight of stainless steel) offers strength at elevated temperatures as well as excellent resistance to corrosion and thermal shock. (Courtesy GTE Wesgo Division.)

crease in engine efficiency and would permit the elimination of radiators, fan belts, cooling system pumps, and coolant lines. Estimated fuel savings could amount to 30% or more.

Additional areas of research include the development of ceramic materials with improved toughness, increased strength, and enhanced corrosion and abrasion resistance. The desire is to be able to produce a low-cost ($3 to $10 per pound), high-strength, high-toughness ceramic with a useful temperature range through 2350°F (1300°C). Potential applications include engines, turbochargers, gas turbines, bearings, pump and valve seals, and other uses involving high-temperature, high-stress environments.

FIGURE 8-10 A variety of components manufactured from silicon nitride, including an exhaust valve and turbine blade. (Courtesy GTE, Wesgo Division.)

Ceramic Cutting Tools. Cutting tool materials have advanced significantly through ceramic technology. Cobalt-bonded tungsten carbide quickly became a popular alternative to high-speed tool steels. Then came the coated carbides, utilizing vapor-deposited coatings of titanium carbide, titanium nitride, and aluminum oxide to reduce wear and friction and to enable faster rates of cutting. Silicon nitride cutting tools now offer even greater tool life, higher cutting speeds, and reduced machine downtime. Aluminum oxide reinforced with 35% silicon carbide whiskers can machine nickel-based superalloys more than 10 times faster than standard carbide materials, the cutting speeds increasing from 200 to nearly 5000 feet per minute. Use of these materials, however, will require additional machine development in the area of high-speed spindles, work-holding devices that can withstand the high centrifugal forces, and chip removal methods that can remove the chips as fast as they are being formed.

A *composite material* is a heterogeneous solid consisting of two or more different materials that are mechanically or metallurgically bonded together. Each of the various components retains its identity in the composite and maintains its characteristic structure and properties. The composite material, however, generally possesses a unique combination of properties, such as stiffness, strength, weight, high-temperature performance, corrosion resistance, hardness, or conductivity, that is not possible with the components by themselves. Analysis of these properties shows that they depend on (1) the properties of the individual components; (2) the relative amounts of the components; (3) the size, shape, and distribution of the discontinuous components; (4) the degree of bonding between components; and (5) the orientation of the various components. The various materials involved can be organics, metals, or ceramics. Hence, sufficient freedom exists so that composite materials can often be designed to provide a specific set of engineering properties and characteristics.

There are many types of composite materials and several methods of classifying them. One such method is based on geometry and consists of three distinct families: *laminar* or layer composites, *particulate* composites, and *fiber-reinforced* composites.

Composite Materials

Laminar or Layer Composites.
Laminar composites are those having alternating layers of material bonded together in some manner and include thin coatings, thicker protective surfaces, claddings, bimetallics, laminates, and sandwiches. Plywood is probably the most common engineering material in this category and is an example of a laminate material. Layers of wood veneer are adhesively bonded with their grain orientations at various angles to one another. Strength and fracture resistance are improved, properties are somewhat uniform within the plane of the sheet, swelling and shrinkage tendencies are minimized, and large-size pieces are now available at reasonable cost. Safety glass is another laminate in which an adhesive layer is placed between two pieces of glass and serves to retain the fragments when the glass is broken. Other laminates are used in decorative items, such as Formica countertops and furniture.

Bimetallic strip consists of two metals with different coefficients of thermal expansion bonded together in a laminate. Changes in temperature produce flexing or curvature in the product, which may be used in thermostat and other heat-sensing applications, as illustrated in Figure 8-11. Still other laminar composites are designed to improve corrosion resistance while retaining low cost, high strength, or light weight. Clad materials fall into this class. Alclad metal, for example, consists of a high-strength, heat-treatable aluminum with an exterior cladding of one of the more corrosion-resistant, non-heat-treatable alloys. U.S. coinage is yet another example, the purpose here being to conserve the more costly high-nickel material while preserving the desired luster and corrosion resistance. Other surface layers can be selected to improve wear resistance or appearance. These can be applied through a variety of means, including plating, hardfacing, and solvent-based coatings.

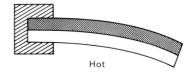

FIGURE 8-11 Schematic of a bimetallic strip where Material A has the greater coefficient of thermal expansion.

Sandwich material is composed of a thick, low-density core placed between thin, high-strength, high-density surfaces. Corrugated cardboard would be an example of a sandwich material. Other engineering sandwiches use cores of polymer foam or honeycomb structure.

Particulate Composites. *Particulate composites* consist of discrete particles of one material surrounded by a matrix of another material. Concrete is a classic example, consisting of sand and gravel particles surrounded by cement. Asphalt consists of a similar aggregate in a matrix of bitumen, a thermoplastic polymer. In these examples, the particles are rather coarse. Other particulate composites involve extremely fine particles. For example, many powder metallurgy products (those with little, if any, interdiffusion between the components) can be considered particulate composites.

One class of particulate-reinforced composite is the dispersion-strengthened materials, in which a small quantity of hard, brittle, fine particles (typically oxides) are dispersed in a softer, more ductile matrix. Pronounced strengthening can be induced, which decreases only gradually as temperature is increased. Creep resistance can be significantly improved. Examples of these materials include sintered aluminum powder (SAP), which consists of an aluminum matrix strengthened by aluminum oxide, and TD-nickel, a nickel alloy containing 1–2 weight percentage of thoria (ThO_2).

Other particulate composites contain large amounts of rather coarse particles and are designed to produce altered combinations of properties, rather than increased strength. Cemented carbides consist of hard ceramic particles, such as tungsten carbide, tantalum carbide, or titanium carbide, dispersed in a metal matrix, generally cobalt. The hard, stiff carbide can withstand the high temperatures of cutting, but it is extremely brittle. Toughness can be imparted, however, by combining the carbide particles and cobalt powder, pressing them into desired shapes, heating to melt the cobalt, and then resolidifying the compacted material. Varying levels of toughness can be imparted by varying the amount of cobalt in the composite.

Grinding and cutting wheels are often formed from alumina (Al_2O_3), silicon carbide (SiC), cubic boron nitride (BN), or diamond bonded in a matrix of glass or polymeric material. As the hard particles wear, they fracture or pull out of the matrix, exposing new cutting edges. Electrical contacts can be designed to have good conductivity and spark-erosion resistance by mixing tungsten powder with powdered silver or copper and processing via powder metallurgy. Many of the additives in plastics can actually be considered particles in a particulate composite. Foundry molds and cores are often made from sand (particles) and an organic or inorganic binder (matrix).

Because of their unique geometry, the properties of particulate composites can be isotropic, that is, uniform in all directions. This quality may be particularly important in many engineering applications.

Fiber-Reinforced Composites—The "String-and-Glue" Materials.

The most popular type of composite material is the *fiber-reinforced* geometry, in which continuous or discontinuous thin fibers of one material are embedded in a matrix of another. The matrix supports and transmits loads to the fibers and provides ductility and toughness, and the fibers carry most of the force. Wood and bamboo are two naturally occurring fiber composites, consisting of cellulose fibers in a lignin matrix. Bricks of straw and mud may well have been the first fabricated material of this variety. Automobile tires now use fibers of nylon, rayon, Kevlar,[1] or steel in various numbers and orientations to reinforce the rubber and to provide added strength and durability. Steel-reinforced concrete is actually a double composite, consisting of a particulate matrix reinforced with steel "fibers."

Glass-fiber-reinforced resins were the first of the modern "fibrous composites" and were developed shortly after World War II in an attempt to produce lightweight materials with high strength and high stiffness. Glass fibers, about 10 microns in diameter, were bonded in a variety of polymers, generally epoxy resins. Limitations were generally related to stiffness, since the glass fibers had a modulus of only 10 to 13 $\times$ 10^6 psi (70,000 to 90,000 MPa).

Further efforts sought to improve the strength and modulus through the development of improved fibers. Boron–tungsten fibers (boron deposited on a tungsten core) have an elastic modulus of 55 $\times$ 10^6 psi (380,000 MPa) with tensile strengths in excess of 400 ksi (2750 MPa), and they can be incorporated into a cast metal matrix if desired. Silicon carbide filaments (SiC on tungsten) have an even higher modulus of elasticity.

Graphite and Kevlar have become some of the most popular reinforcing fibers. Graphite can be either the PAN type, produced by the thermal pyrolysis of synthetic organic fibers, such as viscose rayon or polyacrylonitrile, or the pitch type, made from petroleum pitch. They have low density and a range of high tensile strengths and elastic moduli. The negative coefficient of thermal expansion of graphite can be used to offset the positive values of most matrix materials. Kevlar is an organic aramid fiber with 450 ksi (3100 MPa) tensile strength,

[1]Trade name by DuPont.

TABLE 8.4 **Properties and Characteristics
of Some Common Reinforcing Fibers**

Fiber Material	Specific Strength[a] (10^6 in.)	Specific Stiffness[b] (10^6 in.)	Density (lb/in.^{3})	Melting Temperature[c] (°F)
"E" glass	5.6	114	.092	<3140
High-strength graphite	7.4	742	.054	6690
High-modulus graphite	5.0	1430	.054	6690
Boron	4.7	647	.085	3690
Kevlar	10.1	347	.052	—
Ceramic fiber (Mullite)	1.1	200	.110	5430
Al_2O_3 whiskers	21.0	434	.142	3600
SiC whiskers	26.2	608	.114	4890

[a]Strength divided by density.
[b]Elastic modulus divided by density.
[c]Or maximum temp of use.

a 19×10^6 psi (131,000 MPa) elastic modulus, a density approximately one half that of aluminum, and a negative thermal expansion coefficient. Its additional features of being flame-retardant and transparent to radio signals makes it attractive for a number of military and aerospace applications. Table 8.4 lists some of the significant engineering properties of several of the common reinforcing fibers.

The reinforcing fibers can be arranged in a variety of orientations. Fiberglas, for example, contains short, randomly oriented fibers. Unidirectional fibers can be used to produce highly directional properties, the fiber directions being tailored to the direction of loading. Woven fabrics or tapes can be produced and then layered at various orientations to produce a product. Complex three-dimensional shapes can be woven from fibers and later injected with a matrix material.

The properties of fiber-reinforced composites depend strongly on several characteristics: (1) the properties of the fiber material; (2) the volume fraction of fibers; (3) the aspect ratio of the fibers (i.e., the length-to-diameter ratio); (4) the orientation of the fibers; (5) the degree of bonding between the fiber and the matrix; and (6) the properties of the matrix. The matrix materials are usually tough and ductile, so as to transmit the loads to the fibers and to prevent cracks in the fibers from propagating through the composite. They should also be evaluated for strength, since their strength does contribute to the overall properties of the composite. Finally, the matrix should be selected with consideration for the temperature of operation. Polymeric matrix materials can be used for temperatures below 600°F (315°C). Above this temperature, metal matrices should be considered.

Advanced Composites and Metal Matrix Materials. Advanced composites are materials developed for applications in which exceptionally good combinations of strength, stiffness, and light weight are required. Fiber content generally exceeds 50%, and the modulus of elasticity is typically greater than 16×10^6 (psi). Superior creep and fatigue resistance, low thermal expansion, low friction and wear, vibration-damping characteristics, and environmental stability are all attractive features of these materials.

There are four basic types of advanced composites in which the matrix material is matched to the fiber and application:

1. The *organic or resin matrix* composites frequently use high-strength, high-modulus fibers of boron, graphite, or Kevlar. Properties can be put in desired locations or orientations at about half the weight of aluminum (or one sixth that of steel). Thermal expansions can be designed to be low, or even negative. Table 8.5 compares the properties of some of the common resin-matrix composites to several of the lightweight or low-thermal-expansion metals.

2. *Metal matrix* composites are used where operating temperature is high or extreme strength is desired (the matrix material is also a structural member). The ductile matrix material can be aluminum, titanium, or nickel, and the reinforcing fibers may be of graphite, boron, alumina, or silicon carbide. Fine whiskers of sapphire, silicon carbide, and silicon nitride have also been used to reinforce metallic matrices. Compared to metals, these materials offer higher stiffness and strength (especially at elevated temperatures) and a lower coefficient of thermal expansion. Compared to the organic matrix materials, they offer higher heat resistance, as well as improved electrical and thermal conductivity. Graphite-reinforced aluminum can be designed to have near-zero thermal expansion in the fiber direction. Aluminum-oxide-reinforced aluminum has been

TABLE 8.5 Comparison of the Properties of Fiber Composites (in the Fiber Direction) with Those of Lightweight or Low-Thermal-Expansion Structural Metals

Material	Specific Strength[a] (10^6 in.)	Specific Stiffness[b] (10^6 in.)	Density (lb/in^3)	Thermal Expansion Coefficient [in./(in.-°F)]	Thermal Conductivity [Btu/(hr-ft-°F)]	Raw Material Cost[c] ($/lb)
Graphite-epoxy: high strength (unidirection)	5.4	400	0.056	−0.3	3	50
Graphite-epoxy: high modulus (unidirectional)	2.1	700	0.063	−0.5	75	300
Glass-epoxy (woven cloth)	0.7	45	0.065	6	0.1	10
Kevlar-epoxy (woven cloth)	1	80	0.5	1	0.5	15
Boron-epoxy	3.3	457	0.07	2.2	1.1	300
Aluminum	0.7	100	0.10	13	100	2
Beryllium	1.1	700	0.07	7.5	120	400
Titanium	0.8	100	0.16	5	4	30
Invar[d]	0.2	70	0.29	1	6	22

[a]Strength divided by density.
[b]Elastic modulus divided by density.
[c]As of 1980.
[d]A low-expansion metal containing 36% Ni and 64% Fe.

used in automotive connecting rods to provide stiffness and fatigue resistance with lighter weight. Aluminum reinforced with silicon carbide whiskers has been fabricated into aircraft wing panels, providing a 20% to 40% weight savings. In the future, fiber-reinforced superalloys may well become a preferred material for applications such as turbine blades.

3. *Carbon–carbon* composites (graphite fibers in a carbon matrix) offer the possibility of a heat-resistant material that could operate at temperatures up to 6000°F (3300°C), with a strength 20 times that of conventional graphite and a density that is 30% lighter (1.38 g/cc). Not only does this material withstand high temperatures, it actually gets stronger when heated. For temperatures over 1000°F (540°C), however, the composite requires some form of coating to protect it from oxidizing. Various coatings are used for different temperature ranges. Current applications include the nose cone and leading edge of the space shuttle, racing-car disk brakes (become stronger when hotter), aerospace turbines and jet engine components, rocket nozzles, and surgical implants.

4. In a *ceramic matrix*, the fibers add directional strength and reduce brittleness, while the matrix provides high temperature resistance. A silicon nitride material reinforced with silicon carbide whiskers has 40% more resistance to internal cracking and 25% more resistance to complete fracture.

Hybrid Composites.   *Hybrid composites* involve combinations of filaments set in a common thermoset matrix, the combination of fibers being selected to balance strength and stiffness, to provide dimensional stability, to reduce cost, to reduce weight, or to improve fatigue and fracture resistance. Types of hybrid composites include (1) interply (alternating layers of fibers); (2) intraply (mixed strands in the same layer); (3) interply-intraply (a mixture of 1 and 2); (4) selected placement (using the more costly material only where needed); and (5) interply knitting (stitching the plies together with another fiber).

Design and Fabrication.   The design of composite materials involves such considerations as the selection of components; the determination of the size, shape, distribution, and orientation of the components; and the selection of an appropriate fabrication method. Many of the fabrication methods have been specifically developed for use with composite materials. For example, fibrous composites can be manufactured into useful shapes through simple compression molding, filament winding, pultrusion (in which bundles of coated fibers are drawn through a heated die), cloth laminations, and autoclave curing (in which pressure and elevated temperature are simultaneously applied). Fiber-reinforced plastics have been injection-molded in a process that competes with zinc die castings. Sheets of fiber-reinforced plastic (sheet-molding compounds) can be press-formed to provide lightweight and corrosion-resistant products that can effectively compete with sheet metal. Chapter 20 will present a complete discussion of the fabrication methods that have been developed for composite materials. With the wide variety of materials, geometries, and processes, it is possible for a designer to tailor a composite material specifically to a given application.

Limitations.   A graphite–epoxy composite I-beam has a weight less than one fifth that of steel, one third that of titanium, and one half that of aluminum. Its ultimate tensile strength is equal to or greater than that of the other three materials. In addition, it has an almost infinite fatigue life. The greatest limitations on the use of this and other composites are their relative brittleness and high cost. Graphite fibers, costing between $400 and $500 per pound in the late 1960s, are now available for about $10 per pound, but even that cost is considerably higher than that of many alternative materials.

The manufacture of composites tends to be quite labor-intensive, and there is a lack of trained designers, established design guidelines and data, information about fabrication costs, and methods of quality control and inspection. It is often difficult to predict the interfacial bond strength, the strength of the composite, its response to impacts, and probable modes of failure. Defects can involve delaminations, voids, missing layers, contamination, fiber breakage, and hard-to-detect improperly cured resin. There is often some concern about heat resistance, and many of the composites with polymeric matrices are sensitive to humidity and ultraviolet radiation and tend to cure forever (causing continually changing properties). In addition, most composites have a limited ability to be repaired if damaged, and routine maintenance procedures have not been established. Assembly operations generally require the use of adhesives.

On the positive side, the ability to use a material with a strength and stiffness greater than those of steel and only one fifth the weight may well justify some engineering compromises. In many applications, products can be designed to significantly reduce the number of parts, the number of fasteners, the assembly time, and the cost.

Areas of Application.   Many composite materials are stronger than steel, lighter than aluminum, and stiffer than titanium and offer low thermal conductivity, good heat resistance, good fatigue life, low corrosion rates, and adequate wear resistance. For these reasons, they have become well established in a number of major areas. Aerospace applications are extremely common and may well approach 65% of the weight of an airplane in turn-of-the-century subsonic designs. Figure 8–12 shows a schematic of the U.S. Marine Corps AV-8B Harrier II (built by McDonnell Douglas Corp.). This plane, which can take off on a 1500-foot runway and make vertical landings, is 26% composite material (by weight), primarily graphite-epoxy. Sporting equipment, such as golf club shafts, baseball bats, fishing rods, tennis rackets, bicycle frames, and skis, is now available in a variety of fibrous composites. Figure 8-13 shows several of these applications. Potential automotive uses, in addition to body panels, include drive shafts, springs, and bumpers. Weight savings compared to the cost of existing parts is generally 20% to 25%.

**AV–8B
Composite applications**

Aluminum
Titanium
Composites
Other

Horizontal stabilizer
(full span)
Rudder
Outrigger faring
Wing span
(full span)
Over wing faring
Engine access doors
Flap
Flap slot door
Aileron
Seals
Nose cone
Lid fence and strakes
Sine wave
spars and ribs
Forward fuselage

FIGURE 8-12   Schematic diagram of the U.S. Marine Corps' AV-8B Harrier II fighter plane (built by McDonnell Douglas Corp.) depicting the extensive use of composite materials. Composites account for 26% of the plane's weight and are primarily graphite fiber-reinforced epoxy. (Reprinted with permission of *Metalworking News,* copyright 1987, Fairchild Publications, a Capital Cities/ABC Company.)

FIGURE 8-13   Composite materials are often used in sporting applications to provide light weight, high stiffness, strength, and attractive styling. (Courtesy Fiberite Corporation).

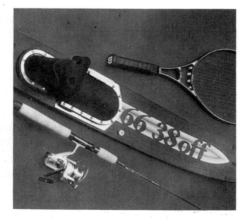

In one recent year, North American truck manufacturers used over 75 million pounds (34 million kilograms) of fiber-reinforced composites for such parts as cab shells and bodies, oil pans, fan shrouds, instrument panels, and engine covers. One prediction estimated that the demand for composite materials will increase from the current 4 billion pounds per year to over 16 billion pounds per year by the year 2000.

Review Questions

1. What are some of the material families that would be classified under the general term of *nonmetallic engineering materials?*
2. How would you describe the current trend in the use of nonmetallic materials?
3. What is the difference between a saturated and an unsaturated molecule?
4. Describe and differentiate the two means of forming polymers, namely, addition and condensation.
5. Describe and differentiate thermoplastic and thermosetting materials.
6. Describe the mechanism by which thermoplastic materials deform.
7. Describe some of the structure modifications that can alter the strength of the thermoplastic polymers.
8. Why are thermosetting polymers characteristically brittle?
9. What are some of the attractive engineering properties of materials that are called *plastics?*
10. What are some of the limiting properties of plastics, and in what general area do they fall?
11. What are some of the primary reasons that additive agents are added to plastics?
12. What are the purposes of a filler material?
13. What is the primary engineering benefit of an oriented plastic?
14. What are some of the properties and characteristics of the "true engineering plastics"?
15. What is the unique mechanical property of elastomeric materials, and what structural feature is responsible for it?
16. How can cross-linking be used to control the engineering properties of elastomers?
17. What are some of the attractive properties of the natural rubber compounds?
18. What attractive features of elastomers make them useful for some tooling applications?
19. What are some of the outstanding physical properties of ceramic materials?
20. In what ways are the structures of crystalline ceramics similar to and different from those of metals?
21. What key property is characteristic of the refractory ceramics?
22. What are cermets, and what properties or combination of properties do they offer?
23. Why do most ceramic materials fail to have their theroretically high tensile strength?
24. Describe the ways in which ceramic materials fail or fracture.
25. What are some of the specific materials that can be classified as structural ceramics?
26. What are some of the attractive and limiting properties of sialon (one of the structural ceramics)?
27. What are some of the attractive possibilities that could accompany the development of a ceramic engine block?
28. What features of ceramic materials make them attractive candidates for cutting tools?
29. What is a composite material?
30. What are the basic structural features of a composite material that influence and determine its properties?
31. What are the three primary geometries of composite materials?
32. What are some of the unique properties or characteristics that can be imparted by the formation of a laminar composite?
33. Describe the various component phases and their contribution to a cemented carbide cutting tool.
34. Which of the three primary composite geometries is most likely to have isotropic properties (i.e., to be uniform in all directions)?
35. What is the primary role of the matrix in a fiber-reinforced composite? Of the fibers?
36. What are some of the possible fiber orientations in a fiber-reinforced composite material?
37. What properties or characteristics are generally sought when one is developing an advanced composite material?
38. In what ways are metal matrix composites superior to straight engineering metals? To the organic matrix composites?
39. What types of composite materials could be fabricated to operate at extremely elevated ranges of temperature?
40. What are some of the considerations and decisions involved in the design of composite materials?
41. What are some of the major limitations on the extensive use of composite materials in engineering applications?

Material Selection

The objective of any practical work dealing with the manufacture of products is to produce components that will adequately perform their designated tasks. Meeting this objective implies the manufacture of components from *selected engineering materials* with the *required geometrical shape and precision* and *companion material structures* that are optimized for the service environment that the components must withstand. The ideal design is one that will just meet all requirements. Anything better tends to waste money or material. Anything worse, and we have failed to manufacture an adequate product.

During recent years, the selection of engineering materials has become extremely important, and the process now requires constant reevaluation. New materials are continually being developed, others may no longer be available, and prices are subject to change. Concerns regarding environmental pollution, recycling, and worker health and safety have imposed new constraints. The desire for weight reduction, energy savings, or improved corrosion resistance might be sufficient to evoke a change in engineering material. Pressures from domestic and foreign competition, increased demand for quality and serviceability, and negative customer feedback can all prompt a reevaluation. Finally, the proliferation of product liability actions, many the result of improper material use, has further emphasized the need for a constant reevaluation of the engineering materials in a product.

The interdependence between materials and their processing must also be recognized. Improper processing of a well-chosen material is likely to result in a defective product. Improvements in processes often dictate a reevaluation of the materials being processed. Likewise, a change in material may well require a change in the manufacturing process. Thus, if design and manufacturing engineers are to achieve satisfactory results at a reasonable cost, they must exercise considerable care in selecting and using engineering materials.

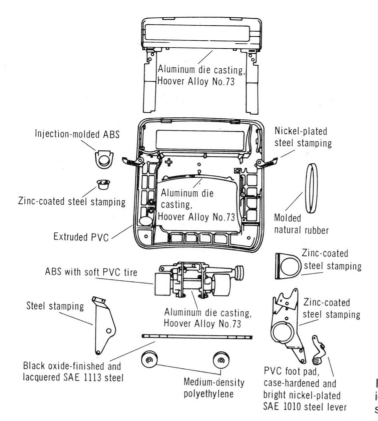

Aluminum die casting, Hoover Alloy No.73

Injection-molded ABS

Zinc-coated steel stamping

Extruded PVC

ABS with soft PVC tire

Steel stamping

Black oxide-finished and lacquered SAE 1113 steel

Aluminum die casting, Hoover Alloy No.73

Aluminum die casting, Hoover Alloy No.73

Medium-density polyethylene

Nickel-plated steel stamping

Molded natural rubber

Zinc-coated steel stamping

Zinc-coated steel stamping

PVC foot pad, case-hardened and bright nickel-plated SAE 1010 steel lever

FIGURE 9-1   Materials used in various parts of a vacuum cleaner assembly. (Courtesy Metals Progress.)

Most modern products are relatively complex and often use a variety of materials. The vacuum cleaner assembly shown in Figure 9–1 is typical: nine different materials were used in the parts that are illustrated. Table 9.1 lists the material changes that were recommended in one particular revision of the entire appliance. The materials for 12 components were changed completely, and that for a thirteenth was modified. Eleven different reasons were given for the changes.

Figure 9–2 shows how the material content of a typical American car has changed from 1977 to 1987. Notice the increased use of lighter-weight materials and high-strength steels, as well as the inclusion of newly developed plastics and composite materials.

If we recognize that the term *engineering materials* includes virtually all metals and alloys, ceramics, glasses, plastics, elastomers, electrical semiconductors, concrete, composite materials, and others, it is not surprising that a single individual will have great difficulty in keeping abreast of, and making the necessary decisions concerning, the materials in even a simple manufactured product. Nevertheless, it is the responsibility of the design engineer, working

TABLE 9.1   Examples of Material Selection and Substitution in a Modern Vacuum Cleaner

Part	Former Material	New Material	Benefits
Bottom plate	Assembly of steel stampings	One-piece aluminum die casting	More convenient servicing
Wheels (carrier and caster)	Molded phenolic	Molded medium-density polyethylene	Reduced noise
Wheel mounting	Screw-machine parts	Preassembled with a cold-headed steel shaft	Simplified replacement, more economical
Agitator brush	Horsehair bristles in a die-cast zinc or aluminum brush back	Nylon bristles stapled to a polyethylene brush back	Nylon bristles last seven times longer and are now cheaper than horsehair
Switch toggle	Bakelite molding	Molded ABS	Breakage eliminated
Handle tube	AISI 1010 lock-seam tubing	Electric seam-welded tubing	Less expensive, better dimensional control
Handle bail	Steel stamping	Die-cast aluminum	Better appearance, allowed lower profile for cleaning under furniture
Motor hood	Molded cellulose acetate (replaced Bakelite)	Molded ABS	Reasonable cost, equal impact strength, much improved heat and moisture resistance: eliminated warpage problems
Extension-tube spring latch	Nickel-plated spring steel, extruded PVC cover	Molded acetal resin	More economical
Crevice tool	Wrapped fiber paper	Molded polyethylene	More flexibility
Rug nozzle	Molded ABS	High-impact styrene	Reduced costs
Hose	PVC-coated wire with a single-ply PVC extruded covering	PVC-coated wire with a two-ply PVC extruded covering separated by a nylon reinforcement	More durability, lower cost
Bellows, cleaning-tool nozzles, cord insulation, bumper strips	Rubber	PVC	More economical, better aging and color, less marking

Source: Metal Progress, by permission.

in conjunction with materials specialists, to select the materials to be used in converting the designs into reality.

The Design Process.   The first step in the manufacture of any product is design, which usually takes place in several distinct stages: (1) conceptual; (2) functional; and (3) production. During the *conceptual-design* stage, the designer is primarily concerned with the functions that the product is to fulfill. Several concepts are often considered, and a decision is made: either the concept is not practical, or it is sound and should be developed further. Here, the only concern about materials is that ones exist that can provide the desired properties.

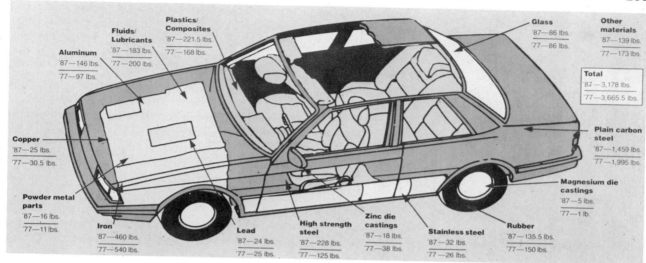

Aluminum
'87—146 lbs.
'77—97 lbs.

Fluids/
Lubricants
'87—183 lbs.
'77—200 lbs.

Plastics/
Composites
'87—221.5 lbs.
'77—168 lbs.

Glass
'87—86 lbs.
'77—86 lbs.

Other
materials
'87—139 lbs.
'77—173 lbs.

Total
'87—3,178 lbs.
'77—3,665.5 lbs.

Copper
'87—25 lbs.
'77—30.5 lbs.

Plain carbon
steel
'87—1,459 lbs.
'77—1,995 lbs.

Magnesium die
castings
'87—5 lbs.
'77—1 lb.

Powder metal
parts
'87—16 lbs.
'77—11 lbs.

Iron
'87—460 lbs.
'77—540 lbs.

Lead
'87—24 lbs.
'77—25 lbs.

High strength
steel
'87—228 lbs.
'77—125 lbs.

Zinc die
castings
'87—18 lbs.
'77—38 lbs.

Stainless steel
'87—32 lbs.
'77—26 lbs.

Rubber
'87—135.5 lbs.
'77—150 lbs.

FIGURE 9-2 The material make-up of a typical U.S. auto and how it has changed from 1977 to 1987. (Reprinted with the permission of Metalworking News, copyright 1987, Fairchild Publications, a Capital Cities/ABC Company.)

If no such materials are available, consideration is given to whether there is a reasonable prospect that new ones could be developed within the limitations of cost and time.

At the *functional-* or *engineering-design* stage, a practical, workable design is developed. Fairly complete drawings are made, and materials are selected and fully specified for the various components. Consideration is given to appearance, cost, reliability, producibility, and serviceability, in addition to the various functional factors. Often, a prototype or working model is constructed to permit a full evaluation of the product. Although it is possible that this testing will show that some changes will have to be made in materials before the product can be advanced to the production-design stage, this possibility should not be taken as an excuse for not doing a thorough job of material selection. In addition, there is much merit in the practice of building all prototypes with the same materials that will be used in production and, where possible, with the same manufacturing techniques. It is of little value to have a perfectly functioning prototype that cannot be manufactured economically in the desired volume, or one that is substantially different from what the production units will be like. A complete job of material analysis, selection, and specification should be done at the development stage of design, rather than leaving these critical aspects to the production-design stage, where decisions may be made by others less knowledgeable about all of the functional aspects of the product.

As the product moves to the *production-design* stage, the materials concerns are now directed to whether the specified materials are compatible with the manufacturing processes and equipment. Can the materials be processed economically, and are they available in the necessary quantities and quality?

As manufacturing progresses, situations often arise that require modifications in the materials and processes that are being used. In most cases, however, any changes made after the tooling and machinery are placed in production tend to be quite costly. Good material selection and thorough product evaluation can do much to eliminate this unnecessary expense.

The availability of new materials and processes often presents possibilities for cost reduction and improved performance. These new materials, however, should be evaluated very carefully to ensure that all of their characteristics are well established. Remember that it is indeed rare for as much to be known about the properties and reliability of a new material as about an established one. Numerous product failures and product liability cases have resulted from new materials being substituted before their long-term properties were fully known.

A Procedure for Material Selection.   The selection of an appropriate material and then converting it into a useful product with the desired shape and properties is a rather complex process. Nearly every engineered item goes through the sequence of activities known as design → material selection → fabrication → evaluation → and possible redesign or modification. Numerous engineering decisions have to be made.

Several methods have been developed for approaching a design and selection problem. The *case-history method* assumes that something has worked successfully before, and that a similar component may be made with the same engineering material and method of manufacture. Although the approach is useful, and many manufacturers evaluate their competitors' products for just this purpose, minor variations in service requirements may well dictate different materials or manufacturing operations. Moreover, this approach tends to preclude the use of new technology, new materials, and other manufacturing advances that have occurred since the formulation of the previous solution. It is equally unwise, however, to totally ignore the potential benefits of past experience.

Other design and selection activities involve the *modification of an existing product,* generally in an effort to reduce cost or to improve quality. Efforts here can begin with an evaluation of the current product and its present method of manufacture. The most frequent pitfall is to lose sight of one of the original design requirements and to recommend a change that proves to be inadequate in the total performance of the product. Several examples are cited in the section on material substitution.

The safest and most thorough approach is to view the task as the *development of an entirely new product.* Here, the full sequence of design, material selection, and the development of a manufacturing sequence is followed.

The *first step* in any material selection problem is to *define the needs of the product.* Without any prior biases about material or method of manufacture, the engineer should form a clear picture of all the characteristics necessary for this part to adequately perform its intended function. These requirements fall into three major areas: (1) shape or geometry considerations; (2) property requirements; and (3) manufacturing concerns.

The area of *shape considerations* primarily influences the selection of the method of manufacture. Although such concerns are often obvious, they may be more complex than one might first imagine. Typical questions include:

1. What is the relative size of the component?
2. How complex is its shape? Are there axes or planes of symmetry? Uniform cross sections? Do you want to consider making it in more than one piece?
3. How many dimensions must be specified?
4. How precise must these dimensions be? Are all precise? How many are restrictive and which ones?
5. How does the component interact with other components?
6. What surface characteristics are needed? Which surfaces need to be smooth? Hard? Which need to be finished? Which do not?
7. How much can a dimension change by wear or corrosion and the part still perform adequately?
8. Could a minor change in the shape significantly improve the suitability of the part (increase strength, reliability, fracture resistance, etc.)?

The defining of *property requirements* is often a far more complex task. Some aspects that should be considered follow.

Mechanical properties:

1. What are the needs with regard to static strength?
2. If overloaded, is the component more likely to fail by deformation or fracture? Do you have a preference?
3. Can you envision impact loadings? If so, of what type and magnitude?
4. Can you envision cyclic loadings? If so, of what type, magnitude, and frequency?
5. Is wear resistance needed? Where? How much? How deep?
6. Over what temperature range must these properties be present?
7. How much can the material deflect, stretch, or compress and still function properly?

Physical properties:

1. Are there any electrical property requirements?
2. Are any magnetic properties desired?
3. Are thermal properties significant? Thermal conductivity? Change of dimensions with change of temperature?
4. Are there any optical requirements?
5. Is weight a significant factor?
6. What about appearance?

Another important area to evaluate is the *service environment* of the product throughout its lifetime:

1. What is the lowest, highest, and normal operating temperature for the component, and how fast is temperature likely to change?

2. Are all of the desired properties required over this range of temperatures?
3. What is the most severe environment anticipated as far as corrosion or deterioration of material properties is concerned?
4. What is the desired service lifetime of the product?
5. What is the anticipated maintenance for this component?
6. What is the potential liability if the product should fail?
7. Should the product be manufactured with recycling in mind?

A final area of concern is to determine various *factors that would influence the method of manufacture:*

1. Have standard components and sizes been specified wherever possible?
2. Has the design addressed the requirements that will facilitate ease of manufacture? Machinability? Weldability? Formability? Hardenability? Castability?
3. How many components are to be made? At what rate?
4. What are the maximum and minimum section thicknesses?
5. What is the desired level of quality compared to that of similar products on the market?
6. What are the anticipated quality control and inspection requirements?
7. Are there any assembly concerns, relationships to mating parts, and so on that should be noted?

Although there is a tendency to want to jump to "the answer," time spent determining requirements will be well rewarded, and it is important that *all* factors be listed and *all* service conditions and uses be considered. Many failures and product liability claims have resulted from simple engineering oversights or the designer's not anticipating reasonable use of a product or conditions outside the specific function for which the product was designed. Figure 9–3 shows such an example. The designer of a large electrical transformer substituted a plastic laminate for maple for the members of the structure that support the heavy copper leads. In service, these members would have been in pure compression, and the laminate would be excellent for this type of loading. The designer failed to recognize, however, that *during shipping* some of the members might be subjected to transverse bending loads, and the material was weak in this regard. As a consequence, several supporting members broke during shipping, and a domino effect produced extensive failure and damage. Construction equipment parts failed frequently during the building of the Alaskan pipeline because they had not been designed to function in subzero temperatures. For other components, the most severe corrosion environment may well be experienced during shipping or storage, as opposed to operational use.

Assuming that the list of required properties has been properly compiled, it is often helpful to indicate the relative importance of the various needs. Some requirements may be "absolutes," whereas others may be relative. Absolutes are those about which there can be no compromise. For example, if ductility is a "must," then gray cast iron would have to be ruled out. If the product must have good electrical conductivity, then plastics would not be appropriate. On

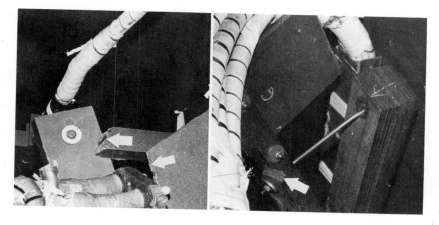

FIGURE 9-3 Failures in the lead-support structure of a large electrical transformer. Material selection failed to consider the stresses experienced during shipping.

the basis of absolutes, many materials can be quickly eliminated from consideration.

In the actual material selection, it is apparent that, in many cases, no one material will emerge as the obvious choice. Often several materials will, to differing degrees, meet the specific requirements. Compromise and opinion enter into the decision making, and it is important that the final choice not overlook a major requirement. The listing and ranking of required properties will go a long way to ensure that the person making the selection has considered all of the necessary factors. Should no material meet the requirements and the compromises appear to be too severe, it may be necessary to redesign the product.

Additional Factors to Consider.   When attempting to evaluate candidate materials, the engineer is often forced to use "handbook-type" data obtained through the use of standardized materials-characterization tests. It is important to note the conditions of these tests in comparison with those of the proposed application. Significant variations in factors such as temperature, rates of loading, and surface finish can lead to major changes in a material's behavior. In addition, one should also keep in mind that the handbook values often represent an average or mean, and that actual material properties vary to either side of that value. Where vital information is missing or the data may not be applicable to the proposed use, one is advised to consult with the various materials producers or qualified materials engineers.

It is also appropriate to introduce *cost* as a selection factor. Cost is not a service requirement and has not been considered up to this point because we have adopted a philosophy that a material must first meet the necessary property requirements to be considered as a viable candidate. Cost, however, is an important part of the selection process, and both material cost and the cost of fabricating the material into the desired component should be considered. Often, the final decision involves a compromise between cost, producibility, and performance or quality. Should a more expensive material be used if it offers improved performance? How much additional expense would be justified to gain ease of fabrication?

Materials availability is another consideration. The selected material may not be available in the size, quantity, or shape desired, or it may not be available in any form at all. Here, one should be prepared to recommend alternatives, provided they are viable materials for the specific use.

Still other factors to be considered when one is making material selections include:

1. What is the possibility of misuse by the user? If the product is to be used by the general public, one should anticipate the worst. Screwdrivers are routinely used as chisels and pry bars (which involve different forms of loading from the intended torsional twist). Scissors may be used as wire cutters. Other products are similarly misused.
2. Have there been any prior failures of this product or similar ones? Vital failure-analysis information often fails to get back into the hands of designers who can benefit from it.
3. Has the selected material (or class of materials) established a favorable or an unfavorable performance record?
4. Can you take advantage of materials standardization, whereby multiple components are manufactured from the same material or use the same manufacturing process? Although you should not sacrifice function, reliability, or appearance for standardization, neither should you overlook the potential for savings and simplification that it has to offer.

Consideration of the Manufacturing Process.   On the basis of the previous considerations, a tentative selection of several candidate materials should be made. If one of these is clearly outstanding, it may be selected at this point. It is more common, however, to carry through several possibilities and to further refine the selection by consideration of the possible fabrication processes and the suitability of each of the "prescreened" materials to each of the processes. Familiarity with the various manufacturing alternatives is a must, along with a knowledge of the associated limitations, economics, product quality, surface finish, precision, and so on. All processes are not compatible with all materials. Steel, for example, cannot be fabricated by die casting. In addition, it is not uncommon for a certain process to be implied in the details of a design. A good engineer, however, should consider all possible methods of manufacture and, if necessary, recommend a change in design to accommodate a more attractive means of production.

The Ultimate Objective.   The real objective of this activity is to arrive at a combination of material and process (or sequence of processes) that is the "best" solution for the manufacture of the desired product. Although the various considerations in arriving at this optimized *manufacturing system* have been presented sequentially, one should be aware of a number of interrelating limitations. Production requirements, such as formability, weldability or machinability, tend to restrict the number of candidate materials. Specific materials, in turn, limit the number of fabrication possibilities. Specific fabrication methods tend to impart characteristic properties to the material, all of which may not be beneficial. Processes designed to alter certain properties (such as heat treatment) may adversely affect others. Economics, environment, energy, efficiency, recycling, inspection, serviceability, and other concerns influence decisions.

Engineers working in this area must be capable of exercising sound judgment. They must understand the product, the materials, the manufacturing processes, and all of the various interrelations. The development of new materials, technological advances in processing methods, increased restrictions in environment and energy, and the demand for products that continue to push the limits of capability further challenge these individuals and ensure their value to employers.

Materials Substitution.   As new technology is developed or market pressures arise, it is not uncommon for a new material to be substituted in an existing design or manufacturing system. Often, the result is improved quality or reduced cost, but it is possible to overlook certain requirements and to cause more harm than good.

Consider the drive to produce a lighter-weight, more-fuel-efficient automobile. The development of high-strength low-alloy (HSLA) steel sheets offered the ability to match the strength of many body panels with thinner-gauge material. As some of the original processing problems were overcome, the substitution appeared to be a natural one. However, the engineer must consider the total picture and be aware of possible compromises. Although strength was increased, corrosion resistance and elastic stiffness were essentially unaltered. Thinner sheets would corrode in a shorter time, and undesirable vibrations would be more of a problem. Design modifications would probably be necessary to accommodate the new material.

Aluminum castings were proposed to replaced cast iron for transmission housings. Corrosion resistance would be enhanced, and weight savings would be substantial. However, even if the mechanical properties were adequate, consideration would have to be directed to aluminum's tendency to transmit noise and vibration. Cast iron is a damping material and eliminates these undesirable features. The use of aluminum would possibly require the addition of some form of sound isolation material. Similarly, cast-iron engine blocks damp out vibrations. When aluminum was used in this application, a companion redesign of the motor mounts was required to accommodate the change.

Although numerous other examples could be cited, it is obvious that the responsible engineer should first consider *all* of the design requirements before authorizing a material substitution. Approaching a design or material modification as thoroughly as a new problem may well prevent costly errors.

The Effect of Product Liability on Material Selection.   Product liability actions, court awards, and rising insurance costs have made it imperative that designers and companies employ the very best procedures in selecting materials.

Although many people would agree that the situation has grown to absurd proportions, there have been many unfortunate instances in which sound procedures were not used in selecting materials and methods of manufacture. No designer or engineer can afford to be negligent.

The five most common causes of product liability losses have been (1) failure to know and use the latest and best information about the materials utilized; (2) failure to foresee and to take into account the *reasonable uses* of the product (where possible, the designer is further advised to foresee and account for potential misuse of the product, as there have been many product liability cases in recent years in which the claimant, injured during misuse of a product, has sued the manufacturer and won); (3) the use of materials about which there were insufficient or uncertain data, particularly about long-term properties; (4) inadequate, and unverified, quality-control procedures; and (5) material selection made by people who were completely unqualified.

An examination of these faults will lead one to conclude that there is no good reason for them to exist. Consideration of each of them, however, is good practice in ensuring the production of a quality product and can greatly reduce the number and magnitude of product liability claims.

Aids to Material Selection.   From the previous discussion in this chapter, it is apparent that those who select materials should have a broad, basic understanding of the nature and properties of materials and their processing characteristics. Providing this background is a primary purpose of this book. However, the number of engineering materials is so great, as is the mass of information that is available and useful in specific situations, that a single book of this type and size cannot be expected to furnish all of the information required. Anyone who does much work in material selection needs to have ready access to multiple sources of data.

One very useful reference is the "Materials Selector" issue of *Materials Engineering*. This annual issue of a monthly magazine provides tabulated data about a number of engineering materials, as well as typical uses, process capabilities, and the compatibility between materials and processes.

Another "must" is the materials-related volumes of the *Metals Handbook*, published by the American Society for Metals (ASM). Volumes 1 through 3 of the ninth edition (1978–1980) deal solely with the properties and selection of engineering metals. The one-volume *Metals Handbook Desk Edition* (1984) provides similar information in a less voluminous, more concise format.

ASM also offers a one-volume *ASM Metals Reference Book* (second edition, 1983), which provides extensive data about metals and metalworking in tabular or graphic form. The new "Materials and Processes Databook" issue of the society's monthly magazine reflects an expanded coverage of a wide range of engineering materials, not just metals. Other privately published handbooks provide voluminous data about the engineering plastics and composite materials.

In addition to these sources, persons selecting materials should have available several of the handbooks published by the various technical societies and trade associations. They may be materials-related (such as the Aluminum Association's *Aluminum Standards and Data* and the Copper Development Association's *Standards Handbook: Copper, Brass, and Bronze*); process-related (such as the *Steel Castings Handbook* by the Steel Founder's Society of America and the *Heat Treater's Guide* by the American Society for Metals); or profession-related (such as the *SAE Handbook* of the Society of Automotive Engineers, the *ASME Handbook* of the American Society for Mechanical Engineers, and the *Tool and Manufacturing Engineers Handbook* of the Society for Manufacturing Engineers). These may be supplemented by a variety of supplier-published references. Although the latter are excellent, low-cost references, the user should recognize that they are clearly focused on the products of the publisher and may not provide an objective viewpoint.

Another area of necessary information is the comparative costs of materials. Since these tend to fluctuate, it may be necessary to use a publication like the *American Metal Market* newspaper. Costs associated with the various processing operations are more difficult to obtain and can vary greatly from one company to another. These costs may be available from within the firm or may have to be estimated from outside sources. One valuable reference would be the *American Machinist Cost Estimator* or the associated computer software.

As one attempts to compare materials and use data from various references, a comparative rating chart, like the one shown in Figure 9–4, may be very useful. The various desired properties are weighted according to their significance, and candidate materials are evaluated on a scale of 1 to 5 relative to each property. One computes a "rating number" by multiplying the property rating by its weighted significance and summing the results. Potential materials can then be compared in a uniform, unbiased manner. Moreover, the method can

RATING CHART FOR SELECTING MATERIALS

Material	Go-No-Go** Screening			Relative Rating Number (†Rating Number x *Weighting Factor)								Material Rating Number
	Corrosion	Weldability	Brazability	Strength (5)*	Toughness (5)	Stiffness (5)	Stability (5)	Fatigue (4)	As-Welded Strength (4)	Thermal Stresses (3)	Cost (1)	$\frac{\Sigma \text{ Rel Rating No.}}{\Sigma \text{ Rating Factors}}$

*Weighting Factor (Range = 1 Lowest to 5 Most Important)
†Range = 1 Poorest to 5 Best
**Code = S = Satisfactory
        U = Unsatisfactory

FIGURE 9-4 Rating chart that may be used for comparing materials for a specific application.

be computerized and run with a large data base of materials information. An additional benefit is that, while making this chart, the designer puts all of the requirements on a single sheet of paper and is then less likely to overlook a major need later in the selection process.

Review Questions

1. In a manufacturing environment, why should the selection and use of engineering materials be a matter of constant reevaluation?
2. What are some of the classes of materials that can be included in the term *engineering materials?*
3. What are the three primary phases of product design, and how does the consideration of materials differ in each?
4. What is the benefit of requiring prototype products to be manufactured from the same materials that will be used in production and by the same manufacturing techniques?
5. What cautions should be exercised when one is considering a new material or process for the manufacture of a product?
6. What are some of the possible pitfalls in the case-history approach to material selection?
7. What is the most frequent problem when one is seeking to improve an existing product?
8. What should be the first step in any material selection problem?
9. Why is it important to resist jumping to ''the answer'' and first perform a thorough evaluation of product needs and requirements?
10. What is the difference between an ''absolute'' and a ''relative'' requirement?
11. What are some possible pitfalls when one is using handbook data to assist in material selection?
12. Why should a consideration of the various fabrication-process possibilities be included in a material selection?
13. Why is it important to consider all of the design and service requirements when considering a material substitution in an existing product?
14. What are some of the most common causes of product liability losses?
15. Why might a single-page comparative rating chart be a useful tool in material selection?
16. Three materials, X, Y, and Z, are available for a certain use.

Any material selected must have good weldability. Tensile strength, stiffness, stability, and fatigue strength are required, fatigue strength being considered the most important and stiffness the least important of these factors. The three materials are rated as follows in these factors:

	X	Y	Z
Weldability	Excellent	Poor	Good
Tensile strength	Good	Excellent	Fair
Stiffness	Good	Good	Good
Stability	Good	Excellent	Good
Fatigue strength	Fair	Good	Excellent

Use a rating chart such as the one in Figure 9–4 to determine which material should be selected.

Special Problem   This problem is proposed as an exercise for correlating the various aspects of material selection.

Select a common home, office, or garage item, such as a pair of scissors, an appliance part, an automobile component, a cooking utensil, a hand tool, or a projector case.

For the selected part, consider and answer the following questions:

a. What is the part supposed to do? What functions must it perform? Under what conditions must it operate?
b. What properties must the material have in order to perform adequately?
c. What material or materials might you suggest for this part? Why?
d. How would you suggest that this part be fabricated? Why?
e. Would the part require heat treatment? What type?
f. Would the part require surface treatment? Why? What type?

PART II

Measurement and Quality Assurance

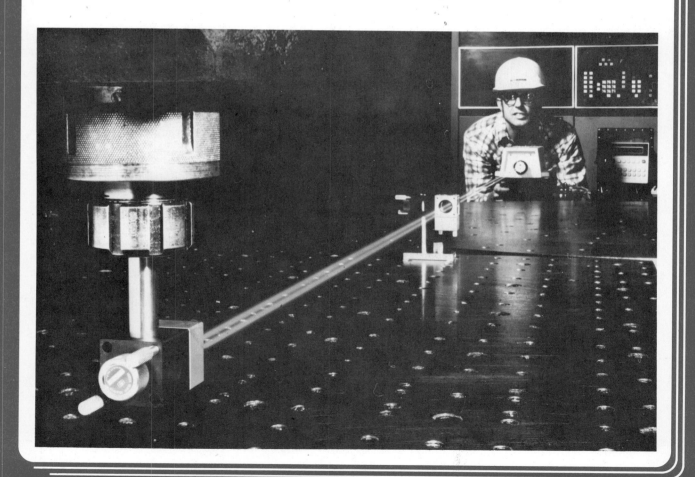

10

Measurement and Inspection

Measurement, that is the act of measuring or being measured, is the fundamental activity of inspection. The intent of inspection is to ensure that what is being manufactured will conform to the specifications of the product. Most products are manufactured to standard sizes and shapes. For example, the base of a 60-watt light bulb has been standardized so that when one bulb burns out, the next will also fit the socket in the lamp. The socket in the lamp has also been designed and made to accept the standard bulb size. Christmas tree light bulbs are made to a different standard size. Standardization is a necessity for interchangeable parts and is also important for economic reasons. A 69-watt light bulb cannot be purchased because that is not a standard wattage. Light bulbs are manufactured only in standard wattages so that they can be produced in large volumes by high-speed, automated equipment. This results in a low unit cost.

Large-scale manufacturing based on the principles of standardization of sizes and interchangeable parts became common practice in the early 1900s. Size control must be built into the machine tools and the workholding devices through the precision manufacture of these machines and their tooling. The output of the machines must then be carefully checked (1) to determine the capability of specific machines and (2) for the control and maintenance of the quality of the product. A designer who specifies the dimensions and tolerances of a part quite often does so to enhance the function of the product, but the designer is also determining the machines and processes needed to make the part. Quite often, it is necessary for the design engineer to alter the design or the specifications to make the product easier or less costly to manufacture, assemble, or inspect, or all of these. Designers should always be prepared to do this provided they are not sacrificing functionality, product reliability, or performance.

Attributes Versus Variables.   The examination of the product either during or after manufacture, either manually or automatically, falls in the province of *inspection*. Basically, inspection of items or products can be done two ways:

1. By attributes, with the use of gages to determine if the product is good or bad resulting in a yes or no, go or no-go decision.
2. By variables, with the use of calibrated instruments to determine the actual dimensions of the product for comparison with the desired size.

In an automobile, a speedometer and an oil pressure gage are variable types of measuring instruments, and an oil pressure light is an attributes type of gage. As is typical of an attributes type of gage, the driver does not know *what* the pressure actually is if the light goes on, only that it is not good.

Measurement is the generally accepted industrial term for inspection by variables. *Gaging* (or *gauging*) is the term for determining whether the dimension or characteristic is larger or smaller than the established standard or range of acceptability. Variable types of inspection generally take more time and are more expensive than attributes inspection, but they give more information because the magnitude of the characteristic is known in some standard unit of measurement.

Standards of Measurement

The four fundamental measures on which all others depend are *length, time, mass,* and *temperature*. Three of these basic measures are defined in terms of material constants, as shown in Table 10.1, along with the original definitions. These four measures, along with the *ampere* and the *candela*, provide the basis for all other units of measurement, as shown in Figure 10-1. Most mechanical measurements involve combinations of units of mass, length, and time. Thus, the newton, a unit of force, is derived from Newton's second law of motion ($f = ma$) and is defined as the force that gives an acceleration of 1 meter per second to a mass of 1 kilogram. A table of metric–English conversions is provided at the end of the chapter (see Table 10.5).

Linear Measurements.   When man first sought a unit of length, he adopted parts of the human body, mainly his hands, arms, or feet. Such tools were not very satisfactory because they were not universally standard in size. Satisfactory measurement and gaging must be based on a reliable, and preferably universal, standard or standards. These have not always existed. For example, although the musket parts made in Eli Whitney's shop were interchangeable, they were not interchangeable with parts made by another contemporary gun maker *from the same drawings,* because the two gunsmiths *had different foot rulers.* Today, the entire industrialized world has adopted the *international meter* as the standard of linear measurement. The inch, used by both the United States and Great Britain, has been officially defined as 2.54 centimeters. Thus, the United States standard inch is 41,929.399 wavelengths of the orange-red light from krypton-86.

TABLE 10.1   The International System of Units: Founded on Seven Base Quantities on Which All Others Depend

Quantity	Name of Base	Symbol	Definition/Comment
Length	Meter (or metre)	Original:	1/10,000,000th of quadrant of earth's meridian passing through Barcelona and Dunkirk.
		Present:	1,650,763.73 wavelengths in vacuum of transition between energy levels $2p_{10}$ and $5d_5$ of krypton-86 atoms, excited at triple point of nitrogen ($-210°C$).
Mass	Kilogram	Original:	Mass of one cubic decimeter (1,000 cubic centimeters) of water at its maximum density (4°C).
		Present:	Mass of Prototype Kilogram No. 1 kept at International Bureau of Weights and Measures at Sèvres, France.
Time	Second	Original:	1/86,400th of mean solar day.
		Present:	9,192,631,770 cycles of frequency associated with transition between two hyperfine levels of isotope cesium-133.
Electric Current	Ampere	Present:	The rate of motion of charge in a circuit is called the *current*. The unit of current is the *ampere*. One ampere exists when the charge flows at a rate of one coulomb per second.
Thermodynamic Temperature	Degree Celsius (Kelvin)	Present:	1/273.16th of the thermodynamic temperature of the triple point of water (0.01°C):
Amount of Substance	Mole	Present:	A mole is an artificially chosen number $N_o = 6.02 \times 10^{23}$ that measures the number of molecules.
Luminous Intensity	Candle	Present:	One lumen per square foot is a footcandle or $I = F/4$.

Although *officially* the United States is committed to conversion to the metric (SI) system of measurement, which uses millimeters for virtually all linear measurements in manufacturing, the English system of feet and inches is still being used by many manufacturing plants, and its use will probably continue for some time.

Length Standards in Industry.   *Gage blocks* provide industry with linear standards of high accuracy that are necessary for everyday use in manufacturing plants. These are small, rectangular, square, or round in cross section and are blocks made from steel or carbide with two very flat and parallel surfaces that are certain specified distances apart (see Figure 10-2). These gage blocks were first conceived by Carl E. Johansson in Sweden just before 1900. By 1911, he was able to produce sets of such blocks on a very limited scale, and they came into limited but significant use during World War I. Shortly after the war, Henry Ford recognized the importance of having such gage blocks generally available. He arranged for Johansson to come to the United States, and through facilities provided by the Ford Motor Company, methods were devised for the large-scale production of gage block sets. Today, gage block sets of excellent quality are produced by a number of companies in this country and abroad.

Gage blocks, made of alloy steel, are hardened to $65R_c$ and carefully heat-treated (*seasoned*) to relieve internal stresses and to minimize subsequent

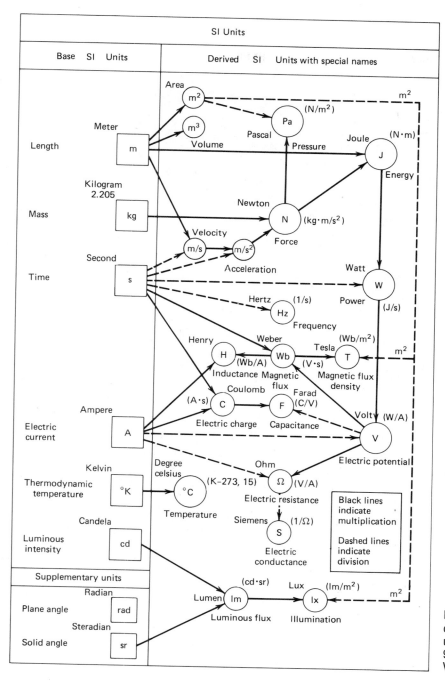

FIGURE 10-1 Relationship of secondary physical quantities to basic SI units. (From NBS Technical Note 938, National Bureau of Standards, Washington, D.C.)

FIGURE 10-2 Standard set of rectangular gage blocks with ±.000050″ accuracy, 3 individual blocks shown below.

Lapped and mirror polished to a very low micro-surface finish.

dimensional change. Carbide gage blocks provide extra wear resistance. The measuring surfaces of each block are surface-ground to approximately the required dimension and are then lapped and mirror-polished to bring the block to the final dimension and to produce a very flat and smooth surface. The surface finish is 0.4 millionths of an inch (0.01 microns).

Gage blocks are commonly made to conform to National Bureau of Standards #731/222 131, having Grades 1, 2, and 3 with tolerances as follows:

Grade	Inches	Millimeters	Recalibration Period
0.5 (laboratory)	±0.000,001	±0.000 03	Annually
1 (laboratory)	±0.000,002	±0.000 05	Annually
2 (precision)	+0.000,004 −0.000,002	+0.000 10 −0.000 05	Monthly to semiannually
3 (working)	+0.000,008 −0.000,004	+0.000 20 −0.000 10	Monthly to quarterly

ANSI/ASME B89.1.OM-1984

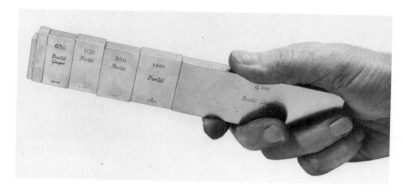

FIGURE 10-3 Seven gage blocks wrung together to build up a desired dimension (Courtesy DoALL Company.)

Blocks up to 1 inch in length have absolute accuracies as stated, whereas the tolerances are per inch of length for blocks larger than 1 inch. Some companies supply blocks in AA quality (which corresponds to Grade 1), and in A + quality (which corresponds to Grade 2).

Grade 0.5 (grand-master) blocks are used as a basic reference standard in calibration laboratories. Grade 1 (laboratory-grade) blocks are used for checking and calibrating other grades of gage blocks. Grade 2 (precision-grade) blocks are used for checking Grade 3 blocks and master gages. Grade 3 (B or working grade) blocks are used to check routine measuring devices, such as micrometers, or in actual gaging operations.

The dimensions of individual blocks are established by light beam interferometry, with which it is possible to calibrate these blocks routinely with an uncertainty as low as 1 part per million.

Gage blocks usually come in sets containing various numbers of blocks of various sizes, such as those shown in Figure 10-2. By wringing the blocks together in various combinations, as shown in Figure 10-3, any desired dimension can be obtained.

For example, here is the breakdown of blocks that one would have in an 81 block set of English #2 gage blocks.

9 Blocks   0.1001 in. through 0.1009 in. in steps of 0.0001 in.
49 Blocks   0.101 in. through 0.149 in in steps of 0.001 in.
19 Blocks   0.050 in. through 0.950 in. in steps of 0.050 in.
4 Blocks   1.000 in. through 4.000 in. in steps of 1.000 in.

Gage blocks are wrung together by sliding one past another using hand pressure. They will adhere to one another with considerable force and must not be left in contact for extended periods of time.

Gage blocks are available in different shapes (squares, angles, rounds, and pins), so that standards of high accuracy can be obtained to fill almost any need. In addition, various auxiliary clamping, scribing, and base block attachments are available that make it possible to form very accurate gaging devices, such as that shown in Figure 10-4.

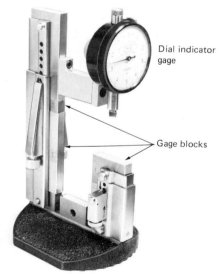

Dial indicator gage

Gage blocks

FIGURE 10-4 Wrung-together gage blocks in a special holder and used with a dial gage to form an accurate comparator. (Courtesy DoALL Company.)

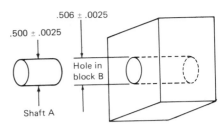

.506 ± .0025

.500 ± .0025

Hole in block B

Shaft A

FIGURE 10-5 When mating parts are designed, each shaft must be smaller than each hole for a clearance fit.

Standard Measuring Temperature. Because all the commonly used metals are affected dimensionally by temperature, a standard measuring temperature of 68°F (20°C) has been adopted for precision-measuring work. All gage blocks, gages, and other precision-measuring instruments are calibrated at this temperature. Consequently, when measurements are to be made to accuracies greater than 0.0001 in. (0.0025 mm), the work should be done in a room in which the temperature is controlled at standard. Although it is true that to some extent both the workpiece and the measuring or gaging device *may* be affected to about the same extent by temperature variations, one should not rely on this. Measurements to even 0.0001 in. (0.0025 mm) should not be relied on if the temperature is very far from 68°F (20°C).

Allowance and Tolerance. Two factors, allowance and tolerance, must be specified if one is to obtain the desired fit between mating parts. *Allowance* is the intentional, desired difference between the dimensions of two mating parts. It is the difference between the dimension of the largest interior-fitting part (shaft) and that of the smallest exterior-fitting part (hole). Figure 10-5 shows Pin A designed to fit into the hole in Block B. This difference thus determines the condition of *tightest* fit between mating parts. Allowance may be specified so that either *clearance* or *interference* exists between the mating parts. With clearance fits, the largest shaft is smaller than the smallest hole, whereas with interference fits, the hole is smaller than the shaft.

Tolerance is an undesirable, but permissible, deviation from a desired dimension in recognition that no part can be made *exactly* to a specified dimension, except by chance, and that such exactness is neither necessary not economical. Consequently, it is necessary to permit the actual dimension to deviate from the desired theoretical dimension (called the *nominal*) and to control the degree of deviation so that satisfactory functioning of the mating parts will still be ensured.

Inspection, by means of measurement techniques, provides feedback information on the actual size of the parts with reference to the size specified by the designer on the part drawing. The conventional drawing shows the nominal diameter of Part A plus a positive tolerance, which is to be less than the smallest hole in Part B. In other words, the minimum diameter of the hole should always be greater than the maximum diameter of the shaft if all parts are to fit or mate together.

The processes that make the shaft are different from those that make the hole, but both the hole and the shaft are subject to deviations in size because of variability in the processes and the materials. Thus, while the designer wishes ideally that all the shafts would be exactly 0.500 and all the holes 0.506, the reality of processing is that there will be deviations in size around these nominal or ideal sizes.

Most manufacturing processes result in products whose measurements of the geometrical features and sizes are distributed normally (see Figure 10-6). That is, most of the measurements are clustered around the average dimension, $\overline{X}$. $\overline{X}$ will be equal to the nominal dimension only if the process is perfectly centered. Parts will be distributed on either side of the average. For normal distributions,

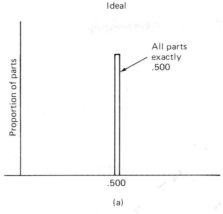

Ideal

All parts
exactly
.500

.500

(a)

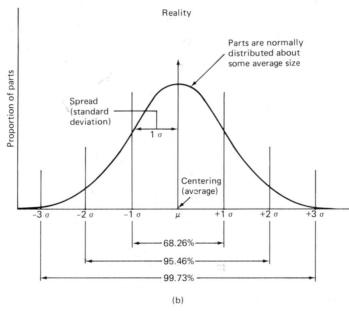

Reality

Parts are normally
distributed about
some average size

Spread
(standard
deviation)

$1\,\sigma$

Centering
(average)

$-3\,\sigma$ $-2\,\sigma$ $-1\,\sigma$ μ $+1\,\sigma$ $+2\,\sigma$ $+3\,\sigma$

68.26%

95.46%

99.73%

(b)

FIGURE 10-6 In the ideal situation,
the process would make all parts exactly the same size. In the real world
of manufacturing, parts have variability in size. The size distribution is
often normal.

99.73% of the measurements will fall within plus or minus 3 standard deviations.
($\pm 3\sigma$), 95.46% in Figure 10-6 will be within $\pm 2\sigma$ and 68.26% within
$\pm 1\sigma$,

where

$$\sigma = \sqrt{\frac{\sum_{i=1}^{n} (X_i - \overline{X})^2}{n}} \qquad (10\text{-}1)$$

and

$$\overline{X} = \frac{\sum_{i=1}^{n} X_i}{n} \quad \text{for } n \text{ items} \qquad (10\text{-}2)$$

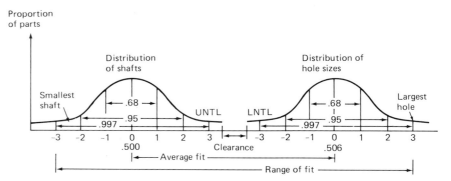

FIGURE 10-7 The manner in which the distributions of the two mating parts interact determines the fit.

In general, the designer applies nominal values to the mating parts according to the desired fit between the parts. Tolerances are added to those nominal values in recognition of the fact that all processes have some natural amount of variability.

Assume the distributions for both the hole and the shaft are normally distributed. As shown in Figure 10-7, the average fit of two mating parts is equal to the difference between the mean of the shaft distribution and the mean of the hole distribution. The "range of fit" will be the difference between the minimum diameter shaft and the maximum diameter hole. The minimum "clearance" would be the difference between the smallest hole and the largest shaft. The tighter the distributions (i.e., the more precise the process), the better the fit between the parts.

If tool wear is considered, the diameter of the shaft will tend to get larger as the tool wears. On the other hand, the diameter of the hole in Part B decreases in size with tool wear. If no corrective action is taken, the distribution means will shift toward each other, and the fit will become increasingly tight (the clearance will be decreased). If the hole and the shaft distributions overlap by 2 standard deviations, as shown in Figure 10-8, 6 parts out of every 10,000 could not be assembled. As the shaft and the hole distributions move closer

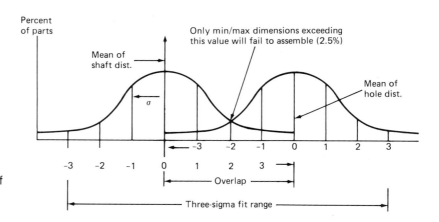

FIGURE 10-8 Shifting the means of the distributions toward each other results in some interference fits.

together, more interference between parts will occur, and the fit will become tighter, eventually becoming an interference fit.

The designer must specify the tolerances and allowances for the mating parts according to the function. Suppose the mating parts are the cap and the body of an ink pen. The cap must fit snugly but must be easily removed by hand. A snug fit would be too tight for a dead bolt in a door lock. A snug fit is also too tight for a high-speed bearing for rotational parts, but not tight enough for permanently mounting a wheel on an axle. The next section introduces the manner in which tolerances and allowances are specified, but design engineers are expected to have a deeper understanding of this topic.

Specifying Tolerances and Allowances.   Tolerance can be specified in three ways: bilateral, unilateral, and limits. *Bilateral* tolerance is specified as a plus or minus deviation from the nominal size, such as 2.000 ± 0.002 in. More modern practice uses the *unilateral* system, where the deviation is in one direction from the basic size, such as

$$2.000 \text{ in.} \begin{matrix} +0.004 \text{ in.} \\ -0.000 \text{ in.} \end{matrix} \quad \text{or} \quad 50.8 \text{ mm} \begin{matrix} +0.1 \text{ mm} \\ -0.0 \text{ mm} \end{matrix} \text{ in metric.}$$

In the first case, that of bilateral tolerance, the dimension of the part could vary between 1.998 and 2.002 in., a total tolerance of 0.004 in. For the example of unilateral tolerance, the dimension could vary between 2.000 and 2.004 in., again a tolerance of 0.004 in. Obviously, in order to obtain the same maximum and minimum dimensions with the two systems, different basic sizes must be used. The maximum and minimum dimensions that result from the application of the designated tolerance are called *limit dimensions,* or *limits.*

There can be no rigid rules for the amount of clearance that should be provided between mating parts; the decision must be made by the designer, who considers how the parts are to function. The American National Standards Institute, Inc. (ANSI), has established eight classes of fits that serve as a useful guide in specifying the allowance and tolerance for typical applications, and that permit the amount of allowance and tolerance to be determined merely by specifying a particular class of fit. These classes are as follows:

Class 1: Loose fit—large allowance. Accuracy is not essential.

Class 2: Free fit—liberal allowance. For running fits where speeds are over 600 RPM and pressures are 600 psi (4.1 MPa) or over.

Class 3: Medium fit—medium allowance. For running fits under 600 RPM and pressures less than 600 psi (4.1 MPa) and for sliding fits.

Class 4: Snug fit—zero allowance. No movement under load is intended, and no shaking is wanted. This is the tightest fit that can be assembled by hand.

Class 5: Wringing fit—zero to negative allowance. Assemblies are selective and not interchangeable.

Class 6: Tight fit—slight negative allowance. An interference fit for parts that must not come apart in service and are not to be disassembled or are to be disassembled only seldom. Light pressure is required for assembly. Not to be used to withstand other than very light loads.

Class 7: Medium force fit—an interference fit requiring considerable pressure to assemble; ordinarily assembled by heating the external member or cooling the internal member to provide expansion or shrinkage. Used for fastening wheels, crank disks, and the like to shafting. The tightest fit that should be used on cast iron external members.

Class 8: Heavy force and shrink fits—considerable negative allowance. Used for permanent shrink fits on steel members.

The allowances and tolerances that are associated with the ANSI classes of fits are determined according to the theoretical relationship shown in Table 10.2. The actual resulting dimensional values for a wide range of basic sizes can be found in tabulations in drafting and machine-design books.

In the ANSI system, the hole size is always considered basic, because the majority of holes are produced through the use of standard-size drills and reamers. The internal member, the shaft, can be made to any one dimension as readily as to another. The allowance and tolerances are applied to the basic hole size to determine the limit dimensions of the mating parts. For example, for a basic hole size of 2 inches and a Class 3 fit, the dimensions would be:

Allowance	0.0014 in.
Tolerance	0.0010 in.
Hole	
Maximum	2.0010 in.
Minimum	2.0000 in.
Shaft	
Maximum	1.9986 in.
Minimum	1.9976 in.

It should be noted that for both clearance and interference fits, the permissible tolerances tend to result in a looser fit.

The *ISO System of Limits and Fits,* shown in Figure 10-9, is widely used in a number of leading metric countries. This system is considerably more complex than the ANSI system just discussed. In this system, each part has a *basic size.* Each limit of size of a part, high and low, is defined by its *deviation* from the

TABLE 10.2   ANSI Recommended Allowances and Tolerances

Class of Fit	Allowance	Average Interference	Hole Tolerance	Shaft Tolerance
1	$0.0025 \sqrt[3]{d^2}$		$+0.0025 \sqrt[3]{d}$	$-0.0025 \sqrt[3]{d}$
2	$0.0014 \sqrt[3]{d^2}$		$+0.0013 \sqrt[3]{d}$	$-0.0013 \sqrt[3]{d}$
3	$0.0009 \sqrt[3]{d^2}$		$+0.0008 \sqrt[3]{d}$	$-0.0008 \sqrt[3]{d}$
4	0		$+0.0006 \sqrt[3]{d}$	$-0.0004 \sqrt[3]{d}$
5		0	$+0.0006 \sqrt[3]{d}$	$+0.0004 \sqrt[3]{d}$
6		$0.00025d$	$+0.0006 \sqrt[3]{d}$	$+0.0006 \sqrt[3]{d}$
7		$0.0005d$	$+0.0006 \sqrt[3]{d}$	$+0.0006 \sqrt[3]{d}$
8		$0.001d$	$+0.0006 \sqrt[3]{d}$	$+0.0006 \sqrt[3]{d}$

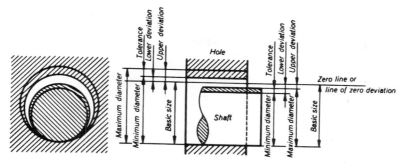

FIGURE 10-9 Basic size, deviation, and tolerance in the ISO system. (By permission from Recommendations R286-1962, System of Limits and Fits, copyright 1962, American National Standards Institute, New York.)

basic size, the magnitude and sign being obtained by subtracting the basic size from the limit in question. The difference between the two limits of size of a part is called *tolerance*, an absolute amount without sign.

There are three classes of fits: (1) *clearance fits;* (2) *transition fits* (the assembly may have either clearance or interference); and (3) *interference fits.*

Either a *shaft-basis system* or a *hole-basis system* may be used, as illustrated in Figure 10-10. For any given basic size, a range of tolerances and deviations may be specified with respect to the line of zero deviation, called the *zero line.* The tolerance is a function of the basic size and is designated by a number symbol, called the *grade* (e.g., the *tolerance grade*). The *position* of the tolerance with respect to the zero line, also a function of the basic size, is indicated by a letter symbol (or two letters), a capital letter for holes, and a lowercase letter for shafts, as illustrated in Figure 10-11.[1] Thus, the specification for a hole and a shaft having a basic size of 45 mm might be 45 H8/g7.

Eighteen standard grades of tolerances are provided, called IT 01, IT 0, and IT 1 thru IT 16, providing numerical values for each nominal diameter, in arbitrary steps up to 500 mm (for example, 0-3, 3-6, 6-10, . . . , 400–500 mm). The value of the tolerance unit, *i*, for Grades 5–16 would be:

$$i = 0.45 \sqrt[3]{d} + 0.001 D$$

where *i* is in micrometers and *D* in millimeters.

Standard shaft and hole deviations are provided by similar sets of formulas. However, for practical application, both tolerances and deviations are provided

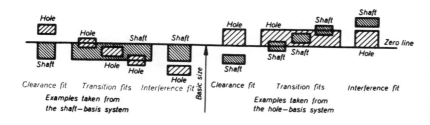

FIGURE 10-10 "Shaft-basis" and "hole-basis" system for specifying fits in the ISO system. (By permission from ISO Recommendation R286-1962, Systems of Limits and Fits, copyright 1962, American National Standards Institute, New York.)

[1]It will be recognized that the "position" in the ISO system essentially provides the "allowance" of the ANSI system.

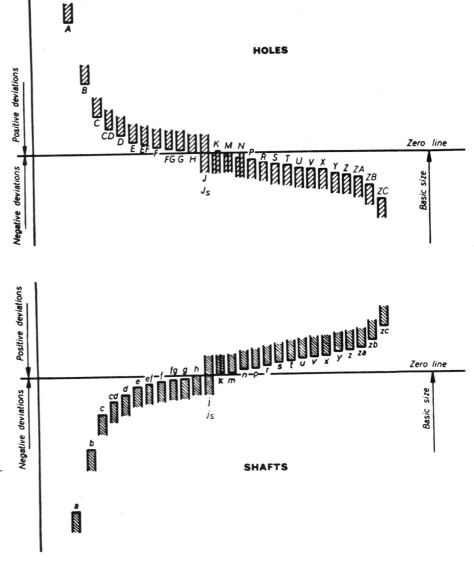

FIGURE 10-11   Positions of the various tolerance zones for a given diameter in the ISO system. (By permission from ISO Recommendation R286-1962, Systems of Limits and Fits, copyright 1962, American National Standards Institute, New York.)

in three sets of rather complex tables. Additional tables give the values for basic sizes above 500 mm, and for "Commonly Used Shafts and Holes" in two categories: "General Purpose" and "Fine Mechanisms and Horology." Horology is the art of making timepieces.

Accuracy Versus Precision in Processes.   It is vitally important that the difference between accuracy and precision be understood. *Accuracy* refers to

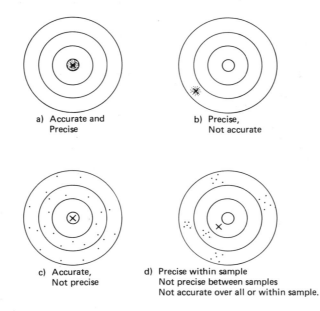

a) Accurate and
   Precise

b) Precise,
   Not accurate

c) Accurate,
   Not precise

d) Precise within sample
   Not precise between samples
   Not accurate over all or within sample.

FIGURE 10-12   Accuracy versus precision. Dots in targets represent location of shots. Cross (X) represents the location of the average position of all shots.

the ability to hit what is aimed at (the bull's-eye), and *precision* refers to the repeatability of the process. Suppose that five sets of five shots are fired at a target from the same gun. Figure 10-12 shows some of the possible outcomes. In Figure 10-12a, the inspection of the target shows that this is a good process—accurate and precise. Figure 10-12b shows precision (repeatability) but poor accuracy. The agreement with a standard is not good. In Figure 10-12c, the process is, "on the average," quite accurate as the X is right in the middle of the bull's-eye, but the process has too much scatter or variability; it does not repeat. Finally, in Figure 10-12d, a failure to repeat accuracy between samples with respect to time is observed; the process is not stable. These four outcomes are typical but not all-inclusive of what may be observed.

The field of metrology, even limited to geometrical or dimensional measurements, is far too great to be covered here. The remainder of this chapter concentrates on basic linear measurements and the attributes and variables devices most commonly found in a company's metrology or quality control facility. At a minimum, such labs would typically contain optical flats; one or two granite measuring tables; an assortment of indicators, calipers, micrometers, and height gages; an optical comparator; a set or two of Grade 1 gage blocks; a coordinate measuring machine; a laser scanning device; a laser interferometer; a toolmaker's microscope; and pieces of equipment specially designed to accommodate the company's products. Table 10.3 provides a summary of inspection methods, listing four basic kinds of devices: air, optical (light and electron), electronic, and mechanical. The variety seems to be endless, but digital electronic readouts connected to any of the measuring devices are becoming the preferred method.

Inspection Methods for Measurement

TABLE 10.3 **Four Basic Kinds of Inspection Method**

Method	Typical Accuracy	Major Applications	Comments
Air	0.5–10 microinches or 2% to 3% of scale range	Gaging holes and shafts using a calibrated difference in air pressure or airflow, with magnifications of 20,000–40,000 to 1; also used for machine control, sorting, and classifying.	High precision and flexibility; can measure out-of-round, taper, concentricity, camber, squareness, parallelism, and clearance between mating parts; noncontact principle good for delicate parts.
Optical light energy	0.5–2 microinches or better with laser interferometry 0.5–1 second of arc in autocollimation optical comparators	Interferometry; checking flatness and size of gage blocks; finding surface flaws; measuring spherical shapes, flatness of surface plates, accuracy of rotary index tables; includes all light microscopes and devices common on plant floor.	Largest variety of measuring equipment; autocollimators are used for making precision angular measurements; lasers are used to make precision in-process measurements; laser scanning.
Electron energy	100 Angstroms (Å)	Precision measurement in scanning electron microscopes of microelectronic circuits and other small precision parts.	Part size restricted by vacuum chamber size; electron beam can be used for processing and part testing of electronic circuits.
Electronic	0.5–10 microinches	Widely used for machine control, on-line inspection, sorting, and classification; OD's, ID's, height, surface, and geometrical relationships, profile tracing for roundness, surface roughness, contours, etc.; most devices are comparators with movement of stylus or spindle producing an electronic signal that is amplified electronically; commonly connected to microprocessors and minicomputers for process adjustment.	Electronic gages come in many forms but usually have a sensory head or detector combined with an amplifier; capable of high magnification with resolution limited by size or geometry of sensory head; readouts commonly have multiple magnification steps; solid-state, digital electronics make these devices small, portable, stable, and extremely flexible, with extremely fast response time.
Mechanical	1–10 microinches	Large variety of external and internal measurements using dial indicators, micrometers, calipers, and the like; commonly used for bench comparators for gage calibration work.	Moderate cost and ease of use make many of these devices the workhorses of the shop floor; highly dependent on workers' skill and often subject to problems of linkages.

The discrete digital readout on a clear LCD display eliminates reading interpretations associated with analog scales and can be directly entered into dedicated microprocessors or computers for permanent record and analysis. The added speed and ease of use for these types of equipment have allowed them to be routinely used on the plant floor instead of in the metrology lab. In summary, the trend toward tighter tolerances (greater precision) and accuracy associated with the need for superior quality and reliability has greatly enhanced the need for improved measurement methods.

Factors in Selecting Inspection Equipment. Many inspection devices with electronic output to communicate directly to microprocessors are being built into the processes themselves and are often computer-aided. In-process inspection generates feedback sensory data from the process or its output to the computer control of the machine, which is the first step in making the processes responsive to changes (Adaptive Control). In addition to in-process inspection, many other quality checks and measurements of parts and assemblies are needed. In general, six factors should be considered when selecting equipment for an inspection job by measurement techniques.

1. *The rule of 10.* The measurement device (or working gage) should be 10 times more precise than the tolerance to be measured. The rule actually applies to all stages in the inspection sequence, as shown in Figure 10-13. The master gage should be precise to 1/10 of that of the inspection device. The reference standard used to check the master gage should be 10 times more precise than the master gage. The application of the rule greatly reduces the probability of rejecting good parts or accepting bad components and performing additional work on them.

2. *Linearity.* This factor refers to the calibration accuracy of the device over its full working range. Is it linear? What is its degree on nonlinearity? Where does it become nonlinear, and what is, therefore, its real linear working region?

3. *Repeat accuracy.* How repeatable is the device in taking the same reading over and over on a given standard?

4. *Stability.* How well does this device retain its calibration over a period of time? Stability is also called *drift*. As devices become more accurate, they often lose stability and become more sensitive to small changes in temperature and humidity.

5. *Magnification.* The amplification of the output portion of the device over the actual input dimension. The more accurate the device, the greater must be its magnification factor, so that the required measurement can be read out (or observed) and compared with the desired standard. Magnification is often confused with resolution, but they are not the same thing.

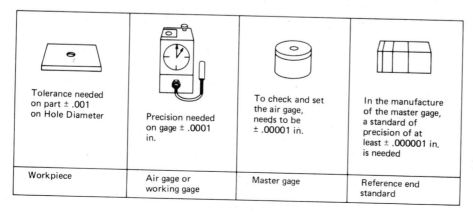

Tolerance needed on part ± .001 on Hole Diameter	Precision needed on gage ± .0001 in.	To check and set the air gage, needs to be ± .00001 in.	In the manufacture of the master gage, a standard of precision of at least ± .000001 in. is needed
Workpiece	Air gage or working gage	Master gage	Reference end standard

FIGURE 10-13 The Rule of 10 states that for reliable measurements each successive step in the inspection sequence should have 10 times the *precision* of the preceding step.

6. *Resolution*. This is sometimes called *sensitivity* and refers to the smallest unit of scale or dimensional input that the device can detect or distinguish. The greater the resolution of the device, the smaller will be the things it can resolve and the greater will be the magnification required to expand these measurements up to the point where they can be observed by the naked eye.

Some other factors of importance in selecting inspection devices include the type of measurement information desired; the range or the span of sizes the device can handle versus the size and geometry of the workpieces; the environment; the cost of the device; and the cost of installing and using the device. The last factor depends on the speed of measurement, the degree to which it can be automated, and the functional life of the device in service.

Measuring Instruments

Because of the great importance of measuring in manufacturing, a great variety of instruments are available that permit measurements to be made routinely, ranging in accuracy from 1/64 to 0.00001 in. and from 0.5 to 0.0003 mm. Machine-mounted measuring devices (probes and lasers) for automatically inspecting the workpiece during manufacturing are beginning to compete with postprocess gaging and inspection, in which the part is inspected, automatically or manually, after it has come off the machine. In-process inspection for automatic size control has been used for some years in grinding to compensate for the relatively rapid wear of the grinding wheel. Today, touch trigger probes, with built-in automatic measuring systems, are being used on machine tools to determine cutting tool offsets and tool wear. These systems are discussed in Chapter 29.

For manually operated analog instruments, the ease of use, precision, and accuracy of measurements can be affected by (1) the least count of the subdivisions on the instrument; (2) line matching; and (3) the parallax in reading the instrument. Elastic deformation of the instrument and workpiece and temperature effects must be considered. Some instruments are more subject to these factors than others. In addition, the skill of the person making the measurements is very important. Digital readout devices in measuring instruments lessen or eliminate the effect of most of these factors, simplify many measuring problems, and lessen the chance of making a math error.

Linear Measuring Instruments.   Linear measuring instruments are of two types: direct-reading and indirect-reading. *Direct-reading instruments* contain a line-graduated scale so that the size of the object being measured can be read directly on this scale. *Indirect-reading instruments* do not contain line graduations and are used to transfer the size of the dimension being measured to a direct-reading scale, thus obtaining the desired size information indirectly.

The simplest, and most common, direct-reading linear measuring instrument is the machinist's rule, shown in Figure 10-14. Metric rules usually have two sets of line graduations on each side: 1/2- and 1-mm divisions; English rules

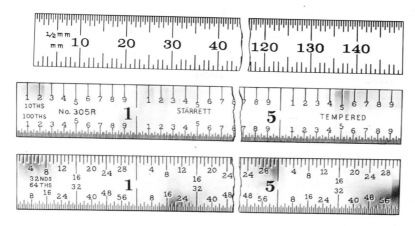

FIGURE 10-14 Machinist's rules. (Top) Metric. (Center and bottom) Inch graduations; 10ths and 100ths on one side, 32nd and 64ths on opposite side. (Courtesy L.S. Starrett Company.)

have four sets: 1/16-, 1/32-, 1/64-, and 1/100-in. divisions. Other combinations can be obtained in each type.

The machinist's rule is an end- or line-matching device; an end and a line, or two lines, must be aligned with the extremities of the object or the distance being measured in order for the desired reading to be obtained. Thus, the accuracy of the resulting reading is a function of the alignment and the magnitude of the smallest scale division. Such scales ordinarily are not used for accuracies greater than 1/64 in. (0.01 in.) or about 1/2 mm.

Several attachments can be added to a machinist's rule to extend its usefulness. Shown in Figure 10-15, the *square head* can be used as a miter or trisquare or to hold the rule in an upright position on a flat surface for making height measurements. It also contains a small bubble-type level so that it can be used by itself as a level. The *bevel protractor* permits the measurement or layout of angles. The *center head* permits the center of cylindrical work to be determined.

The *vernier caliper*, illustrated in Figure 10-16, is an end-measuring instrument, available in various sizes, that can be used to make both outside and inside measurements to theoretical accuracies of 0.001 in. or 0.01 mm. End-measuring instruments are more accurate and somewhat easier to use than line-matching types because their jaws are placed against either end of the object being measured, so that any difficulty in aligning edges or lines is avoided. However, the difficulty remains in obtaining uniform contact pressure, or "feel," between the legs of the instrument and the object being measured.

A major feature of the vernier caliper is the auxiliary scale, shown in Figure 10-17. The caliper shown has a graduated beam with a metric scale on the top, a metric vernier plate, and an English scale on the bottom with an English vernier. The manner in which readings are made is explained in the figure. Figure 10-16 also shows a vernier depth gage for measuring the depth of holes or the length of shoulders on parts and a vernier height gage for making height measurements. Figure 10-18 shows calipers that have a dial indicator or a digital readout that replace the vernier. The latter two calipers are capable of making inside and outside measurements as well as depth measurements, as shown in

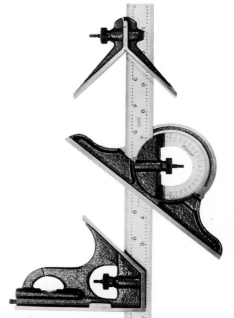

FIGURE 10-15 Combination set. (Courtesy MTI Corporation.)

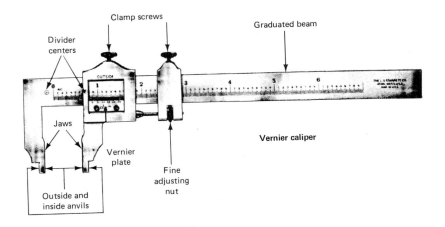

Vernier caliper

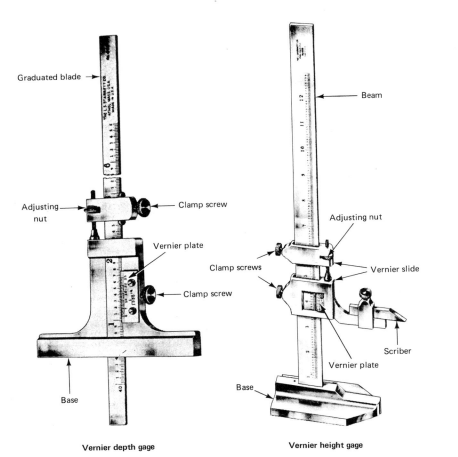

Vernier depth gage

Vernier height gage

FIGURE 10-16 Variations in the vernier caliper design result in other basic gages.

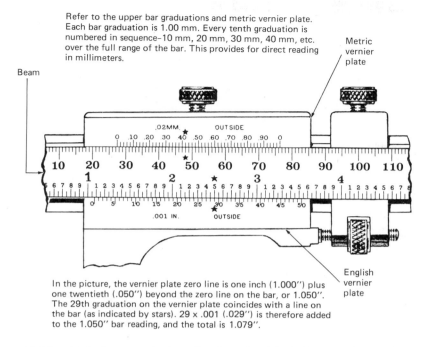

Refer to the upper bar graduations and metric vernier plate. Each bar graduation is 1.00 mm. Every tenth graduation is numbered in sequence–10 mm, 20 mm, 30 mm, 40 mm, etc. over the full range of the bar. This provides for direct reading in millimeters.

Metric vernier plate

Beam

In the picture, the vernier plate zero line is one inch (1.000″) plus one twentieth (.050″) beyond the zero line on the bar, or 1.050″. The 29th graduation on the vernier plate coincides with a line on the bar (as indicated by stars). 29 x .001 (.029″) is therefore added to the 1.050″ bar reading, and the total is 1.079″.

English vernier plate

FIGURE 10-17 Vernier caliper graduated for English and Metric (Direct) reading. The metric reading is 27 + .42 = 27.42 mm.

Figure 10-19, which shows typical applications for this end-measuring device. The digital device can be zeroed at any position, as shown in the left-hand sketch of applications A through D, which greatly speeds the inspection processes and improves reading accuracy (eliminates math errors).

The *micrometer caliper,* more commonly called a *micrometer,* is one of the most widely used measuring devices. Until recently, the type shown in Figure 10-20 was virtually standard. It consists of a fixed anvil and a movable spindle. When the thimble is rotated on the end of the caliper, the spindle is moved away from the anvil by means of an accurate screw thread. On English types, this thread has a lead of 0.025 in., and one revolution of the thimble moves the spindle this distance. The barrel, or sleeve, is calibrated in 0.025-in. divisions, with each 1/10 of an inch being numbered. The circumference at the edge of the thimble is graduated into 25 divisions, each representing 0.001 in.

A major difficulty with this type of micrometer is making the reading of the dimension shown on the instrument. To read the instrument, the division on the thimble that coincides with the longitudinal line on the barrel is added to the largest reading exposed on the barrel.

Micrometers graduated in ten-thousandths of an inch are the same as those graduated in thousandths, except that an additional vernier scale is placed on the sleeve so that a reading of ten-thousandths is obtained and added to the thousandths reading.

The vernier consists of 10 divisions on the sleeve, shown in B, which occupy the same space as 9 divisions on the thimble. Therefore, the difference between the width of 1 of the 10 spaces on the vernier and 1 of the 9 spaces on the

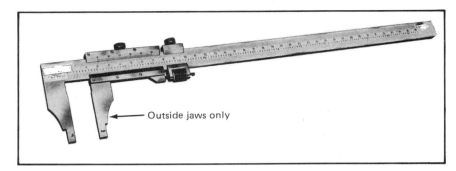

Vernier caliper with inch or metric scales and .001'' accuracy

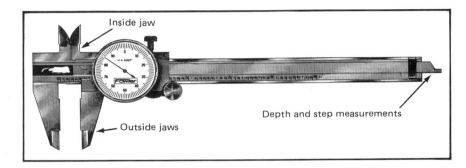

Dial caliper with .001'' accuracy

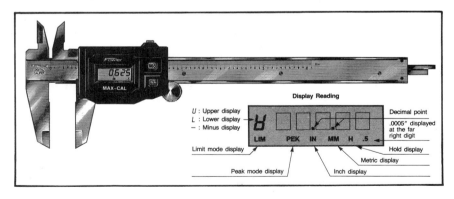

FIGURE 10-18 Three styles of calipers in common use today.

Digital electronic caliper with .001'' (.03 mm) accuracy and .0001'' resolution with inch/metric conversion.

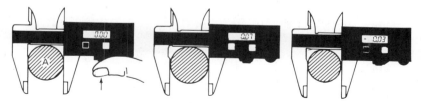

Fig. A. Deviation from reference size. (A = reference size). Use outside jaws.

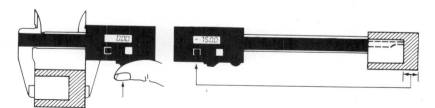

Fig. B. Comparisons, e.g. between plug and hole. Use outside + inside jaws.

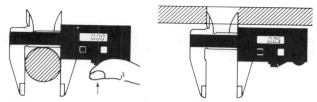

Fig. C. Measurement of wall thickness. Use outside jaws + depth rod.

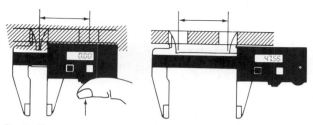

Fig. D. Measurement of center of distance between two identical holes. Use inside jaws.

FIGURE 10-19 Four typical applications of a digital caliper. Reading reset to zero (0.000) in each left-hand figure.

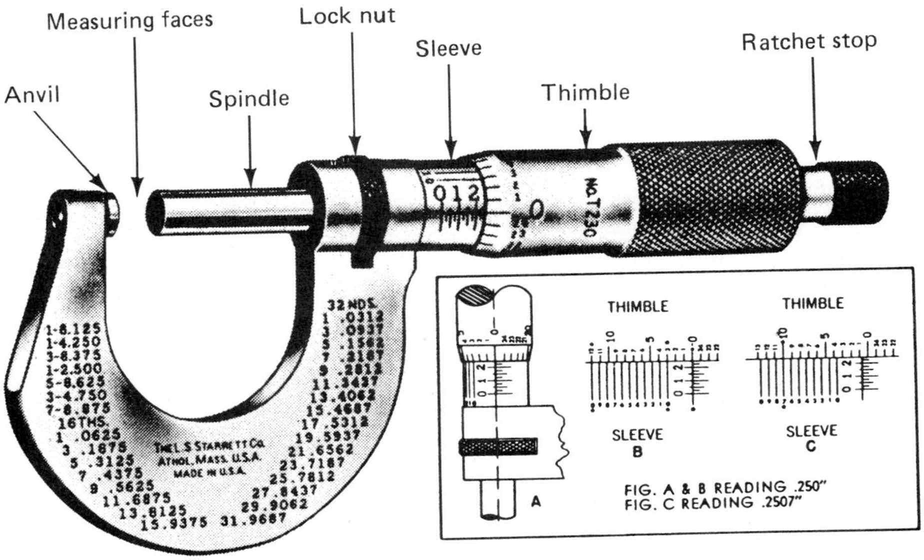

Measuring faces · Lock nut · Sleeve · Thimble · Ratchet stop · Anvil · Spindle

FIG. A & B READING .250″
FIG. C READING .2507″

FIGURE 10-20 Micrometer caliper graduated in ten-thousandths of an inch (Courtesy L.S. Starrett Company.)

thimble is one-tenth of a division on the thimble, or one-tenth of one-thousandth, which is one ten-thousandth. To read a ten-thousandths micrometer, first obtain the thousandths reading, then see which of the lines on the vernier coincides with a line on the thimble. If it is the line marked ''1,'' add one ten-thousandth; if it is the line marked ''2,'' add two ten-thousandths; and so on.

Example: Refer to Insets A and B in Figure 10-20.

The ''2'' line on the sleeve is visible, representing ------------------------------- 0.200 in.
Two additional lines, each representing .025 in. ---------------- 2 × .025″ = 0.050 in.
Line ''0'' on the thimble coincides with the reading line on the
   sleeve, representing --- 0.000 in.
The ''0'' lines on the vernier coincide with lines on the thimble,
   representing --- 0.000 in.
The micrometer reading is ---0.2500 in.

Now you try to read Inset C.

The ''2'' line on the sleeve is visible, representing ------------------------------- 0.200 in.
Two additional lines, each representing 0.025 in. ---------------2 × 0.025″ = 0.050 in.
The reading line on the sleeve lies between the ''0'' and ''1'' on the thimble,
   indicating two-thousandths of an inch are also to be added as read from the vernier.
The ''7'' line on the vernier coincides with a line on the thimble,
   representing 7 × 0.0001″ = 0.0007 in.
The micrometer reading is ---0.2507 in.

However, owing to the lack of pressure control, micrometers can seldom be relied on for accuracy beyond 0.0005 in., and such vernier scales are not used extensively. On metric micrometers, the graduations on the sleeve and thimble are usually 0.5 mm and 0.01 mm, respectively. See the problems at the end of chapter.

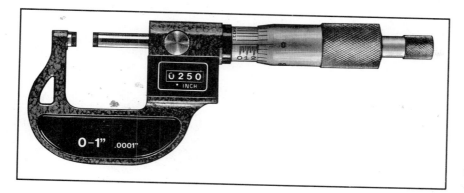

FIGURE 10-21 Digital micrometer for measurements from 0 to 1″, in .0001″ graduations.

Many errors have resulted from the ordinary micrometer being misread, the error being ± 0.025 in. or ± 0.5 mm. Consequently, direct-reading micrometers have been developed. Figure 10-21 shows a digital outside micrometer that reads to 0.001 in. on the digit counter and 0.0001 in. on the vernier on the sleeve. The range of a micrometer is limited to 1 in. Thus, a number of micrometers of various sizes are required to cover a wide range of dimensions. To control the pressure between the anvil, the spindle, and the piece being measured, most micrometers are equipped with a ratchet or a friction device, as shown in Figures 10-20 and 10-21. Calipers that do not have this device may be sprung several thousandths by applying excess torque to the thimble. Micrometer calipers should not usually be relied on for measurements of greater accuracy than 0.01 mm or 0.001 in., unless they are of the new digital design.

Micrometer calipers are available with a variety of specially shaped anvils and/or spindles, such as point, balls, and disks, for measuring special shapes, including screw threads.

Micrometers are also available for inside measurements, and the micrometer principle is also incorporated into a *micrometer depth gage.*

Digital bench micrometers with direct readout to data processors are becoming standard inspection devices on the plant floor (see Figure 10-22). The data processor provides a record of measurement as well as control charts and histograms. Direct gaging height gages, calipers, indicators and micrometers are also available with statistical analysis capability. See Chapter 12 for more discussion on control charts and process capability.

Larger versions of micrometers, called *supermicrometers,* are capable of measuring 0.0001 in. when equipped with an indicator that shows that a selected pressure between the anvils has been obtained. The addition of a digital readout permits the device to measure directly to ± 0.00005 in. (0.001 mm) when it is used in a controlled temperature environment (see Figure 10-23).

The toolmakers' microscope, shown in Figure 10-24, is a versatile instrument that measures by optical means; no pressure is involved. Thus, it is very useful for making accurate measurements on small or delicate parts. The base, on which the special microscope is mounted, has a table that can be moved in two mutually perpendicular, horizontal directions by means of accurate micrometer

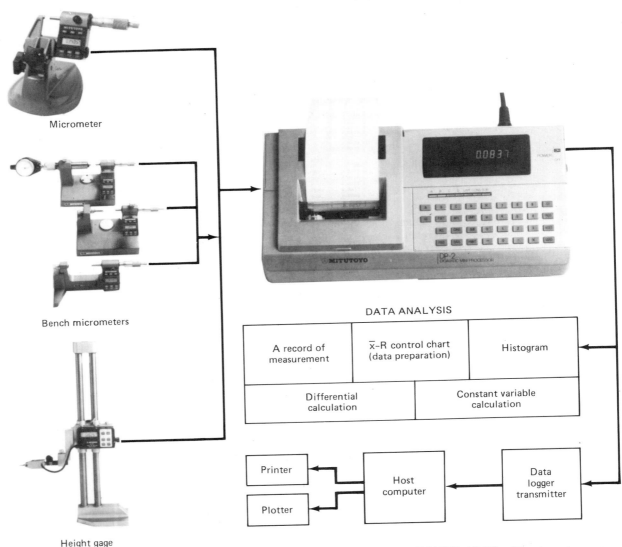

Micrometer

Bench micrometers

Height gage

DATA ANALYSIS

A record of measurement	$\bar{x}$–R control chart (data preparation)	Histogram
Differential calculation		Constant variable calculation

Printer		Host computer		Data logger transmitter
Plotter				

FIGURE 10-22 Direct gaging system for process control and statistical analysis of inspection data. (Courtesy MITUTOYO)

screws that can be read to 0.0001 in., or, if so equipped, by means of the digital readout. Parts to be measured are mounted on the table, the microscope is focused, and one end of the desired part feature is aligned with the crossline in the microscope. The reading is then noted, and the table is moved until the other extremity of the part coincides with the crossline. From the final reading, the desired measurement can be determined. In addition to a wide variety of

FIGURE 10-23 Pratt & Whitney super micrometer modified with a digital electronic readout with accuracy to ± .00005″ (Courtesy CDI)

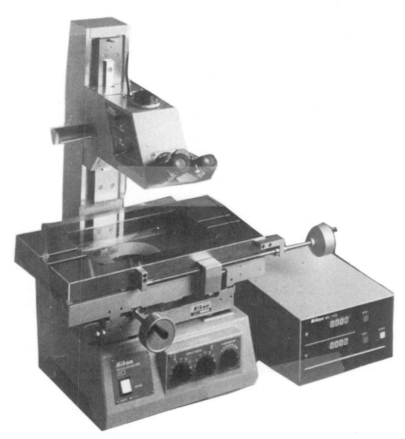

FIGURE 10-24 Toolmakers microscope with digital readouts for X and Y table movements. (Courtesy Nikon)

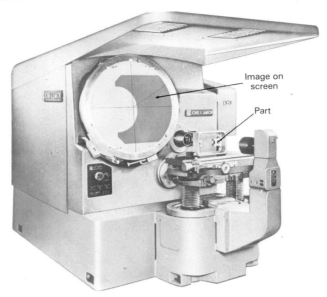

FIGURE 10-25 Optical comparator, measuring the contour on a workpiece. Digital indicators with inch/mm conversions add to the utility of optical comparators.

linear measurements, accurate angular measurements can also be made by means of a special protractor eyepiece. These microscopes are available with digital readouts.

The *optical projector*, or *comparator*, shown in Figure 10-25, is a larger optical device on which both linear and angular measurements can be made. As with the toolmakers' microscope, the part to be measured is mounted on a table that can be moved X and Y directions by accurate micrometer screws. The optical system projects the image of the part on a screen, magnifying it from 5 to more than 100 times. Measurements can be made directly, either by means of the micrometer dials, the digital readouts, or the dial indicators, or on the magnified image on the screen by means of an accurate rule. A very common use for this type of instrument is the checking of parts, such as dies and screws, against a template that is drawn to an enlarged scale, is placed on the screen, and is compared with the projected contour of the part. Some projectors also function as low-power microscopes by providing surface illumination.

Measuring with Lasers.   One of the earliest and most common metrological uses of low-power lasers has been in interferometry. The interferometer is a device that can determine distance and thickness by measuring wavelengths (see Figure 10-26).

First, a beam splitter divides a beam of light into a measurement beam and a reference beam. The measurement beam travels to a reflector (optical glass plate A) resting on the part whose distance is to be measured, while the reference beam is directed at fixed reflector B. Both beams are reflected back through the beam splitter, where they are recombined into a single beam before traveling to the observer. This recombined beam produces interference fringes, depending

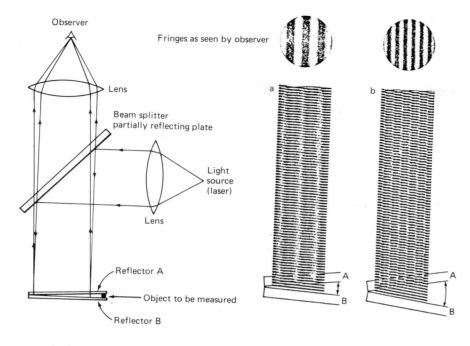

FIGURE 10-26 Interference bands can be used to measure the size of objects to great accuracy. (Based on Michelson interferometer invented in 1882)

on whether the waves of the two returning beams are in phase (called *constructive interference*) or out-of-phase (termed *destructive interference*). In-phase waves produce a series of bright bands, and out-of-phase waves produce dark bands.

The number of fringes can be related to the size of the object, measured in terms of lightwaves of a given frequency. The following example will explain the basics of the method.

To determine the size of object *U* in Figure 10-27, a calibrated reference standard, *S*, plus an optical flat and a toolmaker's flat, is needed along with a monochromatic light source. *Optical flats* are quartz or special glass disks, from 2 to 10 in. (50 to 250 mm) in diameter and about 1/2–1 in. (12 to 25 mm) thick, whose surfaces are very nearly true planes and are nearly parallel. Flats can be obtained with surfaces within 0.000001 in. (0.000 03 mm) of true flatness. It is not essential that both surfaces be accurate or that they be exactly parallel, but one must be certain that only the accurate surface is used in making measurements. A *toolmakers' flat* is similar to an optical flat but is made of steel and usually has only one surface that is accurate. A *monochromatic light source*, light of a single wavelength, must be used. Selenium, helium, or cadmium sources are commonly used along with helium-neon lasers.

The block to be measured is *U*, and the calibrated block is *S*. Distances *a* and *b* must be known but do not have to be measured with great accuracy. By counting the number of interference bands shown on the surface of block *U*, the distance c-d can be determined. Because the difference in the distances between the optical flat and the surface of *U* is 1/2 wavelength, each dark band

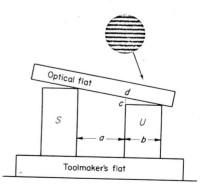

FIGURE 10-27 Method of calibrating gage block by light-wave interference.

indicates a change of 1/2 wavelength in the elevation. If a monochromatic light source having a wavelength of 23.2 microinches (0.589 μm) is used, each interference band represents 11.6 microinches (0.295 μm). Then, by simple geometry, the difference in the heights of the two blocks can be computed. The same method is applicable for making precise measurements of other objects by comparing them with a known gage block.

Light interference also makes it possible to determine easily whether a surface is exactly flat.

The achievement of interference fringes is largely dependent on the coherence of the light used. The availability of highly coherent *laser* light (in-phase light of a single frequency) has made interferometry practical in far less restrictive environments than in the past. The sometimes arduous task of extracting usable data from a close-packed series of interference fringes has been taken over by microprocessors. The laser light has less tendency to diverge (spread out) and is also monochromatic (of the same wavelength). A process that has been largely confined to the optical industry and the metrology lab is now suitable for the factory, where its extremely precise distance-measuring capabilities have been applied to the alignment and calibration of machine tools.

Accurate measurement of distances greater than a few inches was very difficult until the development of laser interferometry, which permits accuracies of ±0.5 parts per million over a distance of 6.1 meters with 0.01 μm resolution. Such equipment is particularly useful in checking the movement of machine tool tables, aligning and checking large assembly jigs, and making measurements of intricate machined parts, such as tire-tread molds.

The Hewlett-Packard laser interferometer, shown in Figure 10-28, uses a helium-neon laser beam split into two beams, each of different frequency and polarized. When the beams are recombined, any relative motion between the optics creates a Doppler shift in the frequency. This shift is then converted into a distance measurement.

The company's first two-frequency interferometer calibration system was introduced in 1970 to overcome workplace contamination by thermal gradients, air turbulence, oil mist, and so on, which affect the intensity of light. Doppler laser interferometers are relatively insensitive to such problems. The system can be used to measure linear distances, velocities, angles, flatness, straightness, squareness, and parallelism in machine tools.

Lasers provide for accurate machine tool alignment. Even larger, modern machine tools can move out of alignment in a matter of months, causing production problems often attributed to the cutting tools, the work holders, the machining conditions, or the numerical control part program.

The most widely used laser technique for inspection and in-process gaging is known as *laser scanning*. At its most basic level, the process consists of placing an object between the source of the laser beam and a receiver containing a photodiode. A microprocessor then computes the object's dimensions based on the shadow that the object casts (see Figure 10-29).

The noncontact nature of laser scanning makes it well suited to in-process measurement, including such difficult tasks as the inspection of hot rolled, or

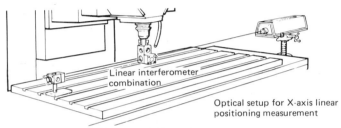

Linear interferometer
combination

Optical setup for X-axis linear
positioning measurement

Linear retroreflector shown
with height adjuster, post and base

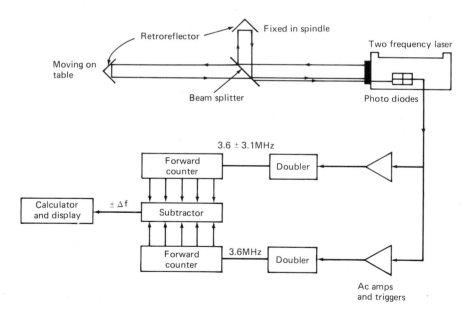

FIGURE 10-28   a) Calibrating the
X-axis linear table displacement of
a vertical spindle milling machine.
b) schematic of optic setup. c) schematic of components of a two frequency laser interferometer. (Courtesy Hewlett-Packard)

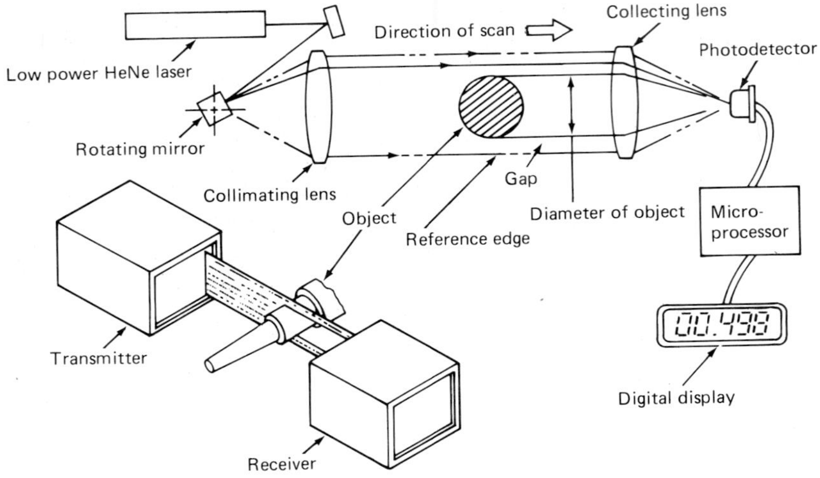

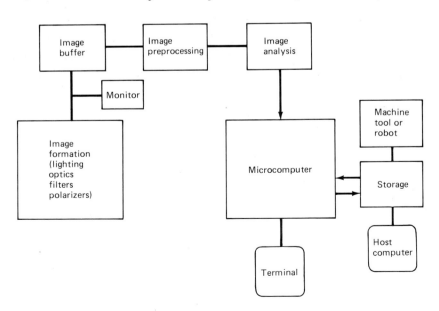

FIGURE 10-29. Scanning laser measurement system.

extruded material. Its comparative simplicity has led to the development of highly portable systems. The bench gage versions can measure to resolutions of 0.0001 mm.

Vision Systems for Measurement

If a picture is worth a thousand words, then vision systems are the tone of inspection methods (see Figure 10-30). Machine vision is used for visual inspection, for guidance and control, or for both. Normal TV image formation on photosensitive surfaces or arrays are used, and the video signals are analyzed information about the object. Each picture frame represents the object at some

FIGURE 10-30   Schematic of element of a machine vision system.

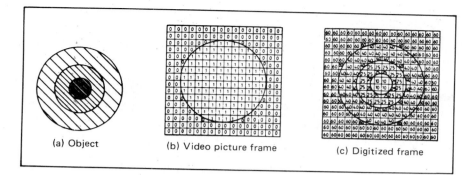

(a) Object   (b) Video picture frame   (c) Digitized frame

FIGURE 10-31 Vision systems use a gray scale to identify objects. a) object with 3 different gray values. b) one frame of object (pixels). c) each pixel assigned a gray scale number.

brief interval of time. Each frame must be dissected into picture elements (called *pixels*). Each pixel is "digitized" (has binary numbers assigned to it) by fixing the brightness or graylevel of each pixel to produce a "bit map" of the object (see Figure 10-31). That is, each pixel is assigned a numerical value based on its shade of gray. Image preprocessing improves the quality of the image data by removing unwanted detail. The bit map is stored in a buffer memory. By analyzing and processing the digitized and stored bit map, the patterns are extracted, edges located, and dimensions determined.

Sophisticated computer algorithms utilizing artificial intelligence have greatly reduced the computer operations needed to achieve a result, but even the most powerful video-based systems currently require 1 to 2 seconds to achieve a measurement. This may be too long a time for many on-line production applications. Table 10.4 provides a comparison of vision systems to laser scanning. Additional material on vision systems will be found in Chapters 29 and 42.

With the recent emphasis on quality and 100% inspection, applications for

TABLE 10.4 Laser Scanning Versus Vision Systems

Variable	Laser-Scanning Systems	Video-Based Systems
Ambient lighting	Independent.	Dependent.
Object motion	Object usually stationary.	Multiple cameras or strobe lighting may be required.
Adaptability to robot systems	Readily adapted; some limitations on robot motion speed or overall system operation.	Readily adapted; image-processing delays may delay system operation.
Signal processing	Simple; computers often not required.	Requires relatively powerful computers with sophisticated software.
Cycle time	Very fast.	Seconds of computer time may be needed.
Applicability to simple tasks	Readily handled; edges and features produce sharp transitions in signal.	Requires extensive use of sophisticated software algorithms to identify edges.
Sizing capability	Can size an object in a single scan per axis.	Can size on horizontal axis in one scan; other dimensions require full-frame processing.
Three-dimensional capability	Limited three dimentionality. Needs ranging capability.	Uses two views or two cameras with sophisticated software or structured light.
Accuracy and precision	Submicrometer 0.001 to 0.0001 in. or better accuracy. Highly repeatable.	Depends on cameras resolution and distance between camera and object. Typically, fractional millimeter 0.004-in. precision with 0.006-in. accuracy tops.

inspection by machine vision have increased markedly. Vision systems can check hundreds of parts per hour for multiple dimensions. Resolutions of ±0.01 in. have been demonstrated but ±0.02 in. is more typical for part location. Machine vision is useful for robot guidance in material handling, welding, and assembly, but nonrobotic inspection and part location applications are still more typical. Their use in inspection, quality control, sorting, and machining tool monitoring will continue to expand. Systems can cost $100,000 or more to install and must be justified on the basis of improved quality rather than labor replacement.

Coordinate Measuring Machines

Precision measurements in three-dimensional Cartesian coordinate space can be made with coordinate measuring machines of the design shown in Figure 10-32. The vertical column rides on a bridge beam and carries a touch-trigger probe. Touching the probe to the surface generates an accurate measurement. Such

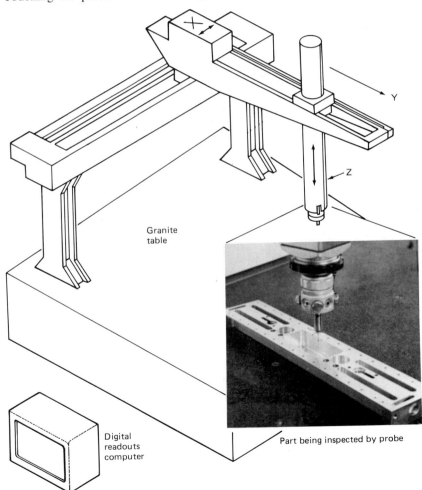

FIGURE 10-32 Coordinate measuring machine with inset showing probe and a part being measured.

machines use digital readouts, air bearings, computer controls, and granite tables to achieve accuracies of the order of 0.0002 to 0.0004 in. over spans of 10 to 30 in. These systems may have computer routines that give the best fit to feature measurements and that provide the means of establishing form tolerances and geometric tolerances. Figures 10-33 and 10-34 give a partial listing of the results one can achieve with these machines.

Straightness

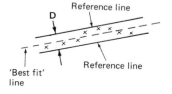

Straightness
Measured or previously calculated points may be used to determine a 'best fit' line. The form routine establishes two reference lines which are parallel to the 'best fit' line, and which just contain all of the measured or calculated points. Straightness is defined as the distance D between these two reference lines.

Flatness

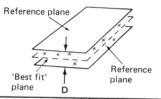

Flatness
Measured or previously calculated points may be used to determine the 'best fit' plane. The form routine establishes two reference planes which are parallel to the 'best fit' plane, and which just contain all of the measured or calculated points. Flatness is defined as the distance D between these two reference planes.

Roundness

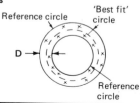

Roundness
Measured or previously calculated points may be used to determine the 'best fit' circle. The form routine establishes two reference circles which are concentric with the 'best fit' circle, and which just contain all of the measured or calculated points. Roundness is defined as the difference D in radius of these two reference circles.

Cylindricity

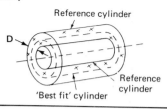

Cylindricity
Measured or previously calculated points may be used to determine the 'best fit' cylinder. The form routine establishes two reference cylinders which are co-axial to the 'best fit' cylinder, and which just contain all of the measured or calculated points. Cylindricity is the difference D in radius of these two reference cylinders. Also applicable to stepped cylinders.

Conicity

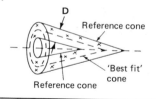

Conicity
Measured or previously calculated points may be used to determine the 'best fit' cone. The form routine establishes two reference cones which are co-axial with and similar to the 'best fit' cone, and which just contain all of the measured or calculated points. Conicity is defined as the distance D between the side of these two reference cones.

FIGURE 10-33 Form tolerances developed by probing surface.

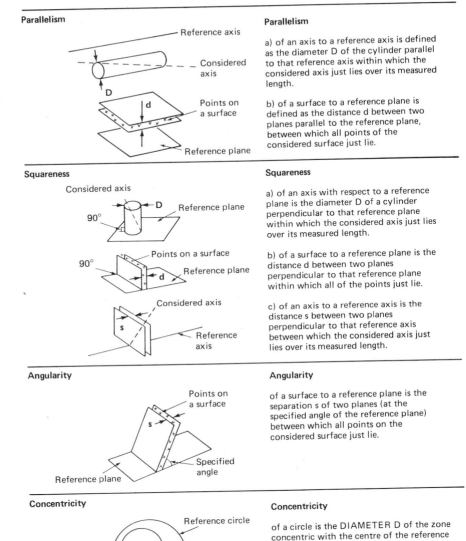

Parallelism

a) of an axis to a reference axis is defined as the diameter D of the cylinder parallel to that reference axis within which the considered axis just lies over its measured length.

b) of a surface to a reference plane is defined as the distance d between two planes parallel to the reference plane, between which all points of the considered surface just lie.

Squareness

a) of an axis with respect to a reference plane is the diameter D of a cylinder perpendicular to that reference plane within which the considered axis just lies over its measured length.

b) of a surface to a reference plane is the distance d between two planes perpendicular to that reference plane within which all of the points just lie.

c) of an axis to a reference axis is the distance s between two planes perpendicular to that reference axis between which the considered axis just lies over its measured length.

Angularity

of a surface to a reference plane is the separation s of two planes (at the specified angle of the reference plane) between which all points on the considered surface just lie.

Concentricity

of a circle is the DIAMETER D of the zone concentric with the centre of the reference circle within which the considered circle centre just lies.

FIGURE 10-34 Geometric tolerances established by CMM.

Angle-Measuring Instruments.   Accurate angle measurements are usually more difficult to make than linear measurements. Angles are measured in degrees (a degree is 1/360 part of a circle) and decimal subdivisions of a degree (or in minutes and seconds of arc). The SI system calls for measurements of plane angles in radians, but degrees are permissible. The use of degrees will continue in manufacturing, but with minutes and seconds of arc possibly being replaced by decimal portions of a degree.

FIGURE 10-35 Measuring an angle on a part with a bevel protractor. (Courtesy Brown & Sharpe Mfg. Co.)

The bevel protractor, illustrated in Figure 10-35, is the most general angle-measuring instrument. The two movable blades are brought into contact with the sides of the angular part, and the angle can be read on the vernier scale to five minutes of arc. A clamping device is provided to lock the blades in any desired position so that the instrument can be used for both direct measurement and layout work.

As indicated previously, an angle attachment on the combination set can also be used to measure angles similarly to the way a bevel protractor is used, but usually with somewhat less accuracy.

The toolmakers' microscope is very satisfactory for making angle measurements, but its use is restricted to small parts. The accuracy obtainable is 5 min of arc. Similarly, angles can be measured on the optical contour projector. Angular measurements can also be made by means of an angular interferometer with the laser system.

A *sine bar* may be used to obtain accurate angle measurements if the physical conditions will permit. This device, as illustrated in Figure 10-36, consists of an accurately ground bar on which two accurately ground pins, of the same diameter, are mounted an exact distance apart. The distances used are usually either 5 or 10 in., and the resulting instrument is called a 5- or 10-in. sine bar. Sine bars also are available with millimeter dimensions. Measurements are made

FIGURE 10-36 Setup to measure an angle on a part using a sine bar. The dial indicator is used to determine when the part surface X is parallel to the surface plate.

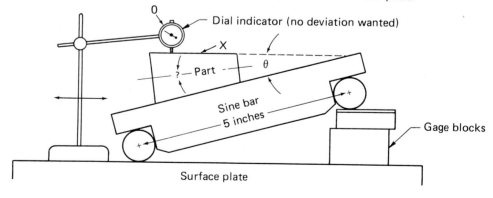

by using the principle that the sine of a given angle is the ratio of the opposite side to the hypotenuse of the right triangle.

The object being measured is attached to the sine bar, and the inclination of the assembly is raised until the top surface is exactly parallel with the surface plate. If a stack of gage blocks is used to elevate one end of the sine bar, as shown in Figure 10-36, the height of the stack directly determines the difference in height of the two pins. The difference in height of the pins can also be determined by a dial indicator gage or any other type of gage. The difference in elevation is then equal to either 5 or 10 times the sine of the angle being measured, depending on whether a 5- or 10-in. bar is being used. Tabulated values of the angles corresponding to any measured elevation difference for 5- or 10-in. sine bars are available in various handbooks. Several types of sine bars are available to suit various requirements.

Accurate measurements of angles to 1 sec of arc can be made by means of *angle gage blocks*. These come in sets of 16 blocks that can be assembled in desired combinations. Angle measurements can also be made to ±0.001 degrees on rotary indexing tables having suitable numerical control.

Gages for Attributes Measuring

In manufacturing, particularly in mass production, it may not be necessary to know the exact dimensions of a part, only that it is within previously established limits. Limits can often be determined more easily than specific dimensions by the use of attributes-type instruments called *gages*. They may be of either fixed type or deviation type, may be used for both linear and angular dimensions, and may be used manually or mechanically (automatically).

Fixed-Type Gages.   Fixed-type gages are designed to gage only one dimension and to indicate whether it is larger or smaller than the previously established standard. They do not determine how much larger or smaller the measured dimension is than the standard. Because such gages fulfill a simple and limited function, they are relatively inexpensive and are easy to use.

Gages of this type are ordinarily made of hardened steel of proper composition and are heat-treated to produce dimensional stability. Hardness is essential to minimize wear and to maintain accuracy. Because steels of high hardness tend to be dimensionally unstable, some fixed gages are made of softer steel with a hard chrome plating on the surface to provide surface hardness. Chrome plating can also be used for reclaiming some worn gages. Where gages are to be subjected to extensive use, they may be made of tungsten carbide at the wear points.

The *plug gage* is one of the most common types of fixed gages. As shown in Figure 10-37, plug gages are accurately ground cylinders used to gage internal dimensions, such as holes. The gaging element of a *plain plug gage* has a single diameter. To control the minimum and maximum limits of a given hole, two plug gages are required. The smaller, or ''go,'' gage controls the minimum because it must go (slide) into any hole that is larger than the required minimum. The larger, or ''not-go,'' gage controls the maximum dimension because it will not go into any hole unless that hole is over the maximum permissible size. As

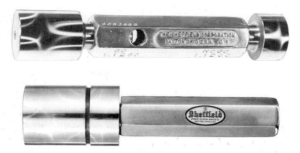

shown in Figure 10-37, the go and not-go plugs are often designed with two gages on a single handle for convenience in use. The not-go plug is usually much shorter than the go plug; it is subjected to little wear because it seldom slides into any holes. *Step-type go–not-go gages* have the go and not-go diameters on the same end of a single plug, the go portion being the outer end. The question is simply, does the gage go or not go into the hole? Such gages require careful use and should never be forced into (or onto) the part.

Plug-type gages are also made for gaging shapes other than cylindrical holes. Three common types are *taper plug gages, thread plug gages,* and *spline gages.* Taper plug gages gage both the angle of the taper and its size. Any deviation from the correct angle is indicated by looseness between the plug and the tapered hole. The size is indicated by the depth to which the plug fits into the hole, the correct depth being denoted by a mark on the plug. Thread plug gages come in go and not-go types. The go gage must screw into the threaded holes, and the not-go gage must not enter.

Ring gages are used to gage shaft or other external round members. These are also made in go and not-go types as shown in Figure 10-38. Go ring gages have plain knurled exteriors, whereas not-go ring gages have a circumferential groove in the knurling, so that they can easily be distinguished. *Ring thread gages* are made to be slightly adjustable because it is almost impossible to make

FIGURE 10-39 Adjustable go and not-go snap gage. (Courtesy Bendix Corporation, Automation and Measurement Division.)

them exactly to the desired size. Thus, they are adjusted to exact, final size after the final grinding and polishing has been completed.

Snap gages are the most common type of fixed gage for measuring external dimensions. As shown in Figure 10-39, they have a rigid, U-shaped frame on which are two or three gaging surfaces, usually made of hardened steel or tungsten carbide. In the adjustable type shown, one gaging surface is fixed, and the other(s) may be adjusted over a small range and locked at the desired position(s). Because in most cases one wishes to control both the maximum and the minimum dimensions, the *progressive* or *step-type snap gage*, shown in Figure 10-39, is used most frequently. These gages have one fixed anvil and two adjustable surfaces to form the outer go and the inner not-go openings, thus eliminating the use of separate go and not-go gages.

Snap gages are available in several types and a wide range of sizes. The gaging surfaces may be round or rectangular. They are set to the desired dimensions with the aid of gage blocks.

Many types of special gages are available or can be constructed for special applications. The *flush-pin gage*, illustrated in Figure 10-40, is an example for gaging the depth of a shoulder. The main section is placed on the higher of the two surfaces with the movable step pin resting on the lower surface. If the depth between the two surfaces is sufficient but not too great, the top of the pin, but not the lower step, will be slightly above the top surface of the gage body. If the depth is too great, the top of the pin will be below the surface. Similarly, if the depth is not great enough, the lower step on the top of the pin will be above the surface of the gage body. When a finger, or fingernail, is run across the top of the pin, the pin's position with respect to the surface of the gage body can readily be determined.

Several types of *form gages* are available for use in checking the *profile* of various objects. Two of the most common types are *radius gages,* shown in Figure 10-41, and *screw-thread pitch gages,* shown in Figure 10-42.

Deviation-Type Gages.   A large amount of gaging, and some measurement, is done through the use of deviation-type gages, which determine the amount by which a measured part deviates, plus or minus, from a standard dimension to which the instrument has been set. In most cases, the deviation is indicated directly in units of measurement, but in some cases, the gage shows only whether the deviation is within a permissible range. A good example of a deviation-type gage is a flashlight battery checker, which shows if the battery is good (green), bad (red), or borderline (yellow), but not how much voltage or current is generated. Such gages use mechanical, electrical, or fluid amplification techniques so that very small linear deviations can be detected. Most are quite rugged, and they are available in a variety of designs, amplifications, and sizes.

Dial indicators, such as shown in Figure 10-4, are a widely used form of deviation-type gage. Movement of the gaging spindle is amplified mechanically through a rack and pinion and a gear train and is indicated by a pointer on a graduated dial. Most dial indicators have a spindle travel equal to about 2½

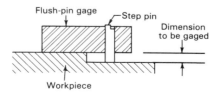

FIGURE 10-40 Flush-pin gage being used to check height of step.

revolutions of the indicating pointer and are read in either 0.001 or 0.0001 in. or 0.02 or 0.002 mm.

The dial can be rotated by means of the knurled bezel ring to align the zero point with any position of the pointer. The indicator is often mounted on an adjustable arm to permit its being brought into proper relationship with the work. It is important that the axis of the spindle be aligned exactly with the dimension being gaged if accuracy is to be achieved. Digital indicators are also readily available, as shown in Figure 10-43.

Dial indicators should be checked occasionally to determine if their accuracy has been lost through wear in the gear train. Also, it should be remembered that the pressure of the spindle on the work varies because of spring pressure as the spindle moves into the gage. This spring pressure normally causes no difficulty unless the spindles are used on soft or flexible parts.

In comparators using electronic magnification, as shown in Figure 10-44, the gaging head is small and readily portable and can be mounted in many ways. The end of the sensing lever is shaped so as to automatically compensate for misalignment in the measuring plane up to about ±15 degrees. The indicator may use either a pointer and graduated scale or a digital readout. Accuracies up to ±0.0001 in. (0.0003 mm) are available, and several ranges can usually be selected by merely turning a knob.

Linear variable-differential transformers (LVDT) are used as sensory elements in many electronic gages, usually with a solid-state diode display, or in automatic inspection setups. These devices can frequently be combined into multiple units for the simultaneous gaging of several dimensions. Ranges and resolutions down to 0.0005 and 0.00001 in. (0.013 and 0.00025 mm), respectively, are available.

Air gages have special characteristics that make them especially suitable for gaging holes or the internal dimensions of various shapes. A typical gage of

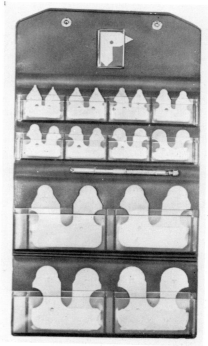

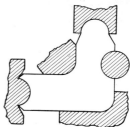

FIGURE 10-41 Set of radius gages, showing how they are used. (Courtesy MTI Corporation.)

FIGURE 10-42 Thread pitch gages. (Courtesy L. S. Starrett Company.)

this type is shown in Figure 10-45. These gages indicate the clearance between the gaging head and the hole by measuring either the volume of air that escapes or the pressure drop resulting from the air flow. The gage is calibrated directly in 0.0001-in. or 0.02-mm. divisions. Air gages have an advantage over mechanical or electronic gages for this purpose in that they detect not only linear size deviations but also out-of-round conditions. Also, they are subject to very little wear because the gaging member is always slightly smaller than the hole and the air flow minimizes rubbing. Special types of air gages can be used for external gaging.

Surface Roughness Measurement

The machining processes to be discussed in Chapters 21 through 32 generate a wide variety of surface patterns. *Lay* is the term used to designate the direction of the predominate surface pattern produced by the machining process. In addition, certain other terms and symbols have been developed and standardized for specifying the surface quality. The most important terms are *surface roughness, waviness,* and *lay,* which are illustrated in Figure 10-46. *Roughness* refers to the finely spaced surface irregularities. It results from machining operations in the case of machined surfaces. *Waviness* is surface irregularity of greater spacing than in roughness. It may be the result of warping, vibration, or the work's being deflecting during machining.

Roughness is measured by the heights of the irregularities with respect to an average line, as illustrated in Figure 10-47. These measurements are usually expressed in micrometers or microinches. In most cases, the arithmetical average (AA) is used. In terms of the measurements indicated in Figure 10-47, the AA would be as follows:

$$AA = \frac{\sum_{i=1}^{n} y_i}{n}$$

FIGURE 10-43 Digital dial indicator with 1″ range and .0001″ accuracy (Courtesy CDI.)

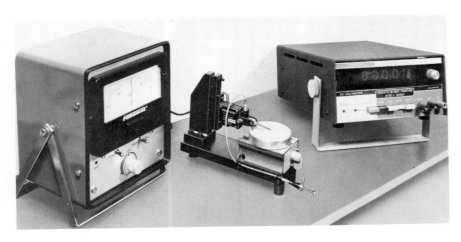

FIGURE 10-44 Electronic-magnification gage being used to gage a computer memory core 0.76 mm (0.030 inch) in diameter and 0.15 mm (0.006 inch) thick. (Courtesy Federal Products Corporation.)

where y_i is a vertical distance from the center line and n is the total number of vertical measurements taken within a specified cutoff distance. This average roughness value is also called R_a. Occasionally used is the *root mean square* (rms) value, which is defined as

$$rms = \sqrt{\frac{\sum_{i=1}^{n} y_i^2}{n}}$$

A variety of instruments are available for measuring surface roughness and surface profiles. The majority of these devices use a diamond stylus that is moved at a constant rate across the surface, perpendicular to the lay pattern. The rise and fall of the stylus is detected electronically (often by an LVDT device), is amplified and recorded on a strip chart, or is processed electronically to produce AA or rms readings for a meter (see Figure 10-48). The unit containing the stylus and the driving motor may be hand-held or supported by the workpiece or some other supporting surface.

The instrument shown in Figure 10-49 is capable of making a series of parallel offset traces on the surface, providing a two-dimensional profile map as shown in the bottom of the figure. Areas of from 0.005×0.005 in.

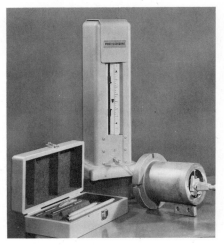

FIGURE 10-45 Gaging the diameter of a hole with an air gage. (Courtesy Bendix Corporation, Automation and Measurement Division.)

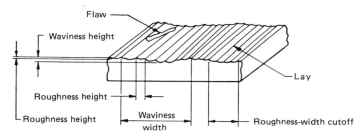

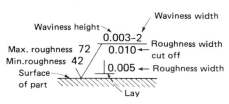

Lay symbols

= Parallel to the boundary line of the nominal surface
⊥ Perpendicular to the boundary line of the nominal surface
X Angular in both directions to the boundary line of the nominal surface
M Multidirectional
C Approximately circular relative to the center
R Approximately radial relative to the center of the nominal surface

FIGURE 10-46 (Top) Terminology used in specifying and measuring surface quality. (Left) Symbols used on drawings by designers with definitions of lay symbols. Quite often, only the desired surface roughness will be specified.

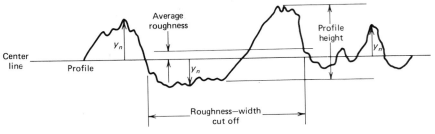

FIGURE 10-47 Schematic of surface profile as produced by a stylus device showing some typical Y values with respect to the center line.

(0.13 × 0.13 mm) up to 2 × 2 in. (50.8 × 50.8 mm), depending on the magnification selected, can be profiled.

The resolution of these devices is determined by the radius or the diameter of the tip of the stylus. When the magnitude of the geometric features begins to approach the magnitude of the tip of the stylus, great caution should be used in interpreting the output from these devices. As a case in point, Figure 10-50 shows a scanning electron micrograph of a face-milled surface, on which has been superimposed (photographically) a scanning electron micrograph of the tip of a diamond stylus (tip radius of 0.0005 in.). Both micrographs have the same final magnification. Surface flaws of the same general size as the roughness created by the machining process are difficult to resolve with the stylus-type device, where both these features are about the same size as the stylus tip.

This example points out the difference between resolution and detection. The stylus tracing devices often can detect the presence of a surface crack, step, or ridge on the part but cannot resolve the geometry of the defect when the defect is of the same order of magnitude as the stylus tip or smaller.

Another problem with these devices is that they produce a reading (a line on the chart) where the stylus tip is not touching the surface, as is demonstrated in Figure 10-51(a), which shows the *s* from the word *trust* on a U.S. dime. The SEM micrograph was made after the topographical map of Figure 10-51(b) had been made. Both figures are at about the same magnification.

The tracks produced by the stylus tip are easily seen in the micrograph. Notice the difference between the features shown in the micrograph and the trace,

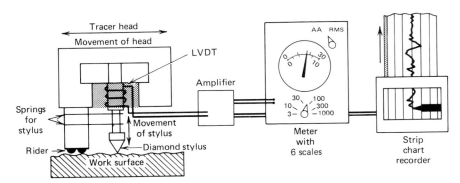

FIGURE 10-48 Schematic of stylus profile device for measuring surface roughness and surface profile with two readout devices shown: a meter for AA or rms values and a strip chart recorder for surface profile.

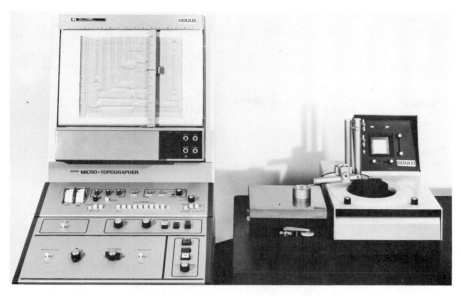

     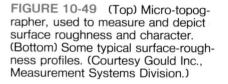

Blanchard ground
X & Y 200X
Z 50X

Milled
X & Y 50X
Z 200X

Ground (Rust
spots). X & Y 50X
Z 200X

Bead blasted
X & Y 50X
Z 200X

EDM machined
X & Y 200X
Z 200X

FIGURE 10-49 (Top) Micro-topographer, used to measure and depict surface roughness and character. (Bottom) Some typical surface-roughness profiles. (Courtesy Gould Inc., Measurement Systems Division.)

FIGURE 10-50 Typical machined steel surface as created by face milling and examined in the SEM. A micrograph (same magnification) of a 0.0005-inch stylus tip has been superimposed at the top.

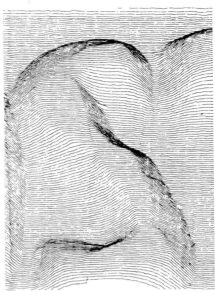

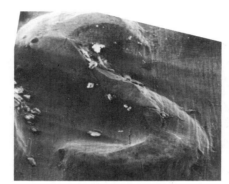

FIGURE 10-51 (a) SEM micrograph of a U.S. dime, showing the "S" in the word TRUST after the region has been traced by a stylus-type machine. (b) Topographical map of the "S" region of the word TRUST from a U.S. dime. Compare to Figure 10-51a.

a)

b)

indicating that the stylus tip was not in contact with the surface many times during its passage over the surface (left no track in the surface) and yet the trace itself is continuous.

Figure 10-52 shows schematics for a new laser-based instrument capable of measuring surface roughness. This instrument has the advantage of not needing to contact the surface to obtain a reading. The parameter S_N is a mean value from the scattered light distribution and has an entirely different meaning from R_a and is not (yet) an established standard. The noncontact feature, coupled with the measurement speed (20 readings per sec), and the ability to read small parts easily are strong advantages.

Surface roughnesses that are typically produced by various manufacturing processes are indicated in Figure 10-53. For the designer's use, sets of *surface-finish blocks*, such as those shown in Figure 10-54, are very useful. It is difficult to visualize a surface having a given roughness, inasmuch as the same value of roughness may reflect different surface characteristics when produced by different processes. Clearly, such systems will be incorporated into machines in the future for in-process control of surface roughness.

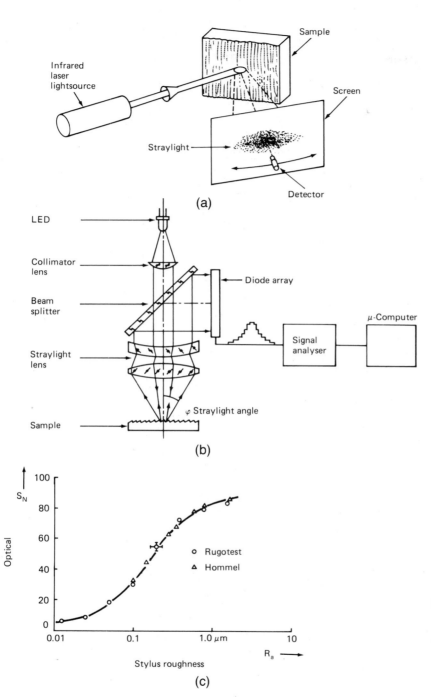

FIGURE 10-52   (a) Laser light scattering on rough surface. (b) Schematic of roughness measuring system. (c) Good correlation between optical and stylus roughness in the low RMS ranges. (Courtesy Rodenstock).

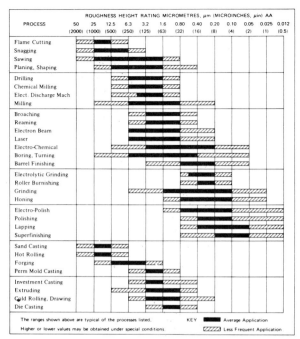

PROCESS	ROUGHNESS HEIGHT RATING MICROMETRES, μm (MICROINCHES, μin) AA												
	50 (2000)	25 (1000)	12.5 (500)	6.3 (250)	3.2 (125)	1.6 (63)	0.80 (32)	0.40 (16)	0.20 (8)	0.10 (4)	0.05 (2)	0.025 (1)	0.012 (0.5)
Flame Cutting													
Snagging													
Sawing													
Planing, Shaping													
Drilling													
Chemical Milling													
Elect. Discharge Mach													
Milling													
Broaching													
Reaming													
Electron Beam													
Laser													
Electro-Chemical													
Boring, Turning													
Barrel Finishing													
Electrolytic Grinding													
Roller Burnishing													
Grinding													
Honing													
Electro-Polish													
Polishing													
Lapping													
Superfinishing													
Sand Casting													
Hot Rolling													
Forging													
Perm Mold Casting													
Investment Casting													
Extruding													
Cold Rolling, Drawing													
Die Casting													

The ranges shown above are typical of the processes listed.

Higher or lower values may be obtained under special conditions.

KEY ■ Average Application ▨ Less Frequent Application

Extracted from General Motors Drafting Standards, June 1973 revision

FIGURE 10-53 Comparison of surface roughness produced by common production processes. (Courtesy *American Machinist.*)

FIGURE 10-54 Set of surface-roughness standards being used in a drafting room. (Courtesy Surface Checking Gage Co.)

TABLE 10.5 Metric—English Conversions

Measurement	Metric Symbol	Metric Unit	English Conversion
Linear dimensions	m	metre*	1 in. = 0.0254 m* and
			1 ft = 0.3048 m*
	cm	centimetre	1 in. = 2.54 cm*
	mm	millimetre	1 in. = 25.4 mm*
	μm	micrometre	1 μin. = 0.0254 μm*
Area	m^2	square metres	1 ft^2 = 0.093 m^2†
	cm^2	square centimetres	1 $in.^2$ = 6.45 cm^2†
	mm^2	square millimetres	1 $in.^2$ = 645.16 mm^2*
Volume and capacity	m^3	cubic metres	1 ft^3 = 0.028 m^3†
	cm^3	cubic centimeters	1 $in.^3$ = 16.39 cm^3†
	mm^3	cubic millimetres	1 $in.^3$ = 16 387 mm^3†
	l	litre	1 ft^3 = 28.32 l† and
			1 U.S. gal = 3.79 l†
Velocity, acceleration, and flow	m/s	metres per second	1 ft/s = 0.3048 m/s*
	m/min	metres per minute	1 ft/min = 0.3048 m/min*
	m/s^2	metres per second squared	1 in/s^2 = 0.0254 m/s^2*
	l/min	litres per minute	1 ft^3/min = 28.3 l/min† and
			1 gallon (U.S. liquid)/min = 3.785 l/min†
Mass	g	gram	1 oz = 28.35 g†
	kg	kilogram	1 lb = 0.45 kg†
	t	metric ton	1 short ton (2000 lb) = 0.9072t†
Force	N	newton	1 lb-force = 4.448 N†
	kN	kilonewton	1 short ton-force (2000 lb) = 8.896 kN†
Bending moment or torque	N · m	newton-metre	1 oz-force/in. = 0.007 N · m†
			1 lb-force/in. = 0.113 N · m†
			1 lb-force/ft = 1.3558 N · m†
Pressure	Pa	pascal	1 lb/ft^2 = 47.88 Pa†
	kPa	kilopascal	1 lb/in^2 = 6.895 kPa†
Energy, work, or quantity of heat	J	joule	1 Btu†† = 1055.056 J†
	kJ	kilojoule	1 Btu†† = 1.055 kJ†
Power	W	watt	1 hp (550 ft-lb/s) = 745.7 W†
			1 hp (electric) = 746 W*
	kW	kilowatt	1 hp (550 ft-lb/s) = 0.7457 kW†
Temperature	C	Celsius	degrees C = $\dfrac{\text{degrees F} - 32}{1.8}$
Frequency	Hz	hertz	1 cycle per second = 1 Hz
	kHz	kilohertz	1000 cycles per second = 1 kHz
	MHz	megahertz	1 000 000 cycles per second = 1 MHz

*Exact   †Approximate   ††International Table

Comments on the metric system:

The *re* spelling of *metre* conforms with the such standards as ANSI Z210.1, ISO Standard 1000, and recommendations of the Society of Manufacturing Engineers Metric Advisory Committee. The use of *metre* also avoids confusion with meter and micrometer measuring instruments.

Metric symbols are usually presented the same way in singular and in plural (1 mm, 100 mm), and periods are not used after symbols, except at the ends of sentences. *Degree* (instead of *radian*) continues to be used for plane angles, but angles are expressed with decimal subdivisions rather than minutes and seconds. Surface finishes are specified in micrometres.

Accuracy of Conversion. Multiplying an English measurement by an exact metric conversion factor often provides an accuracy not intended by the original value. In general, use one less significant digit to the right of the decimal point than was given in the origianl value: 0.032 in. (0.81 mm), 0.0003 in. (0.008 mm), etc. Fractions of an inch are converted to the nearest tenth of a millimetre: ⅛ in. (3.2 mm).

Weight, Mass, and Force. Confusion exists in the use of the term *weight* as a quantity to mean either force or mass. In nontechnical circles, the term *weight* nearly always means mass, and this use will probably persist. Weight is a force generated by mass under the influence of gravity. Mass is expressed in gram (g), kilogram (kg), or metric ton (t) units, and force in newton (N) or kilonewton (kN) units.

Pressure. Kilopascal (kPa) is the recommended unit for fluid pressure. Absolute pressure is specified either by using the identifying phrase "absolute pressure," or by adding the word *absolute* after the unit symbol, separating the two by a comma or a space.

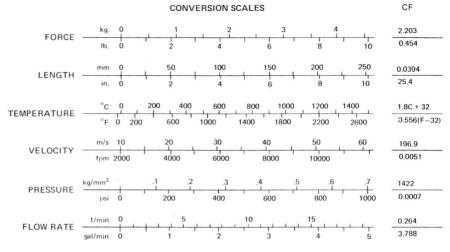

FIGURE 10-55 Conversion scales to convert from English to metric units or vise versa.

CF = *Conversion Factor; example: 10 kg = (2.203)(10)lb., 10 in = 2.54(10)mm.*
1 Newton ≃ 0.1 kg ≃ 0.22 lb; 1 micron = 40 microinches

Review Questions

1. What are some of the advantages to the consumer of standardization and of interchangeable parts?
2. Why is it important to interface the manufacturing engineering requirements with the design phase as early as possible?
3. Explain the difference between attributes and variables inspection.
4. Why have so many variables-type devices in autos been replaced with attributes-type devices?
5. What are the four basic measures on which all others depend?
6. What is a pascal, and how is it made up of the basic measures?
7. What are the different grades of gage blocks, and why do they come in sets?
8. What keeps gage blocks together when they are "wrung together"?
9. What is the difference between tolerance and allowance?
10. What type of fit would describe the following situations?
    a. The cap of a ballpoint pen.
    b. The lead in a mechanical lead pencil, at the tip.
    c. The bullet in the barrel of a gun.
11. List the items you can think of that are assembled with a medium-force fit.
12. Why might you use a shrink fit rather than welding to join two steel parts? What does the word *shrink* imply?

13. Explain the difference between accuracy and precision.
14. Into which of the four basic kinds of inspection does interferometry fall?
15. What factors should be considered in selecting measurement equipment?
16. Explain how you could determine if your ordinary bathroom scale is linear and has good repeat accuracy, assuming the scale is analogue.
17. Design and describe a simple experiment that demonstrates the difference between magnification and resolution.
18. Explain what is meant by the statement that usable magnification is limited by the resolution of the device.
19. What is parallax?
20. What is the rule of 10?
21. What is the principle of vernier calipers?
22. What are the two most likely sources of error in using micrometer calipers?
23. What is the major disadvantage of a micrometer caliper as compared with a vernier caliper? The advantages?
24. What would be the major difficulty in obtaining an accurate measurement with a micrometer depth gage if it were not equipped with a ratchet or friction device for turning the thimble?

25. Suppose you had a 2-ft steel bar in your supermicrometer. Could you detect a length change if the temperature of the bar changed 20°F?
26. Why is the toolmakers' microscope particularly useful for making measurements on delicate parts?
27. In what two ways can linear measurements be made with an optical projector?
28. What type of instrument would you select for checking the accuracy of the linear movement of a machine tool table through a distance of 50 in.?
29. What are the chief disadvantages of using a vision system for measurement compared to laser scanning?
30. What is a coordinate measuring machine?
31. On what principle is a sine bar based?
32. How can the not-go member of a plug gage be easily distinguished from the go member?
33. What is the primary precaution that should be observed in using a dial gage?
34. How does a taper plug gage check both the angle of a taper and its size?

35. Explain how a go–not-go ring gage works for checking a shaft.
36. Why are air gages particularly well suited to gaging the diameter of a hole?
37. Explain the principle of measurement by light-wave interference.
38. How does a toolmakers' flat differ from an optical flat?
39. Why may two surfaces that have the same microinch roughness be quite different in appearance?
40. Why are surface-finish blocks often used for specifying surface finish rather than just microinch values?
41. What limits the resolution of a stylus-type surface-measuring device in finding profiles?
42. What is the general relationship between surface roughness and tolerance? Between tolerance and the cost of producing the surface and/or tolerance?
43. What is the main disadvantage of the laser-based instrument for surface measurement?

Problems

1. Read the 25-division vernier graduated in English; see Figure 10-A.
2. Read the 25-division vernier graduated in metric (direct reading); see Figure 10-B.
3. Convert the larger of the two readings to units of the smaller and subtract.
4. Suppose, that, in Figure 10-36, the height of the gage blocks are 3.2500 inches. What is the angle θ assuming that the dial indicator is reading 0.0 ± 0.001 in.?
5. What is the estimated error in this measurement, given that Grade 3 working blocks are being used?

6. In Figure 10-C, the sleeve–thimble region of three micrometers graduated in thousandths of an inch are shown. What are the readings for these three micrometers? Hint: Think of the various units as if you were making change from a $10 bill. Count the figures on the sleeve as dollars, the vertical lines on the sleeve as quarters, and the divisions on the thimble as cents. Add up your change, and put a decimal point instead of a dollar sign in front of the figures.
7. Figure 10-D shows the sleeve–thimble region of two micrometers graduated in thousandths of an inch with a vernier for an additional ten-thousandths. What are the readings?

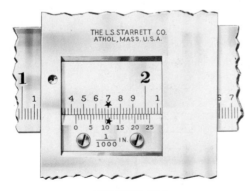

FIGURE 10A

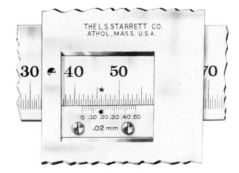

FIGURE 10B

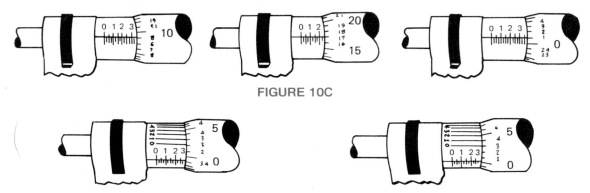

FIGURE 10C

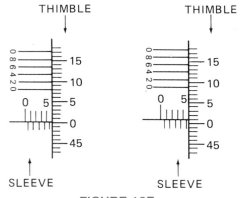

FIGURE 10D

8. In Figure 10-E, two examples of a metric vernier micrometer are shown. The micrometer is graduated in hundredths of a millimetre (0.01 mm), and an additional reading in two-thousandths of a millimetre (0.002 mm) is obtained from the vernier on the sleeve. What are the readings?

9. In checking a 1-in.-square gage block by means of a helium light source, five dark bands were observed. There was a 2-in. distance between the front edges of the two blocks. What was the difference in height between the two blocks?

10. In Figure 10-F, the angle θ on the part needs to be inspected. The setup used is shown in Figure 10-G. (No sine plate was available.) Determine the angle θ from the part drawing and the value of X gage blocks.

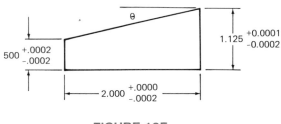

FIGURE 10F

THIMBLE   THIMBLE

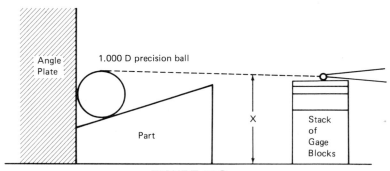

SLEEVE   SLEEVE

FIGURE 10E

FIGURE 10G

11

Nondestructive Inspection and Testing

A major emphasis in any manufacturing operation should be the prevention of defects that may lead to material or product failure. Care in product design, material selection, fabrication of the desired shape, heat treatment, and surface treatment, as well as consideration of all possible service conditions, can do much to ensure the manufacture of a quality product. How do we determine whether these efforts have been successful? How can we be sure that the product is free from any harmful defects or flaws?

Various types of testing can be used to help evaluate the quality of products and ensure the absence of any flaws that might prevent an acceptable service performance. *Destructive testing* subjects selected components to test conditions that intentionally induce failure. By determining the specific conditions under which these components fail, insight can be gained into the general characteristics of a larger production lot. Statistical methods can be used to determine the probability that the entire production run will be good, provided that a certain number of test specimens prove to be satisfactory. Since each of the products tested is destroyed during the evaluation, the cost of such methods must be borne by the remaining products. In addition, there is still some degree of uncertainty about the quality of each of the remaining products, because they have never been individually evaluated.

To provide an increased measure of assurance, *proof testing* may be used. Here, a product is subjected to loads of a determined magnitude, generally equal to or greater than the designed capacity. If the part remains intact, then there is reason to believe that it will perform adequately in the absence of abuse or loads in excess of its rated level. Proof tests can be performed either under laboratory conditions or at the site of installation, as on large manufactured assemblies such as pressure vessels.

In some situations, *hardness tests* can be used to evaluate the quality of a product. Proper material and heat treatment can be reasonably ensured by requiring that all test results fall within a desired range. The tests can be performed quickly, and the surface markings are often small enough so that they can be concealed or easily removed from a product. The results, however, relate only to the surface strength of the product and cannot detect serious defects, such as cracks or voids.

Nondestructive testing is the examination of a product in a manner that will not render it useless for future service. Testing can be performed directly on production items, or even on parts in service, and the only scrap losses are those defective parts that are detected. The entire production lot can be inspected, or representative samples can be taken. Different tests can be applied to the same item, either simultaneously or sequentially, and the same test can be repeated on the same specimen for additional verification. Little or no specimen preparation is required, and the equipment is often portable, permitting on-site testing in most locations.

Various objectives can be sought in nondestructive tests, including the detection of internal or surface flaws, the measurement of dimensions, a determination of the material's structure or chemistry, or the evaluation of a material's physical or mechanical properties. In general, nondestructive tests involve each of the following aspects: (1) a probing medium applied to the product; (2) a possible modification of this probing medium by a flaw, a defect, a material property, or a specimen feature; (3) a sensor to detect the response signal; (4) a device to indicate or record the detector signals; and (5) a means of interpreting the response and of evaluating quality.

How you look at a material or product generally depends on what you are looking at, what you wish to see, and how finely you wish to see it. Each of the various inspection processes has its characteristic advantages and limitations. Some can be performed on only certain types of materials (such as conductors or ferromagnetic materials). Each is limited in the type, size, and orientation of flaws that it can detect. Various degrees of accessibility may be required, and there may be geometric restrictions on part size or complexity. The available or required equipment, the cost of its operation, the need for a skilled operator or technician, and the possibility of producing a permanent test record are all considerations when selecting a test procedure.

Regardless of the specific method, nondestructive testing is a vital element in good manufacturing practice, and its potential value is becoming more widely recognized as productivity demands increase, consumers demand higher-quality products, and product liability continues to be a concern. Rather than being an added manufacturing cost, this practice can actually expand profit by ensuring product reliability and customer satisfaction. While essentially a quality-control operation, nondestructive testing can also be used to aid in product design, to provide on-line control of a manufacturing process, and to reduce overall manufacturing costs.

The remainder of this chapter seeks to provide an overview of the various nondestructive test methods. Each is presented along with its underlying principle, associated advantages and limitations, compatible materials, and typical applications.

Probably the simplest and most widely used nondestructive testing method is simple visual inspection. The human eye is a very discerning instrument, and with training, the brain can readily interpret the signals. Optical aids such as mirrors, magnifying glasses, and microscopes can enhance the capabilities of the system. Digital image analyzers can be used to automate the testing and to make a number of quantitative geometrical evaluations. Borescopes and similar tools can provide accessibility to otherwise inaccessible locations. In the area of limitations, only the surfaces of the product can be examined.

Visual Inspection

TABLE 11.1 Visual Inspection

Method: Visual inspection.

Principle: Illuminate the test specimen and observe the surface with the eye. Use of optical aids or assists is permitted.

Advantages: Simple, easy to use, cheap.

Limitations: Depends on the skill and knowledge of the inspector. Limited to detection of surface flaws.

Material limitations: None.

Geometrical limitations: Any size or shape providing accessibility of surfaces to be viewed.

Permanent record: Photographs or videotapes are possible. Inspectors' reports also provide valuable records.

Liquid Penetrant Inspection

Liquid penetrant testing is a simple method of detecting surface defects in metals and other nonporous material surfaces. The piece to be tested is first subjected to a thorough cleaning, often by means of solvent-type materials, and is dried before the test. Then, a *penetrant*, a liquid material capable of wetting the entire surface and being drawn into fine openings, is applied to the surface of the prepared workpiece by dipping, spraying, or brushing. After a period of time that permits capillary action to draw the penetrant into the surface discontinuities, the excess penetrant liquid is removed. The surface is then coated with a thin film of "developer," an absorbent material capable of drawing traces of penetrant from the defects back onto the surface. Brightly colored dyes or fluorescent materials that radiate in ultraviolet light are generally added to the penetrant to make these traces more visible, and the developer is often selected so as to provide a contrasting background. Radioactive tracers can be added and used in conjunction with photographic paper to produce a permanent image of the defects. Cracks, laps, seams, lack of bonding, pinholes, gouges, and tool marks can all be detected. After inspection, the developer and the residual penetrant are removed by a second cleaning operation. Figure 11-1 shows a schematic representation of the liquid penetrant procedure.

If previous processing involved techniques that could have induced a flow of the surface layers, such as shot peening, honing, burnishing, or various forms

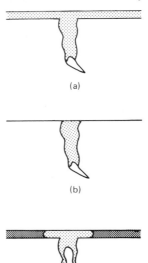

(a)

(b)

(c)

FIGURE 11-1 Liquid penetrant testing: a) application of the penetrant to a clean surface; b) excess penetrant is removed; c) developer is applied and the product inspected.

Magnetic Particle Inspection

of cold working, a chemical etching may be required to first remove any material covering the surface defects. An alternative procedure is to penetrant-test *before* the final surface-finishing operation, when the defects are still open and available for detection.

Penetrant inspection systems can range from a set of aerosol spray cans of cleaner, penetrant, and developer (for portable applications), to automated, mass-production equipment using laser scanners and computer control.

TABLE 11.2   Liquid Penetrant Inspection

Method: Liquid penetrant.
Principle: A liquid penetrant is drawn into surface flaws by capillary action and is subsequently revealed by developer material in conjunction with visual inspection.
Advantages: Simple, inexpensive, versatile, portable, easily interpreted, and applicable to complex shapes.
Limitations: Can detect only flaws that are open to the surface; surfaces must be cleaned before and after inspection; deformed surfaces and surface coatings may prevent detection; and the penetrant may be washed out of large defects.
Material limitations: Must have a nonporous surface.
Geometric limitations: Any size and shape permitting accessibility of the surfaces to be inspected.
Permanent record: Can be made by same techniques as visual inspection.

Magnetic particle inspection is based on the principle that ferromagnetic materials (such as iron, steel, nickel, and cobalt alloys), when magnetized, will have distorted magnetic fields in the vicinity of material defects, as shown in Figure 11-2. Surface and subsurface flaws, such as cracks and inclusions, can produce magnetic anomalies that can be mapped with the aid of magnetic particles on the specimen surface.

As in the previous methods, the specimen must be cleaned before inspection. A suitable magnetic field is then established in the part, produced so as to reveal selected defects. Orientation considerations are quite important since a flaw must produce a significant disturbance of the magnetic field at or near the surface if the flaw is to be detected. For example, placing a bar of steel within an energized coil will produce a magnetic field whose lines of flux travel along the axis of the bar. Any defect perpendicular to this axis will significantly alter the field. If the perturbation is sufficiently large and close enough to the surface, it can be detected by the inspection methods. However, if the flaw is in the form of a crack running down the specimen axis, the defect is oriented in such a way as to produce little perturbation of the lines of flux and is likely to go undetected.

If the same sample is magnetized by passing a current through it to create a circumferential magnetic field, the axial defect now becomes highly detectable, and the defect perpendicular to the axis may go unnoticed. These features are summarized in Figure 11-3. Thus, a series of inspections using different forms of magnetization may be required to fully inspect a product. Passing a current through the product between various points of contact is a popular means of inducing the desired fields. Electromagnetic coils of various shapes and sizes are also used. Alternating current methods are most sensitive to surface flaws. Direct current methods are more capable of detecting subsurface defects, such as nonmetallic inclusions.

Once the specimen is magnetized, magnetic particles are applied to the surface, as either a dry powder or a suspension in a liquid carrier. The particles can be treated with a fluorescent material for observation under ultraviolet light. In addition, they are often made in an elongated form to better reveal the orientation and can be coated with a lubricant to prevent oxidation and enhance their mobility. The resultant distribution of particles is then examined, and any anomalies are interpreted. Figure 11-4 shows the kingpin for a truck front axle: as manufactured, under straight magnetic-particle inspection, and under ultraviolet light with fluorescent particles.

Some residual magnetization will be retained by all parts subjected to magnetic particle inspection. Therefore, it is usually necessary to demagnetize them before further processing or before placing them in use. One common means of demagnetization is to place the parts inside a coil powered by alternating current and then gradually to reduce the current to zero. A final cleaning operation completes the process.

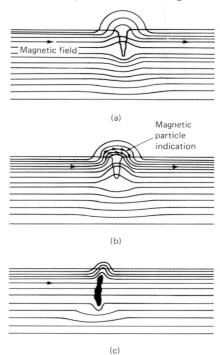

FIGURE 11-2 a) Magnetic field being disrupted by a surface crack; b) Magnetic particles applied and attracted to field leakage; c) Subsurface defects can produce surface-detectable disruptions.

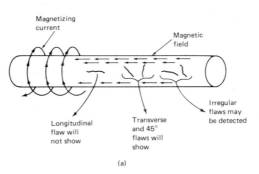

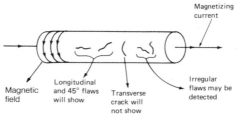

FIGURE 11-3 a) A bar placed within a magnetizing coil will have an axial magnetic field. Defects parallel to this field will go unnoticed while those disrupting the field and sufficiently close to a surface will be detected. b) When magnetized by a current passing through it, the bar has a circumferential magnetic field and the geometries of detectable flaws are reversed.

FIGURE 11-4   A front axle kingpin for a truck. (Left) As manufactured and apparently sound. (Center) Inspected under conventional magnetic particle inspection to reveal numerous grinding-induced cracks. (Right) Fluorescent particles and ultraviolet light make the cracks even more visible. (Courtesy Magnaflux Corporation.)

In addition to in-process and final inspection of parts during manufacture, magnetic particle inspection is extensively used during the maintenance and overhaul of equipment and machinery. The testing equipment ranges from small, lightweight, portable units to heavy, complex, automated systems.

TABLE 11.3   Magnetic Particle Inspection

Method: Magnetic particle.

Principle: When magnetized, ferromagnetic materials have a distorted magnetic field in the vicinity of flaws and defects. Magnetic particles will be strongly attracted to surface regions where the flux is concentrated.

Advantages: Relatively simple; fast; easy to interpret; portable units exist; can reveal subsurface flaws and inclusions (as much as ¼ inch deep) and small, tight cracks.

Limitations: Parts must be relatively clean; alignment of the flaw and the field affects the sensitivity so that multiple inspections with different magnetizations may be required; must demagnetize part after test; high current source is required; some surface processes can mask defects; postcleaning may be required.

Material limitations: Must be ferromagnetic; nonferrous metals such as aluminum, magnesium, copper, lead, tin, titanium, and the ferrous (but not ferromagnetic) austenitic stainless steels cannot be inspected.

Geometric limitations: Size and shape are almost unlimited; most restrictions relate to the ability to induce uniform magnetic fields within the piece.

Permanent record: Can be made by the same techniques as in visual inspection. In addition, the defect pattern can be preserved on the specimen by an application of transparent lacquer or can be transferred to a piece of transparent tape that has been applied to the specimen and peeled off.

Ultrasonic Inspection

Sound has long been used to provide an indication of product quality. A cracked bell will not ring true, but a fine crystal goblet will have a clear ring when tapped lightly. Striking an object and listening to the characteristic ''ring'' is an ancient art, but it is limited to the detection of large defects because the wavelength of audible sound is rather large compared to the size of most defects. By reducing the wavelength of the signal to the ultrasonic range, beyond the range of human hearing, ultrasonic inspection can detect rather small defects and flaws.

Ultrasonic inspection, therefore, involves sending high-frequency vibrations through a material and observing what happens when this ultrasonic beam hits a defect, a surface, or a change of density. At the interface, part of the ultrasonic wave is reflected and part is transmitted. If the incident beam is at an angle to the interface and the material changes across the surface, the transmitted portion of the beam is bent to a new angle by the phenomenon of refraction. By receiving and interpreting the altered signal, ultrasonic inspection can be used to detect flaws within the material, to measure thickness from only one side, and to characterize metallurgical structure.

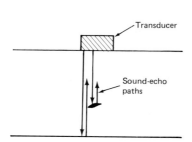

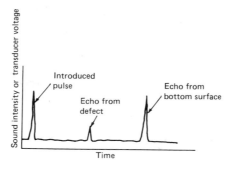

FIGURE 11-5 (Left) Ultrasonic inspection of flat plate with a single transducer. (Right) A plot of sound intensity or transducer voltage versus time depicting the base signal and a secondary peak indicative of an intervening defect.

An ultrasonic inspection system begins with a pulsed oscillator and a transducer, which serves to transform electrical energy into mechanical vibrations. The pulsed oscillator generates a burst of alternating voltage—the principal frequency, duration, profile, and repetition rate being either fixed or variable. This burst is then applied to a sending transducer, which uses a piezoelectric crystal to convert the electrical oscillations into mechanical vibrations. Because air is a poor transmitter of ultrasonic waves, an acoustic coupling medium—generally a liquid such as oil or water—is required to link the transducer to the piece to be inspected and to transmit the vibrations into the part. A receiving transducer is then used to convert the received ultrasonic vibrations back into electrical signals. The receiving transducer can be identical to the sending unit, and the same transducer can even be used for both functions. A receiving unit then amplifies, filters, and processes the signal for display on an oscilloscope, for recording on some form of electromechanical recorder, and for final interpretation. An electronic clock is generally integrated into the system to time the responses and to provide reference signals for comparison purposes.

Depending on the test objectives and the part geometry, several different inspection techniques can be used:

1. In the *pulse-echo* technique, a pulsed ultrasonic beam is introduced into the piece to be inspected, and the echoes from the opposing surfaces and any intervening flaws are monitored by the receiver. The time interval between the initial emitted pulse and the various echoes is displayed on the horizontal axis of an oscilloscope screen. Defects are identified by the position and amplitude of the various echoes. Figure 11-5 shows a schematic of this technique using a single transducer and the companion signal as it would be recorded on an oscilloscope. Figure 11-6a shows the dual-transducer pulse-echo technique.

2. The *through-transmission* technique requires separate sending and receiving transducers. A pulsed beam is sent through the part by the sending transducer, and the transmitted signal is picked up by the receiver, as shown in Figure 11-6b. Flaws in the material decrease the amplitude of the transmitted beam because of back-reflection and scattering.

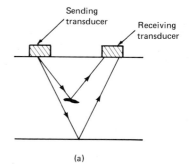

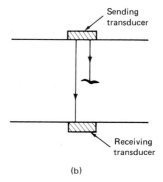

FIGURE 11-6 a) Dual transducer ultrasonic inspection in the pulse-echo mode. b) Dual transducers in through-transmission inspection.

3. *Resonance testing* can be used to determine the thickness of a plate or sheet from one side of the material. Input pulses of varying frequency are fed into the material. When resonance is detected by an increase of the energy at the transducer, the thickness can be calculated from the speed of sound in the material and the time of traverse. Some instruments can be calibrated to provide a direct digital readout of the thickness of the material.

Reference standards, consisting of a series of specimens with known thicknesses or various types and sizes of machined "flaws," are generally used to ensure consistent results and to aid in interpreting any indications of internal discontinuities.

TABLE 11.4   Ultrasonic Inspection

Method: Ultrasonic inspection.

Principle: Sound waves are propagated through a test specimen, and the transmitted or reflected signal is monitored and interpreted.

Advantage: High sensitivity to most cracks and flaws; high-speed test with immediate results; can be automated and recorded; portable; high penetration in most important materials (up to 60 feet in steel); indicates flaw size and location; access to only one side is required; can also be used to measure thickness, Poisson's ratio, or elastic modulus; presents no radiation or safety hazard.

Limitations: Difficult to use with complex shapes; external surfaces and defect orientation can affect the test (may need dual transducer or multiple inspections); a couplant is required; the area of coverage is small (inspection of large areas requires scanning); trained, experienced, and motivated technicians may be required.

Material limitations: Few—can be used on metals, plastics, ceramics, glass, rubber, graphite, and concrete, as well as joints and interfaces between materials.

Geometric limitations: Small, thin, complex-shape parts or parts with rough surfaces and nonhomogeneous structure pose the greatest difficulty.

Permanent record: Ultrasonic signals can be recorded for subsequent playback and analysis. Strip charts can also be used.

Radiography

Radiographic inspection uses the same principles and techniques as medical X rays. In essence, a shadow pattern is created when certain types of radiation (X rays, gamma rays, or neutron beams) penetrate an object and are differentially absorbed because of variations in thickness, density, or chemistry, or because of the presence of defects in the specimen. The transmitted radiation is registered on a photographic film that provides a permanent record and a means of analyzing the component. Fluorescent screens can provide a direct conversion of radiation into visible light and can enable fast and inexpensive viewing without the need for film processing. The image, however, has relatively poor sensitivity compared to that possible by the photographic methods.

X rays are an extremely short-wavelength form of electromagnetic radiation and are capable of penetrating many materials that reflect or absorb visible light. X rays are generated by high-voltage electrical apparatus, the higher the voltage, the shorter the X ray wavelength and the greater the energy and penetrating power of the beam. *Gamma rays* are also electromagnetic radiation, but they are emitted during the disintegration of radioactive nuclei. Various radioactive isotopes can be selected as the radiation source. *Neutron beams* for radiography are generally derived from nuclear reactors, nuclear accelerators, or radio-isotopes. For most applications, it is necessary to moderate the energy and to collimate the beam before use.

The absorption of X rays and gamma rays depends on the thickness, density, and atomic structure of the material being inspected. The higher the atomic number, the greater the attenuation of the beam. Figure 11-7 and the lead photo for Part III of the text show a radiograph of the historic Liberty Bell. Clearly visible is the famous crack, along with the internal spider (installed to support the clapper in 1915) and the steel beam and supports installed in the wooden yoke in 1929. Additional radiographs disclosed shrinkage separations and cracks that were not known to exist, as well as an additional crack in the bell's clapper.

In contrast to X ray absorption, neutron absorption varies widely from atom to atom, with no pattern in terms of atomic number. Unusual contrasts can be obtained that would be impossible with other inspection methods. For example, hydrogen has a high neutron absorption, so the presence of water in regions of a product can be easily detected by neutron radiography. X rays, on the other hand, are readily transmitted through water.

In addition, when the penetrating beam passes through an object, part of the radiation is scattered in all directions, producing an overall "fogging" of the radiograph and a reduction in contrast and image sharpness. The thicker the material, the more troublesome the scattered radiation becomes. Fortunately, measures can be taken to minimize this effect, such as beam filtering, beam collimation, shielding of the specimen, and shielding of the back and sides of the film cassette. Photographic considerations of exposure time and development also affect the quality of the radiographic image.

In order to have a reference for the image densities on a radiograph, it is standard practice to include a standard test piece, or *penetrameter*, in the exposure of the part being inspected. The penetrameters are made of the same material as the specimen or of similar material and contain structural features of known dimensions. The image of the penetrameter then permits direct comparison with the features present in the image of the product being inspected.

Density-sensitive aids have been developed to assist in detecting subtle, but important, details in the image. Television cameras connected to electronic processors with image-enhancing software can do much to accentuate density variations.

Because of the expense of testing, many users recommend extensive use of radiography during the development of a product and a process, followed by spot checks and statistical methods for subsequent production.

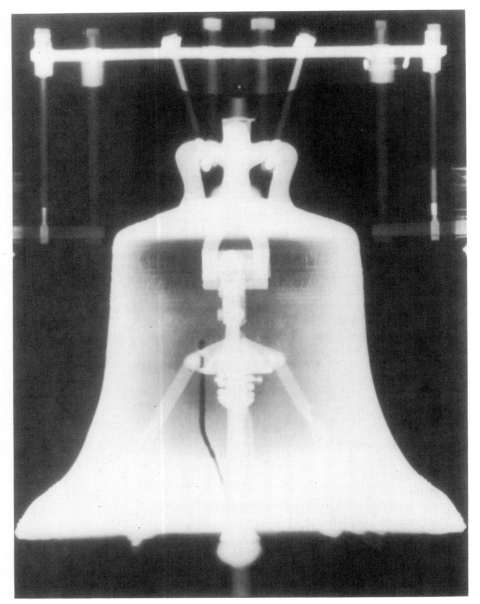

FIGURE 11-7 Full-size radiograph of the Liberty Bell. Photo reveals the famous crack, as well as the iron spider installed in 1915 to support the clapper and the steel beam and supports which were set into the yoke in 1929. (Courtesy Eastman Kodak Company.)

TABLE 11.5 **Radiography**

Method: Radiography.

Principle: Some form of radiation is passed through the material and is differentially absorbed depending on the thickness, the type of material, and the presence of flaws or defects.

Advantages: Probes the internal regions of a material; provides a permanent record of the inspection; can be used to determine thickness of a material; very sensitive to density changes.

Limitations: Most costly of the nondestructive testing methods (involves expensive equipment, film, and processing); radiation precautions are necessary (potentially dangerous to human health); the defect must be at least 2% of the total section thickness to be detected (thin cracks can be missed if oriented perpendicular to the beam); film processing requires time, facilities, and care; complex shapes can present problems; location of an internal defect requires a second inspection at a different angle.

Material limitations: Applicable to most engineering materials.

Geometric limitations: Complex shapes can present problems in setting exposure conditions and obtaining proper orientation of source, specimen, and film. Two-side accessibility is required.

Permanent record: A photographic image is part of the standard test procedure.

When an electrically conductive material is exposed to an alternating magnetic field, such as that generated by a coil of wire carrying an alternating current, small electric currents are generated on or near the surface of the material, as indicated in Figure 11-8. These *eddy currents*, in turn, generate their own magnetic field, which then interacts with the magnetic field of the exciting coil to change its electrical impedance. By measuring the impedance of the exciting coil, or a separate indicating coil, eddy current testing can be used to detect any condition that would affect the current-carrying conditions (or conductivity) of the test specimen, such as that indicated in Figure 11-9.

Eddy current testing can be used to detect surface and near-surface flaws, such as cracks, voids, inclusions, seams, and even stress concentrations. Differences in metal chemistry or heat treatment will affect the magnetic permeability and conductivity of a metal and, hence, the eddy current characteristics. Material mix-ups can be detected. Specimens can be sorted by hardness, case depth, residual stresses, or any other structure-related property. Thicknesses or variations in thickness of platings, coatings, or even corrosion can be detected or measured.

Test equipment can range from simple, portable units with hand-held probes to fully automated systems with computer control and analysis. Each system, however, includes:

1. A source of magnetic field capable of inducing eddy currents in the part being tested. This source generally takes the form of a coil (or coil-containing probe) carrying alternating current. Various coil geometries are used for different-shaped specimens.

Eddy Current Testing

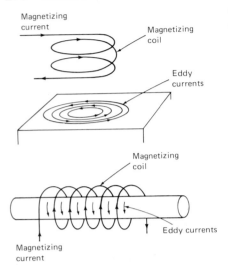

FIGURE 11-8 Relation of the magnetizing coil, magnetizing current, and induced eddy currents. Note: the magnetizing current is actually an alternating current such that the magnetic field forms, collapses, and reforms in the opposite direction. This changing field induces the eddy currents and the changes in the eddy currents induce the secondary magnetic field which interacts with the sensor coil or probe.

TABLE 11.6 Eddy Current Testing

Method: Eddy current testing.

Principle: When an electrically conductive material is brought near an alternating current coil, producing an alternating magnetic field, surface currents (eddy currents) are generated in the material. These surface currents generate their own magnetic field, which interacts with the original, modifying the impedance of the originating coil. Various material properties and defects affect the magnitude and nature of the induced eddy current and can be detected by the electronics.

Advantages: Can detect both surface and near-surface irregularities; applicable to both ferrous and nonferrous metals; versatile—can detect flaws, variations in alloy or heat treatment, variations in plating or coating thickness, wall thickness, and crack depth; intimate contact with the specimen is not required; can be automated; electrical circuitry can be adjusted to select sensitivity and function; pass–fail inspection is easily conducted; high speed; low cost; no final cleanup is required.

Limitations: Response is sensitive to a number of variables, so interpretation may be difficult; sensitivity varies with depth, and depth of inspection depends on the test frequency; reference standards are needed for comparison; trained operators are generally required.

Material limitations: Applicable only to conductive materials, such as metals; some difficulties may be encountered with ferromagnetic materials.

Geometric limitations: Depth of penetration is limited; must have accessibility of coil or probe; constant separation between coils and specimen is required for good results.

Permanent record: Electronic signals can be recorded for permanent record by means of devices such as strip chart recorders.

2. A means of sensing the field changes caused by the interaction of the eddy currents with the original magnetic field. Either the exciting coil itself or a secondary sensing coil can be used to detect the impedance changes. Differential testing can be performed by means of two oppositely wound coils wired in series. In this method, only differences in the signals between the two coils are detected as one or both coils are scanned over the specimen.

3. A means of measuring and interpreting the resulting impedance changes. The simplest method is to measure the induced voltage of the sensing coil, a reading that evaluates the cumulative effect of all variables affecting the

FIGURE 11-9 While eddy currents are constrained to travel within the material, the principles of detectability are quite similar to those of magnetic particle inspection. The eddy current method is more flexible, however, since it can detect features such as differences in heat treatment which simply alter conductivity.

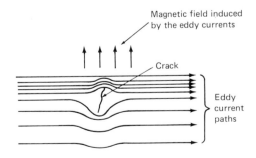

eddy current field. Phase analysis can be used to determine the magnitude and direction of the induced eddy current field. Familiarity with characteristic impedance responses can then be used to identify selected features in the specimen.

According to one comparison, eddy current testing was not as sensitive as penetrant testing in detecting small, open flaws, but the eddy current method requires none of the cleanup operations and is noticeably faster. Likewise, it is not as sensitive to small, subsurface flaws as magnetic particle inspection but can be applied to all metals (ferromagnetic and nonferromagnetic alike). In addition, eddy current testing offers capabilities that cannot be duplicated by the other methods, such as the ability to differentiate chemistries and heat treatments.

Materials undergoing stressing, deformation, or fracture emit sound waves in frequencies as high as 1 megahertz. While these sounds are inaudible to the human ear, they are detectable through the use of sophisticated electronics. Transducers, amplifiers, filters, counters, and microcomputers can be used to isolate and analyze the sonic emissions of cracking or deforming material. Much like the warning sound of ice cracking underneath skates, the acoustic emissions of materials can be used to provide a warning of impending danger. They can detect deformations as small as 10^{-12} in/in that occur in short intervals of time, as well as the delamination of layered materials and fiber failure in composites. In addition, multiple sensors can be coupled to accurately pinpoint the source of these sounds in a triangulation method similar to that used to locate seismic sources (earthquakes) in the earth.

Acoustic Emission Monitoring

TABLE 11.7   Acoustic Emission Monitoring

Method: Acoustic emission monitoring

Principle: Almost all materials emit high-frequency sound (acoustic emissions) when stressed, deformed, or undergoing structural changes, such as the formation or growth of a crack or a defect. These emissions can now be detected and provide an indication of dynamic change within the material.

Advantages: The entire structure can be monitored with near-instantaneous detection and response; only "active" flaws are detected; defects inaccessible to other methods are detected; inspection can be in harsh environments; the location of the emission source can be determined.

Limitations: Only growing flaws can be detected (the mere presence of defects is not detected); background signals may cause difficulty; there is no indication of the size or shape of the flaw; experience is required to interpret the signals.

Material limitations: Virtually unlimited.

Geometric limitations: Requires a continuous sound-transmitting path between the source and the detector; size and shape also determine the relative strengths of the emissions signals that reach the detector.

In essence, acoustic emission monitoring involves simply listening for failure indications. Temporary or transient monitoring can be used to detect the formation of cracks in materials during production, such as in welding operations and the subsequent cooling of the weld region. Monitoring can also be performed during preservice proof testing. Continuous surveillance is often used where the application is particularly critical, such as in bridges and nuclear reactor vessels. The sensing electronics can be coupled with an alarm and a shutdown system to protect and maintain the integrity of the structure.

In contrast to the previous inspection methods, acoustic emission is not a means of detecting an existing defect in a static product. Instead, it is a monitoring technique designed to detect a dynamic change in the material, such as the formation or growth of a crack, a fracture, or a defect, or the onset of plastic deformation.

Other Methods of Nondestructive Testing and Inspection

Leak Testing.   Leak testing is a form of nondestructive testing designed to determine the absence or existence of leak sites and the rate of material loss through the leaks. Various testing methods have been developed, ranging from the rather crude bubble-emission test (pressurize, immerse, and look for bubbles) to advanced techniques involving tracers, detectors, and sophisticated apparatus. Each has its characteristic advantages, limitations, and sensitivity. Selection should be made on the basis of cost, sensitivity, and reliability.

Thermal Methods.   Temperature-based techniques can also be used to evaluate the soundness of engineering materials and components. The parts are heated, and various means are used to detect abnormal temperature distributions, indicative of faults or flaws. Temperature-sensing tools include, thermometers, thermocouples, pyrometers, temperature-sensitive paints and coatings, liquid crystals, infrared scanners, and infrared film. The location of "hot spots" on an operating component can be a valuable means of defect detection and can provide advanced warning of impending failure. Electrical components are frequently inspected by this technique, since the faulty components tend to be hotter than their defect-free counterparts. Composite materials can be subjected to brief pulses of intense heat and can then be inspected to reveal the thermal diffusion pattern. Thermal anomalies appear in the areas of poor bonding between the components.

Strain Sensing.   While primarily used in laboratories as a part of product development, strain-sensing techniques can be used to provide valuable insight into the stresses and the stress distribution within a part. Brittle coatings, photoelastic coatings, or electrical-resistance strain gages are applied to the external surfaces of the part, which is then subjected to an applied stress. The extent and nature of cracking, the photoelastic pattern produced as a result of thinning of the coating, or electrical resistance changes then provide insight into the strain at various locations. X-ray diffraction methods and extensiometers have also been used.

Advanced Optical Methods.   Visual inspection is the simplest of the optical methods and can be assisted by a variety of means, as previously discussed. More recently, we have seen the development of several advanced optical methods. Monochromatic, coherent, high-intensity laser light has been used to detect differences in the backscattered pattern from a part and a "master." The presence or absence of geometrical features, such as holes or gear teeth, is readily detected. Holograms can provide three-dimensional images of an object, and holographic interferometry can be used to detect minute changes in an object under stress.

Resistivity Methods.   The electrical resistivity of a conductive material is a function of its chemistry, its processing history, and its structural soundness. Changes in resistivity from one sample to another can therefore be used for alloy identification, flaw detection, and the assurance of proper processing (such as evaluating heat treatment, the amount of cold work, the integrity of welds, or the depth of case hardening). The development of a sensitive micro-ohm-meter (microhmeter) has greatly expanded the possibilities in this area.

Review Questions

1. What are some of the unattractive features of destructive testing methods?
2. What is a proof test, and what assurance does it provide?
3. How can hardness tests be used to indicate quality?
4. What exactly is nondestructive testing, and what are some attractive features of the approach?
5. What are some of the factors that should be considered when selecting a nondestructive testing method?
6. What are some of the optical aids that may be used during a visual inspection?
7. What are some types of defects that can be detected in a liquid penetrant test?
8. Describe how the orientation of a flaw with respect to a magnetic field can affect its detectability during magnetic particle inspection.
9. Why is part demagnetization a necessary step in magnetic particle inspection?
10. What is the major limitation of "sonic testing," where one listens to the characteristic "ring" of a product in an attempt to detect defects?
11. Why is it necessary to use an acoustic coupling medium to transmit the ultrasonic waves from the transducer into the product?
12. What are some of the types of radiation used in the radiographic inspection of manufactured products?
13. What is a penetrameter, and what is its use in a radiographic inspection?
14. What types of defects or product features can be detected by an eddy current inspection?
15. How does eddy current testing compare with liquid-penetrant and magnetic-particle testing in its ability to detect flaws and in its general usefulness?
16. Why can't acoustic emission methods detect the presence of an existing, but static, defect?
17. How do the various thermal methods of nondestructive testing reveal the presence of defects?
18. What are some of the product features that can be evaluated by the resistivity methods?
19. One manufacturing company routinely uses X ray radiography to ensure the absence of cracks in its products. The primary reason for selecting radiography is the production of a "hard-copy" record of each inspection for use in any possible liability litigation. Do you agree with their selection? If not, which of the other processes might you prefer, and why?
20. For each of the methods listed below, cite one major limitation, in its use:
    —Visual inspection
    —Liquid penetrant inspection
    —Magnetic particle inspection
    —Ultrasonic inspection
    —Radiography
    —Eddy current testing
    —Acoustic emission monitoring

Process Capability and Quality Control

All manufacturing processes display some level of inherent capability or some inherent uniformity or nature. For example, suppose that we view "shooting at a metal target" as a "process" for putting holes in a piece of metal. I hand you the gun and tell you to take five shots at the bull's-eye. Thus, you are the operator of the process. In order to measure process capability (PC)—that is, your ability to consistently hit what you are aiming at, the bull's-eye—the target is inspected after you have finished shooting. So the capability of manufacturing processes is determined by measuring the output of the process. In quality control (QC), the product is examined to determine whether or not the processing accomplished was what was specified in the design. In QC, the objective is to inspect to prevent defects from occurring, not to find the defects (deviations from the ideal or target that will make the products defective) after they have occurred (see Figure 12-1).

PC studies use statistical and analytical tools, as does QC, except that the results are directed at the machines used in the processing rather than the output or products from the processes. Going back to our example, a PC study would be directed toward quantifying the inherent accuracy and precision in the shooting process. The QC program would be geared to root out the problems that can cause defective products during production.

Determining Process Capability

The *nature of the process* refers to both the variability (or inherent uniformity) and the aim of the process. Thus, in the target-shooting example, a perfect process would be capable of placing five shots right in the middle of the bull's-eye, one right on top of the other. The process would display no variability. Such performance would be very unusual in a real industrial process. The variability may have assignable causes and may be correctable if the cause can be found and eliminated. That variability to which no cause can be assigned and

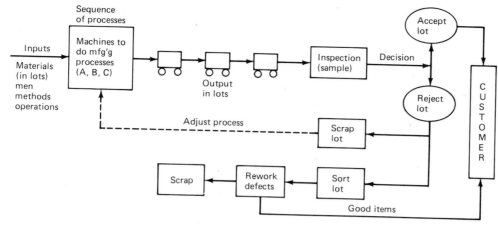

(a) Inspection that finds defects

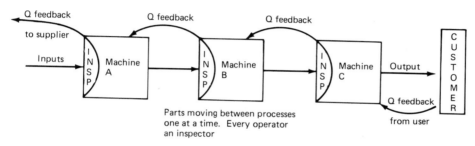

(b) Inspection to prevent defects

FIGURE 12-1   Don't inspect to find defects. Inspect to prevent the defect from happening.

which cannot be eliminated is inherent in the process and is therefore its "nature."

Some examples of assignable causes of variation in processes include multiple machines for the same components, operator blunders, defective materials, or progressive wear in the tools during machining. Sources of inherent variability in the process include variation in material properties, operator variability, vibrations and chatter, and the wear of the sliding components in the machine, perhaps resulting in poorer operation of the machine. These kinds of variations, which occur naturally in processes, usually display a random nature and often cannot be eliminated. In QC terms, these are referred to as *chance causes*. Sometimes, the causes of assignable variation cannot be eliminated because of cost. Almost every process has multiple causes of variability occurring simultaneously, so it is extremely difficult to separate the effects of the different sources of variability during the analysis.

Making PC Studies.   The object of the PC study is to determine the inherent nature of the process as compared to the desired specifications. The output of the process must be examined under normal conditions, or what is typically called *hands-off conditions*. The inputs (e.g., materials, setups, cycle times, temperature, pressure, and operator) are fixed or standardized. The process is allowed to run without tinkering or adjusting while the output (i.e., the product or units or components) is carefully documented with respect to (1) time; (2) source; and (3) order of production. A sufficient number of data have to be taken to ensure confidence in the statistical analysis of the data. The precision of the measurement system should exceed the process capability by at least one order of magnitude (see the rule of 10 in Chapter 10).

Before any data collection, these steps must be taken:

1. Design the PC experiment.
   A. Use "normal" or "hands-off" process conditions; specify machine settings for speed, feed, volume, pressure, material, temperature, operator, and so on. (This is the standard PC study approach.)
   Alt. A. Use specified combinations of all of the input parameters, at various levels, that are believed to influence the quality characteristics being measured. These combinations should be run with the objective of selecting the best. For example, speed levels may be high, normal, and low, or operators may be fast or slow. (This is the Taguchi or factorial approach.)
2. Define the inspection method and the inspection means (the procedure and the instrumentation).
3. Decide how many items will be needed to perform the statistical analysis. (For a Taguchi approach, this means deciding how many replications at each level of combination are desired.)
4. For the standard PC study, use homogeneous input material, and try to contrast it with normal (more variable) input material. (For a Taguchi approach, material *is* an input variable specified at different levels: normal, homogeneous, and highly variable. If a material is not controllable, it is considered a noise factor.)
5. Data sheets must be designed to record date, time, source, order of production, and all the process parameters being used (or measured) while the data are being gathered. (For a Taguchi method, this is usually a simplified experimental design, called an orthogonal array.)
6. Assuming that the standard PC study approach is being used, the process is run, and the parts are made and measured.

Let us assume that the designer specified the part to be 1.000 ± 0.005 in.

After 70 units have been manufactured without any adjustment of the process, each unit is measured, and the data are recorded on the data sheet. A frequency distribution, or histogram, as shown in Figure 12-2, is developed. This histogram shows the raw data and the desired value, along with the upper and lower tolerance limits, where LSL = lower specification limit and USL = upper specification limit. (These are also called the (lower and upper) specification

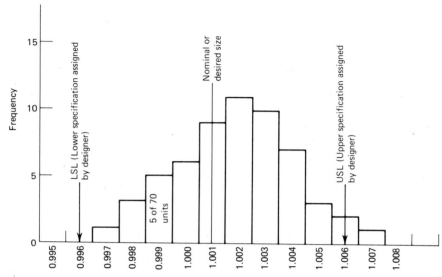

FIGURE 12-2   Histogram of 70 measurements of a parameter. The design specification was 1.001 ± 0.005 inch.

limits.) The statistical data are used to estimate the mean and the standard deviation of this distribution.

The mechanics of this statistical analysis are shown in Figure 12-3. The true mean of the distribution, designated $\bar{X}'$ (X bar prime) is to be compared with

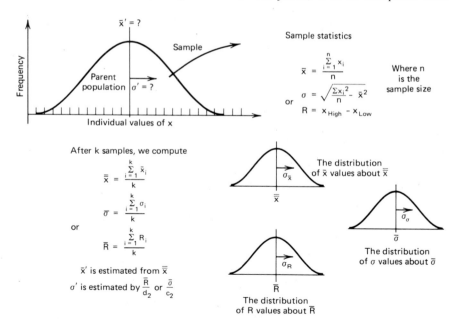

Sample statistics

$$\bar{x} = \frac{\sum\limits_{i=1}^{n} x_i}{n}$$ Where n is the sample size

or $$\sigma = \sqrt{\frac{\sum x_i^2}{n} - \bar{x}^2}$$

$$R = x_{High} - x_{Low}$$

After k samples, we compute

$$\bar{\bar{x}} = \frac{\sum\limits_{i=1}^{k} \bar{x}_i}{k}$$

$$\bar{\sigma} = \frac{\sum\limits_{i=1}^{k} \sigma_i}{k}$$

or

$$\bar{R} = \frac{\sum\limits_{i=1}^{k} R_i}{k}$$

$\bar{x}'$ is estimated from $\bar{\bar{x}}$

σ' is estimated by $\dfrac{\bar{R}}{d_2}$ or $\dfrac{\bar{\sigma}}{c_2}$

The distribution of $\bar{x}$ values about $\bar{\bar{x}}$

The distribution of σ values about $\bar{\sigma}$

The distribution of R values about $\bar{R}$

Each distribution of averages of samples, or standard deviation of samples or ranges of samples must have its own average and standard deviation.

FIGURE 12-3   Calculations needed to obtain estimates of the mean ($\bar{X}'$) and the standard deviation (σ') of the parent population, as in a process capability study.

the nominal value. The estimate of the true standard deviation, designated σ' (sigma prime), is used to determine how the process compares with the desired tolerance. *The purpose of the analysis is to obtain estimates of $\overline{X}'$ and σ' values, the true process parameters, as they are not known.* A sample size of 5 was used in this example, so $n = 5$. Fourteen groups of samples were drawn from the process, so $k = 14$. For each sample, the sample mean $\overline{X}$ and the sample range R are computed. For large samples ($n > 12$), the standard deviation of each sample should be computed, rather than the range. Next, the average of the sample averages, $\overline{\overline{X}}$, is computed. This is sometimes called the *grand average*, and it is used to estimate the mean of the process, $\overline{X}'$. The standard deviation of the process, which is a measure of the spread or variability of the process, is estimated from either the average of the sample ranges, $\overline{R}$ or the average of the sample standard deviations, $\overline{\sigma}$, using either $\dfrac{\overline{R}}{d_2}$ or $\dfrac{\overline{\sigma}}{c_2}$. The factors d_2 and c_2 depend on the sample size n and are given in Table 12.1. The process capability is defined by $\pm 3\sigma'$ or $6\sigma'$. Thus, $\overline{X}' \pm 3\sigma'$ defines the natural capability limits of the process, assuming the process is approximately normally distributed.

Note that a distinction is made between a sample and a population. A sample is of a specified, limited size and is drawn from the population. The population is the large source of items, which can include all the items the process will ever produce under the specified conditions. Our calculations assume that this population was normal or bell-shaped. Figure 10-6 shows a typical normal curve and the areas under the curve as defined by the standard deviation. Other distributions are possible, but the histogram clearly suggested that this process can best be described by a normal probability distribution. Now it remains for the process engineer and the operator to combine their knowledge of the process with the results from the analysis in order to draw conclusions about the ability of this process to meet specifications.

What PC Studies Tell About the Process.   PC studies provide the answers to these questions:

1. Does the process have the ability to meet specifications?
   To answer this question, a process capability index, C_p, is often computed, where

$$C_p = \frac{\text{Tolerance Spread}}{6\sigma'} = \frac{USL - LSL}{6\sigma'} \tag{12-1}$$

   A value of $C_p \geq 1.33$ is considered good.
2. Is the process well centered with respect to the desired nominal specification?
   To answer this question, the degree of bias (β) is computed, where

$$\beta = \frac{\text{Estimated Process Mean} - \text{Nominal}}{1/2(\text{Tolerance Spread})} = \frac{\overline{X}' - \text{Nominal}}{1/2(USL - LSL)} \tag{12-2}$$

TABLE 12.1   Factors for Estimating the Standard Deviation (σ') of a Parent Population from Sample Data for the Average Range ($\bar{R}$) or the Average Sample Standard Deviation ($\bar{\sigma}$).

Sample Size, or the Number of Observations in Subgroup n	Range Factor, for Estimate from $\bar{R}$, $d_2 \equiv \bar{R}/\sigma'$	Standard Deviation Factor, for Estimate from $\bar{\sigma}$, $c_2 \equiv \bar{\sigma}/\sigma'$	Factor Which Relates σ_R, with σ' (σ_R = Std. Dev. of Range Distribution), $d_3 = \sigma_R/\sigma'$
2	1.128	0.5642	0.8525
3	1.693	0.7236	0.8884
4	2.059	0.7979	0.8798
5	2.326	0.8407	0.8641
6	2.534	0.8686	0.8480
7	2.704	0.8882	0.8330
8	2.847	0.9027	0.8200
9	2.970	0.9139	0.8084
10	3.078	0.9227	0.7970
11	3.173	0.9300	0.7870
12	3.258	0.9359	0.7780
13	3.336	0.9410	For $n > 12$, C_2
14	3.407	0.9453	factors should be
15	3.472	0.9490	used with $\bar{\sigma}$
16	3.532	0.9523	
17	3.588	0.9551	
18	3.640	0.9576	
19	3.689	0.9599	
20	3.735	0.9619	
25	3.931	0.9696	
30	4.086	0.9748	

Source: 1950 Manual on Quality Control of Materials; copyright ASTM, 1916 Race St., Philadelphia, PA 19103. Adapted, with permission.

The condition for not producing rejects is

$$\bar{X}' - \text{Nominal} + 3\sigma' < 1/2(USL - LSL) \qquad (12\text{-}3a)$$

or

$$C_p(1 - \beta) > 1 \qquad (12\text{-}3b)$$

Another capability index that combines both factors is

$$C_{pk} = \frac{\text{MIN}\{|USL - \bar{X}'|, |LSL - \bar{X}'|\}}{3\sigma'} \qquad (12\text{-}4)$$

In response to the first question, the width of the histogram is compared with the specifications (see Figure 12-4). In addition, the natural specification limits

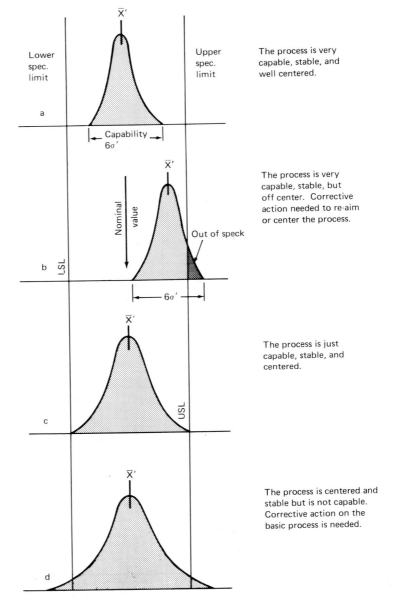

FIGURE 12-4 What P.C. studies tell about the process.

are computed and are then compared with the upper and lower tolerance limits. Three situations exist:

 I. $6\sigma' < USL - LSL$ or $C_p > 1$, or process variability is less than tolerance spread (see Figure 12-4a).

 II. $6\sigma' = USL - LSL$ or $C_p = 1$, or process variability is equal to tolerance spread (see Figure 12-4c).

III. $6\sigma' > USL - LSL$ or $C_p < 1$, or process variability is greater than tolerance spread (see Figure 12-4d).

In Situation I, the machine is capable of meeting the tolerances applied by the designer. Generally speaking, if process capability is on the order of two thirds to three quarters of the design tolerance, there is a high probability that the process will produce all good parts over a long period of time. If the PC is on the order of one half or less of the design tolerance, it may be that the selected process is "too good." That is, the company is making ball bearings when what is called for are marbles. In this case, it may be possible to trade off some precision in this process for looser specifications elsewhere, resulting in an overall economic gain. Quality in well-behaved processes can be maintained by checking the first and last part of a lot or production run. If these parts are good, then the lot is certain to be good. This is called $N = 2$. Naturally if the lot size is 2 or less, this is 100% inspection. Sampling and control charts are also used under these conditions to maintain the process aim and variability.

In Situation III, the process is not capable of meeting the design specifications. There are a variety of alternatives here, including:

a. Shifting this job to another machine with greater process capability.
b. Getting a review of the specifications to see if they may be loosened.
c. Sorting the product to separate the good from the bad. This entails 100% inspection of the product, which may not be a viable economic alternative unless it can be done automatically. Automatic inspection of the product on a 100% basis can ensure near-perfect quality of all the accepted parts. The automated station shown in Figure 12-5 checks parts for the proper diameter with the aid of a linear variable differential transformer (LVDT). As a part approaches the inspection station on a motor-driven conveyor system, a computer-based controller activates a clamping device. Embedded in the clamp is an LVDT position sensor with which the control computer can measure the diameter of the part. Once the measurement

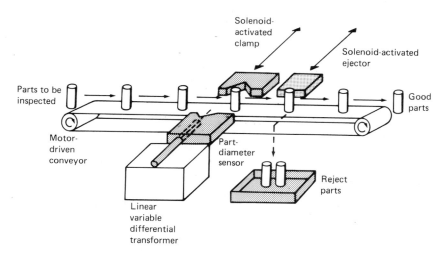

FIGURE 12-5 A linear differential transformer (LVDT) is a key element in an inspection station checking part diameters. Momentarily clamped into the sensor fixture, a part pushed the LVDT armature into the device winding. The LVDT output is proportional to the displacement of the armature. The transformer makes highly accurate measurements over a small displacement range.

has been made, the computer releases the clamp, allowing the part to be carried away. If the diameter of the part is within a given tolerance, a solenoid-actuated gate operated by the computer lets the part pass. Otherwise, the part is ejected into a bin. With the fast-responding LVDT, 100% of manufactured parts can be automatically sorted quickly and economically.

Automated sorting does not determine what caused the defects, so this example is an automated defect-finder. How would one change this inspection system to make it "inspect to prevent" the defect from occurring?

d. Determining whether the precision of the process can be improved by:
   (1) Switching cutting tools, workholding devices or materials.
   (2) Overhauling the existing process and/or developing a preventive maintenance program.
   (3) Finding and eliminating the causes of variability, or
   (4) Combinations of (1), (2), and (3).
   (5) Using a factorial (Taguchi) procedure, that is, designing experiments to reduce the variability of the process.

In Figure 12-4c, the process capability is almost exactly equal to the assigned tolerance spread, so if the process is not perfectly centered, defective products will always result. Thus, Situation II should be treated like Situation III unless the process can be perfectly centered and maintained. Tool wear, which causes the distribution to shift, must be negligible. Then the situation can be treated as that in I, particularly if a small percentage of parts just outside tolerance is acceptable.

The second question deals with the ability of the process to maintain centering so that the average of the distribution comes as close as possible to the desired nominal value (see Figure 12-4b). This process needs to be reaimed or centered so that the mean of the process distribution is at or near the nominal (print) value. Most processes can be reaimed. Poor accuracy is often due to assignable causes, which can be eliminated.

In addition to direct information about the accuracy and the precision of the process, PC studies can also tell the manufacturing engineer how pilot processes compare with production processes, and vice versa. If the source and time of each product are carefully recorded, information about the instantaneous reproducibility can be found and compared with the repeatability of the process with respect to time (time-to-time variability). More important, since almost all processes are duplicated, PC studies generate information about machine-to-machine variability. Suppose, going back to our target-shooting example, that five different guns were used, all of the same make and type. The results would have been different, just as having five marksmen use the same gun would have resulted in yet another outcome. Thus, PC studies generate information about the homogeneity and the differences in multiple machines and operators.

It is quite often the case in such studies that one variable dominates the process. Target shooting viewed as a process is "operator"-dominated in that the outcome is highly dependent on the skill of the "worker." Processes that are not well engineered nor highly automated, or in which the worker is viewed as "highly skilled," are usually operator-dominated. Processes that change or shift uniformly with time but that have good repeatability in the short run are often "machine"-dominated. For example, the mean of a process $(\overline{X}')$ will usually shift after a tool change, but the variability may decrease or remain unchanged. Machines tend to become more precise (to have less variability within a sample) after they have been "broken in" (i.e., the rough contact surfaces have smoothed out because of wear) but will later become less precise (will have less repeatability) because of poor fits between moving elements (called *backlash*) of the machine under varying loads. Other variables that can dominate processes are setup, input components, and even information.

In many machining processes in use today, the task of tool setting has been replaced by an automatic tool positioning capability (see Chapter 29), which means that one source of variability in the process has been eliminated, and the process becomes more repeatable. In the same light, it will be very important in the future for manufacturing engineers to know the process capability of the robots they want to use in the workplace.

The discussion to this point has assumed that the parent population is normally distributed, that is, has the classic bell-shaped distribution in which the percentages shown in Figure 10-6 are dictated by the number of standard deviations from the central value or mean. The shape of the histogram may reveal the nature of the process to be skewed to the left or the right (unsymmetrical), often indicating some natural limit in the process. Drilled holes exhibit such a trend as the drill tends to make the hole oversize. Another possibility is a bimodal distribution (two distinct peaks), often caused by two processes' being mixed together. The possibilities are endless and require a careful recording of all the sources of the data to track down the factors that result in loss of precision and accuracy in the process. Rapid feedback on quality is perhaps the most important factor.

Taguchi Method.   The second approach to making a PC study may involve the Taguchi method, which requires a more sophisticated analysis. The Taguchi approach uses a truncated experimental design (called an *orthogonal array*) to determine which process inputs have the greatest effect on process variability (i.e., precision) and which have the least. Those inputs that have the greatest influence are set at levels that minimize their effect on process variability. Those having the least influence are used to adjust or recenter the process aim. In other words, Taguchi methods seek to minimize or dampen the effect of the causes of variability and thus to reduce the total process variability (see Figure 12-6).

In Chapter 29, process capability is addressed again in the context of machining centers, programmable (NC) machines. In machine tools, accuracy and precision in processing are affected by machine alignment, the setup of the workholder, the design and rigidity (accuracy) of the workholder, the accuracy of the cutting tools, the design of the product, the temperature, and the operating parameters. Taguchi methods provide a means of determining which of the input parameters are most influential in product quality.

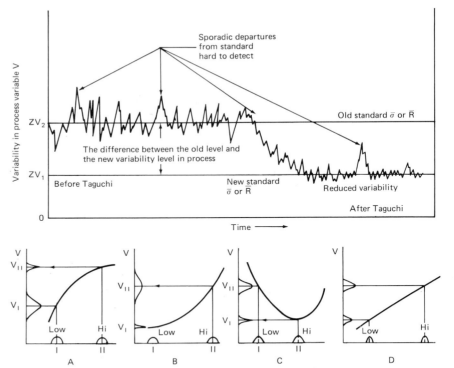

FIGURE 12-6 The Taguchi method can reduce the process variability, as measured by range values (R) or standard deviations (σ).

Inspection and Quality Control

In virtually all manufacturing, it is extremely important that the dimensions and quality of individual parts be known and maintained. This is of particular importance where large quantities of parts, often made in widely separated plants, must be capable of interchangeable assembly. Otherwise, difficulty may be experienced in subsequent assembly or in service, and costly delays and failures may result. In recent years, defective products resulting in death or injury to the user have resulted in expensive litigation and damage awards against manufacturers. Inspection is that function which controls the quality (e.g., the dimensions, the performance, and the color)—manually, by using operators or inspectors, or automatically, by using machines, as discussed previously.

The economics-based question "How much should be inspected?" has three possible answers.

1. *Inspect every item being made*; 100% checking with prompt execution of feedback and immediate corrective action can ensure perfect quality.
2. *Sample*. Inspect some of the product by sampling and make decisions about the quality of the process based on the sample.
3. *None*. Assume that everything made is acceptable or that the product is inspected by the consumer, who will exchange it if it is defective. (This is not a recommended procedure.)

The reasons for not inspecting all of the product (i.e., for sampling) include:

1. The test is destructive.
2. There is too much product for all of it to be inspected.
3. The testing takes too much time or is too complex or too expensive.
4. It is not economically feasible to inspect everything even though the test is simple, cheap, and quick.

Some characteristics are nondissectible; that is, they cannot be measured during the manufacturing process because they do not exist until after a whole series of operations have taken place. The final edge geometry of a razor blade is a good example, as is the yield strength of a rolled bar of steel.

Sampling or looking at some percentage of the whole requires the use of statistical techniques that permit decisions about the acceptability of the whole based on the quality found in the sample. This is known as *statistical process control* (SPC).

Statistical Process Control.   Looking at some (sampling) and deciding about the behavior of the whole (the parent population) is common in industrial inspection operations. The basic SPC techniques are the histogram and control charts. The histogram (see Figure 12-7) could be used to classify the range of shaft diameters after turning. The X axis represents the diameter measurement, and the Y axis represents the frequency with which these measurements were found. The natural process capability limits of the distribution can be compared with the engineering specifications to determine if the process is centered at the print nominal value and to estimate if diameters are being produced within tolerance values (or product specifications).

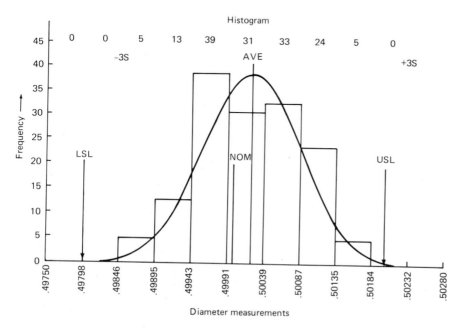

FIGURE 12-7   Histogram of a process to compare measured sizes with designer specifications.

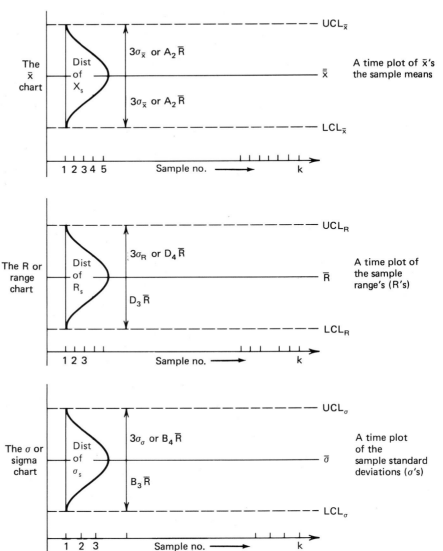

FIGURE 12-8 The basic design of the X chart, the R chart, and the σ chart used in SPC. Values for A_2, D_3, D_4 given in Table 12-2. Also see Figure 12-3.

Control charts for variables are used to monitor the output of a process by sampling (looking at some), by measuring selected quality characteristics, by plotting the sample data on the chart, and then by making decisions about the performance of the process.

Figure 12-8 shows the basic structure of three charts commonly used for variable types of measurements. The $\overline{X}$ chart tracks the aim (accuracy) of the process. The R chart (or σ chart) tracks the precision or variability of the process. Usually, only the $\overline{X}$ chart and the R chart are used, unless the sample size is large, and then σ charts are used in place of R charts. *The data plotted on these charts are sample values, not individual values.* Because some sample statistics tend to be normally distributed about their own mean (see Figure 12-3), $\overline{X}$ values are normally distributed about $\overline{\overline{X}}$, R values are normally distributed about $\overline{R}$, and σ values are normally distributed about $\overline{\sigma}$.

Sampling Errors.   It is important to understand that in sampling, two kinds of decision errors are always possible. Suppose that the process is running perfectly but the sample data indicate that something is wrong. You decide to stop the process to make adjustments. This is a Type I error. Suppose that the process was not running perfectly and was making defective products. However, the sample data did not indicate that anything was wrong. You did not stop the process and set it right. This is a Type II error. Both types of errors are possible in sampling. For a given sample size, reducing the chance of one type of error will increase the chance of the other. Increasing the sample size or the frequency of sampling reduces the probability of errors but increases the cost of inspection. Many companies determine the size of the errors they are willing to accept according to the overall cost of making the errors plus the cost of inspection. If, for example, a Type II error is very expensive in terms of product recalls or legal suits, the company may be willing to make more Type I errors, to sample more, or even to go to 100% inspection on very critical items to ensure that the company is not accepting defective materials as good and passing them on to the customer. The inspection should take place immediately after the processing.

As mentioned earlier, in any continuing manufacturing process, variations from established standards are of two types: (1) *assignable cause variations*, such as those due to malfunctioning equipment or personnel, or to defective material, or due to a worn or broken tool; and (2) normal *chance variations*, resulting from the inherent nonuniformities that exist in materials and in machine motions and operations. Deviations due to assignable causes may vary greatly. Their magnitude and occurrence are unpredictable, and one thus wishes to prevent their occurrence. On the other hand, if the assignable causes of variation are removed from a given operation, the magnitude and frequency of the chance variations can be predicted with great accuracy. Thus, if one can be assured that only chance variations will occur, the quality of the product will be better known, and manufacturing can proceed with assurance about the results. By using statistical process control procedures, one may detect the presence of an assignable-cause variation and remove the cause, often before it causes quality to become unacceptable. Also, the astute application of statistical experimental design methods (Taguchi experiments) can help to identify some assignable causes.

To sum up, use PC analysis and process improvements to get the process in the condition described in Figure 12-4a. Use control charts to keep the process centered ($\overline{X}$ chart) with no change in variability (R chart).

As an example, examine the data in Figure 12-9. This is the frequency distribution of measurements of the diameters of 100 ground pins, which represent the entire population in this case. These pins are supposed to have a nominal diameter of 0.500 in., (12.700 mm) and this was the size appearing most fre-

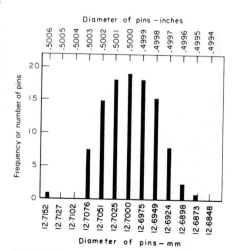

FIGURE 12-9   Frequency distribution of 100 ground pins having a nominal diameter of 12.700 mm (0.5000 inch).

quently (there were 18 of these). The population is assumed to be normal, and only chance variations are occurring. Even so, no pins between 0.5003 and 0.5006 in. were found. The mean for the entire population, μ, is obtained by

$$\mu = \frac{\sum_{i=1}^{n} X_i}{n} = \frac{\sum_{i=1}^{100} X_i}{100} \qquad (12\text{-}5)$$

where X_i is an individual measurement and n is the number of items or measurements. It is a measure of the central tendency around which the individual measurements tend to group. The variability of the individual measurements about the average may be indicated by the standard deviation, σ, where

$$\sigma = \sqrt{\frac{\sum_{i=1}^{n} (X_i - \mu)^2}{n}} \qquad (12\text{-}6)$$

The standard deviation is of particular interest. For normal distributions, 68.26% of all measurement values will lie within the $\pm 1\sigma$ range from the average, 95.46% within the $\pm 2\sigma$ range, and 99.73% within the $\pm 3\sigma$ range (refer to Figure 10-6). If the standard deviation for a population of items is known, then not over 0.07% of the items will fall outside the $\mu \pm 3\sigma$ *as long as only chance variations occur.* For the pin-grinding operation, the average diameter, μ, was 0.5000 in. (12.700 mm), and the standard deviation, σ, was 0.00019 in. (0.0048 mm), making $3\sigma = 0.006$ in. (0.0144 mm). The manufacturing specifications could be set at 0.5006 and 0.4994, with the assurance that fewer than 3 parts in 1000 would fall outside those limits because of chance causes of variation. More likely, a process would be selected that produced a distribution of parts whose $6\sigma'$ value is less than the total part tolerance.

Note that these are not the same values as were obtained earlier in the PC study. Here we are finding the mean and the standard deviation of 100 items, the entire population. In the PC study, the true mean and standard deviation of the population was being *estimated* from sample data, a more typical situation.

Quality Control Charts.   Quality control charts are widely used as aids in maintaining quality and in achieving the objective of detecting trends in quality variation before defective parts are actually produced. These charts are based on the previously discussed concept that if only chance causes of variation are present, the deviation from the specified dimension or attribute will fall within predetermined limits.

When sampling inspection is used, the typical sample sizes are from 3 to about 12 units. Figure 12-10 shows an example of two control charts. The $\overline{X}$ chart tracks the sample averages ($\overline{X}$ values), and the R chart plots the range values (R values). Let us assume that the data shown in Figure 12-9 now represent the results of 20 samples of Size 5 taken from a large number of ground pins, rather than the entire parent population. The sample data are then used to prepare the control charts shown in Figure 12-10.

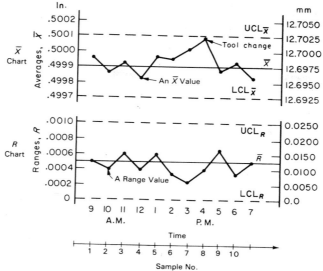

FIGURE 12-10   Statistical quality control charts for 12.700mm (0.5000inch) diameter pins. Note trend in $\overline{X}$ curve before tool change, indicating that the mean was shifting due to tool wear.

The center line of the $\overline{X}$ chart was computed before any actual use of the charts in control work:

$$\overline{\overline{X}} = \sum_{i=1}^{k} \overline{X}_i/k \qquad (12\text{-}7)$$

where $\overline{X}_i$ was the sample average and k was the number of sample averages. The horizontal axis for the charts is time; thus, indicating *when* the sample was taken, $\overline{\overline{X}}$ serves as an estimate for $\overline{X}'$, the true center of the process distribution and the center line of the $\overline{X}$ chart. The upper and lower control limits are commonly based on three standard error units (i.e., standard deviations for $\overline{X}$). Thus,

$$\begin{aligned} UCL_{\overline{x}} &= \text{upper control limit, } \overline{X} \text{ chart} = \overline{\overline{X}} + 3\sigma_{\overline{x}} \\ &= \overline{X}' + A_2\overline{R} \end{aligned}$$

$$(12\text{-}8)$$

$$\begin{aligned} LCL_{\overline{x}} &= \text{lower control limit, } \overline{X} \text{ chart} = \overline{\overline{X}} - 3\sigma_{\overline{x}} \\ &= \overline{X}' - A_2\overline{R} \end{aligned}$$

where $\sigma_{\overline{x}} = \dfrac{\sigma'}{\sqrt{n}}$ (see Table 12.2 for A_2 factor). The standard deviation for the distribution of sample averages, $\sigma_{\overline{x}}$, is determined directly from the estimate of the standard deviation of the parent population, σ'. The $\overline{X}$ chart is used to track the central tendency (aim) of the process. In this example, assume that samples are being taken hourly, and that the average of *each sample* (not individual values) is plotted.

The R chart is used to track the variability or dispersion of the process. A σ chart could also be used. R is computed for each sample ($X_{\text{HIGH}} - X_{\text{LOW}}$). The value of $\overline{R}$ was determined previously in the PC study from

Conditions in control

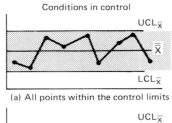

(a) All points within the control limits

(b) Random variation of points

Conditions out of control

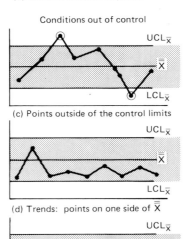

(c) Points outside of the control limits

(d) Trends: points on one side of $\overline{X}$

(e) Cyclic effect

Each point on the chart represents 4 measurements — ie., n = 4.

FIGURE 12-11   Examples of $\overline{X}$ control charts.

$$\overline{R} = \frac{\sum\limits_{i=1}^{k} R}{k} \qquad (12\text{-}9)$$

where $\overline{R}$ represents the average range of k range values. The range values may also be normally distributed about $\overline{R}$, with standard deviation σ_R. To determine the upper and lower control limits for the charts, the following relationships are used.

$$\begin{aligned} UCL_R &= \text{Upper control limit } R \text{ chart, } \overline{R} + 3\sigma_R = D_4\overline{R} \\ &= 2.11\overline{R} \text{ for } n = 5 \\ UCL_R &= \text{Lower control limit, } \overline{R} - 3\sigma_R = D_3\overline{R} = 0 \end{aligned} \qquad (12\text{-}9)$$

where D_4 and D_3 are constants and are given in Table 12.2. For small values of n, the distance between center line R and LCL_R is more than $3\sigma_R$, but LCL_R cannot be negative, as negative range values are not allowed, by definition. Hence, $D_3 = 0$ for values of n up to 6.

After control charts have been established, and the average and range values have been plotted for each sample group, the chart acts as a control indicator for the process. If the process is operating under chance cause conditions, the data will appear random (will have no trends or pattern; see Figure 12-11). If $\overline{X}$ or R values fall outside the control limits or if nonrandom trends (run or cyclic effects) appear, an assignable cause or change may have occurred, and some action should be taken to correct the problem.

As shown in Figures 12-10 and 12-11, trends in the charts often indicate the existence of an assignable cause factor before the process actually produces a point outside the control limit. Thus, $\overline{X}$ values for samples 5, 6, 7, and 8 show a trend toward oversized parts. In this grinding operation, the wheel has worn down (become undersized), so now the parts are becoming oversized, and corrective action should be taken (one redresses and resets the wheel or replaces it with a new wheel). Note that defective parts can be produced even if the points on the charts are in control. That is, it is possible for something to change in the process, causing defective parts to be made, and for the sample point still to be within the control limits. Since no corrective action was suggested by the charts, a Type II error was made. Subsequent operations will then involve performing additional work on products already defective. Thus, the effectiveness of the SPC approach in improving quality is often deterred by the lag in time between the occurrence of the abnormality and its discovery and the subsequent corrective action.

With regard to control charts in general, it should be kept in mind that the charts are capable of indicating only that something has (or has not) happened, and that a certain amount of detective work will be necessary to find out what has occurred to cause a break from the random, normal pattern of sample points on the charts. Keeping careful track of when and where the sample was taken is very helpful in such investigations, but the best procedure is to have the operator do the inspection and run the chart. In this way, quality feedback is very rapid and the causes of defects readily found.

TABLE 12.2 Factors for Determining $\overline{X}$ and R Chart Control Limits, 3 Standard Deviations Assumed

Number of Observations in Subgroup n	Factor for $\overline{X}$ Chart A_2	Factors for R Chart	
		Lower Control Limit D_3	Upper Control Limit D_4
2	1.88	0	3.27
3	1.02	0	2.57
4	0.73	0	2.28
5	0.58	0	2.11
6	0.48	0	2.00
7	0.42	0.08	1.92
8	0.37	0.14	1.86
9	0.34	0.18	1.82
10	0.31	0.22	1.78
11	0.29	0.26	1.74
12	0.27	0.28	1.72
13	0.25	0.31	1.69
14	0.24	0.33	1.67
15	0.22	0.35	1.65
16	0.21	0.36	1.64
17	0.20	0.38	1.62
18	0.19	0.39	1.61
19	0.19	0.40	1.60
20	0.18	0.41	1.59

Upper control limit for $\overline{X} = UCL_x = \overline{X}' + A_2\overline{R}$

Lower control limit for $\overline{X} = LCL_x = \overline{X}' - A_2\overline{R}$

Upper control limit for $R = UCL_R = D_4\overline{R}$

Lower control limit for $R = LCL_R = D_3\overline{R}$

Note that since $\sigma_{\bar{x}} = \dfrac{\sigma'}{\sqrt{n}}$, $3\sigma_{\bar{x}} = \dfrac{3\sigma'}{\sqrt{n}} = \dfrac{3\overline{R}}{\sqrt{n}\,d_2} = A_2\overline{R}$ where $A_2 = \dfrac{3}{\sqrt{n}\,d_2}$

Also $UCL_R = \overline{R} + 3\sigma_R = d_2\sigma' + 3d_3\sigma'$ (see Table 12.1).

Thus $UCL_R = (d_2 + 3\sigma_3)\sigma' = \left(1 + \dfrac{3d_3}{d_2}\right)\overline{R} = D_4\overline{R}$

Similarly, $D_3 = \left(1 - \dfrac{3d_3}{d_2}\right)$

Source: 1950 Manual on Quality Control of Materials; copyright ASTM, 1916 Race St., Philadelphia, PA 19103. Adapted, with permission.

The best way to isolate quality problems quickly is to make everyone an inspector. This means that every worker, foreman, supervisor, engineer, manager, and so on is responsible for making it right the first time and every time. One very helpful tool in this effort is the fishbone chart (see Figure 12-12). The problem (bad welds) can have many causes. Generally, the cause lies in the process, the operators, the materials, or the method, the four main branches on

Determining Causes for Problems in Quality

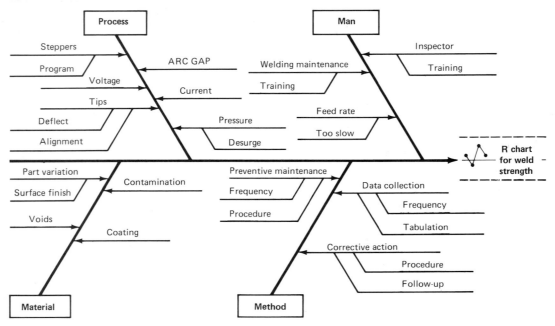

FIGURE 12-12 Fishbone diagram used to determine the causes of quality problems. The problem is bad welds.

the chart. Every time a quality problem is caused by one of these events, it is noted by the observer, and corrective action is taken. As before, experimental design procedures can help to identify the causes that affect performance.

In summary, the data gathered to develop the PC study can be used to prepare the initial control charts from the process after the removal of all assignable causes for variability and the proper setting of the process average. After the charts have been in place for some time, and a large number of data have been obtained from the process output, the PC study can be redone to give better estimates of the natural spread of the process during actual production. On-line or in-process methods, like SPC, and off-line (Taguchi) methods are key elements in a total quality control program.

Total Quality Control (TQC). Quality begins with optimizing the process design. The emphasis is on making it right the first time. Few companies have embarked on implementing Taguchi methods in order to reduce the inherent variability in their processes. A change in company attitude is needed in order to give the responsibility for quality to the worker, along with the authority to stop the process when something goes wrong. An attitude of defect prevention and a habit of constant improvement of quality are fundamental to the Integrated Manufacturing Production System (IMPS) (see Chapter 44). Companies like Toyota have accomplished TQC by extensive education of the workers, giving them the analysis tools they need (control charts with cause-and-effect diagrams) to find and expose the problems. Workers are encouraged to correct their own errors, and 100% inspection (often done automatically) is the rule. Passing defective products on to the next process is not allowed. The goal is perfection. *Quality circles*, now popular in the United States, are just one of the methods used by Japanese industries to achieve perfection.

The designer of the product must have quality in mind during the design phase, seeking the least costly means to ensure the quality of the desired functional characteristics. Major factors that can be handled during the early stages of the product design cycle include temperature, humidity, power variations, and the deterioration of materials and tools. Compensation for these factors is difficult or even impossible to implement after the product is in production. The distinction between superior quality and poor quality products can be seen in their variability in the face of internal and external causes. This is where Taguchi parameter design methods are critical.

The secret to successful process control is putting the control of quality in the hands of the workers. Many companies in this country are currently engaged in SQC (statistical quality control), but they are still inspecting to *find* defects. The number of defects will not be reduced if the inspection stage is merely made better or faster or is automated. Then, one is simply more efficient in discovering defects. The trick is to *inspect* to prevent defects. How can this be done? S. Shingo (1986) suggested these basic ideas: Use source inspection techniques that control quality at the stage where defects originate. Use 100% inspection with immediate feedback, rather than sampling. Make every worker an inspector. Minimize the time it takes to carry out corrective action. Remember that people are human and not infallible, so devise methods and devices that prevent them from making errors. Concentrate on making processes efficient, not on simply making the operators and operations more proficient. The latter approach seems to have been the American stumbling block. Industrial engineers can do operations improvement work, called *operations research*, but not process improvement work. We devise fancy, complex computerized solutions to solve complex manufacturing system problems. Why not simplify the manufacturing system so that the need for complex solutions disappears? For additional reading on this topic, see Chapter 42 on manufacturing systems and Chapter 44 on integrated manufacturing production systems.

Review Questions

1. Define a process capability study in terms of accuracy or precision.
2. What does the phrase *nature of the process* refer to?
3. Suppose you had a process that was accurate and precise, as shown in Figure 12-1a. What might the target look like if, occasionally while shooting, a sharp gust of wind blew left to right?
4. Review the steps required to making a PC study of a process.
5. Why don't standard tables exist detailing the natural variability of a given process, like rolling, extruding, or turning?

6. What are Taguchi or factorial experiments, and how are they related to experimental designs?
7. How does the Taguchi approach differ from the typical "experimental approach"?
8. Why are Taguchi experiments so important compared to classical experiments?
9. Here are some common, everyday processes with which you are familiar. What variable do you think dominates these processes?
   a. Baking a cake (from scratch; from a cake mix).

b. Mowing the lawn.

c. Washing dishes in a dishwasher.

10. Explain why the diameter measurements for holes produced by the process of drilling could have a skewed distribution rather than a normal distribution.

11. What are some common manufactured items that may not receive any final inspection?

12. What are the common reasons for sampling inspection rather than 100% insepction?

13. Fill in this table with one of the following statements:

Type I.

Type II.

No Error.

		In reality, if we looked at everything the process made, we would know that it had	
		Changed	Not Changed
The sample suggested that the process had:	Changed		
	Not Changed		

14. Why, when we sample, can we not avoid making Type I and Type II errors?

15. Which error can lead to legal action from the consumer for a defective product that caused bodily injury?

16. Define and explain the difference between σ', $\sigma_{\bar{x}}$, and σ_R.

17. What is C_p, and why is a value of 0.80 not good? How about a value of (1.00? 1.3?)

18. Explain what is measured by the bias factor.

19. What are some of the alternatives available to you when you have the situation where $6\sigma' > USL - LSL$?

20. C_{pk} is also a process capability index. How does it differ from C_p?

21. In a sigma chart, are σ values for the samples normally distributed about $\bar{\sigma}$? Why or why not?

22. What is an assignable cause, and how is it different from a chance cause?

23. Why is the range statistic used to measure variability when σ is really a better statistic?

24. How is the standard deviation of a distribution of sample means related to the distribution from which the samples were drawn?

25. What is the real secret to superior quality control?

Problems

1. For the items listed below, obtain a quantity of 48. Measure the indicated characteristics and determine the process mean and its standard deviation. Use a sample size of 4, so that 12 samples are produced.

Item	Characteristic(s) You Can Measure
Flat washer	Weight, width, diameter of hole, outside diameter
Paper clip	Length, diameter of wire
Coin (penny, dime)	Diameter, thickness at some point, weight
Your choice	Your choice

2. Perform a process capability study to determine the PC of the process that makes M&M candy. You will need to decide what characteristics you want to measure (e.g., weight, diameter, and thickness), how you will measure them (use rule of 10), and what kind of M&M's you want to inspect (how many bags of M&M's you wish to sample). Take a sample size of 4 ($n = 4$). Make a histogram of the individual data and estimate

$\bar{X}'$ and σ' as outlined in Figure 12-3. If you decide to measure the weight characteristics, you can check your estimate of $\bar{X}'$ by weighing all the M&M's together and dividing by the total number of M&M's.

3. For the data given in Figure 12-2, compute C_p, β, and C_{pk}, making any assumptions needed to perform the calculations.

4. For the data given in Figure 12-7, compute C_p, β, and C_{pk}.

5. Calculate $\bar{\bar{X}}$ and $\bar{R}$ and the control limits for the $\bar{X}$ and R control charts shown in Figure 12-A. The sample mean, $\bar{X}$, and range for the first nine subgroups and the data for each sample are given in the bottom of the figure. There are 25 samples of Size 5. Therefore $k = 25$, $n = 5$. Complete the bottom part of the table, and then compute the control limits for both charts. Construct the charts plotting $\bar{X}$ and $\bar{R}$ as solid lines and control limits as dash lines. The first 8 data points have been plotted. The first 4 data points have been connected. Plot the rest of the data on the charts, and comment on your findings.

6. For the data given in Figure 12-A, estimate the mean and the standard deviation for the process from which these samples were drawn (i.e., the parent population), and discuss the process capability. The USL and LSI for this dimension are .9 and .5, respectively, and the nominal is .7.

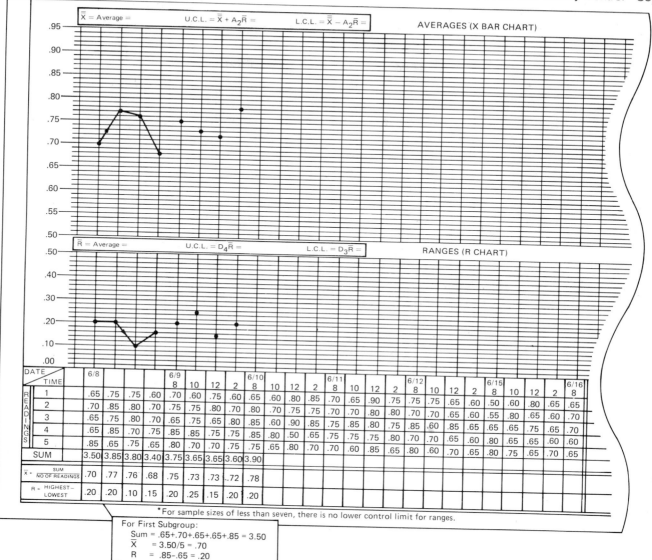

FIGURE 12-A

PART III

Casting Processes

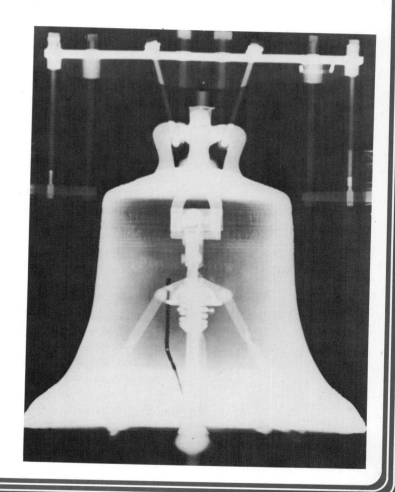

13

Fundamentals of Casting

Materials processing is the science and technology by which a material is converted into a final useful shape, processing the necessary structure and properties for its intended use. More loosely, processing is ''what is done to convert stuff into things.'' Formation of the desired shape is a major portion of processing, and casting is one of the more popular means of achieving this goal.

In casting, a solid is melted, heated to proper temperature, and sometimes treated to modify its chemical composition. The molten material, generally metal, is then poured into a cavity or mold, which contains it in proper shape during solidification. Thus, *in a single step,* simple or complex shapes can be made from any metal that can be melted. The resulting product can have virtually any configuration the designer desires. In addition, the resistance to working stresses can be optimized, directional properties can be controlled, and a pleasing appearance can be produced.

Cast parts range in size from a fraction of an inch and a fraction of an ounce (such as the individual teeth on a zipper), to over 30 feet (10 meters) and many tons (such as the huge propellers and stern frames of ocean liners). Moreover, casting has marked advantages in the production of complex shapes, of parts having hollow sections or internal cavities, of parts that contain irregular curved surfaces (except those made from thin sheet metal), of very large parts, and of parts made from metals that are difficult to machine. Because of these obvious advantages, casting is one of the most important of the manufacturing processes.

Today, it is nearly impossible to design anything that cannot be cast by means of one or more of the available casting processes. However, as in all manufacturing techniques, the best results and economy are achieved if the designer understands the various options and tailors the design to use the most appropriate process in the most efficient manner. The various processes differ primarily in the mold material (whether sand, metal, or other material) and the pouring method (gravity, vacuum, low pressure, or high pressure). All of the processes share the requirement that the material solidify in a manner that would maximize the properties, while simultaneously preventing potential defects, such as shrinkage voids, gas porosity, and trapped inclusions.

Six basic factors are involved in casting processes:

1. A *mold cavity*, having the desired shape and size, must be produced with due allowance for shrinkage of the solidifying metal. Any complexity of shape desired in the finished casting must exist in the cavity. Consequently, the mold material must be able to reproduce the desired detail and also must have a refractory character so that it will not be significantly affected by the molten metal that it must contain. Either a new mold must be prepared for each casting (single-use molds), or the mold must be made from a material that can withstand being used for repeated castings. The latter type are known as *permanent molds*. Since the permanent molds are made of metal or graphite and are quite costly, their use is generally restricted to large production runs. The more economical single-use molds are generally preferred for the production of smaller quantities.

2. A *melting process* must be capable of providing molten material not only at the proper temperature, but also in the desired quantity, with an acceptable quality, and within a reasonable cost.

3. A *pouring technique* must be devised to introduce the molten metal into the mold. Provision should be made to permit the escape of all air or gases in the mold before pouring and those generated by the action of the hot metal entering the mold. The molten metal can then completely fill the cavity, producing a quality casting that is fully dense and free of defects.

4. The *solidification process* should be properly designed and controlled. Provision must be made so that the mold will not cause too much restraint to the shrinkage that accompanies the cooling of solidified metal. Otherwise, the casting will crack when it is still hot and its strength is low. In addition, the design of the casting must be such that solidification and solidification shrinkage can occur without producing internal porosity or voids.

5. It must be possible to remove the casting from the mold (i.e., *mold removal*). When the metal is poured into molds that are broken apart and destroyed after each casting is made, there is no serious difficulty. However, in processes where permanent molds are used, removal of the casting may present a major design problem.

6. After the casting is removed from the mold, various *cleaning, finishing, and inspection* operations may need to be performed. Extraneous material that is attached where the metal entered the cavity, excesses at mold parting lines, and mold material that is attached to the casting surface must all be removed.

These aspects are considered in more detail as we move through the chapter. The fundamentals of solidification, pattern design, gating, and risering are developed. Various defects are also considered, along with their causes and cures.

Casting Terminology. Before we proceed with these fundamentals, it is helpful to first become familiar with a variety of casting terms. Figure 13-1 presents the cross section of a typical two-part sand mold and incorporates many features of the casting process. The casting starts with the construction of a *pattern*, an approximate duplicate of the final casting. The molding material is then packed around the pattern, and the pattern is removed to produce a mold cavity. The *flask* is the box that contains the molding aggregate. In a two-part mold, the *cope* is the top half of the pattern, flask, mold, or core. The *drag* is the bottom half of any of these features. A *core* is a sand shape that is inserted into the mold to produce internal features on a casting, such as holes or passages for water cooling. A *core print* is the region added to the pattern, core, or mold that is used to locate and support the core within the mold. The mold material and the core then combine to form the *mold cavity*, the void into which the molten metal will be poured and solidified to produce the desired casting. A *riser* is an extra void created in the mold that will be filled with molten metal. It provides a reservoir of molten metal that can flow into the mold cavity to

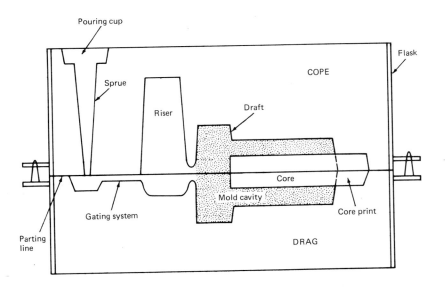

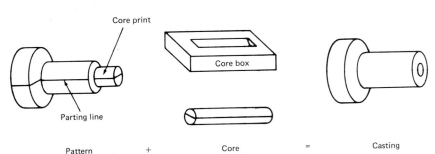

FIGURE 13-1 Cross section of a typical two-part sand mold, indicating various mold components and terminology.

compensate for any material shrinkage that occurs during solidification. Any shrinkage voids should then be in the riser and not in the final casting.

The *gating system* is the network of channels used to deliver the molten metal from outside the mold into the mold cavity. The *pouring cup* is the portion of the gating system that initially receives the molten metal from the pouring vessel and controls its delivery to the rest of the mold. From the pouring cup, the metal travels down the *sprue* (the vertical portion of the gating system), then along horizontal channels (called *runners*), and finally through controlled entrances, or *gates,* into the mold cavity.

The *parting line* or *parting surface* is the interface that separates the cope and drag halves of the mold, flask, or pattern, and the halves of a core during some core-making processes. The *draft* is the taper on a pattern or casting that permits it to be withdrawn from the mold. The mold or die used to produce casting cores is known as a *core box*. Finally, the term *casting* is used to describe both the process and the product when molten metal is poured and solidified in a mold.

Solidification.

Casting is a solidification process, the molten material being poured into a mold and then allowed to freeze into the desired final shape. Many of the structural features that ultimately control product properties are set during solidification. Furthermore, many casting defects, such as gas porosity and solidification shrinkage, are solidification phenomena, and they can be reduced or eliminated by control of the solidification process.

Solidification occurs in two stages, *nucleation* and *growth*, and it is important to control both of these processes. Nucleation occurs when a stable solid particle forms from within the molten liquid. As the material changes state, its internal energy is reduced, since, at lower temperatures, the solid phase is more stable than the liquid. At the same time, however, interface surfaces are created between the newly solidified metal and the parent liquid. Formation of these surfaces requires a positive contribution of energy. As a result, nucleation generally occurs at a temperature somewhat below the equilibrium melting point (the temperature where the internal energies of the liquid and the solid are equal). The difference between the melting point and the temperature of nucleation is known as the amount of *undercooling*.

In most practical situations, the nucleation process prefers existing surfaces, where solidification can begin without creating a full, wrap-around interface. These surfaces are generally in the form of mold or container walls, or solid impurity particles.

If we note that each nucleation event produces a crystal or grain in the final casting and remember that fine-grain materials (many small grains) possess improved strength and mechanical properties, then we conclude that anything that can be done to promote nucleation will be beneficial to the final product. Thus, we may intentionally introduce impurities into the liquid metal before pouring it into the mold. These small particles of solid provide for uniform and widespread nucleation and promote a uniform, fine-grain size in the product. The practice of intentionally introducing impurities is known as *inoculation*, or *grain refining*.

The second step in the solidification process is *growth*, which occurs as the heat of fusion is continually extracted from the liquid material. The direction, rate, and type of growth can be controlled by the way in which heat is extracted. Directional solidification, in which the solidification interface sweeps continuously through the material, can be used to ensure the production of a sound casting. The molten material on the liquid side of the interface can flow into the mold to continuously compensate for the shrinkage that occurs as the material changes from liquid to solid. Faster rates of cooling generally produce products with finer grain size and superior mechanical properties.

Cooling Curves. Cooling curves (as discussed previously, in Chapter 4) can provide one of the most useful tools for studying the solidification process. By inserting thermocouples into a casting and monitoring the temperature versus time, one can obtain valuable insight into what is happening in the various regions.

Figure 13-2 shows a typical cooling curve for a pure or eutectic material and is useful for depicting many of the principal features and terms. The *pouring temperature* is the temperature of the liquid metal when it first enters the mold cavity. *Superheat* is the difference between the pouring temperature and the freezing temperature of the material. The higher the superheat, the more the time allowed for the material to flow into the intricate mold cavity details before it begins to freeze. The *cooling rate* is the rate at which the liquid or solid is cooling and can be viewed as the slope of the cooling curve at any given point.

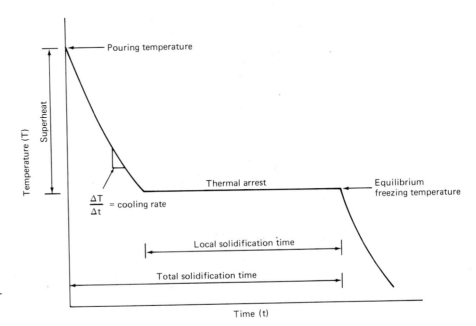

FIGURE 13-2 Cooling curve for a pure metal or eutectic alloy (metals with a distinct freezing point) indicating major casting features related to solidification.

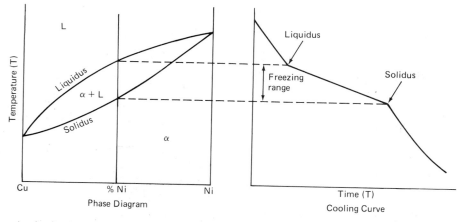

FIGURE 13-3 Cooling curve of an alloy that has a freezing range.

A *thermal arrest* is the plateau in the cooling curve that occurs during the solidification of a material with a fixed melting point. While the material is at this temperature, the heat being removed from the mold comes from the solidification process and does not require a decrease in the material's temperature. The time from the start of pouring to the end of solidification is known as the *total solidification time*. The time from the start of solidification to the end of solidification is called the *local solidification time*.

If an alloy is used that does not have a distinct melting point, the difference between the liquidus and solidus temperatures is known as the *freezing range*, as shown in Figure 13-3. If undercooling is required to induce the initial nucleation, the subsequent solidification can release enough heat to cause an increase in temperature back to the melting point. This increase in temperature is known as *recalescence* and is shown in Figure 13-4.

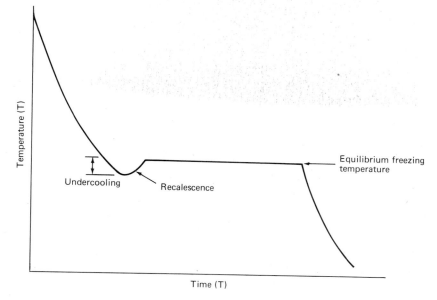

FIGURE 13-4 Cooling curve depicting undercooling and subsequent recalescence.

The specific form of the cooling curve depends on the type of material being poured, the nature of the nucleation process, and the rate and means of heat removal from the mold. Analysis of experimental cooling curves can provide valuable insight into the process and the product. Fast cooling rates and short solidification times lead to finer structures and improved mechanical properties.

Prediction of Solidification Time: Chvorinov's Rule.   The amount of heat that must be removed from a casting to cause it to solidify is directly proportional to the amount of superheating and the amount of metal in the casting, or the casting volume. Conversely, the ability to remove heat from a casting is directly related to the amount of exposed surface area through which the heat can be extracted and the insulating value of the mold. These observations are reflected in Chvorinov's rule,[1] which states that t_s, the total solidification time, can be computed by:

$$t_s = B \, (V/A)^n \text{ where } n = 1.5 \text{ to } 2.0$$

The total solidification time is the time from pouring to the completion of solidification; V is the volume of the casting; A is the surface area; and B is the mold constant, which depends on the characteristics of the metal being cast (its density, heat capacity, and heat of fusion), the mold material (its density, thermal conductivity, and heat capacity), the mold thickness, and the amount of superheat.

Test specimens can be cast to determine B for a given mold material, metal, and condition of casting. This value can then be used to compute the solidification times for other castings made under the same conditions. Since a riser and a casting are both within the same mold and fill with the same metal under the same conditions, Chvorinov's rule can be used to ensure that the casting will solidify before the riser. This is necessary if the liquid within the riser is to effectively feed the casting to compensate for solidification shrinkage. Examples will be given when we discuss risers later in this chapter.

Different cooling rates and solidification times can produce substantial variation in the resulting structure and properties. For instance, die casting, which uses metal molds, has faster cooling and produces higher-strength castings than sand casting, which uses a more insulating mold material. The various types of sands can produce different cooling rates. Sands with high moisture contents extract heat faster than sands with low moisture.

The Cast Structure.   The structure that results when molten metals are poured into molds and permitted to solidify may have as many as three distinct regions or zones. The *chill zone* is a narrow band of randomly oriented crystals that forms on the surface of a casting. Rapid nucleation occurs here because of the presence of the mold walls and the relatively rapid surface cooling. As additional heat is removed from the surfaces, the grains of the chill zone begin to grow inward, and the rate of heat extraction and solidification decreases.

[1]N. Chvorinov, *Proc. Inst. Brit. Found.*, Vol. 32, p. 229 (1938–39).

Since most crystals have directions of rapid growth, a selection process occurs. Those crystals whose rapid growth direction is perpendicular to the casting surface grow fast and shut off adjacent grains whose direction is at some particular angle. The favorably oriented crystals can then grow until all of the liquid has solidified, producing long, thin columnar grains, the region being known as the *columnar zone*. The resulting properties are highly directional, since the selection process has converted the purely random structure of the surface into one of aligned parallel crystals. Figure 13-5 shows a cast structure containing both chill and columnar zones.

In most alloys and some pure metals, new crystals can nucleate in the center of the casting and then grow to produce another region of randomly oriented, spherical-shaped crystals, the *equiaxed zone*. Low pouring temperatures, alloy additions, or the addition of inoculants can be used to promote the formation of this region, which is far more desirable than the columnar-shaped grains. Isotropic properties (uniform in all directions) are observed in this region of the casting.

Molten Metal Problems. Castings begin with molten metal, and many of the reactions that can occur between molten metal and its environment can lead to defects in the casting. Oxygen and molten metal often react to produce metal oxides, which can then be carried with the molten metal during the pouring and filling of the mold. Known as *dross* or *slag*, this material can become trapped in the casting and can impair surface finish, machinability, and mechanical properties.

Control of dross and slag can be achieved through special cautions during melting, pouring, and the design of processes. Fluxes can be used to cover and protect molten metal during melting, or melting and pouring can be performed under a vacuum or a protective atmosphere. Measures can be taken to agglomerate the dross, to cause it to float to the surface of the metal, and then be skimmed off before pouring. Special ladles that pour from beneath the surface can be particularly effective, two such designs being shown in Figure 13-6. Gating systems can often be designed to trap any dross that may enter the mold

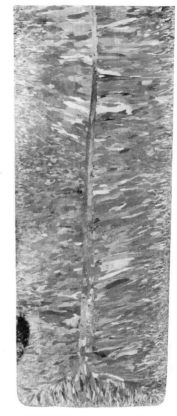

FIGURE 13-5 Cross-sectional structure of a cast metal bar showing the chill zone at the periphery and columnar grains growing toward the center.

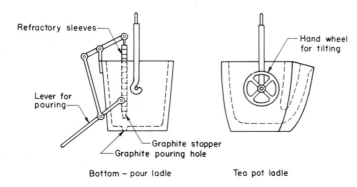

Refractory sleeves

Lever for pouring

Graphite stopper

Graphite pouring hole

Bottom – pour ladle

Hand wheel for tilting

Tea pot ladle

FIGURE 13-6 Two types of ladles used in the pouring of castings. Note how each avoids pouring the impure material from the top of the molten pool.

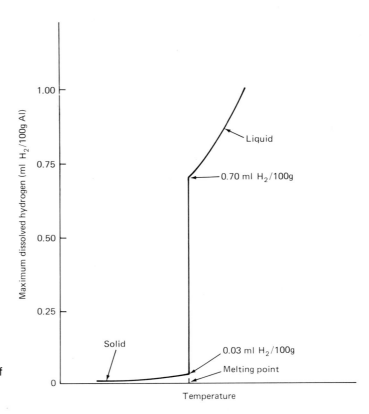

FIGURE 13-7 Maximum solubility of hydrogen in aluminum as a function of temperature.

and keep it from entering the mold cavity, and filters may be inserted in the feeder channels of the mold.

Significant amounts of gas can dissolve in many liquid metals. When these metals solidify, however, the solid structure cannot accommodate the gas, and the rejected gas atoms nucleate bubbles or *gas porosity* within the casting. Figure 13-7 shows the solubility of hydrogen in aluminum as a function of temperature. It is important to note the large change when the material undergoes freezing.

Several different techniques can be applied to prevent the formation of gas porosity. One approach is to prevent the gas from initially dissolving in the molten metal. Melting can be performed under vacuum, in an environment of low-solubility gases, or under a protective flux that excludes contact with the air. Superheat temperatures can be kept low to minimize the extent of solubility. In addition, careful handling and pouring can do much to streamline the flow of molten metal and to minimize turbulence that brings air and molten metal into contact.

Other techniques seek to remove the gas from the molten metal before pouring. *Vacuum degassing* subjects the molten metal to a low-pressure environment. Under such conditions, the amount of dissolved gas reduces as the system seeks to establish an equilibrium with its environment. (See Sievert's law in any basic chemistry text.) Passing small bubbles of inert gas or reactive gas through the melt is also effective, a technique known as *gas flushing*. In seeking to establish equilibrium with this new environment, the dissolved gas enters the flushing gas and is carried away. Nitrogen and chlorine, for example, are particularly effective in removing hydrogen from molten aluminum.

Still another approach is to react the dissolved gas with something to produce a low-density compound. These compounds then float to the surface and can be removed with the dross or slag. Oxygen can be removed from copper by the addition of phosphorus. Steels can be deoxidized with aluminum or silicon. The resulting phosphorus-, aluminum-, or silicon-oxides are then removed from the surface of the molten metal container or are left on the top as the better metal is poured from beneath the surface.

Fluidity.   On pouring, one desires the molten metal to perform two functions: (1) to *flow* into all regions of the mold cavity, and then (2) to *freeze* into this new shape. It is vitally important that these two functions be performed in the proper sequence, for if the metal begins to freeze before it has completely filled the mold, defects known as *misruns* and *cold shuts* are produced. The ability of the metal to flow and fill the mold—its runniness, if you will—is known as *fluidity*. Standard tests to evaluate fluidity include the fluidity spiral, which measures the length to which metal will fill a standard spiral channel, and the Ragone test, where metal is drawn into a vacuum suction tube, and the height is used to determine fluidity.

Pouring Temperature.   While composition, freezing temperature, and freezing range are all important factors affecting fluidity, the most important parameter is the pouring temperature or the amount of superheat. The higher the pouring temperature, the higher the fluidity. Excessive temperatures should be avoided, however. At high pouring temperatures, metal–mold reactions are accelerated, and penetration is possible. Penetration is a defect where the fluidity is so great that the metal not only fills the mold cavity but further fills the small voids between the sand particles in a sand mold. The product surface then consists of sand particles embedded in the casting metal.

The Gating System.   As the metal is poured into the mold, the gating system serves to deliver it to all sections of the mold cavity. The speed or rate of metal movement is important, as is the degree of cooling that occurs while the metal is flowing. Slow filling and high loss of heat result in misruns and cold shuts. Rapid rates of filling can produce erosion of the gating system and the mold cavity and can result in entrapped mold material in the final casting. The cross-sectional areas of the various channels can be selected to regulate flow. The shape and length of the channels are influential in controlling temperature loss. When heat loss is to be minimized, short-length channels with round or square cross sections are the most desirable. Multiple gates and runners can be used to introduce metal to more than one point of a large casting.

Another function of the gating system is to minimize turbulent flow, which can promote excessive absorption of gases and oxidation of the metal, and can

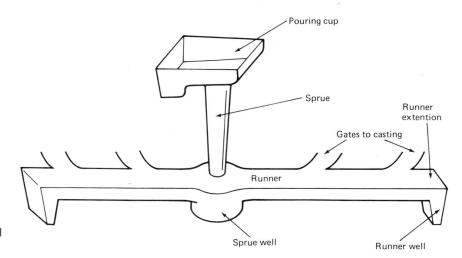

FIGURE 13-8 Typical gating system for a horizontal parting plane mold, showing key components involved in controlling the flow of metal into the mold cavity.

accelerate mold erosion. Figure 13-8 shows a typical gating system for a mold with a horizontal parting line and identifies some of the key components that can be optimized to promote the smooth flow of the metal. Short sprues are desirable to minimize the distance that the metal must fall when entering the mold. Rectangular pouring cups will prevent a vortex or funnel from forming and sucking gas and oxides into the sprue. Tapered sprues can also prevent vortex formation. A large sprue well can serve to dissipate the kinetic energy of the falling stream. Finally, the *choke*, or smallest cross-sectional area, should be located at the base of the sprue. If the choke is at the gates, the metal will flow into the mold cavity with a fountain effect, a very turbulent mode of flow.

The gating system can also be used to trap dross and sand particles and to keep them from entering the mold cavity. Since many foreign particles are of low density, they rise to the top of the molten metal if given sufficient time. Long, flat runners can be beneficial (but these promote cooling of the metal). Also, the first metal to enter the mold is most likely to contain the foreign particles (dross from the top of the pouring ladle, if poured from the top, and loose particles from the walls of the gating system). Runner extensions and wells (see Figure 13-8) can be used to catch and trap this first metal and to keep it from entering the mold cavity. These are particularly effective in aluminum castings, where the aluminum oxide has approximately the same density as molten aluminum. Screens or ceramic filters can also be inserted into the gating system to trap the foreign material.

The specific details of the gating system often depend on the metal being cast. Turbulence-sensitive metals, such as aluminum and magnesium, and alloys with low melting points generally require gating systems that concentrate on eliminating turbulence and trapping dross. Turbulence-insensitive alloys, such as steel, cast iron, most copper alloys, and alloys with a high melting point, generally use short, open gating systems that provide for quick filling of the mold cavity.

Solidification Shrinkage. Once in the mold cavity, most metals and alloys contract during cooling. There are three principal stages during which shrinkage occurs: (1) shrinkage of the liquid; (2) solidification shrinkage as the liquid turns to solid; and (3) solid metal contraction as the solidified material cools to room temperature. The amount of liquid metal contraction depends on the coefficient of thermal contraction (a property of the metal being cast) and the amount of superheat. Liquid contraction, however, is rarely a problem in casting production because the metal in the gating system continues to flow into the mold cavity as the metal in the cavity cools and contracts.

As the metal cools between the liquidus and solidus temperatures and changes state from liquid to solid, significant amounts of shrinkage tend to occur, as indicated by the data in Table 13.1. This table also shows that not all metals contract. Some expand, such as gray cast iron, in which low-density graphite flakes form as part of the solidification structure.

When shrinkage occurs, it is important to know and control the form of the resulting void. Metals and alloys with short freezing ranges, such as pure metals and eutectic alloys, tend to form large cavities or pipes. These can be avoided by designing the casting to have directional solidification. Here, the solidification begins furthest away from the feed gate or riser and moves progressively toward it. As the metal solidifies and shrinks, the shrinkage void is continually filled with liquid metal. Ultimately, the final shrinkage void occurs in the riser or the gating system.

Alloys with large freezing ranges have a wide range of temperatures over which the material is in a mushy state. As the cooler regions complete their solidification, it is almost impossible for additional liquid to feed into the shrinkage voids. Thus, the resultant structure tends to have small, but numerous, shrinkage pores dispersed throughout. This type of shrinkage is far more difficult to prevent by control of the gating and risering, and it may even be necessary to accept the fact that a porous product will result. If a gas- or liquid-tight product is desired, the castings can be impregnated (the pores filled with a resinous material or lower-melting-temperature metal) as a subsequent operation.

After solidification, the casting contracts further as it cools to room temperature. These dimensional changes have to be compensated for when setting the dimensions of the mold cavity or pattern. Additional concern arises when the casting is being made in a rigid mold, such as in die casting where metal molds are used. If the mold provides constraint as the casting tries to contract, tensile forces can be generated within the casting, and cracking can occur. Thus, it may be desirable to eject the castings as soon as possible after the completion of solidification.

Risers and Riser Design. Risers are added reservoirs designed to feed liquid metal to the solidifying casting as a means of compensating for solidification shrinkage. To perform this function, the risers must solidify after the casting. If the reverse were true, liquid metal would flow from the casting into the solidifying riser, and the casting shrinkage would be even greater. Hence, the casting should be designed to produce directional solidification, which

TABLE 13.1 Solidification Shrinkage of Some Common Engineering Metals

Low-carbon steel	2.5%–3.0%
High-carbon steel	4.0%
Pure aluminum	6.6%
Pure copper	4.9%
White cast iron	4.0%–5.5%
Gray cast iron	−1.9%

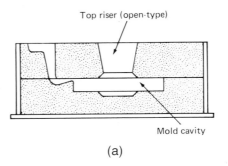

Top riser (open-type)

Mold cavity

(a)

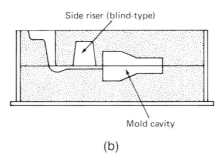

Side riser (blind-type)

Mold cavity

(b)

FIGURE 13-9 Schematic of a sand casting mold, showing a top riser (a) and a side riser (b).

sweeps from the extremities of the mold cavity to the riser. In this way, the riser can continuously feed molten metal and will compensate for the solidification shrinkage of the entire mold cavity. If such solidification is not possible, then multiple risers may be necessary, with various sections of the casting solidifying toward their respective risers.

Finally, risers should be designed to conserve metal. If we define the yield of a casting as the casting weight divided by the total weight of metal poured (sprue, gates, risers, and casting), it is clear that there is a motivation to make the risers as small as possible to still perform their task. This can often be done by proper consideration of riser size, shape, and location, and the nature of the connection between the riser and the casting.

By consideration of Chvorinov's rule, a good shape for a riser would be one that has a long freezing time or a small surface area per unit volume. A sphere would make the most efficient riser but presents considerable difficulty to the patternmaker and moldmaker, who must remove the pattern from the mold. As a result, the most popular shape for a riser is a cylinder, in which the height-to-diameter ratio is varied depending on the nature of the alloy, the location of the riser, the size of the flask, and other variables.

Risers should be located so that directional solidification occurs. Since the thickest regions of a casting are the last to freeze, the risers should be located so as to feed into these heavy sections. Various types of risers are possible. A *top riser*, one that sits on top of a casting, has the advantage of feeding by additional pressure (the weight of the metal), feeding a shorter distance, and occupying less space within the flask, thereby permitting more freedom for the layout of the pattern and the gating system. *Side risers* are located adjacent to the mold cavity in the horizontal direction. Figure 13-9 compares a top and a side riser. If the riser is contained entirely within the mold, it is known as a *blind riser*, while one that is open to the atmosphere is called an *open riser*. Blind risers develop a solid skin because of surface solidification and are generally bigger than open risers, because of the heat lost through the additional surface.

Live (hot) risers receive the last hot metal that enters the mold and generally do so at a time when the metal in the mold cavity has already begun to cool and solidify. Thus, they can be smaller than *dead (cold) risers*, which fill with the colder metal that has already flowed through the mold cavity. Top risers are almost always dead risers. Risers that are part of the gating system are generally live risers.

The minimum size of a riser can be calculated from Chvorinov's rule by setting the total solidification time for the riser to be greater than the total solidification time for the casting. Since both will receive the same metal and are in the same mold, the mold constant, B, will be the same for both regions. Assuming $n = 2$, and a safe difference in solidification time of 25% (the riser takes 25% longer to solidify than the casting), we can write this condition as:

$$t_{\text{riser}} = 1.25 \, t_{\text{casting}}$$
$$(V/A)^2_{\text{riser}} = 1.25 \, (V/A)^2_{\text{casting}}$$

Calculation of the riser size then requires the selection of a riser geometry (generally cylindrical) and specification of a height-to-diameter ratio, so that the riser side of the equation will have only one unknown. For a cylinder of diameter D and height H:

$$V = \pi D^2 H/4$$
$$A = \pi DH + 2\,(\pi D^2/4)$$

Specifying the riser height as a function of the diameter enables the V/A ratio to be written as a simple expression with one unknown, namely, D. The V/A ratio for the casting can be calculated from its particular geometry. Substitution of this information into the above form of Chvorinov's rule will then enable calculation of the required riser size. Note, however, that if the riser and the casting share a surface, as with a top riser, the common surface area should be subtracted from both segments since it will not be a surface of heat loss to either component. While there are a variety of more sophisticated methods for calculating riser size, the Chvorinov's rule method is the only one presented in this text.

A final aspect of riser design is the connection between the riser and the casting. Since it will ultimately be necessary to separate the riser from the casting, it is desirable that the connection area be as small as possible. On the other hand, the connection area should be large enough so that the link does not freeze before solidification of the casting is complete. Short-length connections are most desirable. The adjacent mold material will then receive heat from both the casting and the riser. Therefore, it will heat rapidly and remain hot throughout the cast, retarding solidification of the metal in the channel.

Risering Aids.

Methods have been developed to assist the risers in performing their job and generally have one of two objectives: (1) promoting directional solidification and/or (2) reducing the number and size of the risers, thereby increasing the yield of a casting. Risering aids work in one of two ways: speeding the rate at which a casting solidifies or retarding the solidification of the riser.

External chills are masses of high-heat-capacity, high-thermal-conductivity material that are placed in the mold (adjacent to the casting) to accelerate the cooling of various regions. Chills can effectively promote directional solidification or increase the effective feeding distance of a riser. They can often be used to reduce the number of risers required for a casting.

Internal chills are pieces of metal that are placed within the mold cavity to absorb heat and to promote a more rapid solidification. Since some of this metal often melts and absorbs the heat of fusion as well as heat capacity energy, internal chills are usually more effective than the external variety. However, since internal chills ultimately become part of the final casting, they must be made from the same alloy as that being cast.

Ways to slow the cooling of risers include (1) switching from a blind riser to an open riser; (2) using insulating sleeves and toppings around the riser; and (3) surrounding the sides or top of the riser with exothermic material that supplies added heat to just the riser segment of the mold. These efforts generally seek to reduce the riser size, rather than to promote directional solidification.

Finally, one should note that risers are not always necessary. For alloys with large freezing ranges, the risers would not be particularly effective, and one generally accepts the fine, dispersed porosity. For other processes such as die casting, low-pressure permanent molding, and centrifugal casting, positive pressure provides the necessary feeding action to compensate for the shrinkage.

Patterns. Casting processes can be divided into two basic categories: those for which a new mold must be created for each casting (the expendable-mold processes) and those that use a permanent, reusable mold. Almost all of the expendable-mold processes begin with a permanent, reusable pattern—a duplicate of the part to be cast, modified to reflect the casting process and the material being cast.

The modifications that are incorporated into a pattern are called *allowances,* and the most important of these is the *shrinkage allowance.* Following solidification, the casting continues to contract as it cools, the amount of contraction being as much as 2%, or ¼ inch per foot. Thus, the pattern must be made slightly larger than the desired casting as a means of compensation. The exact allowance depends on the metal that is to be cast. Allowances typical of some common engineering metals are:

Cast iron	0.8%–1.0% (¹⁄₁₀–⅛ in./ft.)
Steel	1.5%–2.0% (³⁄₁₆–¼ in./ft.)
Aluminum	1.0%–1.3% (⅛–⁵⁄₃₂ in./ft.)
Magnesium	1.0%–1.3% (⅛–⁵⁄₃₂ in./ft.)
Brass	1.5% (³⁄₁₆ in./ft.)

The patternmaker often incorporates these allowances into the pattern by using special *shrink rules*, measuring devices that are larger than a standard rule by the desired shrinkage allowance. For example, a shrinkage rule for brass would designate 1 foot at a length that is actually 1 foot ³⁄₁₆ inch. A pattern made to shrinkage rule dimensions would produce a proper-sized casting on cooling.

Some caution should be exercised when using shrinkage rules, however, for thermal contraction may not be the only factor affecting the dimensions after solidification. The various phase transformations discussed in Chapter 4 can often bring about significant expansions or contractions. These include eutectoid reactions, martensitic reactions, and graphitization.

In casting processes where the pattern must be withdrawn from the mold, the mold is generally made in two or more sections. Consideration must be given to the location of the *parting line* or surface where one section of the mold mates with the other section or sections, and to the *draft* or *taper* that must be provided on the pattern to facilitate the pattern's withdrawal. These are illustrated in Figure 13-10. If the surfaces of the pattern that are vertical to the parting line were parallel to the direction of pattern withdrawal, the friction between the pattern and the sand, or any horizontal movement of the pattern during extraction, would tend to cause sand particles to be broken away from the mold. The damage would be particularly severe at the corners where the

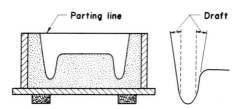

FIGURE 13-10 Relationship of draft to the mold parting plane in a casting.

mold cavity intersects the parting surface. By providing a slight taper on all surfaces parallel to the direction of withdrawal, this difficulty can be minimized. As soon as the pattern is withdrawn a slight amount, it is free from the sand on all surfaces, and it can be withdrawn further without damaging the mold.

The amount of draft is determined by the size and shape of the pattern, the depth of the draw, the method used to withdraw the pattern, the pattern material, the mold material, and the molding procedure. Draft is seldom less than 1° or ⅛ inch per foot, with a minimum of about ¹⁄₁₆ inch on any surface. On interior surfaces where the opening is small, such as a hole in the center of a hub, the draft should be increased to about ½ inch per foot. However, since draft allowances tend to increase the size of a pattern, and thus the size and weight of a casting, it is generally desirable to keep them to the minimum that will permit satisfactory pattern removal. Modern molding procedures, which provide higher strength to the molding sand before the pattern is withdrawn, as well as the use of molding machines that incorporate mechanical pattern withdrawal, have permitted substantial reductions in draft allowances. These improvements have enabled the production of lighter castings with thinner sections, thus saving weight and machining.

When machined surfaces must be provided on castings, it is necessary to provide a *machining allowance*, or *finish allowance* on the pattern. This allowance depends to a great extent on the casting process and the mold material. Ordinary sand castings have rougher surfaces than shell-mold castings. Die castings are sufficiently smooth so that very little or no metal has to be removed, and investment castings frequently require no additional machining. Consequently, the designer should relate the finishing allowance to the casting process and should also remember that draft may provide part or all of the extra metal needed for machining.

There are also some special forms of allowances. If a core is to be used to form a hole or an interior cavity, it, too, must be made oversized to compensate for shrinkage (all of the metal surrounding the hole will contract, forcing the hole to become smaller). However, if a machining allowance is to be included, it should be *subtracted* from the core dimensions because machining tends to increase the size of a hole. If the casting is made directly into a metal mold, all of the previous allowances should be incorporated into the cavity. In addition, the change in mold dimensions caused by the heating of the mold from room temperature to its elevated operating temperature should be included as an additional correction.

Some casting shapes require still another allowance for *distortion*. For example, the arms of a U-shaped section may be restrained by the mold, while the base of the U is free to shrink. This restraint will result in a final casting with outwardly sloping arms. However, by originally designing the arms to slope inward, they will distort to a straight shape on cooling. Long, horizontal sections tend to sag in the center unless adequate support is provided by suitable ribbing. Distortion depends greatly on the particular configuration of the casting, and the designer must use experience and judgment to provide the required distortion allowance.

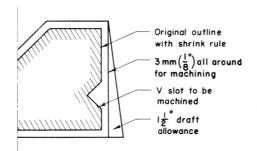

FIGURE 13-11 Various allowances incorporated into a casting pattern.

Figure 13-11 illustrates the manner in which the various allowances are included in the pattern for a simple shape casting. Since allowances tend to increase the weight of a casting and the amount of metal that has to be removed by machining, efforts made to reduce the allowances will be well received.

Design Considerations in Castings

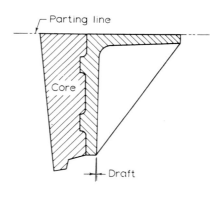

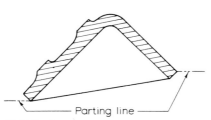

FIGURE 13-12 Elimination of a core by changing the location of the parting plane.

If economy and best results are to be obtained, it is very important that the designer of castings give careful attention to several requirements of the process and, if possible, cooperate closely with the foundry. Frequently, minor and readily permissible changes in design will greatly facilitate and simplify the casting of a component and will reduce the percentage of defects.

One of the first features that must be considered by a designer is the location of the parting plane, an important part of all processes that use segmented or separable molds. The location of the parting plane can affect each of the following: (1) the number of cores: (2) the use of effective and economical gating; (3) the weight of the final casting; (4) the method of supporting the cores; (5) the final dimensional accuracy; and (6) the ease of molding.

In general, it is desirable to minimize the use of cores. Often, a change in the location of the parting plan can assist in this objective, as illustrated in Figure 13-12. Note that the change also reduces the weight of the casting by eliminating the need for draft. Figure 13-13 shows another example of how a simple design change eliminated the need for a core.

The location of the parting plane can also be dictated by certain design features. Figure 13-14 shows how the specification of round edges on a part can restrict the location of the parting plane. The specification of draft can also fix the parting plane, as indicated in Figure 13-15. This figure also shows that considerable freedom can be provided by simply noting the need to provide for a draft or simply letting it be an option of the foundry. Since mold closure may not always be consistent, consideration should also be given to the fact that dimensions across the parting plane are subject to more variation than those that lie within a given segment of the mold.

Controlling the solidification process is of prime importance in obtaining quality castings, and this control is also related to design. Those portions of a casting that have a high ratio of surface area to volume will experience more rapid cooling and will be stronger and harder than the other regions. Heavier sections will cool more slowly and, unless special precautions are observed, may contain shrinkage cavities and porosity or may have large grain-size structures.

Ideally, a casting should have uniform thickness in all directions. In most cases, however, this is not possible. When the section thickness must change, it is best if these changes are gradual, as indicated in the various sections of Figure 13-16.

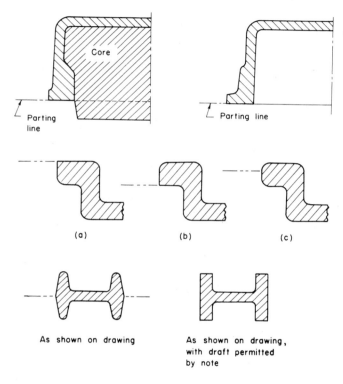

FIGURE 13-13   Elimination of a dry-sand core by a change in part design.

FIGURE 13-14   Effect of rounded edges on the location of the parting plane.

FIGURE 13-15   (top left) Location of the parting plane specified by draft. (Top right) Part with draft unspecified. (Bottom) Various options in producing that part.

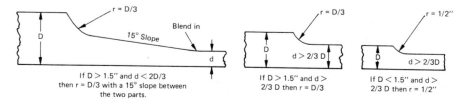

FIGURE 13-16   Guidelines for section changes in castings.

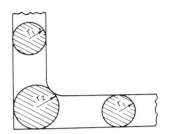

FIGURE 13-17 "Hot spot" at section r_2 caused by intersecting sections.

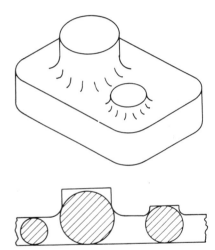

FIGURE 13-18 Hot spot resulting from intersecting sections of various thickness.

When sections of castings intersect, two problems can arise. The first is the possibility of stress concentrators. This problem can be minimized by providing generous fillets (inside radii) at all interior corners. Excessive fillets, however, can cause the second problem, known as *hot spots*. Figure 13-17 shows that localized thick sections tend to exist where sections of castings intersect. These thick sections cool more slowly than the others and tend to be sites of localized, abnormal shrinkage. When the differences in section are large, as illustrated in Figure 13-18, the hot-spot areas are likely to result in serious defects in the form of porosity or shrinkage cavities.

Defects such as voids, porosity, and cracks can be sites of subsequent failure and should be prevented if at all possible. Sometimes cored holes, as illustrated in Figure 13-19, can be used to prevent hot spots. Where heavy sections must exist, an adjacent riser can often be used to feed the section during shrinkage, as in Figure 13-20. If the riser is properly designed, the shrinkage cavity will lie totally within the riser and can be removed when the riser is cut off.

Intersecting ribs can cause shrinkage problems and should be given special consideration by the designer. Where sections intersect to form continuous ribs, contraction occurs in opposite directions as the various ribs contract. As a consequence, cracking frequently occurs during cooling. By staggering the ribs, as shown in Figure 13-21, there is opportunity for slight distortion to occur, thereby ensuring that high stresses are not built up.

Large unsupported areas should be avoided in all types of casting, since such sections tend to warp during cooling. The warpage then disrupts the good, smooth surface appearance that is so often desired. Another appearance consideration is the location of the parting line. Some small amount of fin, or flash, is often present at this location. When the flash is removed, or if it is considered small enough to leave in place, a region of surface imperfection will be present. If this is in the middle of a flat surface, it will be clearly visible. However, if the parting line is placed to coincide with a corner, the "defect" line will go largely unnoticed.

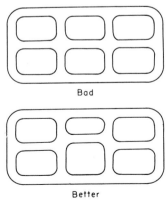

FIGURE 13-21 Method of using staggered ribs to prevent cracking during cooling.

FIGURE 13-19 Method of eliminating unsound metal at the center of heavy sections in castings by using cored holes.

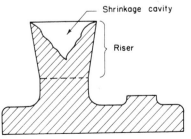

FIGURE 13-20 Use of a riser to keep the shrinkage cavity out of a casting.

TABLE 13.2   **Recommended Minimum Section Thicknesses for Various Engineering Metals and Casting Processes**

Material	Minimum		Desirable		Casting Process
	mm	in.	mm	in.	
Steel	4.76	3/16	6.35	1/4	Sand
Gray iron	3.18	1/8	4.76	3/16	Sand
Malleable iron	3.18	1/8	4.76	3/16	Sand
Aluminum	3.18	1/8	4.76	3/16	Sand
Magnesium	4.76	3/16	6.35	1/4	Sand
Zinc alloys	0.51	0.020	0.76	0.030	Die
Aluminum alloys	1.27	0.050	1.52	0.060	Die
Magnesium alloys	1.27	0.050	1.52	0.060	Die

In designing all types of castings, minimum section thickness must be considered. Exact specifications for economical and practical section thicknesses cannot be given, however, because the recommendations should be based on the shape and size of the casting, the type of metal, the method of casting, and the practice of the individual foundry. Table 13.2, however, seeks to present some reasonable guidelines for common foundry materials.

In conclusion, it is important to note that the design of a casting is sufficiently complex so that it can usually benefit from the input of a skilled and knowledgeable foundryman.

Review Questions

1. What is materials processing?
2. Describe the capabilities of the casting process in terms of the size and shape of the product.
3. What are some of the possible mold materials and pouring methods used in the casting of metals?
4. How might the desired production quantity influence the selection of a single-use or multiple-use molding process?
5. Why is it important to provide a means of venting gases from the mold cavity?
6. What types of problems can arise if the mold material provides too much restraint to the solidifying and cooling metal?
7. What is a casting pattern? A flask? A core? A mold cavity? A riser?
8. What are some of the components that combine to make up the gating system of a mold?
9. What is a parting line or parting surface?
10. What are the two stages of solidification, and what occurs during each?
11. Why is it that most solidification does not begin until the temperature falls somewhat below the equilibrium melting temperature?
12. Why might it be desirable to promote nucleation in a casting?
13. Why might directional solidification be desirable in the production of a cast product?
14. Describe some of the key features observed in the cooling curve of a pure metal, as depicted in Figure 13-2.
15. Discuss the roles of casting volume and surface area as they relate to the total solidification time and Chvorinov's rule.
16. What are some of the characteristics of the specific casting process that are incorporated into the "mold constant," B, in Chvorinov's rule?
17. What is the chill zone of a casting, and why does it form?
18. What is dross or slag, and how can it be prevented from becoming part of a finished casting?
19. What are some of the possible approaches that can be taken to prevent the formation of gas porosity in a metal casting?
20. What is fluidity, and how can it be measured?
21. Why is it important to design the geometry of the gating

system to control the rate of metal flow from the pouring ladle into the mold cavity?

22. What are some of the undesirable consequences that could result from turbulence of the metal in the gating system and the mold cavity?

23. What features of the metal being cast tend to influence whether the gating system is designed to minimize turbulence and reduce dross, or to promote rapid filling to minimize temperature loss?

24. What are the three stages of contraction or shrinkage as a liquid is converted into a finished casting?

25. Why is it more difficult to prevent shrinkage voids from forming in metals or alloys with large freezing ranges?

26. Why is it desirable to design a casting to have directional solidification sweeping from the extremities of the mold to the riser?

27. Based on Chvorinov's rule, what would be an effective shape for a casting riser from an ideal viewpoint? From a practical one?

28. What are some of the assumptions when one uses Chvorinov's rule to calculate the size of a riser in the manner presented in the text?

29. What are two possible objectives of the methods that have been developed to assist the risers in performing their job?

30. Why are internal chills a more effective means of removing heat than external chills?

31. What types of modifications or allowances are generally incorporated into a casting pattern?

32. What is a shrinkage rule, and how does it work?

33. In addition to thermal contraction, what other features can produce a significant change in product dimensions after solidification?

34. What is the purpose of a draft or taper on pattern surfaces?

35. Why is it attractive to make the pattern allowances as small as possible?

36. Fillets are used to reduce the magnitude of stress raisers at intersecting sections. What can occur if too large a fillet is used?

37. Using Chvorinov's rule as presented in the text with $n = 2$, calculate the dimensions of an effective riser for a casting that is a $2'' \times 4'' \times 6''$ rectangular plate. Assume that the casting and the riser are not connected, except through a gate and a runner, and that the riser is a cylinder of height/diameter ratio $H/D = 1.5$. The finished casting is what fraction of the combined weight of the riser and the casting?

38. Reposition the riser in Problem 37 so that it sits directly on top of the flat rectangle, with its bottom circular surface being part of the surface of the casting, and recompute the size and yield fraction. Which approach is more efficient?

39. A rectangular casting having the dimensions $3'' \times 5'' \times 10''$ solidifies completely in 11.5 minutes. Using $n = 2$ in Chvorinov's rule, calculate the mold constant, B. Then compute the solidification time of a $0.5'' \times 8'' \times 8''$ casting poured under the same conditions.

Expendable-Mold Casting Processes

A large number and variety of processes are used to produce molds and castings, and it is helpful to devise some means of classification. One such means breaks the processes down as follows:

1. Single-use molds with multiple-use patterns.
2. Single-use molds with single-use patterns.
3. Multiple-use molds.

This classification will be used in this and the following chapter, grouping Categories 1 and 2 under the general heading of "expendable-mold casting processes."

Although some nonmetals are cast, casting processes are of primary importance in the production of metal products, and the emphasis of these chapters will be on metals casting. The metals most frequently cast are iron, steel, aluminum, brass, bronze, magnesium, and certain zinc alloys. Of these, cast iron is used more than all others because of its low cost, fluidity, low shrinkage, strength, rigidity, ease of control, and wide range of properties.

Sand casting is, by far, the most popular of the casting processes, and uses sand as the primary mold material. The sand grains, mixed with small amounts of other materials to improve moldability and cohesive strength, are packed around a pattern that has the shape of the desired casting. Because the grains pack into thin sections and can be used economically in large quantities, products covering a wide range of sizes and detail can be made by this method. After the sand has been packed around it, the pattern must generally be removed to leave a cavity of the desired shape. If pattern removal is required, the mold must be made in at least two pieces. An opening, called a *sprue hole,* is cut from the top of the mold through the sand and is connected to the mold cavity through a system of channels, called *runners.* The molten metal is poured into the sprue

Sand Casting

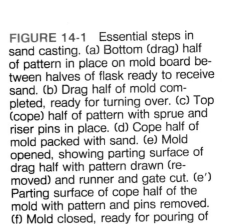

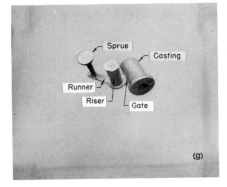

FIGURE 14-1 Essential steps in sand casting. (a) Bottom (drag) half of pattern in place on mold board between halves of flask ready to receive sand. (b) Drag half of mold completed, ready for turning over. (c) Top (cope) half of pattern with sprue and riser pins in place. (d) Cope half of mold packed with sand. (e) Mold opened, showing parting surface of drag half with pattern drawn (removed) and runner and gate cut. (e') Parting surface of cope half of the mold with pattern and pins removed. (f) Mold closed, ready for pouring of the metal. (g) Casting removed from the mold.

hole and enters the cavity through the runners and an opening called a *gate,* which controls the rate of flow. Gravity is usually used to cause the metal to flow into the mold. Because the mold is then destroyed when the finished casting is removed, a new mold must be made for each casting to be poured. Figure 14-1 shows the essential steps and basic components of the sand-casting process.

Patterns. The first step in making a sand casting is the design and construction of a pattern. This is a duplicate of the part to be cast, modified in accordance with the requirements of the casting process, the metal being cast, and the particular molding technique that is being used. The pattern material is determined primarily by the number of castings to be made. Wood patterns are relatively easy to make and are frequently used when small quantities of castings are required. Wood, however, is not very dimensionally stable, as it may warp or swell with changes in humidity, and it tends to wear out fairly rapidly. Metal patterns are more expensive but are more dimensionally stable and longer lasting. Hard plastics, such as urethanes, are being used more frequently, particularly with some of the strong, organically bonded sands that tend to stick to other pattern materials. In the full-mold process, expanded polystyrene (EPS) is used, but here each pattern can be used only once.

Types of Patterns. A number of basic types of patterns are in common use, the particular type being determined primarily by the number of duplicate castings required and the complexity of the part.

One-piece or *solid patterns,* such as the one shown in Figure 14-2, are the simplest and often the least expensive type to make. Essentially, such a pattern is a duplicate of the part to be cast, modified only by provision for the various allowances discussed in Chapter 13 and the possible addition of core prints. While this type of pattern is cheap to construct, the molding process is usually slow. Thus, one-piece patterns are generally used only when one or a few duplicate castings are to be made.

Unless a one-piece pattern is quite simple in shape and contains a flat surface that can be placed on a follow board to form a plane parting surface, it is necessary for the molder to cut an irregular parting surface by hand. This process is time-consuming and requires a skilled worker, but it can often be avoided by using a special follow board that is cut out so that the one-piece pattern sets down into it to the depth of the desired parting line. Here, the follow board determines the parting surface, as illustrated in Figure 14-3.

FIGURE 14-2 Single-piece pattern for a pinion gear.

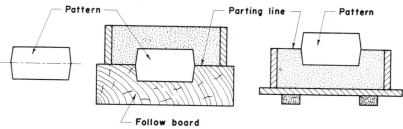

FIGURE 14-3 Method of using a follow-board pattern.

FIGURE 14-4   Split pattern, showing the two sections together and separated. Light-colored portions are core prints.

Split patterns are used where moderate quantities of duplicate castings are to be made; they permit the molding of more complex shapes without resorting to the hand forming of the parting plane or the use of cutout follow boards. The pattern is split into two sections along a single parting plane, which will correspond to the parting plane of the mold. As shown in Figure 14-4, one half of the pattern forms the cavity in the lower, or drag, portion of the mold, and the other half serves a similar function in the upper, or cope, section. Tapered pins in the cope half of the pattern align with corresponding holes in the drag half to hold the pieces in proper position as the mold segments are being made.

Match-plate patterns, such as the one shown in Figure 14-5, are widely used in many foundries because they can be coupled with modern molding machines to produce large quantities of duplicate castings. Here, the cope and drag sections are fastened to the opposite sides of a wood or metal match-plate that is equipped with holes or bushings that mate with pins or guides on the halves of the mold flask. The match-plate is fitted between the two flask sections, and sand is packed into the flask to complete the cope and drag. The mold sections are then separated, and the match-plate pattern is removed. On reclosing of the mold, the cavities in the cope and drag join in proper alignment. This alignment is made possible by the pins and guide holes in the flask segments.

In most cases, the necessary gate and runner system is incorporated on the match-plate. Such a match-plate eliminates the necessity for the molder to cut the gates and runner by hand and also ensures that they will be uniform and of the proper size in each mold, thus reducing the likelihood of defects. The gate and runner system can be seen as the dark center section on the match-plate pattern in Figure 14-5, and on the cope section of Figure 14-6. These patterns also illustrate the common practice of having more than one pattern on a match-plate. Core prints and risers are also included when required.

When large quantities are to be produced, or when the casting is quite large, it may be desirable to have the cope and drag halves of split patterns attached

FIGURE 14-5   Matchplate pattern for molding two parts. (Left) Cope side; (Right) Drag side.

FIGURE 14-6 Cope-and-drag pattern for molding two heavy parts. (Left) Cope section; (Right) Drag section.

to separate match-plates instead of being attached to the opposite sides of a single plate. Large molds can then be handled more easily, or two workers, on two separate machines, can simultaneously produce the two portions of a mold. Such patterns are called *cope-and-drag* patterns, and are illustrated by the pattern in Figure 14-6.

When an object to be cast has protruding sections arranged so that neither a one-piece pattern nor one split along a single parting plane can be removed from the molding sand, a *loose-piece pattern* can sometimes be developed. Loose pieces are held to the remainder of the pattern by beveled grooves or pins, as shown in Figure 14-7. This construction permits all of the pattern except the loose pieces to be withdrawn directly from the sand, after which there is space within the cavity so that the remaining loose pieces may be moved in the necessary direction and the necessary amount to permit their removal. If the loose pieces cannot be held to the main portion of the pattern by grooves or stationary pins, a long, sliding pin may be used. After the sand is compacted, the pin is withdrawn, so that the loose pieces are freed for successive removal. Obviously, loose-piece patterns are expensive to make, require careful maintenance, slow the molding process, and increase molding costs. They do make possible the casting of complex shapes that otherwise could not be cast, except by the full-mold or investment processes. Whenever possible, however, it is desirable to eliminate their necessity by design changes in the casting.

It is very important on all castings that intersecting surfaces be joined by a small radius, called a *fillet*, instead of being permitted to intersect in a line. Fillets prevent shrinkage cracks at such intersections and also eliminate stress concentrations in the finished product. As a general rule, designers should make fillets generous in size, ¼ and ⅛ inch (6.35 and 3.18 mm) radii being common. Fillets can be added to wood patterns by means of wax, leather, or plastic strips of the desired radius, which can be glued to the pattern or pressed into place with a heated fillet tool.

Sand Conditioning and Sand Control. The sand used to make molds must be carefully conditioned and controlled in order to give satisfactory and

FIGURE 14-7 Loose-piece pattern for molding a large worm. After sufficient sand is packed around the pattern halves to hold the pieces in position, the wooden pins are withdrawn. The remaining sand is then rammed around the pattern, the mold is opened, and the pattern is removed.

uniform results. Ordinary silica (SiO_2), zircon, or olivine (forsterite and fayalite) are compounded with additives to meet four requirements:

1. *Refractoriness:* the ability to withstand high temperatures.
2. *Cohesiveness* (also referred to as *bond*): the ability to retain a given shape when packed into a mold.
3. *Permeability:* the ability to permit gases to escape.
4. *Collapsibility:* the ability to permit the metal to shrink after it solidifies and to ultimately free the casting through disintegration of the mold.

Refractoriness is provided by the basic nature of the sand. Cohesiveness, bond, or strength is obtained by coating the sand grains with clays, such as bentonite, kaolinite, or illite, that become cohesive when moistened. Collapsibility is sometimes obtained by adding cereals or other organic materials such as cellulose, which burn out when exposed to the hot metal, thereby reducing the volume of solid bulk and decreasing the strength of the restraining sand. Permeability is a function of the size of the sand particles, the amount and type of clay or other bonding agent, the moisture content, and the compacting pressure.

Good molding sand always represents a compromise between conflicting factors, the size of the sand particles, the amount of bonding agent (such as clay), the moisture content, and the organic matter being selected so as to yield a satisfactory compromise of the four requirements. Sand composition must be carefully controlled to ensure satisfactory and consistent results; a typical green sand mixture consists of 88% silica sand, 9% clay, and 3% water. In addition, it is often desirable to reclaim and reuse the mold material. Mold temperature during the pouring and solidification must be well controlled. If organic materials, such as cereal or wood flour, are used, they burn during the pour, and some of the mold material will have to be discarded and replaced with new.

Each grain of sand should be coated uniformly with the additive agents. To achieve this, the ingredients are put through a *muller,* which provides the necessary mixing. Figure 14-8 shows a modern continuous-type muller, which uses blades and wheels to produce the mixing. The sand is often discharged through an *aerator,* which fluffs the sand so that it does not tend to pack too hard during handling.

FIGURE 14-8 Schematic diagram of sand mullers. (Left) Continuous muller; (Right) conventional batch muller. Plow blades move the sand and the muller wheels mix the components. (From ASM Committee on Sand Molding, "Sand Molding," *Metals Handbook,* Vol. 5, 8th ed., American Society for Metals, Metals Park, Ohio, 1970, p. 163.)

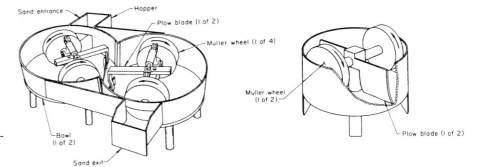

Sand Control.   Although sand control is of little concern to the designer of castings, it is a matter of great concern to the foundry worker, who is expected to deliver castings of good and consistent quality. Standard tests and procedures have been developed to maintain consistent sand quality by evaluating *grain size, moisture content, clay content, compactibility,* and *mold hardness, permeability, and strength.*

Grain size is determined by shaking a known amount of clean, dry sand downward through a set of 11 standard sieves of decreasing mesh size. After shaking for 15 minutes, the amount remaining in each sieve is weighed, and the weights are converted to an AFS (American Foundrymen's Society) grain-fineness number.

Moisture content is most commonly determined by a special device that measures the electrical conductivity of a small sample of sand that is compressed between two prongs. Another method is to measure the direct weight loss from a 50-gram sample after it has been subjected to a temperature of about 230°F (110°C) until all water is driven off.

Clay content can often be determined by washing the clay from a 50-gram sample of molding sand in water that contains sufficient sodium hydroxide to make it alkaline. After several cycles of agitation and washing in such a solution, the clay is fully removed. The remaining sand is dried and then weighed to give the proportion of the original sample that was clay.

Permeability and strength tests are conducted on a standard rammed specimen. A sufficient amount of sand is placed in a 2-inch-diameter steel tube so that after a 14-pound weight is dropped three times from a height of 2 inches, the final height is within $1/32$ inch of 2 inches. Figure 14-9 shows a sand rammer used for the production of standard rammed specimens.

Permeability is a measure of how easily gases can pass through the narrow voids between the sand grains. Air in the mold before pouring, plus steam produced when the hot metal vaporizes the moisture in the sand, must be allowed to escape, rather than be trapped in the casting as porosity or blow holes. For the permeability test, the sample tube containing the rammed sand specimen is placed on a device like that pictured in Figure 14-10 and is subjected to an air pressure of 10 grams per square centimeter. By means of either a flow rate determination or a measurement of the pressure between the orifice and the sand, an AFS permeability number[1] is determined. Most test devices are calibrated to provide a direct readout of the permeability number.

The compressive strength of the sand is determined by removing the rammed

FIGURE 14-9 Sand rammer for preparing a standard rammed foundry sand specimen. (Courtesy Harry W. Dietert Company.)

---

[1]The AFS permeability number can be determined by computing:

$$\text{AFS Number} = (V \times H)/(P \times A \times T), \text{ where}$$

V is the volume of air (2000 cubic centimeters), H is the height of the specimen (5.08 cm), P is the pressure (10 grams per centimeter squared), A is the cross section of the specimen (20.268 cubic centimeters), and T is the time in seconds to pass a flow of 2000 cubic centimeters. Substituting the constants, the permeability number is equal to 3000.2/T where T is in seconds.

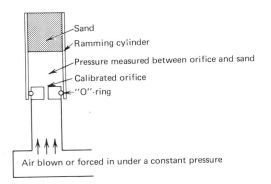

Figure 14-10 (Left) Permeability tester for foundry sand. Standard sample in sleeve is sealed by O-ring on top of unit. (Courtesy Harry W. Dietert Company.) (Right) Schematic of permeability tester in operation.

specimen from the tube and placing it in a Universal sand tester, such as the one shown in Figure 14-11. A compressive load is then applied until the specimen breaks, usually in the range of 10–30 psi (0.07–0.2 MPa). When there is too little moisture in the sand, the grains are poorly bonded and strength is poor. When there is excess moisture, the extra water acts as a lubricant and strength is again poor. Thus, there is a maximum strength, and an optimum water content, that vary with the content of other materials within the sand. A similar optimum also applies to permeability, since unwetted clay blocks passages, as does excess water. Sand coated with a uniform thin film of moist clay is most desired. A ratio of one part water to three parts clay (by weight) generally gives good molding properties.

The hardness to which sand is compacted in a mold is very important because it can quickly provide an indication of mold strength and give insight into the strength–permeability characteristics. It is commonly measured by the instrument shown in Figure 14-12, which measures the resistance of the sand to penetration by a 0.2-inch (5.08-mm) diameter, spring-loaded steel ball.

Compactibility can be determined by sifting sand into a steel cylinder, leveling off the column, striking it three times with a weight (as in the permeability test), and then measuring the change in height. The percent compactibility is determined as the change in height divided by the original height, times 100, and is closely related to the moisture content of the sand. A compactibility of 45% is comparable to the most desirable moisture content. Poorer response is indicative of too little moisture.

Sand Properties and Sand-Related Defects.   The character of the sand grains can be very influential in determining the properties of the molding material. Round sand grains give good permeability and minimize the amount of clay required because of their low surface area. Angular sands give better green strength because of the mechanical interlocking of the grains. Large-size sand grains provide good permeability and better resistance to high-temperature melt-

FIGURE 14-11 Universal strength-testing machine for foundry sand. (Courtesy Harry W. Dietert Company.)

ing and expansion, and fine-grain sands provide good surface finish on the final casting. Uniform-size sands give good permeability, and a wide distribution gives a better surface finish.

When hot metal is poured into a silica sand mold, the silica sand heats up, undergoes one or more phase transformations, and has a large expansion in volume. Only the sand adjacent to the mold cavity expands, however. The remainder stays fairly cool, doesn't expand, and provides resistance to the other expansion. Thus, the sand at the surface of the mold cavity may fail as it undergoes its expansion, the failures taking the form of buckles or folds. Castings having extensive flat surfaces are more prone to sand expansion defects since large amounts of expansion must occur in one fixed direction.

Sand expansion defects can be minimized in a number of ways. Certain sand geometries permit the grains to slide relative to one another, thereby to relieve the expansion stresses. Sands with excess clay and/or volatile materials can absorb the sand expansion. Olivine sand does not undergo the phase transformations on heating, so it expands only about half as much as silica sand. Unfortunately, it is much more expensive. Finally, sand additives, such as cellulose (wood flour), can be added. Wood flour is a volatile material that burns as the sand heats, leaving behind a void that can accommodate sand expansion.

Voids caused by trapped or evolved gas are usually attributed to low sand permeability or to excessive evolved gas, possibly due to high moisture or excessive amounts of volatiles. In addition to the corrective measures related to the above, the cutting of mold vents may be extremely helpful, but this process adds significantly to the mold-making cost.

Penetration of metal between the sand grains can be due to high pouring temperatures (excess fluidity), high metal pressure (possibly due to excess cope height or pouring from too high an elevation above the mold), or the use of coarse, uniform sand particles. Fine-grained materials, such as silica flour, can be used to fill the voids, but permeability is reduced and sand-expansion defects are more likely to form.

FIGURE 14-12 Mold hardness tester. (Courtesy Harry W. Dietert Company.)

FIGURE 14-13   Halves of a tapered cam-latch snap flask, closed at left and open at right.

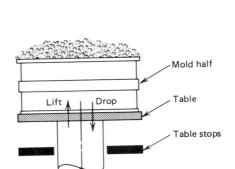

FIGURE 14-14   Jolting a mold half.

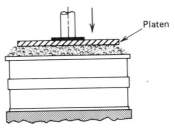

FIGURE 14-15   Squeezing operation.

Hot tears or cracks can form in castings made from alloys with a long freezing range. During solidification, the metal tries to contract but may be restrained by a strong mold or core. Internal stresses can develop in the metal while it is still partially liquid, and if they become great enough, the casting will crack.

One of the problems leading to hot tears is the lack of *collapsibility*, the ability of the sand to break down and crumble after the casting has been poured and solidified. Sand additives such as cellulose can be particularly helpful in providing collapsibility.

The Making of Sand Molds.   Except in very small foundries or when only a few castings of a given design are to be made and *hand ramming* is used, virtually all sand molds are now made with various types of molding machines. These greatly reduce the labor and skill required and produce castings with better dimensional accuracy and consistency.

Molding machines vary in how they pack the sand within the molding flask, whether mechanical assistance is provided for turning and/or handling the mold, and whether a flask is required. Molding usually begins with a pattern, such as the match-plate pattern discussed earlier, and a *flask*. The flasks may be either straight-walled containers with guide pins, or they may be removable jackets, and they are generally constructed of aluminum or magnesium. Figure 14-13 shows a snap-type flask designed to open slightly to permit withdrawal after the mold packing is complete.

The packing of the sand is generally done by the use of one or a combination of several principles. In one, sand is placed on top of the pattern, and the pattern, flask, and sand are then lifted and dropped several times, as shown in Figure 14-14. The kinetic energy of the sand itself produces optimum packing at the pattern. These *jolt-type machines* can be used on the first half of a match-plate pattern or on both halves of a cope-and-drag operation.

Squeezing machines pack the sand by the squeezing action of either an air-operated squeeze head, as indicated in Figure 14-15; a flexible diaphragm; or small, individually activated squeeze heads. Squeezing packs the sand firmly near the squeezing head, but the density diminishes as the distance from the head increases. A high-pressure, flexible-diaphragm squeezing machine, commonly called a *Taccone machine*, can give a more uniform density around all

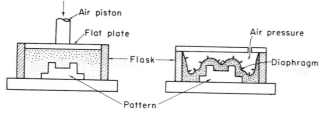

FIGURE 14-16 Schematic diagram showing relative sand densities obtained by flat-plate squeezing machines (left) and by flexible-diaphragm squeezing machine (right).

parts of an irregular pattern. Figure 14-16 compares squeezing with a flat plate and squeezing with a flexible diaphragm.

A combination of jolting and squeezing is commonly used to yield more uniform density throughout the mold depth. Here, the flask, with a match-plate pattern in place, is placed upside down on the table of the machine. After parting dust has been sprinkled on the pattern, the flask is then filled with sand. The table and the flask are then jolted the desired number of times to pack the sand around the pattern. The squeeze head is then lowered, and pressure is applied to pack the sand in the upper portion of the flask. Both cope and drag may be made on the same machine by rolling the flask over and repeating the operations on the cope half, or the cope and drag can be made on separate machines using cope-and-drag patterns.

Except for very small molds, jolt and/or squeeze machines usually provide mechanical assistance for turning over the heavy mold. Some patterns may include the sprue hole, but it is often cut by hand. The pouring basin may be hand-cut, or it may be formed by a protruding shape on the squeeze board. The gate and runner systems are usually included on the pattern, but these may also be hand-cut after molding. The growing trend is to design and construct patterns that will eliminate the majority of handwork in molding.

After the mold is completed, the tapered flask may be removed so that it will not be damaged during pouring. A *slip jacket*—an inexpensive metal band— may be slid around the mold to hold the sand in place. Heavy metal weights are often placed on top of the molds to prevent the sections from becoming separated as the hydrostatic pressure of the molten metal presses upward on the cope. When used, slip jackets and pouring weights are needed only during pouring and for a few minutes afterward, while solidification occurs. They can then be removed and placed on other molds, so the amount of equipment needed in the operation is reduced.

For mass-production molding, a number of machines have been developed that use one or more of the basic compaction methods, but operate automatically at high compaction pressures. These include automatic match-plate machines and automatic cope-and-drag machines. Figure 14-17 depicts a *vertically parted flaskless molding machine*. The halves of two separate molds are made at one time, a complete mold being formed by the joining of two successive mold blocks. This system is unique in that (1) a vertical gating system is now used, and (2) one entire mold cavity is produced per block of sand. Previous techniques required separate cope and drag segments (two molding operations) to produce

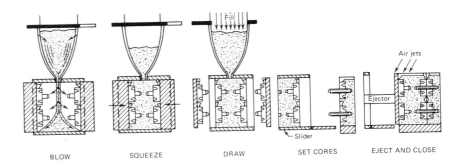

FIGURE 14-17 Method of making molds without the use of a flask. (Courtesy Belens Corporation.)

a mold. If cores are used, they are set in place before the halves are closed. The *H-process* uses a common runner and sprue system to pour a number of vertically parted mold segments.

In *stack molding,* a number of mold sections are piled one on top of the other, each section containing a cope and a drag impression. Metal is poured down a common sprue, which is connected to a gating system at each of the parting planes.

For molds too large to be made either by hand ramming or with one of the molding machines that have been described, large flasks that sit on the foundry floor are used in conjunction with various mechanical aids. A *sand slinger,* a mechanized piece of equipment with impeller blades, can be used to fling sand into the flask at high velocity. If this is skillfully done, uniform compaction of a desired hardness can be achieved. Extra tamping can be done with a pneumatic rammer.

Extremely large molds can be made in pits, as shown in Figure 14-18. Because of the size, complexity, and need for strength, pit molds are frequently built from sections of baked or dried sand. Added binders are generally required to provide the desired strength.

Green-Sand, Dry-Sand, and Skin-Dried Molds. In green-sand molding, the aggregate is sand and the binder is the clay, water, and additives. Green sand, while inexpensive and very common, does create some problems, generally associated with low strength and high moisture. One way to reduce these problems is to dry the mold before pouring. Almost twice as much water is added to the original sand as would be added for a green-sand mold. After molding, a baking operation at or above 300°F is conducted to drive off most of the moisture, reducing the amount of gases that would be evolved when hot metal comes in contact with the sand and producing higher strengths.

Dry-sand molds are not used very extensively because of the time and cost of drying and the availability of alternative methods. An attractive compromise, however, would be to dry only the sand at the mold cavity surface, producing a *skin-dried mold.* Molds in which steel is to be cast are nearly always skin-dried just before pouring because much higher temperatures are involved than in the case of cast iron. Such molds may also be given a high-silica wash before drying to increase the refractoriness of the surface, or zircon sand can be used

FIGURE 14-18 Using dry-sand sections in the construction of a large pit-type mold. (Courtesy Pennsylvania Glass Sand Corporation.)

as a facing. Skin drying to a depth of about one-half inch is then accomplished by means of a torch. Frequently, additional binders, such as molasses, linseed oil, or corn flour, must be added to the facing sand when skin drying is to be performed.

Sodium-Silicate–CO_2 Molding.   In modern foundry practice, many molds and cores are made of a sand that is given increased strength through the addition of 3%–4% sodium silicate, or water glass, a liquid inorganic binder. Sand mixed with the sodium silicate remains soft and moldable until exposed to CO_2 gas, after which it hardens within a few seconds by virtue of the reaction:

$$Na_2SiO_3 + CO_2 \rightarrow Na_2CO_3 + SiO_2 \text{ (colloidal)}$$

Sands of this type can be mixed in a muller and handled in the normal manner, as they are not gassed with the CO_2 until they have been packed around the pattern in the mold flask.

The major advantages of the CO_2–sodium-silicate method are that the CO_2 gas is nontoxic and odorless, and that no heating is required for curing. When hardened, however, these sands have poor collapsibility, making shakeout and core removal difficult. Unlike with other sands, the heating experienced during

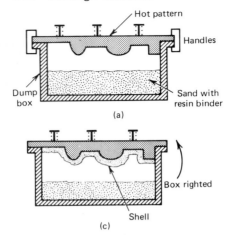

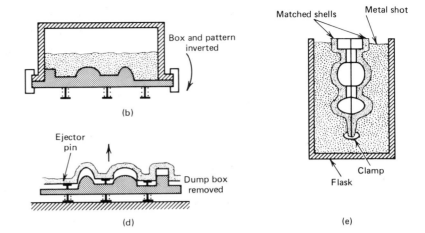

FIGURE 14-19 Schematic of the shell-molding process. A heated pattern is placed over a dump box containing a sand and resin mixture. The box is inverted and a shell partially cures around the pattern. The box is righted, the top is removed, and the shell is further cured and is finally stripped from the pattern. Matched shells are joined and supported in a flask ready for pouring.

the pour actually makes the sand stronger, as in the firing of a ceramic material. Additives that will burn out during the pour are often used to enhance the collapsibility of sodium silicate molds. In addition, care must be taken to prevent the carbon dioxide in the air from hardening the sand before the mold-making process is complete.

A modification of the CO_2 process is used quite extensively in making certain portions of molds where better accuracy of detail and thinner sections are desired than can be achieved with ordinary molding sand. Sand mixed with sodium silicate is packed around the metal pattern to a depth of about 1 inch, followed by regular molding sand as a backing material. After the mold is rammed, CO_2 is introduced through vents in the metal pattern, thereby setting the adjacent sand. The pattern can then be withdrawn with reduced danger of damage to the mold. This procedure is particularly useful where deep draws are necessary.

Another modification of the sodium silicate process is the use of cold-setting additives, such as furfurals and a catalyst, which cause the sodium silicate to set up without external CO_2 gassing. In these *air-set sands*, hardening occurs within a matter of minutes, so the additives must be mixed with the sand just before the molding operation. Molds produced by this method permit thinner sections, deeper draws, and greater accuracy.

Shell Molding.   Large numbers of molds are now made by the *shell-molding process*, which offers better surface finish than can be obtained with ordinary sand molding, better dimensional accuracy, and reduced labor requirements. In many cases, the process can be completely mechanized and adapted for mass production.

As illustrated in Figure 14-19, the shell process involves six basic steps:

1. A mixture of sand and a thermosetting plastic binder is dumped onto a metal pattern that has been heated to 300–450°F (150–230°C) and is held in place for a few minutes. During this interval, heat from the pattern partially cures a layer of the sand–plastic mixture, forming a strong, solid-

bonded region, about ⅛ inch (3.5 mm) thick adjacent to the pattern. The thickness depends on the pattern temperature and the time of contact.

2. The pattern and the sand mixture are then inverted to permit all sand to drop off except the layer of partially cured material adhering to the pattern.
3. The pattern and the partially cured "shell" are then placed in an oven for a few minutes to complete the curing of the shell.
4. The hardened shell is then stripped from the pattern.
5. Two shells are then clamped or glued together to form a complete mold.
6. The bonded shells may be placed in a pouring jacket and backed up with metal shot or sand to provide extra support during the pour.

Because the plastic and the shell sand are molded to a metal pattern and have been compounded to undergo almost no shrinkage, the shell has virtually the same dimensional accuracy as the pattern. Tolerances of 0.003–0.005 inch (0.08–0.13 mm) are readily obtained. The sand used is finer than ordinary foundry sand, and when combined with the plastic resin, it results in a very smooth shell and casting surface. Also, the consistency between castings is superior to that obtained by ordinary green-sand casting.

Figure 14-20 shows a set of patterns, the two shells before clamping, and the resulting casting. Machines for making shell molds vary from simple types for small operations to large, completely automated types for mass production. The cost of a metal pattern is often rather high, and its design must include the gate-and-runner system. Also, fairly large amounts of expensive binder are required. Nevertheless, the amount of material used to form a thin shell is rather small. High productivity, low labor costs, smooth surfaces, and good precision that reduces machining operations—all combine to make the process economical for even moderate quantities. In addition, during pouring, the shell becomes very hot, some of the resin binder burns out, and the collapsibility and shakeout characteristics are excellent.

The V-Process (Vacuum Molding). A relatively new sand-molding process uses no binder at all but attains mold strength through the use of a specially designed vacuum flask. First, a thin sheet of plastic is draped over a special pattern and is drawn tightly to the pattern surface by a vacuum. Then, a vacuum flask is placed over the pattern, the flask is filled with sand, a sprue and a pouring cup are formed, and a second sheet of plastic is placed over the sand. A vacuum is then drawn on the flask, making the sand very hard. The pattern vacuum is released, and the pattern is withdrawn. Mold halves are assembled, and the mold is poured while the mold vaccum is maintained. Figure 14-21 shows this sequence of events.

The V-process has numerous advantages. The castings have no moisture-related defects. No binder is used, so binder cost is eliminated and the sand is immediately reuseable. No fumes (binders burning up) are generated during the pouring operation, and any sand or aggregate can be used. The shakeout characteristics are exceptional, for when the vacuum is released after pouring, the mold virtually collapses. Unfortunately, the process is relatively slow.

FIGURE 14-20   (Bottom) Two halves of a shell-mold pattern. (Top) Shell mold and the resulting casting. (Courtesy Shalco Systems, an Acme-Cleveland Company.)

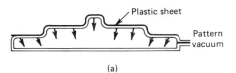

(a)

FIGURE 14-21   Schematic of the V-process or vacuum molding. A vacuum is pulled on a pattern, drawing a plastic sheet tightly against it. A vacuum flask is filled with sand, a second sheet placed on top, and a mold vacuum is pulled. The pattern vacuum is broken, the pattern withdrawn, the mold halves assembled, and the molten metal is poured.

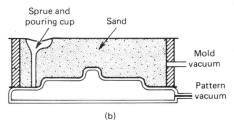

(b)

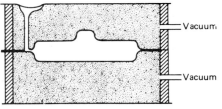

(c)

The Eff-Set Process. In the Eff-set process, frozen water is the binder. Sand with a small amount of clay and quite a bit of water is packed around a pattern, and the pattern is removed. At this point, the mold has just sufficient strength to hold its shape but cannot withstand handling. Liquid nitrogen is then sprayed onto the mold surface, and the ice that forms becomes a firm binder. The mold is assembled, and molten metal is poured while the mold is still frozen. Here again, the process benefits from low binder cost and easy shakeout. In addition, the process has better than normal fluidity since the vaporizing water tends to provide a "cushion" over which the metal flows.

Cores and Core Making

One of the distinct advantages of castings is that internal cavities or reentrant sections can be included with relative ease. Often, such configurations could not be made by any other process. However, to produce such castings, it is necessary to use *cores*. Figure 14-22 shows an example of a product that could not be made by any other process than casting with cores. Of course, cores constitute an added cost, and in many cases, the designer can do much to facilitate and simplify their use.

To develop the use of cores, consider the simple belt pulley shown in Figure 14-23. The shaft hole could be made in a number of ways. First, the pulley could be cast as a solid and the hole made by machining. This procedure would require a substantial amount of costly machining, however, and quite a bit of metal would be wasted. Therefore, it is more economical to cast a pulley with

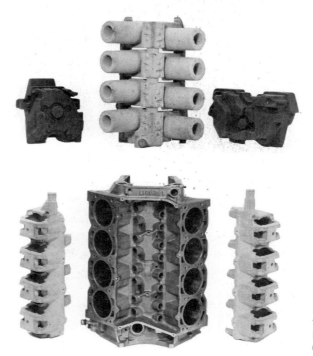

FIGURE 14-22 Dry-sand cores used in the making of a V-8 engine block and the completed casting (bottom center). (Courtesy Central Foundry Division of General Motors Corporation.)

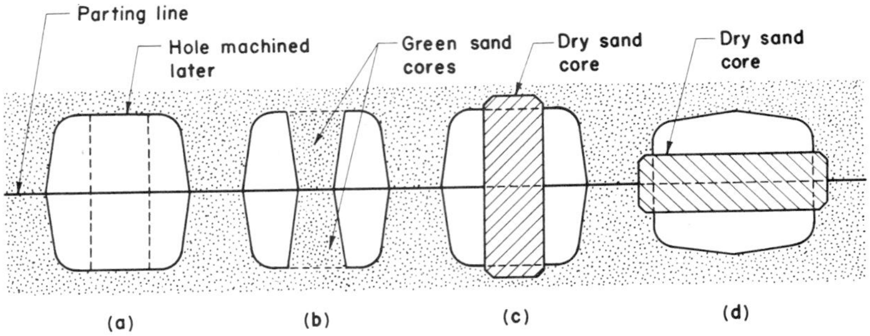

Parting line
Hole machined later
Green sand cores
Dry sand core
Dry sand core

(a)   (b)   (c)   (d)

FIGURE 14-23 Four methods of making a hole in a cast pulley.

a hole in it of the approximate size desired. One procedure would be to use a split pattern, as shown in Figure 14-23b. Each half of the pattern contains a tapered hole into which green molding sand is packed, just as in the remainder of the mold. The sections of sand that protrude into the mold cavity are called *green-sand cores*. Green-sand cores are frequently impractical, however, because they are relatively weak. If they are narrow or long, the pattern cannot be withdrawn without their breaking, or they will have insufficient strength to support their own weight. In addition, the amount of draft that must be provided on the pattern is such that if the core is very long, a considerable amount of machining may have to be done and the advantage of the core is lost. In other cases, the shape is such that a green-sand core cannot be used.

To overcome these difficulties, it is often necessary to use *dry-sand cores*, as shown in Figure 14-23c and d. Dry-sand cores can be made in several ways. In each, the sand, mixed with some form of binding material, is packed into a core box that contains a cavity of the designed shape. The core box, usually made of wood or metal, is analogous to the mold in which a casting is made. The *dump-type core box,* shown in Figure 14-24, is very common. The sand is packed into the box and scraped level with the top surface. A metal plate is

FIGURE 14-24 (Clockwise from upper right) Core box, two core halves ready for baking, and the completed core made by gluing the two halves together.

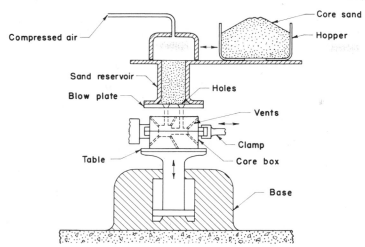

FIGURE 14-25 Method of making cores in a core-blowing machine.

then placed on top of the box, and the box is turned over and lifted upward, leaving the core resting on the plate.

Some cores can be made in a *split core box,* consisting of two halves that are clamped together with an opening in one or both ends, through which the sand is rammed. The halves of the box then separate to permit removal of the core. Cores that have a uniform cross section throughout their length can be made by a core-extruding machine that is similar to the familiar meat grinder. These cores are cut to length as they come from the machine and are placed in core supports for hardening.

In many cases, the sand mix is blown into the core box by equipment similar to that in Figure 14-25. The oldest core-making process uses a vegetable or synthetic oil as a binder. Baking is required to polymerize the oil and produce a strong organic bond between the sand grains. In the *hot-box method,* the core box is heated to around 400°F (200°C) so that the surface of the core polymerizes after the sand is blown in. The core can then be removed and further baked without danger of breakage. In the *cold-box process,* binder-coated sand is first blown into the core box, which is then sealed and gassed with an organic gas that immediately polymerizes the resin at room temperature. Cores can be produced in 30–45 seconds, compared to about 4 minutes for the hot-box process. If the gases are introduced at the surface of the core, hollow cores can be produced, and the uncured sand from the center can be reused. The gases involved may be very toxic, however.

Many cores are now made with airset or "no-bake" binders, which use an organic resin as the binder and a catalyst to cause it to cure at room temperature with no required gassing. The sodium-silicate–CO_2 process and shell molding are other core-making alternatives. Shell-molded cores have excellent permeability since they are generally hollow.

Cores are often molded in halves and assembled after baking or hardening is completed. Rough spots are removed with coarse files or sanding belts, and the

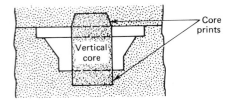

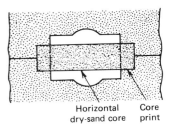

FIGURE 14-26 Mold cavities containing core prints to hold and position dry-sand cores. (Left) Vertical core; (Right) Horizontal core.

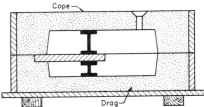

FIGURE 14-27 (Top) Typical chaplets. (Bottom) Method of supporting cores by use of chaplets (relative size of the chaplets is exaggerated).

halves are pasted together, often with hot-glue guns. They may then be given a thin coating to produce a smoother surface or greater refractoriness. Graphite, silica, or mica can be either sprayed on or brushed on.

To function properly, cores must have the following characteristics:

1. Sufficient hardness and strength after baking or hardening to withstand handling and the forces of the molten metal. Compressive strength should be between 100 and 500 psi.
2. Sufficient strength before hardening to permit any required handling.
3. Adequate permeability to permit the escape of gases. Since the core may be largely surrounded by molten metal, cores should possess exceptionally good permeability.
4. Collapsibility to permit the shrinkage that accompanies the cooling of the solidified metal (thereby preventing cracking of the casting), and to permit easy shakeout from the interior of the casting.
5. Adequate refractoriness. Since the cores are largely surrounded by hot metal, they can become quite a bit hotter than the adjacent mold material.
6. A smooth surface.
7. Minimum generation of gases.

Cores are occasionally strengthened by means of wires or rods to give them sufficient strength for handling and to resist the forces that act against them during pouring. In some cases (particularly in steel castings, where considerable shrinkage is encountered), cores are made hollow at their centers or straw is put in the center so they can collapse as the metal cools after solidifying. Failure to provide for free shrinkage may result in cracked castings.

All but very small cores must be vented to enable the gases to escape. Vent holes can often be produced by pushing small wires into the cores. Coke or cinders are sometimes placed in the center of large cores to provide venting.

Since the core material must be subsequently removed, the inside cavities must be connected to the outside of the casting. These connections, or recesses in the mold, are often used to provide support for the cores and to hold them in proper position during mold filling. These recesses in which the ends of cores are placed are called *core prints*. Figure 14-26 shows molds with core prints holding a vertical and a horizontal core.

In some cases, the design of a casting does not permit the core to be supported from the sides of the mold. In such instances, the core can be supported (and prevented from being moved or floated by the molten metal) by means of small metal supports called *chaplets*. Figure 14-27 shows some chaplets and illustrates how they are used. The use of chaplets should be minimized because they become part of the casting and may be a possible source of weakness. When used, chaplets should be of the same, or at least comparable, composition as the casting material. They should be large enough so that they do not completely melt and permit the core to float, but small enough so that the surface melts and fuses with the metal.

Core sections can also be used to facilitate the molding of complex shapes, such as those with reentrant sections, as illustrated in Figures 14-28 and 14-29.

In Figure 14-28, an intermediate flask section, called a *cheek*, is used to contain a separately made green-sand core. Although additional molding operations are required, this procedure may be advantageous when only a few molds are to be made, since it eliminates the need for a special core box.

If any substantial quantity of the pulley were required, the use of a dry-sand core, as illustrated in Figure 14-29, can greatly simplify and speed up the operation. In this case, the pattern is of a different shape, so as to provide a seat for the specially shaped dry-sand core that is set in place to form the reentrant section of the casting. The use of dry-sand cores is particularly advantageous where rapid machine molding is used, and less skill is required of the molders. However, the cost of providing a core box and making the cores must be balanced against the savings in labor cost. Quantity is a primary factor. If modifications can be made in the design of a casting that will eliminate the necessity for dry-sand cores or additional sections and still permit machine molding to be used, this is obviously the best solution.

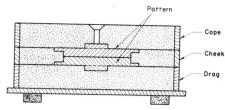

FIGURE 14-28 Method of making a reentrant angle by using a three-piece flask.

Plaster Molding. Plaster of Paris (or gypsum plaster) is the molding material, with small additions of talc, terra alba, or magnesium oxide to prevent cracking and reduce the setting time. Lime or cement may be added to control expansion during baking, and about 25% fiber can be added to improve strength. Sand can also be added as a filler.

The plaster material is mixed with water to produce a fluid mixture that can be poured around a metal pattern. Hydration of the plaster occurs, producing a solid mold, which is then stripped, assembled, and poured. Plaster molds require metal patterns and give excellent surface finish and dimensional accuracy. Unfortunately, they are suitable only for the lower-melting-temperature nonferrous alloys. When iron or steel is poured into a plaster mold, the rapid evolution of the hydration water can cause an explosion.

A variation of plaster molding is the *Antioch process*, in which about 50% each of sand and plaster is mixed with water. After setting and removal of the pattern, the molds are autoclaved in steam. The resulting mold has considerably greater permeability than the normal plaster variety.

The Shaw Process. Another means of obtaining adequate permeability while retaining the supersmooth surface of the plaster casting is the *Shaw process*, also known as *precision molding* or *ceramic molding*. Here, a slurrylike mixture of refractory aggregate, hydrolyzed ethyl silicate, alcohol, and a gelling agent is poured over a reusable pattern. The mixture sets to a rubbery state that permits the pattern to be stripped while retaining the shape and dimensions. The alcohol is then ignited, burns, and fires the ceramic (a furnace firing may also be necessary). Firing makes the mold rigid and hard but, at the same time, forms a network of microscopic cracks. This *microcrazing* produces fissures that provide excellent permeability and good collapsibility to accommodate the shrinkage of the solidifying metal.

Other Single-Use-Mold– Multiple-Use-Pattern Processes

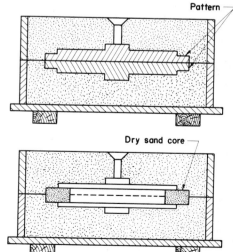

FIGURE 14-29 Molding a reentrant section by using a dry-sand core.

FIGURE 14-31 Method of combining the Shaw process and wax-pattern casting to produce the vanes of an impeller. A wax pattern is added to a metal pattern. After the metal pattern is withdrawn (left), the wax pattern is melted, leaving a cavity as shown (right). (Courtesy Avnet Shaw Division of Avnet, Inc.)

Expendable-Mold Processes Using Single-Use Patterns

The Shaw process can be used for castings of all sizes and all metals, and it produces excellent surface finish, detail, and dimensional accuracy. Dimensional tolerances between 0.002 and 0.010 inch per inch are readily attainable. The molds may be one-piece or multipiece depending on the type and complexity of the pattern. Figure 14-30 shows some of the accurate and complex parts that can be cast by this process. Figure 14-31 shows how the Shaw process can be combined with the lost-wax process (to be discussed shortly) to produce more complex products. Cores can also be used in this technique in much the same way as they are used in sand casting.

Thin sections can be obtained with plaster and ceramic molds because the mold material retards cooling. The thin sections, the better dimensional accuracy, the fine details, the elimination of machining, and the exceptionally fine surface finish are all advantages that must be weighed against the greater cost of the plaster and ceramic molds.

Expendable Graphite Molds. For metals such as titanium that tend to react with many common mold materials, powdered graphite can be combined with cement, starch, and water and compacted around a pattern. The pattern is removed, and the mold is fired at 1800°F (1000°C) to consolidate the graphite into a solid mold. After pouring, the mold is broken for removal of the casting.

Rubber-Mold Casting. Several types of artificial rubbers—usually silicone varieties—are available that can be compounded in liquid form and then poured over patterns to form semirigid molds on hardening. The molds retain sufficient flexibility to permit them to be stripped from the pattern, thereby permitting the casting of intricate shapes with reentrant sections. Unfortunately, these molds are suitable only for small castings of low-melting-point materials. Wax patterns for investment casting can be made, as well as finished castings of plastics and metals which can be cast at temperatures below 500°F (260°C).

Investment Casting. *Investment casting* is actually a very old process. It existed in China for centuries, and Benvenuto Cellini used a form in Italy in the sixteenth century. Dentists have used the process since 1897, but it was not until the end of World War II that it attained industrial importance for making rocket components and jet turbine blades from metals that are not readily machinable. Currently, millions of castings are produced by the process each year.

Its unique characteristics permit the designer almost unlimited freedom in the complexity of shapes that can be produced with extremely close tolerances.

Investment casting uses the same type of binder and molding aggregate as the ceramic molding process (Shaw process) and involves the following steps:

1. *Produce a master pattern.* The pattern may be made from metal, wood, plastic, or some other easily worked material.

2. *From the master pattern, produce a master die.* This die is usually made from low-melting-point metal, steel, or possibly even wood. If low-melting-point metal is used, the die can be cast from the master pattern. Steel dies are often machined directly, eliminating the need for Step 1. Rubber molds can also be made from the master pattern.

3. *Produce the wax patterns.* These are made by pouring molten wax, or injecting it under pressure, into the master die and allowing it to harden. Plastic and frozen mercury have also been used as pattern materials.

4. *Assemble the wax patterns to a common wax sprue.* The individual wax patterns are removed from the master die, and several of them are attached to a central sprue and runners by means of heated tools and melted wax. In some cases, several pattern pieces may be united to form a complex, single pattern that, if made in one piece, could not be withdrawn from a master die. The result of this step is a pattern cluster, or *tree*.

5. *Coat the cluster with a thin layer of investment material.* This step is usually accomplished by dipping the cluster into a thin slurry of finely ground refractory. Fine silicaceous material mixed with a special plaster is often used. A thin but very smooth layer of investment material is deposited adjacent to the wax pattern, ensuring a smooth surface and good detail in the final product.

6. *Produce the final investment around the coated cluster.* The cluster can be dipped repeatedly in the investment material until the desired thickness is obtained, or it can be placed upside down in a flask and the investment material can be poured around it.

7. *Vibrate the flask to remove entrapped air and to settle the investment material around the cluster.* This step is necessary only when the investment material is poured around the cluster.

8. *Allow the investment to harden.*

9. *Melt or dissolve the wax pattern to permit it to run out of the mold.* This step is generally accomplished by placing the molds upside down in an oven where the wax melts and the residue subsequently vaporizes. The wax can be recovered for further use. This step is the most distinctive feature of the process because it permits a complex pattern to be removed from a mold without requiring that the mold be made in two or more sections. Consequently, complicated shapes with reentrant sections can be cast. In the early history of the process, when only small parts were cast, the molds were placed in an oven, and as the wax melted, it was absorbed into the porous investment. Because the wax disappeared from

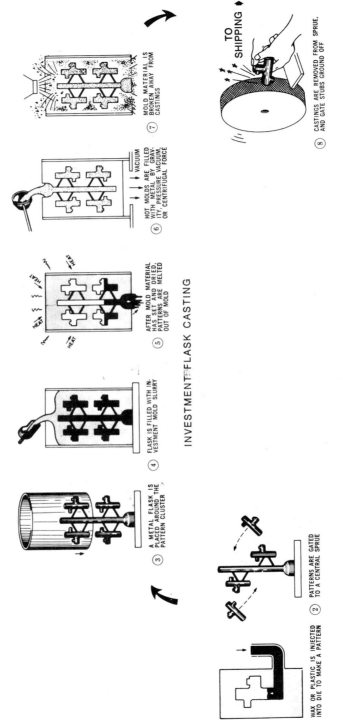

7. MOLD MATERIAL IS BROKEN AWAY FROM CASTINGS

6. HOT MOLDS ARE FILLED WITH METAL BY GRAVITY, PRESSURE, VACUUM, OR CENTRIFUGAL FORCE

VACUUM

5. AFTER MOLD MATERIAL HAS SET AND DRIED, PATTERNS ARE MELTED OUT OF MOLD

HEAT

INVESTMENT-FLASK CASTING

4. FLASK IS FILLED WITH INVESTMENT MOLD SLURRY

3. A METAL FLASK IS PLACED AROUND THE PATTERN CLUSTER

TO SHIPPING

8. CASTINGS ARE REMOVED FROM SPRUE, AND GATE STUBS GROUND OFF

2. PATTERNS ARE GATED TO A CENTRAL SPRUE

1. WAX OR PLASTIC IS INJECTED INTO DIE TO MAKE A PATTERN

FIGURE 14-32 Investment flask-casting procedure. (Courtesy Investment Casting Institute, Dallas, Texas.)

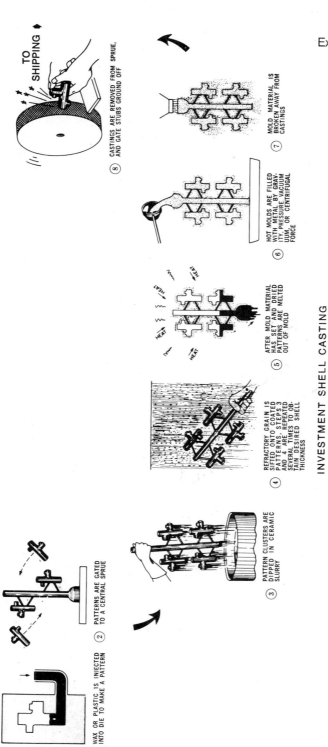

① WAX OR PLASTIC IS INJECTED INTO DIE TO MAKE A PATTERN

② PATTERNS ARE GATED TO A CENTRAL SPRUE

③ PATTERN CLUSTERS ARE DIPPED IN CERAMIC SLURRY

④ REFRACTORY GRAIN IS SIFTED ONTO COATED PATTERNS. STEPS 3 AND 4 ARE REPEATED SEVERAL TIMES TO OBTAIN DESIRED SHELL THICKNESS

⑤ AFTER MOLD MATERIAL HAS SET AND DRIED, PATTERNS ARE MELTED OUT OF MOLD

⑥ HOT MOLDS ARE FILLED WITH METAL BY GRAVITY, PRESSURE, VACUUM, OR CENTRIFUGAL FORCE

⑦ MOLD MATERIAL IS BROKEN AWAY FROM CASTINGS

⑧ CASTINGS ARE REMOVED FROM SPRUE, AND GATE STUBS GROUND OFF

TO SHIPPING

INVESTMENT SHELL CASTING

FIGURE 14-33 Investment shell-casting procedure. (Courtesy Investment Casting Institute, Dallas, Texas.)

sight, the process was called the *lost-wax process*, and the name is still used occasionally.

10. *Preheat the mold in preparation for pouring.* Heating to 1000–2000°F (550–1100°C) ensures that the molten metal will flow more readily to all thin sections. It also gives better dimensional control because the mold and the metal will shrink together during cooling.

11. *Pour the molten metal.* Various methods, beyond simple pouring, are used to ensure complete filling of the mold, especially where complex thin sections are involved. Among these methods are the use of air pressure, evacuation of the air from the mold, and a centrifugal process.

12. *Remove the casting from the mold.* This step is accomplished by breaking the mold away from the casting.

Figure 14-32 schematically presents the investment procedure where the investment material fills the entire flask. Figure 14-33 shows the shell-investment procedure.

Investment casting is obviously a complex process and can be quite expensive. Nevertheless, its unique advantages make it economically feasible in many cases, particularly since many of the steps can be completely automated. Not only can complex shapes be cast that could not be made by other processes, but very thin sections, down to 0.015 inch (0.40 mm), can be produced. The dimensional tolerances are excellent, 0.005–0.010 inch per inch being routinely obtained in combination with very smooth surfaces. Machining can often be completely eliminated or greatly reduced. Where machining is required, allowances of 0.015–0.040 inch (0.4–1 mm) are usually ample. These advantages are especially important where difficult-to-machine metals are involved, for the investment casting process can be applied to all metals.

Although most investment castings are less than 3 inches in size and weigh less than 1 pound, there are no specific size limitations. Castings up to 36 inches and weighing 80 pounds have been produced. Some typical investment castings are shown in Figure 14-34. It should be noted that complexity of shape is the most common characteristic.

The Full-Mold or Lost-Foam Process. The *full-mold process*, also known as the *lost-foam process, evaporative pattern casting*, and the *EPS (expanded polystyrene) technique*, avoids two bothersome restrictions inherent in the making of ordinary sand molds. These are (1) the cost of making a relatively expensive pattern when only a small number of castings are required and (2) the necessity for withdrawal of the pattern from the mold, often requiring some design modification, a complex pattern, or special molding procedures. In the full-mold process, the pattern is made of expanded polystyrene, which *remains in the mold during the pouring of the metal.* When the molten metal is poured, the heat vaporizes the polystyrene pattern almost instantaneously, and the metal fills the space previously occupied by the pattern, as shown in Figure 14-35.

Hard beads of polystyrene are first steam-expanded and dried before injection into a heated metal mold at low pressure, where they further expand, fill the

FIGURE 14-34 Group of typical parts produced by investment casting. (Courtesy Haynes Stellite Company.)

die, and fuse to form a pattern. The dies can be very complex, and large numbers of patterns can be accurately and rapidly produced. Extremely complex patterns can be made by gluing multiple pieces together. Because the pattern is not withdrawn, no draft need be included in the design.

When small quantities are required, the pattern can be hand-cut or machined from pieces of foamed polystyrene, a material that is light in weight (1.2 pounds per cubic foot) and is easily cut. A hand-shaped pouring basin, a sprue, runner segments, and risers can be glued on to form a complete pattern assembly. Small products can be clustered, much as in investment casting.

Several techniques can be used to complete the mold. Green sand can be compacted around the pattern, if one takes care not to crush or distort it during the compaction. Loose sand can be poured around a positioned pattern, and vacuum flasks and plastic sheets can be used, as in the V-process. More frequently, the polystyrene pattern is dipped into a water-based ceramic to form a coating 0.003 inch thick—rigid enough to prevent mold collapse during pouring, but thin enough to be permeable for the escape of the polystyrene gas. The pattern is then positioned in a flask and surrounded by vibrated loose sand, before pouring. Figure 14-36 shows a large styrene pattern with gating system and risers, along with the finished casting ready for machining.

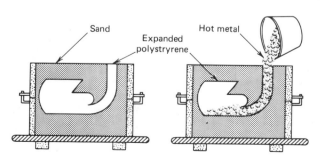

FIGURE 14-35 Schematic of the full-mold process. (Left) Expendable pattern in a sand mold. (Right) Hot metal vaporizes the pattern and fills the resulting cavity.

FIGURE 14-36 Polystyrene pattern (top) used for making the casting (bottom) by the full mold process. Note that the risers, vents, and sprue have been removed in the lower photo. (Courtesy Full Mold Process, Inc.)

The full-mold process can be used for castings of any size, both ferrous and nonferrous (although nonferrous is most common). Because of the reduced pattern cost, it is quite economical where very small quantities are desired. Since the pattern need not be withdrawn, the process is attractive for complex shapes that ordinarily would require costly cores, loose-piece patterns, or extensive finish machining. For high-volume production, mass-produced patterns made in metal molds can directly replace many of the hard patterns and cores used in conventional sand casting. Very accurate castings are produced that often reduce or eliminate machining and finishing operations, and the surface finish is quite good. Cores and parting lines are eliminated, high yields are common, and the backup sand is often directly reusable. For these reasons, full-mold casting is rapidly growing in popularity and use.

Review Questions

1. What are some of the materials used in making casting patterns, and what features would be significant in their selection?
2. What is the benefit of a split pattern over a one-piece or solid pattern? How are the two halves maintained in proper alignment?
3. How is a cope-and-drag pattern different from a match-plate pattern?
4. Under what conditions of product geometry would a loose-piece pattern be attractive?
5. What is a convenient way to add fillets to the pieces of a casting pattern?

6. What are the four requirements of a molding sand, and what feature of the sand material or additive is responsible for filling these requirements?

7. What is a muller, and what function does it perform?

8. What are some of the properties or characteristics of foundry sands that can be evaluated by standard tests?

9. What is a "standard rammed specimen" for evaluating foundry sands, and how is it produced?

10. What is permeability, and why is it important in molding sands?

11. How does the basic size and geometry of the sand grains relate to molding-sand properties?

12. What is the cause of "sand expansion defects"?

13. What features can cause the penetration of molten metal between the grains of the molding sand?

14. Describe the distribution of sand density after compaction by jolting, squeezing, and a jolt–squeeze combination.

15. What is the purpose of the heavy metal weights that are placed on top of the mold sections before pouring?

16. How can the use of vertically parted flaskless molding cut the number of required mold sections in half?

17. What are two of the major sources of problems with green sand?

18. What are some of the prominent advantages and limitations of sodium-silcate–CO_2 molding sands?

19. What material is used to bind the sand grains in the shell-molding process, and what is used to promote the "cure"?

20. What mechanism is used to bond the sand particles in the V-process?

21. Cite some of the prominent advantages and limitations of the V-process.

22. What types of geometric features in a casting require the use of cores?

23. What are some of the limitations of green-sand cores?

24. What is the binder in the hot-box core-making process, and how does it cure?

25. What is one attractive feature of shell-molded cores?

26. Why is it common for greater permeability and collapsibility to be required of cores than of the base molding sand?

27. What are chaplets, and why is it important that they not completely melt during the pouring and solidification of the casting?

28. Why are plaster molds suitable only for the lower-melting-temperature nonferrous metals and alloys?

29. How is permeability imparted to the ceramic mold used in the Shaw process?

30. What are some of the major attractive capabilities of ceramic and plaster molds?

31. What is the major limitation of rubber-mold casting?

32. Why are investment-casting molds generally preheated before pouring?

33. What are some of the benefits of not having to remove the pattern from the mold (as in the lost-foam process)?

34. Cite some of the major attractive features of the full-mold process.

15

Multiple-Use-Mold Casting Processes

The expendable-mold casting processes have several disadvantages, most noticeably the necessity that a separate mold be created for each casting. Variations in mold consistency, mold strength, moisture content, pattern removal, and other factors contribute to dimensional and property variation within a production lot. Consequently, considerable effort has been directed to the development and expansion of the multiple-use-mold casting processes. These processes, however, have other limitations that must be considered by designers. Since the molds are generally made from metal, the processes are often restricted to the casting of the lower-melting-point nonferrous metals and alloys. Some process modifications, such as the use of graphite molds, have been developed to permit the casting of ferrous metals in multiple-use molds.

Permanent-Mold Casting. In the permanent-mold casting process, also known as *gravity die casting*, a reusable mold is machined from gray cast iron, steel, graphite, or other material, the mold segments being hinged to permit rapid and accurate opening and closing. The mold is first preheated, and molten metal is poured in under the action of gravity alone. After solidification, the mold is opened, and the product is removed. The mold is then reclosed, and another casting is poured. Aluminum-, magnesium-, and copper-base alloys are the metals most frequently cast. With graphite molds, iron and steel castings can be made by the permanent-mold process.

Numerous advantages can be cited for the process. The mold is reusable. A good surface finish is obtained if the mold is in good condition. Dimensional accuracy can usually be held within 0.005–0.010 inch (0.13–0.25 mm). By selectively heating or chilling various portions of the mold, or by varying the thickness of the mold wall, directional solidification can be promoted so as to produce sound, defect-free castings with the desired mechanical properties. The faster cooling rates of a metal mold generally produce stronger products than would result from a sand casting. Sand or retractable metal cores can be used to increase the possible complexity of the casting.

From the negative side, the process is generally limited to the lower-melting-point alloys. If it is applied to steels or cast irons, the mold life is extremely short. Even for the low-temperature metals, the mold life is limited because of erosion by the molten metal and thermal fatigue. The actual mold life varies with:

1. The alloy being cast. The higher the melting point, the shorter the mold life.
2. The mold material. Gray cast iron has about the best resistance to thermal fatigue and machines easily. Thus, it is used most frequently for permanent molds.
3. The pouring temperature. Higher pouring temperatures reduce mold life, increase shrinkage problems, and induce longer cycle times.
4. Mold temperature. If the temperature is too low, misruns are produced, and high temperature differences form in the mold. If the temperature is too high, excessive cycle times result, and mold erosion is aggravated.
5. Mold configuration. Differences in section sizes of either the mold or the casting can produce temperature differences and reduce the life of the mold.

Mold complexity is often restricted because the rigid cavity has no collapsibility to compensate for the shrinkage of the casting. As a best alternative, it is common practice to open the mold and remove the casting immediately after solidification, thereby preventing any tearing that may occur on subsequent cool-down.

Permanent molds are usually heated at the beginning of a run and are then maintained at a fairly uniform temperature as a means of controlling the cooling rate of the metal being cast. Since the mold rises in temperature as a casting is poured and sufficient time is permitted for solidification, it may be necessary to

FIGURE 15-1 Truck and car pistons, mass-produced by the millions by permanent mold casting. (Courtesy Central Foundry Division of General Motors Corporation.)

provide a cool-down delay before another casting is poured. Refractory washes are often applied to the mold walls to prevent the casting from sticking and to prolong the mold life. When pouring cast iron, a coating of carbon black is often applied through the use of an acetylene torch.

Since the molds are not permeable, special provision must be made for venting. This is usually accomplished through the slight cracks between mold halves, or by very small vent holes that permit the flow of trapped air but not the passage of molten metal. Since gravity is the only means of inducing metal flow, risers must still be used to compensate for shrinkage.

Mold costs are generally high so that high-volume production is necessary to justify the expense. Mass-production machines are then used to coat the mold, pour the metal, and remove the casting. Figure 15-1 shows a variety of mass-produced car and truck pistons that were manufactured by the permanent-mold process.

Slush Casting.   As a variation of permanent-mold casting, *slush casting* is a process in which the metal is permitted to remain in the mold only until a shell of the desired thickness has formed. The mold is then inverted, and the remaining liquid metal is poured out. The resulting casting is a hollow shape with good surface detail, but variable wall thickness. Low-melting-temperature metals are frequently cast into ornamental objects, such as candlesticks, lamp bases, and statuary, by this technique.

Corthias Casting.   Another variation of the permanent-mold process, *Corthias casting*, uses a plunger that is pushed down into the pouring cup, sealing the sprue hole and displacing some of the molten metal into the outer portions of the mold cavity. The positive pressure produces additional detail and permits thinner sections to be cast successfully..

Low-Pressure Permanent-Mold Casting.   In contrast to the gravity pouring of conventional permanent-mold casting, low-pressure permanent-mold casting (LPPM) forces the molten metal up through a vertical tube by applying low pressure of 5–15 psi to a molten bath below. Figure 15-2 provides an illustration of this process, which can be applied to a wide variety of metals, including steel and cast iron.

Clean metal from the center of the melt is drawn into the mold, never passing through the atmosphere. Thus, the low-pressure process becomes particularly attrractive for rapidly oxidizing metals, such as aluminum. By control of the pressure, the mold can be filled from the bottom in a controlled, nonturbulent manner, and gas porosity and dross are thus minimized. Mold cooling is designed in such a way as to promote directional solidification from the top down. Cycle times may be somewhat slow, but the applied pressure continually feeds molten metal to compensate for shrinkage, and the unused metal in the feed tube simply drops back into the crucible after the pressure is released. Since no risers are used (the pressurized feed tube acts as a riser) and the molten metal in the feed tube can be immediately reused, yields are generally greater than

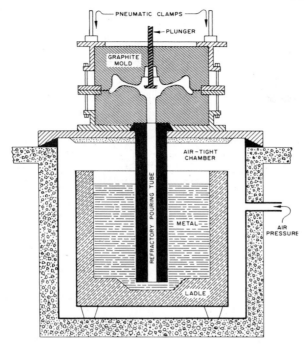

FIGURE 15-2   Schematic of the pressure-pouring process. (Courtesy Amsted Industries.)

85%. Mechanical properties are typically 5% superior to those in conventional permanent-mold castings.

Vacuum Permanent-Mold Casting.   In yet another variation, a vacuum is used to draw material into the permanent mold. All of the benefits and features of the low-pressure process apply here, including the subsurface extraction of molten metal from the melt, the bottom feed to the mold, the minimal metal disturbance during pouring, the self-risering action, and the downward directional solidification. Thin-walled castings can be produced with high metal yield and excellent surface quality. Because of the vacuum being used, the cleanliness of the metal and the dissolved gas content are superior even to those of the low-pressure process. Final castings typically range from 0.4 to 10 pounds and have properties that average 10% to 15% superiority to those of conventional permanent-mold products. Figure 15-3 provides a schematic illustration of vacuum permanent-mold casting. Other variations of the process include (1) incorporating both melting and casting under vacuum conditions and (2) replacing the permanent mold atop the feed tube with expendable-shell or sand molds (the Hitchiner process).

Die Casting.   *Die casting* differs from ordinary permanent-mold casting in that the molten metal is forced into the molds by pressures of thousands of pounds per square inch and is held under this pressure during solidification. Most die castings are made from nonferrous metals and alloys, but ferrous die

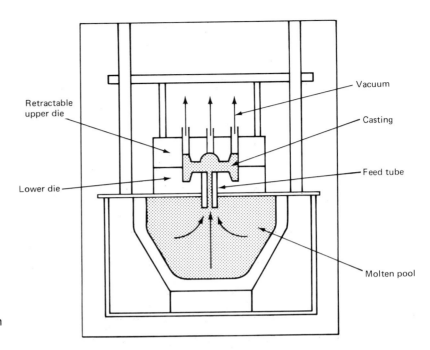

FIGURE 15-3 Schematic illustration of vacuum permanent mold casting.

castings are now being produced. Because of the combination of metal molds or dies and high pressure, fine sections and excellent detail can be achieved, together with long mold life. Special zinc-, copper-, and aluminum-base alloys have been developed that have excellent properties when die cast, further expanding the attractiveness of the process.

Because die-casting dies are usually made from hardened tool steels (cast iron cannot withstand the pressures), they are generally quite expensive. They may be relatively simple, containing only one or two mold cavities, but more often they are quite complex and contain eight or more cavities, as shown in Figure 15-4. In order to be opened to remove the castings, the dies must be made in at least two pieces; but very often they are much more complicated, containing sections that may move in several directions. In addition, the separate die sections must contain water-cooling passages and knock-out pins, which eject the casting. Consequently, such dies usually cost more than $3,000 and often over $10,000.

It follows, then, that the economical use of die casting is closely related to high production rates, the excellent surface qualities that are obtainable, and the almost complete elimination of machining. The size and weight of die castings are increasing constantly, and parts weighing up to about 20 pounds (10 kg) and measuring up to 24 inches (600 mm) can be made routinely.

The die-casting cycle consists of the following steps: (1) closing and locking the dies; (2) forcing the metal into the die and maintaining the pressure; (3) permitting the metal to solidify; (4) opening the die; and (5) ejecting the casting.

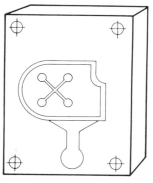

SINGLE CAVITY DIE

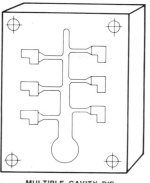

MULTIPLE CAVITY DIE

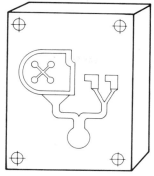

COMBINATION DIE

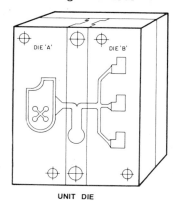

UNIT DIE

FIGURE 15-4 Some common die-casting designs. (Courtesy American Die Casting Institute, Inc., Des Plaines, IL.)

Because the high injection pressures may cause turbulence and air entrapment, the amount of pressure and time of application varies considerably. A recent trend is toward the use of larger gates and lower injection pressure, followed by much higher pressure after the mold has been completely filled and the metal has started to solidify. This approach tends to improve the density and reduce the porosity in the finished casting. Where the shape of the product permits, a very high semiforging pressure may be applied when the casting has cooled to forging temperatures.

Two basic types of die-casting machines are used. Figure 15-5 schematically illustrates the *hot-chamber* (or gooseneck) machine. A "gooseneck" is partially submerged in a pool of molten metal. On each cycle of the machine, a port opens that allows the molten metal to fill the gooseneck. A mechanical plunger or air-injection system then forces the metal out of the gooseneck and into the die, where it rapidly solidifies. This type of machine is fast in operation and has a distinct advantage in that the metal is injected from the same chamber that it is melted in. The process cannot be used for the higher-melting-point metals. When casting aluminum, there is a tendency for molten aluminum to pick up some iron from the casting equipment. Thus, hot-chamber machines are used primarily with zinc- and tin-based alloys.

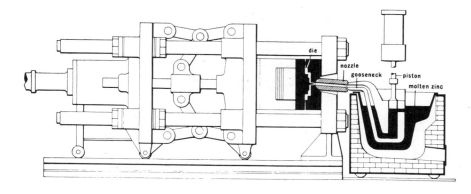

FIGURE 15-5 Schematic of hot-chamber die-casting machine. (Courtesy Zinc Institute, Inc., New York.)

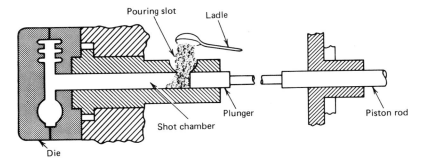

FIGURE 15-6 Schematic diagram of a cold-chamber die-casting machine.

Cold-chamber machines, as illustrated in Figure 15-6 and Figure 15-7, are used for the die casting of higher-melting-point metals. Here, the metal is melted in a separate furnace, then transported to the die-casting machine, where a measured quantity is fed into the shot chamber and is subsequently driven into the die by a hydraulic or mechanical plunger. The process can be used for a variety of alloys and is quite popular for aluminum and magnesium. In addition, the pressure cycle can be controlled to produce high-quality castings. Since the metal must be transferred to the chamber for each cycle, the cold-chamber process has a longer operating cycle than the hot-chamber machines. Nevertheless, productivity is still high, and up to 100 cycles per minute are not uncommon.

A modification of the cold-chamber process, known as the *Acurad process*, uses a double plunger. After the molten metal has been injected into the die and begins to solidify and shrink, a second plunger advances to force more metal into the die cavity and to ensure production of a better casting.

FIGURE 15-7 Cold-chamber die casting machine. Inset shows a close-up of the die and the casting being removed. (Courtesy Reed-Prentice Corporation.)

Die-casting dies fill with metal so fast that there is little time for air in the mold cavity to escape, and the metal mold has no permeability. Trapped-air problems, in the form of blow holes, porosity, or misruns, are quite common. To minimize these, it is crucial that the dies be properly vented, usually by wide, thin (0.005-inch or 0.13-mm) vents at the parting line. The metal filling these escapes must then be trimmed off after the casting has been removed from the mold. This can be done with the trimming dies that also serve to remove the sprues and runners.

No risers are used in die castings since the high pressures help feed metal from the gating system into the casting. However, die castings are generally difficult to make sound because of trapped air and metal turbulence. Fortunately, the porosity tends to be confined to the center of the casting, and the rapidly solidified surface is quite strong and is usually sound and suitable for plating or decorative applications. In the relatively new *pore-free* casting process, air problems are eliminated by introducing oxygen into the mold before each die-casting shot. When the molten metal is injected, it reacts with the oxygen to form fine, dispersed oxide particles, virtually eliminating gas porosity. The products can be welded and heat-treated, and they have greater strength than do conventional die castings. Pore-free casting has been performed on aluminum, zinc, and lead alloys.

Another recent innovation in die casting is the *heated-manifold direct-injection die casting of zinc* (also known as *direct-injection zinc die casting* or *runnerless zinc die casting*). Molten zinc is forced through a heated manifold and heated mininozzles directly into the die cavity, thereby eliminating the need for sprues, gates, and runners. Less cooling of the metal occurs, so the product surface is improved. Energy is conserved, the scrap to be remelted is reduced substantially, and the product quality is increased. Existing die-casting machines can be used with the addition of a manifold and design modifications on the dies.

Sand cores cannot be used in die casting because the high pressures and flow rates cause the cores either to disintegrate or to have excessive metal penetration. As a result, metal cores are used extensively in die castings, and provisions must be made for retracting them, usually before the die is opened for the removal of the casting. Because a close fit must be maintained between the halves of the die and between any cores and the die sections to prevent the metal from flowing into the gap, both construction and maintenance costs are greatly increased when cores are used. It is very important that the direction of the core-retracting motions be either a straight line or a circular arc. As an alternative, loose core pieces can be inserted into the die at the beginning of each cycle and then removed from the casting after it has ejected from the die. Such a procedure permits complex shapes to be cast, such as internal threads, but with considerable reduction in the production rate and with increased cost. Figure 15-8 is an example of a complex die casting that involves the use of several cores.

Because of the method by which they are produced, die castings have certain distinguishing characteristics. The surfaces tend to be harder than the interior as a result of the chilling action of the metal die on the molten metal. The

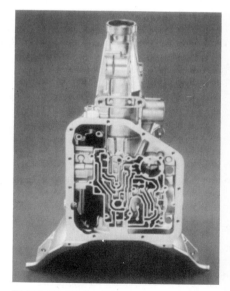

FIGURE 15-8   Die-cast aluminum automatic transmission case, incorporating the bell housing and rear case extension. (Courtesy Central Foundry Division of General Motors Corporation.)

interior regions may be porous because of entrapped air and turbulence. Smooth metal flow, good venting, and proper application of pressure can do much to minimize this problem. Casting design, die design, die maintenance, proper casting procedure, and proper equipment are all concerns in the production of quality die castings.

The die-casting process also permits the incorporation of cast-in inserts. These include prethreaded bosses, electrical heating elements, threaded studs, or high-strength bearing surfaces that are positioned in the die before it is closed. The molten metal is then injected and solidifies around the insert. Suitable recesses must be provided in the die for positioning and support, and the casting cycle tends to be slowed by the additional operations.

Excellent dimensional accuracy can be obtained with die casting. For aluminum-, magnesium-, zinc-, and copper-base alloys, linear tolerances of ± 0.003 inch per inch (± 3 mm per meter) can be maintained. Minimum section thickness and draft depend on the kind of metal, as follows:

Metal	Minimum Section	Minimum Draft
Aluminum alloys	0.035 in. (0.89 mm)	1:100 (0.010 in./in.)
Brass and bronze	0.050 in. (1.27 mm)	1:80 (0.015 in./in.)
Magnesium alloys	0.050 in. (1.27 mm)	1:100 (0.010 in./in.)
Zinc alloys	0.025 in. (0.63 mm)	1:200 (0.005 in./in.)

As a result of the excellent dimensional accuracy and the smooth surfaces that can be obtained, most die castings require no machining except the removal of the small amount of excess metal fin, or flash, around the parting line and possibly the drilling or tapping of holes. Production rates are high, and a set of dies can produce many thousands of castings without a significant change in dimensions

Squeeze Casting (or Liquid-Metal Forging).   In squeeze casting, molten metal is poured into the bottom half of a closed-die set and is allowed to partially solidify. The upper die then descends, applying pressure as the solidification process continues. Gas and shrinkage porosity is substantially reduced; mechanical properties are enhanced; and the process can be applied to either wrought or cast alloys. An added feature is the possibility of producing metal-matrix composites by forcing the pressurized liquid around foam or fiber reinforcements.

Centrifugal Casting.   Hollow-shaped castings can be produced by centrifugal casting, as centrifugal force causes the molten metal to conform to the shape of a rotating mold cavity. In *true centrifugal casting*, the mold rotates about either a horizontal or vertical axis at speeds of from 300 to 3000 rpm as the molten metal is introduced. Either a dry-sand or a metal mold is used, this

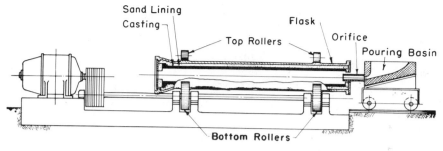

Sand Lining
Casting
Flask
Top Rollers
Orifice
Pouring Basin
Bottom Rollers

FIGURE 15-9 Schematic representation of a horizontal centrifugal casting machine. (Courtesy American Cast Iron Pipe Company.)

choice determining the outer surface finish of the casting. Round shapes are most common, but hexagons and other symmetrical shapes are possible.

For the interior, no core or mold is needed. When the horizontal axis is used, as illustrated in Figure 15-9, the inner surface is always cylindrical. If the vertical axis is used, the inner surface is a section of a parabola, as illustrated in Figure 15-10, the exact shape being a function of the speed of rotation. Wall thickness can be easily varied and depends on how much metal is introduced into the mold.

In true centrifugal casting, the metal is forced against the outer walls of the mold with considerable force, and it solidifies first at the outer surface. Centrifugal force continues to feed molten metal, compensating for shrinkage, so no risers are required. The final product has a strong, dense exterior with all of the lighter impurities tending to be at the inner surface, which can be removed by a light machining cut.

Pipe, pressure vessels, cylinder liners, and brake drums are commonly manufactured by centrifugal casting. The equipment is specialized and can be quite expensive for large castings. For small parts, relatively simple and inexpensive equipment is often available. The permanent molds can also be quite costly, but they do offer a long service life. When ferrous metals are cast, the molds are coated with some type of refractory dust or wash to prolong their life. Since no sprue, gate, or riser is required, yields can be 90% or more. Moreover, it is possible to produce composite castings by spinning a second alloy onto the surface of the first material.

Semicentrifugal Casting. In some cases, the *semicentrifugal casting* principle, illustrated in Figure 15-11, can be used to advantage. In this case, centrifugal force aids the flow of the molten metal from a central reservoir to the extremities of a symmetrical mold, which may be expendable or multiple-use. The rotational speeds are often considerably less than in the case of true centrifugal casting. Frequently, several identical molds are stacked on top of each other, being fed by a common reservoir.

The central pouring reservoir acts as a riser and must be large enough to ensure that it will be the last material to freeze. Since the lighter impurities concentrate in the center, the process is best used for castings from which the central region will be removed by machining. Cores can be used in semicentrifugal casting to increase the complexity of the product.

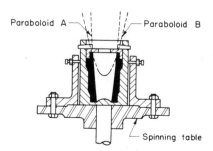

Paraboloid A   Paraboloid B
Spinning table

FIGURE 15-10 Vertical centrifugal casting, showing the effect of rotational speed on the shape of the inner surface. Paraboloid A results from fast spinning; paraboloid B from slow.

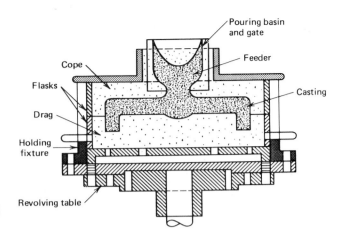

FIGURE 15-11 The semicentrifugal casting process.

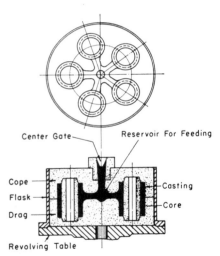

FIGURE 15-12 Method of casting by the centrifuging process. (Courtesy American Cast Iron Pipe Company.)

Centrifuging. Illustrated in Figure 15-12, *centrifuging* is another procedure that can be used to force metal from a central pouring reservoir into thin, intricate mold cavities removed from the axis of rotation. Relatively low rotational speeds are used to produce sound castings with fairly high yields. Investment castings are often poured with centrifuging machines.

In an adaptation of the process, pewter, zinc, or wax can be cast into spinning rubber molds (or stacked molds) to produce products with close tolerances, smooth surfaces, and excellent detail. These can be finished products or patterns for subsequent investment casting.

Continuous Casting. While generally used in the solidification of basic shapes destined for further processing, as shown in Figure 6-5, *continuous casting* can also be used to produce long lengths of special cast-section, as in the product of Figure 15-13. Only a single mold is required to produce a large number of pieces, since each is basically a cut section of a continuous strand. Quality is high as well, since the metal can be protected from contamination during melting and pouring, and there is only a minimum of handling.

Electromagnetic Casting. In electromagnetic casting, the metal is contained and solidified in an electromagnetic field rather than in a conventional mold. This process has recently been adapted for continuous casting. There is no sliding contact with a mold wall, and the metal can be solidified by direct water impingement. The product surface is smooth, and the process can be directly automated. Several large industries are currently using electromagnetic casting in the production of continuously cast aluminum strand.

Melting and Pouring

All of the casting processes require a furnace for melting the metal. Ideally, such a furnace should (1) provide adequate temperature; (2) minimize contamination; (3) make possible holding the metal at the required temperature without harmful effects so that the chemical composition can be altered by alloy addi-

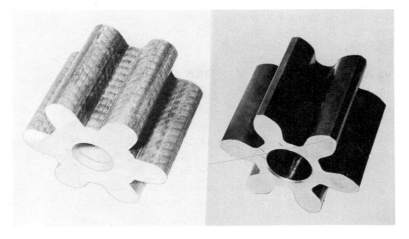

FIGURE 15-13 Gear produced by continuous casting. (Left) Piece cut from continuous casting. (Right) After machining. (Courtesy American Smelting and Refining Company.)

tions; (4) be economical; and (5) be capable of control to prevent pollution of the atmosphere. Except for experimental or very small operations, virtually all foundries use either a cupola, an air furnace, an electric-arc furnace, or an electric-induction furnace. Occasionally, in fully integrated steel mills containing a foundry, steel is taken directly from a basic-oxygen furnace and poured into casting molds, but such a practice is not very common and is reserved for exceptionally large castings. For some experimental work, and in very small foundries, gas-fired crucible furnaces may be used, but these do not have sufficient capacity for most commercial operations.

Selection of the best melting procedure depends on such factors as (1) the temperature needed to melt and superheat the metal; (2) the alloy being melted; (3) the desired melting rate or quantity of metal; (4) the desired quality of the metal; (5) the availability and cost of various fuels; (6) the variety of metals or alloys to be melted; (7) whether melting is to be batch or continuous; (8) the required level of emission control; and (9) the various capital and operating costs.

Cupolas.   A large amount of gray, nodular, and white cast iron is still melted in *cupolas*, although many foundries have converted to electric induction furnaces, and their use is steadily increasing. Basically, a cupola is a refractory-lined, vertical steel shell into which coke, iron (pig iron and/or scrap), limestone or other flux, and possible alloy additions are charged and melted under forced air draft. Operation is similar to that of a blast furnace, the molten metal collecting at the bottom of the cupola to be tapped off at periodic intervals.

Cupolas are simple and economical, can be obtained in a wide range of capacities, and can produce an excellent quality of cast iron if the proper raw materials are used and good control is practiced. Cupolas are used exclusively for the melting of cast iron and can be operated either continuously or as batch-type melters. Temperature and chemistry control is somewhat difficult. The nature of the charged materials and the reactions that occur within the cupola can all affect the product chemistry. Moreover, when the final chemistry is ultimately determined by analysis of the tapped product, a substantial charge of material is already working its way through the furnace. Final chemistry adjustments are often performed in the ladle by means of the various techniques of ladle metallurgy discussed in Chapter 6.

In order to increase the melting rate and give greater economy, a *hot-blast* cupola may be used. Here, the stack gases are put through a heat exchanger to preheat the incoming air to temperatures up to 1200°F (650°C). Water cooling may be incorporated around the shell to prolong the life of the cupola or to enable the elimination of the refractory lining, thereby permitting the use of a wider variety of slags. Oxygen-enriched blasts can be used to increase the temperature and to accelerate the rate of melting. Plasma torches can also be used to melt the iron scrap in a cupola. Here, the reduced air currents in the cupola enable the feeding of uncompacted iron turnings, further reducing the operating costs.

In all modern installations, the large volume of stack gases is passed through dust collectors and various types of pollution-control equipment. Consequently, as indicated in Figure 15-14, a modern cupola installation can be quite complex.

Indirect Fuel-Fired Furnaces. *Indirect fuel-fired furnaces* include crucibles and pot furnaces where the container is heated by an external flame. Their use is typically restricted to the batch-type melting of nonferrous metals. Stirring action, temperature control, and chemistry control are poor, but these furnaces offer low capital cost. The containment crucibles are generally made from clay and graphite, silicon carbide, cast iron, or steel.

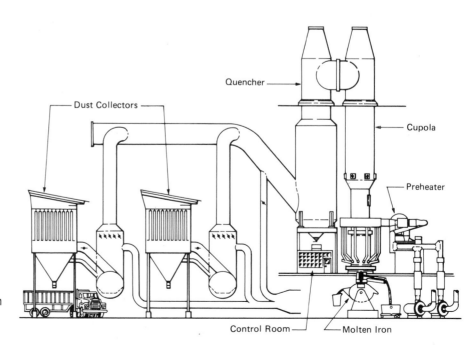

FIGURE 15-14 Modern cupola installation with associated antipollution equipment. (Courtesy *Foundry Management and Technology*.)

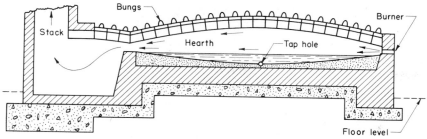

FIGURE 15-15   Section of an air furnace.

Air Furnaces or Direct Fuel-Fired Furnaces.   In *air furnaces or direct fuel-fired furnaces*, illustrated in Figure 15-15, the surface of the metal is heated directly by the burning fuel. These furnaces are similar to small open-hearth furnaces but are somewhat less sophisticated. Their capacity is much larger than that of the crucible furnace, but the operation is still limited to the batch melting of nonferrous metals. (Malleable cast iron can be held and adjusted in these furnaces after melting in a cupola.) The rate of heating and melting, as well as the temperature and composition of the metal, is easily controlled.

Arc Furnaces.   The use of *arc furnaces* in foundries has increased substantially because of (1) their rapid melting rates; (2) their ability to hold the molten metal for any desired period of time to permit alloying; and (3) the greater ease of incorporating adequate pollution control.

The basic features of a *direct-arc* furnace are shown in Figure 15-16. In most types, the top can be lifted or swung off to permit the charge to be introduced. The electrodes are then lowered so that an arc is struck and maintained between the electrodes and the metal charge. The path of the heating current is from one electrode, across the arc to the metal, through the metal, and back through the arc between the metal and another electrode.

Fluxing materials are added to provide a protective cover over the molten metal. Because the metal is thus protected and the metal can be maintained at a given temperature for as long as desired, high-quality metal of any desired composition can be obtained. Furnaces of this type are available in capacities up to 200 tons, but those below 25 tons are most common. Up to 50 tons per hour can be conveniently melted in batch-type operations. Arc furnaces are

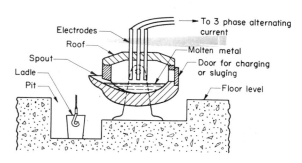

FIGURE 15-16   Schematic diagram of a three-phase electric arc furnace.

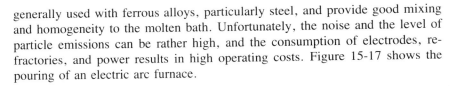

FIGURE 15-17 Electric arc furnace, tilted for pouring. (Courtesy Pittsburgh Lectromelt Furnace Corporation.)

generally used with ferrous alloys, particularly steel, and provide good mixing and homogeneity to the molten bath. Unfortunately, the noise and the level of particle emissions can be rather high, and the consumption of electrodes, refractories, and power results in high operating costs. Figure 15-17 shows the pouring of an electric arc furnace.

Induction Furnaces. Because of their very rapid melting rates and the relative ease of controlling pollution, electric induction furnaces have become very popular means of melting metal. Two basic types exist. The *high-frequency type*, or *coreless* induction unit, shown in a schematic in Figure 15-18, consists of a crucible surrounded by a water-cooled coil of copper tubing. A high-frequency electrical current passes through the coil, establishing an alternating magnetic field. This, in turn, induces secondary currents in the metal being melted, which bring about a rapid rate of heating.

Coreless induction furnaces are used for virtually all common alloys, the maximum temperature being limited only by the refractory and the efficiency of insulating against heat loss. These furnaces provide good control of temperature and composition and are available in capacities through 65 tons. Because there is no contamination from the heat source, they produce very pure metal. Operation is generally on a batch-type basis.

Low-frequency or *channel-type* induction furnaces are also seeing increased use. As indicated in Figure 15-19, only a small channel is surrounded by the alternating-current primary coil. The secondary coil is formed by a loop, or channel, of molten metal, and all metal must circulate through this loop to gain heat. Some molten metal must first be introduced into the furnace to form the

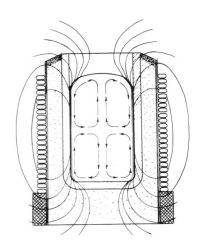

FIGURE 15-18 Schematic showing the basic principle of a coreless induction furnace.

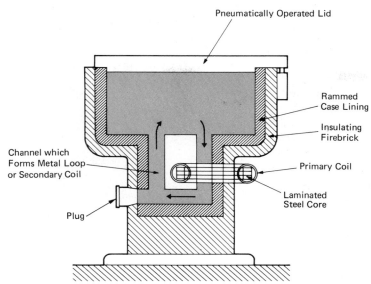

Pneumatically Operated Lid

Rammed Case Lining

Insulating Firebrick

Channel which Forms Metal Loop or Secondary Coil

Primary Coil

Laminated Steel Core

Plug

FIGURE 15-19   Principle of the low-frequency or channel-type induction furnace.

secondary coil, but the subsequent heating rate is very high, and the temperature can be readily controlled. This aspect makes these furnaces very useful as holding furnaces, where it is desirable to maintain molten metal at a constant temperature for an extended period of time, as in die-casting machines or mold-pouring systems. Capacities are available up to 250 tons.

Preheating iron or steel scrap charges to induction or arc furnaces to 600–1000°F (315–540°C) can be used to increase the melting capabilities of a given furnace by as much as 30%.

Pouring Practice.

In order to transfer the metal from the melting furnace into the molds, some type of pouring device, or ladle, must be used. The primary considerations are to maintain the metal at the proper temperature for pouring and to ensure that only quality metal will get into the molds. The type of ladle used is determined largely by the size and number of castings to be poured. In small foundries, the hand-held, shank-type ladle shown in Figure 15-20 is often used. In large foundries, either bottom-pour or teapot-type ladles are used, as illustrated in Figure 13-6. These are often used in conjunction with a conveyor line on which the molds move past the pouring station. When metal is extracted from beneath the surface, slag and oxidized material that floats on top of the melt does not enter the mold.

In modern, mass-production foundries, automatic pouring systems, such as the one shown in Figure 15-21, are used. The molten metal is brought from the main melting furnace to a holding furnace by overhead crane. A programmed amount of molten metal is poured into the individual pouring ladles from the holding furnace and, in turn, is poured automatically into the corresponding molds.

FIGURE 15-20 Pouring a mold from a shank-type ladle. (Courtesy Steel Founders' Society of America, Rocky River, Ohio.)

Vacuum Melting and Pouring.   Just as increasing amounts of metals are being melted in a vacuum to remove gases and ensure higher purity, increasing numbers of castings are being poured in a vacuum so that they retain their purity in the cast form. Vacuum pouring is often done by enclosing both the induction furnace and the mold in a chamber from which the air can be evacuated, and by arranging for the metal to be poured directly from the furnace into the mold.

Another method of vacuum melting and pouring, used when metal of extremely high purity is necessary, is carried out with an arc furnace using consumable metal electrodes of the metal to be melted. When castings are to be

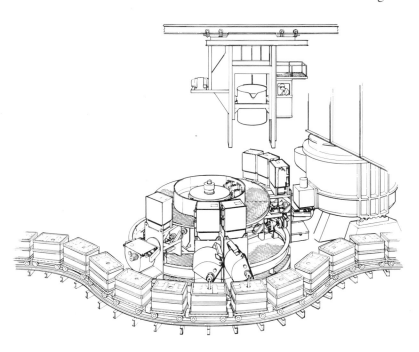

FIGURE 15-21 Machine for automatic pouring of molds on a conveyor line. (Courtesy Roberts Corporation.)

made, the electrode is melted into a crucible, which is then poured into the mold, all of the equipment being operated in a vacuum chamber. Obviously, these vacuum methods are expensive and should be used only when the highest quality is essential.

After solidification and removal from the molds, most castings require some cleaning and finishing operations. These may involve all or several of the following:

Cleaning, Finishing, and Heat-Treating Castings

1. Removing cores.
2. Removing gates and risers.
3. Removing fins, flash, and rough spots from the surface.
4. Cleaning the surface.
5. Repairing any defects.

The required operations are not always done in the same order, and a particular casting process may eliminate some of them. Because cleaning and finishing operations may involve considerable expense, some consideration should be given to them in designing castings and selecting the casting method to be used. Often, substantial savings can be effected. In addition, consideration should be directed toward mechanizing these operations.

Sand cores can usually be removed by shaking. Sometimes, they must be removed by dissolving the core binder. On small castings, gates and risers can sometimes be knocked off. However, on large castings and some small castings, a cutting operation must be performed. On nonferrous and cast iron castings, this is generally done by an abrasive cutoff wheel, a power hacksaw, or a bandsaw. Gates and risers on steel castings are frequently removed by an oxy-acetylene torch. Plasma arc cutting can also be used to remove sprues, gates, and risers.

After the gates and risers are removed, small castings are often put through tumbling barrels, as shown in Figure 15-22, to remove fins, snags, and sand that has adhered to the surface. Tumbling may also be done to remove cores and, in some cases, gates and risers. Frequently, some type of shot or abrasive material is added to the barrel to aid in the cleaning. Larger castings may be passed by conveyor through a cleaning chamber, where they are subjected to blasts of abrasive or cleaning material. Extremely large castings are usually finished manually with pneumatic chisels, portable grinders, and manually di-rected blast hoses.

Although it is desirable that castings contain no defects, it is inevitable that some will occur, particularly in large castings where only one or a few of a particular design will be made. Some types of defects can be repaired readily by means of arc welding. However, it is imperative that the casting be of a material that can be welded satisfactorily, that *all* the defective area be removed down to sound metal by means of chipping or grinding, and that a sound repair weld be made. Surface porosity can be filled with resinous material, such as polyester, by a process known as *impregnation*.

FIGURE 15-22 (Left) Tumbling machine for cleaning castings. (Right) Uncleaned castings in loading hopper (bottom of photo) emerge as cleaned castings at the upper right. (Courtesy Wheelabrator-Frye Inc.)

Heat Treatment and Inspection of Castings.   The heat treatment of castings has become a common way to obtain the full benefit of the alloy additions. Steel castings are almost always given a full anneal to reduce the hardness of rapidly cooled thin sections and to reduce the internal stresses that result from uneven cooling. Nonferrous castings are often heat-treated to a softened, stress-relieved structure before any machining. For final properties, virtually all of the treatments discussed in Chapter 5 can be applied. Ferrous-metal castings often undergo a quench-and-temper treatment, and many nonferrous castings are age-hardened to impart additional strength. The variety of heat treatments is largely responsible for the wide range of properties and characteristics available in cast metal products.

X-ray radiography, liquid penetrant inspection, and magnetic particle inspection are all common means of inspecting cast products. These methods were discussed in detail in Chapter 11.

Robots in Foundry Applications.   Many of the operations that are performed in a foundry are ideally suited to robotic automation in that they are dirty, dangerous, and dull. In the casting process itself, robots have been used to tend stationary, cyclic equipment, such as die-casting machines. If the machines are properly grouped, one robot can often tend two or three machines. In the finishing room, robots have been equipped with plasma cutters or torches to remove sprues, gates, and runners. They can perform grinding and blasting operations, as well as various functions involved in the heat treatment of castings. Robots can dry molds, coat cores, vent molds, and clean or lubricate dies. In the investment-casting process, robots can be used to dip the wax patterns into the refractory slurry and produce the desired molds. Similarly, robots have been used to dip full-mold styrofoam patterns in their refractory coating and hang them on conveyors to dry. In a full implementation of robots, they could also be used to position the pattern, fill the flask with sand, pour the metal, and manipulate a cutting torch to remove the sprue.

Review Questions

1. What are some of the major disadvantages of the expendable-mold casting processes?
2. Cite one possible limitation on the use of multiple-use molds.
3. Describe some of the process advantages of permanent-mold casting.
4. Why are permanent-mold castings generally removed from the mold immediately after solidification has been completed?
5. Are risers used in the permanent-mold process? Can sand cores be used?
6. What types of objects would be possible candidates for manufacture by slush casting?
7. How does low-pressure permanent-mold casting differ from the traditional gravity-pour process?
8. What process feature or features account for the high casting yields in low-pressure permanent-mold casting (typicaly more than 85% of the starting metal ends up in the cast product)?
9. What are some of the cited advantages of vacuum permanent-mold casting?
10. Contrast the feeding pressures on the molten metal in low-pressure permanent molding and die casting.
11. Contrast the materials used to make dies for gravity permanent-mold casting and die casting. Why is there a notable difference?
12. Are risers used in die casting? Can sand cores be used?
13. For what types of materials would a hot-chamber die-casting machine be appropriate?
14. How does the air in the mold cavity escape in the die-casting process?
15. How does the ''pore-free'' process reduce gas porosity problems? What happens to the oxygen that was in the mold cavity?
16. What is the major advantage of the heated-manifold direct injection die casting of zinc?
17. Contrast the structure and properties of the outer and inner surfaces of a true centrifugal casting.
18. What is the purpose of centrifuging when pouring intricate investment castings?
19. What are some of the operating requirements of a casting furnace?
20. In cupola-melting operations, why are many of the chemistry control operations performed in the ladle via ladle metallurgy?
21. What are some of the major pros and cons of indirect fuel-fired furnaces?
22. What are some of the attractive features of arc furnaces in foundry applications?
23. Why are channel-type induction furnaces attractive for metal-holding applications where molten metal must be held at a specified temperature?
24. What are the primary functions of a pouring ladle?
25. What is the benefit of melting and pouring castings in a vacuum?
26. What are some of the typical cleaning and finishing operations performed on castings?
27. How can some defective castings be repaired to permit successful use in their intended applications?
28. What are some of the attractive applications for industrial robots in metal-casting operations?

16

Powder Metallurgy

Powder metallurgy is the name given to a process in which fine powdered materials are blended, pressed into a desired shape (compacted), and then heated (sintered) in a controlled atmosphere at a temperature below the melting point of the major constituent for a sufficient time to bond the contacting surfaces of the particles and establish the desired properties. The process, commonly designated as P/M, readily lends itself to the mass production of small, intricate parts of high precision, often eliminating the need for any additional machining. There is little material waste; unusual materials or mixtures can be utilized; and controlled degrees of porosity or permeability can be produced. Major areas of application tend to be those for which the P/M process has strong economical advantage or those in which the desired properties and characteristics are unique to powder metallurgy.

A crude form of powder metallurgy appears to have existed in Egypt as early as 3000 B.C., using particles of sponge iron. Around 1100 A.D. the Arabs were making high-quality iron swords from iron powder. In the nineteenth century, the process was used to produce platinum and tungsten wires. The first significant use, however, was in the 1920s, when powder metallurgy was used to produce tungsten carbide cutting-tool tips and nonferrous bushings. Lamp filaments, self-lubricating bearings, and metallic filters were other early products. A period of rapid technological development after World War II focused primarily on automotive applications, and iron and steel replaced copper as the dominant P/M material. Accelerated aerospace and nuclear activity spurred the development of applications for refractory and reactive materials. Full-density products emerged in the 1960s and high-performance superalloy components were a highlight of the 1970s. Additional developments in the 1980s have included the commercialization of rapid-solidified and amorphous powders and the development of injection molding technology.

The last three decades have been a period of high development and growth

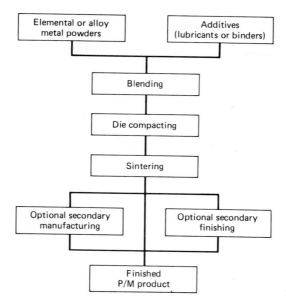

FIGURE 16-1 Simplified flow chart
of the powder metallurgy process.

for the P/M industry. From 1960 to 1980, the consumption of iron powder increased tenfold. P/M has become accepted as a viable, proven method of parts production and is now considered to be an alternative in the manufacture of many components. Most products are still under 2 in. (50 mm) in size, but some have been produced with weights up to 100 pounds (45 kg) with linear dimensions up to 20 in. (500 mm). The current market split for P/M products is 70% automotive, 15% appliances, 12% business machines, and 5% farm and garden equipment. Areas of rapid growth, however, include aerospace applications, advanced composites, electronic components, magnetic materials, and metalworking tools.

The Basic Process. The powder metallurgy process generally consists of four basic steps, namely (1) powder manufacture, (2) mixing or blending, (3) compacting, and (4) sintering. Optional secondary processing often follows to obtain special properties or enhanced precision. Figure 16-1 presents a simplified flowchart of the P/M process.

Powder Manufacture. The properties of powder metallurgy products are highly dependent upon the characteristics of the metal (or material) powders that are used. Important properties and characteristics include chemistry and purity, particle size, size distribution, particle shape, and the surface texture of the particles. Several processes are used to produce metal powders, with each imparting distinct properties and characteristics to the resulting powder and hence to the final product.

The most common means of making metal powder is probably *melt atomization,* and various forms of the process have been developed. In the method

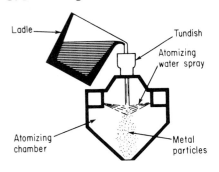

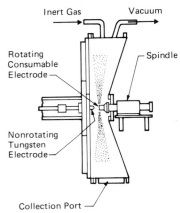

FIGURE 16-2 Two methods for producing metal powders. (Top) Melt atomization. (Bottom) Atomization from a rotating consumable electrode.

illustrated in Figure 16-2 (*top*), molten metal is atomized by a stream of impinging gas or liquid as it emerges from an orifice. The method in Figure 16-2 (*bottom*) utilizes an electric arc impinging on a rapidly rotating electrode, all within a chamber purged with inert gas. Regardless of the particular form, melt atomization processes are particularly useful in producing prealloyed powders. By starting with an alloyed melt or electrode, each powder particle is of the desired alloy composition. Powders of stainless steel, nickel-based alloys (such as Monel), titanium alloys, cobalt-base alloys, and various low-alloy steels are all commercially available. Size and shape of the powder particles can be varied and depend upon such process features as the velocity of the atomizing jets, the speed of electrode rotation, and the time from atomization to solidification.

Other methods of powder manufacture include *reduction of compounds* (generally oxides or ores), *electrolytic deposition from solutions or fused salts, pulverization or grinding, thermal decomposition of hydrides or carbonyls, precipitation from solution,* and *condensation of metal vapors.*

Almost any metal, metal alloy, or nonmetal can be converted to powder form by one or more of the above methods. Prior to further processing, many powders must first undergo a drying operation and possibly heat treatment.

Rapid Solidified Powder (Microcrystalline and Amorphous). Increasing the cooling rate of an atomized liquid can result in the formation of an ultrafine or microcrystalline grain size. If the cooling rate approaches or exceeds 10^6 °C/sec, metals can solidify without becoming crystalline. These amorphous or glassy metals often exhibit unusual or unique properties such as high strength, improved corrosion resistance, and reduced energy to induce and reverse magnetization.

Production of amorphous material, however, requires immensely high cooling rates and therefore ultrasmall dimensions. Atomization with rapid cooling and the "splat quenching" of a metal stream onto a cool surface to produce a continuous ribbon are the two prominent methods. Since much of the ribbon material is further fragmented into powder, powder metallurgy becomes the primary means of fabricating products. Because of their enhanced and often unique properties, considerable effort is now being spent in assessing the possible applications of these materials.

Powder Testing and Evaluation. In addition to evaluation of bulk chemistry, surface chemistry, size, shape, surface texture, and internal structure, metal powders must also be evaluated for their suitability for further processing. *Flow rate,* a measure of the ease by which powder can be fed and distributed into a die, can significantly influence the cycle time of the pressing operation and therefore the rate of production. Poor flow characteristics can also result in a pyramiding effect under nozzles and require the use of multiple feed nozzles or a vibrating operation prior to pressing.

Associated with the flow characteristics is the *apparent density,* a measure of the powder's ability to fill available space without the application of external pressure. *Compressibility tests* evaluate the effectiveness of applied pressure in raising the density of the powder, and *green strength* is used to describe the strength of the pressed powder immediately after compacting. Good green strength is required to maintain smooth surfaces, sharp corners, and intricate details during ejection from the compacting die or tooling and the subsequent transfer to the sintering operation.

Powder Mixing and Blending. It is rare that a single powder will possess all of the characteristics desired in a given process and product. Most commonly, therefore, the starting material is a mixture of various grades or sizes of powder, or powders of different compositions, with additions of lubricants or binders. It is well established that higher product density results in superior mechanical properties, such as strength and fracture resistance. A mixture of various sizes should theoretically flow and compact to higher density, but there is a tendency for the finer sizes to separate or segregate during handling and pouring. Many users, therefore, prefer to use uniform-size powders for their process.

Final product chemistry can often be obtained by combining pure metal or nonmetal powders instead of using prealloyed material. Sufficient interatomic diffusion must then occur during the sintering operation to produce a uniform chemistry and structure. Unique composites can also be produced, such as the intimate distribution of an immiscible reinforcement material in a matrix, or the combination of metals and nonmetals in a single product as in a tungsten–carbide–cobalt-matrix cutting tool.

A powder such as graphite, can even play a dual role, where it serves as a lubricant during compacting and a source of carbon as it alloys with the metal during sintering. Lubricants, such as graphite or stearic acid, improve the flow characteristics and compressibility at the expense of reduced green strength. Binders produce the reverse effect.

Blending or mixing operations can be done either wet or dry, the use of water or other solvents being employed to obtain better mixing, reduce dusting, and lessen explosion hazards. Large lots of powder can be homogenized in regard to both chemistry and distribution of components, sizes, and shapes. Quantities up to 35,000 pounds (16,000 kg) have been blended to ensure uniform behavior during processing and the production of a consistent product.

Compacting. One of the most critical and controlling steps in the P/M process is compacting. Here, loose powder is compressed and densified into a shape known as a *green compact*. High product density and good uniformity of that density throughout the compact are generally desired characteristics. In addition, the compacts should possess sufficient green strength for in-process handling and transport to the sintering furnace.

Most compacting is done with mechanical presses and rigid tools, but hydraulic and hybrid (combinations of mechanical, hydraulic, and pneumatic) presses can also be used. The mechanical presses can be of the eccentric or crank type, toggle or knuckle type, or cam type. Figure 16-3 shows a typical mechanical press for compacting powders and a removable set of compaction tooling that is placed between the moving platens. Compacting pressures gen-

FIGURE 16-3 (Left) Typical press for the compacting of metal powders. The removable dieset (right) allows the machine to be producing parts with one dieset while another is being fitted to produce a second product. (Courtesy Sharples-Stokes Division, Pennwalt Corporation.)

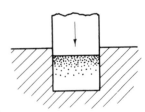

FIGURE 16-4 Compaction with a single punch, showing the resultant nonuniform density.

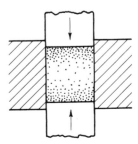

FIGURE 16-5 Density distribution obtained with a double-acting press and two moving punches.

erally range between 5 and 50 tons per square inch (70 to 700 MPa), with 10 to 30 tons per square inch (140 to 415 MPa) being the most common. Most P/M presses have total capacities of less than 100 tons (9×10^5 N), although increasing numbers now have capacities between 200 and 300 tons (18 to 27×10^5 N). Consequently, most powder metallurgy products must have cross sections of less than 3 square inches (2000 mm^2), but with increased press capacity sections up to 10 square inches (6500 mm^2) are becoming increasingly common. At least one production P/M press has a capacity of 3000 tons (27 MN).

In most cases, the prepared powder gravity flows into the die until there is some excess. The excess is scraped off, providing a starting measure based on volume, and the press closes to compact the powder. As an alternative, the amount of powder can be controlled by weighing or a prescribed amount of powder can be preformed into tablets or slugs in a tablet-making machine.

During compacting, the powder particles move primarily in the direction of the applied force. The powder does not flow like a liquid but simply compacts until an equal and opposing force is developed by the friction between the particles and the die surfaces. As illustrated in Figure 16-4, when the pressure is applied by only one punch, maximum density occurs below the punch and decreases as one moves down the column. Thus, it is seldom possible to transmit uniform pressures and produce uniform density throughout a compact. A double-action press, as illustrated in Figure 16-5, enables a more uniform density to be obtained and thicker products to be compacted. Since side-wall friction is a key

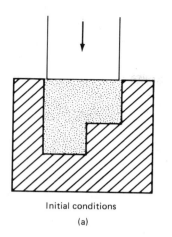

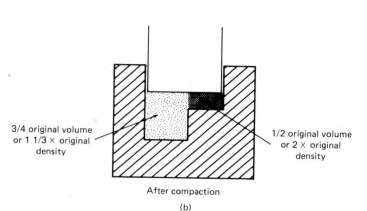

Initial conditions

(a)

After compaction

(b)

3/4 original volume
or 1 1/3 × original
density

1/2 original volume
or 2 × original
density

FIGURE 16-6 A two-thickness part with only one moving punch. (Left) Initial conditions. (Right) After compaction by the upper punch, showing drastic difference in compacted density.

factor in compaction, the resulting density is a strong function of both the thickness and width of the part being pressed. For good uniform compaction, the ratio of thickness and width (t/w) should be kept below 2.0 whenever possible. Products with ratios above 2.0 tend to exhibit considerable variation in density.

Since the product density is a function of thickness, it is impossible to produce uniform density in a multithickness part by compacting with a single punch, as illustrated in Figure 16-6. When nonuniform thickness is desired, more complicated presses or methods must be used. Figure 16-7 illustrates two methods to compact a dual-thickness part. By providing different amounts of motion to the various punches one can produce a uniformly compacted product. When extremely complex shapes are desired, the powder is generally encapsulated in a flexible rubber mold and immersed in a pressurized gas or liquid, a process known as *isostatic* (uniform pressure) *compaction*. Compaction rates are extremely slow, but parts up to several hundred pounds can be effectively compacted. Another way to promote uniform compaction is to increase the amount of lubricant in the powder. This reduces the friction between the powder and the die wall and requires more movement to generate the opposing friction. Unfortunately, additional lubricant may reduce the green strength to the point where it is inadequate for ejection and handling.

Pressing rates vary widely, with about six pieces per minute being the minimum and 100 pieces per minute being quite common. After pressing, the parts are mechanically ejected from the die. By means of mechanisms such as bulk movement of particles, deformation of individual particles, and particle fracture or fragmentation, the density of the powder has been raised to about 80% of an equivalent cast or forged metal. Sufficient strength has been imparted to hold the shape and withstand a reasonable amount of handling. In addition, the nature and uniformity of the remaining porosity has been set.

Dies. Because the powder particles tend to be somewhat abrasive and high pressures are involved, there is considerable wear on the die walls. Conse-

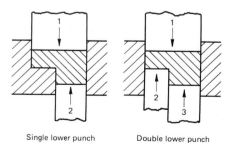

Single lower punch   Double lower punch

FIGURE 16-7 Two methods of compacting two thickness parts to near-uniform density. Both involve the controlled movement of two or more punches.

quently, dies are usually made of hardened tool steel. For particularly abrasive powders, or for high-volume production, cemented carbides may be employed. Die surfaces should be highly polished and the dies should be heavy enough to withstand the high pressing pressures. Lubricants can be used to reduce die wear. If a lubricant is not incorporated with the powder, it can be sprayed onto the die surfaces prior to each filling.

P/M Injection Molding.   A recently developed alternative to conventional compaction is the injection molding of P/M products. Small, complex components that were previously investment-cast or machined directly from metal stock can now be made by mixing metal powder with a thermoplastic material, heating it to a paste-like consistency, and then injecting it into a mold cavity under conditions of pressure and temperature selected to assure die filling. After ejection, the binder material is removed by either solvent extraction or controlled heating to above the volatilization temperature. Parts then undergo conventional sintering, during which they shrink 20% to 25%, and finally obtain the desired density (up to 95% of ideal) and properties.

Removing the binder is currently the most expensive and time-consuming part of the process. Heating rates, temperatures, and debinding times should all be carefully controlled and correlated with part thickness. Up to three days may be required for certain thick-walled sections. A new water-soluble methyl-cellulose binder, however, offers some promise. Water from the binder evaporates early during heating and the methyl-cellulose burns off during sintering.

P/M injection molding is proving to be an attractive means of producing small, intricate parts with thin walls or delicate cross sections that would be impossible to compact in a conventional press. Die design and manufacture are sufficiently costly that large production volumes are generally required to justify the process. However, the relatively high final density (94% to 98% of ideal) and close tolerances (0.3% to 0.5%) make the process attractive for many applications. Current areas of application include sporting goods, dental and medical supplies, office machines, appliances, aerospace components, and both diesel and turbine engines.

Sintering.   In the sintering operation, the compacts are subjected to elevated temperatures in a controlled-atmosphere environment. Most metals are sintered at temperatures of 70% to 80% of their melting point, while certain refractory materials may require temperatures near 90%. When the product is composed of more than one material, the sintering temperature may be above the melting temperature of some of the components. Here, the lower-melting-point materials simply flow into the voids between the higher-melting-point materials.

Most sintering operations involve three stages, and many furnaces employ three distinct zones. The first region, the *burn-off* or *purge chamber*, is designed to combust air, volatilize and remove lubricants or binders that would interfere with good bonding, and slowly raise the temperature of the compacts in a controlled manner. Rapid heating would produce high internal pressure from air entrapped in closed pores and volatilizing lubricants and would result in swelling or fracture of the compacts. If the compacts contain appreciable quantities of volatile materials, their removal will then produce a porous and permeable product. The manufacture of products such as metal filters is designed to take advantage of this feature. The second, or *high-temperature zone,* is the site of actual solid-state diffusion and bonding between the powder particles. The time here must be sufficient to produce the desired density and final properties, and usually varies from 10 minutes to several hours. Finally, a *cooling zone* is required to lower the temperature of the products while maintaining them in a controlled atmosphere. This feature serves to prevent both oxidation that would occur upon direct discharge into air and possible thermal shock. Both batch-type and continuous furnaces can be used.

All three zones of the furnace must operate with a controlled protective atmosphere. This is critical because the fine powder particles still have residual porosity after pressing and large exposed surface areas. At elevated temperatures, rapid oxidation would occur and significantly impair the quality of interparticle bonding. Reducing atmospheres, commonly based on hydrogen, dissociated ammonia, or cracked hydrocarbons, are preferred since they can reduce any oxide already present on the particle surfaces and combust harmful gases that are liberated during the sintering.

During the sintering operation, a number of changes occur in the compact. Metallurgical bonds form between the powder particles as a result of atomic diffusion. If powders of different chemistry are blended, diffusion may also promote the formation of alloys or intermetallic phases. In addition, the increase in density is accompanied by a contraction in product dimensions. Compensation for this final feature will have to be made through the design of an oversized compaction die. Final sintered products generally contain between 10% and 25% porosity.

Presintering.   Powder metallurgy is frequently employed to produce parts from materials that are very difficult to machine. When some machining is desired on these parts, it can often be made easier by employing a *presintering* operation, wherein the compacted parts are heated for a short time at a temperature considerably below the final sintering temperature. This heating imparts sufficient strength to the parts that they can be handled and machined with little difficulty. After machining, they are then given the final sintering, during which little dimensional change occurs. In this manner, machining after final sintering can be reduced to a minimum or even eliminated.

Hot Isostatic Pressing (HIP).   Hot isostatic pressing combines powder compaction and sintering into a single operation. Here the powder is hermetically sealed in a flexible, airtight, evacuated container and then subjected to a high-temperature, high-pressure environment, generally around 10,000 to 15,000 psi (70 to 100 MPa) and about 2300°F (1250°C) for iron and steel. Units capable of 45,000 psi (310 MPa) and 2750°F (1500°C) have been developed to process the nickel-based superalloys, refractory metals, and ceramics. Products emerge at full density with uniform isotropic properties that are often superior to those of other processes. Near-net shapes are possible, thereby eliminating material waste and costly machining operations. Since the powder is totally isolated and compaction and sintering occur simultaneously, the process is attractive for reactive and brittle materials such as beryllium, uranium, zirconium, and titanium. The HIP process has also been employed to densify an existing part (such as one that has been conventionally pressed and sintered), heal internal porosity in castings, and seal internal cracks in various products.

Several aspects of the HIP process make it unattractive for high-volume production. The first is the high cost of "canning" the powder in some flexible isolating medium that can resist the subsequent temperatures and pressures, and then later removing this material from the product. Sheet metal containers are most common, but glass and even ceramic molds have been used. The second problem involves the relatively long time required for the HIP cycle. While new advances have reduced cycle times from 24 hours to 6–8 hours, production is still limited to several loads a day and the capacity per load is limited by the ability to produce and maintain uniform temperature throughout the chamber.

The *sinter–HIP process* is a technique that has been designed to produce full-density products but eliminate the expense of canning and decanning. Conventionally compacted parts are first sintered under vacuum in a HIP chamber for sufficient time to close and isolate all remaining porosity. The vacuum is then broken and high pressure is applied to complete the process. The sealed surface produced durng the vacuum sintering acts as the isolating can during the high-pressure stage. An alternative is first to compact and sinter parts in the conventional manner and then to subject them to the HIP process, but this involves an additional heating and cooling cycle.

Other Techniques to Produce High-Density P/M Products. High-density P/M parts can also be produced by using high-temperature forming methods. Rods, wires, and small billets can be produced by the hot extrusion of encapsulated powder or pressed and sintered slugs. Similarly, forging can be applied to form complex shapes from canned powder or sintered preforms. The products offer the combined benefits of powder metallurgy and the respective forming process, such as complex shape, with the enhanced properties associated with uniform fine-grain size and uniform chemistry.

The newly developed *Ceracon process* was also designed to raise conventional pressed and sintered P/M products to full density, but porous or permeable preforms can now be used. A heated preform is totally surrounded by hot granular material capable of transmitting pressure in a pseudo-uniform manner, and the entire assembly is compacted in a conventional hydraulic press. Because no gas or liquid is involved, encapsulation of the porous or permeable preform is not required. Cycle times are on the order of several seconds, the part and the pressurizing medium separate freely, and the pressure-transmitting granules are reheated and reused.

A unique approach to producing high-density P/M products is the *Osprey process*. Here, molten metal is atomized and propelled toward shaped collector molds by streams of inert or harmless gas (nitrogen or carbon dioxide). Cooling of the droplets is controlled so that they strike the mold in a semisolid state and rapidly freeze. Thus, the product differs from either a full casting or conventional powder metallurgy in that the structure exhibits uniform fine grain size, uniform chemistry, and little, if any, porosity. Deposition rates can be as high as 40 to 50 pounds (18 to 23 kg) per minute.

Secondary Operations.   For many applications, P/M parts are ready to use as they come from the sintering furnace. However, many products now require one or more secondary operations to provide enhanced precision or special characteristics.

A second pressing operation, known as *repressing, coining,* or *sizing,* may be used to restore or improve dimensional precision. The part is placed in a die and subjected to pressures equal to or greater than the initial pressing pressure. A small amount of plastic flow takes place, resulting in a very uniform product with respect to size and sharpness of detail. In addition, coining may also increase part strength by 25% to 50%. Most coining is done cold.

If massive metal deformation takes place in the second pressing, the operation is known as *P/M forging*. Here, the powder metallurgy process is used to produce a preform, which is one forging operation removed from the finished shape. The normal forging operations of producing a billet or bloom, shearing, reheating, and sequentially deforming it to the desired shape are replaced by the manufacture of a comparatively simple-shape preform by powder metallurgy and a final forging operation to produce a more complex shape, add precision, and provide the benefits of metal flow. The forging operation further densifies the preform, often up to 99% of theoretical density, and significantly improves its properties. While atmospheres or coatings must be used to prevent oxidation during hot forging, the process can significantly reduce scrap or waste and extends the capabilities of powder metallurgy by offering improved properties and increased size and complexity. The forged products have no segregation, have a uniform grain size, and can be made from novel alloys or composites. In addition, subsequent machining operations can be reduced or eliminated. Figure 16-8 shows an example of a part made by conventional forging and then by the P/M forge method. The substantial reduction in scrap is clearly illustrated.

The permeability of low-density P/M products opens up two additional possibilities, *impregnation* and *infiltration*. Impregnation refers to the forcing of oil or other liquid into the porous network by either immersing the part in a bath and applying pressure or by a combination vacuum–pressure process. The most common application is that of oil-impregnated bearings. Here the bearing itself contains from 10% to 40% oil by volume, which will be released slowly throughout its lifetime in response to applied loads and temperatures. Similarly, parts may be impregnated with fluorocarbon resin (such as Teflon) to produce products offering a combination of high strength and low friction.

When the porous nature of a P/M part is undesirable, the product may be subjected to metal infiltration. In this process, a molten metal of a melting point lower than that of the P/M constituent is forced into the product under pressure or is absorbed by capillary action. Engineering properties, such as strength and

FIGURE 16-8 Comparison of conventional forging and the forging of a powder metallurgy preform to shape a gear blank (or gear). Moving left to right, the top sequence shows the sheared stock, rough forging, forged blank, and scrap of conventional forging. The finished gear is generally machined from the blank. The bottom pieces are the powder metallurgy preform and finished gear produced without scrap by P/M forging. (Courtesy GKN Forging Limited.)

toughness, are generally comparable to those of solid metal products. Infiltration can also be used to seal pores prior to plating, improve machinability, or make the components gas- or liquid-tight.

Powder metallurgy products may also be subjected to more conventional finishing operations such as heat treatment, machining, and surface treatment. If the part is of high density or has been metal-impregnated, conventional techniques are employed. Special precautions must be taken, however, when processing low-density products. During heat treatment, protective atmospheres must again be maintained and certain liquid quenchants should be avoided. In machining, speeds and feeds must be adjusted and care taken to avoid lubricant pickup and associated problems. Nearly all common methods of surface finishing are applicable, again with some modification for porous or low-density parts.

Properties of P/M Products.   Because the properties of powder metallurgy products depend on so many variables—types and size of powder, amount and type of lubricant, pressing pressure, sintering temperature and time, finishing treatments, and so on—it is difficult to provide generalized information on properties. Products can range all the way from low-density, highly porous parts with tensile strengths as low as 10 ksi (70 MPa) up to high-density pieces with tensile strengths of 180 ksi (1250 MPa) or more. Tensile strengths of 40 to 50 ksi (275 to 350 MPa) are typical of many P/M parts. In general, most mechanical properties show a strong dependence on product density, with the fracture-dependent properties such as toughness, ductility, and fatigue life showing a stronger dependence than strength and hardness. The strength properties of the weaker metals are often equivalent to the same material in wrought form. As alloying elements are added to produce higher-strength powder, the resultant properties tend to fall below those of wrought equivalents by varying but usually substantial amounts. Table 16.1 shows the properties of a few powder metallurgy materials in comparison with those of wrought material of similar composition. As larger presses and processes such as P/M forging and HIP are employed to provide greater density, the strengths of the P/M products approach those of the

TABLE 16.1   Comparison of Properties of Powder Metallurgy Materials and Equivalent Wrought Metals

Material	Form and Composition	Condition	Theoretical Density (%)	Tensile Strength		Elongation in 2 inches (%)
				10^3 psi	MPa	
Iron[a]	Wrought	HR	—	48	331	30
	P/M—49% Fe min	As sintered	89	30	207	9
	P/M—99% Fe min	As sintered	94	40	276	15
Steel[a]	Wrought AISI 1025	HR	—	85	586	25
	P/M—0.25%C, 99.75% Fe	As sintered	84	34	234	2
Stainless[a] steel	Wrought Type 303	Annealed	—	90	621	50
	P/M Type 303	As sintered	82	52	358	2
Aluminum[a]	Wrought 2014	T6	—	70	483	20
	P/M 201 AB	T6	94	48	331	2
Aluminum[a]	Wrought 6061	T6	—	45	310	15
	P/M 601 AB	T6	94	36.5	252	2
Copper[a]	Wrought OFHC	Annealed	—	34	234	50
	P/M Copper	As sintered	89	23	159	8
		Repressed	96	35	241	18
Brass[a]	Wrought 260	Annealed	—	44	303	65
	P/M 70% Cu–30% Zn	As sintered	89	37	255	26

[a]Equivalent wrought metal shown for comparison.
HR = hot rolled; T6 = age hardened.

wrought materials. With full density and fine grain size, the properties of the P/M parts can often exceed those of wrought or cast equivalents. Because the mechanical properties are so dependent upon density, the best approach for powder metallurgy is to design products that will possess the desired properties with the anticipated amount of final porosity.

Physical properties can also be affected by porosity. Corrosion resistance tends to be reduced because of the availability of entrapment pockets and fissures. Electrical, thermal, and magnetic properties all vary with density. On the positive side, the presence of porosity does promote good sound and vibration damping, and many P/M parts are designed to take advantage of this feature.

Design of Powder Metallurgy Parts. Powder metallurgy is an engineering system with the ultimate objective being the economical production of products for specific engineering applications. Success begins with good design and follows through with good material and proper processing. In designing parts that are to be made by powder metallurgy, one must remember that P/M is a special manufacturing process and provision should be made for a number of unique factors. Products that are converted from other manufacturing processes without modification in design rarely perform as well as parts designed specifically for manufacture by powder metallurgy. Some basic rules for the design of P/M parts are

1. The shape of the part must permit ejection from the die.
2. The shape of the part should be such that the powder is not required to flow into small cavities, such as thin walls, narrow splines, or sharp corners.
3. The shape of the part should permit the construction of strong tooling.
4. The shape of the part should make allowance for the thickness to which parts can be compacted.
5. The part should be designed with as few changes in section thickness as possible.
6. Design to take advantage of the fact that certain forms can be produced by P/M which are impossible, impractical, or uneconomical to obtain by any other method.
7. If necessary, the design should be consistent with available equipment. Pressing areas should match press capability, and the number of thicknesses should be consistent with the number of available press actions.
8. Consideration should also be given to product tolerances. Higher precision and repeatability are observed for dimensions in the radial direction (set by the die) than for those in the axial or pressing direction (set by punch movement).
9. Finally, design should consider and compensate for the dimensional changes that will occur after pressing, such as the shrinkage that occurs during sintering.

Since uniform strength requires uniform density, parts should be ideally designed with uniform cross section and a short thickness compared to cross-sectional dimension. Designs should not contain holes whose axes are perpendicular to the direction of pressing. Multiple-stepped diameters, reentrant holes, grooves, and undercuts should be eliminated. Abrupt changes in section, narrow, deep flutes, and internal angles without generous fillets should be avoided. Straight serrations can be molded readily, but diamond knurls cannot. The meeting plane between mold punches should be on a cylindrical or flat surface, never on a sphere. Figure 16-9 illustrates some of these points.

Powder Metallurgy Products.   The products that are commonly produced by powder metallurgy can generally be classified into five groups:

1. *Porous or permeable products, such as bearings, filters, and pressure or flow regulators.* Oil-impregnated bearings, made from either iron or copper alloys, constitute a large volume of P/M products. They are widely used in home appliance and automotive applications since they require no lubrication or maintenance during their service life. P/M filters can be made with pores of almost any size, some as small as 0.0001 in. (0.0025 mm).
2. *Products of complex shapes that would require considerable machining when made by other processes.* Large numbers of small gears are made

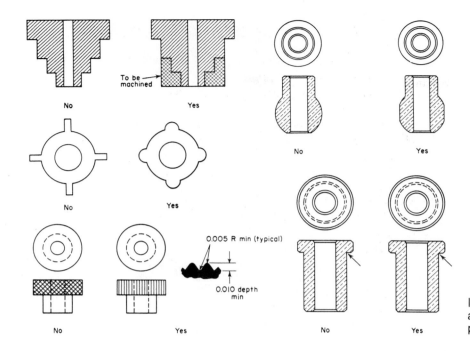

To be machined

No Yes

No Yes

No Yes

0.005 R min (typical)

0.010 depth min

No Yes

No Yes

No Yes

FIGURE 16-9 Examples of poor and good design details for use in powder metallurgy parts.

by the powder metallurgy process. Because of the accuracy and fine finish characteristic of the P/M process, many parts require no further processing and others require only a small amount of finish machining. Other complex shapes, such as pawls, cams, and small activating levers, can often be made quite economically by powder metallurgy.

3. *Products made from materials that are difficult to machine or that have high melting points.* Some of the first modern uses of powder metallurgy were in the production of tungsten lamp filaments and tungsten carbide cutting tools.

4. *Products for which the combined properties of two or more metals (or both metals and nonmetals) are desired.* This unique characteristic of the powder metallurgy process is applied to a number of products. In the electrical industry, copper and graphite are frequently combined in such applications as motor or generator brushes, copper providing the current-carrying capacity and graphite providing lubrication. Similarly, bearings have been made of graphite combined with iron or copper, or of mixtures of two metals, such as tin and copper, where a softer metal is placed in a harder metal matrix. Electrical contacts often combine copper or silver with tungsten, nickel, or molybdenum. The copper or silver provides high conductivity; the material with high melting temperature provides resistance to fusion during the conditions of arcing and subsequent closure.

5. *Products wherein the powder metallurgy process produces clearly superior properties.* The development of processes to produce full-density P/M

FIGURE 16-10 Typical parts produced by the powder metallurgy process. (Courtesy PTX-Pentronix, Inc.)

products has resulted in products that are clearly superior to those produced by competing techniques. In areas of critical importance, such as aerospace applications, the additional cost of the processing may be justified by the enhanced properties of the product. In the production of P/M magnets, a magnetic field can be used to align the particles prior to sintering, thereby producing a high flux density in the product.

Figure 16-10 shows a sample of typical powder metallurgy products.

Advantages and Disadvantages of Powder Metallurgy. Like all other manufacturing processes, powder metallurgy has some distinct advantages and disadvantages that should be considered if the technique is to be economically and successfully employed. Among the important advantages are

1. *Elimination or reduction of machining.* The dimensional accuracy and finish obtainable are such that for many applications machining operations can be totally eliminated. If unusual dimensional accuracy is required, simple coining or sizing operations can often give accuracies equivalent to those obtainable from most production machining.
2. *High production rates.* All steps in the P/M process are simple and readily automated. Labor requirements are low, and product uniformity and reproducibility are among the highest in manufacturing.
3. *Production of complex shapes.* Subject to the limitations discussed above, quite complex shapes can be produced, such as combination gears, cams, and internal keys. It is often possible to produce parts by powder metallurgy that cannot be economically machined or cast.
4. *Possibility of wide variations in compositions.* Parts of very high purity can be readily produced. Metals and ceramics can be intimately mixed.

Immiscible materials can be combined, and solubility limits can be exceeded. In all cases, the homogeneity of the product generally exceeds that of all competing techniques.

5. *Availability of wide variations in properties.* Products can range from low-density parts with controlled permeability to high-density ones with properties that equal or exceed those of equivalent wrought counterparts. Damping of noise and vibration can be tailored into a P/M product. Magnetic properties, wear properties, and others can all be designed to match the needs of a specific application.

6. *Reduction or elimination of scrap.* Powder metallurgy is the only common manufacturing process in which no material is wasted. In casting, machining, and press forming the scrap often exceeds 50% of the starting material. This is particularly important when expensive materials are involved and often makes it possible to use more costly materials without increasing the overall cost of the product. An example of such a product would be the rare-earth magnets.

The major disadvantages of the powder metallurgy process are

1. *Inferior strength properties.* In most cases, powder metallurgy parts have mechanical properties that are less than wrought or cast products of the same material. Their use may be limited when high stresses are involved. However, if the additional expense is justified, the required strength can frequently be obtained by using different materials or by employing alternate or secondary processing techniques.

2. *Relatively high die cost.* Because of the high pressures and severe abrasion involved in the process, the dies must be made of expensive materials and be relatively massive. Production volumes of less than 10,000 identical parts are normally not practical.

3. *High material cost.* On a unit weight basis, powdered metals are considerably more expensive than wrought or cast stock. However, the absence of scrap and the elimination of machining can often offset the higher material cost. Moreover, powder metallurgy is usually employed for rather small parts where the material cost per part is not very great.

4. *Design limitations.* The powder metallurgy process is simply not feasible for many shapes. Parts must be able to be ejected from the die. The thickness-to-diameter ratio is limited. Thin sections are difficult, and the overall size must be within the capacity of available presses.

5. *Property variations produced by variations in density.* The nonuniform product density that results from compacting nonuniform shapes generally results in property variations throughout the part.

Review Questions

1. What are some of the reasons for the rapid growth of the powder metallurgy industry in recent years?
2. What are some of the primary market areas for P/M products?
3. What are some of the application areas that are currently experiencing the greatest growth?
4. What are the four basic steps usually involved in making

products by powder metallurgy?

5. What are some of the important properties and characteristics of metal powders to be used in powder metallurgy?

6. Describe the most common means of producing metal powders and discuss its ability to vary particle size, shape, and chemistry.

7. Why is powder metallurgy a key process in producing products from rapidly solidified material?

8. Why is the flow rate of a powder an important manufacturing-related property?

9. What is green strength and why is it important to the manufacture of quality P/M products?

10. What are some of the objectives of powder mixing or blending?

11. How does the addition of a lubricant affect compressibility? green strength?

12. What are some of the objectives of the compacting operation?

13. In what ways does a metal powder *not* flow and transmit pressure like a liquid?

14. In what ways is double-action pressing more attractive than compaction with a single moving punch?

15. What is isostatic compaction? For what product shapes might it be preferred?

16. What are some of the ways in which a mass of particles can respond to an applied force or pressure and raise its density?

17. What is the role of the thermoplastic material in P/M injection molding?

18. For what types of parts does P/M injection molding extend the capabilities of the powder metallurgy process?

19. What are the three primary stages of the sintering process?

20. Why is it necessary to raise the temperature of P/M compacts slowly to the temperature of sintering?

21. Why is a protective atmosphere required during sintering?

22. What are some of the changes that occur in the compact during sintering?

23. Presintering is generally applied as a preliminary treatment for what manufacturing operation?

24. Since the HIP process can combine compaction and sintering, why is it not the preferred method of manufacture for conventional P/M products?

25. What are some of the attractive properties of hot isostatic pressed products?

26. What is the attractive feature of the sinter–HIP process?

27. What are some of the alternative methods of producing high-density P/M products?

28. What is the purpose of repressing, coining, or sizing operations?

29. Why can the original compaction tooling not be used to shape the product during repressing?

30. What is the difference between impregnation and infiltration?

31. The properties of P/M products are strongly tied to density. Which properties show the strongest dependence?

32. What advice would you want to give to a person who is planning to convert the manufacture of a component from die casting to powder metallurgy?

33. What types of design constraints can be imposed if manufacture must be by existing equipment?

34. Why is powder metallurgy attractive for small parts that would require considerable amounts of machining?

35. Give an example of a product for which two or more materials are mixed to provide two distinct properties.

36. How might you respond to the criticism that P/M parts have inferior strength?

37. Why is P/M not attractive for parts with low production quantities?

38. What features of the P/M process often compensate for the higher cost of the starting material?

Forming Processes

The Fundamentals of Metal Forming

The ultimate goal of a manufacturing engineer is to produce components of a selected material with a required geometrical shape and a structure optimized for the proposed service environment. Of the above aspects, production of the desired shape is a major part of the manufacturing process. Four basic alternatives exist: casting, machining, consolidating smaller pieces (welding, powder metallurgy, mechanical fasteners, epoxy), and deformation processes. Casting processes exploit the fluidity of a liquid as it takes shape and solidifies in a mold. Machining, or, more specifically, material removal processes, provide excellent precision and great flexibility but tend to waste material in the generation of the removed portions. Consolidation processes enable complex shapes to be constructed from simpler components and have a useful domain of application.

Deformation processes, on the other hand, exploit a remarkable property of some materials (usually metals)—their ability to flow plastically in the solid state without deterioration of properties. By simply moving the material to the desired shape, as opposed to removing the unwanted regions, there is little or no waste. The required forces, however, are often high. Machinery and tooling can be quite expensive, and large production quantities may be necessary to justify the process.

The overall usefulness of metals in modern society is largely due to the ease with which they may be formed into useful shapes. Nearly all metal products undergo metal deformation at some stage of their manufacture. Cast ingots, strands, and slabs are reduced in size and converted into basic forms such as sheets, rods, and plates. These forms may then experience further deformation to produce wire or the myriad of finished products formed by processes such as forging, extrusion, and sheet metal forming. The deformation may be bulk flow in three dimensions, simple shearing, simple or compound bending, or any combination of these and other processes. The stresses producing these defor-

TABLE 17.1   Classification of States of Stress

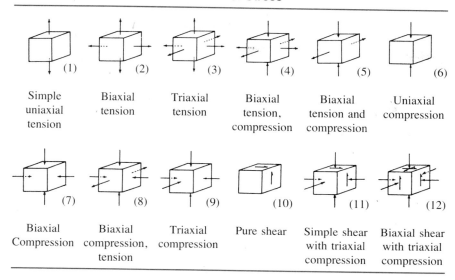

(1)	(2)	(3)	(4)	(5)	(6)
Simple uniaxial tension	Biaxial tension	Triaxial tension	Biaxial tension, compression	Biaxial tension and compression	Uniaxial compression

(7)	(8)	(9)	(10)	(11)	(12)
Biaxial Compression	Biaxial compression, tension	Triaxial compression	Pure shear	Simple shear with triaxial compression	Biaxial shear with triaxial compression

mations can be tension, compression, shear, or various combinations thereof, illustrated in Table 17.1. Speeds, temperatures, tolerances, surface finishes, and deformation amounts span a wide spectrum. The specific processes are numerous and varied, as shown in Table 17.2.

Metal-Forming Processes: Independent Variables.   In general, forming processes tend to be complex systems consisting of independent variables, dependent variables, and independent–dependent interrelations. *Independent variables* are those aspects of the process over which the engineer has direct control and are generally selected or specified when setting up the process. Consider some of the typical independent process variables:

1. *Starting material.* The engineer is free to specify the chemistry and condition of the material to be deformed. In so doing, he is selecting the properties and characteristics of the starting material. These may be chosen for ease in fabrication or may be restricted by the desired final properties of the product.
2. *Starting workpiece geometry.* This may be dictated by previous processing or may be selected by the engineer from a variety of available shapes. Economics often influences the decision.
3. *Tool or die geometry.* This is an area of major significance and has many aspects, such as the diameter of a rolling mill roll, the bend radius in a sheet-forming operation, the die angle in wire drawing or extrusion, and the cavity details in forging. Since the tooling will produce and control the metal flow, success or failure of a process often depends upon tool geometry.

TABLE 17.2 **Classification of Some Forming Operations**

Number	Process	Schematic Diagram	State of Stress in Main Part During Forming[a]
1	Rolling		7
2	Forging		9
3	Extrusion		9
4	Shear spinning		12
5	Tube spinning		9
6	Swaging or kneading		7
7	Deep drawing		In flange of blank, 5 In wall of cup, 1
8	Wire and tube drawing		8

TABLE 17.2 Classification of Some Forming Operations (*Continued*)

Number	Process	Schematic Diagram	State of Stress in Main Part During Forming[a]
9	Stretching		2
10	Straight bending		At bend, 2 and 7
11	Contoured flanging	(a) Convex	At outer flange, 6 At bend, 2 and 7
		(b) Concave	At outer flange, 1 At bend, 2 and 7

[a]See Table 17-1 for key.

4. *Lubrication*. It is not uncommon for friction to account for more than 50% of the power supplied to a deformation process. Since lubricants also act as coolants, thermal barriers, corrosion inhibitors, and parting compounds, their selection is an aspect of great importance.

5. *Starting temperature*. Temperature is one of the most influential of the process variables. Because many material properties vary greatly with temperature, its selection and control may well dictate the success or failure of an operation.

6. *Speed of operation*. Most deformation processing equipment can be operated over a range of speeds. Since speed influences lubricant effectiveness, forces required for deformation (see Figure 2-31), and heat transfer, it is obvious that its selection will be significant.

7. *Amount of deformation*. While this variable may be set by die design, other processes, such as rolling, permit this variable to be set at the discretion of the engineer.

The Dependent Variables.   After the engineer specifies the independent variables, the process then determines the nature of and values for a second set of variables. Known as *dependent variables*, they are the consequences of the dependent variable selection. Examples of dependent variables include:

1. *Force or power requirements.* To deform a selected material from a given starting shape to a specified final shape, with a specified lubricant, tooling geometry, speed, and starting temperature, will require a certain amount of force or power. A change in any of the independent variables will bring about a change in the force or power required. The effect, however, is indirect. The engineer cannot directly specify the force or power; he can only specify the independent variables and then experience the consequences of his selection.

    The ability to predict the forces or powers, however, is extremely important, for only by having this knowledge will the engineer be able to specify or select the equipment for the process, select appropriate tool or die materials, compare various die designs or deformation methods, and ultimately optimize the process.

2. *Material properties of the product.* Although the engineer can specify the properties of the starting material, the combined effects of deformation and temperature as dictated by the process will certainly change them. The customer is not interested in the starting properties but rather in the final properties of the product. Therefore, while it is often desirable to select starting properties based on compatibility with the process, it is also necessary to know or be able to predict how the process will alter them.

3. *Exit temperature.* Deformation generates heat. Hot workpieces cool in cold tooling. Lubricants can break down or decompose when overheated. Engineering properties can be altered by both the mechanical and thermal history of the material. Therefore it is important to know and control the temperature of the material throughout the process.

4. *Surface finish and precision.* Both are product characteristics dependent upon the specific details of the process.

5. *The nature of the material flow.* Deformation processes generally exert external constraints on the material through control and movement of its surfaces. How it flows or deforms internally depends upon the specifics of the material and the process. Since properties depend on deformation history, control here is vital. The customer is satisfied only if the desired geometric shape is produced with the right set of companion properties, and without surface or internal defects.

The Metal Former's Dilemma. As illustrated in Figure 17-1, the problem facing the metal former is quite obvious. On one side are the independent variables, those aspects of the process over which he has direct and immediate control. On the other are the dependent variables, those aspects over which he seeks control but for which his influence is indirect. The *dependent variables are determined by the process,* as consequences of the independent variable selection. If a dependent variable is to be modified, the engineer must determine which independent variable (or variables) is to be changed, in what manner, and by how much. Thus, the engineer must have a knowledge of the *independent–dependent variable interrelations.*

Independent variables		Dependent variables
Starting material		Force or power requirements
Starting geometry	–Experience–	
		Product properties
Tool geometry		
	–Experiment–	Exit temperature
Lubrication		
		Surface finish
Starting temperature	–Theory–	
		Dimensional precision
Speed of deformation		
		Material flow details
Amount of deformation		

FIGURE 17-1 Schematic of the metal forming system showing independent variables, dependent variables, and the various means of relating the two.

The link between independent and dependent variables is the most important area of knowledge for an individual in metal forming. Unfortunately, such links are often difficult to obtain. Metal-forming processes are complex systems composed of the material being deformed, the tooling performing the deformation, lubrication at surfaces and interfaces, and various other process parameters. The number of different forming processes (and variations thereof) is quite large. Various materials often behave differently in the same process. Multitudes of different lubricants exist. Some processes can be viewed as complex systems of 15 or more interacting independent variables.

The ability to predict and control dependent variables can be obtained in three distinct ways:

1. *Experience*. Unfortunately, this requires long-time exposure to the process and is generally limited to the specific materials, equipment, and products encountered in the realm of past contact.
2. *Experiment*. Although possibly the least likely to be in error, direct experiment is both time consuming and costly. Size and speed of deformation are often reduced when conducting laboratory studies. Lubricant performance and heat transfer behave differently at different speeds and sizes, and their effects are generally altered. The most valid experiment, therefore, is one conducted under full-size and full-speed production conditions—generally too costly to consider to any great degree. Laboratory experiments can provide valuable insight, but caution should be exercised in extrapolating their results to more realistic production conditions.
3. *Theory*. Here one attempts to develop a mathematical model of the process into which one can insert numerical values for the various independent variables and compute predictions for the dependent variables. Most techniques rely upon the applied theory of plasticity with various simplifying assumptions. Alternatives vary from crude first-order approximations, such as slab equilibrium or uniform deformation energy calculations, to sophisticated computer-based solutions, such as the finite element and finite difference methods. Solutions may be algebraic equations that describe the process and reveal trends between the variables, or simply a numerical answer based upon the specific input values.

Although the trend is certainly toward mathematical models, both for process design and process computer control, it is important to note that the models can be no better than the accuracy to which the input variables are known. For example, the plasticity behavior (yield strength, ductility, etc.) of the deforming material must be known for the specific conditions of temperature, strain (amount of prior deformation), and strain rate (speed of deformation) being considered. Moreover, the same material may behave differently under the same conditions if its microstructure is different (i.e., a 1040 steel annealed to ferrite and pearlite will behave differently than the same material that has been quenched and tempered to produce tempered martensite). Microstructure and its effects are difficult to describe in quantitative terms that can be handled by a computer. The characterization of material behavior under various process conditions forms the basis of *constitutive relations* and is receiving considerable attention at this time.

Another rather elusive variable is friction. While it is known to be dependent upon pressure, area, surface finish, lubricant, speed, and material, and often varies from location to location within the same process, most models tend to account for its effect with a single variable of constant magnitude.

At first glance, these problems appear to offer a significant barrier to the theoretical approach. However, it must be noted that the same lack of knowledge hinders the individual trying to document, characterize, and extrapolate experience or experimental results. Theory often reveals process aspects that might otherwise go unnoticed, and can be quite useful when extending a process into a previously unknown area or designing a new process or variation.

General Parameters. While much metal-forming knowledge is specific to a given process, there are certain aspects that are common to all processes and will be treated here.

One important area of knowledge is *information about the material being deformed*. What is its strength or resistance to deformation at the relevant conditions of temperature, speed of deformation, and amount of prior straining? What are its formability and fracture characteristics? What is the effect of temperature and variations in temperature? To what extent does the material strain harden? What are its recrystallization kinetics? Does it react with various environments or lubricants? These and many other questions must be answered to determine the suitability of a material to a given metal deformation process. Since material properties are quite varied, such details will not be discussed here. The reader is referred to the various chapters on engineering materials and the additional recommended references, such as those cited in Chapter 9 and the Appendix.

Several other effects are strongly related to the *speed of deformation*. Material behavior can vary markedly with speed. Some materials may shatter or crack if impact-loaded but will deform plastically under slow-speed loadings. Rate-sensitive materials appear stronger when deformed at a faster speed. Thus, more energy is required to produce the same result if we wish to do it faster. Mechanical data obtained from slow-speed tensile tests may not be particularly useful when considering deformation processes at much faster rates of deformation. Moreover, speed sensitivity (the degree to which behavior varies with speed) generally increases when the material is at elevated temperature, as is often the case in metal forming.

In addition to mechanical changes related to speed, the faster speeds also tend to promote improved lubricant efficiency. Also, faster speeds reduce the time for heat transfer and cooling. During hot working, faster speeds allow the workpiece to stay hotter and less heat to be transferred to the tools.

Other general variables include *friction and lubrication* and *temperature*. Both are of sufficient importance that they will be discussed in some detail.

Friction and Lubrication Under Metalworking Conditions.

An important consideration in metal deformation processes is the friction developed between the workpiece and the forming tool or tools. Sometimes more than 50% of the energy supplied by the equipment is spent in overcoming friction. Product quality aspects, such as surface finish and dimensional precision, are directly related to friction. Changes in lubrication can alter the mode of material flow during forming and, in so doing, modify the properties of the final product. Production rates, tool design, tool wear, and process optimization all depend on the ability to determine and control friction.

For most processes, efforts are directed at economically reducing the effects of friction. Other processes, such as rolling, however, can only operate when sufficient friction is present. In all cases, friction effects are hard to measure. Moreover, since they depend on such variables as contact area, speed, and temperature, friction effects are difficult to scale down for testing or scale up to production conditions.

It should be recognized that friction under metalworking conditions is significantly different from the friction encountered in most mechanical devices. The friction conditions of gears, bearings, journals, and similar components generally involve (1) two surfaces of similar material and strength, (2) elastic loads such that neither body undergoes permanent change in shape, (3) wear-in cycles to produce surface compatibility, and (4) generally low to moderate temperatures. Metal-forming operations, on the other hand, involve a hard, nondeforming tool interacting with a soft workpiece at pressures sufficient to cause plastic flow in the weaker material. Only a single pass is involved as the tool shapes the piece, and the workpiece is often at highly elevated temperature.

Figure 17-2 shows the change in frictional resistance with variation in contact pressure. For light, elastic loads, friction is proportional to the pressure normal to the interface, the proportionality constant (often denoted by μ) being known as the *coefficient of friction*. At high pressures, friction becomes independent of contact pressure and is more dependent on the strength of the weaker material.

Understanding of these results can be obtained from modern friction theory, the primary premise of which is that "flat surfaces are not flat" but have some degree of roughness. When two irregular surfaces interact, sufficient contact is established to support the applied load. At the lightest of loads, only three points of contact may be necessary to support the plane. As the load is increased, the

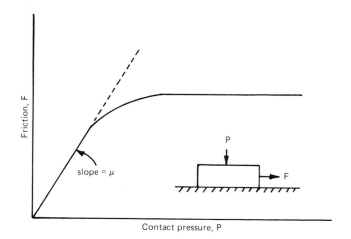

FIGURE 17-2 Effect of contact pressure on the frictional resistance between two surfaces.

contact area increases, initially in a linear fashion. At high enough loads, virtually the entire surface is in physical contact. Additional loads can cause no further change in the interface. If a sliding motion is then introduced, friction can be viewed as the resistance to this motion. From a mechanics viewpoint, this resistance can be in two forms: the force necessary to plow the peaks of the harder material through the softer one, and the force necessary to rip apart any weldments that form at the interface. Since the tears generally form in the weaker material, both components are proportional to the strength of the weaker material and to the actual area of metal-to-metal contact. The curve in Figure 17-2, therefore, can actually be viewed as a plot of actual contact area at the interface versus contact pressure. Unfortunately, Figure 17-2 and the associated theory are for the case of unlubricated contact. The presence of a lubricant and variation in type and amount can significantly alter the friction behavior.

Wear behavior is also a significant concern. Since the workpiece only interacts with the tooling once, any wear experienced by the workpiece is usually not objectionable. In fact, the shiny, fresh-metal surface produced by wear is often desired by customers. Manufacturers who fail to produce enough wear and thereby retain some of the original dull finish may actually be accused of selling old or substandard products. Wear on the tooling, however, is quite the reverse. Tooling is expensive and it is expected to shape many workpieces. Wear here generally means that the dimensions of the workpiece will change. Tolerance control is lost and at some point the tools will have to be replaced. Other consequences of tool wear include increased frictional resistance (increased required power and decreased process efficiency), poor surface finish on the product, and loss of production during tool changes.

Lubrication is of immense importance during metal forming. Lubricants are selected to reduce friction and suppress tool wear. Other considerations in selecting a metalworking lubricant include its ability to act as a thermal barrier, keeping heat in the workpiece and away from the tooling; its ability to act as a coolant and remove heat from the tools; its ability to retard corrosion if left on the formed product; ease of application and removal; lack of toxicity, odor, and flammability; reactivity or lack of reactivity with material surfaces; adaptability over a useful range of pressure, temperature and velocity; surface-wetting characteristics; cost; availability; and its ability to flow or thin and still function. In addition, the behavior of a given lubricant will change with variation in the interface conditions. The exact response will depend on such factors as the surface finish of both surfaces, area of contact, load, speed, temperature, and amount of lubricant.

In view of the important aspects of lubricants, it is amazing how little science has been applied to their development and selection. Most useful lubricants have been developed on the basis of art and experience. Many alternatives exist, often with proprietary additives and unreported chemistries, and are generally marketed under exotic trade names. Selection is often a wasteful, hit-or-miss proposition. Moreover, when a problem is encountered in a process, variation of the lubricant is often the easiest and least expensive change that can be made among the independent process variables.

Nevertheless, the benefits of friction and lubrication control can be great. For example, if one can achieve full-fluid separation between tool and workpiece, the required deformation forces may reduce 30% to 40% and tool wear becomes almost nonexistent. Considerable effort, therefore, has been directed to the measurement of friction for both general metalworking conditions and specific metalforming processes. A scientific base is being established that will ultimately enable the optimum utilization of lubricants in metalworking.

Temperature Concerns.   The importance of temperature in metalworking is every bit as great as that of lubrication. The role of temperature in altering material properties and behavior has already been discussed in Chapter 2. In general, an increase in temperature brings about a decrease in material strength, an increase in ductility, and a decrease in the rate of strain hardening—all effects that would tend to promote ease of deformation.

Forming processes tend to be classified as hot working, cold working, or warm working on the basis of the material being formed and the temperature of forming. *Hot working* is deformation under conditions of temperature and strain rate such that recrystallization is occurring simultaneously with deformation. Generally, the temperature of deformation must exceed 0.6 times the melting point of the material on an absolute temperature scale (Kelvin or Rankine). *Cold working* is deformation under conditions whereby the recovery processes are not active, and it generally requires working temperatures less than 0.3 times the workpiece melting temperature. *Warm working* is deformation under the conditions of transition, that is, a working temperature between 0.3 and 0.6 times the melting point.

Hot Working.   *Hot working* is defined as the plastic deformation of metals above their recrystallization temperature. Here it is important to note that the recrystallization temperature varies greatly with different materials. Lead and tin are hot-worked at room temperature; steels require temperatures near 2000°F;

and tungsten is still in a cold- or warm-working condition at 2000°F. Thus, the term *hot working* does not necessarily imply high or elevated temperature, although such is often the case.

As was discussed in Chapter 3, plastic deformation above the recrystallization temperature does not produce strain hardening. Therefore, hot working does not cause any increase in yield strength or hardness or a corresponding decrease in ductility. In addition, Figures 2-29 and 2-30 show that the yield strength of metals decreases as temperature increases and the ductility improves. The result is a true stress–true strain curve that is essentially flat for strains above the yield point. This is in distinct contrast to the rising curve observed for temperatures below the recrystallization temperature. Thus, hot working can be used to drastically alter the shape of metals without fear of fracture and without requiring excessively high forces. In addition, elevated temperatures promote diffusion that can remove chemical inhomogeneities; pores can be welded shut or reduced in size during deformation; and the metallurgical structure can be altered to improve the final properties. For steels, hot working involves the deformation of face-centered-cubic austenite, which is rather weak and ductile compared to the body-centered-cubic ferrite found at lower temperatures.

From a negative viewpoint, the high temperatures may promote undesirable reactions between the metal and its surroundings. Tolerances are poorer due to thermal contractions and possible nonuniform cooling. The metallurgical structure may also be nonuniform, the final grain size depending upon the reduction, temperature at last deformation, cooling history after deformation, and other factors.

Grain Alteration Through Hot Working.   When metals solidify, particularly in large sections such as ingots or continuously cast strands, coarse dendritic grains form and a certain amount of segregation of impurities occurs. Undesirable grain shapes are quite common such as the columnar grains so visible in Figure 17-3. Small gas cavities or shrinkage porosity can form during solidification. Finally, an as-solidified metal typically has a nonuniform grain structure with a rather large grain size.

Reheating the metal without prior deformation will simply promote grain growth and a concurrent decrease in properties. However, if the metal is deformed sufficiently at temperatures above the recrystallization temperature, the distorted structure is rapidly replaced by new strain-free grains. The metal can then enter into a state of grain growth, be cooled to "freeze-in" the current structure, or be further deformed and recrystallized. The final structure will be that formed by the last recrystallization and the subsequent thermal history. Production of a fine, randomly oriented, spherical-shaped grain structure can result in a net increase not only in strength but also in ductility and toughness.

Another improvement that can be obtained from hot working is the reorientation of inclusions or impurity particles in the metal. With normal melting and cooling, many impurities tend to locate along grain boundary interfaces and, if unfavorably oriented, can assist a crack in its propagation through a metal. When a piece of metal is plastically deformed, the impurity material often distorts and

FIGURE 17-3 Cross section of a cast copper bar (4 in. in diameter) showing the as-cast grain structure.

FIGURE 17-4   "Fiber" structure of a hot-formed (forged) transmission gear blank. (Courtesy Bethlehem Steel Corporation.)

flows along with the metal. This nonmetallic material, however, does not recrystallize with the base metal and often produces an aligned *fiber structure,* as seen in Figure 17-4. Such a structure clearly has directional properties, being stronger in one direction than in another. Moreover, an impurity originally oriented so as to aid crack movement through the metal can be reoriented into a "crack-arrestor" configuration, perpendicular to the direction of crack probagation. Through proper design of the deformation, attractive properties can be produced. Figure 17-5 schematically compares a machined thread and a rolled thread in a threaded fastener. By removing potential failure sites where defects intersect the surface, the rolled thread with its fibered structure possesses improved strength and fracture resistance. Concurrently, however, improperly designed deformation can actually increase the likelihood of failure.

Temperature Variation.   The success or failure of a hot deformation process often depends on the ability to control thermal conditions. Since over 90% of the energy imparted to a deforming workpiece will be converted into heat, it is possible to produce a temperature rise in the workpiece if the deformation is sufficiently rapid. More common, however, is the cooling of the workpiece in its lower temperature environment. Heat is lost through the workpiece surfaces, the bulk of the loss occurring where the workpiece is in contact with lower temperature tooling. As the surfaces cool, nonuniform temperatures are produced in the workpiece. Flow of the hot, weak interior may well result in cracking of the colder, less ductile surfaces. It is desirable, therefore, to maintain temperatures as uniform as possible. Heated dies can reduce the rate of heat transfer at the expense of reduced die life. It is not uncommon, therefore, to see die temperatures in the range of 600° to 800°F (320° to 420°C) when hot-forming steel. While operators would like to use tooling temperatures as high as 1000° to 1200°F (540° to 650°C) to produce improved tolerances and permit longer contact times, tool life drops so rapidly as to make these conditions unattractive. When forming complex shaped products, as in hot forging, thin sections cool faster than thick sections and this factor may further complicate flow behavior. Finally, nonuniform cooling from the temperatures of working

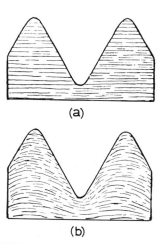

FIGURE 17-5   Schematic comparison of the grain flow characteristics in a machined thread (a) and a rolled thread (b). The rolling operation further deforms the axial structure produced by the previous wire- or rod-forming operations, while machining simply cuts through it.

may serve to introduce significant amounts of residual stress in hot-worked products.

Cold Working.   Plastic deformation of metals below the recrystallization temperature is known as *cold working* and is generally performed at room temperature. In some cases, mildly elevated temperatures may be used to provide increased ductility and reduced strength. From a manufacturing viewpoint, cold working has a number of distinct advantages, and the various cold-working processes have become extremely important. Significant advances in recent years have extended the use of cold forming, and the trend appears likely to continue.

In comparison with hot working, the advantages of cold working include

1. No heating is required.
2. Better surface finish is obtained.
3. Superior dimensional control is achieved; therefore little, if any, secondary machining is required.
4. Products possess better reproducibility and interchangeability.
5. Strength, fatigue, and wear properties are improved.
6. Directional properties can be imparted.
7. Contamination problems are minimized.

Some disadvantages associated with cold-working processes include

1. Higher forces are required for deformation.
2. Heavier and more powerful equipment is required.
3. Less ductility is available.
4. Metal surfaces must be clean and scale-free.
5. Strain hardening occurs (may require intermediate anneals).
6. The imparted directional properties may be detrimental.
7. Undesirable residual stresses may be produced.

If one examines the advantages and disadvantages, it becomes evident that the cold-working processes are particularly well suited for large-volume production where the quantity involved can readily justify the cost of the required equipment and tooling. Considerable effort has been devoted to developing and improving cold-forming machinery. In addition, better and more ductile metals and an improved understanding of basic plastic flow have done much to reduce the difficulties experienced in earlier years. To a very large extent, modern mass production has paralleled, and been made possible by, the development of cold-forming processes. Automated, high-quality production enables the manufacture of low-cost metal products. In addition, most cold-working processes eliminate or minimize the production of waste material and the need for subsequent machining. With increasing efforts in conservation and materials recycling, these benefits become quite significant.

Although the cold-forming processes tend to be better suited to large-scale manufacturing, much effort has been directed to developing methods that enable these processes and their associated equipment to be used economically for quite modest production quantities. A substantial amount of the equipment required is well standardized and not excessively costly.

Relationship of Metal Properties to Cold Working.

The suitability of a metal for cold working is determined primarily by its tensile properties, these being directly influenced by its metallurgical structure. Similarly, the cold working of a metal will alter the tensile properties of the resulting product. Both of these relationships should be considered by the designer when selecting metals that are to be processed by cold working.

No plastic deformation of a metal can occur until the elastic limit is exceeded. Thus, in Figure 17-6, permanent deformation cannot occur until the strain exceeds X_1, the strain associated with the elastic limit, a, on the stress–strain curve. At the other extreme, if the strain exceeds X_4, the metal will rupture. Consequently, from the viewpoint of cold working, two factors are of prime importance: (1) the magnitude of the yield-point stress, which determines the force required to initiate permanent deformation, and (2) the extent of the strain region from $O–X_4$, which indicates the amount of plastic deformation (or ductility) that is available to be utilized. If considerable deformation must be imparted to a metal without rupture, one having tensile properties like those depicted in the left-hand diagram of Figure 17-6 is more desirable than one having properties similar to those of the right-hand diagram. Greater ductility would be available and less force would be required to initiate and continue plastic deformation. A metal having the characteristics of the right-hand diagram would be work-hardened to a greater extent by a given amount of cold work and therefore would not be as suitable for most forming operations. On the other hand, the right-hand metal might be more satisfactory for shearing operations and would be easier to machine, as will be discussed in Chapter 21. Cold-working properties are also affected by grain size, with too large and too small both having undesirable effects.

Springback is an ever-present phenomenon in cold-working operations that can also be explained with the aid of a stress–strain diagram. When a metal is deformed through the application of a load, part of the resulting total deformation is elastic. For example, if a metal is strained to point X_1 in Figure 17-6 and the

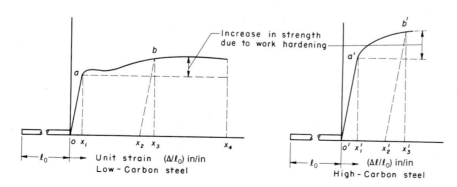

FIGURE 17-6 Use of a stress-strain diagram to reveal the tensile properties of two metals and assess their suitability for cold working.

load is removed, it will return to its original shape and size because all of the deformation is elastic. If, on the other hand, the same metal is strained to X_3, corresponding to point b on the stress–strain curve, the total strain X_3 is made up of two parts, a portion that is elastic and another that is plastic. If the deforming load is removed, the stress reduction will follow line bX_2, and the residual plastic strain will be only X_2. The decrease in strain, X_3-X_2, is known as *elastic springback*. Quite clearly, springback is an important phenomenon in cold working because the deformation must always be carried beyond the desired point by an amount equal to the springback. Moreover, since different materials have different elastic moduli, the amount of springback from a given load will differ from one material to another. Change in material will require changes in the forming process. Fortunately, springback is a design consideration, and most difficulties can be overcome by proper design procedures.

Preparation of Metals for Cold Working.   In order to obtain several of the benefits of cold working, the metal must often receive special treatment prior to processing. First, if better surface finish and dimensional accuracy are to be obtained than those produced by hot working, the starting material must be clean and free of existing scale that might cause abrasion and damage the dies or rolls. Scale is removed by *pickling*, in which the metal is dipped in acid and then washed. Second, to assure good dimensional tolerances in cold-worked parts, it is often necessary to start with a metal that is uniform in thickness and has a smooth surface. For this reason, sheet metal is sometimes given a light cold rolling prior to the major cold working.

The light cold rolling pass can also serve to remove the yield-point phenomenon and associated problems of nonuniform deformation and surface irregularities in the product. Consider the stress–strain curve of Figure 17-7a (a blow-up of the left-hand region of Figure 2-5), which is typical of many low-carbon steels. After loading to the upper yield-point, the material exhibits a "yield-point runout" wherein the material can strain up to several per cent with no additional required force. If a piece of sheet metal were to be formed into an automotive body panel, and a segment of that panel were to receive a total stretch less than the yield-point runout, a stress equal to the yield-point stress would have to be applied. Under this stress, the material is free either not to deform at all, to deform the entire amount of the yield-point runout, or to select some point in between. Often some regions deform the entire amount and thin correspondingly, while adjacent regions resist deformation and retain the original thickness. The resulting ridges and valleys, shown in Figure 17-7b, are referred to as Luders bands, or stretcher strains, and are very difficult to remove or conceal. By first cold rolling the material to a strain near or past the yield-point runout, the subsequent sheet-forming deformation occurs in a region of the curve where a single strain corresponds to each applied stress and the material deforms and thins uniformly.

A third treatment that may be given to a metal prior to cold working is annealing. If the cold working is to involve considerable deformation, it is desirable to have as much starting ductility as possible. In other cases, annealing

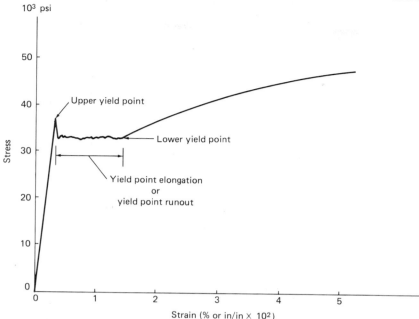

10^3 psi

Stress

Upper yield point

Lower yield point

Yield point elongation
or
yield point runout

Strain (% or in/in × 10^2)

FIGURE 17-7 (Left) The stress-strain curve for a low-carbon steel showing the commonly-observed yield-point runout. (Right) Luders bands or stretcher strains that form when this material is stretched to an amount less than the yield-point runout.

may be performed after the workpiece has been partially shaped by cold working. These anneals are called intermediate anneals and serve to restore sufficient ductility to enable further processing without danger or fracture. If the last anneal is properly positioned in the deformation cycle, the desired shape can be produced with the mechanical properties consistent with that level of cold work imparted since the anneal. For example, if the last anneal were moved earlier in the deformation sequence, a product that used to contain 10% cold work may now contain 40% cold work and possess a stronger, less ductile set of mechanical properties. In all annealing operations, care should be exercised to control the grain size of the resulting material.

Warm Forming.   Deformation produced at temperatures intermediate to hot and cold forming is known as *warm forming*. Compared to cold forming, warm forming offers the advantages of reduced loads on the tooling and equipment, increased material ductility, and a possible reduction in the number of anneals because of the reduction in strain hardening. The use of warm forming can often expand the range of materials and geometries possible for a given forming process or piece of equipment. For high-carbon steels, it may be possible to eliminate the need to spheroidize anneal the material prior to forming. Also, the favorable as-formed properties may well eliminate subsequent heat treatment operations. Compared to hot forming, the warm regimen requires less energy (the decreased energy in heating the workpiece, energy saved through higher precision, and the elimination of some post-forming heat treatments more than offset the increased deformation energy); produces less scaling and decarburization; provides better precision, dimensional control, and surface finish; and generates less scrap. Tools last longer, for, although they must exert 25% to 60% higher forces, there is less thermal shock and thermal fatigue.

Warm forming, however, is still a developing field and there are several barriers to its growth. Material behavior is less well characterized at these previously little-used temperatures (the warm-working temperatures for steel are between 1000° and 1500°F or 550° to 800°C.). Lubricants have not been fully developed for the new conditions of temperature and pressure. Finally, die design technology is not well established for warm working. Nevertheless, the pressures of energy conservation and other benefits strongly favor its continued development.

In general, cold forming is still preferred for small components. Warm forming is now considered to be attractive for larger parts (up to about 10 pounds or 5 kilograms) and steels with more than 0.35% carbon or high alloy content.

Isothermal Forming.   Figure 17-8 shows that some materials, such as titanium and nickel-based superalloys, have yield strengths that are strongly dependent on temperature. Within the realm of typical hot-working temperatures, a cooling of as little as 200°F (100°C) can produce a doubling in strength. The strength variations (cooler, stronger surface and hotter, weaker interior) promote nonuniform deformation and, often, surface cracking.

To adequately deform these materials, the deformation may have to be performed under isothermal conditions. The dies or tooling must be heated to the workpiece temperature, sacrificing die life for product quality. Deformation speeds must be slowed so that the heat generated by deformation can be removed to maintain a uniform and constant temperature. Inert atmospheres are often

FIGURE 17-8 Variation of material strength (as indicated by pressure required to forge a standard specimen) with temperature. Materials with steep curves may have to be isothermally formed. (From *A Study of Forging Variables*, ML-TDR-64-95, March 1964: Courtesy Battelle Columbus Laboratories.)

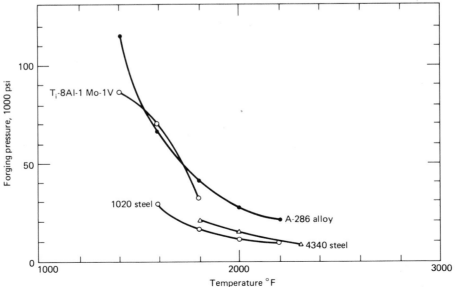

required because of the long times at elevated temperature. From a positive side, the products tend to exhibit close tolerances, low residual stresses, and fairly uniform metal flow. Moreover, while such methods are indeed costly, they may be the only means of producing a satisfactory product from certain materials.

Review Questions

1. What are the four basic classes of processes that are designed to change the shape of engineering materials?
2. Why might large production quantities be necessary to justify metal deformation as a means of manufacture?
3. What is an independent variable in a metal-forming process?
4. Discuss the significance of tool and die geometry in designing a successful metal-forming process.
5. Why is lubrication often a major concern in metal forming?
6. What are some of the secondary effects that may occur when the speed of a metal-forming process is varied?
7. What is a dependent variable in a metal-forming process?
8. Why is it important to be able to predict the forces or powers required to perform specific forming processes?
9. Why is it important to know and control the thermal history of a metal as it undergoes deformation?
10. Why is it often difficult to determine the specific relationship between independent and dependent variables?
11. What are the three distinct ways of determining the interrelation of independent and dependent variables?
12. What features limit the value of laboratory experiments in modelling metal-forming processes?
13. What is a constitutive relation for an engineering material?
14. What information about the material being deformed may be particularly significant to a metal-forming engineer?
15. What are several ways in which the friction conditions during metalworking differ from the friction conditions found in most mechanical equipment?
16. Discuss the significance of wear in metal-forming process as it relates to both the workpiece and the tool.
17. Lubricants are often selected for properties other than their ability to reduce friction. What are some of these additional properties?
18. What are some of the benefits that can be obtained by fully separating the tool and workpiece by an intervening layer of lubricant?
19. If the temperature of a material is increased, what changes in properties might occur that would promote the ease of deformation?
20. Define the various regimens of cold working, warm working, and hot working in terms of the melting point of the material being formed.
21. What is an acceptable definition of *hot working*?
22. What are some of the attractive manufacturing and metallurgical features of hot-working processes?
23. What are some of the negative aspects of hot working?
24. Describe how hot working can be used to improve the grain structure of a metal.
25. If the deformed grains recrystallize during hot working, how can the process impart an oriented or fiber structure to the product?
26. Why are heated dies or tools often employed in hot-working processes?
27. Compared to hot working, what are some of the advantages of cold-working processes?
28. What are some of the disadvantages of cold-forming processes?
29. How can the tensile test properties of a metal be used to assess its suitability for cold forming?
30. Why is elastic springback an important consideration in cold-forming processes?
31. What are Luders bands, or stretcher strains, and what causes them to form?
32. What are some of the advantages of warm forming compared to cold forming? Compared to hot forming?
33. For what types of materials might isothermal forming be required?

Hot-Working Processes

The shaping of metal by deformation is as old as recorded history. The Bible, in the fourth chapter of Genesis, introduces Tubal-Cain and cites his ability as a forger of metal. We do not know what his equipment was, but it is well known that metal forging was commonly practiced before written records. Processes such as rolling and wire drawing were extremely common in the Middle Ages and probably date back much further. In North America, the 1680 Saugus Iron Works near Boston had an operating drop forge, rolling mill, and slitting mill.

Although the basic concepts of many forming processes have remained largely unchanged throughout history, the details and equipment have evolved considerably. Manual processes were converted to machine processes during the Industrial Revolution. The machinery then became bigger, faster, and more powerful. Waterwheel power was replaced by steam and then by electricity. Most recently, computer-controlled, automated operations have become quite common.

Chapters 18 and 19 will present a survey of metal deformation processes. The division, although somewhat artificial, will be made on the basis of temperature. Processes that are normally performed ''hot'' will be presented here and processes normally performed ''cold'' will be deferred to Chapter 19. One should be aware, however, that with increased emphasis on energy conservation, the growth of ''warm working,'' and new advances in technology such as temperature classification is often arbitrary. Processes discussed as hot-working processes are often performed cold, and cold-forming processes can often be aided by some degree of heating.

Hot-Working Processes. The most obvious reason for the popularity of the hot-working processes is that they often provide an attractive means for forming a desired shape. At elevated temperatures, metals weaken and become more ductile. With continual recrystallization, massive deformation can take place without exhausting material plasticity. In steels, hot forming involves the deformation of the weaker austenite structure as opposed to the much stronger, room-temperature ferrite.

Some of the hot-working processes that are of major importance in modern manufacturing are:

1. Rolling.
2. Forging.
3. Extrusion.
4. Hot drawing.
5. Hot spinning.
6. Pipe welding.
7. Piercing.

Figure 18-1 further emphasizes the significance of these processes by schematically depicting many of the hot-forming operations commonly performed by steel makers and suppliers. Hot-worked products often form the starting material for subsequent processing.

Rolling

Rolling is usually the first process in converting a cast material into a finished wrought product. Some finished parts, such as hot-rolled structural shapes, are completed entirely by hot rolling. More often, however, hot-rolled products, such as sheets, plates, bars, and strips, serve as input material for other processes, such as cold forming or machining. From a tonnage viewpoint, hot rolling is predominant among all manufacturing processes, and modern hot-rolling equipment and practices are sufficiently advanced that standardized, uniform-quality products can be produced at low cost. Because the rolls are so massive and costly, hot-rolled products can normally be obtained only in standard shapes and sizes for which there is sufficient demand to permit economical production.

The Basic Rolling Process.

Basically, hot rolling consists of passing heated metal between two rolls that rotate in opposite directions, the space between the rolls being somewhat less than the thickness of the entering metal, as depicted in Figure 18-2. Because the rolls rotate with a surface velocity exceeding the speed of the incoming metal, friction along the contact interface acts to propel the metal forward. The metal is squeezed and elongated with a decrease in cross-sectional area. The amount of deformation that can be achieved in a single pass between a given pair of rolls depends upon the friction conditions along the interface. If too much is demanded, the rolls cannot advance the material and simply skid over its surface. On the other hand, too little deformation per pass results in excessive production cost.

Rolling Temperatures.

In hot rolling, as in all hot working, it is very important that the metal be uniformly heated to the proper temperature before processing. Prolonged exposure to elevated temperatures is usually required, a process known as *soaking*. If the temperature is not uniform, the subsequent deformation will not be uniform. For example, if insufficient soaking has occurred, the hotter exterior will flow in preference to the cooler, stronger interior.

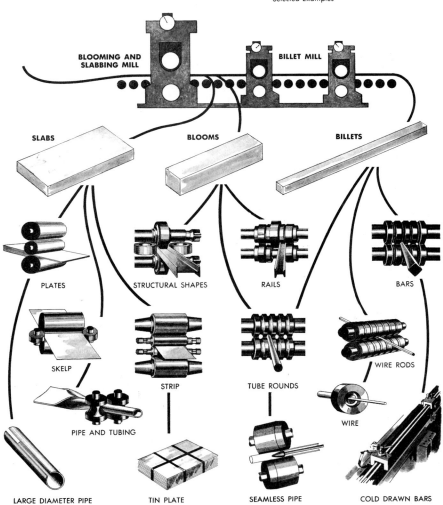

FROM STEEL INGOTS TO
FINISHED PRODUCTS

Selected Examples

BLOOMING AND
SLABBING MILL

BILLET MILL

SLABS

BLOOMS

BILLETS

PLATES

STRUCTURAL SHAPES

RAILS

BARS

SKELP

STRIP

TUBE ROUNDS

WIRE RODS

WIRE

PIPE AND TUBING

LARGE DIAMETER PIPE

TIN PLATE

SEAMLESS PIPE

COLD DRAWN BARS

FIGURE 18-1 Schematic flow chart
for the production of various finished
and semi-finished steel shapes.
(Courtesy American Iron and Steel Institute, Washington, D.C.)

If cooling occurs after removal from soaking or during previous rolling operations, the now cooler surface will resist deformation. Cracking and tearing of
the metal surface may result as the weaker interior tries to deform and elongate.

Much hot rolling is now done in integrated mills where the flow of material
is directed toward specific products. If the starting material is a cooled solid,
such as an ingot, slab, or bloom, it must first be brought to the rolling temperature in gas- or oil-fired soaking pits or furnaces. For plain carbon and low-

alloy steels, a uniform temperature of about 2200°F (1200°C) is required for the start of rolling. More frequently, continuously cast material is solidified and the subsequent cooling is controlled to enable direct insertion into the hot-rolling operation without additional handling or heating. For smaller diameter rods and bars, induction heating coils may be used to heat the material for rolling.

Hot rolling is usually terminated when the temperature falls to about 100° to 200°F (50° to 100°C) above the recrystallization temperature of the material. Maintenance of such a *finishing temperature* assures the production of a uniform fine grain size and prevents the possibility of unwanted strain hardening.

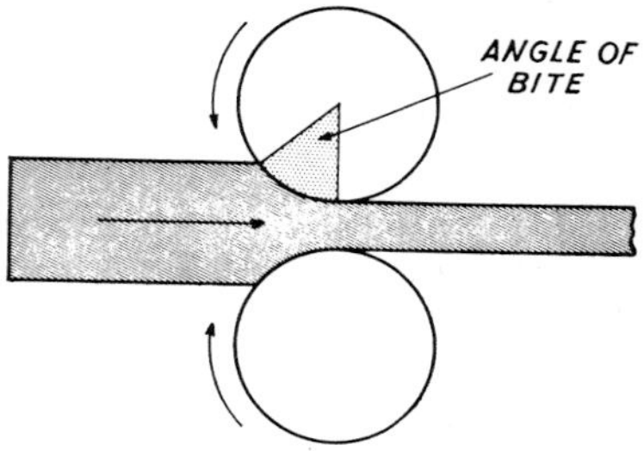

Rolling Mills.
Hot rolling is usually done in stages with a series of rolling mill stands. Thick starting stock is first rolled into *blooms* or *slabs*. Blooms are large bars with a minimum thickness generally greater than 6 in. (150mm) and a square cross section. Slabs have a distinctly rectangular shape. The blooms, in turn, are further reduced in size to form *billets*, and the slabs are rolled into *plate* or *strip*. These products then become the raw material for further hot working or other forming processes.

Rolling mill stands are available in a variety of roll configurations, as illustrated in Figure 18-3. Early reductions, often called primary roughing or breakdown passes, usually employ a two-high or three-high configuration with rolls

FIGURE 18-2   Schematic representation of the hot-rolling process.

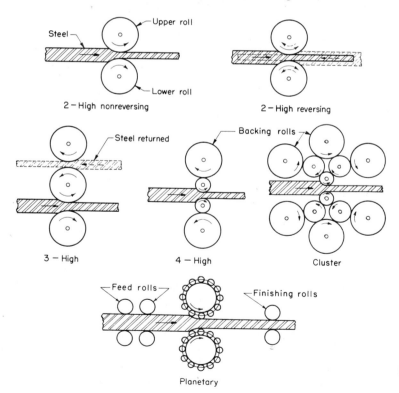

2 – High nonreversing

2 – High reversing

3 – High

4 – High

Cluster

Planetary

FIGURE 18-3   Various roll configurations used in rolling stands.

FIGURE 18-4 Hot ingot entering a three-high blooming mill. (Courtesy Mesta Machine Company.)

of 24- to 55-inch (600 to 1400-mm) diameter. The three-high mill shown in Figure 18-4 is equipped with an elevator on each side of the stand for raising or lowering the bloom and mechanical manipulators for turning the material and shifting it for the various passes as it is rolled back and forth.

Smaller-diameter rolls produce less length of contact for a given reduction and therefore require lower force and less energy for a given change in shape. The smaller cross section, however, provides reduced stiffness, and the rolls are prone to flex elastically as they are supported on the ends and pressed apart by the metal passing through the middle. Four-high and cluster-roll arrangements use backup rolls to provide the necessary support for the smaller work rolls. These configurations are used in the hot rolling of wide plate and sheets, and in cold rolling, where even small deflections in the roll would result in an unacceptable variation in product thickness.

The planetary mill configuration enables extremely large reductions to be performed in a single pass. Each roll consists of a set of planetary rolls carried about a backup or support roll, much like a roller bearing. Because friction cannot provide the necessary propulsion force, external drives are now required to push the material through the mill, and speeds are quite slow. In a typical operation, a 2¼-in. (57 mm) slab can be reduced to 0.10 in. (2.5 mm) sheet in a single pass, but the entering speed is only 6 ft/min (1.8 m/min). Nevertheless, such mills can offer a significant reduction in required capital expenditure, floor space, and operating personnel. Productivity is sacrificed and a small surface

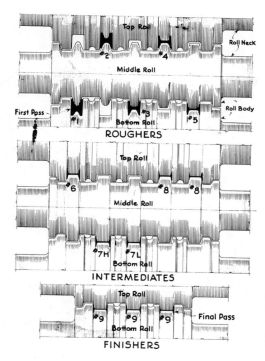

FIGURE 18-5 Rolling sequence used in producing a structural beam. Nine roll passes are required to convert the billet into a beam. (Courtesy American Iron and Steel Institute, Washington, D.C.)

reduction is often required to remove the slight scalloping effect produced by the planetary rolls.

In the rolling of nonflat or shaped products, the sets of rolls are grooved to provide a series of openings which sequentially shape the desired cross section and control the metal flow. Figure 18-5 shows an example of the roll shapes and the pass sequence used in the rolling of an I-beam.

When the volume of a product justifies the investment, finished shapes, such as sheets, bars, plates, and standard structural shapes, may be rolled from billets, blooms, or slabs on continuous rolling mills. These mills, as illustrated in Figure 18-6, consist of a series of nonreversing stands through which the material passes in a continuous fashion. For example, continuous mills for hot-rolling steel strip often consist of a roughing train of approximately four four-high mill stands and a finishing train of six or seven additional four-high stands. In a continuous structural mill, the rolls of each stand contain only one set of shaped grooves, much in contrast to the rolls of Figure 18-5 where the material passes back and forth through the various openings.

In a continuous rolling mill, the rolls of each successive stand must turn faster than those of the preceding one by an amount designed to accommodate the previous reduction in thickness. In essence, the same amount of material must pass through each roll stand in a given time. If the cross-sectional area is smaller, the velocity must be faster. In applications such as a continuous strip mill, the synchronization of six or seven mill stands is a major aspect of mill control.

FIGURE 18-6   A continuous hot-strip finishing mill, viewed from the starting end. (Courtesy Mesta Machine Company.)

This is especially difficult in view of the speeds of current mills where the product may be exiting the final stand at speeds in excess of 70 miles per hour (110 kilometers per hour). Reflecting the high degree of automation and computer control, many current operations now employ continuous casting units directly feeding into continuous rolling mills, such that the time lapse from final solidification to finished rolled product is a matter of a few minutes or less.

Characteristics, Quality, and Tolerances of Hot-Rolled Products. Because they are rolled and finished above the recrystallization temperature, hot-rolled products have little directionality in their properties and are relatively free of residual stresses. These characteristics may vary, however, depending upon the thickness of the product and the presence of complex sections that may induce either nonuniform working in the various directions or nonuniform cooling. Thin sheets often show some definite directional characteristics, whereas thicker plate (such as that above 0.8 in. or 20 mm) will usually have very little. A complex shape, such as an I- or H-beam, will warp substantially if a portion of one flange is cut away because of the high residual stresses (tension) in the edges.

As a result of the hot deformation and the good control that is maintained during processing, hot-rolled products are normally of uniform and dependable quality, and considerable reliance can be placed on them. It is quite unusual to find any voids, seams, or laminations when these products are produced by reliable manufacturers.

Of course, the surfaces of hot-rolled products are usually a bit rough and are originally covered with a tenacious high-temperature oxide known as mill scale. This is usually removed by an acid pickling operation, resulting in a surprisingly smooth surface finish.

The dimensional tolerances of hot-rolled products vary with the kind of metal and the size of the product. For most products produced in reasonably large tonnages, the tolerances are within 2% to 5% of the size (either height or width).

In view of the fact that only standard sizes and shapes are generally available, the variety of hot-rolled products is indeed considerable. In addition, special sizes and shapes can be obtained, but a considerable volume is required to make the costs reasonable. In contrast, flat plate can be economically rolled to specified dimensions in small to medium lots.

Controlled Rolling. As with most deformation processes, rolling is generally viewed as being a means of changing the shape of a material. While heat may be used to reduce forces and promote plasticity, thermal processes to produce or control mechanical properties (heat treatments) are usually performed as subsequent operations. *Thermomechanical processing*, of which *controlled rolling* is an example, consists of simultaneously performing both deformation and controlled thermal processing so as to directly produce the desired levels of strength and toughness in the as-worked product. The heat for the property modification is the same heat used in the rolling operation, and subsequent heat treatment is thereby eliminated.

To achieve this goal requires the design of a closely controlled processing system. The selected material composition must be closely maintained. Then, a time–temperature-deformation system must be developed to achieve the desired objective. Possible goals include production of a uniform fine grain size; controlling the nature, size, and distribution of the various transformation products (such as ferrite, pearlite, bainite, and martensite in steels); controlling the reactions that produce solid solution hardening or precipitation hardening; and producing desired levels of toughness in a rolled product. Starting structure (controlled by composition and prior thermal treatments), deformation details, temperature during the deformation, and the cool-down from the working temperature must all be specified and controlled. Moreover, the attainment of uniform properties requires uniform temperatures and deformations throughout the product. The availability of computer-controlled rolling facilities is almost a necessity for such a process to be performed successfully.

Possible benefits include improved product properties; substantial energy savings (through the elimination of subsequent heat treatment); and the possible substitution of a cheaper, less highly alloyed metal for a highly alloyed one that responds to heat treatment. For these reasons, it is expected that processes such as controlled rolling will assume increased significance in the years to come.

Forging

Forging is the plastic working of metal by means of localized compressive forces exerted by manual or power hammers, presses, or special forging machines. It may be done either hot or cold. However, when it is performed cold, special names are usually given to the processes. Consequently, the term *forging* usually implies hot forging done above the recrystallization temperature.

Forging is the oldest known metal-working process. From the days when prehistoric peoples discovered that they could heat sponge iron and beat it into a useful implement by hammering with a stone, forging has been an effective method of producing many useful shapes. Modern forging has developed from the ancient art practiced by the armor makers and the immortalized village blacksmith. High-powered hammers and mechanical presses have replaced the strong arm, the hammer, and the anvil, and modern metallurgical knowledge supplements the art and skill of the craftsman in controlling the heating and handling of the metal.

A variety of forging processes have been developed that make it economically possible both to forge a single piece and to mass-produce thousands of identical parts. The metal may be (1) *drawn out* to increase its length and decrease its cross section, (2) *upset* to decrease the length and increase the cross section, or (3) *squeezed in closed impression dies* to produce multidirectional flow. As indicated in Table 17-2, the state of stress in the work is primarily uniaxial or multiaxial compression.

The common forging processes are

1. Open-die hammer forging.
2. Impression-die drop forging.
3. Press forging.
4. Upset forging.
5. Automatic hot forging.
6. Roll forging.
7. Swaging.
8. Rotary forging.

Open-Die Hammer Forging. Basically, *open-die hammer forging* is the same type of forging done by the blacksmith of old, but now massive mechanical equipment is used to impart the repeated blows. The metal to be forged is heated throughout to the proper temperature before being placed on the anvil. Gas, oil, or electric furnaces are usually employed, although induction heating has become attractive for many applications. The impact is then delivered by some type of mechanical hammer, the simplest type being a gravity drop or *board hammer*. Here the hammer is attached to the lower end of a hardwood board, which is raised by being gripped between two rotating, roughened rollers, which then separate to release it for free fall. Although some of these are still in use, *steam or air hammers*, which use pressure to both raise and propel the hammer, are far more common. These give higher striking velocities, more control of striking force, easier automation, and the capability of shaping pieces up to several tons. *Computer-controlled hammers* can provide predetermined blows of energy for each of the various stages of an operation. Their use can greatly increase the efficiency of the process and minimize the creation of noise and vibration, which are the most common outlets for the excess energy not absorbed in the deformation of the workpiece.

Figure 18-7 shows a large double-frame steam hammer and a schematic of its operation. Another style is the open-frame design, which allows more room to manipulate the work and more flexibility. The open-frame design is not as strong as the double-frame, however. Open-frame hammers usually range in capacity up to about 5000 pounds (2300 kg) and the double-frame type up to 25,000 pounds (11500 kg).

Open-die forging does not confine the flow of metal, the hammer and anvil often being completely flat. The operator obtains the desired shape by manipulating the workpiece between blows. He may use specially shaped tools or a slightly shaped die between the workpiece and the hammer or anvil to aid in shaping sections (round, concave, or convex), making holes, or performing cutoff operations. Manual manipulators are used to hold and manipulate the larger workpieces, sometimes weighing many tons. Although some finished

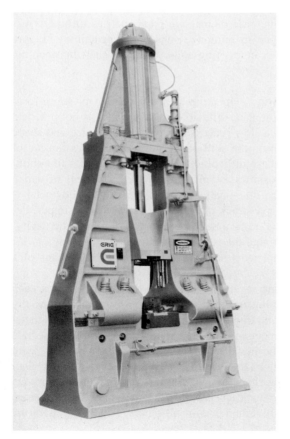

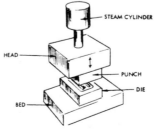

FIGURE 18-7 Double frame steam drop hammer. (Courtesy Erie Press Systems, Erie, Pa.) Inset shows the tooling in schematic.

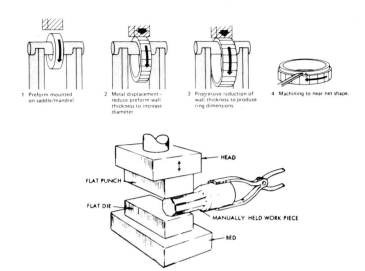

1 Preform mounted on saddle/mandrel.

2 Metal displacement – reduce preform wall thickness to increase diameter.

3 Progressive reduction of wall thickness to produce ring dimensions.

4 Machining to near net shape.

HEAD

FLAT PUNCH

FLAT DIE

MANUALLY HELD WORK PIECE

BED

FIGURE 18-8 (Top) Forging of a seamless ring by the open-die method. (Courtesy Forging Industry Association, Cleveland, Ohio.) (Bottom) Open-die forging of a solid shaft.

parts can be made by this technique, it is most often used to preshape metal for some further operation, as in the case of massive parts such as turbine rotors for which the metal is preshaped to minimize subsequent machining. Figure 18-8 shows the formation of a seamless ring and cylindrical shaft by open-die forging.

Impression-Die Drop Forging. Open-die hammer forging (or smith forging, as it has been called) is a simple and flexible process, but it is not practical for large-scale production. It is a slow process and the resulting size and shape of the workpiece are dependent on the skill of the operator. *Impression-die* or *closed-die forging* overcomes these difficulties by using shaped dies to control the flow of metal. Figure 18-9 shows a typical set of dies, one half of which attaches to the hammer and the other half to the anvil. The heated metal is positioned in the lower cavity and struck one or more blows by the upper die. This hammering causes the metal to flow so as to completely fill the die cavity. Excess metal is squeezed out around the periphery of the cavity to form a *flash*. When the final forging is completed, the flash is trimmed off by means of a trimming die.

Most drop-forging dies contain several cavities, and the deformation in each cavity may take more than one blow of the hammer. The first impression usually is an *edging, fullering,* or *bending* impression for roughly distributing the metal in accordance with the requirements of the later impressions. The intermediate impressions are for *blocking* the metal to approximately its final shape, with generous corner and fillet radii. Final shape and size are imparted in the *final impression*. These steps and the shape of a part at the conclusion of each step are shown in Figure 18-9. Because each part produced is shaped in the same die cavities, each is very closely a duplicate of all others, subject to slight die wear.

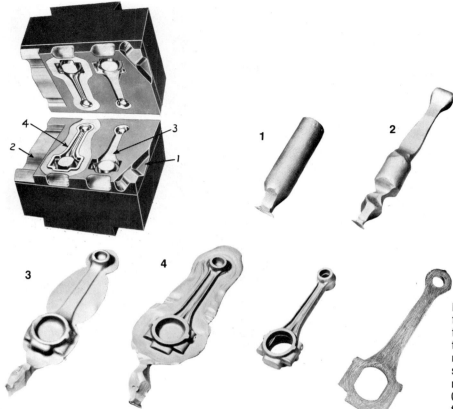

FIGURE 18-9 Impression drop-forging dies and the product resulting from each impression. The flash is trimmed from the finished connecting rod in a separate trimming die. The sectional view shows the grain fiber resulting from the forging process. (Courtesy Forging Industry Association, Cleveland, Ohio.)

The shape of the various cavities causes the metal to flow in desired directions, thereby imparting a desired fiber structure to the product. In addition, the metal may be placed where it is needed to provide the most favorable section modulus to resist the applied loads. These factors, together with the fine-grain structure and the absence of voids, make it possible to obtain about 20% higher strength-to-weight ratios with forgings than with cast or machined parts of the same material.

Board hammers, steam hammers, and air hammers, such as the one shown in Figure 18-10, are all used in impression-die forging. An alternative to the hammer and anvil arrangement is the *counterblow machine,* or *impactor,* like the one illustrated in Figure 18-11. These machines have two horizontal hammers that move together simultaneously to forge a workpiece positioned between them. Because the work is not supported on an anvil, no energy is lost to the machine foundation and a heavy machine base is no longer required. In addition, the machine operates more quietly and with less vibration. Moreover, the operation of the machine can be almost completely automated. The work can be heated by induction coils and mechanically fed into the machine, forged, and removed.

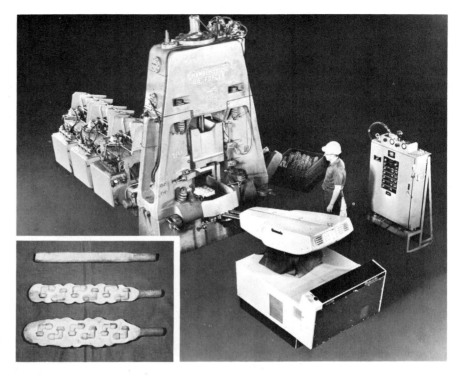

FIGURE 18-10 Fully automated drop-forging system, combining a heating and feeding unit, a pneumatic hammer, and an automatic positioning and handling unit (robot). Inset shows the three stages of the forging process. (Courtesy Chambersburg Engineering Company.)

A recent modification of impression-die forging utilizes a cast preform that is removed from the mold while hot and finish-forged in a die. After forging, the flash is trimmed in the usual manner. In some cases, the four-step process—casting, transfer from the mold to the forging die, forging, and trimming—is completely mechanized. This process, known as the *Autoforge technique*, is used primarily for nonferrous metals.

The Osprey process, described in Chapter 17, has been used to produce preforms from a gas-atomized stream of molten metal. The droplets are sprayed into shaped collector molds, which are essentially one forging operation removed from the final product. The preforms are then removed from the mold, and the final shape and properties are imparted by forging.

Another alternative to impression-die forging is *flashless forging*. This is true closed-die forging, wherein the metal is deformed in a cavity that allows little or no escape of metal. Accurate workpiece sizing is required to assure complete filling of the cavity with no excess material. Accurate workpiece positioning is necessary and die design and lubrication must be such as to control workpiece flow during the operation. The major advantage is the elimination of the scrap generated in the flash area of conventional forgings, which is typically between 20% and 45% of the starting material.

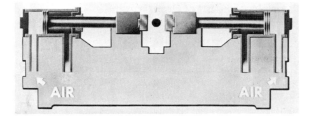

A
CONVENTIONAL
FORGED DISC WITH
PATHS OF FLOW

B
DISC FORMED BY
IMPACTER WITH
PATHS OF FLOW

FIGURE 18-11 Automatic impact forging of aluminum turbine blades. Parts are forged between two air-driven rams at the rate of 40 per minute. Upper diagrams show the equipment schematic and compare the flow of metal in both impact and conventional forging. (Courtesy Chambersburg Engineering Company.)

Die Design Factors.   There are several important factors that must be kept in mind when designing parts that are to be made by closed-die forging. Many of these relate to the design of the forming dies, which are usually made of high alloy or tool steel and are quite costly to construct. Impact resistance, wear resistance, strength at elevated temperature, and the ability to withstand alternating rapid heating and cooling must all be outstanding. Considerable care and maintenance are required to assure a smooth and accurate cavity and parting plane. Better and more economical results will be obtained if the following rules are observed in the design of forgings:

1. The parting of the die should be along a single, flat plane if at all possible. If not, the parting plane should follow the contour of the part.
2. The parting plane should be a plane through the center of the forging and not near an upper or lower edge.

TABLE 18.1   Thickness Tolerances for Steel Drop Forgings

Mass of Forging		Minus		Plus	
lb	kg	in.	mm	in.	mm
1	0.45	0.006	0.15	0.018	0.48
2	0.91	0.008	0.20	0.024	0.61
5	2.27	0.010	0.25	0.030	0.76
10	4.54	0.011	0.28	0.033	0.84
20	9.07	0.013	0.33	0.039	0.99
50	22.68	0.019	0.48	0.057	1.45
100	45.36	0.029	0.74	0.087	2.21

3. Adequate draft should be provided—at least 3° for aluminum and 5° to 7° for steel.
4. Generous fillets and radii should be provided.
5. Ribs should be wide and low.
6. The various sections should be balanced to avoid extreme differences in metal flow.
7. Full advantage should be taken of fiber flow lines.
8. Dimensional tolerances should not be closer than necessary.

Design details, such as the number of intermediate steps and the shape of each, the amount of excess metal required to assure die filling, and the dimensions of the flash at each step, are often a matter of experience. Each component is a new design entity. While computer-aided design has made notable advances, good forging design is still a combination of science and art.

Good dimensional accuracy is a major reason for using impression-die forging. It must be remembered, however, that the dimensions across the parting plane are dependent upon die wear and the thickness of the final flash. With reasonable care, the tolerances shown in Table 18.1 can be consistently maintained in these dimensions. Dimensions contained entirely within a single die cavity can be maintained with considerably greater accuracy. Draft angles approaching zero can also be used in some designs, but such practice is not recommended for general use.

Press Forging.   In hammer or impact forging, the metal flow is a response to the energy in the hammer–workpiece collision. If all of the energy can be dissipated through flow of the surface layers of metal and absorption by the hammer anvil and foundation, the interior regions of the workpiece can remain undeformed. (Consider the deformation of a metal wedge when repeatedly struck by a sledge hammer.) Therefore, when the forging of large sections is required, *press forging* must be employed. Here, the slow squeezing action penetrates through the metal and produces a more uniform metal flow. Problems arise, however, because of the long time of contact between the dies and the hot workpiece. If the workpiece surface cools, it becomes stronger and less ductile, and it may crack during further forming. Heated dies are generally used during press forming to reduce heat loss, promote surface flow, and enable the production of finer details and closer tolerances.

Forging presses are of two basic types, mechanical and hydraulic, and are usually quite massive. Mechanical presses use means such as a flywheel and clutch to provide the motion and are capable of up to 50 strokes per minute. Hydraulic presses, on the other hand, are slower, more massive, and more costly to operate. On the positive side, they usually are much more flexible and have greater capacity. Hydraulic presses with capacities up to 50,000 tons (445 MN) are currently in operation in the United States. Figure 18-12 shows a large press that is capable of forging large sections of aircraft structures as a single piece.

Press forgings usually require somewhat less draft than drop forgings and have higher dimensional accuracy. In addition, press forgings can often be completed in a single closing of the dies. Thus, modern mechanical pressing can take much of the trial and error out of forming and offers improved precision of the product and an increased opportunity for a high degree of automation.

FIGURE 18-12   A 35,000-ton (310-MN) forging press. Foreground shows a 121-inch (3.1-meter) aluminum part weighing 262 pounds (120 kg) that was forged on this press. (Courtesy Wyman-Gordon Company.)

Upset Forging.   *Upset forging* involves increasing the diameter of a material by compressing its length. In terms of the number of pieces produced, it is the most widely used of all forging processes, and its use has increased greatly in recent years. Parts can be upset-forged both hot and cold on special high-speed machines in which the workpiece is rapidly moved from station to station. The starting stock is usually wire or rod, but some machines can forge bars up to 10 inches (250 mm) in diameter.

Upset forging generally employs split dies with several positions or cavities per die, such as the set shown in Figure 18-13. The split dies separate enough for the heated bar to advance between them and move into position. They are then forced together and a heading tool or ram moves longitudinally against the bar, upsetting it into the die cavity. Separation of the die then permits transfer to the next position or removal of the product. By starting a new piece with each die separation and having all cavities perform their operations simultaneously, a finished product can be made with each cycle of the machine. If a shearing operation is performed as the initial piece moves into position, the process can operate with continuous coil or long-length rod as its feedstock.

Upset-forging machines are often used to forge heads on bolts and other fasteners and to shape valves, couplings, and many other small components. The following three rules, illustrated in Figure 18-14, should be followed in designing parts that are to be upset-forged:

1. The limiting length of unsupported metal that can be gathered or upset in one blow without injurious buckling is three times the diameter of the bar.
2. Lengths of stock greater than three times the diameter may be upset successfully provided that the diameter of the cavity is not more than 1½ times the diameter of the bar.
3. In an upset requiring stock with length more than three times the diameter of the bar and when the diameter of the upset is less than 1½ times the diameter of the bar, the length of unsupported metal beyond the face of a die must not exceed the diameter of the bar.

FIGURE 18-13   Set of upset forging dies and punches. The product resulting from each of the four positions is also shown. (Courtesy Ajax Manufacturing Company.)

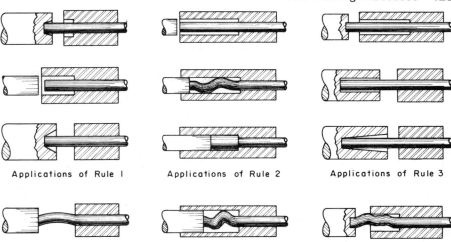

Applications of Rule 1	Applications of Rule 2	Applications of Rule 3

Violation of Rule 1	Violation of Rule 2	Violation of Rule 3

FIGURE 18-14 Schematics illustrating the rules governing upset forging. (Courtesy National Machinery Company.)

Figure 18-15 illustrates the variety of parts that can be produced by upsetting and subsequent piercing, trimming, and machining operations.

Automatic Hot Forging.   Several equipment manufacturers are now offering highly automated upset equipment in which mill-length steel bars (typically 24 feet, or 8 meters, long) are fed in one end at room temperature and hot-forged products emerge from the other end at rates of up to 180 parts per minute (86,400

FIGURE 18-15 Typical parts made by upsetting and related operations. (Courtesy National Machinery Company.)

parts per 8-hour shift). The parts can be solid or hollow, round or symmetrical, up to 12 pounds (6 kg) in weight, and up to 7 in. (180 mm) in diameter.

Consider the normal sequence of activity. The input material is the lowest-cost steel bar stock—hot-rolled and air-cooled carbon or alloy steel. The bar is induction-heated to 2200° to 2350°F (1200° to 1300°C) in under 60 seconds as it passes through induction coils. It is descaled by rolls, sheared into individual blanks, and transferred through several successive forming stages during which it is upset, preformed, final-forged, and finally pierced (if necessary). Small parts can be produced at up to 180 parts per minute, with rates for larger pieces being near 90 per minute.

The process offers numerous advantages. Low-cost input material and high production speeds have already been cited. Minimum labor is required and, since no flash is produced, material savings can be as much as 20% to 30% over conventional forging. With a consistent finishing temperature near 1900°F (1050°C), an air cooling often produces a structure suitable for machining without the need for an additional anneal or normalizing treatment. Tolerances are generally ±0.012 inch (±0.3 mm), surfaces are clean, and draft angles need only be ½° to 1° (as opposed to the conventional 3° to 5°). Tool life is nearly double that of conventional forging because the contact times are only on the order of 6/100 of a second.

To justify the operation, however, large quantities of a given product must be required. A single installation of automatic hot-forging equipment may well require an initial investment in excess of $10 million.

Recently, this technique has been coupled with a high-rate cold-forming operation. Preform shapes are hot-formed at rates that approach 180 parts per minute. These products are then cold-formed to final shape on automatic cold-formers at rates of about 90 parts per minute. The benefits of the combined operations include low production cost coupled with the precision, surface finish, and strain hardening characteristic of a cold-finished product.

Roll Forging.   In *roll forging,* round or flat bar stock is reduced in thickness and increased in length to produce such products as axles, tapered levers, and leaf springs. As illustrated in Figure 18-16, roll forging is performed on machines that have two semicylindrical rolls that are slightly eccentric to the axis of rotation, each roll containing a series of shaped grooves. The heated bar is inserted between the rolls when they are in the open position. As the rolls turn one half-revolution, the bar is progressively shaped and rolled out toward the operator. The piece is then inserted between the next set of smaller grooves and the process is repeated until the desired size and shape are obtained. Figure 18-17 shows the process in schematic.

Swaging.   *Swaging* involves hammering or forcing a tube or rod into a confining die to reduce its diameter, the die itself often playing the role of the hammer. Repeated blows cause the metal to flow inward and take the form of the die. Figure 18-18 illustrates the use of swaging to reduce and form the end of a gas cylinder.

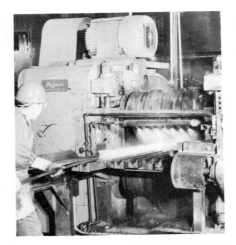

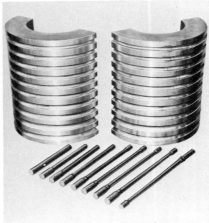

FIGURE 18-16   (Top) Roll forging machine in operation. (Bottom) Rolls from a roll forging machine and the various stages in roll forging a part. (Courtesy Ajax Manufacturing Company.)

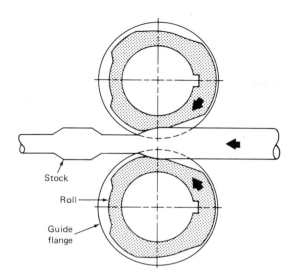

FIGURE 18-17   Schematic of the roll forging process showing the two shaped rolls and the stock being formed. (Courtesy Forging Industry Association, Cleveland, Ohio.)

Stock

Roll

Guide
flange

Rotary Forging.   Rotary forging is a process designed to replace many of the rolling operations performed during the initial breakdown of a cast ingot or strand. Starting material as large as 25½ in. (650 mm) in diameter and several tons in weight is gripped from either end, manipulated, and passed back and forth between four mechanically driven hammers. The hammers surround the workpiece and are capable of delivering up to 175 blows per minute to convert the material into rounds, flats, squares, rectangles, hollows, and multiple-diameter bars that are up to 39 feet (12 meters) long and 3 to 4 inches (75 to 100 mm) thick. Reductions of up to 40% are possible in a single pass, and the deformation-induced heat is often sufficient to maintain a desired forging temperature in the material.

Near Net-Shape Forging Operations.   As much as 80% of the cost of a forged gear can be incurred during machining operations. One aircraft wing spar contained only 4% of the original billet weight, the remainder having been lost through scrap in the forging and subsequent machining operations. As a result, considerable effort is being expended in the development of processes capable of directly forming parts that are close enough to specified dimensions that few or no secondary operations are required. These are known as *net-shape*, or *near net-shape*, operations. Cost savings and increased productivity can be attributed to the reduction in secondary machining operations, reduced quantities of generated scrap, and a decrease in the energy required to produce the product.

Precision or near net-shape forgings can now utilize draft angles of less than 1° (or even zero draft) and can produce complex shapes with such close tolerances that little or no finish machining is required. However, the design and implementation of the process is sufficiently expensive that it is currently reserved for those parts with such a high degree of complexity that they offer a great opportunity for cost reduction.

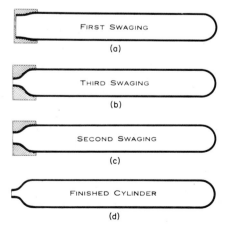

FIRST SWAGING

(a)

THIRD SWAGING

(b)

SECOND SWAGING

(c)

FINISHED CYLINDER

(d)

FIGURE 18-18   Steps in swaging a tube to form the neck of a gas cylinder. (Courtesy USX Corporation.)

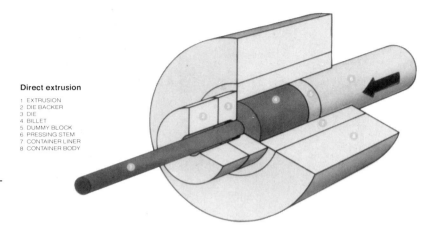

Direct extrusion

1. EXTRUSION
2. DIE BACKER
3. DIE
4. BILLET
5. DUMMY BLOCK
6. PRESSING STEM
7. CONTAINER LINER
8. CONTAINER BODY

FIGURE 18-19 Direct extrusion schematic showing the various equipment components. (Courtesy Wean United, Inc., Hydraulic Machinery Division.)

Extrusion

In the *extrusion process,* metal is compressed and forced to flow through a suitably shaped die to form a product with reduced cross section. Although extrusion may be performed either hot or cold, hot extrusion is commonly employed for many metals to reduce the forces required, eliminate cold-working effects, and reduce directional properties. Basically, the extrusion process is like squeezing toothpaste out of a tube. In the case of metals, a common arrangement is to have a heated billet placed inside a confining chamber. A ram advances from one end, causing the billet to first upset and conform to the confining chamber. As the ram continues to advance, the pressure builds until the material flows plastically through the die, as illustrated in Figure 18-19. The stress state within the material is triaxial compression.

Lead, copper, aluminum, magnesium, and alloys of these metals are commonly extruded, taking advantage of the relatively low yield strengths and low extrusion temperatures. Steels, stainless steels, and nickel-base alloys are far more difficult to extrude. Their yield strengths are high and the metal has the tendency to weld to the walls of the die and confining chamber under the required conditions of temperature and pressure. With the development and use of phosphate-based and molten glass lubricants, however, substantial quantities of extrusions are now produced from the high-strength metals. These lubricants are able to adhere to the billet and prevent metal-to-metal contact throughout the process.

As shown in part of Figure 18-20, almost any cross-sectional shape can be extruded from the nonferrous metals. Size limitations are few because presses are now available that can extrude any shape that can be enclosed within a 30-in. (750-mm) circle. In the case of steels and the other high-strength metals, the shapes and sizes are a bit more limited. As the other segment of Figure 18-20 shows, however, considerable design freedom still exists.

Extrusion is an attractive process for numerous reasons. Many shapes can be produced as extrusions that cannot be achieved by rolling, such as those containing reentrant angles or longitudinal holes. No draft is required, thereby enabling the saving of metal and weight. Since extrusion is compressive in nature, the amount of reduction in a single step is limited only by the capacity

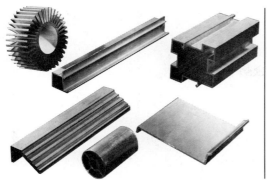

FIGURE 18-20 Typical shaped obtainable by extrusion. (Left) Aluminum products. (Courtesy Aluminum Company of America.) (Right) Steel products. (Courtesy Allegheny Ludlum Steel Corporation.)

of the equipment. Billet-to-product area ratios in excess of 100:1 have become quite common with the weaker metals. In addition, extrusion dies are relatively inexpensive—frequently less than $1,000—and often only one die is required for a given product. Because product changes require only a single die change, small quantities of a desired shape can often be produced economically by extrusion. The major limitation of the process is the requirement that the cross section must be the same for the entire length of the product being extruded.

The dimensional tolerances of extrusions are very good. For most shapes ±0.003 in./in. (0.003 mm/mm) or a minimum of ±0.003 in. (0.07 mm) is easily attainable. Grain structure is typical of other hot-worked metals, but strong directional properties (longitudinal versus transverse) are usually observed. Standard product lengths are about 20 to 24 feet (7 to 8 meters). Lengths in excess of 40 feet (13 meters) have been produced.

Extrusion Methods. Three basic methods are employed to produce extrusions. Hot extrusion is usually done by either *direct extrusion* or *indirect extrusion,* both processes being illustrated in Figure 18-21. Although the indirect configuration reduces friction between the billet and chamber wall (the billet does not have to advance with the ram) and is more energy-efficient, the added equipment complexity and restricted length of product tend to favor the direct method. Figure 18-22 shows a large aluminum extrusion press and the inset shows an extrusion emerging from the die. Extrusion speeds are often rather fast to minimize the cooling of the billet within the chamber. On the other hand, the speed may be restricted by the large amounts of heat that are generated by the massive deformation and the associated rise in temperature. The third type of extrusion, *impact extrusion,* is usually performed cold and will be discussed in Chapter 19.

Extrusion of Hollow Shapes. Hollow shapes can be extruded by several methods. For tubular products, the stationary or moving mandrel processes of Figure 18-23 are often used. For more complex internal cavities, a spider mandrel or torpedo die, as in Figure 18-24, is usually employed. As the hot metal flows beyond the spider, further reduction between the die and the mandrel

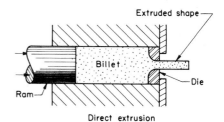

FIGURE 18-21 Direct and indirect extrusion.

FIGURE 18-22 A 14,000-ton (125-MN) extrusion press. Inset shows an aluminum extrusion emerging from the die. (Courtesy Aluminum Company of America.)

forces the seams to close and weld back together. Since the metal has never been exposed to contamination, perfect welds result, and the location of the spider arms is almost impossible to find in the product. Obviously, the cost for hollow extrusions will be greater than that for solid ones, but a wide variety of shapes can be produced that cannot be economically made by any other process.

Metal Flow in Extrusion.   The flow of metal during extrusion is often rather complex and some care must be exercised to prevent cracks and defects from forming. In direct extrusion, metal near the center of the chamber can often pass through the die with little distortion, while metal near the surface undergoes considerable shearing. In addition, friction between the forward moving billet and the chamber and die serves to further impede surface flow. The result is a final deformation pattern such as the one shown in Figure 18-25. If the surface regions of the billet undergo excessive cooling, the deformation is even further

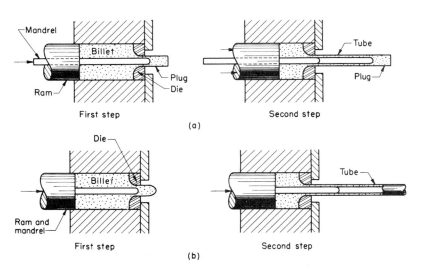

FIGURE 18-23 Two methods of extruding hollow shapes using internal mandrels.

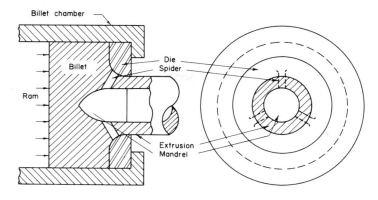

FIGURE 18-24 Extrusion of a hollow shape using a spider mandrel.

impeded and cracks tend to form on the product surface. If quality extrusions are to be produced, process control must be exercised in the areas of design, lubrication, extrusion speed, and temperature control.

Hot Drawing of Sheet and Plate

Drawing is a plastic deformation process in which a flat sheet or plate is formed into a recessed, three-dimensional part with a depth several times the thickness of the metal. As a punch descends into a die (or the die moves upward over a punch), the metal assumes the configuration of the mating punch and die tooling. Hot drawing is used for forming relatively thick-walled parts of simple geometry, usually cylindrical. There is often considerable thinning of the material as it passes through the dies. In contrast, cold drawing uses relatively thin metal, changes the thickness very little or not at all, and produces parts in a variety of shapes.

Hot drawing is illustrated schematically in Figure 18-26. A heated flat disc is positioned over a female die. The punch then descends, pushing the metal through the die, converting the flat blank to a cylindrical cup. If the difference

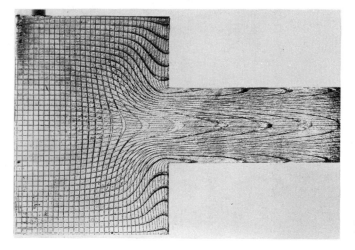

FIGURE 18-25 Grid pattern showing the metal flow in a direct extrusion. The billet was sectioned and the grid pattern was engraved prior to extrusion.

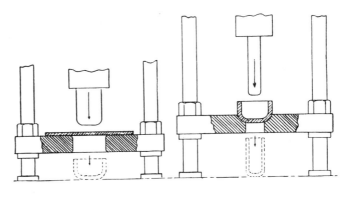

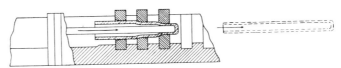

FIGURE 18-26 Methods of cup forming or hot drawing. (Upper left) First draw. (Upper right) Redraw operation. (Lower) Multiple-die drawing. (Courtesy USX Corporation.)

between the punch diameter and die opening is less than twice the thickness of the material being drawn, the cup wall is simultaneously thinned and elongated (a process often called *ironing* or *wall ironing*). As indicated in Table 17-2, the stress state during drawing is biaxial compression (in the outer flange) and uniaxial tension (in the cup wall).

Further reduction in diameter can be obtained by redrawing with a smaller punch and die set, as shown in the upper right hand segment of Figure 18-26.

FIGURE 18-27 Hot drawing of a large tank segment from ½-inch (13-mm) plate on a 2000-ton (17.9-MN) press. The dies for this operation weigh over 70 tons (63,500 kg). (Courtesy Lukens Steel Company.)

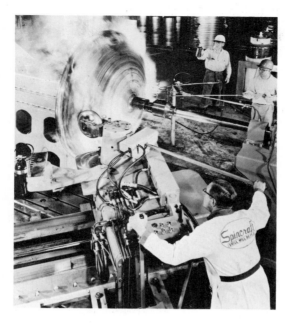

FIGURE 18-28 Hot spinning of a large workpiece using a machine equipped with power-assist controls. (Courtesy Spincraft, Inc.)

The lower segment of this figure illustrates still another alternative wherein the cup is pushed through a series of dies with a single punch.

Hot drawing is used primarily for the forming of thick-walled cylindrical components such as oxygen tanks and large artillery shells. The process can also be used for shaping more complex parts, but here the female die is closed and pressure is applied through a mating male shape. Figure 18-27 shows a large press using this technique to produce a shallow-drawn rectangular product from ½-in. thick (13 mm) plate.

In the *spinning* process, localized pressure is applied to one side of a flat rotating disk, forcing the metal to flow against a rotating male form held against the other side. Most spinning is done cold with thin sheets of metal, and the primary discussion of the process will be deferred until Chapter 19. Figure 19-57 in that chapter presents a schematic of the basic process.

Hot spinning is used when thicker plates of metal, usually steel, are to be formed into axisymmetric shapes. Metal up to 6 in. (150 mm) thick can be routinely spun into dished pressure vessel and tank heads. Thinner plates of hard-to-form metals, such as titanium, are also shaped by hot spinning. The basic theory of spinning is the same whether hot or cold, but, as with other hot processes, hot spinning enables the forming of greater thicknesses of metal with no strain hardening. Simple, hand-held wooden or metal tools are often employed in cold spinning. In hot spinning, however, heavy rollers must be used with mechanical holding and control mechanisms. Figure 18-28 shows a typical hot spinning operation.

Hot Spinning

FIGURE 18-29 Method of making butt-welded pipe from continuous skelp. (Courtesy American Iron and Steel Institute, Washington, D.C.)

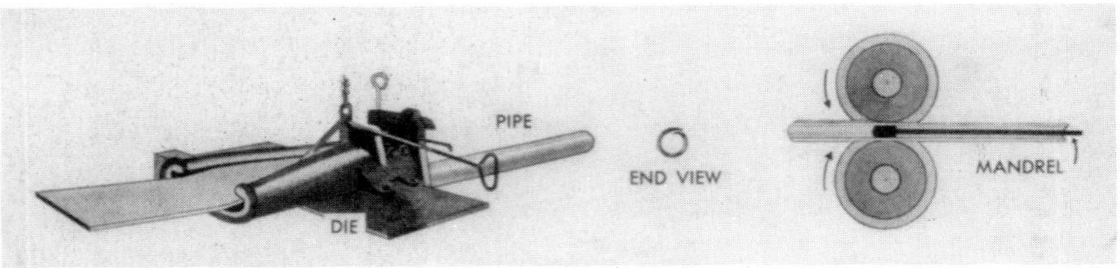

FIGURE 18-30 Method of making lap-welded pipe from continuous skelp. (Courtesy American Iron and Steel Institute, Washington, D.C.)

Pipe Welding

Large quantities of small-diameter steel pipe are made by two processes that use the hot forming of steel strip and deformation-induced welding of its free edges. Both of these processes, *butt welding of pipe* and *lap welding of pipe*, utilize steel in the form of *skelp*—long, narrow strips of the desired thickness. Because the skelp has been previously hot-rolled and the welding process produces further compressive working and recrystallization, pipe welded by these processes is very uniform in quality.

Butt-Welded Pipe. Figure 18-29 schematically illustrates the butt-welding process for making pipe. Steel skelp is unwound from a coil and is heated to forging temperature as it passes through a furnace. Upon leaving the furnace, it is pulled through forming rolls that shape it into a cylinder. The pressure exerted between the edges of the skelp as it passes through the rolls is sufficient to upset the metal and produce a welded seam. Additional sets of rollers then size and shape the pipe and it is cut to standard, preset lengths. Typical product diameters range from ⅛ in. (3 mm) to 3 in. (75 mm). Single production units can operate at rates as high as 30,000 feet per hour (10,000 meters per hour).

Lap-Welded Pipe. As illustrated in Figure 18-30, the lap-welding process for making pipe differs from the butt-welding technique in that the skelp now has beveled edges and a mandrel is used in conjunction with the rollers to produce the weld. This process is used primarily for larger sizes of pipe, from about 2 in. (50 mm) to 14 in. (400 mm) in diameter. Because of the necessity of supporting and removing the mandrel, product lengths are limited to about 20 feet (7 meters).

Piercing

Thick-walled, seamless tubing is made by the *piercing process,* the principle of which is illustrated in Figure 18-31. A heated round billet, with its leading end center-punched, is pushed longitudinally into the gap between two large convex-

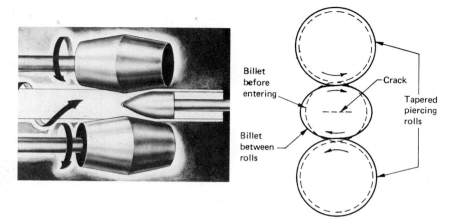

FIGURE 18-31   (Left) Principle of the Mannesmann process of producing seamless tubing. (Courtesy American Brass Company.) (Right) Mechanism of crack formation in the Mannesmann process.

tapered rolls. These rolls are powered to rotate in the same direction and have their roll axes at opposite angles of about 6° to the axis of the billet. The clearance between the rolls is set to be somewhat less than the diameter of the billet. As the billet is caught by the rolls, it too is rotated and the inclination causes the material to be propelled forward. The reduced clearance between the rolls forces the billet to deform into an elliptical shape. As this elliptical section rotates, the metal shears about the major axis, causing a crack to open along the center axis of the billet. As the crack forms, the billet is forced over a pointed mandrel that enlarges and shapes the opening to form a seamless tube. After the short-length billet has been converted into tube, the piercer point drops off and the support bar is removed. The tube can then be passed through reeler and sizing rolls, as shown in Figure 18-32, to straighten it and reduce it to the desired size. Figure 18-33 shows such a mill in operation with a thick-walled tube emerging in the direction of the viewer. If required, the seamless tube can also be put through a plug mill, where it is expanded over a larger mandrel to increase its diameter and reduce the wall thickness.

The *Mannesmann-type mills* commonly used in hot piercing (and illustrated in the previous figures) can be used to produce tubing up to 12 in. (300 mm) in diameter. Tubes of larger diameter can be produced on *Stiefel piercing mills*, which use the same principle of piercing but replace the convex rolls of the Mannesmann mill with larger-diameter conical discs.

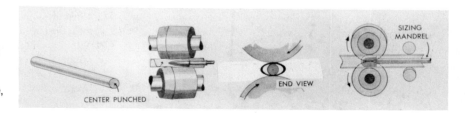

FIGURE 18-32   Schematic diagram of the production of seamless tubing by the Mannesmann process. (Courtesy American Iron and Steel Institute, Washington, D.C.)

FIGURE 18-33   Pierced tube emerging from a Mannesmann mill. (Courtesy The Timkin Company.)

Review Questions

1. Why is the division of forming processes into hot working and cold working a somewhat artificial classification?
2. What are some features of material at elevated temperature that make hot forming an attractive technique to shape metal?
3. Why should an individual working with sheet metal or wire be knowledgeable about hot rolling?
4. Why are hot-rolled products generally limited to standard shapes and sizes?
5. Why is it important to establish uniform elevated temperature in a material to be hot-rolled?
6. Discuss the relative advantages and typical uses of two-high rolling mills with large diameter rolls, three-high mills, and four-high mills.
7. Why is speed synchronization of the various rolls so vitally important in a continuous or multistand rolling mill?
8. Describe the surface quality and product precision of hot-rolled metal products.
9. What is thermomechanical processing, and what are some of its possible advantages?
10. What are some of the possible goals or objectives of controlled rolling?
11. Why are steam or air hammers more attractive than board hammers for hammer forging?
12. What is the difference between open-die and impression-die forging?
13. Why is open-die forging not a practical technique for large-scale production of identical products?
14. What structural features make it possible for forgings to have a higher strength-to-weight ratio than cast or machined products?
15. Describe the sequence of activities in the Autoforge process.
16. What is the major advantage of flashless forging?
17. What are some of the metallurgical properties and characteristics that would be desirable in a forging die material?
18. Why are different tolerances usually applied to dimensions within a die cavity and dimensions across the parting plane?
19. Why are heated dies generally employed in press-forging operations?

20. What is upset forging?
21. What are some of the typical products produced by upset-forging operations?
22. What are some of the attractive advantages of automatic hot forging?
23. How does roll forging differ from a conventional rolling operation?
24. What operation or series of operations can be replaced by a rotary forging installation?
25. What is the objective of near net-shape forging?
26. For what type of products would the near net-shape forging process be attractive?
27. What are some of the attractive features of the extrusion process?
28. What is the primary shape limitation of the extrusion process?
29. What is meant by the term *ironing* in hot drawing operations?
30. Most spinning is done cold. What types of products would require hot spinning?
31. What two hot-forming operations can be used to produce pipe from steel strip?
32. What limits the length of seamless pipe that can be produced by hot-piercing operations?

Cold-Working Processes

As discussed in Chapter 17, a number of attractive features are associated with the cold deformation of a metal. Since strain hardening serves to increase the strength of a material, cheaper, weaker starting materials may be selected or costly heat treatment of the finished product may be eliminated. By combining annealing and cold working, properties can be tailored to the specific needs of the product. Surface finish and precision are quite good. Energy savings are possible, since heating is not required to bring the workpiece to the forming temperature.

As a result of these features, numerous cold-working processes have been developed to perform a variety of deformations. These processes can be classified into four basic categories: *squeezing operations, bending operations, shearing operations,* and *drawing operations.* Table 19.1 lists a number of cold-working processes by their primary form of deformation.

Squeezing Processes

Most of the cold-working processes that employ squeezing have identical hot-working counterparts or are extensions of hot-working processes. The primary reason for deforming cold rather than hot is to obtain better dimensional accuracy and surface finish. In many cases, the equipment is basically the same, except that it must be more powerful to deform the higher-strength starting material and overcome the additional resistance caused by strain hardening. If power is limited, compromise may have to be made in the size of the workpiece or the amount of deformation.

Cold Rolling. *Cold rolling* accounts for the greatest tonnage of cold-worked products. Sheets, strips, bars, and rods are cold-rolled to obtain products that have smooth surfaces and accurate dimensions. Most cold rolling is done on four-high, cluster, or planetary rolling mills.

TABLE 19.1   Classification of the Major Cold-Working Operations

Squeezing		Bending	
1. Rolling	7. Staking	1. Angle	5. Seaming
2. Swaging	8. Coining	2. Roll	6. Flanging
3. Cold forging	9. Peening	3. Draw and	7. Straightening
4. Extrusion	10. Burnishing	compression	
5. Sizing	11. Die hobbing (hubbing)	4. Roll-forming	
6. Riveting	12. Thread rolling		

Shearing		Drawing	
1. Shearing;	4. Notching;	1. Bar and tube	5. Stretch forming
slitting	Nibbling	drawing	6. Shell drawing
2. Blanking	5. Shaving	2. Wire drawing	7. Ironing
3. Piercing;	6. Trimming	3. Spinning	8. Superplastic
lancing;	7. Cutoff	4. Embossing	forming
perforating	8. Dinking		

Cold-rolled sheet and strip are obtainable in four conditions: *skin-rolled, quarter-hard, half-hard,* and *full hard*. Skin-rolled metal is given only a ½% to 1% reduction to produce a smooth surface and uniform thickness and to remove the yield-point phenomenon (that is, to prevent formation of Luders bands upon further forming). This material is well suited for further cold-working operations when good ductility is required. Quarter-hard, half-hard, and full hard sheet and strip have experienced greater amounts of cold reduction, up to 50%. Consequently, their yield points have been increased, they have acquired definite directional properties, and they have correspondingly decreased ductility. Quarter-hard steel can be bent back on itself across the grain without breaking. Half-hard and full hard can be bent back 90° and 45°, respectively, about a radius equal to the material thickness.

Cold-Rolled Shapes.   If the product has a uniform (or near-uniform) cross section and relatively small transverse dimensions (less than about 2 in. or 5 cm), cold rolling of rod or bar may be an attractive alternative to extrusion or machining. Strain hardening can be employed to provide up to 20% additional strength to the material, and the process offers smooth surfaces and good precision. Like the hot forming of structural shapes discussed in Chapter 18, the cold rolling of shapes generally requires a series of shaping operations. The production of a given part may require separate passes (and roll grooves) for sizing, breakdown, roughing, semiroughing, semifinishing, and finishing. A minimum order of 3–5 tons of product may be required to justify the expense of tooling and make the process economical.

Swaging.   *Swaging* is a process of reducing the diameter, tapering, or pointing round bars or tubes by external hammering. If a shaped mandrel is inserted into

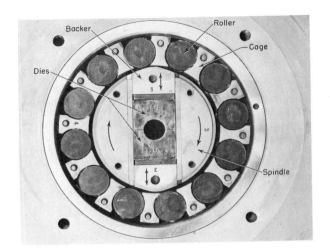

a tube before swaging, the metal can be collapsed around it to simultaneously shape and size the interior and exterior of a product.

Cold swaging is performed by a rotary machine, as shown in Figures 19-1 and 19-2. High-speed rotation of the spindle causes the die segments and backer blocks to separate in response to the action of centrifugal force. As the spindle rotates, however, the backer blocks encounter opposing rollers that are mounted in a massive machine housing. Sustained rotation, usually motivated by a massive flywheel or motor drive, forces the dies tightly together, allowing the backer blocks to pass under the rollers. Once cleared, the dies again separate and the cycle repeats. Thus, the workpiece is cyclically squeezed from various external angles (the parting line of the dies constantly changing as the spindle rotates).

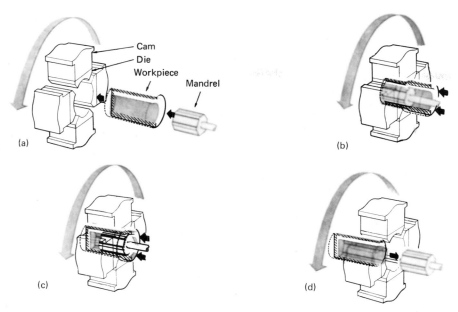

FIGURE 19-3   Method of forming internal details by swaging over a mandrel. (Courtesy Cincinnati Milacron, Inc.)

The operator simply inserts a bar or tube between the dies and gradually advances it until the desired length of material has been swaged. As the diameter is reduced, the product elongates.

Figure 19-3 shows the simultaneous formation of an internal shape. An open- or closed-end workpiece is placed over a shaped mandrel and the assembly is inserted between the rotating swaging dies. As the dies then reciprocate and

FIGURE 19-4   Typical parts swaged with internal details. (Courtesy Cincinnati Milacron, Inc.)

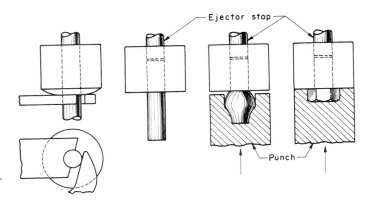

FIGURE 19-5 Steps in a cold-heading operation.

rotate, they shape the exterior of the workpiece and force the interior to conform to the shape of the mandrel, imparting an accurate internal shape. The operation requires only a few seconds, and the accuracy and surface finish are excellent. When a long product is desired, the workpiece can be fed over a short, stationary mandrel. Internal mandrels can be used to form internal gears, splines, recesses, and sockets. Figure 19-4 shows a variety of swaged products that contain shaped internal details.

Cold Forging. Large quantities of products are now being made by *cold forging,* in which metal is squeezed into a die cavity that imparts the desired shape. *Cold heading,* illustrated schematically in Figure 19-5, is used for making enlarged sections on the ends of a piece of rod or wire, such as the heads of nails, bolts, rivets, or other fasteners. Two variations of the process are common. In the sequence illustrated, a piece of rod is first cut to length and transferred to a holder–ejector assembly. Heading punches then strike one or more blows on the exposed end to perform the upsetting. If intermediate shapes are required, the various heading punches are rotated into position between strokes. When heading is completed, the ejector stop advances to expel the product from the holding die.

In the second variation, a continuous rod is fed forward, clamped, and the head is formed. The rod is further advanced, cut to length, and the cycle repeats. This procedure is particularly attractive for making nails when the point is formed by the cutoff operation.

Enlarged sections at locations other than the ends of a rod or wire can be made by the *upsetting process* illustrated in Figure 19-6. In this procedure, however, ejector pins must be provided in both the punch and the die.

Through the use of various types of dies and by combining heading, upsetting, extrusion, bending, coining, thread rolling, and knurling, a wide variety of relatively complex parts can be rapidly cold-formed to close tolerances. Figure 18-15 presented a display of typical products, the larger ones being hot-formed and machined and the smaller ones being cold-formed. Cold-forming is a chip-

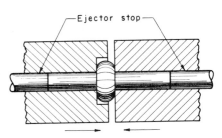

FIGURE 19-6 Method of upsetting the center portion of a rod. Both dies grip the stock during upsetting.

CUTTING (74% WASTE)   COLD FORMING (6% WASTE)

FIGURE 19-7   Manufacturing of a spark plug body. (Left) By machining from hexagonal bar stock. (Right) By cold forming. (Courtesy National Machinery Co.)

less manufacturing process, capable of producing parts that would otherwise be machined from bar stock or hot forgings. Thus, material waste can be substantially reduced.

Figure 19-7 compares the manufacture of a spark-plug body machined from hexagonal bar stock to manufacture by cold forming. Material is saved, machining time and cost are reduced, and the product is stronger, as illustrated by the flow lines in Figure 19-8. These features are often sufficient to pay for the cost of the forming equipment. By converting from screw-machining to cold-forming, a manufacturer of cruise-control housings reduced his material usage by 65% (from 197 pounds per 1000 parts to 67 pounds per 1000 parts) and increased production rate from 6–8 per minute to over 40 per minute.

While generally associated with small parts of nonferrous metal, cold forming is now used extensively on steel, in parts up to 100 pounds (45 kg) in weight and 7 in. (180 mm) in diameter. Shapes are usually axisymmetric or those with relatively small departures from symmetry. As with the other cold-forming processes, production rates are high, dimensional tolerances and surface finish are excellent, and machining can be reduced (material savings). Strain hardening can provide additional strength, and favorable grain flow can be imparted. The combination of tooling cost and production speeds, however, generally requires large volume production.

Small electronic components are also being produced by cold forming. By using forming instead of machining, the productivity of a single machine is doubled and significant material savings are realized by the elimination of chips. The tolerances obtainable in these *microforming* operations are typically 0.0002 inch (0.005 mm) or less.

Extrusion.   Great advances have been made in *cold extrusion* in recent years, as well as in combining cold extrusion and cold heading. Figure 19-9 illustrates the basic principles of *forward* and *backward* cold extrusion using open and closed dies. This process is often called *impact extrusion* and was first used only with the low-strength metals such as lead, tin, zinc, and aluminum to produce such items as collapsible tubes for toothpaste, medications, and other creams;

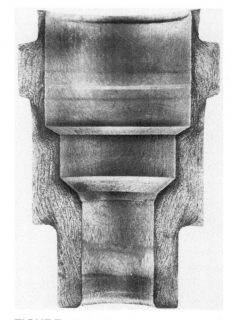

FIGURE 19-8   Section of the cold-formed spark plug body of Figure 19-7, etched to reveal the flow lines. The cold-formed structure produces an 18% increase in strength over the machined product. (Courtesy National Machinery Co.)

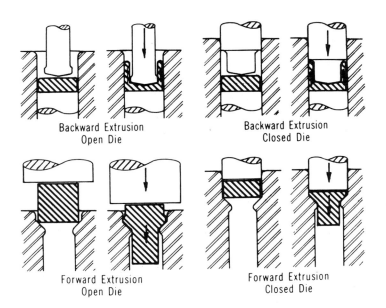

FIGURE 19-9 Methods of cold extrusion.

small "cans" for shielding electronic components; cases for flashlight batteries; and larger cans for food and beverages. In recent years, cold extrusion has been used for forming mild-steel parts, often in combination with cold heading. When heading alone is used, there is a definite limit to the ratio of the head and stock diameters (as presented in Figure 18-13 and related discussion). As can be seen in Figure 19-10, the combination of extrusion and cold heading can be used to overcome this difficulty. Considerable metal can be saved by not having to machine parts from the larger-diameter stock. Compared to a multiple-step heading operation on small-diameter rod, the combined process enables the use of an intermediate-size rod as a starting material. This is often cheaper, since one generally has to pay for the additional drawing operations required to produce the smaller stock.

FIGURE 19-10 Steps in the forming of a bolt by extrusion, cold heading, and thread rolling. (Courtesy National Machinery Co.)

FIGURE 19-11   Cold-forming sequence involving cutoff, squaring, two extrusions, an upset, and a trimming operation. Also shown is the finished part and the trimmed scrap. (Courtesy National Machinery Co.)

Figure 19-11 illustrates still another cold-forming operation, this time involving two extrusion operations and a central upset.

Another type of cold extrusion, known as *hydrostatic extrusion,* utilizes high fluid pressure to extrude a billet through a die, the product emerging either into atmospheric pressure or a lower-pressure fluid chamber. The pressure-to-pressure process, illustrated in Figure 19-12, can be used to extrude relatively brittle materials, such as molybdenum, beryllium, and tungsten. Billet-chamber friction is eliminated, billet-die lubrication is enhanced by the pressure, and the surrounding pressurized environment suppresses crack initiation and growth.

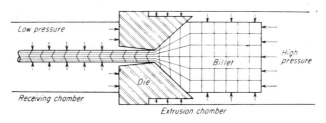

FIGURE 19-12   Hydrostatic extrusion method for the extrusion of relatively brittle materials using differential hydrostatic pressures. (Courtesy American Machinist.)

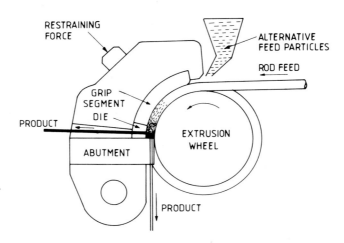

FIGURE 19-13 Schematic of the Conform continuous extrusion process showing various types of feeds and alternate locations of product withdrawal. (Courtesy United Kingdom Atomic Energy Authority, Springfields Nuclear Laboratories.)

Continuous Extrusion.   Conventional extrusion begins with a finite-length billet or workpiece and extrudes a finite-length product. However, if the pushing force could be exerted on the periphery of the feedstock, instead of on the back, continuous feedstock could be converted to continuous product in a process known as *continuous extrusion*. A number of techniques have been developed and have met with varying degrees of success. Probably the most significant, in terms of commercial manufacturing, is the *Conform process,* illustrated schematically in Figure 19-13. Continuous feedstock is inserted into a grooved wheel and is friction-propelled into a mating die shoe, where it is upset to conform to the chamber, pressurized, and finally extruded through a die opening.

Since friction is the propulsion force, the feedstock can be solid rod, metal powder, other particulates, or even punchouts or swarf from other manufacturing operations. Metallic and nonmetallic powders can be intimately mixed and extruded. Rapidly solidified materials can be formed to shape without exposure to the elevated temperatures that would harm their properties. Polymeric materials and even fiber-reinforced plastics have been extruded. The primary feedstock, however, is nonferrous metal, particularly aluminum and copper alloys.

Continuous extrusion appears to be an attractive means of producing nonferrous products with small, but uniform, cross sections. Extrusion can perform massive reductions through a single die, such that one Conform extrusion can produce deformation equivalent to 10 conventional drawing or cold-rolling passes. In addition, sufficient heat is generated that if the process is run fast, the product emerges in an annealed condition, ready for further processing without any intermediate heat treatment.

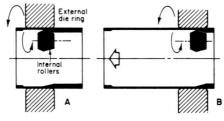

FIGURE 19-14 Roll extrusion process with internal rollers. (Courtesy *Materials Engineering.*)

Roll Extrusion.   Thin-wall cylinders can be produced from thicker-wall material by the *roll-extrusion process* depicted in Figure 19-14. The squeezing action of a rotating roller forces the material to flow forward between the roller and the external confining ring. Although cylinders from 0.75 in. (19 mm) to 156 in. (4000 mm) in diameter have been made by this process, it is most commonly used for those in the range of 3 in. (75 mm) to 20 in. (500 mm).

An alternative procedure is to use an internal mandrel and an external roller, as in Figure 19-65. However, in the latter process, provision must be made for the extraction of the mandrel.

Sizing. *Sizing* involves squeezing areas of forgings or ductile castings to a desired thickness or precision. It is used principally on bosses and flats, with only enough deformation occurring to bring the region to a desired dimension. By using this process, designers can make the general tolerances of a part more liberal, often enabling the use of less costly production methods. The few close dimensions are then obtained by one or two simple and inexpensive sizing operations. Sizing is usually performed between simple dies on a mechanically driven press, thereby assuring precise dimensional control.

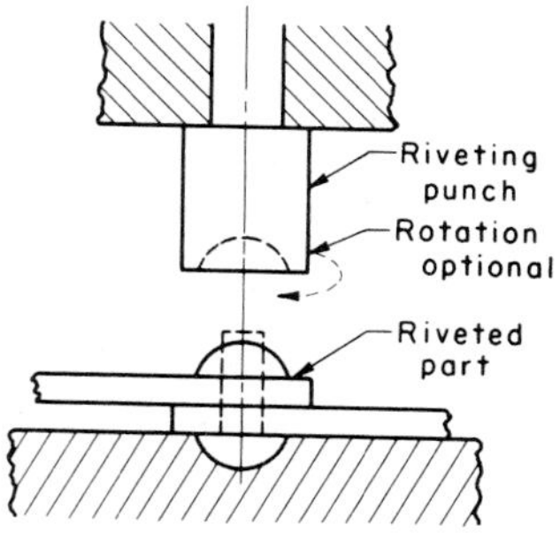

FIGURE 19-15   Method of fastening by riveting.

Riveting.   In *riveting,* a head is formed on the shank end of a fastener to provide a permanent method of joining sheets or plates of material. Although riveting is usually done hot in structural applications, in manufacturing it is almost always done cold. Quite commonly, where there is access to both sides of the work, the method illustrated in Figure 19-15 is used. The shaped punch may be held and advanced by a press or contained in a special hand-held riveting hammer. When a press is used, the rivet is usually headed in a single squeezing action, although sometimes the heading punch rotates to shape the head progressively (orbital forming). Special riveting machines, such as those used in aircraft assembly, punch the hole for the rivet, place the rivet in position, and perform the heading operation in about 1 second.

It is often desirable to use riveting in situations in which there is access to only one side of the assembly. Several types of rivets are available for these applications, two of which are illustrated in Figure 19-16. Both involve cold working. The explosive type is activated by the application of a heated tool to the rivet head, causing the charge to explode and expand the shank into a retaining head. The pull type, or pop rivet, mechanically expands the shank, after which the pull pin is cut off flush with the head.

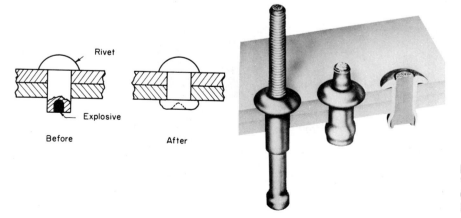

FIGURE 19-16   Rivets for use in "blind" riveting. (Left) Explosive type; (Right) Shank-type pull-up. (Courtesy Huck Manufacturing Company.)

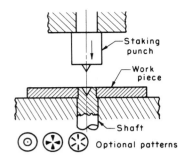

FIGURE 19-17 Fastening by staking.

Staking. *Staking* is a commonly used method of permanently joining two parts together whereby one part protrudes through a hole in the other. It is sufficiently simple in both method and final appearance that it is often overlooked by product designers. As shown in Figure 19-17, a shaped punch is driven into the protruding piece, deforming it sufficiently to cause it to squeeze outward against a second piece and lock the two together. Because the staking punch is simple and the operation can be completed with a single stroke of the press, it is a convenient and economical fastening method when permanence is desired and the appearance of the punch mark is not objectionable. Figure 19-17 illustrates some of the decorative punch designs that are in common use.

Coining. *Coining* involves cold working by means of a positive displacement punch while the metal is completely confined within a set of dies. The process, illustrated schematically in Figure 19-18, is used to produce coins, medals, and other products for which exact size and fine detail are required in a variable thickness product. Because of the confinement of the metal and the positive displacement of the punch, there is no possibility for excess metal to flow from the die and very high pressures are required. Pressures as high as 200 ksi (1400 MPa) are commonly used. Accurate volumetric measurement of the metal put into the die is essential to avoid breakage of the dies or press.

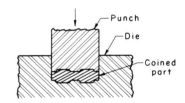

FIGURE 19-18 The coining process.

Hobbing. *Hobbing*[1] is a cold-working process that is used to form cavities in various types of female dies, such as those used for molding plastics. As shown in Figure 19-19, a male *hob* is made with the contour of the part that will ultimately be formed by the die. The hob is hardened and is then pressed

[1]*Hobbing* is also the name of a machining process used for cutting gears. See Chapter 31.

FIGURE 19-19 Hobbing a die block in a hydraulic press. Inset shows close-up of the hardened hob and the impression in the die block. The die block is contained in a reinforcing ring. The outer surface of the die block is then machined flat to remove the bulged metal.

into an annealed die block (usually by a hydraulic press) until the desired impression is produced. Flow of the metal in the die block can often be aided and controlled by machining away some of the metal where large amounts of plastic flow would occur. The die block is usually round and is reinforced by a heavy steel ring during the hobbing operation. After hobbing is completed, the die is removed from the reinforcing ring, excess metal displaced during the deformation, is machined away, and the piece is hardened by heat treatment.

Because one hob may be used to form a number of identical cavities and because it is generally easier to machine a male shape as opposed to a female shape, hobbing is frequently more economical than producing dies by conventional die sinking (machining). The hobbing process may also be referred to as *hubbing*.

Surface Improvement by Cold Working.

Cold-working processes are also used extensively for improving or altering the surface of metal products. *Peening* involves subjecting the surface to repeated blows by impelled shot or a round-nose tool. The highly localized blows deform and tend to stretch the metal surface. Because the surface deformation is resisted by the metal underneath, the result is a surface layer under residual compressive stress. This condition is favorable to resisting cracking under fatigue conditions, such as repeated bending, because the compressive stresses subtract from the applied tensile loads. For this reason, shafting, crankshafts, connecting rods, gear teeth, and other cyclic-loaded components are frequently peened.

In most manufacturing, peening is done by means of shot that is impelled by the type of mechanism shown in Figure 19-20. When weldments are peened to reduce distortion and prevent cracking, manual or pneumatic hammers are commonly used.

Burnishing involves rubbing a smooth, hard object (under considerable pressure) over the minute surface protrusions that are formed on a metal surface during machining or shearing, thereby reducing their depth and sharpness

FIGURE 19-20   Device for peening with impelled shot. Inset shows a steel surface that has been peened with steel shot, 25X. (Courtesy Wheelabrator-Frye, Inc.)

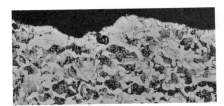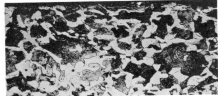

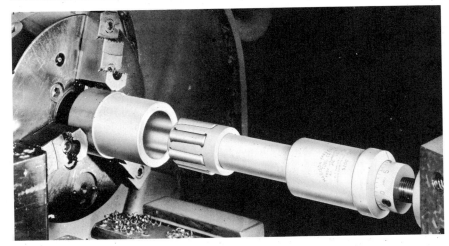

FIGURE 19-21 Tool for roller burnishing. The burnishing rolls are moved outward by means of a taper. (Top left) Section of the surface before burnishing. (Top right) Surface after burnishing. (Courtesy Madison Industries, Inc.)

through plastic flow. The edge surfaces of sheet metal stampings can be burnished by pushing the stamped parts through a slightly tapered die having its entrance end a little larger than the workpiece and its exit a slight amount smaller than the workpiece. The rubbing against the sides of the die occurs under sufficient pressure to smooth the slightly rough edges that are produced during the blanking of the part (see Figure 19-37).

Roller burnishing is used to improve the size and finish of internal or external cylindrical or conical surfaces, as illustrated in Figure 19-21. The hardened rolls of the tool press against the surface and roll the protrusions of the machined surface to a more nearly flat geometry. Being cold-worked and in residual compression, the surface also possesses improved wear and fatigue resistance.

Bending

Bending is the plastic deformation of metals about a linear axis with little or no change in the surface area. When multiple bends are made simultaneously with the use of a die, the process is sometimes called *forming*. The various bend axes can be at angles to each other, but each axis must be linear and independent of the others for the process to be classified as a true bending operation and be treatable by simple bending theory. If the axes of deformation are not linear or are not independent, the process becomes one of drawing and/or stretching, not bending; these operations will be treated later.

As shown in Figure 19-22, bending causes the metal on the outside to be stretched while that on the inside is compressed. That location that is neither stretched nor compressed is known as the neutral axis of the bend. Since the yield strength of metals in compression is somewhat higher than the yield strength in tension, the metal on the outer side yields first, and the neutral axis is displaced from the center of the two surfaces. In fact, the neutral axis is generally located between one third and one half of the way from the inner surface, the precise location depending upon the bend radius and the material. Because of the preferred tensile deformation, the metal is thinned somewhat at the bend, the thinning being more pronounced in the center of the sheet where the material cannot freely pull in along the axis of the bend. Both of these effects are shown schematically in Figure 19-22.

Looking to the inner side of the bend, it is possible for the compressive forces to induce upsetting, which would cause the material to become longer in the direction parallel to the bend axis. This effect can become quite pronounced in the bending of thick, narrow pieces.

Still another consequence of the condition of combined tension and compression is the tendency of the metal to unbend somewhat after forming, a phenomenon known as *springback*. To form a desired angle, metals must be overbent in such a way that upon springback, the material assumes the desired shape of the product.

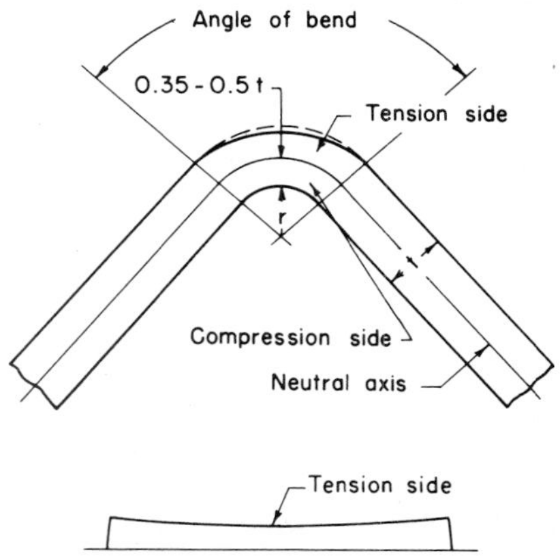

Angle Bending.   A *bar folder*, as shown in Figure 19-23, can be used to make angle bends up to 150° in sheet metal under $\frac{1}{16}$ in. (1.5 mm) thick. The sheet of metal is inserted under the folding leaf and moved to proper position. Raising the handle then actuates a cam, causing the leaf to clamp the sheet. Further motion of the handle then bends the metal to the desired angle. Bar folders are manually operated and are usually less than 12 feet (4 meters) long.

FIGURE 19-22   (Top) Nature of a bend in sheet metal. (Bottom) Cross section (exaggerated) of the tension side of a bent sheet or bar, showing the thickness variation due to restraint at the edges.

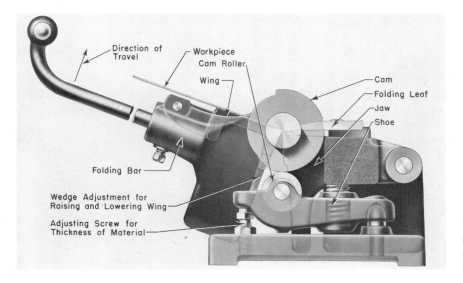

FIGURE 19-23   Phantom section of a bar folder. (Courtesy Niagara Machine and Tool Works.)

FIGURE 19-24 (Left) Modern press brake with CNC gaging system. (Courtesy DiAcro Division, Houdaille Industries, Inc.) (Right) Close-up view of press brake dies. (Courtesy Cincinnati Incorporated.)

Bends in heavier sheet or more complex bends in thin material are generally made on *press brakes,* such as the one shown in Figure 19-24. These are mechanically or hydraulically driven presses with a long, narrow bed and relatively slow, short, adjustable strokes. The metal is bent between interchangeable dies that are attached to the bed and the ram. As illustrated in Figures 19-24 and 19-25, different dies can be used to produce many types of bends. The metal can be fed inward between successive strokes to produce various repeated bends, such as corrugations. Figure 19-26 shows how a complex bend can be formed with repeated strokes, proper positioning, and more than one set of tooling. Seaming, embossing, punching, and other operations can also be performed by inserting suitable dies into press brakes, but these operations can usually be done more efficiently on other types of equipment when the volume is sufficient to justify their use.

Design for Bending. Several factors must be considered in designing parts that are to be shaped by bending. Of primary importance is determining the smallest bend radius that can be formed without metal cracking, the *minimum bend radius.* This is directly dependent upon the ductility of the metal and can be related to the percent reduction in area observed in a standard tensile test.

FIGURE 19-25 Several types of dies used on press brakes for forming angles and rounds. (Courtesy Cincinnati Incorporated.)

     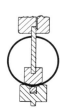

FIGURE 19-26 Dies and stages used in forming a roll bead on a press brake. (Courtesy Cincinnati Incorporated.)

Figure 19-27 shows the relationship between the ratio of the minimum bend radius R to the thickness of the material t for a range of materials with varying ductility. As can be noted, it is seldom feasible to produce a bend with radius less than the thickness of the metal. In general, bends should be designed with the largest possible radii. This permits easier forming and allows the designer to select from a wider variety of engineering materials.

If the metal has been previously cold-worked or has marked directional properties, these characteristics can have considerable effect on its bending behavior. Whenever possible, it is best to make the bend axis perpendicular to the direction of previous working. If two perpendicular bends are involved, it is preferable to orient the bend axes at angles of 45° to the rolling direction of the metal.

A second matter of design concern is determining the length of a flat blank that will produce a bent part of the desired dimensions. The fact that the neutral axis is not at the center line of the metal causes metal to thin and lengthen during bending. The amount of lengthening is a function of the stock thickness and the bend radius. Figure 19-28 illustrates one method that has been found to give satisfactory results in determining the blank length for bent products.

A third important design factor is determining the minimum length of a leg that can be successfully bent. In most cases, the length of the protruding leg should be at least 1½ times the thickness of the metal plus the bend radius.

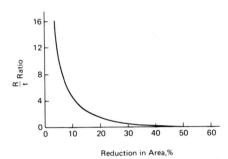

FIGURE 19-27 Curve relating the minimum bend radius (relative to thickness) to the ductility of the metal as measured by the reduction in area in a uniaxial tensile test.

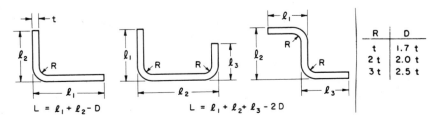

R	D
t	1.7 t
2 t	2.0 t
3 t	2.5 t

FIGURE 19-28 A method of determining the blank size for bending operations. Because of thinning, the product will lengthen during forming.

FIGURE 19-29   Cold roll bending of structural shapes. (Courtesy Buffalo Forge Company.)

Whenever possible, the tolerance on bent parts should not be less than $\frac{1}{32}$ in. (or 0.8 mm). Ninety-degree bends should not be specified without first determining whether the bending method will permit a full right angle to be obtained. Multiple bend parts should be designed with most (or all) bends of the same radius to reduce setup and tooling costs. Consideration should also be given to providing regions for adequate clamping or handling during manufacture. Bending near the edge of a material will distort the edge. If an undistorted edge is required, additional material must be included and a trimming operation performed after bending.

Roll Bending.   Plates, sheets, and rolled shapes can be bent to a desired curvature on forming rolls of the type shown in Figure 19-29. These machines usually have three rolls in the form of a pyramid, with the two lower rolls being driven and the position of the upper roll being adjustable to control the degree of curvature. When the rolls are supported by a frame on each end, one of the supports can often be swung clear to permit the removal of closed shapes from the rolls. Roll-bending machines are available in a wide range of sizes, some being capable of bending plate up to 6 in. (150 mm) thick.

Draw Bending and Compression Bending.   Many modern bending machines utilize a clamp and pressure tool to produce bending about a form block. *Draw bending,* illustrated in Figure 19-30, is perhaps the most versatile and accurate means of bending. The workpiece is clamped against a bending form and the entire assembly rotates to draw the workpiece across a pressure tool. In *compression bending,* also shown in Figure 19-30, the bending form remains stationary and the pressure tool traverses along the workpiece.

Cold Roll Forming.   The *cold roll forming* of flat strip into complex sections has become a highly developed forming technique. As depicted in Figures 19-31 and 19-32, the process involves the progressive bending of metal strip as it passes through a series of forming rolls. Various moldings, channeling, gutters

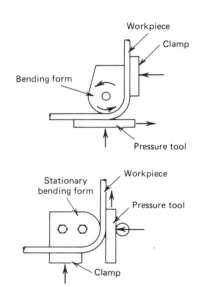

FIGURE 19-30   (Top) Draw bending, in which the form block rotates. (Bottom) Compression bending, in which a moving tool compresses the workpiece against a stationary form.

STAGES OF FORMING

PROFILES OF TOP, BOTTOM AND SIDE ROLLS

THIS SHOWS FINISHED SEC- TION WHICH CAN BE CUT OFF IN ANY DE- SIRED LENGTHS

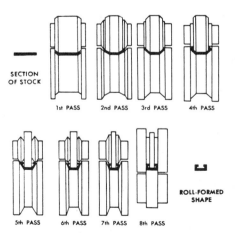

FIGURE 19-31 Schematic representation of the cold roll-forming process. Inset shows typical shapes formed by this process. (Courtesy Van Huffel Tube Corporation.)

and downspouts, automobile bumpers, and other shapes of uniform cross section can be formed on machines that can produce up to 10,000 feet (3000 meters) of product a day. By changing the rolls, a single machine can produce a wide variety of different shapes. Since changeover, setup, and adjustment may take several hours, however, a production run of at least 10,000 feet (3000 meters) of product is usually required. When tubes or pipe are desired, a resistance welding unit or seaming operation is usually combined with the roll forming.

Seaming.   *Seaming* is used to join the ends of sheet metal to form containers such as cans, pails, and drums. The seams are formed by a series of small rollers on seaming machines that range from small hand-operated types to large automatic units capable of producing hundreds of seams per minute in the mass production of products. Figure 19-33 shows several of the more common seam designs.

Flanging.   *Flanges* can be rolled on sheet metal in essentially the same manner as seaming. In many cases, however, the forming of flanges and seams is a drawing operation, since the bending must be performed along a curved axis.

Straightening.   *Straightening* or *flattening* has as its objective the opposite of bending and is often performed before cold-forming operations to assure the use of flat or straight material. Two different techniques are quite common. *Roll*

SECTION OF STOCK

1st PASS   2nd PASS   3rd PASS   4th PASS

5th PASS   6th PASS   7th PASS   8th PASS

ROLL-FORMED SHAPE

FIGURE 19-32 Eight-roll sequence for forming a box channel. (Courtesy Aluminum Association, New York.)

FIGURE 19-33 Various types of seams used on sheet metal.

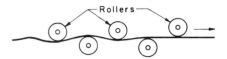

FIGURE 19-34 Method of straightening rod or sheet by passing it through a set of straightening rolls. For rods, another set of rolls is used to provide straightening in the transverse direction.

straightening or *roller leveling*, illustrated in Figure 19-34, subjects the material to a series of reverse bends. The rod, sheet, or wire is passed through a series of rolls with progressively decreased offsets from a straight line. As the metal is bent back and forth, it is stressed slightly beyond its elastic limit, thereby replacing any permanent set with a flat or straight profile.

Sheet may also be straightened by a process called *stretcher leveling*. As shown in Figure 19-35, the sheets are gripped mechanically and stretched slightly beyond the elastic limit to produce uniform length and the desired flatness.

Shearing

Shearing is the mechanical cutting of materials without the formation of chips or the use of burning or melting. When the two cutting blades are straight, the process is called shearing. When the blade geometry is curved, as in the edges of punches and dies, the processes have special names, such as *blanking, piercing, notching, shaving,* and *trimming.* All of these are essentially shearing operations, however.

A simple type of shearing process is illustrated in Figure 19-36. As the punch (or upper blade) descends against the workpiece, the metal is first deformed plastically into the die (or over the lower blade). Because the clearance between

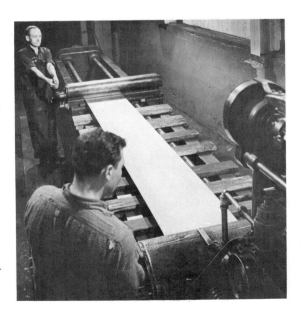

FIGURE 19-35 Straightening sheets of brass by stretching. This process is sometimes called stretcher leveling. (Courtesy Scovill Manufacturing Company.)

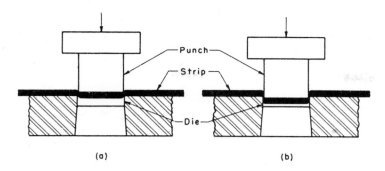

FIGURE 19-36 A simple type of shearing process: blanking with a punch and die.

the two tools is only 5% to 10% of the thickness of the metal being cut, the deformation is highly localized. The punch penetrates into the metal, the material flows into the die, and the opposite surface bulges slightly. When the penetration reaches about 15% to 60% of the thickness of the metal, the amount depending upon the material ductility and strength, the applied stress exceeds the shear strength and the metal suddenly shears or ruptures through the remainder of its thickness. These two stages of the shearing process, deformation and fracture, can often be seen on the edges of sheared parts and are clearly visible in Figure 19-37.

Because of the normal nonhomogeneities in a metal and the possibility of nonuniform clearance between the shear blades, the final shearing does not occur uniformly. Fracture and tearing start at the weakest points and proceed progressively and intermittently to the next stronger locations. The result is a rough and ragged sheared edge.

If the punch and die (or shearing blades) have proper clearance and are maintained in good condition, sheared edges may be produced that are sufficiently smooth to use without further finishing. The quality of the sheared edge can be further improved if the strip stock is clamped firmly against the die from above,

FIGURE 19-37 Conventionally sheared surface showing the two distinct regions of deformation and fracture. (Courtesy American Feintool, Inc.)

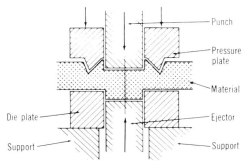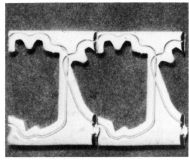

FIGURE 19-38 (Left) Method of obtaining a smooth edge in shearing by using a shaped pressure plate to put the metal into localized compression and an ejector descending in unison with the punch. (Courtesy *Metal Progress.*) (Right) Stock skeleton after shearing, showing the compression indentation. (Courtesy Clark Metal Products Company.)

the punch and die are maintained with proper clearance and alignment, and the movement of the piece through the die is restrained by an opposing plunger or rubber die cushion applying pressure from below the workpiece. These measures cause the shearing to take place uniformly around the edge rather than randomly at the weakest points.

If a compressive pressure is superimposed on the shearing process, fracture is suppressed and the relative amount of smooth edge (produced by deformation) is increased. Above a certain threshold pressure, a 100% smooth edge can be obtained. Figure 19-38 schematically depicts one production process that has been designed to produce smooth and square sheared edges, a process known as *fineblanking*. A V-shaped protrusion is incorporated into the holddown or pressure plate at a location slightly external to the contour of the cut. As the holddown is pressed against the material to be sheared, the protrusion penetrates the surface and places the region to be cut into a localized state of compression. Matching upper and lower punches then grip the material and descend in unison to produce a smooth and square sheared edge, as shown in Figure 19-39. A reduced clearance between the punch and the die is used in fineblanking, and a triple action press is generally required. Fineblanked parts are usually less than ¼ in. (6 mm) in thickness and are characterized by complex peripheries and close tolerances. The production of holes, slots, bends, and semipierced projections is often incorporated into the fineblanking operation.

A similar technique using the same principle is illustrated in Figure 19-40.

FIGURE 19-39 Fine-blanked surface of the same component as shown in Figure 19-37. (Courtesy American Feintool, Inc.)

Incoming bar stock is pressed against the closed end of a feed hole, putting the stock in a state of compression. A transverse punch then shears the material into burr-free slugs for further processing.

When sheets of metal are to be sheared along a straight line, *squaring shears* are frequently used, such as the one illustrated in Figure 19-41. As the upper ram descends, a clamping bar or set of clamping fingers presses the sheet of metal against the machine table to hold it firmly in position. The moving blade then comes down across the fixed blade and shears the metal. On larger shears, the moving blade is usually set at an angle or "rocks" as it descends, making the cut in a progressive fashion from one end of the material to the other. This action significantly reduces the amount of cutting force required, although the total energy expended is the same. (If work is force times distance, the low force-long stroke operation can be used in place of the high force-short stroke approach.)

Slitting is the shearing process used to cut rolls of sheet metal into several rolls of narrower width. Here the shearing blades are in the form of circumferential mating grooves on cylindrical rolls, the raised ribs of one roll matching the recessed grooves on the other. The process is continuous and can be performed rapidly and economically. Moreover, since the distance between adjacent sets of shearing edges is fixed, the width of the slit strips is very accurate and constant, more consistent than that obtained from alternative cutting processes.

Piercing and Blanking.   *Piercing* and *blanking* are shearing operations wherein the shear blades take the form of closed, curved lines on the edges of a punch and die. They both involve the same basic cutting action, with the difference being primarily one of definition. Figure 19-42 shows that in blanking, the piece being punched out is the desired workpiece and any major burrs or undesirable features should be left on the remaining strip. In piercing, the piece being punched out is the scrap and the remainder of the strip becomes the

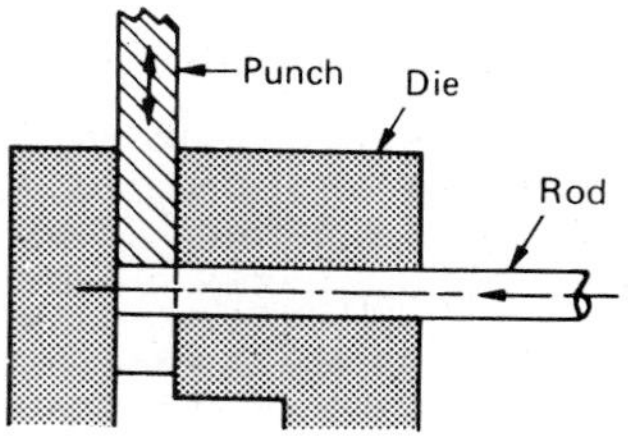

FIGURE 19-40   Method of smooth shearing rod by putting it into compression during shearing.

FIGURE 19-41   A 10-foot (3-meter) power shear for ¼-inch (6.5 mm) steel. (Courtesy Cincinnati Incorporated.)

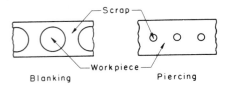

FIGURE 19-42 Schematic showing the difference between piercing and blanking.

desired workpiece. Piercing and blanking are usually done on some form of mechanical press.

Several variations of piercing and blanking have been developed and have come to acquire specific names. *Lancing* is a piercing operation that may take the form of a slit in the metal or an actual hole, as shown in Figure 19-43. The purpose of lancing is to permit the adjacent metal to flow more readily in subsequent forming operations. In the case illustrated, the lancing makes it easier to form the grooves, which were shaped before the ashtray was blanked from the strip stock and drawn to final shape.

Perforating consists of piercing a large number of closely spaced holes.

Notching is essentially the same as piercing, except that the edge of the sheet of metal forms a portion of the piece that is punched out. It is used to form notches of any desired shape along the edge of a sheet.

Nibbling is a variation of notching in which a special machine makes a series of overlapping notches, each further into the sheet of metal. As can be seen in Figure 19-44, the already sheared edge often forms one end of the notch being cut. By repetitive shearing, a desired shape can be cut from sheets of metal up to ¼ in. (6.5 mm) thick. If the operation is started from a punched or drilled hole, interior segments of a sheet can be cut free.

Shaving is a finishing operation in which a very small amount of metal is sheared away from the edge of a blanked part. Its primary use is to obtain greater dimensional accuracy, but it may also be employed to produce a square or smoother edge. Because only a small amount of metal is to be removed, the punches and dies must be made with very little clearance. Parts such as small gears can be shaved to tolerances of 0.001 in. (0.025 mm) after blanking.

A *cutoff* operation is one in which a stamping is removed from a strip of stock by means of a punch and die. Frequently, an irregular shape is imparted to the cutoff operation such that it forms part or all of the desired periphery on the workpiece.

Dinking is a modified shearing operation that is used to blank shapes from low-strength materials, such as rubber, fiber, or cloth. The procedure is illustrated in Figure 19-45. The shank of a die is struck with a hammer or mallet or the die is operated by some form of mechanical press.

Tools and Dies for Piercing and Blanking. The basic components of piercing and blanking die sets, as shown in Figure 19-46, include a *punch*, a

FIGURE 19-43 Steps in making a simple ashtray. (Left to right) Piercing and lancing, blanking, and the final formed ashtray.

FIGURE 19-44   Shearing operations being performed on a nibbling machine. (Courtesy Tech-Pacific.)

die, and a *stripper plate.* Theoretically, the punch should be of dimensions such that it would just fit within the die with a uniform clearance approaching zero. On its downward stroke, it should not enter the die but should stop as its base aligns with the top surface of the die. In general practice, however, the clearance is generally between 5% to 7% of the stock thickness and the punch enters the die by a small amount.

It is also essential that the punch and die be in proper alignment such that a uniform clearance is maintained around the entire periphery. The die is usually attached to the bolster plate of the press, which, in turn, is attached to the main press frame. The punch is attached to the movable ram, moving in and out of the die with each stroke of the press. Frequently, the punch and die are mounted on a *punch holder* and *die shoe* to form a die set as shown in Figure 19-47. The holder and shoe are permanently aligned and guided by two or more guide pins. Once a punch and die are correctly aligned and fastened to the die set, the unit can be inserted directly into the press without having to set and check the tool alignment. This can significantly reduce the amount of production time lost during tool change. Moreover, when a given punch and die are no longer needed, they can be removed and new tools attached to the shoe and holder assembly.

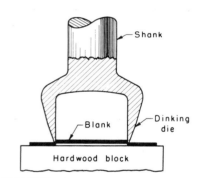

FIGURE 19-45   The dinking process.

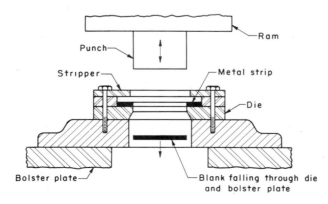

FIGURE 19-46   The basic components of piercing and blanking dies.

FIGURE 19-47 Typical die set having two guideposts. (Courtesy Danly Machine Specialties, Inc.)

The punch face may be ground normal to the axis of motion, or it may have a slight angle, a feature known as *shear*. Shear reduces the maximum cutting force since the entire periphery of a cut is no longer made at the same time. The amount of shear is limited to the thickness of the metal and is frequently less. A shear of one half the thickness reduces the cutting load by about 25%. Shear equal to the full thickness can reduce the load by 50%. Since the cutting distance is proportionately increased, the total energy required to make the cut remains unchanged.

A *stripper plate* is often attached above the die to prevent the material from riding upward with the punch on its upward travel. It should be located a sufficient distance above the die to permit easy movement and positioning of the metal sheet. The hole in the stripper plate should be larger than the punch so that there will be no restricting friction between them.

In most cases, the punch holder is attached directly to the ram of the press such that motion of the ram both raises and lowers the punch. On small die sets, however, it is common practice for springs to be incorporated into the die set to provide the upward motion. The press ram simply contacts the top of the punch holder and forces it downward. As the ram retracts, the springs raise the punch. This form of construction makes the die set fully self-contained so that it can be inserted and removed from a press quite rapidly, thereby reducing setup time.

A wide variety of standardized, self-contained die sets have been developed. Known as *subpress dies,* they can be assembled and combined on the bed of a press to pierce or blank large parts that would otherwise require large and costly die sets. Figure 19-48 shows one such setup.

Punches and dies are usually made of nondeforming, or air-hardening, tool steel so that they can be hardened after machining without danger of warpage. Beyond a depth of about ⅛ in. (3 mm) from the face, the die is usually provided

FIGURE 19-48 Typical setup for piercing and blanking using self-contained subpress tool units. (Courtesy Strippit Division, Houdaille Industries, Inc.)

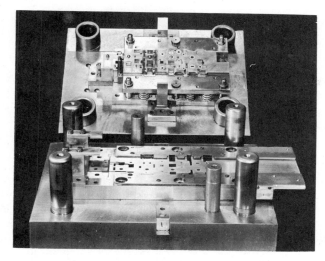

FIGURE 19-49 Progressive piercing, forming, and cutoff die set built up mostly from standard components. The part produced is shown at bottom. (Courtesy Oak Manufacturing Company.)

with an angular clearance or back relief (see Figure 19-46) to reduce friction between the part and the die and to permit the part to fall freely from the die after being sheared. The ⅛ in. land provides adequate strength and sufficient metal so that the die can be resharpened by grinding a few thousandths of an inch from its face.

Dies may be made in a single piece, which results in a basically simple but costly die, or made in sections that are fastened together on the die shoe. The latter procedure usually simplifies making the die and repairing it in case of damage, since only the damaged section must be replaced. Many standardized punch and die components are available from which complex die sets can often be constructed at greatly reduced cost. Figure 19-49 shows such a die set. Substantial savings can often be achieved if designers would determine what standard die components are available and modify the design of parts that are to be pierced and blanked so that these components can be utilized. An added advantage is that when the die set is no longer needed the components can be removed and used to construct tooling for another product.

Another technique that can be used to cut metal up to about ½ inch (13 mm) thick is to use "steel-rule" dies, as illustrated in Figure 19-50. Here the die is made from hardened, relatively thin steel strips that are mounted and supported on edge. A second die plate is used in place of the upper punch and holds the steel strips. Neoprene rubber pads are inserted between the strips to replace the stripper plate. As the ram ascends, the rubber expands to push the blank free of the "steel-rule" cavity. Such die sets are usually much less expensive to construct than solid dies and are therefore more attractive for small quantities of parts.

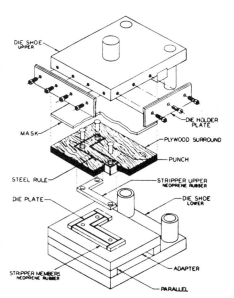

FIGURE 19-50 Construction of a steel-rule die set. (Courtesy J.J. Raphael.)

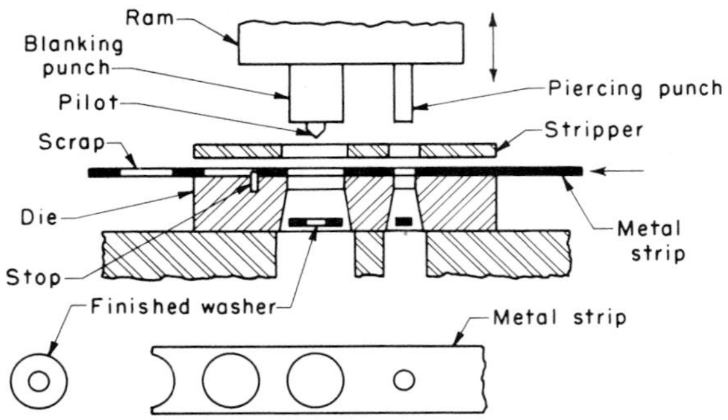

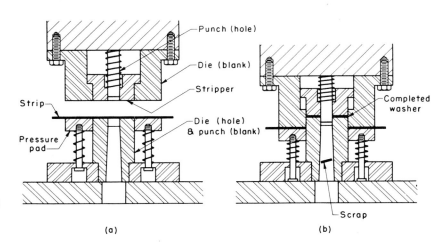

FIGURE 19-51   Progressive piercing and blanking die for making a washer.

Only simple piercing or blanking can be performed with the type of punch and die illustrated in Figure 19-46. Unfortunately, many parts require both piercing and blanking, and it is desirable to perform both with a single cycle of a press. This can be accomplished with the types of dies shown in Figures 19-51 and 19-52. For simplicity, their operation will be discussed in terms of manufacturing simple, flat washers from a continuous strip of metal.

The *progressive die set,* as depicted in Figure 19-51, is the simpler of the two types. Basically, it consists of two or more sets of punches and dies mounted in tandem. The strip stock is fed into the first die, where a hole is pierced as the ram descends on its first stroke. When the ram rises, the stock is moved into position under the blanking punch. Positioning is accomplished automatically by a stop mechanism that engages a previously punched hole. As the ram descends on the second stroke, the pilot on the bottom of the blanking punch enters the hole that was pierced on the previous stroke to ensure accurate alignment. Further descent of the punch blanks the completed washer from the strip and at the same time the first punch pierces the hole for the next washer. Thus, a finished part is completed with each stroke of the press.

Progressive dies can be used for many combinations of piercing, blanking, forming, lancing, and drawing, as shown in Figures 19-43 and 19-53. They

FIGURE 19-52   Method for making a simple washer in a compound piercing and blanking die. Part is blanked (a) and subsequently pierced (b). The blanking punch contains the die for piercing.

have the added advantage of being fairly simple to construct and are economical to maintain and repair, since a defective punch or die does not necessitate the replacement of the entire set. However, if accurate alignment of the various operations is required, they are not quite as effective as compound dies.

In compound dies, such as the one shown in Figure 19-52, piercing and blanking or other combinations of operations occur sequentially during a single stroke of the ram while the strip is held firmly in position. While dies of this type are more accurate, they are usually more expensive to construct and more subject to breakage.

Numerically controlled, turret-type punch presses can be used to punch large numbers of holes in sheet metal components. In these, as many as 60 separate punches and dies are contained in coordinated turrets that can be quickly rotated into operation as desired. Between punches, the workpiece is rapidly positioned through X–Y movements of the worktable.

Designing for Piercing and Drawing.   The construction, operation, and maintenance of piercing and blanking dies can be greatly facilitated if designers of the parts to be fabricated keep a few simple rules in mind:

1. If solid dies are to be used, blank corners should be radiused wherever possible. Square corners are preferred for sectional dies.
2. The width of any projection or slot should be at least 1½ times the metal thickness and never less than ³⁄₃₂ inch (2.5 mm)
3. Diameters of pierced holes should not be less than the thickness of the metal or a minimum of 0.025 in. (0.65 mm). Smaller holes can be made, but with difficulty.
4. The minimum distance between holes or between a hole and the edge of the stock should be at least equal to the metal thickness.
5. Tolerances should be kept as great as possible. Tolerances below about ±0.003 inch (±0.08 mm) mean that shaving will be required.
6. The pattern of parts on the strip should be arranged to minimize scrap.

Drawing

Cold drawing is a term that can refer to two somewhat different operations. If the starting stock is in the form of sheet metal, cold drawing is the forming of parts wherein plastic flow occurs over a curved axis. This is one of the most important of all cold-working operations because a wide range of shapes, from small cups to large automobile body panels, can be drawn in a few seconds each. Cold drawing is similar to hot drawing, but the higher deformation forces,

FIGURE 19-54 Cold drawing of rods on a chain-driven multiple-die draw bench. (Courtesy Scovill Manufacturing Company.) Inset shows a schematic of the operation that produces finite lengths of straight rod or tube. (Courtesy Wean United, Inc.)

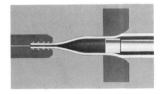

FIGURE 19-55 Method of cold-drawing smaller tubing from larger tubing. The die sets the outer dimension while the mandrel sizes the inner diameter. (Courtesy Copperweld Tubing Group.)

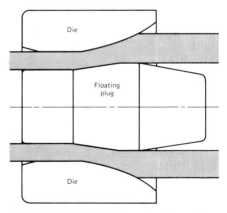

FIGURE 19-56 Tube drawing with a floating plug.

thinner metal, limited ductility, and closer dimensional tolerances create some distinctive problems.

If the starting stock is wire, rod, or tubing, cold drawing refers to the process of reducing the cross section of the material by pulling it through a die, a tensile equivalent to extrusion.

Rod, Bar, and Tube Drawing. One of the simplest cold-drawing operations is *rod* or *bar drawing,* illustrated in Figure 19-54. One end of a bar is reduced or pointed, inserted through a die of somewhat smaller cross section than the original bar, gripped and pulled in tension, drawing the remainder of the bar through the die. The bars reduce in section, elongate, and strain-harden. Since the product cannot be conveniently bent or coiled, straight-pull draw benches are employed on finite length stock. The reduction in area per pass is usually restricted to 20% to 50% to avoid tensile fracture. Therefore, multiple draws through a series of dies may be required to produce a desired final size. Intermediate annealing may be required to restore ductility and enable further working.

Tube drawing, depicted in Figure 19-55, is used to produce high quality tubing where the product can benefit from the smoother surfaces, thinner walls, more accurate dimensions, and strain hardening that accompany cold forming. Internal mandrels are used to control the inside diameter of tubes from about ½ in. (12.5 mm) to 10 in. (250 mm) in diameter. Heavy-walled tubes and those less than ½ in. (12.5 mm) in diameter are often drawn without a mandrel in a process known as *tube sinking.* Precise control of the inner diameter is sacrificed for process simplicity and the ability to draw extremely long lengths of product. If a controlled internal diameter must be produced in a long-length product, the manufacturer may replace the internal mandrel with a "floating plug" as shown in Figure 19-56. If properly designed to correspond to the material, reduction, and friction conditions, the plug will assume a stable position within the die. The plug then sizes the internal diameter while the external die shapes the outside of the tube.

Products with shaped cross section can also be produced by the drawing of

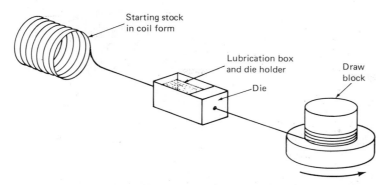

FIGURE 19-57 Schematic of wire drawing on a continuous draw block. The rotating motor on the draw block provides a continuous pull on the incoming wire.

bar stock. Precise dimensions can be produced with little or no material waste or finish machining. Steels, copper alloys, and aluminum alloys have all been formed in this manner.

Wire Drawing. *Wire drawing* is essentially the same as bar drawing except that it involves smaller-diameter material that can be coiled and is generally performed as a continuous operation on draw blocks, such as illustrated in Figure 19-57. The drawing of steel wire begins when large coils of hot-rolled material approximately ⅜ in. (9 mm) in diameter are first descaled by either mechanical flexing or acid pickling and rinsing. The cleaned product is then further processed by immersion in a lime bath or other procedure to provide neutralization of the remaining acid, corrosion protection, and a carrier for surface lubrication. The drawing of nonferrous wire begins with essentially the same starting geometry but undergoes different preparation treatment.

After the initial preparation, one end of the coil is pointed, fed through a die, and gripped, and the drawing process begins. Wire dies generally have the configuration shown in Figure 19-58 and are usually made of tungsten carbide or polycrystalline, man-made diamond. Single-crystal natural diamonds are frequently used for the drawing of very fine wire. Lubrication boxes often precede the individual dies to assure uniform coating of the material.

Small-diameter wire is usually drawn on tandem machines of the type shown in Figure 19-59. These contain from 3 to 12 dies, each held in water-cooled die blocks. The reduction in each die is controlled at each station to avoid bunch-up or excessive tension in the wire. If extensive deformation is required, intermediate anneals must be performed between the various stages of drawing.

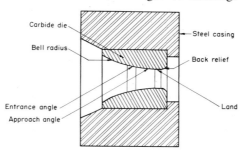

FIGURE 19-58 Cross section through a typical carbide wire drawing die showing the characteristic regions of the contour.

FIGURE 19-59 Wire being drawn on a machine having eight synchronized draw blocks. (Courtesy Vaughn Machinery Company.)

Wires of a variety of tempers can be produced by controlling the placement of the last anneal in the process cycle. Dead-soft wire is annealed in controlled-atmosphere furnaces after the final draw.

Spinning. *Spinning* is a rather fascinating cold-forming operation in which a rotating disk of sheet metal is drawn over a male form. Localized pressure is applied through a simple, round-ended wooden or metal tool or small roller, which traverses the entire surface of the part. Figure 19-60 shows the progressive forming of a spinning operation.

The form, or *chuck*, is rotated on a rapidly rotating spindle, often the drive section of a simple lathe. The disk of metal is centered and held against the small end of the form by a follower attached to the tailstock of the lathe. As the disc and form rotate, the operator applies localized pressure against the metal, causing it to flow against the form, as shown in Figure 19-61. Because the final diameter of the part is less than that of the initial disc, leading to a reduction in circumference, the operator must stretch the metal radially a corresponding amount to prevent circumferential buckling. (In essence, this is a shrink forming operation, as will be explained later, and is illustrated in Figure 19-69.) Considerable skill is required on the part of the operator. While there is usually some thinning of the metal, it can usually be kept to a minimum by a skilled worker.

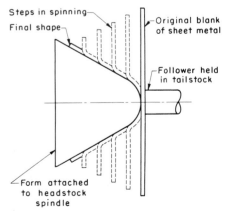

FIGURE 19-60 Progressive stages in the forming of sheet metal by spinning.

FIGURE 19-61   Two stages in spinning a metal reflector. (Courtesy Spincraft, Inc.)

Inasmuch as the metal is not pulled across it under pressure, the form block can often be made of hardwood or even plastic. Its major requirement is to present a smooth surface, since any irregularities will transfer to the finished part. Thus, the tooling cost for spinning is extremely low, making the process quite attractive for small production quantities. The process, however, is also used for many high-production applications, such as the manufacture of lamp reflectors, cooking utensils, bowls, and the bells of some musical instruments. If large quantities of identical parts are to be spun, a metal form block is generally used.

Spinning is best suited to shapes that can be withdrawn directly from a one-piece chuck, but other shapes can be spun over multipiece or offset chucks. Complex shapes have also been made on form blocks of frozen water, which is simply melted out after spinning.

FIGURE 19-62   Machine-spinning of heavy-gage sheet metal. (Courtesy Cincinnati Milacron, Inc.)

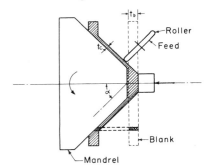

FIGURE 19-63 The basic shear-forming process.

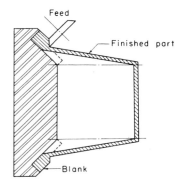

FIGURE 19-64 Forming a conical part by reverse shear forming.

To retain the advantages of spinning without requiring high operator skill, spinning machines such as that in Figure 19-62 have been developed. The action of the spinning rollers is controlled automatically and is programmed for each particular part. Each successive part then undergoes the same deformation sequence, resulting in better consistency and fewer rejects. Spinning machines also enable the spinning of thicker metal.

Shear Forming. *Shear forming* or *flow turning,* illustrated in Figure 19-63, is a simplified version of the spinning process in which each element of the blank maintains its distance from the axis of rotation. Because of this fact, the metal flow is entirely in shear and no radial stretch has to take place to compensate for the circumferential shrinkage. As a consequence, conical, hemispherical, cylindrical, and similar shapes can often be formed more readily by this process than by normal spinning, and shear forming lends itself to a high degree of automation. Wall thickness will vary with the angle of the region. As shown in Figure 19-63, the relationship between the wall thickness of the product and that of the blank for conical parts is $t_c = t_b \sin \alpha$. If α is less than $30°$, it may be necessary to complete the forming in two stages with an intermediate anneal between. Reductions in wall thickness as high as 8:1 are possible, but the limit is usually set at about 5:1 or an 80% reduction.

Conical shapes are usually shear-formed by the *direct process* depicted in Figure 19-63. They can also be formed, however, by the *reverse process* illustrated in Figure 19-64. By varying the direction of feed of the forming rollers in the reverse process, it is possible to form concave or convex parts without having to have a matching mandrel.

Shear-formed cylinders can be made by either the direct or indirect processes, as shown in Figure 19-65. The reverse process has the advantage that a cylinder can be formed that is longer than the mandrel. When the flow is induced by a roller and a mandrel, the process is also known as roll extrusion.

When long, thin-walled cylinders are desired, a "flo-reform" process, illustrated in Figure 19-66, is used. The process is a combination of shear forming and conventional spinning. The first steps involve no change in the diameter of the workpiece (shear forming), while the latter stages involve a pulling-in (spinning).

Stretch Forming. *Stretch forming,* the principle of which is illustrated in Figure 19-67, was developed by the aircraft industry to enable sheet metal parts,

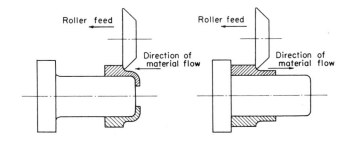

FIGURE 19-65 Shear forming a cylinder by the direct process (left) and the reverse process (right).

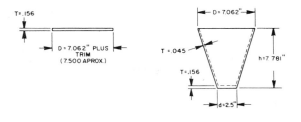

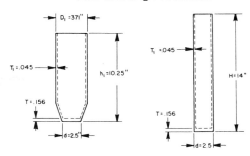

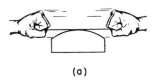

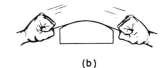

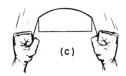

FIGURE 19-66 Flo-reform process for forming long cylinders in four steps. (Courtesy Lodge and Shipley Company.)

particularly large ones, to be formed economically in small quantities. As shown, only a single male form block is required. The sheet of metal is gripped by two or more sets of jaws that stretch it and wrap it around the form block as the latter rises upward. Various combinations of stretching, wrapping, and motion of the block or grips are employed, depending on the shape of the part. Figure 19-68 shows a large part being formed by this process.

Through proper control of the stretching, most or all of the compressive stresses that often accompany bending or forming are eliminated. Consequently, there is very little springback, and the workpiece conforms very closely to the shape of the form block. Because the form block is almost completely in compression, it can be made of wood, Kirksite (a zinc alloy that can be cast at low temperature), or even plastic.

Stretch forming, or *stretch–wrap forming* as it is often called, is frequently used to form cowlings, wing tips, scoops, and other large aircraft panels from aluminum and stainless steel. Automobile and truck panels can be stretch-formed from low-carbon steel.

If mating male and female dies are used to shape the metal while it is being stretched, the process is known as *stretch–draw forming*.

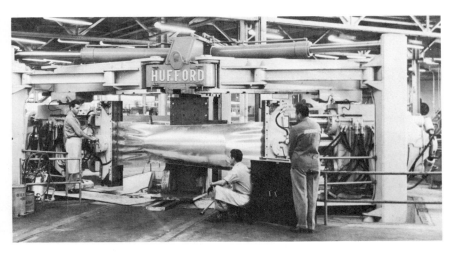

FIGURE 19-67 Schematic representation of the motions and steps involved in stretch–wrap forming.

FIGURE 19-68 Large aircraft part being formed by stretch–wrap forming. (Courtesy Hufford Machine Works.)

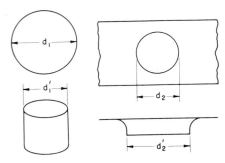

Shrink forming   Stretch forming

FIGURE 19-69 The two basic types of flow during drawing.

Deep Drawing or Shell Drawing.   The drawing of closed cylindrical or rectangular containers, or similar shapes the depth of which is greater than the smallest dimension of the opening, is one of the most important and widely used manufacturing processes. Because the process had its earliest uses in manufacturing artillery shells and cartridge cases, it is sometimes called *shell drawing*. When the depth of the product is less than its diameter (or smallest surface dimension), the process is considered to be *shallow drawing*. When the depth is greater than the diameter, it is known as *deep drawing*.

There are basically two types of drawing, as illustrated in Figure 19-69, and both involve multiaxial or curved-axis bending. In *shrink forming,* there is circumferential compression resulting from the decrease in diameter from d_1 to d_2, and the metal may compensate by thickening. Since the metal is thin, however, an alternative response is to relieve the circumferential compression by buckling or wrinkling. A frequent means to suppress this is to compress the sheet between the die and a hold-down surface during forming. By resisting the formation of wrinkles, the hold-down forces the metal to compensate by increasing in either radial length or thickness.

In *stretch forming* there is a circumferential stretching that occurs and a corresponding thinning of the metal. This thinning can be the cause of tearing during forming, or service failures, such as premature failure of the part under corrosive conditions.

Many drawn products contain regions of both shrink and stretch forming. Effective design and control can be a complex problem. In general, it should be recognized that regions with large amounts of shrink or stretch will probably cause some degree of difficulty.

Computer techniques have done much to assist the design of complex-shaped drawn products, but their success is still not a guaranteed certainty. Successful drawing is still a combination of science, experience, empirical data, and experimentation. The body of existing knowledge is quite substantial, however, and is being used with outstanding results.

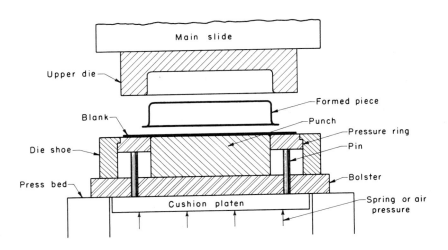

FIGURE 19-70 The method of deep drawing on a single-action press.

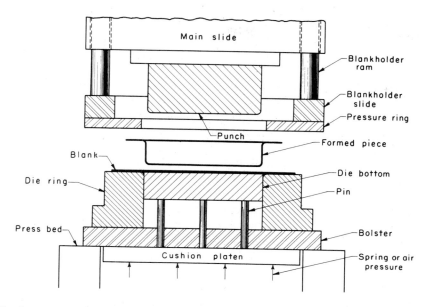

FIGURE 19-71   Drawing dies for use on a double-action press.

To draw a part without wrinkling, tearing, or undesirable variation in thickness, the flow of metal must be controlled. This is usually accomplished with some form of pressure ring or pad. In single-action presses, as shown in Figure 19-70, where there is only one movement of the slide, spring or air pressure is used to control the flow of metal between the upper die and pressure ring. When double-action presses are available, with two or more independent rams, as illustrated in Figure 19-71, the force applied to the pressure ring is controlled independently of the position of the main slide. This permits the pressure to be varied as needed during the drawing operation. For this reason, double-action presses are usually specified for drawing more complex parts, while single-action presses can be used for the simpler types of operations.

Because the flow of metal is generally not uniform throughout the workpiece, many drawn parts have to be trimmed after forming to remove excess or undesired material. Figure 19-72 shows a typical part before and after trimming. Obviously, trimming adds to the production cost because it converts some of

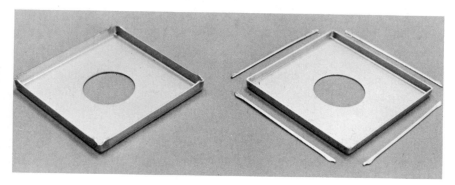

FIGURE 19-72   Pierced, blanked, and formed part before and after trimming.

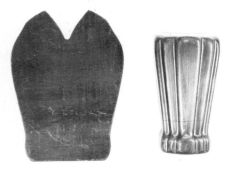

FIGURE 19-73 Irregularly shaped blank and finished drawn stove leg. No trimming is necessary.

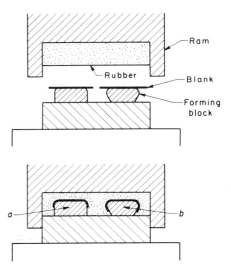

FIGURE 19-74 The Guerin process for forming sheet metal.

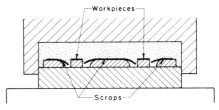

FIGURE 19-75 Method of blanking sheet metal using the Guerin process.

the starting material to scrap and adds another operation to the manufacturing process. In addition, most trimming operations require either accurate positioning and manipulation of the workpiece or the construction of a separate trimming die.

In many cases, trimming after drawing can be avoided by using a blank that has been cut to a special complex shape before drawing. Figure 19-73 illustrates this procedure, which also requires close control of lubrication and hold-down pressure. The use of a preshaped blank requires a special-shaped blanking die; but since the starting material for most drawing operations is first formed by blanking, the added cost may be minimal. The choice of method is often dependent upon the complexity of the part and the quantity to be produced.

When the mating tools for blanking and drawing are made of special steels, they are generally quite expensive. Consequently, numerous methods have been developed to permit these operations to be performed with less expensive tooling or to enable more extensive deformation to be performed with fewer sets of dies or fewer in-process anneals. While most of these methods have limitations as to the complexity of shapes or types of metal that can be formed, they are certainly very useful when they can be used. Some of these methods will now be presented.

Forming with Rubber Tooling or Fluid Pressure. Several methods of forming employ rubber or fluid pressure to produce the desired deformation, thereby eliminating either the male or female member of the die set. The *Guerin process,* depicted in Figure 19-74, utilizes the phenomenon that rubber of the proper consistency, *when totally confined,* acts as a fluid and will transmit pressure uniformly in all directions. Blanks of sheet metal are placed on top of form blocks, which can be made of wood, Bakelite, polyurethane, epoxy, or low-melting-point zinc alloys (such as Kirksite). The upper ram contains a pad of rubber 8 to 10 in. (200 to 250 mm) thick encased in a steel container. As the ram descends, the rubber pad is confined and transmits force to the metal, causing it to bend to the desired shape. Since no female die is used and inexpensive form blocks replace the male die, tooling cost is quite low. There are no mating tools to align, process flexibility is quite high (different shapes can even be formed at the same time), wear on the material and tooling is low, and the surface quality of the workpiece is easily maintained. When reentrant sections are formed, as in part *b* of Figure 19-74, it must be possible to slide the parts lengthwise from the form blocks or to disassemble the form blocks from within the product.

The Guerin process was developed in the aircraft industry, where the production of small numbers of duplicate parts clearly favors the low cost of tooling. It can be used on aluminum sheet up to 1/8 in. (3 mm) thick and on 1/16 in. (1.6 mm) stainless steel. Magnesium can also be formed if it is heated and shaped over heated form blocks.

Most forming done with the Guerin process is multiple-axis bending, but some shallow drawing can also be performed. The process can also be used for piercing and blanking thin gages of aluminum, as illustrated in Figure 19-75.

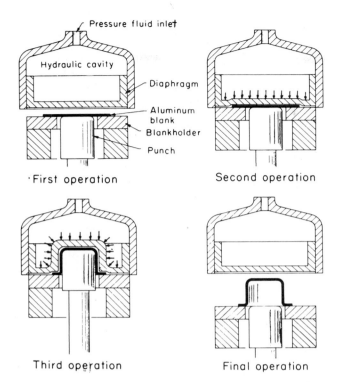

First operation

Second operation

Third operation

Final operation

FIGURE 19-76 Hydroform process showing (1) blank in place, no pressure in the cavity; (2) press closed and cavity pressurized; (3) ram advanced with cavity maintaining fluid pressure; and (4) pressure released and ram retracted. (Courtesy Aluminum Association, New York.)

For this application, the blanking blocks, shaped the same as the desired workpiece, have a face, or edge, made of hardened steel. Round-edge supporting blocks are positioned a short distance from the blanking blocks to support the scrap skeleton and permit the metal to bend away from the shearing edges.

In the *Hydroform process* (also known as "rubber bag forming" or Flexforming), the rubber pad is replaced by a flexible diaphragm backed by controlled hydraulic pressures up to about 20,000 psi (140 MPa). Deeper parts can be formed with truly uniform fluid pressure, as illustrated in Figure 19-76. Cycle times are on the order of 1–3 minutes, but the reduced tool costs make the process attractive for prototype manufacturing and low volume production (up to about 10,000 identical parts).

In *bulging*, oil or rubber is used to transmit the pressure necessary to expand a metal blank or tube outward against a female mold or die. Complex, multiple-piece male members are not required and split female dies facilitate easy removal of the product. For simple shapes, rubber tooling can be inserted, compressed, and then easily removed, as shown in Figure 19-77. For more complicated shapes, fluid pressure is used to form the bulge. Complex equipment is now required to provide the necessary seals yet enable the easy insertion and removal of material required for mass production.

Drawing on a Drop Hammer. When small quantities of shallow-drawn parts are required, they can often be made most economically through the use

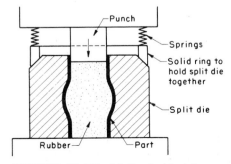

FIGURE 19-77 Method of bulging tubes with rubber tooling.

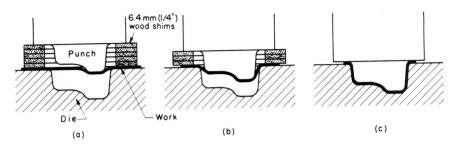

FIGURE 19-78 Method of deep drawing on a drop hammer using Kirksite dies and a book of shims.

of Kirksite (zinc alloy) dies and a drop hammer. The dies can be directly cast, eliminating the expense of costly machined steel dies, and they can be remelted and cast into other shapes when no longer needed. Often, a stack of shims, made of thin plywood, is placed on top of the sheet of metal to permit the ram to descend only partway, as shown in Figure 19-78. These shims further act as a pressure pad to restrict and control the flow of metal and thus inhibit wrinkling. One shim is withdrawn after each stroke of the ram, permitting the ram to fall further and deepen the draw. Wrinkles that form can be hammered out between strokes with a hand mallet.

Drawing on a drop hammer is considerably cruder than drawing with steel dies, but it is often the most economical method for small quantities. It is most suitable for aluminum alloys, but thin carbon and stainless sheets can also be drawn successfully.

High-Energy-Rate Forming.   A number of methods have been developed to form metals through the release and application of large amounts of energy in a very short time interval. These processes are known as the *high-energy-rate forming processes,* often abbreviated as HERF. Many metals tend to deform more readily under the ultrarapid load application rates used in these processes, a phenomenon apparently related to the relative rates of load application and the movement of dislocations through the metal. As a consequence, HERF makes it possible to form large workpieces and difficult to form metals with less expensive equipment and tooling than would otherwise be required.

Another advantage of HERF is that there is less difficulty from springback. This is probably associated with two factors: (1) high compressive stresses are

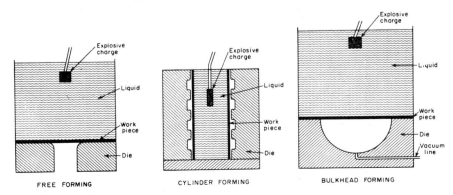

FIGURE 19-79 Three methods of high-energy-rate forming with explosive charges. (Courtesy Materials Engineering.)

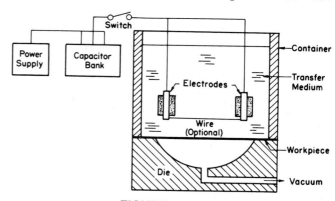

FIGURE 19-81 Components in the spark-discharge method of forming.

FIGURE 19-80 Explosively formed elliptical dome 10 feet (3 meters) in diameter being removed from a forming die. (Courtesy NASA.)

set up in the metal when it is forced against the die, and (2) some slight elastic deformation of the die occurs under the high pressure. The latter results in the workpiece being overformed and makes it appear that no springback has occurred.

The high energy-release rates are obtained by five methods: (1) underwater explosions, (2) underwater spark discharge (electrohydraulic techniques), (3) pneumatic-mechanical means, (4) internal combustion of gaseous mixtures, and (5) the use of rapidly formed magnetic fields (electromagnetic techniques).

Figure 19-79 illustrates three commonly used procedures involving the use of *explosive charge:* free forming, cylinder forming, and bulkhead forming. Although these procedures can be used for a wide range of products, they are particularly suitable for parts of thick material as in the 10-foot (3-meter) diameter elliptical dome of Figure 19-80. Only a tank of water in the ground is required, with about 6 feet (2 meters) of water above the workpiece. The female die can be made of inexpensive materials such as wood, plastic, or Kirksite.

The *spark-discharge method,* shown in Figure 19-81, uses the energy of an electrical discharge to shape the metal. Electrical energy is stored in large capacitor banks and then released in a controlled discharge, either between two electrodes or across an exploding bridgewire. High-energy shockwaves propagate through a pressure-transmitting medium and deform the metal. The initiating wire can be preshaped, and shockwave reflectors can be used to adapt the process to a variety of components. The space between the blank and the die is usually evacuated before the discharge occurs to prevent the possibility of puckering due to entrapped air.

The spark-discharge methods are most often used for bulging operations in small parts, such as those shown in Figure 19-82, but parts up to 50 in. (1.3

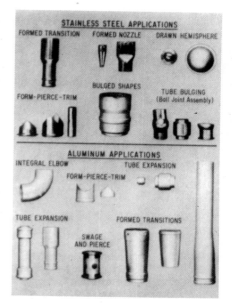

FIGURE 19-82 Some typical operations performed by spark-discharge techniques. (Courtesy General Dynamics, Fort Worth Division.)

FIGURE 19-83 Typical parts manufactured by the pneumatic-mechanical HERF technique. (Courtesy Interstate Drop Forge Inc.)

meters) in diameter have been formed. Compared to explosive forming, the discharge techniques are much easier and safer, use smaller tanks, and do not have to be performed in remote areas.

To make the HERF techniques adaptable for mass production within a plant, the *pneumatic-mechanical* and *internal combustion* presses were developed. In the pneumatic-mechanical presses, one portion of the forming die is attached to the stationary bolster of the press bed and the other to a piston rod. Low-pressure gas acts on the entire area of the piston, holding it up against a seal. A small area on the other side of the piston is exposed to high-pressure gas. When the pressure of the high-pressure gas is raised above a certain value, the seal is broken. The entire area of the piston is then exposed to high pressure and the piston is driven downward very rapidly, bringing the dies into contact. Figure 19-83 shows some typical parts made by this method.

The internal combustion presses operate on the same principle as an automobile engine. A gaseous mixture is exploded within a cylinder, causing a piston to be driven downward very rapidly. The upper segment of the forming die is attached to the bottom of the piston. Internal combustion presses can produce ram velocities up to 50 feet per second (15 meters per second) and operate at up to 60 strokes per minute. Either single or repeated blows can be used to form the part.

Electromagnetic forming is based on the principle that the electromagnetic field of an induced current always opposes the electromagnetic field of the inducing current. A capacitor is discharged through a coiled conductor that is within or surrounding a cylinder or adjacent to a flat sheet of metal that is to be formed. The discharge induces a current in the workpiece, causing it to be repelled from the coil and formed against a die or mating workpiece. The process is very rapid and is used primarily in expanding or contracting tubing to various shapes, or permanently assembling component parts. Coining, forming and swaging have also been performed with electromagnetic forces. Figure 19-84 shows some typical electromagnetically formed components.

Ironing. *Ironing* is the name given to the process of thinning the walls of a drawn cylinder by passing it between a punch and die where the separation is

FIGURE 19-84 Two parts shaped by electromagnetic forming. Both are approximately 5 in. (125 mm) in diameter. (Courtesy General Atomic Division of General Dynamics Corporation.)

less than the original wall thickness. The walls are elongated and thinned while the base remains unchanged, as shown in Figure 19-85. The most common example of an ironed product is the thin-walled beverage can.

Embossing.   *Embossing,* shown in Figure 19-86, is a method of producing lettering or other designs in thin sheet metal. Basically, it is a very shallow drawing operation with the depth of the draw being from one to three times the thickness of the metal.

Superplastic Sheet Forming.   The commercial development of superplastic metals (materials with more than 100% elongation at selected strain rates and temperatures, and possibly as high as 2000% to 3000%) has made possible the economical production of large, complex-shaped products, often in very limited production quantities. Deep or complex shapes can now be made as one-piece, single-operation pressings, rather than multistep conventional pressings or multipiece assemblies. The elevated temperatures required to promote superplasticity can reduce the required forces such that many of the forming techniques are adaptations of processes used to form thermoplastics (to be discussed in Chapter 20), and the tooling is relatively inexpensive. Because of the low forming pressures, form blocks can often be used in place of die sets. Thermoforming, in which a vacuum or pneumatic pressure forces the sheet to conform to a heated male or female die, is quite popular. Blow forming, vacuum forming, deep drawing, and combined superplastic forming and diffusion bonding are other possibilities. Precision is excellent and fine details or textures can be reproduced. Products have a fine, uniform grain size.

The major disadvantage to superplastic forming is the inherently low forming rate necessary to maintain superplastic behavior. Typical cycle times may be from 5 minutes to as much as 40 minutes per part, rather than the several seconds that is typical of conventional presswork. As some compensation, many of the applications of superplastic forming serve to eliminate a considerable number of subsequent assembly (joining) operations. The weight of products can be reduced, there are fewer fastener holes to initiate fatigue cracks, tooling costs are reduced, and there is a shorter production lead time.

Design Aids for Sheet Metal Forming.   A majority of sheet metal failures are the result of either excessive thinning or fracture, both being the result of excessive deformation in a given region. A fast and economical means of evaluating the severity of deformation in formed parts is the use of *strain analysis* and *forming limit diagrams.* A pattern or grid, such as the one in Figure 19-87, is placed on the surface of the sheet by scribing, printing, or etching. The sheet is then deformed and the distorted pattern is measured and evaluated. Regions where the area has expanded are locations of sheet thinning and possible failure. Areas that have contracted have undergone sheet thickening and may be sites of possible buckling or wrinkles. The major strain (strain in the direction of the largest radius or diameter) and the minor strain (strain 90° from the major) can be measured for various locations, and the values can be plotted on a forming-

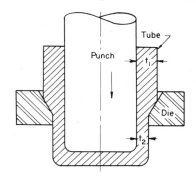

FIGURE 19-85   Schematic of the ironing process.

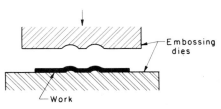

FIGURE 19-86   Embossing.

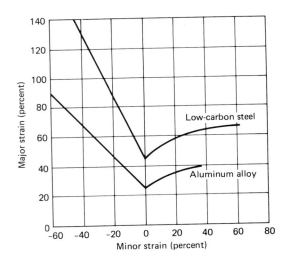

FIGURE 19-87 (Left) Typical pattern for sheet metal deformation analysis. (Right) Forming limit diagram used to determine whether a metal can be shaped without risk of fracture.

limit diagram, as shown in Figure 19-87. If both major and minor strains are positive (right-hand side of the diagram), the deformation is known as stretching and the sheet metal will decrease in thickness. If the minor strain is negative, the stretching in the direction of the major strain is partially or wholly compensated for by a contraction in the minor strain direction. The deformation is known as drawing and the thickness may decrease, increase, or stay the same depending upon the relative magnitudes of the two strains.

Strains that fall above the forming limit line indicate areas of probable fracture. These undesirable deformation characteristics should then be changed by means such as lubrication modifications, die design changes, or variation in the clamping or hold-down pressure. A strain analysis can also be used to orient parts relative to the prior rolling direction, evaluate various lubricants as to their effectiveness, and assist in the design of dies for complex-shaped products.

Alternatives to Deformation Methods

Electroforming. The *electroforming* process produces metal parts by electroplating metal (as discussed in Chapter 40) onto an accurately made mandrel that has the inverse contour and surface finish of the finished product. When metal has been deposited to the desired thickness, the workpiece is separated from the mandrel, the exact method of separation depending upon the mandrel material and shape.

The mandrels are made from a variety of materials, including plastics, glass, Pyrex, and various metals (most commonly aluminum or stainless steels). If the material is nonconductive, a conductive coating must first be applied in order to perform the electroplating. The electroformed products are typically made from nickel, iron, copper, or silver, in thicknesses up to ⅝ in. (16 mm).

A wide variety of parts and shapes can be made by electroforming, the principal limitation being the need to remove the product from the mandrel. The electroforming process often imparts distinct properties to the metals. Dimensional tolerances are good, often up to 0.0001 in. (0.0025 mm) and surface finishes of 2 microinches (0.05 μm) can be obtained quite readily if the mandrel is sufficiently smooth. For most applications, the wall thickness is selected to provide adequate strength. However, where additional strength is required, the electroformed part can be reinforced with other materials, often cast directly into the shell.

Electroforming is particularly useful for high-cost metals and for low production quantities (because of the low cost of tooling). Some care must be exercised to minimize residual stresses, which may range from 2000 to 5000 psi, or higher if plating temperatures and current densities are not properly controlled.

Plasma Spray Forming.   Parts can also be formed by injecting powder into an intensely hot stream of ionized gas (a plasma with temperatures of up to 20,000°F or 11,000°C) where it melts and is propelled onto a shaped mandrel or form. The droplets then undergo rapid solidification to form a fine-grain-size product. Layer after layer is deposited to build up a desired size, shape, and thickness. The mandrel or form is then removed by machining or chemical etching. Most applications involve the fabrication of specialized products from difficult-to-form or ultra high melting-point materials.

Classification of Presses.   Many types of presses have been developed to perform the variety of cold-working operations. When selecting a press for a given application, consideration should be given to the capacity required, the question of whether the power source should be *hydraulic* or *mechanical,* and the method of transmitting the power to the ram (the type of drive). Table 19.2 lists the various types of presses available and Figure 19-88 illustrates the more important drive mechanisms. In general, mechanical drives provide faster action and more positive displacement control, whereas greater forces and more flexibility can be obtained with hydraulic drives.

Presses

TABLE 19.2   Classification of the Various Drive Mechanisms of Commercial Presses

Foot	Mechanical	Hydraulic
Kick presses	Crank	Single-slide
	Single	Multiple-slide
	Double	
	Eccentric	
	Cam	
	Knuckle joint	
	Toggle	
	Screw	
	Rack and pinion	

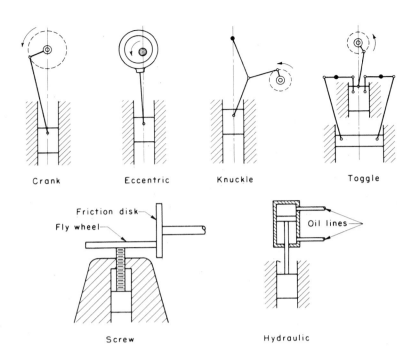

FIGURE 19-88 Schematic representation of the various drive mechanisms for presses.

Foot-operated presses, commonly called *kick-presses,* are only used for very light work. *Crank-driven presses* are the most common type because of their simplicity. They are used for most piercing and blanking operations and for simple drawing. Double-crank presses provide a method of actuating the blank holders or operating multiple-action dies. *Eccentric* or *cam drives* are used when only a short ram stroke is required. Cam action can provide a dwell at the bottom of the stroke and is often the preferred method of actuating the blank holder in deep-drawing processes. *Knuckle-joint drives* provide a very high mechanical advantage along with fast action. They are often used in coining, sizing, and Guerin forming. *Toggle mechanisms* are used principally in drawing presses to actuate the blank holder.

Hydraulic presses are available in many varieties and sizes. Because almost unlimited capacities can be provided, most large drawing presses are of this type. By using several hydraulic cylinders, programmed loads can be applied to the ram and a separate force and timing can be used on the blank holder. Although most hydraulic presses tend to be slow, some are available that can provide up to 600 strokes per minute in a high-speed blanking operation.

Types of Press Frame. Another matter of importance in selecting a press is determining the type of frame, because this factor often imposes limitations on the size and type of work that can be done. Table 19.3 presents a classification of basic frame types.

Arch-frame presses, having their frames in the shape of an arch, are seldom used today, except with screw drives for coining operations.

TABLE 19.3 Classification of Presses According to Type of Frame

Arch	Gap	Straight-Sided
Crank or eccentric	Foot	Many variations, but all
Percussion	Bench	with straight-sided frames
	Vertical	
	Inclinable	
	Inclinable	
	Open back	
	Horn	
	Turret	

Gap-frame presses, having a *C*-shaped frame as shown in Figure 19-89, provide good clearance for the dies and permit large stock to be fed into the press. They are made in a wide variety of sizes, with capacities ranging from 1 to 250 tons (10 to 2200 kN). As shown in Figure 19-89, they can often be inclined to permit the completed work to drop out of the back side of the press. Also included on that press is a sliding bolster, a common feature that permits a second die to be set up on the press while another is in operation. Die change-over then requires only a few minutes to unclamp the punch segment of one die set from the ram, move the second die set into position, clamp the upper segment of the new set to the press ram, and start a new operation.

Bench presses are small, inclinable, gap-frame presses, 1 to 8 tons (10 to 70 kN) in capacity, that are made to be mounted on a bench. *Open-back presses* are gap-frame presses that are not inclinable. *Horn presses* are upright, open-back presses that have a heavy cylindrical shaft or "horn" in place of the usual bed. This permits curved or cylindrical workpieces to be placed over the horn for such operations as seaming, punching, riveting, and so forth, as illustrated in Figure 19-90. On some presses, both a horn and a bed are provided, with provision for swinging the horn aside when it is not needed.

Turret presses use a modified gap-frame construction but have upper and lower turrets that carry a number of punches and dies. The two turrets are geared together so that any desired set can be quickly rotated into position. Another type uses a single turret on which subpress die sets are mounted.

Straight-sided presses are available in a wide variety of sizes and designs. This type of frame is used on most hydraulic presses and for larger-sized and specialized mechanically driven presses. A typical example is shown in Figure 19-91.

Special Types of Presses. A number of special presses are available, designed to perform specific types of operations. Two interesting examples are the *transfer press* and the *multislide press*. The transfer press, illustrated in Figure 19-92, has a single long slide with the provision for mounting a number of die sets side by side. Each die set can be adjusted individually. Stock is then

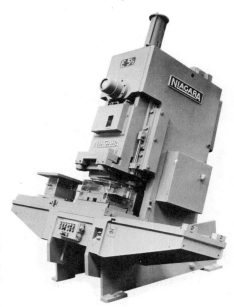

FIGURE 19-89 Inclinable gap-frame press with sliding bolster to accommodate two die sets for rapid change of tooling. (Courtesy Niagara Machine & Tool Works.)

FIGURE 19-90 Making a seam on a horn press. (Courtesy Niagara Machine & Tool Works.)

FIGURE 19-91 A 200-ton (1800-kN) straight-side press. (Courtesy Rousselle Corporation.)

FIGURE 19-92 Typical transfer-type press. Inset shows parts at several of the eight stations. (Courtesy Verson Allsteel Press Company.)

FIGURE 19-93 Transfer mechanism used in transfer presses. (Courtesy Verson Allsteel Press Company.)

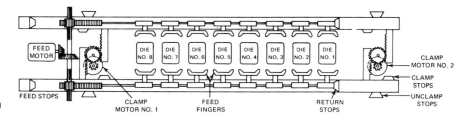

FIGURE 19-94 Possible operations which may be performed during the production of stamped and drawn parts on a transfer press. (Courtesy U.S. Baird Corporation, Stratford, Conn.)

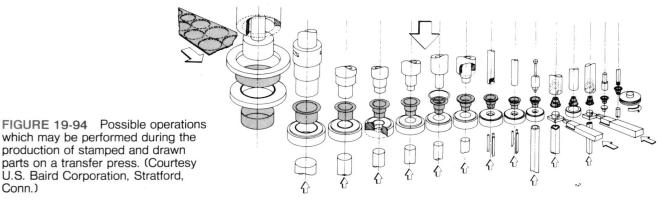

FIGURE 19-95   Multislide machine viewed from the "back" side. (Courtesy U.S. Baird Corporation, Stratford, Conn.)

fed through the various stations. After the completion of each stroke of the ram, the part is automatically and progressively transferred to the next station by the mechanism shown in Figure 19-93. Such presses have high production rates, up to 1500 parts per hour, and are also very flexible. Figure 19-94 illustrates the variety of operations that can be sequentially performed in the production of stamped and drawn sheet metal parts.

The multislide machine, shown in Figures 19-95 and 19-96, contains a series of bending or forming slides that move horizontally. As indicated on the sche-

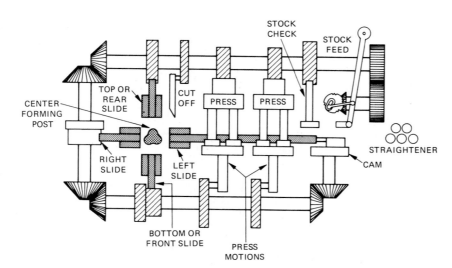

FIGURE 19-96   Schematic of the operating mechanism of a multislide machine. (Courtesy U.S. Baird Corporation, Stratford, Conn.)

FIGURE 19-97 Example of the piercing, blanking, and forming of a part on a multislide machine. (Courtesy U.S. Baird Corporation, Stratford, Conn.)

matic diagram, these slides are driven by cams on four shafts that are located on four sides of the machine and are driven by miter gears. Coiled strip stock is fed into the machine and is progressively pierced, notched, bent, and cut off at the various slide stations, as indicated in Figure 19-97. Strip stock up to about 3 in. (75 mm) wide and 3/32 in. (2.5 mm) thick and wires up to about 1/8 in. (3 mm) in diameter are commonly processed. Products such as hinges, links, clips, and razor blades can be formed at very high rates on these machines.

Press Feeding Devices. Although hand feeding is still used in many press operations, operator safety and the desire to increase productivity have motivated a strong shift to feeding by some form of mechanical device. Dial-feed mechanisms, such as the one shown in Figure 19-98, enable the operator to place the workpieces into the front holes of the dial. The dial then indexes with each stroke of the press to move the parts into proper position between the punch and die.

When continuous strip is used it can be fed automatically by press-driven roll feeds mounted on the side of the press. Lightweight parts can be fed by suction-cup mechanisms, vibratory-bed feeders, and similar devices. Robots are being used in increasing numbers to place parts into presses and remove them after forming. In general, the technology and equipment exist to replace manual feeding in almost all cases if such a transition is truly desired.

FIGURE 19-98 Dial feed device being used on a punch press. (Courtesy E.W. Bliss Company.)

Review Questions

1. What are some of the attractive features of cold-working processes?
2. What is the purpose of subjecting sheet metal to a skin-rolling pass?
3. Why should the cold rolling of shaped product from bar not be considered an attractive means of producing a small amount of prototype product?
4. Describe how the swaging process can impart different sizes and shapes to an interior cavity and the exterior of a product.
5. How can cold forging be used to reduce material waste substantially?
6. What are some of the geometric limitations of cold forging?
7. What are some of the attractive properties or characteristics of cold forging?
8. If a product contains a large-diameter head and a small-diameter shank, how can the processes of cold extrusion and cold heading be combined to save metal?
9. What are some of the unique capabilities and attractive features of hydrostatic extrusion?
10. What are some of the possible feedstock forms for continuous extrusion processes?
11. What type or geometry of product is typically produced by roll extrusion?
12. How might the use of sizing operations actually provide economic savings?
13. What types of rivets can be used when there is access to only one side of a joint?
14. What type of geometry is necessary to join two materials by staking?
15. Why might hobbing be an attractive means of producing a number of identical die cavities?
16. How does the peening operation improve the fracture resistance of a product?
17. What is burnishing?
18. Why does a metal usually thin in the region of a bend?
19. What is springback, and why is it a concern during bending?
20. What types of operations can be performed with press brakes?
21. Why is it generally desirable to specify the largest possible bend radius?
22. If a right-angle bend is to be made in a cold-rolled sheet, should it be made with the bend lying along or perpendicular to the direction of previous rolling?
23. From a manufacturing viewpoint, why is it desirable for all bends in a product (or component) to have the same radius?
24. What is the primary difference between draw bending and compression bending?
25. What types or geometries of products are generally produced by cold roll forming?
26. What are the two primary methods of straightening or flattening rod or sheet?
27. Why are sheared or blanked edges generally not smooth?
28. What simple measures can often be employed to improve the quality of a sheared edge?
29. Why are fineblanking presses noticeably more complex than those used in conventional blanking?
30. What are the similarities and differences between piercing and blanking?
31. What are some types of blanking or piercing operations that have come to acquire specific names?
32. Why is it essential that a blanking punch and die be in proper alignment?
33. What is the major benefit of mounting punches and dies on independent die sets?
34. What is the purpose of grinding an angle, or shear, on the surface of a punch?
35. What economical benefit can often be attained by making dies in sections?
36. What is the major benefit of assembling a complex die set from standard subpress dies?
37. What is a progressive die set?
38. How do compound dies differ from progressive dies?
39. What two distinctly different processes are often referred to as cold drawing?
40. What is the difference between tube drawing and tube sinking?
41. Why are rods generally drawn on draw benches, while wire is drawn on draw block machines?
42. What is the primary benefit of tube drawing with a floating plug?
43. How does the operator in a spinning process compensate for the decreasing circumference of the material?
44. Why is the tooling cost for a spinning operation relatively low?
45. How is shear forming different from spinning?
46. For what types of products would stretch forming be an attractive means of manufacture?
47. What types of defects or failures are generally associated with a shrink-forming mode of deformation? With the stretch-forming mode?
48. What is the function of the pressure-ring or hold-down in a deep-drawing operation?
49. How does the Guerin process reduce the cost of tooling? The hydroform process?

50. What are some of the basic methods that have been used to achieve the high energy-release rates needed in the HERF processes?
51. Why is springback rather minimal in high-energy-rate forming?
52. What types of operations are generally performed by electromagnetic forming?
53. What is the major limitation of the superplastic forming of sheet metal?
54. What benefits can be obtained through superplastic forming?
55. How can strain analysis be used to determine locations of possible defects or failure in sheet metal components?
56. Describe two alternative methods of producing thin, complex-shape products without employing metal deformation techniques.
57. What are the two primary types of power (or drives) used in commercial presses and the attractive features of each?
58. For what types of products would a horn press be useful?
59. Describe how multiple operations are preformed simultaneously in a transfer press.

Fabrication of Plastics, Ceramics, and Composites

Plastics, ceramics, and composites are considerably different from metals, and as a result, the principles of material selection and design and the processes of fabrication are often different. Some similarities certainly exist. For example, the specific material is still selected for its ability to provide the required properties or characteristics. The fabrication process is still selected for its ability to produce the desired shape economically and practically. However, certain differences should be noted. Plastics, ceramics, and composites are often used closer to their design limits than metals. Many fabrication processes can take the raw material to a finished product in a single operation, as opposed to the sequence of activities generally used on metals. Large, complex shapes can be formed as a single unit, often eliminating the need for multipart assembly operations. When necessary, however, the joining and fastening operations are quite different from those used on metals. Since these materials can offer integral color and the processes used to manufacture the shapes may provide the required surface finish and precision directly, finishing operations can often be eliminated. Altering of the final dimensions and finish by machining-type operations is often difficult and costly.

The properties of the resultant product are strongly dependent on the process used to fabricate the shape. Thus, providing the desired characteristics involves not just the selection of an appropriate material, but the determination of a material and processing system that, together, will attain the necessary properties, precision, and finish.

Not only is there a large number of plastic or polymeric materials available, there is also a large number of different processes by which a plastic can be converted into a desired shape or product. The specific method or methods used depend to a large extent on the nature of the polymer, specifically whether it is

Fabrication of Plastics

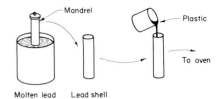

FIGURE 20-1   Steps in the casting of plastic parts.

thermoplastic or thermosetting. Thermoplastic polymers can be heated to temperatures at or near the melting temperature so that the material becomes either a formable solid or a liquid. The polymer is then cast, injected into a mold, or forced through a die to produce the desired shape. Thermosetting polymers allow far fewer options because, once the polymerization has occurred, the framework structure is established and no further deformation can occur. Thus, the polymerization reaction and the shape-forming process must be accomplished simultaneously.

Casting, blow molding, hot-compression molding, transfer molding, cold molding, injection molding, extrusion, and *vacuum forming* are all used extensively to shape polymers. Each process has certain advantages and limitations relating to part design, compatible materials, and production cost. Since it is desirable to convert the material into a finished product in a single operation, it is important that the selected material–process system be capable of producing both the shape and the desired properties.

Casting.   Casting is the simplest of the polymer-forming processes because no fillers are used and no pressure is required as the liquid plastic is poured into the mold. Several varieties of the process have been developed. Plate glass molds can be used to cast thick plastic sheets. Moving belts of stainless steel can contain and cool the resins to produce thinner sheets. Molten plastic can be spun against a mold wall (centrifugal casting) to produce hollow shapes. Small products can be cast from the thermoset resins by a technique that involves a strippable metal mold, as depicted in Figure 20-1. A steel pattern is dipped into molten lead, removed, and allowed to cool. The thin lead sheath that remains when the pattern is removed becomes the mold for the plastic resin. The resin is then cured in the mold by heating for long times at low temperatures (150–200°F, or 65–95°C). On removal from the curing oven, the lead sheaths are stripped from the finished product and reused. Other types of plastics can be cured at room temperature.

Because cast plastics contain no fillers, they have a distinctive lustrous appearance. The process is inexpensive because no expensive tooling or equipment is involved. Typical products include sheets and small objects, such as jewelry, ornamental shapes, gears, and lenses. Precision can be quite high.

Blow Molding.   In blow molding, a hot tube of polymer, called a *parison,* is introduced into a mold; the mold closes; and air or gas pressure expands the tube outward against the walls of the mold. The mold then opens to release the product, which is a hollow shape, such as a bottle or container. Figure 20-2 depicts a schematic of this process.

In a modification of this process, a sheet of heated plastic is placed between two cavities, the bottom one being the shape of the desired product. Both cavities are pressurized to 300–600 pounds per square inch (psi) with a nonreactive gas, such as argon. When the pressure in the lower segment is vented, the material is "blown" to final shape.

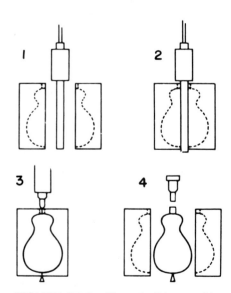

FIGURE 20-2   Steps in "blow molding" plastic parts. (1) a tube of heated plastic is placed in the open mold, (2) the mold closes over the extruded tube, (3) air forces the tube against the sides of the mold, and (4) the mold opens to release the product.

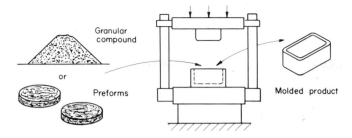

FIGURE 20-3 Schematic representation of the production of plastic parts by the hot-compression molding process.

Hot-Compression Molding or Compression Molding. In hot-compression molding, illustrated schematically in Figure 20-3, solid granules or preformed tablets of the raw, unpolymerized plastic are loaded into the cavity of an open, heated mold. A heated plunger (the male member of the mold), usually attached to the upper portion of a press, descends to close the mold and creates sufficient pressure to force the plastic, as it melts and becomes fluid, into all portions of the cavity. After continued exposure to heat has allowed the material to set (also called *cure* or *polymerize*), the mold is opened, and the part is removed. Multiple cavities can be contained within a single mold to produce more than one part in a single pressing. The process is simple, but it is usually restricted to the thermosetting polymers. Alternate heating and cooling of the mold would be required for the thermoplastic materials, and this is not economical.

In order to contain the material within the die and to enable pressure to be built up, some type of seal is required on the hot-compression molding dies. Three types of seals are common, and each is illustrated in Figure 20-4. In the *flash-type,* excess material is provided that is squeezed out of the cavity during the final stages of mold closing. The resulting *flash* must be removed from the product and usually requires an additional manufacturing operation. However, this type of mold is relatively inexpensive to make and it is not necessary to control precisely the amount of raw material introduced into the cavity. The dimensions of the product in the direction of mold opening and the product density will tend to vary.

In the *straight plunger mold,* no material can escape from the cavity. Close dimensional control requires that the raw material be precisely measured. Pressure control can then be used to control the density of the product. The *landed-plunger mold* is the most common, providing both good pressure and a definite cutoff to ensure accurate dimensions.

Transfer Molding. In order to prevent the turbulence and uneven flow that often results from the nonuniform, high pressures of hot-compression molding, transfer molding is sometimes used. A double-chamber mold is used to form the thermosetting polymer. Unpolymerized raw material is placed within the plunger cavity, where it is heated until molten. The plunger then descends,

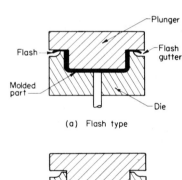

(a) Flash type

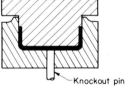

(b) Straight plunger type

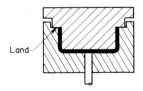

(c) Landed plunger type

FIGURE 20-4 Three types of molds used in the hot-compression molding process.

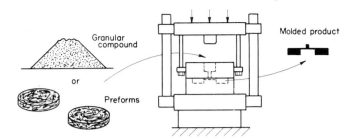

FIGURE 20-5 Schematic diagram
of the transfer molding process.

forcing the molten plastic into the adjoining die cavities. Because the material
enters the cavities as a liquid, there is little pressure until the cavity is completely
filled. Thin sections, excellent detail, and good tolerances and finish are all
characteristics of the process. In addition, transfer molding is quite useful when
inserts are to be incorporated into the product. These are positioned in the cavity
and are maintained in place as the liquid resin is introduced around them. Figure
20-5 provides an illustration of this process, which combines elements of both
compression and injection molding and permits some of the advantages of injection molding to be used with the thermosetting polymers.

Cold Molding.   In the cold molding process, depicted in Figure 20-6, the raw
thermosetting material is pressed to shape while cold and is then removed from
the mold and cured in an oven. The process is quite economical, but the resulting
products do not have very good surface finish or dimensional precision.

Injection Molding.   Illustrated schematically in Figure 20-7, injection molding is used to produce more thermoplastic products than any other process.
Granules of raw material are fed by gravity from a hopper into a pressure
chamber ahead of a plunger. As the plunger advances, the plastic is forced
through a heated chamber, where it is preheated. From the preheating segment,
it is forced through the *torpedo* section, where it is melted and superheated to
400–600°F (200–300°C). It then leaves the torpedo section through a nozzle
that seats against the mold and allows the molten plastic to enter the closed-die
cavities through suitable gates and runners. The die remains cool, so the plastic
solidifies almost as soon as the mold is filled. To ensure proper filling of the
cavity, the material must be forced into the mold rapidly under considerable

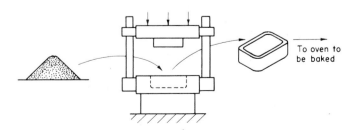

FIGURE 20-6 Schematic diagram
showing the steps in cold molding.

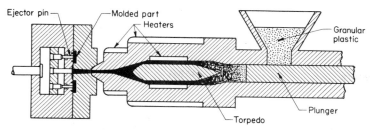

Ejector pin — Molded part
Heaters
Granular plastic
Plunger
Torpedo

FIGURE 20-7 Schematic diagram of the injection-molding process.

pressure, typically 5–20 ksi (35–140 MPa). Premature solidification would cause defective products. While the mold is being opened, the part ejected, and the mold reclosed, the material for the next part is being heated in the torpedo. The complete molding cycle takes typically 1–30 sec and is very similar to the die casting of molten metals. Figure 20-8 shows a typical injection-molding machine.

Because thermosetting plastics must be held at elevated temperature and pressure for sufficient time to permit curing, the injection molding process must be modified for this type of polymer. In the *jet molding* process, shown schematically in Figure 20-9, the polymer is preheated in the feed chamber to about 200°F (95°C) and then is further heated to the temperature of polymerization as it passes through the nozzle. Additional time in the heated mold completes the curing process. Care must be exercised to prevent the material in the nozzle from curing during this time and clogging material flow. Water cooling is intro-

FIGURE 20-8 A plastic injection-molding machine. Inset shows a plastic part being removed from the mold. (Courtesy Pennwalt Chemicals Corporation.)

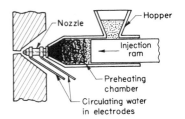

FIGURE 20-9 Jet molding process for the injection molding of thermosetting plastics.

duced to the nozzle area as soon as the cavity is nearly filled. The water cools the material in this region and retards the hardening reaction. Because of the relatively long cycle time, little injection molding of thermosets is performed. The properties can often compete with die-cast metals, provided the lower rigidity of the polymer is not objectionable.

Reaction Injection Molding. In reaction injection molding (RIM), two monomers are mixed together as they are injected into a mold. A chemical reaction takes place between the two components to form a plastic polymer. Heating of the material is not required. Rather, the chemical reaction gives off energy that must be removed. Production rates are determined primarily by the curing time of the polymer, which is often less than 1 min.

Extrusion. Plastic products with long, uniform cross sections are readily produced by the *extrusion* process depicted in Figure 20-10. Thermoplastic granules are fed through a hopper into the barrel chamber of a screw extruder. The rotating screw conveys the material through a preheating section, where it is heated, homogenized, and compressed, and then forces it through a heated die and onto a conveyor belt. As the plastic passes onto the belt, it is cooled by air or water sprays, which harden it sufficiently to preserve the shape imparted to it by the die. It continues to cool as it passes along the belt and is then either cut into lengths or coiled, depending on whether the material is rigid or flexible and what the customer desires. The process is continuous and provides a cheap and rapid method of molding. Common production shapes include solid forms (including those with reentrant angles), tubes, pipes, and even coated wires and cables. If the emerging tube is blown up by air pressure, allowed to cool, and then rolled, the product can be a double layer of sheet or film.

Vacuum Forming. Vacuum forming or thermoforming is used extensively to form shapes from thermoplastic materials. As illustrated in Figure 20-11, a sheet of plastic is placed over a die or pattern and heated until it becomes soft. A vacuum is then drawn between the sheet and the form so that the plastic draws down to the desired shape. The material is then cooled, the vacuum dropped, and the part removed from the cavity. The entire cycle requires only a few minutes, is quite economical, and is used for a wide variety of products ranging from panels for light fixtures to pages of Braille text for the blind. Moving tools (dies) can also be used to assist in the forming.

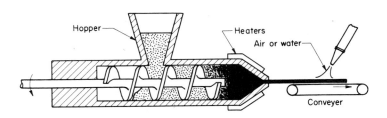

FIGURE 20-10 Extrusion process for producing plastic parts.

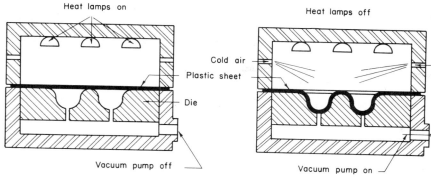

Heat lamps off

Cold air

Plastic sheet

Die

Vacuum pump off

Vacuum pump on

FIGURE 20-11   Method of molding plastic shapes by heat and vacuum (thermoforming).

Foamed Plastics.   Foamed plastics have become an important and widely used form of product. A foaming agent is mixed with the plastic resin and releases gas or volatilizes when the combination is heated during molding. The resulting products have very low densities, ranging from 2 to 40 lb/ft³ (32–641 kg/m³). "Open-cell" foams have interconnected bubbles that permit the permeability of gas or liquid. "Closed-cell" foams have the property of being gas- or liquid-tight.

Both rigid and flexible foamed plastics have been produced from both thermoplastic and thermosetting material. The rigid type is useful for structural applications, for packaging and shipping containers, as patterns for the full-mold casting process (see Chapter 14), and for providing rigidity to thin-skinned metal components, such as aircraft fins and stabilizers, by producing the foam directly between the skins. Flexible foams are used primarily for cushioning.

It is also possible to produce products that have a solid outer skin and a rigid foam core, but the process generally requires the connection of multiple injection devices to a single mold. Figure 20-12 shows the cross section of a plastic gear made with a rigid foamed core and a solid outer skin.

FIGURE 20-12   Plastic gear having a solid outer skin and a rigid foam core. (Courtesy *American Machinist.*)

Other Plastic-Forming Processes. In the *calendering* process, molten plastic is poured onto a set of rolls with a small opening. As the rolls turn, they squeeze out a thin sheet or film of polymer, which is then cooled for subsequent use.

Conventional *drawing* and *rolling* can be performed to produce fibers or to change the shapes of plastic extrusions. In addition to changing the product dimensions, these processes can also serve to induce crystallization or to produce a preferred orientation of the chains in thermoplastic polymers.

Filaments, fibers, and yarns can be produced by *spinning*, a modified form of extrusion. Molten thermoplastic polymer is forced through a die containing many small holes. Where multistrand yarns or cables are desired, the dies can rotate or spin to produce the twists and wraps.

Designing for Fabrication. The primary design consideration in any manufacturing operation is that the part satisfy its functional requirements. A material must be selected that has the required properties with respect to tensile strength, impact strength, and so on. Plastics offer a number of unique material properties, including light weight, corrosion resistance, good thermal and electrical insulation, and the possibility of integral color. At the same time, consideration should be given to possible limitations in terms of elevated temperatures of operation, poor dimensional stability, or the deterioration of properties with age. Likewise, each of the various fabrication processes has unique advantages and limitations. Designers and fabricators should work together to select the details that best use the unique features of the fabrication process to be employed.

Since molding-type processes are the most common means of plastic fabrication, they are used here as an example. In every casting process in which a fluid or semifluid material is introduced into a mold cavity and allowed to solidify to a desired shape, certain basic problems are encountered. First, the proper amount of material must be introduced and caused to flow in such a way as to completely fill the cavity. Second, any entrapped material within the cavity, usually air, must be removed. Third, any shrinkage that occurs during solidification and/or cooling must be taken into account. Heat transfer should be considered to ensure controlled cooling and solidification within the mold. Finally, it must be possible to remove the part from the mold after it has solidified. Part appearance, the resultant engineering properties, and the economics of production are all closely dependent on good design of the molding process.

In all cast products, it is important to provide adequate fillets between adjacent sections to ensure a smooth flow of the plastic into all sections of the mold and also to eliminate stress concentrations at sharp interior corners. Such fillets also make the mold less expensive to produce and lessen the danger of mold breakage where thin, delicate mold sections are encountered. It is even desirable to round exterior edges slightly where permissible. A radius of 0.010–0.015 in. (0.25–0.40 mm) is scarcely noticeable but will do much to prevent an edge from chipping. Where plastics are used for electrical applications, sharp corners should be avoided because they increase voltage gradients and may lead to product failure.

Wall section thickness is also very important, since the curing time is determined by the thickest section. If possible, sections should be kept nearly uniform in thickness, since nonuniform wall thickness can lead to serious warpage and dimensional control problems. As a general rule, one should use the minimum wall thickness that will provide satisfactory end-use performance of the part. Minimum wall thickness is determined primarily by the size of the part and, to some extent, by the type of plastic used:

Minimum recommended	0.025 in. (0.65 mm)
Small parts	0.050 in. (1.25 mm)
Average-sized parts	0.085 in. (2.15 mm)
Large parts	0.125 in. (3.20 mm)

Thick corners should be avoided because they are likely to lead to gas pockets, undercuring, or cracking. Where extra strength is desired at corners, it can usually be obtained by ribbing.

Economical production is also facilitated by appropriate dimensional tolerances. A minimum tolerance of ± 0.003 in. (0.08 mm) should be allowed in the direction parallel to the parting line of the mold. In the direction perpendicular to the parting line, a minimum tolerance of ± 0.010 in. (0.25 mm) is desirable. In both cases, increasing these tolerances by 50% can reduce manufacturing difficulties and costs appreciably.

Since most molds are reusable, careful attention should also be given to the problem of removing the part from the mold. The rigid metal molds should be designed so that they can be easily opened and closed. A small amount of unidirectional taper should be provided on each side of the mold's parting plane to facilitate the withdrawal of the part. Undercuts should be avoided wherever possible, since they prevent part removal unless special mold sections are used. Such special sections must move at right angles to the opening motion of the major mold, making the molds costly to produce and maintain and also slowing production.

The overall molding cycle can be as short as 2 sec or as long as several minutes, and one part to several dozen may be formed in each cycle. The actual cycle time is generally determined by the heat transfer capabilities of the mold.

Inserts. Because of the difficulty of molding threads in plastic parts and the fact that cut threads tend to chip, tapped or threaded inserts are generally used where considerable strength is required and frequent disassembly of the parts may occur. Several types of inserts are shown in Figure 20-13, the use of which requires attention to design detail if satisfactory results are to be obtained. Inserts are usually made of brass or steel and are held in the plastic only by a mechanical

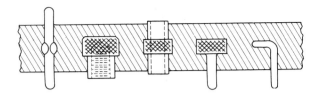

FIGURE 20-13 Typical metal inserts for use in plastic parts.

bond. Therefore, it is necessary to provide suitable knurling or grooving so that the insert may be gripped firmly and will not become loose in service. A medium or coarse knurl is quite satisfactory for resisting torsional loads and moderate axial loads. A groove is excellent for axial loads but offers little resistance to torsional stresses. Consideration should also be given to minimizing the stresses around a cast-in insert as a result of the differential contraction during cooling. Preheating the inserts is often beneficial; the cast plastic can be selected to minimize shrinking; and the use of selected nonmetallic inserts should be considered.

If an insert is to act as a boss for mounting or is an electrical terminal, it should protrude slightly above the surface of the plastic in which it is embedded. This protrusion permits a firm connection to be made without creating an axial load that would tend to pull the insert from the compound. On the other hand, if the insert is used to hold two mating parts closely together, the insert should be flush with the surface. The parts can then be held together snugly without danger of a loosening of the insert. Where it is necessary to keep the surface of an insert entirely free of any plastic, a shouldered design is most satisfactory. However, if an insert is used to fasten mating parts that must fit closely, a depression must be made in the mating part to provide clearance for the shoulder. Similarly, a depression has to be provided in the mold. Both operations add to the cost.

Inserts must have adequate support. The wall thickness of the surrounding plastic must be sufficient to support any load that may be transmitted by the insert. For small inserts, the wall thickness should be at least half the diameter of the insert. Above ½ in. (13 mm) in diameter, the wall thickness should be at least ¼ in. (6.5 mm).

Design Factors Related to Finishing. In the designing of plastic parts, a prime objective should be to eliminate any necessity for machining after molding. This is especially important where machined areas would be exposed, since these are poor in appearance and may also absorb moisture. The parting surfaces of molds are difficult to maintain in perfect condition so that they mate properly. Radii or curved surfaces where parting lines meet make it even more difficult to maintain perfect mating and should be avoided. The result of poor parting-line fit is a small fin or "flash," as illustrated in Figure 20-14. When fillers are used, as they are in most plastics, the exterior surface is a thin film of pure resin without any filler. This film provides the high luster that is characteristic of plastic products. If the flash is trimmed off, a line of exposed filler is produced,

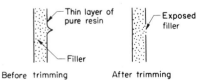

Before trimming   After trimming

FIGURE 20-14 Effect of trimming the flash from a plastic part containing filler.

which may be objectionable. If the parting lines are located at sharp corners, it is easier to maintain satisfactory mating of the mold sections, and any exposed filler resulting from the removal of a fin will be confined to a corner, where it will be less noticeable.

Because plastics have low moduli of elasticity, large flat areas are not rigid and should be avoided whenever possible. Ribbing or doming, as illustrated in Figure 20-15, is helpful in providing the required stiffness. Flat surfaces also tend to reveal flow marks from molding and scratches that occur from service. External ribbing can then serve the dual function of increasing strength and rigidity and also preventing these surface marks from being obvious. Dimpled, or textured, surfaces often provide a pleasing appearance and do not reveal scratches.

Holes that must be formed by pins in the mold should be given special consideration. In compression-type molding, such pins are subjected to considerable bending during the mold closing. Where these pins are supported only at one end, their length should not exceed twice the diameter. In transfer-type molds, the length can be five times the diameter without the problems becoming excessive.

Holes that are to be threaded after molding, or that are to be used for self-tapping screws, should be countersunk slightly. This countersink facilitates starting the tap or screw and prevents chipping at the outer edge of the hole. If the threaded hole is to be less than ¼ in. (6.35 mm) in diameter, it is best to cut the thread after molding by means of a thread tap. For diameters above ¼ in. it is usually better to mold the thread or to use an insert. If the threads are molded, however, special provisions must be made to remove the part from the mold. These measures are generally uneconomical because they force mold delays and reduce productivity.

Machining of Plastics.

Plastics can be machined, sawed, drilled, and threaded much as metals can, but their properties vary so greatly that it is impossible to give instructions that are exactly correct for all. It is very important, however, to remember some of the general characteristics that affect their machinability. First, all plastics are poor heat conductors. Consequently, little of the heat that results from chip formation will be conducted away through the material or will be carried away in the chips. As a result, cutting tools run very hot and may fail more rapidly than when they cut metal. Carbide tools are frequently more economical to use than high-speed tool steels if the cuts are of moderately long duration or if high-speed cutting is to be performed.

Second, because considerable heat and high temperatures do develop at the point of cutting, thermoplastics tend to soften and swell and to bind or clog the cutting tool. The elastic flexing of plastics, coupled with the softening and related effects, results in less precise final dimensions. Thermosetting plastics give less trouble in this regard.

Third, cutting tools should be kept very sharp at all times. Drilling is best done by means of straight-flute drills, or by *dubbing* the cutting edge of a regular twist drill to produce a zero rake angle. These drills are shown in Figure 20-16.

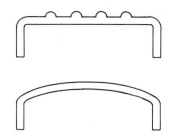

FIGURE 20-15 Method of providing stiffness in large surfaces of plastic parts by the use of ribbing and doming.

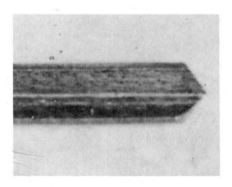

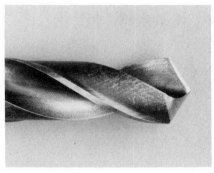

FIGURE 20-16 Straight-flute drill (top) and "dubbed" drill (bottom) used for drilling plastics.

Rotary files and burrs, saws, and milling cutters should be run at high speeds so as to improve cooling, but with the feed carefully adjusted to prevent jamming of the gullets. In some cases, coolants can be used advantageously if they do not discolor the plastic or cause gumming. Water, soluble oil and water, and weak solutions of sodium silicate in water have been used.

Fourth, filled and laminated plastics are usually quite abrasive, and the fine dust produced during machining can often be a health hazard.

Responding to the listed limitations, laser machining of plastics can be quite attractive. Because it vaporizes the material to be removed instead of forming chips, precise cuts can be achieved. Minute holes can be drilled, such as those in the nozzles of the popular aerosol cans.

Finishing of Plastic Parts.   Special attention must often be given to appearance details because plastics are frequently used for products where consumer acceptance is of great importance. In the majority of cases, however, plastic parts can be designed to require very little finishing or decorative treatment, thus promoting economy of manufacture. In some cases, fins and rough spots can be removed, and smoothing and polishing can be performed by barrel tumbling with suitable abrasives or polishing agents. Required decoration or lettering can be obtained by etching the mold. These procedures can produce letters or designs that protrude from the surface of the plastic only about 0.004 in. (0.01 mm). When higher relief is desired, the mold must be engraved, which adds significantly to mold cost.

Whenever possible, depressed letters or designs should be avoided. Such features must be raised above the surrounding surface of the mold, requiring the entire remaining mold surface to be cut away at considerable expense. This cost can frequently be reduced by setting the letters in a small area raised above the main plastic surface. With this technique, only a small amount of the die area will need to be undercut.

Other finishing methods for plastics include printing, hot stamping, vacuum metallizing, electroplating, and painting. Some of these are discussed in Chapter 40.

Welding and Joining of Plastics.   Many plastics can be joined by heat, which can be applied in the form of a stream of hot gases or through a tool, such as a soldering iron. Heat can also be generated through ultrasonic vibrations. The joining of plastics is developed further in Chapter 36.

Processing of Rubber and Elastomers

Rubber products can be produced by several different processes. In the simplest of these, they are formed from a liquid preparation or compound, and the products are known as *latex products*. A form is repeatedly immersed in the latex compound, removed, and allowed to dry. This process causes a certain amount of the liquid to adhere to the surface with each dipping, building up to a final desired thickness. After vulcanization, usually in steam, the products can be stripped from the molds.

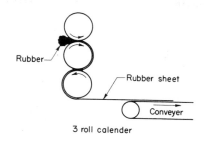

FIGURE 20-17 (Left) Three roll calender used for producing rubber or plastic sheet. (Courtesy Farrel-Birmingham Company, Inc.) (Right) Schematic diagram showing the method of making sheets of rubber in a three-roll calender.

Dipping can be accelerated by the *anode process*, a technique that uses negative electrical charges on the latex particles. A coagulant is deposited on the form or mold and releases positively charged ions when dipped into the latex. These ions neutralize the charges on the adjacent latex particles and cause them to be deposited on the form. The process goes on continuously, so any desired thickness can be deposited in a single immersion.

When products are to be made of solid elastomers, the first step is the compounding of the elastomers, vulcanizers, fillers, antioxidants, accelerators, and other pigments. This compounding is usually done in some form of mixer, which breaks down the elastomer and permits the addition of the other components to form a homogeneous mass. The mix is then put on a mill, in which chilled iron rolls rotate toward each other at different speeds. Water circulating through the interior of the rolls removes the heat generated during the milling action and prevents the start of vulcanization. Sulfur and accelerators are added in this operation.

Rubber compounds and plastics are made into sheet form on *calenders*, such as that shown in Figure 20-17. The sheet coming from the calender is often rolled with a fabric liner to prevent the adjacent layers from sticking together. The covering of cord or woven fabric with rubber or elastomer is also done on a three- or four-roll calender. In the three-roll geometry, only one side of the fabric can be coated in each pass. The four-roll arrangement, shown schematically in Figure 20-18, permits the coating of both sides in a single pass.

Products such as inner tubes, garden hose, tubing, and moldings are produced by extrusion. The compounded material from the mill is forced through a die by a screw device similar to that described previously for plastic material.

Excellent adhesives have been developed that permit the bonding of rubber or artificial elastomers to metal, usually brass or steel. Tanks fabricated in this manner can be used to transport and store a wide variety of corrosive liquids. Only moderate pressures and temperatures are required for excellent adhesion.

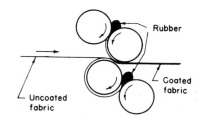

FIGURE 20-18 Arrangement of the rolls, fabric, and coating material for coating both sides of a fabric in a Z-type four-roll calender.

Processing of Ceramics

The processing of ceramic materials generally falls into two distinct classes, based on the properties of the material. *Glasses* are manufactured into useful articles by first producing a viscous liquid, then shaping the material under viscous flow conditions, and finally cooling the material to produce a solid product. *Crystalline ceramics* have a characteristically brittle behavior and are normally manufactured into useful components by first preparing a shape by pressing moist aggregates or powder, followed by drying, and then bonding by one of a variety of mechanisms: chemical reaction, vitrification, or sintering.

Fabrication Techniques for Glasses. Glass is generally shaped at elevated temperatures where the viscosity can be controlled. Sheet and plate glass is formed at or above the melting range by processes such as rolling through water-cooled rolls or floating on a bath of molten tin. Other glass shapes can be produced by casting the molten material into a mold. Mating male and female die members can be used to press controlled masses of viscous material into desired shapes, as in Figure 20-19. A process similar to the blow molding of plastics can be used to force prepressed (cut-shaped) pieces of viscous material against the outside of prepared mold cavities, as in Figure 20-20. Finally, fibers or rods can be produced by drawing the viscous material through prepared dies.

Fabrication of Crystalline Ceramics. Crystalline ceramics are generally processed in the solid state, beginning with particles or aggregates. Many of the processes are quite similar to those used in powder metallurgy and are summarized in Table 20.1. Dry powders can be pressed by uniaxial or isostatic methods. Hot pressing techniques can also be employed.

In *slip casting*, the ceramic powder is mixed into a liquid slurry, which is then cast into a mold containing very fine pores. The liquid is removed from the slurry by capillary action, allowing the ceramic particles to arrange into a "green" body with sufficient strength for subsequent handling. Hollow shapes can be produced by pouring out the remaining slurry once a desired thickness of solid has formed on the mold walls.

Plastic forming involves the blending of the ceramic with additives to make the mixture formable under pressure and heat. Injection molding can be performed with a mixture of ceramic particles and plastic resins. Extrusion and other forming processes can be conducted with a ceramic–binder mixture. The additive material is subsequently removed by controlled heating before the fusion of the remaining ceramic.

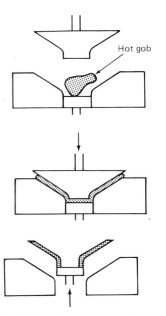

FIGURE 20-19 The pressing of viscous glass with mating male and female die members.

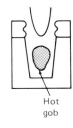 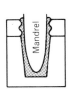 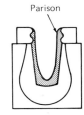 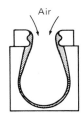

FIGURE 20-20 The forming of a thin-walled glass shape by use of combined pressing and blow molding.

TABLE 20.1   Processes Used to Form Crystalline Ceramics

Process	Dry Axial Pressing	Isostatic Processing	Slip Casting	Injection Molding	Forming Processes (Ex.: Extrusion)
Starting material	Dry powder	Dry powder	Slurry	Ceramic–plastic blend	Ceramic–binder blend
Advantages	Low cost Can be automated	Uniform density Variable cross sections Can be automated	Large sizes Complex shapes Low tooling cost	Complex cross sections Fast Can be automated High volume	Low cost Variable shapes (such as long lengths)
Limitations	Limited cross sections Density gradients	Long cycle times Small number of products per cycle	Long cycle times Labor-intensive	Binder must be removed High tool cost	Binder must be removed Particles oriented by flow

Firing of Ceramic Products.   After a drying process, designed to control the dimensional changes and to minimize the stresses, distortion, and cracking, the ceramic shape is generally subjected to a firing operation to fuse the particles together and produce the desired level of strength. Control of the temperature and the time can control the diffusion process and produce products with controlled grain size, pore size, and pore shape.

In some firing operations surface melting, or vitrification, can occur. This surface liquid flows and produces a glassy bond between the ceramic particles.

An alternative to the firing process is *cementation*, in which the ceramic particles are joined by the use of a binder that does not require firing or sintering. A liquid material first coats the ceramic particles, and a chemical reaction then converts that liquid to a solid, forming strong, rigid bonds.

Machining of Ceramics.   Some of the engineering ceramics can be shaped to precision by machining operations, but these operations are generally of the nonconventional form. Diamond abrasives, lasers, electron beams, and chemical etchants have all been used to machine ceramics, and ceramic machining is becoming an important manufacturing process.

Fabrication of Composite Materials

As presented in Chapter 8, composite materials can offer a number of attractive properties and have begun to earn a reputation as "the material of the future." Such growth and usage can be made possible, however, only if there is a concurrent development in the associated manufacturing processes. Faster production speeds, a reduction in hand labor through the increased use of automation, reduced variability, and integrated quality control will all be keys to the continued growth of the composite industry.

Fabrication of Particulate Composites.   Particulate composites rarely involve processes unique to the composite. Instead, the particles are dispersed in the matrix either through introduction into a liquid melt or slurry, or through mixing various components in the powder metallurgy technique. Subsequent processing generally follows conventional processing methods or ones that have been developed for powder metallurgy, and these need not be described separately in this section.

Fabrication of Laminar Composites.

The production of laminar composites involves processes designed to form a quality bond between distinct layers of different materials. Many laminar composites, such as claddings and bimetallics, are produced by hot or cold *roll bonding*. Sheets of the various materials are simultaneously passed through the rolls of a conventional rolling mill. If the amount of deformation is great enough, surface oxides and contaminants are broken up and dispersed, metal-to-metal contact is established, and the two surfaces become joined by a solid-state bond. U.S. coinage is a common example of a roll-bonded material.

Explosive bonding is another practical means of bonding layered materials. A sheet of explosive material progressively detonates above the layers to be joined, and a pressure wave sweeps across the interface. A small open angle is maintained between the two surfaces. As the pressure wave propagates, any surface films are liquefied and are jetted out the open interface. Clean metal surfaces are then forced together at high pressures, forming a solid-state bond with a characteristically wavy configuration at the interface. Wide plates (too wide to conveniently roll-bond) and dissimilar materials with large differences in mechanical properties are attractive candidates for explosive bonding.

Lamination processes provide still another alternative. These are discussed in detail later in this chapter.

Fabrication of Fiber-Reinforced Composites.

A number of processes have been developed for fabricating products from fiber-reinforced composites. Variations are based primarily on the orientation of the fibers, the length of the continuous filaments, and the geometry of the final product. Each process seeks to embed the fiber in a selected matrix with the alignment and spacing necessary to produce the desired final properties. Discontinuous fibers can be combined with the matrix to provide either a random or a preferred orientation. Normally, continuous fibers are unidirectionally aligned as rods or tapes, woven into fabric layers, wound around a mandrel, or woven into a three-dimensional shape. Various processes have been developed to combine the fiber and the matrix into a unified form. Other processes convert these basic forms into selected products with specific shapes and properties.

Processes Designed to Combine Fibers and a Matrix.

If the matrix material can be liquefied and the temperature is not harmful to the fibers, *casting* processes can be an attractive means of coating the reinforcing fibers. The pouring of concrete around steel reinforcing rods is a crude example of this method. In the case of the high-tech fiber-reinforced plastics and metals, the liquid is placed between the fibers by means of capillary action, vacuum infiltration, or pressure casting. Another alternative is to draw the fibers through a bath of molten matrix material and then form them into aligned bundles before the liquid solidifies.

Prepregs involve the formation of a woven fabric that is then infiltrated with the matrix material. In the case of a polymeric matrix, the infiltration is conducted under conditions in which the resin undergoes only a partial cure. Later fabrication then involves stacking the prepreg layers and subjecting them to heat and pressure to further cure the resin into a continuous solid matrix. The prepreg layers can be combined in various orientations to provide the desired directional properties.

Fibers can also be wound around a mandrel, prepregged with matrix material, and then removed to produce *tapes* containing unidirectionally aligned filaments.

Individual filaments can be coated with the matrix by drawing through a molten bath, plasma spraying, vapor deposition, electrodeposition, and other techniques. These coated fibers can then be assembed and bonded to produce composite products.

Where the temperature of a molten matrix could be objectionable, the matrix can be bonded to the fibers by means of either diffusion bonding or deformation bonding (hot pressing or rolling), often using layers of foil and woven fibers. Loosely woven fibers can also be infiltrated with a particulate matrix, which is then compacted at high pressures and sintered to form a solid mass.

Sheet-molding compounds (SMCs) are comprised of chopped fibers and resin in sheets approximately 0.1 in. (2.5 mm) thick. With strengths of 5–10 ksi (35–70 MPa) and the ability to be press-formed in heated dies, SMCs are a viable alternative to sheet metal in applications where light weight, corrosion resistance, and integral color are attractive features.

Pultrusion.

Pultrusion is a continuous process that closely resembles the extrusion of a metal. Bundles of continuous reinforcing fibers are drawn through a resin bath and then through a preformer to produce the desired cross-sectional shape. This material is then pulled through a series of heated dies, which further shape the product and cure the resin. On emergence from the heated dies, the product is cooled by air or water and is cut to length to form continuous-reinforced shapes with a constant cross section.

Filament Winding.

The availability of plastic-coated, high-strength, continuous filaments or tapes of various materials, such as glass, graphite, and boron, has made it possible to produce container-type shapes that have exceptional strength-to-weight ratios. The filaments are wound over a form, using longitudinal, circumferential, or helical patterns, or a combination of these, designed to take advantage of the highly directional strength properties. By adjusting the density of the filaments in various locations and selecting the orientation of the wraps, products can be designed to have strength where needed and lighter weight in the less critical regions. After wrapping, the resin (often of an epoxy type) is cured to provide a continuous polymeric matrix that binds the adjacent fibers together, and transmits the stress from one fiber to another. After curing, the product is stripped from the form.

FIGURE 20-21 A large tank being made by filament winding. (Courtesy Rohr Corporation.)

Figure 20-21 shows a large tank being made by this process. Such tanks can be made in virtually any size, some being as large as 15 ft (4.6 m) in diameter and 65 ft (20 m) long. Moderate production quantities are feasible, and because the process can be highly mechanized, uniform quality can be maintained. The special tooling for a new size or design (i.e., a new form block) is relatively inexpensive, so the process offers tremendous potential for cost savings and high flexibility. In converting the production of a 12-ft-diameter fuselage section from an aluminum assembly to a filament winding, one airline manufacturer cited a 30% reduction in labor to produce the part, a 65% reduction in total manufacturing time, a reduction in the number of component parts and assembly operations, and a substantial reduction in tooling costs.

Lamination. In the laminating process, prepreg sheets of partially cured resin and aligned or woven fibers are stacked to a desired thickness and are cured under pressure and heat. The resulting products have unusually high strength properties as a result of the integral fiber reinforcement. Because the surface is a thin layer of pure resin, laminates usually possess a smooth, attractive appearance. By the use of transparent resins, the fiber material can be made visible and can produce a variety of decorative effects.

Laminated materials can be produced as sheets, tubes, and rods. Flat sheets are produced as illustrated in Figure 20-22. Figure 20-23 illustrates the method that is used to produce laminated tubing. The impregnated stock is wound on a

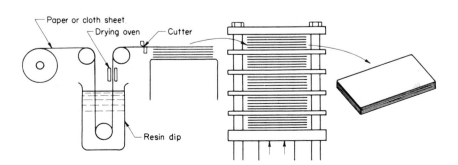

FIGURE 20-22 Method of producing flat laminated plastic sheets.

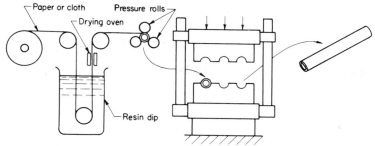

FIGURE 20-23 Method of producing laminated plastic tubing.

mandrel of the proper diameter. After winding, the assembly is cured in a molding press, and the mandrel is removed. Rods are produced in a similar manner by the use of a small mandrel and its removal before curing.

Because of their excellent strength properties, plastic laminates find a wide variety of uses. Some sheets can be easily cut by blanking and piercing. Gears machined from thick laminated sheets have unusually quiet operating characteristics when matched with metal gears.

Many laminated products are not flat, but contain relatively simple curves and contours. Boats, automobile body panels, safety helmets, and similar products can be manufactured by processes that require zero to moderate pressure and low temperatures, often provided by simple heat lamps. In one technique, only a female mold is required, and this can be made from metal, hardwood, or even particle board. The laminating material, generally in the form of fabric dipped in resin, is placed in the mold in successive layers and various orientations until the desired thickness is obtained. Caution is taken to ensure that no impurities, such as oil, dirt, or other contaminants, will be trapped between the layers. The entire mold assembly is then placed in a bag, from which all air is evacuated (*vacuum-bag molding*). External air pressure then holds the laminate against the mold during the curing operation, which may be done by subjecting the material to an environment of live steam. Matched metal dies can be used in place of the mold and bag when production quantities are large and part quality needs to be high. Curing is accomplished during the compression stage with heated dies (*compression molding*).

If the polymer can be cured at room temperature or under simple heat lamps, the vacuum bag can often be eliminated. The pliable resin-coated cloth is simply placed in the mold or over a form (*hand lay-up*) and is cured in the desired shape. If prepreg layers are not used, a layer of woven fiber can be put in place, the resin painted on, and the process repeated to build the desired shape. Tooling costs are sufficiently low so that single items or small quantities are economically feasible. Tool modifications are rather inexpensive, and manufacturing lead-time can be rather short. Large parts can be produced as single units, considerably reducing the number of assembly operations.

Where the use of continuous or woven fibers can be sacrificed, parts have been produced by mixing chopped fibers and resin and spraying the combination onto a mold form (*spray molding*).

Injection Molding. The injection molding of fiber-reinforced plastics is a process designed to compete with metal die castings and offers comparable properties at considerably reduced weight. In the simplest variation, chopped or continuous fibers are placed in a mold cavity, which is then closed and injected with resin. An improved method uses chopped fibers, up to ¼ in. (6 mm) in length, which are premixed with thermoplastic resin (often nylon) and simultaneously injected. Another alternative uses a feedstock of discrete pellets that have been manufactured by slicing continuous-fiber pultruded rods. Both glass and carbon fibers have been used, and the product tolerances are quite good. The benefits compared to those of conventional plastic molding include increased rigidity and impact strength, less likelihood of brittle failure during impact, better dimensional stability at elevated temperatures and in humid environments, greater abrasion resistance, improved surface finish (less dimensional contraction and less severe sink marks), and more isotropic properties.

Braiding, or Three-Dimensional Weaving. In the braiding processes, the Kevlar, graphite, fiberglass, silicon carbide, or other high-strength reinforcing fibers are first braided or woven into a three-dimensional preform. Resin is injected into the assembly, and the resultant product is cured for use. Complex shapes can be produced, the fiber orientations being selected and altered for their optimum properties. Moreover, because of the three-dimensional inter-weaving of the fibers, products of this process lack the possible delamination (layer-separation) planes present in many composite materials. Computers can be used to design and control the weaving process, thereby reducing the high cost of the more labor-intensive fabrication methods.

Fabrication of Metal-Matrix Composites. Continuous-fiber metal-matrix composites can be produced by filament winding, extrusion, or pultrusion, or by shaping and bonding discrete fiber-reinforced sheets that have been produced by electroplating, plasma-spraying, deposition-coating or vapor-depositing metal onto a fabric or mesh. Diffusion bonding of foil–fabric sandwiches, vacuum infiltration with metal, roll bonding, and coextrusion are all alternative means of producing fiber-reinforced metal stock. Products with discontinuous fibers are often produced by powder metallurgy techniques followed by some form of thermomechanical processing.

Graphite-reinforced aluminum has been shown to be twice as stiff as steel and one third to one fourth the weight, with practically zero thermal expansion. Magnesium, copper, and titanium alloys have also been used as the metal matrix.

Secondary Processing and Finishing of Fiber Composites. Fiber-reinforced composites can often be processed with conventional equipment (sawed, drilled, routed, tapped, threaded, turned, milled, sanded, and sheared), but special precautions should be exercised. Cutting some materials may be like cutting multilayer cloth, and precautions should be used to prevent the formation of splinters, cracks, and frayed or delaminated edges. Sharp tools, high speeds, and low feeds are generally required. The cutting debris should be quickly removed to prevent the cutters from becoming clogged.

In addition, many of the reinforcing fibers are extremely abrasive in nature and quickly dull most conventional cutting tools. Diamond, or polycrystalline diamond, tooling may be required to achieve realistic tool life. Abrasive slurries can be used in conjunction with rigid tooling to ensure the production of smooth surfaces. Lasers and water jets (streams of water as small as 0.006 in. in diameter, flowing under pressures up to 60,000 psi, and often containing an abrasive material in a slurry) are alternative cutting tools. However, lasers may burn or carbonize the material or may produce undesirable heat-affected zones. Water jets can create moisture problems with plastic resins.

Where fiber-reinforced materials must be joined, the major concern is the lack of continuity of the fibers in the joint area. Thermoplastics can be welded. Thermoset materials must use mechanical joints or adhesives, each method having its characteristic advantages and limitations.

Review Questions

1. How are the fabrication processes applied to plastics, ceramics, and composites different from those applied to metals?
2. How does the fabrication of a thermoplastic polymer differ from the processing of a thermosetting polymer?
3. Why do cast plastic resins typically have a lustrous appearance?
4. Why is hot-compression molding generally not an attractive fabrication process for thermoplastic polymers?
5. Why must the amount of material introduced into a straight-plunger compression mold be precisely controlled?
6. What are some of the attractive features of the transfer-molding process?
7. What is the most popular fabrication process for thermoplastic materials (in terms of the number of parts produced)?
8. How long is a typical molding cycle in the injection molding of thermoplastics?
9. Why is the cycle time for the injection molding of thermosetting polymers signficantly longer than that for the thermoplastics?
10. What is the distinctive feature of the reaction-injection molding process?
11. What are some of the typical production shapes that are produced by the extrusion of plastics?
12. What is the difference between open-cell and closed-cell foamed plastics?
13. What are some typical applications for rigid-type foamed plastics?

14. What are some of the attractive design properties of plastics?
15. What are some of the design limitations of plastics?
16. Why should adequate fillets be included between the adjacent sections of a mold?
17. Why is it most desirable to have uniform wall thickness in plastic products?
18. Why are product dimensions less precise perpendicular to the parting line?
19. What is the primary purpose of incorporating metal inserts into molded plastic products?
20. Why is it so important to properly locate the parting line in the design of plastic-molded products?
21. Why are heat-related problems common in the machining of plastics?
22. When designing a decorative surface (design or lettering) on a plastic product, why is it desirable that the details be raised on the product, rather than depressed?
23. What are some of the types of additives mixed with elastomers when they are used to produce solid products?
24. What are the two basic types of ceramic materials, and how does their processing differ?
25. Discuss the role of additives in the plastic forming of ceramics? Do the additives become part of the finished product?
26. What changes or development will have to occur in the manufacturing processes associated with composites if the materials are to realize their full potential?
27. What are some of the processes that can be used to establish a quality bond between the layers of a laminar composite?

28. What is a prepreg in the manufacture of fiber-reinforced composites?
29. What are sheet-molding compounds (SMCs)?
30. In what way is pultrusion similar to the extrusion of metal?
31. What are some of the ways in which filament winding can bring about a cost savings in the manufacture of small quantities of large parts?
32. What types of geometries can be produced with laminated sheets of woven fibers?

33. Discuss several ways in which reinforcing fibers and resin can be mixed and shaped by injection molding.
34. What are some of the attractive features of the three-dimensional weaving of fiber-reinforced composites?
35. What is the major design concern when the joining of fiber-reinforced composites is being considered?

Material Removal Processes

Chip-Type Machining Processes (Metal Cutting)

Introduction

Metal cutting, commonly called *machining,* is the removal of the unwanted metal from a workpiece in the form of chips so as to obtain a finished product of desired size, shape, and finish. United States industries annually spend $60 billion to perform metal removal operations because the vast majority of manufactured products require machining at some stage in their production, ranging from relatively rough or nonprecision work, such as cleanup of castings or forgings, to high-precision work involving tolerances of 0.0001 inch or less. Thus, machining undoubtedly is the most important of the basic manufacturing processes.

Over the past 80 years, the process has been the object of considerable research and experimentation that have led to improved understanding of the nature of both the process itself and the surfaces produced by it. While this research effort has led to improvements in machining productivity, the complexity of the process has resulted in slow progress in obtaining a complete theory of chip formation. Most theories have ignored the plastic deformation properties of the work material and/or have not properly characterized the interactions at the contact zone between the tool and the chip. Metal cutting, an unconstrained deformation process (like tensile testing), is a very large strain (plastic deformation) process operating at exceptionally high strain rates, making it unique. The problem is further complicated by tool geometry variations, the wide variety of tool materials used in the process, temperature or heat problems, and the great variation in operating conditions of the machines performing this process. In addition, the environment in which the process is performed has a significant influence on the outcome. The objective of this chapter is to put all this in perspective for the practicing engineer.

Basic Chip Formation Processes. There are seven basic chip formation processes: shaping, turning, milling, drilling, sawing, broaching, and abrasive machining (grinding). Grinding will be treated separately in Chapter 27. For all metal-cutting processes, it is necessary to distinguish between speed, feed, and depth of cut. The shaping process will be used to introduce these terms. See Figure 21-1. In general, speed (V) is the primary cutting motion, which relates the velocity of the cutting tool relative to the work piece. It is generally given in units of surface feet per minute (sfpm), inches per minute (ipm), or meters per minute (m/m) or meters per second (m/s). Speed (V) is shown in Figure 21-1a with the heavy dark arrow. Feed (f) is the amount of material removed per revolution or per pass of the tool over the workpiece. Feed units are inches/revolution, inches per cycle, inches per minute, or inches per tooth, depending on the process. Feed is shown with dashed arrows. In shaping, feed is in inches/stroke and the workpiece feeds perpendicular to the cutting stroke after the tool is retracted by the ram. See Figure 21-1b and 21-1c. The depth

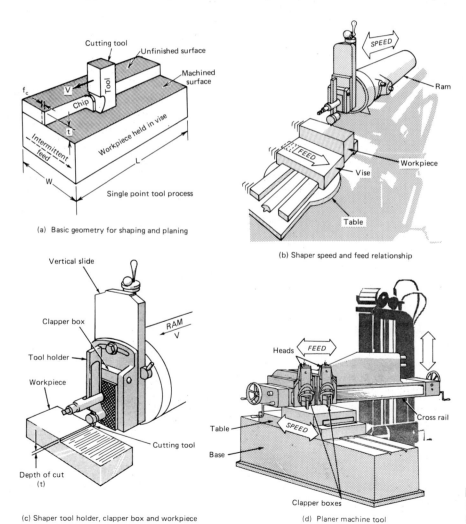

(a) Basic geometry for shaping and planing

(b) Shaper speed and feed relationship

(c) Shaper tool holder, clapper box and workpiece

(d) Planer machine tool

FIGURE 21-1 Basics of single point machining, using shaping/planing as an example.

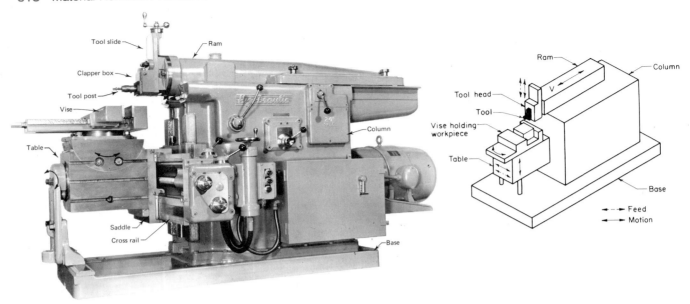

FIGURE 21-2 Elements of a hydraulic-drive, horizontal push-cut shaper with a universal table. (Courtesy Rockford Machine Tool Company.)

of cut reflects the third dimension in these processes and is indicated in the schematics by the letter *t*. See Figures 21-1a and 21-1c.

Machines for metal cutting are called *machine tools*. The machine tool for shaping is shown in Figure 21-1b and 21-2. *Workpieces* are held in the workholding device, in this example a vise. A *cutting tool* is used to machine the metal. Machine tools, cutting tools, and workholding devices are usually manufactured by separate companies and the tooling can cost as much or more than the machine tool.

To process different metals, the input parameters to the machine tools must be determined.

Again using shaping as an example, the cutting tool, held in the tool post located in the ram, reciprocates over the work with a forward stroke, cutting at velocity (V) and a quick return stroke at velocity (V_R). The RPM of drive crank (N_s) determines the velocity of the operation. See Figure 21-3.

The stroke ratio, $R_s = \dfrac{\text{Cutting stroke angle}}{360 \text{ degrees}} = \dfrac{200}{360} = \dfrac{5}{9}$ or $\dfrac{5}{9}$ gives the percentage of time the tool is advancing.

The number of strokes per minute is N_s, the RPM of the drive crank.

Feed f_c is in inches per stroke and is at right angles to the cutting direction.

The length of stroke, ℓ, must be greater than the block length (or length of cut), L, since velocity is position variant. Let ℓ = twice the length of the block being cut or $2L$. The cut velocity, V, is assumed to be twice the average forward velocity V, of the ram. The general relationship between cutting speed and RPM is

$$V = \frac{\pi D N_s}{12} \text{ in feet/minute} \qquad (21\text{-}1)$$

where D = diameter (of rotational member) in inches.

For shaping, cutting speed

$$V = \frac{2\,\ell N_s}{12\,R_s} \qquad (21\text{-}2)$$

in feet/minute.

Cutting speed (V) is selected (by the process engineer or the machinist) depending on the cutting tool material and the work material. Handbooks are available to help in this selection. Table 21.1 shows an example of such a table for shaping and planing. This table has suggested speeds for high speed steel and carbide tool materials.

Once a cutting speed (V) is selected, the RPM, N_s, of the machine can be calculated. Table 21.1 also provides suggested feed values, f_c, in inches per stroke (or cycle) and also recommends depths of cut. The maximum depth of cut is based on the horsepower available to form the chips. This calculation requires that the metal removal rate (MRR) be known. The MRR is the volume of metal removed per unit time or

MRR = L × W × t / CT (cubic in./min.) (21-3)

where

$$\left.\begin{array}{l} W = \text{width of block being cut} \\ L = \text{length of block being cut,} \\ t = \text{depth of cut} \end{array}\right\} \text{ so volume of cut } = W\,L\,T$$

and CT is the time in minutes to cut that volume.

In general, CT is the total length of the cut divided by the feed rate. For shaping, it is the width of block divided by the feed rate of the tool across the width. Thus, for shaping

$$CT = \frac{W}{N_s \times f_c} \text{ (min.)}$$

$$\text{Also } CT = S/N_s$$

where the number of strokes for the job is $S = \dfrac{W}{f_c}$ for a surface of width W.

Substituting for N_s from equation 21-2, we have

$$CT = 2\,\ell\,s\,/\,2\,R_s V \qquad (21\text{-}4)$$

Similarly, the equation for MRR can be written in terms of shaping parameters as MRR = $L\,t\,N_s f_c$.

The basic equations for the cutting time per piece (CT) in minutes and the rate of metal removal (MRR) in volume per minute vary somewhat from process to process. The shaping equations apply to planing (see Figure 21-1d) which differs from shaping in that in planing the work reciprocates (instead of the tool) and the tool feeds. These basic equations are as fundamental as the processes

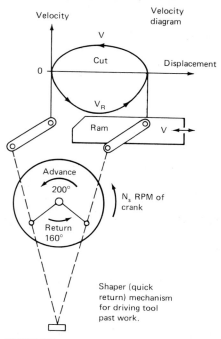

FIGURE 21-3 The ram of the shaper carries the cutting tool and reciprocates it back and forth across the workpiece.

TABLE 21.1 Example of a Table of Recommended Cutting Speeds, Feeds, and Depth of Cut

Material	Hardness (Bhn)	Condition	Depth of Cut in	Depth of Cut mm	HSS Speed fpm	HSS Speed m/min	HSS Feed in/stroke	HSS Feed mm/stroke	Tool Material AISI	Tool Material ISO	Carbide Speed fpm	Carbide Speed m/min	Carbide Feed in/stroke	Carbide Feed mm/stroke	Carbide Grade C	Carbide Grade ISO
Free machining alloy steels, wrought; Medium carbon resulfurized; 4140 4142Te 4147Te; 4140Se 4145Se 4150	150 to 200	Hot rolled, normalized, annealed or cold drawn	.005	0.1	45	14	*	*	M2, M3	S4, S5	250	76	*	*	C-6	P10
			.100	2.5	65	20	.050	1.25	M2, M3	S4, S5	300	90	.080	2.05	C-6	P20
			.500	12	40	12	.060	1.50	M2, M3	S4, S5	225	69	.060	1.50	C-6	P30
	200 to 250	Hot rolled, normalized, annealed or cold drawn	.005	0.1	40	12	*	*	M2, M3	S4, S5	250	76	*	*	C-6	P10
			.100	2.5	55	17	.040	1.0	M2, M3	S4, S5	300	90	.060	1.50	C-6	P20
			.500	12	35	11	.050	1.25	M2, M3	S4, S5	225	69	.050	1.25	C-6	P30
	275 to 325	Quenched and tempered	.005	0.1	35	11	*	*	M2, M3	S4, S5	220	67	*	*	C-6	P10
			.100	2.5	45	14	.030	.75	M2, M3	S4, S5	250	76	.060	1.50	C-6	P20
			.500	12	30	9	.045	1.15	M2, M3	S4, S5	170	52	.050	1.25	C-6	P30
	325 to 375	Quenched and tempered	.005	0.1	30	9	*	*	M2, M3	S4, S5	200	60	*	*	C-6	P10
			.100	2.5	40	12	.030	.75	M2, M3	S4, S5	225	69	.060	1.50	C-6	P20
			.500	12	25	8	.045	1.15	M2, M3	S4, S5	150	46	.050	1.25	C-6	P30
Medium and high carbon leaded; 41L30 41L47 51L32 86L40; 41L40 41L50 52L100; 41L45 43L40 86L20	150 to 200	Hot rolled, normalized, annealed or cold drawn	.005	0.1	55	17	*	*	M2, M3	S4, S5	280	85	*	*	C-6	P10
			.100	2.5	75	23	.050	1.25	M2, M3	S4, S5	300	90	.080	2.05	C-6	P20
			.500	12	45	14	.060	1.50	M2, M3	S4, S5	260	79	.060	1.50	C-6	P30

Material	Brinell hardness	Condition	Depth of cut	Speed	Feed	Tool material	Speed	Feed	Tool material
	200 to 250	Hot rolled, normalized, annealed or cold drawn	.005	45	*	M2, M3	260	*	C-6
			.100	65	.040	M2, M3	290	.060	C-6
			.500	40	.050	M2, M3	225	.050	C-6
			0.1	14	*	S4, S5	79	*	P10
			2.5	20	1.0	S4, S5	88	1.50	P20
			12	12	1.25	S4, S5	69	1.25	P30
	275 to 325	Quenched and tempered	.005	40	*	M2, M3	230	*	C-6
			.100	50	.030	M2, M3	260	.060	C-6
			.500	35	.045	M2, M3	200	.050	C-6
			0.1	12	*	S4, S5	70	*	P10
			2.5	15	.75	S4, S5	79	1.50	P20
			12	11	1.15	S4, S5	60	1.25	P30
	325 to 375	Quenched and tempered	.005	35	*	M2, M3	210	*	C-6
			.100	45	.030	M2, M3	240	.060	C-6
			.500	30	.045	M2, M3	175	.050	C-6
			0.1	11	*	S4, S5	64	*	P10
			2.5	14	.75	S4, S5	73	1.50	P20
			12	9	1.15	S4, S5	53	1.25	P30
Alloy steels, wrought Low carbon 4012 4615 4817 8617 4023 4617 4820 8620 4024 4620 5015 8622 4118 4621 5115 8822 4320 4718 5120 9310 4419 4720 6118 94B15 4422 4815 8115 94B17	125 to 175	Hot rolled, annealed or cold drawn	.005	40	*	M2, M3	275	*	C-6
			.100	60	.050	M2, M3	300	.080	C-6
			.500	35	.060	M2, M3	250	.060	C-6
			0.1	12	*	S4, S5	84	*	P10
			2.5	18	1.25	S4, S5	90	2.05	P20
			12	11	1.50	S4, S5	76	1.50	P30
	175 to 225	Hot rolled, annealed or cold drawn	.005	35	*	M2, M3	230	*	C-6
			.100	55	.050	M2, M3	250	.080	C-6
			.500	30	.060	M2, M3	225	.060	C-6
			0.1	11	*	S4, S5	70	*	P10
			2.5	17	1.25	S4, S5	76	2.05	P20
			12	9	1.50	S4, S5	69	1.50	P30

*Feed = ¾ the width of square nose finishing tool.
Source: *Machining Data Handbook*, Vol. 1, 3rd edition, Machinability Data Center, Cincinnati, Ohio.

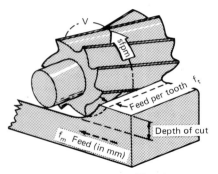

FIGURE 21-4 Slab milling is an example of a multiple tooth-cutting process which creates a flat surface.

themselves. In the main, if one keeps track of the units and visualizes the process, the equations are, for the most part, straightforward.

Shaping and planing are examples of single-point tool cutting processes. The process of milling is an example of a multiple-tool process. In milling, there are two feeds: the amount of metal an individual tooth removes, called the feed per tooth, f_t and the rate at which the table is translated past the rotating cutting tool. This feed is in inches per minute, f_m. Figure 21-4 shows the typical arrangement for milling where the multiple cutting tool is rotated while the workpiece feeds past the cutter. The feed per tooth is

$$f_t = \frac{f_m}{N_s\, n} \tag{21-5}$$

where N_s is the rpm of the cutter, $n =$ number of teeth in the cutter and f_m is the table feed in inches per minute.

Another way to think about the processes is according to the type of surface generated: flat, internal cylindrical, external cylindrical, or contoured. Figures 21-5 and 21-6 add further clarification to the relationship between the workpiece and the tool, showing the similarity between linear and cylindrical processes. Table 21.2 summarizes process information for the basic machining processes, noting typical sizes of parts that are manufactured for these processes as well as their typical production rates, tolerances, and surface finishes. Additional specific information on these processes is given in Chapters 23–30, including equations for cutting time and metal removal rate.

Operation	Block Diagram	Most Commonly Used Machines	Machines Less Frequently Used	Machines Seldom Used
Turning	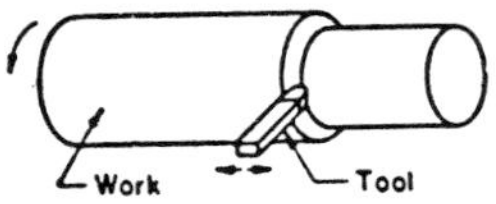	Lathe	Boring mill	Vertical shaper Milling machine
Grinding	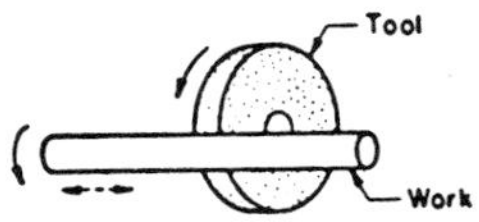	Cylindrical grinder		Lathe (with special attachment)
Sawing (of plates)	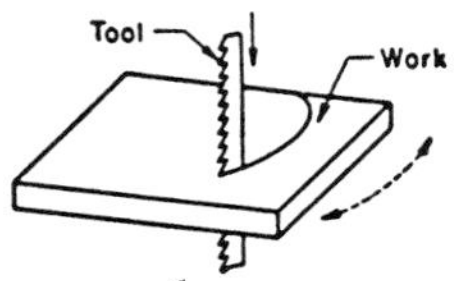	Contour or band saw	Flame cutting or plasma arc	
Drilling	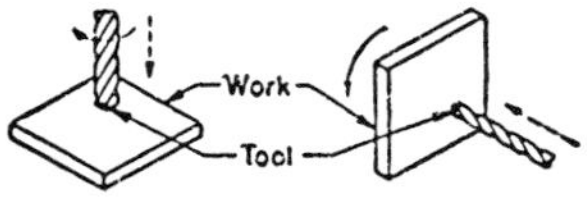	Drill press Machining center Vert. milling machine	Lathe Horizontal boring machine	Horizontal milling machine Boring mill

FIGURE 21-5 Operations and machines for machining flat surfaces.

Operation	Block Diagram	Most Commonly Used Machines	Machines Less Frequently Used	Machines Seldom Used
Boring	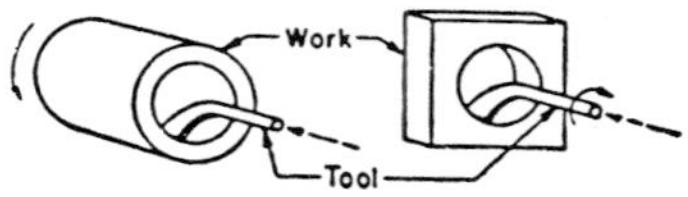	Lathe Boring mill Horizontal boring machine Machining center		Milling machine Drill press
Reaming	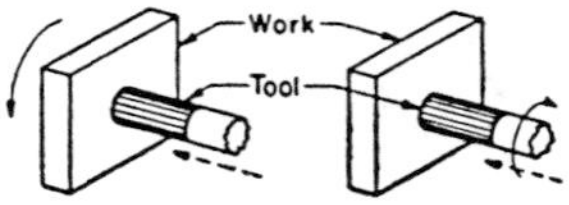	Lathe Drill press Boring mill Horizontal boring machine Machining center	Milling machine	
Grinding	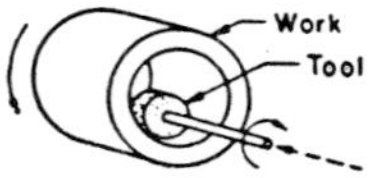	Cylindrical grinder		Lathe (with special attachment)
Sawing	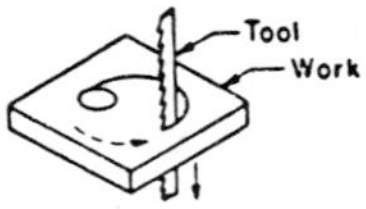	Contour or band saw		
Broaching	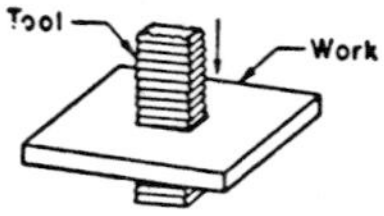	Broaching machine	Arbor Press (keyway broaching)	
Shaping	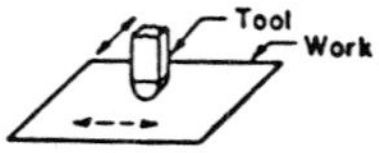	Horizontal shaper	Vertical shaper	
Planing		Planer		
Milling	slab milling 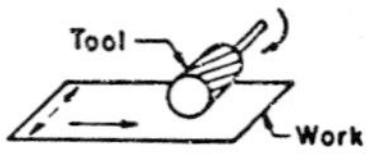	Milling machine		Lathe (with special attachment)
	face milling 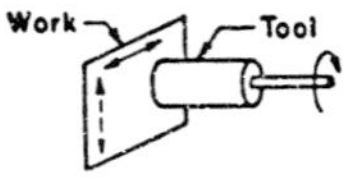	Milling machine Machining center		Drill press (light cuts)

FIGURE 21.5 (continued)

Operation	Block Diagram	Most Commonly Used Machines	Machines Less Frequently Used	Machines Seldom Used
Facing		Lathe	Boring mill	
Broaching		Broaching machine		
Grinding		Surface grinder		Lathe (with special attachment)
Sawing		Cutoff saw	Contour saw	

FIGURE 21.5 (continued)

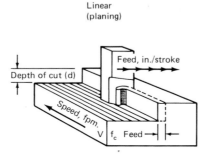

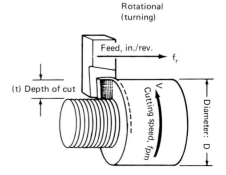

FIGURE 21-6 Relationship between linear and rotational single point machining.

Speed = relative motion between tool and work, stated in surface feet per minute (fpm). Reciprocating motions may be crank-powered, hydraulically, or direct-reversible, electrically driven. There will always be accelerations for a portion of the stroke. Cutting speed is assumed to be the average speed. Feed with single-point tools is the amount the tool or work table is indexed.

Speed, stated in surface feet per minute, (sfpm) is the peripheral speed at the cutting edge. To convert rpm into sfpm, use the following:

$$V = \frac{\pi D N_s}{12} \text{ (converting D to ft.)}$$ This applies to milling, drilling, turning, and all rotary operations. Feed per revolution in turning (and drilling) is a geared feed driven from the main spindle.

TABLE 21.2 Summary of Basic Machining Processes

Applicable Process	Raw Material Form	Size		Typical Production Rate	Material Choice	Typical Tolerance	Typical Surface Roughness
		Maximum	Minimum				
Shaping	Bar Plate Casing	3 ft × 6 ft	Limited usually by ability to hold part	1–4 parts/hour	Low- to medium-carbon steels and nonferrous metals best; no hardened parts	±0.001–±0.002 in. (larger parts) ±0.0001–±0.0005 in. (small-medium parts)	63–250
Planing	Bar Plate Casting	42 ft wide × 18 ft high × 76 ft long	Parts too large for sharper work	1 part/hour	Low- to medium-carbon steels or nonferrous materials best	±0.001–±0.005 in.	63–125
Gear shaping	Blanks	120-in.-dia. gears 6-in. face width	1 in. dia.	1–60 parts/hour	Any material with good machinability rating	±0.001 in. or better at 200 D.P. to 0.0065 in. at 30 D.P.	63
Turning (engine lathes)	Cylinders Preforms Castings Forgings	78 in. dia. × 73 in. long	1/64 in. typical	1–10 parts/hour	All ferrous and nonferrous material considered machinable	±0.002 in. on dia. common; ±0.001 in. obtainable	125–250
Turning (turret lathe)	Bar Rod Tube Preforms	36 in. dia. × 93 in. long	1/64 in. dia.	1 part/minute	Any material with good machinability rate	±0.003 in. on dia. where needed; ±0.010 in. common	125 average
Turning (automatic screw machine)	Bar Rod	Generally 2 in. dia. × 6 in. long	1/16 in. dia. and less, weight less than 1 ounce	10–30 parts/minute	Any material with good machinability rating	±0.0005 in. possible; ±0.001 to ±0.003 in. common	63 average
Turning (Swiss automatic machining)	Rod	Collets adapt to 1/2 in. dia.	Collets adapt to less than 1/2 in.	12–30 parts/minute	Any material with good machinability rating	±0.0002 in. to ±0.001 in. common	63 and better
Boring (vertical)	Casting Preforms	98 in. × 72 in.	2 in. × 12 in.	2–20 hrs/pc	All ferrous and nonferrous	±0.0005 in.	90–250
Milling	Bar Plate Rod Tube	4–6 ft long	Limited usually by ability to hold part	1–100 parts/hour	Any material with good machinability rating	±0.0005 in. possible; ±0.001 in. common	63–250
Hobbing (milling gears)	Blanks Preforms Rods	10-ft-dia. gears 14-in. face width	0.100 in. dia.	1 part/minute	Any material with good machinability rating	±0.001 in. or better	63

TABLE 21.2   *Continued*

Applicable Process	Raw Material Form	Size		Typical Production Rate	Material Choice	Typical Tolerance	Typical Surface Roughness
		Maximum	**Minimum**				
Drilling	Plate Bar Preforms	3½-in.-dia. drills, (1 in. dia. normal)	0.002 in. drill dia.	2–20 sec/hole after setup	Any unhardened material; carbides needed for some case-hardened parts	±0.002–±0.010 in. common; ±0.001 in. possible	63–250
Sawing	Bar Plate Sheet	2 in. armor plate (1/2 in. is preferred)	0.010 in. thick	3–30 parts/hour	Any nonhardened material	±0.015 in. possible	250–1000
Broaching	Tube Rod Bar Plate	74 in. long	1 in.	300–400 parts/minute	Any material with good machinability rating	±0.0005–±0.001 in.	32–125
Grinding	Plate Rod Bars	36 in. wide × 7 in. dia.	0.020 in. dia.	1–1000 pieces/hour	Nearly all metallic materials plus many nonmetallic	0.0001 in. and less	16

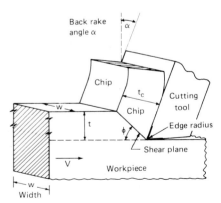

FIGURE 21-7   Schematic of orthogonal machining. The cutting edge of the tool is perpendicular to the direction of motion (V). The back rake angle is α. The shear angle is ϕ.

Fundamental Structure of Chip Formation.   In order to understand this complex process, the tool geometry is simplified from the three-dimensional (oblique) geometry, which typifies most processes, to a two-dimensional (orthogonal) geometry. The workpiece is a flat plate as shown in Figure 21-7. This model is sufficient to allow us to consider the behavior of the work material during chip formation, the influence of the most critical elements of the tool geometry (the edge radius of the cutting tool and the back rake angle, α), and the interactions that occur between the tool and the freshly generated surfaces of the chip against the rake face and the new surface as rubbed by the flank of the tool.

Basically the chip is formed by a localized shear process which takes place over very narrow regions. This large-strain, high-strain-rate, plastic deformation evolves out of a radial compression zone which travels ahead of the tool as it passes over the workpiece (see Figure 21-8). This radial compression zone has, like all plastic deformations, an elastic compression region which becomes the plastic compression region as the field boundary is crossed. The plastic compression generates dense dislocation tangles and networks in annealed metals. When this work-hardening reaches a saturated condition (fully work-hardened), the

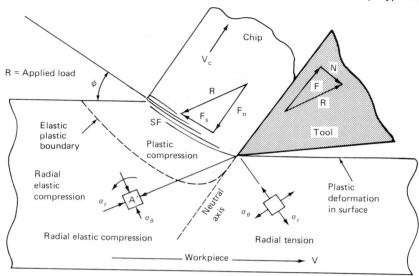

FIGURE 21-8 Machining process produces a radial compression ahead of the shear process. The stress reverses from compression to tension across the neutral axis (NA).

material has no recourse but to shear. The shear process is always nonhomogeneous (discontinuous), composed of a series of shear fronts or narrow bands which produce a lamellar structure in the chips (see Figure 21-9). *This is the fundamental structure that occurs on the microscale in all metals when machined.* Through the use of special metallurgical preparations, it can also be observed developing on the side of the work, as shown in Figure 21-10.

In Figure 21-9, an oxide particle in the surface can be observed. This is a hard second-phase particle which acts as a barrier to the shear front dislocations, which cannot penetrate the particle. The dislocations create voids around the particles in the shear front layers. If there are enough particles of the right size and shape, the chip will fracture through the shear zone, forming segmented chips. "Free-machining steels," which have small percentages of hard second-phase particles added to them, use this metallurgical phenomenon to break up the chips for easier chip handling.

In Figure 21-9 a pattern in the lamella lying perpendicular to the shear fronts can be observed. This pattern was produced by the cutting edge of the tool during the previous pass of the tool over the workpiece, subsequently sheared on the next pass. Thus this pattern actually reflects the microgeometry of the cutting edge of the tool.

In Figure 21-10, the narrow shear fronts on the side of the chip are observed to develop at the tool tip and progress toward the free surface, producing ultimately the lamellar structure seen in Figure 21-9 on the top of the chip. Shear fronts are very narrow (100 to 500 angstroms) compared to the thickness of the lamella (2 to 4 μm) and account for the large strain rates. The micrographs of this figure were made in a high-resolution scanning electron microscope.

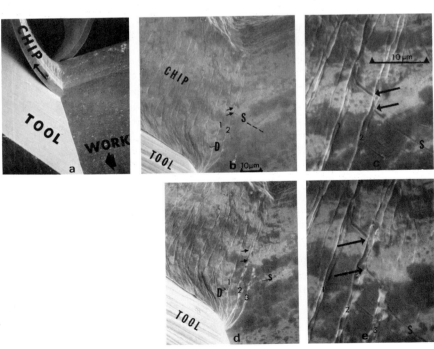

FIGURE 21-9 SEM micrograph of the top of copper chip, clearly showing the lamella structure produced by the shear process in metal cutting.

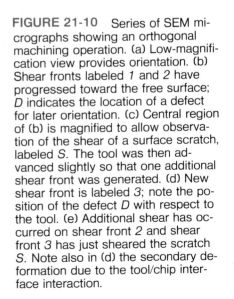

FIGURE 21-10 Series of SEM micrographs showing an orthogonal machining operation. (a) Low-magnification view provides orientation. (b) Shear fronts labeled *1* and *2* have progressed toward the free surface; *D* indicates the location of a defect for later orientation. (c) Central region of (b) is magnified to allow observation of the shear of a surface scratch, labeled *S*. The tool was then advanced slightly so that one additional shear front was generated. (d) New shear front is labeled *3*; note the position of the defect *D* with respect to the tool. (e) Additional shear has occurred on shear front *2* and shear front *3* has just sheared the scratch *S*. Note also in (d) the secondary deformation due to the tool/chip interface interaction.

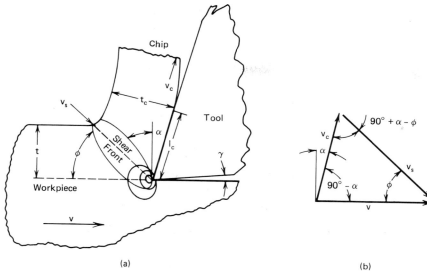

(a)

(b)

FIGURE 21-11 (a) Schematic of orthogonal machining process. (b) Velocity diagram associated with orthogonal machining.

Orthogonal machining is done to test machining mechanics and theory. Orthogonal machining (measuring two forces) can be obtained in practice by

1. Machining a plate as shown in Figure 21-7.
2. End-cutting a tube wall in a turning setup. (See Chapter 23).

In *oblique* machining, the cutting edge and the cutting motion are not perpendicular to each other as in shaping, drilling, milling, and single-point turning. Since the orthogonal case is more easily modeled, it will be used here to describe the process further.

For the purpose of modeling chip formation, assume that the shear process takes place on a single narrow plane rather than on a set of shear fronts which actually comprise a narrow shear zone. Further, assume that the tool cutting edge is perfectly sharp and no contact is being made between the flank of the tool and the new surface. The workpiece passes the tool with velocity V, the cutting speed. See Figure 21-11a. The depth of cut is t. Ignoring the plastic compression, chips having thickness t_c and velocity V_c are formed by the shear process. The shear process then has velocity V_s and occurs at shear angle ϕ. The tool geometry is given by the back rake angle α and the clearance angle γ. The velocity triangle for V, V_c, and V_s is also shown (Figure 21-11b). The chip makes contact with the rake face of the tool over length l_c. The plate thickness is called w. In order to compute the shear angle the equation for chip thickness ratio, r_c, defined as t/t_c can be derived:

$$r_c = \frac{t}{t_c} = \frac{AB \sin \phi}{AB \cos (\phi - \alpha)} \tag{21-6}$$

where AB is the length of the shear plane (see Figure 21-12).

Orthogonal Machining

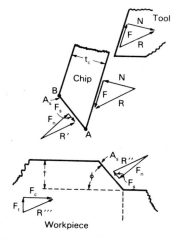

FIGURE 21-12 Free-body diagram of orthogonal chip formation process, showing equilibrium condition between resultant forces R and R'.

Equation (21-5) may be solved for the shear angle ϕ as a function of the measurable chip thickness ratio by expanding the cosine term and simplifying.

$$\tan \phi = \frac{r_c \cos \alpha}{1 - r_c \sin \alpha} \qquad (21\text{-}7)$$

There are numerous other ways to measure chip ratios and obtain shear angles both during (dynamically) and after (statically) the cutting process. For example, the ratio of the length of the chip, L_c to the length of the cut, L, can be used to determine r_c. Many researchers use the chip compression ratio, which is the reciprocal of r_c, as a parameter. See problem 2 at the end of the chapter for another method. The shear angle can be measured statically by instantaneously interrupting the cut through the use of "quick-stop devices." These devices disengage the cutting tool from the workpiece while cutting is in progress, leaving the chip attached to the workpiece. Optical and scanning electron microscopy is then used to observe the direction of shear. High-speed motion pictures have also been used to observe the process at frame rates as high as 30,000 frames per second. More recently, machining stages have been built that allow the process to be performed inside a scanning electron microscope and recorded on video tapes for high-resolution, high-magnification examination of the deformation process. Figure 21-10 was made this way. Using sophisticated electronics and slow-motion playback, this technique can be used to measure (for the first time) the shear velocity. See Figure 21-11. The vector sum of V_s and V_c equals V.

For consistency of volume, we observe that

$$r_c = \frac{t}{t_c} = \frac{\sin \phi}{\cos (\phi - \alpha)} = \frac{V_c}{V} \qquad (21\text{-}6\,\text{alt.})$$

indicating that the chip ratio (and therefore the shear angle) can be determined dynamically if a reliable means to measure V_c can be found. The ratio of V_s to V is

$$\frac{V_s}{V} = \frac{\cos \alpha}{\cos (\phi - \alpha)} \qquad (21\text{-}8)$$

These velocities are important in power calculations, heat determinations, and vibration analysis associated with chip formation.

During the cutting, the chip undergoes a shear strain of

$$\epsilon_c = \frac{\cos \alpha}{\sin \phi \cos (\phi - \alpha)} \qquad (21\text{-}9)$$

which shows that the shear strain is dependent on the rake angle α and the shear direction, ϕ. Generally speaking, metal cutting strains are quite large compared to other plastic deformation processes, being on the order of 2 to 4 in./inch.

This large strain occurs, however, over very narrow regions (the shear fronts) which results in extremely high shear strain rates, $\dot{\epsilon}_c$, which typically are in the range of 10^4 to 10^8 in./in./sec. The strain rate in metal cutting can be estimated by

$$\dot{\epsilon}_c = \frac{V}{t_{sf}} \qquad (21\text{-}10)$$

where t_{sf} is the thickness of the shear front.

It is this combination of large strains and high strain rates operating within a process constrained only by the rake face of the tool that results in great difficulties in theoretical analysis of this process.

Effects of Work Material Properties. As noted previously, the properties of the work material are important in chip formation. High-strength materials require larger forces than materials of lower strength, causing greater tool and work deflection and increased friction, heat generation, and operating temperatures, and requiring greater work input. The structure and composition also influence metal cutting. Hard or abrasive constituents, such as carbides in steel, accelerate tool wear.

Work material ductility is an important factor. Highly ductile materials not only permit extensive plastic deformation of the chip during cutting, which increases work, heat generation, and temperature, but also result in longer, "continuous" chips that remain in contact longer with the tool face, thus causing more frictional heat. Chips of this type are severely deformed and have a characteristic curl. On the other hand, some materials, such as gray cast iron, lack the ductility necessary for appreciable plastic deformation. Consequently, the compressed material ahead of the tool fails in brittle fracture, sometimes along the shear front, producing small fragments. Such chips are termed *discontinuous* or *segmented* (see Figure 21-13).

A variation of the continuous chip, often encountered in machining ductile materials, is associated with a "built-up" edge (BUE) formation on the cutting tool. The local high temperature and extreme pressure in the cutting zone cause the work material to adhere or pressure-weld to the cutting edge of the tool forming the built-up edge, rather like a dead metal zone in the extrusion process.

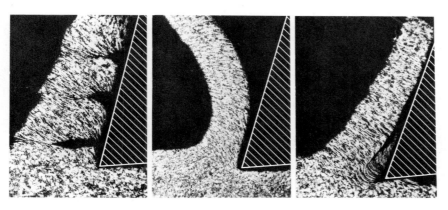

FIGURE 21-13 Three characteristic types of chips. (Left to right) Discontinuous, continuous, and continuous with built-up edge. Chip samples produced by "quick stop" techniques. (Courtesy Cincinnati Milacron, Inc.)

Although this material protects the cutting edge from wear, it modifies the geometry of the tool. BUE's are not stable and will break off periodically, adhering to the chip or passing under the tool and remaining on the machined surface. Built-up edge formation can be eliminated or minimized by reducing the depth of cut, increasing the cutting speed, using positive rake tools, applying a coolant, or changing cutting tool materials.

Mechanics of Machining. Orthogonal machining has been defined as a two-force system. Consider Figure 21-12, which shows a free-body diagram of a chip that has been separated at a shear plane. It is assumed that the resultant force R acting on the back of the chip is equal and opposite to the resultant force R' acting on the shear plane. The resultant R is composed of the friction force F and the normal force N acting on the tool/chip interface contact area. The resultant force R'' is composed of a shear force F_s and normal force F_n acting on the shear plane area A_s. Since neither of these two sets of forces can usually be measured, a third set is needed which can be measured, using a dynamometer (force transducer), mounted either in the workholder or the tool holder. Note that this set has resultant R''', which is equal in magnitude to all the other resultant forces in the diagram. The resultant force R''' is composed of a cutting force F_c and a tangential (normal) force F_t. Now it is necessary to express the desired forces (F_s, F_n, F, N) in terms of the measured dynamometer components, F_c and F_t, and appropriate angles. In order to do this, a circular force diagram is developed in which all six forces are collected in the same force circle. This is shown in Figure 21-14. The only symbol in this figure as yet undefined is β, which is the angle between the normal force N and the resultant R. It is used to describe the friction coefficient μ on the tool/chip interface area, which is defined as F/N, so that

$$\beta = \tan^{-1} \mu = \tan^{-1} \frac{F}{N} \tag{21-11}$$

The friction force F and its normal N can be shown to be

$$F = F_c \sin \alpha + F_t \cos \alpha \tag{21-12}$$
$$N = F_c \cos \alpha - F_t \sin \alpha \tag{21-13}$$

where

$$R = \sqrt{F_c^2 + F_t^2} \tag{21-14}$$

Notice that in the special situation where the back rake angle is zero, $F = F_t$ and $N = F_c$, so that, in this orientation, the friction force and its normal can be directly measured by the dynamometer.

The forces parallel and perpendicular to the shear plane can be shown (from the force circle diagram in Figure 21-14) to be

$$F_s = F_c \cos \phi - F_t \sin \phi \tag{21-15}$$
$$F_n = F_c \sin \phi + F_t \cos \phi \tag{21-16}$$

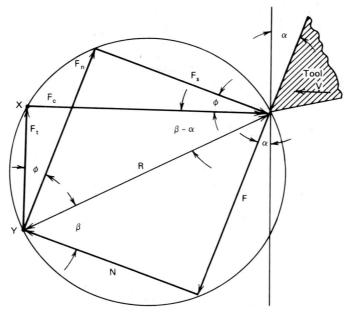

FIGURE 21-14 Circular force diagram used to derive equations for F_s, F_n, F, and N as functions of F_c, F_t, ϕ, α, and β.

F_s is of particular interest as it is used to compute the shear stress on the shear plane. This shear stress is defined as

$$\tau_s = \frac{F_s}{A_s} \qquad (21\text{-}17)$$

where

$$A_s = \frac{tw}{\sin \phi} \qquad (21\text{-}18)$$

recalling that t was the depth of cut and w was the width of the workpiece. The shear stress is therefore

$$\tau_s = \frac{F_c \sin \phi \cos \phi - F_t \sin^2 \phi}{tw} \text{ psi} \qquad (21\text{-}19)$$

For a given polycrystalline metal, this shear stress is a material constant, not sensitive to variations in cutting parameters, tool material, or the cutting environment. Figure 21-15 gives some typical values for the flow stress for a variety of metals, plotted against hardness.

It is the objective of some metal-cutting researchers to be able to derive (predict) the shear stress τ_s and the shear direction, ϕ, from dislocation theory. Correlations of the shear stress with metallurgical measures like hardness or dislocation stacking fault energy have been useful in these efforts.

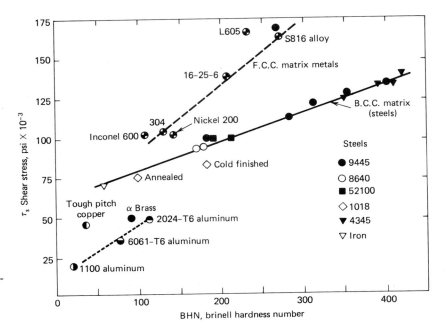

FIGURE 21-15 Shear stress, τ_s variation with the Brinell hardness number for a group of steels and aerospace alloys. Data of some selected fcc metals are also included. (Adapted with permission from S. Ramalingham and K. J. Trigger, Advances in Machine Tool Design and Research, 1971, Pergamon Press.)

Energy and Power in Machining. The cutting force system in a conventional, oblique chip formation process, shown schematically in Figure 21-16, has three components:

1. F_c = primary cutting force acting in the direction of the cutting velocity vector. This force is generally the largest force and accounts for 99% of the power required by the process.
2. F_f = feed force acting in the direction of the tool feed. This force is usually about 50% of F_c but accounts for only a small percentage of the power required because feed rates are usually small compared to cutting speeds.
3. F_r = radial or thrust force acting perpendicular to the machined surface. This force is typically about 50% of F_f and contributes very little to power requirements because velocity in the radial direction is negligible.

Figure 21-16 also shows the general relationship between these forces and speed, feed, and depth of cut. These figures cannot be used to determine forces for a specific process, as the following discussion will explain.

The energy per unit time, P or power required for cutting is

$$P = F_c V \text{ ft-lb/min} \qquad (21\text{-}20)$$

The horsepower at the spindle of the machine is therefore

$$\text{HP} = \frac{F_c V}{33,000} \qquad (21\text{-}21)$$

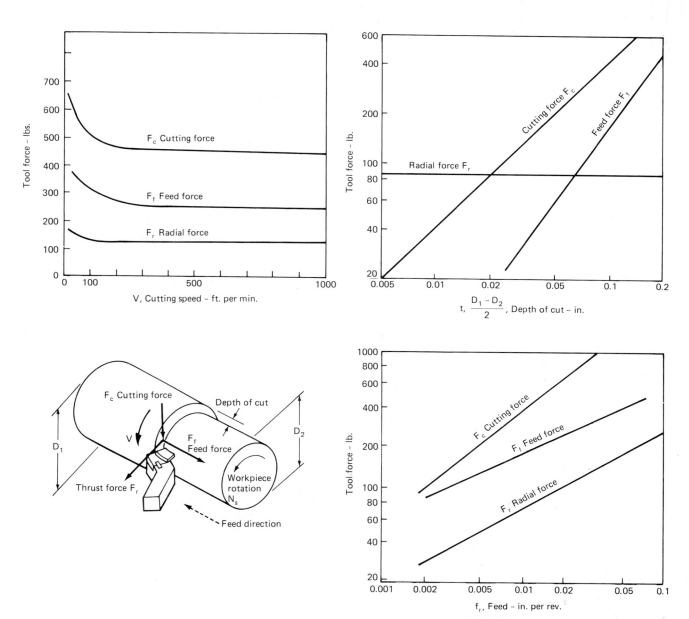

FIGURE 21-16 Three measurable components of forces acting on a single point tool (oblique machining) with effects on the forces of varying speed, depth of cut, and feed.

In metal cutting a very useful parameter is called the unit or specific horsepower HP_s which is defined as

$$HP_s = \frac{HP}{MRR} \text{ (hp/in}^3\text{/min)} \tag{21-22}$$

In turning, for example, where $MRR = 12Vf_rt$,

$$HP_s = \frac{F_c}{396{,}000\ f_rt} \tag{21-23}$$

Thus this term represents the approximate power needed at the spindle to remove a cubic inch of metal per minute. Specific power factors for some common materials are given in Table 21-3. Specific horsepower is related to and correlates well with shear stress for a given metal. Notice the similarity between equations 21-19 and 21-23. The major difference is that unit power is sensitive to material properties (like hardness) and rake angle, depth of cut, and feed. τ_s is sensitive to material properties only.

Essentially, the majority of the energy is consumed in shear and tool/chip interface friction.

$$U = U_s + U_f \tag{21-24}$$

Where total energy is

$$U = \frac{F_c}{f_rt} \tag{21-25}$$

and specific shear energy is

$$U_s = \frac{F_sV_s}{f_rtV} = \frac{F_s \cos \alpha}{f_rV \cos (\phi - \alpha)} \tag{21-26}$$

Specific friction energy is

$$U_f = \frac{FV_c}{f_rVt} = \frac{Fr_c}{f_rt} \tag{21-27}$$

Usually 30% to 40% of the total energy goes into friction and 60% to 70% into the shear process.

Specific power can be used in a number of ways. First, it can be used to estimate the motor horsepower required to perform a machining operation for a given material by multiplying HP_s values from the table by the approximate MRR for the process. The motor horsepower, HP_m, is then

$$HP_m = \frac{HP_s \times MRR \times CF}{E} \tag{21-28}$$

E is the efficiency of the machine. This factor accounts for the power needed to overcome friction and inertia in the machine and drive moving parts. Usually 80% is used. Correction factors (CF) may also be used to account for variations

TABLE 21.3 Values for Specific or Unit HP$_s$ For Various Metals During Metal Removal

Material	Hardness Bhn or R	HP$_s$ hp/in^3/min	kW/cm^3/min	Comments
Steels, including plain carbon, alloy, tool hot or cold rolled or cast	85–200	1.1	0.050	Values assume normal feed ranges and sharp tools. Multiply value by 1.25 for a dull tool.
	35–40R$_c$	1.4	.064	
	40–50R$_c$	1.5	.068	
	50–55R$_c$	2.0	.091	
	55–58R$_c$	3.4	.155	
Cast iron	100–190	0.7–1.0	0.03–0.045	Add 10% for milling to table value.
	190–300	1.4–1.6	0.05–0.07	
Stainless steels	150–450	1.2–1.4	0.05–0.068	
Iron base alloys	180–320	1.2–1.6	0.055–0.073	High temperature alloys
Nickel alloys	80–360	1.8–2.0	0.82–0.091	
Nickel/Cobalt base alloys	200–360	2.0–2.5	0.09–0.11	High temperature materials
Aluminum				
2014-T6, 2017-T4	30–150 at 500 kg	0.25 to	.014 to	
6064-T6		0.34	.016	
Pure-108	50 ft	0.16	.007	
Hard (rolled)	Hard	0.33	.015	
Magnesium alloys	40–90 at 500 kg	0.16	.007	
Copper	50 R$_B$	0.9–1.0	0.041–0.046	
Copper alloys	10–80 R$_B$	0.5–0.6	0.022–0.030	
	80–100 R$_B$	0.8–1.0	0.036–0.046	
Titanium	250–375	1.8–2.0	0.82–0.091	
Tungsten, Tantalum	210–320	2.6–2.8	0.12–0.13	

Sources: Authors' data; Monarch Machine Tool Company; GE Company (Handbooks for High-Efficiency Metal Cutting; METCUT (*Machining Data Handbook*).

in cutting speed, feed, and rake angle. There is usually a tool wear correction factor of 1.25 used to account for the fact that dull tools use more power than sharp tools.

The primary cutting force F_c can be roughly estimated according to

$$F_c \cong \frac{HP_s \times MRR \times 33,000}{V} \qquad (21\text{-}29)$$

This type of estimate of the major force F_c is useful in analysis of deflection and vibration problems in machining and in the proper design of work-holding devices, as these devices must be able to resist movement and deflection of the part during the process.

In general, increasing the speed, the feed, or the depth of cut will increase the power requirement. Doubling the speed doubles the HP directly. Doubling the feed or the depth of cut doubles the cutting force F_c. In general, increasing the speed does not increase the cutting force F_c, which has always been a puzzle. However, recent research has shown that the periodicity or spacing of the shear fronts remains constant for a given material regardless of the cutting speed. Doubling the velocity essentially doubles the number of shear fronts produced in a given amount of time. Put another way, if speed is doubled, chip length is doubled for the same amount of cutting time. For constant shear front spacing (constant lamella size) this means twice as many shear fronts. Therefore energy is doubled. Cutting force F_c, on the other hand, reflects a change in F_s, where $F_s = \tau_s A_s$, τ_s = shear stress, and A_s = shear area. Neither of these two quantities was changed by the change in speed. Thus F_c remains constant. Since a change in feed or depth of cut changes area, A_s directly, changes in f_r or t directly effect F_c. However, speed had a strong effect on tool life because most of the input energy is converted into heat, which raises the temperature of the chip, the work, and the tool, to the latter's detriment.

Equation 21-28 can be used to estimate the maximum depth of cut, t, for a process as limited by the available power.

$$t_{max} = \frac{HP_m \times E}{12\, HP_s\, V f_r (CF)} \qquad (21\text{-}30)$$

Heating and Temperature in Metal Cutting.   In metal cutting, the power put into the process $(F_c V)$ is largely converted to heat, elevating the temperatures of the chip, the workpiece, and the tool. These three elements of the process, along with the environment (which includes the cutting fluid), act as the heat sinks. Figure 21-17 shows the distribution of the heat to these three sinks as a function of cutting speed.

There are three main *sources* of heat. Listed in order of their heat-generating capacity, they are

1. The shear zone itself, where plastic deformation results in the major heat source. Most of this heat stays in the chip.

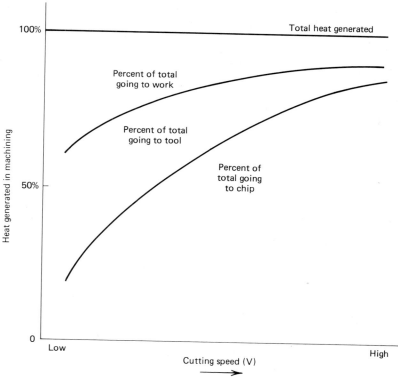

FIGURE 21-17   Distribution of heat generated in machining to the chip, tool, and workpiece. Heat going to the environment is not shown.

2. The tool/chip interface contact region, where additional plastic deformation takes place in the chip and there is considerable heat generated due to sliding friction.

3. The flank of the tool, where the freshly produced workpiece surface rubs the tool.

There have been numerous experimental techniques developed to measure cutting temperatures and some excellent theoretical analysis of this "moving" multiple-heat-source problem. Space does not permit us to explore this problem in depth. Figure 21-18a shows the effect of cutting speed on the tool/chip interface temperature. The rate of wear of the tool at the interface can be shown to be directly related to temperature (see Figure 21-18b). Because cutting forces are concentrated on small areas near the cutting edge, these forces produce large pressures. The tool material must be hard to resist wear and tough to resist cracking and chipping. Tools used in interrupted cutting, like milling, must be able to resist impact loading as well. Tool materials must sustain these properties at elevated temperatures, as shown in Figure 21-19, which relates the hardness of various tool materials to temperature.

The challenge to manufacturers of cutting tools has always been to find materials that satisfy these severe conditions. Cutting tool materials that do not lose

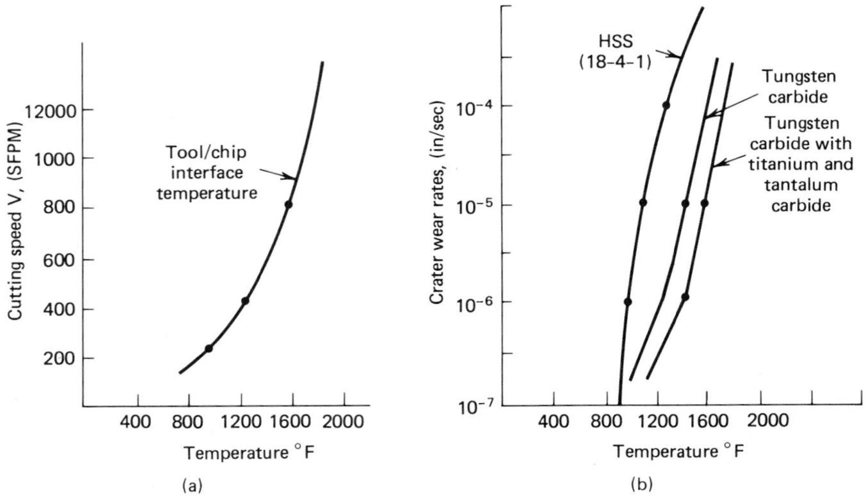

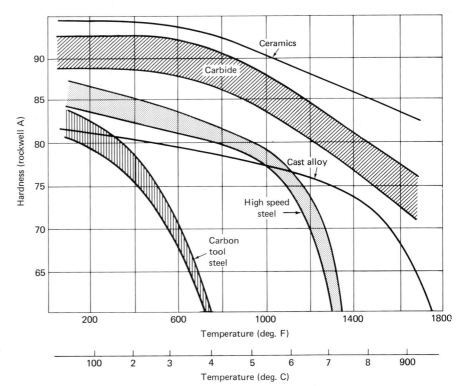

FIGURE 21-18 Typical relationships of temperature to cutting speed and crater wear for machining steels.

(a)

(b)

FIGURE 21-19 Hardness of tool materials decreasing with increasing temperature. Some materials display a more rapid drop in hardness above some temperatures. (From *Metal Cutting Principles*, 2nd ed; courtesy Ingersoll Cutting Tool Company.)

hardness at the high temperatures associated with high speeds are said to have "hot hardness," but obtaining this property usually requires a trade-off in toughness, as hardness and toughness are generally opposing properties. The next chapter will address cutting tool materials in more depth.

The process of shaping has been used in this chapter to introduce the metal-cutting processes. From a consideration of the relative motions between the tool and the workpiece, shaping and planing are the simplest of all machining operations, and the machines that do the operations are among the simplest of all machine tools. Straight-line cutting motion with a single-point cutting tool is used to generate a flat surface. The tool is fed at right angles to the cutting motion between successive strokes. In addition to plain flat surfaces, the shapes most commonly produced on the shaper and planer are those illustrated in Figure 21-20. Relatively skilled workers are required to operate shapers and planers, and most of the shapes that can be produced on them also can be made by much more productive processes, such as milling, broaching, or grinding. Consequently, except for certain special types, planers that will do only planing have become obsolete, and shapers are used mainly in tool and die work, in low volume production, or in the manufacture of gear teeth (see Chapter 31).

Shapers, as machine tools, usually are classified according to their general design features as follows:

1. Horizontal.
    a. Push-cut.
    b. Pull-cut or draw cut shaper.
2. Vertical
    a. Regular or slotters.
    b. Keyseaters.
3. Special.

They are also designed as to the type of drive employed—*mechanical drive* or *hydraulic drive*. Most shapers are of the *horizontal push-out* type (see Figure 21-2), where cutting occurs as the ram *pushes* the tool across the work.

Workholding Devices for Shapers.   On horizontal push-cut shapers, the work is usually held in a heavy vise mounted on the top side of the table. Shaper vises have a very heavy movable jaw, because the vise must often be turned so that the cutting forces are directed against this jaw.

In clamping work in a shaper vise, care must be exercised to make sure that it rests solidly against the bottom of the vise (on parallel bars) so that it will not be deflected by the cutting force, and that it is held securely yet not distorted by the clamping pressure.

Figure 21-21 illustrates the use of parallel bars for raising work to the proper height in the vise jaws and also several methods of clamping rough and irregularly shaped work. These latter procedures prevent the clamping action from

The Shaping and Planing Processes

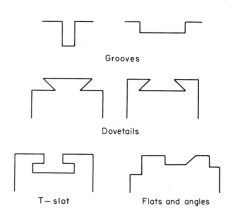

Grooves

Dovetails

T–slot   Flats and angles

FIGURE 21-20   Types of surfaces commonly machined by shaping and planing.

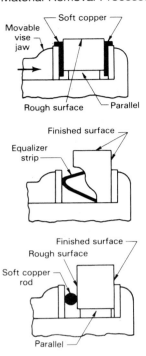

FIGURE 21-21 Methods of clamping workpieces in a shaper vise.

tilting the work in the vise. Work that cannot be held conveniently in a vise can be clamped directly to the top or sides of the table, using T-slots in the table.

Cutting Tools for Shaping. Most shaping is done with simple high-speed steel or carbide-tipped cutting tool bits held in a heavy, forged tool holder, as shown in Figure 21-1c. Although shapers are versatile tools, the precision of the work done on them is greatly dependent on the operator. Feed dials on shapers nearly always are graduated in 0.001-in. divisions, and work is seldom done to greater precision than this. A tolerance of 0.002 to 0.003 in. is desirable on parts that are to be machined on a shaper, because this gives some provision for errors due to clamping, possible looseness or deflection of the table, and deflection of the tool and ram during cutting.

Planing. Planing can be used to produce horizontal, vertical, or inclined flat surfaces on workpieces that are too large to be accommodated on shapers. However, planing is much less efficient than other basic machining processes that will produce such surfaces. Consequently, planing and planers have largely been replaced by planer-type milling machines or machines that can do both milling and planing. These will be discussed in Chapter 25.

Figures 21-1d and 21-22 show the basic components and motions of planers. In most planing, the action is opposite to that of shaping. The work is moved past one or more stationary, single-point cutting tools. Because a large and heavy workpiece and table must be reciprocated at relatively low speeds, several tool heads are provided, often with multiple tools in each head. In addition, many planers are provided with tool heads arranged so that cuts occur on both directions of the table movement. However, because only single-point cutting tools are used and the cutting speeds are quite low, planers are low in productivity as compared with some other types of machine tools.

Planers are made in four basic types. Figure 21-22 depicts the most common, double-housing type. It has a closed housing structure, spanning the reciprocating worktable, with a crossrail supported at each end on a vertical column and carrying two toolheads. An additional toolhead usually is mounted on each column, so that four tools (or four sets) can cut during each stroke of the table. The closed-frame structure of this type of planer limits the size of the work that can be machined. Open-side planers have the crossrail supported on a single column. This provides unrestricted access to one side of the table to permit wider workpieces to be accommodated. Some open-side planers are convertible, in that a second column can be attached to the bed when desired so as to provide added support for the crossrail.

Workholding and Setup on Planers. Workpieces in planers are usually large and heavy. They must be securely clamped to resist large cutting forces and the high inertia forces that result from the rapid velocity changes at the ends of the strokes. Special stops are provided at each end of the workpiece to prevent it from shifting.

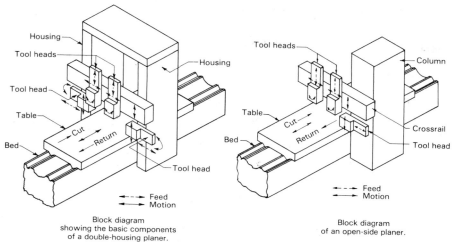

Block diagram
showing the basic components
of a double-housing planer.

Block diagram
of an open-side planer.

Interchangeable multiple tool
holder for use on planers. (*Courtesy
Gebr. Boehringer GMBH.*)

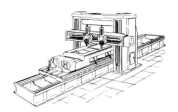

Double Housing Planer

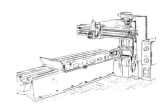

Open Side Planer

FIGURE 21-22 Elements of double-housing and open-side planers. (Schematics from *Manufacturing Producibility Handbook*; courtesy General Electric Company. Photograph, courtesy Gebr. Boehringer GMBH.)

Considerable time usually is required to set up the planer, thus reducing the time the machine is available for producing chips. Sometimes special setup plates are used for quick setup of the workpiece. Another procedure is to use two tables. Work is setup on one table while another workpiece is being machined on the other. The tables can be fastened together for machining long workpieces.

The large workpieces can usually support heavy cutting forces, so large depths of cut are recommended. This helps reduce the cost. Consequently, planer tools usually are quite massive and can sustain the large cutting forces. Usually, the main shank of the tools is made of plain carbon steel, and a high-speed or carbide tip is clamped or brazed to it. Chip breakers should be used to avoid long and dangerous chips in ductile materials.

Theoretically, planers have about the same precision as shapers. The feed and other dimension-controlling dials usually are graduated in 0.001-in. divisions. However, because larger and heavier workpieces are usually involved, and much longer beds and tables, the working tolerances for planer work should be somewhat greater than for shaping.

Summary. In this chapter the basics of machining have been presented. Chapters 23 through 31 will elaborate on the machine tools that have been developed to perform specific machining operations. Often a machine can perform several of the basic processes. As described in Chapter 1 and Chapter 29, machining centers are now widely used. Figure 21-5 and Table 21.2 should be carefully studied; they summarize the relationships between the basic processes and the machine tools that can be used to perform these processes. Generally speaking, these machines will be of the A(2) or A(3) level of automation. Machining centers, which are numerical control machines, are A(4). Machining centers have automatic tool change capability and are usually capable of milling, drilling, boring, reaming, tapping (hole threading), and other minor machining processes. For particular machines, you will need to become familiar with new terminology, but in general all machining processes will need inputs concerning rpm (given that you selected the cutting speed), feeds, and depths of cut. Note also from these tables that the same process can be performed on two or more different machine tools. There are many ways to produce flat surfaces, internal and external cylindrical surfaces, and special geometries in parts. Generally, the quantity to be made is the driving factor in the selection of processes, as we shall see in later discussions.

Review Questions

1. Why has the metal-cutting process resisted theoretical solution for so many years?
2. What are the variables that must be considered in understanding a machining process?
3. Which of the basic chip formation processes are single point and which are multiple point?
4. Is feed related to speed in all machining operations?
5. Milling has two feeds. What are they, and which one does the operator need to know about?
6. What is the fundamental mechanism of chip formation?
7. What is the main implication of Figure 21-11, given that this series of micrographs was made at very low speed?
8. What is the difference between oblique machining and orthogonal machining?
9. Suppose you are shaping a block 4" × 7" × 2". Why should the block be cut in the 4-in. direction rather than in the 7-in. direction?
10. If the job described in problem 9 had been done on a planer rather than a shaper, is it likely that the setup would have been made to cut the block in the 7-in. direction?

11. How does the magnitude of the strain and strain rate values of metal cutting compare to tensile testing?

12. Why is titanium such a difficult metal to machine? (Note its high value of HP$_s$).

13. Explain why you get segmented or discontinuous chips when you machine cast iron.

14. Why is shear stress so important?

15. Which of the three cutting forces in oblique cutting consumes most of the power? How is that power divided and where does the energy consumed ultimately go?

16. What would you have to know about a process in order to estimate the primary cutting force?

17. How is cutting speed related to tool wear?

18. What is the relationship between hardness and temperature in metal-cutting tool materials?

19. What are the benefits in machining of having a large shear angle?

20. What is the effect on ϕ, the shear angle, of having a negative rake tool versus a positive rake tool?

21. Why are not all tools made with large positive rakes?

22. What is the basic difference between a shaper and a planer?

23. For what type of work are shapers best suited?

24. How is shaper feed expressed?

25. Why is the shaper seldom used for production work?

26. Why is it seldom necessary to use carbide cutting tools on a shaper?

27. Why is it difficult to use high cutting speeds on a planer?

28. How does the planer feed differ from shaper feed?

29. Why are such large tools needed in planers?

30. Why is it so desirable to make more than one cut simultaneously on a planer?

Problems

1. Suppose you have the following data obtained from a metal-cutting experiment (orthogonal machining). Compute the shear angle, the shear stress, the specific energy, the shear strain and the coefficient of friction at tool/chip interface. How do your HP$_s$ and τ_s values compare with the values found in Chapter 21?

Machining data for 1020 steel, as received, machined in air with a K3H carbide tool, orthogonally (tube cutting on lathe) with tube OD = 2.875. The cutting speed was 530 fpm. The tube wall thickness was .200 inches. The back rake angle was zero for all cuts.

Data

Run number	F_c	F_t	feed ipr × 1/1000	chip ratio r_c	ϕ	τ	μ	HP$_s$
1	330	295	4.89	.331				
2	308	280	4.89	.381				
3	410	330	7.35	.426				
4	420	340	7.35	.426				
5	510	350	9.81	.458				
6	540	395	9.81	.453				

2. Suppose you weighted a short chip fragment. You measured the length of the short chip fragment, 1_c, with $1_c \times W_c \times t_c$ = volume of chip fragment. The density of this metal is ρ, lbs/in.^{3} Can you obtain the chip thickness ratio, r_c?

3. For the data in problem 1, determine the specific shear energy and the specific friction energy.

4. Suppose you want to shape a block of metal 7 in. wide and 4 in. long, using a shaper as set up in Figure 21-1. You have determined for this metal that the cutting speed should be 25 sfpm, the depth of cut needed here for roughing is 0.25 in., and the feed will be 0.1 in. per stroke. Determine the approximate crank rpm and then estimate the cutting time and the MRR.

5. Could you have saved any time in problem 4 by cutting the block in the 7-in. direction? Redo with $L = 7$ and $W = 4$ inches.

6. Derive equations for F and N using the circular force diagram. (*Hint:* Make a copy of the diagram. Extend a line from point X intersecting force F perpendicularly. Extend a line from point Y intersecting the previous line perpendicularly. Find the angle α made by these constructions.)

7. Derive equations for F_s and F_n using the circular force diagram. *Hint:* Construct a line through X parallel to vector F_n. Extend vector F_s to intersect this line. Construct a line from X perpendicular to F_n. Construct a line through point Y perpendicular to the line through X.

8. Derive equation (21-2), the shaping cutting speed.

9. How many strokes per minute would be required to obtain a cutting speed of 36.6 m (120 feet) per minute on a typical mechanical-drive shaper if a 254-mm (10-in.) stroke is used?

10. How much time would be required to shape a flat surface 254 mm (10 in.) wide and 203 mm (8 in.) long on a hydraulic-

drive shaper, using a cutting speed of 45.7 m (150 feet) per minute, a feed of 0.51 mm (0.020 in.) per stroke, and an overrun of 12.7 mm (½ in.) at each end of the cut?

11. What is the metal-removal rate in problem 10 if the depth of cut is 6.35 mm (¼ in.)?

12. If the work material in problem is 11 gray cast iron, what will be the power required?

13. A hydraulic-drive shaper has a 5.6-kW (7 ½-hp) motor, and 75% of the motor output is available at the cutting tool. The specific power for cutting a certain metal is 0.03 W/mm³ (.67 hp/in.³/min). What is the maximum depth of cut that can be taken in shaping a surface in this material if the surface is 305 × 305 mm (12 × 12 in.), the feed is 0.64 mm (0.025 in.) per stroke, and the cutting speed is 54.9 mm (180 feet) per minute?

Cutting Tools for Machining

Success in metal cutting depends upon the selection of the proper cutting tool (material and geometry) for a given work material. A wide range of cutting tool materials is available with a variety of properties, performance capabilities, and cost. These include high carbon steels and low/medium alloy steels, high-speed steels, cast cobalt alloys, cemented carbides, cast carbides, coated carbides, coated high-speed steels, ceramics, sintered polycrystalline cubic boron nitride (CBN), sintered polycrystalline diamond, and single-crystal natural diamond. Figure 22-1 presents a chronological rating of cutting tool materials, showing the rapid advances that have occurred in this field in the last two decades. The tool materials are rated by their permissible cutting speed in machining steel

Introduction

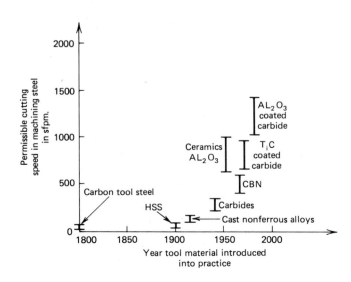

FIGURE 22-1 The acceleration of cutting tool material technology measured by permissible cutting speed (sfpm) for machining steel.

materials. Figure 22-2 outlines the input variables that influence the tool material selection decision.

The cutting tool is the most critical part of the machining system. The cutting tool material, cutting parameters, and tool geometry selected directly influence the productivity of the machining operation. The elements influencing the decision are

- Work material characteristics (chemical and metallurgical state).
- Part characteristics (geometry, accuracy, finish, and surface-integrity requirements).
- Machine tool characteristics including the workholders (adequate rigidity with high horsepower, and wide speed and feed ranges).
- Support systems (operator's ability, sensors, controls, methods of lubrication, and chip removal).

Tool material technology is advancing rapidly, enabling many difficult-to-machine materials to be machined at higher removal rates and/or cutting speeds

FIGURE 22-2 The selection of the cutting tool (material, geometry) and the cutting conditions for a given application depends upon many variables.

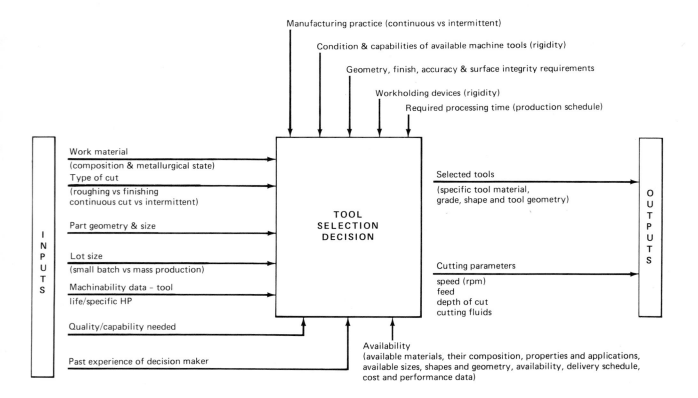

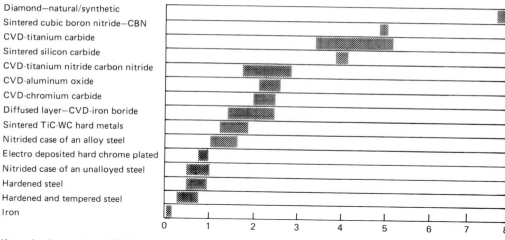

Knoop hardness scale — 1,000 Kp/mm$_2$

FIGURE 22-3 Vickers hardness ranges for various cutting tool materials.

with greater performance reliability. Higher speed and/or removal rates usually improve productivity. Predictable tool performance is essential when machine tools are computer-controlled and have minimal operator interaction. Long tool life is desirable when machines are placed in cellular manufacturing systems.

The cutting tool is subjected to severe conditions. Tool temperatures of 1000° Celsius, severe friction, and high local stresses require that the tool have these characteristics.

1. High hardness (Figure 22-3).
2. Resistance to abrasion, wear, chipping of the cutting edge.
3. High toughness (impact strength).
4. High hot hardness (Figure 21-19).
5. Strength to resist bulk deformation.
6. Good chemical stability (inertness or negligible affinity with the work material).
7. Adequate thermal properties.
8. High elastic modulus (stiffness).
9. Consistent tool life.
10. Correct geometry and surface finish.

Table 22.1 compares these properties for various cutting tool materials. Figure 22-3 compares various tool materials on the basis of hardness, the most critical characteristic.

Naturally, it would be most convenient if these materials were also easy to fabricate, readily available, and inexpensive, since cutting tools are routinely replaced. Obviously many of the requirements conflict, and therefore, tool selection will always require trade-offs.

TABLE 22.1 Salient Properties of Cutting Tool Materials[a]

	Carbon and Low/Medium Alloy Steels	High-speed Steels	Sintered (demented) Carbides	Coated HSS	Coated Carbides	Ceramics	Polycrystalline CBN	Diamond
Toughness	← decreasing →							
Hot hardness	← increasing →							
Impact strength	← decreasing →							
Wear resistance	← increasing →							
Chipping resistance	← decreasing →							
Cutting speed	← increasing →							
Depth of cut	light to medium	light to heavy	light to heavy	light to heavy	light to heavy	light to heavy	light to heavy	very light for single crystal diamond
Finish obtainable	rough	rough	good	good	good	very good	very good	excellent
Method of manufacture	wrought	wrought cast, HIP sintering	cold pressing and sintering, PM	PVD[b] after forming	CVD[c]	cold pressing and sintering or HIP sintering	high pressure– high temperature sintering	high pressure– high temperature sintering
Fabrication	machining and grinding	machining and grinding	grinding	machining and grinding, coating	grinding before coating	grinding	grinding and polishing	grinding and polishing
Thermal shock resistance	← increasing →							
Tool material cost	← increasing →							

[a] Overlapping characteristics exist in many cases. Exceptions to the rule are very common. In many classes of tool materials a wide range of composition and properties is obtainable.
[b] Physical vapor deposition
[c] Chemical vapor disposition

Cutting Tool Materials

In nearly all machining operations, cutting speed and feed are limited by the capability of the tool material. Speeds and feeds must be kept low enough to provide for an acceptable tool life. If not, the time lost changing tools may outweigh the productivity gains from increasing cutting speed.

High-speed steels and cemented carbides (coated and uncoated) are currently the most extensively used tool materials. Diamond and CBN are used for special applications in which, in spite of higher cost, their use is justified. Cast cobalt alloys are being phased out because of the high raw material cost and the increasing availability of alternate tool materials. New ceramic materials are being introduced that will have significant impact on future manufacturing productivity.

Tool requirements for other processes that use noncontacting tools, as in electro-discharge machining (EDM) and electro-chemical machining (ECM), or *no tools at all* (as in laser machining), are discussed in Chapter 32. Grinding abrasives will be discussed in Chapter 27.

Tool Steels.   Carbon steels and low/medium alloy steels, called *tool steels*, were once the most common cutting tool materials. Plain carbon steels of 0.90% to 1.30% carbon when hardened and tempered have good hardness and strength and adequate toughness and can be given a keen cutting edge. However, tool steels lose hardness at temperatures above 400°F because of tempering (refer to Figure 21-19) and have largely been replaced by other materials for metal cutting.

Low/medium alloy steels have alloying elements such as Mo and Cr, which improve hardenability, and W and Mo, which improve wear resistance. These tool materials also lose their hardness rapidly when heated to about their tempering temperature of 300°–650°F, and they have limited abrasion resistance. Consequently, low/medium alloy steels are used in relatively inexpensive cutting tools (such as drills, taps, dies, reamers, broaches, and chasers) for certain low-speed cutting applications when the heat generated is not high enough to reduce their hardness significantly. High-speed steels, cemented carbides, and coated tools are also used extensively to make these kinds of cutting tools. Though more expensive, they have longer tool life and improved performance.

High-Speed Steel.   First introduced in 1900 by Taylor and White, high-speed steel is superior to tool steel in that it retains its cutting ability at temperatures up to 1100°F, exhibiting good "red hardness." Compared with tool steel, it can operate at about double the cutting speed with equal life, resulting in its name *high-speed steel,* often abbreviated HSS.

High-speed steels contain significant amounts of W, Mo, Co, V, and Cr besides Fe and C. W, Mo, Cr, and Co in the ferrite as a solid solution provide strengthening of the matrix beyond the tempering temperature, thus increasing the hot hardness. Vanadium (V), along with W, Mo, and Cr, improves hardness and wear resistance. Extensive solid solutioning of the matrix also ensures good hardenability of these steels.

Although many formulations are used, (see Table 22.2), a typical composition is that of the 18–4–1 type (tungsten 18%, chromium 4%, vanadium 1%), called T1. Comparable performance can also be obtained by the substitution of approximately 8% molybdenum for the tungsten, referred to as a tungsten equivalent (W_{eq}). High-speed steel is still widely used for drills and many types of

TABLE 22.2   Nominal Percentage of Chemical Compositions of High-Speed Steels

AISI Tool Steel Type	Percentage of Chemical Composition[a]							
	C	Cr	V	W	Mo	Co	Cb	W_{eq}
Tungsten high-speed steel								
T1[b]	0.70	4.0	1.0	18.0				18.0
T2[b]	0.85	4.0	2.0	18.0				18.0
T3	1.0	4.0	3.0	18.0	0.60			18.8
T4	0.75	4.0	1.0	18.0	0.60	5.0		19.2
T5	0.80	4.25	1.0	18.0	0.90	8.0		19.8
T6	0.80	4.25	1.5	2.0	0.90	12.0		21.8
T7	0.80	4.0	2.0	14.0				14.0
T8	0.80	4.0	2.0	14.0	0.90	5.0		15.8
T9	1.20	4.0	4.0	18.0				18.0
T15	1.55	4.50	5.0	12.0	0.60	5.0		13.2
Molybdenum high-speed steels								
M1[b]	0.80	4.0	1.00	1.5	8.0			17.5
M2[b]	0.85	4.0	2.00	6.0	5.0			16.0
M3	1.00	4.0	2.75	6.0	5.0			16.0
M4	1.30	4.0	4.00	5.5	4.5			14.5
M6	0.80	4.0	1.50	4.0	5.0	12.0		14.0
M7[b]	1.00	4.0	2.00	1.75	8.75			19.25
M8	0.80	4.0	1.50	5.0	5.0		1.25	15.0
M10	0.85	4.0	2.00		8.0			16.0
High-hardness (molydbenum base) cobalt high-speed steels								
M30	0.85	4.0	1.25	2.0	8.0	5.0		18.0
M34	0.85	4.0	2.00	2.0	8.0	8.0		18.0
M35	0.85	4.0	2.00	6.0	5.0	5.0		16.0
M36	0.85	4.0	2.00	6.0	5.0	8.0		16.0
M41	1.10	4.25	2.00	6.75	3.75	5.0		14.25
M42	1.10	3.75	1.15	1.50	9.50	8.25		20.5
M43	1.20	3.75	1.60	2.75	8.00	8.25		18.75
M44	1.55	4.25	2.00	5.25	6.50	12.00		18.25
M45	1.25	4.25	1.60	8.25	5.0	5.50		18.25
M46	1.25	4.00	3.20	2.00	8.25	8.25		18.0

[a]Normal ranges of manganese, silicon, phosphorus, and sulfur are assumed. Balance Fe in all cases.
[b]Widely available

$$W_{eq} = 2(\%Mo) + \%W$$

general-purpose milling cutters and in single-point tools used in general machining. It has been almost completely replaced for high-production machining by carbides, coated carbides, and coated HSS.

High-speed steel tools are fabricated by three methods: cast, wrought, and sintered (using the powder metallurgy technique). Improper processing of cast and wrought products can result in carbide segregation, formation of large carbide particles and significant variation of carbide size, and nonuniform distribution of carbides in the matrix. The material will be difficult to grind to shape and will cause wide fluctuations of properties, inconsistent tool performance, distortion and cracking.

To overcome some of these problems, a powder metallurgy technique has been developed which uses the hot isostatic pressing (HIP) process on atomized, prealloyed tool steel mixtures. Because the various constituents of the PM alloys are ''locked'' in place by the compacting procedure, the end product is a more homogeneous alloy. PM high-speed-steel cutting tools exhibit better grindability, greater toughness, better wear resistance, and higher red (or hot) hardness, and they perform more consistently. They are about double the cost of regular HSS.

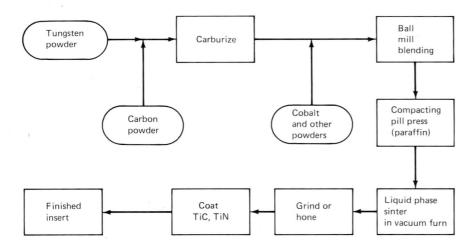

Tungsten is carburized in a high-temperature furnace, mixed with cobalt and blended in large ball mills. After ball milling, the powder is screened and dried. Paraffin is added to hold the mixture together for compacting. Carbide inserts are compacted using a pill press. The compacted powder is sintered in a high-temperature vacuum furnace. The solid cobalt dissolves some tungsten carbide, then melts and fills the space between adjacent tungsten carbide grains. As the mixture is cooled, most of the dissolved tungsten carbide precipitates onto the surface of existing grains. After cooling, inserts are finish ground and honed or used in the pressed condition.

FIGURE 22-4   The PM process for making cemented carbide insert tools.

Cast Cobalt Alloys.   Cast cobalt alloys, popularly known as *Stellite tools,* are cobalt-rich, chromium–tungsten–carbide cast alloys having properties and applications in the intermediate range between high-speed steel and cemented carbides. Though comparable in room temperature hardness to high-speed steel tools, cast cobalt alloy tools retain their hardness to a much higher temperature. Consequently, they can be used at higher cutting speeds than HSS tools (@ 25% higher). Cast cobalt alloys are hard as cast and cannot be softened or heat-treated.

Cast cobalt alloys contain a primary phase of Co-rich solid solution strengthened by Cr and W and dispersion-hardened by complex, hard, refractory carbides of W and Cr. Other elements added include V, B, Ni, and Ta. The casting provides a tough core and elongated grains normal to the surface. The structure is not, however, homogeneous.

Tools of cast cobalt alloys are generally cast to shape and finished to size by grinding. They are available only in simple shapes, such as single-point tools and saw blades, because of limitations in the casting process and expense involved in the final shaping (grinding). The high cost of fabrication is primarily due to the high hardness of the material in the as-cast condition.

Materials machinable with this tool material include plain carbon steels, alloy steels, nonferrous alloys, and cast iron.

Cast cobalt alloys are currently being phased out for cutting-tool applications because of increasing costs, shortages of strategic raw materials (Co, W, and Cr), and the development of other, superior tool materials at lower cost.

Carbides or Sintered Carbides.   These nonferrous alloys are called sintered (or cemented) carbides because they are manufactured by powder metallurgy techniques. See Figure 22-4. These materials became popular during World War II as they afforded a four- or five-fold increase in cutting speeds. The early versions, which are still widely used, had tungsten carbide as the major constituent, with a cobalt binder in amounts of 3% to 13%. Most carbide tools in use today are WC-based (either straight WC or multicarbides of W-Ti or W-Ti-Ta, depending upon the work material to be machined with cobalt as the binder. These tool materials are much harder, and chemically more stable; they have better hot hardness, high stiffness, and lower friction; and they operate at higher cutting speeds than HSS and cobalt alloys. They are more brittle and more expensive and use strategic metals (W, Ta, Co) more extensively.

Cemented carbide tool materials based on TiC have been developed primarily for auto industry applications using predominantly Ni and Mo as a binder. These are used for higher speed ($>$1000 ft./min.) finish-machining of steels, and some malleable cast irons.

Cemented carbide tools are available in insert form in many different shapes: squares, triangles, diamonds, and rounds. They can be either brazed or mechanically clamped onto the tool shank. The latter is more popular (see Figure 22-5) because when one edge becomes dull, the insert is rotated or turned over for a new edge. Brazed tools are more accurate than insert tools because the tip

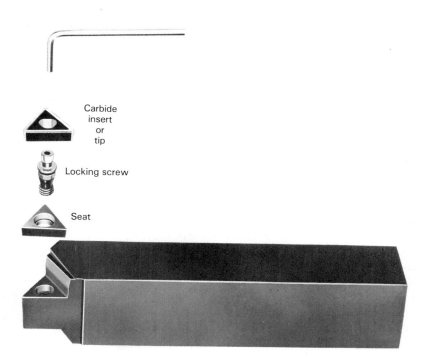

Carbide
insert
or
tip

Locking screw

Seat

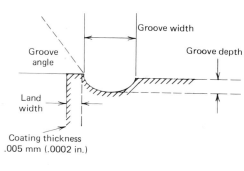

Groove width

Groove depth

Groove angle

Land width

Coating thickness
.005 mm (.0002 in.)

SECTION AA

FIGURE 22-5 (Top) Examples of throwaway carbide cutting tool tips with grooves on rake face. (Left) Components of a typical mounting holder. (Section AA) groove design on coated tool to reduce forces and breakup chips. (Courtesy General Electric.)

is permanent. Inserts can be purchased in the "as pressed" state or the insert can be ground to closer tolerances. Naturally, precision-ground inserts cost more. Any part tolerance less than ±.003 in. normally cannot be manufactured without radial adjustment of the cutting tool, even with ground inserts. If no radial adjustment is performed, precision-ground inserts should be used only when the part tolerance is between ±.003 and ±.006. Pressed inserts have an application advantage as the cutting edge is unground and thus does not leave grinding marks on the part after machining. Ground inserts can break under heavy cutting loads because the grinding marks on the insert produce stress concentrations that result in brittle fracture. Diamond grinding is used to finish carbide tools. Abusive grinding can lead to thermal cracks and premature (early) failure of the tool.

Since cemented carbide tools are relatively brittle, a 90-degree corner angle at the cutting edge is desired. To strengthen the edge and prevent edge chipping, it is rounded off by honing, or an appropriate chamfer or a negative land on the rake face is provided. A *chip groove* (see section AA of Figure 22-5) with a positive rake angle at the tool tip may also be used to reduce cutting forces without reducing the overall strength of the insert significantly. The groove also breaks up the chips for easier disposal by causing them to curl tightly.

For very low speed cutting operations, the chips tend to weld to the tool face and cause subsequent microchipping of the cutting edge. Cutting speeds are generally in the 150 to 600 ft./min. range. Higher speeds (>1000 ft/min) are recommended for certain less difficult-to-machine materials (such as aluminum alloys), and much lower speeds (100 ft/min) for more difficult-to-machine materials (such as titanium alloys). In interrupted cutting applications, it is important to prevent edge chipping by choosing the appropriate cutter geometry and cutter position with respect to the workpiece. For interrupted cutting, finer grain size and higher cobalt content improve toughness in straight WC-Co grades.

After use, carbide inserts (called disposable or "throwaway" inserts) are generally recycled by separating Ta, WC, and Co. This recycling not only conserves strategic materials but also reduces costs. A new trend is to regrind these tools for future use when the actual size of the insert is not of critical concern.

Ceramics.   Ceramics are made of pure aluminum oxide, Al_2O_3. Very fine particles are PM formed into cutting tips under a pressure of 20 to 28 tons/in^2 (267 to 386 MPa) and sintered at about 1800°F (1000°C). Unlike the case with ordinary ceramics, sintering occurs without a vitreous phase.

Ceramics usually are in the form of disposable tips. They can be operated at from two to three times the cutting speeds of tungsten carbide, almost completely resist cratering, usually require no coolant, and have about the same tool life at their higher speeds as tungsten carbide does at lower speeds. As shown in Table 22.3, ceramics are usually as hard as carbides but are more brittle (lower bend strength) and therefore require more rigid tool holders and machines in order to take advantage of their capabilities. Their hardness and chemical inertness make ceramics a good material for high-speed finishing and/or high-removal-rate

TABLE 22.3 **Properties of Cutting Tool Materials Compared for Carbides, Ceramics, HSS, and Cast Cobalt**

	Hardness Rockwell A or C	Transverse Rupture (bend) Strength ($\times 10^3$ psi)	Compressive Strength ($\times 10^3$ psi)	Modulus of Elasticity (E) ($\times 10^6$ psi)
Carbide C1–C4	90–95 R_A	250–320	750–860	89–93
Carbide C5–C8	91–93 R_A	100–250	710–840	66–81
High Speed Steel	86 R_A	600	600–650	30
Ceramic (oxide)	92–94 R_A	100–125	400–650	50–60
Cast cobalt	46–62 R_C	80–120	220–335	40

Exact properties depend upon materials, grain size, bonder content, volume fraction of each constituent, and method of fabrication.

machining applications of super alloys, hard-chill cast iron, and high-strength steels. Because ceramics have poor thermal and mechanical shock resistance, interrupted cuts and interrupted application of coolants can lead to premature tool failure. Edge chipping is usually the dominant mode of tool failure. Ceramics are not suitable for aluminum, titanium, and other materials that react chemically with alumina-based ceramics.

Coated Carbide Tools. In cutting tools, material requirements at the surface of the tool need to be abrasion-resistant, hard, and chemically inert to prevent the tool and the work material from interacting chemically with each other during cutting. A thin, chemically stable, hard refractory coating of TiC, TiN, or Al_2O_3 accomplishes this objective. The bulk of the tool is a tough, shock-resistant carbide which can withstand high-temperature plastic deformation and resist breakage. The result is a composite tool as shown in Figure 22-6. Surface treatments for cutting tools are summarized in Table 22.4.

To be effective, the coatings should be hard, refractory, chemically stable, and chemically inert to shield the constituents of the tool and the workpiece from interacting chemically under cutting conditions. The coatings must be fine-grained, free of binders and porosity. Naturally, the coatings must be metallurgically bonded to the substrate. Interface coatings are graded to match the properties of the coating and the substrate. The coatings must be thick enough to prolong tool life but thin enough to prevent brittleness.

Coatings should have a low coefficient of friction so the chips do not adhere to the rake face. TiC-coated tools were introduced in 1969. Coating materials now include single coatings of TiC, TiN, Al_2O_3, HfN, or HfC, and multiple coatings of Al_2O_3 or TiN on top of Al_2O_3 or TiC. Chemical vapor deposition (CVD) is the technique commonly used in the coating process. See Figure 22-7. The coatings are formed by chemical reactions that take place only on or near the substrate. Like electroplating, chemical vapor deposition is a process in which the deposit is built up atom by atom. It is therefore capable of producing

Titanium carbide remains as the basic material covering the substrate for strength and wear resistance. The second layer is aluminum oxide which has proven chemical stability at high temperatures and resists abrasive wear. The third layer is a thin coating of titanium nitride to give the insert a lower coefficient of friction and to reduce edge build-up.

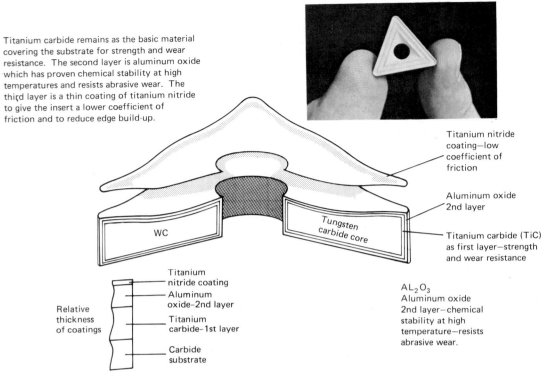

Titanium nitride coating—low coefficient of friction

Aluminum oxide 2nd layer

Titanium carbide (TiC) as first layer—strength and wear resistance

WC

Tungsten carbide core

Relative thickness of coatings

Titanium nitride coating
Aluminum oxide–2nd layer
Titanium carbide–1st layer
Carbide substrate

AL_2O_3 Aluminum oxide 2nd layer—chemical stability at high temperature—resists abrasive wear.

FIGURE 22-6 Triple-coated carbide tools provide resistance to wear and plastic deformation in machining of steel, abrasive wear in cast iron, and built-up edge formation.

TABLE 22.4 Surface Treatments for Cutting Tools

Process	Method	Hardness and Depth	Advantages	Limitations
Black Oxide	HSS cutting tools are oxidized in a steam atmosphere at 1000°	No change in prior steel hardness	Prevents built-up edge formation in machining of steel	Strictly for HSS tools
Nitriding Case Hardening	Steel surface is coated with nitride layer by use of cyanide salt at 900°F to 1600°F, or ammonia, gas or N_2 ions	To 72 R_C^a Case depth: 0.0001 in. to 0.100 in.	High production rates with bulk handling. High surface hardness. Diffuses into the steel surfaces. Simulates strain hardening	Can only be applied to steel. Process has embrittling effect because of greater hardness. Post-heat treatment needed for some alloys
Electrolytic Electroplating	The part is the cathode in a chromic acid solution; anode is lead. Hard chrome plating is the most common process for wear resistance	70–72 R_C^a 0.0001 to 0.100 in.	Low friction coefficient, anti-galling. Corrosion resistance. High hardness	Moderate production; pieces must be fixtured. Part must be very clean. Coating does not diffuse into surface, which can affect impact properties

TABLE 22.4 *Continued*

Process	Method	Hardness and Depth	Advantages	Limitations
Vapor Deposition Chemical Vapor Deposition (CVD)	Deposition of coating materials by chemical reactions in the gaseous phase. Reactive gases replace a protective atmosphere in a vacuum chamber. At temperatures of 1800°F to 1200°F, a thin diffusion zone is created between the base metal and the coating	To 84 R_C^a 0.0002 to 0.0004 in.	Large quantities per batch. Short reaction times reduce substrate stresses. Excellent adhesion, recommended for forming tools. Multiple coatings can be applied (TiN, TiC, AL_2O_3). Line of sight not a problem	High temperatures can affect substrate metallurgy, requiring post heat treatment, which can cause dimensional distortion (except when coating sintered carbides). Necessary to reduce effects of hydrogen chloride on material properties, such as impact strength. Usually not diffused. Tolerances of +.001 required for HSS tools
Physical Vapor Desposition (PVD sputtering)	Plasma is generated in a vacuum chamber by ion bombardment to dislodge particles from a target made of the coating material. Metal is evaporated and is condensed or attracted to substrate surfaces	To 84 R_C^a To 0.0002 in. thick	A useful experimental procedure for developing wear surfaces. Can coat substrates with metals, alloys, compounds, and refractories. Applicable for all tooling.	Not a high production method. Requires care in cleaning. Usually not diffused
PVD (Electron beam)	A plasma is generated in vacuum by evaporation from a molten pool which is heated by an electron beam gun	To 84 R_C^a To 0.0002 in. thick	Can coat reasonable quantities per batch cycle. Coating materials are metals, compounds, alloys, and refactories. Substrate metallurgy is preserved. Very good adhesion. Fine particle deposition. Applicable for all tooling	Parts require fixturing and orientation in line-of-sight process. Ultra cleanliness required
PVD/ARC	Titanium is evaporated in a vacuum and reacted with nitrogen gas. Resulting titanium nitride plasma is ionized and electrically attracted to the substrate surface. A high energy process with multiple plasma guns	To 84 R_C^a To 0.0002 in. thick	Process at 900°F preserves substrate metallurgy. Excellent coating adhesion. Controllable deposition of grain size and growth. Dimensions, surface finish, and sharp edges are preserved. Can coat all high speed steels without distortion	Parts must be fixtured for line-of-sight process. Parts must be very clean. No byproducts formed in reaction. Usually only minor diffusion

[a]Rockwell hardness values above 68 are estimates.

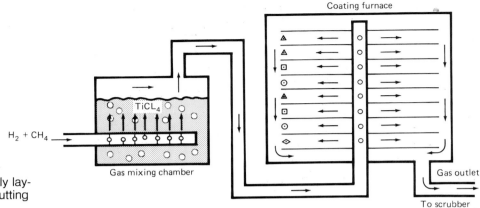

FIGURE 22-7 Chemical vapor deposition (CVD) is used to apply layers, (TiC, TiN, etc.) to carbide cutting tools.

deposits of maximum density and of closely reproducing fine detail on the substrate surface. Figure 22-7 illustrates a typical TiC-coating apparatus. A mixture of hydrogen, methane, and titanium tetrachloride gases is formed in the gas-mixing chamber. The gas mixture is routed into the coating chamber where the carbide tool blanks are heated to a temperature of about 1000°C. The tungsten carbide catalyzes the reaction which produces the TiC coating:

$$TiCl_4 + CH_4 \rightarrow TiC + 4HCl \uparrow \qquad (22\text{-}1)$$

Control of critical variables such as temperature, gas concentration, and flow pattern is required to assure adhesion of the coating to the substrate. The coating-to-subtrate adhesion must be better for cutting tool inserts than for most other coatings applications to survive the cutting pressure and temperature conditions without flaking off.

The purpose of multiple coatings is to tailor the coating thickness for prolonged tool life. Multiple coatings allow a stronger metallurgical bond between the coating and the substrate and provide a variety of protection processes for machining different work materials, thus offering a more general-purpose tool material grade. Figure 22-8 shows micrographs of multicoated WC tools. A very thin final coat of TiN coating (5 μm) can effectively reduce crater formation on the tool face by one to two orders of magnitude relative to uncoated tools.

Coated inserts of carbides are finding wide acceptance in many metal-cutting applications. Coated tools have two or three times the wear resistance of the best uncoated tools with the same breakage resistance. This results in a 50% to 100% increase in speed for the same tool life. Because most coated inserts cover a broader application range, fewer grades are needed, and therefore inventory costs are lower. Aluminum oxide coatings have demonstrated excellent crater wear resistance by providing a chemical/diffusion reaction barrier at the tool/chip interface, permitting a 90% increase in cutting speeds in machining some steels.

Coated carbide tools have progressed to the place where in the United States about 50% of the carbide tools used in metalworking are coated.

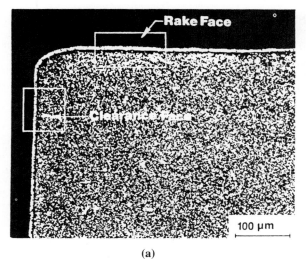

(a)

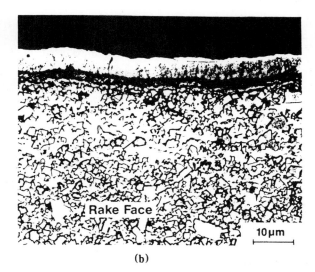

(b)

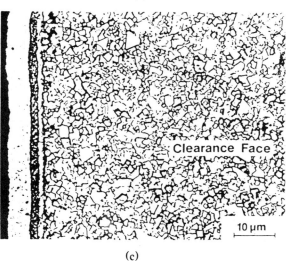

(c)

FIGURE 22-8 Micrographs of a multicoating [TiN-Ti (C,N)-TiC] over an engineered cemented tungsten carbide substrate (note that the top layer is TiN).

a. A low magnification micrograph showing the areas around the rake face and the clearance face of the tool.

b. Micrograph at higher magnification of an area around the rake face showing a thin cobalt-enriched straight WC layer near the rake face for increased toughness and edge strength.

c. Micrograph at higher magnification of an area around the clearance face showing a thin layer of cobalt-depleted multicarbide near the clearance face for increased wear resistance and high-temperature deformation resistance. (Courtesy R. Komanduri & Desai, J.D., *Tool Materials for Machining*, G.E. TIS Report No. 82CRD220, August 1982.)

TiN-Coated High-Speed Steel. Probably the most dynamic area for new developments in cutting tools is the application of coatings to high-speed steels. First introduced in 1980 for gear cutters (hobs) and in 1981 for drills, TiN-coated HSS tools have demonstrated their ability to more than pay for the extra cost of the coating process.

In addition to hobs, gear-shaper cutters, and drills, HSS tooling coated by TiN now includes reamers, taps, chasers, spade-drill blades, broaches, bandsaw and circular saw blades, insert tooling, form tools, end mills, and an assortment of other milling cutters.

Physical vapor deposition (PVD) has proved to be the most viable process for coating HSS, primarily because it is a relatively low-temperature process that does not exceed the tempering point of HSS. Therefore, no subsequent heat treatment of the cutting tool is required. Films of .0001 in. to .0002 in. in thickness adhere well and withstand minor elastic, plastic, and thermal loads. Thicker coatings tend to fracture under the typical thermo-mechanical stresses of machining.

The advantages of TiN-coated HSS tooling include reduced tool wear. Less tool wear results in less stock removal during tool regrinding, thus allowing individual tools to be reground more times. For example, a TiN hob can cut 300 gears per sharpening; the uncoated tool would cut only 75 parts per sharpening. Therefore the cost per gear is reduced from 20 cents to 2 cents. Naturally, reduced tool wear means longer tool life as shown in Figure 22-9.

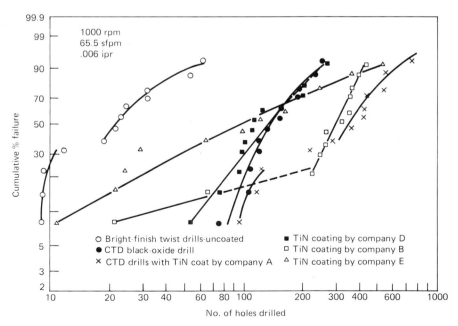

FIGURE 22-9 TiN-coated HSS drills outperform uncoated drills.

Drill performance based on the number of holes drilled with 1/4 inch diam. drills in T-1 structural steel.

Higher hardness, with typical values for the thin coatings, "equivalent" to R_c 80-85, vs. R_c 65-70 for hardened HSS, means reduced abrasion wear.

Relative inertness (that is, TiN does not react significantly with most workpiece materials) results in greater tool life through a reduction in adhesion. Tenfold increases in tool life have been reported with MRRs doubled.

TiN coatings have a low coefficient of friction. This results in an increase in the shear angle, which reduces the cutting forces, spindle power, and heat generated by the deformation processes.

In the PVD method, evaporation methods (sputtering) are used to eject atoms, ions, and/or neutral or charged clusters of atoms from a titanium source in a vacuum chamber. The titanium particles travel in more or less straight lines, depending upon chamber pressure, and condense on the parts to be coated. Because the PVD processes are primarily line-of-sight, the parts are rotated during coating to obtain uniform coating.

The PVD process depends on gas pressure (nitrogen), transit distance, and rate of removal from the source. Argon is generally used to support the discharge on sputtering sources. Total vacuum pressures in the range of 1 to 30 $\times$ 10^{-3} Torr are used for coating.

PVD processes are carried out with the workpieces heated to temperatures in the range of 400°–900°F. Substrate heating enhances coating adhesion and film structure.

Because surface pretreatment is critical in PVD processing, tools to be coated are subject to a vigorous cleaning process. Precleaning methods typically involve degreasing, ultrasonic cleaning, and freon drying. Deburring, honing, and more active cleaning methods are also used.

Diamonds.   *Diamond* is the hardest material known. Industrial diamonds are now available in the form of polycrystalline compacts, which are finding industrial application in the machining of aluminum, bronze, and plastics, greatly reducing the cutting forces as compared to carbides. Diamond machining is done at high speeds with fine feeds for finishing and produces excellent finishes. Single-crystal diamonds, with a cutting-edge radius of 100 angstroms or less, are being used for precision machining of large mirrored surfaces. Single-crystal diamonds have been used for years to machine brass watch faces, thus eliminating polishing. They have also been used to slice biological materials into thin films for viewing in transmission electron microscopes. (This process is known as ultramicrotomy and is one of the few industrial versions of orthogonal machining in common practice.)

The salient features of diamond tools include high hardness, good thermal conductivity, the ability to form a sharp edge of cleavage (single-crystal, natural diamond), very low friction, nonadherence to most materials, the ability to maintain a sharp edge for a long period of time especially in machining soft materials like copper and aluminum, and good wear resistance.

To be weighed against these advantages are some shortcomings which include a tendency to interact chemically with elements of Group IVB to Group VIII of the periodic table. In addition, diamond wears rapidly when machining or grinding mild steel. It wears less rapidly with high-carbon alloy steels than with low-carbon steel, and has occasionally machined gray cast iron (which has high carbon content) with long life. Diamond has a tendency to revert at high temperatures (700°C) to graphite and/or to oxidize in air. Diamond is very brittle and is difficult and costly to shape into cutting tools.

Limited supply, increasing demand, and high cost of natural diamonds have led to the ultra-high-pressure (@ 50Kbar), high-temperature (@ 1500°C) synthesis of diamond from graphite at the General Electric Company in the mid 1950s, and the subsequent development of polycrystalline sintered diamond tools in the late 1960s.

Polycrystalline diamond tools consist of a thin layer (@ 0.5 to 1.5 mm) of fine-grain-size diamond particles sintered together and metallurgically bonded to a cemented carbide substrate. A high-temperature/high-pressure process, using conditions close to those used for the initial synthesis of diamond, is needed. Fine diamond powder (1 to 30 μm) is first packed on a support base of cemented carbide in the press. At the appropriate sintering conditions of pressure and temperature in the diamond stable region, complete consolidation and extensive diamond-to-diamond bonding take place. Sintered diamond tools are then finished to shape, size, and accuracy by laser cutting and grinding. See Figure 22-10. The cemented carbide provides the necessary elastic support for the hard and brittle diamond layer above it. The main advantages of sintered polycrystalline tools over natural single-crystal tools are better quality, greater toughness, and improved wear resistance, resulting from the random orientation of the diamond grains and the lack of large cleavage planes.

Sintered polycrystalline diamond tools are more expensive (an order of magnitude or more) than the conventional cemented carbides or ceramic tools because of the high cost of both the processing technique and the finishing methods. This is also the case with sintered polycrystalline cubic boron nitride, described in the next section. In spite of the higher cost, many applications have been found for these tools because they increase productivity, provide long tool life, and can be economical on the basis of overall cost per part. Diamond tools offer dramatic performance improvements over carbides. Tool life is often greatly improved as is control over part size, finish, and surface integrity.

Positive rake tooling is recommended for the vast majority of diamond tooling applications. If BUE is a problem, increasing cutting speed and using more positive rake angles may eliminate it. If edge breakage and chipping is a problem, one can reduce the feed rate.

Coolants are not generally used in diamond machining unless, as in the machining of plastics, it is necessary to reduce airborne dust particles.

Diamond tools can be reground.

Cubic Boron Nitrides.

CBN is a man-made tool material, developed by General Electric, (called Borazon). It is made in a compact form for tools by a process quite similar to that used for sintered polycrystalline diamonds. It retains its hardness at elevated temperatures (Knoop 4700 at 20°C, 4000 at 1000°C)

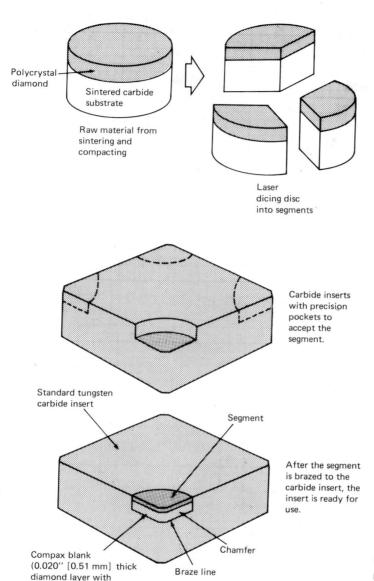

Polycrystal diamond

Sintered carbide substrate

Raw material from sintering and compacting

Laser dicing disc into segments

Carbide inserts with precision pockets to accept the segment.

Standard tungsten carbide insert

Segment

After the segment is brazed to the carbide insert, the insert is ready for use.

Compax blank (0.020'' [0.51 mm] thick diamond layer with carbide substrate)

Braze line

Chamfer

FIGURE 22-10 Polycrystalline diamond tools are carbides coated with diamonds.

and has low chemical reactivity at the tool/chip interface. This material can be used to machine hard aerospace materials like Inconel 718 and René 95 as well as chilled cast iron.

Cubic boron nitride, though not as hard as diamond, is less reactive with materials like hardened steels, hard-chill cast iron, and nickel-base and cobalt-base super alloys. CBN can be used efficiently and economically to machine such difficult-to-machine materials at higher speeds (@ 5 times) and with a higher removal rate (@ 5 times) than cemented carbide, and with superior accuracy, finish, and surface integrity. CBN tools are available in basically the same sizes and shapes as sintered diamond and are made by the same process. Their costs are somewhat higher than either cemented carbide or ceramic tools.

The two predominant wear modes of CBN tools are notching at the depth of cut line (DCL) and microchipping. In some cases, the tool will exhibit flank wear of the cutting edge. These tools have been used successfully for heavy interrupted cutting and for milling white cast iron and hardened steels using negative lands and honed cutting edges.

Since diamond and CBN are extremely hard but brittle materials, new demands are being placed on the machine tools and on machining practice in order to take full advantage of the potential of these tool materials.

These demands include

- Use of more rigid machine tools, and machining practices involving gentle entry and exit of the cut in order to prevent microchipping.
- Use of high-precision machine tools, because these tools are capable of producing high finish and accuracy.
- Use of machine tools with higher power, because these tools are capable of higher removal rates.

Tool Geometry

For cutting tools, geometry depends mainly on the properties of the tool material and the work material. The standard terminology is shown in Figure 22-11 for single-point tools. The most important angles are the rake angles and the end and side relief angles.

The *back rake angle* affects the ability of the tool to shear the work material and form the chip. It can be positive or negative. Positive rake angles reduce the cutting forces resulting in smaller deflections of the workpiece, toolholder, and machine.

If the back rake angle is too large, the strength of the tool is reduced as well as its capacity to conduct heat. In machining hard work materials, the back rake angle must be small, even negative for carbide and diamond tools. The higher the hardness, the smaller the back rake angle. For high-speed steels, back rake angle is normally chosen in the positive range, depending on the type of tool (turning, planing, end milling, face milling, drilling, and so on), and the work material.

For carbide tools, inserts for different work materials and toolholders can be supplied with several standard values of back rake angle: −6 degrees to + 6 degrees. The side rake angle and the back rake angle combine to form the effective rake angle. See Figure 22-11. This is also called the true rake angle or resultant rake angle of the tool.

True rake inclination of a cutting tool has a major effect in determining the amount of chip compression and the shear angle. A small rake angle causes high compression, tool forces, and friction, resulting in a thick, highly deformed, hot chip. Increased rake angle reduces the compression, the forces, and the friction, yielding a thinner, less-deformed and cooler chip. Unfortunately,

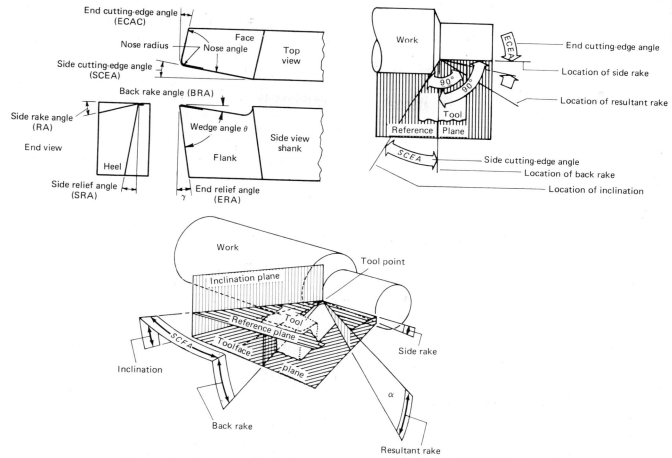

Back and side rake angles combine to produce a resultant or effective rake for cutting. Inclination is the angle between the cutting edge and its projection on the reference plane. All angles are positive.

FIGURE 22-11 Standard terminology to describe the geometry of single-point tools.

it is difficult to take advantage of these desirable effects of larger positive rake angles, since they are offset by the reduced strength of the cutting tool, due to the reduced tool section, and by its greatly reduced capacity to conduct heat away from the cutting edge.

In order to provide greater strength at the cutting edge and better heat conductivity, zero or negative rake angles commonly are employed on carbide, ceramic, polydiamond, and CBN cutting tools. These materials tend to be brittle, but their ability to hold their superior hardness at high temperatures results in their selection for high-speed and continuous machining operations. The negative rake angle increases tool forces, but this is necessary to provide the added support to the cutting edge. This is particularly important in making intermittent cuts and in absorbing the impact during the initial engagement of the tool and work.

In general, the power consumption is reduced by approximately 1% for each 1 degree in α. γ is the end relief angle. The wedge angle θ determines the strength of the tool and its capacity to conduct heat and depends on the values of α and γ.

The relief angles mainly affect the tool life and the surface quality of the workpiece. To reduce the deflections of the tool and the workpiece and to provide good surface quality, larger relief values are required. For high-speed steel, relief angles in the range of 5–10 degrees are normal, with smaller values being for the harder work materials. For carbides, the relief angles are lower to give added strength to the tool.

The side and end cutting-edge angles define the nose angle and characterize the tool design. The nose radius has a major influence on surface finish. Increasing the nose radius usually decreases tool wear and improves surface finish.

Tool nomenclature varies with different cutting tools, manufacturers, and users. Many terms are still not standard because of all this variety. The most common tool terms will be used in later chapters to describe specific cutting tools.

The introduction of coated tools has spurred the development of improved tool geometries. Specifically, *low-force groove* (LFG) geometries have been developed which reduce the total energy consumed and break up the chips into shorter segments (see Figure 22-5). The effect of these grooves is effectively to increase the rake angle, which increases the shear angle and lowers the cutting force and power. This means that higher cutting speeds or lower cutting temperatures (and better tool lives) are possible.

As a chip breaker, the groove deflects the chip at a sharp angle and causes it to break into short pieces that are easier to remove and are not so likely to become tangled in the machine and possibly cause damage to personnel. This is particularly important on high-speed, mass-production machines.

The shapes of cutting tools as used for various operations and materials are compromises, resulting from experience and research so as to provide good overall performance. Table 22.5 gives representative rake angles and suggested cutting speeds.

Tool Failure and Tool Life

In metal cutting, the failure of the cutting tool can be classified into two broad categories, according to the failure mechanisms that caused the tool to die (or fail):

1. *Slow-death mechanisms:* gradual tool wear on the flank(s) of the tool or on the rake face of tool (called *crater wear*) or both.
2. *Sudden-death mechanisms:* rapid, usually unpredictable and often catastrophic failures resulting from abrupt, premature death of a tool.

Figure 22-12 shows a sketch of a ''worn'' tool, showing crater and flank wear, along with wear of the tool nose radius and an outer diameter groove (the DCL groove). As the tool wears, its geometry changes. This geometry change will influence the cutting forces, the power being consumed, the surface finish obtained, the dimensional accuracy, and even the dynamic stability of the

TABLE 22.5 Representative Machining Conditions for Various Work and Tool–Material Combinations

Work Material	Tool	Rake Angles (degrees)		Cutting Speed	
		Back	Side	m/min	fpm
B1112 steel	HSS	16	22	69	225
	WC	0	3	168	550
	Ceramic	− 5	− 5	427	1400
4140 steel	HSS	12	14	40	130
	WC	0	3	91	300
	Ceramic	− 5	− 5	274	900
8620 steel	HSS	uncoated			100
	WC	uncoated			400
	WC	coated with TiC			600
	WC	coated with AL_2O_3			1100
	WC, AL_2O_3 with LFG				1300
18–8 steel (stainless)	HSS	8	14	27	90
	WC	4	8	84	275
	Ceramic	− 5	− 5	152	500
Gray cast iron (medium)	HSS	5	12	34	110
	WC	0–4	2–4	69	225
	Ceramic	− 5	− 5	244	800
Brass (free-machining)	HSS	0	0	76	250
	WC	0	4	221	725
Aluminum alloys	HSS	35	15	91	300 plus
	WC	10–20	10–20	122	400 plus
Magnesium alloys	HSS	0	10	91	300 plus
	WC	10	10	213	700 plus
Titanium (turning)	WC	0	5	46	150

Table Valve Typical for
  Lathe turning operation
  Single-point tool

Feed: 0.38 mm/rev (0.015 ipr)
Depth: 3.18 mm (0.125 in.)

process, as worn tools often chatter in processes that are usually relatively free of vibration. The actual wear mechanisms active in this high-temperature environment are abrasive, adhesion, diffusion, or chemical interactions. It appears that in metal cutting, any or all of these mechanisms may be operative at a given time in a given process.

The sudden-death mechanisms are more straightforward but less predictable. These mechanisms are categorized as plastic deformation, brittle fracture, fatigue fracture, or edge chipping. Here again it is difficult to predict which mechanism will dominate and result in a tool failure in a particular situation. What can be said is that tools, like people, die (or fail) from a great variety of causes under widely varying conditions. Therefore tool life should be treated as a random variable, or probabilistically, and not as a deterministic quantity.

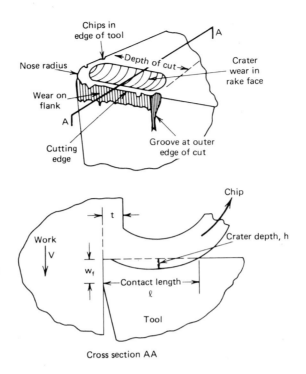

FIGURE 22-12 Sketch of a worn tool showing various wear elements resulting during oblique cutting.

During machining, the tool is performing in a hostile environment wherein high contact stresses and high temperatures are commonplace, and therefore tool wear is always an unavoidable consequence. In Figure 22-13 four characteristic tool wear curves (average values) are shown for four different cutting speeds, V_1 through V_4. V_4 is the fastest cutting speed and therefore generates the highest interface temperatures and the fastest wear rates. Such curves often have three general regions as shown in the figure. The central region is a steady-state region (or the region of secondary wear). This is the normal operating region for the tool. Such curves are typical for both flank wear and crater wear. When the amount of wear reaches the value W_f, the permissible tool wear on the flank, the tool is said to be "worn out." W_f is typically 0.025 to 0.030 in. For crater wear, the depth of the crater is used to determine tool failure.

Suppose that the tool wear experiment was repeated 15 times, without changing any of the input parameters. The result would look like Figure 22-14, which depicts the variable nature of tool wear and shows why tool wear must be treated as a random variable. In Figure 22-14 the average time is denoted as μ_t and the standard deviation as σ_t, where the wear limit criterion was 0.25 in. At a given time during the test, 35 minutes, the tool displayed flank wear ranging from 0.13 to 0.21 in. with an average of $\mu_w = .0175$ in. with standard deviation σ_w.

Other criteria that can be used to define tool death in addition to wear limits are surface finish, failure to conform to size (tolerances), increases in cutting

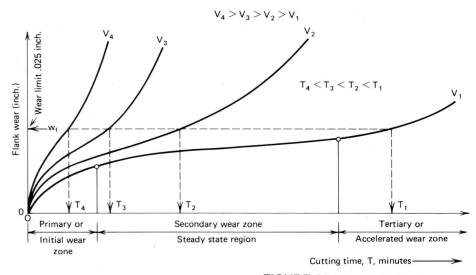

FIGURE 22-13 Typical tool wear curves for flank wear at different velocities.

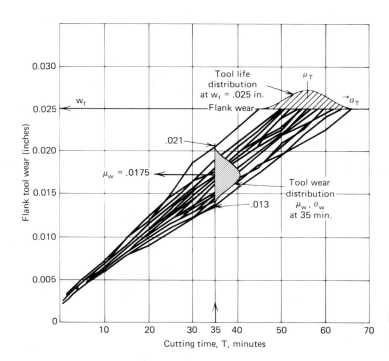

FIGURE 22-14 Tool wear on the flank displays a random nature as does tool life.

forces and power, time required to drill a hole under a given load, or complete failure of the tool. In automated processes, it is very beneficial to be able to monitor the tool wear on-line, so that the tool can be replaced prior to failure wherein defective products may also result. The feed force has been shown to be a good indirect measure of tool wear. That is, as the tool wears and dulls, the feed force increases more than the cutting force increases.

Once criteria for failure have been established, tool life is that time elapsed between start and finish of the cut, in minutes. Other ways to express tool life, other than time, include

1. Volume of metal removed between regrinds or replacement of tool.
2. Number of holes drilled with a given tool. See Figure 22-9.
3. Number of pieces machined per tool.

Taylor's Tool Life Model.   F. W. Taylor, in 1907, published his now famous "Taylor tool life equation" wherein tool life (T) was related to cutting speed V and feed (f). These equations had the form

$$T = \frac{\text{constant}}{f_x V_y}$$

which over the years took the more widely published form

$$VT^n = C \tag{22-2}$$

where n = exponent which depends mostly on tool material but is affected by work material, cutting conditions, and environment and

   C = constant which depends on all the input parameters, including feed.

This equation was obtained from empirical log-log plots, like Figure 22-15, which used the values of T (time in minutes) associated with V (cutting speed)

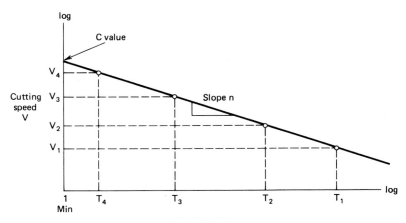

FIGURE 22-15   Construction of the Taylor tool life curve from deterministic tool wear plots like those of Figure 22-13.

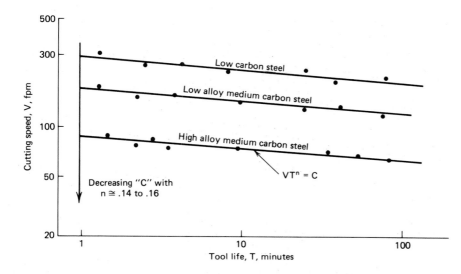

FIGURE 22-16   Log–log tool life
plots for three steel work materials
cut with HSS tool material.

for a given amount of tool wear, W_f. (See the dashed-line construction in Figure 22-13.) Figure 22-16 shows typical tool life curves for one tool material and three work materials. Notice that all three plots have about the same slope, n. Typical values for n are 0.14 to 0.16 for HSS, 0.21 to 0.25 for uncoated carbides, 0.30 for TiC insets, 0.33 for polydiamonds, 0.35 for TiN inserts, and 0.40 for ceramic-coated inserts.

It takes a great deal of experimental effort to obtain the constants for the Taylor equation, as each combination of tool and work material will have different constants. Note that for a tool life of 1 minute, $C = V$ or the cutting speed that yields about 1 minute of tool life for this tool. A great deal of research has gone into developing more sophisticated versions of the Taylor equation wherein constants for other input parameters (typically feed, depth of cut, and work material hardness) are experimentally determined. For example:

$$V T^{\,n} f^{\,m} d^{p} = K' \tag{22-3}$$

where n, m, and p are exponents and K' is a constant. Equations of this form are also deterministic.

The problem has been approached probabilistically in the following way. Since T depends upon speed, feed, materials, and so on, one writes

$$T = \frac{C^{1/n}}{V^{1/n}} = \frac{K}{V_m} \tag{22-4}$$

where K is now a random variable which represents the effects of all unmeasured factors and V_m is an input variable.

The sources of tool life variability include factors such as:

1. Variation in work material hardness (from part to part and within a part).
2. Variability in cutting tool materials, geometry, and preparation.

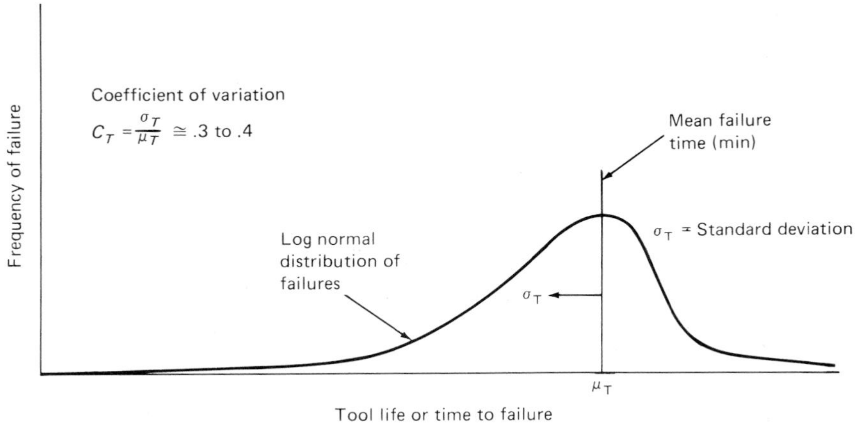

FIGURE 22-17 Tool life viewed as random variable has a log normal distribution with a large coefficient of variation.

3. Vibrations in machine tool, including rigidity of work- and tool-holding devices.
4. Changing surface characteristics of workpieces.

The examination of the data from a large number of tool life studies, wherein a variety of steels were machined, has shown that, regardless of the tool material or process, tool life distributions are usually log normal and typically have a large standard deviation. See Figure 22-17.

Reconditioning Cutting Tools. In the reconditioning of tools by sharpening and recoating, care must be taken in grinding the tool's surfaces. The following guidelines should be observed:

1. Resharpen to original tool geometry specifications. Restoring the original tool geometry will help the tool achieve consistent results on subsequent uses. Computer numerical control (CNC) grinding machines for tool resharpening have made it easier to restore a tool's original geometry.
2. Grind cutting edges and surfaces to a fine finish. Rough finishes left by poor and abusive regrinding hinder the performance of resharpened tools. For coated tools, tops of ridges left by rough grinding will break away in early tool use, leaving uncoated and unprotected surfaces that will cause premature tool failure.
3. Remove all burrs on resharpened cutting edges. If a tool with a burr is coated, premature failure can occur because the burr will break away in the first cut, leaving an uncoated surface exposed to wear.
4. Avoid resharpening practices that overheat and burn or melt (called *glazed over*) the tool surfaces, as this will cause problems in coating adhesion. Polishing or wire brushing of tools causes similar problems.

The cost of each recoating is about 1/5th the cost of purchasing a new tool. By recoating, the tooling cost per workpiece can be cut by between 20% and 30%, depending on the number of parts being machined.

The cutting speed has such a great influence on the tool life compared to the feed or the depth of cut that it greatly influences the overall economics of the machining process. For a given combination of work material and tool material, a 50% increase in speed results in a 90% decrease in tool life while a 50% increase in feed results in a 60% decrease in tool life. A 50% increase in depth of cut produces only a 15% decrease in tool life. This says that in limited horsepower situations, depth of cut and then feed should be maximized while speed is held constant and horsepower consumed is maintained within limits. As cutting speed is increased, the machining time decreases but the tools wear out faster and must be changed more often. In terms of costs, the situation is as shown in Figure 22-18, which shows the effect of cutting speed on the cost per piece.

The total cost per operation is comprised of four individual costs: machining costs, tool costs, tool changing costs, and the handling costs. The machining cost is observed to decrease with increased cutting speed because the cutting time decreases. Cutting time is proportional to the machining costs. Both the tool costs and the tool changing costs increase with increases in cutting speeds. The handling costs are independent of cutting speed. Adding up each of the individual costs results in a total unit cost curve which is observed to go through a minimum point. For a turning operation, the total cost per piece, C, equals

$$C = C_1 + C_2 + C_3 + C_4$$
$$= \text{Machining cost} + \text{tool cost} + \text{tool changing cost} \quad (22\text{-}5)$$
$$+ \text{ handling cost}$$

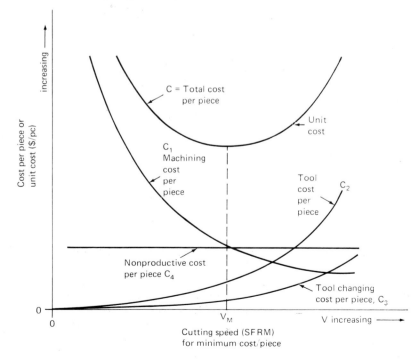

FIGURE 22-18 The costs of a machining process versus cutting speed.

Expressing each of these cost terms as a function of cutting velocity will permit the summation of all the costs.

$$C_1 = CT \times C_o \qquad \text{where } C_o = \text{operating cost (\$/min) and}$$
$$CT = \text{cutting time (min/piece)}$$

$$C_2 = \frac{C_t \times CT}{T} \qquad \text{where } T = \text{tool life (min)}$$
$$\text{and } C_t = \text{initial cost of tool (\$)}$$

$$C_3 = \frac{T_c \times C_o \times CT}{T} \qquad \text{where } T_c = \text{time to change tool (min)}$$

C_4 = labor, overhead, and machine tool costs consumed while part is being loaded, unloaded, tools advanced, machine broken down, etc.

Since $CT = L / Nf_r$ for turning
$$= \pi DL / 12Vf_r$$
and $T = (K/V)^{1/n}$ rewriting equation (22-2)

C can be expressed in terms of V.

$$C = \frac{L\pi DC_o}{12Vf_r} + \frac{C_t V^{1/n}}{tK^{1/n}} + \frac{t_c C_o V^{1/n}}{t\,K^{1/n}} + C_4 \qquad (22\text{-}6)$$

To find the minimum, take $dC/dV = 0$ and solve for V

$$V_m = \left(\frac{1}{n} - 1\right)\left(\frac{C_t + (C_o \times t_c)}{C_o}\right) \qquad (22\text{-}7)$$

A word of caution here is appropriate. Note that this derivation was totally dependent upon the Taylor tool life equation. Such data may not be available because they are expensive and time-consuming to obtain. Even when the tool life data are available, this procedure assumes that the tool fails only by whichever wear mechanism (flank or crater) was described by this equation and by no other failure mechanism. Recall that tool life had a very large coefficient of variation and was probabilistic in nature. This derivation assumes that for a given V there is one T, and this simply is not the case. The model also assumes that the workpiece material is homogeneous, the tool geometry is preselected, the depth of cut and feed rate are known and remain unchanged during the entire process, sufficient horsepower is available for the cut at the economic cutting conditions, and the cost of operating time is the same whether the machine is cutting or not cutting.

Table 22.6 shows an example of the analysis of tooling economics, where a comparison is being made between four different tools, all used for turning hot rolled 8620 steel with triangular inserts. Operating costs for the machine tool were $60 per hour. The low force groove insert has only three cutting edges available instead of six. It takes three minutes to change inserts and ½ minute to unload a finished part and load in a new 6-inch-diameter bar stock. The length of cut was about 24 inches. This table should be carefully analyzed and studied so that each line is understood. Note that the cutting tool cost per piece was

TABLE 22.6 Cost Comparison of Four Tool Materials, Based on Equal Tool Life

	Uncoated	TiC-coated	Al_2O_3-coated	Al_2O_3 LFG
Cutting speed (surface ft/min)	400	640	1100	1320
Feed (in/rev)	0.020	0.022	0.024	0.028
Cutting edges available per insert	6	6	6	3
Cost of an insert ($)	4.80	5.52	6.72	6.72
Tool life (pieces/cutting edge)	40	40	40	40
Tool life (min/cutting edge)	192	108	60	40
Tool change time per piece (min)	0.075	0.075	0.075	0.075
Nonproductive cost per piece ($)	0.50	0.50	0.50	0.50
Machining cost per piece ($)	4.8	2.7	1.50	1.00
Machining time per piece (min)	4.8	2.7	1.5	1.0
Tool change cost per piece ($)	0.08	0.08	0.08	0.08
Cutting tool cost per piece ($)	0.02	0.02	0.03	0.06
Total cost per piece ($)	5.40	3.30	2.11	1.64
Cost savings (%)	0	39	61	70
Production rate (pieces/hr)	11	18	29	38
Improvement in productivity based on pieces/hr (%)	0	64	164	245

Source: Data from T. E. Hale et al., "High Productivity Approaches to Metal Removal," *Materials Technology*, Spring 1980, p. 25.

three times higher for the low force groove tool over the carbide but really of no consequence since the major cost per piece comes from two sources: the machining cost per piece and the nonproductive cost per piece.

Machinability

Machinability is a widely used and much maligned term which has many different meanings but generally refers to the ease with which a metal can be machined to an acceptable surface finish. The principal definitions of the term are entirely different, the first based on material properties, the second based on tool life, and the third based on cutting speed.

1. Machinability is defined by the ease or difficulty with which the metal can be machined. In this light, specific energy, horsepower, or shear stress are used as measures and, in general, the larger the shear stress or specific power values, the more difficult the material is to machine, requiring greater forces and lower speeds. In this definition, the material is the key.

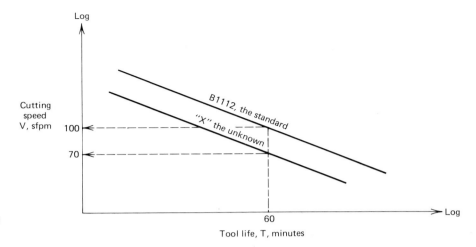

Log

Cutting
speed
V, sfpm

100

70

B1112, the standard

"X" the unknown

60

Log

Tool life, T, minutes

FIGURE 22-19   Machinability ratings defined by deterministic tool life curves.

2. Machinability is defined by the relative cutting speed for a given tool life of the tool cutting some material compared to a standard material cut with the same tool material. As shown in Figure 22-19, tool life curves are used to develop machinability ratings. The material chosen for the standard material was B1112 steel, which has a tool life of 60 minutes at a cutting speed of 100 sfpm. Material "X" has a 70% rating, which implies that steel "X" has a cutting speed of 70% of B1112 for equal tool lives. Note that this definition assumes that the tool fails when machining "X" by whatever mechanism dominated the tool failure when machining the B1112. There is no guarantee that this will be the case. ISO standard 3685 has machinability index numbers based on 30 minutes of tool life with flank wear of .33mm.

3. Cutting speed is measured by the maximum speed at which a tool can provide satisfactory performance for a specified time under specified conditions. See ASTM standard E 618-81: Evaluating Machining Performance of Ferrous Metals Using an Automatic Screw/Bar Machine.

4. Other definitions of machinability are based on the ease of removal of the chips (chip disposal), the quality of the surface finish of the part itself, the dimensional stability of the process, or the cost to remove a given volume of metal.

Further definitions are being developed based on the probabilistic nature of the tool failure, in which machinability is defined by a tool reliability index. Using such indexes, various tool replacement strategies can be examined and optimum cutting rates obtained. These approaches account for the tool life variability by developing coefficients of variation for common cutting tool/work material combinations. The coefficient of variation is the ratio of the standard deviation of the tool life distribution divided by the mean of that distribution. See Figure 22-17.

The results to date are very promising. One thing is clear, however, from this sort of research. While many manufacturers of tools have worked at developing materials that have greater tool life at higher speeds, few have worked to develop tools that have less variability in tool life at all speeds.

The reduction in variability is fundamental to achieving smaller coefficients of variation, which typically are of the order of 0.3 to 0.4. This means that a tool with a 100-minute average tool life has a standard deviation of 30 to 40 minutes and so there is a good probability that the tool will fail early. In automated equipment, where early unpredicted tool failures are extremely costly, reduction of the tool life variability will pay great benefits in improved productivity and reduced costs.

Cutting Fluids

From the day that Frederick W. Taylor demonstrated that a heavy stream of water flowing directly on the cutting process allowed the cutting speeds to be doubled or tripled, cutting fluids have flourished in use and variety and are employed in virtually every machining process. The cutting fluid acts primarily as a coolant and secondly as a lubricant, reducing the friction effects at the tool/chip interface and the work/flank regions. The cutting fluids also carry away the chips and provide friction (and force) reductions in regions where the bodies of the tools rub against the workpiece. Thus, in processes like drilling, sawing, tapping, and reaming, portions of the tool apart from the cutting edges come in contact with the work, and these (sliding friction) contacts greatly increase the power needed to perform the process unless properly lubricated.

The reduction in temperature greatly aids in retaining the hardness of the tool, thereby extending the tool life or permitting increased cutting speed with equal tool life. In addition, the removal of heat from the cutting zone reduces thermal distortion of the work and permits better dimensional control. Coolant effectiveness is closely related to the thermal capacity and conductivity of the fluid used. Water is very effective in this respect but presents a rust hazard to both the work and tools and also is ineffective as a lubricant. Oils offer less effective coolant capacity but do not cause rust and have some lubricant value. In practice, straight cutting oils or emulsion combinations of oil and water or wax and water are frequently used. Various chemicals can also be added to serve as wetting agents or detergents, rust inhibitors, or polarizing agents to promote formation of a protective oil film on the work. The extent to which the flow of a cutting fluid washes the very hot chips away from the cutting area is an important factor in heat removal. Thus the application of a coolant should be copious and of some velocity.

The possibility of a cutting fluid provides lubrication between the chip and the tool face is an attractive one. An effective lubricant can modify the geometry of chip formation so as to yield a thinner, less-deformed, and cooler chip. Such action could discourage the formation of a built-up edge on the tool and thus promote improved surface finish. However, the extreme pressure at the tool/chip interface and the rapid movement of the chip away from the cutting edge make it virtually impossible to maintain a conventional hydrodynamic lubricat-

TABLE 22.7   **Cutting Fluid Contaminants**

Category	Contaminants	Effects
I. Solids	Metallic fines, chips	Scratch product's surface
	Grease and sludge	Plug coolant lines
	Debris and trash	Wear on tools and machines
II. Tramp fluids	Hydraulic oils (coolant)	Decrease cooling efficiency
	Water (oils)	Cause smoking
		Clog paper filters
		Grow bacteria faster
III. Biologicals (coolants)	Bacteria	Acidify coolant
	Fungi	Break down emulsions
	Mold	Cause rancidity, dermatitis
		Require toxic biocides

ing film at the tool/chip interface. Consequently, any lubrication action is associated primarily with the formation of solid chemical compounds of low shear strength on the freshly cut chip face, thereby reducing chip–tool shear forces or friction. For example, carbon tetrachloride is very effective in reducing friction in machining several different metals and yet would hardly be classified as a good lubricant in the usual sense. Chemically active compounds, such as chlorinated or sulfurized oils, can be added to cutting fluids to achieve such a lubrication effect. Extreme-pressure lubricants are especially valuable in severe operations, such as internal threading (tapping) where the extensive tool–work contact results in high friction with limited access for a fluid.

FIGURE 22-20   A well-designed recycling system for coolants will return more than 99% of the fluid for reuse.

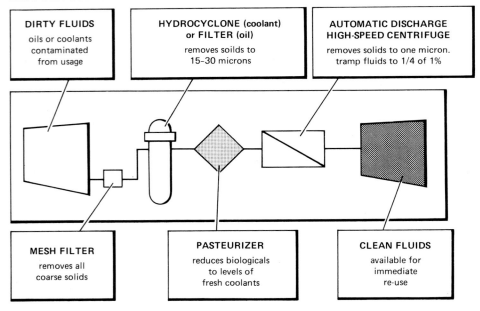

DIRTY FLUIDS	HYDROCYCLONE (coolant) or FILTER (oil)	AUTOMATIC DISCHARGE HIGH-SPEED CENTRIFUGE
oils or coolants contaminated from usage	removes soilds to 15–30 microns	removes solids to one micron. tramp fluids to 1/4 of 1%

MESH FILTER	PASTEURIZER	CLEAN FLUIDS
removes all coarse solids	reduces biologicals to levels of fresh coolants	available for immediate re-use

In addition to functional effectiveness as coolant and lubricant, cutting fluids should be stable in use and storage, noncorrosive to work and machines, and nontoxic to operating personnel. The cutting fluid should also be restorable by using a closed recycling system that will purify the used coolant and cutting oils. Cutting fluids become contaminated in three ways (see Table 22.7). All these contaminants can be eliminated by filtering, hydrocycloning, pasteurizing, and centrifuging. Coolant restoration eliminates 99% of the cost of disposal and 80% or more of new fluid purchases. See Figure 22-20 for a schematic of a coolant recycling system.

Review Questions

1. For metal-cutting tools, what is the most important material property—that is, the most critical characteristic? Why?
2. What is hot hardness?
3. What is impact strength and how is it measured? Why is it an important property in cutting tools?
4. What is HIP and how is it used?
5. What are the primary considerations in tool selection?
6. What is the general strategy behind coated tools?
7. What is meant by the statement, "Tool life is a random variable"?
8. How is a CBN tool manufactured?
9. F. W. Taylor was one of the discoverers of high-speed steel. For what else is he well known?
10. What casting process do you think was used to fabricate cast cobalt alloys?
11. What does the sintering step do in the manufacture of carbides?
12. What does *cemented* mean in the manufacture of carbides?
13. What advantage do ground carbide inserts have over pressed carbide inserts?
14. What is a chip groove?
15. What is the DCL?
16. Suppose you made four beams out of carbide, HSS, ceramic, and cobalt. The beams are identical in size and shape, differing only in material. Which beam would
    a. deflect the most, under the same amount of weight?
    b. resist penetration the most?
    c. bend the farthest without breaking?
    d. support the greatest compressive load?
17. Why are multiple coats or layers put on the carbide base for coated tools?
18. What are the most common surface treatments for
    a. high-speed steels?
    b. carbides?
    c. ceramics?
19. What makes the process which makes TiC coatings for tools a problem? See equation 22-1.
20. Why does a TiN-coated tool consume less power than an uncoated HSS under exactly the same cutting conditions?
21. Why is CBN better for machining steel than diamond?
22. What is a coefficient of variation?
23. What is the typical value of a coefficient of variation in metal-cutting tool life distributions?
24. How is machinability defined?
25. What are the chief functions of cutting fluids?

Problems

1. Figure 22A contains data for cutting speed and tool life. Determine the constants for the Taylor tool life equation for these data. What do you think the tool material might have been?
2. Suppose you have a turning operation using a tool with zero back rake and 5° end relief. The insert flank has a wear land on it of .020 in. How much has the diameter of the workpiece grown (increased) due to this flank wear, assuming that the tool has not been reset to compensate for the flank wear?
3. In Figure 22B, a single point tool is shown. Fill in the blanks for the tool nomenclature:

   A = _____ F = _____
   B = _____ G = _____
   C = _____ and E = _____
   D = _____

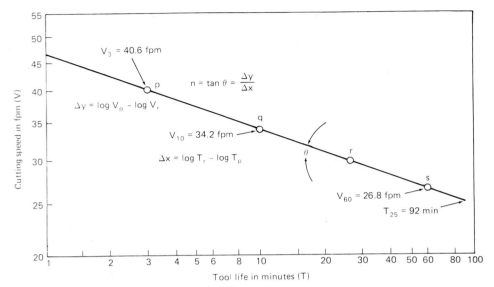

FIGURE 22A

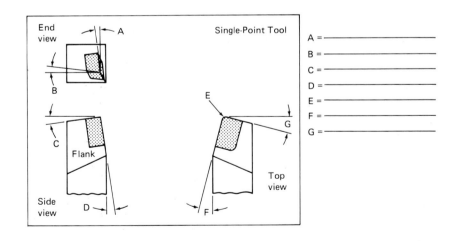

A Side rake
B Side relief
C End relief
D Back rake
E Nose radius
F Side cutting edge
G End cutting edge

FIGURE 22B

4. The following data have been obtained for machining AA390 aluminum, a Si-Al alloy:

Workpiece Material	Tool Material	Cutting speed for 20 minutes	Tool Life of 30 minutes	60 minutes
Sand casting	Diamond polycrystal	731m/min	642m/min	514m/min
Permanent mold casting	Diamond polycrystal	591m/min	517m/min	411m/min
PMC with flood cooling	Diamond polycrystal	608m/min	554m/min	472m/min
Sand casting	WC-K-20	175m/min	161m/min	139m/min

Compute the C and n values for the Taylor tool life equation. How do these n values compare to the typical values?

5. In Figure 22C, the insert at the top is set with a zero side cutting edge angle. The insert at the bottom is set so that the edge contact length is increased from .250 in. depth-of-cut to .289 in. The feed was .010 ipr.

a. Determine the side cutting edge angle for the offset tool.
b. What is the uncut chip thickness in the offset position?
c. What effect will this have on the process?

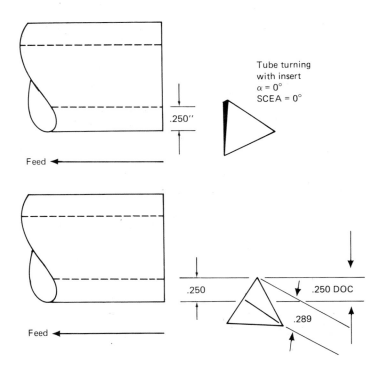

Tube turning with insert
$\alpha = 0°$
SCEA = 0°

FIGURE 22C

Turning and Boring and Related Processes

Introduction

Turning is the process of machining external cylindrical and conical surfaces. It is usually performed on a lathe. As indicated in Figure 23-1, relatively simple work and tool movements are involved in turning a cylindrical surface. The workpiece is rotated into a longitudinally fed, single-point cutting tool. If the tool is fed at an angle to the axis of rotation, an external conical surface results. This is called *taper turning*. If the tool is fed at 90° to the axis of rotation, using a tool that is wider than the width of the cut, the operation is called *facing,* and a flat surface is produced.

By using a tool having a specific form or shape and feeding it inward against the work, external cylindrical, conical, and irregular surfaces of limited length can also be turned. The shape of the resulting surface is determined by the shape and size of the cutting tool. Such machining is called *form turning*. If the tool is fed to the axis of the workpiece, it will be cut in two. This is called *parting* or *cutoff* and a simple, thin tool is used. A similar tool is used for *necking* or *partial cutoff*.

Boring is a variation of turning. Essentially it is internal turning. Boring can use single-point cutting tools to produce internal cylindrical or conical surfaces. It does not create the hole but, rather, bores the hole to a specific size. Boring can be done on most machine tools that can do turning. However, boring also can be done using a rotating tool with the workpiece remaining stationary. Also, specialized machine tools have been developed that will do boring, drilling, and reaming but will not do turning. Other operations, like *threading* and *knurling,* can be done on machines used for turning. In addition, drilling, reaming, and tapping can be done on the rotation axis of the work.

584

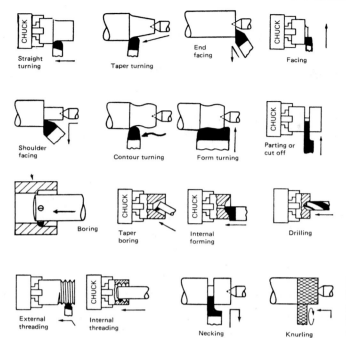

FIGURE 23-1 Turning, facing, boring, and related processes typically done on a lathe. The arrows indicate the motion of the tool relative to the work.

Turning constitutes the majority of lathe work. The cutting forces, resulting from feeding the tool from right to left, should be directed toward the headstock to force the workpiece against the workholder and thus provide better work support.

If good finish and accurate size are desired, one or more roughing cuts usually are followed by one or more finishing cuts. Roughing cuts may be as heavy as proper chip thickness, tool life, lathe horsepower and the workpiece permit. Large depths of cut and smaller feeds are preferred to the reverse procedure, because fewer cuts are required and less time is lost in reversing the carriage and resetting the tool for the following cut.

On workpieces that have a hard surface, such as castings or hot-rolled materials containing mill scale, the initial roughing cut should be deep enough to penetrate the hard materials. Otherwise, the entire cutting edge operates in hard, abrasive material throughout the cut, and the tool will dull rapidly. If the surface is unusually hard, the cutting speed on the first roughing cut should be reduced accordingly.

Finishing cuts are light, usually being less than 0.015 in. in depth, with the feed as fine as necessary to give the desired finish. Sometimes a special finishing tool is used, but often the same tool is used for both roughing and finishing cuts. In most cases one finishing cut is all that is required. However, when exceptional accuracy is required, two finishing cuts may be made. If the diameter is controlled manually, a short finishing cut (¼ in. long) is made and the diameter

Fundamentals of Turning, Boring, and Facing

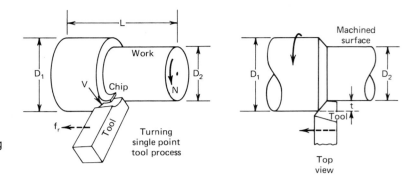

FIGURE 23-2 Basics of the turning process, normally performed in a lathe.

checked before the cut is completed. Because the previous micrometer measurements were made on a rougher surface, it may be necessary to reset the tool in order to have the final measurement, made on a smoother surface, check exactly.

In turning, the primary cutting motion is rotational with the tool feeding parallel to the axis of rotation (Figure 23-2). The rpm of the rotating workpiece, N, establishes the cutting velocity, V, according to $V = \pi DN/12$ surface feet per minute (sfpm). The feed, f_r, is given in inches per revolution (ipr). The depth of cut is t where

$$t = (D_1 - D_2)/2 \text{ in inches} \qquad (23\text{-}1)$$

The length of cut is the distance traveled parallel to the axis, L, plus some overrun, A, to allow the tool to enter and/or exit the cut. Once the cutting speed, feed, and depth of cut have been selected for a given material being cut with a tool of known cutting tool material, the rpm for the machine tool can be determined.

$$N = 12 \ V/\pi D_1 \text{ (using the larger diameter).} \qquad (23\text{-}2)$$

The cutting time $= \text{CT} = (L + A)/f_r N$ where A is overrun allowance. $\qquad (23\text{-}3)$

The MRR $= \dfrac{\text{volume removed}}{\text{time}} = \dfrac{(\pi D_1^2 - \pi D_2^2)L}{4L/f_r N}$ (omitting allowance term).

By rearranging and subbing for N, we obtain

$$\text{MRR} = \frac{12 \ (D_1^2 - D_2^2)f_r V}{4D_1} \qquad (23\text{-}4)$$

$$\text{also} \qquad \text{MRR} \cong 12 \ Vf_r t \qquad (22\text{-}5)$$

Note that equation (23-5) is an approximate equation which assumes that the depth of cut, t, is small compared to the uncut diameter D_1. See problem 7 in this chapter.

Boring. *Boring* always involves the enlarging of an existing hole, which may have been made by a drill or may be the result of a core in a casting. An equally important, and concurrent, purpose of boring may be to make the hole concentric with the axis of rotation of the workpiece and thus correct any eccentricity that may have resulted from the drill's having drifted off the center line. Concentricity is an important attribute of bored holes.

When boring is done in a lathe, the work usually is held in a chuck or on a face plate. Holes may be bored straight, tapered, or to irregular contours. Figure 23-3a shows the relationship of the tool and the workpiece for boring. Boring is essentially internal turning while feeding the tool parallel to the rotation axis of the workpiece.

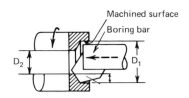

(a) Boring a drilled hole

Given V and f_r, for a cut of length L, the cutting time is

$$CT = (L + A)/Nf_r, \tag{23-6}$$

where $N = 12V/\pi D_1$ for D_1 = diameter of bore, where A is overrun allowance.

$$\text{The MRR} = \frac{(\pi D_1^2 L - \pi D_2^2 L)/4}{L/f_r N}$$

$$\cong 12\, Vf_r t \text{ (omitting allowance term)} \tag{23-7}$$

where D_2 = original hole diameter

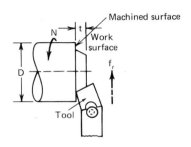

(b) Facing from the cross slide

Facing. *Facing* is the producing of a flat surface as the result of a tool's being fed across the end of the rotating workpiece, as shown in Figure 23-3b. Unless the work is held on a mandrel, if both ends of the work are to be faced, it must be turned end for end after the first end is completed and the facing operation repeated.

The cutting speed should be determined from the largest diameter of the surface to be faced. Facing may be done either from the outside inward or from the center outward. In either case, the point of the tool must be set exactly at the height of the center of rotation. Because the cutting force tends to push the tool away from the work, it is usually desirable to clamp the carriage to the lathe bed during each facing cut to prevent it from moving slightly and thus producing a surface that is not flat.

In the facing of castings or other materials that have a hard surface, the depth of the first cut should be sufficient to penetrate the hard material to avoid excessive tool wear.

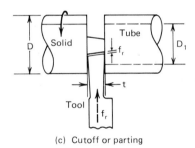

(c) Cutoff or parting

FIGURE 23-3 Basic movements of boring, facing, and cutoff (or parting) processes.

In facing, the tool feeds perpendicular to the axis of the rotating workpiece. Because the rpm is constant, the speed is continually decreasing as the axis is approached. The length, L, is $D/2$ or $(D - D_1)/2$ for a tube.

The cutting time = $CT = (L + A)/f_r N$ in minutes.

$$\text{The MRR} = \text{VOL/CT} = \frac{\pi D^2 t f_r N}{4L} = 6\, Vf_r t, \text{ in in.}^3/\text{min.} \tag{23-8}$$

for $L = D/2$.

Parting. *Parting* is the operation by which one section of a workpiece is severed from the remainder by means of a cutoff tool as shown in Figure

23-3c. Because parting tools are quite thin and must have considerable overhang, this process is less accurate and more difficult. The tool should be set exactly at the height of the axis of rotation, be kept sharp, have proper clearance angles, and be fed into the workpiece at a proper and uniform feed rate.

In parting or cutoff work, the tool is fed (plunged) perpendicular to the rotational axis, as it was in facing. The length of cut for solid bars is $D/2$.

$$\text{For tubes, } L = \frac{D - D_1}{2}$$

In cutoff operations, the width of the tool is t in inches.

The equations for CT and MRR are then basically the same as for facing.

In boring, facing, and cutoff operations, the speeds and feeds selected are generally less than those recommended for turning because of the large overhang of the tool often needed to complete the cuts. Recalling the basic equation for deflection of a cantilever beam, modifying for machining,

$$\delta = \frac{Pl^3}{3EI} = \frac{F_c l^3}{3EI} \qquad (23\text{-}9)$$

the overhang l greatly affects the deflection δ.

The reduction of the feed (or depth of cut) reduces the forces operating on the tools and the reduction of the speed usually reduces the probability of chatter and vibration.

The feed force, F_f, will deflect the tool to the side, resulting in loss of accuracy in cutoff lengths. At the outset, the forces will be balanced if there is no side rake on the tool. As the cutoff tool reaches the axis of the rotating part, the feed force will deflect the tool away from the spindle, resulting in a change in the length of the part.

Large holes may be precision-bored using the lathe setup shown in Figure 23-4, where a pilot bushing is placed in the spindle to mate with the hardened ground pilot of the boring bar. This setup eliminates the cantilever problems common to boring.

In most respects the same principles are used for boring as for turning. Again, the tool should be set exactly at the same height as the axis of rotation. Slightly larger end clearance angles sometimes have to be used to prevent the heel of the tool from rubbing on the inner surface of the hole. Because the tool overhang will be greater, feeds and depths of cut may be somewhat less than for turning to prevent tool vibration and chatter. In some cases the boring bar may be made of tungsten carbide because of this material's greater stiffness.

There always is a tendency for bored holes to be slightly bell-mouthed because of the tool's springing away from the work as it progresses into the hole. This usually can be corrected by repeating the cut with the same tool setting.

Because the rotational relationship between the work and the tool is a simple one and is employed on several types of machine tools, such as lathes, drilling machines, and milling machines, boring very frequently is done on such ma-

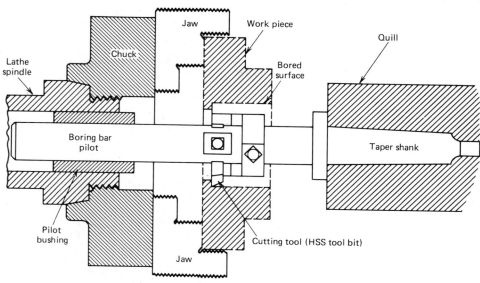

FIGURE 23-4 Pilot boring bar mounted in tailstock of lathe for precision-boring large hole in casting. The size of the hole is controlled by the rotation diameter of the cutting tool.

chines. However, several machine tools have been developed primarily for boring, especially in cases involving large workpieces or for large-volume boring of smaller parts. Such machines are also capable of performing other operations, such as milling and turning. Because boring frequently follows drilling, many boring machines also can do drilling, permitting both operations to be done with a single setup of the work.

Drilling. Most *drilling* on lathes is done with the drill held in the tailstock quill and fed against a workpiece that is rotated in a chuck. Drills with taper shanks are mounted directly in the quill hole. See Figure 23-5. Feeding is by hand by means of the handwheel on the outer end of the tailstock assembly.

It also is possible to drill on a lathe with the drill bit mounted and rotated in the spindle while the work remains stationary, supported on the tailstock.

Usual speeds are used for drilling in a lathe. Because the feed is manual, care must be exercised, particularly in drilling small holes. Coolants should be used where required. In drilling deep holes, the drill should be withdrawn occasionally to clear chips from the hole and to aid in getting coolant to the cutting edges.

Reaming. *Reaming* on a lathe involves no special precautions. Reamers are held in the tailstock quill, taper-shank types being mounted directly and straight-shank types by means of a drill chuck. Rose-chucking reamers usually are used. (See Chapter 24.) Fluted-chucking reamers also may be used, but these should be held in some type of holder that will permit the reamer to float.

Knurling. *Knurling* produces a regularly shaped, roughened surface on a workpiece. Although knurling also can be done on other machine tools, even

Face plate

FIGURE 23-5   Drilling and boring in a lathe using fixture (fix) mounted on a face plate (fp) to hold the workpiece (w).

on flat surfaces, in most cases it is done on external cylindrical surfaces using lathes. Knurling is a chipless, cold-forming process, using a tool of the type shown in Figure 23-6. The two hardened rolls are pressed against the rotating workpiece with sufficient force to cause a slight outward and lateral displacement of the metal so as to form the knurling in a raised, diamond pattern. Another type of knurling tool produces the knurled pattern by cutting chips. Because it involves less pressure and thus does not tend to bend the workpiece, this method is often preferred for workpieces of small diameter and for use on automatic or semiautomatic machines.

Turning and Boring Tapers.   The turning and boring of uniform tapers are common lathe operations. Such tapers can be specified either in degrees of included angle between the sides or as the change in diameter per unit of length—millimeters per millimeter or inches per foot.

Four methods are available for turning external tapers on a lathe, and three for boring internal tapers. The simplest method employs the compound rest; this method is suitable for both external and internal tapers. However, because the length of travel of the compound rest is quite limited—seldom over a few inches—only short tapers can be turned or bored by this method. This rest is particularly useful for steep tapers. The compound rest is swiveled to the desired angle and locked in position. The compound slide is then fed manually to produce the desired taper. The tool should be set at exactly the height of the axis of rotation of the workpiece in all taper turning and boring. See Figures 23-1 and 23-11.

Because the graduated scale on the base of the compound rest usually is calibrated only to one-degree divisions, it is difficult to make the angle setting with accuracy. If accuracy is required, tapers made by this method are checked by means of plug or ring gages, readjusting the setting of the compound rest until the gage fits perfectly. Also, the compound rest cannot be set directly to the correct angle if the taper is dimensioned in millimeters per millimeter or inches per foot.

Both external and internal tapers can be made on a lathe by using a *taper attachment,* such as shown in Figure 23-7. In this device there is an *extension* bolted to the rear of the carriage. When the carriage is moved, the cross slide is caused to move transversely.

A raised *guide bar* is pivoted to any desired angle (within its limits). A *guide shoe* slides on the guide bar, so that, when the carriage is moved longitudinally, the guide follows the guide bar and moves the cross slide and tool post transversely to provide the proper tool angle.

Graduations of taper in mm/mm or in./ft. are provided at one end (degrees at the other end), so the attachment can be set to the desired taper. While taper attachments provide an excellent and convenient method of cutting tapers, they ordinarily can be used only for tapers of less than 0.5 mm/mm or 6 in./ft.

(a)

Two forming rolls

(b)

FIGURE 23-6 (a) Knurling in a lathe, using a forming-type tool, and showing the resulting pattern on the workpiece. (b) Knurling tool with forming rolls. (Courtesy Armstrong Bros. Tool Co.)

(rear view)

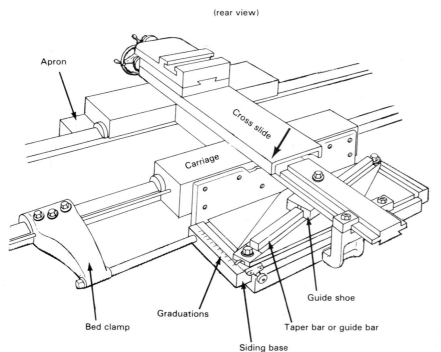

Apron

Cross slide

Carriage

Bed clamp

Graduations

Siding base

Guide shoe

Taper bar or guide bar

FIGURE 23-7 Taper attachment moves cross slide transversely when carriage moves, but only if the bed clamp is fastened. Taper bar is set for the angle to be machined.

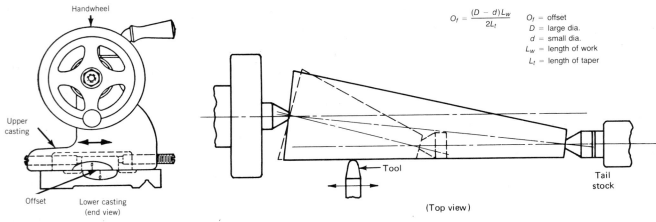

$$O_f = \frac{(D - d)L_w}{2L_t}$$

O_f = offset
D = large dia.
d = small dia.
L_w = length of work
L_t = length of taper

(Top view)

FIGURE 23-8 Method of turning tapers by setting the tailstock off center.

External tapers also can be turned on workpieces that are mounted between centers by *setting over the tailstock*. This method is illustrated in Figure 23-8. The tailstock is moved out of line with the headstock spindle. The set-off distance from the center line is given by the formula in the figure. This method is limited to small tapers and is seldom used.

In specifying tapers on drawings, one must remember that it is difficult for the machinist to measure the smaller diameter of a taper accurately if it is the end of a workpiece.

Both internal and external tapers can be programmed on a numerically controlled lathe. The movement of two perpendicular axes can be programmed, resulting in a tapered surface.

Milling. *Milling* can be done on a lathe but it requires a special attachment. The *milling attachment* is a special vise that attaches to the cross slide to hold work. The milling cutter is mounted and rotated by the spindle. The work is fed by means of the cross-slide screw.

Tool-post grinders are often used to permit grinding to be done on a lathe.

Duplicating attachments are available that, guided by a template, will automatically control the tool movements for turning irregularly shaped parts. In some cases, the first piece, produced in the normal manner, may serve as the template for duplicate parts. To a large extent, duplicating lathes using templates have been replaced by numerically controlled lathes.

Dimensional Accuracy. *Dimensional accuracy* in turning operations is controlled by the wear at the nose of the tool. Precision is influenced by deflection due to the cutting forces and surface roughness. Tool wear causes the workpiece dimension to change from the initial diameter when the tool is sharp to the diameter obtained after the tool has worn. The cutting forces increase as the tool wears, resulting in increased deflection between the workpiece and the cutting tool. Built-up edge (BUE) may form at the tip of the cutting tool. The BUE has the tendency to change the actual diameter of the workpiece. Thus, to hold close tolerances, the size of the wear land, the magnitude of the thrust force, and the elimination of the BUE should be taken into account.

Dimensional accuracy will also be influenced by the workpiece shape and material (rigidity of all elements, surface finish, and vibration). An example of this is dimension accuracy in deep boring operations, where boring bar rigidity is a problem.

Turned surfaces display characteristic turning grooves that are produced by the feed and the tool tip corner radius as shown in Figure 23-9. The roughness resulting from feed marks from a round-nosed tool can be approximated by the formula:

$$y = CR - \sqrt{CR^2 - f^2/4} \cong \frac{f^2}{8CR} \tag{23-10}$$

where
y = roughness
CR = corner radius of insert
f = feed, ipr

To improve the surface finish, reduce feed and increase the corner radius.

Other factors like built-up edge formations, cutting-edge sharpness and tool-wear grooves in the flank wear area also affect the surface finish in turning. Flank wear and BUE can combine to affect both surface finish and accuracy. Wear on the corner radius may cause grooves and nicks, which produce additional surface roughness on the finish-turned surfaces. Thus, to hold the surface roughness within specified limits, minimize tool wear and use small feeds and large-corner-radius tools. To minimize BUE formation, employ cutting speeds higher than those used in rough turning operations.

Lathe Design and Terminology

Knowing the terminology of a machine tool is fundamental to understanding how it performs the basic processes, how the workholding devices are interchanged, and how the cutting tools are mounted and interfaced to the work.

Lathes are machine tools designed primarily to do turning, facing, and boring. Very little turning is done on other types of machine tools, and none can do it with equal facility. Because lathes also can do facing, drilling, and reaming, their versatility permits several operations to be done with a single setup of the workpiece. Consequently, the lathe is the most common machine tool.

Lathes in various forms have existed for more than 2,000 years, but modern lathes date from about 1797, when Henry Maudsley developed one with a lead-screw, providing controlled mechanical feed of the tool. This ingenious Englishman also developed a change-gear system that could connect the motions of the spindle and leadscrew and thus enable threads to be cut (see Chapter 1).

Lathe Design.   The essential components of an engine lathe are depicted in Figure 23-10. These are the bed, headstock assembly, tailstock assembly, carriage assembly, quick-change gearbox, and the leadscrew and feed rod.

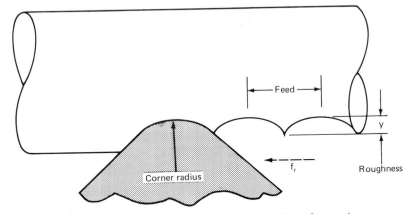

The feed and the corner radius of the cutting tool influence the surface roughness.

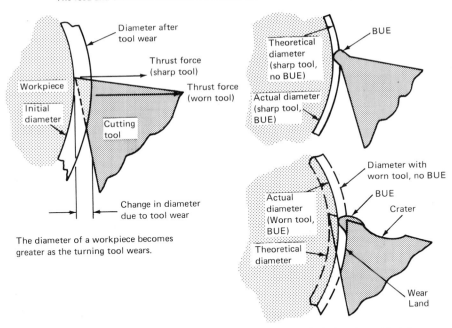

The diameter of a workpiece becomes greater as the turning tool wears.

Regardless of whether the tool is dull or sharp, a built-up edge (BUE) causes the diameter of the workpiece to be smaller than desired.

FIGURE 23-9 Accuracy and precision in turning is a function of many factors.

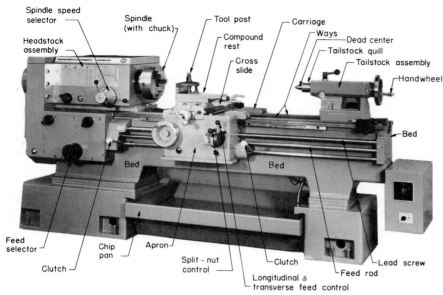

Spindle speed selector

Headstock assembly

Spindle (with chuck)

Tool post

Compound rest

Cross slide

Carriage

Ways

Dead center

Tailstock quill

Tailstock assembly

Handwheel

Bed

Bed

Bed

Feed selector

Chip pan

Apron

Clutch

Split - nut control

Longitudinal & transverse feed control

Clutch

Feed rod

Lead screw

FIGURE 23-10   Modern engine lathe, with the names of the principal parts. (Courtesy Heidenreich & Harbeck.)

The *bed* is the base and backbone of a lathe. It usually is made of well-normalized or aged gray or nodular cast iron and provides a heavy, rigid frame on which all the other basic components are mounted. Two sets of parallel, longitudinal *ways,* inner and outer, are contained on the bed. On modern lathes, the ways are surface-hardened and precision-machined, and care should be taken to assure that the ways are not damaged. Any inaccuracy in them usually means that the accuracy of the entire lathe is destroyed.

The *headstock,* mounted in a fixed position on the inner ways, provides powered means to rotate the work at various rpms. Essentially, it consists of a hollow spindle, mounted in accurate bearings, and a set of transmission gears—similar to a truck transmission—through which the spindle can be rotated at a number of speeds. Most lathes provide from 8 to 18 rpms. On modern lathes all the rpms can be obtained merely by moving from two to four levers. An increasing trend is to provide a continuously variable speed range through electrical or mechanical drives.

The accuracy of a lathe is greatly dependent on the *spindle.* It carries the workholders and is mounted on heavy bearings, usually preloaded tapered roller or ball types. The spindle has a hole extending through its length, through which long bar stock can be fed. The size of this hole is an important dimension of a lathe because it determines the maximum size of bar stock that can be machined when the materials must be fed through the spindle.

The spindle protrudes from the gearbox and contains means for mounting various types of workholding devices (chucks, face plates, and dog plates).

Power is supplied to the spindle from an electric motor through a V-belt or silent-chain drive. Most modern lathes have motors of from 5 to 15 horsepower to provide adequate power for carbide and ceramic tools at the high cutting speeds.

The tailstock assembly consists, essentially, of three parts. A lower casting fits on the inner ways of the bed, can slide longitudinally, and can be clamped in any desired location. An upper casting fits on the lower one and can be moved transversely upon it, on some type of keyed ways, to permit aligning the tailstock and headstock spindles (for turning tapers). The third major component of the assembly is the *tailstock quill*. This is a hollow steel cylinder, usually about 2 to 3 in. in diameter, that can be moved longitudinally in and out of the upper casting by means of a handwheel and screw. The open end of the quill hole has a Morse taper. Tools or a lathe center (See Figure 23-33) are held on the quill. A graduated scale usually is engraved on the outside of the quill to aid in controlling its motion in and out of the upper casting. A locking device permits clamping the quill in any desired position.

The *carriage assembly,* shown in Figure 23-11, together with the apron, provides the means for mounting and moving cutting tools. The *carriage,* a relatively flat H-shaped casting, rides on the outer set of ways on the bed. The *cross slide* is mounted on the carriage and can be moved by means of a feed screw that is controlled by a small handwheel and a graduated dial. The cross slide thus provides a means for moving the lathe tool in the facing or cutoff direction.

On most lathes the tool post is mounted on a *compound rest.* The compound rest can rotate and translate with respect to the cross slide, permitting further positioning of the tool with respect to the work.

The *apron,* attached to the front of the carriage, has the controls for providing manual and powered motion for the carriage and powered motion for the cross slide. Figure 23-11 shows front and rear views of a typical apron. The carriage

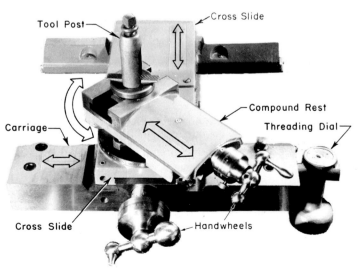

FIGURE 23-11   Lathe carriage assembly. (*Courtesy Sheldon Machine Company, Inc.*)

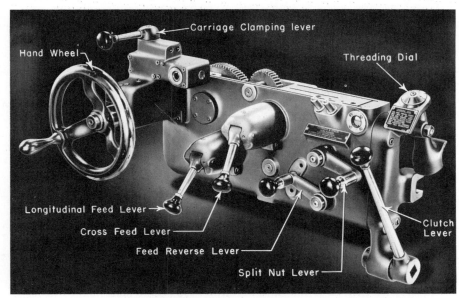

Front side of a lathe apron. (*Courtesy American Tool Works Company.*)

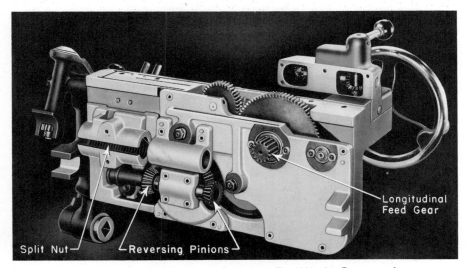

Back side of a lathe apron. (*Courtesy American Tool Works Company.*)

FIGURE 23-11 (continued)   The carriage and apron assembly for an engine lathe.

is moved parallel to the ways by turning a handwheel on the front of the apron, which is geared to a pinion on the back side. This pinion engages a rack that is attached beneath the upper front edge of the bed in an inverted position.

Powered movement of the carriage and cross slide is provided by a rotating *feed rod,* shown in Figure 23-10. The feed rod, which contains a keyway, passes through the two reversing bevel pinions shown in Figure 23-11 and is keyed to them. Either pinion can be activated by means of the feed reverse lever, thus providing "forward" or "reverse" power to the carriage. Suitable clutches connect either the rack pinion or the cross-slide screw to provide longitudinal motion of the carriage or transverse motion of the cross slide.

For cutting threads, a *lead screw,* shown in Figure 23-10, is used. When a friction clutch is used to drive the carriage, motion through the lead screw is by a direct, mechanical connection between the apron and the lead screw. A *split nut,* shown in Figure 23-11, is closed around the lead screw by means of a lever on the front of the apron directly driving the carriage without any slippage.

Modern lathes have *quick-change gearboxes,* driven by the spindle and connect feed rod and lead screw. The associated gearing, the lead screw, and feed rod connect the carriage to the spindle, so the cutting tool can be made to move a specific distance, either longitudinally or transversely, for each revolution of the spindle. Typical lathes may provide as many as 48 feeds, ranging from 0.002 to 0.118 in. (0.05 to 3 mm) per revolution of the spindle, and, through the lead screw, leads from 1½ to 92 threads per inch.

Size Designation of Lathes.

The size of a lathe is designated by two dimensions. The first is known as the *swing.* This is the maximum diameter of work that can be rotated on a lathe. Swing is approximately twice the distance between the line connecting the lathe centers and the nearest point on the ways. The maximum diameter of a workpiece that can be mounted between centers is somewhat less than the swing diameter because the workpiece must clear the carriage assembly as well as the ways. The second size dimension is the *maximum distance between centers.* The swing thus indicates the maximum workpiece diameter that can be turned in the lathe, while the distance between centers indicates the maximum length of workpiece that can be mounted between centers.

Types of Lathes.

Lathes used in manufacturing can be classified as speed, engine, toolroom, turret, automatics, tracer, and numerical control.

Speed lathes usually have only a headstock, tailstock, and a simple tool post mounted on a light bed. They ordinarily have only three or four speeds and are used primarily for wood turning, polishing, or metal spinning. Spindle speeds up to about 4,000 rpm are common.

Engine lathes are the type most frequently used in manufacturing. Figure 23-10 is an example of this type. They are heavy-duty machine tools with all the components described previously and have power drive for all tool movements except on the compound rest. They commonly range in size from 12 to 24 inches swing and from 24 to 48 inches center distance, but swings up to 50 inches and center distances up to 12 feet are not uncommon. Binns Machinery, in Cincinnati, Ohio, builds very large lathes (36- to 60-foot-long beds) which are extremely rigid and therefore capable of performing roughing cuts in iron and steel at depths of cut of ½ to 2 inches; cutting speed 50–200 sfpm with WC tools run at .010 to .100 ipr feeds. To perform heavy cuts requires rigidity in the machine tool, the cutting tools, the workholder, and the workpiece (using steady-rests and other supports).

Most engine lathes are equipped with chip pans and a built-in coolant circulating system. Smaller engine lathes, with swings usually not over 13 inches, also are available in *bench type,* designed for the bed to be mounted on a bench or table.

Toolroom lathes have somewhat greater accuracy and, usually, a wider range of speeds and feeds than ordinary engine lathes. Designed to have greater versatility to meet the requirements of tool and die work, they often have a continuously variable spindle speed range and shorter beds than ordinary engine lathes of comparable swing, since they are generally used for machining relatively small parts. They may be either bench or pedestal type.

Several types of special-purpose lathes are made to accommodate specific types of work. On a *gap-bed lathe,* for example, a section of the bed, adjacent to the headstock, can be removed to permit work of unusually large diameter to be swung. Another example is the *wheel lathe,* which is designed to permit the turning of railroad-car wheel-and-axle assemblies.

Although engine lathes are versatile and very useful, the time required for changing and setting tools and for making measurements on the workpiece is often a large percentage of the cycle time. Often the actual chip-production time is less than 30% of the total cycle time. Methods to reduce setup and tool change time sharply are discussed in Chapter 28. Much of the operator's time is consumed by simple, repetitious adjustments and in watching chips being made. The placement of such machines in cells, as discussed in Chapter 42, greatly increases the productivity of the workers because they can run more than one machine. Turret lathes, screw machines, and other types of semiautomatic and automatic lathes have been highly developed and are widely used in manufacturing as another means to improve cutting productivity.

Turret Lathes.   The basic components of a *turret lathe* are depicted in Figure 23-12. Basically, a longitudinally feedable, hexagon turret replaces the tailstock. The turret, on which six tools can be mounted, can be rotated about a vertical axis to bring each tool into operating position, and the entire unit can be moved longitudinally, either manually or by power, to provide feed for the tools. When the turret assembly is backed away from the spindle by means of a capstan wheel, the turret indexes automatically at the end of its movement, thus bringing each of the six tools into operating position.

The square turret on the cross slide can be rotated manually about a vertical axis to bring each of the four tools into operating position. On most machines, the turret can be moved transversely, either manually or by power, by means of the cross slide, and longitudinally through power or manual operation of the carriage. In most cases, a fixed tool holder also is added to the back end of the cross slide; this often carries a parting tool.

Through these basic features of a turret lathe, a number of tools can be set up on the machine and then quickly be brought successively into working po-

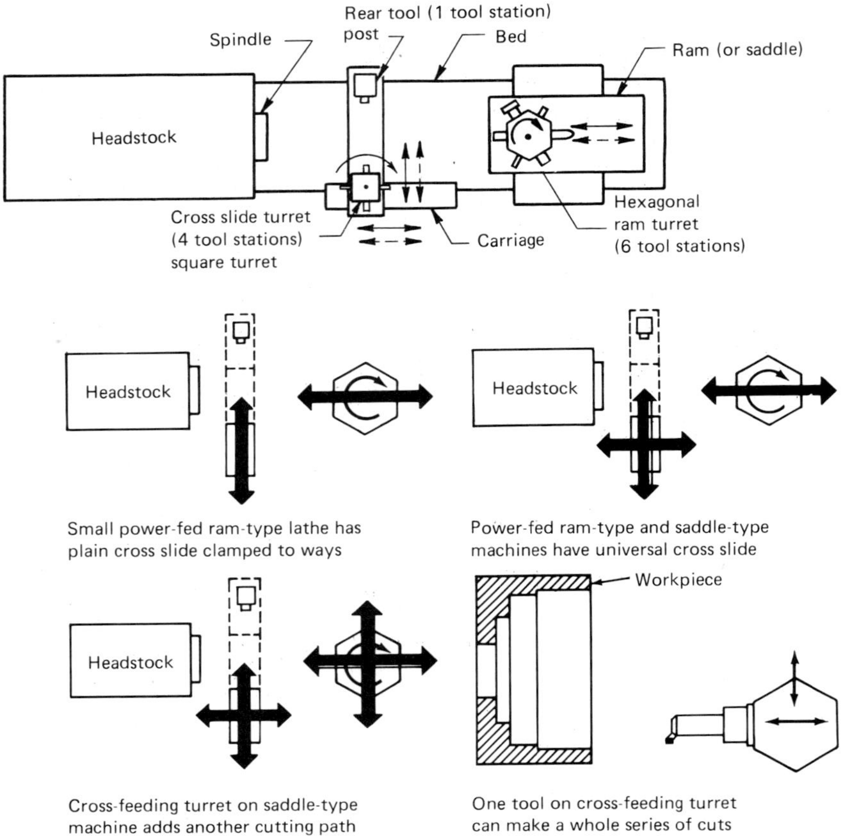

FIGURE 23-12 Block diagrams (top views) of ram- and saddle-type turret lathes.

sition so that a complete part can be machined without the necessity for further adjusting, changing tools, or making measurements.

The two basic types of turret lathes are the ram-type turret lathe and the saddle-type turret lathe.

The *saddle-type turret lathe* provides a more rugged mounting for the hexagon turret than can be obtained by the ram-type mounting. In the *ram-type turret lathe* the ram and turret are moved up to the cutting position by means of the capstan wheel, and the power feed is then engaged. As the ram is moved toward the headstock, the turret is automatically locked into position so that rigid tool support is obtained. A set of rotary stopscrews, such as those shown in Figure 23-13, control the forward travel of the ram, one stop being provided for each face on the turret. The proper stop is brought into operating position automatically when the turret is indexed. A similar set of stops usually is provided to limit movement of the cross slide. In *saddle-type lathes,* the main turret is mounted directly on the saddle, and the entire saddle and turret assembly reciprocates. Larger turret lathes usually have this type of mounting. However, because the saddle-turret assembly is rather heavy, this type of mounting provides less rapid turret reciprocation. When such lathes are used with heavy tooling

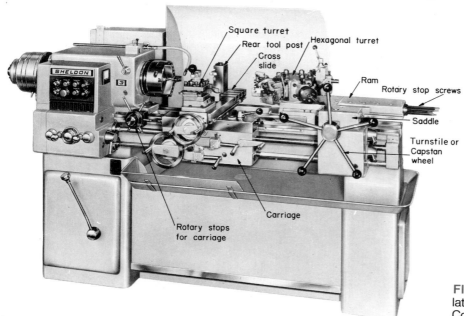

FIGURE 23-13   Ram-type turret lathe. (Courtesy Sheldon Machine Company, Inc.)

for making heavy or multiple cuts, a *pilot arm* attached to the headstock engages a pilot hole attached to one or more faces of the turret to give additional rigidity. Such a device is shown in Figure 23-14.

Turret lathe headstocks have two features not found on ordinary engine lathes: rapid shifting between at least two spindle speeds, with a brake to stop the spindle very rapidly, and automatic stock-feeding for feeding bar stock through the spindle hole.

If the work is to be held in a chuck, some type of air-operated chuck, or special clamping fixture, frequently is employed to reduce the setup time to a minimum.

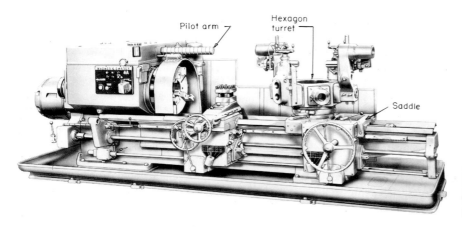

FIGURE 23-14   Saddle-type turret lathe, having a side-hung carriage. (Courtesy Warner & Swassey Company.)

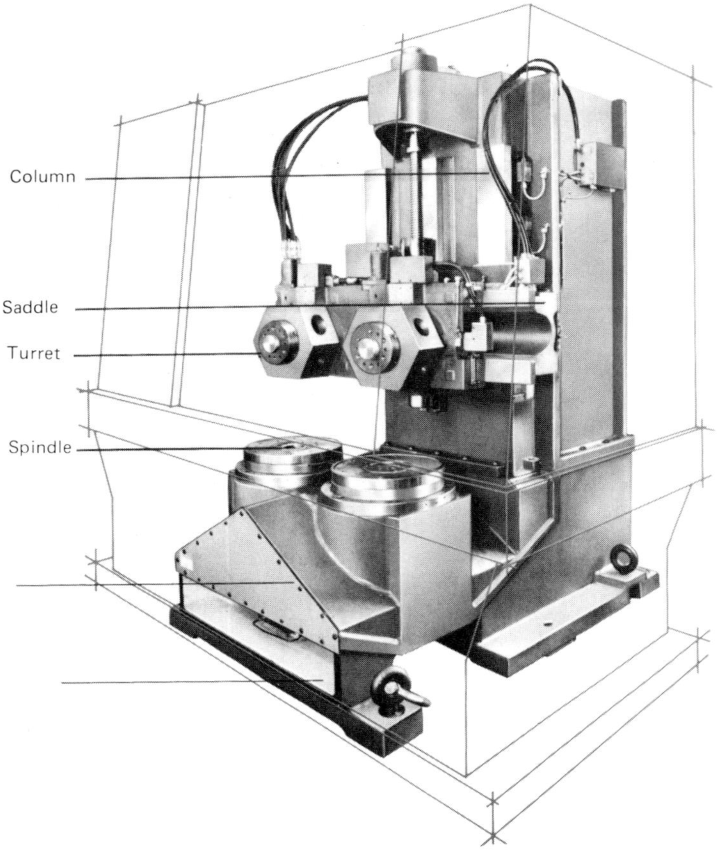

FIGURE 23-15 Basic structure of a twin spindle vertical turret lathe (bullard Aku-Turn), typically NC- or CNC-controlled.

Vertical Turret Lathes. In machining large and/or heavy parts, such as pinions, couplings, and ring gears, vertical turret lathes are used. These are essentially regular turret lathes turned on end. Their rotary work tables commonly range from 24 to 48 inches in diameter and are equipped with both removable chuck jaws and T-slots for clamping the work. The saddle carries one or two five- or six-sided turrets. Usually, each motion of successive tools can be controlled by means of stops so that duplicate workpieces can be machined with one tooling setup. Figure 23-15 shows a twin spindle vertical turret lathe with the guards and covers removed.

Automatic Turret Lathes. After a turret lathe is tooled, the skill required of the operator is very low, and the motions are simple and repetitive. As a result, several types of automatic turret lathes have been developed that require no operator. One type uses buttons and knobs on a control panel to define (program) machine motions. A second type has a turret, the movement of which is controlled by setting trip blocks and pins.

Ordinary turret lathes use the ten-station tooling setups for complete machining of a piece and minimize machine-controlling time. However, an operator is required to control the machine and to feed the work into machining position. Automatic turret lathes can eliminate these last two functions. They usually have provision for manual operation and are not quite so productive as *screw machines*, which are lathes designed for completely automatic operation. Screw machines originally were designed for machining small parts, such as screws, bolts, bushings, and so on, from bar stock—hence the name ''screw machines.'' Now they are used for producing a wide variety of parts, covering a considerable range of sizes, and are even used for some chucking-type work.

Single-Spindle Automatic Screw Machines.

There are two common types of *single-spindle screw machines*. One, an American development and commonly called the turret type (Brown & Sharpe), is shown in Figure 23-16. The other is of Swiss origin and is referred to as the Swiss type. The *Brown & Sharpe screw machine* is essentially a small automatic turret lathe, designed for bar stock, with the main turret mounted in a vertical plane on a ram. Front and rear toolholders can be mounted on the cross slide. All motions of the turret, cross slide, spindle, chuck, and stock-feed mechanism are controlled by cams. The turret cam is essentially a program that defines the movement of the turret during a cycle. These machines usually are equipped with an automatic rod-feeding magazine that feeds a new length of bar stock into the collet as soon as one rod is completely used.

Often screw machines of the Brown & Sharpe type are equipped with a transfer or ''picking'' attachment. This device picks up the workpiece from the spindle as it is cut off and carries it to a position where a secondary operation

FIGURE 23-16 On the turret-type single-spindle automatic, the tools must take turns to make cuts.

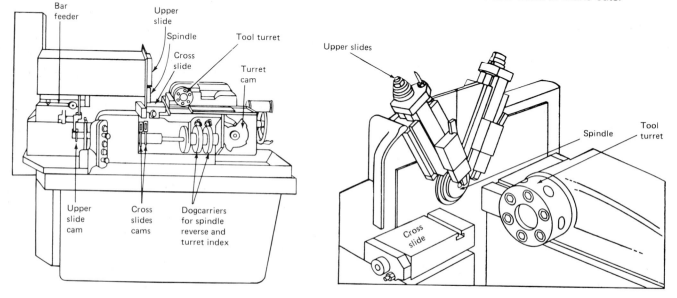

FIGURE 23-17 Close-up view of a Swiss-type screw machine, showing the tooling and radial tool sides, actuated by rocker arms. (Courtesy George Gorton Machine Corporation.)

is performed by a small, auxiliary power head. In this manner screwdriver slots are put in screw heads, small flats are milled parallel with the axis of the workpiece, or holes are drilled normal to the axis.

On the *Swiss-type automatic screw machine*, the cutting tools are held and moved in radial slides as shown in Figure 23-17. Disc cams move the tools into cutting position and provide feed into the work in a radial direction only; they provide any required longitudinal feed by reciprocating the headstock.

Most machining on Swiss-type screw machines is done with single-point cutting tools. Because they are located close to the spindle collet, the workpiece is not subjected to much deflection. Consequently, these machines are particularly well suited for machining very small parts and are used primarily for such work.

Both types of single-spindle screw machines can produce work to close tolerances, the Swiss-type probably being somewhat superior for very small work. Tolerances of 0.0002 to 0.0005 in. are not uncommon. The time required for setting up the machine is usually an hour or two and can be much less. One person can tend many machines, once they are properly tooled. They have short cycle times, frequently less than 30 seconds per piece.

Multiple-Spindle Automatic Screw Machines. Single-spindle screw machines utilize only one or two tooling positions at any given time. Thus the total cycle time per workpiece is the sum of the individual machining and tool-positioning times. On *multiple-spindle screw machines*, sufficient spindles, usually four, six, or eight, are provided so that all tools cut simultaneously. Thus the cycle time per piece is equal to the maximum cutting time of a single tool position plus the time required to index the spindles from one position to the next.

The two distinctive features of multiple-spindle screw machines are shown in Figures 23-18. First, the multiple spindles are carried in a rotatable drum that indexes in order to bring each spindle into a different working position. Second, a nonrotating tool slide contains the same number of tool holders as there are spindles and thus provides and positions a cutting tool (or tools) for each spindle. Tools are fed by longitudinal reciprocating motion. Most machines have a cross slide at each spindle position so that an additional tool can be fed from the side for facing, grooving, knurling, beveling, and cutoff operations. These slides also are shown in Figure 23-18. All motions are controlled automatically.

With a tool position available on the end tool slide for each spindle (except for a stock-feed stop at position 6), when the slide moves forward, these tools cut essentially simultaneously. At the same time, the tools in the cross slides move inward and make their cuts. When the forward cutting motion of the end tool slide is completed, it moves away from the work, accompanied by the outward movement of the radial slides. The spindles are indexed one position, by rotation of the spindle carrier, to position each part for the next operation to be performed. At spindle position 5, finished pieces are cut off. Bar stock is fed to correct length for the beginning of the next operation. Thus a piece is completed each time the tool slide moves forward and back.

Multiple-spindle screw machines are made in a considerable range of sizes, determined by the diameter of the stock that can be accommodated in the spindles. There may be four, five, six or eight spindles. The operating cycle of the end tool slide is determined by the operation that requires the longest time.

The only attention a multiple-spindle screw machine requires is to keep the bar stock feed rack supplied and to check the finished products occasionally to make sure that they are within desired tolerances. One operator usually services many machines.

Most multiple-spindle screw machines utilize specially shaped cams to control the motions. Setting up the cams and the tooling for a given job may require from 2 to 20 hours. However, once such a machine is set up, the cycle time is very short. Often a piece may be completed every 10 seconds. Typically, from 2000 to 5000 parts are required in a lot to justify setting up and tooling a multiple-spindle automatic screw machine.

The precision of multiple-spindle screw machines is good but seldom as good as that of single-spindle machines. However, tolerances from 0.0005 to 0.001 in. on the diameter are typical.

Multiple-station lathes, available for work that must be held in chucks, are essentially vertical, multiple-spindle screw machines. A number of chuck-equipped spindles are mounted in a rotary indexing table and are indexed successively under a series of vertical rams in which the cutting tools are mounted. One chuck position remains at rest and has no tool ram; therefore workpieces

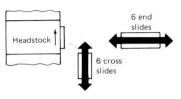

Headstock

6 end slides

6 cross slides

All spindles on multiple-spindle automatic have the same tool path

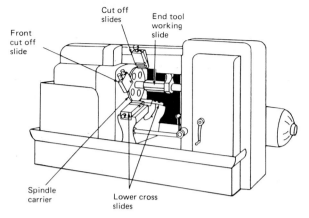

Cut off slides

Front cut off slide

End tool working slide

Spindle carrier

Lower cross slides

The six spindle automatic

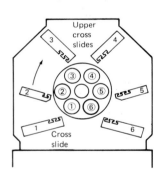

Upper cross slides

3   4

3   4

2   5

1   6

1   Cross slide   6

Spindle arrangement for 6 spindle automatic. The shaded circle shows the position where the barstock is usually fed. The cutoff position is the one preceding the bar feed position.

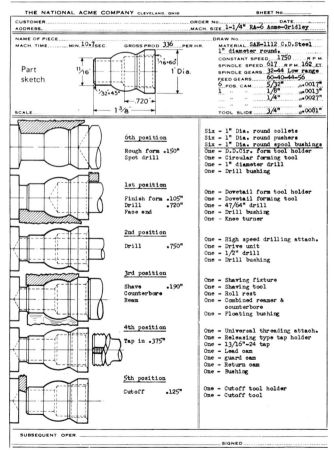

THE NATIONAL ACME COMPANY CLEVELAND, OHIO   SHEET NO.

CUSTOMER   ORDER NO.   DATE
ADDRESS.   MACH. SIZE 1-1/4" RA-6 Acme-Gridley

NAME OF PIECE   DRAW NO.
MACH. TIME   MIN.10.7 SEC.   GROSS PROD. 336 PER HR.   MATERIAL SAE-1112 C.D.Steel
                                                         1" diameter round.
                                            CONSTANT SPEED 1750 R P M
Part sketch   SPINDLE SPEED 617 R P M 162 F.T.
                                            SPINDLE GEARS 32-44 Low range
11/16   16x60   FEED GEARS 60-40-44-56
         1 Dia.   6 POS. CAM   5/32"   .0017"
                                            1   1/8"   .0013"
1/32x45   "   "   1/4"   .0027"
       .720   "   "   6
SCALE   TOOL SLIDE 3/4"   .0081"
        1 3/8

6th position   Six - 1" Dia. round collets
                      Six - 1" Dia. round pushers
Rough form .150"   Six - 1" Dia. round spool bushings
Spot drill   One - D.D.Cir. form tool holder
                      One - Circular forming tool
                      One - 1" diameter drill
                      One - Drill bushing

1st position   One - Dovetail form tool holder
                      One - Dovetail forming tool
Finish form .105"   One - 47/64" drill
Drill   .720"   One - Drill bushing
Face end   One - Knee turner

2nd position   One - High speed drilling attach.
                      One - Drive unit
Drill   .750"   One - 1/2" drill
                      One - Drill bushing

3rd position   One - Shaving fixture
                      One - Shaving tool
Shave   .190"   One - Roll rest
Counterbore   One - Combined reamer &
Ream   counterbore
                      One - Floating bushing

4th position   One - Universal threading attach.
                      One - Releasing type tap holder
Tap in .375"   One - 13/16"-24 tap
                      One - Lead cam
                      One - guard cam
                      One - Return cam
                      One - Bushing

5th position   One - Cutoff tool holder
                      One - Cutoff tool
Cutoff   .125"

SUBSEQUENT OPER.

SIGNED

Tooling sheet for making a part on a six-spindle.

FIGURE 23-18 The multiple spindle automatic makes all cuts simultaneously and then performs the noncutting functions (tool withdrawal, index, bar feed) at high speed.

can be loaded and unloaded at this position while machining takes place at all the others.

Although screw machines and automatic turret lathes are automatic *types* of lathes, the term *automatic lathe* generally is applied to a lathe that is semiautomatic and makes simultaneous cuts using ''massed'' tooling (Figure 23-19) but does not use turret or screw-machine principles. The tools are fed and retracted automatically by means of cam-controlled mechanisms. In most cases, an operator is required to load and unload the machine, so they do not repeat cycles automatically. The majority of automatic lathes have only a single spindle, but some specialized multispindle machines are also used.

In a typical *single-spindle automatic lathe*, the massed cutting tools are held in *tool blocks*, or *slides*, which are power-actuated and controlled to move the tools into position and feed them along the work. Sometimes the front block provides only radial motion for facing, form cutting, and cutoff operations. The rear block has both radial and longitudinal motions, which are controlled by a plate cam, as illustrated in Figure 23-19, which shows the motions of the tool blocks and the portion of the metal removed by each tool. On some machines a third overhead tool block is also provided.

In most cases the work is held between centers, utilizing various types of power-actuated chucks, collets, and tailstocks so that the work-handling time is minimized. In some cases the work is fed into the machine from a hopper-type feeding device and clamped and discharged automatically.

The total machining cycle on an automatic lathe is usually very short, often less than one minute. Sometimes a part is put successively into two to four automatic lathes to complete its machining. Because they are fairly flexible, quite a variety of shapes and sizes can be handled in one lathe by changing the tooling setup.

Tracer and NC lathes will be discussed in Chapter 29.

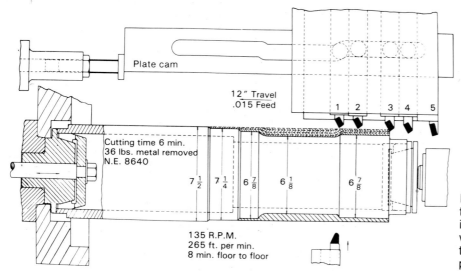

FIGURE 23-19 Movements of the tool blocks for the machining process in a single-spindle automatic lathe, with the metal to be removed by each tool indicated. (Courtesy Gisholt Corporation.)

Types of Boring Machines

Vertical Boring and Turning Machines. Figure 23-20 shows the basic elements of a vertical boring and turning machine. These machine tools are structurally similar to double-housing planers, except that the table rotates instead of reciprocating. Functionally, a vertical boring machine essentially is the same as a vertical turret lathe, but it usually has two main tool heads instead of a turret. Thus turning, facing, and usually boring (but not milling) are done on *vertical boring mills* or machines.

Vertical boring machines come with tables ranging from about 3 to 40 feet in diameter. Toolheads have both horizontal and vertical feed and therefore can be used for boring and for facing cuts. Usually, one or both can also be swiveled about a horizontal axis to permit boring at an angle. Most machines also have a side toolhead, sometimes provided with a four-sided turret. This toolhead has vertical and horizontal feed and is used primarily for turning. Single-point tools customarily are used, and turning, facing, and boring.

Many modern boring machines are numerically controlled (see Chapter 29). This permits the operator to make tool settings merely by setting dials and also to preset the adjustment for a cut while one is being made. The tool can be moved very quickly to the proper position for the next cut, reducing the amount

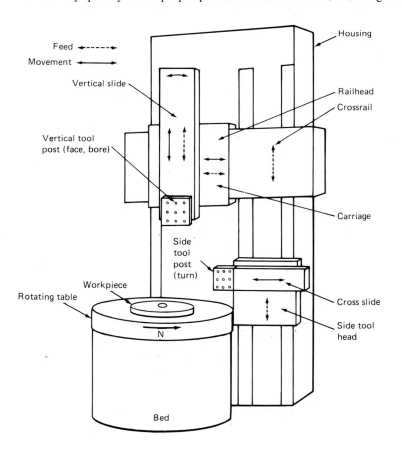

FIGURE 23-20 Block diagram of a vertical boring and turning machine.

of machine-controlling time and increasing the productivity of these large and costly machines.

Horizontal Boring (Drilling and Milling) Machines.

Horizontal boring machines are very versatile and thus particularly useful in machining large parts. The basic components of these machines are indicated in Figure 23-21. The essential features are

1. A rotating spindle that can be fed horizontally (tool rotates).
2. A table that can be moved and fed in two directions in a horizontal plane.
3. A headstock that can be moved vertically.
4. An outboard bearing support for a long boring bar.

The spindle will accept both drills and milling cutters. A wide range of rpms is provided, and heavy bearings are incorporated that will absorb thrust in all directions. The spindle is also provided with longitudinal power feed so that drilling and boring can be done through a considerable distance without the table's being moved.

Boring on this type of machine is done by means of a rotating single-point tool. The tool can be mounted in either a stub-type bar, held only in the spindle,

FIGURE 23-21 Boring a weldment on a horizontal boring, drilling, and milling machine. A line-type boring bar is being used with an outboard bearing support. (Courtesy Lucas Machine Division, The New Britain Machine Company.)

FIGURE 23-22 Adjusting boring bar, using the offset-radius principle.

as shown in Figure 23-22 and Figure 23-23, or in a long line-type bar that has its outer end supported in a bearing on the outboard column, as shown in Figure 23-21. The outboard bearing provides rigid support for the boring bar and permits very accurate work to be done. However, because of the flexibility inherent in a long boring bar and offset tool holder, horizontal boring machines are used primarily for boring holes less than 12 inches in diameter, or for long holes, or for a series of in-line holes. Unless they are very long or the shape of the workpiece prevents it, larger holes usually are bored on a vertical boring mill.

Mass-Production Boring Machines. Special boring machines are built for machining specific parts in mass production. The workpiece usually remains stationary, and boring is done by one or more rotating boring tools, typically carried in a reciprocating powerhead, such as is shown in Figure 23-24.

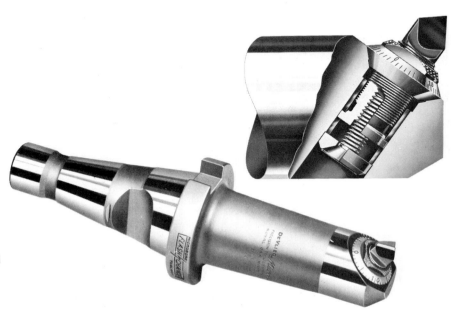

FIGURE 23-23 Adjustable boring tool. Extension of the single-point tool from the bar is adjustable, as shown in sectional view. (Courtesy DeVlieg Machine Company.)

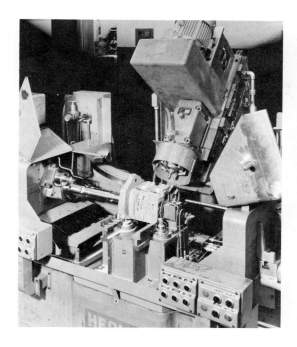

FIGURE 23-24   (Left) Production-type boring machine, having multiple heads, that completes a part in 51 seconds. (Above) Close-up view of one multiple-spindle boring head on a production-type machine. (Courtesy Healt Machine Co.)

In most cases the operation is automatic once the workpiece is placed in the workholding device. Such machines usually are very accurate and often are equipped with automatic gaging and sizing controls. (See discussion of transfer lines in Chapter 42.)

With rotating workpieces, the size of the hole is controlled by transverse movement of the tool holder. When boring is done with a rotating tool, size is controlled by changing the offset radius of the cutting-tool tip with respect to the axis of rotation. A general-purpose type of adjustable boring bar is shown in Figure 23-22. The type shown in Figure 23-23 has more precise control and is used on larger-scale manufacturing. Two or more adjustable cutting tools can be built into a single bar, thus permitting more than one diameter to be bored simultaneously. For boring relatively long holes, the type of boring bar shown in Figure 23-25 has a special advantage. As shown, the smaller, forward bit corrects misalignment of the original hole and provides a guide hole for the nose cone. The nose cone then provides good alignment and support for the rear bit, which bores the final hole to size.

Boring Machine Precision.   As with turning, the precision obtainable in boring depends considerably on the rigidity of the tool support. On specialized, production-type boring machines, tolerances are readily held to within 0.0005 in. on small diameters, whereas on general-purpose machines tolerances of 0.001 in. are typical unless the boring bar overhang becomes excessive.

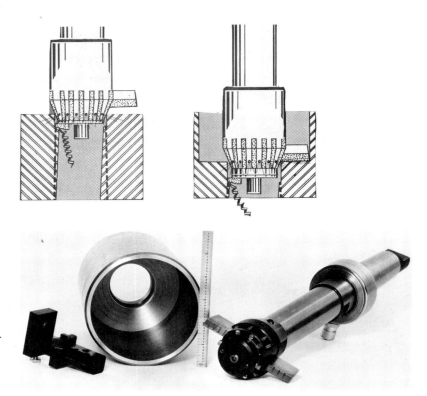

FIGURE 23-25   Boring tool employing a centering tool and conical guide, for boring large holes in a single operation. (Courtesy Vernon Devices, Inc.)

Jig Borers.   *Jig borers* are very precise vertical-type boring machines designed for use in making *jigs* and *fixtures*. From the viewpoint of boring operations, they contain no unusual features, except that the spindle and spindle bearings are constructed with very high precision. Their unique features are in the design of the worktable controls, which permits very precise movement and control, thus making them especially useful in layout work. Modern NC machining centers have eliminated the need for jig borers.

Cutting Tools for Lathes

Lathe Cutting Tools.   Most lathe operations are done with relatively simple, single-point cutting tools, such as those illustrated in Figure 23-26. On right-hand and left-hand turning and facing tools, the cutting takes place on the side of the tool; therefore the side rake angle is of primary importance and deep cuts can be made. On the round-nose turning tools, cutoff tools, finishing tools, and some threading tools, cutting takes place on or near the end of the tool, and the back rake is therefore of importance. Such tools are used with relatively light depths of cut.

Because tool materials are expensive, it is desirable to use as little as possible. It is essential, at the same time, that the cutting tool be supported in a strong,

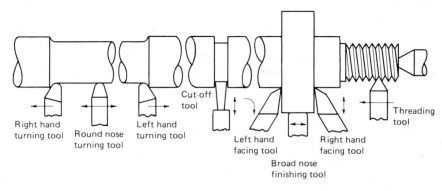

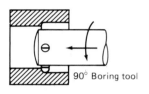

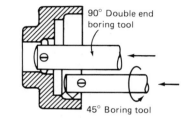

FIGURE 23-26 Shapes and uses of common single-point lathe tools.

rigid manner to minimize deflection and possible vibration. Consequently, lathe tools are supported in various types of heavy, forged steel tool holders, as shown in Figure 23-27. The tool bit should be clamped in the tool holder with minimum overhang. Otherwise, tool chatter and a poor surface finish may result.

In the use of carbide, ceramic, or coated carbides for mass production work, throwaway inserts are used; these can be purchased in a great variety of shapes, geometrics (nose radius, tool angles, and groove geometry), and sizes (see Figure 23-28).

When several different operations on a lathe are performed repeatedly in sequence, the time required for changing and setting tools may constitute as much as 50% of the total time. As a consequence, quick-change tool holders, such as shown in Figure 23-29, are being used increasingly. The individual tools, preset in their holders, can be interchanged in the special tool post in a few seconds. With some systems a second tool may be set in the tool post while a cut is being made with the first tool and can then be brought into proper position by rotating the post.

In lathe work, the nose of the tool should be set exactly at the same height as the axis of rotation of the work. However, because any setting below the axis causes the work to tend to ''climb'' up on the tool, most machinists set their tools a few thousandths of an inch above the axis, except for cutoff, threading, and some facing operations.

Drill chucks are used on lathes for holding drills, center drills, and reamers mounted in the tailstock quill. Large drills are mounted directly in the quill hole by means of their taper shanks.

a.

b.   H.S.S. Bit

c.

d.

Key: a. cutoff
      b. boring bars
      c. R.H. facing
      d. L.H. turning
      e. threading tool

e.

FIGURE 23-27 Common types of forged tool holders. (Courtesy Armstrong Bros. Tool Company.) See Figure 23-26.

FIGURE 23-28 Typical insert shapes, available cutting edges per insert, and insert holders. (Adapted from *Turning Handbook of High Efficiency Metal Cutting.* Courtesy General Electric Company.)

Insert shape	Available cutting edges	Typical insert holder
Round ◯	4–10 on a side 8–20 total	15° Square insert
80°/100° diamond ▱	4 on a side 8 total	
Square ▭	4 on a side 8 total	Triangular insert 0°
Triangle △	3 on a side 6 total	
55° diamond ◇	2 on a side 4 total	35° diamond
35° diamond ◇	2 on a side 4 total	5°

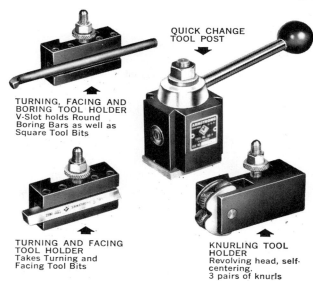

QUICK CHANGE
TOOL POST

TURNING, FACING AND
BORING TOOL HOLDER
V-Slot holds Round
Boring Bars as well as
Square Tool Bits

TURNING AND FACING
TOOL HOLDER
Takes Turning and
Facing Tool Bits

KNURLING TOOL
HOLDER
Revolving head, self-
centering.
3 pairs of knurls

FIGURE 23-29   Quick-change tool post and accompanying tool holders. (Courtesy Armstrong Bros. Tool Company.)

Form Tools.   Form tools, made by grinding the inverse of the desired work contour, are extensively used on lathes. See Figure 23-18. A threading tool is often a form tool. Although form tools are relatively expensive to manufacture, they make it possible to machine a fairly complex surface with a single inward feeding of one tool. For mass-production work, adjustable form tools of either flat or rotary types, such as are shown in Figure 23-30, are used. These, of course, are expensive to make initially but can be resharpened by merely grinding a small amount off the face and then raising or rotating the cutting edge to the correct position.

The use of form tools is limited by the difficulty of grinding adequate rake angles for all points along the cutting edge. A rigid setup is needed to resist the large cutting forces that develop with these tools.

Turret-Lathe Tools.   In turret lathes, the work is generally held in collets and the correct amount of bar stock is fed into the machine to make one part. The tools are arranged in sequence at the tool stations with depths of cut all preset. The following factors should be considered when setting up a turret lathe.

1. *Setup time*: time required to set the tooling and set stops. Standard tool-holders and tools should be used as much as possible to minimize setup time. Setup time can be greatly reduced by eliminating adjustment in the setup. (See Chapter 44.)
2. *Workholding time*: time to load and unload parts and/or stock.
3. *Machine-controlling time*: time required to manipulate the turrets. Can be reduced by combining operations where possible. Dependent on the sequence of operations established by the design of the setup.

FIGURE 23-30   Circular and block types of form tools. (Courtesy Speedi Tool Company, Inc.)

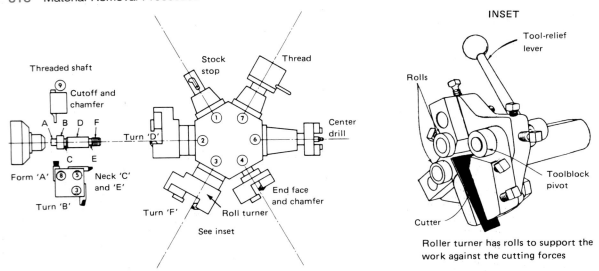

INSET

Tool-relief lever

Rolls

Toolblock pivot

Cutter

Roller turner has rolls to support the work against the cutting forces

FIGURE 23-31 Turret lathe tooling setup for producing part shown. Numbers in circles indicate the sequence of operation from 1 to 9. Operation 3 is a combined operation. The roll turner is turning surface F while tool 3 on the square post is turning surface B.

4. *Cutting time*: time during which chips are being produced; should be as short as is economically practical and represent the greatest percentage of the total cycle time possible.

5. *Cost*: cost of the tool, setup labor cost, lathe operator labor cost, and the number of pieces to be made.

There are essentially 11 tooling stations, as shown in Figure 23-31, with six in the turret, four in the indexable tool post, and one in the rear tool post. The tooling is more rugged in turret lathes because heavy, simultaneous cuts are often made. Tools mounted in the hex turret that are used for turning are often equipped with pressure rollers set on the opposite side of the rotating workpiece from the tool to counter the cutting forces.

Turret lathes are most economical in producing lots too large for engine lathes but too small for automatic screw machines or automatic lathes. However, in recent years much of this work has been assumed by numerical control lathes, which will be discussed in Chapter 29. Numerical control lathes represent a higher level of automation, A(4), than turret lathes. See the Yardstick for Automation in Chapter 1.

Workholding in Lathes

Workholding Devices for Lathes. Five methods are commonly used for supporting workpieces in lathes:

1. Held between centers.
2. Held in a chuck.
3. Held in a collet.
4. Mounted on a face plate.
5. Mounted on the carriage.

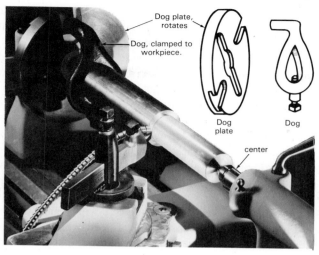

FIGURE 23-32 Work being turned between centers in a lathe, showing the use of dog and dog plate. (Courtesy South Bend Lathe.)

In the first four of these methods the workpiece is rotated during machining. In the fifth method, which is not used extensively, the tool rotates while the workpiece is fed into the tool.

Lathe Centers. Workpieces that are relatively long with respect to their diameters usually are machined between centers. See Figure 23-32. Two *lathe centers* are used, one in the spindle hole and the other in the hole in the tailstock quill. Two types are used, called *dead* and *live* by the user. Dead centers are *solid*, that is, made of hardened steel with a Morse taper on one end so that it will fit into the spindle hole. See Figure 23-33. The other end is ground to a 60-degree taper. Sometimes the tip of this taper is made of tungsten carbide to provide better wear resistance. Before a center is placed in position, the spindle hole should be carefully wiped clean. The presence of foreign material will prevent the center from seating properly and it will not be aligned accurately.

Live centers of the type, shown in Figure 23-34, are designed so that the end that fits into the workpiece is mounted on ball or roller bearings. It is free to rotate. No lubrication is required. Live centers may not be as accurate as the plain type and therefore are not often used for precision work.

Before a workpiece can be mounted between lathe centers, a 60-degree center hole must be drilled in each end. This can be done in a drill press, or in a lathe by holding the work in a chuck. A combination center drill and countersink

FIGURE 23-33 Solid or "dead" lathe center. (Courtesy Chicago-Latrobe Twist Drill Works.)

FIGURE 23-34 "Live" type of lathe center. (Courtesy Motor Tool Mfg. Co.)

ordinarily is used, with care taken that the center hole is deep enough so that it will not be machined away in any facing operation and yet is not drilled to the full depth of the tapered portion of the center drill.

Because the work and the center of the headstock end rotate together, no lubricant is needed in the center hole at this end. Because the center in the tailstock quill does not rotate, adequate lubrication must be provided. A mixture of white lead and oil is often used. Failure to provide proper lubrication at all times will result in scoring of the workpiece center hole and the center, and inaccuracy and serious damage may occur. Live centers are often used in the tailstock to overcome these problems.

The workpiece must rotate freely, yet no looseness should exist. Looseness usually will be manifested in chattering of the workpiece during cutting. The setting of the centers should be checked after cutting for a short time. Heating and thermal expansion of the workpiece will reduce the clearances in the setup.

A mechanical connection must be provided between the spindle and the workpiece to cause it to rotate. This is accomplished by some type of *lathe dog*. The lathe dog and *dog plate* provide a mechanical connection between the workpiece and the spindle. The dog is clamped to the work. The *tail* of the dog enters a slot in the dog plate, which is attached to the lathe spindle in the same manner as a lathe chuck. For work that has a finished surface, a piece of soft metal, such as copper or aluminum, can be placed between the work and the dog setscrew clamp to avoid marring.

Mandrels.   Workpieces that must be machined on both ends or are disc-shaped are often mounted on mandrels for turning between centers. Three common types of mandrels are shown in Figure 23-35. *Solid mandrels* usually vary from 4 to 12 in. in length and are accurately ground with a 1:2000 taper (0.006 in. per foot). After the workpiece is drilled and/or bored, it is pressed on the mandrel. The mandrel should be mounted between centers so that the cutting force tends to tighten the work on the mandrel taper. Solid mandrels permit the work to be machined on both ends as well as on the cylindrical surface. They are available in stock sizes but can be made to any desired size.

Gang (or disc) mandrels are used for production-type work because the workpieces do not have to be pressed on and thus can be put in position and removed more rapidly. However, only the cylindrical surface of the workpiece can be machined when this type of mandrel is used.

Cone mandrels have the advantage that they can be used to center workpieces having a range of hole sizes.

Lathe Chucks.   Lathe chucks are used to support a wider variety of workpiece shapes and to permit more operations to be performed than can be accomplished when the work is held between centers. Two basic types of chucks are used. There are illustrated in Figure 23-36.

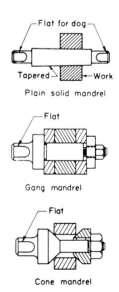

FIGURE 23-35   Three types of mandrels.

FIGURE 23-36 Lathe chucks. (Left) Four-jaw independent. (Right) Three-jaw, self-centering. (Courtesy Cushman Industries, Inc.)

Three-jaw self-centering chucks are used for work that has a round or hexagonal cross section. The three jaws are moved inward or outward simultaneously by the rotation of a spiral cam, which is operated by means of a special wrench through a bevel gear. If they are not abused, these chucks will provide automatic centering to within about 0.001 in. However, they can be damaged through use and will then be considerably less accurate.

Each jaw in a *four-jaw independent chuck* can be moved inward and outward independent of the others by means of a chuck wrench. Thus they can be used to support a wide variety of work shapes. A series of concentric circles engraved on the chuck face aid in adjusting the jaws to fit a given workpiece. Four-jaw chucks are heavier and more rugged than the three-jaw type and, because undue pressure on one jaw does not destroy the accuracy of the chuck, they should be used for all heavy work. The jaws on both three- and four-jaw chucks can be reversed to facilitate gripping either the inside or the outside of workpieces.

Combination four-jaw chucks are available in which each jaw can be moved independently or can be moved simultaneously by means of a spiral cam. Two-jaw chucks are also available. For mass-production work, special chucks often are used in which the jaws are actuated by air or hydraulic pressure, permitting very rapid clamping of the work. The rapid exchange of chucks is discussed in Chapters 28 and 44.

Collets. By the use of *collets,* smooth cold-rolled bar stock or machined workpieces can be held more accurately than with regular chucks. As shown in Figure 23-37, collets are relatively thin tubular steel bushings that are split into three longitudinal segments over about two-thirds of their length. At the split end, the smooth internal surface is shaped to fit the piece of stock that is to be held. The external surface of the collet is a taper that mates with an internal taper of a collet sleeve. When the collet is pulled inward into the spindle (by

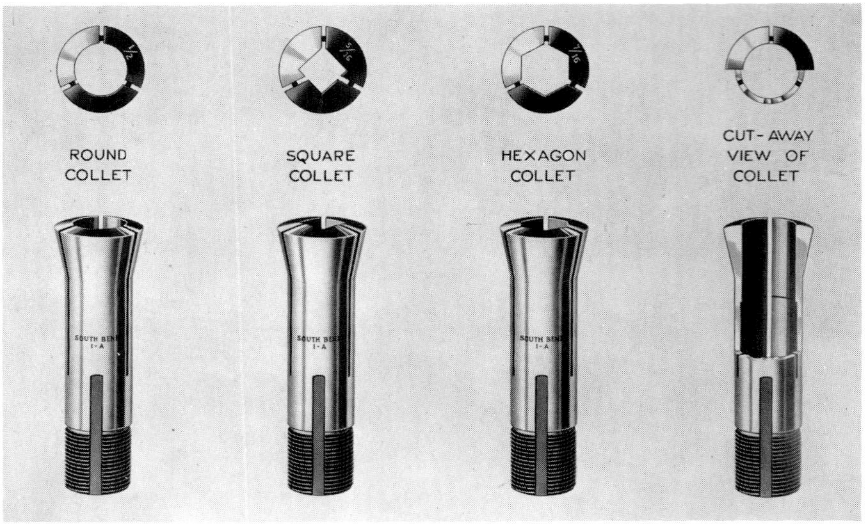

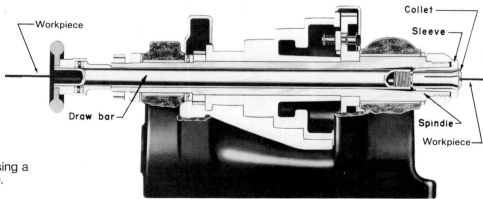

FIGURE 23-37 Several types of lathe collets. (Courtesy South Bend Lathe.)

means of the draw bar), the action of the two mating tapers squeezes the collet segments together, causing them to grip the workpiece. See Figure 23-38.

Collets are made to fit a variety of symmetrical shapes. If the stock surface is smooth and accurate, good collets will provide very accurate centering, runout less than 0.0005 in. However, the work should be no more than 0.002 in. larger or 0.005 in. smaller than the nominal size of the collet. Consequently, collets are used only on drill-rod, cold-rolled, extruded, or previously machined stock.

Collets that can open automatically and feed bar stock forward to a stop mechanism are commonly used on turret lathes. Another type of collet similar to a Jacobs drill chuck has a greater size range than ordinary collets; therefore fewer are required.

FIGURE 23-38 Method of using a draw-in collet in a lathe spindle. (Courtesy South Bend Lathe.)

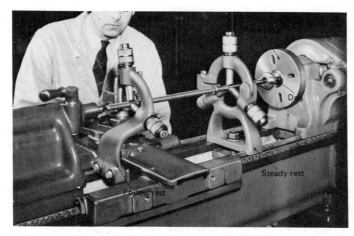

FIGURE 23-39 Cutting a thread on a long slender workpiece, using a follow rest (left) and a steady rest (right). (Courtesy South Bend Lathe.)

Face Plates. *Face plates* are used to support irregularly shaped work that cannot be gripped easily in chucks or collets. The work can be bolted or clamped directly on the face plate or can be supported on an auxiliary fixture that is attached to the face plate. The latter procedure is time-saving when identical pieces are to be machined.

Mounting Work on the Carriage. When no other means is available, boring occasionally is done on a lathe by mounting the work on the carriage, with the boring bar mounted between centers and driven by means of a dog.

Steady and Follow Rests. If one attempts to turn a long slender piece between centers, the radial force exerted by the cutting tool, or the weight of the workpiece itself, may cause it to be deflected out of line. *Steady rests and follow rests,* shown in Figure 23-39, provide means for supporting such work between the headstock and the tailstock. The steady rest is clamped to the lathe ways and has three movable fingers that are adjusted to contact the work and align it. A light cut should be taken before adjusting the fingers to provide a smooth contact-surface area.

A steady rest also can be used in place of the tailstock as a means of supporting the end of long pieces, pieces having too large an internal hole to permit using a regular dead center, or work where the end must be open for boring. In such cases the headstock end of the work must be held in a chuck to prevent longitudinal movement. Tool feed should be toward the headstock.

The follow rest is bolted to the lathe carriage. It has two contact fingers that are adjusted to bear against the workpiece, opposite the cutting tool, in order to prevent the work from being deflected away from the cutting tool by the cutting forces.

Review Questions

1. What is the tool–work relationship in turning?
2. What different kinds of surfaces can be produced by turning?
3. How does form turning differ from ordinary turning?
4. What is the basic difference between facing and a cutoff operation?
5. Name six different machining operations that can be done on a lathe.
6. Why is it difficult to make heavy cuts if a form turning tool is complex in shape?
7. What is the "swing" of a lathe?
8. Why is the spindle of the lathe hollow?
9. What function does a lathe carriage have?
10. How is feed specified on a lathe?
11. What function is provided by the lead screw on a lathe that is not provided by the feed rod?
12. How can work be held and supported in a lathe?
13. How is rotation provided to a workpiece that is mounted between centers on a lathe?
14. What will happen to the workpiece when turned if it is held between centers and the centers are not exactly in line?
15. Why is it not advisable to hold hot-rolled steel stock in a collet?
16. How does a steady rest differ from a follow rest?
17. What are the advantages and disadvantages of a four-jaw independent chuck over a three-jaw chuck?
18. Why should the overhang distance of a lathe tool be minimized?
19. What occurs if a lathe tool is set below the center line of the workpiece in turning?
20. How can a tapered part be turned on a lathe?
21. Why is it desirable to use a heavy depth and a light feed in turning rather than the opposite?
22. On what diameter is the rpm based for a facing cut, assuming given work and tool materials?
23. Why is it usually necessary to take relatively light cuts when boring on a lathe?
24. What are two basic ways knurling is done?
25. Why are saddle-type turret lathes used much less frequently than ram-type lathes?
26. What important factor must be kept in mind in tooling multiple-spindle screw machines that does not have to be considered in single-spindle machines?
27. What are the two objectives of boring?
28. Why does boring assure concentricity between the hole axis and the axis of rotation of the workpiece (for boring tool), whereas drilling does not?
29. Why are vertical boring mills better suited than a lathe for machining large workpieces?
30. What is the principal advantage of a horizontal boring machine over a vertical boring machine for large workpieces?
31. Why is a horizontal boring, drilling, and milling machine such an important tool for machining very large workpieces?
32. Where in Chapter 23 is a workpiece being held in a fixture mounted on a faceplate?
33. How is the workpiece in Figure 23-6 being held?
34. In what three figures in Chapter 23 is a dead center shown?
35. In what two figures in Chapter 23 is a live center shown?
36. In which figures showing setups do you find
    a. a three-jaw chuck
    b. a collet
    c. a faceplate
    d. a four-jaw chuck
    being used as a workholding device?
37. How many form tools are being utilized in the process shown in Figure 23-18 to machine the part?

Problems

1. At what speed should a 3-in.-diameter bar be rotated to provide a cutting speed of 200 feet per minute?
2. Assume that the workpiece in problem 1 above is 8 in. (203.2mm) long and a feed of 0.020 in. (0.51mm) per revolution is used. Compute the cutting time across its entire length. Do not forget to add an allowance.
3. If the depth of cut in problem 2 is ⅛ in., what is the metal removal rate (MRR)?
4. The following data apply for machining a part on a turret lathe and on an engine lathe:

	Engine Lathe	Turret Lathe
Times in minutes to machine part	30	5
Cost of special tooling	0	300
Time to setup	30 minutes	3 hours
Labor rates	8 $/hr	8 $/hr
Machine rates	10 $/hr	12 $/hr

a. How many pieces would have to be made for the cost of the engine lathe to just equal the cost of the turret lathe? This is the BEQ.

b. What is the cost per unit at the BEQ?

5. A hole 89 mm in diameter is to be drilled and bored through a piece of 1340 steel that is 200 mm long, using a horizontal boring, drilling, and milling machine. High-speed tools will be used. The job will be done by center drilling, drilling with an 18-mm drill, followed by a 76-mm drill, then bored to size in one cut, using a feed of 0.50 mm/rev. Drilling feeds will be 0.25 mm/rev for the smaller drill and 0.64 mm/rev for the larger drill. The center drilling operation requires .5 minute. To set or change any given tool and set the proper machine speed and feed requires 1 minute. Select the initial cutting speeds, and compute the total time required for doing the job. (Neglect setup time for the workpiece.)

6. In Figure 23A below are three plots of unit production cost ($/unit) versus production volume (Q = build quantity). Note that this plot is made on log–log paper. Cost per unit for a particular process decreases with increased volume. For a particular process there is no minimum cost but rather production volumes within which particular processes are most economical.

a. Each of these curves is a plot of the equation for total cost per unit, which means each is the sum of the fixed cost per unit (setup, tooling, overhead) and the variable costs per unit (direct labor, direct material). Estimate these costs for the NC lathe from the data in curves.

b. For what build quantities is the NC lathe most economical (approximately)?

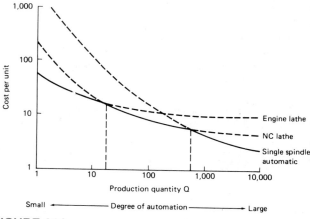

FIGURE 23A

c. What cost per unit does the NC lathe approach as the build quantity becomes very large?

d. What happens to these plots if you plot them on regular Cartesian coordinates? Try it and comment on what you find.

e. Many manufacturers have found innovative ways to eliminate setup time in many of their processes. What is the impact of this on these kinds of plots, on cost per unit economics, and on job shop inventories?

7. The derivation of the approximate equation 23-5 for the MRR for turning process requires an assumption regarding the diameters of the parts being turned. Derive equation 23-5 from 23-4 and state the assumption.

Drilling and Related Hole-Making Processes

In manufacturing, it is probable that more holes are produced than any other shape, and a large proportion of these are made by drilling. Consequently, drilling is a very important process. Although drilling appears to be a relatively simple process, certain aspects of it cause considerable difficulty. Most drilling is done with a tool having two cutting edges. These edges are at the end of a relatively flexible tool. Cutting action takes place inside the workpiece. The only exit for the chips is the hole that is filled by the drill. Friction results in heat that is additional to that due to chip formation. The counterflow of the chips makes lubrication and cooling difficult.

In recent years, new drill point geometries and TiN coatings have resulted in improved hole accuracy, longer life, self-centering action, and increased-feed-rate capabilities. However, virtually 99% of the drills manufactured have the conventional point and geometry shown in Figure 24-1. If the drill is reground, the original drill geometry may be lost. Drill accuracy and precision will also depend upon the drilling machine tool, the workholding device, the drill holder, and the surface of the workpiece. Poor surface conditions (sand pockets and/or chilled hard spots on castings, hard oxide scale on hot rolled metal) can accelerate early tool failure and degrade the hole-drilling process.

Fundamentals of the Drilling Process. The process of drilling creates two chips. A conventional two-flute drill, with drill of diameter D, has two principal cutting edges rotating at an rpm of N and feeding axially. The rpm of the drill is established by the selected cutting velocity, V, where

$$N = \frac{12\,V}{\pi\,D} \qquad (24\text{-}1)$$

with V in surface feet per minute and D in inches.

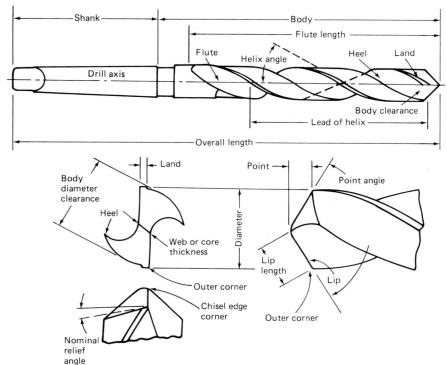

FIGURE 24-1 Nomenclature and geometry of conventional twist drill.

This equation assumes that V is the cutting speed at the outer edges of the cutting lip (point X in Figure 24-2).

The feed, f_r, is given in inches per revolution. The depth of cut in drilling is equal to half the feed rate or $t = f_r/2$. The feed rate in inches per minute, f_m, is $f_r N$. The length of cut in drilling equals the depth of the hole, L, plus an allowance for approach and for the tip of drill, usually $A = D/2$.

Given a selected cutting speed and feed for drilling a hole in a certain metal with a drill of known tool material, the rpm of the spindle of the machine is determined from equation (24-1), the maximum velocity occurring at the extreme ends of the drill lips. The velocity is very small near the center of the chisel end of the drill.

For drilling, cutting time is $CT = (L + A)/f_r N = \dfrac{L + A}{f_m}$ (24-2)

The metal removal rate is MRR $= \dfrac{\text{Volume}}{\text{Cut}}$ (24-3)

$= \dfrac{\pi D^2 L/4}{L/f_r N}$ (omitting allowances)

which reduces to MRR $= (\pi D^2/4)f_r N$ (cubic inches per minute).   (24-4)

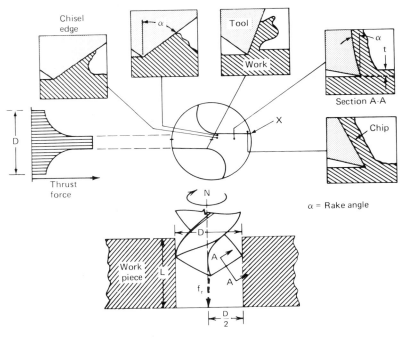

FIGURE 24-2 Conventional drill geometry and symbols with cutting cross sections and thrust force.

An approximate form of MRR $= 3DVf_r$ is often used. (24-5)

Note that the units here are not consistent.

Summary of Drilling Parameters

1. Cutting speed at drill outside diameter $= (\pi/12) \times$ (Drill diameter) $\times$ (rpm)
2. Feed rate (ipm) = Feed (ipr) $\times$ (rpm)
3. Maximum chip load = Feed (ipr)/2. . . .for two-fluted drill
4. Material removal rate (in.^{3}/min.) $= (\pi/4) \times$ (Drill diameter)$^2 \times$ Feed rate (ipm)
5. Drilling time/hole $= \dfrac{\text{Length drilled } + \text{ Air cut}}{\text{Feed rate}}$

   $+ \dfrac{\text{Rapid traverse length, including withdrawal}}{\text{Rapid traverse rate}}$

   $+$ Prorated downtime to change drill per hole
6. Cost/Hole = (Drilling time/hole) $\times$ (Labor $+$ Machine rate)
   $+$ Prorated cost of purchasing and regrinding drill/hole
7. Prorated downtime to change drill/hole $= \dfrac{\text{Drill change downtime}}{\text{Holes drilled per drill regrind}}$
8. Prorated cost of purchasing and regrinding drill/hole
   $= \dfrac{(\text{Purchase cost } + \text{ Cost per regrind}) \times (\text{Number of regrinds})}{\text{Number of holes drilled}}$

Drill Designs.   The most common type of drills are *twist drills*. These have three basic parts: the *body*, the *point*, and the *shank*, shown in Figure 24-1. The body contains two or more spiral or helical grooves, called *flutes*, separated by *lands*. To reduce the friction between the drill and the hole, each land is reduced in diameter except at the leading edge, leaving a narrow *margin* of full diameter to aid in supporting and guiding the drill and thus aiding in obtaining an accurate hole. The lands terminate in the point, with the leading edge of each land forming a cutting edge.

The principal rake angles behind the cutting edges are formed by the relation of the flute helix angle to the work. This means that the rake angle of a drill varies along the cutting edges (or lips), being 0 degrees close to the point and equal to the helix angle out at the lip. Because the helix angle is built into the drill, the primary rake angle cannot be changed by normal grinding. The helix angle of most drills is 24 degrees, but drills with larger helix angles—often above 30 degrees—are used for materials that can be drilled very rapidly, resulting in a large volume of chips. Helix angles ranging from 0 to 20 degrees are used for soft materials, such as plastics and copper. Straight-flute drills (zero helix and rake angles) are also used for drilling thin sheets of soft materials. It is possible to change the rake angle adjacent to the cutting edge by a special grinding procedure called *dubbing*.

The cone-shaped point on a drill contains the cutting edges and the various clearance angles. This cone angle affects the direction of flow of the chips across the tool face and into the flute. The 118 degrees cone angle that is used most often has been found to provide good cutting conditions and reasonable tool life when drilling mild steel, thus making it suitable for much general-purpose drilling. Smaller cone angles—from 90 to 118 degrees—are sometimes used for drilling more brittle materials, such as gray cast iron and magnesium alloys. Cone angles from 118 to 135 degrees are often used for the more ductile materials, such as aluminum alloys. Cone angles less than 90 degrees frequently are used for drilling plastics. Many methods of grinding drills have been developed that produce point angles other than 118 degrees.

The flutes serve as channels through which the chips exit the hole and coolant gets to the cutting edges. Although most drills have two flutes, some, as shown in Figure 24-3, have three.

The relatively thin *web* between the flutes forms a metal column or backbone. If a plain conical point is ground on the drill, the intersection of the web and the cone produces a straight-line *chisel end,* which can be seen in the end view of Figure 24-1. The chisel point, which also must act as a cutting edge, forms a 56-degree negative rake angle with the conical surface. Such a large negative rake angle does not cut efficiently, causing excessive deformation of the metal. This results in high thrust forces and excessive heat being developed at the point. In addition, the cutting speed at the drill center is low, approaching zero. As a consequence, drill failure on a standard drill occurs both at the center, where the cutting speed is lowest, and at the outer tips of the cutting edges, where the speed is highest.

When the rotating, straight-line chisel point comes in contact with the work-

FIGURE 24-3 Types of twist drills and shanks. (Left to right) Straight-shank, three-flute core drill; taper-shank; straight-shank; bit-shank; straight-shank, high-helix-angle; straight-shank, straight-flute; taper-shank, subland drill.

piece, it has a tendency to slide or ''walk'' along the surface, thus moving the drill away from the desired location. The conventional point drill, when used on machining centers or high-speed automatics, will require additional supporting operations like center drilling, burr removal, and tool change, all of which increase total production time and reduce productivity.

Special methods of grinding drill points have been developed to eliminate or minimize the difficulties caused by the chisel point and to obtain better cutting action and tool life (see Figure 24-4). *Web thinning* uses a narrow grinding wheel to remove a portion of the web near the point of the drill. Such methods have had varying degrees of success, and they require special drill-grinding equipment.

The center core or slot point drill is relatively new to the marketplace. This drill has twin carbide tips brazed on a steel shank. The work material in the slot is not machined but rather fractured away. All rake angles of the cutting edge are positive, which further reduces the cutting force. This drill operates at about 30–50% less thrust than conventional drills.

The conventional point also has a tendency to produce a burr on the exit side of a hole. Some type of chip breaker often is incorporated into drills. One procedure is to grind a small groove in the rake face, parallel with and a short distance back from the cutting edge. Drills with a special chip-breaker rib as an

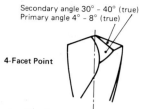

Secondary angle 30° – 40° (true)
Primary angle 4° – 8° (true)

4-Facet Point

Four-Facet
Good self-centering ability
Breaks up chips for deep-hole drilling
Can be generated in a single grinding operation
Eliminate center drilling in NC

130°

"S" form chisel

S-Point

Helical (S-shape chisel point)
Can eliminate center drilling on NC machining centers
Excellent hole geometry
Close relationship between drill size and hole size
Increased tool life
Lower thrust requirements
Leaves burr on breakthrough

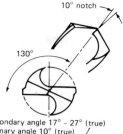

10° notch

130°

Secondary angle 17° – 27° (true)
Primary angle 10° (true)

135°

Split Point

Split point or crankshaft drill
Good self-centering ability
Breaks up chips for deep-hole drilling

Bickford
Combination of helical and Racon point features
Self-centering and reduced burrs
Excellent hole geometry
Increased tool life

Racon (radiused conventional point)
Increased feed rates
Increased tool life (8-10 times in C.I.)
Reduced burrs at breakthrough
Not self centering

Relieved helical
Reduces thrust force
Eliminate chisel edge
Equil. rake angle

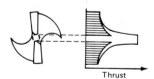

Body

Carbide tip

20°

Convex rake face

Noncutting zone

Margin

Twisted flute

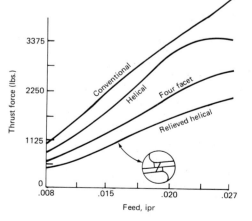

Thrust force (lbs.)

3375

2250

1125

0

Conventional
Helical
Four facet
Relieved helical

.008   .015   .020   .027

Feed, ipr

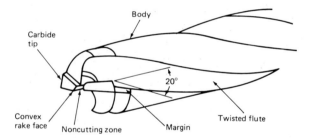

Center core drill or
Slot point drill
Greatly reduced thrust
Center core removed by ductile fracture (tension)

Thrust

FIGURE 24-4 Variations to the conventional drill geometry are usually aimed at reducing the thrust force at the chisel end of the drill.

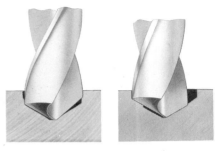

FIGURE 24-5 Effects of improper drill grinding. (Left) Angles of two tips are different. (Right) Lengths of the lips are not equal. (Courtesy Cleveland Twist Drill Company.)

integral part of the flute are available. The rib interrupts the flow of the chip, causing it to break into short lengths.

The original drill point produced by the manufacturer lasts only until the first regrind; thereafter performance and life depend upon the quality of regrind.

Proper regrinding (reconditioning) of a drill is a complex and important operation. If satisfactory cutting and hole size are to be achieved, it is essential that the point angle, lip clearance, lip length, and web thinning be correct. As illustrated in Figure 24-5, incorrect sharpening often results in unbalanced cutting forces at the tip, causing misalignment and oversized holes. Drills, even small drills, should always be machine-ground, never hand-ground. Drill grinders, often computer-controlled, should be used to assure exact reproduction of the geometry established by the manufacturer of the drill. This is extremely important when drills are used on mass-production or numerically controlled machines. Companies invest huge sums in NC machining centers but overlook the value of a top-quality drill-grinding machine.

Drill shanks are made in several types. The two most common types are the straight and the taper. *Straight-shank* drills are usually used for sizes up to ½ inch and must be held in some type of drill chuck. *Taper shanks* are available on drills from ⅛ to ½ inch and are common on drills above ½ inch. Morse tapers, having a taper of approximately ⅝ inch per foot, are used on taper-shank drills, ranging from a number 1 taper on ⅛-in. drills to a number 6 on a 3½-in. drill.

Taper-shank drills are held in a female taper in the end of the machine tool spindle. If the taper on the drill is different from the spindle taper, adapter sleeves are available. The taper assures the drill's being accurately centered in the spindle. The *tang* at the end of the taper shank fits loosely in a slot at the end of the tapered hole in the spindle. The drill may be loosened for removal by driving a metal wedge called a *drift* through a hole in the side of the spindle and against the end of the tang. It also acts as a safety device to prevent the drill from rotating in the spindle hole under heavy loads. However, if the tapers on the drill and in the spindle are proper, no slipping should occur. The driving force to the drill is carried by the friction between the two tapered members.

Standard drills are available in four size series, the size indicating the diameter of the drill body:

- *Millimeter series:* 0.01- to 0.50-mm increments, according to size, in diameters from 0.015 mm.
- *Numerical series:* No. 80 to No. 1 (0.0135 to 0.228 in.).
- *Lettered series:* A to Z (0.234 to 0.413 in.).
- *Fractional series:* ¹⁄₆₄ to 4 in. (and over) by 64ths.

TiN coating of conventional drills greatly improves drilling performance. The increase in tool life of TiN-coated drills over uncoated drills in machining steel is over 200–1000%.

The bores of rifle barrels were once drilled using conventional drills. Today, *deep-hole drills,* or *gun drills,* are used when deep holes are to be drilled. A deep hole is one in which the depth of the hole is 3 or more times the diameter.

Coolants can be fed internally through these drills to the cutting edges. The coolants flush the chips out the flutes. The special design of these drills reduces the tendency of the drill to drift, thus producing a more accurately aligned hole. The ones shown in Figure 24-6 are designed for single-lip end cutting of a hole in a single pass. Solid deep-hole drills have alloy-steel shanks with a carbide-edged tip that is mechanically fixed to it. The cutting edge cuts through the center on one side of the hole, leaving no area of material to be extruded. The cutting is done by the outer and inner cutting angles which meet at a point. Theoretically the depth of the hole has no limit, but practically it is restricted by the torsional rigidity of the shank.

Gun drills have a single-lip cutting action. Bearing areas and lifting forces generated by the coolant pressure counteract the radial and tangential loads. The single-lip construction forces the edge to cut in a true circular pattern. The tip thus follows the direction of its own axis. The *trepanning gun drill* leaves a solid core.

Hole straightness is affected by variables such as diameter, depth, uniformity of the workpiece material, condition of the machine, sharpness of cutting edges,

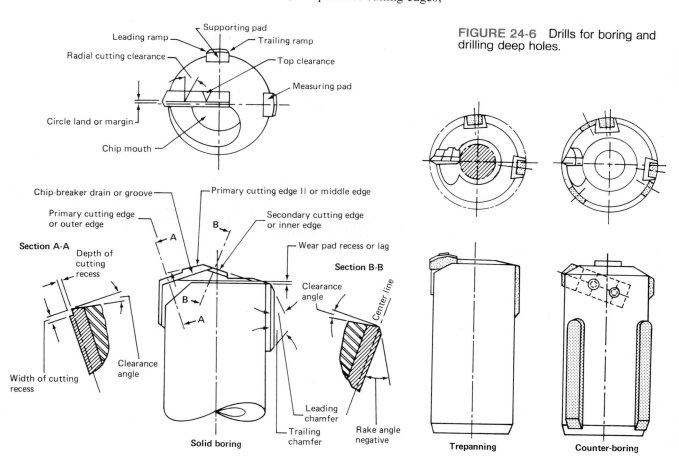

FIGURE 24-6 Drills for boring and drilling deep holes.

TABLE 24.1 **Comparison of Drilling Processes**

Type of Drill	Twist Drill	Gun Drill	Spade Drill	Ejector Drill	Trepanning	Solid Boring Drill
			Diameter [mm]			
Typical size range	0.5 to 50	1.4 to 25	25 to 150	18 to 60	42 to 250	12 to 200
			Hole depth/diameter ratio			
Minimum practical	no minimum	<1	no minimum	1	10	1
Common maximum	5–10	100	>40	50	100	100
Ultimate	>50	200	10(vertical) >100(horizontal)	>50	>100	>100

FIGURE 24-7 Hole cutter used for thin sheets. (Courtesy Armstrong-Blum Manufacturing Company.)

FIGURE 24-8 Combination center drill and countersink. (Courtesy Chicago-Latrobe Twist Drill Works.)

feeds and speeds used, and whether the tool or the workpiece is rotated or counter-rotated.

Two-flute drills are available that have holes extending throughout the length of each land to permit coolant to be supplied, under pressure, to the point adjacent to each cutting edge. These are helpful in providing cooling and also in promoting chip removal from the hole in drilling to moderate depths. They require special fittings through which the coolant can be supplied to the rotating drill and are used primarily on automatic and semiautomatic machines. See Table 24.1 for comparison of drilling processes.

Larger holes in thin material may be made with a *hole cutter,* shown in Figure 24-7, whereby the main hole is produced by the thin-walled, multiple tooth cutter with saw teeth. Hole cutters are often called hole saws.

When starting to drill a hole, a drill can deflect rather easily because of the "walking" action of the chisel point. Consequently, to assure that a hole is started accurately, a *center drill and countersink,* illustrated in Figure 24-8, is used prior to a regular chisel-point twist drill. The center drill and countersink has a short, straight drill section extending beyond a 60-degree taper portion. The heavy, short body provides rigidity so that a hole can be started with little possibility of tool deflection. The hole should be drilled only partway up on the tapered section of countersink. The conical portion of the hole serves to guide the drill being used to make the main hole. Combination center drills are made in four sizes to provide the proper-size starting hole for any drill. If the drill is sufficiently large in diameter, or if it is sufficiently short, satisfactory accuracy often may be obtained without center drilling. Special drill holders are available that permit drills to be held with only a very short length protruding.

Because of its flexibility, a drill may drift off centerline during drilling. The use of a center (start) drill helps to assure that a drill starts drilling at the desired location. Nonhomogeneities in the workpiece and imperfect drill geometries may also cause the hole to be oversize or off-line. If accuracy in these respects is desired, it is necessary to follow center drilling and drilling by boring and reaming, as illustrated in Figure 24-9. Boring trues the hole alignment, whereas reaming brings the hole to accurate size and improves the surface finish.

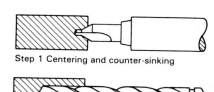

Step 1 Centering and counter-sinking

Step 3 Truing hole with boring cutter

Step 2 Drilling

Step 4 Final sizing with reamer

FIGURE 24-9 Steps required to obtain a hole that is accurate as to size and aligned on center.

Special *combination drills* are made that can drill two or more diameters, or drill and countersink and/or counterbore, in a single operation. Some of these are illustrated in Figure 24-10. A *step drill* has a single set of flutes and is ground to two or more diameters. *Subland drills* have a separate set of flutes, on a single body, for each diameter or operation; they provide better chip flow, and the cutting edges can be ground to give proper cutting conditions for each operation. Combination drills are expensive and may be difficult to regrind but can be economical for production-type operations if they reduce work handling, setups, or separate machines and operations.

Spade drills, such as shown in Figure 24-11, are widely used for making holes one inch or larger in diameter at low speeds/high feeds. See Table 24.2. The workpiece usually has an existing hole, but a spade drill can drill deep holes in solids or stacked materials. Spade drills are less expensive because the long supporting bar can be made of ordinary steel. The drill point can be ground with a minimum chisel point. The main body can be made more rigid because no flutes are required. The main body can be provided with a central hole through which a fluid can be circulated to aid in cooling and in chip removal. The cutting blade is easier to sharpen; only the blades need to be TiN-coated.

Spade drills are often used to machine a shallow locating cone for a subsequent smaller drill and at the same time to provide a small bevel around the hole to facilitate later tapping or assembly operations. Such a bevel also frequently eliminates the need for deburring. This practice is particularly useful on mass-production and numerically controlled machines.

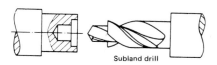

Subland drill

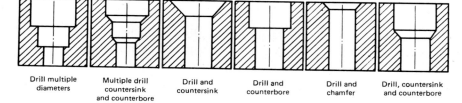

Drill multiple diameters

Multiple drill countersink and counterbore

Drill and countersink

Drill and counterbore

Drill and chamfer

Drill, countersink and counterbore

FIGURE 24-10 Special-purpose subland drill (above), and some of the operations possible with such drills (below).

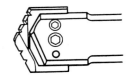

Regular spade drill

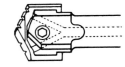

Spade drill with oil holes

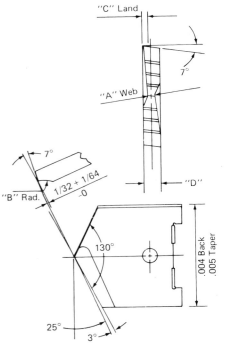

FIGURE 24-11 (Top) Regular spade drill. (Middle) Spade drill with oil holes. (Bottom) Spade drill geometry, nomenclature.

TABLE 24.2 Recommended Surface Speeds and Feeds with Spade Drills for Various Materials

Material	Surface Speed (ft per min)
Mild machinery steel .2 to .3 carbon	65–110
Steel, annealed .4 to .5 carbon	55– 80
Tool steel, 1.2 carbon	45– 60
Steel forging	35– 50
Alloy steel	45– 70
Stainless steel, free machining	50– 70
Stainless steel, hard	25– 40
Cast iron, soft	80–150
Cast iron, medium hard	55–100
Cast iron, hard, chilled	25– 40
Malleable iron	70– 90
Brass and bronze, ordinary	200–300
Bronze, high tensile	70–150
Monel metal	35– 50
Aluminum and its alloys	200–300
Magnesium and its alloys	250–400

Feed Rates for Drilling (inches per revolution)

Drill Size (inches)	Cast Iron Malleable Iron Brass Bronze	Medium Steel Stainless Steel Monel Metal Drop-Forged Alloys Tool Steel (annealed)	Tough Steel Drop Forging Aluminum
1 to 1¼	.010–.020	.008–.014	.006–.012
1¼ to 1¾	.010–.024	.008–.018	.008–.017
1¾ to 2½	.010–.030	.010–.024	.010–.017
2½ to 4	.012–.032	.012–.030	.010–.017
4 to 6	.012–.032	.010–.024	.008–.017

Source: Waukesha Cutting Tools, Inc.

Carbide-tipped drills and drills with indexable inserts are also available (see Figure 24-12), with one-piece and two-piece inserts for drilling shallow holes in solid workpieces. Notice the central round insert in the larger drill.

Indexable insert drills can produce a hole four times faster than a spade drill because they run at high speeds/low feeds and are really more of a boring operation than a drilling process.

Tool Holders for Drills. Straight-shank drills must be held in some type of drill chuck (see Figure 24-13). Chucks are adjustable over a considerable size range and have radial steel fingers. When the chuck is tightened by means of a chuck key, these fingers are forced inward against the drill. On smaller drill presses, the chuck often is permanently attached to the machine spindle, whereas

Insert

FIGURE 24-12   One- and two-tipped insert drills. (Courtesy Waukesha.)

on larger drilling machines the chucks have a tapered shank that fits into the female Morse taper of the machine spindle.

Special types of chucks in semiautomatic or fully automatic machines permit quite a wide range of sizes of drills to be held in a single chuck (see Figure 24-14).

Chucks using chuck keys require that the machine spindle be stopped in order to change a drill. To reduce the downtime when drills must be changed frequently, *quick-change chucks* are used. Each drill is fastened in a simple round collet that can be inserted into the chuck hole while it is turning by merely raising and lowering a ring on the chuck body. With the use of this type of chuck, center drills, drills, counterbores, reamers, and so on can be used in quick succession.

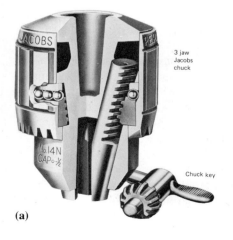

3 jaw
Jacobs
chuck

Chuck key

(a)

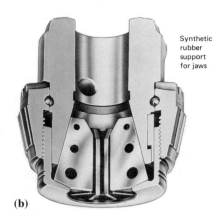

Synthetic
rubber
support
for jaws

(b)

FIGURE 24-13   Two of the most commonly used types of drill chucks. (Courtesy Jacobs Manufacturing Company.)

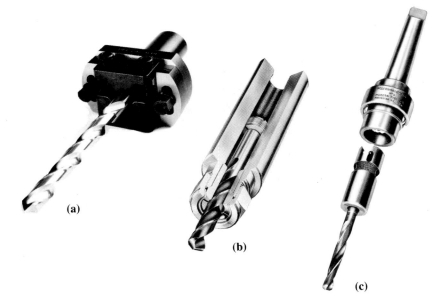

(a)

(b)

(c)

FIGURE 24-14 Three types of drill chucks used on automatic equipment. (left) Universal type. (Courtesy Brookfield Tool Co.) (center) Collet type. (Courtesy Erickson Tool Company.) (right) Quick-change drill chuck. (Courtesy Consolidated Machine Tool Division, Farrel Birmingham Company, Inc.)

Work Holding for Drilling.   Work that is to be drilled is ordinarily held in a vise or a special jig or fixture (see Chapter 28). Even in light drilling the work should not be held on the table by hand unless entirely adequate leverage is available. This is a dangerous practice and can lead to serious accidents, because the drill has a tendency to catch on the workpiece and cause it to rotate. Drilling vises and jigs frequently are made in such a way that they can be flipped onto a second side to permit drilling to be done on two faces of the work with a single clamping. These are called *tumble jigs*.

Work that is too large to be held in a vise can be clamped directly to the machine table using suitable bolts and clamps and the slots or holes in the table.

Machine Tools for Drilling

The basic work and tool motions required for drilling—relative rotation between the workpiece and the tool, with relative longitudinal feeding—also occurs in a number of other machining operations. Thus, drilling can be done on a variety of machine tools such as lathes, vertical milling machines, boring machines, and, of course, machining centers. This chapter will consider only those machines that are designed, constructed, and used primarily for drilling. Drill presses consist of a *base,* a *column* that supports a *powerhead,* a *spindle,* and a *worktable*. On small machines the base rests on a work bench, whereas on larger machines it rests on the floor (see Figure 24-15). The column may be either round or of box-type construction, the latter being used on larger, heavy-duty machines, except in radial types. The powerhead contains an electric motor and means for driving the spindle in rotation at several speeds. On small drilling

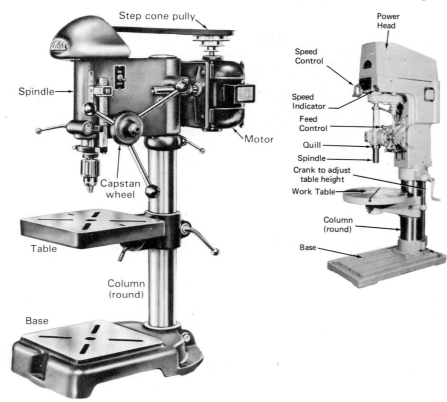

Step cone pully

Spindle

Motor

Capstan
wheel

Table

Column
(round)

Base

Power
Head

Speed
Control

Speed
Indicator

Feed
Control

Quill

Spindle

Crank to adjust
table height

Work Table

Column
(round)

Base

FIGURE 24-15 (left) Fifteen-inch bench-type drill press, usually set on a table. (Courtesy Altas Press Company.) (right) Upright drilling machine. (Courtesy Buffalo Forge Company.)

machines this may be accomplished by shifting a belt on a step-cone pulley, but on larger machines a geared transmission is used. See Figure 24-16 for block diagrams.

The heart of any drilling machine is its spindle. In order to drill satisfactorily, the spindle must rotate accurately and also resist whatever side forces result from the drilling. In virtually all machines the spindle rotates in preloaded ball or taper-roller bearings. In addition to powered rotation, provision is made so that the spindle can be moved axially to feed the drill into the work. On small machines the spindle is fed by hand, whereas on larger machines power feed is provided. Except on some small bench types, the spindle contains a hole with a Morse taper in its lower end into which taper-shank drills or drill chucks can be inserted.

The worktables on drilling machines may be moved up and down on the column to accommodate work of various sizes. On round-column machines the table usually can also be rotated out of the way so that workpieces can be mounted directly on the base. On some box-column machines the table is mounted on a subbase so that it can be moved in two directions in a horizontal plane by means of feed screws.

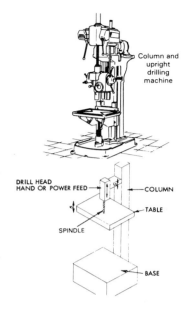

Column and
upright
drilling
machine

DRILL HEAD
HAND OR POWER FEED — COLUMN

— TABLE

SPINDLE

— BASE

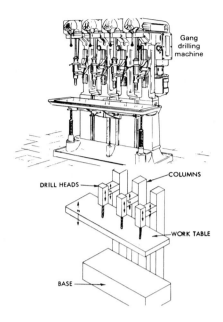

Gang
drilling
machine

DRILL HEADS

COLUMNS

WORK TABLE

BASE

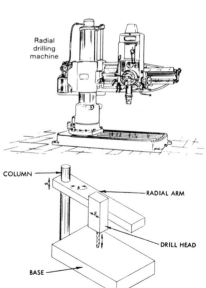

Radial
drilling
machine

COLUMN —

— RADIAL ARM

— DRILL HEAD

BASE

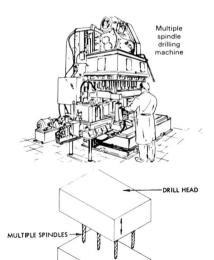

Multiple
spindle
drilling
machine

— DRILL HEAD

MULTIPLE SPINDLES —

WORK PIECE —

FIGURE 24-16 Four principal types of drilling machines. (From Manufacturing Producibility Handbook; courtesy General Electric Company.)

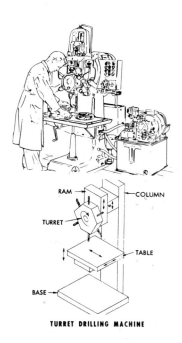

FIGURE 24-17   Turret drilling machines (left) and NC turret drilling machines (right) have largely replaced the gang-drilling machines in many production facilities. (From Manufacturing Producibility Handbook, courtesy General Electric Company.)

Types of Drilling Machines.   Drilling machines usually are classified in the following manner:

1. Bench (Figure 24-15).
   a. Plain.
   b. Sensitive.
2. Upright (Figure 24-15).
   a. Single-spindle (Figure 24-16).
   b. Turret (Figure 24-17).
   c. NC turret.
3. Radial (Figure 24-16).
   a. Plain.
   b. Semiuniversal.
   c. Universal.
4. Gang (Figure 24-16).
5. Multispindle (Figure 24-16).
6. Deep-hole (Figure 24-18).
   a. Vertical.
   b. Horizontal.
7. Transfer.

With bench drill presses, holes up to ½ inch in diameter can be drilled. The same type of machine can be obtained with a long column so that it can stand on the floor. The size of bench and upright drilling machines is designated by *twice* the distance from the center line of the spindle to the nearest point on the

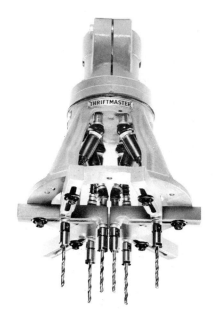

FIGURE 24-18   Multiple-spindle drill head attachment head, showing six spindles. (Courtesy Thriftmaster Products Incorporated.)

column, this being an indication of the maximum size of the work that can be drilled in the machines. For example, a 15-inch drill press will permit a hole to be drilled at the center of a workpiece 15 inches in diameter.

Sensitive drilling machines are essentially smaller plain bench-type machines with more accurate spindles and bearings. They are capable of operating at higher speeds—up to 30,000 rpm. Very sensitive, hand-operated feeding mechanisms are provided for use in drilling small holes. Such machines are used for tool and die work and for drilling very small holes, often less than a few thousandths of an inch in diameter, when high spindle speeds are necessary to obtain proper cutting speed and sensitive feel in order to provide delicate feeding to avoid breakage of the very small drills.

Upright drilling machines usually have spindle speeds ranges from 60 to 3500 rpm and power feed rates, in from 4 to 12 steps, from about 0.004 to 0.025 in. per revolution. Most modern machines use a single-speed motor and a geared transmission to provide the range of speeds and feeds. The feed clutch disengages automatically when the spindle reaches a preset depth.

Worktables on most upright drilling machines contain holes and slots for use in clamping work and nearly always have a channel around the edges to collect cutting fluid, when it is used. On box-column machines, the table is mounted on vertical ways on the front of the column and can be raised or lowered by means of a crank-operated elevating screw.

In mass production, *gang-drilling machines* are often used when several related operations, such as holes of different sizes, reaming, or counterboring, must be done on a single part. These consist essentially of several independent columns, heads, and spindles mounted on a common base and having a single table. The work can be slid into position for the operation at each spindle. They are available with or without power feed. One or several operators may be used. This machine would be an example of a simple small cell except that the machines are usually not single-cycle automatics.

Turret-type, upright drilling machines, such as shown in Figure 24-17, are used when a series of holes of different size, or a series of operations (such as center drilling, drilling, reaming, and spot facing), must be done repeatedly in succession. The selected tools are mounted in the turret. Each tool can quickly be brought into position merely by rotation of the turret. These machines automatically provide individual feed rates for each spindle and are often numerically controlled. *Radial drilling machine tools* are used on large workpieces that cannot easily be handled manually. As shown in Figure 24-17, these machines have a large, heavy, round, vertical column supported on a large base. The column supports a radial arm that can be raised and lowered by power and rotated over the base. The spindle head, with its speed- and feed-changing mechanism, is mounted on the radial arm. It can be moved horizontally to any desired position on the arm. Thus, the spindle can quickly be properly positioned for drilling holes at any desired point on a large workpiece mounted either on the base of the machine or even sitting on the floor.

Plain radial drilling machines provide only a vertical spindle motion. On *semiuniversal machines*, the spindle head can be pivoted at an angle to a vertical plane. On *universal machines,* the radial arm is rotated about a horizontal axis to permit drilling at any angle.

Radial drilling machines are designated by the radius of the largest disc in which a center hole can be drilled when the spindle head is at its outermost position. Sizes from 3 to 12 feet are available. Radial drilling machines have a wide range of speeds and feeds, can do boring, and include provisions for tapping (internal threading). See Chapter 30.

Multiple-spindle drilling machines, shown in Figure 24-16, are mass-production machines with as many as 50 spindles driven by a single powerhead and fed simultaneously into the work. Figure 24-18 shows a multiple-spindle head for a drill press and Figure 24-19 shows the methods of driving and positioning the spindles, which permits them to be adjusted so that holes can be drilled at any location within the overall capacity of the head. A special drill jig is made

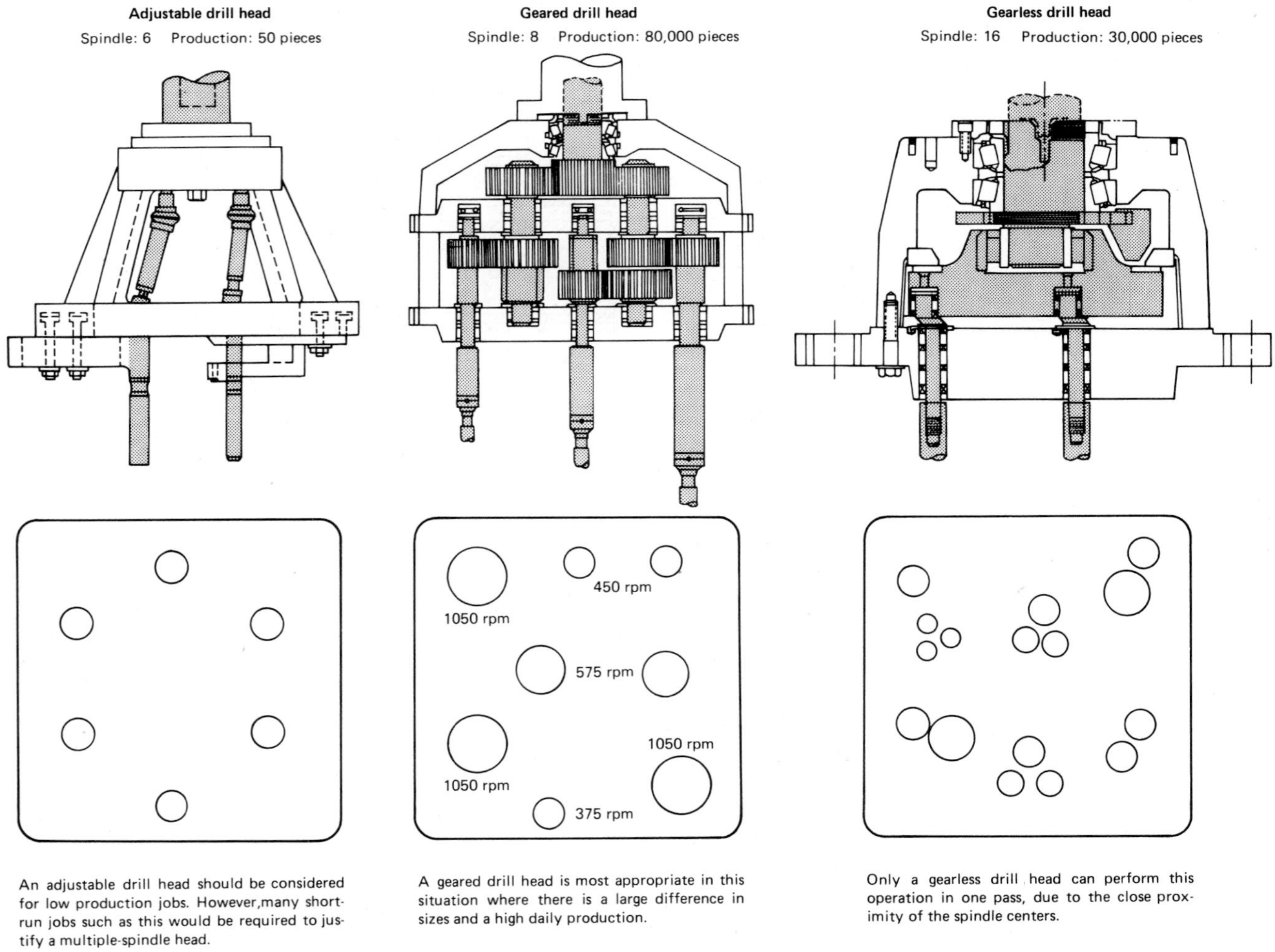

FIGURE 24-19 The three basic types of multiple-spindle drill heads. (left) Adjustable. (middle) Geared. (right) Gearless. (Courtesy Zagar Incorporated.)

Adjustable drill head

Spindle: 6   Production: 50 pieces

Geared drill head

Spindle: 8   Production: 80,000 pieces

Gearless drill head

Spindle: 16   Production: 30,000 pieces

450 rpm

1050 rpm

575 rpm

1050 rpm

1050 rpm

375 rpm

An adjustable drill head should be considered for low production jobs. However, many short-run jobs such as this would be required to justify a multiple-spindle head.

A geared drill head is most appropriate in this situation where there is a large difference in sizes and a high daily production.

Only a gearless drill head can perform this operation in one pass, due to the close proximity of the spindle centers.

for each job to provide accurate guidance for each drill. Although such machines and workholders are quite costly, they can be cost-justified when the quantity to be produced will justify the setup cost and the cost of the jig. Reducing setup on these machines is difficult. See Chapter 44.

Numerically controlled drill presses, other than turret drill presses, are not common because drilling and all its related processes can be done on vertical or horizontal NC machining centers equipped with automatic tool changers. See Chapter 29.

Special machines are used for drilling long (deep) holes, such as are found in rifle barrels, connecting rods, and long spindles. High cutting speeds, very light feeds, and a copious flow of cutting fluid assure rapid chip removal. Adequate support for the long, slender drills is required. In most cases horizontal machines are used (see Figure 24-20). The work is rotated in a chuck with steady rests providing support along its length, as required. The drill does not rotate and is fed into the work. Vertical machines are also available for shorter work pieces.

FIGURE 24-20 Horizontal deep-hole drilling machine with enlargement (above) of cutting tool. (Azad and Chandrashekar, *Mechanical Engineering*, Sept. 1985, p. 62, 63.)

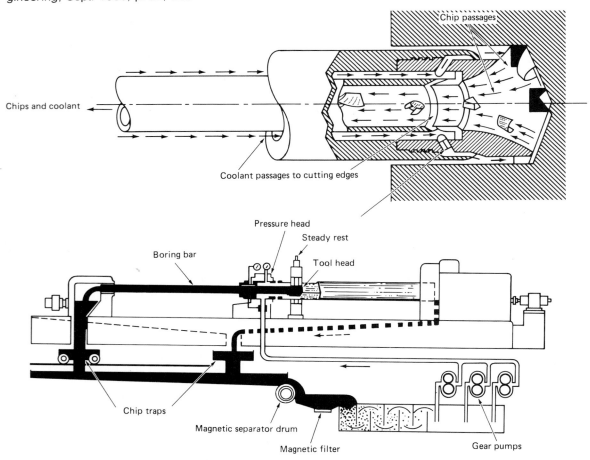

TABLE 24.3   **Drill Feeds Versus Drill Size**

Diameter of drill	mmpr	ipr
Less than 3 mm (⅛ in.)	0.03–0.05	0.001–0.002
3 to 6.4 mm·(⅛ to ¼ in.)	0.05–0.10	0.002–0.004
6.4 to 12.7 mm (¼ to ½ in.)	0.10–0.18	0.004–0.007
12.7 to 25.4 mm (½ to 1 in.)	0.18–0.38	0.007–0.015
Over 25.4 mm (1 in.)	0.38–0.64	0.015–0.025

Drilling Speeds and Feeds.   Cutting speeds specified for drilling are the surface speeds at the outside of the drill. These surface speeds are used to compute the rotational speed of the drill. Also, consider how heat will be conducted away from the cutting edges: by conduction through the workpiece, by cutting fluids, or by rapid chip flow. In drilling deep holes, or in drilling holes in materials that do not conduct heat rapidly, cutting speeds may have to be reduced unless an ample supply of coolant can be provided at the cutting edges. Coating drills with TiN allows marked increases in cutting speeds and tool life in drills.

Drilling feeds are expressed as *millimeters per revolution* or *inches per revolution* (mmpr or ipr). Starting or recommended feeds are given in standard reference books as a function of work material and tool material. As the diameter of the drill increases, the feed is also increased, as shown in Table 24.3.

Cutting Fluids in Drilling.   For shallow holes, the general rules relating to cutting fluids, as given in Chapter 28, are applicable. When the depth of the hole exceeds one diameter, it is desirable to increase the lubricating quality of the fluid because of the rubbing between the drill margins and the wall of the hole. The effectiveness of a cutting fluid as a coolant is quite variable in drilling. While the rapid exit of the chips is a primary factor in heat removal, this action also tends to restrict entry of the cutting fluid. This is of particular importance in drilling materials that have poor heat conductivity. Cutting fluids for drilling are given in Table 24.4.

If the hole depth exceeds two or three diameters, it is usually advantageous to withdraw the drill each time it has drilled about one diameter of depth, in order to clear chips from the hole. Some machines are equipped to provide this "pecking" action automatically.

Where cooling is desired, the fluid should be applied copiously. For severe conditions, drills containing coolant holes have a considerable advantage; not only is the fluid supplied near the cutting edges, but the coolant flow aids in flushing the chips from the hole. Where feasible, drilling horizontally has distinct advantages over drilling vertically downward.

Counterboring, Countersinking, and Spot Facing.   Drilling often is followed by *counterboring, countersinking,* or *spot facing*. As shown in Figure

TABLE 24.4 Cutting Fluids for Drilling

Work material	Cutting fluid
Aluminum and its alloys	Soluble oil, kerosene, and lard-oil compounds; light, non-viscous neutral oil; kerosene and soluble oil mixtures
Brass	Dry, soluble oil; kerosene and lard-oil compounds; light, nonviscous neutral oil
Copper	Soluble oil, strained lard oil, oleic-acid compounds
Cast iron	Dry or with a jet of compressed air for cooling
Malleable iron	Soluble oil, nonviscous neutral oil
Monel metal	Soluble oil, sulfurized mineral oil
Stainless steel	Soluble oil, sulfurized mineral oil
Steel, ordinary	Soluble oil, sulfurized oil, high extreme-pressure-value mineral oil
Steel, very hard	Soluble oil, sulfurized oil, turpentine
Wrought iron	Soluble oil, sulfurized oil, mineral–animal oil compound

24-21, each provides a bearing surface at one end of a drilled hole. They are usually done with a special tool having from three to six cutting edges.

Counterboring provides an enlarged cylindrical hole with a flat bottom so that a bolt head, or a nut, will have a smooth bearing surface that is normal to the axis of the hole; the depth may be sufficient so that the entire bolt head or nut will be below the surface of the part (see Figure 24-22). The pilot on the end of the tool fits into the drilled hole and helps to assure concentricity with the original hole. Two or more diameters may be produced in a single counterboring operation. Counterboring also can be done with a single-point tool, although this method ordinarily is used only on large holes and essentially is a boring operation.

Countersinking makes a beveled section at the end of a drilled hole to provide a proper seat for a flat-head screw or rivet. The most common angles are 60, 82, and 90 degrees. Countersinking tools are similar to counterboring tools except that the cutting edges are elements of a cone, and they usually do not have a pilot because the bevel of the tool causes them to be self-centering.

Spot facing is done to provide a smooth bearing area on an otherwise rough surface at the opening of a hole and normal to its axis. Machining is limited to minimum depth that will provide a smooth, uniform surface. Spot faces thus are somewhat easier and more economical to produce than counterbores. They usually are made with a multiedged end-cutting tool that does not have a pilot, although counterboring tools frequently are used.

FIGURE 24-21 Surfaces produced by counterboring, countersinking, and spot facing.

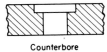

Counterbore

Countersink

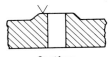

Spotface

FIGURE 24-22 Counterboring tools. (Bottom to top) Interchangeable counterbore; solid, taper-shank counterbore with integral pilot; replaceable counterbore and pilot; replaceable counterbore, disassembled. (Courtesy Ex-Cell-O Corporation and Chicago-Latrobe Twist Drill Works.)

The drilling machine should be rigid and have sufficient strength and power to withstand the cutting forces. A lack of rigidity in the tool, the workpiece, or the machine permits the affected members to deflect under cutting forces and cause chatter. The load builds up, the material being cut is sheared, the load is suddenly released, and the deflected member springs back to its normal position. The cycle rapidly repeats itself with the result that the cutting lips have a vibrating action against the work. As part of the routine preventive maintenance effort, backlash in the feed mechanism should be minimized to reduce strain on the drill when it breaks through the bottom of the hole. Jigs and workholding devices on indexing machines must be free from play and firmly seated. The tapers in the machine spindle, drill sleeves, adaptors, and shanks should be free from burrs and scratches that may cause drill runout. Spindle overhang should be minimized. Spindle bearings should be in good condition to prevent run out and end play.

Table 24.5 gives some typical drilling problems along with their probable causes and cures. One very simple way to check drill performance is to examine the chips. They should be the same from both flutes. If the chips are not of the same size, the point has been improperly ground and one lip is doing most of the cutting.

Drilling Practice

TABLE 24.5 Causes/Cures in Drilling Problems

Problem	Cause/Cure
Outer corners of drill break down	Cutting speed too high; hard spots in material; no cutting fluid at tips; flutes clogged with chips; insufficient feed in materials like stainless steel
Chipping in cutting edge	Feed too high; lip relief too large
Checks or cracks in cutting edges	Overheating; poor cooling; drill cooled too rapidly while sharpening or drilling
Chipped drill margin	Wrong size drill bushing; oversize, allowing vibration; undersize, causing interference
Broken drill body Fracture in hole	Point improperly ground; feed too heavy for work material; backlash or looseness in spindle, workholder, or workpiece; dulling of drill, increasing cutting forces; flutes clogged with chips
Broken tang	Bad fit between taper shank and socket caused by chips, burrs, dirt, wear, etc.
Drill splits up center	Lip relief angles too small or feed too large
Drill will not enter	Drill is worn, dull; web too thick; lip relief angle too small
Walls of hole rough	Bad (unequal) point geometry; poor lubrication at point; wrong cutting fluid; feed too large; insufficient rigidity in workholder or spindle
Hole oversize	Unequal rake angles on cutting edges; one edge duller than the other; unequal edge length; loose spindle

FIGURE 24-23 Standard nomenclature for hand and chucking reamers.

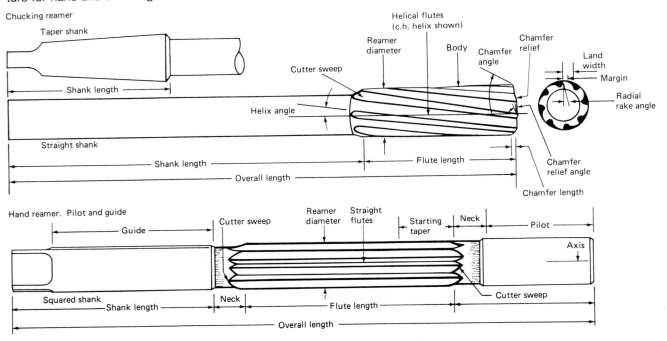

Reaming removes a small amount of material from the surface of holes. It is done for two purposes: to bring holes to a more exact size and to improve the finish of an existing hole. Multiedge cutting tools are used as shown in Figure 24-23. No special machines are built for reaming. The same machine that was employed for drilling the hole can be used for reaming by changing the cutting tool.

In order to obtain proper results, only a minimum amount of materials should be left for removal by reaming. As little as 0.005 in. is desirable, and in no case should the amount exceed 0.015 in. A properly reamed hole will be within 0.001 in. of correct size and have a fine finish.

Types of Reamers. The principal types of reamers, shown in Figure 24-24, are

1. Hand reamers.
    a. Straight.
    b. Taper.
2. Machine or chucking reamers.
    a. Rose.
    b. Fluted.
3. Shell reamers.
4. Expansion reamers.
5. Adjustable reamers.

Hand reamers are intended to be turned and fed by hand and to remove only a few thousandths of an inch of metal. They have a straight shank with a square

Reaming

FIGURE 24-24 Types of reamers. (Top to bottom) Straight-fluted rose reamer; straight-fluted chucking reamer; straight-fluted taper reamer; straight-fluted hand reamer; expansion reamer; shell reamer; adjustable insert-blade reamer.

tang for a wrench. They can have straight or spiral flutes and be solid or expandable. The teeth have relief along their edges and thus may cut along their entire length. However, the reamer is tapered from 0.005 to 0.010 in. in the first third of its length to assist in starting it in the hole, and most of the cutting therefore takes place in this portion.

Machine or *chucking reamers* are for use with various machine tools at slow speeds. The best feed is usually 2 to 3 times the drilling feed. Machine reamers have chamfers which carry the forward cutting edges. The chamfer causes the reamer to seat firmly and concentrically in the drilled hole, allowing the reamer to cut at full diameter. The longitudinal cutting edges do little or no cutting. Chamfer angles are usually 45 degrees. They have straight or tapered shanks and either straight or spiral flutes. *Rose chucking reamers* are ground cylindrical and have no relief behind the outer edges of the teeth. All cutting is done on the beveled ends of the teeth. *Fluted chucking reamers*, on the other hand, have relief behind the edges of the teeth as well as beveled ends. They can therefore cut on all portions of the teeth. Their flutes are relatively short and they are intended for light finishing cuts. For best results they should not be held rigidly but permitted to float and be aligned by the hole.

Shell reamers often are used for sizes over ¾ inch in order to save cutting-tool material. The shell, made of tool steel for smaller sizes and with carbide edges for larger sizes or for mass-production work, is held on an arbor that is made of ordinary steel. One arbor may be used with any number of shells. Only the shell is subject to wear and need be replaced when worn. They may be ground as rose or fluted reamers.

Expansion reamers can be adjusted over a few thousandths of an inch to compensate for wear or to permit some variation in hole size to be obtained. They are available in both hand and machine types.

Adjustable reamers have cutting edges in the form of blades that are locked in a body. The blades can be adjusted over a greater range than expansion reamers. This permits adjustment for size and to compensate for regrinding. When the blades become too small from regrinding, they can be replaced. Both tool steel and carbide blades are used.

Taper reamers are used for finishing holes to an exact taper. They may have up to eight straight or spiral flutes. Standard tapers, such as Morse, Jarno, or Brown & Sharpe, come in sets of two. The *roughing reamer* has nicks along the cutting edge to break up the heavy chips that result as a cylindrical hole is cut to a taper. The *finishing reamer* has smooth cutting edges.

Reaming Practice.

If the material to be removed is free-cutting, reamers of fairly light construction will give satisfactory results. However, if the material is hard and tough, solid-type reamers are recommended, even for fairly large holes.

To meet quality requirements including both finish and accuracy (tolerances on diameter, roundness, straightness, and absence of bell-mouth at ends of holes), reamers must have adequate support for the cutting edges, and reamer deflection must be minimal.

Reaming speed is usually two thirds the speed for drilling the same materials. However, for close tolerances and fine finish, speeds should be slower.

Feeds are usually much higher than those for drilling and depend upon material. A feed of between 0.0015 and 0.004 in. per flute per revolution is recommended as a starting point. Use the highest feed that will still produce the required finish and accuracy. Recommended cutting fluids are the same as those for drilling.

Reamers, like drills, should not be allowed to become dull. The chamfer must be reground long before it exhibits excessive wear. Sharpening is usually restricted to the starting taper or chamfer. Each flute must be ground exactly even, or the tool will cut oversize.

Reamers tend to chatter when not held securely, when the work or workholder is loose, or when the reamer is not properly ground. Irregularly spaced teeth may help reduce chatter. Other cures for chatter in reaming are to reduce the speed, vary the feed rate, chamfer the hole opening, use a piloted reamer, reduce the relief angle on the chamfer, or change cutting fluid. Any misalignment between the workpiece and the reamer will cause chatter and improper reaming.

Review Questions

1. What functions are performed by the flutes on a drill?
2. What determines the rake angle of a drill? See Figure 24-2.
3. Basically, what determines what helix angle a drill should have?
4. When a large-diameter hole is to be drilled, why is a small-diameter hole often drilled first?
5. Equation 24-4 for the MRR for drilling can be thought of as _____ times _____ where $f_r N$ is the feed rate of the drill bit.
6. Are the recommended surface speeds (feeds) given in Table 24.2 typically higher or lower than those recommended for twist drills?
7. What can happen when an improperly ground drill is used to drill a hole?
8. Why are most drilled holes oversize with respect to the nominally specified diameter?
9. What are the two primary functions of a combination center drill?
10. What is the function of the margins on a twist drill?
11. What factors tend to cause a drill to "drift" off the center line of a hole?
12. For what types of holes are drills having coolant passages in the flutes advantageous?
13. In drilling, the deeper the hole, the greater the torque. Why?
14. Why do cutting fluids for drilling usually have more lubricating qualities than those for most other machining operations?
15. How is feed expressed in drilling?
16. How does a gang-drilling machine differ from a multiple-spindle drilling machine?
17. For what type of work is a quick-change drill chuck advantageous?
18. What may result from holding the workpiece by hand when drilling?
19. What is the rationale behind the operation sequence shown in Figure 24-9?
20. In terms of thrust, what is unusual about the slot-point drill compared to other drills?
21. What is the purpose of spot facing?
22. How does the purpose of counterboring differ from that of spot facing?
23. What are the primary purposes of reaming?
24. What are the advantages of shell reamers?
25. A drill that operated satisfactorily for drilling cast iron gave very short life when used for drilling a plastic. What might be the reason for this?
26. What precautionary procedures should one use when drilling a deep, vertical hole in mild steel when using an ordinary twist drill?
27. What is the advantage of a spade drill? Is it really a drill?
28. What is a "pecking" action in drilling?
29. Why does drill feed increase with drill size?
30. Suppose you specified a feed that was too large. What kinds of problems do you think this might cause? See Table 24-5 for help.

Problems

1. How much time will be required to drill a 1.5-in.-diameter hole through a piece of 1020 cold rolled steel that is 2 in. thick, using a high-speed drill? (Values for feed and cutting speed need to be specified by student.)

2. What is the metal-removal rate when a 1½-in.-diameter hole, 2 in. deep, is drilled in 1020 steel at a cutting speed of 120 fpm with a feed of 0.020 ipr?

3. If the specific horsepower for the steel in problem 2 is 0.9, what horsepower would be required?

4. If the specific power of AISI steel is 0.7, and 75% of the output of the 1.5-kW motor of a drilling machine is available at the tool, what is the maximum feed that can be used in drilling a 2-in.-diameter hole with a HSS drill? (Use the cutting speed suggested in problem 2.) ·

5. Show how the approximate equation for 24-5 for MRR in drilling was obtained. What assumption was needed?

6. An indexable-insert drill can produce a hole four times faster than a spade-blade drill but, with inserts, may cost 50% to 75% more than the equivalent spade blade and holder. For making only a couple of holes, the extra cost is not justified.

   Manufacturer's charts will help determine the best feed, speed, and rpm to run the drills. For example a 1½-in. hole is to be drilled in 4140 steel annealed to Bhn 275. For the spade blade, speed is 80 sfm, feed 0.009 ipr, and spindle rotation 204 rpm. For the indexable-insert drill, speed is 358 sfm, feed 0.007 ipr, and spindle rotation 891 rpm. Determine the number of holes needed to justify the indexable-insert drill, using the cost data below.

   Ignore tool life and assume that the blades and the indexable drills make about the same number of holes. Assume the holes are 3 in. deep, with no allowance needed. Cost of drills:

Spade blade	Indexable-insert drill
$139.00 holder	$273.00 drill
21.90 blade	12.80 two inserts
$160.90	$285.80

   Assume, for this example, a $45/hr machine rate includes the cost of labor and machine burden.

7. Let us assume that you are drilling eight holes, equally spaced in a bolt-hole circle. That is, there would be holes at 12, 3, 6, and 9 o'clock and four more holes equally spaced between them. The diameter of the bolt hole circle is 6 in. The designer says that the holes must be 45 ± 1 degree from each other around the circle.

   a. Compute the tolerance between hole centers.

   b. Do you think a typical multiple-spindle drill setup could be used to make this bolt circle—using eight drills all at once? Why or why not?

   c. Do you think that the use of a jig may help improve the situation?

Milling

Milling is a basic machining process by which a surface is generated progressively by the removal of chips from a workpiece fed into a rotating cutter in a direction perpendicular to the axis of the cutter. Sometimes the workpiece remains stationary, and the cutter is fed to the work. In nearly all cases, a multiple-tooth cutter is used so that the material removal rate is high. Often the desired surface is obtained in a single pass of the cutter or work and, because very good surface finish can be obtained, milling is particularly well suited to and widely used for mass-production work. Several types of milling machines are used, ranging from relatively simple and versatile machines that are used for general-purpose machining in job shops and tool-and-die work to highly specialized machines for mass production. Unquestionably, more flat surfaces are produced by milling than by any other machining process.

The cutting tool used in milling is known as a *milling cutter*. Equally spaced peripheral teeth will intermittently engage and machine the workpiece. This is called *interrupted cutting*.

Milling operations can be classified into two broad categories called peripheral milling and face milling. Each has many variations.

In *peripheral milling* the surface is generated by teeth located on the periphery of the cutter body. See Figure 25-1. The surface is parallel with the axis of rotation of the cutter. Both flat and formed surfaces can be produced by this method, the cross section of the resulting surface corresponding to the axial contour of the cutter. This process is often called *slab milling* and is usually performed on horizontal spindle machines. In slab milling, the tool rotates (mills) at some rpm (N) while the work feeds past the tool at a rate f_m.

Introduction

Fundamentals of Milling Processes

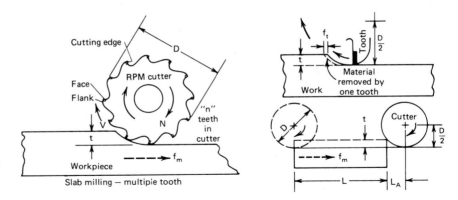

FIGURE 25-1 Basics of the milling process—Peripheral or slab milling. Also see Figure 25-8.

The surface cutting speed is established by the cutter of diameter (D) according to

$$V = \pi DN/12. \qquad (25\text{-}1)$$

in surface feet per minute where D is in inches. The depth of cut is t in inches. The width of cut is the width of the cutter or the work in inches and is given the symbol W. The length of the cut, L, is the length of the work plus some allowance, L_A, for approach and overtravel.

The feed of the table, f_m, in inches per minute is related to the amount of metal each tool removes during a revolution (this is called the feed per tooth), f_t, according to

$$f_m = f_t Nn \qquad (25\text{-}2)$$

where n is the number of teeth in the cutter (teeth/rev.). Given a selected cutting speed, V, and feed, f_t, for a given work material and tool material combination, the spindle rpm is computed from equation 25-1. The cutting velocity is that which occurs at the cutting edges of the milling cutter.

The cutting time $= \mathrm{CT} = (L + L_A)/f_m$ in minutes, where $\qquad$ (25-3)

the length of approach $= L_A = \sqrt{\dfrac{D^2}{4} - \left(\dfrac{D}{2} - t\right)^2} = \sqrt{t\,(D - t)} \qquad$ (25-4)

The MRR $= \mathrm{Vol}/\mathrm{CT} = LWt/\mathrm{CT} = Wtf_m$ in.^{3}/min., ignoring L_A. $\qquad$ (25-5)

Values for f_t are given in Table 25.1 along with recommended cutting speeds.

In *face milling* and *end milling*, the generated surface is at right angles to the cutter axis. Most of the cutting is done by the peripheral portions of the teeth, with the face portions providing some finishing action. Face milling is done on both horizontal-spindle and vertical-spindle machines.

The tool rotates (face mills) at some rpm (N) while the work feeds past the tool. See Figure 25-2. The surface cutting speed is related to the cutting diameter D according to equation 25-1 in surface feet per minute. The depth of cut is t in

TABLE 25.1 Suggested Starting Feeds and Speeds Using High-Speed Steel and Carbide Cutters

Material	Feed (in/tooth) / Speed (fpm)	Carbide Cutters						High-speed Steel Cutters					
		Face Mills	Slab Mills	End Mills	Full and Half Side Mills	Saws	Form Mills	Face Mills	Slab Mills	End Mills	Full and Half Side Mills	Saws	Form Mills
Malleable iron Soft/Hard	Feed per tooth	.005–.015	.005–.015	.005–.010	.005–.010	.003–.004	.005–.010	.005–.015	.005–.015	.003–.010	.006–.012	.003–.006	.005–.010
	Feet per min.	200–300	200–300	200–350	200–300	200–350	175–275	60–100	60–90	60–100	60–100	60–100	60–80
Cast steel Soft/Hard	Feed per tooth	.008–.015	.005–.015	.003–.010	.005–.010	.002–.004	.005–.010	.010–.015	.010–.015	.005–.010	.005–.010	.002–.005	.008–.012
	Feet per min.	150–350	150–350	150–350	150–350	150–300	150–300	40–60	40–60	40–60	40–60	40–60	40–60
100–150 BHN Steel	Feed per tooth	.010–.015	.008–.015	.005–.010	.008–.012	.003–.006	.004–.010	.015–.030	.008–.015	.003–.010	.010–.020	.003–.006	.008–.010
	Feet per min.	450–800	450–600	450–600	450–800	350–600	350–600	80–130	80–130	80–140	80–130	70–100	70–100
150–250 BHN Steel	Feed per tooth	.010–.015	.008–.015	.005–.010	.007–.012	.003–.006	.004–.010	.010–.020	.008–.015	.003–.010	.010–.015	.003–.006	.006–.010
	Feet per min.	300–450	300–450	300–450	300–450	300–450	300–450	50–70	50–70	60–80	50–70	50–70	50–70
250–350 BHN Steel	Feed per tooth	.008–.015	.007–.012	.005–.010	.005–.012	.002–.005	.003–.008	.005–.010	.005–.010	.003–.010	.005–.010	.002–.005	.005–.010
	Feet per min.	180–300	150–300	150–300	160–300	150–300	150–300	35–60	35–50	40–60	35–50	35–50	35–50
350–450 BHN Steel	Feed per tooth	.008–.015	.007–.012	.004–.008	.005–.012	.001–.004	.003–.008	.003–.008	.005–.008	.003–.101	.003–.008	.001–.004	.003–.008
	Feet per min.	125–180	100–150	100–150	125–180	100–150	100–150	20–35	20–35	20–40	20–35	20–35	20–35
Cast Iron Hard BHN 225–350	Feed per tooth	.005–.010	.005–.010	.003–.008	.003–.010	.002–.003	.005–.010	.005–.012	.005–.010	.003–.008	.005–.010	.002–.004	.005–.010
	Feet per min.	125–200	100–175	125–200	125–200	125–200	100–175	40–60	35–50	40–60	40–60	35–60	35–50
Cast Iron Medium BHN 180–225	Feed per tooth	.008–.015	.008–.015	.005–.010	.005–.012	.003–.004	.006–.012	.010–.020	.008–.015	.003–.010	.008–.015	.003–.005	.008–.012
	Feet per min.	200–275	175–250	200–275	200–275	200–250	175–250	60–80	50–70	60–90	60–80	60–70	50–60
Cast Iron Soft BHN 150–180	Feed per tooth	.015–.025	.010–.020	.005–.012	.008–.015	.003–.004	.008–.015	.015–.030	.010–.025	.004–.010	.010–.020	.002–.005	.010–.015
	Feet per min.	275–400	250–350	275–400	275–400	250–350	250–350	80–120	70–110	80–120	80–120	70–110	60–80
Bronze Soft/Hard	Feed per tooth	.010–.020	.010–.020	.005–.010	.008–.012	.003–.004	.008–.015	.010–.025	.008–.020	.003–.010	.008–.015	.003–.005	.008–.015
	Feet per min.	300–100	300–800	300–1000	300–1000	300–1000	200–800	50–225	50–200	50–250	50–225	50–250	50–200
Brass Soft/Hard	Feed per tooth	.010–.020	.010–.020	.005–.010	.008–.012	.003–.004	.008–.015	.010–.025	.008–.020	.005–.015	.008–.015	.003–.005	.008–.015
	Feet per min.	500–1500	500–1500	500–1500	500–1500	500–1500	500–1500	150–300	100–300	150–350	150–350	150–300	100–300
Aluminum Alloy Soft/Hard	Feed per tooth	.010–.040	.010–.030	.003–.015	.008–.025	.003–.006	.008–.015	.010–.040	.015–.040	.005–.020	.010–.030	.004–.008	.010–.020
	Feet per min.	2000 UP	2000 UP	2000 UP	2000 UP	2000 UP	2000 UP	300–1200	300–1200	300–1200	300–1200	300–1000	300–1200

Generally, lower end of range used for inserted blade cutters, higher end of range for indexable insert cutters.

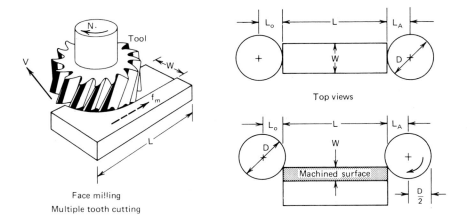

FIGURE 25-2 Basics of face and end milling processes.

Face milling
Multiple tooth cutting

inches. The width of cut is W in inches and may be width of the workpiece or width of the cutter depending upon the setup. The length of cut is the length of the workpiece, L, plus an allowance for approach, L_A and overtravel, L_o in inches. The feed of the table, f_m, in inches per minute is related to the amount of metal each tooth removes during a pass over the work and this is called the feed per tooth, f_t, where $f_m = f_t N n$. The number of teeth in the cutter is n.

The cutting time, $CT = (L + L_A + L_o)/f_m$ in minutes. $\qquad$ (25-6)

The MRR $= \text{Vol}/CT = Lwt/CT = Wtf_m$ in in.^{3}/min (ignore L_o and L_A). See Figure 25-3. The length of approach is usually equal to the length of overtravel which usually equals $D/2$ inches.

For a setup in which the tool does not completely pass over the workpiece,

$$L_o = L_A = \sqrt{W(D-W)} \text{ for } W < \frac{D}{2} \text{ or} \qquad (25\text{-}7)$$

$$L_o = L_A = D/2 \text{ for } W \geq D/2. \qquad (25\text{-}8)$$

In milling, surfaces can be generated by two distinctly different methods, illustrated in Figure 25-4. *Up milling* is the traditional way to mill and is called *conventional* milling. The cutter rotates against the direction of feed of the workpiece. In *climb* or *down milling*, the rotation is in the same direction as the feed. The method of chip formation is completely different in the two cases. In up milling the chip is very thin at the beginning, where the tooth first contacts the work, and increases in thickness, becoming a maximum where the tooth leaves the work. The cutter tends to push the work along and lift it upward from the table. This action tends to eliminate any effect of looseness in the feed screw and nut of the milling machine table and results in a smooth cut. However, the action also tends to loosen the work from the clamping device, therefore greater clamping forces must be employed. In addition, the smoothness of the generated surface depends greatly on the sharpness of the cutting edges. If milling

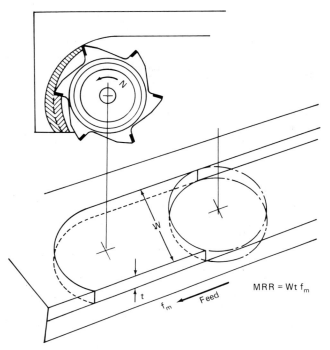

$$MRR = W t \, f_m$$

FIGURE 25-3 Face milling MRR in up milling.

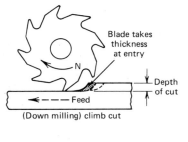

Blade takes thickness at entry

Depth of cut

Feed

(Down milling) climb cut

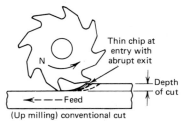

Thin chip at entry with abrupt exit

Depth of cut

Feed

(Up milling) conventional cut

Peripherial or slab milling

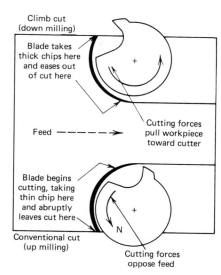

Climb cut (down milling)

Blade takes thick chips here and eases out of cut here

Cutting forces pull workpiece toward cutter

Feed

Blade begins cutting, taking thin chip here and abruptly leaves cut here

Conventional cut (up milling)

Cutting forces oppose feed

Face or end milling

FIGURE 25-4 Climb cut (or down) milling versus conventional cut (or up) milling for slab or end milling.

conditions create a built-up edge on the cutting edge, the BUE will not affect the surface in climb milling.

In down milling, maximum chip thickness occurs close to the point at which the tooth contacts the work. Because the relative motion tends to pull the workpiece into the cutter, any possibility of looseness in the table feed screw must be eliminated if down milling is to be used. It should never be attempted on machines that are not designed for this type of milling. Virtually all modern milling machines are capable of doing down milling. Because the material yields in approximately a tangential direction at the end of the tooth engagement, there is less tendency for the machined surface to show toothmarks (than when up milling is used) and the cutting process is smoother with less chatter. Another advantage of down milling is that the cutting force tends to hold the work against the machine table, permitting lower clamping forces. However, the fact that the cutter teeth strike against the surface of the work at the beginning of each chip can be a disadvantage if the workpiece has a hard surface, as castings sometimes do. This may cause the teeth to dull rapidly. Metals that readily work-harden should be climb milled.

Milling is an interrupted cutting process wherein entering and leaving the cut subjects the tool to impact loading, cyclic heating, and cyclic cutting forces. As shown in Figure 25-5, the cutting force, F_c, builds rapidly as the tool enters the work at Ⓐ and progresses to Ⓑ, peaks as the blade crosses the direction of feed at Ⓒ, decreases to Ⓓ, and then drops to zero abruptly upon exit. The diagram does not indicate the impulse loads caused by impacts. The interrupted-cut phenomenon explains in large part why milling cutter teeth are designed to have small positive or negative rakes, particularly when the tool material is carbide or ceramic. These brittle materials tend to be very strong in compression, and negative rake results in the cutting edges' being placed in compression by the cutting forces rather than tension. Cutters made from HSS are made with positive rakes, in the main, but must be run at lower speeds. Positive rake tends to lift the workpiece while negative rakes compress the workpiece and allow heavier cuts to be made. Table 25.2 summarizes some additional milling problems.

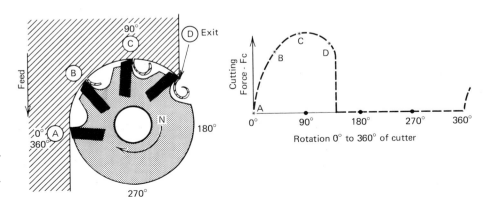

FIGURE 25-5   Conventional face milling (left) with cutting force diagram for F_c (right), showing the interrupted nature of the process. (From *Metal Cutting Principles*, 2nd ed., Ingersoll Cutting Tool Company.)

TABLE 25.2 Probable Causes of Milling Problems

Problem	Probable Cause	Cures
Chatter (vibration)	1. Lack of rigidity in machine, fixtures, arbor, or workpiece	Use larger arbors
	2. Cutting load too great	Decrease feed/tooth or number of teeth in contact with work
	3. Dull cutter	Sharpen or replace
	4. Poor lubrication	Flood coolant
	5. Straight-tooth cutter	Use helical cutter
	6. Radial relief too great	
	7. Rubbing, insufficient clearance	Check tool angles
Loss of accuracy (cannot hold size)	1. High cutting load causing deflection	Decrease number of teeth in contact with work
	2. Chip packing	Adjust cutting fluid to wash chips out of teeth
	3. Chips not cleaned away before mounting new piece of work	
Cutter rapidly dulls	1. Cutting load too great	Decrease feed/tooth or number of teeth in contact
	2. Insufficient coolant	Add blending oil to coolant
Poor surface finish	1. Feed too high	Check to see if all teeth are set at same height
	2. Tool dull	
	3. Speed too low	
	4. Not enough cutter teeth	
Cutter digs in (hogs into work)	1. Radial relief too great	Check to see that workpiece is not deflecting and is securely clamped
	2. Rake angle too large	
	3. Improper speed	
Work burnishing	1. Cut is too light	Enlarge feed
		Sharpen cutter
	2. Insufficient radial relief	
	3. Land too wide	
Cutter burns	1. Not enough lubricant	Add sulfur-based oil
	2. Speed too high	Reduce cutting speed
		Flood coolant
Teeth breaking	1. Feed too high	Decrease feed/tooth
		Use cutter with more teeth
		Reduce table feed

Milling cutters can be classified according to the way the cutter is mounted in the machine tool. *Arbor cutters* have a center hole so they can be mounted on an arbor. *Shank cutters* have either a tapered or straight integral shank. Those with tapered shanks can be mounted directly in the milling machine spindle, whereas straight-shank cutters are held in a chuck. *Facing cutters* usually are bolted to the end of a stub arbor. Common types of milling cutters, classified in this manner, are as follows:

Milling Cutters

Arbor Cutters	Shank Cutters
Plain	End mills
Side	Solids
Staggered-tooth	Inserted-tooth
Slitting saws	Shell
Angle	Hollow
Inserted-tooth	T-slot
Form	Woodruff key seat
Fly	Fly

Figures 25-6 and 25-7 show several types of arbor-type and shank-type milling cutters, respectively.

Another method of classification applies only to face and end-mill cutters and relates to the direction of rotation. A *right-hand cutter* must rotate counterclockwise when viewed from the front end of the machine spindle. Similarly, a *left-hand cutter* must rotate clockwise. All other cutters can be reversed on the arbor

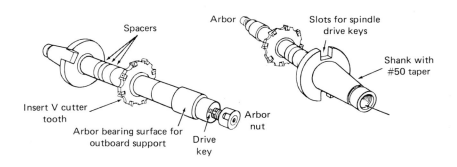

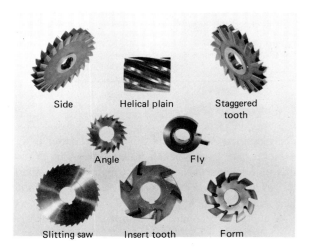

FIGURE 25-6 Arbor (2 views) and arbor-type milling cutters.

FIGURE 25-7 Shank-type milling cutters. (Left to right) T-slot, shell and mill. Woodruff key seater or hollow end mill, solid end mill.

to change them from one hand to the other. Positive rake angles are used on general-purpose HSS milling cutters. Negative rake angles are commonly used on carbide- and ceramic-tipped cutters employed in mass-production milling in order to obtain the greater strength and cooling capacity which they provide. TiN coating of these tools is quite common, resulting in significant increases in tool life.

Types of Milling Cutters.

Plain milling cutters used for plain or slab milling have straight or helical teeth on the periphery and are used for milling flat surfaces. *Helical mills* (Figure 25-8) engage the work gradually, and usually more than one tooth cuts at a given time. This reduces shock and chattering tendencies and promotes a smoother surface. Consequently, this type of cutter usually is preferred over one with straight teeth.

Side milling cutters are similar to plain milling cutters except that the teeth extend radially part way across one or both ends of the cylinder toward the

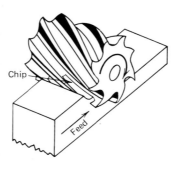

FIGURE 25-8 Manner in which chips are formed progressively by the teeth of a plain helical-tooth milling cutter in up milling. (Courtesy Cincinnati Milacron, Inc.)

center. The teeth may be either straight or helical. Frequently, these cutters are relatively narrow, being disklike in shape. Two or more side milling cutters often are spaced on an arbor to straddle the workpiece (called *straddle milling*) and machine two parallel surfaces at once.

Interlocking slotting cutters consist of two cutters similar to side mills but made to operate as a unit for milling slots. The two cutters are adjusted to the desired width by inserting shims between them.

Staggered-tooth milling cutters are narrow cylindrical cutters having staggered teeth, and with alternate teeth having opposite helix angles. They are ground to cut only on the periphery, but each tooth also has chip clearance ground on the protruding side. These cutters have a free cutting action that makes them particularly effective in milling deep slots. See Figure 25-9.

Slitting saws are thin, plain milling cutters, usually from $1/32$ to $3/16$ in. thick, which have their sides slightly "dished" to provide clearance and prevent binding. They usually have more teeth per unit of diameter than ordinary plain milling cutters and are used for milling deep narrow slots and cutting-off operations.

Angle milling cutters are made in two types: single-angle and double-angle. Angle cutters are used for milling slots of various angles or for milling the edges of workpieces to a desired angle. *Single-angle cutters* have teeth on the conical surface, usually at an angle of 45 to 60 degrees to the plane face. *Double-angle cutters* have V-shaped teeth, with both conical surfaces at an angle to the end faces, but not necessarily at the same angle. The V-angle usually is 45, 60, or 90 degrees.

Most larger-sized milling cutters are of the *inserted-tooth type*. The cutter body is made of steel, with the teeth made of high-speed steel, carbides, or TiN carbides, fastened to the body by various methods. An insert tooth cutter using indexable carbide or ceramic inserts is shown in Figure 25-10. This type of construction reduces the amount of costly material that is required and can be used for any type of cutter but most often is used with face mills.

Form milling cutters have the teeth ground to a special shape—usually an irregular contour—to produce a surface having a desired transverse contour. They must be sharpened by grinding only the tooth face, thereby retaining the original contour as long as the plane of the face remains unchanged with respect to the axis of rotation. Convex, concave, corner-rounding, and gear-tooth cutters are common examples. See Figure 25-9.

End mills are shank-type cutters having teeth on the circumferential surface and one end. They thus can be used for facing, profiling, and end milling. The teeth may be either straight or helical, but the latter is more common. Small end mills have straight shanks, whereas taper shanks are used on larger sizes.

Plain end mills have multiple teeth that extend only about halfway toward the center on the end. They are used in milling slots, profiling, and facing narrow surfaces. *Two-lip mills* have two straight or helical teeth that extend to the center. Thus they may be sunk into material, like a drill, and then fed lengthwise to form a groove.

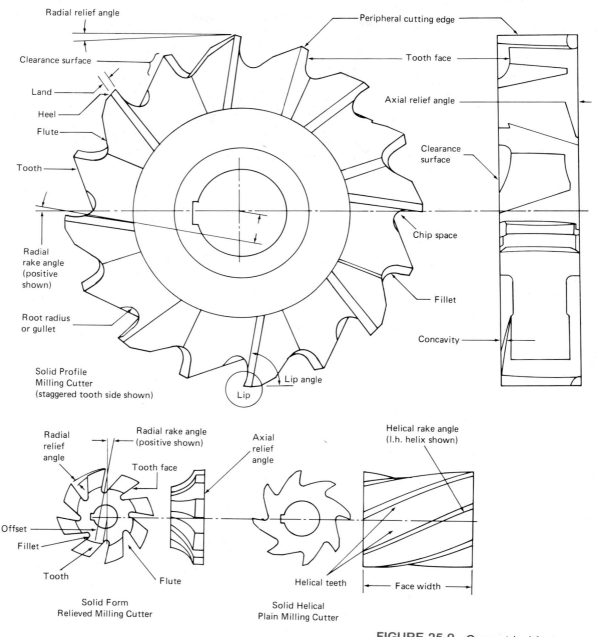

Radial relief angle

Clearance surface

Land

Heel

Flute

Tooth

Radial
rake angle
(positive
shown)

Root radius
or gullet

Solid Profile
Milling Cutter
(staggered tooth side shown)

Lip angle

Lip

Peripheral cutting edge

Tooth face

Axial relief angle

Clearance
surface

Chip space

Fillet

Concavity

Radial
relief
angle

Radial rake angle
(positive shown)

Tooth face

Offset

Fillet

Tooth

Flute

Axial
relief
angle

Solid Form
Relieved Milling Cutter

Helical rake angle
(l.h. helix shown)

Helical teeth

Face width

Solid Helical
Plain Milling Cutter

FIGURE 25-9 Geometrical features
of solid (HSS) arbor cutters for
milling.

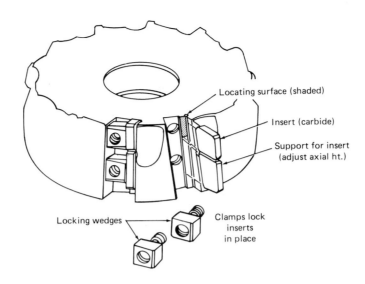

FIGURE 25-10 Insert-tooth face milling cutter.

Shell end mills are solid-type, multiple-tooth cutters, similar to plain end mills but without a shank. The center of the face is recessed to receive a screw head or nut for mounting the cutter on a separate shank or a stub arbor. One shank can hold any of several cutters and thus provides great economy for larger-sized end mills.

Hollow end mills are tubular in cross section, with teeth only on the end but having internal clearance. They are used primarily on automatic screw machines for sizing cylindrical stock, producing a short cylindrical surface of accurate diameter.

T-slot cutters are integral-shank cutters with teeth on the periphery and *both* sides. They are used for milling the wide groove of a T-slot. In order to use them, the vertical groove must first be made with a slotting mill or an end mill to provide clearance for the shank. Because the T-slot cutter cuts on five surfaces simultaneously, it must be fed with care.

Woodruff keyseat cutters are made for the single purpose of milling the semi-cylindrical seats required in shafts for Woodruff keys. They come in standard sizes corresponding to Woodruff key sizes. Those below 2 inches in diameter have integral shanks; the larger sizes may be arbor-mounted.

Occasionally, *fly cutters* may be used for face milling or boring, or both operations may be done with a single tool at one setup. A single-point cutting tool is attached to a special shank, usually with provision for adjusting the effective radius of the cutting tool with respect to the axis of rotation. The cutting edge can be made in any desired shape and, because it is a single-point tool, is very easy to grind.

Milling Machines

Because the milling process is versatile and highly productive, a variety of machines have been developed to employ the milling principle. The basic, general-purpose milling machines provide a high degree of flexibility. Another type, *duplicators,* is used exclusively for reproducing parts from templates or patterns. A third type encompasses *special-purpose machines* that are used in mass-production manufacturing. Machines of the fourth type are numerically controlled and do other basic machining operations in addition to milling. These commonly are called *machining centers* and will be discussed in Chapter 29.

Milling machines have an accurate, rugged, rotating spindle. They can also be used for other machining operations, such as drilling and boring. The common types may be classified according to their general characteristics as follows:

1. Column-and-knee type (general purpose).
   a. Plain or horizontal.
      (1) Power table feed.
      (2) Hand table feed.
   b. Universal.
   c. Vertical.
   d. Turret-type universal.
2. Bed type (manufacturing).
   a. Simplex.
   b. Universal.
   c. Triplex.
3. Planer type (large work only).
4. Special.
   a. Rotary table.
   b. Drum type.
   c. Profilers.
   d. Duplicators.

Basic Milling-Machine Construction.   Most basic milling machines are of column-and-knee construction, employing the components and motions shown in Figure 25-11. The column, mounted on the base, is the main supporting frame for all the other parts and contains the spindle with its driving mechanism. This construction provides controlled motion of the worktable in three mutually perpendicular directions: (1) through the *knee* moving vertically on ways on the front of the column, (2) through the *saddle* moving transversely on ways on the knee, and (3) through the *table* moving longitudinally on ways on the saddle. All these motions can be imparted either by manual or powered means. In most cases, a powered rapid traverse is provided in addition to the regular feed rates for use in setting up work and in returning the table at the end of a cut.

Milling machines having only the three mutually perpendicular table motions are called *plain column-and-knee type*. These are available with both horizontal and vertical spindles as shown in Figure 25-11. On the horizontal type, an adjustable over-arm provides an outboard bearing support for the end of the cutter arbor. In some vertical-spindle machines the spindle can be fed up and down, either by power or by hand. Vertical-spindle machines are especially well suited for face- and end-milling operations. They also are very useful for drilling

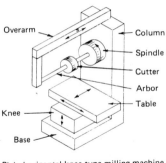

Plain horizontal knee type milling machine

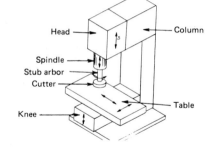

Vertical knee type milling machine

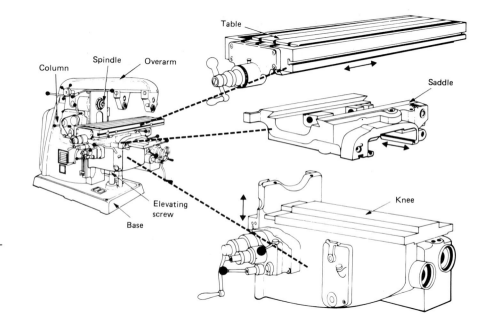

FIGURE 25-11 Major components of a plain column-and-knee type milling machine, which can have horizontal spindle or vertical spindle, (Top, from Manufacturing Producibility Handbook; courtesy General Electric Company. Bottom, courtesy Cincinnati Milacron, Inc.)

and boring, particularly when holes must be spaced accurately in a horizontal plane, because of the controlled table motion.

Universal Column-and-Knee Milling Machines. *Universal column-and-knee milling machines* differ from plain column-and-knee machines in that the table is mounted on a housing that can be swiveled in a horizontal plane, thereby increasing its flexibility. Helices, as found in twist drills, milling cutters, and helical gear teeth, can be milled on universal machines.

Turret-Type Milling Machines. *Turret-type column-and-knee milling machines,* as shown in Figure 25-12, have dual heads that can be swiveled about a horizontal axis on the end of a horizontally adjustable ram. This permits milling

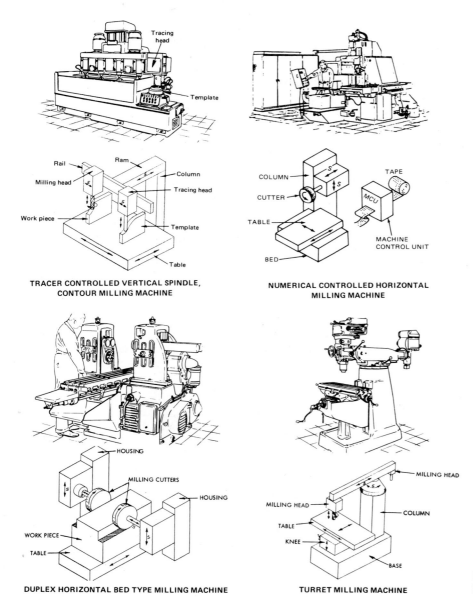

TRACER CONTROLLED VERTICAL SPINDLE,
CONTOUR MILLING MACHINE

NUMERICAL CONTROLLED HORIZONTAL
MILLING MACHINE

DUPLEX HORIZONTAL BED TYPE MILLING MACHINE

TURRET MILLING MACHINE

FIGURE 25-12 Four types of milling machine tools.

to be done horizontally, vertically, or at any angle. This added flexibility is advantageous when a variety of work has to be done, as in tool and die or experimental shops. They are available with either plain or universal tables.

Bed-Type Milling Machines.   In production manufacturing operations, ruggedness and the capability of making heavy cuts are of more importance than versatility. *Bed-type milling machines,* such as shown in Figure 25-12, are made for these conditions. The table is mounted directly on the bed and has only longitudinal motion. The spindle head can be moved vertically in order to set up the machine for a given operation. Normally, once the setup is completed, the spindle head is clamped in position and no further motion of it occurs during machining. However, on some machines vertical motion of the spindle occurs during each cycle.

After such milling machines are set up, little skill is required to operate them, permitting the use of lesser-skilled operators. Some machines of this type are equipped with automatic controls so that all the operator has to do is load and unload workpieces into the fixture and set the machine into operation. Often a fixture is provided at each end of the table so that one workpiece can be unloaded while another is being machined.

Bed-type milling machines with single spindles sometimes are called *simplex milling machines;* they are made with both horizontal and vertical spindles. Bed-type machines also are made in *duplex* and *triplex* types, having two or three spindles respectively. These permit the simultaneous milling of two or three surfaces at a single pass.

Planer-Type Milling Machines. *Planer-type milling machines,* as illustrated in Figure 25-13, utilize several milling heads, which can remove large amounts of metal while permitting the table and workpiece to feed quite slowly. Often only a single pass of the workpiece past the cutters is required. Through the use of different types of milling heads and cutters, a wide variety of surfaces

FIGURE 25-13 Large planer-type milling machine. Inset shows 90-degree head being used. (Courtesy Cosa Corporation.)

can be machined with a single setup of the workpiece. This is an advantage when heavy workpieces are involved.

Rotary-Table Milling Machines.

Some types of face milling in mass-production manufacturing are often done on *rotary-table milling machines*. Roughing and finishing cuts can be made in succession as the workpieces are moved past the several milling cutters while held in fixtures on the rotating table. The operator can load and unload the work without stopping the machine.

Profilers and Duplicators.

Milling machines that can duplicate external or internal geometries in two dimensions are called *profilers* or tracer-controlled machines. As shown in Figure 25-12, a tracing probe follows a two-dimensional template and, through electronic or air-actuated mechanisms, controls the cutting spindles in two mutually perpendicular directions. See Figure 25-14.

Hydraulic tracer control is not new, but a great variety of control systems for contouring and straight-line milling have become available during the past few years. These range from simple single-axis control systems for die sinking to control for profiling and three-axis contouring.

Hydraulic tracers have transformed standard and special milling machines into production tools with virtually unlimited capabilities for contouring and straight-line milling. They can produce shapes that are impossible to produce manually, such as free-form dies and molds.

All hydraulic tracers work basically the same way, in that they utilize a stylus connected to a precision servomechanism for each axis of control. Figure 25-14 illustrates a stylus that is linked mechanically to the servos in a two-axis tracer valve. The servos are connected to hydraulic actuators on the machine

FIGURE 25-14   Small knee-type vertical-spindle milling machine modified with tracer control system for X and Y control.

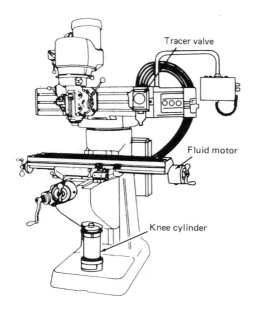

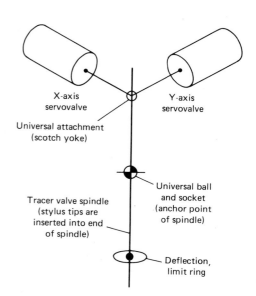

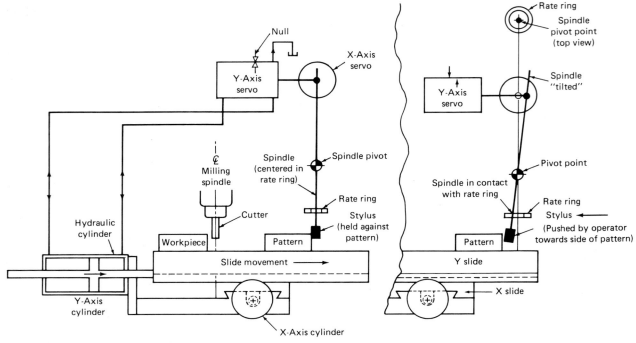

FIGURE 25-15 Principle of operation of a hydraulic tracer.

slides. As the stylus traces a template, the servos control the motion of the slides so that the milling cutter duplicates the template shape onto the workpiece. See Figure 25-15 for details of the principle of operation.

As the tracing stylus is tilted toward the pattern, the servovalve is shifted off null, resulting in Y slide motion opposite to that of the stylus direction. When the stylus is in contact with the pattern the servovalve is at its null position and Y slide motion is stopped.

Duplicators produce forms in three dimensions. A tracing probe follows a three-dimensional master. Often the probe does not actually contact the master, a variation in the length of a spark between the probe and the master controlling the drives to the quill and the table, thereby avoiding wear on the master or possible deflection of the probe. On some machines, the ratio between the movements of the probe and cutter can be varied.

Duplicators are widely used to machine molds and dies and sometimes are called *die-sinking machines*. They are used extensively in the aerospace industry to machine parts from wrought plate or bar stock as substitutes for forgings when the small number of parts required would make the cost of forging dies uneconomical.

Accessories for Milling Machines. The usefulness of ordinary milling machines is greatly extended by employing various accessories.

The *vertical milling attachment*, shown in Figure 25-16, is used on a horizontal milling machine to permit vertical milling to be done. Ordinarily, heavy work cannot be done with such an attachment.

The *universal milling attachment,* shown in Figure 25-17, is similar to the vertical attachment but can be swiveled about both the axis of the milling machine spindle and a second, perpendicular axis to permit milling to be done at any angle.

The *slotting attachment* permits adapting a horizontal milling machine to the work of a vertical shaper. Although not used extensively, it is useful for cutting small keyways.

The *universal dividing head* is by far the most widely used milling machine accessory, providing a means for holding and indexing work through any desired arc of rotation. The work may be mounted between centers, as shown in Figure 25-17, or held in a chuck that is mounted in the spindle hole of the dividing head. The spindle can be tilted from about 5 degrees below horizontal to beyond the vertical position.

Basically, a dividing head is a rugged, accurate, 40:1 worm-gear reduction unit. The spindle of the dividing head is rotated one revolution by turning the input crank 40 turns. An index plate, mounted beneath the crank, contains a number of holes, arranged in concentric circles and equally spaced, with each circle having a different number of holes. A plunger pin on the crank handle can be adjusted to engage the holes of any circle. This permits the crank to be turned an accurate, fractional part of a complete circle as represented by the increment between any two holes of a given circle on the index plate. Utilizing the 40:1 gear ratio and the proper hole circle on the index plate, the spindle can be rotated a precise amount by the application of either of the following rules:

Rule 1. The number of turns of crank $= \dfrac{40}{\text{cuts per revolution of work}}$

Rule 2. The holes to be indexed $= \dfrac{40 \times \text{holes in index circle}}{\text{cuts per revolution of work}}$

FIGURE 25-16   Vertical milling attachment for horizontal milling machine. (Courtesy Brown & Sharpe Manufacturing Company.)

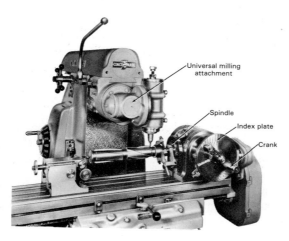

FIGURE 25-17   Milling a helical groove using a universal dividing head and a universal milling attachment. (Courtesy Cincinnati Milacron, Inc.)

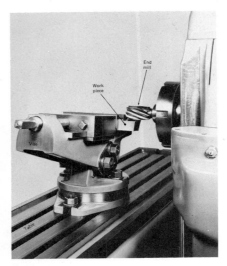

FIGURE 25-18 Universal vise on a horizontal-spindle milling machine. (Courtesy Cincinnati Milacron, Inc.)

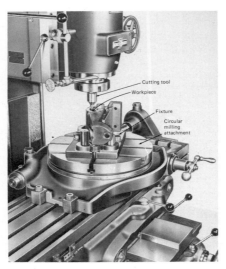

FIGURE 25-19 Milling a circular slot using a fixture mounted on a cir-cular-milling attachment in a vertical-spindle milling machine. (Courtesy Cincinnati Milacron, Inc.)

If the first rule is used, an index circle must be selected that has the proper number of holes to be divisible by the denominator of any resulting fractional portion of a turn of the crank. In using the second rule, the number of holes in the index circle must be such that the numerator of the fraction is an even multiple of the denominator. For example, if 24 cuts are to be taken about the circumference of a workpiece, the number of turns of the crank required would be 1⅔. An index circle having 12 holes could be used with one full turn plus eight additional holes. The second rule would give the same result. Adjustable *sector arms* are provided on the index plate that can be set to a desired number of holes, less than a full turn, so that fractional turns can be made readily without the necessity for counting holes each time.

Dividing heads are made having ratios other than 40:1; the ratio should be checked before using.

Because each full turn of the crank on a standard dividing head represents 360/40, or 9 degrees of rotation of the spindle, indexing to a fraction of a degree can be obtained. For example, the space between two adjacent holes on a 36-hole circle represents ¼ degree.

Indexing can be done in three ways. *Plain indexing* is done solely by the use of the 40:1 ratio in the dividing head. In *compound indexing,* the index plate is moved forward or backward a number of hole spaces each time the crank handle is advanced. For *differential indexing* the spindle and the index plate are con-nected by suitable gearing so that, as the spindle is turned by means of the crank, the index plate is rotated a proportional amount.

The dividing head also can be connected to the feed screw of the milling-machine table by means of gearing. This procedure is used to provide a definite rotation of the workpiece with respect to the longitudinal movement of the table, as in cutting helical gears. This procedure is illustrated in Chapter 31.

Workholding Devices in Milling. Although T-slots are provided on milling machine tables so that workpieces can be clamped directly to the table, more often various workholding devices, called vises or fixtures, are utilized. Smaller workpieces usually are held in a vise mounted on the table. A universal vise, shown in Figure 25-18, is particularly useful in tool-and-die work. Fixtures are used for larger volumes as shown in Figure 25-19. Fixtures reduce machine-loading time and assure proper part location with respect to the cutting tools. Fixtures provide clamping, which counteracts the cutting forces. Workholding devices are discussed in Chapter 28.

Milling Tolerances. As with all tooling applications, the tolerances that can be maintained in milling are dependent upon the rigidity of the workpiece, the accuracy and rigidity of the machine spindle, the precision and accuracy of the workholding device, and the quality of the cutting tool itself. Milling produces forces that contribute to chatter and vibration because of the intermittent cutting action. Soft materials tend to adhere to the cutter teeth and make it more difficult to hold tolerances. Fine-grain materials such as cast iron are easy to mill, as is hardened aluminum.

Within these criteria, properly maintained cutters used in rigid spindles on properly fixtured workpieces can expect to machine within tolerances ± .0005 in. with surface flatness tolerances of .001 in. per foot. Such tolerances are also possible on slotting operations with milling cutters, but ±.001 in. to ±.002 in. is more probable.

Flatness specifications are more difficult to maintain in steel, easier to maintain in some types of aluminum, cast iron, and other nonferrous material.

Milling Surface Finish. The average surface finishes that can be expected on free machining materials range from 60 to 150 microinches. Conditions can exist, however, that can produce wide variations on either side of these ranges.

Review Questions

1. Why is milling better suited than shaping for producing flat surfaces in mass-production machining?
2. How does face milling differ basically from peripheral milling?
3. What type of milling (up or down) uses the least power? Explain why.
4. Why may down milling dull the cutter more rapidly than up milling in machining sand castings?
5. What are two common ways of classifying milling cutters?
6. What kind of milling do you have if the diameter D of the cutter shown in Figure 25-2 is less than the width W of the workpiece?
7. What is the advantage of a helical-tooth cutter over a straight-tooth cutter for slab milling?
8. What would the cutting force diagram for F_c look like if all eight teeth are considered? See right side of Figure 25-5.
9. Explain what steps are required to produce a T-slot by milling.
10. In a typical solid arbor milling cutter, shown in Figure 25-9, why are the teeth staggered?
11. Why would a plain column-and-knee milling machine not be suitable for milling the flutes on a large twist drill?
12. Why is a turret-type milling machine often preferred for tool-and-die work?
13. What is the distinctive feature of a universal milling machine?
14. Explain how controlled movements of the work in three mutually perpendicular directions are obtained in column-and-knee-type milling machines.
15. What is the function of the rate ring in the tracer milling machine?

16. Why are bed-type milling machines preferred over column-and-knee types for production milling?
17. How does a duplicator differ from a profiler?
18. Why have planer-type milling machines replaced ordinary planers?
19. What are two common milling machine accessories?
20. What is the basic principle of a universal dividing head?
21. The input end of a universal dividing head can be connected to the feed screw of the milling machine table. For what purpose?
22. What is the purpose of the hole-circle plate on a universal dividing head?
23. Explain how a standard universal dividing head, having hole circles of 21, 24, 27, 30, and 32 holes, would be operated to cut an 18-tooth gear.
24. Explain how table feed (ipm) and spindle rpm are specified or computed for a milling machine.
25. Why must the number of teeth on the cutter be known when calculating milling machine feed, that is, the table feed in inches/minute?
26. In Figure 25-12, duplex milling is shown. Is the cutter performing up or down milling, and why?
27. Why is the question of up or down milling more critical in horizontal slab milling than in vertical spindle (end or face) milling?
28. Suppose you wanted to machine a cast iron with BHN of 275. The process to be used was face milling and a HSS cutter was going to be used. What feed and speed values would you select?

Problems

1. In question 28, you were asked to select a feed per tooth and a cutting speed for a milling process. Reasonable values for feed and speed are 0.005 to 0.010 in. per tooth and 200 sfpm. Compute the input values for the machine tool. Use the information in problem 6 as input where needed.

2. How much time will be required to face-mill an AISI 1020 steel surface that is 12 in. long and 5 in. wide, using a 6-in. diameter, eight-tooth tungsten carbide inserted-tooth cutter? Select values of feed per tooth and cutting speed from Table 25.1.

3. If the depth of cut is 0.35 in., what is the metal-removal rate in problem 2?

4. Estimate the power required for the operation of problem 3. Do not forget to consider Figure 25-5.

5. If all the flat surfaces on the bearing block shown in Figure 28-5 must be machined, would you machine them in the same sequence on a shaper as on a milling machine? Explain why or why not.

6. A gray cast iron surface 6 in. wide and 18 in. long may be machined on either a vertical milling machine, using an 8-in.-diameter cutter have 10 inserted HSS teeth, or on a hydraulic shaper with a HSS tool. If milling is used, the feed per tooth is 0.01 in. For shaping, a feed of 0.015 in. per stroke would be used. Setup time for the milling machine would be 60 minutes and for the shaper would be 10 minutes. The time required to load and unload either machine is 2 minutes. Machine-hour charges for the milling machine and shaper would be $24.50 and $16.50, respectively. Labor cost would be $8.75 per hour in each case. Which machine would be more economical for this job?

7. In problem 6, what percentage of the cycle time for the job is consumed in non-metal-removal activities (setup time and part loading/unloading), assuming that 10 parts are to be made?

8. In Figure 21-4, the feed per tooth is .006 in. per tooth. The cutter is rotating at an rpm which will produce a surface cutting speed of 125 sfpm. The cutter diameter is 3 in. The depth of cut is .5 in. The block is 2 in. wide. What is the feed per minute of the milling machine table?

9. What is the MRR for this situation described in problem 8?

10. Suppose you want to do the job described in problem 2 by slab milling. You have selected a 6-in.-diameter cutter with eight carbide insert teeth. The cutting speed will be 500 sfpm and the feed per tooth will be 0.010 ipt. Determine the input parameters for the machine (rpm of arbor and table feed), the CT, and MRR. Compare these answers with what you got previously.

Broaching, Sawing, Filing

The process of *broaching,* as illustrated in Figure 26-1, is one of the most productive of the basic machining processes. Broaching competes economically with milling and boring and is capable of producing precision-machined surfaces. Broaching is similar to shaping, but a broach finishes an entire surface in a single pass whereas a shaper requires many strokes. Broaches are used in production to finish holes, splines, and flat surfaces. Feed per tooth in broaching is the change in height of successive teeth. Broaching is similar to sawing except that the saw makes many passes through the cut. The heart of this process lies

Broaching

FIGURE 26-1 Basic shape and nomenclature for conventional pull (hole) broach. Section A–A' shows the cross section of a roughing tooth.

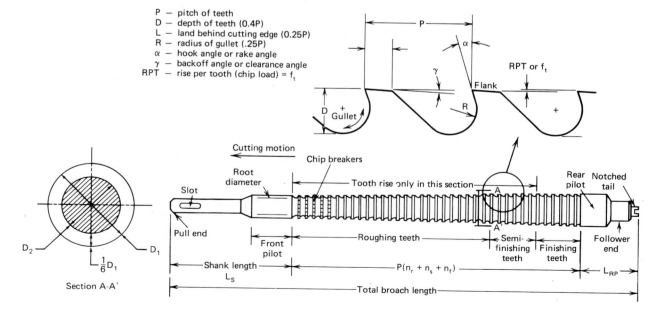

P — pitch of teeth
D — depth of teeth (0.4P)
L — land behind cutting edge (0.25P)
R — radius of gullet (.25P)
α — hook angle or rake angle
γ — backoff angle or clearance angle
RPT — rise per tooth (chip load) = f_t

in the broaching tool, in which roughing, semifinishing, and finishing teeth are combined into one tool. Broaching is unique in that it is the only one of the basic machining processes in which *feed,* which determines the chip thickness, is built into the tool, called a *broach*. The machined surface is always the inverse of the profile of the broach, and, in most cases, it is produced with a single linear stroke of the tool across the workpiece (or the workpiece across the broach).

A broach is composed of a series of teeth, each tooth standing slightly higher than the last. This rise per tooth, RPT, also know as *step,* determines the depth of cut by each tooth (chip thickness), so that no feeding of the broaching tool is required. The frontal contour of the teeth determines the shape of the resulting machined surface. As the result of these conditions built into the tool, no complex motion of the tool relative to the workpiece is required and the need for highly skilled machine operators is minimized.

Figure 26-2 shows a pull type of broach in a vertical pull-down broaching machine. The pull end of the broach is passed through the part and a key mates to the slot. The broach is pulled through the part. The broach is retracted (pulled

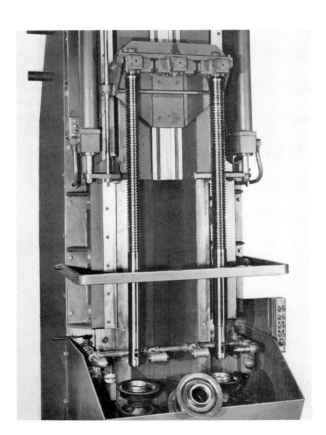

FIGURE 26-2 Two pull broaches in a vertical pull-down broaching machine.

up) out of the part. The part is transferred from the left fixture to the right fixture. One finished part is completed in every time cycle.

In broaching, the tool (or work) is translated past the work (or tool) with a single stroke of velocity V. The feed is provided by a gradual increase in height of successive teeth. The rise per tooth varies depending on whether the tooth is for roughing (t_r), semifinishing (t_s), or final sizing or finishing (t_f). In a typical broach there are 3 to 5 semifinishing and finishing teeth specified. The number of roughing teeth must be determined so that broach length, which is needed to estimate the cutting time, can be calculated. Other lengths needed for a typical pull broach are shown in Figure 26-1. The chip-breakers in first section of roughing teeth may be extended to more teeth if the cut is heavy or the material difficult to machine.

The pitch or distance between teeth is P.

$$P \cong .35\sqrt{L_w} \qquad (26\text{-}1)$$

where length of cut usually equals L_w as shown in Figure 26-3.

The number of roughing teeth is $n_r = \dfrac{[d - (n_f t_f) - (n_s t_s)]}{t_r}$

where d is the total amount of metal to be removed and t_r = rise per tooth.

The length of the broach $L_B = (n_r + n_s + n_f)P + L_s + L_{RP}$ for pull broach.

The length of stroke $L = L_B - L_W$ if broach moves past work in inches or $L_W + L_B$ if work moves past broach.

The cutting time, $\text{CT} = L/12V$, $\qquad (26\text{-}2)$

where V is the cutting speed, in surface feet per minute.

The metal removal rate depends upon the number of teeth (roughing) contracting the work.

$\text{MRR (per tooth)} = 12\, t_r WV\ \text{in}^3/\text{min per roughing tooth}, \qquad (26\text{-}3)$

where W is the width of broach tooth.

The maximum number of roughing teeth in contact with part $n = L_W/P$ for broach larger than part.

$\text{MRR (for process)} = 12\, t_r WVn\ \text{in}^3/\text{min}$ where n is usually rounded off to next largest whole number.

The pull broach must be strong enough so that it will not be pulled apart. The strength of a pull broach is determined by its minimum cross section which occurs either at the root of the first tooth or at the pull end.

$$\text{Allowable pull} = \frac{\text{Area of minimum section} \times \text{Y.S. of broach material}}{\text{Factor of safety}}$$

$$(26\text{-}4)$$

The push broach must be strong enough so that it will not buckle. If length-

Fundamentals of Broaching

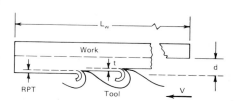

FIGURE 26-3 The feed in broaching depends upon the rise per tooth (RPT). The sum of t gives the depth of cut, d.

to-diameter ratio, L/D_r, is greater than 25, the broach must be considered a long column which can buckle if overloaded.

$$\text{Let } L = \text{length from push end to first tooth}$$
$$D_r = \text{root diameter at } \tfrac{1}{2}\,L \qquad\qquad (25\text{-}5)$$
$$S = \text{factor of safety}$$
$$\text{Allowance load} = \frac{13.5 \times 10^6 \times D_r^4}{SL^2}$$

For L/D_r less than 25, normal broach loads are not critical.

Calculations of the total push or pull load depends upon the number of teeth engaged, n_B; the width of the cut, w; the RPT per tooth engaged, t; and the shear strength of the metal being machined.

Let F_{CB} be the broach pull.

$$F_{CB} \cong 5\tau_s twn_B \qquad\qquad (26\text{-}6)$$

τ_s is found from Figure 21-15 depending upon the BHN for the metal. This force estimate can be used to estimate the horsepower needed for the broaching machine. The number of teeth actually in contact, n_B, is estimated from L_w/P.

The Advantages and Limitations of Broaching

Because of the features built into a broach, it is a simple and rapid method of machining. There is a close relationship between the contour of the surface to be produced, the amount of material that must be removed, and the design of the broach. For example, the total depth of the material to be removed cannot exceed the total step provided in the broach, and the step of each tooth must be sufficient to provide proper chip thickness for the type of material to be machined. Consequently, either a special broach must be made for each job, or the workpiece must be designed so that a standard broach can be used. Broaching is widely used and particularly well suited for mass production because the volume can easily justify the cost of a rather expensive tool. It is also used for certain simple and standardized shapes, such as keyways, where stock broaches can be used.

Broaching originally was developed for machining internal keyways. However, its obvious advantages quickly led to its development for mass-production machining of various surfaces, such as flat, interior or exterior cylindrical and semicylindrical, and many irregular surfaces. Because there are few limitations as to the contour form that broach teeth may have, there is almost no limitation in the shape of surfaces that can be produced by broaching. The only physical limitations are that there must be no obstruction to interfere with the passage of the entire tool over the surface to be machined and that the workpiece must be strong enough to withstand the forces involved. In internal broaching, a hole must exist in the workpiece into which the broach may enter. Such a hole can be made by drilling, boring, or coring.

Broaching usually produces better accuracy and finish than can be obtained by milling or reaming. Although the relative motion between the broaching tool and the work usually is a simple linear one, a rotational motion can be added to permit the broaching of spiral splines or gun-barrel rifling.

Classification of Broaches.   Broaches commonly are classified as follows:

Purpose	Motion	Construction	Function
Single	Push	Solid	Roughing
Combination	Pull	Built-up	Sizing
	Stationary		Burnishing

Broach Design.   Figure 26-1 shows the principal components of a broach and the shape and arrangement of the teeth. Each tooth is essentially a single-edge cutting tool, arranged much like the teeth on a saw except for the step, which determines the depth cut by each tooth. The depth of cut varies from about 0.006 in. for roughing teeth in machining free-cutting steel to a minimum 0.001 in. for finishing teeth. The exact amount depends on several factors. Too-large cuts impose undue stresses on the teeth and the work; too-small cuts result in rubbing rather than cutting action. The strength and ductility of the metal being cut are the primary factors.

Where it is desirable for each tooth to take a deep cut, as in broaching castings or forgings that have a hard, abrasive surface layer, *rotor-cut* or *jump-cut* tooth design, shown in Figure 26-4, may be used. In this design, two or three teeth in succession have the same diameter, or height, but each tooth of the group is notched or cut away so that it cuts only a portion of the circumference or width. This permits deeper but narrower cuts by each tooth without increasing the total load per tooth. This tooth design also reduces the forces and the power requirements.

Tooth loads and cutting forces also can be reduced by using the *double-cut* construction shown in Figure 26-5. Pairs of teeth have the same size, but the first has extra-wide chip-breaker notches and removes metal over only a part of its width, while the smooth second tooth completes the cut. Chip-breaker notches are used on round broaches to break up the chips as shown in Figure 26-6.

FIGURE 26-4   Rotor or jump tooth broach design.

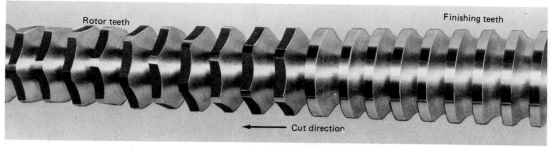

Rotor teeth

Finishing teeth

← Cut direction

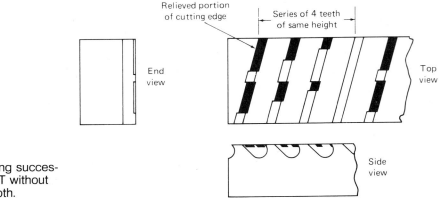

Relieved portion
of cutting edge

Series of 4 teeth
of same height

End
view

Top
view

Side
view

FIGURE 26-5 Overlapping successive teeth permit large RPT without increasing the load per tooth.

FIGURE 26-6 Round, push-type broach with chip-breaker grooves on alternate teeth except at the finishing end. (Courtesy duMont Corp.)

A third construction for reducing tooth loads utilizes the principle illustrated in Figure 26-7. Employed primarily for broaching wide, flat surfaces, the first few teeth in *progressive* broaches completely machine the center, while succeeding teeth are offset in two groups to complete the remainder of the surface. Rotor, double-cut, and progressive designs require the broach to be made longer than if normal teeth were used, and they therefore can be used only on a machine having adequate stroke length.

The faces of the teeth on surface broaches may be either normal to the direction of motion or at an angle of from 5 to 20 degrees. The latter, *shear-cut,* broaches provide smoother cutting action with less tendency to vibrate. Other shapes that can be broached are shown in Figure 26-8 along with push- or pull-type broaches used for the job. The pitch of the teeth and the gullet between them must be sufficient to provide ample room for chip clearance. All chips produced by a given tooth during its passage over the full length of the workpiece must be contained in the space between successive teeth. At the same time, it

FIGURE 26-7 Progressive surface broach. (Courtesy Detroit Broach & Machine Company.)

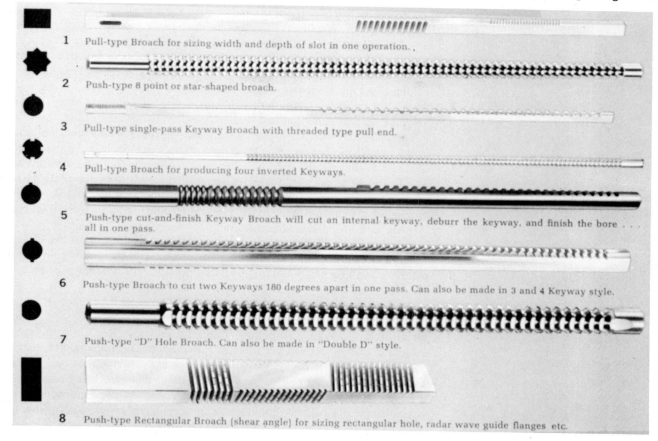

1   Pull-type Broach for sizing width and depth of slot in one operation.

2   Push-type 8 point or star-shaped broach.

3   Pull-type single-pass Keyway Broach with threaded type pull end.

4   Pull-type Broach for producing four inverted Keyways.

5   Push-type cut-and-finish Keyway Broach will cut an internal keyway, deburr the keyway, and finish the bore . . . all in one pass.

6   Push-type Broach to cut two Keyways 180 degrees apart in one pass. Can also be made in 3 and 4 Keyway style.

7   Push-type "D" Hole Broach. Can also be made in "Double D" style.

8   Push-type Rectangular Broach (shear angle) for sizing rectangular hole, radar wave guide flanges etc.

FIGURE 26-8   Examples of push-type and pull-type broaches. (Courtesy DuMont Corp.)

is desirable to have the pitch sufficiently small so that at least two or three teeth are cutting at all times. See Figure 26-9.

The *hook* determines the primary rake angle and is a function of the material being cut. It is 15 to 20 degrees for steel and 6 to 8 degrees for cast iron. *Back-off* or end clearance angles are from 1 to 3 degrees to prevent rubbing.

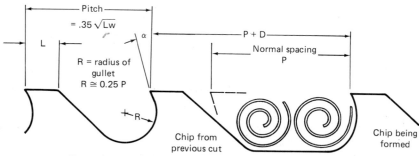

Extra-wide spacing may be used when chip disposal is a problem
Chip adhering to broach tooth will be displaced by the next chip formed

FIGURE 26-9   The gullet area provides room for the chips.

Most of the metal removal is done by the *roughing teeth. Semifinishing teeth* provide surface smoothness, whereas *finishing teeth* produce exact size. On a new broach all the finishing teeth usually are made the same size. As the first finishing teeth become worn, those behind continue the sizing function. On some round broaches, *burnishing teeth* are provided for finishing. These have no cutting edges but are rounded disks that are from 0.001 to 0.003 in. larger than the size of the hole. The resulting rubbing action smooths and sizes the hole. They are used primarily on cast iron and nonferrous metals.

The *pull end* of a broach provides a means of quickly attaching the broach to the pulling mechanism. The *front pilot* aligns the broach in the hole before it begins to cut, and the *rear pilot* keeps the tool square with the finished hole as it leaves the workpiece. *Shank length* must be sufficient to permit the broach to pass through the workpiece and be attached to the puller before the first roughing tooth engages the work. If a broach is to be used on a vertical machine that has a tool-handling mechanism, a *tail* is necessary.

A broach should not be used to remove a greater depth of metal than that for which it is designed—the sum of the steps of all the teeth. In designing workpieces, a minimum of 0.020 in. should be provided on surfaces that are to be broached, and about ¼ in. is the practical maximum.

Broaching Speeds.   Broaching speeds are relatively low (25 to 20 sfpm), seldom exceeding 50 feet per minute. However, because a surface usually is completed in a single stroke, the productivity is high. A complete cycle usually requires only from 5 to 30 seconds, with most of that time being taken up by the return stroke, broach handling, and workpiece loading and unloading. Such cutting conditions facilitate cooling and lubrication and result in very slow tool wear which reduces the necessity for frequent resharpening and prolongs the life of the expensive broaching tool.

For a given cutting speed and material, the force required to pull or push a broach is a function of the tooth width, the step, and the number of teeth cutting. Consequently, it is necessary to design or specify a broach within the stroke length and power limitations of the machine on which it is to be used.

Broach Materials and Construction.   Because of the low cutting speeds employed, most broaches are made of alloy or high-speed tool steel, even with some mass-production work. TiN coatings of HSS broaches is becoming more common, greatly prolonging the life of broaches. When they are used in continuous mass-production machines, particularly in surface broaching, tungsten carbide teeth may be used, permitting them to be used for long periods of time without resharpening.

Most internal broaches are of solid construction. Quite often, however, they are made of *shells* mounted on an arbor (Figure 26-10). When the broach (or a section of it) is subject to rapid wear, a single shell can be replaced. This will

FIGURE 26-10 Shell construction for a pull broach.

be much cheaper than replacing an entire solid broach. Shell construction, however, is initially more expensive than a solid broach of comparable size.

Small surface broaches may be of solid construction, but larger ones are usually built up in sections, as shown in Figure 26-11. Sectional construction makes the broach easier and cheaper to construct and sharpen. It also often provides some degree of interchangeability of the sections.

Sharpening Broaches.   Most broaches are resharpened by grinding the hook faces of the teeth. The lands of internal broaches must not be reground because this would change the size of the broach. Lands of flat surface broaches sometimes are ground, in which case all of them must be ground to maintain their proper relationship.

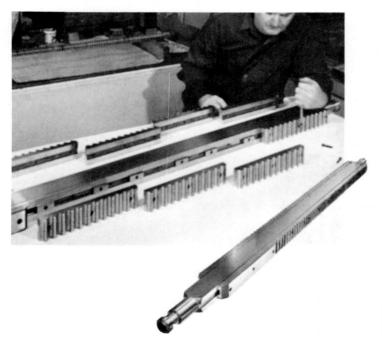

FIGURE 26-11 Eighty-inch-long broaches constructed from toothed inserts are cheaper to build and sharpen.

Broaching Machines

Because all the factors that determine the shape of the machined surface and that determine all cutting conditions except speed are built into the broaching tool, broaching machines are relatively simple. Their primary functions are to impart plain reciprocating motion to the broach and to provide a means for automatically handling the broach.

Most broaching machines are driven hydraulically, although mechanical drive is used in a few special types. The major classification relates to whether the motion of the broach is vertical or horizontal, as follows:

Vertical	Horizontal	Rotary
Broaching presses (push broaching)	Pull	Special types
Pull down (Figure 26-2)	Surface	
Pull up	Continuous	
Surface		

The choice between vertical and horizontal machines is determined primarily by the length of the stroke required and the available floor space. Vertical machines seldom have strokes greater than 60 inches because of height limitations. Horizontal machines can have almost any length of stroke, but they require greater floor space.

Broaching Presses. As shown in Figure 26-12, *broaching presses* essentially are arbor presses with a guided ram. They are used with push broaches, have a capacity of from 5 to 50 tons, and are used only for internal broaching. The forward guide of the broach is inserted through the hole in the workpiece as it rests on the press table, often in a fixture. As the ram descends, it engages the upper end of the broach and pushes it through the work.

Broaching presses are relatively slow, in comparison with other broaching machines, but they are inexpensive, flexible, and can be used for other types of operations, such as bending and staking.

Vertical Pull-Down Machines. The major components of vertical pull-down machines are a worktable, usually having a spherical-seated workholder, a broach elevator above the table, and a pulling mechanism below the table. As shown in Figure 26-2, when the elevator raises the broach above the table the work can be placed into position. The elevator then lowers the pilot end of the broach through the hole in the workpiece, where it is engaged by the puller. The elevator then releases the upper end of the broach, and it is pulled through the workpiece. The workpieces are removed from the table, and the broach is raised upward to be engaged by the elevator mechanism. In some cases wherein machines have two rams, they are arranged so that one broach is being pulled down while the work is being unloaded and the broach raised at the other station.

FIGURE 26-12   Arbor press used to broach keyway in a gear.

In Figure 26-2, the part is being broached in two passes, first on the left, then on the right.

Vertical Pull-Up Machines.   In *vertical pull-up machines,* the pulling ram is above the worktable and the broach-handling mechanism below it. The work is placed in position, above the pilot, while the broach is lowered. The handling mechanism then raises the broach until it engages the puller head. As the broach is pulled upward, the work comes to rest against the underside of the table, where it is held until the broach has been pulled through. The work then falls free, often sliding down a chute into a tote bin.

Pull-up machines may have up to eight rams. Because the workpieces need only be placed in the machines, and the broach handling and work removal are automatic, they are highly productive. For certain types of work, automatic feeding can be provided.

Vertical Surface-Broaching Machines.   On *vertical surface-broaching machines* the broaches usually are mounted on guided slides to provide support against lateral thrust. Because there is no need for handling the broach, they are simpler but much heavier than pull- or push-broaching machines. Many have two or more slides so that work can be loaded at one while another part is machined at the other. The operating cycle is very short as there is no handling of the broach. Slide or rotary-indexing fixtures are usually used to hold the work. This reduces the work-handling time and minimizes the total cycle time.

Horizontal Broaching Machines.   The primary reason for employing a *horizontal* configuration for pull- and surface-broaching machines is to make possible longer strokes and the use of longer broaches than can be conveniently accommodated in vertical machines. Horizontal pull-broaching machines are vertical machines turned on their side. See Figure 26-13. When internal surfaces are to be broached, such as holes, the broach must have a diameter-to-length ratio large enough to make it self-supporting without appreciable deflection. Consequently, horizontal machines are seldom used for small holes. In surface broaching, the broach is always supported in guides and therefore no such limitation is encountered. See Figure 26-14. Broaching that requires rotation of the broach, as in rifling and spiral splines, usually is done on horizontal machines.

Continuous Surface-Broaching Machines.   In *continuous surface-broaching machines,* the broaches usually are stationary, and the work is pulled past the cutters by means of an endless conveyor. Fixtures are usually attached to the conveyor chain so that the workpieces can be placed in them at one end of the machine and removed at the other, sometimes automatically. Such machines are being used increasingly in mass production (see Figure 26-14).

Rotary Broaching Machines.   In *rotary broaching machines,* occasionally used in mass production, the broaches are stationary, and the work is passed

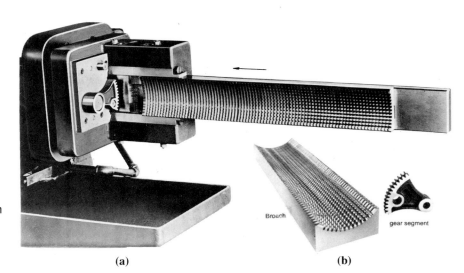

FIGURE 26-13 Broaching the teeth in a gear segment by horizontal surface broaching in one pass. (Courtesy Apex Broach and Machine Company.)

(a)   (b)

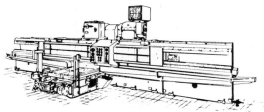

HORIZONTAL SURFACE BROACHING MACHINE

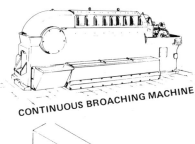

CONTINUOUS BROACHING MACHINE

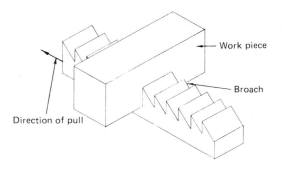

Work piece

Broach

Direction of pull

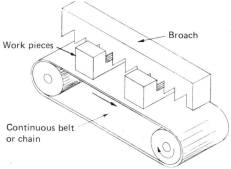

Work pieces

Broach

Continuous belt
or chain

FIGURE 26-14   Large horizontal surface broaching machine on left and continuous horizontal surface broacher on right. (Courtesy General Electric Company.)

beneath or between them. The work is held in fixtures on a rotary table. The advantage of these machines is that there is no lost time due to noncutting, reciprocating strokes.

Sawing

Sawing is a basic machining process in which chips are produced by a succession of small cutting edges, or *teeth,* arranged in a narrow line on a saw "blade." As shown in Figure 26-15, each tooth forms a chip progressively as it passes through the workpiece, and the chip is contained within the space between two successive teeth until these teeth pass from the work. Because sections of considerable size can be severed from the workpiece with the removal of only a small amount of the material in the form of chips, sawing is probably the most

V

Blade

Chips

Feed

Work material chips
(plastic)

FIGURE 26-15   Formation of chips in sawing. (Courtesy DoALL Company.)

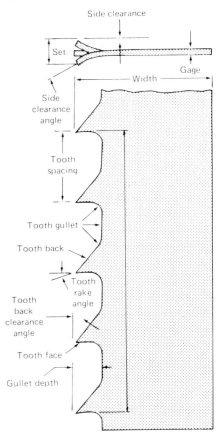

FIGURE 26-16   Standard nomenclature for a saw blade.

economical of the basic machining processes with respect to the waste of material and power consumption, and in many cases with respect to labor.

In recent years vast improvements have been made in saw blades and sawing machines resulting in improved accuracy and precision of the process. Most sawing is done to sever bar stock and shapes into desired lengths for use in other operations. There are many cases in which sawing is used to produce desired shapes. Frequently, and especially for producing only a few parts, contour sawing may be more economical than any other machining process.

Saw Blades.   Figure 26-16 gives the standard nomenclature for a saw blade. Saw blades are made in three basic configurations. The first, commonly called a *hacksaw* blade, is straight, relatively rigid, and of limited length with teeth on one edge. The second is sufficiently flexible so that a long length can be formed into a continuous band with teeth on one edge; these are known as *band-saw blades*. The third form is a rigid disk having teeth on the periphery; these are *circular saws*.

All saw blades have certain common and basic features. These are (1) material, (2) tooth form, (3) tooth spacing, (4) tooth set, and (5) blade thickness. Small hacksaw blades usually are made entirely of tungsten or molybdenum high-speed steel. Blades for power-operated hacksaws often are made with teeth cut from a strip of high-speed steel that has been electron-beam-welded to the heavy main portion of the blade, which is made from a tougher and cheaper alloy steel. Band-saw blades frequently are made with this same type of construction, as shown in Figure 26-17, but with the main portion of the blade made of relatively thin, high-tensile-strength alloy steel to provide the required flexibility. Band-saw blades also are available with tungsten carbide teeth and TiN coatings.

Three common *tooth forms* are shown in Figure 26-18.

Tooth spacing is very important in all sawing because it determines three factors. First, it controls the size of the teeth. From the viewpoint of strength,

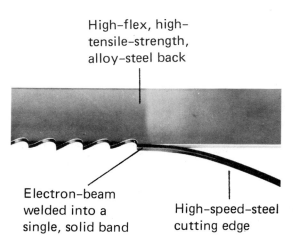

High–flex, high–tensile–strength, alloy–steel back

Electron–beam welded into a single, solid band

High–speed–steel cutting edge

FIGURE 26-17   Method of providing HSS teeth on a softer steel band. (Courtesy DoALL Company.)

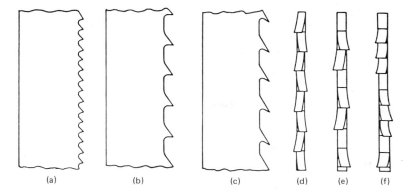

(a)   Regular or standard tooth for ferrous metal, general purpose cutting
(b)   Skip tooth with large gullets for machining softer, nonferrous metals
(c)   Hook (10° positive rache) tooth for harder nonferrous alloys
(d)   Symmetrical tooth straight set saw for brass & plastic
(e)   Raker tooth set for general purpose sawing, uniform thickness
(f)   Wavy tooth set for thin, or non uniform thickness of stock.

FIGURE 26-18   Types of blades used in sawing.

large teeth are desirable. Second, tooth spacing determines the space *(gullet)* available to contain the chip that is formed. As shown in Figure 26-15, the chip cannot drop from this space until it emerges from the slot, or *kerf*. The space must be such that there is no crowding of the chip and no tendency for chips to become wedged between the teeth and not drop out. Third, tooth spacing determines how many teeth will bear against the work. This is very important in cutting thin material, such as tubing. At least two teeth should be in contact with the work at all times. If the teeth are too coarse, only one tooth rests on the work at a given time, permitting the saw to rock, and the teeth may be stripped from the saw. Hand hacksaw blades have 14 to 32 teeth per inch. In order to make it easier to start a cut, some hand hacksaw blades are made with a short section at the forward end having teeth of a special form with negative rake angles. Tooth spacing for power hacksaw blades ranges from 4 to 18 teeth per inch.

Tooth set, illustrated in Figure 26-18, refers to the manner in which the teeth are offset from the center line in order to make a cut (kerf) wider than the thickness of the back portions of the blade. This permits the saw to move more freely in the kerf and reduces friction and heating. *Raker-tooth saws* are used in cutting most steel and iron. *Straight-set teeth* are used for sawing brass, copper, and plastics. Saws with *wavy-set teeth* are used primarily for cutting thin sheets and thin-walled tubing.

The *blade thickness* of nearly all hand hacksaw blades is 0.025 in. Saw blades for power hacksaws vary in thickness from 0.050 to 0.100 in.

Hand hacksaw blades come in two standard lengths, 10 and 12 in. All are ½ in. wide. Blades for power hacksaws vary in length from 12 to 24 in. and in width from 1 to 2 in. Wider and thicker blades are desirable for heavy-duty work. The blade should be at least twice as long as the maximum length of cut that is to be made.

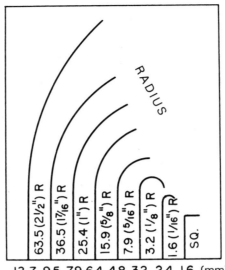

FIGURE 26-19 Relationship of bandsaw width to the minimum radius that can be cut.

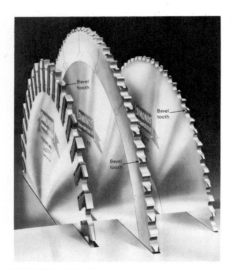

FIGURE 26-20 (Left to right) The inserted tooth, segmental-tooth, and integral-tooth forms of saw constructions. (Courtesy Simonds Saw and Steel Company.)

Band-saw blades are available in straight, raker, wave, or combination sets. In order to reduce the noise from high-speed bandsawing, it is becoming increasingly common to use blades that have more than one pitch, size of teeth, and type of set. Blade width is very important in bandsawing because it determines the minimum radius that can be cut. This relationship is illustrated in Figure 26-19. The most common widths are from ¹⁄₁₆ to ½ in., although wider blades can be obtained. Because wider blades are stronger, as wide a blade as possible should be used. Consequently, cutting small radii requires a narrower and weaker blade and is more time-consuming.

Band-saw blades come in tooth spacings from 2 to 32 teeth per inch.

Disc or *circular saws* necessarily differ somewhat from straight blade forms. Because they must be relatively large in comparison with the work, only the sizes up to about 18 inches in diameter have teeth that are cut into the disc. See Figure 26-20. Larger saws use either *segmented* or *inserted* teeth. The teeth are made of high-speed steel, or tungsten carbide. The remainder of the disc is made of ordinary, less expensive, and tougher steel. *Segmental* blades are composed of segments mounted around the periphery of the disc, usually fitted with a tongue and groove and fastened by means of screws or rivets. Each segment contains several teeth. If a single tooth is broken, only one segment need be replaced to restore the saw to operating conditions.

Figure 26-20 shows a common tooth form used in circular saws, in which every other tooth is beveled on both sides. Sometimes the first tooth is beveled on the left side, the second tooth on both sides, the third tooth on the right side, the fourth tooth on the left side, and so forth. Another method is to bevel the opposite sides of successive teeth. Beveling is done to produce a smoother cut. Precision circular saws made from carbide are becoming available which are very thin (.03 in.) and have high cutting-off accuracy, around ± .0008 in. with negligible burrs.

Circular saws for cutting metal are often called *cold saws* to distinguish them from friction-type disc saws that heat the metal to the melting temperature at the point of metal removal. Cold saws cut rapidly and produce surfaces that are comparable in smoothness and accuracy with surfaces made by slitting saws in a milling machine or by a cutoff tool in a lathe.

Types of Sawing Machines. Metal-sawing machines may be classified as follows:

1. Reciprocating saw.
   a. Manual hacksaw.
   b. Power hacksaw.
2. Band saw.
   a. Vertical cutoff.
   b. Horizontal cutoff.
   c. Combination cutoff and contour.
   d. Friction.
3. Circular saw.
   a. Cold saw.
   b. Steel friction disc.
   c. Abrasive disc.

Power Hacksaws. As the name implies, power hacksaws are machines that mechanically reciprocate a large hacksaw blade. As shown in Figure 26-21, they consist of a bed, a work-holding frame, a power mechanism for reciprocating the saw frame, and some type of feeding mechanism. Because of the inherent inefficiency of cutting in only one stroke direction, they have often been replaced by more efficient, horizontal band-sawing machines.

Band-Sawing Machines. The earliest metal-cutting band-sawing machines were direct adaptations from wood-cutting band saws. Modern machines of this type are much more sophisticated and versatile and have been developed specifically for metal cutting. To a large degree, they were made possible by the development of vastly better and more flexible band-saw blades and simple flash-welding equipment, which can weld the two ends of a strip of band-saw blade together to form a band of any desired length. Three basic types of band-sawing machines are in common use.

Upright cutoff band-sawing machines, such as shown in Figure 26-22 are designed primarily for cutoff work on single stationary workpieces that can be held on a table. On many machines, the blade mechanism can be tilted to about

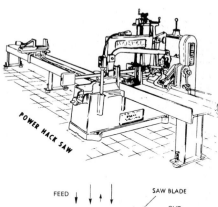

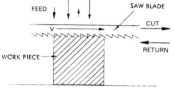

FIGURE 26-21 Power hacksaw with reciprocating blade. (From *Manufacturing Producibility Handbook*; courtesy General Electric Company.)

FIGURE 26-22 Cutting a pipe at a 45-degree angle on an upright cutoff bandsawing machine. (Courtesy Armstrong-Blum Mfg. Co.)

45 degrees, as shown, to permit cutting at an angle. They usually have automatic power feed of the blade into the work, automatic stops, and provision for supplying coolant.

Horizontal metal-cutting band-sawing machines were developed to combine the flexibility of reciprocating power hacksaws and the continuous cutting action of vertical band saws. These heavy-duty automatic bandsaws feed the saw vertically by a hydraulic mechanism and have automatic stock feed that can be set to feed the stock laterally any desired distance after a cut is completed and automatically clamp it for the next cut. Such machines can be arranged to hold, clamp, and cut several bars of material simultaneously. CNC bandsaws are available with automatic storage and retrieval systems for the bar stock. Smaller and less expensive types have swing-frame construction, with the band-saw head

Blade installation

(b)

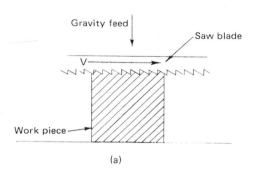

Gravity feed

Saw blade

V

Work piece

(a)

FIGURE 26-23 (Left) Horizontal band-sawing machine sawing an I-beam. (Right) Horizontal band saw with machine blade guards removed for blade installation.

mounted in a pivot on the rear of the machine. Feed is accomplished by gravity through rotation of the head about the pivot point. Because of their continuous cutting action, horizontal band-sawing machines are very efficient (see Figure 26-23).

Combination cutoff and contour band-sawing machines, such as shown in Figure 26-24, can be used not only for cutoff work but also for contour sawing. They are widely used for cutting irregular shapes in connection with making dies and the production of small numbers of parts and are often equipped with rotary tables. Additional features on these machines include a table that pivots so that it can be tilted to any angle up to 45 degrees. Second, a small flash welder is provided, on the vertical column, so that a straight length of band-saw blade can be welded quickly into a continuous band. A small grinding wheel is located beneath the welder so that the flash can be ground from the weld to provide a smooth joint that will pass through the saw guides. This welding and grinding unit makes it possible to cut internal openings by drilling a hole, inserting one end of the saw blade through the hole, and then butt-welding the two ends together. When the cut is finished, the band is cut apart and removed from the opening. Third, the speed of the saw blade can be varied continuously over a wide range to provide correct operating conditions for any

FIGURE 26-24 Contour band sawing on vertical band-sawing machine. Schematic of bandsaw shown in inset.

material. Fourth, a method of power feeding the work is provided, sometimes gravity-actuated.

Contour-sawing machines are made in a wide range of sizes, the principal size dimension being the throat depth. Sizes from 12 to 72 in. are available. The speeds available on most machines range from about 50 to 2000 feet per minute. Modern horizontal band saws are accurate to ± .002 in. per vertical inch of cut but have feeding accuracy of only ± .005 in. subject to the size of the stock and the rate of the feed. Repeatability from one feed to the next may be ± .010 to ± .020 in.

Sawing centers with microprocessor controls have opened up new automation aspects for sawing. Such control systems can improve accuracy to within ± .005 in. over entire cuts by controlling saw speed, blade feed pressure, and feed rate.

Special band-sawing machines are available with very high speed ranges, up to 14,000 feet per minute. These are known as *friction bandsawing machines*. Material is not cut by chip formation. Instead, the friction between the rapidly moving saw and the work is sufficient to raise the temperature of the material at the end of the kerf to or just below the melting point where its strength is very low. The saw blade then pulls the molten, or weakened, material out of the kerf. Consequently, the blades do not need to be sharp; they frequently have no teeth, only occasional notches in the blade to aid in removing the metal.

Almost any material, including ceramics, can be cut by friction sawing. Because only a small portion of the blade is in contact with the work for an instant and then is cooled by its passage through the air, it remains cool. Usually the major portion of the work, away from contact with the saw blade, also remains quite cool. The metal adjacent to the kerf is heat-affected, recast, and sometimes harder than the bulk metal. Friction sawing is also a very rapid method for trimming the flash from sheet-metal parts and castings.

Circular-Blade Sawing Machines. Machines employing rotating circular saw blades are used exclusively for cutoff work. These range from small, simple types, in which the saw is fed manually, to very large saws having power feed and built-in coolant systems, commonly used for cutting off hot-rolled shapes as they come from a rolling mill, such as shown in Figure 26-25. In some cases friction saws are used for this purpose, having discs up to 6 feet in diameter and operating at surface speeds up to 25,000 feet per minute. Steel sections up to 24 inches can be cut in less than one minute by this technique.

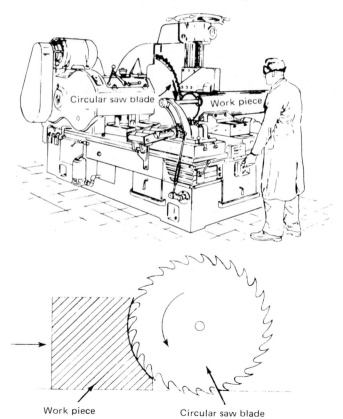

FIGURE 26-25 Circular or cold sawing cutting off large tubing.

Although technically not a sawing operation, cutoff work up to about 6 inches is often done utilizing thin *abrasive* discs. The equipment used is the same as for sawing. It has the advantage that very hard materials that would be very difficult to saw can be cut readily. A thin rubber- or resinoid-bonded abrasive wheel is used. Usually a somewhat smoother surface is produced.

Cutting Fluids.   Cutting fluids should be used for all bandsawing with the exception that cast iron is always cut dry. Commercially available oils or light cutting oils will give good results in cutting ferrous materials. Beeswax or paraffin are common lubricants for cutting aluminum and aluminum alloys.

Feed and Speed.   Because of the many different types of feed involved in bandsawing, it is not practical to provide tabular feed or pressure data. Under general conditions, however, an even pressure, without forcing the work, gives best results.

A free-cut curl in the chips indicates ideal feed pressure. Burned or discolored chips indicate excessive pressure, which can cause tooth breakage and premature wear.

In general, HSS blades are run at 200 to 300 fpm when cutting 1-in. thick low- and medium-carbon steels. For high-carbon steels, alloy steels, and tool and die steels, the range is from 150 to 225 fpm, and most stainless steels are cut at 100 to 125 fpm.

Filing

Basically, the metal-removing action in filing is the same as in sawing, in that chips are removed by cutting teeth that are arranged in succession along the same plane on the surface of a tool, called a *file*. There are two differences: (1) the chips are very small and therefore the cutting action is slow and easily controlled, and (2) the cutting teeth are much wider. Consequently, fine and accurate work can be done.

Types of Files.   Files are classified according to the following:

1. The type, or *cut*, of the teeth.
2. The degree of coarseness of the teeth.
3. Construction.
   a. Single solid units for hand use or in die-filing machines.
   b. Band segments, for use in band-filing machines.
   c. Discs, for use in disc-filing machines.

Four types of *cuts* are available. *Single-cut files* have rows of parallel teeth that extend across the entire width of the file at the angle of from 65 to 85 degrees. *Double-cut files* have two series of parallel teeth that extend across the width of the file. One series is cut at an angle of 40 to 45 degrees. The other series is coarser and is cut at an opposite angle that varies from about 10 to 80 degrees. A *vixen-cut file* has a series of parallel curved teeth, each extending

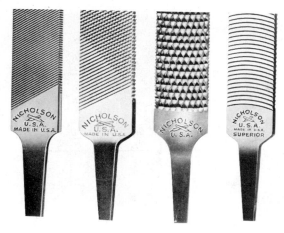

FIGURE 26-26   Four cuts of files. (Left to right) Single, double, rasp, and curved (Vixen). (Courtesy Nicholson File Company.)

across the file face. On a *rasp-cut* file, each tooth is short and is raised out of the surface by means of a punch. These four types of cuts are shown in Figure 26-26.

The coarseness of files is designated by the following terms, arranged in order of increasing coarseness: *dead smooth, smooth, second cut, bastard, coarse,* and *rough*. There is also a series of finer Swiss pattern files, designated by numbers from 00 to 8.

Files are available in a number of cross-sectional shapes: *flat, round, square, triangular,* and *half-round*. Flat files can be obtained with no teeth on one or both narrow edges, known as *safe edges,* so as to prevent material from being removed from a surface that is normal to the one being filed.

Most files for hand filing are from 10 to 14 in. in length and have a pointed *tang* at one end on which a wood or metal handle can be fitted for easy grasping.

Filing Machines.   Although an experienced machinist can do very accurate work by hand filing, it is a slow and tiresome task. Consequently, three types of filing machines have been developed that permit quite accurate results to be obtained rapidly and with much less effort. *Die-filing machines,* shown in Figure 26-27, hold and reciprocate a file that extends upward through the worktable. The file rides against a roller guide at its upper end, and cutting occurs on the downward stroke; therefore the cutting force tends to hold the work against the table. The table can be tilted to any desired angle. Such machines operate at from 300 to 500 strokes per minute, and the resulting surface tends to be at a uniform angle with respect to the table. Quite accurate work can be done. Because of the reciprocating action, approximately 50% of the operating time is nonproductive.

Band-filing machines provide continuous cutting action. Most band filing is done on contour band-sawing machines by means of a special band file that is substituted for the usual band-saw blade.

FIGURE 26-27   Die-filing machine.

(a)

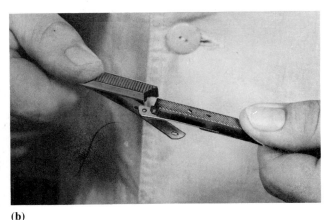

(b)

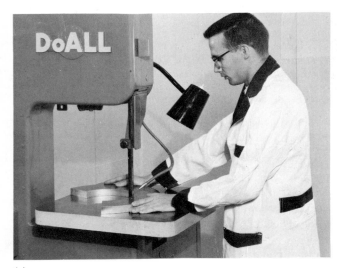

(c)

FIGURE 26-28   Band file segments (top) are joined together to form a continuous band which runs on a band-filing machine. (Courtesy DoALL Co.)

The principle of a band file is shown in Figure 26-28. Rigid, straight file segments, about 3 in. long, are riveted to a flexible steel band near their leading ends. One end of the steel band contains a slot that can be hooked over a pin in the other end to form a continuous band. As the band passes over the drive and idler wheels of the machine, it flexes so that the ends of adjacent file segments move apart. When the band becomes straight, the ends of adjacent segments move together and interlock to form a continuous straight file. Where the file passes through the worktable, it is guided and supported by a grooved guide, which provides the necessary support to resist the pressure of the work against the file.

Band files are available in most of the standard cuts and in several widths and shapes. Operating speeds range from about 50 to 250 feet per minute.

Although band filing is considerably more rapid than can be done on a die-filing machine, it usually is not quite as accurate. Frequently, band filing may be followed by some finish filing on a die-filing machine.

Some *disc-filing machines* are used, having files in the form of discs, as shown in Figure 26-29. These are even simpler than die-filing machines and provide continuous cutting action. However, it is difficult to obtain accurate results by their use.

FIGURE 26-29   (Above) Disc-type filing machine. (Left) Some of the available types of disc files. (Courtesy Jersey Manufacturing Company.)

Review Questions

1. What is unique about broaching, as compared with the other basic machining processes?
2. Why can a thick saw blade not be used as a broach?
3. Broaching machines are simpler in a basic design than most other machine tools. Why is this?
4. Why is broaching particularly well suited for mass production?
5. Explain how internal spiral grooves are produced by broaching.
6. Why is it necessary to relate the design of a broach to the specific workpiece that is to be machined?
7. What two methods can be utilized to reduce the force and power requirements for a particular broaching cut?
8. For a given job, how would a broach having rotor-tooth design compare in length with one having regular, full-width teeth?
9. Why are the pitch and radius between teeth on a broach of importance?
10. Why are broaching speeds usually relatively low, as compared with other machining operations?
11. What are the advantages of shell-type broach construction?
12. Why are most broaches made from alloy or high-speed steel rather than from tungsten carbide?

13. What are the advantages of TiN-coated broaching tools?
14. For mass-production operations, which process is preferred, pull-up broaching or pull-down broaching?
15. Can continuous broaching machines be used for broaching holes? Why or why not?
16. The sides of a square, dead-end hole must be machined all the way to the bottom. The hole is drilled to full depth and the bottom end milled flat. Is it possible to machine the hole square by broaching? Why or why not?
17. The interior flat surfaces of socket wrenches, which have one ''closed'' end, often are finished to size by broaching. By examining one of these, determine what design modification was incorporated to make this operation possible.
18. Why is sawing one of the most efficient of the chip-forming processes?
19. Explain why tooth spacing is important in sawing.
20. What is the tooth gullet?
21. Explain what is meant by the ''set'' of the teeth on a saw blade.
22. Why can a band-saw blade not be hardened throughout the entire width of the band?
23. What is the general relationshp between the width of a band-saw blade and the radius of a cut that can be made with it?

24. What are the advantages and disadvantages of circular saws?
25. Why have band-sawing machines largely replaced those using reciprocating saws?
26. Explain how the hole shown in Figure 26-24 is made on a band-sawing machine.
27. How does friction sawing differ from ordinary band sawing?
28. What is the disadvantage of using gravity to feed a saw in cutting round bar stock?
29. To what extent is filing different from sawing?
30. What is a safe edge on a file?
31. Why is an end-filing machine more efficient than a die-filing machine?
32. How does a rasp-cut file differ from other types of files?

Problems

1. A surface 12 in. long is to be machined with a flat, solid broach that has a rise per tooth of 0.0047 in. What is the minimum cross-section area that must be provided in the chip gullet between adjacent teeth?
2. The pitch of the teeth on a simple surface broach can be determined by equation 26-1. If a broach is to remove 0.25 in. of material from a gray iron casting that is 3 in. wide and 17.75 in. long, and if each tooth has a rise per tooth of .004 in., what will be the length of the roughing section of the broach?
3. Estimate the (approximate maximum) horsepower needed to accomplish the operation described in problem #2 at a cutting speed of 10 meters per minute. (*Hint:* First find the HP used per tooth and determine the maximum number of teeth engaged at any time.)
4. Estimate the approximate force acting in the forward direction during cutting for the conditions stated in problems #2 and 3.
5. In making a 6-in. cut in a piece of AISI 1020 CR steel that is one inch thick, the material is fed to a bandsaw blade with teeth having a pitch of 1.27mm (20 pitch) at the rate of 0.0001 in. per tooth. Estimate the cutting time for the cut.
6. The strength of a pull broach is determined by its minimum cross section, which usually occurs either at the root of the first tooth or at the pull end. See Figure 26-1. Suppose the minimum root diameter is D_r, the pull end diameter is D_p and the width of the pull slot is W. Write an equation for the allowable pull in psi using 200,000 as the yield strength for the broach material.

Abrasive Machining Processes

Abrasive machining is the basic process in which chips are formed by the small cutting edges on abrasive particles, or abrasive grits. Unquestionably, abrasive machining is the oldest of the basic machining processes. Museums abound with examples of utensils, tools, and weapons that ancient man produced by rubbing hard stones against softer materials to abrade away unwanted portions, leaving desired shapes. For centuries, only natural abrasives were available, and abrasive machining was surpassed in importance and use by more modern, basic machining processes, which were developed around superior cutting materials. However, the development of man-made abrasives and a better fundamental understanding of the abrasive machining process has resulted in placing abrasive machining and its variations among the most important of all the basic machining processes.

The results that can be obtained by abrasive machining range from the finest and smoothest surfaces produced by any machining process, in which very little material is removed, to rough, coarse surfaces that accompany high material-removal rates.

The abrasive particles may be (1) free (see Chapter 32); (2) mounted in resin on a belt (called *coated product*); or (3) close packed into wheels or stones, with abrasives held together by bonding material (called *bonded product*). See Figure 27-1.

The metal-removal process is basically the same in all three cases but with important differences due to spacing of active grains (grains in contact with work) and the rigidity and degree of fixation of the grains. Table 27.1 summarizes the primary abrasive processes.

Abrasive machining processes have two unique characteristics. First, each cutting edge is very small and many of these edges can cut simultaneously. When suitable machines are employed, very fine cuts are possible, and fine

Introduction

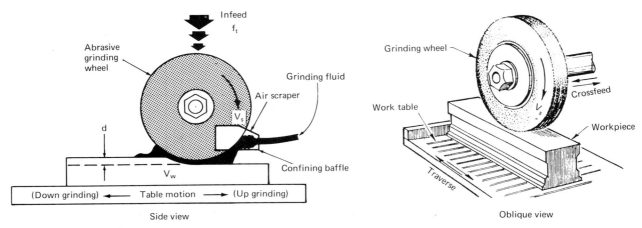

FIGURE 27-1 Schematic of a surface grinding process.

surfaces and close dimensional control can be obtained. Second, because extremely hard abrasive particles can be produced, very hard materials, such as hardened steel, glass, carbides, and ceramics, can be machined very readily. As a result, the abrasive machining processes are not only important as manufacturing processes; they are indeed essential. Many of our modern products, such as modern machine tools, automobiles, space vehicles, and aircraft, could not be manufactured without these processes.

Abrasives

An *abrasive* is a hard material that can cut or abrade other substances. Certain natural abrasives have existed from the earliest times. Sandstone was used by ancient peoples to sharpen tools and weapons. Early grinding wheels were cut

TABLE 27.1   Abrasive Machining Processes

Process	Particle Mounting	Features
Grinding	Bonded	Uses wheels, accurate sizing, finishing, low MRR. Can be done at high speeds (over 12,000 sfpm)
Creep feed grinding	Bonded open, soft	Uses wheels with long cutting arc, very slow feed and large depth of cut
Abrasive machining[a]	Bonded	High MRR, to obtain desired shapes and approximate sizes
Snagging	Bonded belted	High MRR, rough rapid technique to clean up and deburr castings, forgings
Honing	Bonded	"Stones" containing fine abrasives; primarily a hole-finishing process.
Lapping	Free	Fine particles embedded in soft metal or cloth; primarily a surface-finishing process

[a]The term *abrasive machining* applied to one particular form of the grinding process is unfortunate, because all these processes involve machining with abrasives.

from slabs of sandstone, but, because they were not uniform throughout, they wore unevenly and did not produce consistent results. Emery, a mixture of alumina (Al_2O_3) and magnetite (Fe_3O_4), is a natural abrasive used on coated paper and cloth. Corundum (natural Al_2O_3) and diamonds are other naturally occurring abrasive materials. However, the development of artificial abrasives having known uniform properties has permitted abrasive processes to become precision manufacturing processes.

Today, the only natural abrasives that have commercial importance are quartz, sand, garnets, and diamonds. Quartz sand is used primarily in coated abrasives and in air blasting (which will be discussed in Chapter 40), but artificial abrasives are also making inroads in these applications.

Hardness, the ability to resist penetration, is an important property for an abrasive. Table 27.2 lists the primary abrasives and their approximate Knoop hardness (kg/mm^2).

Two other properties are significant in abrasive grits. *Attrition* or fine, abrasive wear action of the grits results in dulled edges, grit flattening, and wheel glazing. *Friability* refers to the fracture of the grits and is the opposite of toughness. In grinding, it is important that grits be able to fracture to expose new, sharp edges.

Diamonds are the hardest of all materials. Those that are used for abrasives are either natural, off-color stones (called garnets) that are not suitable for gems, or small, synthetic stones that are produced specifically for abrasive purposes. Man-made stones appear to be somewhat more friable and thus tend to cut faster and cooler. They do not perform as satisfactorily in metal-bonded wheels. Diamond abrasive wheels are used extensively for sharpening carbide and ceramic cutting tools. Diamonds also are used for truing and dressing other types of abrasive wheels. Obviously, because of their cost, diamonds are used only when cheaper abrasives will not produce the desired results.

Garnets are used primarily in the form of very finely crushed and graded powders for fine polishing.

Artificial abrasives date from 1891, when E. G. Acheson, while attempting to produce precious gems, discovered how to make *silicon carbide* (SiC). Silicon

TABLE 27.2   Knoop Hardness Values for Common Abrasives

Abrasive Material	Year of Discovery	Hardness (Knoop)	Comment/Uses
Quartz (SiO_2)	?	320	Sand blasting
Aluminum oxide	1893	2100	Softer and tougher than silicon carbide; used on steel, iron, brass
Silicon carbide	1891	2400	Used for brass, bronze, aluminum, stainless steel, and cast iron
Borazon (cubic boron nitride)	1962	4700	For grinding hard, tough tool steels, stainless steel, aerospace metals, hard coatings
Diamond (synthetic)	1955	7000	Used to grind tungsten carbide and some die steels

FIGURE 27-2 Loose abrasive grains at high magnification, showing their irregular, sharp cutting edges. (Courtesy Norton Company.)

carbide is made by charging an electric furnace with silica sand, petroleum coke, salt, and sawdust. By passing large amounts of current through the charge, a temperature of over 4000°F is maintained for several hours, and a solid mass of silicon carbide crystals results. After the furnace has cooled, the mass of crystals is removed, crushed, and graded to various desired sizes. As can be seen in Figure 27-2, the resulting crystals, or grains, are irregular in shape with cutting edges having every possible rake angle.

Silicon carbide crystals are very hard, friable, and rather brittle. This limits their use. Silicon carbide is sold under the trade names Carborundum and Crystolon.

Aluminum oxide (Al_2O_3) is the most widely used artificial abrasive. Also produced in an arc furnace, from bauxite, iron filings, and small amounts of coke, it contains aluminum hydroxide, ferric oxide, silica, and other impurities. The mass of aluminum oxide that is formed is crushed, and the particles are graded to size. Common trade names for aluminum oxide abrasives are Alundum and Aloxite.

Although aluminum oxide is softer than silicon carbide, it is considerably tougher. Consequently, it is a better general-purpose abrasive.

Cubic boron nitride (CBN) is not found in nature. It is produced by a combination of intensive heat and pressure in the presence of a catalyst. CBN is extremely hard, registering at 4700 on the Knoop scale. It is the second hardest substance created by man or nature. Hardness, however, is not everything.

CBN far surpasses diamond in the important characteristic of thermal resistance. At temperatures of 650°C, at which diamond may begin to revert to plain carbon dioxide, CBN continues to maintain its hardness and chemical integrity. When the temperature of 1400°C is reached, CBN changes from its cubic form to a hexagonal form and loses hardness. CBN can be used successfully in grinding iron, steel, alloys of iron, Ni-based alloys, and other materials. CBN works very effectively (long wheel life, high G-ratio, good surface quality, no burn or chatter, low scrap rate, and overall increase in parts/shift) on hardened materials (R_c50 or higher). It can also be used for soft steel under selected situations. CBN does well at conventional grinding speeds (6,000 to 12,000 fpm), resulting in lower total grinding cost/piece in conventional equipment. CBN can also perform well at high grinding speeds (12,000 fpm and higher) and will enhance the benefits from future machine tools. CBN can solve difficult-to-grind jobs, but it also generates cost benefits in many production grinding operations despite its higher cost. CBN is manufactured by the General Electric Company under the trade name of Borazon.

Abrasive Grain Size and Geometry. To enhance the process capability of grinding, abrasive grains are sorted into sizes by mechanical sieving machines. The number of openings per linear inch in a sieve (or screen) through which most of the particles of a particular size can pass determines the grain size. See Figure 27-3.

A 24 grit would pass through a standard screen having 24 openings per inch but would not pass through one having 30 openings per inch. These numbers

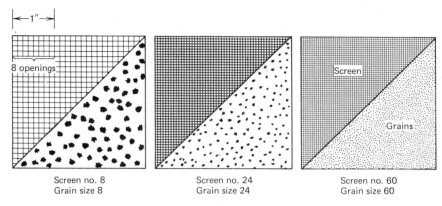

Screen no. 8
Grain size 8

Screen no. 24
Grain size 24

Screen no. 60
Grain size 60

FIGURE 27-3 Typical screens for sifting abrasives into sizes. (Courtesy Carborundum Company.)

have since been specified in terms of millimeters and micrometers (See ANSI B74.12 for details). Commercial practice commonly designates grain sizes from 4 to 24, inclusive, as *coarse;* 30 to 60, inclusive, as *medium;* and 70 to 600, inclusive, as *fine.* Silicon carbide is obtainable in grit sizes ranging from 2 to 240 and aluminum oxide in sizes from 4 to 240. Sizes from 240 to 600 are designated as *flour* sizes. These are used primarily for lapping, or in fine honing stones.

The grain diameter can be estimated from the screen number (S), which corresponds to the number of openings per inch. The mean diameter of the grain (g) is related to the screen number by $g \cong 0.7/S$.

Regardless of the size of the grain, only a small percentage (2% to 5%) of the surface of the grain is operative at any one time. That is, the depth of cut for an individual grain (the actual feed per grit) with respect to the grain diameter is very small. Thus, the chips are small. As the grain diameter decreases, the number of active grains per unit area increases. The cuts become finer because grain size is the controlling factor for surface finish (roughness). Of course, the MRR also decreases.

The grain shape is also important because it determines the tool geometry— that is, the back rake angle and the clearance angle at the cutting edge of the grit. See Figure 27-4.

FIGURE 27-4 The rake angle for abrasive particles can be positive, zero or negative. The cavity must be large enough to hold all the chips during the cut.

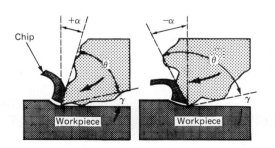

Chip

Workpiece

Workpiece

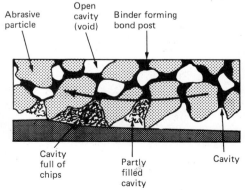

Abrasive particle

Open cavity (void)

Binder forming bond post

Cavity full of chips

Partly filled cavity

Cavity

In Figure 27-4, γ is the clearance angle, θ is the wedge angle, and α is the rake angle. The cavities between the grits provide space for the chips. The volume of the cavities must be greater than the volume of the chips.

Obviously, there is no specific rake angle but rather a distribution of angles. Thus, a grinding wheel can present to the surface rake angles ranging from $+45$ degrees to negative 60 degrees or greater. Grits with large negative rake angles or rounded cutting edges do not form chips but rather *plow* a groove in the surface (see Figures 27-5 and 27-6) or just *rub* it. Thus, abrasive machining is a mixture of *cutting, plowing, and rubbing* with the percentage of each being highly dependent on the geometry of the grit. As the grits are continuously abraded, fractured, or dislodged from the bond, new grits are exposed and the mixture of cutting, plowing, and rubbing is continuously changing. One hundred percent of the energy of rubbing and plowing goes into the workpiece. In cutting, 95–98% of the energy (the heat) goes into the chip.

Grinding

Grinding, wherein the abrasives are bonded together into a wheel, is the most common abrasive machining process. The performance of grinding wheels is greatly affected by the bonding material and the spatial arrangement of the particles, known as the structure.

Grinding Wheel Structure. The spacing of the abrasive particles with respect to each other is called *structure*. Close-packed grains have dense structure; open structure means widely spaced grains. Open-structure wheels have larger chip cavities but fewer cutting edges per unit area (see Figure 27-7).

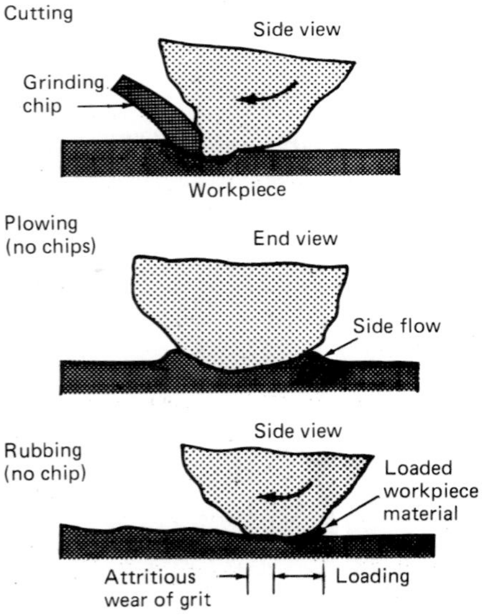

In grinding, the chips are small but are formed by the same basic mechanism of compression and shear as discussed in Chapter 21 for regular metal cutting. Figure 27-8 shows steel chips from a grinding process, at high magnification. They show the same shear front—lamella structure as chips from other machining processes. The chips often have sufficient heat energy to burn or melt in the atmosphere. Burning chips are the sparks observed during grinding with no cutting fluid. The feeds and depths of cut in grinding are small while the cutting speeds are high, resulting in high specific horsepower numbers. Because cutting is obviously more efficient than plowing or rubbing, grain fracture and grain pullout are natural phenomena used to keep the grains sharp. As the grains become dull, cutting forces increase and there is an increased tendency for the grains to fracture or break free from the bonding material. The latter action can be controlled by varying the strength of the bond, known as the *grade*. Grade thus is a measure of how strongly the grains are held in the wheel. It really is dependent on two factors: the strength of the bonding materials and the amount of the bonding agent connecting the grains. This latter factor is illustrated in Figure 27-9. Abrasive wheels are usually porous. The grains are held together with "posts" of bonding material. If these posts are large in cross section, the force required to break a grain free from the wheel is greater than when the posts are small. If a high dislodging force is required, the bond is said to be *hard*. If only a small force is required, the bond is said to be *soft*. Wheels

FIGURE 27-5 The grits interact with the surface three ways: cutting, plowing, and rubbing.

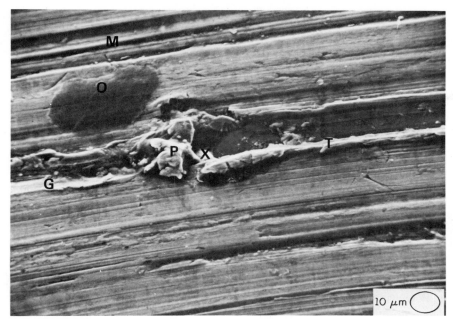

FIGURE 27-6 SEM micrograph of a ground steel surface showing a plowed track (T) in the middle and a machined track (M) above. The grit fractured, leaving a portion of the grit in the surface (X), a prow formation (P), and a groove (G) where the fractured portion was pushed further across the surface. The area marked (O) is an oil deposit.

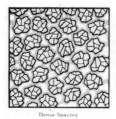
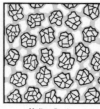
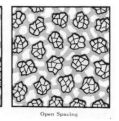

Dense Spacing   Medium Spacing   Open Spacing

FIGURE 27-7 Meaning of grid wheel "structure". (Courtesy Carborundum Company.)

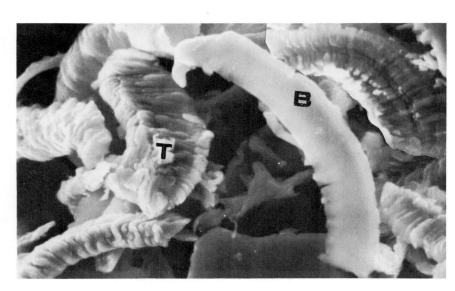

FIGURE 27-8 SEM micrograph of stainless steel chips from a grinding process. The tops (T) of the chips have the typical shear-front-lamella structure while the bottoms (B) are smooth; 4800x.

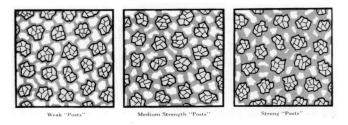

Weak "Posts"   Medium Strength "Posts"   Strong "Posts"

FIGURE 27-9 The grade of a grinding wheel depends on the amount of bonding agent that is holding the abrasive grains in the wheel. (Courtesy Carborundum Company.)

commonly are referred to as hard or soft, referring to the net strength of the bond resulting from both the strength of the bonding material and its disposition between the grains.

The loss of grains from the wheel means that the wheel is changing size. The grinding ratio or G ratio is defined as the cubic inches of stock removed divided by the cubic inches of wheel lost. In conventional grinding, G ratio is in the range of 20:1 to 80:1. The G ratio is a measure of grinding production and reflects the amount of work a wheel can do during its useful life.

A typical vitrified grinding wheel will consist of (in volume percentages) 50% abrasive particles, 10% bond, and 40% cavities. The manner in which the wheel performs is influenced by

1. The mean force required to dislodge a grain from the surface (the grade of the wheel).
2. The cavity size and distribution (the structure).
3. The mean spacing of active grains in the wheel surface (grain size).
4. The properties of the grain (hardness and friability).
5. The geometry of the cutting edges of the grains (rake angles and cutting edge radius compared to depth of cut).
6. The process parameters (speeds, feeds, cutting fluids) and type of grinding (surface, cylindrical).

Bonding Materials for Grinding Wheels.   Bonding material is a very important factor to be considered in selecting a grinding wheel. It determines the strength of the wheel, thus establishing the maximum operating speed. It determines the elastic behavior or deflection of the grits in the wheel during grinding. The wheel can be hard or rigid or it can be flexible. Finally, the bond determines the force required to dislodge an abrasive particle from the wheel and thus plays a major role in the cutting action. Bond materials are formulated so that the ratio of bond wear matches the rate of wear of the abrasive grits. Bonding materials in common use are

- *Vitrified bonds.* They are composed of clays and other ceramic substances. The abrasive particles are mixed with the wet clays so that each grain is coated. Wheels are formed from the mix, usually by pressing, and then dried. They are then fired in a kiln, which results in the bonding material's becoming hard and strong, having properties similar to glass. Vitrified wheels are porous, strong, rigid, and unaffected by oils, water, or temperature over the ranges usually encountered in metal cutting. The operating speed range in most cases is 5500 to 6500 feet per minute, but some wheels now operate at surface speeds up to 16,000 feet per minute.

- *Resinoid,* or phenolic resins. Because plastics can be compounded to have a wide range of properties, such wheels can be obtained to cover a variety of work conditions. They have, to a considerable extent, replaced shellac and rubber wheels.

    Composite materials are being used in rubber-bonded or resinoid-bonded wheels that are to have some degree of flexibility or are to receive considerable abuse and side loading. Various natural and synthetic fabrics and fibers, glass fibers, and nonferrous wire mesh are used for this purpose.

- *Silicate* wheels use silicate of soda (waterglass) as the bond material. The wheels are formed and then baked at about 500°F for a day or more. Because they are more brittle and not as strong as vitrified wheels, the abrasive grains are released more readily. Consequently, they machine at lower surface temperatures than vitrified wheels and are useful in grinding tools when heat must be kept to a minimum.

- *Shellac-bonded* wheels are made by mixing the abrasive grains with shellac in a heated mixture, pressing or rolling into the desired shapes, and baking for several hours at about 300°F. This type of bond is used primarily for strong, thin wheels having some elasticity. They tend to produce a high polish and thus have been used in grinding such parts as camshafts and mill rolls.

- *Rubber* bonding is used to produce wheels that can operate at high speeds but must have a considerable degree of flexibility so as to resist side thrust. Rubber, sulfur, and other vulcanizing agents are mixed with the abrasive grains. The mixture then is rolled out into sheets of the desired thickness, and the wheels are cut from these sheets and vulcanized. Rubber-bonded wheels can be operated at speeds up to 16,000 feet per minute. They commonly are used for snagging work in foundries and for thin cutoff wheels.

Abrasive Machining.   The condition wherein very rapid metal removal can be achieved by grinding is the one to which some have applied the term *abrasive machining*. The metal-removal rates are compared with, or exceed, those obtainable by milling or turning or broaching, and the size tolerances are comparable. It obviously is just a special type of grinding, using abrasive grains as cutting tools, as do all other types of abrasive machining. Abrasive grinding will produce sufficient localized plastic deformation and heat in the surface so as to develop tensile residual stresses, layers of overtempered martensite (in steels), and even microcracks because this process is quite abusive.

Low-Stress Grinding.   Conventional grinding should be replaced by procedures that develop lower surface stresses in those applications where service failures due to fatigue or stress corrosion are possible (see Figure 27-10). This is accomplished by employing softer grades of grinding wheels, reducing the grinding speeds and infeed rates, and using chemically active cutting fluids, like

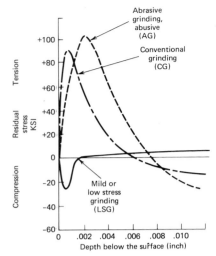

FIGURE 27-10 Typical residual stress distributions produced by surface grinding with different conditions. Material 4340 steel (From M. Field and W.P. Koster, "Surface Integrity in Grinding," in New Developments in Grinding, Carnegie Mellon University Press, Pittsburgh, Pa., 1972, p. 666.)

Grinding conditions

	Abusive AG	Conventional CG	Low stress LSG
Wheel	A46MV	A46KV	A46HV or A60IV
Wheel speed ft/min.	6,000 to 18,000	4,500 to 6,500	2500-3000
Down feed in/pass	.002 to .004	.001 to .003	.0002 to .005
Cross feed in/pass	.040 – .060	.040 – .060	.040-.060
Table speed ft/min	40 to 100	40 to 100	40 to 100
Fluid	Dry	Sol oil (1:20)	Sulfurized oil

highly sulfurized oil or KNO_2 in water. These procedures may require the addition of a variable-speed drive to the grinding machine. Generally, only about 0.005 to 0.010 in. of surface stock needs to be finish-ground in this way, as the depth of the surface damage due to conventional grinding or abusive grinding is 0.005 to 0.007 in. High-strength steels, high-temperature nickel, and cobalt-base alloys and titanium alloys are particularly sensitive to surface deformation and cracking problems from grinding. Other postprocessing processes, like polishing, honing, and chemical milling plus peening, can be used to remove the deformed layers in critically stressed parts. It is strongly recommended, however, that testing programs be used along with service experience on critical parts before these procedures are employed in production. See discussion of residual stresses in Chapter 40.

Dressing and Truing.   As the wheel is used, there is a tendency for the wheel to become *loaded* (metal chips become lodged in the cavities between the grains) and the grains get dull (grits wear, flatten, and polish). Unless the wheel is cleaned and sharpened (or *dressed*), the wheel will not cut as well and will tend to plow and rub more. The dulled grains cause the cutting forces on the grains to increase, ideally resulting in the grains' fracturing or being pulled out of the bond, thus providing a continuous exposure of sharp cutting edges. Such a continuous action ordinarily will not occur for light feeds and depths of cut. For heavier cuts, grinding wheels do become somewhat self-dressing, but workpiece burn occurs before the wheel reaches a fully dressed condition. This results in the scrapping of several workpieces before parts of good quality are ground. Figure 27-11 shows an arrangement for stick dressing a grinding wheel.

Grinding wheels lose their geometry during use. *Truing* restores the original shape. A single-point diamond tool can be used to *true* the wheel while fracturing abrasive grains to expose new grains and new cutting edges on worn, glazed

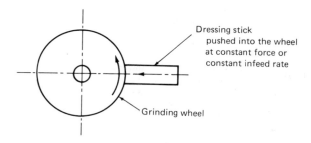

Dressing stick
pushed into the wheel
at constant force or
constant infeed rate

Grinding wheel

FIGURE 27-11   Schematic arrangement of stick dressing.

grains (see Figure 27-12). Truing can also be accomplished by grinding the grinding wheel with controlled-path or powered rotary devices using conventional abrasive wheels. The precision in generating a trued wheel surface by these methods is poorer than by the method described earlier.

Resin-bonded wheels can be trued by grinding with hard ceramics such as tungsten carbide. The procedure for truing and dressing a CBN wheel in a surface grinder might be as follows: Use .0002-in. downfeed per pass and cross-feed slightly more than half the wheel thickness at moderate table speeds. The wheel speed is the same as the grinding speed. The grinding power will gradually increase, as the wheel is getting dull, while being trued. When the power exceeds normal power drawn during workpiece grinding, stop the truing operation. Dress the wheel face open using a J-grade stick, with abrasive one grit size smaller than CBN. Continue the truing. Repeat this cycle until the wheel is completely trued.

Modern grinding machines are equipped so that the wheel can be dressed and/or trued continuously or intermittently while grinding continues. A common way to do this is by *crush dressing*. Crush dressing consists of forcing a hard steel roll, having the same contour as the part to be ground, against the grinding wheel while it is revolving—usually quite slowly. The crushing action fractures and dislodges some of the abrasive grains, exposing fresh sharp edges. This procedure usually is employed to produce and maintain a special contour to the abrasive wheel, as in form grinding. Crush dressing is a very rapid method of dressing grinding wheels and, because it fractures abrasive grains, results in free cutting and somewhat cooler grinding. The resulting surfaces may be slightly rougher than when diamond dressing is used.

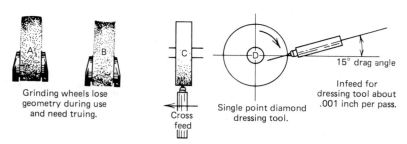

Grinding wheels lose
geometry during use
and need truing.

Cross
feed

Single point diamond
dressing tool.

15° drag angle

Infeed for
dressing tool about
.001 inch per pass.

FIGURE 27-12   Diamond nibs may be used for truing wheels in batch operations.

STANDARD MARKING SYSTEM CHART
ANSI STANDARD B74.13 – 1970

Sequence:	1	2	3	4	5	6
Prefix	Abrasive Type	Grain Size	Grade	Structure	Bond Type	Manufacturer's Record

51 — A — 36 — L — 5 — V — 23

PREFIX	ABRASIVE TYPE	ABRASIVE (GRAIN) SIZE				GRADE			STRUCTURE		BOND TYPE	MANUFACTURER'S RECORD
		COARSE	MEDIUM	FINE	VERY FINE	SOFT	MEDIUM	HARD	DENSE TO OPEN			
Manufacturer's symbol indicating exact kind of abrasive (use optional)	A - Aluminum Oxide C - Silicon Carbide	8 10 12 14 16 20 24	30 36 46 54 60	70 80 90 100 120 150 180	220 240 280 320 400 500 600	A E B F C G D H	I M J N K O L P	Q V R W S X T Y U Z	1 2 3 4 5 6 7 8	9 10 11 12 13 14 15 16 etc.	B - Resinoid BF - Resinoid Reinforced E - Shellac O - Oxychloride R - Rubber RF - Rubber Reinforced S - Silicate V - Vitrified	Manufacturer's private marking to identify wheel (use optional)
									(use optional)			

FIGURE 27-13 Standard marking system for grinding wheels.

Grinding-Wheel Identification.   Most grinding wheels are identified by a standard marking system that has been established by the American National Standards Institute, Inc. This system is illustrated and explained in Figure 27-13. The first and last symbols in the marking are left to the discretion of the manufacturer.

Grinding-Wheel Geometry.   The shape and size of the wheel are critical selection factors. Obviously, the shape must permit proper contact between the wheel and all of the surface that must be ground. Grinding-wheel shapes have been standardized and eight of the most commonly used types are shown in Figure 27-14. Types 1, 2, and 5 are used primarily for grinding external or internal cylindrical surfaces and for plain surface grinding. Type 2 can be mounted for grinding either on the periphery or the side of the wheel. Type 4 is used with tapered safety flanges so that if the wheel breaks during rough grinding, such as snagging, these flanges will prevent the pieces of the wheel from flying and causing damage. Type 6, the straight cup, is used primarily for surface grinding but can also be used for certain types of offhand grinding. The flaring-cup type of wheel is used for tool grinding. Dish-type wheels are used for grinding tools and saws.

Type 1, the straight grinding wheels, can be obtained with a variety of standard faces. Some of these are shown in Figure 27-15.

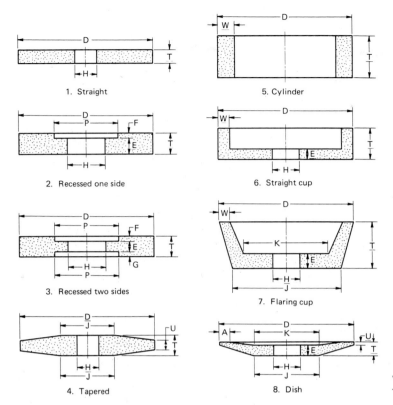

FIGURE 27-14 Standard grinding-wheel shapes commonly used. (Courtesy Carborundum Company.)

The size of the wheel to be used is determined, primarily, by the spindle rpms available on the grinding machine and the proper cutting speed for the wheel, as dictated by the type of bond. For most grinding operations the cutting speed is about 4500 to 6500 feet per minute, but different types and grades of bond often justify considerable deviation from this average speed. For certain types of work using special wheels and machines, as in thread grinding and "abrasive machining," much higher speeds are used.

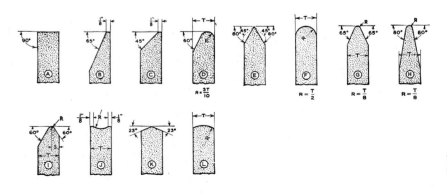

FIGURE 27-15 Standard face contours for straight grinding wheels. (Courtesy Carborundum Company.)

The operation for which the abrasive wheel is intended will also influence the wheel shape and size. The major use categories are

1. *Cutting off:* for slicing and slotting parts; use thin wheel, organic bond.
2. *Cylindrical between centers:* grinding outside diameters of cylindrical workpieces.
3. *Cylindrical, centerless:* grinding outside diameters with work rotated by regulating wheel.
4. *Internal cylindrical:* grinding bores and large holes.
5. *Snagging:* removing large amounts of metal without regard to surface finish or tolerances.
6. *Surface grinding:* grinding flat workpieces.
7. *Tool grinding:* for grinding cutting edges on tools like drills, milling cutters, taps, reamers, and single-point, high-speed steel tools.
8. *Off-hand grinding:* work or the grinding tool is hand-held.

In many cases, the classification of processes coincides with the classification of machines that do the process.

Other factors that will influence the choice of wheel to be selected include the workpiece material, the amount of stock to be removed, the shape of the workpiece, and the accuracy and surface finish desired.

Workpiece material has a great impact on choice of the wheel. Hard, high-strength metals (tool steels, alloy steels) are generally ground with aluminum oxide wheels or cubic boron nitride wheels. Silicon carbide and CBN are employed in grinding brittle materials (cast iron and ceramics) as well as softer, low-strength metals like aluminum, brass, copper, and bronze. Diamonds have taken over the cutting of tungsten carbides, and CBN is used for precision grinding of tool and die steel, alloy steels, stainless, and other very hard materials. There are so many factors that affect the cutting action that there are no hard and fast rules with regard to abrasive selection.

Selection of grain size is determined by whether coarse or fine cutting and finish are desired. Coarse grains take larger depths of cut and cut more rapidly. Hard wheels with fine grains leave smaller tracks and therefore usually are selected for finishing cuts. If there is a tendency for the work material to load the wheel, larger grains with more open structure may be used for finishing.

Balancing Grinding Wheels.   Because of the high rotation speeds involved, grinding wheels must never be used unless they are in good balance. A slight imbalance will produce vibrations that will cause waviness in the work surface. It may cause a wheel to break, with the probability of serious damage and injury. The wheel should be mounted with proper bushings so that it fits snugly on the spindle of the machine. Rings of blotting paper should be placed between the wheel and the flanges to assure that the clamping pressure is evenly distributed. Most grinding wheels will run in good balance if they are mounted properly and trued. Most machines have provision for compensating for a small amount of wheel imbalance by attaching weights to one mounting flange. Some have provision for semiautomatic balancing with weights that are permanently attached to the machine spindle.

Safety in Grinding. Because the rotational speeds are quite high, and the strength of grinding wheels usually is much less than the materials being ground, serious accidents occur much too frequently in connection with the use of grinding wheels. Virtually all such accidents could be avoided and are due to one or more of four causes. First, grinding wheels occasionally are operated at unsafe and improper speeds. All grinding wheels are clearly marked with the maximum rpm at which they should be rotated. They are all tested to considerably above the designated rpm and are safe at the specified speed *unless abused. They should never, under any condition, be operated above the rated speed.* Second, a most common form of abuse, frequently accidental, is dropping the wheel or striking it against a hard object. This can cause a crack (which may not be readily visible), resulting in subsequent failure of the wheel while rotating at high speed under load. If a wheel is dropped or struck against a hard object, it should be discarded and never used unless tested at above the rated speed in a properly designed test stand. A third common cause of grinding wheel failure is improper use, such as grinding against the side of a wheel that was designed for grinding only on its periphery. The fourth and most common cause of injury from grinding is the absence of a proper safety guard over the wheel and/or over the eyes or face of the operator. The frequency with which operators will remove safety guards from grinding equipment or fail to use safety goggles or face shields is amazing and inexcusable.

The Use of Cutting Fluids in Grinding. Because grinding involves cutting, the selection and use of a cutting fluid is governed by the basic principles discussed in Chapter 20. If a fluid is used, it should be applied in sufficient quantities and in a manner that will assure that the chips are washed away and not trapped between the wheel and the work. This is of particular importance in grinding horizontal surfaces. In hardened steel, the use of a fluid can help to prevent fine microcracks that result from highly localized heating. The air scraper shown in Figure 27-1 permits the cutting fluid (lubricant) to get onto the face of the wheel.

Much snagging and off-hand grinding is done dry. On some types of material, dry grinding produces a better finish than can be obtained by wet grinding. Grinding fluids strongly influence the performance of CBN wheels. Straight, sulfurized, or sulfochlorinated oils can enhance performance considerably.

Grinding machines commonly are classified according to the type of surface they produce. Table 27.3 presents such a classification, with further subdivision to indicate characteristic features of different types of machines within each classification

Grinding Machines

Grinding on all machines is done in three ways. In the first, the depth of cut (f_r) is obtained by *infeed*—moving the wheel into the work, or the work into the wheel. See Figure 27-1. The desired surface is then produced by traversing the

TABLE 27.3 Grinding Machines

Type of Machine	Type of Surface	Specific Types or Features
Cylindrical external	External surface on rotating, usually cylindrical part	Work rotated between centers Centerless Chucking Tool post Crankshaft, cam, etc.
Cylindrical internal	Internal diameters of holes	Chucking Planetary (work stationary) Centerless
Surface conventional	Flat surfaces	Reciprocating table or rotating table Horizontal or vertical spindle
Creep feed	Deep slots, profiles	Rigid, chatter-free, creep feed Continuous dressing Heavy coolant flows NC or CNC control Variable speed wheel
Tool grinders	Tool angles and geometries	Universal Special
Other	Special or any of the above	Disc, contour, thread, flexible shaft, swing frame, snag, pedestal, bench

wheel across (cross feed) the workpiece, or vice versa. In the second method, known as *plunge-cut* grinding, the basic movement is of the wheel being fed radially into the work while the latter revolves on centers. It is similar to form cutting on a lathe; usually a formed grinding wheel is used. See Figure 27-16. In the third method, the work is fed very slowly past the wheel and the total

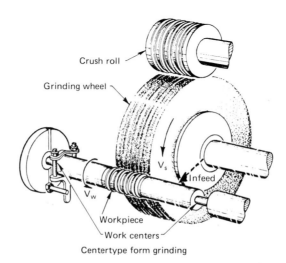

FIGURE 27-16 Plunge cut grinding of cylinder held between centers.

Centertype form grinding

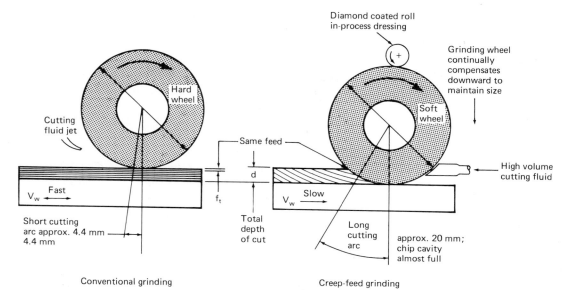

FIGURE 27-17 Conventional grinding contrasted to creep feed grinding.

depth (*d*) is accomplished in a single pass. See Figure 27-17. This is called creep feed grinding. See Table 27.4.

Grinding machines that are used for precision work have certain important characteristics that permit them to produce parts having close dimensional tolerances. They are constructed very accurately, with heavy, rigid frames to assure permanency of alignment. Rotating parts are accurately balanced to avoid vibration. Spindles are mounted in very accurate bearings, usually of the preloaded ball-bearing type. Controls are provided so that all movements that determine dimensions of the workpiece can be made with accuracy—usually 0.001 or .0001 in.

TABLE 27.4 Starting Conditions for CBN Grinding

Grinding Variable	Conventional grinding	Creep Feed Grinding	High-speed Grinding
Wheel speed (fpm)	5500–9500 versus 4500 to 6500 vitrified	5000–9000 versus 3000–5000	12000 to 25000
Table speed (fpm)	80–150	0.5–5	5–20
Feed (*f_t*) in/pass	0.0005 to 0.0015	0.100 to 0.250	.250 to .500
Grinding fluids	10% heavy duty soluble oil or 3–5% light duty soluble for light feeds	Sulfurized or sulfochlorinated straight grinding oil applied at 80 to 100 gallons per minute at 100 psi or more	

The abrasive dust that results from grinding must be prevented from entering between moving parts. All ways and bearings must be fully covered or protected by seals. If this is not done, the abrasive dust between moving parts becomes embedded in the softer of the two, causing it to act as lap and abrade the harder of the two surfaces, resulting in permanent loss of accuracy.

These special characteristics add considerably to the cost of these machines and require that they be operated by skilled personnel. Production-type grinders are more fully automated and have higher metal-removal rates and excellent dimensional accuracy. Fine surface finish can be obtained very economically.

Cylindrical Grinding.   *Center-type grinding* is commonly used for producing external cylindrical surfaces. Figures 27-16 and 27-18 show the basic principles and motions of this process. The grinding wheel revolves at an ordinary cutting speed, and the workpiece rotates on centers at a much slower speed, usually

A. Grinding wheel
B. Grinding face
C. Wheel spindle
D. Work piece
E. Work centers
F. Face plate
G. Dog

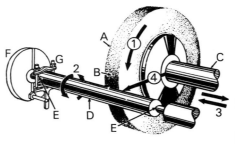

Movements

1. Wheel   2. Work (rotates)
3. Traverse   4. Infeed

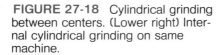

FIGURE 27-18 Cylindrical grinding between centers. (Lower right) Internal cylindrical grinding on same machine.

from 75 to 125 feet per minute. The grinding wheel and the workpiece move in opposite directions at their point of contact. The depth of cut is determined by infeed of the wheel or workpiece. Because this motion also determines the finished diameter of the workpiece, accurate control of this movement is required. Provision is made to traverse the workpiece with respect to the wheel, usually by reciprocating the work. However, in very large grinders, the wheel is reciprocated because of the massiveness of the work. For form grinding, the detail of the wheel is maintained by periodic crush roll dressing.

A *plain center-type cylindrical grinder* is shown in Figure 27-18. On this type the work is mounted between headstock and tailstock centers. Solid dead centers are always used in the tailstock, and provision usually is made so that the headstock center can be operated either dead or alive. High-precision work usually is ground with a dead headstock center, because this eliminates any possibility that the workpiece will run out of round due to any eccentricity in the headstock.

The table assembly can be reciprocated, in most cases, by using a hydraulic drive. The speed can be varied, and the length of the movement can be controlled by means of adjustable trip dogs.

Infeed is provided by movement of the wheelhead at right angles to the longitudinal axis of the table. The spindle is driven by an electric motor that is also mounted on the wheelhead. If the infeed movement is controlled manually by some type of vernier drive to provide control to 0.001 in. or less, the machine is usually equipped with digital readout equipment to show the exact size being produced. Most production-type grinders have automatic infeed with retraction when the desired size has been obtained. Such machines are usually equipped with an automatic diamond wheel-truing device that dresses the wheel and resets the measuring element before grinding is started on each piece.

The longitudinal traverse should be about one fourth to three fourths of the wheel width for each revolution of the work. For light machines and fine finishes, it should be held to the smaller end of this range.

The depth of cut (infeed) varies with the purpose of the grinding operation and the finish desired. When grinding is done to obtain accurate size, infeeds of 0.002 to 0.004 in. commonly are used for roughing cuts. For finishing, the infeed is reduced to 0.00025 to 0.0005 in. The design allowance for grinding should be from 0.005 to 0.010 in. on short parts and on parts that are not to be hardened. On long or large parts and on work that is to be hardened, a grinding allowance of from 0.015 to 0.030 in. is desirable. When grinding is used primarily for metal removal (so-called abrasive machining), infeeds are much higher, 0.020 to 0.040 in. being common. Continuous downfeed often is used, rates up to 0.100 per minute being common.

Grinding machines are available in which the workpiece is held in a chuck for grinding both external and internal cylindrical surfaces. *Chucking-type external grinders* are production-type machines for use in rapid grinding of relatively short parts, such as ball-bearing races. Both chucks and collets are used for holding the work, the means dictated by the shape of the workpiece and rapid loading and removal.

In *chucking-type internal grinding machines* the chuck-held workpiece revolves, and a relatively small, high-speed grinding wheel is rotated on a spindle arranged so that it can be reciprocated in and out of the workpiece. Infeed movement of the wheelhead is normal to the axis of rotation of the work. See Figure 27-18.

Centerless Grinding. *Centerless grinding* makes it possible to grind both external and internal cylindrical surfaces without requiring the workpiece to be mounted between centers or in a chuck. This eliminates the requirement of center holes in some workpieces and the necessity for mounting the workpiece, thereby reducing the cycle time.

The principle of *centerless external grinding* is illustrated in Figure 27-19. Two wheels are used. The larger one operates at regular grinding speeds and does the actual grinding. The smaller wheel is the *regulating* wheel. It is mounted at an angle to the plane of the grinding wheel. Revolving at a much slower surface speed—usually 50 to 200 feet per minute—it controls the rotation and longitudinal motion of the workpiece. It usually is a plastic- or rubber-bonded wheel with a fairly wide face.

The workpiece is held against the work-rest blade by the cutting forces exerted by the grinding wheel and rotates at approximately the same surface speed as that of the regulating wheel. This axial feed is calculated approximatley by the equation

$$F = dN \sin \theta \qquad (27\text{-}1)$$

where F = feed, millimeters or inches per minute
d = diameter of the regulating wheel, millimeters or inches
N = revolutions per minute of the regulating wheel
θ = angle of inclination of the regulating wheel

Centerless grinding has several important advantages:

1. It is very rapid; infeed centerless grinding is almost continuous.
2. Very little skill is required of the operator.
3. It can often be made automatic.

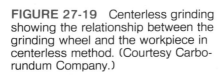

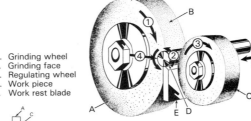

A. Grinding wheel
B. Grinding face
C. Regulating wheel
D. Work piece
E. Work rest blade

θ = Angle of tilt of regulating wheel.

Movements
1. Grinding wheel   2. Work
3. Regulating wheel 4. Infeed
5. Traverse

4. Where the cutting occurs, the work is fully supported by the work rest and the regulating wheel. This permits heavy cuts to be made.
5. Because there is no distortion of the workpiece, accurate size control is easily achieved.
6. Large grinding wheels can be used, thereby minimizing wheel wear.

Thus centerless grinding is ideally suited to certain types of mass-production operations.

The major disadvantages are as follows:

1. Special machines are required that can do no other type of work.
2. The work must be round—no flats, such as keyways, can be present.
3. Its use on work having more than one diameter or on curved parts is limited.
4. In grinding tubes, there is no guarantee that the OD and ID are concentric.

Special centerless grinding machines are available for grinding balls and tapered workpieces. The centerless grinding principle can also be applied to internal grinding, but the external surface of the cylinder must be finished accurately before the internal operation is started. However, it assures that the internal and external surfaces will be concentric. The operation is easily mechanized for many applications.

Tool-Post Grinders. *Tool-post grinders,* illustrated in Figure 27-20, are used on lathes, occasionally, to grind cylindrical parts. The wheelhead is either a high-speed electric or air motor with the grinding wheel often mounted directly on the motor shaft. The entire mechanism is mounted either on the tool post or on the compound rest. The lathe spindle provides rotation for the workpiece, and the lathe carriage is used to reciprocate the wheelhead.

Although tool-post grinders are versatile and useful, care should be taken to

FIGURE 27-20 Grinding a bearing seat by means of a tool-post grinder on a lathe. (Courtesy Dunore Company.)

cover the ways of the lathe with a closely woven cloth to provide protection from the abrasive dust that can become entrapped between the moving parts.

Surface Grinding. *Surface grinding* is used primarily to grind flat surfaces. However, formed, irregular surfaces can be produced on some types of surface grinders by use of a formed wheel.

There are four basic types of surface-grinding machines, differing in the movement of their tables and the orientation of the grinding wheel spindles:

1. Horizontal spindle and reciprocating table.
2. Vertical spindle and reciprocating table.
3. Horizontal spindle and rotary table.
4. Vertical spindle and rotary table.

These machines are illustrated in Figure 27-21.

Surface-Grinding Machines. The most common type of surface-grinding machine has a reciprocating table and horizontal spindle, such as shown in Figure 27-22. The table can be reciprocated longitudinally either by handwheel or by hydraulic power. The wheelhead is given transverse motion at the end of each table motion, again either by handwheel or by hydraulic power feed. Both the longitudinal and transverse motions can be controlled by limit switches. Infeed on such grinders is controlled by handwheels or automatically.

The size of such machines is designated by the size of the surface that can be ground.

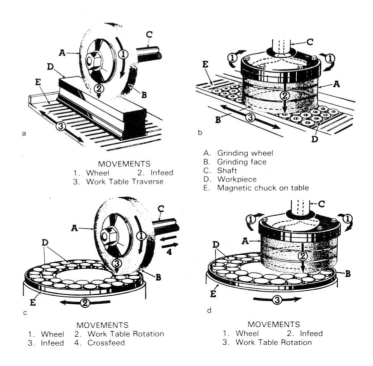

FIGURE 27-21 Surface grinding. (a) Horizontal surface grinding with reciprocating table. (b) Vertical spindle with reciprocating table. (c) and (d) Both horizontal and vertical spindle machines can have rotary tables. (Courtesy Carborundum Company.)

MOVEMENTS
1. Wheel 2. Infeed
3. Work Table Traverse

A. Grinding wheel
B. Grinding face
C. Shaft
D. Workpiece
E. Magnetic chuck on table

MOVEMENTS
1. Wheel 2. Work Table Rotation
3. Infeed 4. Crossfeed

MOVEMENTS
1. Wheel 2. Infeed
3. Work Table Rotation

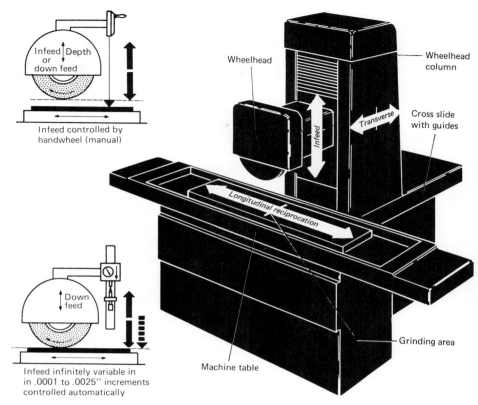

Infeed | Depth
or |
down feed |

Infeed controlled by
handwheel (manual)

Down
feed

Infeed infinitely variable in
in .0001 to .0025" increments
controlled automatically

Wheelhead

Wheelhead
column

Transverse

Cross slide
with guides

Infeed

Longitudinal reciprocation

Grinding area

Machine table

FIGURE 27-22 Horizontal spindle
surface grinder.

In using such machines, the wheel should overtravel the work at both ends of the table reciprocation, so as to prevent the wheel from grinding in one spot while the table is being reversed. The transverse motion should be one fourth to three fourths of the wheel width between each stroke.

Vertical-spindle reciprocating-table surface grinders differ basically from those with horizontal spindles only in that their spindle are vertical and that the wheel diameter must exceed the width of the surface to be ground. Usually, no traverse motion of either the table or the wheelhead is provided. Such machines can produce very flat surfaces.

Rotary-table surface grinders are of two types, but those with horizontal spindles are limited in the type of work they will accommodate and therefore are not used to a great extent.

Vertical-spindle rotary-table surface grinders are primarily production-type machines. They frequently have two or more grinding heads, and therefore both rough grinding and finish grinding are accomplished in one rotation of the work-piece. The work can be held either on a magnetic chuck or in special fixtures attached to the table.

By using special rotary feeding mechanisms, machines of this type often are made automatic. Parts are dumped on the rotary feeding table and fed automatically onto work-holding devices and moved past the grinding wheels. After they pass the last grinding head, they are automatically unloaded.

Creep Feed Grinders. This is a grinding method, often done in the surface-grinding mode, that is markedly different from conventional surface grinding. See Figure 27-17. In contrast to conventional techniques, the depth of cut is increased 1000 to 10,000 times and the work feed is decreased in the same proportion; hence the name "creep feed grinding." See Table 27-4. The machine tools to perform this type of grinding must be specially designed with high static and dynamic stability, stick-slip free ways, adequate damping, increased horsepower, infinitely variable spindle speed, variable but extremely consistent table feed (especially in the low ranges), high-pressure cooling systems, integrated devices for dressing the grinding wheels, and specially designed (soft with open structure) grinding wheels. The process is mainly being applied to grinding deep slots with straight parallel sides or to grinding complex profiles in difficult-to-grind materials. The process is capable of producing extreme precision at relatively high metal removal rates. Because the process can operate at relatively low surface temperatures, the surface integrity of the metals being ground is good.

The grinding wheels maintain their initial profile much longer. Creep feed grinding eliminates preparatory operations like milling or broaching since profiles are ground into the solid workpiece. This process is greatly enhanced by continuous dressing (form-truing and dressing the grinding wheel throughout the process rather than between cycles). This technique results in higher MRRs, improved dimensional accuracy and form tolerance, reduced grinding forces (and power), and reduced thermal effects while sacrificing wheel wear.

Disc Grinders. *Disc grinders* have relatively large side-mounted abrasive discs. The work is held against one side of the disc for grinding. Both single and double discs grinders are used; in the latter type the work is passed between the two disc and is ground on both sides simultaneously. On these machines, the work is always held and fed automatically. On small, single-disc grinders the work can be held and fed by hand while resting on a supporting table. Although manual disc grinding is not very precise, flat surfaces can be obtained quite rapidly with little or no tooling cost. On specialized, production-type machines, excellent accuracy can be obtained very economically.

Tool and Cutter Grinders. Simple, single-point tools often are sharpened by hand on bench or pedestal grinders *(offhand grinding)*. More complex tools, such as milling cutters, reamers, hobs, and single-point tools for production-type operations, require more sophisticated grinding machines, commonly called *universal tool and cutter grinders*. These machines are similar to small universal cylindrical center-type grinders, but they differ in four important respects:

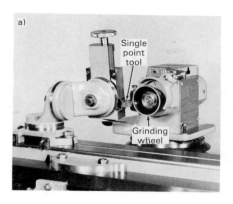

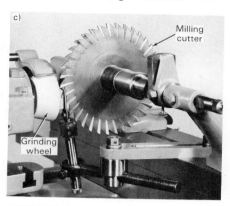

FIGURE 27-23 Three typical setups for grinding single- and multiple-edge tools on a universal tool and cutter grinder. (a) Single-point tool is held in a device that permits all possible angles to be ground. (b) Edges of a large hand reamer are being ground. (c) Milling cutter is sharpened with a cupped Al_2O_3 grinding wheel.

1. The headstock is not motorized.
2. The headstock can be swiveled about a horizontal as well as a vertical axis.
3. The wheelhead can be raised and lowered and can be swiveled through at 360-degree rotation about a vertical axis.
4. All table motions are manual, no power feeds being provided.

Specific rake and clearance angles must be created, often repeatedly, on a given tool or on duplicate tools. Tool and cutter grinders have a high degree of flexibility built into them so that the required relationships between the tool and the grinding wheel can be established for almost any type of tool. Although setting up such a grinder is quite complicated and requires a highly skilled worker, after the setup is made for a particular job, the actual grinding is accomplished rather easily. Figure 27-23 shows several typical setups on a tool and cutter grinder.

Hand-ground cutting tools are not accurate enough for automated machining processes. Many NC machine tools have been sold on the premise that they can position to very close tolerances—within plus or minus 0.0001 to 0.0002 in.—only to have the initial workpieces produced by those machines out of tolerance by as much as 0.015 to 0.020 in. In most instances, the culprit was a poorly ground tool. For example, a twist drill with a point ground 0.005 in. off center can "walk" as much as 0.015 in., thus causing poor hole location. Many companies are turning to CNC grinders to handle the regrinding of their cutting tools. A six-axis CNC grinder is capable of restoring the proper tool angles (rake and clearance), concentricity, cutting edges, and dimensional size.

Snagging. *Snagging* is a type of rough manual grinding that is done to remove fins, gates, risers, and rough spots from castings or flash from forging, preparatory to further machining. The primary objective is to remove substantial amounts of metal rapidly without much regard for accuracy. Pedestal-type or *swing grinders* ordinarily are used. Portable electric or hand air grinders also are used for this purpose and for miscellaneous grinding in connection with welding.

FIGURE 27-24 Mounted abrasive wheels and points. (Courtesy Norton Company.)

FIGURE 27-25 Various forms of coated abrasives. (Courtesy Carborundum Company.)

Control of the Grinding Process

Mounted Wheels and Points. *Mounted wheels and points* are small grinding wheels of various shapes that are permanently attached to metal shanks that can be inserted in the chucks of portable, high-speed electric or air motors. They are operated at speeds up to 100,000 rpm, depending on their diameters, and are used primarily for deburring and finishing in mold and die work. Several types are shown in Figure 27-24.

Coated Abrasives. *Coated abrasives* are being used increasingly in finishing both metal and nonmetal products. These are made by gluing abrasive grains onto a cloth or paper backing. Synthetic abrasives—Carborundum and Alundum—are used most commonly, but some natural abrasives—sand, flint, garnet, and emery—also are employed. Various types of glues are utilized to attach the abrasive grains to the backing, usually compounded to allow the finished product to have some flexibility.

Coated abrasives are available in sheets, rolls, endless belts, and disks of various sizes. Some of the available forms are shown in Figure 27-25. Although the cutting action of coated abrasives basically is the same as with grinding wheels, there is one major difference: they have little tendency to be self-sharpened when dull grains are pulled from the backing. Consequently, when the abrasive particles become dull or the belt loaded, the belt must be replaced.

Both force sensors and power monitors can be used to control the grinding process. According to R. P. Lindsay of the Norton Company, measuring power is a way to measure normal force (F_n), provided wheel speed is considered. It is difficult to measure normal force but easy to measure power. He argues as follows:

Figure 27-26 is a generalized representation of all grinding systems. The wheel and workpiece are connected to the machine slide and bed by system springs. Once wheel and work contact is made, further infeed of the slide builds up the normal (F_n) and tangential (F_t) forces. As the forces increase, the work removal rate (v_w) and wheelwear rate (v_s) increase. When these rates equal the slide infeed rate (v_f), the steady state is reached and F_n and F_t will be constant. For steady state: $(v_w) + (v_s) = (v_r)$ and F_n and $F_t = $ constant values.

Thus the infeed rate (v_r) is the input to the grinding system but F_n is input to the grinding process.

Figure 27-27 shows metal removal rate (MRR') per unit width of wheel versus normal force (F_n') per unit width for two wheel grades. The linear slope for

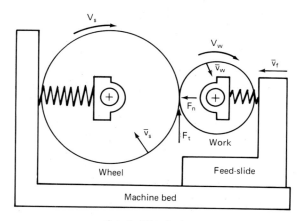

Grinding Terminology

F_n = force normal to contact surfaces per inch of wheel width	$\bar{v}_w$ = rate of radial wheel wear
	Z_s = volumetric rate of wheel wear
F_t = force tangent to contact surfaces per inch of wheel width	Z_w = volumetric rate of stock removal or material removal rate (MRR)
f_t = workpiece infeed rate	
F_{bd} = threshold force intensity, wheel	Λ_s = wheel-wear removal parameter
F_{th} = threshold force intensity, workpiece	Λ_w = metal removal parameter
	a = depth of grind per pass
$\bar{v}_f$ = feed rate of cross slide	b = width of grind
V_s = surface speed of grinding wheel	b_c = crossfeed rate (surface grinding)
$\bar{v}_s$ = rate of radial workpiece removal	D_s = diameter of wheel
	D_w = diameter of workpiece
V_w = surface speed of workpiece	

FIGURE 27-26   Generalized model for grinding processes.

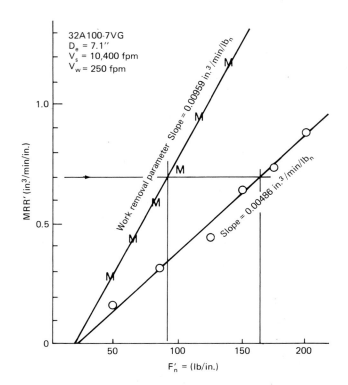

FIGURE 27-27 Normal force per unit width versus metal removal rate per unit width for M and O wheel grades.

each is the work removal parameter having units of cubic inches per minute being removed per pound of normal force. The M grade cuts about twice as fast (0.00959 versus 0.00486) as the 0 grade.

Figure 27-28 shows power per unit width versus normal force per unit width for these wheels. Notice that there is no difference in slope here. Some power is generated at some normal force regardless of wheel grade. Therefore, if power is measured, it is a direct measurement of normal force.

Figure 27-29 shows power′ versus MRR′ for both wheel grades. The constant slope here is called the specific power having the units horsepower required for one cubic inch per minute removal rate. Notice the fast-cutting M grade had a lower specific power (14.35 versus 25.75) than the 0 grade. Power, though, is tangential force times wheel speed:

$$\text{Power} = \frac{F_t \times V_s}{33,000} \frac{\text{lb/ft/min}}{\left(\dfrac{\text{ft/lb}}{\text{min/hp}}\right)} \tag{27-2}$$

Measuring power is a way to measure normal force provided the change in wheel speed is considered. It is very hard to measure normal force in production but fairly easy to measure power. Normal force is better, but power is an acceptable, easier alternative for normal force, especially in production grinding. Then Figure 27-27 can be used in place of Figure 27-27.

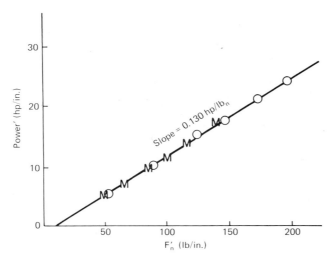

FIGURE 27-28 Power per unit width versus normal force per unit width.

Figure 27-30 shows the *G* ratio for an internal grinding operation where the wheel wear is not linear but the metal removal parameter is. Generally speaking, the surface finish would be getting rougher with increased F_n.

Design Considerations in Grinding.   Almost any shape and size of work can be finished on modern grinding equipment, including flat surfaces, straight or tapered cylinders, irregular external and internal surfaces, cams, antifriction bearing races, threads, and gears. The most accurate threads are formed from solid cylindrical blanks on special thread-grinding machines. Gears that must operate without play are hardened and then finish-ground to close tolerances.

Two important design recommendations are to reduce the area to be ground and to keep all surfaces that are to be ground in the same or parallel planes, as shown in Figure 27-31.

Abrasive machining can remove scale as well as parent metal. Large allow-

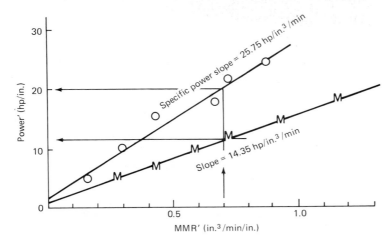

FIGURE 27-29 Power per unit width versus metal removal rate per unit width.

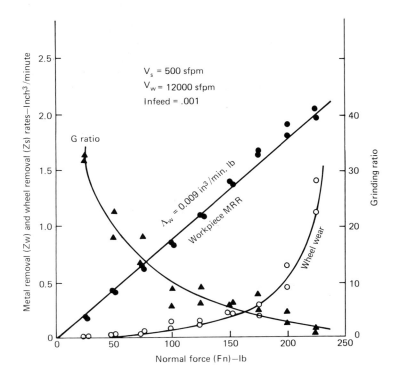

FIGURE 27-30 Grinding ratio, wheel wear and metal removal parameters for grinding AISI 52100 steel.

Original design of base plate

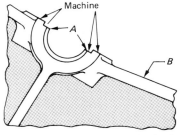

Original design of crankshaft bearing bracket

Redesigned to reduce weight and grinding time

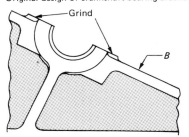

Redesign eliminated shoulders and made part suitable for grinding in single setup

FIGURE 27-31 Reducing area to be ground and keeping all surfaces to be ground in the same or parallel planes are two important design recommendations. (*Machine Design*, June 1, 1972, p. 87.)

ances of material, needed to permit conventional metal-cutting tools to cut below hard or abrasive inclusions, are not necessary for abrasive machining. An allowance of 0.015 in. is adequate, assuming, of course, that the part is not warped or out of round. This small allowance requirement results in savings in machining time, in material (often 60% less metal is removed), and in shipping of unfinished parts.

Honing is a stock removal process that uses fine abrasive stones to remove very small amounts of metal. Cutting speed is much lower than that of grinding. The process is used to size and finish bored holes, remove common errors left by boring (taper, waviness, and tool marks) or remove the tool marks left by grinding. The amount of metal removed is typically about 0.005 in. or less. Although honing occasionally is done by hand, as in finishing the face of a cutting tool, it usually is done with special equipment. Most honing is done on internal cylindrical surfaces, such as automobile cylinder walls. The honing stones usually are held in a honing head, with the stones being held against the work with controlled light pressure. The honing head is not guided externally but, instead, *floats* in the hole, being guided by the work surface.

The stones are given a complex motion so as to prevent a single grit from repeating its path over the work surface. Rotation is combined with an oscillatory axial motion. For external and flat surfaces, varying oscillatory motions are used. The length of the motions should be such that the stones extend beyond the work surface at the end of each stroke. A cutting fluid is used in virtually all honing operations. The critical process parameters are rotational speed, V_r, oscillation speed, V_o, the length and position of stroke, and the honing stick pressure. See Figure 27-32. V_c and the inclination angle are both products of V_o and V_r.

Honing

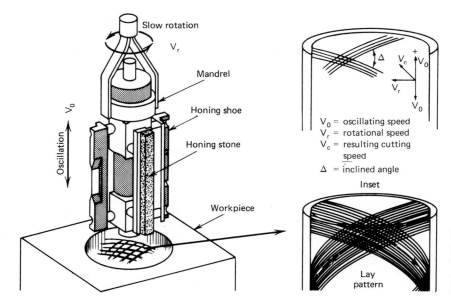

V_0 = oscillating speed
V_r = rotational speed
V_c = resulting cutting speed
Δ = inclined angle

Inset

FIGURE 27-32 Schematic of honing head showing the manner in which the stones are held. Rotary, oscillatory motion results in a cross-hatched lay pattern.

TABLE 27.5 **Honing Reference Values**

For	Honing Parameters	Conventional Abrasives	Diamonds	CBN
High MRR	$V_c \left(\dfrac{m}{min} \right)$	20–30	40–70	35–90
	$P_s \left(\dfrac{N}{mm^2} \right)$	1–2	2–8	2–4
Best Quality Surface	$V_c \left(\dfrac{m}{min} \right)$	5–30	40–70	20–60
	$P_s \left(\dfrac{N}{mm^2} \right)$	0.5–1.5	1.0–3.0	1.0–2.0

Honing Stones. Virtually all honing is done with stones made by bonding together various fine artificial abrasives. *Honing stones* differ from grinding wheels in that additional materials, such as sulfur, resin, or wax, are often added to the bonding agent to modify the cutting action. The abrasive grains range in size from 80 to 600 grit. The stones are equally spaced about the periphery of the tool. Table 27.5 gives some reference values for V_c and honing stick pressure, P_s, for various abrasives.

Single- and multiple-spindle honing machines are available in both horizontal and vertical types. Some are equipped with special, sensitive measuring devices that collapse the honing head when the desired size has been reached.

For honing single, small, internal cylindrical surfaces a procedure is often used wherein the workpiece is manually held and reciprocated over a rotating hone.

If the volume of work is sufficient, honing is a fairly inexpensive process. A complete honing cycle, including loading and unloading the work, is often less than 1 minute. Size control within 0.0003 in. is achieved routinely.

Superfinishing is a variation of honing that uses

1. Very light, controlled pressure, 10 to 40 psi.
2. Rapid (over 400 per minute), short strokes—less than ¼ in.
3. Stroke paths controlled so that a single grit never traverses the same path twice.
4. Copious amounts of low-viscosity lubricant-coolant flooded over the work surface.

This procedure, illustrated in Figure 27-33, results in surfaces of very uniform, repeatable smoothness.

Superfinishing is based on the phenomenon that a lubricant of a given viscosity will establish and maintain a separating, lubricating film between two mating surfaces if their roughness does not exceed a certain value and if a certain critical pressure, holding them apart, is not exceeded. Consequently, as the minute

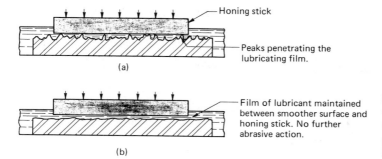

(a)

Honing stick

Peaks penetrating the lubricating film.

(b)

Film of lubricant maintained between smoother surface and honing stick. No further abrasive action.

FIGURE 27-33 Manner in which a film of lubricant is established between the work and the abrasive stone in superfinishing as the work becomes smoother.

peaks on a surface are cut away by the honing stone, applied with a controlled pressure, a certain degree of smoothness is achieved. The lubricant establishes a continuous film between the stone and the workpiece and separates them so that no further cutting action occurs. Thus, with a given pressure, lubricant, and honing stone, each workpiece is honed to the same degree of smoothness.

Superfinishing is applied to both cylindrical and plane surfaces. The amount of metal removed usually is less than 0.002 in., most of it being the peaks of the surface roughness. Copious amounts of lubricant-coolant maintain the work at a uniform temperature and wash away all abraded metal particles to prevent scratching.

Lapping

Lapping is an abrasive surface-finishing process wherein fine abrasive particles are "charged" (caused to become embedded) into a soft material, called a *lap*. The material of the lap may range from cloth to cast iron or copper, but it is always softer than the material to be finished, being only a holder for the hard abrasive particles. Lapping is applied to both metals and nonmetals.

As the charged lap is rubbed against a surface, the abrasive particles in the surface of the lap remove small amounts of material from the harder surface. Thus it is the abrasive that does the cutting, and the soft lap is not worn away because the abrasive particles become embedded in its surface instead of moving across it. This action always occurs when two materials rub together in the presence of a fine abrasive; the softer one forms a lap, and the harder one is abraded away.

In lapping, the abrasive is usually carried between the lap and the work surface in some sort of a vehicle, such as grease, oil, or water. The abrasive particles are from 120 grit up to the finest powder sizes. As a result, only very small amounts of metal are removed, usually considerably less than 0.001 in. Because it is such a slow metal-removing process, lapping is used only to remove scratch marks left by grinding or honing, or to obtain very flat or smooth surfaces, such as are required on gage blocks or for liquid-tight seals where high pressures are involved.

Materials of almost any hardness can be lapped. However, it is difficult to lap soft materials because the abrasive tends to become embedded. The most common lap material is fine-grained cast iron. Copper is used quite often and is the common material for lapping diamonds. For lapping hardened metals for metallographic examination, cloth laps are used.

Lapping can be done either by hand or by special machines. In hand lapping, the lap is flat, similar to a surface plate. Grooves usually are cut across the surface of a lap to collect the excess abrasive and chips. The work is moved across the surface of the lap, using an irregular, rotary motion, and is turned frequently to obtain a uniform cutting action.

In lapping machines for obtaining flat surfaces, workpieces are placed loosely in holders and are held against the rotating lap by means of floating heads. The holders, rotating slowly, move the workpieces in an irregular path. When two parallel surfaces are to be produced, two laps may be employed, one rotating below and the other above the workpieces.

Various types of lapping machines are available for lapping round surfaces. A special type of centerless lapping machine is used for lapping small cylindrical parts, such as piston pins and ball-bearing races.

Because the demand for surfaces having only a few micrometers of roughness on hardened materials has become quite common, the use of lapping has increased greatly. However, it is a very slow method of removing metal, obviously costly compared with other methods, and should not be specified unless such a surface is absolutely necessary.

Review Questions

1. What machining processes use abrasive particles for cutting tools?
2. What is attrition in an abrasive grit?
3. Why is friability an important grit property?
4. What is the relationship between grit size and surface finish?
5. Why is aluminum oxide used more frequently than silicon carbide as an abrasive?
6. Why is CBN superior to silicon carbide as an abrasive in some applications?
7. What materials commonly are used as bonding agents in grinding wheels?
8. Why is the grade of a bond in a grinding wheel important, and how does grade differ from structure?
9. What is crush dressing?
10. What are the common causes of grinding accidents?
11. How does loading differ from glazing?
12. What is meant by the statement that grinding is a mixture of processes?
13. What is accomplished in dressing a grinding wheel?
14. How does abrasive machining differ from ordinary grinding?
15. What is a grinding ratio?
16. How is the feed of the workpiece controlled in centerless grinding?
17. Why is grain spacing important in grinding wheels?
18. How is the machining time in surface grinding calculated?
19. Why should a cutting fluid be used in copious quantities in doing wet grinding?
20. What is plunge-cut grinding?
21. If grinding machines are placed among other machine tools, what precautions must be taken?
22. How is the machining time for creep feed surface grinding calculated?
23. What is the purpose of low-stress grinding, and how is it done?
24. The number of grains per square inch that actively contact and cut a surface decreases with increasing grain diameter. Why is this so?
25. Is the grain diameter equal to the grain-size number?
26. Why are centerless grinders so popular in industry compared to center-type grinders?
27. How is the through-feed rate varied in a centerless grinder? Could you do this while the process is in operation?
28. Why are vacuum chucks used in surface grinding and not in milling? See Chapter 29.
29. How does creep feed grinding differ from conventional surface grinding?

30. Why does a lap not wear, since it is softer than the material being lapped?
31. How do honing stones differ from grinding wheels?
32. What is meant by "charging" a lap?
33. Why is a honing head permitted to float in a hole that is being honed?
34. In what respect does the cutting action of a coated abrasive differ from that of an abrasive wheel?
35. What is the relationship of wheel wear to metal removal rate?
36. What is the inclined angle in honing and what determines it?
37. Perhaps you have observed the following wear phenomenon: A set of marble stairs shows wear on the treads in the regions where people step when they climb (or descend) the stairs. The higher up the stairs, the less the wear on the tread. Given that soles of shoes (leather, rubber) are far softer than marble or granite:
    a. Why do the stairs wear?
    b. Why are the lower stairs more worn than the upper stairs?
38. Explain why it is that a small particle of a material can be used to abrade a surface made of the same material—that is, why does the small particle act harder or stronger than the bulk material?
39. What other machine tool does a surface grinder resemble?
40. What determines the actual "cutting speed" in grinding? Use surface grinding in your discussion.

Workholding Devices

Introduction

In previous chapters attention has been directed repeatedly to the manner in which workpieces are mounted and held in the various machine tools. Workholding devices—that is, jigs and fixtures—are critical components in the manufacturing of interchangeable parts. With workholders, tolerance levels can be achieved which otherwise would be impossible with a given combination of cutting tools and machine tools. In this chapter, workholding devices (jigs and fixtures) will be considered as important production tools or adjuncts, with primary attention being directed toward their functional characteristics, their relationship to the machine tools and the manufacturing processes, and the manner in which workholding devices can be designed to be more flexible—that is, able to accommodate more than one part. Flexible workholders are a critical element in manufacturing cells, particularly in unmanned cells. For the cell to be flexible, workholding devices should be able to accommodate all the parts within the *family of parts* (see Chapters 42 and 44). This design requirement has added significantly to the complexity of conventional jig and fixture design.

Many different schemes are being proposed to provide workholder flexibility. Programmable clamps using air-activated plungers; part encapsulation with a low melting point alloy; and NC-controlled clamping machines are some of the more recently developed systems. Despite their flexibilities, these clamping systems have some significant drawbacks. They are expensive, and the individual systems are not integrated into a single machine tool unit. Therefore, for every machine tool, an additional clamping system must be developed that requires more money and more floor space. Let us begin with the basics of workholding devices.

Conventional Fixture Design

In the conventional method of fixture design, tool designers rely on their experience and intuition to design "single-purpose" fixtures for specific machining operations, using a trial-and-error method until the workholders perform satisfactorily. Designers do not usually calculate the clamping forces or stress distributions in the fixturing elements to determine the loads that will deform the fixtures or the workpieces elastically or plastically.

Workholding devices have two primary functions: locating and clamping.

Locating refers to orienting and positioning the part in the machine tool with respect to the cutting tools to achieve the required specifications.

Clamping refers to holding or maintaining the part in that location during the operations.

Jigs and *fixtures* are specially designed and built workholding devices which hold the work during machining or assembly operations. In addition, a jig determines a location dimension that is produced by machining or fastening. For example, location dimensions determine the position of a hole on a plate. Consider the subject of dimensioning as used in drafting practice. Dimensions are of two types: *size* and *location*. Size dimensions denote the size of geometrical shapes—holes, cubes, parallelopipeds, and so forth—of which objects are composed. Location dimensions, on the other hand, determine the position or location of these geometrical shapes *with respect to each other*. Thus, in Figure 28-1, *a* and *c* are location dimensions, while *e* and *g* are size dimensions. With location dimensions in mind, one can precisely define a jig as follows: *A jig is a special device that, through built-in features, determines location dimensions that are produced by machining or fastening operations.* The key requirement of a jig is that it determine a location dimension. Thus jigs automatically accomplish layout.

In establishing location dimensions, jigs may do a number of other things. They frequently guide tools, as in drill jigs, and thus determine the location of a component geometrical shape. However, they do not always guide tools. In the case of welding jigs, component parts are held (located) in a desired relationship with respect to each other while an unguided tool accomplishes the fastening. The guiding of a tool is not a necessary requirement of a jig.

Similarly, jigs usually hold the work that is to be machined, fastened, or assembled. However, in certain cases, the work actually supports the jig. Thus, although a jig *may* incidentally perform other functions, the basic requirement is that, through qualities that are built into it, certain critical dimensions of the workpiece are determined.

A *fixture* is *a special device that holds work during machining or assembly operations.* The key characteristic is that it is a *special* workholding device,

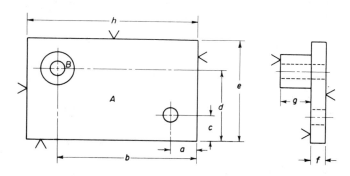

FIGURE 28-1 Drawing of a plate showing locating dimensions versus sizing dimensions.

designed and constructed for a particular part or shape. A general-purpose device, such as a vise, a chuck or a clamp, is not a fixture. Thus a fixture has as its specific objective the facilitating of *setup,* or making holding easier.

Modular fixtures employ a collection of general-purpose components, like clamps, blocks, and T-slotted plates which can be built up into a fixture for a specific job. The versatile pieces can be disassembled and recombined for other tasks. This is an "erector-set" approach to the problem, in contrast to reliance on hard fixturing and permanent dedicated fixtures. Modular fixturing applies not only to machining applications but also to fixtures used in gaging, inspection, assembly, welding, and other manufacturing processes.

Because many jigs hold the work while determining critical dimensions, they usually meet all the requirements of a fixture. It is equally evident that fixtures never determine dimensions of parts that they hold—a basic requirement of a jig.

Design Criteria for Workholders

To meet all the design criteria for workholders is impossible. Compromise is inevitable. Still, it is useful to list the optimal design objectives to illustrate the positioning, holding, and supporting functions that fixtures must fulfill.

Positive Location.   A fixture must, above all else, hold the workpiece precisely in space to prevent each of 12 kinds of degrees of freedom—linear movement in either direction along the X, Y, and Z axes and rotational movement in either direction about each axis.

Repeatability.   Identical workpiece specimens should be located by the workholder in precisely the same space on repeated loading and unloading cycles. It should be impossible to load the workpiece incorrectly.

Adequate Clamping Forces.   The workholder must hold the workpiece immobile against the forces of gravity, centrifugal forces, inertial forces, and cutting forces. Milling and broaching operations, in particular, tend to pull the workpiece out of the fixture, and the designer must calculate these machining forces against the fixture's holding capacity. The device must be rigid.

Reliability.   The clamping forces must be maintained during machine operation every time the device is used. The mechanism must be easy to maintain and lubricate.

Ruggedness.   Workholders usually receive more punishment during the loading and unloading cycle than during the machining operation. The device must endure impact and abrasion for at least the life of the job. Elements of a device that are subject to damage and wear should be easily replaceable.

Design and Construction Ease.   Workholders should use standard elements as much as possible to allow the engineer to concentrate on function rather than on construction details. Modular fixtures epitomize this design rule as the entire workholder is made from standard elements, permitting a bolt-together approach for substantial time and cost savings over custom work-holders.

Low Profile.   Workholder elements must be clear of the machine path. Designing lugs on the part for clamping can simplify the fixture and allow proper tool clearance.

Workpiece Accommodation.   Surface contours of castings or forgings vary from one part to the next. The device should tolerate these variations without sacrificing positive location or other design objectives.

Rapid, Easy Operation.   The workholder elements should not obstruct the loading or unloading of workpieces. In manual operations, the operator should not have to reach past the tool to load or unload parts.

Freedom from Part Distortion.   Parts being machined can be distorted by gravity, the machining forces, or the clamping forces. Once clamped into the device, the part must be unstressed or, at least, undistorted. Otherwise, the newly machined surfaces take on any distortions caused by the clamping forces.

The classical design of a workholder (for example, a drill jig) involves the following steps:

Design Steps

1. Analyze the drawing of the workpiece and determine the machining operations required to machine it. Note the critical dimensions and tolerances.
2. Determine the orientation of the workpiece in relation to the cutting tools and their movements.
3. Perform an analysis to estimate the magnitude and direction of the cutting forces.
4. Study the standard devices available for workholders and for the clamping functions. Can an off-the-shelf device be modified? What standard elements can be used?
5. Form a mental picture of the workpiece in position in the workholder in the machine tool with the cutting tools performing the required operation(s).
6. Make a three-dimensional sketch of the workpiece in the workholder in its required position to determine the location of all the elements—clamps, locator buttons, bushings, and so forth.

After determining the workpiece's orientation in the workholder, the next step is to locate it in that position. This location is also used for all similar workpieces. The designer must select or design locating devices (supports) that insure that every workpiece placed in the device occupies the same position with respect

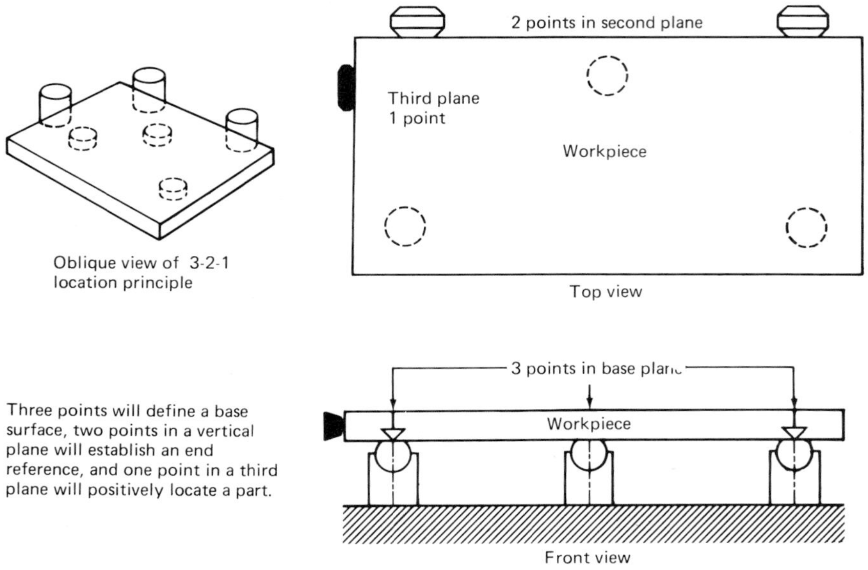

Oblique view of 3-2-1
location principle

2 points in second plane

Third plane
1 point

Workpiece

Top view

Three points will define a base
surface, two points in a vertical
plane will establish an end
reference, and one point in a third
plane will positively locate a part.

3 points in base plane

Workpiece

Front view

FIGURE 28-2 Workpiece location is based on the 3–2–1 principle.

to the cutting tools. Thus, when the machining operation is performed, the workpieces are processed identically. In locating the workpiece, the basic 3–2–1 principle of location is used. See Figure 28-2. For positive location, the fixture must position the workpiece in each of three perpendicular planes. Positioning processes can vary greatly, but workholder design always begins by defining the first plane of reference with three points. Once the object is defined in a single plane, supported at three points (like a three-legged stool on a floor), a second plane can be assigned that is perpendicular to the first. To do this, the object is brought up against any two points in the second plane. To continue the example, the stool is slid along the floor until two legs touch a wall.

A third plane, perpendicular to each of the other two, is then defined by designating one point on it. As long as an object is in contact with three points on the first plane, two points on the second, and a single point on the third, it is positively located in space. The location points within each plane should be selected as far apart as possible for maximum stability.

In practice, it is often necessary to support a workpiece on more points than this 3–2–1 formula dictates. The machining of a large rectangular plate, for example, typically requires support at four or more points. However, any extra points must be carefully established to support the workpiece in a plane defined by three—and only three—points.

Appropriate clamping devices are selected so that the clamping forces hold the workpiece in the proper location and resist the cutting forces, centrifugal forces, and vibrations. If possible, the machining forces should act into the location points, not into the clamps, so that smaller clamps can be used. In reality, the worker often determines clamping force when the part is loaded into the workholder.

TABLE 28.1 Twenty Principles of Jig and Fixture Design

1. Determine the critical surfaces or points for the part.
2. Decide on locating points and clamping arrangements.
3. For mating parts, use corresponding locating points or surfaces to ensure proper alignment when assembled.
4. Try to use 3–2–1 location with three assigned to largest surface. Additional points should be adjustable.
5. Locating points should be visible so that the operator can see if they are clean. Can they be replaced if worn?
6. Provide clamps that are as quick-acting and easy-to-use as is economically justifiable.
7. Clamps should not require undue effort by operator to close or to open, nor should they harm hands or fingers during use.
8. Clamps should be integral parts of device. Avoid loose parts that can get lost.
9. Avoid complicated clamping arrangements or combinations that can wear out or malfunction.
10. Locate clamps opposite locaters (if possible) to avoid deflection/distortion during machining and springback afterward.
11. Take the thrust of the cutting forces on the locaters (if possible) and not on the clamps.
12. Arrange the workholder so that the workpiece can be easily loaded and unloaded from the device and so that it can be loaded only in the correct manner (mistake-proof) and in such a way that the location can be found quickly.
13. Consistent with strength and rigidity, make the workholder as light as possible.
14. Provide ample room for chip clearance and removal.
15. Provide accessibility for cleaning.
16. Provide for entrance and exit of cutting fluid (which may carry off chips) if one is to be used.
17. Provide four feet on all movable workholders.
18. Provide hold-down lugs on all fixed workholders.
19. Provide keys to align fixtures on machine tables.
20. Do not sacrifice safety for production.

Fixtures are usually fastened to the table of the machine tool. Though used mostly on milling and broaching machines, fixtures are also designed and used to hold workpieces for various operations on most of the standard machine tools. Some additional design rules for fixture design are given in Table 28.1.

Clamping Considerations

Clamping of the work is closely related to support of the work. Clamping *stresses* should be kept low. Any clamping, of course, induces some stresses that tend to cause some distortion of the workpiece, usually elastic. If this distortion is measurable, it will cause some inaccuracy in final dimensions, as illustrated in an exaggerated manner in Figure 28-3. The obvious solution is to spread the clamping forces over a sufficient area to reduce the stresses to a level that will not produce appreciable distortion.

The clamping forces should direct the work against the points of location and work support. Clamped surfaces often have some irregularities that may produce

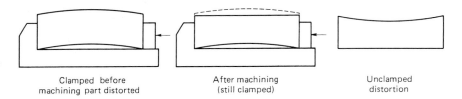

FIGURE 28-3 Exaggerated illustration of the manner in which excessive clamping forces can affect the final dimensions of a workpiece.

Clamped before
machining part distorted

After machining
(still clamped)

Unclamped
distortion

force components in an undesired direction. Consequently, clamping forces should be applied in directions that will assure that the work will remain in the desired position.

Whenever possible, jigs and fixtures should be designed so that the forces induced by the cutting process act to hold the workpiece in position against the supports. These forces are predictable, and proper utilization of them can materially aid in reducing the magnitude of the required clamping stresses. In addition to locating the work properly, the stops or work-supporting areas must be arranged so as to provide adequate support against the cutting forces. As shown in Figure 28-4a, having the cutting force act against a fixed portion of the jig or fixture and not against a movable section permits lower clamping forces to be used. Figure 28-4b illustrates the principle of keeping the points of clamping as nearly as possible in line with the action forces of the cutting tool so as to reduce their tendency to pull the work from the clamping jaws. Compliance with this principle results both in lower clamping stresses and less massive clamping devices. The location points should be as far apart as possible but positioned so as to not allow the cutting force to distort the work. The

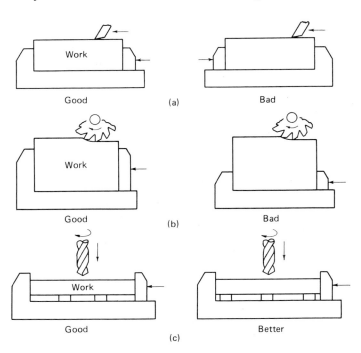

Good   (a)   Bad

Good   (b)   Bad

Good   (c)   Better

FIGURE 28-4 Proper work support to resist the forces imposed by cutting tools. In (c), three buttons form triangle for work to rest on.

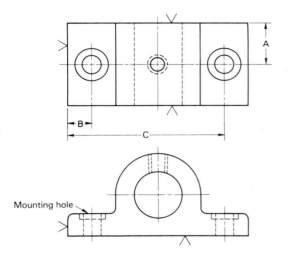

FIGURE 28-5 (left) Drawing of a bearing block, with two mounting holes. (right) Box-type drill jig for drilling the holes.

cutting forces may distort the work, with resulting inaccuracy, or broken tools. These design suggestions materially reduce vibration and chatter during the cutting process.

As many operations as possible and practical should be performed with each clamping of the workpiece. This principle has both physical and economic aspects. Because some stresses result from each clamping, with the possibility of accompanying distortion, greater accuracy is achieved if multiple operations are performed with each clamping. From the economic viewpoint, if the number of jigs or fixtures is reduced, less capital will be required and less time will be spent handling the workpiece loading and unloading.

An Example of Jig Design

The principles of work location and tool guidance are illustrated in Figure 28-5. The two mounting holes in the base of the bearing block are to be located and drilled. The dimensions, A, B, and C, are determined by the jig. There also is one other location dimension that must be controlled. The axes of the mounting holes must be at right angles to the bottom surface of the block.

The way in which the part dimensions are obtained in the finished workpiece is as follows: The surfaces marked with a V are reference (or location) surfaces and are finished (machined) prior to insertion of the part into the drill jig. The part rests in the jig on four buttons marked X in Figure 28-5. These buttons, made of hardened steel, are set into the bottom plate of the jig and are accurately ground so that their surfaces are in a single plane. The left-hand end of the part is held against another button Y in Figure 28-5. This locating button is built into the jig so that its surface is at right angles to the plane of the X buttons. When the block is placed in the jig, its rear surface rests against three more buttons marked Z. These buttons are located and ground so that their surfaces lie in a plane that is at right angles to the planes of both the X and Y buttons. The part is held in its located position by the two clamps marked C.

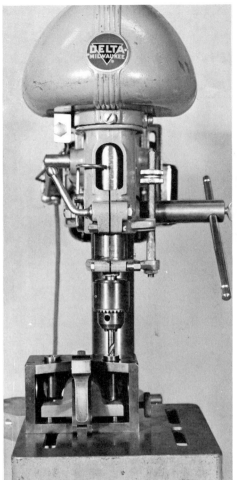

FIGURE 28-6 (Left) Drilling the mounting holes in a bearing block, using the drill jig shown in Figure 28-5. (Right) Close-up view, showing the block in the jig, resting on the location buttons, and the drill being guided by the drill bushing.

The use of four buttons on the bottom of this jig (X buttons) appears not to adhere to the 3–2–1 principle stated previously. However, although only two X buttons would have been required for complete location, the use of only two buttons would not have provided adequate support during drilling. The thrust from the drills would have dislodged the part from the locators. Thus the 3–2–1 principle is a *minimum* concept and often must be exceeded.

To assure that the mounting holes are drilled in their proper locations, the drill must be located and then guided during the drilling process. This is accomplished by the two drill bushings, marked K in Figure 28-5. Such drill bushings are accurately made of hardened steel with their inner and outer cylindrical surfaces concentric. The inner diameter is made slightly larger than the drill—usually 0.005 to 0.002 in.—so that the drill can turn freely but not shift appreciably. The bushings are accurately mounted in the upper plate of the jig and positioned so that their axes are exactly perpendicular to the plane of the X

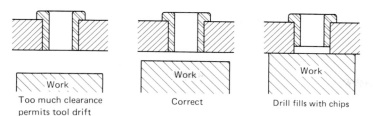

Too much clearance permits tool drift

Correct

Drill fills with chips

FIGURE 28-7 Proper clearance between drill bushing and workpiece.

buttons, at a distance *A* from the *Z* buttons and at distances *B* and *C*, respectively, from the plane of the *Y* button. Note that the bushings are sufficiently long that the drill is guided close to the surface where it will start drilling. Consequently, when the workpiece is properly placed and clamped in the jig, the drill will be located and guided by the bushings so that the critical dimension on the workpiece will be correct. Figure 28-6 shows the right hole being drilled. The box construction is rigid but open.

When jigs or fixtures are used in connection with chip-making operations, adequate provision must be made for the easy removal of the chips. This is essential for several reasons. First, if chips become packed around the tool, heat will not be carried away and tool life can be decreased. Figure 28-7 illustrates how insufficient clearance between the end of a drill bushing and the workpiece can prevent the chips from escaping, whereas too much clearance may not provide accurate drill guidance and can result in broken drills.

A second reason why chips must be removed is so that they do not interfere with the proper seating of the work in the jig or fixture. This is illustrated in Figure 28-8. Even though chips and dirt always have to be cleaned from the locating and supporting surfaces by a worker or by automatic means, such as an air blast, the design details should be such that chips and other debris will not readily adhere to, or be caught in or on, the locating surfaces, corners, or overhanging elements and thereby prevent the work from seating properly. Such a condition results in distortion, high clamping stresses, and incorrect finish dimensions.

The cost of the workholders must be justified by the quantities of production involved, and their primary purpose is to increase productivity and quality. While work is being put into or being taken out of jigs and fixtures, the machines with which they are used are not making chips. The load/unload time plus the machining time (also called the run time) plus any delay times equals the cycle time for a part. The load/unload time is greatly influenced by the choice of clamps.

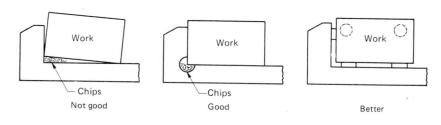

Not good

Good

Better

FIGURE 28-8 Methods of providing chip clearance to assure proper seating of the work.

TABLE 28.2 **Effects of Reducing Load/Unload Time**

	Load/Unload Time (min)	Machining Time (min)	Total Cycle Time (min)	Cost per Unit	Cost Saving per Unit (%)
A					
Labor: $10.00/hr	3	10	13	$7.25	7.8
Machine: $23.50/hr	2	10	12	$6.69	
B					
Labor: $8.00/hr	3	5	8	$4.53	12.6
Machine: $26.00/hr	2	5	7	$3.96	

In Table 28.2, the effect of reducing change time by one minute is shown for two conditions—one in which the machining time is 10 minutes and the other in which it is only 5 minutes. In the first case, A, an operator receives $10.00 per hour, and the cost for the machine is $23.50 per hour, which includes all overhead and interest costs. In the second case, B, the operator is paid $8.00 per hour because the machine is more nearly automatic, but this results in a cost for the machine of $26.00 per hour. The results of decreasing the load/unload time by a constant amount are very evident. Although the cost saving in each case happens to be the same ($0.57 per unit), the percentage savings is much greater in the case of the more productive machine. Many machines are equipped with matched (or multiple) fixtures so one part can be loaded while another part is being machined. The initial cost for the workholders is doubled and the savings must come from decreased load/unload time. In Chapter 44, the cycle time problem is approached from a systems aspect.

There are several ways in which jigs and fixtures can be made easier to load/unload. Some clamping methods can be operated more readily than others. For example, in the drill jig shown in Figure 28-5, a *knurled clamping screw* is used to hold the block against the buttons at the end of the jig. To clamp or unclamp the block in this direction requires several motions. On the other hand, a *cam latch* is used to close the jig and hold the workpiece against the rear locating buttons. This type of latch can be operated with a single motion. Certainly the device should be designed so that the part cannot be loaded incorrectly. This "mistake proofing" is often accomplished by the clamping device. If the part is not loaded properly, it cannot be clamped.

Ease of operation of workholders not only directly increases the productivity of such equipment but also results indirectly in better quality and fewer lost-time accidents.

As will be pointed out in Chapter 29, the workholder is as critical as the machine tool and the cutting tool to the final quality of the part. See Figure 29-22. The use of the workholder eliminates manual layout of the desired features of the part on the raw material. Manual layout requires a highly skilled worker and is very time-consuming. The workholder permits a lower-skilled person to achieve quality and repeatable production with far greater efficiency.

Every part coming out of the workholder should be the same, resulting in interchangeable parts. But what about the first part? What about the initial setup of the workholder into the machine? In many cases, this setup operation takes hours and the machine is not producing anything during this time. Rapid exchanges in tooling (RETAD) will be discussed as a key technique in IMPS. See Chapter 44. Reducing setup times permits shorter production runs (smaller lot sizes). Do not confuse initial setup (of workholders) with part loading and unloading or tool changing. The trick with initial setup is to do it quickly and to get the first part out of the process as a good part, with no adjustment of the machine, the tooling, or the workholder.

Types of Jigs

Jigs are made in several basic forms and carry names that are descriptive of their general configurations or predominant features. Several of these are illustrated in Figure 28-9.

A *plate jig* is one of the simplest types, consisting only of a plate that contains the drill bushings and a simple means of clamping the work in the jig, or the

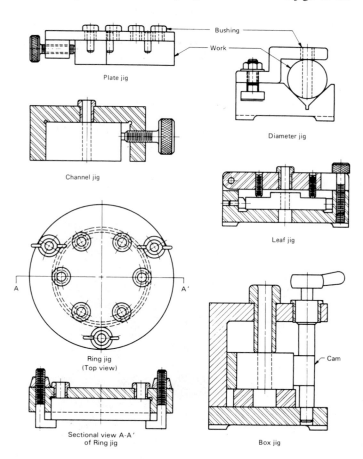

FIGURE 28-9 Examples of some common types of workholders—jigs.

jig to the work. In the latter case, wherein the jig is clamped to the work, the device is sometimes called a *clamp-on jig*. Such jigs frequently are used on large parts, where it is necessary to drill one or more holes that must be spaced accurately with respect to each other, or to a corner of the part, but that need not have an exact relationship with other portions of the work.

Channel jigs also are simple and derive their name from the cross-sectional shape of the main member. They can be used only with parts having fairly simple shapes.

Ring jigs are used only for drilling round parts, such as pipe flanges. The clamping force must be sufficient to prevent the part from rotating in the jig.

Diameter jigs provide a means of locating a drilled hole exactly on a diameter of a cylindrical or spherical piece.

Leaf jigs derive their name from the hinged leaf or cover that can be swung open to permit the workpiece to be inserted and then closed to clamp the work in position. Drill bushings may be located in the leaf as well as in the body of the jig to permit locating and drilling holes on more than one side of the workpiece. Such jigs are called *rollover jigs* or *tumble jigs* when they require turning to permit drilling from more than one side.

Box jigs are very common, deriving their name from their boxlike construction. They have five fixed sides and a hinged cover or leaf, or, as shown in Figure 28-9, a cam that locks the workpiece in place. Usually, the drill bushings are located in the fixed sides to assure retention of their accuracy. The fixed sides of the box usually are fastened by means of dowel pins and screws so that they can be taken apart and reassembled without loss of accuracy. Because of their more complex construction, box jigs are costly, but their inherent accuracy and strength can be justified when there is sufficient volume of production. They have two obvious disadvantages: (1) it usually is more difficult to put work into them than into simpler types, and (2) there is a greater tendency for chips to accumulate within them. Figure 28-5 shows a box-type jig.

FIGURE 28-10 (Left) Two types of universal jigs, manual (top) and power-actuated (bottom). (Right) Completed jig, made from unit shown at lower left, (Courtesy Cleveland Universal Jig Division, The Industrial Machine Company.)

Because jigs must be constructed very accurately and be made sufficiently rugged so as to maintain their accuracy despite the abuse to which they inevitably are subjected in use, they are expensive. Consequently, several methods have been devised to aid in lowering the cost of manufacturing jigs. One way to reduce this cost is to use simple, standardized plate and clamping mechanisms called *universal jigs,* such as shown in Figure 28-10. These can easily be equipped with suitable locating buttons and drill bushings to construct a jig for a particular job. Such universal jigs are available in a variety of configurations and sizes and, because they can be produced in quantities, their cost is relatively low. However, the variety of work that can be accommodated by such jigs obviously is limited.

Many examples of conventional fixtures have appeared in the text. Production milling, broaching, and boring processes as performed on NC machines, conventional equipment, or machining centers, routinely use fixtures to locate and hold the part properly with respect to the cutting tools on the machine tool. Like cutting tools, fixtures are sold separately and are not usually supplied by the machine tool builder. Traditionally, beginning with Eli Whitney, manufacturers have designed and built custom-made, dedicated fixtures. See Figures 25-19, and 26-13 for two examples. Because of the pressure of shorter production runs and smaller lot sizes, many companies are turning to modular fixturing approaches. The greatest advantage of these systems is that the fixture can be quickly constructed.

Perhaps the most common fixture uses the vise as its base element. Figure 28-11 shows a typical setup using a vise as a fixture. The vise jaws are readily modified to conform to the 3–2–1 location principle and provide adequate clamp-

Examples of Conventional Fixtures

FIGURE 28-11   (Left) Standard vise with removable jaw plates can be modified and used as a fixture for milling (right). (Courtesy Kurt Manufacturing Company).

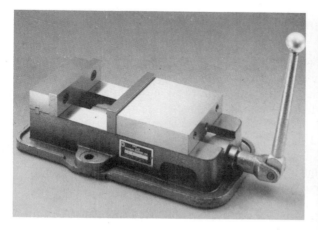

1. Pre-assembled Mini-System and top jaws.

2. Assembly being inserted into the Master Jaw

3. Quickly retighten cap screws.

4. All 3 jaws changed in 5 minutes or less.

FIGURE 28-12 Quick-changing of the top jaws on a three-jaw chuck. (Courtesy Huron Machine Products.)

ing forces for almost every machining operation. The four vises shown in Figure 28-11 are mounted on a subplate for rapid insertion on the machine.

The chucks used in lathes are really general-purpose fixtures for rotational parts. Newer chuck designs have greatly improved their flexibility (the range of diameters the chuck can accommodate in a given setup and speed of setup). Figure 28-12 shows a complete change of top jaws for a three-jaw chuck being done in less than 5 minutes. The normal time for this part of the setup might exceed 15 minutes. New quick-change insert top jaws may even snap in by hand with no jaw nuts, keys, screws, or tools. Such jaws are changeable by robots.

Figure 28-13 shows an example of the *intermediate jig concept,* also discussed in Chapter 44, applied to lathes and chucks. An adaptor or intermediate fixture is bolted to the machine spindle and is a permanent part of the machine tool. The intermediate fixture will accept mating chucks that have been preset for the workpiece prior to insertion. Different chuck designs mount interchangeably on

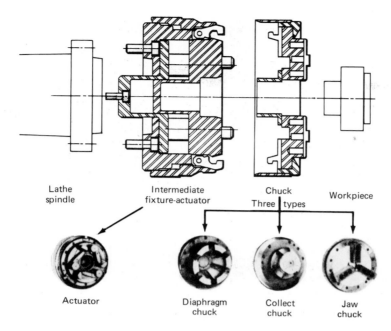

Lathe spindle · Intermediate fixture-actuator · Chuck Three types · Workpiece

Actuator · Diaphragm chuck · Collect chuck · Jaw chuck

FIGURE 28-13 Example of the intermediate jig concept applied to lathe chucks. (Courtesy Sheffer Collet Company.)

the common actuator. This method greatly reduces setup time and permits the operator to perform chuck maintenance and retooling (setup) while the machine is running. These chucks can be exchanged automatically.

Most producers of chucks use some variation of equation 28-1 to compute the maximum rpm at which the chuck can run:

$$S_m^2 = \frac{F_m}{3 \times (2.84 \times 10^{-5}) \times W \times D} \qquad (28\text{-}1)$$

where S_m = maximum rpm at which gripping force equals $1/3$ F_m
$\quad F_m$ = maximum rated gripping force, at rest (lbs)
$\quad W$ = combined weight of jaws (lbs)
$\quad D$ = distance from spindle centerline to center of jaw mass (in.)

Thus, with this equation, a 10 in. power chuck with a published rating, F_m, of 13,200 lbs would retain one third of its initial gripping force at 2507 rpm. (Check this calculation using $W = 8$ lbs, $D = 3.1$ in.). The greater the rpm, the greater the centrifugal force factor. This is an important factor in high-speed machining operations in which the part is rotating.

In Figure 28-14 some typical types of clamps that are used in fixtures are shown. The *strap clamp* comes in many forms and sizes. The force can be applied by a hand knob, a cam, or a wrench turning down a nut. Figure 28-15 shows some examples of power-actuated clamps. *Extending clamps* operate in a manner similar to that of a manual clamp–strap assembly. They extend forward horizontally, then clamp down. *Edge* clamps have a very low profile. They simultaneously clamp down and forward.

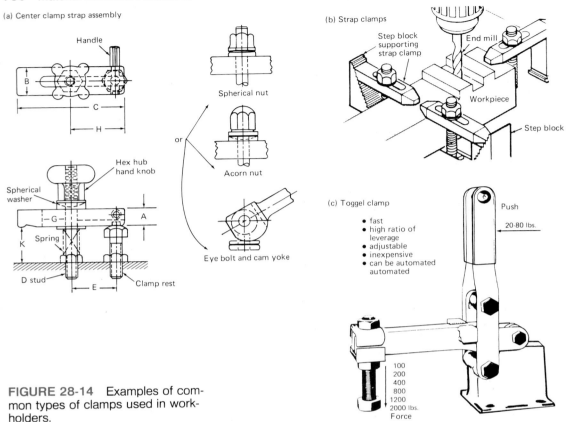

(a) Center clamp strap assembly

Handle

B

C

H

Hex hub hand knob

Spherical washer

G

Spring

A

K

D stud

E

Clamp rest

or

Spherical nut

Acorn nut

Eye bolt and cam yoke

(b) Strap clamps

Step block supporting strap clamp

End mill

Workpiece

Step block

(c) Toggel clamp

Push

20-80 lbs.

- fast
- high ratio of leverage
- adjustable
- inexpensive
- can be automated automated

100
200
400
800
1200
2000 lbs.
Force

FIGURE 28-14 Examples of common types of clamps used in workholders.

Modular Fixturing

Modular fixtures have all the same design criteria as conventional fixtures plus one more—*versatility*. Modular fixture elements must be useful for a variety of machining applications and easily adaptable to different workpiece geometries. Individual fixture designs can be photographed or entered into a CAD library for future reference. After the job is done, the fixture itself can be dismantled and the elements returned to the toolroom.

The erector-set approach uses either tee-slot or dowel-pin designs. Figures 28-16 and 28-17 show two examples of modular fixturing. The designs begin with base plates. Elements for locating and clamping are added to the subplate. Rectangular, square, and round are the typical patterns for the subplates. Also shown are the typical components for modular fixturing systems used for mounting points, locators, attachments, and so on. The standard elements needed to construct the fixture include riser blocks, vee blocks, angle plates, cubes, box parallels, and the like. Smaller elements like locator pins, supports, pads, and clamps are added to the subplate on the larger structural elements. Mechanical clamping devices are shown, but power-assisted clamps are available. The base

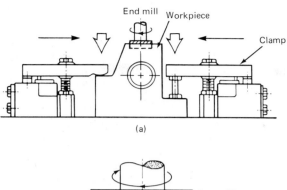

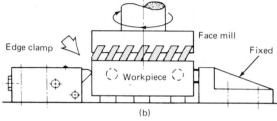

FIGURE 28-15 Examples of power clamping devices: (a) Extending clamp. (b) Edge clamp.

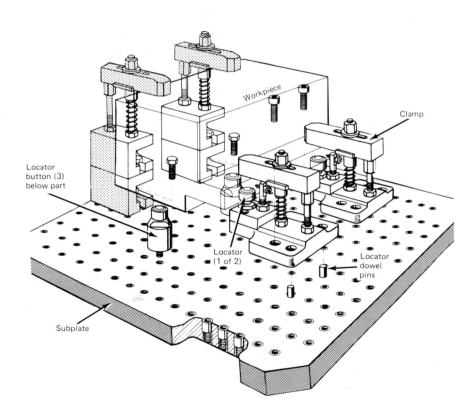

FIGURE 28-16 Modular fixturing begins with a subplate (grid base) and adds locators and clamps.

FIGURE 28-17 The dedicated fixture on left versus the modular fixture on the right. (*Manufacturing Engineering,* January 1984.)

and fixturing elements are made to tolerances of plus or minus .0002 to .0004 in. in flatness, parallelism, and size. Figure 28-17 shows a part in a dedicated fixture compared to a modular fixture. The dedicated fixture represents a capital investment that must be absorbed by the job and must be maintained after the job is complete. The modular fixture is disassembled and the elements reused later in fixtures for other parts. Modular fixtures are commonly used for prototype tooling and small batch production runs. They are being incorporated more frequently into regular production as users gain confidence in this approach.

The Group Jig/Fixture

The concept of group jigs and fixtures originated from the group technology (GT) concept. GT will be discussed in Chapter 42 as a method to form cellular manufacturing systems by determining a family of parts. For a cell to be flexible, the workholding devices should be able to accommodate all the parts within the parts family. For unmanned cells, the workholders will also have to compensate for variation in cutting forces, centrifugal forces, and so forth. Group workholders are designed to accept every part family member, with adapters that accommodate minor part variations. Figure 28-18 shows a group jig for processing a part family. Drilling the holes of different parts in this part family requires only one group jig. This jig has different auxiliary adapters to accommodate minor differences in hole sizes, numbers, and locations, and in part sizes and shapes. Therefore, instead of designing, fabricating, and using individual drill jigs as is done in a conventional production method, only one group jig and adapter or bushing plate is required. The jig never leaves the machine and therefore never gets lost.

Within group technology is the composite part concept, which is used solely as an aid in designing group fixtures and jigs and group tooling setups. As

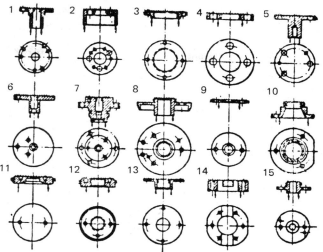

The family of parts is composed of 15 parts

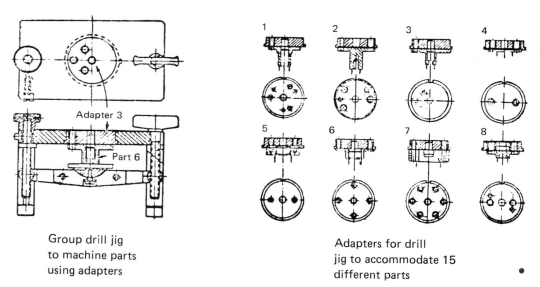

Group drill jig
to machine parts
using adapters

Adapters for drill
jig to accommodate 15
different parts

FIGURE 28-18   Example of a master drill jig which accommodates adapters so that 15 parts in the family can be drilled with very rapid changeover and savings in tooling costs.

Figure 28-19 shows, a group of similar parts (that is, similar in design) is represented by the composite part that possesses all the part family's shape characteristics and machining features. In other words, the composite part is an envelope, the shape of which encompasses the shapes of all the parts in the family. The theory is that if the tooling is designed for the composite part, any part that fits within the envelope could be machined without any tooling changes.

These GT approaches have been shown to be effective in the reduction of setup time.

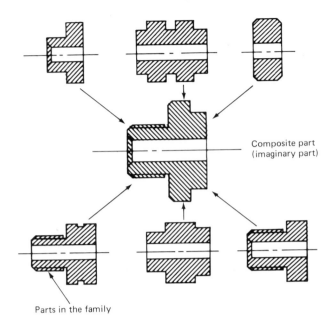

Composite part
(imaginary part)

Parts in the family

FIGURE 28-19 Example of composite part for a part family with six parts.

Assembly Jigs. Because *assembly jigs* usually must provide for the introduction of several component parts and the use of some type of fastening equipment, such as welding or riveting, they commonly are of the open-frame type. Such jigs are widely used in the automobile and aircraft industries. Large jigs of this type are shown in Figure 28-20 for the assembly of aircraft wings. This jig is constructed mainly of reinforced concrete.

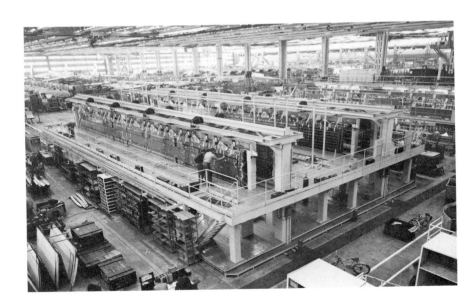

FIGURE 28-20 Example of large assembly jig for an airplane wing.

Other Workholding Devices

Magnetic Workholders. Workpieces usually are held in a different manner on surface grinders than on other machine tools. To obtain high accuracy, it is desirable to reduce clamping forces and distribute them over the entire area of the workpiece. Also, grinding is frequently done on quite thin or relatively delicate workpieces, which would be difficult to clamp by normal methods. In addition, there often is the problem of grinding a number of small, duplicate workpieces. Magnetic chucks solve all these problems very satisfactorily.

Magnetic chucks are available in disc or rectangular shapes as shown in Figure 28-21. Dry-disc rectifiers are used to provide the necessary direct-current power. Some older types of magnetic chucks utilize permanent magnets.

Magnetic chucks provide an excellent means of holding workpieces provided the cutting or inertial forces are not too great. The holding force is distributed over the entire contact surface of the work, the clamping stresses are low, and therefore there is little tendency for the work to be distorted. Consequently, pieces can be held and ground accurately. Also, a number of small pieces can be mounted on a chuck and ground at the same time. Magnetic chucks provide great part-to-part repeatability because the holding power from one part to the next is the same. Initial setup is usually fast, simple, and relatively inexpensive. Parts loading and unloading is also relatively easy.

It often is necessary to demagnetize work that has been held on a magnetic chuck. Some electrically powered chucks provide satisfactory demagnetization by reversing the direct current briefly when the power is shut off.

Electrostatic Workholders. Magnetic chucks can be used only with ferromagnetic materials. Electrostatic chucks can be used with any electrically conductive material. This principle, indicated in Figure 28-22, directs that the work be held by mutually attracting electrostatic fields in the chuck and the workpiece. These provide a holding force of up to 20 psi (21,000 Pa). Nonmetal parts usually can be held if they are flashed (that is, coated) with a thin layer of metal. These chucks have the added advantage of not inducing residual magnetism in the work.

Vacuum Chucks. Vacuum chucks are also available. In one type, illustrated in Figure 28-23, the holes in the work plate are connected to a vacuum pump and can be opened or closed by means of valve screws. The valves are opened in the area on which the work is to rest. The other type has a porous plate on which the work rests. The workpiece and plate are covered with a polyethylene sheet. When the vacuum is turned on, the film forms around the workpiece, covering and sealing the holes not covered by the workpiece and thus producing a seal. The film covering the workpiece is removed or the first cut removes the film covering the workpiece. Vacuum chucks have the advantage that they can be used on both nonmetals and metals and can provide an easily variable force. Magnetic, electrostatic, and vacuum chucks are used for some light milling and turning operations.

LOW PROFILE INTERLOC CHUCK

CERAMAX CHUCK

SINE PLATES

ROTARY ELECTROMAGNETIC CHUCKS

FIGURE 28-21 Examples of magnetic chucks (Courtesy O. S. Walker).

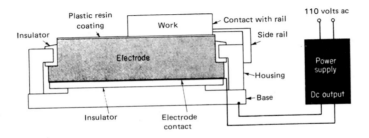

FIGURE 28-22 Principle of electrostatic chuck.

Economic Justification of Jigs and Fixtures

As discussed previously, workholders are expensive, even when designed and constructed by using standard components. Obviously, their cost is a part of the total cost of production, and one must determine whether they can be justified economically by the savings in labor and machine cost and improvements in quality that will result from their use. Often it is only through the use of such devices that the design specifications can be met and sustained from part to part.

In order to determine the economic justification of any special tooling, the following factors must be considered:

1. The cost of the tooling.
2. Interest or profit charges on the tooling cost.
3. The savings resulting from the use of the tooling. The savings can result from reduced cycle times, improved quality, or use of lower-skilled labor.
4. The savings in machine cost due to increased productivity.
5. The number of units that will be produced using the tooling.

Considering the savings due to use of lower-skilled labor, the economic relationship between these factors can be expressed in the following manner:

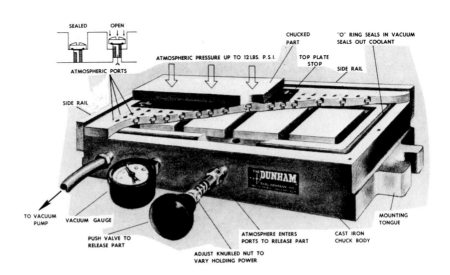

FIGURE 28-23 Cutaway view of a vacuum chuck. (Courtesy Dunham Tool Company, Inc.)

savings per piece (exclusive of tooling costs) $\geq$ additional cost per piece

$$
\underbrace{
\begin{array}{c}
\text{total cost per piece} \\
\text{without tooling}
\end{array}
\quad - \quad
\begin{array}{c}
\text{total cost per} \\
\text{piece using tooling} \\
\text{(exclusive of tooling costs)}
\end{array}
}
\quad \geq \quad \text{tooling cost per piece}
$$

$$
\left[
\begin{array}{c}
\text{labor cost} \\
\text{per piece} \\
\text{without} \\
\text{tooling}
\end{array}
+
\begin{array}{c}
\text{machine and} \\
\text{overhead cost} \\
\text{per piece} \\
\text{without} \\
\text{tooling}
\end{array}
\right]
-
\left[
\begin{array}{c}
\text{labor cost} \\
\text{per piece} \\
\text{with} \\
\text{tooling}
\end{array}
+
\begin{array}{c}
\text{machine and} \\
\text{overhead cost} \\
\text{per piece} \\
\text{with tooling}
\end{array}
\right]
\geq
\begin{array}{c}
\text{cost} \\
\text{of} \\
\text{tooling}
\end{array}
+
\begin{array}{c}
\text{interest on} \\
\text{tooling cost}
\end{array}
$$

$$
[(R)(t) + (R_m)(t)] - [(R_t)(t_t) + (R_m)(t_t)] \geq \frac{C_t + (C_t/2)(n)(i)}{N} \tag{28-2}
$$

where R = labor rate per hour, without tooling
R_t = labor rate per hour, using tooling
t = hours per piece, without tooling
t_t = hours per piece, using tooling
R_m = machine cost per hour, including all overhead
C_t = cost of the special tooling
n = number of years tooling will be used
i = interest rate (or what invested capital is worth)
N = number of pieces that will be produced with the tooling

This equation can be expressed in a simpler form:

$$
(R + R_m) t - (R_t + R_m)t_t \geq \frac{C_t}{N}\left(1 + \frac{n \times i}{2}\right) \tag{28-3}
$$

This equation assumes straight-line depreciation and computes interest on the average amount of capital invested throughout the life of the tooling.[1] When the time over which the tooling is to be used is less than one year, companies often do not include an interest cost. If this factor is neglected, the right-hand term of equation (28-3) reduces to C_t/N.

The equations assume that the material cost will be the same regardless of whether special tooling is used or not. This is not always true. Although these equations are not completely accurate for all cases, they are satisfactory for determining tooling justification in most cases, because the life of such tooling seldom exceeds five years and more frequently does not exceed two years. The equation does not include the cost of poor quality. This can be included by estimating the decrease in the number of defective parts when the workholder is used versus when it is not used.[2]

[1] For the use of more sophisticated economic analysis, see E. P. DeGarmo, J.R. Canada, and W. G. Sullivan, *Engineering Economy,* 6th ed., Macmillan Publishing Company, Inc., New York, 1979.
[2] For a discussion of new measures and methods on cost accounting, see C. S. Park, "Counting the Costs," Jan. 1987, *Mechanical Engineering,* p. 66.

The following example illustrates the use of equation (28-3) to determine tooling justification. In drilling a series of holes on a radial drill, the use of a drill jig will reduce the time from ½ hour per piece to 12 minutes per piece. If a jig is not used, a machinist, whose hourly rate is $8.00, must be used. If the jig is used, the job can be done by a machinist whose rate is $6.50 per hour. The hourly rate for the radial drill is $8.75 per hour.

The cost of making the jig would include $250 for design, $75 for material, and 50 hours of toolmaker's labor, which is charged at the rate of $12.00 per hour to include all machine and overhead costs in the toolmaking department.

Investment capital is worth 16% to the company. It is estimated that the jig would last 3 years and that it would be used for the production of 300 parts over this period. Is the jig justified? How many parts would have to be produced with the jig to "break even"—that is, increased costs just equal savings?

The cost of the jig, C_t, is estimated to be $925.00.

$$C_t = \$250 + \$75 + \$12 \times \$50 = \$925$$

Substituting the values given in equation (28-3) yields:

$$(8.00 + 8.75)0.5 - (6.50 + 8.87)\,0.2 \geqslant \frac{\$925}{300}\left[1 + \frac{3 \times 0.16}{2}\right]$$

$$\text{or } 5.33 \geqslant 3.82$$

Thus, use of the jig is justified. By omitting the value 300 in the solution above and solving for N, it is found that at least 216 pieces would have to be produced with the jig for it to break even or pay out.

It should be noted that equations (28-2) and (28-3) assume that the time (of the people and machines) saved by the use of the special tooling can be used for other operations. If this is not the case, the cost analysis should be altered to take this important fact into account. Otherwise, the tooling justification may be substantially in error.

The concepts of group technology and NC machines, discussed in Chapters 29 and 42, may eliminate the need for designing and building a new jig or fixture every time a new part is designed. New measures of manufacturing performance that include terms for quality and flexibility are being developed.

Review Questions

1. What are the two primary functions of a workholding device?
2. What distinguishes a jig from a fixture?
3. An early treatise defined a jig as "a device that holds the work and guides a tool." Why was this definition incorrect?
4. Why would an ordinary vise not be considered to be a fixture?
5. What basic criteria should be considered in designing jigs and fixtures?
6. What difficulties can result from not keeping clamping stresses low in designing jigs and fixtures?
7. Explain the 3–2–1 concept for workpiece location.
8. Which of the basic design principles relating to jigs and fixtures would most likely be in conflict with the 3–2–1 location concept?
9. Why would you want a drill bushing to be readily removable?

10. What are two reasons for not having drill bushings actually touching the workpiece?

11. Why does the use of down milling often make it easier to design a milling fixture than it would be if up milling were used?

12. Explain what is meant by the expression, "A jig is a skill-transfer unit."

13. A large assembly jig for an airplane-wing component gave difficulty when it rested on four points of support but was satisfactory when only three supporting points were used. The assembled wing components were not consistant in shape. Why?

14. Explain why the use of a given fixture may not be economical when used with one machine tool but may be economical when used in conjunction with another machine tool.

15. What are roll-over jigs and what advantages do they offer?

16. In the clamps shown in Figure 28-14, what purpose has the spherical washer?

17. What are other common types of clamps?

18. What is the purpose of dimensioning the strap–clamp assembly in Figure 28-14 with letters?

19. Figure 28-2 showed three types of locator buttons. Identify them.

20. In Figure 28-5, why are there not three points put on the X plane, two points on the Z plane, and one point on the Y plane?

21. Which set of locators in Figure 28-5 establishes the A dimension?

22. To prepare the workpiece shown in Figure 28-5, which surface would you have milled first, the bottom or the back or the front?

23. For the part shown in Figure 28-5, why do you not drill the holes first, then mill? Why mill at all?

24. The holes are countersunk after they are drilled. How must the jig be designed in order to put the countersinks on the mounting holes while the part is in the jig?

Problems

1. Using the following values, determine the number of pieces that would have to be made to justify the use of a jig costing $3,000:

$$R = \$5.75$$
$$R_t = \$4.50$$
$$t_t = 1\ 1/4$$
$$t = 2\ 1/2$$
$$i = 10\%$$
$$R_m = \$4.50$$
$$n = 3$$

2. Suppose, in the sample problem given in the text, modular fixturing is used, which reduces the toolmaker's labor to four hours and the design cost to $100 (4 hours at $25 per hour). The material cost (modular elements) for the subplate structural elements, clamps, etc. was $600. What is the break-even quantity for a modular fixture? Note: The modular fixture is used for the job, then disassembled and returned to the tool room. The parts are reused in other workholders.

Machining Centers

Introduction

Machining, as a fundamental method of material removal, has evolved from individual machines which, with the aid of man, performed individual processes to machines capable of performing many processes. The processes of turning (boring), drilling (and related processes of reaming and tapping), milling, shaping (and planing), broaching, and grinding have been discussed in earlier chapters. Processes are what this book is all about. However, in the early 1950s, it became possible to control the movement of machine tools by numbers, and soon numerically controlled (NC) milling machines, NC drilling machines, and NC boring machines were commercially available. In 1968, Kearney & Trecker marketed a NC machine which could automatically change tools so that many different processes could be done on one machine. Such a machine became known as a *machining center*—a machine that can perform a variety of processes and change tools automatically while under programmable control. See Figure 29-1. Programmable control means that the workpiece (or the tool) can be *moved,* using software, to specific locations and can even machine multiple surfaces.

The study of *machining centers* begins with the history of numerical control, often abbreviated as NC. NC is programmable automation in which certain functions of the machine tools are controlled by numbers or letters. The program of instruction defines how a particular part is to be made. If the design changes (or the method), it is necessary to change the program instructions. Nowadays, writing software programs for instruction/control of machines is fairly common but in 1955, before the advent of NC machines, it was unheard of.

Numerical Control. The advent and wide-scale adoption of numerically (tape- and computer-) controlled machine tools has been the most significant

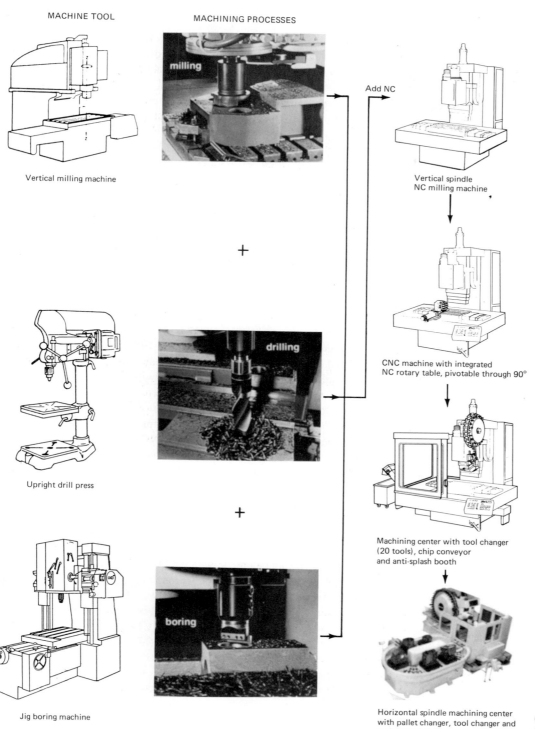

MACHINE TOOL

MACHINING PROCESSES

milling

Vertical milling machine

Upright drill press

Jig boring machine

drilling

boring

+

+

Add NC

Vertical spindle
NC milling machine

CNC machine with integrated
NC rotary table, pivotable through 90°

Machining center with tool changer
(20 tools), chip conveyor
and anti-splash booth

Horizontal spindle machining center
with pallet changer, tool changer and
anti-splash booth

FIGURE 29-1 The
evolution of the
machining center.

development in manufacturing during the past 30 years. These machines raised automation to a new level (A4) by providing positional feedback as well as programmable flexibility. Numerical control of machine tools created entirely new concepts in manufacturing. Certain operations are now routine that previously were very difficult if not impossible to accomplish. In earlier years highly trained NC programmers were required. The development of low-cost, solid-state microprocessing chips has resulted in machines that can be programmed in a very short time, by personnel having only a few hours of training, using only simple machine shop language. As a consequence, there are few manufacturing facilities today, from the largest down to the smallest job shops, that do not have, and routinely use, one or more numerically controlled machine tools.

NC came into being to fill a need. The United States Air Force and the airframe industry were seeking a means to manufacture complex contoured aircraft components to close tolerances on a highly repeatable basis. John Parsons of the Parsons Corporation of Traverse, Michigan, had been working on a project for developing equipment that would machine templates to be used for inspecting helicopter blades. He conceived of a machine controlled by numerical data to make these templates and took his proposal to the USAF. Parsons convinced the USAF to fund the development of a machine. The Massachusetts Institute of Technology (MIT) was subcontracted to build a prototype machine. The prototype was a conventional two-axis tracer mill retrofitted with servomechanisms. As luck would have it, the servomechanism lab was located next to a lab where one of the very first digital computers (Whirlwind) was being developed. This computer generated the digital numerical data for the servomechanisms.

By 1962, NC machines accounted for about 10% of total dollar shipments in machine tools. Today over 75% of the money spent for drill presses, milling machines, lathes, and machining centers goes for NC equipment.

The early machines were continuous-path or contouring machines where the entire path of the tool was controlled with close accuracy in regard to position and velocity. Today, milling machines, machining centers, and lathes are the most popular applications of continuous-path control requiring feedback control. Next, point-to-point machines were produced where the path taken between operations is relatively unimportant and therefore not continuously monitored. Point-to-point machines are chiefly used for drilling, milling straight cuts, cutoff, and punching. Automatic tool changers, which require that the tools be precisely set to a given length prior to installation in the machines, permitted the merging of many processes into one machine. See Figure 29-2. The two- (or four-sided) tombstone fixture shown has multiple mounting and locating holes for attaching part-dedicated fixture plates, which greatly extends the utility of a horizontal machining center.

During the early days of NC, programming was done manually or with a computer assist for complex workpiece geometries. The chief problems in NC programming were tool radius compensation and tool path interpolation which will be discussed later. Computer software capability to perform linear, circular, parabolic, and other kinds of interpolations had to be developed. Such software is now routinely available on NC machines.

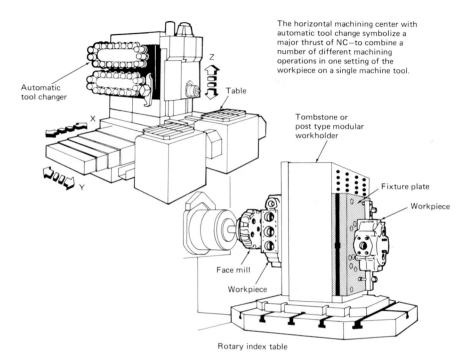

The horizontal machining center with automatic tool change symbolize a major thrust of NC—to combine a number of different machining operations in one setting of the workpiece on a single machine tool.

Automatic tool changer

Table

Z

X

Y

Tombstone or post type modular workholder

Fixture plate

Workpiece

Face mill

Workpiece

Rotary index table

FIGURE 29-2 Horizontal machining center with three axes of simultaneous control typifies the state of art in machining tools. (Adapted from 1987 NC/CIM Guide Book, Modern Machine Shop.)

The USAF sponsored additional research at MIT to develop a computer language that would use simple English-like statements to produce an output that would control the NC machines. This language, called *APT* (Automatically Programmed Tools), was announced by MIT in 1959. It required a computer with a relatively large memory. It was widely adopted and is still used extensively in industry today. See Figure 29-3. *Postprocessing* takes the output from the APT program and converts it to input for a particular machine. Traditional postprocessing yields NC workpiece programs that are not exchangeable. To machine an identical workpiece on another machine, the program must be postprocessed again unless the machine and control are exactly the same.

It was envisioned that a large computer could be used directly to control, in real time, a number of machines. A limited number of direct numerical control (DNC) systems were developed, with the idea that the programs were to be sent directly to the machines (eliminating paper tape handling). The main-frame computer would be shared on a real-time basis by many machine tools. The machine operator would have access to the main computer through a remote terminal at the machine while management would have up-to-the-minute data on production status and machine utilization. This version of DNC had very few takers. Instead, NC machines became computer numerical control (*CNC*) machines through the development of small inexpensive computers, microprocessors with large memories, and programmable computers. See Figure 29-4. Now functions like program storage, tool offset and tool compensation, program-editing capability, various degrees of computation, and the ability to send

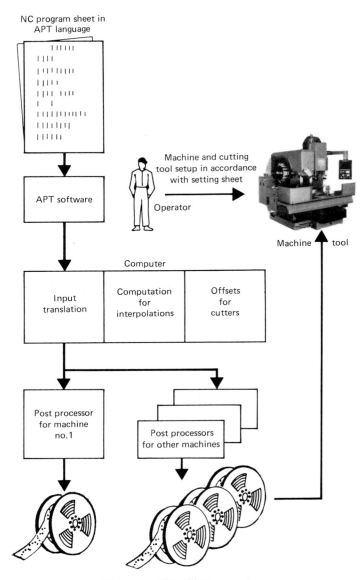

NC program sheet in
APT language

Machine and cutting
tool setup in accordance
with setting sheet

Operator

Machine tool

APT software

Computer

| Input translation | Computation for interpolations | Offsets for cutters |

Post processor
for machine
no. 1

Post processors
for other machines

NC tapes for NC machines

FIGURE 29-3 Steps in numerical control (NC) tape preparation, using APT computer software.

and receive data from a variety of sources, including remote locations, are routinely available through the onboard computer. The computer can store multiple-part programs, recalling them as needed for different parts. Immediately it was found that the machine tool operator could readily learn how to program these machines (manually) for many component parts, often eliminating the need for a part programmer.

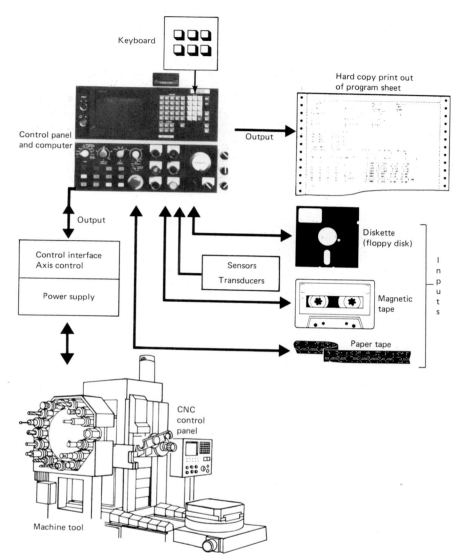

Keyboard

Hard copy print out
of program sheet

Control panel
and computer

Output

Control interface
Axis control

Power supply

Output

Sensors
Transducers

Diskette
(floppy disk)

Magnetic
tape

Paper tape

Inputs

CNC
control
panel

Machine tool

FIGURE 29-4   Computer numerical
control (CNC) inputs to control panel
and onboard computer provide output
to the machine tool.

In recent years, the DNC concept has been revived with the small computers at the machines being networked to a larger computer to provide enhanced memory and computer computational capacity. Minicomputers, supervising and controlling a flexible manufacturing system (FMS) or manufacturing cell, are networked to a large central computer. This DNC now means *distributed NC* with the distribution of NC programs by a central computer to individual CNC units.

The FMS, a special type of manufacturing system, will be discussed in detail in Chapter 42. Historically speaking, the first examples of FMS systems appeared in the late 1960s, but few companies adopted them because of their high initial cost.

The next level of automation, the A(5) level, requires that the control system perform an evaluation function of the process. Figure 29-5 shows diagrams for A(3) open-loop and three schemes for A(4) closed-loop control. The closer the feedback device is to the tool, the better the control scheme, but, usually, the more expensive. Figure 29-6 compares block diagrams for A(4) closed-loop control (with encoder) to A(5) adaptive control. A(5) requires feedback from the process or the product and seeks to optimize the process. CNC machines, with their onboard computers, are potentially capable of the A(5) level of automation. In the standard versions of NC and CNC machines in use today, speed

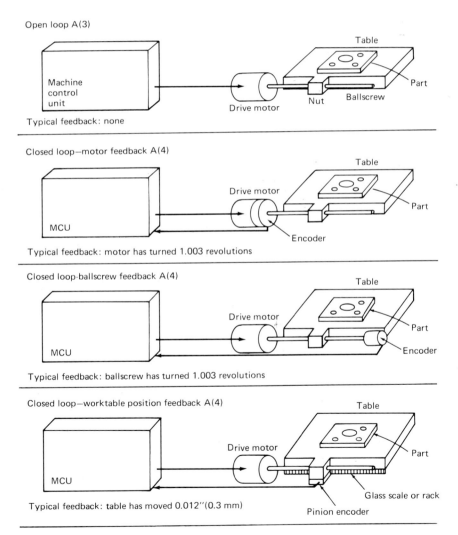

FIGURE 29-5 Axis position control schemes for NC and CNC machine tools.

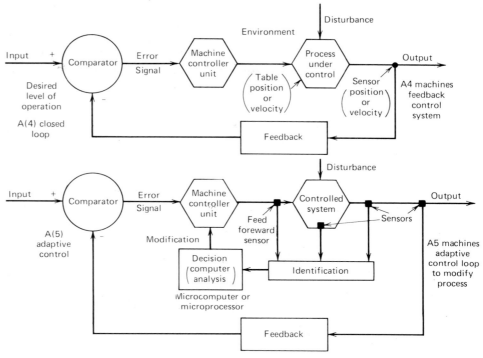

FIGURE 29-6 Block diagrams of A(4) and A(5) level of control.

and feed are fixed in the program unless the operator overrides them at the machine. If either speed or feed is too high, the result can be rapid tool failure, poor surface quality, or damaged parts. If the speed and feed are too low, production time is greater than desired for best productivity. An *adaptive control* (AC) system that can sense deflection force, heat, torque, and the like will use these measurements to make decisions about how the input parameters might be altered to *optimize* the process. This means that the computer must have *mathematical models* in its software that describe how this process behaves and mathematical functions that state what is to be optimized (cost per piece, surface quality, MRR, power consumed, and so on). The models require theory, and herein lies the problem. The theory of shape-generated processes (metal cutting and metal forming) is segregated and incomplete. With incomplete theory, the models become suspect and the systems unreliable. In addition, current versions of AC systems have been very costly to develop, suffer from inadequate data banks, and produce variable cycle times. However, AC systems have the potential of markedly improving productivity, quality, and machine utilization. Reaching this level of automation on a routine shop floor basis requires great expenditures of time and money. Most systems monitor parameters such as spindle deflection, horsepower, or cutting forces and override feed rates while trying to optimize metal removal rates.

Advantages of Numerical Control.   Through the use of either tape or computer control, there can be good assurance that consecutive parts are duplicated and that a part made at some later date will be the same as one made today. Thus, repeatability and quality are improved. Workholding devices can be made more universal and setup time reduced, along with tool change time, thus making programmable machines economical for producing small lots or even a single piece. When combined with the managerial and organizational strategies of group technology (GT), programmable machines lead to tremendous improvements in productivity. See Chapters 42 and 44. GT basically leads to the creation of families of parts made in machining cells containing flexible/programmable machines. The compatibility of the components (similarity in process and sequences of processes) greatly enhances the productivity (utility) of the programmable equipment.

A side result has been the decrease in the non-chip-producing time of machine tools. The operator was relieved of the jobs of setting speeds and feeds and locating the tool relative to the work. Even simple forms of NC and digital readout equipment have provided both greater productivity and increased accuracy. Most of the early NC machine tools were developed for special types of work where accuracies of as much as 0.00005 in. were required, and many NC machines are built to provide accuracies of at least 0.0001 in., whether it was needed or not. This forced many machine tool builders to redesign their machines because the operator was not available to compensate for the machine (positioning) error. While most NC machine tools today will provide greater accuracy than is required for most jobs, the tendency today is toward greater accuracy and precision (that is, better quality) but at no increase in cost. Therefore, NC machines will continue to be the very backbone of the machine tool business. Hopefully, all builders will one day arrive at a common language so that if one can program one machine, then one can program them all.

Basic Principles of Numerical Control.   As the name implies, *numerical control* is a method of controlling the motion of machine components by means of numbers or coded instructions. Assume that three 1-inch holes in the part shown in Figure 29-7 are to be drilled and bored on the vertical spindle machining center shown in Figure 29-8. The centers of these holes must be located relative to each other and with respect to the left-hand edge of the workpiece (X direction) and on the bottom edge (Y direction). The depth of the hole will be controlled by the Z (or W) axis.

The holes will be produced by center drilling, hole drilling, boring, reaming, and counter boring (five tool changes). If this were done conventionally or in manned cells, three or four different machines might be required. The movements of the table are controlled by coordinate systems. The accurate positioning of the cutting tool with respect to the work is established by zero points. The center of the table or a point along the edge of the traverse range is commonly used. See Figure 29-8. The workpiece is positioned on the table with respect to this zero point. For our example, the lower left-hand corner of the part is placed at the zero point.

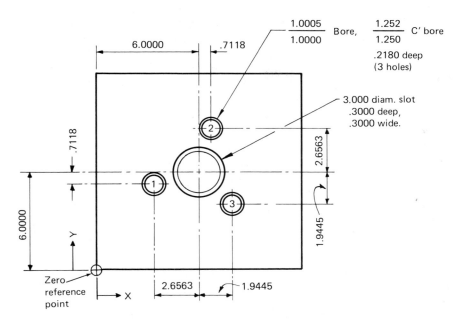

FIGURE 29-7 Part to be machined on a numerically controlled point-to-point machine tool.

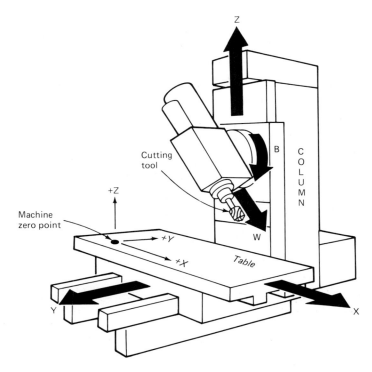

FIGURE 29-8 Five-axis vertical-spindle machining center showing location of machine zero point on table.

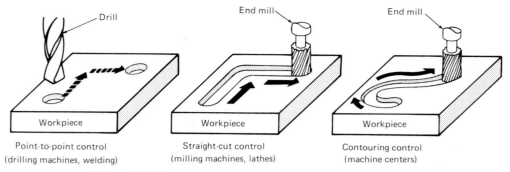

Point-to-point control
(drilling machines, welding)

Straight-cut control
(milling machines, lathes)

Contouring control
(machine centers)

FIGURE 29-9 NC and CNC systems are subdivided into three basic categories; point-to-point controls, straight-cut controls, and contouring controls.

NC and CNC machines can be subdivided into three types, as shown in Figure 29-9. In *point-to-point machines*, the tool path is not controlled. *Straight-line* controls permit axially parallel traverses at desired table feed rates, but only one axis drive is operated at a time. *Contouring* permits two or three axes to be controlled simultaneously, permitting two-dimensional or three-dimensional geometries to be generated. The point-to-point machines are usually *open-loop* rather than *closed-loop*, as was shown in Figure 29-6.

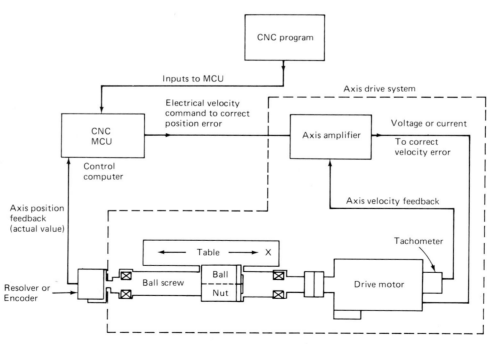

FIGURE 29-10 A CNC control system includes a velocity loop within an axis drive system and a position loop external to the axis drive system.

The X axis of this five-axis vertical-spindle CNC machine tool shown in Figure 29-8 will be used to explain how the closed-loop positional control works. The CNC control system, shown schematically in Figure 29-10, uses a resolver or encoder to provide axis position feedback to the machine control unit (MCU). That is a closed-loop control which requires a transducer or sensing device to detect machine table position (and velocity for contouring) and transmit that information back to the control unit to compare the current status with the desired state. If they are different, the control unit produces a signal to the drive motors to move the table, reducing the error signal and ultimately moving the table to the desired position at the desired velocity. For most NC controls, the feedback signals are supplied by transducers actuated either by the feed screw or by the actual movement of the component. The transducers may provide either *digital* or *analogue* information (signals). The resolver, shown in Figure 29-11, measures (indirectly) the movement of the table by the rotation of the ball screw. A pulse disc on the end of the ball screw converts analogue movement back into digital pulses which are used to calculate table movement. The tachometer on the drive motor measures the table velocity.

Two basic types of digital transducers are used. One supplies *incremental* information and tells how much motion of the input shaft or table has occurred since the last time. The information supplied is similar to telling a newspaper carrier that papers are to be delivered to the first, fourth, and eighth houses from a given corner on one side of a block. To follow the instructions, the carrier would need a means of counting the houses (pulses) as he passed them. He would deliver papers as he has counted 1, 4, and 8. The second type of digital information is *absolute* in character, with each pulse corresponding to a specific location of the machine components. To continue the carrier analogy, this would correspond to telling the carrier to deliver papers to the houses having house numbers 2400, 2406, and 2414. In this case it would only be necessary for the carrier (machine component) to be able to read the house numbers (addresses) and stop and deliver a paper when arriving at a proper address. This ''address'' system is a common one in numerical-control systems, because it provides absolute location information relative to a machine zero point.

When analogue information is used, the signal is usually in the form of an electric voltage that varies as the input shaft is rotated or the machine component

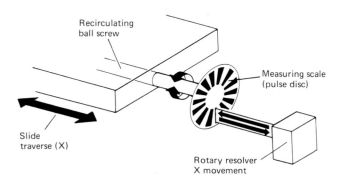

FIGURE 29-11   Table positions can be measured using a resolver.

is moved, the variable output being a function of movement. The movement is evaluated by measuring, or matching, the voltage, or by measuring the ratios between the applied and feedback voltages; this eliminates the effect of supply-voltage variations.

The input information, that is, the location of the holes, is given in binary form to the machine control unit in the form of a punched tape, magnetic tape in a cassette, or a floppy disc, or it is given directly. This command signal is converted into pulses by the machine control unit (MCU), which in turn drives the servomotor or stepper motor.

AC (alternating current) servomotors are rapidly replacing DC (direct current) motors on new CNC machine tools. The reasons for change are better reliability, better performance-to-weight ratio, and lower power consumption. See Figure 29-12. The magnets are mounted on the rotor of the AC servomotor while the windings are located in the stator. On the DC servomotor, the magnets are mounted in the stator while the windings are located in the rotor.

With the AC servomotor, the armature current is synchronized to the magnetic field by sensing the position of the magnetic poles through electrical commutation and control of the current in the stator winding. Electronic communication avoids the wear caused by spark formation between the DC motor's brushes and commutator bar and by friction of the brushes. It also avoids spot overheating of the commutator during long standstill periods under load.

The AC servomotor's high-gain, closed-loop positioning system also has attained positioning and holding accuracy equivalent to that of the DC motor. These benefits, along with the higher speed capabilities and torque-to-inertia

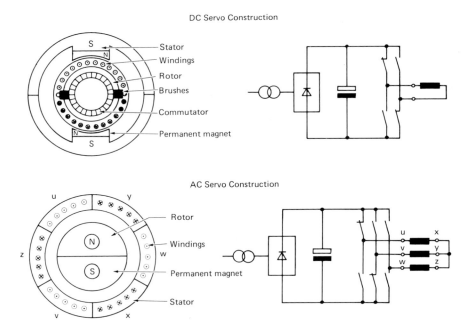

FIGURE 29-12   DC servomotor construction versus AC servo construction. (*Modern Machine Shop,* July 1983, p. 72.)

ratio, are available without increasing the complexity of the system with additional electronics. When the command counter reaches zero, the correct number of pulses has been sent to move the table to the desired position. In the closed-loop system, a comparator is used to compare feedback pulses with the original value, generating an error signal. Thus, when the machine control unit receives a signal to execute this command, the table is moved to the specified location, with the actual position being monitored by the feedback transducer. Table motion ceases when the error signal has been reduced to zero and the function (drilling a hole) takes place. Closed-loop systems tend to have greater accuracy and respond faster to input signals but may exhibit stability problems (oscillating about a desired value instead of achieving it) not found in open-loop machines.

If the system is point-to-point (or positioning), the control system disregards the paths between points. Some positioning systems provide for control of straight cuts along the machine axes and produce diagonal paths at 45 degrees to the axes by maintaining one-to-one relationships between the motions of perpendicular axes. Contouring systems generate paths between points by interpolating intermediate coordinate positions. As many of these systems as desired can be combined to provide control in several axes; two- and three-axis controls are most common, but some machines have as many as seven. In many, conversion to either English or metric measurement is available by merely throwing a switch.

The components required for such a numerical control system now are standardized items of hardware. In most cases, the drive motor is electric, but hydraulic systems are also used. They are usually capable of moving the machine elements, such as tables, at high rates of speed, up to 200 inches per minute being common. Thus exact positioning can be achieved more rapidly than by manual means. The transducer can be placed on the drive motor or connected directly to the lead screw, with special precautions being taken, such as the use of extra-large screws and ball nuts, to avoid backlash and to assure accuracy. In other systems, the encoder is attached to the machine table, providing direct measurement of the table position. Various degrees of accuracy are obtainable. Guaranteed positioning accuracies of 0.001 in. or 0.0001 in. are common, but greater accuracies can be obtained at higher cost. Most NC systems are built into the machines, but they can be retrofitted to some machine tools.

Initially, NC machines provided only tool setting and speed and feed values with the depth of cut being the positioning of the work relative to the tool. The remaining functions were controlled by the operator. Gradually the functions were incorporated into the control system so that the machines could change the tools automatically, change the speeds and feeds as needed for different operations, position the work relative to the tools, control the cutter path and velocity, reposition the tool rapidly between operations, and start and stop the sequence as needed.

Contouring requires directional changes at controlled velocity. This is illustrated in Figure 29-13, where a cutter is milling a circular slot for the part in Figure 29-7. Any lost motion in the system will distort the shape of the circle. The positional feedback transducer for the X axis is placed on the lead screw of

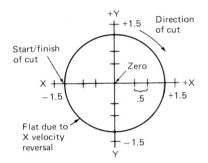

X and Y position to machine a circle

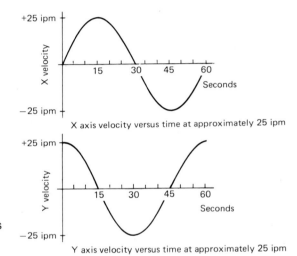

X axis velocity versus time at approximately 25 ipm

Y axis velocity versus time at approximately 25 ipm

FIGURE 29-13 Contouring requires velocity feedback as well as positional feedback.

the table. The velocity transducer (a tachometer) is placed on the rear of the motor. This arrangement provides for minimal distortion (flats on the circle at the points of axis velocity reversal) in the contoured circle. Recirculating ball screw drives, of the type shown in Figure 29-14, greatly reduces the backlash in the drive systems, helping to eliminate problems of servo loop oscillation and machine instability. Some of the functions in programmable machines re-

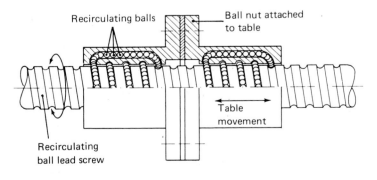

FIGURE 29-14 The ball lead screw adds to the accuracy and precision of NC and CNC machines.

quire feed-forward or preset loops. The machine must know in advance the rough dimensions of a casting so that it can determine how many roughing cuts are needed prior to the finishing cut. NC machines are manufactured with greater accuracy and repeatability and more rapid table movements than conventional machines.

For most CNC work, the operator (or part programmer) still plans the sequence of operations, selects the cutting tools and workholding devices, and selects the speeds, feeds, and depths of cut. Common machining routines like pocket milling or peck-drilling have been preprogrammed into many CNC machines. These are called "canned cycles." As shown in Figure 29-15, the operator merely supplies the information requested by the control menu and the machine fills in the necessary data to the previously written program in order to perform the desired machining routines.

To insure accurate machining of a workpiece on a CNC machine, the control system has to "know" certain dimensions of the tools. These *tool dimensions*

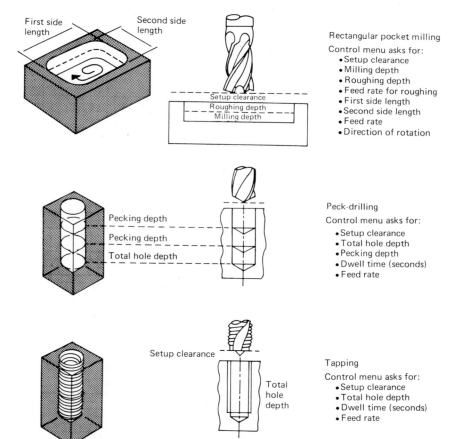

Rectangular pocket milling

Control menu asks for:
- Setup clearance
- Milling depth
- Roughing depth
- Feed rate for roughing
- First side length
- Second side length
- Feed rate
- Direction of rotation

Peck-drilling

Control menu asks for:
- Setup clearance
- Total hole depth
- Pecking depth
- Dwell time (seconds)
- Feed rate

Tapping

Control menu asks for:
- Setup clearance
- Total hole depth
- Dwell time (seconds)
- Feed rate

Source: Heidenhain Corp. (Elk Grove Village, Ill.)

FIGURE 29-15 Canned or preprogrammed machining routines greatly simplify the programming of CNC machines.

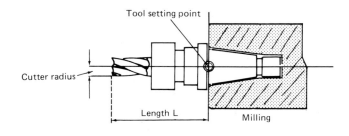

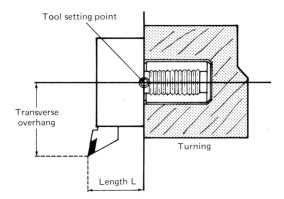

FIGURE 29-16 In NC machines, the tool dimensions must be accurately set and known to the control system.

are referenced to a fixed *setting point* on the tool holder, as shown in Figure 29-16. For the milling cutter the dimensions are length L and cutter radius. For the turning tool the dimensions are length L and transverse overhang. These dimensions are part of the information given to the operator on the setting sheet. See Figure 29-3. The program assumes that the tools will have the specified dimensions.

Part Programming.   Obviously, the preparation of the tape (or other input media) for use in NC is a critical step. In most cases, simple standard languages and programs have been developed. Manually prepared tapes can be punched on a typewriter-like machine, or on devices designed specifically for punching tapes.

The basic steps can be illustrated by reference to the part shown in Figure 29-7.

The first step is to modify the drawing to establish the zero reference axes, the X, Y, and Z directions. The zero reference point was the lower left-hand corner of the part. The hole labeled 1 would be $+3.3437$ in the X direction and $+5.2882$ in the Y direction. The part should be dimensioned with respect to the reference zero. The setup instructions given to the operator establish the workpiece properly on the machine table with respect to the tool. Obviously, this step can be avoided if the original drawing is made in the desired form.

The second step is to make a *part program*. The program (1) defines the sequence of operations required to fabricate the part; (2) gives the *X, Y,* and *Z* coordinate positions of the operations; (3) specifies the spindle traverse that determines the depth of the cut, the spindle speed, and feed and determines also whether the same tool can continue the next operation or whether a tool change is required. The last four items are specified by code symbols or what are called *NC words.* The NC words are put together in a specified order to define a *block* of information needed to execute an operation. By convention, the data are usually arranged in blocks in the following sequential order. See Table 29.1.

After the program has been prepared, it is used to prepare the tape or may be directly entered into the control panel of a CNC machine. Usually longer programs are entered by tape or disc and short programs entered manually. The tape uses a binary code to convert the numerical and instructional information into commands for the machine. Table 29.2 shows the Arabic versus the binary numbers. Binary numbers facilitate the rapid and accurate operation of control systems and computers. The conversion from Arabic to binary is done by the control system or computer, not by the programmer.

Punched tape is not a recent idea; the old player-piano roll was a form of tape control, and punched cards had been used for many years for controlling complicated weaving and business machines. Thus tape control of machine tools is an extension of an existing basic concept in which holes, representing information that has been punched into the tape, are "read" by sensing devices and used to actuate relays or other devices that control various electrical or mechanical mechanisms.

Tape-controlled machine tools use a one-inch-wide paper or Mylar tape containing eight information channels. Figure 29-17 shows an example of a block of such tape, punched with the information required for one operation on a turret-type drilling machine. Either of two codes are used, EIA244A code or ASCII, which is used in computer and telecommunications work as well as in

TABLE 29.1 Common NC Words, Defined

NC Word	Use
N	*sequence number:* identifies the block of information.
G	*preparatory function:* requests different control functions, including preprogrammed machining routines.
X,Y,Z,A,B,C	*dimensional coordinate data:* linear and angular motion commands for the axis of the machine.
F	*feed function:* set feed rate for this operation.
S	*speed function:* set cutting speed for this operation.
T	*tool function:* tells the machine the location of the tool in the tool holder or tool turret.
M	*miscellaneous function:* turn coolant on or off, open spindle, reverse spindle, tool change, etc.
EOB	*end of block:* indicates to the MCU that a full block of information has been transmitted and the block can be executed.

TABLE 29.2 Arabic and Binary Numbers

Arabic	Binary	Powers of 2
0	0	
1	1	2^0
2	10	2^1
3	11	
4	100	2^2
5	101	
6	110	
7	111	
8	1000	2^3
9	1001	
10	1010	
11	1011	
12	1100	
13	1101	
14	1110	
15	1111	
16	10000	2^4
17	10001	
18	10010	
19	10011	
20	10100	
21	10101	
22	10110	
23	10111	
24	11000	
25	11001	
26	11010	
27	11011	
28	11100	
29	11101	
30	11110	
31	11111	
32	100000	2^5
64	1000000	2^6
128	10000000	2^7

Value of 182 expressed in
Arabic numbers

$$1 \times 10^2 = 100$$
$$8 \times 10^1 = 80$$
$$2 \times 10^0 = \underline{2}$$

1 8 2   182

Value of 182 expressed in
binary numbers

$$1 \times 2^7 = 128$$
$$0 \times 2^6 = 0$$
$$1 \times 2^5 = 32$$
$$1 \times 2^4 = 16$$

Remember: $0 \times 2^3 = 0$
Any number raised to the $1 \times 2^2 = 4$
zero power equals 1 $1 \times 2^1 = 2$
                                 $0 \times 2^0 = \underline{0}$

1 0 1 1 0 1 1 0   182

MACHINE INFORMATION

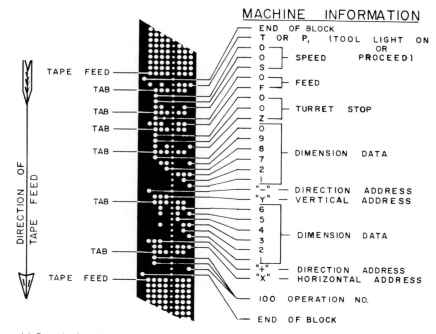

END OF BLOCK
T OR P, (TOOL LIGHT ON OR PROCEED)
O
O — SPEED
S
O — FEED
F
O
O — TURRET STOP
Z
O
9
8
7 — DIMENSION DATA
2
I
"−" DIRECTION ADDRESS
"Y" VERTICAL ADDRESS
6
5
4 — DIMENSION DATA
3
2
"+" DIRECTION ADDRESS
"X" HORIZONTAL ADDRESS
IOO OPERATION NO.
END OF BLOCK

TAPE FEED
TAB
TAB
TAB
TAB
TAB
TAB
TAB
TAPE FEED

DIRECTION OF TAPE FEED

(a) Example of one block of control tape

FIGURE 29-17 (a) The *X, Y, Z*, positions and other instructions comprise a block of machine information. (b) Standard NC tape format defined by Electronics Industries Association. (c) Original standard for tape coding as defined by EIA was RS-244. The ASCII subset is now EIA standard RS-358. Both codes are used.

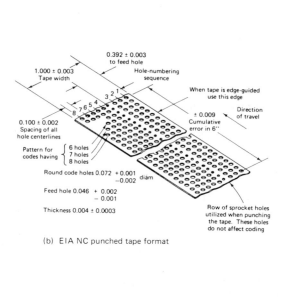

0.392 ± 0.003 to feed hole
1.000 ± 0.003 Tape width
Hole-numbering sequence
When tape is edge-guided use this edge
± 0.009 Cumulative error in 6″
Direction of travel
0.100 ± 0.002 Spacing of all hole centerlines
Pattern for codes having { 6 holes 7 holes 8 holes }
Round code holes 0.072 +0.001 −0.002 diam
Feed hole 0.046 +0.002 −0.001
Thickness 0.004 ± 0.0003
Row of sprocket holes utilized when punching the tape. These holes do not affect coding

(b) EIA NC punched tape format

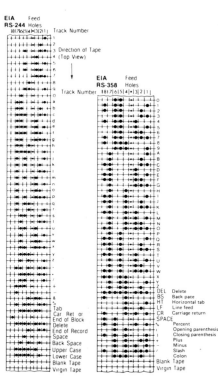

(c)

779

NC. The codes are not interchangeable even though both are based on binary numbering. The binary base 2 system is used because electronic circuitry responds to either of two conditions: on or off, or zero or one. Thus, all the numbers, symbols, and letters that are needed to control the machine are communicated to the machine by the presence or absence of holes in the eight tracks of the tape.

Four basic types of tape format are used for NC input to communicate dimensional and nondimensional information. These are *fixed-sequential format, work-address format, tab-sequential format, and word-address format.* Most new NC or CNC systems use the word-address format, which allows the words to be presented in any order and is the most flexible.

After the program is converted to tape or disc, but before it is used, it is *verified,* or checked, to make sure it is correct. The verification step can use a special NC plotting machine that will trace out all the tool–work paths as they would occur on the machine tool. A sample part is machined in plastic or wax for checking the part specifications from the drawing against the real part. Today CNC machines permit the user to program the interface between the machine tool and the control, greatly reducing the number of machining system components and interconnections. Current CNCs have extensive self-diagnostics and performance-monitoring systems. The greatest advances in CNC technology, however, are in part programming, where easy-to-use, menu-driven software makes programming almost as simple as setting up the machine manually. In NC machines, the operator may override the tape when necessary but cannot reprogram the machine unless a new tape is prepared. The CNC machine has the capability of reading a program into its computer memory, and the program can be modified at the machine like any other computer program.

On CNC machines, the machine tool operator may perform all the programming steps right at the console of the machine, programming the processing steps for the part directly into the computer memory. The program can be saved by having the machine print out a copy of the program, which can be used later for reorders of the same part. Features such as program edit, canned routines, program storage, diagnostics, constant surface speed, and tape punch are common on today's CNC machines.

As more and more design work is done on the computer (CAD) using data bases and software that are comparable to the machine tools, there will be less dependence on tape for program storage and more utilization of floppy discs, hard discs, and other typical computer storage means. For example, a machining cell composed of NC machine tools designed for a family of ten component parts may be able to do the ten different parts without needing retooling or refixturing, but it will still have ten different programs for these parts for each machine. If the programs are computer-stored, they can be readily accessed, but if they are stored on tape, delays will occur in dumping the different programs in or out of the control computer.

Contouring in NC and CNC Machines.   Although the majority of NC and tape-controlled machine tools do not provide for machining contoured surfaces,

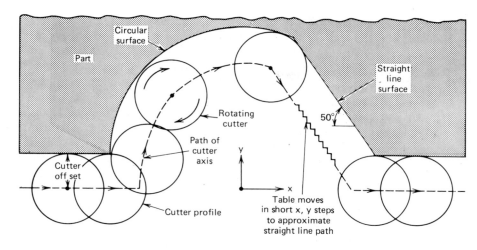

FIGURE 29-18 Two classic problems in NC programming are the determination of cutter offset and interpolation of cutter paths.

many do. Most CNC and DNC machines provide this feature. The required curves and contours are generated approximately by a series of very short, straight lines or segments of some type of regular curves, such as hyperbolas. Consequently, the program fed to the machine is arranged to approximate the required curve within the desired accuracy. Figure 29-18 illustrates how a desired straight line or curved surface can be approximated by short line segments or straight-line interpolation. Interpolation refers to the fact that curved surfaces as generated by machine tools must be approximated by a series of very short, straight-line movements in the X, Y, and Z directions. The length of the segment must be varied in accordance with the deviation permitted. Most machine tools with contouring capability will produce a surface that is within 0.001 in. of the one desired, and many will provide considerably better performance. Most contouring machines have either two- or three-axis capability, but a good many have up to five-axis capability.

Obviously, contour machining (and interpolation) requires that complex information be entered into the MCU because the number of straight-line or curved segments may be quite large. Manual programming of the tape can be quite laborious. Computer programming can translate simple commands into the complex information required by the machine.

Computer Languages for NC Control. The control mechanisms on NC machines do not understand ordinary shop language that people use to describe what machining, or machine-work movements, must take place. In addition, the effects of various sizes and types of cutting tools, requiring different offsets as illustrated in Figure 29-18, must be accounted for. Tool radius offset accounts for the fact that the center of tool rotation (as held in the spindle) must be offset from the workpiece surface to be generated. The path of the cutter center line will have different dimensions than that of the surface. The capabilities of the machine tool regarding available power, speeds, feeds, table travel, and so on must be taken into account. This barrier has been diminished by the use of computers and special simplified programming languages which the computer can understand and convert into the commands required by NC machine controls. One such language is APT II (Automatically Programmed Tools), a revised version of a language developed by MIT. APT is the most widely used language in the United States and is used for both positioning and continuous-path programming. APT is designed to run on main-frame computers with large memories. ADAPT, a language developed by IBM under an Air Force contract, has many of the features of APT but is designed for smaller computers. Another APT version, for smaller computers like the PDP 11/70, is UNIAPT, which delivers an EIA tape with virtually no sacrifice of features and capabilities compared to APT. AUTOSPOT was developed by IBM for point-to-point positioning but today's version can be used for contouring. Other languages (like SPLIT and COMPACT II) have been developed on a proprietary basis for specific machine tool systems and are available on a lease basis. The basic sequence of operations for computer-assisted part programming is as follows (see Figure 29-3):

The part programmer prepares a manuscript which specifies the part geometry, the tool path, and the operations needed for their sequence, using English-like statements. Auxiliary machine functions, like "coolant on," are also programmed. The program is written on a computer terminal. The computer performs an input translation, followed by the required arithmetic calculations, including things like cutter offset computations, in order to obtain the coordinate points the cutter must follow. The individualities of the specific machine tools, relative to the available speeds, feeds, accelerations, and so forth, are provided for by a postprocessor program, the output from which is the needed NC tape or disc or direct program feed for the machine tool.

The workpiece, no matter how complex, can be conceived of as a compilation of points, straight lines, planes, circles, cylinders, and other mathematically defined geometries. Historically, the part programmer's job has been to translate the component parts into basic geometric elements, defining each geometric element in terms of workpiece dimensions. In recent years, much of this work has become routine in CAD systems. However, NC programming languages and CAD software are often incompatible, which has hindered the CAD to CAM step.

An example of how the APT language would be used on a part is shown in Figure 29-19. Essentially, it is necessary to tell the computer the location of a center of the bolt hole circle or one of the holes in the circle with respect to a zero point, the number of holes in the circle, and the radius of the circle. Numerals 18, 10, and 1 (457.2, 254, 25.4) are X, Y, and Z coordinates for table and tool. Numeral 40 is table movement rate. Numeral 12 is feed rate for drill. Numeral 45 is 45° of rotation. Numeral 7 is the instruction for seven duplicate holes to be drilled. The trend to small computers, programmable controllers, and CNC machines will continue to lessen the need for and use of main-frame and minicomputer APT. Many machines, NC and conventional, are being retrofitted to CNC at a cost of $15,000 to $40,000.

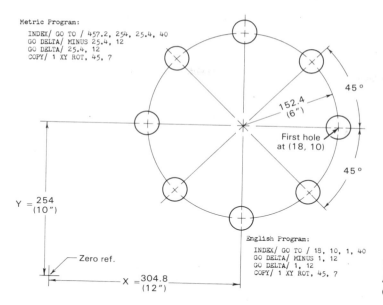

```
Metric Program:
  INDEX/ GO TO / 457.2, 254, 25.4, 40
  GO DELTA/ MINUS 25.4, 12
  GO DELTA/ 25.4, 12
  COPY/ 1 XY ROT, 45, 7
```

45°

152.4
(6")

First hole
at (18, 10)

45°

Y = 254
    (10")

Zero ref.

X = 304.8
    (12")

```
English Program:
  INDEX/ GO TO / 18, 10, 1, 40
  GO DELTA/ MINUS 1, 12
  GO DELTA/ 1, 12
  COPY/ 1 XY ROT, 45, 7
```

FIGURE 29-19   Bolt hole circle and APT programs for machining the eight holes.

Machining Center Features and Trends.

Computer and numerical control is used on a wide variety of machines. These range from single-spindle drilling machines, which often have only two-axis control and can be obtained for about $10,000, to machining centers, such as shown in Figure 29-20. This machining center can do drilling, boring, milling, tapping, and so forth, with four-axis control. It can automatically select and change 32 preset tools. The table can move left/right or in/out and the spindle can move up/down or in/out,

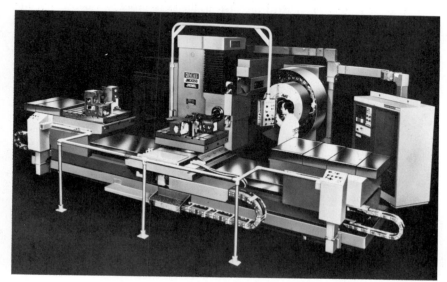

FIGURE 29-20   Horizontal-spindle, four-axis CNC machining center. (Courtesy DeVlieg Machine Company.)

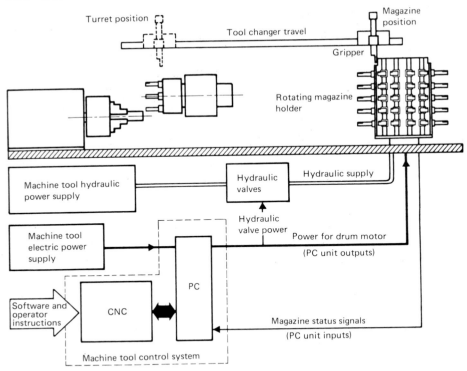

FIGURE 29-21 The concept of automatic tool changing has been extended to lathes. (Courtesy Sandvik.)

with positioning accuracy in the range of 0.0003 in. in 40 in. of travel. The machine has automatic tool change and automatic work transfer so that workpieces can be loaded/unloaded while machining is in process. Such a machine can cost over $200,000. Between these extremes are numerous machine tools that do less varied work than the highly sophisticated machining centers but which combine high output, minimum setup time in changing from one job to another, and remarkable flexibility because of the number of tool motions that are provided.

The concept of automatic tool changing has been extended to CNC lathes. See Figure 29-21. The tools are held on a rotating tool magazine and a gantry-type tool changer is used to change the tools. Each magazine holds one type of cutting tool. This is an example of the trend of providing greater versatility along with high productivity in lathes. The versatility is being further increased by combining both rotary-work and rotary-tool operations—turning and milling—in a single machine. There are numerous tape-controlled machines that provide four- and five-axis contouring capability. Tools are changed in six seconds or less. It is also common to provide two or more worktables, permitting work to be set up while machining is done on the workpiece in the machine, with the tables being interchanged automatically. Consequently, the productivity of such machines can be very high, the chip-producing time often approaching 50% of the total.

Numerical control has been applied to a wide variety of other production processes. NC turret punches with *X–Y* control on the table, NC wire EDM machines, laser welders, flame cutters, and many other machines are readily available.

Two new trends are observed in the development of machining centers. One is the growing interest in smaller, more compact machining centers and the other is the emphasis on extended-shift or even unmanned operations. Modern machining centers have contributed significantly to improved productivity in many companies. They have eliminated the time lost in moving workpieces from machine to machine and the time needed for workpiece loading and unloading for separate operations. In addition, they have minimized the time lost in changing tools, carrying out gaging operations, and aligning workpieces on the machine.

The latest generation of machining centers is aimed at improving utilization still further by reducing the time when machines are stopped, either during pauses in a shift or between shifts. Delays are caused by tool breakage, unforeseen tool wear, limited number of tools or an inadequate number of available workpieces. Machines are fitted with tool breakage monitors, tool wear compensating devices, and means for increasing the number of tools and workpieces available.

Probes on NC machines can greatly improve the process capability of the machine tool. There is a big difference between the claimed program resolution for a NC machine and the accuracy and precision (the process capability) in the actual parts. As shown in Figure 29-22, true positioning accuracy and precision are affected by machine alignment, machine and fixture setup, variations in the workholding device, raw material variations, workpiece location in the fixture variations, and cutting tool tolerances. Thus, the finished workpiece may be unacceptable even though the machine is more than capable of producing the

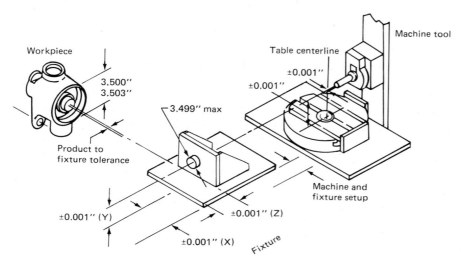

FIGURE 29-22 Process capability in NC machines is affected by many factors.

FIGURE 29-23 (Above) Probe from the tool changer can be mounted in the spindle (right) for accurate location of part features.

part to the design specifications. The part program has no assurance that the part is properly located in the fixture or that the fixture is properly located on the table of the machine. A probe, carried in the tool storage magazine and mounted when needed in the spindle like a cutting tool, can establish the location of the surface features relative to each other and to the spindle axis within 0.0005 in. See Figure 29-23. The machine controller, using the probe data, will then shift the program reference accordingly. The probe can be used to determine the amount of material on a rough casting, locate a corner of a part, define the center of a hole, or check for the presence or absence of a feature. All of the variability described in Figure 29-22 can be compensated for except for variations in the cutting tool geometry. These can be handled by a probe mounted on the machine tool. The CNC turret lathe shown in Figure 29-24 has a retractable probe mounted in the headstock. This probe can be used to check the tool location, tool dimensions (see Figure 29-16), and even tool wear, automatically updating tool-offset data in the control computer. A second probe is mounted on the tool turret for checking part setup and alignment as well as inprocess inspection. Thus, the machine tool can function like a coordinate measuring machine. By comparing the actual touched location with the programmed location, the measuring routine determines appropriate compensation.

Many of the new machining centers are equipped with novel automatic pallet changers and workpiece loading and unloading devices. Robots are increasingly being applied for workpiece handling in machine groups, including manufacturing cells. In some cases the robot is also used for tool-changing functions. See Chapter 42.

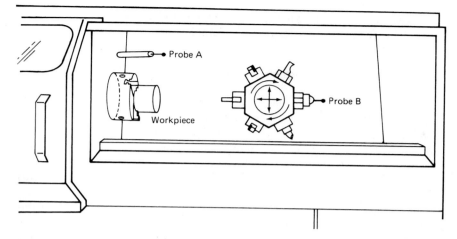

FIGURE 29-24 CNC turret lathe with retractable probe A in the head-stock and probe B mounted on the turret.

Economic Considerations in Tape and Numerical Control. NC and CNC machines are costly, but their use usually can be justified economically in from one to three years, primarily through substantial savings in setup and machining time (particularly when the parts being processed are a GT family of parts), and significant improvements in quality. The NC machine may be the key machine in a manufacturing cell with parts being loaded and unloaded with a robot.

Advantages and Disadvantages. The advantages of programmable machine tools can be summarized as follows:

1. *Flexibility.* It is easy to change from one part to another or to change part design.
2. *Superior process capability.* Greater accuracy and precision is built into the machines, resulting in better quality and a high order of repeatability.
3. *High production rates.* Optimum feeds and speeds for each operation, less time spent in noncutting functions.
4. *Lower tooling costs.* Expensive jigs and templates are not needed.
5. *Less lead time.* Programs can be prepared in less time than conventional jigs and templates, and less setup time is required.
6. *Fewer setups per workpiece.* More operations can be done at each setup of the workpiece.
7. *Better machine utilization.* There is less machine idle time, owing to more efficient table or tool movement between successive operations and fewer setups. Cycle time is easily altered.
8. *Reduced inventory.* Less inventory needs to be carried because parts can be run economically in smaller quantities.
9. *Reduction in space required.* Greater productivity and reduced tooling lessen floor and storage space, and smaller economic lot sizes reduce storage space required for inventories.
10. *Less scrap.* Operator errors are substantially reduced. The first part off the machine can be a good part.

FIGURE 29-25 Spiral groove with variable depth machined in copper plate. (Courtesy Robert Tidmore, NASA Marshall Space Flight Center Test Laboratory, Fabrication Division.)

11. *Less skill required of the operator.* Program planning in preparing tapes reduces the necessity for operator decisions.

12. *Manufacture of unique geometries.*

Through the use of computer programming of NC or CNC machines, one can generate surfaces and geometrical configurations that are not possible to make by any other method, at least not economically. Such a part is seen in Figure 29-25. This is the copper base of a cooling device. The spiral groove has constant width but constantly varying depth from start to finish. The APT programming language was used and the part machined on a three-axis CNC continuous-path vertical milling machine.

The major disadvantage of NC machine tools is their high initial cost. This means that they must be justified from an economic viewpoint. Modern approaches to engineering economic analysis are strongly recommended. The control equipment now is virtually all made of solid-state modules, and reliability is excellent. Programming has been greatly simplified but a staff of part programmers may be needed for complex parts. NC and CNC machines can have high maintenance costs. Overall, NC machines have provided a much needed solution for small- and medium-quantity production, and it is easy to see why they have been so widely adopted.

Review Questions

1. How is a machining center different from a vertical-spindle milling machine?
2. Who is credited with marketing the first machining center?
3. What was DNC as first practiced, and what does DNC usually mean today?
4. How does as adaptive control system differ from a numerical control system?
5. In the decision analysis portion of the AC system, optimization of the process is required. What does this imply about the process?
6. How does feed forward differ from feedback in these systems?
7. What are some examples of everyday devices or machines that employ feedback in their control systems?
8. Why was it necessary for machine tool builders to improve the feed screws on their machines when they made them into NC machines?
9. Explain the problems of cutter offset and interpolation in NC programming.
10. Some of the functions performed by the operator in piece-part manufacturing are very difficult to complete automatically. Name the functions and explain why.
11. The first NC machines were closed-loop control; later NC machines were open-loop. What change did NC require on the part of the machine tool builders?

12. Can a continuous-path NC machine be open-loop? Why or why not?
13. Why are there no NC shapers or broaches?
14. Why is manual programming not used for continuous-path NC? What about for straight-cut or point-to-point programming?
15. What is the name of the modular-type workholder shown in Chapter 29? Modular workholders were described in Chapter 28.
16. What are the three basic closed-loop feedback (A4) schemes?
17. What is the difference between the zero reference point and the machine zero point?
18. What is an encoder?
19. Why does contouring require both velocity and positional feedback?
20. What is a peck-drilling subroutine for a CNC machine?
21. What is pocket milling and what kind of milling cutter is usually used to perform it?
22. What is the tool setting point on a NC or CNC machine?
23. How are probes used in CNC machines to improve process capability?
24. What are G words used for in NC?
25. How can an NC program be verified?
26. What is APT?

Problems

1. What are the X and Y dimensions for the center position of holes 2 and 3 from the part shown in Figure 29-7?

2. Configurations obtained from continuous-path machining are the result of a series of straight-line, parabolic-span, or higher-order curves. The degree to which curved surfaces correspond to their design depends on how many lines or spans are used. Four equal chords in a circle describe a square, as shown in Figure 29-A. Six make a hexagon, and as the number increases the lines themselves come closer to a perfect circle. Number of lines needed is determined by a maximum tolerance allowed between the design of the curved section and the actual chord programmed. The program for a parabolic-span type of control unit requires enough spans for any deviation to stay within an acceptable tolerance. For a tolerance of $T = 0.001$ in., how long should the span be for a curve with a 5-in. radius? Assume that the arc is part of a circle. What is the span angle here, in degrees?

3. Suppose the acceptable tolerance was 0.0001 inc. Determine the span angle.

4. For the bolt hole circle shown in Figure 29-19, compute the X, Y dimensions for the hole 45 degrees above the first hole.

5. Write the dimension for the first hole at (18,10) in Figure 29-19 in binary-coded decimals.

6. Suppose the plate shown in Figure 29-7 was to be profile-milled around the periphery with a 1″ diameter end milling cutter, as shown in Figure 29B. The dashed line is the cutter path. The programmer must calculate an offset path to allow for cutter diameter. Since the programmed points are followed by the cutter centerline and the profile is made at the tool's periphery, the programmer called for a half-inch cutter offset. Working with computer assistance, the programmer would describe the part profile to be machined and specify the cutter. The computer would generate the cutter path. Complete the table above to specify the cutter path, starting with the origin at the zero reference point. Move the tool around the plate counterclockwise.

Programmed point locations

PT	X	Y
1		
2		
3		
4		
5		
1		

7. Suppose surface finish is very important for the profile milling job described in problem 6. Therefore down milling is going to be used. Rewrite the NC program points to accommodate this requirement. Show the new path on a sketch like that in Figure 29B.

8. The circular slot in the plate in Figure 29-7 can be end milled. This is a contouring cut. See Figure 29-13. Compute the cutter's X and Y velocity components as it is passing the number 3 hole, 225 degrees around the slot from the start of the cut, using 25 ipm as the peak velocity value.

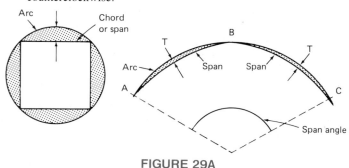

FIGURE 29A

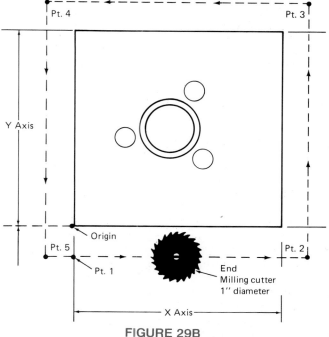

FIGURE 29B

30

Thread Manufacturing

Introduction

Screw threads probably are the most important of all the machine elements. *Threading* or *thread cutting* refers to the manufacture of threads on external diameters. *Tapping* refers to machining threads in (drilled) holes. Without these processes our present technological society would come to a grinding halt. More screw threads are made each year than any other machined element. They range in size from those used in small watches to threaded shafts 10 inches in diameter. They are made in quantities ranging from one to several million duplicate threads. Their precision varies from that of cheap dime-store screws to that of lead screws for the most precise machine tools. Consequently, it is not surprising that many very different procedures have been developed for making screw threads and that the production cost by the various methods varies greatly. Fortunately, some of the most economical methods can provide very accurate results. However, as in the design of most products, the designer can greatly affect the ease and cost of producing specified screw threads. Thus, understanding thread-making processes permits the designer to specify and incorporate screw threads into designs while avoiding needless and excessive cost.

A screw thread is a ridge of uniform section in the form of a helix on the external or internal surface of a cylinder, or in the form of a conical spiral on the external or internal surface of a frustum of a cone. These are called *straight* or *tapered* threads, respectively. Tapered threads are used on pipe joints or other applications where liquid-tight joints are required. Straight threads, on the other hand, are used in a wide variety of applications, most commonly on fastening devices, such as bolts, screws, and nuts, and as integral elements on parts that are to be fastened together. But, as mentioned previously, they find very important applications in transmitting controlled motion, as in lead screws and precision measuring equipment.

Three basic methods are used to produce threads: *cutting, rolling,* and *casting.* Although both external and internal threads can be cast, relatively few are made

790

in this manner, primarily in connection with die casting, investment casting, or the molding of plastics. Today by far the largest number of threads are made by rolling. Both external and internal threads can be made by rolling, but the material must be ductile; because this is a less flexible process than thread cutting, it essentially is restricted to standardized and simple parts. Consequently, large numbers of threads still are, and will continue to be, made by cutting processes, including grinding.

Screw-Thread Standardization.   Starting with Sir Joseph Whitworth in England in 1841 and William Sellers in the United States in 1864, much time and effort have been devoted to screw-thread standardization. In 1948, representatives of the United States, Canada, and Great Britain adopted the Unified and American Screw Thread Standards, based on the form shown in Figure 30-1. In 1968 the International Organization for Standards (ISO) recommended the adoption of a set of metric standards, based on the basic thread profile shown in Figure 30-2. It appears likely that both types of threads will continue to be used for some time to come.

Screw-Thread Nomenclature.   The standard nomenclature for screw-thread components is illustrated in Figure 30-3. In both the *Unified* and *ISO systems,* the crests of external threads may be flat or rounded. The root usually is made rounded to minimize stress concentration at this critical area. The in-

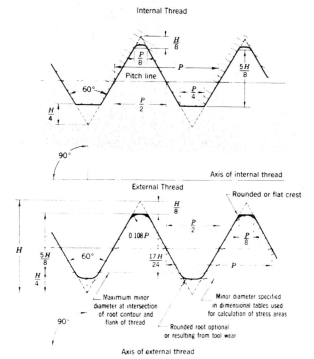

FIGURE 30-1 Unified and American screw-thread form.

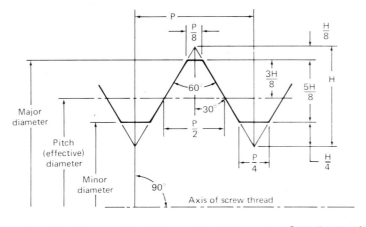

D = Major diameter of thread of nut ⎫ Nominal
d = Major diameter of Screw ⎬ Diameter
D₁ = Minor diameter of thread of nut ⎭
d₁ = Minor diameter of thread of screw
D₂ = Pitch diameter of thread of nut

d_2 = Pitch diameter of thread of screw
H = Height of the complete theoretical thread profile
H_1 = Engagement
P = Pitch

$$H = \frac{\sqrt{3}}{2} P$$

FIGURE 30-2 Basic profile of metric general purpose screw thread per 1SO/R68-1969.

ternal thread has a flat crest in order to mate with either a rounded or V-root of the external thread. A small round is used at the root to provide clearance for the flat crest of the external thread.

In the metric system, the *pitch* always is expressed in millimeters, whereas in the American (Unified) system, it is a fraction having as the numerator 1 and as the denominator the number of threads per inch—thus ¹⁄₁₆ pitch being ¹⁄₁₆ of an inch. Consequently, in the Unified system threads more commonly are described in terms of threads per inch rather than by the pitch.

While all elements of the thread form are based on the *pitch diameter*, screw-thread sizes are expressed in terms of the *outside,* or *major* diameter and the *pitch* or *number of threads per inch*. In threaded elements, *lead* refers to the

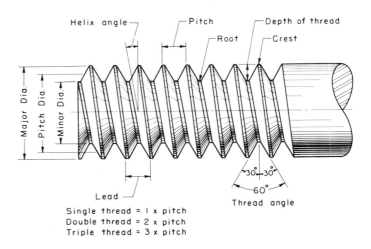

FIGURE 30-3 Standard screw-thread nomenclature.

axial advance of the element during one revolution; therefore lead equals pitch on a single-thread screw.

Types of Screw Threads. Eleven types, or series, of threads are of commercial importance, several having equivalent series in the metric system and Unified systems:

1. *Coarse-thread series* (UNC and NC). For general use where not subjected to vibration.
2. *Fine-thread series* (UNF and NF). For most automotive and aircraft work.
3. *Extra-fine-thread series* (UNEF and NEF). For use with thin-walled material or where a maximum number of threads are required in a given length.
4. *Eight-thread series* (8UN and 8N). Eight threads per inch for all diameters from 1 through 6 in. It is used primarily for bolts on pipe flanges and cylinder-head studs where an initial tension must be set up to resist steam or air pressures.
5. *Twelve-thread series* (12UN and 12N). Twelve threads per inch for diameters from ½ through 6 in. It is not used extensively.
6. *Sixteen-thread series* (16UN and 16N). Sixteen threads per inch for diameters from ¾ through 6 inc. It is used for a wide variety of applications that require a fine thread.
7. *American Acme thread*. See Figure 30-4.
8. *Buttress thread*.
9. *Square thread*.
10. *29° Worm Thread*. These last four of the threads are used primarily in transmitting power and motion.
11. *American standard pipe thread*. This thread, shown in Figure 30-4, is the standard tapered thread used on pipe joints in this country. The taper on all pipe threads is ¾ in. per foot.

As has been indicated, the Unified threads are available in a coarse (UNC and NC), fine (UNF and NF), extra-fine (UNEF and NEF), and three-"pitch" (8, 12, and 16) series, the number of threads per inch being according to an arbitrary determination based on the major diameter.

Many nations have now adopted ISO threads into their national standards. Besides metric ISO threads, there are also inch-based ISO threads, namely the UN series with which people in the United States, Canada, and Great Britain are familiar. ISO offers a wide range of metric sizes. Individual countries have the choice of accepting all or a selection of the ISO offerings.

The size listings of metric threads starts with "M" and continue with the outside diameter in millimeters. Most ISO metric thread sizes come in coarse, medium, and fine pitches. When a coarse thread is designated it is not necessary to spell out the pitch. Example: A coarse 10 mm OD thread is called out as "M 12." This thread has a pitch of 1.75 mm, but the pitch may be omitted from the call-out. A fine 12 mm OD thread is available. It has a 1.25 mm pitch

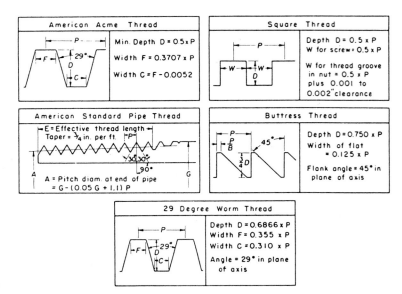

FIGURE 30-4   Special thread forms.

and must be designated "M 12 × 1.25." An extra-fine 12 mm OD thread having 0.75 mm pitch would receive the designation "M 12 × 0.75." See Table 30.1 for a comparison of Unified and ISO thread sizes.

The sign "×" is not employed as a multiplication symbol in metric practice but rather is used to relate these two attributes of the threads. The full description of a thread fastener obviously includes information beyond the thread specification. Head type, length, length of thread, design of end, thread runout, heat treatment, applied finishes, and other data may be needed fully to specify a bolt besides the designation of the thread. The "×" sign should not be used to

TABLE 30.1   Comparison Between Selection Unified and ISO Threads

	Unified			ISO	
Diameter		Threads per inch		(Threads per inch)	
Number Inches	mm	UNC	UNF	Coarse	Fine
#2	2.18	56	64	M2 × 0.4 (63.5)	
#4	2.84	40	48	M2 × 0.45 (56.4)	
#8	4.17	32	36	M4 × 0.7 (36.3)	
#10	4.82	24	28	M5 × 0.8 (31.8)	
¼ in.	6.35	20	28	M6 × 1.0 (25.4)	
½ in.	12.7	13	20	M12 × 1.75 (14.5)	M12 × 1.25 (20.3)
¾ in.	10.05	10	16	M20 × 2.5 (10.2)	M20 × 1.5 (16.9)
1 in.	25.4	8	14	M24 × 3 (8.47)	M24 × 2 (12.7)

separate any of the other characteristics. The availability of fasteners, particularly nuts, containing plastic inserts to make them self-locking and thus able to resist loosening due to vibration, and the use of special coatings that serve the same purpose, have resulted in less use of finer-thread-series fasteners in mass production. Coarser-thread fasteners are easier to assemble and less subject to cross-threading (binding).

Thread Classes.   In the Unified system, manufacturing tolerances are specified by three classes. Class 1 is for ordnance and other special applications. Class 2 threads are the normal production grade, and Class 3 threads have minimum tolerances where tight fits are required. The letters A and B are added after the class numerals to indicate external and internal threads, respectively.

In the ISO system, tolerances are applied to "positions" and "grades." Tolerance positions denote the limits of pitch and crest diameters, using "e" (large), "g" (small), and "H" (no allowance) for internal threads. The grade is expressed by numerals 3 through 9. Grade 6 is roughly equivalent to U.S. grades 2A and B, medium quality, general purpose threads. Below 6 is fine quality and/or short engagement. Above 6 is coarse quality and/or long length of engagement.

Thread Designation.   In the Unified system, screw threads are designated by symbols as follows:

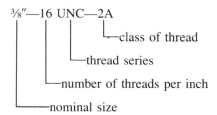

$\frac{3}{8}$″—16 UNC—2A
- class of thread
- thread series
- number of threads per inch
- nominal size

This type of designation applies to right-hand threads. For left-hand threads, the letters LH are added after the thread class symbol.

In the ISO system, threads are designated as follows:

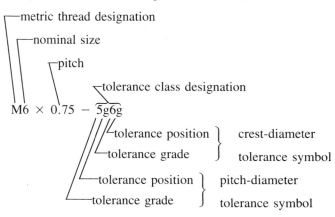

- metric thread designation
- nominal size
- pitch
- tolerance class designation

M6 × 0.75 − 5g6g
- tolerance position } crest-diameter
- tolerance grade   } tolerance symbol
- tolerance position } pitch-diameter
- tolerance grade   } tolerance symbol

Thread Cutting

Threads can be cut by the methods shown in Table 30.2.

TABLE 30.2   Thread-Cutting Methods

External	Internal
Threading (on an engine lathe)	Threading (on an engine lathe)
With a die held in a stock (manual)	With a tap and holder (manual, semiautomatic, or automatic)
With an automatic die (turret lathe or screw machine)	With a collapsible tap (turret lathe, screw machine, or special threading machine)
By milling	By milling
By grinding	

Cutting Threads on a Lathe.   Lathes provided the first method for cutting threads by machine. Although most threads are now produced by other methods, lathes still provide the most versatile and fundamentally simple method. Consequently, they often are used for cutting threads on special workpieces where the configuration or nonstandard size does not permit them to be made by less costly methods.

There are two basic requirements for thread cutting. An accurately shaped and properly mounted tool is needed because thread cutting is a form-cutting operation. The resulting thread profile is determined by the shape of the tool and its position relative to the workpiece. The second requirement is that the tool must move longitudinally in a specific relationship to the rotation of the workpiece, because this determines the *lead* of the thread. This requirement is met through the use of the *lead screw* and the *split nut*, which provide positive motion of the carriage relative to the rotation of the spindle.

External threads can be cut with the work mounted either between centers (Figure 30-5) or held in a chuck. For internal threads, the work must be held in a chuck. The cutting tool usually is checked for shape and alignment by means of a thread template, as indicated in Figure 30-6. Figure 30-7 illustrates two methods of feeding the tool into the work. If the tool is fed radially, cutting takes place simultaneously on both sides of the tool. With this true form-cutting procedure, no rake should be ground on the tool, and the top of the tool must be horizontal and be set exactly in line with the axis of rotation of the work, as shown in Figure 30-8; otherwise, the resulting thread profile will not be correct. An obvious disadvantage of this method is that the absence of side and back rake results in poor cutting (except on cast iron or brass). The surface on steel usually will be rough. Consequently, the second method commonly is used, with the compound swiveled 20 degrees. The cutting then occurs primarily on the left-hand edge of the tool, and some side rake can be provided.

Proper speed ratio between the spindle and the lead screw is set by means of the gear-change box. Modern industrial lathes have ranges of ratios available so that nearly all standard threads can be cut merely by setting the proper levers on the quick-change gear box.

To cut a thread, it also is essential that a constant positional relationship be maintained between the workpiece, the cutting tool, and the lead screw. If this is not done, on successive cuts the tool will not be positioned correctly in the thread space. Correct relationship is obtained by means of a *threading dial,* shown in Figure 30-5, which is driven directly by the lead screw through a worm gear. Because the workpiece and the lead screw are directly connected, the threading dial provides a means for establishing the desired positional relationship between the workpiece and the cutting tool.

The threading dial is graduated into an even number of major and half divisions. If the feed mechanism is engaged in accordance with the following rules, correct positioning of the tool will result:

1. *For even-number threads:* at any line on the dial.

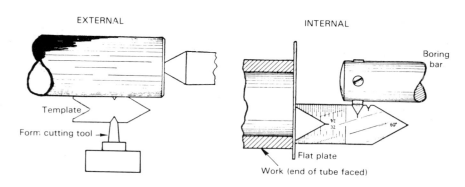

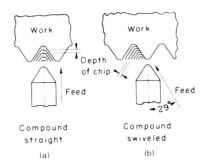

FIGURE 30-7 Top view of two methods of feeding the tool into the work in cutting threads on a lathe: (a) Radial feed, (b) half thread angle feed.

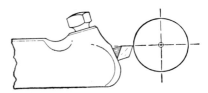

FIGURE 30-8 Proper relationship of the thread-cutting tool to the workpiece center line. (Courtesy South Bend Lathe.)

2. *For odd-number threads:* at any numbered line on the dial.

3. *For threads involving ½ numbers:* at any odd-numbered line on the dial.

4. *For ¼ or ⅛ threads:* return to the original starting line on the dial.

To start cutting a thread, the tool usually is fed inward until it just scratches the work, and the cross-slide dial reading is then noted or set at zero. The split nut is engaged and the tool permitted to run over the desired thread length. When the tool reaches the end of the thread, it is quickly withdrawn by means of the cross-slide control. The split nut is then disengaged and the carriage returned to the starting position, where the tool is clear of the workpiece. At this point the future thread will be indicated by a fine scratch line. This permits the operator to check the thread lead by means of a scale or thread gage to assure that all settings have been made correctly.

Next, the tool is returned to its initial zero depth position by returning the cross slide to the zero setting. By using the compound rest, the tool can be moved inward the proper depth for the first cut. A depth of 0.010 to 0.025 in. usually is used for the first cut and smaller amounts on each successive cut, until the final cut is made with a depth of only 0.001 to 0.003 in. to produce a good finish. When the thread has been cut nearly to its full depth, it is checked for size by means of a mating nut or thread gage. Cutting is continued until a proper fit is obtained.

To cut right-hand threads, the tool is moved from right to left. For left-hand threads the tool must be moved from left to right. Otherwise, the procedure is essentially the same. Internal threads are cut in the same basic manner except that the tool is held in a boring bar. Tapered threads can be cut either with a taper attachment or by setting the tailstock off center. It should be remembered that in cutting a tapered thread the tool must be set normal to the axis of rotation of the workpiece, not the tapered surface

Cutting screw threads on a lathe is a slow, repetitious process that requires considerable operator skill. The cutting speeds usually employed are from one third to one half of regular speeds to enable the operator to have time to manipulate the controls and to ensure better cutting. The cost per part can be high, which explains why other methods are used whenever possible.

Cutting Threads with Dies.   Straight and tapered external threads up to about 1½ in. in diameter can be cut quickly by means of threading dies, such as are shown in Figure 30-9. Basically, these are similar to hardened, threaded nuts with multiple cutting edges. The cutting edges at the starting end are beveled to aid in starting the dies on the workpiece. As a consequence, a few threads at the inner end of the workpiece are not cut to full depth. Such threading dies are made of carbon or high-speed tool steel.

Solid-type dies are seldom used in manufacturing because they have no provision compensating for wear. The solid-adjustable type, shown in Figure 30-9(b), is split and can be adjusted over a small range by means of a screw

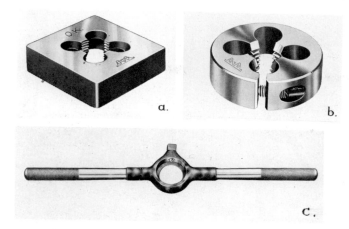

FIGURE 30-9 (a) Solid threading die. (b) Solid-adjustable threading die. (c) Threading-die stock for round die (die removed). (Courtesy TRW-Greenfield Tap & Die.)

to compensate for wear or to provide a variation in the fit of the resulting screw thread.

These types of threading dies usually are held in a *stock* for hand rotation. A suitable lubricant is desirable to produce a smoother thread and to prolong the life of the die, since there is extensive friction during the cutting process.

Self-Opening Die Heads.

A major disadvantage of solid-type threading dies is that they must be unscrewed from the workpiece to remove them. They are therefore not suitable for use on high-speed, production-type machines and *self-opening die heads* are used instead on turret lathes, screw machines, and special threading machines for cutting external threads.

There are three types of self-opening die heads, all having four sets of adjustable, multiple-point cutters that can be removed for sharpening or for interchanging for different thread sizes. This permits one head to be used for a range of thread sizes (see Figure 30-10). The cutters can be positioned radially or tangentially, resulting in less tool flank contact and friction rubbing. In some self-opening die heads, the cutters are circular, with an interruption in the circular form to provide an easily sharpened cutting face. The cutters are mounted on the holder at an angle equal to the helix angle of the thread.

As the name implies, the cutters in self-opening die heads are arranged to open automatically when the thread has been cut to the desired length, thereby permitting the die head to be quickly withdrawn from the workpiece. On die heads used on turret lathes, the operator usually must reset the cutters in the closed position before making the next thread. The die heads used on screw machines and automatic threading machines are provided with a mechanism that automatically closes the cutters after the heads are withdrawn.

Cutting threads by means of self-opening die heads frequently is called *thread chasing*. However, some people apply this term to other methods of thread cutting, even to cutting a thread in a lathe.

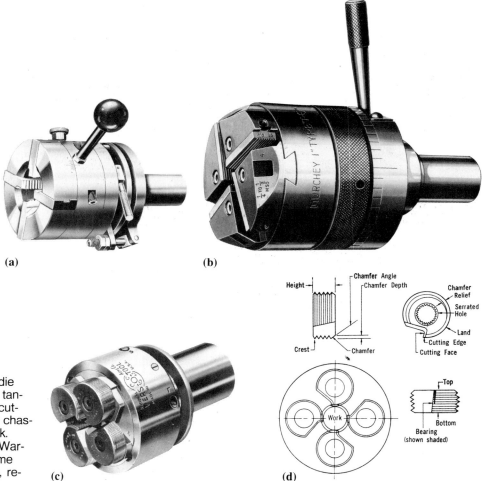

FIGURE 30-10 Self-opening die heads, with (a) radial cutter, (b) tangential cutters, and (c) circular cutters. (d) Terminology of circular chasers and their relation to the work. (Courtesy Geometric Tool Co., Warner & Swasey Co., National Acme Co., TRW-Greenfield Tap & Die, respectively.)

Thread Tapping.   The cutting of an internal thread by means of a multiple-point tool is called *thread tapping,* and the tool is called a *tap.* A hole of diameter slightly larger than the minor diameter of the thread must already exist, made by drilling/reaming, boring, or die casting.

For small holes, solid *hand taps,* such as shown in Figure 30-11, are usually used. The flutes create cutting edges on the thread profile and provide space for the chips and the passage of cutting fluid. Such taps are made of either carbon or high-speed steel and are now routinely coated with TiN. The flutes can be either straight, helical, or spiral.

Hand taps, shown in Figure 30-11, have square shanks and are made in three types, usually in sets. The tapered end of *taper taps* will enter the hole a sufficient distance to help align the tap. In addition, the threads increase gradually to full depth, and therefore this type of tap requires less torque to use.

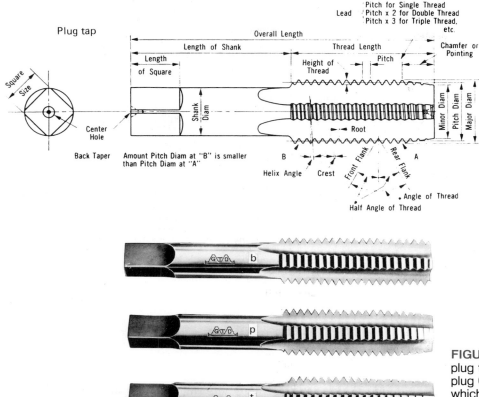

FIGURE 30-11 Terminology for a plug tap with photographs of taper (t), plug (p), and bottoming (b) taps which are used serially in threading holes. (Courtesy TRW-Greenfield Tap & Die.)

However, only a through hole can be threaded completely with a taper tap because it cuts to full depth only behind the tapered portion. A blind hole can be threaded to the bottom using three types of taps in succession. After the taper tap has the thread started in proper alignment, a *plug tap*, which has only a few tapered threads to provide gradual cutting of the threads to depth, is used to cut the threads as deep into the hole as its shape will permit. A *bottoming tap*, having no tapered threads, is used to finish the few remaining threads at the bottom of the hole to full depth. Obviously, producing threads to the full depth of a blind hole is time-consuming, and it also frequently results in broken taps and defective workpieces. Such configurations usually can be avoided if designers will give reasonable thought to the matter.

Taps operate under very severe conditions because of the heavy friction (high torque) involved and the difficulty of chip removal. Also, taps are relatively fragile. *Spiral-fluted taps,* illustrated in Figure 30-12, provide better removal of chips from a hole, particularly in tapping materials that produce long, curling chips. They also are helpful in tapping holes where the cutting action is interrupted by slots or keyways. The *spiral point* cuts the thread with a shearing

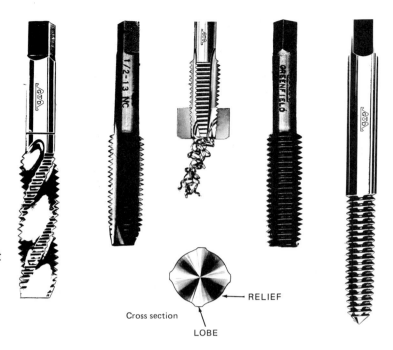

FIGURE 30-12 (Left to right) Spiral fluted tap; spiral point tap; spiral point tap cutting chips; fluteless bottoming tap and fluteless plug tap for cold-forming internal threads. Inset: Cross section of fluteless forming tap. (Courtesy TRW-Greenfield Tap & Die.)

RELIEF

Cross section

LOBE

action that pushes the chips ahead of the tap so that they do not interfere with the cutting action and the flow of cutting fluid into the hole.

Collapsing Taps.   *Collapsing taps* are similar to self-opening die heads in that the cutting elements collapse inward automatically when the thread is completed. This permits withdrawing the tap from the workpiece without the necessity of unscrewing it from the thread. They can either be self-setting, for use on automatic machines, or require manual setting for each cycle. Figure 30-13 shows some of the types available.

Hole Preparation.   Drilling is the most common method of preparing holes for tapping, and when close control over size is required, reaming may also be necessary. The drill size determines the final thread contour and the drilling torque. Unless otherwise specified, the tap drill size for most materials should produce approximately 75% thread, that is, 75% of full thread depth.

Machine Tapping.   Solid taps also are used in tapping operations on machine tools, such as lathes, drill presses, and special tapping machines. In tapping on a drill press, a tapping attachment often is used. These devices rotate the tap slowly when the drill press spindle is fed downward against the work. When the tapping is completed and the spindle raised, the tap is automatically driven in the reverse direction at a higher speed to reduce the time required to back the tap out of the hole. Some modern machine tools provide for extremely fast spindle reversal for backing taps out of holes.

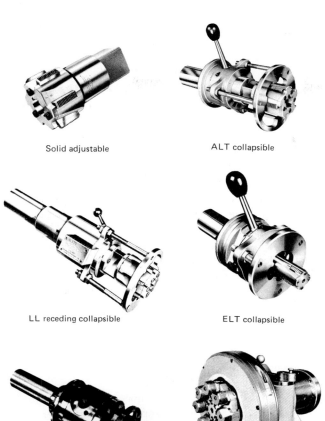

Solid adjustable

ALT collapsible

LL receding collapsible

ELT collapsible

2LLS receding collapsible

CBLM receding collapsible

FIGURE 30-13   Solid adjustable and collapsible taps (Courtesy Teledyne Landis Machine).

When solid taps are used on a screw machine or turret, the tap is prevented from turning while it is being fed into the work. As the tap reaches the end of the hole, the tap is free to rotate with the work. The work is then reversed and the tap, again prevented from rotating, is backed out of the hole.

The machine should have adequate power, rigidity, speed and feed ranges, cutting fluid supply, and positive drive action. Chucks, tap holders, and collets should be checked regularly for signs of wear or damage. Accurate alignment of the tap holder, machine spindle, and workpiece is vital to avoid broken taps or bell-mouthed, tapered, or oversized holes.

Tapping Cutting Time.   The equation to calculate the cutting for tapping is:

$$CT \cong \pi D L n / 8 V \qquad (30\text{-}1)$$

where CT = cutting time (min.)
     D = tap diameter (in.)

L = depth of tapped hole (in.)

n = number of threads per inch (tpi)

V = cutting speed (sfpm)

Special Threading and Tapping Machines.   Special machines are available for production threading and tapping. Threading machines usually have one or more spindles on which a self-opening die head is mounted, with suitable means for clamping and feeding the workpiece. A typical machine of this type is shown in Figure 30-14.

Some special tapping machines are similar in construction, with self-collapsing taps substituted for the threading dies. More commonly, tapping machines resemble drill presses, modified to provide spindle feeds both upward and downward, with the speed and feed more rapid on the upward motion.

Thread Milling.   Highly accurate threads, particularly in larger sizes, are often form-milled. Either a single- or a multiple-form cutter may be used.

A single-form cutter has a single annular row of teeth. As shown in Figure 30-15, the cutter is tilted at an angle equal to the helix angle of the thread and is fed inward radially to full depth while the work is stationary. The workpiece then is rotated slowly, and the cutter simultaneously is moved longitudinally, parallel with the axis of the work (or vice versa), by means of a lead screw, until the thread is completed. The thread can be completed in a single cut, or roughing and finish cuts can be used. This process is used primarily for large-lead or multiple-lead threads.

Some threads can be milled more quickly by using a multiple-form cutter having multiple rows of teeth set perpendicular to the cutter axis (the rows having no lead). The cutter must be slightly larger than the thread to be cut. It is set parallel with the axis of the workpiece and fed inward to full-thread depth while the work is stationary. The work then is rotated slowly for a little over one revolution, and the rotating cutter is simultaneously moved longitudinally with respect to the workpiece (or vice versa) according to the thread lead. When the work has revolved one revolution, the thread is complete. This process cannot be used on threads having a helix angle greater than about 3 degrees because clearance between the sides of the threads and the cutter depends on the cutter diameter's being substantially less than that of the workpiece. Thus, although the process is rapid, its use is restricted to threads of substantial diameter and not more than about 2 in. long.

Common Tapping Problems.   Tap overloading is often caused by poor lubrication, lands that are too wide, chips packed in the flutes, or tap wear. Surface roughness in the threads has many causes. A negative grind on the heel may keep the tap from tearing threads when backing out.

When a tap loses speed or needs more power, it generally indicates that the tap is dull (or improperly ground), or the chips are packed in flutes (loaded). The flutes may be too shallow or the lands too deep.

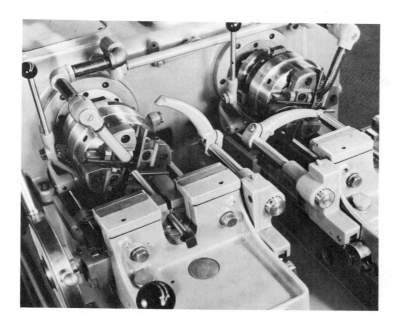

FIGURE 30-14   Two-spindle, automatic threading machine. (Courtesy Landis Machine Company.)

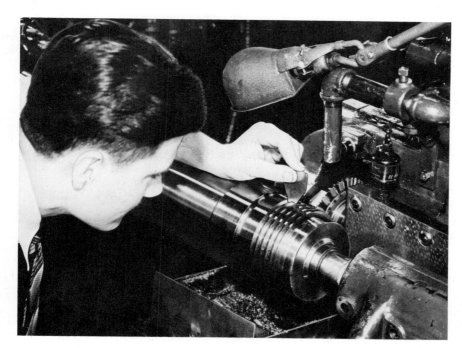

FIGURE 30-15   Milling a large thread with a single-form cutter. The cutter can be seen behind the thread. (Courtesy Lees-Bradner Company.)

When tapping soft ductile metals, loading can usually be overcome by polishing the tap before usage.

Improper hole size due to drill wear increases the percentage of threads being cut. Dull tools can also produce a rough finish or work-harden the hole surface and cause the tap to dull more quickly. Check to see that the axis of the hole and tap are aligned. If the tap cuts when backing out, check to see if the hole is oversize.

Cutting Fluids. Cutting fluids should be kept as clean as possible and should be supplied in copious quantities to reduce heat and friction and to aid in chip removal. Long tap life has been reported to result from routing high pressure coolants through the top to flush out the chips and cool the cutting edges. Recommended cutting fluids are listed in Table 30.3.

Thread Grinding. Grinding can produce very accurate threads, and it also permits threads to be made on hardened materials. Three basic methods are used. *Center-type grinding with axial feed* is the most common method, being similar to cutting a thread on a lathe. A shaped grinding wheel replaces the single-point tool. Usually, a single-ribbed grinding wheel is employed, but multiple-ribbed wheels are used occasionally. The grinding wheels are shaped by special diamond dressers or by crush dressing and must be inclined to the helix

TABLE 30.3 Cutting Fluids for Tapping (HSS Tools)

Work material	Cutting fluid
Aluminum	Kerosene and lard oil; Kerosene and light-base oil
Brass	Soluble oil or light-base oil
Naval brass	Mineral oil with lard or light-base oil
Manganese bronze	Mineral oil with lard or light-base oil
Phosphor bronze	Mineral oil with lard or light-base oil
Copper	Mineral oil with lard or light-base oil
Iron, cast	Dry or soluble oil
malleable	Soluble oil or sulfur-base oil
Magnesium	Light-base oil diluted with kerosene
Monel metal	Sulfur-base oil
Steels:	
up to 0.25 C	Sulfur-base or soluble oil
free machining	Sulfur-base or soluble oil
0.30–0.60 C annealed	Sulfur-base oil
0.30–0.60 C heat treated	Chlorinated sulfur-base oil
tool, high-carbon, HSS	Chlorinated sulfur-base oil
stainless	Chlorinated sulfur-base oil
Titanium	Chlorinated sulfur-base oil
Zinc die castings	Kerosene and lard oil

angle of the thread. Wheel speeds are in the high range. Several passes usually are required to complete the thread.

Center-type infeed thread grinding is similar to multiple-form milling in that a multiple-ribbed wheel, as wide as the length of the desired thread, is used. The wheel is fed inward radially to full thread depth, and the thread blank is then turned through about 1½ turns as the grinding wheel is fed axially a little more than the width of one thread.

Centerless thread grinding, illustrated in Figure 30-16, is used for making headless set screws. The blanks are hopper-fed to position *A*. The regulating wheel causes them to traverse the grinding wheel face, from which they emerge at position *B* in completed form. A production rate of 60 to 70 screws of ½-inch length per minute is possible.

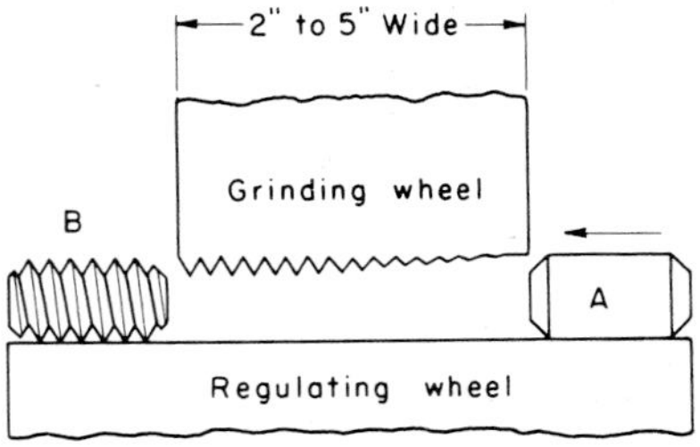

FIGURE 30-16 Principle of centerless thread grinding.

Thread Rolling. *Thread rolling* is used to produce threads in substantial quantities. This is a cold-forming process operation in which the threads are formed by rolling a thread blank between hardened dies that cause the metal to flow radially into the desired shape. Because no metal is removed in the form of chips, less material is required, resulting in substantial savings. In addition, because of cold working, the threads have greater strength than cut threads, and a smoother, harder, and more wear-resistant surface is obtained. In addition, the process is fast, with production rates of one per second being common. The quality of cold-rolled (or fluteless-tapped) products is consistently good and tap life is greater than that of HSS machine taps. Chipless operations are cleaner and there is a savings in material (15% to 20% savings in blank stock weight is typical).

Thread rolling is done by four basic methods. The simplest of these employs one fixed and one movable flat rolling die, as illustrated in Figure 30-17. After

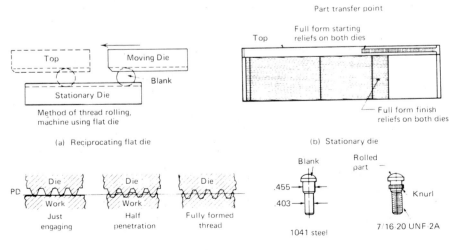

(a) Reciprocating flat die

(b) Stationary die

(c) Action of die in forming thread

FIGURE 30-17 Combination threads rolling and knurling of wheel bolt at 70 per minute by flat die roll threading.

the blank is placed in position on the stationary die, movement of the moving die causes the blank to be rolled between the two dies and the metal in the blank is displaced to form the threads. As the blank rolls, it moves across the die parallel with its longitudinal axis. Prior to the end of the stroke of the moving die, the blank rolls off the end of the stationary die, its thread being completed.

One obvious characteristic of a rolled thread is that its major diameter always is greater than the diameter of the blank. When an accurate class of fit is desired, the diameter of the blank is made about 0.002 in. larger than the thread-pitch diameter. If it is desired to have the body of a bolt larger than the outside diameter of the rolled thread, the blank for the thread is made smaller than the body.

Thread rolling can be done with cylindrical dies. Figure 30-18 illustrates the three-roll method commonly employed on turret lathes and screw machines. Two variations are used. In one, the rolls are retracted while the blank is placed in position. They then move inward radially, while rotating, to form the thread. More commonly the three rolls are contained in a self-opening die head similar to the conventional type used for cutting external threads. The die head is fed onto the blank longitudinally and forms the thread progressively as the blank rotates. With this procedure, as in the case of cut threads, the innermost 1½ to 2 threads are not formed to full depth because of the progressive action of the rollers.

The two-roll method is commonly employed for automatically producing large quantities of externally threaded parts up to 6 in. in diameter and 20 in. in length. The planetary type machine is for mass production of rolled threads on diameters up to 1 in.

Not only is thread rolling very economical, but the threads are excellent as to form and strength. The cold working contributes to increased strength, particularly at the critical root areas. There is less likelihood of surface defects (produced by machining) which can act as stress raisers.

Large numbers of threads are rolled on thin, tubular products. In this case external and internal rolls are used. The threads on electric lamp bases and sockets are examples of this type of thread.

Chipless Tapping.   Unfortunately, most internal threads cannot be made by rolling; there is insufficient space within the hole to permit the required rolls to be arranged and supported, and the required forces are too high. However, many internal threads, up to about ½ in. in diameter, are cold-formed in holes in ductile metals by means of *fluteless taps*. Such a tap and its special cross section are shown in Figure 30-12. The forming action is essentially the same as in rolling external threads. Because of the forming involved and the high friction, the torque required is about double that for cutting taps. Also, the hole diameter must be controlled carefully to obtain full thread depth without excessive torque. However, fluteless taps produce somewhat better accuracy than cutting taps. A lubricating fluid should be used, water-soluble oils being quite effective.

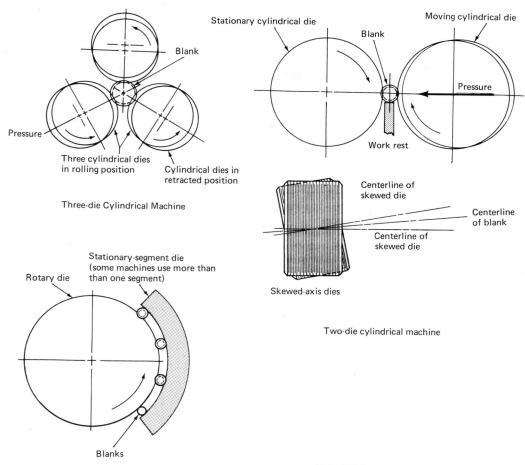

FIGURE 30-18   Methods for roll-forming threads using cylindrical dies.

Fluteless taps are especially suitable for forming threads in dead-end holes because no chips are produced. They come in both plug and bottoming types.

Machining versus Rolling Threads.   Threads are cut when full thread depth is needed (more than one pass necessary); for short production runs; when the blanks are not very accurate; when proximity to the shoulder in end threading is needed; for tapered threads; or when the workpiece material is not adaptable for rolling. Table 30.4 provides recommended drill sizes for chipless tap sizes 0 through ¾ in.

TABLE 30.4 Drill Selector Chart For Chipless Tapping

Tap Size		75% Thread Theoretical Hole Core Size	75% Nearest Drill Size	75% Decimal Equivalent	70% Thread Theoretical Hole Core Size	70% Nearest Drill Size	70% Decimal Equivalent	65% Thread Theoretical Hole Core Size	65% Nearest Drill Size	65% Decimal Equivalent
0	80	.0536	1.35 mm	.0531	.0540	1.35 mm	.0531	.0545		
1	64	.0650	1.65 mm	.0650	.0655	1.65 mm	.0650	.0661		
	72	.0659	1.65 mm	.0650	.0663			.0669	1.7 mm	.0669
2	56	.0769	1.95 mm	.0768	.0774	1.95 mm	.0768	.0781	5/64	.0781
	64	.0780	5/64	.0781	.0785	47	.0785	.0791	2.0 mm	.0787
3	48	.0884	2.25 mm	.0886	.0890	43	.0890	.0898	43	.0890
	56	.0899	43	.0890	.0904			.0911	2.3 mm	.0906
4	40	.0993	2.5 mm	.0984	.1000	39	.0995	.1010	39	.0995
	48	.1014	38	.1015	.1020	38	.1015	.1028	2.6 mm	.1024
5	40	.1123	34	.1110	.1130	33	.1130	.1140	33	.1130
	44	.1134	33	.1130	.1141	2.9 mm	.1142	.1150	2.9 mm	.1142
6	32	.1221	3.1 mm	.1220	.1230	3.1 mm	.1220	1243		
	40	.1253	1/8	.1250	.1260	3.2 mm	.1260	.1270	3.2 mm	.1260
8	32	.1481	3.75 mm	.1476	.1490			.1503	25	.1495
	36	.1498	25	.1495	.1507	3.8 mm	.1496	.1518	24	.1520
10	24	.1688			.1700	18	.1695	.1717	11/64	.1719
	32	.1741	17	.1730	.1750			.1763		
12	24	.1948	10	.1935	.1960	9	.1960	.1977	5.0 mm	.1968
	28	.1978	5.0 mm	.1968	.1989	8	.1990	.2003	8	.1990
1/4	20	.2245	5.7 mm	.2244	.2260			.2280	1	.2280
	28	.2318			.2329	5.9 mm	.2323	.2343	A	.2340
5/16	18	.2842	7.2 mm	.2835	.2861	7.25 mm	.2854	.2879	7.3 mm	.2874
	24	.2912	7.4 mm	.2913	.2927			.2941	M	.2950
3/8	16	.3431	11/32	.3437	.3452	8.75 mm	.3445	.3474	S	.3480
	24	.3537	9.0 mm	.3543	.3552	9.0 mm	.3543	.3566		
7/16	14	.4011			.4035	Y		.4059	13/32	
	20	.4120	Z		.4137	10.5 mm		.4154		
1/2	13	.4608			.4634			.4660		
	20	.4745			.4762			.4779		
9/16	12	.5200			.5229			.5257		
	18	.5342	13.5 mm	.5315	.5361			.5380		
5/8	11	.5787	37/64	.5781	.5817	37/64	.5781	.5848		
	18	.5967	19/32	.5937	.5986			.6004		
3/4	10	.6990			.7024			.7058	45/64	.7031
	16	.7181	23/32	.7187	.7202	23/32	.7187	.7224		

Source: James Abbott, ''Chipless Tapping Comes of Age,'' Modern Machine Shop, June 1984.

TABLE 30.4 Drill Selector Chart For Chipless Tapping (*continued*)

60% Thread			55% Thread			50% Thread			
Theoretical Hole Core Size	Nearest Drill Size	Decimal Equivalent	Theoretical Hole Core Size	Nearest Drill Size	Decimal Equivalent	Theoretical Hole Core Size	Nearest Drill Size	Decimal Equivalent	Tap Size
.0549	54	.0550	.0554	54	.0550	.0558	1.4 mm	.0551	80
									0
.0666			.0672	51	.0670	.0677	51	.0670	64
.0673	51	.0670	.0679	51	.067	.0683			72 1
.0787	47	.0785	.0794	2.0 mm	.0787	.0799			56
.0796	2.0 mm	.0787	.0802			.0807	2.05 mm	.0808	64 2
.0905	2.3 mm	.0906	.0913	2.3 mm	.0906	.0919			48
.0917	2.3 mm	.0906	.0924	2.35 mm	.0925	.0929	2.35 mm	.0925	56 3
.1018	38	.1015	.1028	2.8 mm	.1024	.1035	2.6 mm	.1024	40
.1035	2.6 mm	.1024	.1043	37	.1040	.1049	37	.1040	48 4
.1148	2.9 mm	.1142	.1158	32	.1160	.1165	32	.1160	40
.1157			.1166	32	.1160	.1173	32	.1160	44 5
.1252	1/8	.1250	.1264	3.2 mm	.1260	.1274			32
.1278	3.25 mm	.1280	.1288	30	.1285	.1295	30	.1285	40 6
.1512	3.8 mm	.1496	.1524	24	.1520	.1534	3.9 mm	.1535	32
.1526	24	.1520	.1537	3.9 mm	.1535	.1546	23	.1540	36 8
.1729	11/64	.1719	.1746	17	.1730	.1758			24
.1772	16	.1770	.1784	4.5 mm	.1772	.1794			32 10
.1989	8	.1990	.2006	5.1 mm	.2008	.2018	7	.2010	24
.2014	7	.2010	.2028			.2039	13/64	.2031	28 12
.2295	1	.2280	.2315			.2330	5.9 mm	.2323	20
.2354	15/64	.2344	.2368	6.0 mm	.2362	.2379	B	.2380	28 1/4
.2898	L	.2900	.2917	7.4 mm	.2913	.2936			18
.2955	7.5 mm	.2953	.2969	19/64	.2969	.2983	7.6 mm	.2992	24 5/16
.3495	8.9 mm	.3504	.3516			.3537	9.0 mm	.3543	16
.3580	T	.3580	.3594	23/64	.3594	.3608			24 3/8
.4084			.4108			.4132	Z		14
.4171			.4188			.4205			20 7/16
.4686	15/32		.4712	12.0 mm		.4738	12.0 mm		13
.4796			.4813			.4830	31/64		20 1/2
.5285			.5313	17/32	.5312	.5342	17/32	.5312	12
.5398			.5417			.5436	35/64	.5469	18 9/16
.5879			.5910	15 mm	.5906	.5941	19/32	.5937	11
.6023			.6042			.6061			18 5/8
.7092	18 mm	.7087	.7126			.7160			10
.7245			.7266			.7287	18.5 mm	.7283	16 3/4

Review Questions

1. How does the pitch diameter differ from the major diameter for a standard screw thread?
2. For what types of threads are the pitch and the lead the same?
3. Why are pipe threads tapered?
4. By what three basic methods can external threads be produced?
5. Explain the meaning of ¼″-20 UNC-3A.
6. What is meant by the designation M20 × 2.5-6g6g? (What does the ''x'' mean?)
7. Why are fine-series threads being used less now than in former years?
8. In cutting a thread on a lathe, how is the pitch controlled?
9. When possible, parts should be designed so that any required threads can be made by methods other than cutting on a lathe with a single-point tool. Explain why and what a designer should do to avoid this process.
10. What is the function of a threading dial on a lathe?
11. What controls the lead of a thread when it is cut by a threading die?
12. What is the basic purpose of a self-opening die head?
13. Why are cutting fluids for tapping usually oil base rather than water base?
14. What is the reason for using a taper tap before a plug tap in tapping a hole?
15. What difficulties are encountered if full threads are specified to the bottom of a dead-end hole?
16. Can a fluteless tap be used for threading a hole in gray cast iron? Why or why not?
17. What provisions should a designer make so that a dead-end hole can be threaded?
18. What is the major advantage of a spiral-point tap?
19. Can a fluteless tap be used for threading to the bottom of a dead-end hole? Why or why not?
20. Is it desirable for a tapping fluid to have lubricating qualities? Why?
21. How does thread milling differ when using single-form versus double-form cutters?
22. What are the advantages of making threads by grinding?
23. Why has thread rolling become the most commonly used method for making threads?
24. How may you determine whether a thread has been produced by rolling rather than by cutting?
25. What is a fluteless tap?
26. In Chapter 23, what figure(s) showed a threading dial?

Problems

1. Calculate the cutting time needed to tap a ¾ × 2 in. hole using a cutting speed of 30 sfpm. The tap has 10 tpi.
2. The new manufacturer manager (a recent MBA graduate) has recommended to you that chipless tapping be adopted for tapping holes in the deep dead-end holes on the 2-cylinder engine blocks that the company makes. Chipless tapping can run at twice the speed of the conventional tapping process, works well on deep holes, and provides better quality and finish with longer tool life. The tapping process is the bottleneck process in machining all the cast iron engine blocks, and furthermore about 10% of the blocks have to be scrapped due to broken taps. What do you recommend?

Gear Manufacturing

31

Gears transmit power or motion mechanically between parallel, intersecting, or nonintersecting shafts. Although usually hidden from sight, gears are one of the most important mechanical elements in our civilization, possibly even surpassing the wheel, since most wheels would not be turning were power not being applied to them through gears. They operate at almost unlimited speeds under a wide variety of conditions. Millions are produced each year in sizes from a few millimeters up to more than 30 feet in diameter. Often the requirements that must be, and routinely are, met in their manufacture are amazingly precise. Consequently, the machines and processes that have been developed for producing gears are among the most ingenious we have. In order to understand the functional requirements of these machines and processes, it is helpful briefly to consider the basic theory of gears and their operation.

Gear Theory and Terminology.   Gears, basically, are modifications of wheels with teeth being added to prevent slipping and to assure that their relative motions are constant. However, it should be noted that the relative surface velocities of the wheels (and shafts) are determined by the diameters of the wheels.

Although wooden teeth or pegs were attached to discs to make gears in ancient times, the teeth of modern gears are produced by machining or forming teeth on the outer portion of the wheel. The *pitch circle,* shown in Figure 31-1 and Figure 31-2, corresponds to the diameter of the wheel. Thus the angular velocity of a gear is determined by the diameter of this imaginary pitch circle. All design calculations relating to gear performance are based on the pitch-circle diameter or, more simply, the *pitch diameter (PD)*.

For two gears to operate properly, their pitch circles must be tangential to each other. The point at which the two pitch circles are tangent, at which they intersect the center line connecting their centers of rotation, is called the *pitch*

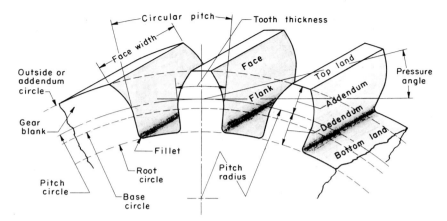

FIGURE 31-1   Gear-tooth nomenclature.

point. The common normal at the point of contact of mating teeth must pass through the pitch point. This condition is illustrated in Figure 31-2.

To minimize friction and wear, and thus increase their life and efficiency, gears are designed to have rolling motion between mating teeth rather than sliding motion. To achieve this condition, most gears utilize a tooth form that is based on an *involute curve*. This is the curve that is generated by a *point* on a straight line when the line rolls around a *base circle*. A somewhat simpler method of developing an involute curve is that shown in Figure 31-3, by unwinding a tautly held string from a base circle; point *A* generates an involute curve.

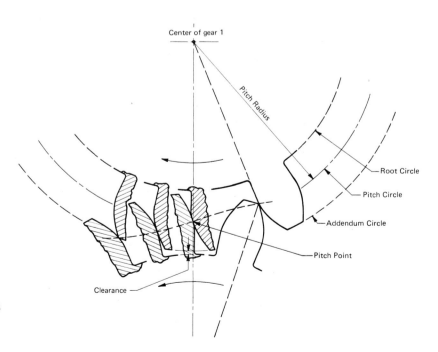

FIGURE 31-2   Tangent pitch circles between two gears produce a pitch point.

There are four reasons for using the involute form for gear teeth. First, such a tooth form provides the desired pure rolling action. Second, even if a pair of involute gears is operated with the distance between the centers slightly too large or too small, the common normal at the point of contact between mating teeth will always pass through the pitch point. Obviously, the theoretical pitch circles in such cases will be increased or decreased slightly. Third, the *line of action,* or *path of contact,* that is, the locus of the points of contact of mating teeth, is a straight line that passes through the pitch point and is tangent to the base circles of the two gears. The fourth reason is that a true involute tooth form can be produced by a cutting tool that has straight-sided teeth. This permits a very accurate tooth profile to be obtained through the use of a simple and easily made cutting tool.

The basic size of gear teeth may be expressed in two ways. The common practice, especially in the United States and England, is to express the dimensions as a function of the *diametral pitch (DP)*. *DP is the number of teeth (N) per unit of pitch diameter (PD);* thus $DP = N/PD$. Dimensionally, DP involves inches in the English system and millimeters in the SI system, and it is a measure of tooth size. The second method for specifying gear tooth size is by means of the *module (M),* defined as *the pitch diameter divided by the number of teeth,* or $M = PD/N$. It thus is the reciprocal of diametrical pitch and is expressed in inches or millimeters. Any two gears having the same diametral pitch or module will mesh properly if they are mounted so as to have the correct distances and relationship.

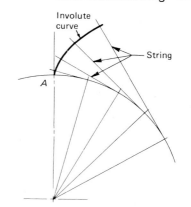

FIGURE 31-3 Method of generating an involute curve by unwinding a string from a cylinder.

The important tooth elements can be specified in terms of the diametral pitch or the module and are as follows:

1. *Addendum:* the radial distance from the pitch circle to the outside diameter.
2. *Dedendum:* the radial distance from the pitch circle to the root circle. It is equal to the addendum plus the *clearance,* which is provided to prevent the outer corner of a tooth from touching against the bottom of the tooth space.
3. *Circular pitch:* the distance between corresponding points of adjacent teeth, measured along the pitch circle. It is numerically equal to π/diametral pitch.
4. *Tooth thickness:* the thickness of a tooth, measured along the pitch circle. When tooth thickness and the corresponding *tooth space* are equal, no *backlash* exists in a pair of mating gears.
5. *Face width:* the length of the gear teeth in an axial plane.
6. *Tooth face:* the mating surface between the pitch circle and the addendum circle.
7. *Tooth flank:* the mating surface between the pitch circle and the root circle.

Four shapes of involute gear teeth are used in the United States:

1. 14½-degree pressure angle, full-depth (used most frequently).
2. 14½-degree pressure angle, composite (seldom used).
3. 20-degree pressure angle, full-depth (seldom used).
4. 20-degree pressure angle, stub-tooth (second most common).

In the 14½-degree full-depth system, the tooth profile outside the base circle is an involute curve. Inward from the base circle the profile is a straight radial line that is joined with the bottom land by a small fillet. With this system, the teeth of the basic rack have straight sides.

The 14½-degree composite system and the 20-degree full-depth system provide somewhat stronger teeth. However, with the 20-degree full-depth system considerable undercutting occurs in the dedendum area, and therefore, stub teeth often are used. The addendum is shortened by 20%, thus permitting the dedendum to be shortened a similar amount. This results in very strong teeth without undercutting.

Table 31.1 gives the formulas for computing the dimensions of gear teeth in the 14½-degree full-depth and 20-degree stub-tooth systems.

Physical Requirements of Gears. A consideration of gear theory leads to five requirements that must be met in order for gears to operate satisfactorily:

1. The actual tooth profile must be the same as the theoretical profile.
2. Tooth spacing must be uniform and correct.
3. The *actual* and theoretical pitch circles must be coincident and be concentric with the axis of rotation of the gear.
4. The face and flank surfaces must be smooth and sufficiently hard to resist wear and prevent noisy operation.
5. Adequate shafts and bearings must be provided so that desired center-to-center distances are retained under operational loads.

The first four of these requirements are determined by the material selection and manufacturing process. The various methods of manufacture that are used rep-

TABLE 31.1   Standard Dimensions for Involute Gear Teeth

	14½ Degree, Full Depth	20 Degree Stub Tooth
Pitch diameter	$\dfrac{N}{DP}$	$\dfrac{N}{DP}$
Addendum	$\dfrac{1}{DP}$	$\dfrac{0.8}{DP}$
Dedendum	$\dfrac{1.157}{DP}$	$\dfrac{1}{DP}$
Outside diameter	$\dfrac{N+2}{DP}$	$\dfrac{N+1.6}{DP}$
Clearance	$\dfrac{0.157}{DP}$	$\dfrac{0.2}{DP}$
Tooth thickness	$\dfrac{1.5708}{DP}$	$\dfrac{1.5708}{DP}$

FIGURE 31-4 Types of gears. (Top) Spur gear and rack, worm and worm gear, continuous herringbone gears. (Center) Spiral bevel gear; helical gears; crown gear. (Bottom) Straight, Zerol, and hypoid bevel gears. (Courtesy Gleason Works.)

resent attempts to meet these requirements to varying degrees with minimum cost, and their effectiveness must be measured in terms of the extent to which the resulting gears embody these requirements.

The more common types of gears are shown in Figure 31-4. *Spur gears* have straight teeth and are used to connect parallel shafts. They are the most easily made and the cheapest of all types.

The teeth on *helical gears* lie along a helix, the angle of the helix being the angle between the helix and a pitch cylinder element parallel with the gear shaft. Helical gears can connect either parallel or nonparallel nonintersecting shafts. Such gears are stronger and quieter than spur gears because the contact between mating teeth increases more gradually and more teeth are in contact at a given time. Although they usually are slightly more expensive to make than spur gears, they can be manufactured in several ways and are produced in large numbers.

Helical gears have one disadvantage. When they are loaded a side thrust is created that must be absorbed in the bearings. *Herringbone gears* neutralize this side thrust by having, in effect, two helical-gear halves, one having a right-hand and the other a left-hand helix. The *continuous* herringbone type is rather difficult to machine but is very strong. A modified herringbone type is made by machining a groove, or gap, around the gear blank where the two sets of teeth would come together. This provides a runout space for the cutting tool in making each set of teeth.

A *rack* is a gear with infinite radius, having teeth that lie on a straight line on a plane. The teeth may be normal to the axis of the rack or helical so as to mate with spur or helical gears, respectively.

A *worm* is similar to a screw. It may have one or more threads, the multiple-thread type being very common. Worms usually are used in conjunction with a *worm gear*. High gear ratios are easily obtainable with this combination. The axes of the worm and worm gear are nonintersecting and usually are at right angles. If the worm has a small helix angle, it cannot be driven by the mating worm gear. This principle frequently is employed to obtain nonreversible drives. Worm gears usually are made with the top land concave to permit greater area of contact between the worm and the gear. A similar effect can be achieved by using a *conical worm,* in which the helical teeth are cut on a double-conical blank, thus producing a worm that has an hourglass shape.

Bevel gears, teeth on a cone, are used to transmit motion between intersecting shafts. The teeth are cut on the surface of a truncated cone. Several types of bevel gears are made, the types varying as to whether the teeth are straight or curved and whether the axes of the mating gears intersect. On *straight-tooth* bevel gears the teeth are straight, and if extended all would pass through a common apex. *Spiral-tooth* bevel gears have teeth that are segments of spirals. Like helical gears, this design provides tooth overlap so that more teeth are engaged at a given time and the engagement is progressive. *Hypoid* bevel gears also have a curved-tooth shape but are designed to operate with nonintersecting axes. They are used in the rear axles of most automobiles so that the drive shaft axis can be below the axis of the axle and thus permit a lower floor height. *Zerol* bevel gears have teeth that are circular arcs, providing somewhat stronger teeth than can be obtained in a comparable straight-tooth gear. They are not used extensively. When a pair of bevel gears are the same size and have their shafts at right angles, they are termed *miter gears.*

A *crown gear* is a special form of bevel gear having a 180-degree cone apex angle. In effect, it is a disk with the teeth on the side of the disk. It also may be thought of as a rack that has been bent into a circle so that its teeth lie in a plane. The teeth may be straight or curved. On straight-tooth crown gears the teeth are radial. Crown gears seldom are used, but they have the important quality that they will mesh properly with a bevel gear of any cone angle, provided that the bevel gear has the same tooth form and diametral pitch. This important principle is incorporated in the design and operation of two very important types of gear-generating machines that will be discussed later.

Most gears are of the external type, the teeth forming the outer periphery of the gear. Internal gears have the teeth on the inside of a solid ring, pointing toward the center of the gear.

Gears are made in very large numbers both by machining and by cold-roll forming. In addition, significant quantities are made by extrusion, by blanking, by casting, and some by powder metallurgy and by a forging process. However, it is only by machining that all types of gears can be made in all sizes, and although roll-formed gears can be made with accuracy sufficient for most applications—even for automobile transmissions—machining still is unsurpassed for gears that must have very high accuracy. Also, roll forming can be used only on ductile metals.

Basic Methods for Machining Gears.   Four basic methods are employed for machining gears, each having certain advantages and limitations as to quality, flexibility, and cost.

Form cutting utilizes the principle illustrated in Figure 31-5, the cutter having the same form as the *space* between adjacent teeth. Usually a multiple-tooth form cutter (see Figure 31-6) is used as shown in Figure 31-5. The tool is fed radially toward the center of the gear blank to the desired tooth depth, then across the tooth face to obtain the required tooth width. When one tooth has been completed, the tool is withdrawn, the gear blank is indexed, and the cutting of the next tooth space is started. Basically, form cutting is a simple and flexible method of machining gears. The equipment and cutters required are relatively

Gear Manufacturing

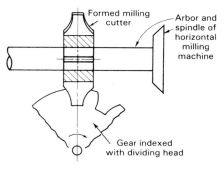

FIGURE 31-5   Basic method of machining a gear by form milling or form cutting.

FIGURE 31-6   Form cutter (left) and stocking cutter for machining gear teeth. (Courtesy Brown Sharpe Mfg. Co.)

FIGURE 31-7 Blind spline broaching machine with automated pick-and-place arms for load/unload. (Courtesy Apex Broach and Machine Company.)

simple, and standard machine tools (milling machines) often are used. However, in most cases the procedure is quite slow, and considerable care is required on the part of the operator; therefore it usually is employed where only one or a few gears are to be made.

Another way to form cut teeth is to cut all the tooth spaces simultaneously by gear broaching. The circular table in Figure 31-7 holds ten sets of progressive tooling. The table rotates moving one set of tooling at a time under two workpieces. The arms load and unload a set of parts every 15 seconds. Excellent gears can be made by *broaching*. However, a separate broach must be provided for each size of gear. The tooling tends to be expensive, a restriction for this method, even though the cycle time can be very fast.

Template machining utilizes a simple, single-point cutting tool that is guided by a template. By using a template that is several times larger than the gear tooth that is to be cut, good accuracy can be achieved. However, the equipment is specialized, and the method is seldom used except for making large bevel gears.

Most high-quality gears that are made by machining are made by the *generating process*. This process is based on the principle that any two involute gears, or any gear and a rack, of the same diametral pitch will mesh together properly. Utilizing this principle, one of the gears (or the rack) is made into a cutter by proper sharpening. It can be used to cut into a mating gear blank and thus generate teeth on the blank. Two principal methods for gear generating are *shaping* and *hobbing*. Shaping is explained in Figure 31-8 and Figure 31-11.

To carry out the process, the cutter and the gear blank must be attached rigidly to their respective shafts, and the two shafts must be interconnected by suitable gearing so that the cutter and the blank rotate positively with respect to each other and with the same pitch-line velocities. To start cutting the gear, the cutter is reciprocated and is fed radially into the blank between successive strokes. When the desired tooth depth has been obtained, the cutter and blank

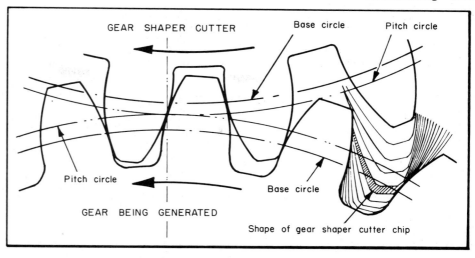

FIGURE 31-8 (Top) Generating action of a Fellows gear-shaper cutter. (Bottom) Series of photographs showing various stages in generating one tooth in a gear by means of a gear shaper, action taking place from right to left, corresponding to the diagram above. One tooth of the cutter was painted white. (Top Courtesy Fellows Gear Shaper Co.)

are then slightly indexed after each cutting stroke. The resulting generating action is indicated schematically in the upper diagram of Figure 31-8 and shown in the cutting of an actual gear tooth in the photographs in the lower portion of the same figure.

Machines for Form-Cutting Gears. In machining gears by the form-cutting process, the form cutter is mounted on the machine spindle, and the gear blank is mounted on a mandrel held between the centers of some type of indexing device. Figure 31-9 shows the arrangement that is employed when, as is often the case, the work is done on a universal milling machine; the cutter is mounted on an arbor, and a dividing head is used to index the gear blank. When a helical gear is to be cut, as in the case shown, the table must be set at an angle equal to the helix angle, and the dividing head is geared to the longitudinal feed screw of the table so that the gear blank will rotate as it moves longitudinally.

Standard cutters usually are employed in form-cutting gears. In the United States, these come in eight sizes for each diametral pitch and will cut gears having the number of teeth indicated in Table 31.2.

A single cutter will not produce a theoretically perfect tooth profile for all sizes of gears in the range for which it is intended. However, the change in tooth profile over the range covered by each cutter is very slight, and most of the time, satisfactory results can be achieved. When greater accuracy is required, half-number cutters (such as 3½) can be obtained. Typical cutters are shown in

FIGURE 31-9 Form-cutting a helical gear on a universal milling machine, using an indexing head. Inset shows closeup of gear and cutter.

TABLE 31.2

Cutter Number	Gear Tooth Range
1	135 teeth to rack
2	55–134
3	35–54
4	26–34
5	21–25
6	17–20
7	14–16
8	12–13

Figures 31-6, 31-9, and 31-10. Cutters are available for all common diametral pitches and 14½-degree and 20-degree pressure angles. If the amount of metal that must be removed to form a tooth space is large, roughing cuts may be taken with a *stocking cutter*. The stepped sides of the stocking cutter remove most of the metal and leave only a small amount to be removed subsequently by the regular form cutter in a finish cut.

Straight-tooth bevel gears can be form-cut on a milling machine, but this is seldom done. Because the tooth profile in bevel gears varies from one end of the tooth to the other, after one cut is taken to form the correct tooth profile at the smaller end, the relationship between the cutter and the blank must be altered. Shaving cuts then are taken on the side of each tooth to form the correct profile throughout the entire tooth length.

Although the form cutting of gears on a milling machine is a flexible process and is suitable for gears that are not to be operated at high speeds or that need not operate with extreme quietness, the process is slow and requires skilled labor.

Semiautomatic machines are available for making gears by the form-cutting process. Such a machine is shown in Figure 31-10. The procedure utilized is essentially the same as on a milling machine, except that, after setup, the various operations are completed automatically. Gears made on such machines are no more accurate than those produced on a milling machine, but the possibility of error is less, and they are much cheaper because of reduced labor requirements. For large quantities, however, form cutting is not used.

Machines for Gear Generating. *Gear shapers* generate gears by a reciprocating tool motion. Figure 31-11 shows a machine called a gear shaper and Figure 31-12 shows the design of the cutter. The gear blank is mounted on the

FIGURE 31-10 Cutting a gear on a semiautomatic gear-cutting machine with cutters making simultaneous roughing and finish cuts. (Courtesy Brown & Sharpe Mfg. Co.)

FIGURE 31-11 Fellows gear shaper, showing the motions of the gear blank and the cutter. Inset shows a close-up view of the cutter and teeth. (Courtesy Fellows Gear Shaper Company.)

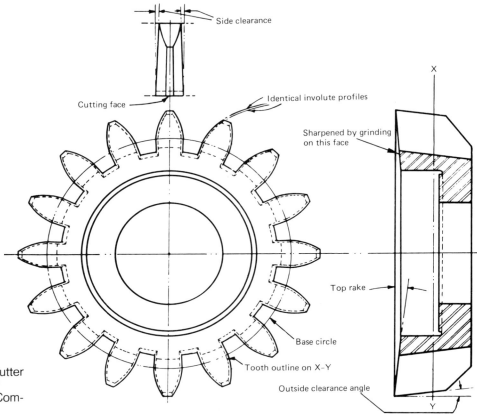

FIGURE 31-12 Details of the cutter used on the Fellows gear shaper (Courtesy Fellows Gear Shaper Company.)

rotating table (or vertical spindle) and the cutter on the end of a vertical, reciprocating spindle. The spindle and the table are connected by means of gears so that the cutter and gear blank revolve with the same pitch-line velocity. Cutting occurs on the down stroke (sometimes on the up stroke). At the end of each cutting stroke, the spindle carrying the blank retracts slightly to provide clearance between work and tool on the return stroke. Because of the reciprocating action of the cutter, these machines are commonly called *gear shapers*.

To start cutting a gear, the cutter is fed radially inward before each cutting stroke, as it and the blank rotate. When the proper depth is reached, the inward feed stops and the cutter and blank continue their rotation until all the teeth have been machined by the generation process.

Either straight- or helical-tooth gears can be cut on gear shapers. To cut helical teeth, both the cutter and the blank are given an oscillating rotational motion during each stroke of the cutter, turning in one direction during the cutting stroke and in the opposite direction during the return stroke. Because the cutting stroke can be adjusted to end at any desired point, gear shapers are particularly useful for cutting cluster gears. Some machines can be equipped with two cutters simultaneously to cut two gears, often of different diameters. Gear shapers also can be adapted for cutting internal gears.

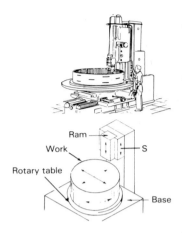

Ram

Work

Rotary table

S

Base

T = Table   W = Work   C = Cutting tool

FIGURE 31-13   (Left) Schematic of vertical shaper. (From *Manufacturing Producibility Handbook*; courtesy General Electric Company.) (Right) Typical job setup in vertical shaper. (Courtesy Colt Industries.)

Special types of gear shapers have been developed for mass-production purposes. The *rotary gear shaper* essentially is ten shaper units mounted on a rotating base and having a single drive mechanism. Nine gears are cut simultaneously while a finished gear is removed and a new blank is put in place on the tenth unit. *Planetary gear shapers* holding six gear blanks move in planetary motion about a large, central gear cutter. The cutter has no teeth in one portion to provide a space where the gear can be removed and a new blank placed on the empty spindle.

CNC gear shapers are now available with hydromechanical stroking systems which produce a uniform cutting velocity during the cutting portion of the down stroke. These machines can operate at 500–1700 rpm and use TiN-coated cutters to enhance tool life.

Vertical shapers for gear generating, shown in Figure 31-13, have a vertical ram and a round table that can be rotated in a horizontal plane by either manual or power feed. These machine tools are sometimes called *slotters*. Usually, the ram is pivoted near the top so that it can be swung outward from the column through an arc of about 10 degrees.

Because one circular and two straight-line motions and feeds are available, vertical shapers are very versatile tools and thus find considerable use in one-of-a-kind manufacturing. Not only can vertical and inclined flat surfaces be machined, but external and internal cylindrical surfaces can be generated by circular feeding of the table between strokes. This may be cheaper than turning or boring for very small lot sizes. A vertical shaper can be used for generating gears or machining curved surfaces, interior surfaces, and arcs by using a stationary tool and rotating the workpiece. A *keyseater* is a special type of vertical shaper designed and used exclusively for machining keyways on the inside of wheel and gear hubs.

For machining continuous herringbone gears, a *Sykes gear-generating machine* is used.

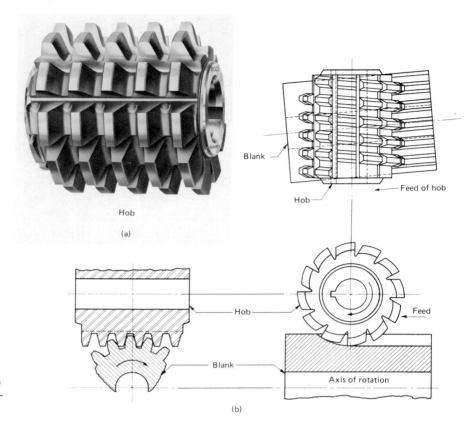

Hob

(a)

Blank

Feed of hob

Hob

Hob

Blank

Feed

Axis of rotation

(b)

FIGURE 31-14 Relationship of hob (upper left) to the gear blank in machining a spur gear by hobbing.

Gear-Hobbing Machines.   Involute gear teeth could be generated by a cutter that has the form of a rack. Such a cutter would be simple to make but has two major disadvantages. First, the cutter (or the blank) would have to reciprocate, with cutting occurring only during one stroke direction. Second, because the rack would have to move longitudinally as the blank rotated, the rack would need to be very long (or the gear very small) or the two would not be in mesh after a few teeth were cut. A *hob* overcomes the preceding two difficulties. As shown in Figure 31-14, a hob can be thought of, basically, as one long rack tooth that has been wrapped around a cylinder in the form of a helix and fluted at intervals to provide a number of cutting edges. Relief is provided behind each of the teeth. The cross section of each tooth, normal to the helix, is the same as that of a rack tooth. A hob can also be thought of as a gashed worm.

The action of a hobbing machine cutting a spur gear is illustrated in Figure 31-15. To cut a spur gear, the axis of the hob must be set off from the normal to the rotational axis of the blank by the helix angle of the hob. In cutting helical gears, the hob must be set over an additional amount equal to the helix angle of the gear. The cutting of a gear by means of a hob is a continuous action. The hob and the blank are connected by proper gearing so they rotate in mesh.

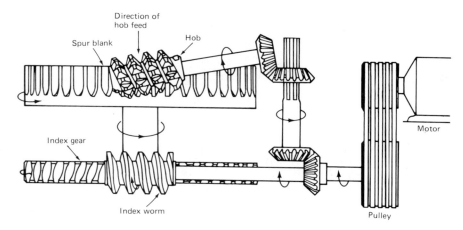

FIGURE 31-15   Schematic of gear-hobbing machine.

To start cutting a gear, the rotating hob is fed inward until the proper setting for tooth depth is obtained. The hob is then fed in a direction parallel with the axis of rotation of the blank. As the gear blank rotates, the teeth are generated and the feed of the hob across the face of the blank extends the teeth to the desired tooth face width.

Hobbing is rapid and economical. More gears are cut by this process than by any other. The process produces excellent gears and can also be used for splines and sprockets. Single- double-, and triple-thread hobs are used. Multiple-thread types increase the production rate but do not produce accuracy as high as single-thread hobs.

Gear-hobbing machines are made in a wide range of sizes. Machines for cutting accurate large gears frequently are housed in temperature-controlled rooms, and the temperature of the cutting fluid is controlled to avoid dimensional change due to variations in temperature.

Bevel Gear-Generating Machines.   The machines used to generate the teeth on various types of bevel gears are among the most ingenious of all machine tools. Two basic types utilize the principle that a crown gear will mate properly with any bevel gear having the same diametral pitch and tooth form. One is used for cutting straight-tooth bevel gears and the other for cutting curved-tooth bevel gears.

The basic principle of these machines is indicated in Figure 31-16. The blank and a connecting gear, having the same cone angle and diametral pitch, are mounted on a common shaft so that the connecting gear meshes with the regular teeth on the crown gear, which also has a reciprocating cutter to generate the teeth on the gear blank. As the connecting gear rolls on the crown gear, the reciprocating cutter generates a tooth in the gear blank. Obviously, with only one cutter tooth, only a single tooth space would be cut in the gear blank. However, this limitation is overcome in actual machines by two modifications. First, by indexing the connecting gear and blank unit, the process is repeated as often as required to cut all the teeth in the blank. Second, instead of a single

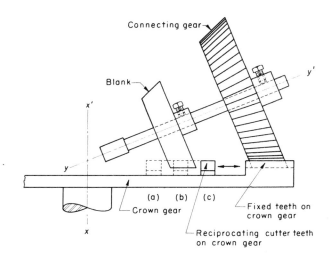

FIGURE 31-16   Principle of bevel gear-generating machines.

reciprocating cutter, two half-tooth cutters are approximating the inner sides of two adjacent teeth, so that they cut both faces of one tooth on the blank.

The application of this modified basic principle in actual machines is shown in Figures 31-17 and 31-18. The cradle acts as a crown gear and carries the two cutters, which reciprocate in slides simultaneously in opposite directions. It can be noted in Figure 31-18 that straight lines through the faces of the cutters, the axis of rotation of the gear blank, and elements of the pitch cone of the gear blank all meet at a common point that is in the plane of the imaginary crown gear and on the axis of the cradle, thus fulfilling the requirements shown in Figure 31-16.

In operation, while the cradle is at the extreme upward position, the gear blank is fed inward toward the cutter until the position for nearly full tooth depth is reached. The cradle then starts to roll downward, and a tooth is generated during the downward roll. When the cradle reaches the full down position, the blank is automatically fed inward a small amount—usually 0.020 to 0.015 in.—and a finish cut is made during the upward roll of the cradle. At the completion

FIGURE 31-17   (Top) Schematic showing the roll of the gear blank and cutters during the cutting of one tooth on a bevel gear-generating machine, as seen from the front. (Bottom) Photographs showing the same roll action, as seen from the back of the machine. (Courtesy Heidenreich & Harbeck.)

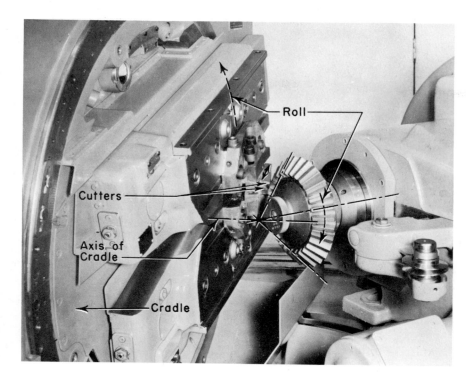

FIGURE 31-18 Cutters and gear blank on a Gleason straight-tooth gear-generating machine, showing the relationship between the axis of rotation of the blank and the axis of the cradle, which, with the cutter, simulates a crown gear. (Courtesy Gleason Works.)

of the upward roll, the blank is withdrawn, indexed, and moved inward automatically, ready to start the next tooth. These machines can cut bevel gears of different cone angles.

Provision is made so that the spindle on which the gear blank is mounted can be swung through an angle to accommodate bevel gears of different cone angles. This spindle is connected to the cradle by means of suitable internal change gears to provide the required positive rolling motion between the two.

For cutting small straight-tooth bevel gears, a machine employing two revolving disc-type cutters is used. These cutters reciprocate on the cradle slides as they rotate. Because of their multiple cutting edges and continuous rotary, rather than a reciprocating cutting action, they are much more productive. See Figure 31-19.

Generating machines for straight-tooth bevel gears often are provided with a mechanism that produces a slight crown on the teeth. Such teeth are slightly thicker at the middle than at the ends, to avoid having applied loads concentrated at the tooth ends where they are the weakest.

Machines used for generating spiral, Zerol, and hypoid bevel gears employ the same basic crown-gear principle, but a multitooth rotating cutter is used. This cutter has its axis of rotation parallel with the axis of roll of the cradle. Figure 31-20 shows the relationship of the gear blank to the axis of the theoretical crown gear (the cradle) and of the cutter teeth to the theoretical mating bevel gear.

FIGURE 31-19 Bevel gear-generating machine, utilizing two disc-type cutters. (Courtesy Gleason Works.)

FIGURE 31-20   Relationship of the gear blank to the axis of the theoretical crown gear (the cradle) and of the cutter teeth to the theoretical mating bevel gear. (Courtesy Gleason Works.)

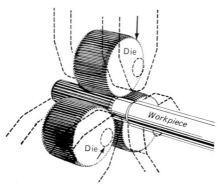

FIGURE 31-21   Method for forming gear teeth and spline by cold forming.

Cold Roll Forming of Gears.   The manufacture of gears by *cold roll-forming* has been highly developed and widely adopted in recent years. Currently, millions of high-quality gears are produced annually by this process; many of the gears in automobile transmissions are made this way. As indicated in Figure 31-21, the process is basically the same as that by which screw threads are roll-formed, except that in most cases the teeth cannot be formed in a single rotation of the forming rolls; the rolls are fed inward gradually during several revolutions.

Because of the metal flow that occurs, the top lands of roll-formed teeth are not smooth and perfect in shape; a depressed line between two slight protrusions can often be seen, as shown encircled in Figure 31-22. However, because the top land plays no part in gear-tooth action, if there is sufficient clearance in the mating gear, this causes no difficulty. Where desired, a light turning cut is used to provide a smooth top land and correct addendum diameter.

The hardened forming rolls are very accurately made, and the roll-formed gear teeth usually have excellent accuracy. In addition, because the severe cold working produces tooth faces that are much smoother and harder than those on ordinary machined gears, they seldom require hardening or further finishing, and they have excellent wear characteristics.

The process is rapid (up to 50 times faster than gear machining) and easily mechanized. No chips are made and thus less material is needed. Less skilled labor is required. Small gears often are made by rolling a length of shaft and then slicing off the individual gear blanks. Usually soft steel is required and four to five inches in diameter is about the limit, with fewer than six teeth, coarser than 12 diametral pitch, and no pressure angle less than 20 degrees.

FIGURE 31-22 (Top) Worm being rolled by means of rotating rolling tools. (Bottom left) Typical worm made by rolling, with enlarged view of end of one tooth. (Right) Gear made by rolling. (Courtesy Landis Machine Company.)

Other Gear-Making Processes.   Gears can be made by the various casting processes. *Sand-cast gears* have rough surfaces and are not accurate dimensionally. They are used only for services where the gear moves slowly and where noise and inaccuracy of motion can be tolerated. Gears made by *die casting* are fairly accurate and have fair surface finish. They can be used to transmit light loads at moderate speeds. Gears made by *investment casting* may be accurate and have good surface characteristics. They can be made of strong materials to permit their use in transmitting heavy loads. In many instances, gears that are to be finished by machining are made from cast blanks, and in some larger gear the teeth can be cast to approximate shape to reduce the amount of machining.

Large quantities of gears are produced by *blanking* in a punch press. The thickness of such gears usually does not exceed about 1/16 in. By shaving the gears after they are blanked, excellent accuracy can be achieved. Such gears are used in clocks, watches, meters, and calculating machines. *Fine blanking* is also used to produce thin, flat gears of good quality.

High-quality gears, both as to dimensional accuracy and surface quality, can be made by the *powder metallurgy process*. Usually, this process is employed only for small sizes, ordinarily less than 1 in. in diameter. However, larger and excellent gears are made by forging powder metallurgy preforms. This results in a product of much greater density and strength than usually can be obtained by ordinary powder metallurgy methods, and the resulting gears give excellent service at reduced cost. Gears made by this process often require little or no finishing.

Large quantities of plastic gears are made by *plastic molding*. The quality of such gears is only fair, and they are suitable only for light loads. Accurate gears suitable for heavy loads frequently are machined out of laminated plastic materials. When such gears are mated with metal gears, they have the quality of reducing noise.

Quite accurate small-sized gears can be made by the *extrusion* process. Typically, long lengths of rod, having the cross section of the desired gear, are extruded. The individual gears are then sliced from this rod. Materials suitable for this process are brass, bronze, aluminum alloys, magnesium alloys, and, occasionally, steel.

Flame machining (oxyacetylene cutting) can be used to produce gears that are to be used for slow-moving applications wherein accuracy is not required.

A few gears are made by the hot roll-forming process. In this process a cold master gear is pressed into a hot blank as the two are rolled together.

Gear Finishing.   In order to operate efficiently and have satisfactory life, gears must have accurate tooth profiles and the faces of the teeth must be smooth and hard. These qualities are particularly important when gears must operate quietly at high speeds. When they are produced rapidly and economically by most of the processes except cold-roll forming, the tooth profiles may not be as accurate as desired, and the surfaces are somewhat rough and subject to rapid wear. Also, it is difficult to cut gear teeth in a hardened gear blank, and therefore economy dictates that the gear be cut in a relatively soft blank and subsequently be heat-treated to obtain greater hardness. Such heat treatment usually results in some slight distortion and surface roughness. Although most roll-formed gears have sufficiently accurate profiles, and the tooth faces are adequately smooth and frequently have sufficient hardness, this process is feasible only for relatively small gears. Consequently, a large proportion of high-quality gears are given some type of finishing operation after they have received primary machining or after heat treatment. Most of these finishing operations can be done quite economically because only minute amounts of metal are removed.

Gear shaving is the most commonly used method for finishing spur and helical gear teeth prior to hardening. The gear is run, at high speed, in contact with a shaving tool, usually of the type shown in Figure 31-23. Such a tool is a very

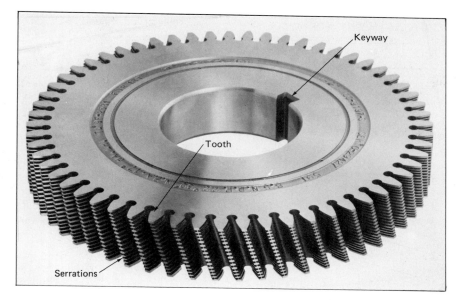

Keyway

Tooth

Serrations

FIGURE 31-23   Rotary gear shaving cutter.

accurate, hardened, and ground gear that contains a number of peripheral serrations, thus forming a series of sharp cutting edges on each tooth. The gear and shaving cutter are run in mesh with their axes crossed at a small angle, usually about 10 degrees. See Figure 31-24. As they rotate, the gear is reciprocated longitudinally across the shaving tool (or vice versa). During this action, which usually requires less than a minute, very fine chips are *shaved* from the gear-tooth faces, thus eliminating any high spots and producing a very accurate tooth profile.

Rack-type shaving cutters sometimes are used for shaving small gears—the cutter reciprocating lengthwise, causing the gear to roll along it, as it is moved sideways across the cutter and fed inward.

Although shaving cutters are costly, they have a relatively long life because only a very small amount of metal is removed, usually 0.001 to 0.004 in. Some gear-shaving machines produce a slight crown on the gear teeth during shaving. Most gears are not hardened prior to shaving, although it is possible to remove very small amounts of metal from hardened gears if they are not too hard. However, modern heat-treating equipment makes it possible to harden gears after shaving without harmful effects, and therefore this practice is usually followed.

Roll finishing is a cold-forming process that is used to finish helical gears. The unhardened gear is rolled with two hardened, accurately formed rolling dies. The center distance between the dies is reduced to cold-work the surfaces and produce highly accurate tooth forms. High points on the unhardened gear are plastically deformed so that a smoother surface and more accurate tooth form are achieved. Because the operation is one of localized cold working, some

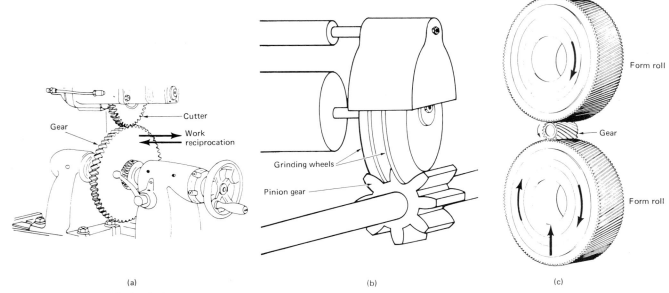

(a)   (b)   (c)

FIGURE 31-24 Methods for gear finishing by shaving, grinding, and roll forming or finishing.

undesirable effect may accrue, such as localized residual stresses and nonuniform surface characteristics. Surface finishes of 6 to 8 microinches have been achieved. If roll finishing is to be used, attention must be paid to the prerolled geometry. Designers should consult the manufacturers of gear-rolling machines for specific recommendations.

Grinding is used to obtain very accurate teeth on hardened gears. Two methods are used. One employs a formed grinding wheel that is trued to the exact form of a tooth by means of diamonds mounted on a special holder and guided by a large template. The other method is involute-generation grinding which uses straight-sided grinding wheels which simulate one side of a rack tooth. The surface of the gear tooth is ground as the gear rolls (and reciprocates) past the grinding wheels. Grinding produces very accurate gears, but because it is slow and expensive, it is used only on the highest-quality hardened gears.

Lapping also can be used for finishing hardened gears. The gear to be finished is run in contact with one or more cast-iron lapping gears under a flow of very fine abrasive in oil. Because lapping removes only a very small amount of metal, it usually is employed on gears that previously have been shaved and hardened. This combination of processes produces gears that are nearly equal to ground gears in quality but at considerably lower cost.

Gear Inspection. As with all manufactured products, gears must be checked to determine whether the resulting product meets the design specifications and requirements. Because of their irregular shape and the number of factors that must be measured, inspection of gears is somewhat difficult. Among the factors to be checked are the linear tooth dimensions—thickness, spacing, depth, and so on—tooth profile, surface roughness, and noise. Several special devices, most of them automatic or semiautomatic, are used for such inspection.

Gear-tooth vernier calipers can be used to measure the thickness of gear teeth on the pitch circle, as shown in Figure 31-25. CNC gear inspection machines, such as shown in Figure 31-26, can quickly check several factors, including variations in circular pitch, involute profile, lead, tooth spacing, and variations in pressure angle. The gear usually is mounted between centers. The probe is moved to the gear through *X, Y, Z* translations while the gear can be rotated (C). The inset in Figure 31-26 shows a typical display for involute profile.

Because noise level is important in many applications, not only from the viewpoint of noise pollution but also as an indicator of probable gear life, special equipment for its measurement is quite widely used, sometimes integrated into mass-production assembly lines.

FIGURE 31-25 Using gear-tooth vernier calipers to check the tooth thickness at the pitch circle.

FIGURE 31-26 CNC gear inspection machine. (Courtesy Illitron.)

Review Questions

1. Why can the relative angular velocities of two mating spur gears not be determined by their outside diameters alone?
2. Why is the involute form used for gear teeth?
3. What is the diametral pitch of a gear?
4. What is the relationship between the diametral pitch and the module of a gear?
5. On a sketch of a gear, indicate the pitch circle, addendum circle, dedendum circle, and circular pitch.
6. What five requirements must be met in order for gears to operate satisfactorily? Which of these are determined by the manufacturing process?
7. What are the advantages of helical gears, compared with spur gears?
8. What is the principal disadvantage of helical gears?
9. A gear that has a pitch diameter of 6 in. has a diametral pitch of 4. What number of form cutter would be used in cutting it?
10. What difficulty would be encountered in hobbing a herringbone gear?
11. What is the only type of machine on which full-herringbone gears can be cut?
12. What modification in design is made to herringbone gears to permit them to be cut by hobbing?
13. Why are not more gears made by broaching?
14. What is the most important property of a crown gear?
15. What are three basic processes for machining gears?
16. Which basic gear-machining process is utilized in a Fellows gear shaper?
17. Could a helical gear be machined on a plain milling machine? Why or why not?
18. Explain how the gear blank and the machine table are interconnected when a helical gear is machined on a milling machine.
19. What is the relationship between a crown gear and Gleason gear generators?
20. Why is a gear-hobbing machine much more productive than a gear shaper?
21. Why is a gear shaper more likely to be used for machining cluster gears than a hobbing machine?
22. What are the advantages of cold roll forming for making gears?
23. Assume that 10,000 spur gears, $1\frac{1}{8}$ in. in diameter and $\frac{3}{8}$ in. thick are to be made of 70–30 brass. What manufacturing methods would you consider?
24. If only three gears of the sort described in question 23 were to be made, what process would you select?
25. Why is cold roll forming not suitable for making gray cast iron gears?
26. Under what conditions can shaving not be used for finishing gears?
27. What inherent property accrues from cold roll forming of gears that may result in improved gear life?
28. Can lapping be used to finish cast iron gears?
29. What factors usually are checked in inspecting gears?
30. What are the basic methods for gear finishing?

Problems

1. A single-thread hob that has a pitch diameter of 76.2 mm is used to cut a gear having 36 teeth. If a cutting speed of 27.4 meters per minute is used, what will be the rpm of the gear blank?
2. If the gear in problem 1 has a face width of 76.2 mm, a feed of 1.9 mm per revolution of the workpiece is used, and the approach and overtravel distances of the hob are 38 mm, how much time will be required to hob the gear?
3. In Figure 31-5 a form-milling operation is shown. The cutter is shown in Figure 31-6. The gear is to be made from 4340 steel, R_c50 prior to heat treat and final grind. Select the proper speeds and feeds for the job (the cutter is 4 in. in diameter) and then compute the cutting time to mill this gear.)
4. Compare the cutting time for problem 3 with that of gear shaping.
5. A gear-broaching machine of the type shown in Figure 31-7 could do the gear in 15 seconds (~240 parts per hour). How many additional gears per year are needed to cover the broaching tooling cost if each broach on the machine cost $250? Do you think the broach tool life is sufficient to handle that number of parts? What about TiN-coating the broaches (cost: $100 per broach)?

Nontraditional Machining Processes

Machining processes that involve compression-shear chip formation have a number of inherently adverse characteristics and limitations. Though often necessary, chip formation is an expensive, difficult process. Large amounts of energy are utilized in producing an unwanted product—chips. Further expenditure of energy and money is required to remove these chips and to dispose of, or recycle, them. A large amount of energy ends up as undesirable heat that often produces problems of distortion and surface cracking. Cutting forces create problems in holding the work and sometimes cause distortion. Undesirable deformation and residual stresses in the workpiece often require further processing to remove the effects. Finally, there are definite limitations in regard to the delicacy of the work that can be machined. For example, the production of the semiconductor "chip," shown in Figure 32-1, would not be possible with any of the chip-making processes. In view of these adverse and limiting characteristics, it is not surprising that in recent years substantial effort has been devoted to developing and perfecting material-removal processes that replace conventional machining. Nontraditional machining processes (NTM) is one designation for this diverse family of unconventional processes, which are generally nonmechanical, do not produce chips or a lay pattern in the surface, and often involve new energy modes.

For the purposes of our discussion, these NTM processes can be divided into four basic groups:

1. *Chemical.* Chemical reaction, sometimes enhanced by electrical or thermal energy, is the dominant mode of material removal.
2. *Electrochemical.* Electrolytic dissolution dominates the material-removal process.
3. *Mechanical.* Multipoint cutting or erosion dominates the removal process.
4. *Thermal.* High temperatures in very localized regions to melt and vaporize material dominate the removal process.

Introduction

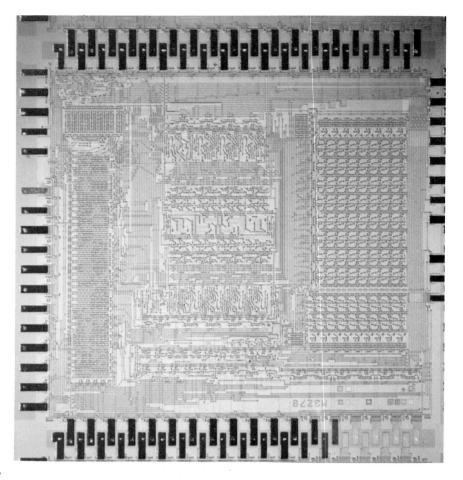

FIGURE 32-1   (Right) Enlarged view of one portion of a microprocessor chip. This chip, shown in full size above, measures only 5 mm on a side and contains over 3000 transistors. (Courtesy Bell Laboratories.)

Tables 32.1 through 32.4 give a summary of the characteristics for these procedures. Not included in this grouping are the new forming processes, like hot isostatic pressing or HERF, discussed in earlier chapters, which do not remove material. When examining these tables, recall that conventional turning has these typical values: surface finish, 32 to 250μ in. AA; MRR, 10 to 200 in.^{3}/min; HP$_s$, 0.5 to 2 HP/in.^{3}/min; V, 100 to 1000 ft/min; Accuracy $\cong$ .002 in. NTM processes typically have low metal-removal rates compared to machining and very high specific horsepowers. They typically have better accuracy, usually at slow rates of processing, which often results in less subsurface damage than conventional processing. These processes are usually employed when conventional machining or grinding cannot be used, often because the materials are too hard. There are numerous hybrid forms of all these processes, generally developed for special applications. Only the main ones will be described here because of space limitations.

TABLE 32.1 Summary of Chemical NTM Processes

Process	Typical Surface Finish AA (μ in.)	Typical Metal Removal Rate	Typical Specific Horsepower (hp/in.³/min)	Typical Penetration Rate (ipm) or Cutting Speed (sfpm)	Typical Accuracy (in.)	Comments
Chemical machining	63–250, but can go as low as 8	30 in³/min	Chemical energy	0.001–0.002 ipm	0.001–0.006; material and process dependent	Almost all materials possible; depth of cut limited to ½ in.; no burrs; no surface stresses; tooling low cost
Electropolishing	4–32, but can go as low as 2 or 1 or better	Very slow	50–200 amperes per square foot	0.0005–0.0015 ipm	NA[a]; process used to obtain finish	High quality, no stress surface; removes residual stresses; makes corrosion-resistant surfaces; may be considered to be an electrochemical process
Photochemical machining (blanking)	63–250, but can go as low as 8	Same as chemical milling	DC power	0.0004–0.0020 ipm	10% of sheet thickness or 0.001–0.002 inch	Limited to thin material; burr-free blanking of brittle material; tooling low cost; used in microelectronics
Thermochemical machining (combustion machining for deburring)	Burr-free	Minute with rapid cycle time	NA[a]	15 to 50 sec cycle times	NA[a]	Vaporizes burrs and fins on cast or machined parts; deburrs steel gears automatically

[a]NA, not applicable for this process.

TABLE 32.2 Summary of Electrochemical NTM Processes

Process	Typical Surface Finish AA (μin.)	Typical Removal Rates (in.³/1000 Amp-Min)	Typical Specific Horsepower (hp/in.³/min)	Typical Penetration Rate (ipm) or Cutting Speed (sfpm)	Typical Accuracy (in.)	Comments
Electrochemical machining (ECM)	16–63	0.06 in W, Mo 0.16 in CI 0.13 in steel, Al 0.60 in Cu	160	.1 to .5 ipm	0.0005–0.005 ≅ 0.002 in cavities	Stress-free metal removal in hard to machine metals; tool design expensive; disposal of chemicals a problem; MRR independent of hardness; deep cuts will have tapered walls
Electrochemical grinding (ECG)[a]	8–32	0.010 in Chromium 0.126 Al 0.268 Cu 0.135 Fe, Ti 0.060 W	High	Cutting rates about same as grinding; wheel speeds, 4000–6000	0.001–0.0005	Special form of ECM; grinding with ECM assist; good for grinding hard conductive materials like tungsten carbide tool bits; no heat damage, burrs, or residual stresses
Electrolytic hole machining (Electrostream)[b]	16–63	NA[c]	NA[c]	0.060–0.120 ipm	≅ 0.001 or 5% of diameter of hole	Special version of ECM for hole-drilling small round or shaped holes; multiple-hole drilling; typical holes 0.004 to 0.03 in. in diameter with depth-to-diameter ratio of 50:1

[a]Honing can also be done with EC assistance which can quadruple the MRR over conventional honing and yield 2 μ in. AA finish.
[b]Trademark of General Electric Company.
[c]NA, not applicable for this process.

TABLE 32.3 Summary of Mechanical NTM Processes

Process	Typical Surface Finish AA (μ/in.)	Typical Metal Removal Rate (in.³/min)	Typical Specific Horsepower (hp/in.³min)	Typical Penetration Rate (ipm) or Cutting Speed (sfpm)	Typical Accuracy (in.)	Comments
Abrasive flow machining	30–300; can go as low as 2	Low	NA[a]	Low	0.001–0.002	Typically used to finish inaccessible integral passages; often used to remove recast layer produced by EDM; used for burr removal; (cannot do blind holes)
Fluid or water jet machining	50–100	Very low	NA[a]	Depends on material	±0.010 at 3 to 4 inches standoff	Used on wood, nonmetals; pressures of 55,000 psi and jet velocity of 1700 to 3000 ft/sec
Abrasive waterjet machining	50–75 for 0.003" to .020" diameter stream	Very low; fine finishing process, 0.001	NA[a]	Very low, .6 ipm to 100 ipm	≅ 0.005 typical ± .020 in. cut line tolerance	Use in heat-sensitive or brittle materials, glass, titanium, and composites and nonmetals; produces tapered walls in deep cuts; burrless
Hydrodynamic machining	Generally 30–100	Depends on material	NA[a]	Depends on material	0.001 possible	Used for soft nonmetalic slitting; no heat-affected zone; produces narrow kerfs (0.001–0.020 in.); high noise levels
Ultrasonic machining (impact grinding)	16–63; as low as 6 to 10 with 9 micron abrasive	Slow, 0.05 typical	200	0.02–0.150 ipm	0.001–0.0005	Most effective in hard materials, R_c >40; tool wear and taper limit hole depth to width at 2.5 to 1; tool also wears

[a]NA, not applicable for this process.

TABLE 32.4 Summary of Thermal NTM Processes

Process	Typical Surface Finish AA (μ/in.)	Typical Metal Removal Rate (in.³/min)	Typical Specific Horsepower (hp/in.³/min)	Typical Penetration Rate (ipm) or Cutting Speed (sfpm)	Typical Accuracy (in.)	Comments
Electron beam machining (EBM)	32–250	0.0005 maximum; extremely low	10,000	200 sfpm; 6 ipm	0.001–0.0002	Micromachining of thin materials and hole-drilling minute holes with 100:1 depth-to-diameter ratios; work must be placed in vacuum but suitable for automatic control; beam can be used for processing and inspection; used widely in microelectronics
Laser beam machining (LBM)	32–250	0.0003; extremely low	60,000	4 ipm	0.005–0.0005	Can drill 0.005 to 0.050 in. diameter holes in materials 0.100 in. thick in seconds; same equipment can weld, surface heat-treat, engrave, trim, blank, etc.; has heat-affected zone and recast layers which may need to be removed
Electrical discharge machining (EDM)	32–105	0.3	40	0.5 ipm	0.002–0.00015 possible	Oldest of NTM processes; widely used and automated; tools and dies expensive; cuts any conductive material regardless of hardness; delicate, burr-free parts possible; always forms recast layer
Electrical discharge wire cutting	32–64	0.10–0.3	40	4 ipm to 10 ipm	≅ 0.0002 to ±0.0001 using molybdenum wire	Special form of EDM using traveling wire; cuts straight narrow kerfs in metals 0.001 to 3 in. thick; wire diameters of 0.002 to 0.010 used; N/C machines allow for complex shapes
Plasma beam machining (PBM)	25–500	10	20	50 sfpm; 10 ipm; 120 ipm in steel	± 0.02 to ± 0.125	Clean, rapid cuts and profiles in almost all plates up to 8 in. thick with 5° to 10° taper; other names: plasma arc, arc cutting

Basically, *chemical machining* is the simplest and oldest of the NTM chemical processes. It has been used for many years in the production of engraved plates for printing and in making small name plates. However, in its use as a machining process it is applied to parts ranging from very small electronic circuits, such as that shown in Figure 32-1, to very large parts up to 15 meters (50 feet) long.

In chemical machining, material is removed from selected areas of a workpiece by immersing it in a chemical reagent. Material is removed by microscopic electrochemical cell action, as occurs in corrosion or chemical dissolution of a metal; no external electrical circuit is involved. This controlled chemical dissolution will simultaneously etch all exposed surfaces, which enhances the productivity even though the penetration rates of the etch may be only 0.0005 to 0.0030 in. per minute. The basic process takes many forms: *chemical milling,* for pockets, contours, and overall metal removal, so named because in its earliest use it replaced milling; *chemical blanking,* for etching through thin sheets; *photochemical machining,* for etching photosensitive resists in microelectronics— see Chapter 41; *gel milling,* using reagent in gel form; *chemical or electrochemical polishing,* where weak chemical reagents are used, sometimes with remote electric assist, for polishing or deburring; *chemical jet machining,* using a single chemically active jet.

Chemical Machining (CHM)

Chemical Milling or Blanking. The material-removal processing steps in chemical milling and chemical blanking are:

1. *Prepare.* Degrease, clean, rinse, pickle, or preclean to provide for good adhesion for masking material.
2. *Mask.* Coating or protecting areas not to be etched.
3. *Etch.* Chemical dissolution by spray or dip followed by rinse.
4. *Remove mask.* Strip or demask, clean and desmut as necessary.
5. *Finish.* Post treatments and finish inspection.

After cleaning and masking, the part is immersed or sprayed with the proper etchant and is permitted to remain in the reagent. With the spray technique, this is not required; therefore this procedure usually is preferred when the size and shape of the workpiece permit. The major complexity in the process involves providing a maskant on the surface of the workpiece so that etching will occur only on desired areas. Usually the entire surface is covered with maskant, portions of which are manually removed from those areas to be etched. A variation of CHM is *photochemical machining* wherein photographic techniques are used to apply the maskant. Figure 32-2 shows the steps that are involved when chemical machining is done through the use of photosensitive resists. These are as follows:

1. Clean the workpiece.
2. Coat the workpiece with a light-sensitive emulsion, usually by dipping or spraying. The emulsion, or resist, is then dried, usually in an oven.
3. Prepare the ''art work.'' An accurate drawing of the workpiece is made, usually on polyester drafting film or glass and up to 50 or more times the

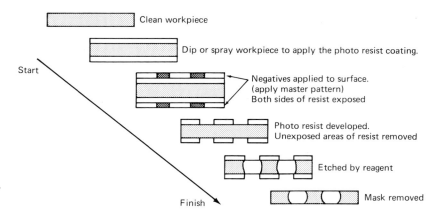

Start

Clean workpiece

Dip or spray workpiece to apply the photo resist coating.

Negatives applied to surface.
(apply master pattern)
Both sides of resist exposed

Photo resist developed.
Unexposed areas of resist removed

Etched by reagent

Finish

Mask removed

FIGURE 32-2 Basic steps in photo-chemical machining. (PCM).

FIGURE 32-3 Typical parts produced by chemical machining (Courtesy Chemcut Corporation.)

size of the final part. With such magnifications, an accuracy of 0.025 in. in the original drawing will permit 0.0005 in. to be achieved in the workpiece. By special procedures, lines of 2μm can be made.

4. Reduce the original drawing by photographic means to obtain a negative that is the master pattern, exactly the size of the finished part. This reduction may require several steps, using industrial photographic equipment.

5. Apply the master pattern to the workpiece using a vacuum frame to ensure good contact. Precise registry of duplicate negatives on each side of the workpiece is essential for accurately blanked parts. Expose to blue light, passing the light through the negative. Mercury-vapor lamps commonly are employed as the light source. Exposure to the light hardens the selected areas of the resist so that it will not be washed away in the subsequent developing.

6. Remove the negatives and develop the workpiece. This removes, or dissolves away, the unexposed areas of the resist, thereby exposing the areas of the workpiece that are to be acted upon by the chemical reagent. The final developing step is to rinse away all residual material.

7. Spray the workpiece with (or immerse it in) the reagent.

8. Remove the remaining maskant.

Chemical machining with the aid of photosensitive resists has been widely used for the production of small, complex parts, such as electronic circuit boards, and very thin parts that are too small to be blanked by ordinary blanking dies. See Figure 32-3. Chemical machining is often followed by plating, sputtering, and vacuum deposition (see Chapter 40) to deposit metallic films of controlled thickness. The process has been the basic technology of the microelectronics industry; see Chapter 41.

The Use of Scribed Maskants. Although photosensitive resists are used in the majority of chemical machining operations, there are some cases in which scribed-and-peeled maskants are employed: (1) when the workpiece is not flat,

(2) when it is very large, and (3) for low-volume work when the several steps required in using photosensitive resists are not economically justified. In this procedure the maskant is applied to the entire surface of the workpiece, usually by dipping or spraying. It is then removed from those areas where metal removal is desired by scribing through the maskant with a knife and peeling away the desired portions. When volume permits, scribing templates can be used.

Chemical Machining to Multiple Depths.   If all areas are to be machined to the same depth, only a single masking, or resist application sequence, and immersing are required. Machining to two or more depths, called *step mach-ining,* can be accomplished by removing the maskant from additional areas after the original immersion. Figure 32-4 illustrates the steps required for stepped chemical machining.

Parts having either uniformly or variably tapered cross sections can be produced by chemical machining by withdrawing them from the etch bath at controlled rates, in the vertical position. In this way, different areas are exposed to the chemical action for differing amounts of time.

Design Factors in Chemical Machining.   When designing parts that are to be made by chemical machining, several unique factors related to the process must be kept in mind. First, dimensional variations can occur through size changes in the art work due to temperature and humidity changes. These usually

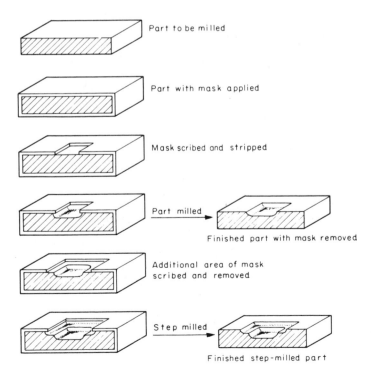

Part to be milled

Part with mask applied

Mask scribed and stripped

Part milled

Finished part with mask removed

Additional area of mask scribed and removed

Step milled

Finished step-milled part

FIGURE 32-4   Steps required to produce a stepped contour by chemical machining.

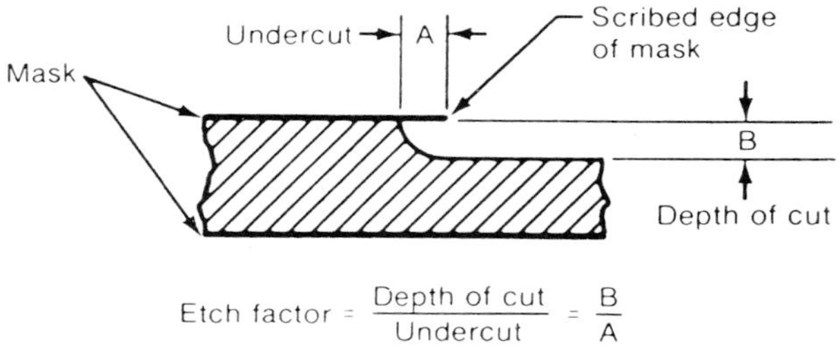

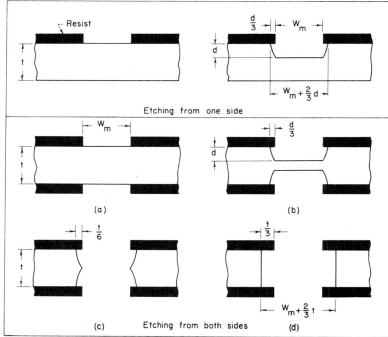

FIGURE 32-5 Etch factor in photo-chemical machining (Note: Inverse of that used in chemical machining.)

$$\text{Etch factor} = \frac{\text{Depth of cut}}{\text{Undercut}} = \frac{B}{A}$$

can be eliminated or controlled by drawing the art work on thicker polyester films or on glass. If very accurate dimensions must be held, the room temperature and humidity should be controlled. The photographic film used in making the master negative also can be affected to some degree by temperature and humidity, but control of handling and processing conditions can eliminate this difficulty.

The second item that must be considered is the *etch factor* or *etch radius,* which describes the undercutting of the maskant. The etchant acts on whatever surface is exposed. Areas that are exposed longer will have more metal removed from them. Consequently, as the depth of etch increases, there is a tendency to undercut or etch under the maskant, as illustrated in Figure 32-5. When the etch depth is only a few hundredths of a millimeter, as often is the case, this causes little or no difficulty. But when the depth is substantial, whether etching from only one or both sides, and when doing chemical blanking, the conditions shown in Figure 32-6 result. In making grooves, the width of the opening in the maskant

FIGURE 32-6 Effect of "etch factor" in chemical machining.

must be reduced by an amount sufficient to compensate for the etch radius. This radius varies from about one fourth to three fourths of the *depth* of the etch, depending on the type of material and to some extent on the depth of the etch. Consequently, it is difficult to produce narrow grooves except when the etch depth is quite small.

An allowance for the etch factor must be taken into account in designing the part and the original art work or scribing template. The values indicated in Figure 32-6 are *minimum* values; it has been found that results will vary between etching machines, and actual etch allowances will have to be somewhat greater and adapted to the specific conditions.

In chemical blanking, with etching occurring from both sides, a sharp edge remains along the line at which breakthrough occurs, as in Figure 32-6c. Because such an edge is usually objectionable, etching ordinarily is continued to produce the straight-wall condition shown in Figure 32-6d.

Etching from both sides, of course, requires the preparation of two maskant patterns and careful registration of them on the two sides of the workpiece.

If the bath is not agitated properly, the "overhang" condition depicted in Figure 32-7 may result, particularly on deep cuts. Not only is the resulting dimension of the opening incorrect, but a very sharp edge may be produced. Other common defects, like *islands* and *dishing,* and processing problems like selective etching at the grain boundaries, microcracks, and pitting, caused by selective etching (etching at unequal etch rates) of the workpiece, are to be expected.

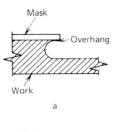

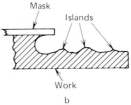

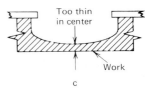

FIGURE 32-7 Typical chemical milling defects. (a) Overhang: deep cuts with improper agitation. (b) Islands: Isolated high spots from dirt, residual maskant, or work material inhomogeneity. (c) Dishing: thinning in center due to improper agitation or stacking of parts in tank.

Advantages and Disadvantages of Chemical Machining.
Chemical machining has a number of distinct advantages. Except for the preparation of the art work and master negative, or a scribing template, the process is relatively simple, does not require highly skilled labor, induces no stresses or cold working in the metal, and can be applied to almost any metal—aluminum, magnesium, titanium, and steel being most common. Large areas can be machined; tanks for parts up to 12 × 50 feet are available. Machining can be done on parts of virtually any shape. Thin sections, such as honeycomb, can be machined because there are no mechanical forces involved. Consequently, chemical machining is very useful and economical for weight reduction. Figure 32-8 shows a typical large, thin part that has been chemically milled.

The tolerances expected with chemical machining range from ±0.0005 in. with care on small etch depth to ±0.004 in. in routine production involving substantial depths. The surface finish is good, seldom having a roughness greater than 0.0001 in.

In using chemical machining, some disadvantages and limitations should be kept in mind. The metal removal rate is slow *in terms of unit area exposed,* being about 0.2 to 0.04 pound per minute per square foot exposed in the case of steel. However, because large areas can be exposed all at once, the overall removal rate may compare favorably with other metal removal processes, particularly when the workpiece metal is thin and unable to sustain large cutting forces.

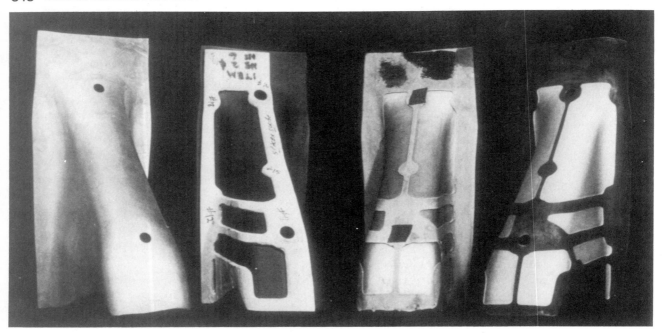

FIGURE 32-8 Iconel 718 aircraft engine parts. These sheet metal parts for a jet fighter engine, are chemically milled to remove weight. (left) As-formed workpiece; (middle left) workpiece coated with liquid rubber, fiberglass scribing template in place; (middle right) scribed workpiece; (right) finished part. About 0.035″ of stock is removed with the 0.070″ thick workpieces. Tolerances are held to +/− 0.004″ (*Manufacturing Engineering*, August 1983, p. 55.)

The soundness and homogeneity of the metal are very important. Wrought materials should be uniformly heat-treated and stress-relieved prior to processing. Although chemical machining induces no stresses, it may release existing residual stresses in the metal and thus cause warping. Castings can be chemically machined provided they are not porous and have uniform grain size. Lack of the latter can cause difficulty. Because of the different grain structures that exist near welds, weldments usually are not suitable for chemical machining.

Chemical milling and conventional machining often can be combined advantageously for producing parts, a fact that often is overlooked.

The tolerance in chemical milling increases with the depth of cut and with faster etch rates and varies for different metals, as shown in Figure 32-9.

Thermochemical machining (TCM) has been developed for the removal of burrs and fins by exposing the workpiece to hot corrosive gases for a short period of time. See Figure 32-10. The workpiece remains unaffected and relatively cool because of its low surface-to-mass ratio and the short exposure time. Alternatively, fine burrs can be removed quickly by exposing the parts to a suitable chemical spray and at much less cost than if it were done by hand. Of course, smaller amounts of metal are removed from all exposed surfaces, and this must be permissible if the process is to be used. Consequently, the procedure usually can be used only for removing small burrs.

The hot gases are formed by detonating explosive mixtures of oxygen, hydrogen, and natural gas in a chamber with the parts. A thermal shock wave vaporizes the burrs found on gears, die castings, valves, and so on, in a few

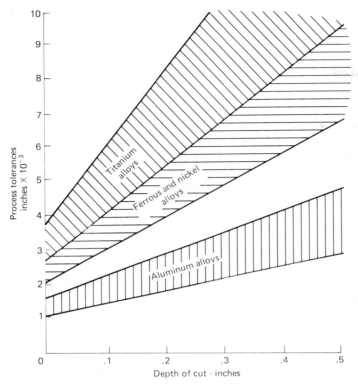

FIGURE 32-9 Chemical milling tolerance bands showing variation with respect to depth of cut and different metals. (Adapted from *Chemical Machining: Production with Chemistry*, MDC 82-1-2, Metcut Research Associates, Inc., Machinability Data Center, Cincinnati, Ohio, 1982.)

milliseconds. The process has been automated and cycle times of 15 to 50 seconds are typical.

TCM will remove burrs or fins from a wide range of materials, but it is particularly effective with materials of low thermal conductivity. It will deburr thermosetting plastics, but not thermoplastic materials. Any workpiece of modest size requiring manual deburring or flash removal should be considered a candidate for thermal deburring. Die castings, gears, valves, rifle bolts, and

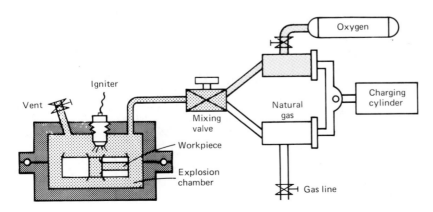

FIGURE 32-10 Thermochemical machining process for the removal of burrs and fins. (Courtesy *Machining Data Handbook*, Vol. 2.)

similar small parts are deburred readily, including blind, internal, and intersecting holes in inaccessible locations. Carburetor parts are processed in automated equipment. Maximum burr thickness should be about $\frac{1}{15}$ of the thinnest feature on the workpiece. Uniformity of results and greater quality assurance over hand deburring is a special advantage of TCM.

Electrochemical Machining (ECM)

Electrochemical machining, commonly designated ECM, removes material by anodic dissolution with a rapidly flowing electrolyte. It is basically a *deplating* process in which the tool is a cathode and the workpiece is the anode; both must be electrically conductive. The electrolyte, which can be pumped rapidly through or around the tool, sweeps away the waste product (sludge) and captures it by settling in filters. The shape of the cavity is the mirror image of the tool, which is advanced by means of a servomechanism which controls the gap (0.003 to 0.030 in., with 0.010 in. typical) between the electrodes. The tool advances into the work at a constant feed rate which matches the dissolution and deplating rates of the workpiece. The electrolytes are highly conductive solutions of inorganic salts, usually NaCl, KCl, $NaNO_3$ (or other proprietary mixtures), and are operated at about 90° to 125°F with flow rates ranging from 50 to 200 feet per second (fps). Tools are usually made of copper or brass and sometimes stainless steel. The process is shown schematically in Figure 32-11.

As shown in Figure 32-12, the metal-removal rate, in terms of penetration of the tool into the workpiece, is primarily a function of the current density. Current densities from 1500 to 2000 amperes per square inch are used and, in suitable applications, ECM provides metal removal rates on the order of 0.1 $in^3/min/1000$ A. The cutting rate is solely a function of the ion-exchange rate and is not affected by the hardness or toughness of the work material. Cutting

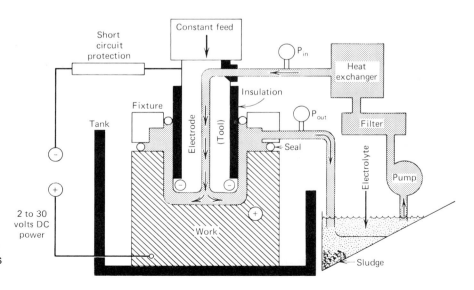

FIGURE 32-11 Schematic diagram of electrochemical machining process (ECM).

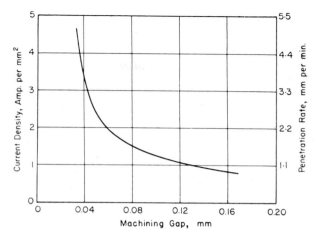

FIGURE 32-12   Relationship of current density, penetration rate, and machining gap in electrochemical machining.

rates up to 0.1 in. of depth per minute are obtained routinely in Waspalloy, a very hard metal alloy.

ECM is well suited for mass production of complex shapes in difficult-to-machine but conductive materials. The principal tooling cost is for the preparation of the tool electrode, which can be time-consuming and costly, requiring several "cut and try" efforts, except for simple shapes. There is no wear of the tool during actual cutting as the tool is protected cathodically. The process produces a stress-free surface. The ability to cut the entire cavity simultaneously is very productive. Process control must be exact to obtain tight tolerances, and the tools must be designed to compensate for the variable current densities produced by electrode geometries or electrolyte variations. For example, corners in cavities are automatically rounded because of the concentration of the current density at the edge of the tool.

The ECM process has a number of modified forms. Electrochemical polishing operates essentially the same as ECM but the feed is halted. Lower current densities and slower electrolyte flow rates greatly reduce the metal removal rates so that the surface develops a fine finish, 10 to 12 μ in. AA being typical.

Electrochemical hole-drilling processes have been developed for drilling very small holes using high voltages and acid electrolytes. The tool is a drawn glass nozzle with an internal electrode. Multiple sets of glass tubes are employed and over 50 holes per stroke can be done. This technique was developed to drill the cooling holes in turbine blades for jet engines. Stress-free holes from 0.004 to 0.030 in. in diameter with 50:1 depth-to-diameter ratios are routinely accomplished in nickel and cobalt alloys. Acid is used so that the dissolved metals go into solution instead of forming a sludge.

The process can be used to drill shaped holes in difficult-to-machine, conductive metals. Holes up to 24 in. in depth with diameters ranging from 0.020 to 0.050 in. are possible. The major differences between this process and the hole-drilling process described above are the reduced voltage levels (5 to 10 volts DC) and special electrodes, which are long, straight, acid-resistant tubes coated with an enamel insulation. The acid is pressure-fed through the tube and returns through the gap (0.001 to 0.002 in.) between the insulated tube wall and the hole wall.

Electrochemical Grinding.   *Electrochemical grinding,* commonly designated ECG, is a variant of electrochemical machining. In ECG, the tool electrode is a rotating, metal-bonded, diamond grit grinding wheel. The setup shown in Figure 32-13 for grinding a tool is typical. As the electric current flows between the workpiece and the wheel, through the electrolyte, the surface metal is changed to a metal oxide, which is ground away by the abrasives. As the oxide film is removed, new surface metal is oxidized and removed. ECG is a low-voltage, high-current electrical process. The MRR is dependent on many variables. Table 32.5 gives typical values. The metal bond of the wheel is the cathode.

The wheels used in ECG must be electrically conductive abrasive wheels. For most metals, resin-bonded aluminum oxide wheels are recommended. The resin bond is loaded with copper to provide for electrical conductivity. Electrical resistance must be negligible. The wheels are dressable, using a variety of wheel-dressing measures, and can then be used for precision form-grinding operations.

The purpose of the abrasive is to increase the efficiency of the ECG process and permit the continuance of the process. The abrasive particles are always nonconductive material such as aluminum oxide, diamond, or borazon (CBN). Thus they act as an insulating spacer maintaining a separation of from .0005 to .003 in. (.012mm to .050mm) between the electrodes. A dead short would result if the insulating spacer were absent. The particles also serve to wipe away the residue and to cut chips if the wheel should contact the workpiece, particularly in the event of a power failure. With proper operation, less than 5% of the material is removed by normal chip forming. The process is used for shaping and sharpening carbide cutting tools, which cause high wear rates on expensive diamond wheels in normal grinding. Electrochemical grinding greatly reduces this wheel wear. Fragile parts (honeycomb structures), surgical needles, and tips of assembled turbine blades have been ECG-processed successfully. The lack

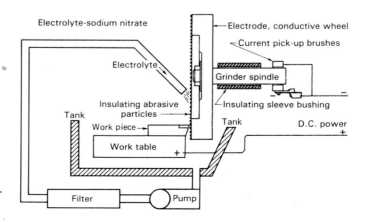

FIGURE 32-13 Equipment setup and electrical circuit for electrochemical grinding.

TABLE 32.5 Metal Removal Rates for ECG for Various Metals

Metal	Valency	Density lb in.³	Density g cm³	Metal-removal Rate at 1000A lb/h	Metal-removal Rate at 1000A in.³/min	Metal-removal Rate at 1000A cm³/min
Aluminum	3	0.098	2.67	0.74	0.126	2.06
Beryllium	2	0.067	1.85	0.37	0.092	1.50
Chromium	2	0.260	7.19	2.14	0.137	2.25
	3			1.43	0.092	1.51
	6			0.71	0.046	0.75
Cobalt	2	0.322	8.85	2.42	0.125	2.05
	3			1.62	0.084	1.38
Niobium	3	0.310	8.57	2.55	0.132	2.16
(Columbium)	4			1.92	0.103	1.69
	5			1.53	0.082	1.34
Copper	1	0.324	8.96	5.22	0.268	4.39
	2			2.61	0.134	2.20
Iron	2	0.284	7.86	2.30	0.135	2.21
	3			1.53	0.090	1.47
Magnesium	2	0.063	1.74	1.00	0.265	4.34
Manganese	2	0.270	7.43	2.26	0.139	2.28
	4			1.13	0.070	1.15
	7			0.65	0.040	0.66
Molybdenum	3	0.369	10.22	2.63	0.119	1.95
	4			1.97	0.090	1.47
	6			1.32	0.060	0.98
Nickel	2	0.322	8.90	2.41	0.129	2.11
	3			1.61	0.083	1.36
Silicon	4	0.084	2.33	0.58	0.114	1.87
Silver	1	0.379	10.49	8.87	0.390	6.39
Tin	2	0.264	7.30	4.88	0.308	5.05
	4			2.44	0.154	2.52
Titanium	3	0.163	4.51	1.31	0.134	2.19
	4			0.99	0.101	1.65
Tungsten	6	0.697	19.3	2.52	0.060	0.98
	8			1.89	0.045	0.74
Uranium	4	0.689	19.1	4.90	0.117	1.92
	6			3.27	0.078	1.29
Vanadium	3	0.220	6.1	1.40	0.106	1.74
	5			0.84	0.064	1.05
Zinc	2	0.258	7.13	2.69	0.174	2.85

Source: 1985 SCTE Conference Proceedings, ASM, 1986.

of heat damage, burrs, and residual stresses is very beneficial, particularly when coupled with MRRs that are competitive with conventional grinding but with far less wheel wear.

Electrochemical Deburring. Limited use is made of the ECM principle for removing burrs from parts. The work is put into a rotating, electrically insulated drum that contains two current-carrying electrodes that are insulated from the drum. Small graphite spheres, added to the electrolyte, receive an inductive charge from the electrodes. The potential gradient across the sphere-to-workpiece gap is sufficient to cause electrochemical machining to occur as the spheres move randomly over the workpiece. Because the current density is higher at the protrusions of the burrs than at smooth areas on the workpiece, they are preferentially removed. As in chemical deburring, there is a slight dimensional change throughout the workpiece, in this case due to the general ECM action and to the natural abrasive character of the graphite spheres.

Mechanical NTM Processes

Ultrasonic machining, sometimes called *impact grinding,* employs an ultrasonically vibrating tool to impel the abrasive in a slurry against the workpiece. The tool forms a reverse image in the workpiece as the abrasive-loaded slurry abrades (machines) the material. Boron carbide, aluminum oxide, and silicon carbide are the most commonly used grit materials. The process can cut virtually any material but is most effective on materials with hardness greater than $R_c = 40$. Figure 32-14 shows a simple schematic of this process.

Ultrasonic machining uses a transducer to impart high-frequency vibrations to the toolholder. Abrasive particles in the slurry are accelerated to great speed

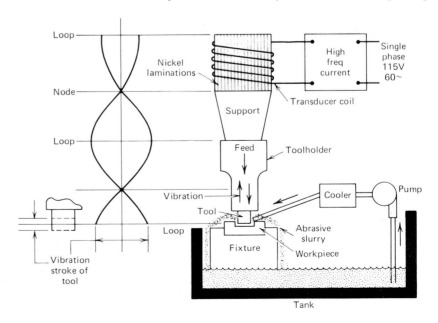

FIGURE 32-14 Sinking a hole in a workpiece by ultrasonic machining.

by the vibrating tool and perform the actual cutting. The tool materials are usually brass, carbide, mild steel, or tool steel and will vary in tool wear depending upon their hardness. Wear ratios of 1:1 or 100:1 (material removed versus tool lost to wear) are possible. The tool must be strong enough to resist fatigue failure.

The cut will be oversize by about twice the size of the abrasive grit being used, and holes will be tapered, usually limiting the hole depth to a diameter ratio of about 3:1. Surface roughness is controlled by the size of the grit (finer finish with smaller grits). Holes, slots, or shaped cavities can be readily eroded in any material, conductive or nonconductive, metallic, ceramic, or composite.

A variation of ultrasonic machining, *rotary ultrasonic machining,* uses a rotating diamond tool vibrating at 20,000 Hz during rotation, to drill, thread, grind, or mill hard, brittle materials. It does not use an abrasive slurry.

Ultrasonics have also been employed for coining, lapping, deburring, and broaching. Plastics can be welded using ultrasonics.

Three other mechanical NTM processes are listed in Table 32.3.

Waterjet machining uses a high-velocity fluid jet (Mach 2) impinging the workpiece, performing a slitting operation. See Figure 32-15. A long-chain polymer may be added to the water of the jet to make the jet coherent (that is, to not come out of the nozzle as a mist) under the 10,000-to 60,000-psi nozzle pressure. Alternate fluids (alcohol, glycerine, cooking oils) have been used in processing meats, baked goods, and frozen foods. The jet is typically 0.003 to 0.020 in. in diameter and exits the orifice at velocities up to 3000 fps. This process has historically been used to cut soft nonmetalics like acoustic tile, plastics, paperboard, asbestos (with no dust), leather, rubber, and fiberglass. More recently, a full range of ferrous and nonferrous metals have been machined. The majority of the metal-working applications require the addition of an abrasive to the waterjet stream. The abrasives are added after the stream has left the orifice. See inset in Figure 32-15. Generally, the kerf of the cut is about 0.001 in. greater than the diameter of the jet and is not sensitive to dwell.

FIGURE 32-15 Schematic of hydrodynamic jet machining. The intensifier elevates the fluid to the desired nozzle pressure while the accumulator smooths out the pulses in the fluid jet.

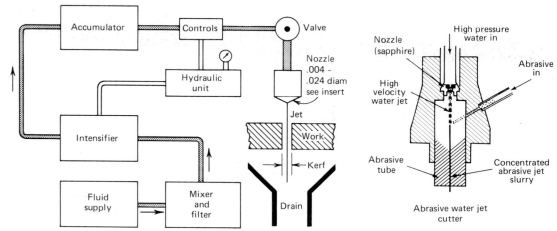

Cutting rates vary from 250 fpm for acoustic tile up to 6000 fpm for paper products.

Abrasive jet cutting adds these additional parameters to waterjet machining: abrasive material (density, hardness, shape); abrasive size or grit; flow rate (pounds per minute); abrasive feed method (pressurized or suction); types of nozzle and mixing chamber; abrasive orifice diameter size; abrasive orifice or accelerator tube material.

Typical abrasive jet cutting systems operate under the following conditions: water pressure of 30,000–50,000 psi; water orifice sizes of .010–.022 in. diameter; standoff distances .020–.060 in.; abrasive grit sizes of 60, 80, 100, 120; abrasive material of garnet, silica, or aluminum oxide. Flow rate of abrasive can be from one half to three pounds per minute.

The inside diameters of the abrasive nozzles or accelerator tubes are from .040–.125 of an in. in diameter. These tubes are normally made of carbide.

TABLE 32.6   Typical Cutting Rates and Depths of Cut for Abrasive Water-jet Machining

Material	Inch Depth of Cut	Cut Rate ipm
Aluminum	2.00	2.0– 6.0
Chrome/Vanadium 58R/C	.69	2.0– 3.5
Glass	.75	24.0– 40.0
Glass mirror	.25	42.0–100.0
Graphite/Epoxy	.38	24.0– 50.0
Graphite/Epoxy	.44	20.0– 45.0
Graphite/Epoxy	1.31	15.0– 20.0
Graphite/Epoxy	2.00	5.0– 7.0
Graphite/Epoxy (al honeycomb)	.75	24.0– 35.0
Hastelloy	.06	19.0– 26.0
HY-80	.38	4.0– 5.0
Inconel	1.50	1.0– 1.5
Inconel	1.75dia	1.0– 1.5
Kevlar	.88	24.0– 36.0
Stainless Steel 316	3.00dia	.5– 2.0
Stainless Steel 15-5 PH	2.50	.5– 1.0
Stainless Steel 15-5 PH	.13	9.0– 15.0
Titanium	1.00	1.5– 2.5
Zircalloy	.44	4.5– 6.5

Process parameters	
Water pressure	40,000–50,000 psi
Water orifice	.014″ diameter
Abrasive flow	1.8#/min. 60 grit garnet
Water flow	approx 1 gallon/min
Standoff from part	.030/.050 inch

Source: 1985 SCTE Proceedings in Nontraditional Machining, ASM, 1987.

Abrasive water-jet cutting can be used to cut any material by using the appropriate abrasive, water-jet pressure, and feed rate.

These processes are good on composites as the cutting rates are reasonable and they do not delaminate the layered material. See Table 32.6.

Abrasive flow machining (AFM) uses a slurry of abrasive material flowing over or through a part, under pressure, to perform edge finishing, deburring, radiusing, polishing, or minor surface machining. Aluminum oxide, silicon carbide, boron carbide, and diamonds are used as abrasives in grit sizes ranging from No. 8 to No. 700. This process is useful for polishing and deburring inaccessible internal passageways and works for most materials.

The AFM process should be considered when workpieces have needed laborious hand finishing or have complex shapes or internal edges that are difficult to reach, or when bulk methods (tumbling, vibrating) are not satisfactory. The process is effective in removing the recast layers produced by the EDM or laser beam machining discussed next.

Electrodischarge machining (EDM) cuts metal by discharging electric current stored in a capacitor bank across a thin gap between the tool (cathode) and the workpiece (anode). Literally thousands of sparks per second are generated and each spark produces a tiny crater by melting and vaporization, thus eroding the shape of the tool into the workpiece. The dielectric fluid (kerosene) flushes out the "chips" and confines the spark. Of all the exotic metalworking processes, none has gained greater industry-wide acceptance than EDM. Figure 32-16 shows a schematic of the process.

In EDM, each spark contains a discrete, measured, and controllable amount of energy; therefore MRR and surface finish can be predicted while size is carefully controlled. The heat generated by the spark melts the metal, and the impact of the spark causes the metal to be ejected, possibly vaporized, and

Thermal Processes

FIGURE 32-16 EDM or spark erosion machining of metal, using high-frequency spark discharges in a dielectric, between the shaped tool (cathode) and the work (anode). The table can make X–Y movements.

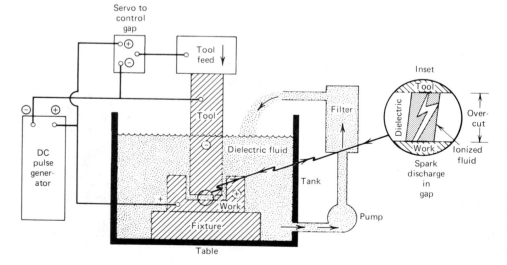

recast in the dielectric as spheres. Materials of any hardness can be cut as long as the material can conduct electricity. Any shape that can be cut into the tool can be reproduced in the workpiece. Spark reversals and unflushed chips cause sparking on the tool, producing unwanted tool wear and taper; therefore for production runs tools are usually made in duplicate sets by numerical control machining. About 80% to 90% of EDM work is in the manufacture of tool and die sets for production casting, forging, stamping, and extrusions.

The absence of almost all mechanical forces makes it possible to machine fragile parts without distortion. In addition, fragile tools, even wires, can be used. The controllability, versatility, and accuracy of this method usually result in superior design flexibility. Significant cost reductions in the manufacture of tools and dies from steels of any hardness and carbides can be achieved.

EDM is slow compared to conventional methods and produces a matte surface finish composed of many small craters. Figure 32-17 shows a scanning electron micrograph of an EDM surface on top of a ground surface. Note the small sphere in the lower right corner attached to the surface. In EDM, surface finish varies widely as a function of the spark frequency, voltage, and current, parameters which, of course, also control the MRR.

The distance between the surface of the electrode and the surface of the workpiece represents the *overcut* and is equal to the length of the spark, which is essentially constant over all areas of the electrode, regardless of size or shape. Typical overcut values range from 0.0005 to 0.020 in. Overcut depends on the gap voltage plus the chip size, which varies with the amperage. EDM equipment manufacturers publish overcut charts for the different power supplies for their machines, but these values should be used mainly as guides for the tool designer. The dimensions of the electrode (the tool) are basically equal to the desired dimensions of the part less the overcut values.

While different materials are used for electrodes, graphite has emerged as the best EDM tool material. The choice depends on the application, considering how easily the electrode material can be machined, how fast it wears (this is called *spark erosion*), how fast it cuts, what kind of quality of finish it can

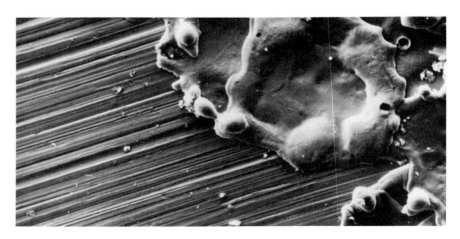

FIGURE 32-17 EDM surface on top of a ground surface in steel, spheroidal nature of debris from the surface in evidence around the craters (750X).

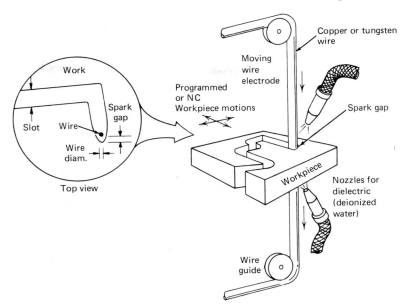

FIGURE 32-18 Schematic diagram of equipment for electrodischarge machining using a moving wire electrode.

produce, what type of power supply is being used, and how much the material costs. Copper, brass, copper-tungsten, aluminum, 70/30 zinc tin, and other alloys are used for electrode material when graphite is not selected.

The dielectric fluid has four main functions: insulation between tool and work, spark conductor, coolant, and flushing medium. This fluid must ionize to provide a channel for the spark and deionize quickly to become an insulator. Polar compounds, like glycerine–water (90:10) with triethylene oil as an additive, have been shown to improve the MRR and decrease the tool wear when compared with traditional cutting fluids like kerosene.

A special form of EDM is shown in Figure 32-18 wherein the electrode is a continuously moving conductive wire. The tensioned wire of copper, brass, or tungsten is used only once, traveling from a take-off spool to a take-up spool while being "guided" to produce a straight narrow kerf in plates up to 3 in. thick. The wire diameter ranges from 0.002 to 0.010 in. with positioning accuracy up to plus/minus 0.0002 in. in machines with NC or tracer control. The dielectric is usually deionized water. This process is widely used for the manufacture of punches, dies, and stripper plates, with modern machines capable of cutting die relief, intricate openings, tight radius contours, and corners routinely. See Figure 32-19 for examples of wire EDM products.

Advantages and Disadvantages of EDM. Electrodischarge machining is applicable to all materials that are fairly good electrical conductors, including metals, alloys, and most carbides. The melting point, hardness, toughness, or brittleness of the material imposes no limitations. Thus, it provides a relatively simple method for making holes of any desired cross section in materials that are too hard or brittle to be machined by most other methods. Because the forces

FIGURE 32-19   Examples of wire EDM workpieces made on NC machine. (Hatachi.)

between the tool and the workpiece are virtually zero, very delicate work can be done. The process leaves no burrs on the edges. It is widely used to produce forming dies for sheet metal in the automotive industry.

On most materials, the process produces a thin, hard recast surface, which may be an advantage or a disadvantage, depending on the use. When the workpiece material is one that tends to be brittle at room temperatures, the surface may contain fine cracks caused by the thermally induced stresses. Consequently, some other finishing process often is used subsequent to EDM to remove a thin surface layer, particularly if the product will be used in a fatigue environment.

Electron Beam Machining.   As a metals-processing tool, the electron beam is used mainly for welding, to some extent for surface hardening, and occasionally for cutting, (mainly drilling).

Electron beam machining (EBM) is a thermal NTM process which uses a beam of high-energy electrons focused on the workpiece to melt and vaporize metal. This micromachining process is performed in a vacuum chamber (10^{-5}mm of mercury). Magnetic lenses are used to focus the beam. Deflection coils control the position of the beam. The desired beam path can be programmed with a computer to produce any desired pattern in the work. The spot size diameters are on the order of 0.0005 to 0.001 in. and holes or narrow slits with

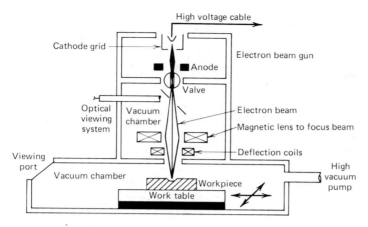

FIGURE 32-20 Electron beam machining uses high-energy electron beam (10^9 watts/in^2) to melt and vaporize metal in 0.001-in.-diameter spots.

depth-to-width ratios of 100:1 can be "machined" with great precision in a short time in any material. The interaction of the beam with the surface produces dangerous X-rays; therefore shielding is necessary. The layer of recast material and the depth of heat damage is very small. For micromachining, processing speeds can exceed that of EDM or ECM. Typical tolerances are about 10% of the hole diameter or slot width. These machines require high voltages (50 to 200 kilovolts) to accelerate the electrons to speeds of 0.5 to 0.8 the speed of light and should be operated by fully trained personnel. See Figure 32-20 for a schematic.

Laser Beam Machining.   *Laser beam machining (LBM)* is a thermal NTM process which uses a laser to melt and vaporize materials. See Figure 32-21. The beam can be focused down to 0.005 in. in diameter for drilling microholes through metals as thick as 0.100 in., but hole depth-to-diameter ratios of 10:1 are more typical. High-energy solid-state and gas lasers are needed, with the optical characteristics of the workpiece determining the wavelength of light energy that should be used. The range is 0.4880 μm for argon lasers up to 10.6

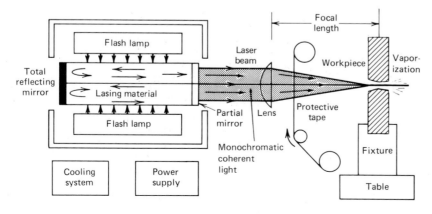

FIGURE 32-21 Schematic diagram of a laser beam machine, a thermal NTM process that can micromachine any material.

μm for carbon dioxide gas lasers. See Table 32.7 for a listing of commercially available lasers for material processing. The small beam divergence, high peak powers (in pulsed lasers), and single frequency provide power densities of the order of 10^5 to 10^{10} watts per square inch, allowing holes 0.020 in. in diameter to be drilled in milliseconds with accuracy of $\pm$ 0.001 in. However, this is not a mass metal-removal process and is limited to rather thin stock (.10 to .20 in.) as the cutting speed drops off rapidly with thickness. Hole geometry is irregular, and there will be a recast layer and a heat-affected zone which can be detrimental to material properties. High-precision systems can hold line cuts to $\pm$ .005 in. positional accuracy and some to $\pm$ .001 in. per foot of travel. Lasers are used to cut, weld, heat-treat, and trim by varying the power density along with

TABLE 32.7   Commercial Lasers Available for Machining, Welding, and Trimming

Laser Type	Wave Length μm	Mode of Operation	Power W	Pulses per Second	Pulse Length Time	Application	Comments
Argon	0.4880 0.5145	Repetitively pulsed	20 peak; 0.005 average	60	50 μs	Scribing thin films	Power low
Ruby	0.6943	Normal pulse	2×10^5 peak	Low (5 to 10)	0.2 to 7 ms	Large material removal in one pulse, drilling diamond dies, spot welding	Often uneconomical
Nd-Glass	1.06	Normal pulse	2×10^6 peak	Low (0.2)	0.5 to 10 ms	Large material removal in one pulse	Often uneconomical
Nd-YAG[a]	1.06	Continuous	1,000	—	—	Welding	Compact; economical at low powers
Nd-YAG	1.06	Repetitively Q-switched	3×10^5 peak; 30 average	1 to 24,000 300	50 to 250 ns 50 ns	Resistor trimming, electronic circuit fabrication	Compact and economical
Nd-YAG	1.06	Normal pulsed	400	300	0.5 to 7 ms	Spot weld, drill	
CO_2[b]	10.6	Continuous	15,000	—	—	Cutting organic materials, oxygen-assisted metal cutting	Very bulky at high powers
CO_2	10.6	Repetitively Q-switched	75,000 peak 1.5 average	400	50 to 200 ns	Resistor trimming	Bulky but economical
CO_2	10.6	Superpulsed	100 average	100	100 μs and up	Welding, hole production, cutting	Bulky but economical

Source: Modified from J. F. Ready, Selecting a laser for material working, *Laser Focus* (March 1970), p. 40.
[a]Neodymium-yttrium aluminum garnet.
[b]CO_2 plus He plus N_2 mixture.

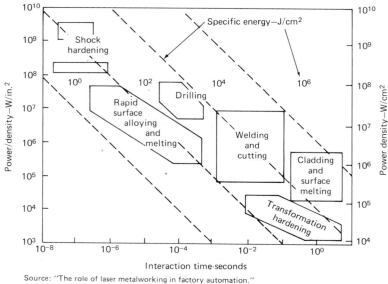

Source: "The role of laser metalworking in factory automation."
Westinghouse R&D center (Pittsburgh).

FIGURE 32-22 The power density input and the percentage of surface heat absorbed determines whether the laser cuts, welds, or heat-treats.

appropriate adjustments in output beam intensity, focus, and duration (see Figure 32-22).

Off-the-shelf laser systems are now available with NC controls and are being used for applications ranging from cigarette paper cutting to drilling microholes in turbine engine blades. Protective materials are absolutely necessary when working around laser equipment because of the potential damage to eyesight from either direct or scattered laser light.

Plasma Arc Cutting.

Plasma arc cutting (PAC) uses a superheated stream of electrically ionized gas to melt and remove material. See Figure 32-23. The 20,000° to 50,000°F plasma is created inside a water-cooled nozzle by electrically ionizing a suitable gas such as nitrogen, hydrogen, argon, or mixtures of these gases. The process can be used on almost any conductive metal. The plasma arc is a mixture of free electrons, positively charged ions and neutral atoms. The arc is initiated in a confined gas-filled chamber by a high-frequency spark. The high-voltage, direct-current power sustains the arc, which exits from the nozzle at near sonic velocity. The workpiece is electrically positive. The high-velocity gases melt and blow away the molten metal "chips." Dual-flow torches use a secondary gas or water shield to assist in blowing the molten metal out of the kerf, giving a cleaner cut. A water shield or injection is sometimes used to assist in confining the arc, blasting away the scale, and reducing smoke. The main advantage of PAC is speed. Mild steel ¼ in. thick can be cut at 125 ipm. Speed decreases with thickness. Greater nozzle life and faster cutting speeds accompany the use of water-injection-type torches. Control of nozzle standoff from the workpiece is important. One electrode size can be used to

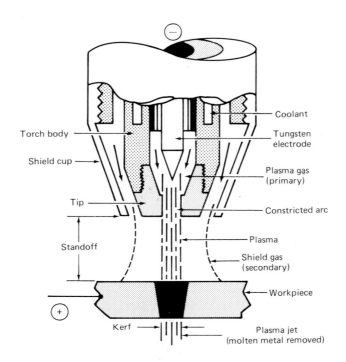

FIGURE 32-23 Plasma arc machining.

machine a wide range of materials and thicknesses by suitable adjustments to the power level, gas type, gas flow rate, traverse speed, and flame angle. PAC is sometimes called plasma beam machining.

PAC can machine exotic metals at high rates. Profile cutting of metals, particularly of stainless steel and aluminum, has been the most prominent commercial application. However, mild steel, alloy steel, titanium, bronze, and most metals can be cut cleanly and rapidly. Multiple-torch cuts are possible on programmed or tracer-controlled cutting tables on plates up to 6 in. thick in stainless steel. Smooth cuts free from contaminants are a PAC advantage. Well-attached dross on the underside of the cut can be a problem and there will be a heat-affected zone (HAZ). The depth of the HAZ is a function of the metal, its thickness, and the cutting speed. Surface heat treatment and metal joining are beginning to use the plasma torch.

Review Questions

1. What are the four basic types of NTM processes?
2. Why might chipless machining processes have greater importance in the future?
3. How does the MRR for AWM compare to conventional metal cutting?
4. What are the steps in chemical machining using photosensitive resists? See also Chapter 41.
5. Why is it preferable in chemical machining to apply the etchant by spraying instead of immersion?
6. What are the advantages of chemical blanking over regular blanking using punch and die methods?
7. Explain how multiple depths can be produced by chemical machining.

8. Would it be feasible to produce a groove 2 mm wide and 3 mm deep by chemical machining?

9. A drawing calls for making a groove 23 mm wide and 3 mm deep by chemical machining. What should be the width of the opening in the maskant?

10. Could an ordinary steel weldment be chemically machined? Why or why not?

11. How would you produce a tapered section by chemical machining?

12. What is the principal application of thermochemical machining?

13. Is ECM related to chemical machining?

14. What effect does work-material hardness have on the metal-removal rate in ECM?

15. Explain the basic principle involved in electrochemical deburring.

16. What is the principal cause of tool wear in ECM?

17. Basically, what type of process is electrochemical grinding?

18. Would electrochemical grinding be a suitable process for sharpening ceramic tools? Why or why not? What about using ultrasonics?

19. Upon what factors does the metal-removal rate depend in ECM?

20. Why is the tool insulated in the ECM schematic?

21. Is ultrasonic machining really a chipless process?

22. Where, in the ultrasonic stroke, is the acceleration and deceleration of the tool the greatest? Given that $F = ma$, what does this mean about the force given to the grits in the slurry?

23. What is the nature of the surface obtained by electrodischarge machining?

24. What is the principal advantage of using a moving-wire electrode in electrodischarge machining?

25. What effect would increasing the voltage have on the metal-removal rate in electrodischarge machining? Why?

26. If the metal from which a part is to be made is quite brittle and the part will be subjected to repeated tensile loads, would you select ECM or electrodischarge machining for making it? Why?

27. If you had to make several holes in a large number of duplicate parts, would you prefer ECM, EDM, EBM, or LBM? Why?

28. What process would you recommend to make many small holes in a very hard alloy when the holes will be used for cooling and venting?

29. Why are the specific power values so large for LBM?

30. Explain (using a little physics and metallurgy) why the chips in thermal processes are often hollow spheres.

PART VI

Joining Processes

Gas Flame Processes: Welding, Cutting, and Straightening

Welding is a process in which two materials, usually metals, are permanently joined together by coalescence, the coalescence resulting from a combination of temperature, pressure, and metallurgical conditions. The particular combination of temperature and pressure can range from high temperature with no pressure to high pressure with no increase in temperature. Thus, welding can be accomplished under a wide variety of conditions, and numerous welding processes have been developed and are used routinely in manufacturing. Nevertheless, the average person has little concept of the importance of welding as a manufacturing process. If it were not for the use of welding, a large portion of our metal products would have to be drastically modified, would be considerably more costly, or could not perform as efficiently.

To obtain coalescence between two metals, there must be sufficient proximity and activity between the atoms of the pieces being joined to cause the formation of common metallic crystals. The ideal metallurgical bond requires (1) perfectly smooth, flat or matching surfaces; (2) clean surfaces, free from oxides, absorbed gases, grease, and other contaminants; (3) metals with no internal impurities; and (4) two metals that are both single crystals with identical crystallographic structure and orientation. Obviously, these conditions would be difficult to obtain, even under ideal laboratory conditions, and are virtually impossible in normal production. Consequently, the various joining methods have been designed to overcome or compensate for this impossibility. Surface roughness is overcome either by force, causing plastic deformation of the asperities, or by melting the two surfaces so that fusion occurs. In solid-state welding, contaminated layers are removed by mechanical or chemical cleaning prior to welding or by causing sufficient metal flow along the interface so that they are squeezed out of the weld. In fusion welding, where a pool of molten metal exists, the contaminants are removed by the use of fluxing agents. If welding is performed in a vacuum, either by solid-state or fusion processes, the contaminants are removed much more easily and coalescence is established with considerable ease. Thus, in outer space mating parts may weld under extremely light loads, even when such was not intended.

The various welding processes differ not only in the way in which temperature and pressure are combined and achieved but also with regard to the attention that must be given to the cleanliness of the metal surfaces prior to welding and to possible oxidation or contamination of the metal during the welding process. If high temperatures are used, most metals are more adversely affected by the surrounding environment. If actual melting occurs, serious modification of the metal may result. The metallurgical structure and quality of the metal can be affected, often adversely, by the heating and cooling cycle of the welding process. These effects and their possible consequences should be considered in the design of a product and in selecting the joining process.

To summarize the above, the production of a quality weld requires (1) a satisfactory heat and/or pressure source, (2) a means of protecting or cleaning the metal, and (3) caution to avoid, or compensate for, harmful metallurgical effects.

Classification of Welding Processes.   The American Welding Society has defined and classified the various welding processes in the manner presented in Figure 33-1 and has assigned short letter symbols to facilitate their designa-

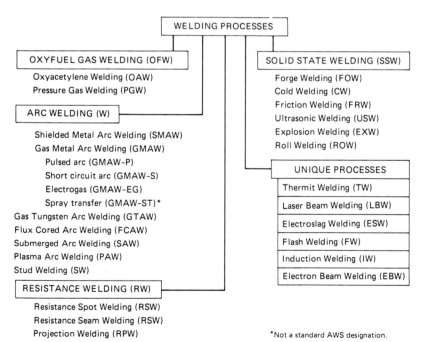

FIGURE 33-1   Classification of common welding processes with their AWS (American Welding Society) designations.

*Not a standard AWS designation.

tion. These processes provide a variety of ways of meeting the three requirements stated above and make it possible to achieve effective and economical
welds in nearly all metals and combinations of metals. As a result, welding has
replaced other types of permanent fastening to such a degree that a large portion
of manufactured products contain one or more welds.

Oxyfuel Gas Welding[1]

Oxyfuel gas welding (OFW) covers a group of welding processes that use as
their heat source the flame produced by the combustion of a fuel gas and oxygen.
The oxygen in this case is usually supplied in relatively pure form, but in rare
cases air can be used.

It was the development of a practical torch to burn acetylene and oxygen
shortly after 1900 that brought welding out of the blacksmith's shop, demonstrated its potential, and started its development as a manufacturing process.
While gas-flame welding has largely been replaced by other processes in the
manufacturing environment, it is still popular for many applications because of
its portability and the low capital investment required. Acetylene is still the
principal fuel gas employed in the process.

The combustion of oxygen and acetylene (C_2H_2) by means of a welding torch
of the type shown in Figure 33-2 produces a temperature of about 6300°F
(3500°C) in a two-stage reaction. In the first stage, the oxygen and acetylene
react to produce carbon monoxide and hydrogen:

$$C_2H_2 + O_2 \rightarrow 2\,CO + H_2$$

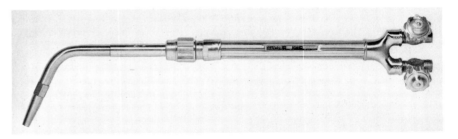

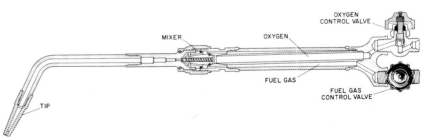

FIGURE 33-2   Oxyacetylene welding torch and schematic. (Courtesy
Victor Equipment Company.)

[1]The American Welding Society has selected this term to identify ''gas-oxygen flame
welding,'' and their terminology will be used throughout this section.

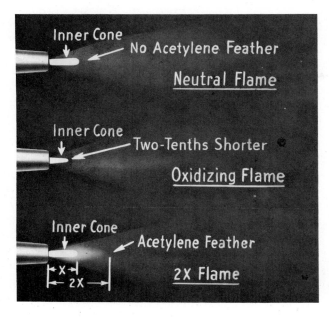

FIGURE 33-3   Neutral, oxidizing, and carburizing flame characteristics. (Courtesy Linde Division, Union Carbide Corporation.)

This reaction occurs near the tip of the torch. The second stage of the reaction involves the combustion of the CO and H_2 and occurs just beyond the first combustion zone. The specific reactions of stage two are:

$$2CO + O_2 \rightarrow 2CO_2$$
$$H_2 + \tfrac{1}{2} O_2 \rightarrow H_2O$$

The oxygen for these secondary reactions is generally obtained from the atmosphere.

The two-stage combustion process produces a flame having two distinct zones, as shown in Figure 33-3. The maximum temperature occurs at the end of the inner cone, where the first stage of combustion is complete. Most welding should be performed with the torch positioned so that the point of maximum temperature is just above the metal being welded. The outer zone of the flame serves to preheat the metal and, at the same time, provides shielding from oxidation, since some of the oxygen from the surrounding air is used in the secondary combustion.

As illustrated in Figure 33-3, three types of flames can be obtained by varying the oxygen-to-acetylene (or oxygen-to-other fuel gas) ratio. If the ratio is about 1:1 to 1.15:1, all reactions are carried to completion and a *neutral flame* is produced. Most welding is done with a neutral flame, since such a flame has minimum chemical effect on most heated metals.

A higher ratio produces an *oxidizing flame,* quite similar in appearance but possessing an excess of oxygen. Such flames are used only in welding copper and copper alloys and as a decarburizing flame for steels, the excess oxygen reacting with the carbon in the steel.

Excess fuel produces a flame that is *carburizing*. The excess fuel decomposes to carbon and hydrogen, and the flame temperature is not so great. Flames with a slight excess of fuel are reducing flames. No carburization occurs, but the metal is well protected from oxidation. Flames of this type are used in welding Monel (a nickel–copper alloy), low-carbon steels, and some alloy steels and in applying some hard-facing materials.

For welding purposes, acetylene is usually obtained in portable storage tanks that hold up to 300 cubic feet (8.5 cubic meters) at 250-psi (1.7 MPa) pressure. Because acetylene is not safe when stored as a gas at pressures above 15 psi (0.1 MPa), it is usually dissolved in acetone. The storage cylinders are filled with a porous filler, such as balsa-wood chips or infusorial earth. Acetone is absorbed into the voids in the filler material and serves as a medium for dissolving the acetylene.

Stabilized methylacetylene propadiene, best known by the trade name of MAPP gas, has become a competitor of acetylene, particularly when portability is important. While flame temperature is slightly lower, the gas is more dense, thus providing more energy for a given volume. In addition, it can be stored in ordinary pressure tanks. The oxygen for gas-flame welding almost always comes from pressurized tanks.

The pressures used in gas-flame welding range from 1 to 15 psi (7 to 105 MPa) and are controlled by pressure regulators on each tank. Because mixtures of acetylene and oxygen or air are highly explosive, precautions must be taken to avoid mixing the gases improperly or by accident. All acetylene fittings have left-hand threads, whereas those for oxygen are equipped with right-hand threads. This prevents the making of any improper connections.

The tip size (orifice diameter) of the torch can be varied to control the shape of the inner cone, the flow rate of the gases, and the size of the material that can be welded. Larger tips permit greater flow of gases, resulting in greater heat input without requiring higher gas velocities that might blow the molten metal from the weld puddle. Thicker metal requires larger torch tips, and larger tips operate with higher gas pressure.

Uses, Advantages, and Limitations.   Almost all oxyfuel gas welding is *fusion welding*. The metals to be joined are simply melted where welding is desired, and no pressure is involved. Because a slight gap usually exists between the pieces being joined, filler metal is often added in the form of a wire or rod that is melted in the flame or in the pool of weld metal. Welding rods come in standard sizes, with diameters from $\frac{1}{16}$ to $\frac{3}{8}$ in. (1.5 to 9.5 mm) and lengths from 24 to 36 in. (0.6 to 0.9 meters). They are available in standard grades to provide specified minimum tensile strengths or in compositions to match the base metal. *Fluxes* are often used to clean the surfaces to be joined and free them from contaminating oxide, thereby promoting the formation of a better bond. The fluxes can be added as a powder, or the welding rod can be dipped in a flux paste or be precoated.

FIGURE 33-4 Pressure gas welding of pipe. (Courtesy Linde Division, Union Carbide Corporation.)

Good-quality welds can be obtained by the OFW processes if proper technique and care are used. The temperature of the work can be easily controlled. However, exposure of the heated and molten metal to the various gases in the flame and atmosphere makes it difficult to prevent contamination. Since the heat source is not concentrated, considerable areas of the metal are heated and distortion is likely to occur. Thus, in production applications, the flame welding processes have largely been replaced by arc welding. Nevertheless, flame welding is still a common means of making repairs on in-service equipment and fabricating small quantities of specialized products.

Pressure Gas Welding. *Pressure gas welding* (PGW) is a process used to make butt joints between the ends of objects such as pipe and railroad rail. The ends are heated with a gas flame to a temperature below the melting point and then forced together under considerable pressure. Thus, the process is a form of solid-state welding. Figure 33-4 shows the pressure gas welding of pipe.

For many years, metal sheets and plates have been cut by means of oxyfuel torches and electric arc equipment. Developed originally for use in salvage and repair work, then used to prepare plates for welding, these processes are now widely used to cut sheets and plates into desired shapes for assembly and other processing operations. In recent years, the development of laser and electron beam equipment has made possible the cutting of both metals and nonmetals at speeds of up to 1000 in. per minute (25 meters per minute). Accuracies of up to 0.01 in. (0.25 mm) are readily attainable and speeds of 50 in. per minute (1.3 meters per minute) are quite common. Figure 33-5 shows a chart of the commonly used torch and arc cutting processes with their AWS designations.

Cutting Processes

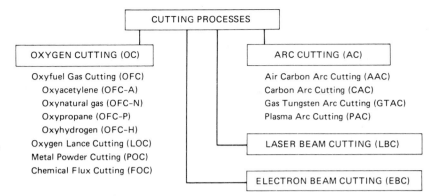

FIGURE 33-5 Classification of common cutting processes with their AWS (American Welding Society) designations.

Oxygen Torch Cutting

By far, the majority of all thermal cutting is done by *oxyfuel gas cutting* (OFC). In a few cases, primarily when the metal is nonferrous, the metal is merely melted by the flame of the oxyfuel gas torch and blown away to form a gap, or *kerf,* in the metal. However, when ferrous metal is being cut, the process is one of rapid oxidation (burning) of iron at high temperatures according to the chemical equation

$$3 \text{ Fe} + 2 \text{ O}_2 \rightarrow \text{Fe}_3\text{O}_4 + \text{heat}$$

Because this reaction does not occur until the metal is at approximately 1600°F (870°C), an oxyfuel flame is first used to raise the metal to the temperature at which burning will commence. Then a stream of pure oxygen is added to the torch (or the oxygen content of the oxyfuel mixture is increased) to oxidize the iron. The liquid iron oxide is then expelled from the joint by the kinetic energy of the oxygen–gas stream.

Theoretically, no further heat is required, but usually additional heat must be supplied to compensate for losses to the atmosphere and the surrounding metal. In addition, the additional heat can be used to assure that the reaction progresses in the desired direction. Under ideal conditions, when the area of burning is sufficiently confined so as to conserve the heat of combustion, no supplemental heating is required and a supply of oxygen through a small pipe is sufficient to keep the cut progressing. This is known as *oxygen lance cutting* (OLC). A temperature of about 2200°F (1200°C) has to be achieved in order for this procedure to be effective.

Fuel Gases for Oxyfuel Gas Cutting. Acetylene is by far the most common fuel used in oxyfuel gas cutting; thus, the process is often called oxyacetylene cutting (OFC-A). The type of torch commonly used is shown in Figure 33-6. The tip contains a circular array of small holes through which the oxygen–acetylene mixture is supplied for the heating flame. A larger hole in the center supplies a stream of oxygen controlled by a lever valve. The rapid flow of the cutting oxygen not only promotes the rapid oxidation but also serves to blow the formed oxides from the cut.

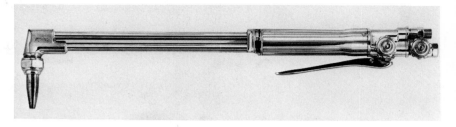

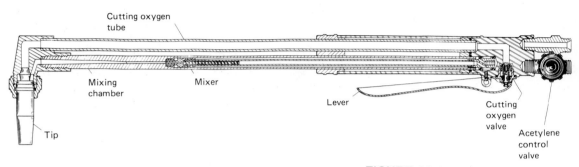

Cutting oxygen
tube

Mixing
chamber

Mixer

Lever

Cutting
oxygen
valve

Acetylene
control
valve

Tip

FIGURE 33-6   Oxyacetylene cutting
torch and schematic. (Courtesy Victor
Equipment Company.)

If the torch is adjusted and manipulated properly, a smooth cut can be produced, as shown at the top of Figure 33-7. As indicated in the remaining photographs of that figure, quality cutting requires careful selection of the preheat condition, oxygen flow rate, and cutting speed. Oxygen purity over 99.5% is required for the most efficient cutting.

Cutting torches are often manipulated manually. However, in most manufacturing applications, the traverse of the desired path is controlled by mechanical or programmable means. Figure 33-8 shows a portable, electrically driven carriage with a mounted cutting torch. For straight cuts, the device travels along a section of portable track. More complex cuts such as circles or arcs can also be cut. When large numbers of duplicate shapes are to be cut, a template-controlled flame cutting machine, such as the one in Figure 33-9, can be used. Most recently, the marriage of computer numerically controlled (CNC) machines and cutting torches has proved to be quite popular. This approach, along with the use of robot-mounted torches, provides great flexibility with good precision and control. Accuracies of ± 0.015 in. (± 0.40 mm) are possible, but ± 0.03 to 0.04 in. (± 0.75 to 1.0 mm) are more common.

Fuel gases other than acetylene are also used for oxyfuel gas cutting, the most common being natural gas (OFC-N) and propane (OFC-P). Their use is generally a matter of economics and gas availability. For certain special work, hydrogen may be used (OFC-H).

In preparing plate edges for subsequent welding, two or three simultaneous cuts are often made to establish the desired edge geometry. Figure 33-10 shows an example of a three-torch cut.

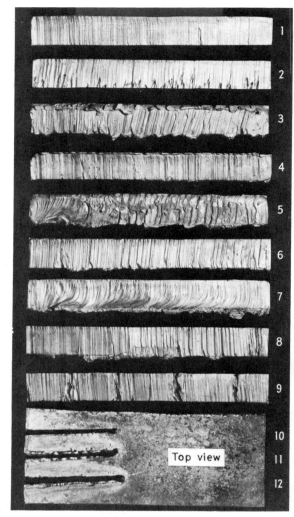

(1) **Correct Procedure**
Compare this cut in 1-in. plate with those below. The edge is square, the drag lines are vertical and not too pronounced.

(2) **Preheat Flames Too Small**
They are only about ⅛ in. long. Result: cutting speed was too slow, causing bad gouging effect at bottom.

(3) **Preheat Flames Too Long**
They are about ½ in. long. Result: surface has melted over, cut edge is irregular, and there is too much adhering slag.

(4) **Oxygen Pressure Too Low**
Result: top edge has melted over because of too slow cutting speed.

(5) **Oxygen Pressure Too High**
Nozzle size also too small. Result: entire control of the cut has been lost.

(6) **Cutting Speed Too Slow**
Result: irregularities of drag lines are emphasized.

(7) **Cutting Speed Too High**
Result: a pronounced rake to the drag lines and irregularities on the cut edge.

(8) **Blowpipe Travel Unsteady**
Result: the cut edge is wavy and irregular.

(9) **Lost Cut Not Properly Restarted**
Result: bad gouges where cut was restarted.

(10) **Good Kerf**
Compare this good kerf (viewed from the top of the plate) with those below.

(11) **Too Much Preheat**
Nozzle also is too close to plate. Result: bad melting of the top edges.

(12) **Too Little Preheat**
Flames also are too far from the plate. Result: heat spread has opened up kerf at top. Kerf is tapered and too wide.

FIGURE 33-7 The cut edge of metal that has been properly and improperly cut by the oxyacetylene process. (Courtesy Linde Division, Union Carbide Corporation.)

Stack Cutting. In order to cut a stack of thin sheets of steel successfully, two precautions must be observed. First, the sheets should be flat, smooth, and free of scale. Second, they should be clamped together tightly so that there are no intervening gaps that could interrupt uniform oxidation or permit slag and molten metal to be entrapped.

Stack cutting is a useful technique when a modest number of duplicate parts is required but not a sufficient number to justify the construction of a blanking die. Obviously, the accuracy of stack cutting is noticeably poorer than that obtained with a blanking die.

Metal Powder Cutting and Chemical Flux Cutting. Hard-to-cut materials can often be cut by modified torch techniques. *Metal powder cutting* (POC)

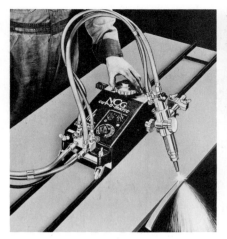

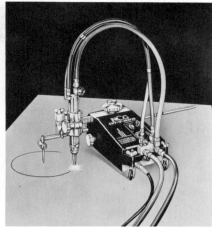

FIGURE 33-8   Oxyacetylene cutting with the torch being carried by a small, portable, electrically-driven carriage. (Courtesy National Cylinder Gas Company.)

FIGURE 33-9   Three parts being cut simultaneously from heavy steel plate using a tracer-pantograph machine that is guided by the template at the bottom. (Courtesy Linde Division, Union Carbide Corporation.)

FIGURE 33-10   Plate edge being prepared for welding. Three simultaneous cuts are being made with oxyacetylene cutting torches. (Courtesy Linde Division, Union Carbide Corporation.)

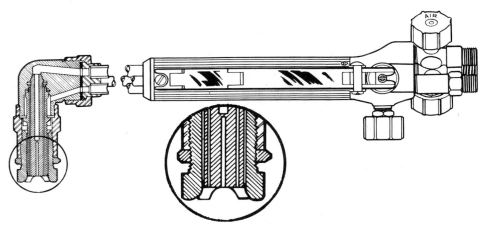

FIGURE 33-11 Underwater cutting torch. Note the extra set of gas openings in the nozzle to permit the flow of compressed air and the extra control valve. (Courtesy Bastian-Blessing Company.)

injects iron powder into the flame to raise its cutting temperature. *Chemical flux cutting* (FOC) adds a fine stream of special flux to the cutting oxygen to increase the fluidity of the high-melting-point oxides. Both methods have been replaced to a considerable extent by plasma arc torches (PAC), which will be discussed in Chapter 34.

Underwater Torch Cutting.   Steel can be cut underwater by use of a specially designed torch such as the one shown in Figure 33-11. An auxiliary skirt surrounds the main tip and an additional set of gas passages supplies a flow of compressed air to provide secondary oxygen for the oxyacetylene flame and to expel water from the zone where the burning of metal occurs. The torch is either ignited in the usual manner before descent or by an electric spark device after being submerged.

Acetylene gas is used for depths up to about 25 feet (7.5 meters). For greater depths, hydrogen is used, since the environmental pressure is too great for the safe use of acetylene.

Flame Straightening

Flame straightening is basically the creation of a controlled, localized upsetting in order to straighten warped or buckled plates. The theory of the process is illustrated in Figure 33-12. If a straight piece of metal is heated in a localized area, as indicated by the curved path in the upper diagram, the metal adjacent to *b* will be upset as it softens and tries to expand against the restraining cool metal. When the upset portion cools, it will contract and the resulting piece will be shorter on edge *b* and will bend to the shape in the lower diagram.

If the starting material is bent or warped, as in the lower segment of Figure 33-12, the upper surface can be heated. Upsetting and contraction will shorten the upper surface at *a'*, bringing the plate back to a straight or flat condition. Such a procedure can often be used in the repair of structures that have been bent in an accident, such as automobile frames.

A similar process can be used to flatten metal plates that have become dished due to buckling. Localized spots about 2 in. (50mm) in diameter are quickly heated to the upsetting temperature so that the surrounding metal remains cool. Cool water is then sprayed on the plate and the contraction of the upset spot brings the buckle into an improved degree of flatness. To remove large buckles, the process may have to be repeated at several spots within the area.

Several cautions should be noted. First, in straightening steel, consideration should be given to the possible phase transformations that could occur during the heating and cooling and the associated consequences. Since rapid cooling is used in the process and martensite may form, a subsequent tempering operation may be required. Secondly, the flame-straightening process should not be attempted with thin material. For the process to work, the metal adjacent to the heated area must have sufficient rigidity to promote upsetting. If this is not the case, localized heating and cooling will simply transfer the buckle from one area to another.

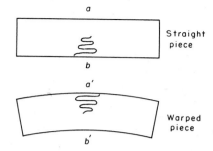

FIGURE 33-12   The theory of flame straightening.

Review Questions

1. What is an acceptable definition of *welding*?
2. What are the four conditions required for an "ideal metallurgical bond"?
3. What are some of the ways in which welding processes overcome or compensate for the inability to meet the ideal bond conditions?
4. What are some of the problems that might occur when high temperatures are used in welding?
5. What are the two stages in the combustion of oxygen and acetylene?
6. What is the location of the maximum temperature in an oxyacetylene flame?
7. What function or functions are served by the outer zone of the welding flame?
8. What three types of flame can be produced by varying the oxygen–fuel ratio?
9. What are some of the attractive features of MAPP gas?
10. Why might a welder want to change the tip size (orifice diameter) in an oxyacetylene torch?
11. Why is added filler metal often required in oxyacetylene welding?

12. In what way does the torch cutting of ferrous metals differ from the cutting of nonferrous alloys?
13. Why might continuously cast steel strands be effectively cut by an oxygen lance as they emerge from the casting operation?
14. How does an oxyacetylene cutting torch differ from an oxyacetylene welding torch?
15. What are some of the ways in which cutting torches can be mechanically manipulated?
16. What two precautions must be observed in order to cut a stack of thin sheets successfully?
17. What modification must be incorporated into a cutting torch to permit it to cut metal underwater?
18. If a curved plate is to be straightened by flame straightening, should the heat be applied to the longer or shorter surface of the arc?
19. Why does the flame-straightening process not work for thin sheets of metal?

Arc Processes: Welding and Cutting

Arc Welding

Almost from the time when electricity became a commercial reality, it was recognized that an electric arc between two electrodes was a concentrated heat source, approaching 7000°F (3900°C) in temperature. As early as 1881, various persons attempted to use an arc between a carbon electrode and a metal workpiece as the heat source for fusion welding. The basic circuit is shown in Figure 34-1. As in gas-flame welding, filler metal was added in the form of a metallic wire. In a later modification, bare metal wire replaced carbon as the electrode. As it melted in the arc, it automatically supplied the necessary filler metal.

The results of these early efforts were very uncertain. Because of the instability of the arc, a great amount of skill was required to maintain it, and contamination and oxidation of the weld resulted from its exposure to the atmosphere at such high temperatures. Furthermore, there was little or no understanding of the metallurgical effects and requirements of such a process. Consequently, although the great potential was recognized, very little use was made of the process until after World War I. About 1920, shielded metal electrodes were developed. These electrodes provided for a stable arc by shielding it from the atmosphere and gave some fluxing action to the molten pool. Having

FIGURE 34-1 The basic circuit for arc welding.

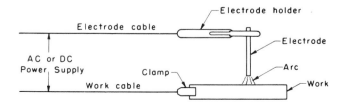

thus overcome the major problems of arc welding, the process began to expand rapidly. Today, a considerable variety of arc-welding processes are available.

All arc-welding processes employ the same basic circuit depicted in Figure 34-1, except that alternating current is used at least as frequently as direct current. If the work is made positive (the anode of the circuit) in the direct current mode, and the electrode is made negative, *straight polarity* (spdc) is said to be employed. When the work is negative and the electrode is positive, the process is using *reverse polarity* (rpdc). When bare electrodes are used, greater heat is liberated at the anode and spdc conditions are preferred. Certain shielded electrodes, however, change the heat conditions and are used with reverse polarity.

All arc welding is done with metal electrodes. In one family of processes, the electrode is consumed *(consumable electrode processes)* and thus supplies the needed filler metal to fill the voids in the joint. Consumable electrodes have a melting temperature below the temperature of the arc. Small droplets are melted from the end of the electrode and pass to the workpiece. The size of these droplets varies greatly and the mechanism of the transfer varies with different types of electrodes and processes. Figure 34-2 depicts metal transfer by the globular, spray, and short-circuit modes. As the electrode melts, the arc length and the resistance of the arc path vary. This requires the electrode to be moved toward the work to maintain a stable arc and satisfactory welding conditions.

Manual arc welding is no longer performed with bare electrodes. Shielded (covered) electrodes are always used. Continuous bare-metal wire is often used as the electrode in automatic or semiautomatic arc welding. However, this is always in conjunction with a separate shielding and arc-stabilizing medium and automatic feed-controlling devices that maintain the proper arc length.

In the other family of arc-welding processes, the electrode is made of tungsten, which is not consumed by the arc, except by relatively slow vaporization. In these *nonconsumable electrode processes,* a separate filler metal wire must be used to supply the necessary metal.

In summary, the various arc-welding processes require selection or specification of the welding voltage, welding current, arc polarity (straight polarity, reversed polarity, or alternating polarity), arc length, welding speed (how fast

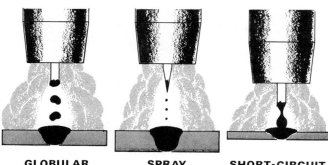

GLOBULAR　　**SPRAY**　　**SHORT-CIRCUIT**

FIGURE 34-2　Modes of metal transfer during arc welding. (Courtesy Republic Steel Corporation.)

the electrode is moved across the workpiece), arc atmosphere, electrode or filler material, and flux. The filler material is generally selected to match the base metal with respect to properties and/or alloy content (chemistry).

Shielded Metal Arc Welding. *Shielded metal arc welding* (SMAW) uses electrodes that consist of metal wire, usually from $\frac{1}{16}$ to $\frac{3}{8}$ in. (1.6 to 9.5 mm) in diameter, upon which is extruded a coating containing chemical components that act to accomplish a number of desirable objectives, including all or a number of the following:

1. Provide a protective atmosphere.
2. Stabilize the arc.
3. Act as a flux to remove impurities from the molten metal.
4. Provide a protective slag to accumulate impurities, prevent oxidation, and slow the cooling of the weld metal.
5. Reduce weld-metal spatter and increase the efficiency of deposition.
6. Add alloying elements.
7. Affect arc penetration (the depth of melting in the workpiece).
8. Influence the shape of the weld bead.
9. Add additional filler metal.

Coated electrodes are classified by the tensile strength of the deposited weld metal, the welding position in which they may be used, the type of current and polarity (if direct current), and the type of covering. A four- or five-digit system of designation is used, as presented in Figure 34-3. As an example, type E7016 is a low-alloy steel electrode that will provide a deposit with a minimum tensile strength of 70,000 psi in the non-stress-relieved condition; it can be used in all positions, with either alternating or reverse-polarity direct current; and it has a low-hydrogen plus potassium coating.

In general, the cellulosic coatings contain about 50% SiO_2; 15% TiO_2; small amounts of FeO, MgO, and Na_2O; and about 30% volatile matter. The titania coatings have about 30% SiO_2; 50% TiO_2; small amounts of FeO, MgO, Na_2O,

FIGURE 34-3 Designation system for arc-welding electrodes.

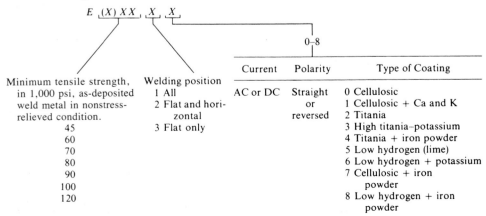

$E\ (X)\,X\,X\ \ X\ \ X$

		0–8
	Current	Polarity	Type of Coating

Minimum tensile strength, in 1,000 psi, as-deposited weld metal in nonstress-relieved condition.
45
60
70
80
90
100
120

Welding position
1 All
2 Flat and horizontal
3 Flat only

Current: AC or DC

Polarity: Straight or reversed

Type of Coating
0 Cellulosic
1 Cellulosic + Ca and K
2 Titania
3 High titania–potassium
4 Titania + iron powder
5 Low hydrogen (lime)
6 Low hydrogen + potassium
7 Cellulosic + iron powder
8 Low hydrogen + iron powder

and Al_2O_3; and about 5% volatile material. The low-hydrogen coatings have various compositions designed to prevent hydrogen from dissolving in the weld metal. The presence of dissolved hydrogen is closely associated with the formation of microcracks in the weld. To be effective, the electrodes must also be baked just prior to use to remove all moisture (another source of hydrogen) from the coating.

All electrodes are marked with colors in accordance with a standard established by the National Electrical Manufacturers Association so that the type can be readily identified. Electrode selection consists of determining electrode coating, coating thickness, electrode composition, and electrode diameter.

As the coating on the electrode melts and vaporizes, it forms a protective atmosphere that stabilizes the arc and protects the molten and hot metal from contamination. Fluxing constituents unite with any impurities in the molten metal and float them to the surface to be entrapped in the slag coating that forms over the weld. The slag coating protects the cooling metal from oxidation and slows down the cooling rate to prevent the formation of hard, brittle structures. The slag is then easily chipped from the weld when it has cooled. Figure 34-4 provides a schematic of metal deposition from a shielded electrode.

Electrodes having iron powder in the coating significantly increase the amount of metal that can be deposited with a given-size electrode wire and current, and are used extensively in production-type welding. Other electrodes possess coatings designed to melt more slowly than the filler wire, such that if the electrode is dragged along the work, the center wire will be recessed by the proper arc length. These are called *contact* or *drag electrodes*.

Carbon steels, alloy steels, stainless steels, and cast irons are commonly welded by the shielded metal arc process. Reverse-polarity DC is used to obtain deep penetration, with alternate modes being employed in welding thin sheet. The mode of metal transfer is either globular or short circuit, and the arc temperatures are rather low (9000°F or 5000°C). Typical welding voltages are 15 to 45 volts with currents between 10 and 500 amps.

In order to provide electrical contact with the center, filler-metal wire, most SMAW electrodes are finite-length "sticks." Length is limited since the current must be supplied near the arc or the electrode will overheat and ruin the coating. Therefore, shielded metal arc welding with stick electrodes is used most in job

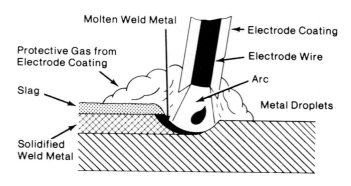

FIGURE 34-4 Schematic diagram of shielded metal arc welding (SMAW). (Courtesy American Iron and Steel Institute, Washington, D.C.)

shops and repair welding operations. Although some techniques have been developed to provide continuous shielded metal arc welding, the general trend is to use alternative processes for heavy-use production welding.

Gas Tungsten Arc Welding.   *Gas tungsten arc welding* (GTAW), formerly known as TIG welding (for tungsten inert-gas), was one of the first developments away from the use of ordinary shielded electrodes. A tungsten electrode is positioned in a special holder through which inert gas flows to form an inert shield around the arc and pool of molten metal. Argon, helium, or a mixture thereof is used to isolate this region from the atmosphere.

Under the inert atmosphere, the tungsten electrode is not consumed at the temperatures of the arc. The arc length remains constant and the arc is stable and easy to maintain. The tungsten electrodes are often treated with thorium or zirconium to provide better current-carrying and electron-emission characteristics. A high-frequency, high-voltage current may be superimposed on the regular AC or DC welding current to make it easier to start and maintain the arc. Figure 34-5 shows a typical water-cooled GTAW torch.

For applications in which a close fit exists, no filler metal may be needed. If filler metal is required, it must be supplied as a separate wire, as indicated in Figure 34-6. The filler metal is generally selected to match the chemistry of the metal being welded. When high deposition rates are desired, a separate electrical circuit is provided to preheat the filler wire. As shown in Figure 34-7, the deposition rate of heated wire can be several times that of a cold wire. Moreover,

FIGURE 34-5   Welding torch used in nonconsumable-electrode, gas tungsten arc welding (GTAW). (Courtesy Linde Division, Union Carbide Corporation.)

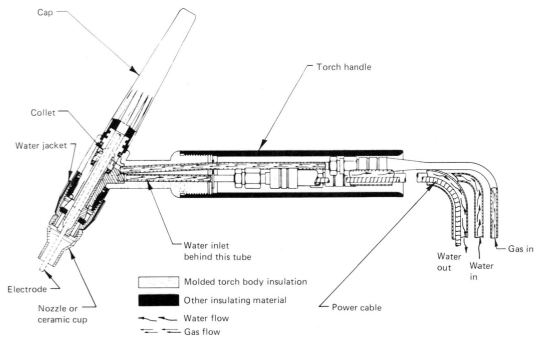

Cap

Torch handle

Collet

Water jacket

Electrode

Nozzle or ceramic cup

Water inlet behind this tube

Molded torch body insulation

Other insulating material

Water flow

Gas flow

Power cable

Water out   Water in   Gas in

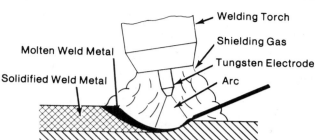

FIGURE 34-6 Schematic diagram of gas tungsten arc welding (GTAW). (Courtesy American Iron and Steel Institute, Washington, D.C.)

the deposition rate can be further increased by oscillating the filler wire from side to side when making a weld pass. The hot-wire process is not practical when welding copper or aluminum, however, because of the low resistivities of the filler wire.

With skilled operators, gas tungsten arc welding can produce welds that are scarcely visible. In addition, the process produces very clean welds. Since no flux is employed, no special cleaning or slag removal is required. However, the surfaces to be welded must be clean and free of oil, grease, paint, and rust, because the inert gas does not provide any cleaning or fluxing action.

All metals and alloys can be welded by this process. Maximum penetration is obtained with straight polarity DC conditions, although AC may be specified to break up surface oxides (as when welding aluminum). Reverse polarity is rarely used because it tends to melt the tungsten electrode. Weld voltage is typically 20 to 40 volts and weld current varies from less than 125 amperes for rpdc to 1000 amperes for spdc.

Gas Tungsten Arc Spot Welding. A variation of gas tungsten arc welding is employed for making spot welds between two pieces of metal without requiring access to both sides of the joint. The basic procedure is illustrated in

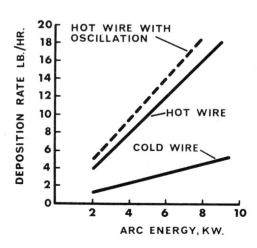

FIGURE 34-7 Comparison of the metal deposition rates with hot and cold filler wire in GTAW welding. (Courtesy *Welding Journal*.)

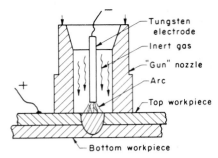

FIGURE 34-8 Schematic diagram of the method of making spot welds with the inert-gas-shielded tungsten arc process.

FIGURE 34-9 Making a spot weld by the inert-gas-shielded tungsten arc process. (Courtesy Air Reduction Company, Inc.)

Figure 34-8. A modified inert-gas tungsten arc gun is used with a vented nozzle on the end. The nozzle is pressed firmly against one of the two pieces of the joint to hold them in reasonably good contact. (The workpieces must be sufficiently rigid to sustain the contact pressure.) Inert gas, usually argon or helium, flows through the nozzle to provide a shielding atmosphere. Automatic controls advance the electrode to make the momentary contact with the workpiece necessary to start the arc, then withdraw it to the correct distance for stabilized arcing. The duration of arcing is automatically timed so that the two workpieces are heated sufficiently to form an acceptable spot weld. The depth and size of the weld nugget are controlled by the amperage, time, and type of shielding gas.

Because access to only one side of the work is required, this type of spot welding has an advantage over the more conventional resistance process in certain applications. One such example is the fastening of sheet metal to heavier framework, as illustrated in Figure 34-9.

Gas Metal Arc Welding. *Gas metal arc welding* (GMAW), formerly known as MIG welding (for metal inert-gas), was a logical outgrowth of gas tungsten arc welding. The process is similar, but the arc is now maintained between the workpiece and an automatically fed, consumable wire electrode. As shown in the schematic of Figure 34-10, the consumable electrode provides the filler metal, so no additional feed is required.

Figure 34-11 shows a complete schematic of GMAW equipment, showing the electrical circuits and mechanisms for water cooling and flow of the shielding gas. Although argon, helium, and mixtures of the two can be used for welding virtually any metal, they are used primarily with the nonferrous metals. In welding steel, some O_2 or CO_2 is usually added to improve the arc stability and reduce weld spatter. The cheaper CO_2 can be used alone in welding steel, provided that a deoxidizing electrode wire is employed.

The specific shielding gases can have considerable effect on the nature of metal transfer from the electrode to the work and also affect the heat transfer behavior, penetration, and tendency for undercutting (weld pool extending laterally beneath the surface of the base metal). Several types of electronic controls can be used to alter the wave form of the current. This makes it possible to

FIGURE 34-10 Schematic diagram of gas metal arc welding (GMAW). (Courtesy American Iron and Steel Institute, Washington, D.C.)

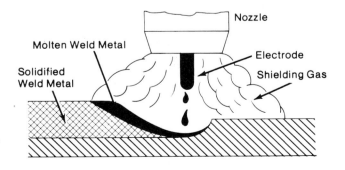

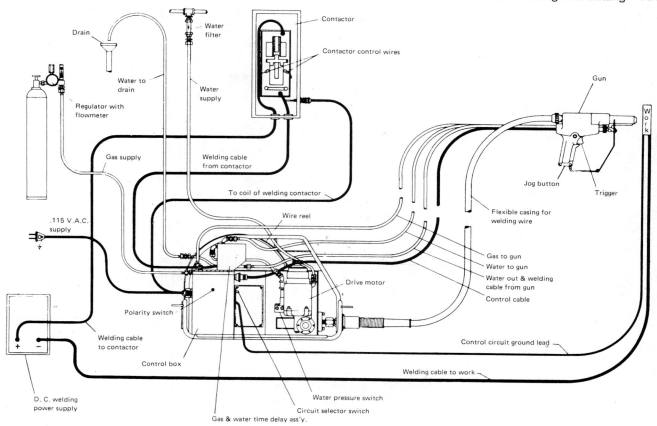

Drain

Water filter

Water to drain

Water supply

Regulator with flowmeter

Contactor

Contactor control wires

Gun

Gas supply

Welding cable from contactor

To coil of welding contactor

Wire reel

Jog button

Trigger

Flexible casing for welding wire

.115 V.A.C. supply

Gas to gun

Water to gun

Water out & welding cable from gun

Control cable

Drive motor

Polarity switch

Welding cable to contactor

Control box

Control circuit ground lead

Welding cable to work

D. C. welding power supply

Water pressure switch

Circuit selector switch

Gas & water time delay ass'y.

Work

control the mechanism of metal transfer, from drops, to spray, to short-circuiting drops. Some of these variations include *pulsed arc welding* (GMAW-P), *short circuiting arc welding* (GMAW-S), and *spray transfer welding* (GMAW-ST). *Buried arc welding* (GMAW-B) is another variation in which carbon dioxide-rich gas is used and the arc is buried in its own crater.

Gas metal arc welding is fast and economical because there is no frequent changing of electrodes, as with stick-type electodes. In addition, there is no slag formed over the weld, the process can be readily automated, and, if done manually, the welding head is relatively light and compact, as shown in Figure 34-12. A reverse-polarity DC arc is generally used because of its deep penetration, spray transfer, and smooth welds with good profile. Process variables include type of current, current magnitude, shielding gas, type of metal transfer, electrode diameter, electrode composition, electrode stickout (extension beyond the gun), welding speed, welding voltage, and arc length.

A number of industrial robots are now availabe to perform gas metal arc welding. To function properly, however, the computer electronics of these robots must be shielded from the high-frequency interference of the welding process.

FIGURE 34-11 Schematic diagram of the equipment used for gas metal arc welding (GMAW). (Courtesy Air Products and Chemicals, Inc.)

FIGURE 34-12   Welding with a manually-held gas metal arc welding gun. (Courtesy Air Products and Chemicals, Inc.)

Pulsed Arc Gas Metal Arc Welding.   Pulsed arc welding (GMAW-P), has developed into an extremely attractive welding technique. High-current capacitor discharges can be used to join both similar and dissimilar metals in a period of milliseconds. Because of the short duration of arcing, temperatures are reduced and a number of benefits can be obtained. Thinner material can be welded; distortion is reduced or eliminated; workpiece discoloration is minimized; heat-sensitive parts can be welded; high-conductivity metals can be joined; electrode life is extended; electrode cooling techniques may not be required; and fine microstructures are produced in the weld pool. In addition, the use of pulsed power lowers spattering and improves the safety of the process. The high speed of the process is attractive for productivity and the energy or power required to produce a weld is considerably lower than with other methods (reduced cost). Controls can be adjusted to alter the shape of the weld pool and vary the penetration.

Flux-Cored Arc Welding.   *Flux-cored arc welding* (FCAW) utilizes a continuous, hollow electrode wire filled with a granular flux, as shown in Figure 34-13. It can be viewed as an adaptation of the shielded metal arc process wherein the electrode is now continuous and less bulky, since no binder is required to hold the flux onto the electrode. A protective atmosphere is provided by the vaporized flux, and further protection results from the slag overlayer.

For enhanced weld properties, an externally supplied shielding gas, generally CO_2, may be employed, essentially creating a gas metal arc process with a flux-

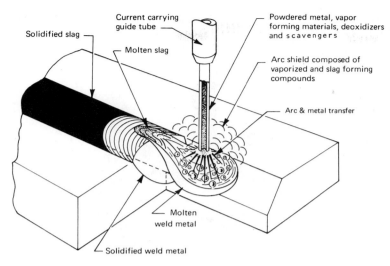

Solidified slag

Current carrying
guide tube

Molten slag

Powdered metal, vapor
forming materials, deoxidizers
and scavengers

Arc shield composed of
vaporized and slag forming
compounds

Arc & metal transfer

Molten
weld metal

Solidified weld metal

FIGURE 34-13 Schematic representation of the flux-cored arc welding process (FCAW). (Courtesy The American Welding Society, New York.)

cored electrode. This process is used primarily for welding ferrous material, almost always with a reverse-polarity DC power source.

Submerged Arc Welding.   In *submerged arc welding* (SAW), as illustrated in Figure 34-14, the arc is maintained beneath a blanket of granular flux. The flux is deposited just ahead of the electrode, which is in the form of a coiled wire, copper-coated to provide good electrical contact. Because the arc is completely submerged in the flux, only a few small flames are visible. The granular flux provides excellent shielding of the molten metal. Because the pool of molten metal is relatively large, good fluxing action occurs to remove any impurities. A portion of the flux is melted and solidifies into a glasslike covering over the weld. This, along with the flux that is not melted, provides a thermal coating that slows the cooling of the weld area and helps to produce a soft, ductile weld. The solidified flux then cracks loose from the weld upon cooling (because of the differential thermal contraction) and is easily removed. Surplus unmelted flux is recovered by a vacuum system and reused.

Thus, submerged arc welding can be used to produce welds of extremely high quality. Either AC or DC current can be used as the power source.

Submerged arc welding is most suitable for making flat butt or fillet welds in low-carbon steel (<0.3% carbon). With some preheat and postheat precautions, medium-carbon and alloy steels and some cast irons, copper alloys, and nickel alloys can be welded. The process is not suitable for high-carbon steels, tool steels, aluminum, magnesium, titanium, lead, or zinc. Several reasons may be responsible for the incompatibility of these metals and submerged arc welding. The fluxes required for these metals may not be suitable for use at the high temperatures of the submerged arc. Some of the metals are extremely reactive at elevated temperatures and others have low sublimation temperatures.

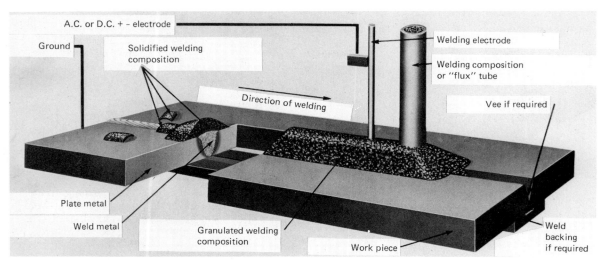

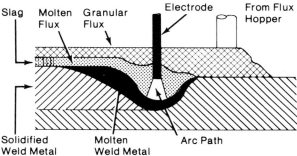

FIGURE 34-14   (Top) Basic features of the submerged arc welding process (SAW). (Courtesy Linde Division, Union Carbide Corporation.); (Bottom) Cut-away schematic of submerged arc welding. (Courtesy American Iron and Steel Institute, Washington, D.C.)

High welding speeds, high deposition rates, deep penetration, and high cleanliness (because of the flux action) are all characteristic of submerged arc welding. Welding speeds of 30 in. (750 mm) per minute in 1-in. (25 mm)-thick steel plate or 12 in. (300 mm) per minute in 1½-in. (40 mm) plate are common. Single-pass welds 1½ in. (40 mm) deep can be made and almost any thickness of base metal can be joined. Because the metal is deposited in fewer passes than with alternative processes, there is less possibility of entrapped slag or voids, and weld quality is further enhanced. When higher deposition rates are desired, multiple electrode wires can be employed. This technique is widely used in large-volume welding, as in the building of ships or the manufacture of large-diameter steel pipe or tanks.

Limitations to the process include the need for extensive flux handling, possible contamination of the flux by moisture (leading to porosity), the large volume of slag that must be removed, the restriction to flat welding because of the flux and slag, the high heat inputs that promote large-grain-size structures, and the slow cooling rate (which permits segregation and possible hot cracking). In addition, chemical control is quite important, since the electrode material often comprises over 70% of the molten weld region.

The electrodes are classified by composition and are available in diameters from 0.045 in. (1.1 mm) to ⅜ in. (9.5 mm). Larger electrodes carry higher currents and provide for rapid deposition rates, but penetration is shallower. The wire may be solid alloy material, plain steel with the alloy additions coming from the flux, or tubular metal with an alloy-element core. The fluxes are classified according to the properties of the weld metal and are designed to have low melting temperatures and good fluidity at high temperatures and to be brittle after cooling.

Submerged arc equipment may be semiautomatic, with the operator controlling the speed, or fully automatic. They may be manual (as in Figure 34-15), portable (wherein the welder traverses a stationary workpiece), or stationary (wherein the workpiece passes under the arc).

In a modification of the process known as *bulk welding*, iron powder is deposited in the prepared gap between the plates to be joined (ahead of the flux and on top of a backing strip) as a means of increasing deposition rate. A single pass can then apply 110 lb/hr (50 kg/hr) of weld metal, equivalent to seven or eight conventional submerged-arc passes.

By using the arrangement shown in Figure 34-16, vertical welds can be made by the submerged arc process. Stationary copper side molds are employed to contain the weld pool. Flux is continuously added through the wire guide to replace the flux that solidifies at the copper–weld interface. Good quality welds can be made in plates up to 4 in. (100 mm) thick with this procedure, but for plates thicker than about 2 in. (50 mm), the electroslag process (to be discussed in Chapter 35) is usually more economical.

FIGURE 34-15 Manual submerged arc welding. (Courtesy Lincoln Electric Company.)

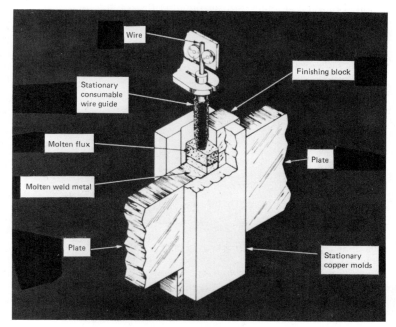

FIGURE 34-16 Setup for making vertical welds by the submerged arc process. (Courtesy *Welding Journal.*)

Plasma Arc Welding. In *plasma arc welding* (PAW), the arc is struck between a nonconsumable electrode and either the welding gun (nontransferred arc) or the workpiece (transferred arc), as shown in Figure 34-17. The arc column is constricted within a small-diameter nozzle, through which an inert gas is directed. Since the arc fills most of the nozzle opening, the gas must flow through the arc, where it is heated to a high temperature and forms a plasma. It is the flow of hot gases that actually transfers heat to the workpiece and melts the metal. Plasma arc welding provides greater energy concentration, fast welding speeds, deep penetration, a narrow heat-affected zone, reduced distortion, less demand for filler metal, higher temperatures, and a process that is insensitive to arc length. In addition, nearly all metals and alloys can be welded by this technique.

With a low-pressure plasma, the metal is simply "melted in" and a filler material is often used. At higher pressures, a "keyhole" effect occurs, wherein the plasma arc forms a hole completely through the sheet (up to ¼ in. or 6 mm thick) that is surrounded by molten metal. As the arc travels, the liquid metal flows to fill the keyhole. If the pressure is increased further, the molten metal is expelled from the region and the process becomes one of plasma cutting, which will be discussed later in this chapter.

Many plasma torches employ a small nontransferred arc within the torch to heat the orifice gas and ionize it. The ionized gas then forms a good conductive path for the main transferred arc. This dual-arc technique permits instant ignition of a low-current arc, which can be lower in magnitude, more stable, and more readily controlled than that of an ordinary plasma torch. Separate DC power supplies are used for the pilot and main arcs. An inert shielding gas is frequently supplied through an outer cup surrounding the torch.

Stud Welding. *Stud welding* (SW) is an arc-welding process by which fasteners can be welded into place instead of having to use riveting or drilling and

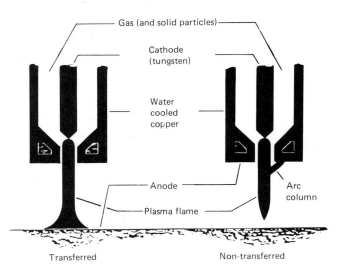

FIGURE 34-17 Two types of plasma arc torches, transferred-arc and nontransferred-arc. (Courtesy Linde Division, Union Carbide Corporation.)

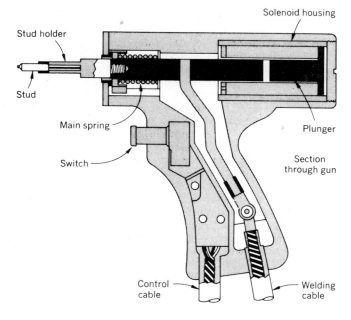

Stud holder

Solenoid housing

Stud

Main spring

Switch

Plunger

Section through gun

Control cable

Welding cable

FIGURE 34-18   Schematic diagram of a stud welding gun. (Courtesy *American Machinist.*)

tapping. A special gun is used, such as the one shown in Figure 34-18, into which the stud is inserted. A DC arc is established between the end of the stud and the workpiece, and a small amount of metal is melted. The two pieces are then brought together under light pressure and allowed to solidify. Automatic equipment controls the establishing of the arc, its duration, and the application of pressure to the stud.

Figure 34-19 shows some of the wide variety of studs that are specially made for this process. Each contains a recessed end that is filled with flux. A ceramic ferrule, such as the one shown in the center photo of Figure 34-19, is usually placed over the end of the stud before it is positioned in the gun. During the arc, the ferrule serves to concentrate the heat and isolate the hot metal from the atmosphere. It also confines the molten or plastic metal to the weld area and shapes it around the base of the stud, as shown in the right-hand photo of Figure 34-19. After the weld has cooled, the brittle ferrule is broken from the stud. Burnoff or melting reduces the length of the stud. The original design of the stud should compensate for this change.

Stud welding requires almost no skill on the part of the operator. Once the stud and ferrule are placed in the gun and the gun positioned on the work, all the operator has to do is pull the trigger. The remainder of the cycle is fully automatic and consumes less than one second. Thus, the process is well suited to manufacturing, serving to eliminate the drilling and tapping of many special holes. Production-type stud welders can produce over 1000 welds per hour.

Advantages and Disadvantages of Arc Welding.

Because of its great flexibility and the wide variety of processes that are available, arc welding has

FIGURE 34-19 (Left) Some of the types of studs used for stud welding. (Center) Stud and ceramic ferrule. (Right) Stud after welding, and a section through a welded stud. (Courtesy Nelson Stud Welding Co.)

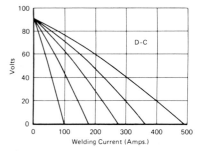

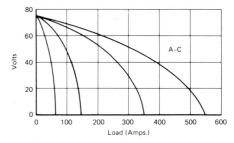

FIGURE 34-20 Drooping-voltage characteristics of typical arc-welding power supplies. (Top) Direct current; (Bottom) Alternating current.

become an extremely useful, versatile, and widely used process. However, except for gas-shielded tungsten arc spot welding, stud welding, and, to some degree, submerged arc welding, the various arc-welding processes have one common disadvantage—the quality of the weld depends upon the skill and integrity of the individual who does the welding. While automation and robotics are helping to reduce this problem, the selection, training, and supervision of welding personnel are still of great importance. In addition, the numerous nondestructive inspection techniques, described in Chapter 11, should be used liberally to assure weld quality.

Power Sources for Arc Welding. Arc welding requires a large amount of current that does not change in magnitude as the voltage varies over a considerable range. The load voltage is usually from 30 to 40 volts, although the actual voltage across the arc varies from about 12 to 30 volts, depending primarily upon the arc length. Both DC and AC sources are available and generally have "drooping voltage" characteristics, as shown in Figure 34-20, and current capacities from 150 to 1000 amperes. These characteristics assure that the current does not vary greatly as the voltage fluctuates over the usual operating range.

In earlier years, most direct current for welding was provided by motor–generator sets. Today, however, solid-state transformer–rectifier machines, such as the one shown in Figure 34-21, are the most frequently used power source. Using a three-phase power supply, these machines can usually provide both AC and DC output. When welding is to be performed where electric power is not available, gasoline-driven DC generators are used.

If only AC welding is to be done, relatively simple transformer-type power supplies can be used. These are usually single-phase devices with low power factors. However, when several machines are to be operated, as in a production-type shop, they may be connected to the various phases of a three-phase supply to help balance the load.

FIGURE 34-21 Rectifier-type ac and dc welding power supply with 300 amperes capacity. The cover has been removed to show the interior. (Courtesy Lincoln Electric Company.)

Jigs and Positioners.   Jigs or fixtures (also called positioners) are frequently used to hold the work in production welding. By positioning and manipulating the workpiece, the welding operation can often be performed in a more favorable position. Figure 34-22 shows a large weldment being fabricated on a positioner of moderate size. Special positioners have been designed to hold large sections of ships which weigh many tons. Parts can also be mounted on numerically controlled (N/C) tables that manipulate and position the workpiece with respect to the welding tool.

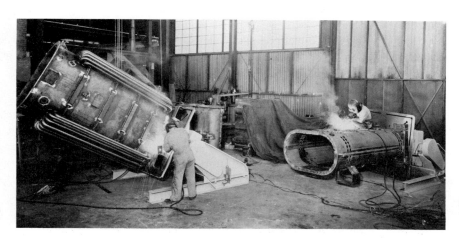

FIGURE 34-22 Transformer cases being welded while held on a welding positioner. (Courtesy Panjiris Weldment Company.)

Arc Cutting

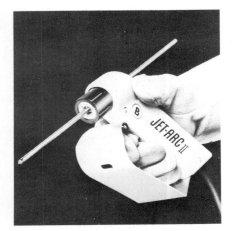

FIGURE 34-23 Gun used in the arc-air cutting process. Note the air holes surrounding the electrode in the holder. (Courtesy Jackson Products.)

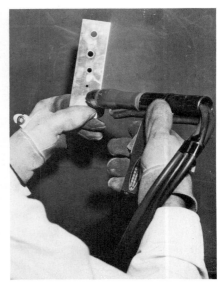

FIGURE 34-24 Making holes in sheet metal by the inert-gas arc process. (Courtesy Hobart Brothers Company.)

Virtually all metals can be cut by electric arc procedures, wherein the material is melted by the intense heat of the arc and then permitted, or forced, to flow from the region of the slit or notch (kerf). Most of the processes are simply adaptations of the arc-welding procedures discussed previously in this chapter and are listed in the presentation of Figure 33-5.

Carbon-Arc (CAC) and Shielded Metal Arc Cutting (SMAC). These methods use the arc to melt the metal, which is then removed from the cut by gravity or the force of the arc. Use is generally limited to small shops, garages, and homes where the equipment investment is small.

Oxygen-Arc Cutting (AOC). In this process, an electric arc and a stream of oxygen are employed to make the cut. The electrode is a coated ferrous-metal tube, the tube serving as the conductor for maintaining the arc, while the bore of the tube conveys and directs oxygen to the area of incandescence. In easily oxidizable metals, such as steel, the arc simply preheats the base metal, which then reacts with oxygen, is liquified, and is expelled by the oxygen stream.

Air Carbon-Arc Cutting. Here an arc is maintained between a carbon electrode and the workpiece, and high-velocity jets of air are directed at the molten metal from holes in the electrode holder, as shown in Figure 34-23. While there is some oxidation, the primary function of the air stream is to blow the molten material from the cut. Thus, the process can be used on metal that does not readily oxidize.

Air carbon-arc cutting is particularly effective for cutting cast iron and for gouging steel plates in preparation for welding. Speeds up to 24 in. (600 mm) per minute are readily attained. From a negative viewpoint, however, the process is quite noisy and the hot metal particles tend to be blown out over a substantial area. For cutting stainless steel and nonferrous metals, the plasma arc cutting process tends to be more efficient.

Gas Metal-Arc Cutting (GMAC). If the wire feed rate and other variables of gas metal-arc welding (MIG welding) are adjusted so that the electrode penetrates completely through the workpiece, then cutting rather than welding will occur. Wire feed rate controls the quality of the cut and the voltage determines the width of the slit.

Gas Tungsten-Arc Cutting (GTAC). The same basic circuit and shielding gas are used in gas tungsten arc cutting as in gas tungsten arc welding; however, a high-velocity jet of gas now passes through the nozzle to expel the molten metal. Figure 34-24 shows a modification of gas tungsten arc spot welding being used to create holes, up to ⅜ in. (9.5 mm) in diameter, in sheet metal.

Plasma Arc Cutting (PAC). The torches used in *plasma arc cutting* produce the highest temperature available from any practical source. Figure 34-17 presented the two types of plasma torches, and their operation was discussed briefly in a previous section of this chapter. With the nontransferred-type torch, the arc column is completely within the nozzle, and a temperature of about 30,000°F (16,650°C) is obtained. With the transfer-type torch, the arc column is between the electrode and the workpiece, and the temperatures can be up to 60,000°F (33,300°C). Obviously, such high temperatures provide a means of rapidly cutting any material by simply melting it and blowing it from the cut, as shown in Figure 34-25.

Early efforts to employ this technique showed that the speed, versatility, and operating cost were far superior to those of the oxyfuel cutting methods. However, these early systems could not constrict the arc to produce a cut of sufficient quality to meet the demands of manufacturing. Therefore, plasma arc cutting was generally used to cut only stainless steel and nonferrous metals that could not be cut by the oxidation-type cutting techniques.

Radial impingement of water on the arc was found to provide the desired constriction and produces an intense, highly focused arc. Water-injected torches can now cut virtually any metal in an economical fashion. Compared to the oxyfuel technique, plasma cutting is more economical (cost per unit length of cut is a fraction of that of oxyfuel), more versatile (it can cut all metals as easily as mild steel), and much faster (typically 5 to 8 times faster than oxyfuel). Cutting speeds up to 300 in. (7.6 meters) per minute have been obtained in ¼-in (6.35 mm) aluminum, and up to 100 in. (2.5 meters) per minute in ½-in. (12.7 mm) steel. The combination of extremely high temperature and jetlike action of the plasma produces narrow kerfs and remarkably smooth surfaces, nearly as smooth as can be obtained by sawing. Cut surfaces are often within 2 degrees of vertical. Surface oxidation is nearly eliminated by the cooling effect of the water spray. In addition, the heat-affected zone in the metal is only one-third to one-quarter as large as that produced by oxyfuel cutting. Heat-related distortion is virtually eliminated.

Integration with CNC (computer numerically controlled) machines provides for fast, clean, accurate cutting. Transferred-type torches are usually used for cutting metals, while the nontransferred type must be used for cutting the poorer-conductivity nonmetals.

Inexpensive nitrogen is the primary gas used to cut all types of metal. For cutting thick pieces, 5 to 6 in. (125 to 150 mm) thick, an argon–hydrogen mixture is used to provide a deeper-penetrating arc.

FIGURE 34-25 Cutting sheet metal with a plasma torch. (Courtesy GTE Sylvania.)

Metallurgical and Heat Considerations in Flame and Arc Cutting

Flame and arc processes expose metals to localized high temperatures and could produce harmful metallurgical effects. In most cases, little or no difficulty is experienced. In others, however, definite steps should be taken to avoid or overcome the harmful consequences.

In low-carbon steels, with less than 0.25% carbon, oxyacetylene cutting usually causes no serious metallurgical effects. Although there is often some minor hardening in a thin zone near the cut (the heat-affected zone), and a small amount of grain growth, these effects are usually eliminated if any subsequent welding is done along the cut edges. However, in steels of higher carbon content, these effects can be quite serious, and preheating and/or postheating may be required. For alloy steels, additional consideration should be given to the effects of the various alloy elements. Chromium, molybdenum, and tungsten can be particularly detrimental to cutting.

From a heat-effect viewpoint, arc air cutting is about the same as arc welding. If welding is to follow cutting, the welding heat effects will be of major consequence and no special cutting precautions need be employed. However, if no subsequent welding is to be performed, consideration should be given to the potential heat effects of the process and their consequence with regard to the anticipated loads and stresses.

Plasma arc cutting is so rapid and the heat is so specifically localized that the heat-affected zone is usually less than $3/32$ in. (2.4 mm), and the original properties of the metal are generally modified only within $1/16$ in. (1.6 mm) of the cut.

All of these processes produce some residual stresses, with the cut surface generally in tension. Except in the case of thin sheet, gas or arc cutting should not produce warping of the product. However, if subsequent machining removes only a portion of the cut surface, or an insufficient depth from the cut, the resulting imbalance in residual stresses can cause the material to warp. Thus, if subsequent machining is to be done, it may be necessary to remove all cut surfaces to a substantial depth to assure dimensional stability. Machining cuts should be sufficiently deep to get below the hardened surface in one pass and avoid dulling the tools.

All flame- or arc-cut edges are rough to varying degrees and thus contain geometrical notches that can act as stress raisers and reduce the endurance or fracture strength. If cut edges are to be subjected to high or repeated tensile stressing, the cut surface and the heat-affected zone should be machined away, or at least given a stress-relief heat treatment.

Review Questions

1. What sorts of problems plagued the early attempts to develop arc welding?
2. What are the three basic types of current and polarity that are used in arc welding?
3. What is the difference between a consumable electrode and a nonconsumable electrode? For which processes does a filler metal have to be added by a separate mechanism?
4. What are the three modes of metal transfer that can occur during arc welding?
5. What are some of the process variables that must be specified when setting up an arc-welding process?
6. What are some of the roles of the electrode coatings in shielded metal arc welding?

7. How are welding electrodes commonly classified?
8. Why are shielded metal arc electrodes commonly baked just prior to welding?
9. What is the function of the slag coating that forms over a shielded metal arc weld?
10. What benefit is obtained by placing iron powder in the coating of a welding electrode to be used in welding ferrous metals?
11. Why are shielded metal arc electrodes generally limited in length and the process restricted to intermittent operation?
12. What are some of the commonly used shielding gases in the gas tungsten arc process?
13. What methods can be employed to increase the rate at which filler metal is deposited during gas tungsten arc welding?

14. What are some of the attractive features of gas tungsten arc welding?

15. What is the primary attraction of gas tungsten arc spot welding?

16. Why might gas metal arc welding be preferred over shielded metal arc for production welding?

17. What are some of the benefits that can be obtained by the reduced heat input and temperatures of the pulsed arc technique?

18. What is the advantage of placing the flux in the center of an electrode (flux-cored arc welding) as opposed to the outside (shielded metal arc welding)?

19. What are some of the functions of the flux in submerged arc welding?

20. From a production viewpoint, what are some of the attractive characteristics of submerged arc welding? Major limitations?

21. What is the primary objective in bulk welding?

22. How is the heating of the workpiece different in plasma arc welding compared to the other arc welding techniques?

23. What are some of the attractive features of plasma arc welding?

24. What is the primary difference between plasma arc welding and plasma arc cutting?

25. Why is stud welding a rather specific-use arc-welding technique?

26. What is the function of the ceramic ferrule placed over the end of the stud in stud welding?

27. What is the function of jigs and positioners in welding?

28. Why might a metal such as stainless steel be difficult to cut by oxygen-arc cutting?

29. How can gas metal arc welding and gas tungsten arc welding be converted to cutting?

30. What technique is used to constrict the arc in plasma arc cutting so as to produce a narrow, controlled cut?

31. In what ways is plasma arc cutting more attractive than oxy-fuel cutting?

32. Why is plasma arc cutting an attractive means of cutting metals that do not readily oxidize?

33. Why are nontransferred-type torches required when using plasma arc cutting to cut nonmetallic materials?

34. Why is there little concern about heat-related effects on cut edges that will be subsequently welded?

35. in what way might the residual stresses induced during cutting operations become objectionable?

36. Why might it be wise to finish-machine the cut edge of a highly stressed machine part?

Resistance Welding

In resistance welding, both heat and pressure are used to effect coalescence. The heat is the consequence of the electrical resistance of the workpieces and the interface between them. The pressure is varied throughout the weld cycle. A certain amount of pressure is applied initially to hold the workpieces in contact and thereby control the electrical resistance at the interface. When the proper temperature is attained, the pressure is increased to facilitate coalescence. Because pressure is utilized, coalescence occurs at a lower temperature than that required for oxyfuel gas or arc welding. In fact, in many resistance welding operations, melting of the base metal does not occur. Therefore, resistance welding processes could well be considered as a form of solid-state welding, though they are not officially classified as such by the American Welding Society.

In some of the resistance welding processes, additional pressure can be applied immediately after coalescence to provide a certain amount of forging action, with some accompanying grain refinement. Also, some additional heating can be employed after welding to provide tempering and/or stress relief.

Usually the required temperature can be attained and coalescence achieved in a few seconds or less. Consequently, resistance welding is a very rapid and economical process, extremely well suited to automated manufacturing.

Heating. The heat for resistance welding is obtained by passing a large electrical current through the workpieces for a short period of time. The amount of heat input can be determined by the basic relationship:

$$H = I^2RT$$

where H is the total heat input, I is the current, R is the electrical resistance of the circuit, and T is the length of time in which current is flowing. It is important

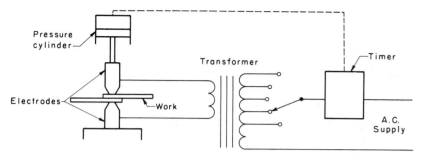

FIGURE 35-1   The fundamental resistance-welding circuit.

to note that the workpieces form part of the electrical circuit, as illustrated in Figure 35-1, and that the total resistance between the electrodes consists of three components:

1. the resistance of the workpieces.
2. the contact resistance between the electrodes and the work.
3. the resistance between the surfaces to be joined, known as the *faying* surfaces.

Because it is desirable to have the maximum temperature occur where the weld is to be made, it is essential to keep resistances (1) and (2) as low as possible with respect to resistance (3). For materials having low electrical resistance, such as aluminum and copper, this condition is difficult to achieve, and they require much larger currents and more attention to interface conditions than does steel.

The resistance of the workpieces is determined by the type and thickness of the metal. It is usually much less than the other two resistances because of the larger area involved and the relatively high electrical conductivity of most metals. The resistance between the work and the electrodes can be minimized by using electrode materials that are excellent electrical conductors, by controlling the shape and size of the electrodes, and by using proper pressure between the work and the electrodes. However, since any change in the pressure between the work and the electrodes also changes the pressure between the faying surfaces, only limited control of the electrode-to-work resistance can be obtained in this manner.

Finally, the resistance between the faying surfaces is a function of (1) the quality of the surfaces; (2) the presence of nonconductive scale, dirt, or other contaminants; (3) the pressure; and (4) the contact area. These factors must all be controlled to obtain uniform results.

As indicated in Figure 35-2, the objective of resistance welding is to simultaneously bring both of the faying surfaces to the proper temperature while keeping the remaining material and the electrodes relatively cool. The electrodes are usually water-cooled to keep their temperature low and to aid in keeping them in proper condition. When metals of different thickness or different conductivity are to be welded, they can generally be brought to the proper temperature in a simultaneous manner by using a larger electrode or one with higher conductivity against the thicker, or higher-conductivity, material.

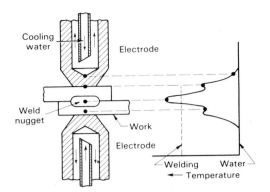

FIGURE 35-2　The desired temperature distribution across the electrodes and the workpieces in lap resistance welding.

Pressure.　Because the pressure in resistance welding promotes a forging action, resistance welds can be made at lower temperatures than welds made by other processes. However, the control of both the magnitude and timing of the pressure is very important. If too little pressure is used, the contact resistance is high and surface burning or pitting of the electrodes may result. On the other hand, if excessive pressure is applied, molten or softened metal may be expelled from between the faying surfaces or the work may be indented by the electrodes. Ideally, a moderate pressure should be applied prior to and during the passage of the welding current to establish the proper resistance at the interface. The pressure should then be increased considerably just as the proper welding heat is attained. This completes the coalescence and forges the weld to produce a fine grain structure.

On small, foot-operated machines, only a single spring-controlled pressure is used. On larger, production-type welders, the pressure is generally applied through air or hydraulic cylinders that are controlled and timed automatically.

Current Control.　With the surface conditions held constant and the pressure controlled, the temperature in resistance welding is then regulated by controlling the magnitude and timing of the welding current. Very precise and sophisticated controls are available for this purpose.

The welding current is usually obtained from a "step-down" transformer. On small machines, the magnitude is controlled through taps on the primary coil of the transformer or by an autotransformer that varies the primary voltage supplied to the main transformer. On larger machines, several other methods are used. In *phase-shift control*, the magnitude and wave shape of the primary current are altered. With *slope control*, the current is permitted to rise gradually to full magnitude in from 3 to 25 cycles.

In large, production-type welders, the magnitude, duration and timing of both current and pressure are carefully programmed. Figure 35-3 shows a relatively simple current and pressure cycle for resistance welding that includes forging and postheating operations.

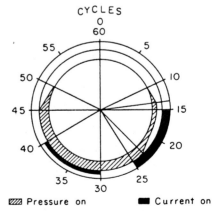

FIGURE 35-3　Typical current and pressure cycle for resistance welding. The cycle includes forging and postheating operations.

Power Supply.   The magnitude of the current required for resistance welding is so great (up to 100,000 amps) that special types of circuits are employed in most machines to reduce the load on power lines. Single-phase circuits are generally used only in smaller machines. Larger machines employ three-phase circuits. In one type, three-phase power is rectified and the energy stored in a special transformer. When the DC flow through the transformer is interrupted, the collapse of the field provides the high current flow in the secondary circuit. Many resistance welders use DC welding current, obtained through solid-state rectification of three-phase power. Such machines reduce the current demand per phase, give a balanced load, and produce excellent welds.

Resistance Spot Welding.   *Resistance spot welding* (RSW) is the simplest and most widely used form of resistance welding. As shown in Figure 35-4, the overlapping work is positioned between water-cooled electrodes, which have reduced areas at the tips to produce welds that are usually from ¹⁄₁₆ to ½ in. (1.5 to 13 mm) in diameter. After the electrodes are closed on the work, the controlled cycle of pressure and current is applied, producing a weld at the metal interface. The electrodes then open and the work is removed.

A satisfactory spot weld, such as the one shown in Figure 35-5, consists of a *nugget* of coalesced metal formed between the faying surfaces. There should be little indentation of the metal under the electrodes. The strength of the welds should be such that, in a tensile or tear test, the weld will remain intact and failure will occur in the heat-affected zone surrounding the nugget, as illustrated in Figure 35-6. If proper current density and timing, electrode shape, electrode pressure, and surface conditions are maintained, sound spot welds can be obtained with excellent consistency.

Spot-Welding Machines.   A variety of spot-welding equipment is available to meet the various needs of production operations. For light-production work for which complex current–pressure cycles are not required, the simple *rocker-arm machine,* shown in Figure 35-7, is often used. The lower electrode arm is stationary, and the upper electrode, mounted on a pivot arm, is brought down into contact with the work by means of a spring-loaded foot pedal. On larger machines, or machines for larger-volume work, the electrode motion can be

Resistance Welding Processes

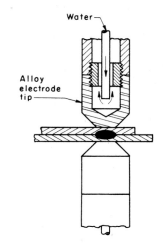

FIGURE 35-4   The arrangement of the electrodes and the work in spot welding.

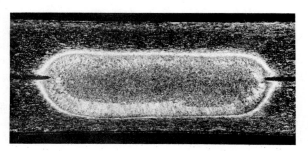

FIGURE 35-5   A spot-weld nugget between two sheets of 0.05 in. (1.3 mm) aluminum alloy. The radius of the upper electrode was greater than that of the lower electrode. (Courtesy Lockheed Aircraft Corporation.)

FIGURE 35-6 Tear test of a satis-factory spot weld, showing how fail-ure occurs outside of the weld.

FIGURE 35-7 Foot-operated rocker-arm, spot-welding machine. (Courtesy Sciaky Bros., Inc.)

obtained through an air cylinder or an electric motor. Rocker-arm machines are available with throat depths up to about 48 in. (1200 mm) and with transformer capacities up to 50 kVa. They are used primarily on steel.

Most large spot welders, and those used at high production rates, are of the *press-type,* as shown in Figure 35-8. On these machines, the movable electrode has a straight-line motion, provided by an air or hydraulic cylinder, and complex pressure cycles can be programmed and controlled. Capacities up to 500 kVa and 60-inch (1500 mm) throat depth are common. Special-purpose press-type spot welders employ multiple welding heads and can be used in high-volume mass production. Some, such as the one shown in Figure 35-9, can make up to 200 individual spot welds in under 6 seconds.

The application of spot welding is greatly extended through the use of *portable spot-welding guns,* such as those shown in Figure 35-10. Each gun is connected to the power supply and control unit by flexible air hoses, electrical cables, and water hoses for cooling (where required). Such equipment permits the welding unit to be brought to the work, greatly extending the use of spot welding in applications wherein the work is too large to be positioned on a welding machine.

Portable welding guns are frequently installed on industrial robots, which can be programmed to position the gun at the desired (three-dimensional) location and produce spot welds automatically, without the need for an operator. Such an application is now common in the automotive industry.

Spot-Welding Metals.   One of the greatest advantages of spot welding is the fact that virtually all of the commercial metals can be spot-welded, and most of them can be spot-welded to each other. In only a few cases do the welds tend to be brittle. Table 35.1 shows some of the combinations of metals that can be spot-welded satisfactorily.

Although the majority of spot welding is done on wrought sheet, other forms of metal can also be spot-welded. Sheets can be spot-welded to rolled shapes and steel castings, and some types of die castings can be welded without diffi-culty. Except for aluminum, most metals require no special preparation, except to be sure that the surface is free of corrosion and is not badly pitted. For best results, aluminum and magnesium should be cleaned immediately prior to weld-ing by mechanical or relatively simple chemical means. Metals that have high electrical conductivity require clean surfaces to assure that the electrode-to-metal resistance is low enough for adequate temperature to be developed in the metal itself. Silver and copper are difficult to weld because of their high thermal conductivity. However, by the use of proper procedures, many copper alloys can be readily resistance-welded. Water cooling adjacent to the spot-weld area can be used to assure that adequate welding temperature is obtained only in the desired spot area.

The practical limit of thicknesses that can be spot-welded by ordinary processes is about 1/8 in. (3 mm), where each piece is of the same thickness. When the thicknesses vary, a thin piece can be easily welded to another piece

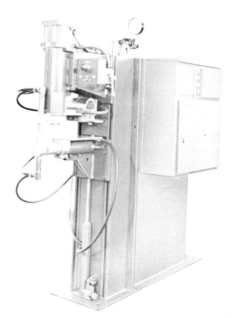

FIGURE 35-8 Single phase, air-operated, press-type resistance welder with microprocessor control. (Courtesy Sciaky Bros., Inc.)

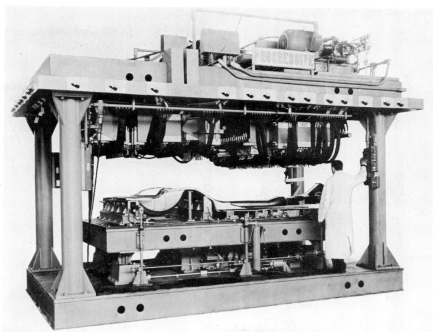

FIGURE 35-9 A large press-type spot welder that has 50 transformers and makes 200 spot welds on an automobile underbody in less than 6 seconds. (Courtesy Progressive Welder Sales Company.)

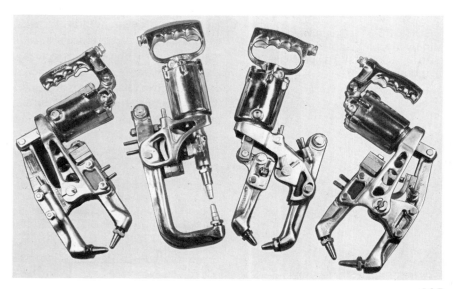

FIGURE 35-10 Some forms of portable spot-welding guns. (Courtesy Progressive Machinery Corporation.)

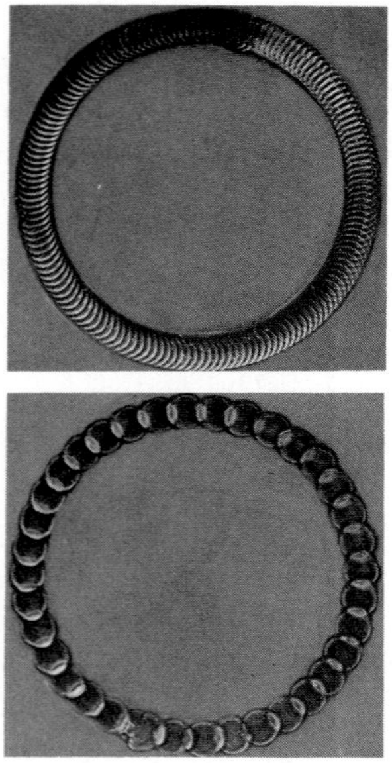

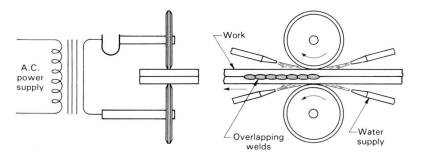

FIGURE 35-11 Seam welds made with overlapping spots of varied spacing. (Courtesy Taylor-Winfield Corporation.)

TABLE 35.1 **Metal Combinations That Can Be Spot-Welded**

Metal	Aluminum	Brass	Copper	Galvanized Iron	Iron (Wrought)	Monel	Nickel	Nickel Silver	Nichrome	Steel	Tin Plate	Zinc
Aluminum	×										×	×
Brass		×	×	×	×	×	×	×	×	×	×	×
Copper		×	×	×	×	×	×	×	×	×	×	×
Galvanized iron		×	×	×	×	×	×	×	×	×	×	
Iron (wrought)		×	×	×	×	×	×	×	×	×	×	
Monel		×	×	×	×	×	×	×	×	×	×	
Nickel		×	×	×	×	×	×	×	×	×	×	
Nickel silver		×	×	×	×	×	×	×	×	×	×	
Nichrome		×	×	×	×	×	×	×	×	×	×	
Steel		×	×	×	×	×	×	×	×	×	×	
Tin plate	×	×	×	×	×	×	×	×	×	×	×	
Zinc	×	×	×									×

that is much thicker than ⅛ in. Pushing the limits of the process, two ½-in. (12.7 mm) steel plates have been successfully spot-welded in an effort to replace large amounts of riveting.

Resistance Seam Welding. *Resistance seam welds* (RSEW) are made by two distinctly different processes. In most cases, when the weld is between two sheets of metal, the seam is actually a series of overlapping spot welds, as shown in Figure 35-11. The basic equipment is the same as for spot welds, except that two rotating discs are used as electrodes, arranged as shown in Figure 35-12. The metal passes between the electrodes, and timed pulses of current pass through it to form the overlapping elliptical welds. The timing of the welds

FIGURE 35-12 Schematic representation of seam welding.

and the movement of the work must be adjusted so that the workpieces do not get too hot. External cooling of the work, by air or water, is often employed. In a variation of the process, a continuous seam can often be produced by passing a continuous current through the rotating electrodes.

Roll-spot welding is used primarily in the production of liquid- or gas-tight sheet-metal vessels, such as gasoline tanks, automobile mufflers, and heat exchangers. Figure 35-13 shows a typical seam-welding machine. Figure 35-14 shows a complex cycle of electrode force and current that can be programmed into such a machine.

The second type of resistance seam welding is used to make butt welds between metal plates. In this process, the electrical resistance of the abutting metal(s) is utilized, but high-frequency current (up to 450 kHz) is employed in the heating. This confines the heat to the surfaces to be joined and their immediate surroundings. (An alternative method employs high-frequency induction heating.) As they attain the welding temperature, the heated surfaces are pressed together to form the weld.

The most extensive use of resistance butt welding is probably in the manufacture of pipe and tube, as illustrated in Figure 35-15. The process can also be used to produce structural shapes from flat stock, as shown in Figure 35-16. Material from 0.005 in. (0.13 mm) to more than 3.4 in. (19 mm) in thickness can be welded at speeds up to 250 feet (82 meters) per minute. The combination of high-frequency current and high welding speed produces a very narrow heat-affected zone. Almost any type of metal can be welded, including dissimilar metals. The process is particularly attractive for high-conductivity metals, such as aluminum and copper.

Projection Welding.   Two disadvantages of spot welding are that electrode maintenance is a considerable problem and usually only one spot weld is made at a time. If more strength or bonding is required than can be provided by a single spot, then several welds must be made. *Projection welding* (RPW) provides a means of overcoming both of these disadvantages and has become particularly attractive to mass production.

The principle of projection welding is illustrated in Figure 35-17. A *dimple* is embossed on one of the workpieces at the location where a weld is desired.

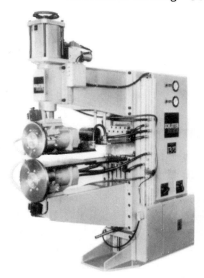

FIGURE 35-13   A typical commercial seam welder. (Courtesy H.A. Schlatter AG.)

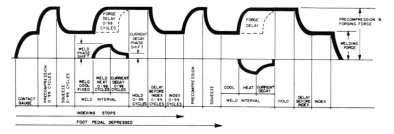

VARIABLE ELECTRODE FORCE — INTERMITTENT DRIVE — SINGLE IMPULSE WITH CURRENT DECAY FOR WELDING LIGHT ALLOYS

FIGURE 35-14   A complex cycle of electrode force and current used in spot and seam welding.

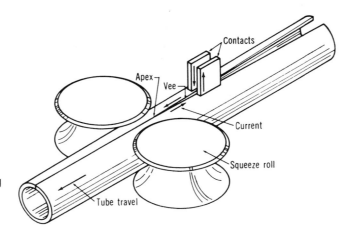

FIGURE 35-15 A method of making a seam weld in pipe by means of a high-frequency current as the heat source.

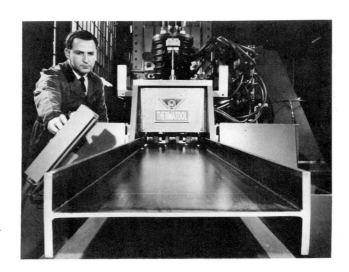

FIGURE 35-16 Fabricating an I-beam from three plates by simultaneous high-frequency resistance welding. (Courtesy AMF Thermatool, Inc.)

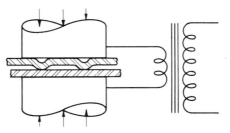

FIGURE 35-17 The principle of projection welding.

The workpieces are then placed between plain, large-area electrodes in a projection welding machine, and pressure and current are applied as in spot welding. Because the contact area between the work and the electrodes is much greater than the area on the end of the dimple, nearly all of the resistance of the circuit is in the dimple, and the heating is concentrated where the weld is desired. As the metal heats and becomes plastic, pressure causes the dimple to flatten and form a weld.

Several projection welds can be made at one time by placing several dimples between the electrodes. The number of projections is limited only by the ability of the machine to provide the required current and pressure. In addition, many spot-welding machines can be converted to projection welding by simply changing the electrodes.

Because the projections are press-formed, they can often be produced during other blanking and forming operations with virtually no additional cost. Another important advantage is that the dimples, or projections, can be made in almost any shape—such as round, oval, or circular—to produce welds of shapes to suit various design purposes. They should be designed, however, so that the weld forms outward from the center of the projection.

Bolts and nuts can be attached to other metal parts by projection welding. Contact is made at a projection that is formed on the bolt or nut; current is applied, and the pieces are pressed together.

Advantages and Disadvantages of Resistance Welding. Resistance welding processes have a number of distinct advantages that account for their wide use, particularly in mass production:

1. They are very rapid.
2. The equipment is semiautomatic or fully automated.
3. They conserve material; no filler metal is required.
4. Skilled operators are not required.
5. Dissimilar metals can be easily joined.
6. A high degree of reliability and reproducibility can be achieved.

Resistance welding also has some disadvantages, the principal ones being

1. The equipment has a high initial cost.
2. There are limitations on the type of joints that can be made (mostly lap joints).
3. Skilled maintenance personnel are required to service the control equipment.
4. For some materials, the surfaces must receive special preparation prior to welding.

Resistance welding is one of the most common methods of high-volume joining. However, because of the rapid heat inputs, short welding times, and rapid quenching by both the base metal and the electrodes, the cooling rates in spot and seam welds can be extremely high. Martensite can form in steels containing more than 0.15% carbon. For these materials, a post-weld heating is generally required to temper the weld.

Review Questions

1. What are the two major roles of applied pressure in resistance welding?
2. What are the three components that comprise the total resistance between the electrodes?
3. What measures can be taken to reduce the resistance between the electrodes and the workpieces?
4. What design features can be altered to permit the joining of different thicknesses or metals of different conductivity?
5. What are the possible consequences of too little pressure during the cycle?
6. What problems can occur if the weld pressure is too great?
7. What is the simplest and most widely used form of resistance welding?
8. What is the typical size of a spot-weld nugget?
9. What are the two basic types of stationary spot-welding machines?

10. What is the major advantage of spot-welding guns as opposed to the press-type machines?

11. What special measures should be taken when resistance-welding aluminum?

12. What is the practical limit of the thicknesses of material that can be readily spot-welded?

13. What is the difference between roll-spot welding and continuous seam welding?

14. What is the benefit of using high-frequency current for producing butt-welded joints?

15. What two disadvantages of spot welding can be overcome by using the projection approach?

16. What are some of the attractive features of resistance welding when viewed from a manufacturing standpoint?

17. What problems might be encountered when spot welding medium- or high-carbon steels?

18. What limitations exist on the type of joints that can be made by resistance welding?

Other Welding and Related Processes

As indicated in the listings of Figures 33-1 and 33-5, there are a number of very useful welding and cutting processes that utilize sources of heat other than an oxyfuel flame, electric arc, or electrical resistance. Although some are quite old, others are among the newest of the manufacturing processes.

Forge Welding. *Forge welding* (FOW) is the most ancient of the welding processes, and a review of it is of both historical and practical value as it helps us to understand how and why the modern welding practices were developed. The armor makers of ancient times occupied a position of prominence in their society, largely because of their ability to join two pieces of metal into a single strong product. More recently, the proverbial village blacksmith was the master of the art of forge welding. With his hammer and anvil, and a high degree of skill, he could join pieces of metal to form a wide variety of products.

The blacksmith used a charcoal forge as his source of heat. Pieces that were to be welded were heated to their forging temperature and the ends were scarfed by hammering to permit them to be fitted together without excessive thickness. The ends were again heated to elevated temperature and were dipped into some borax, which acted as a flux. Heating was then continued until the blacksmith judged by the color that the workpieces were at the proper temperature for welding. They were then withdrawn from the heat and struck (either on the anvil or by the hammer) to knock off any scale and impurities. The ends to be joined were overlapped on the anvil and hammered to the degree necessary to produce an acceptable weld.

Thus, the competent blacksmith was able to perform a type of solid-state welding and could often produce a welded joint that was fully as strong as the original metal. However, because of the crudeness of his heat source, the un-

Solid-State Welding Processes

certainty of the temperature, and the difficulty of maintaining metal cleanliness, a great amount of skill was required and the results were somewhat variable. The quality of deformation welds clearly depends upon the material temperature, surface cleanliness, and the amount of deformation.

Forge-Seam Welding.   Although forge welding, as performed by the blacksmith, is seldom done today, a large amount of *forge-seam welding* is used in the manufacture of pipe, as discussed in Chapter 18. A heated strip of steel is formed into a cylinder and the edges are welded together in either a lap or butt joint. Welding is the result of pressure and deformation as the metal is pulled through a conical welding bell or passed between welding rolls. Figures 18-28 and 18-29 show schematic diagrams of these modern-day forge-welding processes.

Cold Welding.   *Cold welding* is a unique variation of forge welding that uses *no heating* and only a single blow or application of pressure. The surfaces to be joined are first cleaned, usually by wire brushing. After being placed in contact, they are subjected to localized pressures sufficient to cause about 30% to 50% localized cold working. A solid-state bond is produced, as shown in Figure 36-1. While some heating of the metal is certain to occur due to the severe, localized deformation, the high localized pressure and deformation is the primary factor in producing coalescence. The use of the process is generally confined to the joining of small parts, such as the electrical connections shown in Figure 34-1.

Roll Welding or Roll Bonding.   In the *roll welding* or *roll bonding* (ROW) process, two or more sheets of metal are joined by simultaneously passing them

FIGURE 36-1   (Left) Section through a cold weld. (Right) Small parts joined by cold welding. (Courtesy Koldweld Corporation.)

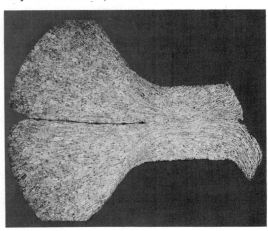

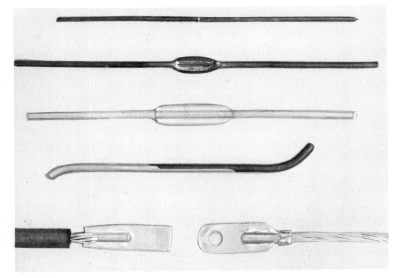

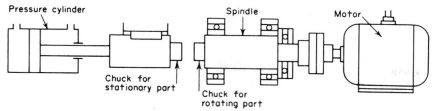

FIGURE 36-2   Schematic diagram of the equipment used for friction welding. (Courtesy Materials Engineering.)

through a rolling mill. The process can be performed either hot or cold and can produce extremely strong bonds between similar or dissimilar metals, as evidenced by its use in producing the material for standard United States coinage.

By coating portions of one surface with a material that prevents bonding, rolling can be used to produce sheets of material that are bonded (or not bonded) in selected regions. A common example of the use of this technique is in refrigerator freezer panels (or solar-energy collector panels), where the no-bond regions are expanded to produce the flow path for the coolant (or heat-transmitting fluid). In this way, inexpensive sheet metal can be used to fabricate components that once required the use of more costly tubing.

Friction Welding and Inertia Welding.

The heat for *friction welding* (FRW) comes from mechanical friction between two abutting pieces of metal that are held together while one rotates and the other is stationary. In the friction-welding process, the moving part is held in a motor-driven collet while the stationary part is held against it at a selected level of pressure. Friction quickly generates enough heat to raise the abutting surfaces to the welding temperature. As soon as this temperature is reached, rotation is stopped and the pressure is maintained or increased to complete the weld. Figure 36-2 shows a schematic of friction welding.

Inertia welding differs from friction welding in that the moving piece is now attached to a rotating flywheel, as shown in Figure 36-3. The flywheel is brought to a specified rotational speed and is then separated from the driving motor. The rotating assembly is then pressed against the stationary member and the kinetic energy of the flywheel is converted into frictional heat. The weld is formed as the flywheel stops its motion and the pieces remain pressed together. This method produces extremely repeatable conditions (consistent welds) and can be readily automated. Figure 36-4 shows an inertia welding machine and a plot of rotational speed (surface velocity), torque, and degree of upset throughout the process.

In both friction and inertia welding, the total cycle time to produce a weld is usually less than 25 seconds and the actual time of heating and welding is about 2 seconds. No material is melted; the process is totally solid state. Because of the short period of heating and the limited time for heat to flow away from the joint, the weld and heat-affected zones are very narrow. Surface impurities on the joint faces tend to be displaced radially into a small upset flash that can be removed after welding if desired. Because virtually all of the energy is converted to heat, the process is very efficient, and it can be used to join many metals or

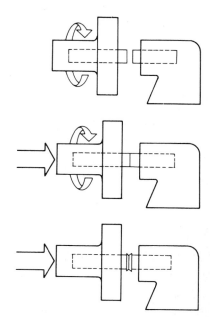

FIGURE 36-3   Schematic representation of the three steps in inertia welding.

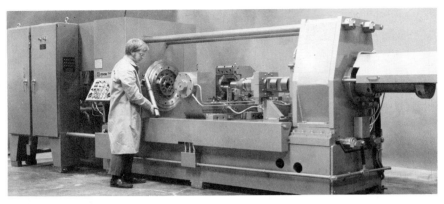

FIGURE 36-4 (Top) Inertia-type friction-welding machine and welded part. (Bottom) Relationship between surface velocity (speed), torque, and upset throughout the inertia welding process. (Courtesy Production Technology, Inc.)

combinations of dissimilar metals. Since grain size is refined during hot working, the strength of the weld is almost the same as the base metal.

Unfortunately, the processes are restricted to joining round bars or tubes of the same size, or connecting bars or tubes to flat surfaces. The ends of the workpieces must be cut true and be fairly smooth. Figure 36-5 shows some typical applications.

Ultrasonic Welding. *Ultrasonic welding* (USW) is a solid-state process in which coalescence is produced by the localized application of high-frequency

FIGURE 36-5 Some typical friction-welded parts. (Left) Impeller made by joining a chrome-moly steel shaft to a nickel-steel casting. (Center) Stud plate with two mild steel studs joined to a square plate. (Right) Tube component where a turned segment is joined to medium-carbon steel tubing. (Courtesy Newcor Bay City, Div. of Newcor, Inc.)

(10,000 to 200,000 cps) vibratory energy to surfaces that are held together under pressure. Although there is some increase in temperature at the faying surfaces, it is always far below the melting point of the materials. Instead, it appears that the rapid reversals of stress along the contact interface facilitate coalescence by breaking up and dispersing the surface contaminants.

Figure 36-6 depicts the basic components of the ultrasonic welding process. The ultrasonic transducer is essentially the same as that employed in ultrasonic machining, depicted schematically in Figure 32-14. It is coupled to a force-sensitive system that contains a welding tip on one end. The pieces to be welded are placed between this tip and a reflecting anvil, thereby concentrating the vibratory energy within the work. Either stationary tips (for spot welds) or rotating discs (for seam welds) can be employed.

Ultrasonic welding is restricted to the joining of thin materials—sheet, foil, and wire—or the attaching of thin sheets to heavier structural members. The maximum thickness is about 0.1 in. (2.5 mm) for aluminum and 0.04 in. (1.0 mm) for harder metals. As indicated in Table 36.1, the process is particularly valuable because of the number of dissimilar metals that can be readily joined. Because the temperatures are low and no arcing or current flow is involved, the process can be applied to sensitive electronic components. Intermetallic compounds seldom form and there is no contamination of the weld or surrounding area. The equipment is simple and reliable, and only moderate skill is required of the operator. The required surface preparation is less than for most competing processes (such as resistance welding) and less energy is needed to produce a weld. Typical applications include joining the dissimilar metal in bimetallics, making microcircuit electrical contacts, welding refractory or reactive metals, bonding ultrathin metal, and encapsulating explosives or chemicals.

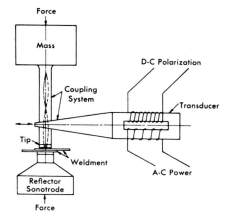

FIGURE 36-6 Schematic diagram of the equipment used in ultrasonic welding.

TABLE 36.1 Metal Combinations Weldable by Ultrasonic Welding

Metal	Aluminum	Copper	Germanium	Gold	Molybdenum	Nickel	Platinum	Silicon	Steel	Zirconium
Aluminum	×	×	×	×	×	×	×	×	×	×
Copper		×		×		×	×		×	×
Germanium			×	×		×	×	×		
Gold				×		×	×	×		
Molybdenum					×	×			×	×
Nickel						×	×		×	×
Platinum							×		×	
Silicon										
Steel									×	×
Zirconium										×

Diffusion Welding.   *Diffusion welding* (DFW) or *diffusion bonding* occurs when properly prepared surfaces are maintained in contact under sufficient pressure and time at elevated temperature. In contrast to the deformation welding methods, plastic flow is limited and the principal bonding mechanism is diffusion. Under low pressure and elevated temperature, a well-prepared interface can be viewed as a planar grain boundary with intervening voids and impurities. Atomic diffusion (the result of time at elevated temperature) then promotes the necessary void shrinkage and grain boundary migration to form a metallurgical bond.

The diffusion-welding process can be controlled through the variation of surface condition and preparation, temperature, time at temperature, pressure, and the possible use of intermediate material layers which can promote diffusion or prevent the formation of undesirable intermetallic compounds. Some intermediate layers melt to form a temporary liquid which significantly accelerates the rate of diffusion.

Explosion Welding.   *Explosion welding* (EXW) is used primarily for bonding sheets of corrosion-resistant metals to heavier plates of base metal (a *cladding* operation), particularly when large areas are involved. An explosive material, usually in the form of a sheet, is placed on top of the two layers of metal and detonated in a progressive fashion. A compressive stress wave, on the order of hundreds of thousands of pounds per square inch (thousands of megapascals), progresses across the surface of the plates, so that a small open angle is formed between the two colliding surfaces. Surface films are liquified or scarfed off of the metals and are jetted out of the interface. The clean metal surfaces then coalesce under the high pressure. The result is a high-strength cold weld with a wavy configuration at the interface. Like many of the other solid-state welding processes, numerous combinations of dissimilar metals can be joined.

Other Welding and Cutting Processes

Thermit Welding.   The heating and coalescence in *thermit welding* (TW) is produced by superheated molten metal and slag obtained from the reaction between iron oxide and aluminum. The term *thermit* is used to refer to a mechanical mixture of about one part finely divided aluminum and three parts iron oxide. When this mixture is ignited by a magnesium fuse (the ignition temperature is about 2100°F or 1150°C), it reacts according to the following chemical equation:

$$8 \; Al \; + \; 3 \; Fe_3O_4 \rightarrow 9 \; Fe \; + \; 4 \; Al_2O_3 \; + \; heat$$

A temperature of over 5000°F (2750°C) is produced in about 30 seconds, superheating the molten iron, which flows into a prepared joint to provide both heat and filler metal. Runners and risers must be provided, as in a casting, to channel the molten metal and compensate for solidification shrinkage.

The thermit process is extremely old and has been replaced to a large degree by alternative methods. Nevertheless, it is still very effective and can be used to join thick sections of ferrous material, particularly in remote locations or when more sophisticated welding equipment is not available. One such application is the field repair of large steel castings which have broken or cracked.

Electroslag Welding.   *Electroslag welding* (ESW), illustrated in Figure 36-7, is a very effective process for welding thick sections of steel plate. Heat is derived from the passage of electrical current through a liquid slag. Resistance heating within the slag raises the temperature to around 3200°F (1760°C). Because there is no arc involved, the process is entirely different from submerged arc welding, and the electrical resistance of the metal being welded plays no part in producing the heat. Instead, the molten slag melts the edges of the pieces that are being joined, as well as the continuously fed electrodes that supply the filler metal. Multiple electrodes are often required to provide an adequate supply of filler and maintain the molten pool. Under normal operating conditions, there is a 2½-in. (65 mm)-deep layer of molten slag, which serves to protect and cleanse the underlying ½- to ¾-in. (12 to 20 mm)-deep pool of molten metal. These liquids are confined to the gap by means of sliding water-cooled plates.

Since the easiest conditions for maintaining a deep slag bath exist in vertical joints, the process is most frequently used for this application. Circumferential joints can also be produced in large pipe by using special curved slag-holder plates and rotating the pipe to maintain the area where welding is occurring in a vertical position.

Because very large amounts of weld metal and heat can be supplied, electroslag welding is the best of all the welding processes for making welds in thick plates. The thickness of the plates can vary from ½ to 36 in. (13 to 900 mm) and the length of the weld (amount of vertical travel) is almost unlimited. Edge preparation is minimal, requiring only squared edges separated by 1 to 1½ in. (25 to 35 mm). Applications have included building construction, machine manufacture, heavy pressure vessels, and the joining of large castings and forgings.

Control of the solidification is vitally important to obtaining a good weld, since slow cooling tends to produce a coarse grain structure. The associated cracking tendencies can be suppressed by promoting a wide shallow pool of molten metal. This can be accomplished through control of the current, voltage, slag depth, number of electrodes, and electrode extension. A large heat-affected zone and extensive grain growth are also common features of the process. The long thermal cycle, however, serves to minimize residual stresses, distortion, and cracking in the heat-affected zone. If good fracture resistance is desired, subsequent heat treatment of the welded structure may be necessary.

Electron Beam Welding.   *Electron beam welding* (EBW) is a fusion welding process in which heating results from the impingement of a beam of high-velocity electrons on the metal to be welded. Originally developed for obtaining ultra-high-purity welds in reactive and refractory metals, its unique qualities have led to substantial use in numerous applications.

The electron optical system for the process is shown in Figure 36-8. A high-voltage current heats a tungsten filament to about 4000°F (2200°C), causing it to emit high-velocity electrons. By means of a control grid, accelerating anode, and focusing coils, the electrons are collected into a concentrated beam and focused onto the workpiece in a spot from 1/32 to 1/8 in. (0.8 to 3.2 mm) in

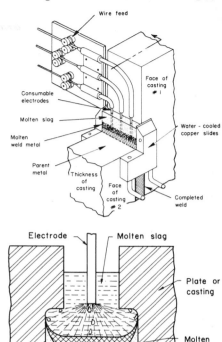

FIGURE 36-7   (Top) Arrangement of equipment and workpieces for making a vertical weld by the electroslag process. (Bottom) Section through the workpieces and the weld during the making of an electroslag weld.

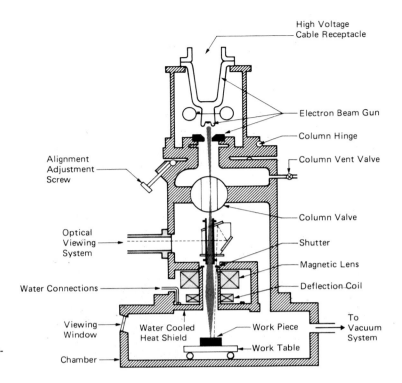

FIGURE 36-8 Schematic diagram of the electron beam welding process. (Courtesy *American Machinist.*)

diameter. Since electrons cannot travel well through air, the beam must be generated and focused in a very high vacuum, typically at pressures of 1×10^{-4} mm of Hg (.01 Pa) or less.

In many operations, the workpiece is also enclosed in the high-vacuum chamber and must be positioned and manipulated in this vacuum. When welding under these conditions, the vacuum assures degasification and decontamination of the molten weld metal, and very high quality welds are obtained. However, the size of the vacuum chamber tends to impose serious limitations on the size of the workpiece that can be accommodated, and the need to break and reestablish the high vacuum as pieces are inserted and removed places a considerable restriction on productivity. As a consequence, electron beam welding machines have been developed to operate at pressures considerably higher than those required for beam generation. Some permit the workpiece to remain outside of the vacuum chamber entirely, the beam emerging through a small orifice in the vacuum chamber to strike the adjacent workpiece. High-capacity vacuum pumps compensate for the leakage through the orifice. Although these machines offer more production freedom, the penetration of the beam and the depth-to-width ratio of the molten region are considerably reduced as the pressure increases (that is, they produce shallower, wider welds).

In general, two ranges of voltage are employed in electron beam welding. High-voltage equipment employs 50 to 100 kilovolts and provides a smaller spot size and greater penetration than does the lower-voltage type, which uses from 10 to 30 kilovolts. Because of their high electron velocities, the high-voltage units emit considerable quantities of harmful X-rays and thus require expensive shielding and indirect viewing systems for observing the work. The X-rays produced by the low-voltage machines are sufficiently soft that they are absorbed by the walls of the vacuum chamber. They are less critical in adjustment and the work can be viewed directly through viewing ports.

Materials that are difficult to weld by other processes, such as zirconium, beryllium, and tungsten, can be welded successfully by electron beam welding, but the weld configuration should be simple and preferably flat. As shown in Figure 36-9, very narrow welds can be obtained with remarkable penetrations. The high power and heat concentrations can produce fusion zones with depth-to-width ratios of 25:1 with low total heat input, low distortion, and a very narrow heat-affected zone. Heat-sensitive materials can be welded without damage to the base metal. High welding speeds are common; no filler metal is required; the process can be performed in all positions; and preheating or post-heating is generally unnecessary.

On the negative side, the equipment is quite expensive and extensive joint preparation is required. Because of the deep and narrow weld profile, joints must be precisely aligned over the entire length of the weld. Machining and fixturing tolerances are often quite demanding. In addition, the vacuum requirements tend to limit production rate, and the size of the vacuum chamber may restrict the size of the workpiece that can be welded.

The electron beam process is best employed when extremely high quality welds are required or where other processes will not produce the desired results. Nevertheless, its unique capabilities have resulted in its routine use in a number of applications, particularly in the automotive and aerospace industries.

FIGURE 36-9 (Left to right) Electron beam welds in 7079 aluminum, thick stainless steel, and a multiple-tier weld in stainless steel tubing. (Courtesy Hamilton Standard Division of United Aircraft Corp.)

Laser Beam Welding. The heat source in *laser beam welding* (LBW) is a focused laser beam, usually providing power intensities in excess of 10 kilowatts per square centimeter. The high-intensity beam provides a very thin column of vaporized metal with a surrounding liquid pool. As the laser advances, the liquid flows into the channel to produce a weld with depth-to-width ratio generally greater than 4:1. Laser beam welding is most effective for simple fusion welds without filler metal, but filler metal can be added if necessary.

The deep-penetration welds produced by lasers are similar to electron beam welds, but the laser beam technique offers several distinct advantages:

1. A vacuum environment is not required.
2. No X-rays are generated.
3. The laser beam is easily shaped and directed with reflective optics.
4. Because only a light beam is involved, there does not need to be any physical contact between the workpieces and the welding equipment. The beam can pass through transparent materials, permitting welds to be made inside transparent containers.

Because the laser beams are highly concentrated sources of energy, laser welds are usually small, often less than 0.001 in. (0.025 mm). While the power intensity is quite high, the weld time is extremely small and the total heat input is often in the range of 0.1 to 10 joules. Thus, laser beam welds are quite attractive to the electronics industry for applications such as connecting leads on small electronic components and integrated circuits. Lap, butt, tee, and cross-wire configurations can all be used. It is even possible to weld wires without removing the polyurethane insulation. The laser simply evaporates the insulation and completes the weld with the internal wire. Figure 36-10 shows several types of laser beam welds.

The equipment required for laser beam welding is quite costly but is generally designed for operation by semiskilled workers. Most industrial lasers are of the CO_2 variety and consume considerable amounts of power. In addition, reflected or scattered laser beams can be quite dangerous to human eyes, even at great distances from the welding site. Eye protection is a must.

FIGURE 36-10 (Left and center) Small electronic welds made by laser welding. (Courtesy Linde Division, Union Carbide Corporation.), (Right) Laser butt weld of 0.125 in. (3mm) stainless steel, made at 60 in. per minute (1.5 meters per minute) with a 1250 watt laser. (Courtesy Coherent, Inc.)

Laser Beam Cutting. *Laser beam cutting* (LBC) uses the intense heat from a laser beam to melt and/or evaporate the material being cut. Any known material can be cut by this process. For some of the nonmetallic materials, the mechanism is purely evaporation, but for most metals the material is simply melted and interacts with a flow of gas—inert gas simply to blow away the molten metal and provide a smooth, clean kerf, or oxygen to speed the process through oxidation. The temperature may well be in excess of 20,000°F (11,093°C). Cutting speeds as high as 1000 in. (25 meters) per minute are not uncommon in nonmetals, and even tough-to-cut steels can be cut at 20 in. (500 mm) per minute or more.

Clean, accurate cuts are characteristic of the process, and the kerf and heat-affected zone are narrower than with any other thermal cutting process. No postcut finishing is required in many applications, even though the process does produce a thin recast surface. Figure 36-11 shows a laser cut of 0.25 inch (6 mm) thick carbon steel, cut at 70 in. per minute (1.8 meters per minute) with a 1250 watt laser.

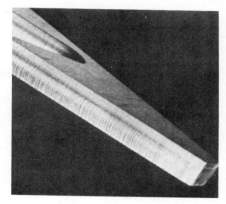

FIGURE 36-11 Surface of ¼ in. (6 mm) thick carbon steel cut with a 1250 watt laser at 70 in. per minute (1.8 meters per minute). (Courtesy Coherent, Inc.)

Virtually all composite materials can be laser-cut, and the process is achieving prominence in this area. The more uniform the thermal characteristics of the components, the better the cut and the less thermal damage to the material. Thus, Kevlar reinforced epoxy cuts most easily and gives a narrow heat-affected zone. Glass-reinforced epoxy is more difficult because of the greater thermal differences. Graphite-reinforced epoxy is even worse, because of the high dissociation temperature and thermal conductivity of the graphite. By the time the graphite has absorbed sufficient cutting heat, a large amount has been conducted into the composite, decomposing the epoxy to a greater depth.

Through use of a fiber-optic cable, laser energy can be piped to the end of a robot arm, eliminating the need to mount and maneuver a heavy, bulky laser. Cutting, drilling, welding, and heat treating can then be performed with multiple axes of motion that can be programmed for specific parts. Fiber-optic cables can transmit a steady power of 400 watts, and up to 10,000 watts in the pulsed mode of operation. Lasers have also been mounted on traditional machine tools, including CNC-type machines. The use of lasers for machining of metal is discussed in Chapter 32.

Laser Spot Welding. Lasers have also been used to produce spot welds in a process that offers unique advantages over the conventional resistance methods. A small clamping force is applied to assure contact of the workpieces and a fine-focused beam scans the area of the weld. Welding is performed in the keyhole mode, wherein the laser produces a small hole through the molten puddle. As the beam is moved, molten metal flows into the hole and solidifies, forming a fusion-type nugget.

Laser spot welding is a noncontact process and produces no indentations. Because no electrodes are involved, electrode wear is no longer a production problem. Weld quality is independent of surface resistance and electrode condition, and no water cooling is required. Heat input is reduced, and therefore the size of heat-affected zones and the magnitude of related problems are reduced

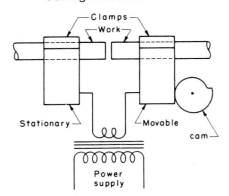

FIGURE 36-12 Schematic diagram of the flash welding process.

as well. Speed of welding and strength of the resulting joint are comparable to resistance welding.

Flash Welding. In *flash welding* (FW), two pieces of metal are mounted in a machine and lightly touched together. An electric current may be passed through the joint to provide preheat (optional), after which the pieces are withdrawn slightly. A flashing action (arc) then occurs, which melts the interface and expels the liquid and oxides. The pieces are then forced together under high pressure to upset the joint and form the weld. The force is maintained until solidification is complete, after which the product is removed from the machine and the upset portion is machined away. Figure 36-12 provides a process schematic.

The flashing action must be long enough to provide heat for melting and to lower the strength of the metal to allow for plastic deformation. Sufficient upsetting should occur that all impure metal is squeezed out into the *flash* and only sound metal remains in the weld. Figure 36-13 shows parts before and after flash welding.

Flash welding can be employed for butt welding of solid or tubular metals and is widely used in the manufacture of tubular metal furniture, pipe, and other such products. The equipment required is generally rather large and expensive, but excellent welds can be made at high production rates. Therefore the process is well suited for mass production operations. Although the resulting flash must be subsequently removed (and this may be a problem on the inside of a tube), in most cases no preliminary surface preparation is required.

Percussion welding is a similar process whereby the heating is obtained by an arc produced by a rapid discharge of stored electrical energy, and is followed by a rapid application of force to expel the metal and produce the joint. In percussion welding, the arc duration is only 1 to 10 milliseconds. While the heat is intense, it is also highly concentrated. Thus, the heat-affected zone is quite small, and the process is attractive in applications in which the adjacent components are heat-sensitive, as in the electronics industry.

Welding of Plastics

Mechanical fasteners, adhesives, and welding processes can all be employed to form joints between engineering plastics. Fasteners are quick and are suitable for most materials, but their use may be expensive, they generally do not provide leak-tight joints, and the localized stresses may cause them to pull free of the polymeric material. Threaded metal inserts may have to be incorporated into the plastic components, further increasing the product cost. Adhesives can provide excellent properties and fully sound joints, but they are often difficult to handle and relatively slow to cure. In addition, considerable attention is required in the areas of joint preparation and surface cleanliness. Welding can produce bonded joints with mechanical properties that approach those of the parent material. Unfortunatley, only the thermoplastic polymers can be welded, since these materials can be melted or softened by heat without degradation. The thermosetting polymers do not soften with heat but tend only to char or burn.

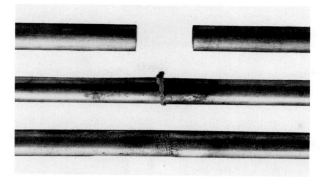

FIGURE 36-13   Flash-welded parts. (Top) Before welding; (Center) Welded, with flash; (Bottom) With flash removed by grinding.

The processes used to weld plastics can be effectively divided into two groups: (1) those that utilize mechanical movement and friction to generate the heat, such as ultrasonic welding, friction welding, and vibration welding, and (2) those that require external heat sources, such as hot plate welding, hot gas welding, and resistive and inductive implant welding.

Ultrasonic welding of plastics uses high-frequency mechanical vibrations to create the bond. Parts are held together and are subjected to ultrasonic vibrations (20 to 40 kilohertz) perpendicular to the area of contact. The high-frequency stresses generate heat at the interface sufficient to produce a quality weld in a period of ½ to 1½ seconds. The process can be readily automated, but the tools are quite expensive and large production runs are generally required. In addition, welding is usually restricted to small components with weld lengths not to exceed a few inches or centimeters.

The *friction welding of plastics* (also called spin welding) is essentially the same as the friction welding of metals, but melting now occurs at the joint interface. High-quality welds are produced with good reproducibility, and little end preparation is necessary. The major limitation is that at least one of the components must exhibit circular symmetry, and the axis of rotation must be perpendicular to the mating surface. Weld strengths vary from 50% to 95% of the parent material in bonds of the same plastic. Joints between dissimilar materials generally give poorer strengths.

In *vibration welding,* frictional heat is again generated by relative movement between the two parts, but the direction of movement is now parallel to the interface (as opposed to the ultrasonic case in which the direction is perpendicular), and the frequencies are considerably less, on the order of 100 to 240 Hertz. Molten material is produced, the vibration is stopped, parts are aligned, and the weld region cools and solidifies. The entire process takes about 1 to 5 seconds. Long-length, complex joints can be produced at rather high production rates. Nearly all thermoplastics can be joined, independent of whether their prior processing was by injection molding, extrusion, blow molding, thermoforming, foaming, or stamping.

Hot plate welding is probably the simplest of the mass-production techniques to join plastics. The parts to be joined are held in fixtures that press them against

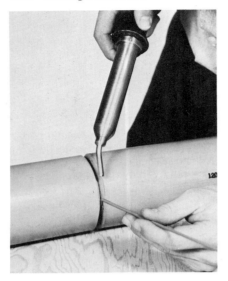

FIGURE 36-14 Using a hot-gas torch to make a weld in plastic pipe.

an electrically heated tool. Contact is maintained until the surface has melted and the remaining material has softened to a specified distance from the interface. The tool is then removed and the two prepared surfaces are pressed together in proper orientation and alignment and allowed to cool. Contaminated surface material is usually displaced into a flash region. Weld times are comparatively slow, ranging from 10 seconds to several minutes. The joint strength is usually equal to that of the parent material, but the joint design is limited to a square-butt configuration. If the joint surface has a nonflat profile, shaped heating tools can be employed. Heated-tool welding can also be used to produce lap seam welds in flexible plastic sheets. Pressure is applied by rollers after the material has passed over the heater.

The *hot gas welding of plastics* is similar to the oxyacetylene welding of metals. Compressed air, nitrogen, hydrogen, oxygen, or carbon dioxide is heated by an electrical coil as it passes through the welding gun, such as the one shown in Figure 36-14. The hot gas stream emerges from the gun at 400° to 570°F (200° to 300°C) and impinges upon the joint area and selected filler material. V-groove or fillet welds are the most common joint configurations used with this process. Because the plastic does not melt and flow, additional filler material usually has to be added. As illustrated in Figure 36-14, thin rods of plastic material are heated simultaneously with the workpiece and are then forced into the softened joint area, providing both filler material and the pressure needed to produce coalescence. The process is usually slow and the results are generally dependent upon operator skill. Therefore it seldom sees production application but is a popular process for repair of components.

In the *implant welding of plastics,* metal inserts are placed between the parts to be joined and are then heated by means of induction or resistance heating. (*Note:* The resistance method requires that wires be placed along the joint to carry current to the implants; this is not required for induction heating). The thermoplastic material adjacent to the implant melts and flows to form a joint. Since a weld forms only in the vicinity of the implants, the process resembles spot welding and produces joints that are considerably weaker than those formed by processes that bond the entire contact area. When bonding is desired over larger areas, iron oxide paste or metal tapes can be used to concentrate the induction heating at the interface.

Welding-Related Processes

Surfacing. *Surfacing* is the process of depositing a layer of metal or other material of one composition upon the surface or edge of a base material of a different composition. The usual objectives are to obtain improved resistance to wear, abrasion, or chemical attack. The process is often called *hard facing,* because the deposited surfaces are usually harder than the base metal. This does not always have to be true, however. In some cases, a softer metal, such as bronze, is applied to a harder base metal.

Surfacing Materials. The materials most commonly used for surfacing include (1) carbon and low-alloy steels; (2) high-alloy steels and irons; (3) cobalt-base alloys; (4) nickel-base alloys, such as Monel, Nichrome, and Hastelloy; (5) copper-base alloys; (6) stainless steels; and (7) ceramic and refractory carbides, oxides, borides, silicides, and similar compounds.

Surfacing Methods and Application.

Surfacing materials can be deposited by nearly all of the gas-flame or arc welding methods, including: oxyfuel gas, shielded metal arc, gas metal arc, gas tungsten arc, submerged arc, and plasma arc. Arc welding is frequently used for the deposition of high-melting-point alloys. Submerged arc welding is used when large areas are to be surfaced or a large amount of surfacing material is to be added, as in Figure 36-15. The plasma arc process further extends the process capabilities because of its extreme temperatures. To obtain true fusion of the surfacing material, a transferred arc is used and the surfacing material is injected in the form of a powder. If a nontransferred arc is used, only a mechanical bond is produced. This becomes a form of metallizing, which will be discussed next. Laser hard facing has also been performed.

Metallizing or Metal Spraying.

In *metallizing* or *metal spraying,* surfacing materials are melted and atomized in a special torch or gun and sprayed onto a base metal. Figure 36-16 shows a typical oxyacetylene metal-spraying gun. A wire of the surfacing metal is fed automatically through the center of the flame, where it melts, is atomized by a stream of compressed air, and is blown onto the base metal. Almost any kind of material that can be made into a wire can be deposited by this procedure.

An alternative type of gun uses material in the form of powders, which are then blown through the flame. This technique has the added advantage that brittle materials, such as cermets, oxides, and carbides can also be sprayed.

The most sophisticated of the metallizing techniques is the plasma spray process, illustrated in Figure 36-17. Since plasma temperatures can reach 30,000°F (16,500°C), the process can be used to spray materials with extremely

FIGURE 36-15 Applying a hard surfacing layer to a cylindrical drum with the submerged arc process. (Courtesy Linde Division, Union Carbide Corporation.)

FIGURE 36-16 Schematic diagram of an oxyacetylene metal-spraying gun. (Courtesy Metallizing Engineering Company, Inc.)

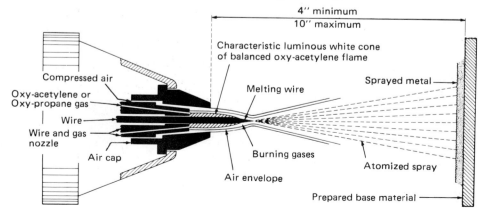

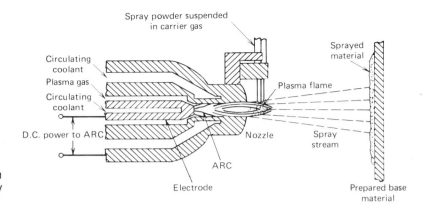

FIGURE 36-17 Schematic diagram of a plasma-arc spray gun. (Courtesy METCO, Inc.)

high melting points, as high as 6000°F (3300°C). Metals, alloys, ceramics, carbides, cermets, intermetallics, and plastic-base powders have all been successfully deposited.

Metallizing guns can be either hand-held or machine-mounted and mechanically driven, as shown in Figure 36-18. A stand-off distance of 6 to 10 inches (150 to 250 mm) is usually maintained between the spray nozzle and the workpiece.

Surface Preparation for Metallizing. Because metallizing relies on a purely mechanical bond between the deposited and base metals, it is essential that the base be properly prepared to promote good mechanical interlocking. To begin, the base metal must first be clean and free of oil. The surface is then roughened by one of a variety of methods. Grit blasting with a sharp, abrasive grit is the most common technique. For cylindrical surfaces that can be rotated in a lathe, an alternative method is to turn very rough threads and then roll the crests over slightly with a knurling tool. An alternate method that can be used on flat surfaces is to cut a series of parallel grooves, using a rounded grooving

FIGURE 36-18 Martensitic stainless being sprayed onto a shaft to improve the wear resistance. The shaft is being rotated in a lathe and the spray gun is using a metal wire feed. Other types of metallizing guns use powder material feed-stock. (Courtesy Wall Colmonoy Corporation.)

tool, and then roll over the lands between the grooves, as indicated in Figure 36-19. Figure 36-20 shows a comparison of surfaces prepared by grit blasting and rough machining. If the sprayed surface is to be subjected to subsequent machining, the base should be prepared by either rough machining or grooving to provide the maximum interlocking.

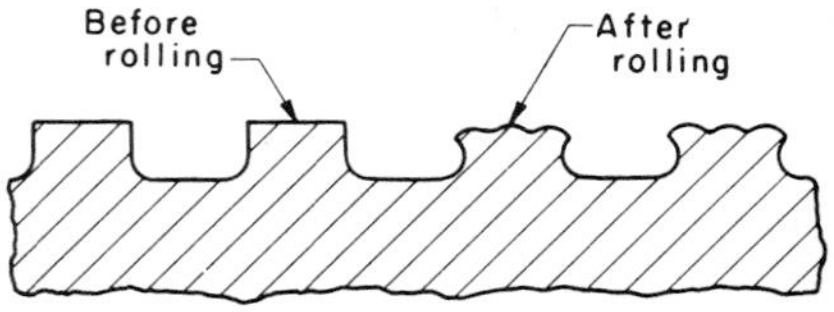

Characteristics of Sprayed Metals.   During deposition, the sprayed material is broken up into fine, molten particles, mixed with air, and then cooled rapidly upon impacting the base metal. As a result, such coatings tend to contain particles of oxidized metal and are harder, more porous, and more brittle than in the conventional wrought state. While these characteristics would be detrimental to many applications, they actually serve to make the sprayed metals more resistant to abrasion and wear and better able to serve as bearing surfaces. The characteristic porosity retains lubricants, thus adding to their ability to resist wear.

Although sprayed coatings have only about 85% to 90% of the density and 30% to 50% of the strength of wrought metals, their electrical conductivity is nearly as good.

Applications of Metal Spraying.   Metal spraying has a number of applications, including

1. *Protective coatings.* Zinc and aluminum are sprayed on iron and steel to provide corrosion resistance.
2. *Building up worn surfaces.* Worn parts can often be salvaged by adding metal to the depleted regions.
3. *Hard surfacing.* Although metal spraying should not be compared to hard facing deposits applied by welding techniques, it can be useful when thin coatings are adequate.

FIGURE 36-19   Method of preparing surfaces for metal spraying by machining grooves and rolling the edges.

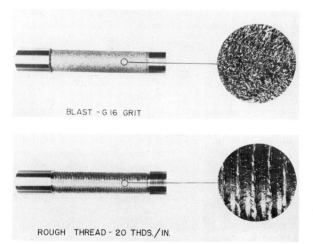

BLAST - G 16 GRIT

ROUGH THREAD - 20 THDS./IN.

FIGURE 36-20   Surfaces prepared for metal spraying by grit blasting and rough turning. (Courtesy Metallizing Engineering Company, Inc.)

4. *Applying coatings of expensive metals.* Metal spraying provides a simple method for applying thin coatings of noble metals to small areas when conventional plating would not be economical.

5. *Electrical conductivity.* Because metal can be sprayed on almost any surface, metal or nonmetal, it can be used to apply a conductive surface to an otherwise poor conductor or a nonconductor. Copper or silver is frequently sprayed on glass or plastics for this purpose.

6. *Reflecting surfaces.* Aluminum, sprayed on the back of glass by a special fusion process, makes an excellent reflecting surface.

7. *Decorative effects.* One of the earliest and still important uses of metal spraying was to obtain decorative effects. Because sprayed metal can be treated in a variety of ways, such as buffed, wire-brushed, or left in the as-sprayed condition, it is frequently specified as a decorative device for both manufactured products and architectural materials.

Review Questions

1. What were some of the limitations that made the forge welds of a blacksmith somewhat variable in nature and highly dependent upon the skill of the individual?
2. What is the primary feature promoting coalescence in cold welding?
3. Describe how sheet metal can be used to fabricate products with fluid-flow channels that once required the use of metal tubing.
4. What is the source of heat in friction welding?
5. How does inertia welding differ from friction welding?
6. How are surface impurities removed in the friction and inertia welding processes?
7. What are some of the geometric limitations of friction and inertia welding?
8. What are some of the geometric limitations of ultrasonic welding?
9. What are the conditions necessary to promote diffusion welding?
10. If the interface of a weld is viewed in cross section, what is the distinctive geometric feature of an explosive weld?
11. In what ways is a thermit weld similar to the production of a casting?
12. What is the source of the welding heat in thermit welding?
13. What is the source of the welding heat in electroslag welding?
14. What are some of the various functions of the slag in electroslag welding?
15. Why is a high vacuum required in the electron beam chamber of an electron beam welding machine?
16. What types of production limitations are imposed by the high vacuum requirements of electron beam welding? What compromises are made when welding is performed on pieces outside of the vacuum chamber?
17. What are the major drawbacks of high-voltage electron beam welding equipment?
18. What sort of manufacturing difficulties can arise from the deep and narrow weld profile of electron beam and laser welding?
19. What are some of the ways in which laser beam welding is more attractive than electron beam welding?
20. Why is laser beam welding an attractive process for use on electronic components?
21. Why can laser beam cutting be performed at such high cutting rates?
22. Why are laser welding and laser cutting attractive processes for integration with industrial robots? (*Hint:* Weight on the extended arm of a robot produces elastic flexing of the components, which in turn affects the accuracy of positioning.)
23. In the flash welding process, what features determine the minimum duration of arcing and the amount of upsetting required?
24. Why would it be difficult to produce a weld in thermosetting polymers but not difficult with the thermoplastics?
25. What are some of the alternative methods by which surfacing materials can be deposited onto a metal substrate?
26. What are some of the attractive features of plasma spray metallizing?
27. What are some of the ways in which a surface can be prepared to receive a metallizing treatment?
28. How can metal spraying be used to salvage worn parts?

Brazing and Soldering

There are many joining or assembly operations for which welding may not be the best choice. Perhaps the heat of welding is objectionable, or the materials possess poor weldability, or alternate processes are less expensive. In such cases, low-temperature joining methods may be preferred. These include brazing, soldering, adhesive joining, and the use of mechanical fasteners. In brazing and soldering, a low-melting-point material is melted, drawn into the space between two solid surfaces by capillary action, and allowed to solidify. Adhesive bonding utilizes a polymerizable resin which fills the space between the surfaces to be joined. Variations in surface finish and fit are more tolerable, since capillary action is not required, and oxides may not be a problem since the adhesive may actually adhere better to a tight oxide layer. Mechanical fasteners span a wide spectrum, including rivets, bolts, screws, nails, and others. While some forms may be permanent, others offer the advantage of easy disassembly for service or replacement of components.

Chapter 37 will discuss brazing and soldering. Adhesive bonding and mechanical fasteners will be presented in Chapter 38.

Brazing

Brazing is the joining of metals through the use of heat and a filler metal for which the melting temperature is above 840°F (450°C)[1] but below the melting point of the metals being joined. In comparison with welding, the brazing process is different in a number of ways:

1. The *composition* of the brazing alloy is significantly different from the base metal.
2. The *strength* of the brazing alloy is substantially lower than the base metal.

[1]The temperature is an arbitrary one, set to distinguish brazing from soldering.

3. The *melting point* of the brazing alloy is lower than that of the base metal, so the base metal is not melted.
4. Bonding requires *capillary action*.

Because of these differences, the brazing process has several distinct advantages:

1. Virtually all metals can be joined by some type of brazing metal.
2. The process is ideally suited for dissimilar metals, such as the joining of ferrous to nonferrous, or metals with widely different melting points.
3. Since less heating is required than for welding, the process can be performed quickly and economically.
4. The lower temperatures reduce problems associated with heat-affected zones, warping, or distortion. Thinner and more complex assemblies can be joined successfully.
5. Brazing is highly adaptable to automation and performs well in mass production of delicate assemblies. A strong permanent joint is formed.

The major disadvantage of brazing is that subsequent heating can cause inadvertent melting of the braze metal, weakening or destroying the joint. Too often brazed joints fail when people apply heat in an attempt to straighten or repair damaged assemblies. While this is certainly not the result of defective brazing, it can lead to most unfortunate consequences. Therefore, if brazing is used in the manufacture of products that may be subject to service or repair, adequate warning should be provided.

The Nature and Strength of Brazed Joints.   Just as in welding, brazing forms a strong metallurgical bond at the interfaces. The bonding is enhanced by clean surfaces, proper clearance, good wetting, and good fluidity. The resulting strength can be quite high, certainly higher than the strength of the brazing alloy and possibly higher than the strength of the metal being brazed. Attainment of this high strength, however, requires optimum processing and design.

Bond strength is a strong function of the clearance between the parts to be joined. There must be sufficient clearance so that the braze metal will wet the joint and flow into it under the force of capillary action. As the gap is increased beyond this optimum value, however, the joint strength decreases rapidly, dropping off to that of the braze metal itself.

Proper clearance varies considerably, depending primarily upon the type of braze metal being used. Copper requires virtually no clearance when heated in a hydrogen environment. Silver-alloy brazing metals require about 0.0015 to 0.002 in. (0.04 to 0.05 mm) and a clearance of 0.02 to 0.03 in. (0.5 to 0.75 mm) should be specified when using 60–40 brass to braze iron and copper. It should be noted that these are the clearances that should exist *at the temperature of the brazing process.* Any effects of thermal expansion should be compensated for in specifying the dimensions of the basic components.

Wettability is a strong function of the surface tensions between the braze metal and the base alloy. Generally, the wettability is good when the two metals can form intermediate diffused alloys. Sometimes, the wettability can be improved, as is done when steel is tin plated to accept a lead–tin solder.

Fluidity is a measure of the flow characteristics of the molten braze metal and is a function of the metal, its temperature, surface cleanliness, and clearance.

Brazing Metals.   The most commonly used brazing metals are copper and copper alloys, silver and silver alloys, and aluminum alloys. Table 37.1 lists some of the frequently used brazing metals, compatible processes, and the materials that are commonly brazed.

Copper is used primarily for brazing steel and other high-melting-point materials, such as high-speed steel and tungsten carbide. Its use is confined almost exclusively to furnace operations conducted in a protective hydrogen atmosphere, in which the copper is extremely fluid and requires no flux. Copper brazing is used extensively for assemblies composed of low-carbon steel stamp-

TABLE 37.1   Commonly Used Brazing Metals and Their Uses

Braze Metal	Composition	Brazing Process	Base Metals
Brazing brass	60% Cu, 40% Zn	Torch Furnace Dip Flow	Steel, copper, high-copper alloys, nickel, nickel alloys, stainless steel
Manganese bronze	58.5% Cu, 1% Sn, 1% Fe, 0.25% Mn, 39.5% Zn	Torch	Steel, copper, high-copper alloys, nickel, nickel alloys, stainless steel
Nickel silver	18% Ni, 55–65% Cu, 27–17% Zn	Torch Induction	Steel, nickel, nickel alloys
Copper silicon	1.5% Si, 0.25% Mn, 98.25% Cu, 1.5% Si, 1.00% Zn, 97.5% Cu	Torch	Steel
Silver alloys (no phosphorus)	5–80% Ag, 15–52% Cu, balance Zn + Sn + Cd	Torch Furnace Induction Resistance Dip	Steel, copper, copper alloys, nickel, nickel alloys, stainless steel
Silver alloys (with phosphorus)	15% Ag, 5% P, 80% Cu	Torch Furnace Induction Resistance Dip	Copper, copper alloys
Copper phosphorus	93% Cu, 7% P	Torch Furnace Induction Resistance	Copper, copper alloys

ings, screw-machine parts, and tubing—materials common to many mass-produced products.

Copper alloys were the earliest brazing materials. Today, copper–zinc alloys are used extensively for brazing steel, cast irons, and copper. Copper–phosphorus alloys are used for the fluxless brazing of copper since the phosphorus can reduce the copper oxide film. Manganese bronzes are also used as brazing material.

Pure silver is used in brazing titanium. *Silver solders*, silver–copper–zinc alloys, are used in joining steels, copper, brass, and nickel. Although these brazing alloys are quite expensive, such a small amount is required that the cost per joint is really quite low. The silver alloys are also used in brazing stainless steels. However, since the brazing temperatures are in the range of carbide precipitation (sensitization), only stabilized or low-carbon stainless steels should be brazed with these alloys if good corrosion resistance is required in the product.

Aluminum–silicon alloys, containing about 6% to 12% silicon, are used for brazing aluminum and aluminum alloys. By using a braze metal that is not greatly unlike the base metal, the possibility of galvanic corrosion is reduced. However, since these brazing alloys have melting points of about 1130°F (610°C), and the melting temperature of commonly brazed aluminum alloys such as 3003 is around 1290°F (670°C), the control of the brazing temperature is quite critical. In brazing aluminum, proper fluxing action, surface cleaning, and/or the use of a controlled-atmosphere or vacuum environment is required to assure adequate flow of the braze metal.

A commonly used procedure when brazing aluminum is to use sheets of metal that have one or both surfaces precoated with the brazing alloy to a thickness of about 10% of the total sheet. Bonds are made by simply coating the joint area with a suitable flux, placing the materials in contact, and heating. The ''brazing sheets'' have sufficient coating to form adequate fillets.

Amorphous alloy brazing sheets have been produced by cooling metal at rates in excess of 1 million degrees Centigrade per second. The resulting metal foils are extremely thin, 0.0015 in. (0.04 mm) being typical, and exhibit excellent ductility and flexibility in the amorphous condition (even with alloys that are characteristically brittle in their crystalline condition). Shaped inserts can be cut or stamped from the foil, inserted into the joint, and heated. The resulting braze exhibits uniform melting and penetration. Since the braze material is fully dense when inserted, no shrinkage or movement is observed during the brazing operation. A variety of brazing alloys are currently available in the form of amorphous foils.

A nickel–chromium–iron–boron brazing alloy is frequently used to braze heat-resistant alloys that are to be used at high temperatures. Although the service temperature of the product will be above the melting point of the braze alloy, the braze material will not melt. During the brazing operation, the boron in the brazing alloy diffuses into the base metal, raising the melting point of the remaining alloy to above the subsequent service temperature.

Cast iron is not readily wettable because of the graphite. Consequently, before cast iron can be brazed, the graphite must be removed by etching.

Fluxes.

Fluxes play a very important part in brazing by (1) dissolving oxides that may be on the surface prior to heating, (2) preventing the formation of oxides during heating, and (3) lowering the surface tension of the molten brazing metal and thus promoting its flow into the joint.

One of the primary factors affecting the quality and uniformity of brazed joints is cleanliness. Although fluxes will dissolve modest amounts of oxides, *they are not cleaners*. Before a flux is applied, dirt and oil should be removed from the surfaces that are to be brazed. The less the flux has to do prior to heating, the more effective it will be during the brazing operation.

Borax has been commonly used as a brazing flux. *Fused borax* should be used because the water in ordinary borax causes bubbling when the flux is heated. Alcohol can be mixed with fused borax when a paste consistency is desired.

A number of modern fluxes are also available that have melting temperatures lower than borax and are somewhat more effective in removing oxidation. The particular flux should be selected for compatibility with the base metal being brazed and the particular process being used. Paste fluxes are utilized for furnace, induction, and dip brazing and are usually applied by brushing. Either paste or powdered fluxes can be used with the torch brazing process. Application is usually done by dipping the heated end of the filler wire into the flux.

Fluxes for aluminum are usually mixtures of metallic halide salts, with sodium and potassium chlorides comprising 15% to 85% of the mixture. Activators such as fluorides and lithium compounds are also added. The aluminum brazing fluxes *do not* dissolve the surface oxide on aluminum.

Since most brazing fluxes are corrosive, the residue should be removed from the work immediately after brazing is completed. This is particularly important in the case of aluminum, to which the chlorides are particularly detrimental. Considerable effort has been directed to developing fluxless procedures for brazing aluminum, as will be discussed later.

Applying the Brazing Metal.

Brazing metal can be applied to joints in several ways. The oldest (and a common technique used when torch brazing) uses brazing metal in the form of a rod or wire. After the joint area has been heated to a temperature at which the braze metal will melt, the wire or rod is melted by the torch and capillary action draws it into the prepared gap. While the base metal is hot enough to melt the braze alloy and assure its remaining molten and flowing into the joint, the actual melting should still be done with the torch.

This method of braze metal application requires considerable labor, and care is necessary to assure that the filler metal has flowed to the inner portions of the joint. To avoid these difficulties, the braze metal is often applied to the joint prior to heating, usually in the form of wires or shims. In cases where it can be

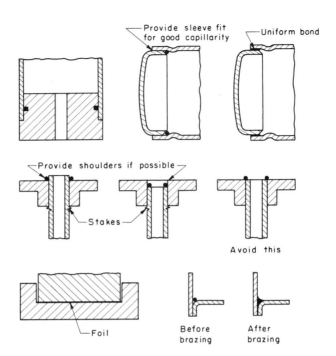

FIGURE 37-1 Methods of applying braze metal in sheet or wire form to assure proper flow into the joint.

done, rings or shims of braze metal are fitted into internal grooves in the joint before the parts are assembled, as shown in Figure 37-1. When this procedure is employed, the parts generally must be held together by press fits, riveting, staking, tack welding, or a jig to maintain their proper alignment. In such preloaded joints, care must also be exercised to assure that the filler metal is not drawn away from the intended surface by the capillary action of another surface of contact. Capillary action will always pull the molten braze metal into the smallest clearance, whether or not such was intended.

Another precaution that must be observed is that the flow of filler metal not be cut off by the absence of required clearances or by the presence of entrapped air. Also, fillets and grooves within the joint may act as reservoirs and trap the filler metal.

Special brazing jigs and fixtures are often used to hold the parts that are to be brazed during the heating. These assure proper alignment of components and are particularly useful when producing complex assemblies. When these are used, however, it is usually necessary to provide springs that will compensate for thermal expansion, particularly when two or more dissimilar metals are being joined. Figure 37–2 shows an excellent example of this procedure.

Heating Methods Used in Brazing. A common source of heat for brazing is a gas-flame torch. In this *torch-brazing* procedure, oxyacetylene, oxyhydrogen, or other gas-flame combinations can be used. Most repair brazing is done in this manner because of its flexibility and simplicity, but the process is also

FIGURE 37-2   Brazing fixture, with provision for expansion caused by the brazing heat. (Courtesy Aluminum Company of America.)

widely used in production applications, as illustrated in Figure 37-3. Its major drawbacks are the difficulty in controlling the temperature, difficulty in maintaining uniform heating, and meeting the cost of the skilled labor. In production-type torch brazing, specially shaped torches are often used to speed the heating and aid in reducing the amount of skill required.

Numerous brazing operations are performed in controlled-atmosphere or vacuum furnaces. In *furnace brazing,* the brazing metal must be preloaded into the

FIGURE 37-3   Manual torch brazing of large heat exchangers using pre-placed rings of brazing alloy. (Reprinted with permission of Handy & Harman, from *The Brazing Book.*)

FIGURE 37-4 Typical furnace-brazed assemblies. (Courtesy Pacific Metals Company.)

work. If the work is not of such a nature that its preassembly will maintain the parts in proper alignment, brazing jigs or fixtures must be used. Fortunately, assemblies that are to be furnace-brazed can usually be designed so that jigs and fixtures are not needed; a light press fit will often suffice. Figure 37-4 shows a number of typical furnace-brazed assemblies.

Because excellent control of the furnace-brazing temperatures can be obtained and no skilled labor is required, furnace brazing is particularly well suited for mass production operations. Either box- or continuous-type furnaces can be used, the latter being more suitable for mass production work. If the furnace atmosphere can reduce oxide films, flux may not be required. If reactive materials are to be brazed, vacuum furnaces are often used.

The third form of heating is *salt-bath brazing,* wherein the parts are dipped in a bath of molten salt that is maintained at a temperature slightly above the melting point of the brazing metal. This process offers several distinct advantages: (1) the work heats very rapidly because it is in complete contact with the heating medium, (2) the salt bath acts as a protective medium to prevent oxidation, and (3) thin pieces can easily be attached to thicker pieces without danger of overheating because the salt bath maintains a uniform temperature that is less than the melting point of the parent metal. This latter feature makes the process well suited for brazing aluminum, for which precise temperature control is required.

Here, again, the parts must be held in jigs or fixtures (or be prefastened in some manner), and the brazing metal must be preloaded into the work. To assure that the bath remains at the desired temperature, its volume must be substantially larger than that of the assemblies to be brazed.

In *dip brazing*, the assemblies are immersed in a bath of molten brazing metal. The bath thus provides both the heat and braze metal for the joint. However, since the braze metal will usually coat the entire workpiece, it is a wasteful process and is usually employed only for small products.

Induction brazing utilizes high-frequency induction currents for heating. This process offers the following advantages, which account for its extensive use:

1. The heating is very rapid. Usually only a few seconds are required for the complete cycle.
2. The operation can be made semiautomatic, and therefore only semiskilled labor is required.
3. Heating can be confined to only the area of the joint through use of specially designed coils and short heating times. This reduces problems associated with scale, discoloration, and distortion.
4. Uniform results are readily obtained.
5. By making new and relatively simple heating coils, a wide variety of work can be performed with a single power supply.

The high-frequency power supplies are available in both large and small capacities at very modest cost. The only other equipment necessary to perform induction brazing is a simple heating coil designed to fit around the joint. These are generally formed of copper tubing, designed to carry a supply of cooling water. Although the filler material can be added to the joint manually after it is heated, the usual practice is to use preloaded joints to speed the operation and produce more uniform bonds. In addition, induction brazing is so rapid that it can often be used to braze parts with high surface finishes, such as silver plating, without affecting the surface. Figure 37-5 shows a typical induction-brazing operation.

Some *resistance brazing* is also performed, in which the parts to be joined are pressed between two electrodes as a current is passed through. Unlike resistance welding, however, most of the resistance is provided by the electrodes, which are made of carbon or graphite. Thus, most of the heating is by means of conduction from the hot electrodes. The resistance process is used primarily to braze electrical components, such as conductors, cable connectors, and similar devices. Equipment is generally an adaptation of conventional resistance welders.

Flux Removal. Although not all brazing fluxes are corrosive, most of them are. Thus, flux residues should be completely removed from the product. Since many of the commonly used fluxes are soluble in hot water, their removal is not very difficult. Immersion in a hot water tank for a few minutes will usually give satisfactory results, provided that the water is sufficiently hot. In addition, flux removal is usually easier if the process is performed when the flux is still hot.

FIGURE 37-5 Typical induction brazing operation. (Courtesy Lepel High Frequency Laboratories, Inc.)

Blasting with grit or sand is another effective method of flux removal, but this procedure cannot be used if the surface finish is to be maintained. Fortunately, such drastic treatment is seldom necessary.

Fluxless Brazing. Both the application and removal of brazing flux involves significant costs, particularly when complex joints and assemblies are involved. Consequently, a large amount of work has been devoted to the development of procedures in which a flux is not required. Much of this work has been directed to the brazing of aluminum. Because of its light weight and good thermal conductivity, aluminum is particularly attractive for applications such as automobile radiators, when weight reduction is a desired objective.

Several characteristics of aluminum make the metal particularly difficult to braze: its low melting point, the high galvanic potential, and the presence of a refractory oxide film. Nevertheless, successful fluxless brazing has been achieved, but with rather complicated vacuum furnace techniques. Vacuums up to 1×10^5 torr (0.0013 Pa) are required and a "getter" metal is often necessary to aid in absorbing the small amounts of oxygen, nitrogen, and occluded gases that remain in the "vacuum" or are evolved from the aluminum being brazed. The aluminum must be carefully cleaned and degreased prior to brazing. In addition, proper joint design is quite critical, with sharp V-edged joints giving the best results.

Some success has been achieved in the fluxless induction brazing of aluminum in air, using an aluminum braze alloy containing about 7% silicon and 2.5% magnesium. The resulting magnesium vapor appears to reduce some of the oxide

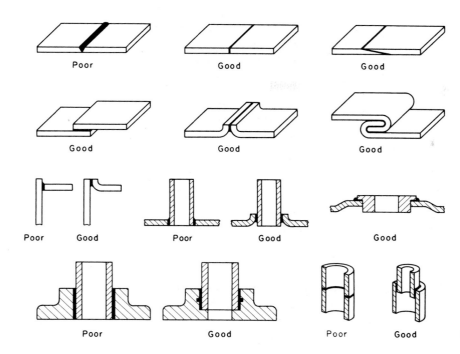

FIGURE 37-6 Examples of good and bad joint design for brazing.

on the surface of aluminum and permits the braze metal to flow and cover the metal surface.

Design of Brazed Joints.   Three types of brazed joints are used: *butt, scarf,* and *lap* or *shear*. These, together with some examples of good and poor joint design, are shown in Figure 37-6. Because the basic strength of a brazed joint is generally less than that of the parent metals, the desired strength is often obtained by specifying sufficient joint area. Often, some type of lap joint is employed when maximum strength is required. If the joints are made very carefully, a lap of 1 to 1¼ times the thickness of the metal can develop strength as great as the parent metal. For joints that are made in routine production, however, it is best to use a lap equal to three times the material thickness. Full electrical conductivity can usually be obtained with a lap of about 1½ times the material thickness.

If maximum joint strength is required, it is important that pressure be applied to the parts during heating and maintained until the braze metal has cooled sufficiently to attain most of its strength. In many cases, however, the needed pressure can be obtained simply through proper joint selection and design.

Finally, in designing brazed assemblies, one must be sure that no gases can be trapped within the joint. As the trapped gases expand upon heating, they can act to prevent the filler metal from flowing throughout the joint.

Braze Welding.   Braze welding differs from straight brazing in that capillary action is not required to distribute the filler metal. Here the molten filler is simply deposited by gravity, as in oxyacetylene gas welding. Because relatively low temperatures are required and warping is minimized, braze welding is very effective for the repair of steel products and ferrous castings. It is also attractive for joining cast irons since the low heat does not alter the graphite shape and the process does not require good wetting characteristics. Strength is determined by the braze metal being used and the amount applied. Considerable buildup may be required if full strength is to be restored to the repaired part.

Braze welding is almost always done with an oxyacetylene torch. The surfaces are first "tinned" with a thin coating of the brazing metal, and the remainder of the filler metal is then added. Figure 37-7 shows a large casting being repaired by braze welding.

Soldering

By definition, *soldering* is a brazing-type operation in which the filler metal has a melting temperature below 840°F (450°C). Bond strength is relatively low, the bonding being the result of adhesion between the solder and the parent metal.

Solder Metals.   Most solders are alloys of lead and tin with the addition of a very small amount of antimony, usually less than 0.5%. The three most commonly used alloys contain 60%, 50%, and 40% tin and all melt below 465°F (240°C). Because tin is expensive, those alloys having higher proportions of tin are used only when the high fluidity is required. For *wiped* joints and for filling dents and seams, as in automobile body work in which little strength is required, solders containing only 20% to 30% tin are used.

Other soldering alloys are used for special purposes. Tin–antimony alloys are useful in electrical applications. Bismuth alloys have very low melting points. Aluminum is often soldered with tin–zinc, cadmium–zinc, or aluminum–zinc alloys. Lead–silver or cadmium–silver alloys can be used for high-temperature service, and indium–tin alloys can be used to join metal to glass.

Soldering Fluxes.   As in brazing, soldering requires that the metal surfaces be clean and free of oxide. Fluxes are used for this purpose, but it is essential that all dirt, oil, and grease be removed before the flux is applied. Soldering fluxes are not intended to remove any appreciable amount of surface contamination.

Soldering fluxes are classified as *corrosive* and *noncorrosive*. A common noncorrosive flux is rosin (the residue after distilling turpentine) in alcohol. This is suitable for copper and brass and for tin-, cadmium-, and silver-plated surfaces, provided that the surfaces have been previously cleaned. Aniline phosphate is a more active noncorrosive flux, but it has limited use because it emits toxic gases when heated. It is suitable for use with copper and brass, aluminum, zinc, steel, and nickel.

The two most commonly used corrosive-type fluxes are muriatic acid and a mixture of zinc and ammonium chlorides. Acid fluxes are very active but are highly corrosive. Chloride fluxes are effective on aluminum, copper, brass, bronze, steel, and nickel, provided that no oil is present on the surface.

FIGURE 37-7   Large casting being repaired by braze welding. Four hours were required to prepare the joint and three hours for welding. Twenty pounds (9 kg) of braze metal were used. (Courtesy Anaconda Brass Company.)

Heating for Soldering. Although any method of heating that is suitable for brazing can be used for soldering, furnace and salt-bath heating are seldom used. Dip soldering is used extensively for joining wire ends (particularly in electronics work), soldering automobile radiators, and tinning. Induction heating is used when large numbers of identical parts are to be soldered. Nevertheless, most soldering is still done with electric soldering irons or guns, the primary requirements being only a source of sufficient heat and adequate heat transfer to the metals being joined. For low-melting-point solders, infrared heat sources can be employed.

A unique source of soldering heat is used in the process of *vapor-phase reflow soldering*. Fixtured parts with preloaded joints are exposed to boiling fluorinated hydrocarbon vapor. The vapors condense on the part surfaces, releasing a high latent heat sufficient to melt and reflow the lead–tin solder. Since the entire workpiece surface is heated uniformly, identical results are produced at all locations. High production rates can be achieved; the results are independent of product geometry or complexity; and costs are usually less than those for conventional soldering. Most applications, such as the joining of components to printed circuit boards, focus on the ability to maintain close temperature control and contamination-free environments.

In the various processes, the joints can be preloaded with solder, or the filler metal can be supplied from a wire. The particular method of heating usually dictates which procedure is used.

Design and Strength of Soldered Joints. Soldered joints will seldom develop shear strengths in excess of 250 psi (1.72 MPa). Consequently, if appreciable strength is required, soldered joints should not be used or some form of mechanical joint, such as a rolled-seam lock joint, should be made prior to soldering. Butt joints should never be used, and designs in which peeling action is possible should be avoided.

In making soldered joints, the part should be held firmly so that no movement can occur until the solder has cooled to well below the solidification temperature. Otherwise, the resulting joint will be full of cracks and have very little strength.

Flux Removal. Soldering flux should generally be removed from the finished joints, either to prevent corrosion or for the sake of appearance. Flux removal is rarely difficult, provided that the type of solvent in the flux is known. Water-soluble fluxes can be removed with hot water and a brush. Alcohol will remove most rosin fluxes. However, when the flux contains some form of grease, as in most paste fluxes, a grease solvent must be used, followed by a hot water rinse.

Review Questions

1. What are some of the alternative low-temperature joining processes?
2. How does brazing differ from welding?

3. Why is brazing an appropriate method of joining dissimilar metals with widely different melting points?
4. What features are necessary if the strength of a brazed joint

is to exceed the strength of the metal being brazed?

5. What are some of the most commonly used brazing metals?

6. What is a silver solder and how can its cost be justified in a manufacturing application?

7. What cautions should be exercised when one is using silver solder to braze stainless steels?

8. Why is it preferable to use a braze metal with chemistry similar to that of the base metal?

9. What are the three functions of a flux in brazing?

10. Why should flux residue be removed immediately following the brazing operation?

11. What are some of the necessary considerations when designing preloaded joints for brazing?

12. What is the purpose of a brazing jig or fixture, and how are they different from jigs or fixtures used for other joining operations?

13. What are the primary limitations of the torch-brazing process?

14. Why might reducing atmospheres or a vacuum be specified when using a furnace-brazing operation?

15. What are some of the attractive features of salt-bath brazing?

16. Why is dip brazing usually restricted to use with small parts?

17. What are some of the attractive features of induction brazing?

18. What is the major advantage of fluxless brazing?

19. What techniques can be used to compensate for the reduced strength of the brazed region in a brazed assembly?

20. How does braze welding differ from brazing?

21. How does soldering differ from brazing?

22. What alloy system contains the majority of soldering alloys?

23. What is vapor-phase reflow soldering?

24. Why should soldering not be specified when appreciable strength is required in a joint?

Adhesive Bonding and Mechanical Fasteners

Tremendous advances have been made in recent years in the development, use, and reliability of adhesive bonding. The use of structural adhesives (wherein the adhesive is a load-transmitting part of the product) has increased rapidly, even in such quality- and durability-conscious areas as the automotive and aircraft industries. Both metals and nonmetals can be bonded, and the expanding role of plastics and composites has brought added impetus to their use.

Adhesive Bonding

Adhesive Materials and Their Properties. Structural adhesives span a wide range of material types and forms, including thermoplastic resins, thermosetting resins, artificial elastomers, and even some ceramics. They can be applied as drops, beads, pellets, tapes, or coatings (films) and are available in the form of liquids, pastes, gels, and solids. Curing can be performed by the use of heat, radiation or light (photoinitiation), moisture, activators, catalysts, multiple-component reactions, or combinations thereof.

Commonly used structural adhesives include

1. *Epoxies.* Epoxies are the oldest, most common, and most diverse of the adhesive systems. They are strong, versatile adhesives that can be designed to offer high adhesion, good tensile and shear strength, high rigidity, creep resistance, easy curing, and good tolerance to elevated temperatures. Single-component epoxies use heat as the curing agent. Most, however, are two-component blends involving a resin and a curing agent, plus possible additives such as elastomers, nitrile, vinyls, or nylon that serve to increase flexibility, peel resistance, impact resistance, and other characteristics. After curing at room temperature, these materials can develop shear strengths as high as 5000 to 10,000 psi (35 to 70 MPa).

943

2. *Cyanoacrylates*. Cyanoacrylates are liquid monomers that polymerize when spread into a thin film between two surfaces. Trace amounts of moisture on the surfaces can promote curing at amazing speeds, often as little as two seconds. Thus, the cyanoacrylates offer a one-component adhesive system that cures at room temperature with no external impetus. Commonly known as "super glues" (most of the commercial varieties are of this type), this family of adhesives is now available in the form of liquids, gels, toughened versions designed to overcome brittleness, and even nonfrosting varieties. In addition to their other properties, the cyanoacrylates provide excellent adhesion to most commercial plastics.

3. *Anaerobics*. These one-component, room-temperature-curing polyester acrylics remain liquid when exposed to air. When confined to small spaces and shut off from oxygen, as in a joint to be bonded, the polymer becomes unstable. In the presence of iron or copper, it polymerizes into a bonding-type resin. Additives can reduce odor, flammability, and toxicity and speed the curing operation. The anaerobics are extremely versatile and can bond almost anything, including oily surfaces. The joints resist vibrations and offer good sealing to moisture and other environmental influences. Unfortunately, they are somewhat brittle.

4. *Acrylics*. The acrylic-based adhesives are relatively new. They offer good strength, toughness, and versatility and are often able to bond a variety of materials, including plastics, metals, ceramics, and composites. Most involve application systems whereby a catalyst primer (curing agent) is applied to one of the surfaces to be joined and the adhesive is applied to the other. The pretreated parts can be stored separately for weeks without damage. Upon assembly, the components cure to a strong thermoset bond at room temperature. Heat can often accelerate the curing and at least one variety cures with ultraviolet light. In comparison to other varieties of adhesives, the acrylics offer strengths comparable to the epoxies, with the added advantages of room temperature curing and a no-mix application system.

5. *Urethanes*. Urethane adhesives are a large and diverse family of adhesives generally targeted for applications that involve temperatures under 150°F (65°C) and components that can undergo great elongation. Both one-part thermoplastic and two-part thermosetting systems are available. In general, they cure quickly to handling strength but are slow to reach the full-cure condition. Two minutes to handling with 24 hours to complete cure is common at room temperature.

6. *Silicones*. Silicone adhesives cure from the moisture in the air. They form low-strength structural joints and are usually selected when much expansion and contraction is expected in the joint, flexibility is desired (as in sheet metal parts), or good sealing is required.

7. *High-temperature adhesives*. When strength must be retained at temperatures over 500°F (290°C), high-temperature structural adhesives should be specified. These include epoxy phenolics, modified silicones or phenolics, polyamides, and some ceramics.

8. *Hot melts.* Although generally not considered to be true structural adhesives, the hot melts are being used increasingly to transmit loads between pieces. These are thermoplastic materials which are solid at room temperature but melt when heated into the range of 200° to 300°F (100° to 150°C). They are generally applied as heated liquids and form a bond as the molten adhesive then cools. Another alternative is to position the adhesive prior to operations such as the paint bake process in automobile manufacture. During the baking, the adhesive melts, flows into seams and crevices, and seals them against the entry of corrosive moisture.

Table 38.1 presents a listing of popular structural adhesives along with their service and curing temperatures and expected strengths.

TABLE 38.1 Some Common Structural Adhesives, Their Cure Temperatures, Maximum Service Temperatures, and Strengths Under Various Types of Loadings

Chemical Type	Cure Temperature [(°F) °C]	Service Temperature [(°F) °C]	Lap Shear Strength[a] [MPa at °C (psi at °F)]	Peel Strength at Room Temperature [(lb/in.) N/cm]
Butyral-phenolic	(275–350) 135–177	(−60–175) −51–79	6.9 at 79 (1000 at 175) 17.2(2500) at RT	(10) 17.5
Epoxy (room-temperature cure)	(60–90) 16–32	(−60–180) −51–82	10.3 at 82 (1500 at 180) 17.2(2500) at RT	(4) 7.0
Epoxy (elevated-temperature cure)	(200–350) 93–177	(−60–350) −51–177	10.3 at 177 (1500 at 350) 17.2(2500) at RT	(5) 8.8
Epoxy-nylon	(250–350) 121–177	(−420–180) −251–82	13.8 at 82 (2000 at 180) 41.4(6000) at RT	(70) 122.6
Epoxy-phenolic	(250–350) 121–177	(−420–500) −251–260	6.9 at 79 (1000 at 175) 17.2(2500) at RT	(10) 17.5
Neoprene-phenolic	(275–350) 135–177	(−60–180) −51–82	6.9 at 82 (1000 at 180) 13.8(2000) at RT	(15) 26.3
Nitrile-phenolic	(275–350) 135–177	(−60–250) −51–121	13.8 at 121 (2000 at 250) 27.6(4000) at RT	(60) 105.1
Polyimide	(550–650) 288–343	(−420–1000) −251–538	6.9 at 538 (1000 at 1000) 17.2(2500) at RT	(3) 5.3
Urethane	(75–250) 24–121	(−420–175) −251–79	6.9 at 79 (1000 at 175) 17.2(2500) at RT	(50) 87.6

[a]RT, room temperature.

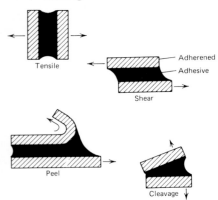

FIGURE 38-1 Types of stresses in adhesive-bonded joints.

Joint Design and Preparation.   Adhesive-bonded joints are often classified as either continuous surface or core to face. In *continuous-surface bonds,* both of the adhering surface areas are relatively large and are of the same size and shape. *Core to face bonds* have one adhered area that is very small compared to the other, as in the bonding of lightweight honeycomb core structures or a corrugated layer to the face sheets. In designing a bonded joint, one must certainly consider the types of stress to which the joint will be subjected. As shown in Figure 38-1, these stresses are tension, shear, cleavage, and peel. As noted in Table 38-1, most of the adhesives are much weaker in peel and cleavage. Therefore, it is important that joints be designed so that as much of the stress as possible is in shear or tension. In these modes, all of the bonded area shares equally in bearing the load.

Looking further, one will note that the shear strengths of common structural adhesives range from 2000 to 6000 psi (14 to 40 MPa) at room temperature, while the tensile strengths are only 600 to 1200 psi (4 to 8 MPa). Therefore, the best adhesive-bonded joints are those that are designed to utilize the superior lap shear strengths. Figure 38-2 shows some of the commonly used joint designs and indicates their relative effectiveness. If additional strength is desired, it is generally obtained by increasing the bond area.

To obtain satisfactory and consistent quality in adhesive-bonded joints, it is essential that the surfaces be properly prepared. A standard procedure should be established, and frequent and adequate checks should be made to assure that it is being followed. A commonly used four-step procedure to prepare surfaces includes

1. *Cleaning.* All contaminants and grease must be removed from the surface. Proper wetting of the surfaces requires that they be meticulously clean.
2. *Etching.* The surface should be made chemically receptive to the adhesive primer and should be sufficiently rough to promote good adhesion. Etching further promotes good wetting characteristics.
3. *Rinsing.*
4. *Drying.*

A low-viscosity primer may be applied to the surfaces by spraying or brushing. After the primer has dried, the adhesive is applied, usually in liquid or paste form. If the adhesive contains a solvent, most of it must be removed before the joint is closed.

When elevated temperatures are required for curing, as with the thermoset adhesives, heat lamps, ovens, heated-platen presses, and autoclaves are used, depending upon the specific conditions of manufacture.

If more strength is required than can be provided by the structural adhesive, spot welds can be combined with adhesive bonding.

Advantages and Limitations of Adhesive Bonding.   Adhesive bonding has a number of obvious advantages. Almost all materials or combinations of materials can be joined. For most adhesives, the curing temperatures are quite low, seldom exceeding 350°F (180°C). A substantial number cure at room tem-

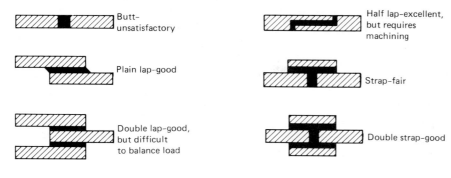

FIGURE 38-2 Possible designs of adhesive-bonded joints and a rating of their performance in service.

perature or slightly above and can provide adequate strength for many applications. Very thin or delicate materials, such as foils, can be joined to each other or to heavier sections. Heat-sensitive materials can be joined without damage and heat-affected zones are not present in the product. When joining dissimilar metals, the adhesives provide a bond that can tolerate the thermal stresses of differential expansion and contraction.

Because adhesives bond the entire joint area, good load distribution and fatigue resistance are obtained and stress concentrations are avoided. Similarly, because of the large amount of contact area that can usually be obtained, the total joint strength compares favorably with that produced by alternative methods of joining or attachment. Smooth contours are obtainable, and no holes have to be made, as with rivets or bolts.

Adhesives are inexpensive and generally weigh less than the fasteners needed to produce a joint of comparable strength. In addition, the adhesive can also serve to provide thermal and electrical insulation; act as a damper to noise, shock, and vibration; and provide protection against galvanic corrosion where dissimilar metals are joined. By providing both a joint and a seal against moisture, gases, and fluids, adhesive-bonded assemblies often offer improved corrosion resistance throughout their useful lifetime.

From a manufacturing viewpoint, surface preparation may be reduced since bonding can occur with an oxide film in place and rough surfaces are actually beneficial because of the increased contact area. Tolerances are less critical since the adhesives are more forgiving than alternative methods of bonding. Additional cost savings can result from simplified machining and assembly, reduced finishing requirements, elimination of mechanical fasteners, and the absence of highly skilled labor. Bonding can be achieved at locations that would prevent access to many types of welding apparatus. In addition, the bonds can be made extremely fast, some curing mechanisms taking as little as 2 to 3 seconds. Robotic dispensing systems can be utilized to provide a high degree of automation.

The major disadvantages of adhesive bonding are:

1. Most of the adhesives are not stable above 350°F (180°C), although a few can be used at temperatures up to 500°F (260°C).
2. It is difficult to determine the quality of an adhesive-bonded joint by nondestructive means, although some methods have been developed that give good results for certain types of joints.
3. Surface preparation, adhesive preparation, and curing procedures are quite critical if good and consistent results are to be obtained.
4. Life expectancy of the joint is hard to predict.
5. Assembly time may be greater than for alternative methods, depending upon the curing mechanism.
6. Some adhesives contain objectionable chemicals or solvents.

Nevertheless, the extensive and successful use of adhesive bonding provides proof that these factors can be overcome if adequate quality-control procedures are adopted and followed.

One should note that while the unit area strengths of adhesives are relatively low, this is not a major disadvantage or limitation, since sufficient area can generally be provided in a properly designed joint.

Mechanical Fasteners

Assembly in manufacturing is often performed by some means of mechanical fastening, a classification that includes a wide variety of techniques and fasteners designed to suit the individual requirements of a multitude of joints and assemblies. Included within this family are integral fasteners, threaded discrete fasteners (to include screws, bolts, studs, inserts, and other types), nonthreaded discrete fasteners (such as rivets, pins, retaining rings, staples, and wire stitches), special-purpose fasteners (such as the quick-release and tamper-resistant types), shrink and expansion fits, press fits, and others. Selection of the specific fastener or fastening method depends primarily on the materials to be joined, the function of the joint, strength and reliability requirements, weight limitations, dimensions of the components, and environmental conditions. Other considerations include cost, installation equipment and accessibility, appearance, and the need or desirability for disassembly. When disassembly and reassembly are desired, threaded fasteners or other styles that can be removed quickly and easily should be specified. Such fasteners should not have a tendency to loosen after installation, however. If disassembly is not necessary, permanent fasteners are often preferred.

Strength is imparted to the joint by mechanical interlocking and interference of the surfaces, and no fusion or adhesion of the surfaces is required. Fasteners and fastening processes should be selected to provide the desired strength and properties, reflecting the nature and magnitude of subsequent loading. Consideration should also be given to the presence of vibrations and/or cyclic stresses that would tend to promote loosening over a period of time. Weight considerations are significant in many applications, particularly aerospace and automotive. The need to withstand corrosive environments, operate at high or low temperatures, or face other severe conditions is often very influential in selecting fasteners or fastening processes.

Mechanical fastening has the additional advantage of being able to join similar or different materials in a wide variety of sizes, shapes, and joint designs.

Integral fasteners are formed areas of a component that interfere or interlock with other components of the assembly and are most commonly found in sheet metal products. Examples include lanced or shear-formed tabs, extruded hole flanges, embossed protrusions, edge seams, and crimps. Figure 38-3 shows some of these techniques, all of which involve some form of metal shearing and/or forming. The common beverage can includes several of these joints, namely, an edge seam to join the top of the can to the body and an embossed protrusion that is subsequently flattened as a means of attaching the opener-tab.

Discrete fasteners are separate pieces the function of which is to join the primary components. These include bolts, screws, nuts (with accessory washers, etc.), rivets, quick-release fasteners, staples, and wire stitches. Over 150 billion discrete fasteners are consumed annually, 27 billion by the automotive industry alone. A typical railroad box car contains 1200 mechanical fasteners; a numerically controlled turret lathe, 1700; and a standard telephone, 70. The variety here is so immense that the major challenge is to select an appropriate fastener for the task at hand and, if possible, an optimum fastener. This task is made even more difficult by the inconsistent nomenclature and identification schemes. While some are classed or identified by the specific product or application, others are classed by the material from which they are made, their size, their shape, or their primary operational features. (Discussion of the primary terms used in identifying discrete fasteners can be found in ANSI Standard B18.12.) On the positive side of the problem is the commercial availability of a wide range of standard and special types, sizes, materials, and strengths, such that an appropriate fastener can generally be found for almost all joining needs. The fasteners are easy to install, easy to remove, and easy to replace. In addition, most standard varieties are interchangeable. Various finishes and coatings can be applied to withstand a multitude of service conditions.

Shrink and expansion fits form another major class of mechanical joining. Here, a dimensional change is introduced to one or both of the components by heating or cooling (heating one part only, heating one and cooling the other, or cooling one). Assembly is then performed and a strong interference fit is established when temperature uniformity is restored. Joint strength is exceptionally high. In addition, the procedures can be used to produce a prestressed condition in a weak, low-cost material, often enabling it to replace a more costly, stronger one. Similarly, a corrosion-resistant cladding or lining can be easily provided to a less costly bulk material.

Press fits are quite similar to shrink and expansion fits, with the same results being obtained by mechanical force instead of differential temperatures.

Problems with Mechanical Fastening.

When a product is assembled with fastened joints, the fasteners become extremely vulnerable sites for failure. The cause of such failure generally relates to one of four areas: (1) the design of the fastener and the manufacturing techniques used to make it, (2) the material from which the fastener is made, (3) joint design, or (4) the means and details of

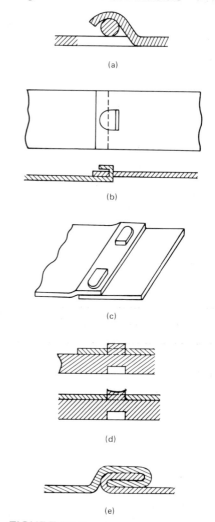

FIGURE 38-3 Several types of integral fasteners: a) lanced tab to fasten wires or cables to sheet or plate; b) and c) assembly through folded tabs and slots for different types of loading; d) use of a flattened embossed protrusion; and e) single-lock seam.

installation. Fasteners may have insufficient strength or corrosion resistance, or they may be subject to hydrogen embrittlement. Installation may have imparted too much or too little preload (too tight or too loose). The joint surfaces may not be flat or parallel and the area under the fastener head may be insufficient to bear the load. Vibration loosening and fastener fatigue are common causes of failure. In addition, joint design should consider the particular stress distribution, since much of the load will be concentrated on the fasteners (in contrast to the previously discussed adhesive joints that distribute the load uniformly over the entire joint area).

Nearly all fastener failures can be avoided by proper design and fastener selection. Consideration should be given to the operating environment, required strength, and magnitude and frequency of vibration. The need for weight savings and the desirability for disassembly and reassembly will influence selection. Standard fasteners should be used whenever possible and as little variety as necessary should be specified.

Fastener design should incorporate a shank-to-head fillet whenever possible. Rolled threads offer superior strength and corrosion-resistant coatings offer enhanced performance. Joint design should seek to avoid such features as offset or oversized holes. Proper installation and tightening are critical to good performance.

Review Questions

1. What are some of the common forms in which structural adhesives can be applied?
2. What are some of the various ways in which structural adhesives can be cured?
3. What general magnitude of shear strength could be expected from a high-strength epoxy?
4. Describe the curing mechanism of the anaerobic adhesives.
5. From a manufacturing viewpoint, what is an attractive feature of the acrylic adhesives?
6. Why might hot-melt adhesives be attractive to the automotive industry for applications involving sealing of joints?
7. Why is it desirable that adhesive joints be designed so the adhesive is loaded in pure shear?
8. What are some of the popular means of applying heat when elevated temperatures are used in curing?
9. Why are the structural adhesives attractive when joining dissimilar metals or materials?
10. In view of the relatively low strengths of the structural adhesives, how can adhesively bonded joints attain strengths comparable to other methods of joining?
11. What additional engineering properties or characteristics can be offered by the structural adhesives in addition to strength?
12. Why are adhesive joints not attractive for applications that involve exposure to elevated temperature?
13. What factors would influence the selection of a specific fastener or fastening method?
14. What types of fasteners should be selected if there is a need to disassemble and reassemble the product?
15. How do press fits differ from shrink or expansion fits? How are they similar?
16. How is the stress distribution different for adhesive joints and joints formed with discrete fasteners?
17. From a manufacturing viewpoint, why is it desirable to use standard fasteners and minimize the variety of fasteners within a given product?

Manufacturing Concerns in Welding and Joining

The majority of engineers enter industry with little basic understanding of the problems inherent in welding and joining and of the considerations that are necessary in designing a joint and selecting a joining technique. Welding is a unique process and should be used only when proper consideration has been given to its particular characteristics and requirements. Proper joint design is critical to the successful application of welding. Process selection is a complex procedure because of the large number of available processes, the variety of possible joint configurations, and the numerous parameters that must be specified for each operation. Heating, melting, and resolidification can cause numerous problems, including drastic changes in the material properties. Weld metal properties can be further changed by dilution of the filler by melted base metal, vaporization of various alloy elements, or a variety of gas–metal reactions.

Various types of weld defects can also be produced. These include: cracks in a variety of forms, cavities (both gas and shrinkage), inclusions (slag, flux, and oxides), incomplete fusion between the weld and base metals, incomplete penetration (insufficient weld depth), unacceptable weld shape or contour, arc strikes, spatter, undesirable metallurgical changes (aging, grain growth, or transformations), and excessive distortion.

Types of Fusion Welds and Types of Joints

There are four basic types of fusion welds, as illustrated in Figure 39-1. *Bead welds* require no edge preparation. However, because the weld is made on a flat surface and the penetration is limited, bead welds are suitable only for joining thin sheets of metal, building up surfaces, and depositing hard facing (wear-resistant) materials.

Groove welds are used when full-thickness strength is desired on thicker materials. Some sort of edge preparation is required to form a groove between the abutting edges. V, double V, U, and J (one-sided V) configurations are most

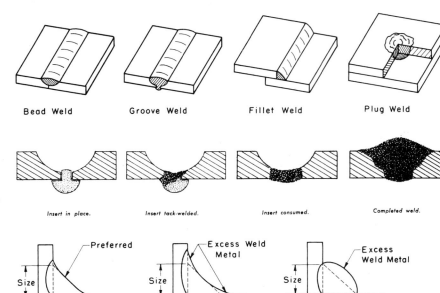

FIGURE 39-1   The four basic types of fusion welds.

Bead Weld   Groove Weld   Fillet Weld   Plug Weld

FIGURE 39-2   The use of a consumable backup insert in making fusion welds. (Courtesy Arcos Corporation.)

Insert in place.   Insert tack-welded.   Insert consumed.   Completed weld.

FIGURE 39-3   Preferred shape and the method of measuring the size of fillet welds.

common and are usually produced by oxyacetylene flame cutting. The specific type of groove usually depends upon the thickness of the work, the welding process to be employed, and the position of the work. The objective of the process is to obtain a sound weld throughout the full thickness with a minimum deposit of weld metal. If possible, single-pass welding is preferred, but multiple passes may be required, depending upon the thickness of the material and the welding process being used. As shown in Figure 39-2, special consumable inserts are often used to assure proper spacing between the mating edges and to assure proper quality in the root pass. These are especially useful in pipeline welding, particularly in field conditions and other applications in which the welding must be done from only one side of the work.

Fillet welds are used for tee, lap, and corner joints. The size of the fillet is measured by the leg of the largest 45° right triangle that can be inscribed within the contour of the weld cross section. This is shown in Figure 39-3, which also depicts the proper shape for fillet welds to avoid excess metal deposition and reduce stress concentration. Fillet welds require no special edge preparation. They may be continuous or intermittent, with spaces being left between short lengths of weld.

Plug welds are used to attach one part on top of another, replacing rivets or bolts. A hole is made in the top plate, and welding is started at the bottom of this hole.

Figure 39-4 shows the five basic types of joints that can be made with the use of bead, groove, and fillet welds. Figure 39-5 shows several methods to make these joints. In selecting the type of weld joint to be used, the primary consideration should be the type of loading that will be applied. Too often, this

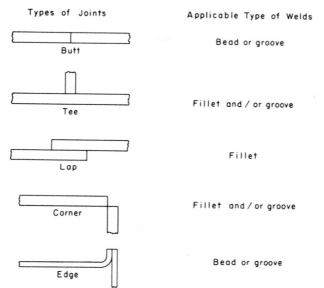

Types of Joints — Applicable Type of Welds

Butt — Bead or groove

Tee — Fillet and / or groove

Lap — Fillet

Corner — Fillet and / or groove

Edge — Bead or groove

FIGURE 39-4   Basic types of fusion weld joints, and the types of welds used in making them.

basic principle is neglected, and a large proportion of what are erroneously called ''welding failures'' are actually the result of such oversight. Cost and accessibility for welding are also important factors in joint selection but should be viewed as secondary to loading considerations. Cost is affected by the required edge preparation, the amount of weld metal that must be deposited, the type of process and equipment that must be used, and the speed and ease with which the welding can be accomplished. Accessibility will obviously have considerable influence on several of these factors.

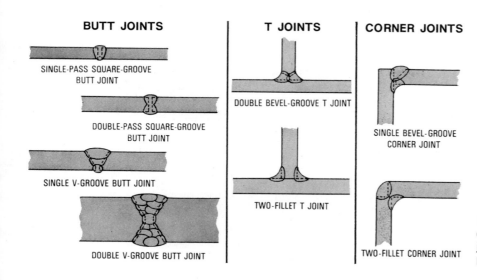

BUTT JOINTS

SINGLE-PASS SQUARE-GROOVE BUTT JOINT

DOUBLE-PASS SQUARE-GROOVE BUTT JOINT

SINGLE V-GROOVE BUTT JOINT

DOUBLE V-GROOVE BUTT JOINT

T JOINTS

DOUBLE BEVEL-GROOVE T JOINT

TWO-FILLET T JOINT

CORNER JOINTS

SINGLE BEVEL-GROOVE CORNER JOINT

TWO-FILLET CORNER JOINT

FIGURE 39-5   Various weld procedures used to form several common joints. (Courtesy Republic Steel Corporation.)

Design Considerations

Welding is a unique process that cannot be substituted directly for other methods of fastening without proper consideration being given to its particular characteristics and requirements. Unfortunately, welding is so easy and convenient to use that these facts are often overlooked. Since the cause of many "welding failures" can often be traced to such negligence, proper design considerations should be employed in all welding operations.

One very important point that must be kept in mind in the use of welding is that it produces monolithic, or one-piece, structures. When two pieces are welded together, they become one piece. This factor can cause significant complications if not properly taken into account. For example, a crack in one piece of a multipiece structure may not be serious because it will seldom progress beyond the single piece in which it occurs. In contrast, when a large structure, such as a ship hull, pipeline, storage tank, or pressure vessel, consists of many pieces welded together, a crack that starts in a single plate or weld can propagate for a great distance and cause complete failure. Obviously, such a fracture is not the fault of the welding process but simply a reflection of the monolithic nature of the product.

Another factor that relates to design is that a given material in small pieces may not behave the same in a large piece, as is often produced by welding. The importance of this fact is illustrated vividly in Figure 39-6, which shows the relationship between energy absorption and temperature for the same steel when tested as a Charpy impact specimen and as a large, welded structure. In the form of a Charpy bar, the material exhibited ductile behavior and good energy absorption at temperatures down to 25°F (−4°C). When it was welded into a large structure, however, brittle behavior was observed at temperatures as high as 110°F (43°C). Thus, the notch-ductility characteristics of steels used in large welded structures may be of great importance. More than one welded structure has failed because the designer did not take this fact into consideration.

Another common error is to make structures too rigid by welding, thereby restricting their ability to redistribute high stresses and avoid failure. Considerable thought may be required to design structures and joints that permit sufficient flexibility. However, the multitude of successful welded structures attests to the fact that such designs are indeed possible.

Accessibility of the joint for welding, welding position, component matchup, and the specific nature of the joint are all important considerations in welding design.

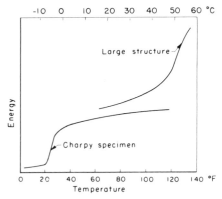

FIGURE 39-6 Effect of size on the transition temperature and energy-absorbing ability of a certain steel.

Heat Effects

Welding Metallurgy. Heating and cooling are essential and integral components of all welding processes (except cold welding) and tend to produce metallurgical changes that are usually undesirable. In fusion welding, the heating is sufficient to produce some melting of the base metal and is then followed by rapid cooling. The thermal effects tend to be most pronounced for this type of welding but also exist to a lesser degree in other types of welding or thermal joining in which the heating–cooling cycle is less severe. If these thermal effects

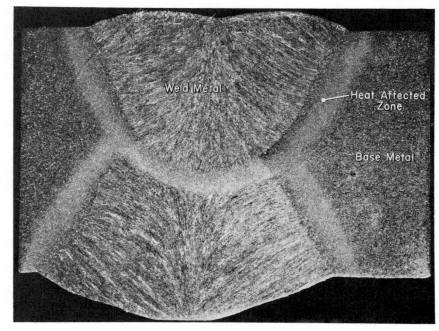

FIGURE 39-7 Grain structure and various zones in a fusion weld.

are properly considered, the adverse results can be avoided and excellent service results can be obtained. If they are overlooked, the results can be disastrous—the failure being that of the designer or fabricator, not of the welding process.

Because such a wide range of metals are welded and such a variety of processes are used, welding metallurgy is an extensive subject. Nevertheless, a few basic considerations should help in understanding the effects that occur. In fusion welding a pool of molten metal is created, the molten metal coming from the parent plate or a mixture of parent and filler material. This molten pool is contained in a metal mold formed by the surrounding plate and is generally very small compared to the surrounding metal. Thus, fusion welding involves *casting* a small amount of molten metal *into a metal mold*. The resultant structure and its properties can be best understood by first analyzing the casting and then considering the effects of the associated heat treatment on the adjacent base metal.

Figure 39-7 shows a typical microstructure produced by a fusion weld. In the center of the weld is a zone made up primarily of weld metal that has solidified from the molten state. Actually, it is a mixture of parent metal and electrode or filler metal, the ratio depending upon the process used, the type of joint, and the edge preparation. This zone is cast metal with a microstructure reflecting the cooling rate in the weld. As such, it cannot be expected to have the same properties and characteristics as the *wrought* parent metal. It can achieve equivalent mechanical properties only through the addition of filler metal that promotes high strength *in the as-cast condition.* Thus, in making quality fusion welds, one should select filler rods or electrodes that, *in the as-deposited condition,* will have properties that equal or exceed those of the parent metal in its wrought condition. Meeting this requirement is the basis for several American Welding Society specifications for electrodes and filler rods.

The grain structure of the weld metal zone may be fine or coarse, equiaxed or dendritic, depending upon the type and volume of weld metal and the rate of cooling. Most electrode and filler rod compositions tend to produce fine, equiaxed grains, but the volume of weld metal and variations in cooling rate can easily defeat these objectives.

Fusion welds are, therefore, prone to all of the problems and defects associated with metal casting, such as gas porosity, inclusions, blowholes, cracks, and shrinkage. Because the amount of metal that is molten is usually small compared to the total mass of the workpiece, rapid solidification and rapid cooling of the solidified metal are quite common. Associated with these conditions may be the inability to expel dissolved gases, chemical segregation, grain size variation, grain shape problems, and orientation effects.

Adjacent to the weld metal is the ever-present, and generally undesirable, *heat-affected zone (HAZ).* In this region, the parent metal is not melted but is subjected to elevated temperatures for a brief period of time. Since the temperature and its duration vary widely with location, fusion welding might be more appropriately described as "a casting and an abnormal, widely varying heat treatment." The adjacent metal may well experience enough heat to bring about structure and property changes, such as phase transformations, grain growth, precipitation, embrittlement, or even cracking. The variation in thermal history produces a variety of microstructures and a range of properties. In steels, the structures can range from hard, brittle martensite all the way through coarse pearlite and ferrite.

The heat-affected zone may no longer possess the desirable properties of the parent material, and, since it was not molten, it cannot assume the properties of the selected weld metal. Consequently, it is often the weakest area in the as-welded material. Except when there are obvious defects in the weld deposit, most welding failures originate in the heat-affected zone. Outside of this region, one finds base metal that has not been affected by the heat of the welding process.

It is apparent, therefore, that the structure and properties of a weld are complex and varied. Associated problems, however, can be reduced or eliminated in several ways. First, consideration should be given to the thermal characteristics of the various processes. Table 39.1 classifies some of the more common welding processes with regard to their *rate* of heat input. Processes with low rates of heat input (slow heating) tend to produce high total heat content within the metal, slow cooling rates, and large heat-affected zones. High-heat-input processes, on the other hand, have low total heats, fast cooling rates, and small heat-affected zones. The size of the heat-affected zone will also increase with increased starting temperature, decreased welding speed, increased thermal conductivity of the base metal, and a decrease in base metal thickness. Weld ge-

TABLE 39.1 Classification of Common Welding Processes by Rate of Heat Input

Low rate of heat input	*High rate of heat input*
Oxyfuel welding	Plasma-arc welding
Electroslag welding	Electron beam welding
Flash welding	Laser welding
	Spot and seam resistance welding
Moderate rate of heat input	Percussion welding
Shielded metal-arc welding	
Flux cored-arc welding	
Gas metal-arc welding	
Submerged-arc welding	
Gas tungsten-arc welding	

ometry is also important, fillet welds producing smaller heat-affected zones than butt welds.

When the as-welded results are unacceptable, the entire piece can be heat-treated after welding. Much of the variation can be reduced or eliminated, but the results are limited to those structures that can be produced by heat treatment. Structures and properties associated with cold working, for example, could not be achieved. Additional problems may occur as one seeks to produce controlled heating and cooling (heat treatment) with the large, complex-shaped structures commonly produced by welding. Finally, furnaces, quench tanks, and related equipment may not be available to handle the full size of the welded assembly.

Another procedure that can reduce the variation in microstructure, particularly the sharpness of the variation, is to preheat the base metal adjacent to the weld just prior to welding. For plain carbon steels, a temperature of 200° to 400°F (100° to 200°C) is usually adequate. This heating serves to reduce the cooling rate of the weld deposit and also of the immediately adjacent metal in the heat-affected zone. The reduced cooling rate results in the production of a more gradual change in microstructures and the elimination of a metallurgical stress raiser.

If the carbon content of plain carbon steels is greater than about 0.3%, the cooling rates encountered in normal welding are sufficient to produce hardening (untempered martensite) and the associated loss of ductility. This is also characteristic of most alloy steels because of their high hardenability. In welding these steels, special prewelding and postwelding heat cycles must be used. In contrast, the weldable low-alloy steels have gained high acceptance because they can be welded without the necessity of preheating or postheating.

In processes wherein little or no melting occurs and there is considerable pressure applied to the heated metal, as in forge or resistance welding, the weld region may actually retain some of the characteristics of the wrought metal.

While the discussion of metallurgical effects has largely focused on steels, other metals also exhibit heat-related changes. The exact effects of the heating and cooling associated with welding will depend upon the specific transformations and changes that can occur within the alloy being joined.

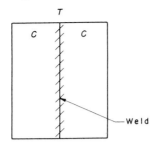

FIGURE 39-8 Schematic of the residual stresses in a fusion weld.

Thermally Induced Stresses. Another effect of the heating and cooling that accompany welding is the introduction of residual stresses. In welding, these may be of two types and are most pronounced in fusion welding where maximum heating occurs. Their effects can be observed in the form of dimensional changes, distortion, or cracking.

Residual welding stresses are the result of restraint to thermal expansion and contraction *offered by the pieces being welded.* They exist independent of whether the pieces are attached to other portions of the structure or are restrained in any manner. The way in which these residual stresses occur can be explained with the help of Figure 39-8. As the weld is made, the liquid region conforms to the shape of the "mold" and the adjacent metal becomes hot and expands. Expansion perpendicular to the weldment can be absorbed by flow of the molten pool, but expansions parallel to the weld line tend to be restrained by the cooler, stronger parent metal adjacent to this region. This resistance can often be sufficient to induce plastic deformation of the heat-affected zone. The material is *upset,* accommodating the thermal expansion by becoming thicker instead of longer.

After the weld metal solidifies, the weld pool and adjacent heat-affected region cool and contract. This contraction is also resisted by the cooler surrounding metal. The cooling region wants to contract but is restrained and must remain in a "stretched" condition, known as *residual tension* (region *T*). Similarly, the forces exerted by this contracting region try to squeeze the adjacent material and produce residual compression (regions *C*). While the net force is zero, in keeping with the basic laws of physics and mechanics, the localized variations can be substantial.

Thermal contractions occur both parallel (longitudinal) and perpendicular (transverse) to the weld, as shown in Figure 39-9. Generally, the transverse contractions are compensated for by movement of the base metal. That is, thermal contraction simply results in the welded assembly being shorter than the components were at the time of welding. If the base metals are restrained from movement, however, significant stresses can be induced. These are the second type of residual stresses and are known as *reaction stresses*. The magnitude of the reaction stresses is an inverse function of the length between the weld joint and the point of maximum rigidity. They can never exceed the yield strength of the parent metal (yielding would occur to relieve them) and are generally less than the stresses parallel to the weld. In the welding of steel, the actual values seldom exceed 15,000 psi (10.3 MPa).

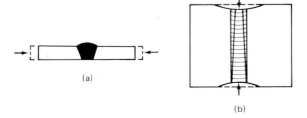

(a)

(b)

FIGURE 39-9 Shrinkage of a typical butt weld in the transverse (left) and longitudinal (right) directions.

As depicted in Figure 39-10, the longitudinal residual stresses are generally greatest in the weld metal and are tensile in nature. These decrease rapidly as one moves away from the weld, becoming compressive stresses of low magnitude and finally returning to zero. The transverse stresses within the weld metal are of uniform tension, except at the ends where they can be relieved by a pulling-in of the edges. These stresses are balanced by compression of the adjacent base metal (not depicted in Figure 39-10).

Effects of Thermal Stresses.

The most apparent result of the thermal stresses induced by welding is a distortion or warping as the material seeks to reduce the imbalance. Figure 39-11 shows the distortion effects of several types of basic weldments. Unfortunately, no fixed rules can be given to avoid warping because the possible conditions that cause it can be widely varied. Several procedures, however, have proved to be rather effective.

Welds should be made with the least amount of weld metal necessary to form the joint. Faster welding speeds reduce the welding time and reduce the volume of metal that is heated. Welding sequences should be designed to use as few weld passes and permit the plates to have as much freedom as possible. It is beneficial to weld toward the point of greatest freedom, such as from the center to the edge.

In another alternative, the components can be oriented out of position, such that the resulting distortion moves them to the desired final shape. Another common procedure is completely to restrain the components during welding, thereby forcing some plastic flow in the material and/or in the cooling weld metal. This procedure is used most effectively on small weldments in which the high reaction stresses are not likely to cause cracking. Still another procedure is to balance the resulting thermal stresses by depositing the weld metal in

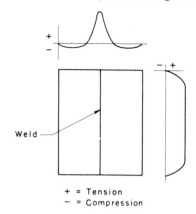

+ = Tension
− = Compression

FIGURE 39-10 Longitudinal and transverse residual stresses in a butt-welded joint.

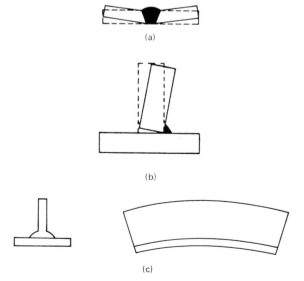

(a)

(b)

(c)

FIGURE 39-11 Distortions observed in several welding operations: (a) V-groove butt weld; (b) One-side fillet weld in a T-joint; (c) Two-fillet weld T-joint with a high vertical web.

predetermined patterns or areas, such as short lengths or on alternating sides of a plate. Warping can also be reduced by the use of *peening*. As the weld bead surface is hammered with the peening tool, the metal is flattened and tries to spread. Being held back by the underlying material, the surface becomes compressed or squeezed. The resulting compressive residual stresses serve to offset the tensile stresses induced by welding. Each pass of multipass welds, except the first and the last, is frequently hammered with a peening tool to promote the needed movement.

With regard to the residual stresses themselves, there is no substantial evidence that they have a harmful effect on the strength performance of weldments, *except in the presence of notches or in very rigid structures in which no plastic flow can occur.* These are two conditions that should not exist in weldments *if they have been properly designed and proper workmanship has been employed.* Unfortunately, welding makes it easy to inadvertently join heavy sections and produce rigid configurations that will not permit the small amounts of elastic or plastic movement required to reduce concentrated stresses. All too often, geometric notches, such as sharp interior corners, are incorporated into welded structures. Other harmful ''notches'' such as gas pockets, rough beads, porosity, and arc ''strikes'' can serve as initiation sites for weld failures, but these too can be avoided by proper welding procedures, good workmanship, and adequate supervision and inspection.

Residual stresses can cause additional warpage when weldments are machined so as to unbalance the stress equilibrium. Consequently, weldments that are to undergo appreciable machining are frequently given a stress-relief heat treatment prior to machining.

The reaction stresses can also cause distortion as they are superimposed on the applied loads of the system. Their most significant effect, however, is the tendency to cause cracking during or immediately following welding (as the weld is cooling). This is particularly likely when welds are made under conditions in which there is great restraint to the normal shrinkage that occurs transverse to the direction of welding. When a multipass weld is being made, such a crack often occurs in one of the early beads when there is insufficient metal to withstand the shrinkage stresses. This condition may be quite serious if the crack goes undetected and is not chipped out and repaired or melted and rewelded during subsequent passes.

A 1983 National Bureau of Standards study determined that fracture and fracture prevention through overdesign cost the United States about $120 billion per year. Since many of the structural failures occurred at weldments, it is important to take measures to reduce the tendency for cracking. Joints should be designed to keep restraint to a minimum, and the metals and alloys should be selected with welding in mind (more problems exist with higher carbon contents, higher alloy contents, and higher strength materials). The thicker the material being welded, the more likely it is that cracking will occur.

Crack-prevention efforts should also be directed at maintaining the proper size and shape of the weld bead. A concave fillet profile has a greater tendency to crack upon cooling, since contraction actually increases the length of the surface.

Convex fillets can shrink while cooling and not put the surface into tension. Welds with high penetration (high depth–width ratio) are more prone to cracking. Other methods to prevent cracking focus on slowing the cooling rate or promoting plasticity of the metals. The metals to be welded may be preheated and additional heat applied between the welding passes to retard cooling. Some welding codes also require weldments to be stress-relieved after welding but prior to use. Finally, efforts should be directed toward reducing the hydrogen pickup during welding. Slower welding and cooling will allow the hydrogen to escape, and the use of low-hydrogen electrodes and low-moisture fluxes will reduce its likelihood.

Summary. If the potential benefits of the welding process are to be obtained and harmful side effects are to be avoided, the effects of heating and cooling should be fully considered in designing and producing weldments. Unfortunately, there are numerous failures in welded structures that demonstrate that designers and fabricators often overlook these important factors.

In a recent attempt to assure quality, an infrared camera, computer, and control equipment were used to monitor and control fusion welding. In high-strength steel weldments, the cooling rate of the weld metal determines the metallurgical structure and therefore the strength and toughness. By monitoring the cooling rate these properties can be properly controlled. Other aspects of the system can be used to monitor the heat-affected zone, determine penetration of the weld, and assure proper weld puddle geometry.

Since flame and arc cutting operations also involve localized heating and cooling, they can also experience many of the same problems. In general, however, the products undergo further fabrication (often welding) and the problems of cutting are less severe and less extensive.

In summary, welding is an excellent fabricating process, but if satisfactory results are to be obtained, it cannot be used thoughtlessly, any more than the nuts can be left off the bolts in a bolted joint. Before welding is adopted as the joining technique in any manufacturing enterprise, one must make sure that the product or structure has been properly designed for welding.

Review Questions

1. Why might it be difficult to select the best procedure for making a given production-type weld?
2. What are some of the common types of weld defects?
3. What are the four basic types of fusion welds?
4. What types of weld joints commonly employ fillet welds?
5. What are some of the factors that influence the cost of making a weldment?
6. Why is it important to consider welded products as monolithic structures?
7. How is the "brittleness" of a large piece of metal different from that of a small piece?
8. How might excessive rigidity actually be a liability in a welded structure?
9. In what way is the weld pool segment of a fusion weld like a metal casting?
10. Why is it not uncommon for the selected filler metal to have a chemical composition that is different from the material being welded?

11. What are some of the defects or problems that can occur in the molten metal region of a fusion weld?

12. Why do properties vary widely in most welding heat-affected zones?

13. What types of structure and property modifications can occur in welding heat-affected zones?

14. Why do most welding failures occur in the heat-affected zone?

15. Describe the cooling rates and size of the heat-affected zone for processes with low rates of heat input.

16. What are some of the difficulties or limitations encountered in heat treating large structures after welding?

17. What is the purpose of prewelding and postwelding heating operations?

18. What are the two types of residual stresses that occur in weldments and what are their causes?

19. What is the typical nature of residual stresses that remain in the weld region, tension or compression? On the basis of your answer, what problems might you expect in this region?

20. How are reaction stresses affected by the distance between the weld and the point of fixed constraint?

21. What are some of the observed effects of the thermally induced stresses?

22. What are some of the techniques that can reduce the amount of distortion in a welded structure?

23. Why might a welded structure warp if the structure is machined after welding?

24. What are some of the techniques that can be employed to reduce the likelihood of cracking in a welded structure?

25. Two pieces of AISI 1025 steel are being shielded-metal arc-welded with E6012 electrodes. Some difficulty is being experienced with cracking in the weld beads and in the heat-affected zones. What possible corrective measures might you suggest?

26. A base for a special machine tool will weigh 1400 pounds (635 kg) if made as a gray iron casting. Pattern cost would be $450 and the foundry has quoted a price of $0.60 per pound ($1.32 per kilogram) for making the casting. If the part is made as a weldment, it will require 800 pounds (363 kg) of steel costing $0.14 per pound ($0.31 per kilogram). Cutting, edge preparation, and setup time will require 30 hours at a rate of $10 per hour for labor and overhead. Welding time will be 55 hours at an hourly rate of $9.50. Two hundred pounds (91 kilograms) of electrode will be required, costing $0.17 per pound ($0.37 per kilogram).

  a. Which method of fabrication will be more economical if only one part is required?

  b. What number of parts would be required for welding and casting to break even?

  c. Since the base of the machine tool will now be made of steel rather than cast iron, what special property of the cast iron will be lost, and why might this be sufficient to dictate manufacturing process regardless of economics?

VII

Processes and Techniques Related to Manufacturing

Surface Treatments and Finishing

Introduction

A large portion of manufactured products must be given some type of surface treatment before they can be sold or used. These processes can be for decoration and eye appeal, for protection, or for the removal of surface defects (scratches, pores, burrs, fins, and blemishes) produced by processing and handling which detract from the appearance of the product or present possible hazards to users. Many commonly used materials, such as most irons and steels, do not inherently possess the colors that customers want in products, especially in large-volume consumer goods. Materials often are not adequately resistant to the environments in which they will be used. As materials become scarce and more costly, the temptation to substitute alternative materials with modified surfaces increases. As a consequence, after achieving their desired shape, manufactured products are often given additional treatments to clean, protect, color, or alter surface properties.

These important decorative and protective surface treatments add to the cost of manufactured products. Further, as with other manufacturing processes, there is often a definite relationship between design and the effectiveness of the finishing process. Mechanical finishing of parts in bulk using mass-finishing processes is among the oldest known process, dating back to the Roman empire. Mass finishing usually involves loading parts into a container usually with some media (the material that finishes the work) and a chemical compound. The container is moved and the contents rub against each other. Today, bulk finishing processes enable manufacturers to alter properties and achieve high-quality finishes in mass quantities at low unit cost. Consequently, designers should be knowledgeable about them. Through proper coordination of design and the shape-producing processes, finishing costs can be reduced, improved, or eliminated.

Abrasive Cleaning.   One of the most common steps preliminary to the application of decorative or protective surface treatments is the removal of sand or scale from metal parts. Sand often adheres to certain types of castings, and scale often results when metal has been processed at elevated temperatures. In some cases, sand may be removable by simple vibratory shaking, but often some type of *abrasive cleaning* is employed to remove such foreign materials. Some type of abrasive, usually sand, steel grit, or shot, is impelled against the surface to be cleaned. Fine glass shot may be used for some materials.

In *shot blasting,* a high-velocity air blast is used as the impelling agent. Air pressures from 60 to 100 psi (0.4 to 0.69 MPa) are used for ferrous metals and from 10 to 60 psi (0.07 to 0.4 MPa) for nonferrous metals. A nozzle with about ³⁄₈-in. diameter is often used. For large parts or for only a few parts, the blast may be directed by hand, as shown in Figure 40-1.

Small parts are usually sand- or shot-blasted inside an enclosed hood, with the parts traveling past stationary nozzles. When this is done manually, protective clothing and breathing equipment must be provided and precautions taken to prevent the resulting dust from being spread. Ordinarily, a separate, well-vented room or booth is used with suitable collection equipment so as to avoid air pollution.

Equipment that impels the abrasive particles by mechanical means using the centrifugal principle is more economical in the use of energy and less polluting.

If sand is used, it should be clean, sharp-edged silica sand. Steel grit will clean more rapidly than sand and causes much less dust. However, sand has a lower cost and somewhat greater flexibility.

Obviously, abrasive cleaning is effective only if the abrasive can reach all the areas of the surface that must be cleaned, which may be difficult if the surface is very complex. Abrasive cleaning cannot be used if sharp edges and corners must be maintained on the part because the process tends to round the edges. If the nozzles are directed manually, the labor cost tends to become high.

Barrel Finishing or Tumbling.   *Rotary barrel tumbling* has been used for centuries. In the Middle Ages, casks filled with abrasive stones and metal parts (pieces of armor) were rolled about the courtyard for days until the desired finishing was obtained. Today *barrel finishing* is used to deburr, radius, descale, remove rust, polish, brighten, surface-harden, and prepare parts for other finishing processes or assembly. The rate of stock removal can range from heavy grinding-type stock removal (.001 to .005 in.) to fine finishing to tolerances as close as .0001 in.

Parts are placed in a special barrel or drum until it is nearly full. Occasionally, the loaded barrel is rotated without the addition of any abrasive agents. However, in most cases, a media of metal slugs or jacks, or some abrasive, such as sand, granite chips, slag, or aluminum oxide pellets, are added. The rotation of the barrel produces a natural "landslide" action of media and parts as the barrel rotates. See Figure 40-2. During rotation, the media and workpieces rise in the barrel until gravity causes the top layer to tumble or slide downward. Only a small fraction of the load is finished at any time, resulting in long process times.

Mechanical Cleaning and Finishing

FIGURE 40-1   Manual shot blasting. (Courtesy Norton Company.)

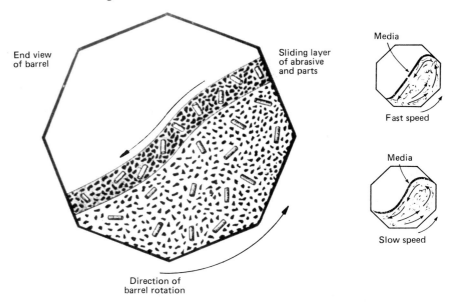

End view
of barrel

Sliding layer
of abrasive
and parts

Media

Fast speed

Media

Slow speed

Direction of
barrel rotation

FIGURE 40-2 Tumbling or barrel finishing provides low cost, mass finishing.

Increasing the rotation speed improves the processing time but the workpieces may be damaged. The sliding action produces a cutting action that usually will remove fins, flashes, scale, and sand. The process can be used only on parts that are sufficiently rugged to withstand the tumbling action. However, by a suitable selection of abrasives, fillers, barrel size, and speeds, and by careful packing of the barrel, an amazing range of parts can be tumbled successfully. Delicate parts should not shift loosely during tumbling. In some cases such parts must be attached to racks within the barrel so that they will not strike each other.

Tumbling is an inexpensive cleaning method. Various shapes of slug materials are used. Several shapes often are mixed in a given load so that some will reach into all sections and corners to be cleaned. Tumbling usually is done dry, but it can also be done wet. Obviously, tumbled parts will have rounded edges and corners. The equipment can be arranged so that loading and unloading are accomplished quickly and so that the slug material is separated from the workpieces by falling through suitable grid tables.

Barrel finishing is another term for tumbling. The finishing action abrades the surface and is controlled by the ratio of the volume of work to abrasive, the speed of rotation, the part geometry, and the shape of the abrasive pellets. The development of synthetically shaped media has greatly enhanced the process. Media come in a variety of shapes and sizes as shown in Figure 40-3. The different shapes and sizes permit the finishing of complex parts with irregular openings. Steel media in the shape of balls, shot, or rounded-end pins can be used to provide a burnishing action; no cutting is involved. Different compounds are added to the media depending on the objectives (cleaning, deburring, burnishing, descaling, rust inhibiting, abrasion). The primary disadvantage of barrel

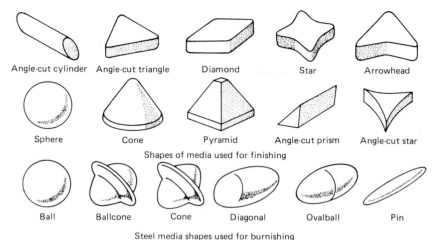

Angle-cut cylinder   Angle-cut triangle   Diamond   Star   Arrowhead

Sphere   Cone   Pyramid   Angle-cut prism   Angle-cut star

Shapes of media used for finishing

Ball   Ballcone   Cone   Diagonal   Ovalball   Pin

Steel media shapes used for burnishing

FIGURE 40-3 The development of synthetic shaped media has greatly improved the reliability of mass finishing.

finishing is that the finishing occurs equally on all surfaces and cannot be limited to one area. The processing cycle time is often slow, the process is noisy; the space requirements are high, and, while the process can be automated, other mass-finishing processes *(vibrating finishing)* are more easily automated.

In contrast to tumbling, *vibrating finishing* is done in open containers. See Figure 40-4. Tubs are either rectangular or round and are also referred to as bowls. The container is vibrated at 900 to 3600 cycles per minute. Vibratory finishing has shorter cycle time than barrel finishing because the entire load is under constant agitation. The process is less noisy and can be controlled and automated more readily; and open tubs allow inspection during processing.

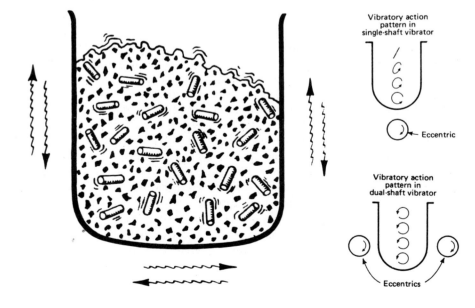

Vibratory action pattern in single-shaft vibrator

Eccentric

Vibratory action pattern in dual-shaft vibrator

Eccentrics

FIGURE 40-4 Vibratory finishing is performed with either single eccentric shaft or dual-shaft vibrator for faster processing times.

Vibrating finishing also does a better job on deburring internal recesses and on smoothing surfaces. The control of vibration frequency and amplitude depend on the part size, part shape, part weight, part material, media, and compound.

In *barrel burnishing,* no cutting action is involved. Instead, the slug material produces peening and rubbing actions, reducing the minute irregularities and producing an even surface. Since burnishing will not remove visible scratches or pits, in most cases the parts should be rolled or vibrated with a fine abrasive. However, the resulting surface is smooth, uniform, and free of porosity.

Barrel burnishing is normally done wet, using water to which some lubricating or cleaning agent has been added, such as soap or cream of tartar. Because the rubbing action between the work and the media is very important, the barrel should not be loaded more than half full with work and media, and the volume ratio of media to work should be about 2 to 1. The ratio should be such that the workpieces do not rub against each other. The speed of rotation of the barrel should be adjusted so that the workpieces will not be thrown out of the mass as they reach the top position and roll down the inclined surfaces. Balls from ⅛ to ¼ inch in diameter, pins, jacks, and ball cones commonly are used.

Centrifugal barrel tumbling is similar to barrel tumbling except that the barrel is placed at the end of a rotating arm. This adds a centrifugal force to the weight of the parts in the barrel and can make the speed of the process 25 to 50 times that of conventional barrel finishing.

Spindle finishing is also similar to barrel finishing except that the workpiece is attached to a rotating shaft and placed in the barrel with media moving in the opposite direction to that of part rotation. The abrasive removes burrs and produces a smooth radius on the edges. Deburring takes only 1 to 2 minutes per part.

Media. *Media* selection is critical to the success of any mass-finishing process. Table 40.1 gives the typical ratios of media to parts by volume. Synthetic media are usually preferred over natural media, which are random in shape and usually not as hard. The primary function of the media is to keep the parts from impinging on each other (cushioning) while cleaning and finishing. Some type of filler is often added to act as a carrying agent for the media and to prevent impingement. Scrap punchings, minerals, leather scraps, and sawdust are typical fillers. Natural abrasives include slag, cinders, sand, corundum, granite chips, limestone, and hardwood pegs, cylinders and cubes. Popular manufactured media use ceramics, polyesters, or resin plastics for a matrix with an abrasive. AL_2O_3 (ceramic), emery, flint, and silicon carbide are typical abrasives and represent 50%–70% by weight in the media. Media are usually made by casting. Steel media with no abrasive are used for burnishing and light deburring.

The need to prevent lodging in part recesses led to the development of many sizes and shapes of media. If media are not to lodge, they must be of definite, uniform shape, equilateral or angle-cut triangles, for example, or a variation, such as arrowheads, stars, rectangles, and diamond shapes. Cylinders, angle-cut cylinders, rods, or spheres are also available. Whatever the shape, the media must fit into holes and recesses without lodging. See Figure 40-5. All of these

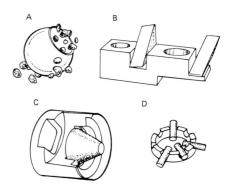

FIGURE 40-5 The selection of geometry of the media depends upon the part geometry.

TABLE 40.1   **Typical Media-to-Part Ratios for Mass Finishing**

Media/Part Ratio by Volume	Typical Application
0:1	Part-on-part processing or burr removal without media
1:1	Produces very rough surfaces and is suitable for parts in which part-on-part damage is not a problem
2:1	Somewhat less severe part-on-part damage, but more action from less media
3:1	May be acceptable for very small parts and very small media. Part-on-part contact is likely on larger and heavier parts
4:1	In general, a good average ratio for many parts. A good ratio for evaluating a new deburring process
5:1	Better for nonferrous parts subject to part-on-part damage
6:1	Suitable for nonferrous parts, especially preplate surfaces on zinc parts with resin-bonded media
8:1	For improved preplate surfaces with resin-bonded media
10:1	Produces very fine finishes

Source: American Machinist, August 1983.

shapes are produced in a variety of sizes (1/8 in. to more than 2 in.). Different combinations of abrasives and shapes perform tasks that range from light deburring with a very fine finish to fast cutting for a rough finish.

During finishing operations the matrix of the media, which is softer than the abrasive materials, is eroded, leaving the abrasive exposed on the surface to perform the desired work. Although ceramic abrasives normally produce a matte-type surface finish, the process parameters—especially the compound—can be changed to provide a very smooth finish.

Compounds.   The importance of the selection of the *compound* for the mass-finishing process is often underrated. The compound is a combination of compatible materials influenced by many factors. Compounds are liquid or dry, abrasive or nonabrasive, acidic, neutral, or alkaline.

Compound types include cleaning, deburring, burnishing, or special-purpose types, such as descaling, rust inhibitors, and abrasive compounds.

Cleaning compounds like dilute acids or soaps remove excessive soils from parts and media and are often used for parts coated with heavy oil and grease.

Deburring compounds are used for keeping parts and media clean and for inhibiting corrosion.

Burnishing compounds can range from mildly acidic to strongly alkaline. With steel media in a vibratory machine, mildly acidic compounds are normally used because they develop brighter colors. In addition, burnishing compounds can control part color and readily brighten surfaces.

Another function of the compound is to inhibit corrosion of both ferrous and nonferrous metals. Corrosion-inhibiting compounds are particularly important when steel media are used.

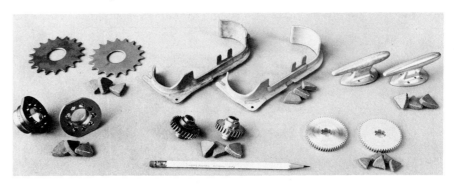

FIGURE 40-6   Parts before and after barrel finishing using triangular media. (Courtesy Norton Company).

In deburring and finishing operations, many small particles are abraded from both the media and the workpiece and must be suspended in the compound solution to prevent them from adhering to the parts.

One of the most important functions of the compound is to condition the water. Because water varies considerably in hardness, or metallic-ion content, depending on geographic locality, water-conditioning agents are generally used to isolate metallic ions and thus make the water ''soft.'' Consistent water conditions must be maintained to ensure uniform and repeatable finishing results.

The compound's cooling function is also important during this period. Steel media, for example, become hot and must be cooled.

Figure 40-6 shows several examples of parts before and after barrel finishing. Finishing time varies from 10 minutes for soft nonferrous parts to 2 or more hours for steel parts. Although they are batch processes, barrel and vibratory finishing are quite simple and economical. Sometimes the parts may be put through more than one barrel, using increasingly fine abrasives. Despite recent advances, mass finishing remains an art, not a science. Selection of media, equipment, and compounds is usually made by trial and error, with various methods tested until the desired result is achieved. Maintaining consistent results is even more difficult.

Because all of the factors—workpiece, equipment media, and compound— are interrelated and therefore interact, the change in one can cause the entire process to fail. Locating the source of the problem will be difficult.

Buffing.   *Buffing* is a polishing operation in which the workpiece is brought in contact with a revolving cloth buffing wheel that has been *charged* with a very fine abrasive, such as polishing *rouge*. Obviously, buffing is closely related to *lapping* (see Chapter 27) in that the cloth buffing wheel is a vehicle for the abrasive. The abrasive removes minute amounts of metal from the workpiece, thus eliminating fine scratch marks and producing a very smooth surface. When softer metals are buffed, there is some indication that a small amount of metal flow may occur that helps to reduce high spots and produce a high polish.

Buffing wheels are made of discs of linen, cotton, broadcloth, or canvas that are made more or less firm by the amount of stitching used to fasten the layers of cloth together. Buffing wheels for very soft polishing or for polishing into

FIGURE 40-7   Parts being buffed on an automatic machine. (Courtesy Murray-Way Corporation).

interior corners may have no stitching, the cloth layers being kept in proper position by the centrifugal force resulting from the rotation of the wheel. Various types of polishing rouges are available, most of them being primarily ferric oxide in some type of binder.

Buffing should be used only to remove very fine scratches or to remove oxide or similar surface coatings. If it is done manually, holding the work against the rotating buffing wheel, it is quite expensive because of the labor cost. However, semiautomatic buffing machines are available, such as the one shown in Figure 40-7 in which the workpieces are held in fixtures on a rotating circular worktable and moved past a series of individually driven buffing wheels. The fixtures can also be mounted on conveyors (rails), which increases the number of buffing stations. Such machines are usually housed in airtight rooms to reduce air pollution. The buffing wheels can be adjusted to desired positions so as to buff different portions of specific workpieces. If the workpieces are not too complex, very good results can be obtained quite economically with such equipment. Obviously, part design plays an important role when automatic buffing is to be employed.

Wire Brushing.   A high-speed, rotary wire brush sometimes is used to clean surfaces; it also does a minor amount of smoothing. *Wire brushing* can be done by hand application of the workpiece to the brush, but more commonly automatic machines are used in which the parts are moved past a series of rotating brushes similar to the procedure shown in Figure 40-7. Wire brushing normally removes very little metal except fine, sharp high spots. It produces a surface composed of fairly uniform fine scratches that, for some purposes, is satisfactory as a final finish or that can easily be removed by barrel finishing or buffing. Abrasive brushing can also be performed with plastic or fiber wheels.

Belt Sanding.   A simple and common method for obtaining smooth surfaces is by *belt sanding*. The workpieces are held against a moving abrasive belt until the desired degree of finish is obtained. A series of belts of varying degrees of fineness can be used. If flat surfaces are desired, the belt passes over a flat table at the point where the work is held against it.

Belt sanding, as used for finishing, cannot be considered a sizing operation because only sufficient sanding is done to remove high spots and thus produce a smoother surface. The resulting surface is composed of very fine scratches, their fineness depending on the grit of the belt.

Although greatly improved sanding belts now are available and fairly smooth surfaces can be obtained, belt sanding is a hand operation and therefore is quite labor-intensive. Consequently, it should not be used except when more economical methods cannot be utilized. It has the further disadvantage of not being effective where recesses or interior corners are involved. As a result, it is used primarily for delicate parts or where small quantities are involved.

Electropolishing.   *Electropolishing,* the reverse of electroplating, sometimes is used for polishing metal parts. The workpiece is made the anode in an electrolyte with a cathode added to complete the electrical circuit. In the resulting deplating, material is removed most rapidly from raised rough spots, producing a very smooth polished surface. Because it usually is not economical to remove more than about 0.001 in. of material, the process is used primarily to produce mirrorlike surfaces, and the initial surface must be quite smooth. A final finish of less than 2 microinches can be obtained if the initial roughness does not exceed 7 to 8 microinches rms.

Electropolishing was originated for polishing metallurgical specimens and later was adapted for polishing stainless steel sheets and parts. It is particularly useful for polishing irregular shapes that would be difficult to buff. For best results the metal should be fine-grained and free of surface defects.

Chemical Cleaning

At some stage in the finishing of virtually all metal products it is necessary to employ *chemical cleaning* to remove oil, dirt, scale, or other foreign material that may adhere to the surface so that subsequent painting or plating can be done successfully. One or more of three cleaning processes is used.

Alkaline Cleaning.   *Alkaline cleaning* employs such agents as sodium metasilicate or caustic soda with some type of soap to aid in emulsification. Wetting agents are often added to assist in obtaining thorough cleaning. The cleaning action is by emulsification of the oils and greases. Thus the solution must penetrate any dirt that covers them. It also is necessary to rinse thoroughly the cleansing solution from the work surface so that no residue is left.

The cleansing bath must be controlled to maintain a constant and proper pH value. Too high as well as too low pH levels can produce poor results. For steel, a pH value of 9 to 14 is typical.

Solvent Cleaning.   Solvents clean by dissolving soils, oils, fats, and the like in solvent derived from coal or petroleum. There are three main types:

1. Petroleum solvents such as kerosene and naphtha applied by wiping or immersion.
2. Chlorinated solvents such as trichloroethylene used in vapor degreasers.
3. Water-solvent mixtures called emulsions or diphase cleaners (separate layers of a solvent and an aqueous phase).

Emulsifiable solvents are formed by combining an organic solvent with a hydrocarbon-soluble emulsifying agent such as sulfonated castor oil with water, or blending a soap and an organic solvent such as kerosene with a small amount of water. An emulsion forms when the soil-and-solvent combination is flushed with water.

The work is dipped in the solvent solution and then rinsed once or twice. If the work is to be electroplated, it should have a subsequent treatment in an alkaline cleaner to remove any organic matter that remains on the metal. As a preparation for painting, solvent cleaning and rinsing are usually adequate. Solvent cleaning is used extensively for metals such as aluminum, lead, and zinc, which are chemically active and might be attacked by alkaline cleaners.

Vapor Degreasing. *Vapor degreasing* is widely used to remove oil from ferrous parts and from such metals as aluminum and zinc alloys, which would be attacked by alkaline cleaners. A nonflammable solvent, such as trichloroethylene, is heated to its boiling point, and the parts to be cleaned are hung in its vapors. The vapor condenses on the work and washes off the grease and oil. Excess vapor is condensed by cooling coils in the top of the vapor chamber. The grease and oil from the work are washed off into the liquid solvent, causing the bath to become dirty. Because they are only slightly volatile at the boiling temperature of the solvent, the vapor remains relatively clean at all times and so continues to clean effectively.

It must be remembered that vapor degreasing is effective only if the vapor condenses on the work. Thus the work must remain relatively cool. This offers no difficulty except in the case of thin sheets that contain considerable amounts of oil and do not have sufficient heat capacity to remain cool and thus condense enough vapor to bring about satisfactory cleaning.

Vapor degreasing is a rapid process and has almost no visible effect on the surface. Its major disadvantages are that vapor alone does not remove solid dirt and that it frequently must be followed by alkaline cleaning to remove remaining organic matter. If the surface has substantial solid dirt in addition to oil, this may be removed by passing the work through a boiling liquid, thereby removing most of the dirt and some of the oil, then through cold liquid to cool the work, and finally through hot vapor to remove the remaining oil.

Pickling. *Pickling* involves dipping metal parts in dilute acid solutions to remove the oxides and dirt that are left on the surface by various processing operations. The most commonly used pickling solution is a 10% sulfuric acid bath at temperatures from 150° to 185°F. Muriatic acid also is used, either cold or hot. When used cold, pickling baths have approximately equal parts of acid and water. At temperatures ranging from $100°$ to $150°$ more dilute solutions are used.

It is very important that parts be thoroughly cleaned prior to pickling. The pickling solution does not act as a cleaner, and any dirt or oil on the surface will result in an uneven removal of the oxides. Alkaline cleaning is usually employed for this purpose. Pickling *inhibitors,* which decrease the attack of the acid on the metal but do not interfere with the action of the acid on the oxides, are frequently added to the pickling bath.

After the parts are removed from a pickling bath, they should be rinsed thoroughly to remove all traces of acid and then dipped in a slightly alkaline bath to prevent rusting. When it will not interfere with further processing, a dip in cold *milk of lime* often is used. Parts should not be overpickled as this may result in roughening of the surface.

Ultrasonic Cleaning.   *Ultrasonic cleaning* is used extensively when very high quality cleaning is required for relatively small parts. In this process the parts are suspended or placed in wire baskets in a cleaning bath, such as *freon,* that contains an ultrasonic transducer operating at a frequency that causes *cavitation* in the liquid. Excellent results can be obtained in from 60 to 200 seconds in most cases. It usually is best to remove gross dirt, grease, and oil before doing ultrasonic cleaning.

Painting.   *Paints* are by far the most widely used finish on manufactured products. A great variety is available to meet a wide range of requirements. Today, most paints and enamels are synthetic organic compounds that dry by polymerization or by a combination of polymerization and adsorption of oxygen. Water is frequently the carrying vehicle for the pigments. Moderate amounts of heat can be used to accelerate the drying, but many synthetic paints and enamels will dry in less than one hour without the use of heat. The older oil-based paints and enamels usually have a long drying time and require excessive environmental protection measures and are therefore seldom are used.

Table 40.2 lists the more commonly used organic finishes and their important characteristics. *Nitrocellulose lacquers,* although very fast drying and capable of producing very beautiful finishes, are not sufficiently durable for most commercial applications. The *alkyds* are general-purpose paints but do not have sufficient durability for hard service conditions. The *acrylic enamels* are widely used for automobile finishes. *Silicones* and *fluoropolymers* are specialty finishes; their high cost is justified only when their special properties are important.

Asphaltic paints, which are solutions of asphalt in some type of solvent such as benzine or toluol, are still used extensively, especially in the electrical industry, when resistance to corrosion is required but appearance is not of prime importance.

Paint Application.   In manufacturing, almost all painting is done by one of four methods: *dipping, hand spraying, automatic spraying, or electrostatic spray finishing.* In most cases at least two coats of paint are required. The first

TABLE 40.2 **Commonly Used Organic Finishes and Their Qualities**

Material	Durability (Scale of 1–10)	Relative Cost (Scale of 1–10)	Characteristics
Nitrocellulose lacquers	1	2	Fast drying; low durability
Epoxy esters	1	2	Good chemical resistance
Alkyd-amine	2	1	Versatile; low adhesion
Acrylic lacquers	4	1.7	Good color retention; low adhesion
Acrylic enamels	4	1.3	Good color retention; tough; high baking temperature
Vinyl solutions	4	2	Flexible; good chemical resistance; low solids
Silicones	4–7	5	Good gloss retention; low flexibility
Fluoropolymers	10	10	Excellent durability; difficult to apply

(or prime) coat serves primarily to (1) assure adhesion, (2) provide a leveling effect by filling in minor porosity and other surface blemishes, and (3) improve corrosion resistance and thus prevent later coatings from being dislodged in service. These properties are less easily obtainable in the more highly pigmented paints that are used for final coats because of their better color and appearance. In using multiple coats, one must be sure that the carrying vehicles in the final coats do not unduly soften the previous coats.

Paint application by *dipping* is used extensively. The parts are either dipped manually into the paint or are passed down into the paint while on a conveyor. Obviously, all of the workpiece is coated, and thus it is a very simple and generally economical technique when all surfaces require painting. Consequently, it is used for prime coats and for small parts, which, if spray painted, would have significant loss of paint due to overspray. On the other hand, the unnecessary amount of paint used can make the process uneconomical if only some surfaces actually require painting or in cases where very thin, uniform coatings of some of the modern primers are adequate, particularly on large objects such as automobile bodies. Other difficulties with dipping are the tendency of the paint to run, thus producing a wavy surface, and the final drop of paint that usually is left at the lowest drip point. It also is essential that the paint in the dip tanks be kept stirred at all times and be of uniform viscosity.

Spray painting is probably the most widely used painting process because of its versatility and economy in the use of paint. The paint is atomized by three methods: by air, by mechanical pressure, or electrostatically. Either manual or automatic application is used, but high production of similar parts is almost always done with an automatic system. When hand spraying by air or mechanical means, the worker must exercise considerable skill in obtaining proper coverage without allowing the paint to "run" or "drape." Consequently, only a very thin film can be deposited at one time—usually not over 0.001 in.—if conventional methods are used. As a result, several coats must usually be applied with some intervening time for drying. Somewhat thicker coatings can be applied in one operation by using a *hot spray* method. In this procedure, the paint is sprayed on while hot.

Obviously, spray painting by hand is costly from a labor viewpoint and is often replaced by automatic methods. The simplest type of automatic equipment consists of a chain conveyor on which the parts are moved past a series of spray heads. However, if regular spray heads are used, results are often not satisfactory. A large amount of the paint may be wasted and it is difficult to get uniform coverage. Robotic spray-finishing systems use robots capable of mimicking the movement of a human painter. Robots are being used in monotonous and repetitious situations wherein a complete and uniform application of material is required. Robotic spray painting also removes the human from an unpleasant, even unhealthy environment.

Good results are obtained from either manual or automatic spray painting by using the *electrostatic principle*. An air spray gun atomizes the paint, giving the particles a negative electrostatic charge and considerable velocity. The atomized particles behave like tiny magnets. They are attracted to and deposited on the work, which is grounded electrically. When electrostatically charged paint droplets enter the field of the grounded target, they follow the lines of the force field to the target. The higher the DC voltage, the greater the attraction and wrap-around affect.

Electrostatic finishing can result in a 60%–80% savings in coating materials compared to other techniques in which a great deal of paint is lost by overspraying or blow-by. Electrostatic air-spray systems provide the finest possible finish (for refrigerators and automobiles), great material savings, and reduced emissions. The disadvantage of the process is that part edges and holes receive a heavier buildup due to the concentration of force lines on any sharp edge. Because only a minimal amount of paint will reach recessed areas, a separate conventional spray process may be needed to perform manual touch ups.

In *airless electrostatic painting,* the paint is fed onto the interior of a rapidly rotating cone or disc that is one electrode of a high-potential electrostatic circuit. The rotation of the cone or disc causes the paint to flow outward to the edge by centrifugal force. As the thin film of paint reaches the edge and then is spun off, the particles are charged electrostatically and atomized without the need for any air pressure. With the workpiece being the other electrode of the circuit, the paint is transferred, as in the previously described method. The primary advantage of this method, which is illustrated in Figure 40-8, is that because no air pressure is used for atomization, there is less spray loss, resulting in a higher efficiency in the use of paint (as high as 99%). Furthermore, fumes are much less of a problem.

Airless systems can also use hydraulic pressure to atomize the paints by discharging them through a small orifice at 500 to 4500 psi. The atomized particles have sufficient momentum to be carried to the target.

FIGURE 40-8 Electrostatically finishing aluminum extrusions, using two reciprocating disks. (Courtesy Ransburg Corporation.)

Airless electrostatic spray painting is used for machinery, lawn mowers, earth-moving equipment, outdoor furniture, and other applications for which over-spray must be kept to a minimum and a coating buildup of .002 to .003 in. is desired.

Electrocoating is the most recent basic development in paint application. It permits the economy of ordinary dip painting to be achieved but overcomes its disadvantages while permitting thinner and more uniform coatings and superior coverage in interior recesses. The principle of the process is shown in Figure 40-9. The paint particles, in a water solvent, are given an electrostatic charge by applying a DC voltage between the tank (cathode) and the workpiece (anode). As the workpiece enters and passes through the tank, the paint particles are attracted to and deposited on it in a uniform, thin coating from 0.0008 to 0.0015 in. thick. When the coating reaches the desired thickness, determined by controlling the conditions, no more paint is deposited. The water in the deposited film is drawn away by electroosmosis, leaving a coating that is composed of more than 90% resins and pigments. The workpiece is removed from the dip tank, rinsed by a water spray, and baked for 10 to 20 minutes at about 375°F, depending upon the particular paint.

Electrocoating is especially suitable for applying the prime coat to complex metal structures, such as automobile bodies, when good corrosion resistance is necessary. The flow of paint to hard-to-reach areas can be improved by placing

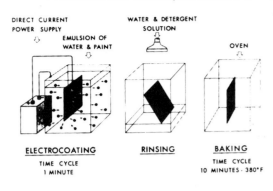

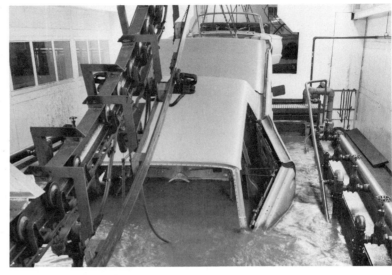

FIGURE 40-9 (Left) Steps in the electrocoating process. (Right) Electrocoating automobile body at Ford Motor plant.

electrodes in the workpiece at strategic locations. In addition, because the solvent is water, there is no fire danger as exists when large-area tanks are employed with regular dipping primers. As shown in Figure 40-9, electrocoating is readily adapted to conveyor-line production.

A new development is the application of paint in powder form by an electrostatic spray process. Several coats, such as primer and finish, can be applied and then followed by a single baking, instead of baking after each coat, as in conventional spray processes.

Most paints and enamels used in manufacturing require from 2 to 24 hours to dry at normal room temperatures. This obviously is not practical. They can be dried satisfactorily in from 10 minutes to 1 hour at temperatures of from 275 to 450°F. Consequently, some drying at elevated temperatures is usually done, using either a baking oven or more often a tunnel or panel of infrared heat lamps. The heat lamps involve relatively low investment, do not require much floor space, and are very flexible.

Although drying at elevated temperatures can be accomplished without difficulty on metal parts, this is not the case with wooden products. The temperatures are high enough to expand the gases, moisture, and sap that are in the wood (even though the wood is quite dry) and forces them to the surface after the paint has started to harden. They form small bubbles that roughen the surface or, if they break, leave small holes in the painted surface.

Metal/Oxide Coatings and Platings

Hot-Dip Coatings. Large quantities of metal parts are given corrosion-resistant coatings by being dipped into certain molten metals. Those most commonly used are zinc, tin, and an alloy of lead and tin.

Hot-dip galvanizing is the most widely used method of providing steel with a protective coating. After the parts, or sheets, have been cleaned, they are *fluxed* by dipping them into a solution of zinc chloride and hydrochloric acid. Next they are dipped into a molten zinc bath. The resulting zinc coating is complex, consisting of a layer of $FeZn_2$ at the metal surface, an intermediate layer of $FeZn_2$, and an outer layer of pure zinc. Hot-dip galvanizing provides corrosion resistance.

The coating thickness should be controlled. Coatings that are too thick crack and peel. A wide variety of "spangle" patterns can be obtained by proper processing. When done properly, bending and forming processes can follow galvanizing without damaging the coating. However, rimmed steel should not be galvanized.

Tin plating is a hot-dip coating process. After the steel has been cleaned, it is dipped into a molten tin bath. The bath has a layer of zinc chloride floating on top of the molten tin. In this manner the work passes through the zinc chloride before entering the molten tin. As the work leaves the tin bath, it passes through rollers immersed in palm oil, thus removing the excess tin. Most tin plate is now produced by an electroplating process that gives more uniform coating with less tin being required.

Terne coating is similar to hot-dip coating, but an alloy of 15% to 20% tin and the remainder lead is used in place of pure tin. This process is therefore cheaper than tin coating and provides satisfactory corrosion resistance for some purposes.

Phosphate Coatings. Two phosphate coating processes are used to provide corrosion resistance, usually to steel. In these processes the surface of the metal is converted into an insoluble crystalline phosphate by treatment with a dilute acid phosphate solution.

Parkerizing produces a fairly corrosion-resistant coating from 0.00015 to 0.0003 in. The treatment requires about 45 minutes and provides quite good corrosion resistance for parts that are to be painted. *Bonderizing* is similar to Parkerizing, but its primary purpose is not to give corrosion resistance but to form a surface to which paint will adhere tightly. The coating is thinner than that obtained by Parkerizing, but it reduces the chemical activity of the metal surface so that corrosion at the paint–metal interface is retarded. As a result, if the paint coat is scratched, there is less likelihood of rust starting and progressing and thereby causing the paint adjacent to the scratch to loosen.

Blackening. Many steel parts are treated to produce a black, lustrous surface that will be resistant to rusting when handled. Such coatings usually are obtained by converting the surface into black iron oxide. One method is to heat the parts in a closed box of spent carburizing compound at 1200°F for about 1½ hours and then quench them in oil. Another method consists of immersing the parts in special blackening salts at 300°F for about 15 minutes. A third method is to heat the parts in a rotary retort furnace to about 750°F. A small quantity of linseed or fish oil is then added. After a few minutes the parts are removed from the furnace, spread out, and allowed to cool. When they have cooled they are dipped into an oil that helps retard rusting.

A *gun-metal finish* is obtained by heating the parts in a retort with a small amount of charred bone to 750°. When the parts are oxidized, they are allowed to cool to about 650°F. A mixture of bone and some carbonic oil is then added and the heating continued for several hours. The work is then removed from the furnace and dipped in sperm oil.

Large quantities of both metal and plastic parts are *electroplated* to provide corrosion or wear resistance, improved appearance (such as color or luster), or an increase in dimensions. Plating is applied to virtually all base metals—copper, brass, nickel–brass, aluminum, steel, and zinc-base die castings—and also to plastics. If plastics are to be plated, they must first be coated with some electrically conductive material.

The most common plating metals are tin, cadmium, chromium, copper, gold, platinum, silver, and zinc. Except for the making of tin plate for the container industry, chromium is by far the most common metal used for plating. However, in most cases a thin layer of both copper and nickel is deposited before the chromium. Gold, silver, and platinum are very important plating metals in the jewelry and electronics industries. Although the methods may vary somewhat in details, all electroplating processes are essentially the same. The basic process is indicated in Figure 40-10. The parts to be plated become the cathodes and are suspended in a solution that contains dissolved salts of the metal to be deposited. The anode is the metal to be deposited. Other materials may be added to the electrolyte to increase its conductivity. When a DC voltage is applied, metallic ions migrate to the cathode and, upon losing their charges, are deposited as metal upon it.

Successful plating depends greatly on (1) the preparation of the surface, (2) the ability of the bath to produce coatings in recessed areas, and (3) the crystalline character of the deposited metal. The actual deposition of the plating metal is governed by the bath composition and concentration, the bath temperature, and the current density. These are interdependent and must be carefully controlled to obtain satisfactory and consistent results.

Surfaces that are to be electroplated must be prepared properly for satisfactory results. All defects, pin holes, and scratches must be removed if a smooth, lustrous finish is desired. The surface must be chemically clean. Proper combinations of degreasing, cleaning, and pickling are used to assure a clean surface to which the plating material will adhere.

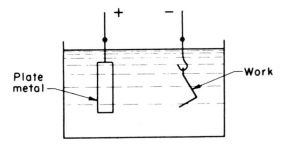

FIGURE 40-10   Basic electric circuit for electroplating.

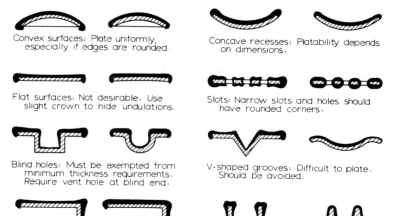

Convex surfaces: Plate uniformly, especially if edges are rounded.

Concave recesses: Platability depends on dimensions.

Flat surfaces: Not desirable. Use slight crown to hide undulations.

Slots: Narrow slots and holes should have rounded corners.

Blind holes: Must be exempted from minimum thickness requirements. Require vent hole at blind end.

V-shaped grooves: Difficult to plate. Should be avoided.

Sharply angled edges: Plating is thinner in center areas. Round all areas.

Fins: Increase plating time and costs. Reduce durability of finish.

FIGURE 40-11   Design details for electroplating.

Plating solutions are chosen on the basis of their ability to deposit metal in recesses. Cyanide solutions have better ''throwing power'' than acid solutions and are therefore commonly used, although they are more dangerous to handle.

The plating metal tends to be attracted to and build up on corners and protrusions (see Figure 40-11). This makes it difficult to obtain uniform plating thickness on parts of irregular shape containing recesses and interior corners. Improved results can often be obtained by using several properly spaced anodes, or anodes having a shape similar to that of the workpiece.

Nickel plating provides good corrosion resistance but does not retain its luster and is expensive. Consequently, it has largely been replaced as an outer coating by chromium when appearance is important and by cadmium for many applications for which appearance is not of much importance and only moderate corrosion resistance is required. Chromium seldom is used alone. In modern practice, the first layer usually is bright acid copper, which produces a leveling effect and makes it possible to reduce the thickness of the nickel layer that follows. The nickel layer need not exceed 0.0003 to 0.0006 in. Chromium is then plated as the final layer to provide both protection and appearance.

Electroplating is typically done as a continuous process. The parts to be plated are hung from conveyors and are lowered into successive plating, washing, and fixing tanks wherein the various operations are performed. Such methods make it possible to obtain economical plating when the volume of work is high. Ordinarily, only one type of workpiece can be plated at a time because the solutions, timing, and conditions of current density must be changed when different sizes and shapes are to be processed.

Hard chromium plating is used to build up worn parts to larger dimensions, to coat tools to reduce friction and wear, and to resist wear and corrosion. Hard chromium coatings are always applied directly to the base material and usually are much thicker than ordinary chrome plating, commonly from 0.003 and 0.010 in. thick. However, greater thicknesses—up to 0.030 in.—are used for such items as diesel cylinder liners. The hardness of hard chrome plating typically is from 66 to 70R_c. Hard chrome plating does not have a leveling effect, and therefore defects or roughness in the base surface will be amplified and made more apparent. Very smooth surfaces can be obtained by subsequent grinding and polishing processes.

Cadmium-titanium plating is used successfully to provide corrosion resistance for high-strength steels that are subject to hydrogen embrittlement when plated with either zinc or cadmium alone.

Anodizing.   *Anodizing* is a process widely used to provide corrosion-resistant and decorative finishes to aluminum. It is somewhat the reverse of electroplating in that (1) the work is made the anode in an electrolytic circuit and (2) instead of a layer of material being added to the surface, the reaction progresses inward, increasing the thickness of the highly protective but thin aluminum oxide layer that normally exists on aluminum.

One of the common forms of anodizing uses a 3% solution of chromic acid as the electrolyte at a temperature of about 100°F. The voltage is raised from 0 to about 40 at the rate of about 8 volts per minute and then is maintained at full voltage for about 30 to 60 minutes with a current density of about 1 to 3 amp/ft^2. This treatment produces a converted layer that is about 0.00005 to 0.0001 in. thick. It is used primarily on aircraft materials. Because the coating is integral to the part, the metal can be subjected to quite severe forming and drawing operations without destroying the coating or reducing its protective qualities. Parts that are anodized in this manner usually have a grayish-green color resulting from the presence of chromium in the coating. Other colors can be obtained by the use of suitable dye materials. The anodized surface also provides a good paint base.

More complex anodizing treatments are often used, some of which are *Alumilite* finishes. The most common of these uses a solution containing 15% to 25% sulfuric acid. This produces a transparent coating on pure aluminum and an opaque coating on alloys. These coatings are submicroscopically porous. A wide variety of colors can be obtained by the use of suitable dyes that penetrate these pores. Some of the colors will not resist sunlight, but it is possible to use a special type of dye and a sealing process that will produce good sun resistance.

Inasmuch as anodyzing does not add to the dimensions, there is no necessity for providing any dimensional allowance, as must be done when electroplating is used.

Electroless Nickel Plating.   Because it is almost impossible to obtain a uniform plating thickness on even moderately complex shapes with electroplating, and because of the large amounts of energy consumed, extensive work has been done in developing *electroless plating,* with major success in plating with nickel. Basically, electroless plating is accomplished by autocatalytic reduction of the metallic ion of the plating metal in an aqueous solution in which the workpiece acts as the catalyst. In the case of electroless nickel plating, sodium hypophosphite acts as the reducing agent, reducing nickel salts to nickel metal and, incidentally, supplying a small amount of phosphorus so that the resulting plating material is a solid solution of phosphorus (about 8%) nickel.

In addition to having as good corrosion resistance as electroplated nickel, electroless nickel has an as-deposited hardness of about 49 to $55R_c$, which can be increased to as high as $80R_c$ by suitable heat treatment. Also, because this is purely a chemical process, the coatings obtained are uniform in thickness, not being affected by part complexity.

Nickel-Carbide Plating. A very useful electroless plating process has been developed wherein minute particles of silicon carbide are codeposited in a nickel-alloy matrix. As shown in Figure 40-12, the particles of silicon carbide are 0.000,04 to 0.0001 in. in size and constitute about 25% of the volume of the deposit. The coating is as corrosion-resistant as nickel. Because of the very high hardness of the carbide particles (about 4500 on the Vickers scale, where tungsten carbide is 1300 and hardened steel about 900, or $62R_c$), the resistance of such coatings to wear and abrasion is outstanding. As with electroless nickel plating, the thickness of the coating is not affected by part shape.

Coating thicknesses typically are up to 0.008 in. The process has a wide range

FIGURE 40-12   (Left) Photomicrograph of nickel-carbide coating produced by electroless plating showing uniform thickness of deposit on irregularly shaped part (Right) High magnification section. (Courtesy Electro-Coatings Inc.)

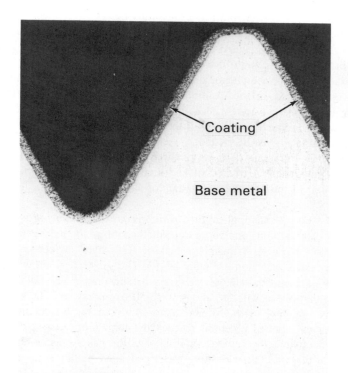

Coating

Base metal

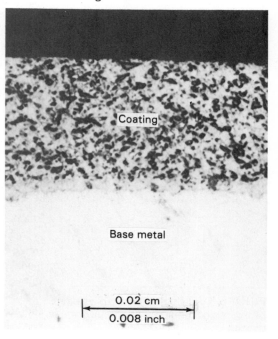

Coating

Base metal

0.02 cm

0.008 inch

of applications, being of outstanding value for coating molding dies for plastics that contain substantial amounts of abrasive-filled materials, such as glass fibers.

A modification of this process utilizes minute artificial polycrystalline diamond particles in place of silicon carbide. Such coatings usually do not exceed 0.001 in. in thickness, and they have outstanding wear resistance.

Impact plating, obtained by tumbling the parts in a tumbling barrel that contains a water slurry of very fine powder particles in the plating metal, glass spheres, and a "promotor" chemical, can be used to obtain a thin coating that is satisfactory for some purposes. The small glass balls peen the fine powder particles onto the workpieces, producing some cold welding. The deposited coatings are lamellar in structure and quite uniform in thickness. Any metal that can be obtained in a very fine powder form can be used as a plating material. One advantage of the process is that there is no danger of hydrogen embrittlement; therefore, it can be used on hardened steel.

Vaporized Metal Coatings

Vacuum Coating.   *Vacuum deposition or coating* is widely used to deposit thin films of metal and metal compounds on various substrate materials. The process involves the evaporation of the metal or the compound in a high vacuum and the subsequent condensation of the vapor on the cool workpiece. A pressure (vacuum) of from 0.012 to 1.33 Pa usually is required. Such coatings are used as electrical conductors and resistors in the electronics industry, as decorative coatings, and as reflective surfaces. Virtually any metal can be deposited by the vacuum-coating process, aluminum, chromium, gold, nickel, silver, germanium, and platinum being very common. The coatings are usually less than 0.00002 in. in thickness and often are as little as 1 microinch.

Vaporization is a surface phenomenon and does not constitute boiling. As the metal leaves the heated surface in atomic form, it travels in line-of-sight direction to the surface of the substrate. Thus, if an entire surface of a shape is to be coated, it must be rotated to expose all the surfaces. Fortunately, in most cases only a single surface must be coated.

Because of the very thin deposits required, very little metal is needed. Expensive materials, such as gold or silver, often can be used economically over inexpensive part materials, such as plastics or steel.

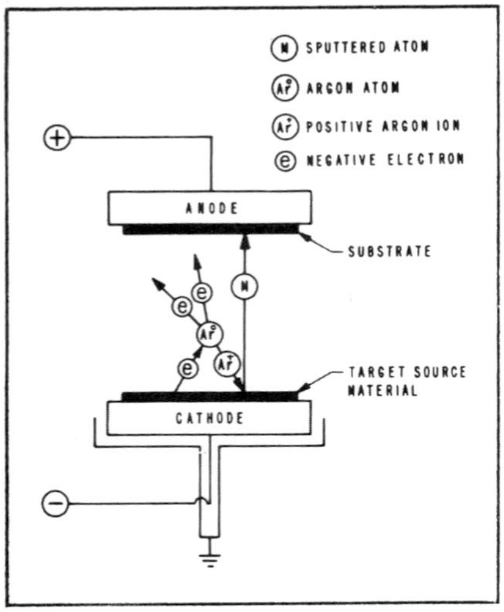

Sputtering.   For certain applications, *radio-frequency sputtering* is used as a substitute for electroplating and vapor-deposited coatings. Its most extensive use is for depositing thin films of metals in making solid-state devices and circuits. The basic process is indicated in Figure 40-13. The substrate upon which metal is to be deposited and the source metal are arranged in a gas-filled chamber (often argon) that is evacuated to about 10 to 50 μm pressure. The substrate is made positive, relative to the source material, by a radio-frequency power source. When the applied potential reaches the ionization energy of the gas, electrons, generated at the cathode, collide with the gas atoms, ionizing them and creating a plasma. These positively charged ions, having high kinetic energy, are accelerated toward the cathode target, overcoming the binding energy

FIGURE 40-13   Schematic diagram of the radio-frequency sputtering process.

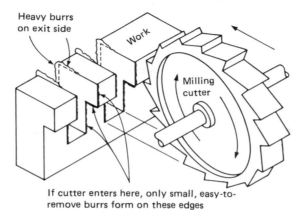

Heavy burrs
on exit side

Work

Milling
cutter

If cutter enters here, only small, easy-to-
remove burrs form on these edges

FIGURE 40-14 Heavy burrs will
form on the exit side of a milled slot.
(Source: L. X. Gillespie, *American
Machinist*, November 1985.)

of the target material, and dislodge atoms that then travel across the electrode gap and are deposited on the substrate. Because of the energy of these atoms, usually between 15 and 50 electron volts, their adherence to the substrate is considerably better than if they were deposited by ordinary vacuum evaporation. This technique has been used to coat the edges of razor blades with chromium to prevent corrosion on the blade edges.

Burr Removal

Burrs form on the edge of all machined and stamped parts and must, in most cases, be removed. For example, the milling process leaves burrs on the exit side of the cut, as can be seen in Figure 40-14. Burrs are generally small, sometimes flexible projections adhering strongly to the edge of a workpiece. Burrs are typically only .003 in. thick and .001 to .005 in. high. If they are not removed, they can cause assembly failures, short circuits, injuries to workers, and even fatigue failures.

Processes previously described that are used for deburring are grinding, chamfering, barrel tumbling, vibratory finishing, centrifugal and spindle finishing, abrasive jet, water jet, wire brushing, belt sanding, chemical machining, electropolishing, ECM, filing, ultrasonic machining, and abrasive machining. Some other deburring processes are liquid-hone deburring using a 60 grit abrasive and thermal energy deburring. Thermal energy deburring uses a high temperature wave front—produced by igniting natural gas in a closed container—to vaporize the burrs. The short duration wavefront heats the burr to 6000°F while the part is exposed to only 200°F. Up to 80 parts per hour can be deburred by this process.

Of all the processes, tumbling and vibratory finishing are usually the most economical. In sufficient quantities, deburring by these processes adds 1 to 6 cents per part. Most deburring processes remove metal from exposed surfaces while removing burrs. Because burr size is not controllable it is important to allow some tolerances on part dimensions and edge radii for deburring. Table 40.3 provides a listing of many deburring processes and suggests allowances

TABLE 40.3   Allowances Recommended for Deburring Processes[a]

Process	Edge Radius, mm (in.)	Stock Loss, mm (in.)	Surface Finish, μm AA (μin. AA[b])
Barrel tumbling	0.08–0.5 (0.003–0.020)	0–0.0025 (0–0.001)	1.5–0.5 (60–20)
Vibratory deburring	0.08–0.5 (0.003–0.020)	0–0.025 (0–0.001)	1.8–0.9 (70–35)
Centrifugal barrel tumbling	0.08–0.5 (0.003–0.020)	0–0.025 (0–0.001)	1.8–0.5 (70–20)
Spindle finishing	0.08–0.5 (0.003–0.020)	0–0.025 (0–0.001)	1.8–0.5 (70–20)
Abrasive-jet deburring	0.08–0.25 (0.003–0.010)	0–0.05 (0–0.002)[c]	0.8–1.3 (30–50)
Water-jet deburring	0–0.13 (0–0.005)(p)	0(p)	
Liquid hone deburring	0–0.13 (0–0.005)	0–0.013 (0–0.0005)	
Abrasive-flow deburring	0.025–0.5 (0.001–0.020)	0.025–0.13 (0.001–0.005)[d]	1.8–0.5 (70–20)
Chemical deburring	0–0.5 (0–0.002)	0–0.025 (0–0.001)	1.3–0.5 (50–20)
Ultrasonic deburring	0–0.05 (0–0.002)	0–0.025 (0–0.001)	0.5–0.4 (20–15)
Electrochemical deburring	0.05–0.25 (0.002–0.010)	0.025–0.08 (0.001–0.003)[e]	
Electropolish deburring	0–0.25 (0–0.010)	0.025–0.08 (0.001–0.003)[e]	0.8–0.4 (30–15)
Thermal-energy deburring	0.05–0.5 (0.002–0.020)	0	1.5–1.3(p) (60–50)
Power brushing	0.08–0.5 (0.003–0.020)	0–0.013 (0–0.0005)	
Power sanding	0.08–0.8 (0.003–0.030)[f]	0.013–0.08 (0.0005–0.003)	1.0–08 (40–30)
Mechanical deburring	0.08–1.5 (0.003–0.060)		
Manual deburring	0.05–0.4 (0.002–0.015)[g]		

[a]Based on a burr 0.08 mm (0.003 in.) thick and 0.13 mm (0.005 in.) high in steel. Thinner burrs can generally be removed much more rapidly. Values shown are typical. Stock-loss values are for overall thickness or diameter. Location A implies that loss occurs over external surfaces, B that loss occurs over all surfaces, and C that loss occurs only near edge. (p) indicates best estimate.
[b]Values shown indicate typical before and after measurements in a deburring cycle.
[c]Abrasive is assumed to contact all surfaces.
[d]Stock loss occurs only at surfaces over which medium flows.
[e]Some additional stray etching occurs on some surfaces.
[f]Flat sanding produces a small burr and no radius.
[g]Chamfer is generally produced with a small burr.
Source: L. X. Gillespie, *American Machinist,* November 1985.

for that process, assuming that the part has a "typical burr" of 0.003 in. of thickness.

Designing to Eliminate the Need for Burring.
Knowing how and where burrs are likely to form is valuable information for the engineer who designs parts so that the burr is easy to remove or even to eliminate. The addition of a recess or a groove can eliminate the need for deburring as shown in Figure 40-15. The burr produced by a cutoff tool or slot milling cutter is contained below the surface. This approach presumes that it is cheaper to perform an additional operation (undercutting or grooving) than it is to remove the burr.

Chamfers added to sharp corners often eliminate the need to deburr. The chamfering tool removes the large burrs formed by facing, turning, and boring and produces a relief for mating parts. The burrs produced by chamfers are small and can be easily removed if necessary. Edge breaks should be specified to allow for either a chamfer or a radius. This allows manufacturing the freedom to determine whether a machining or a deburring process provides the most economical solution. A typical corner break has a 1/64 in. radius and an included angle of 45 degrees. Radii should not be specified larger than 0.020 in. or smaller than .003 in.

Influence of Processing on Subsurface State.
It is important to understand that the processes themselves impart certain properties to the materials which will influence their performance when put into service. While satisfactory performance of manufactured products is obviously dependent on the design, the manufacturing, and the assembly, failures of component parts in service (often resulting in product liability claims) usually arise from a combination of factors. Here is how some typical processes affect the workpiece properties.

Various machining processes produce widely varied surface textures (roughness, waviness, and lay) in the workpieces. In addition, different processes

FIGURE 40-15 Designing extra recesses and grooves into a part may eliminate the need to deburr. (Source: L. X. Gillespie, *American Machinist*, November 1985.)

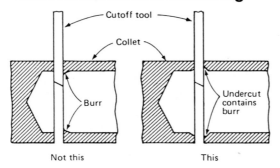

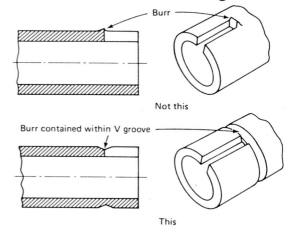

produce changes in the physical or metallurgical properties in the near surfaces of the component parts. For the most part, these changes take place, in the subsurface, to a depth of .005 to .050 in. below the surface. The effect of these changes can be beneficial or detrimental, depending on the process used to create the surface. The process selected is often dictated by the material selected and the functional design.

Machining processes (both chip-forming and chipless) cause plastic deformation. Surfaces, when cut, generally are left with a tensile residual stress, microcracks, and a hardness different from that of the bulk metal. The EDM process leaves a layer of hard, recast metal on the surface which usually contains microcracks. Ground surfaces can have either residual tension or residual compression stresses, depending on the mix between chip formation and plowing and rubbing during the grinding. In some materials ground at high speed, phase transformations can occur in the subsurface.

Roller burnishing, discussed in Chapter 19, is often used as a finishing process. The rolls impact a smooth surface with a residual compressive stress. Shot peening or tumbling processes can impart a residual compressive stress. Welding processes leave residual stresses in the metal near the weld joint due to shrinkage of the molten weld metal as it cools and contracts. Similar shrinkage problems create residual stresses in castings. In summary, the principal causes of alterations in the subsurface state, due to material processing operations, are:

1. Plastic strain or plastic deformation.
2. Large temperature gradients or high temperatures.
3. Chemical reactions.
4. Differential shrinkage.

To understand the subtle nature of this problem, look at Figure 40-16, which shows the depth of subsurface damage due to machining as a function of back rake angle of the tool. Suppose production control called for an increase in the production rate which, in turn, forced the manufacturing engineer to change

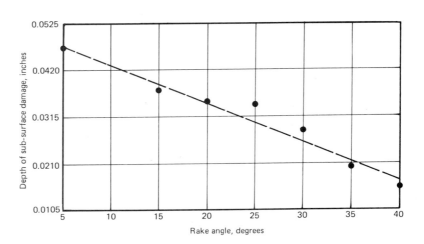

FIGURE 40-16 The depth of damage in a machined part surface increases with decreasing rake angles.

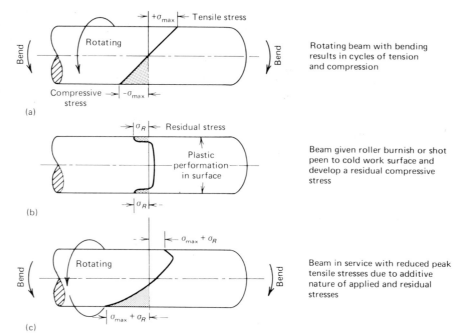

Rotating beam with bending results in cycles of tension and compression

(a)

Beam given roller burnish or shot peen to cold work surface and develop a residual compressive stress

(b)

Beam in service with reduced peak tensile stresses due to additive nature of applied and residual stresses

(c)

FIGURE 40-17 Lowering of peak tensile stresses in a rotating beam with bending by adding residual compressive stress during manufacture.

tool materials in order to be able to increase cutting speed, to meet the new production rates. Thus, he changed from a high-speed steel tool with a large rake (30 degrees) to a carbide tool with a small rake (5 degrees) but still got the same surface finish. The result was a doubling of the depth of surface damage, possibly causing the part to fail in service, whereas before it had performed quite admirably.

Residual stresses are often the product of nonuniform or localized plastic deformation. Since this describes most metalworking or machining processes, most components will have some residual stresses left in them after processing. These internal or ''locked-in'' stresses can be detrimental or beneficial in terms of fatigue behavior or corrosion resistance in the parts.

Suppose a round beam has been machined. When placed in service, the beam has a load on it so that it is bent while rotating. This results in a cyclic fatigue situation shown in Figure 40-17. The residual compressive stress has the effect of lowering the peak tensile stress in the surface, thereby enhancing the fatigue life of the part. What is important to understand here is that the final process can have a significant influence on the component's performance. For example, Table 40.4 shows some test results for fatigue strength in reverse cantilever bending as a function of eight different surface finishes. Notice that for a life of 10^7 cycles, the same part will sustain almost five times the load if the surface is prepared by ultrasonic machining rather than by EDM.

Figure 40-18 shows the results of another study wherein specimens were prepared by milling and turning and then either polished, shot-peened, or roller-burnished. Only the average line is shown for clarity. Suppose that the average

TABLE 40.4 Fatigue Strength in Reverse Cantilever Bending of Ti—5Al–2.5Sn as a Function of Surface Generation Method for a Life of 10^7 Cycles

Surface Generation Method	Strength[a] (psi)
Ultrasonic	98,000
Slab mill	86,000
Chemical mill + vacuum anneal	77,000
Shot peen	76,000
As rolled and received	61,000
Chemical mill	59,000
Ground	52,000
Electric discharge machined	21,000

[a]Average values.
Source: Data from Rooney, WADC report.

applied stress in this situation was around 41,000 to 42,000 psi. The difference in life between a milled specimen and one that has been milled and roller-burnished is 610,000 cycles (190,000 to 700,000 cycles). To put it another way, roller burnishing demonstrated the ability to improve the life of the part sevenfold! Similar results have been obtained in the area of stress corrosion resistance. Both the design engineer and the manufacturing engineer should always be aware of the functional behavior of the component part in that the processes can greatly influence the life of the part in service. Quite often the worst environment that a part sees in its lifetime occurs during its manufacture.

FIGURE 40-18 Variation in fatigue life of rotary bend 2024-T4 aluminum specimens as a function of surface-finishing processes.

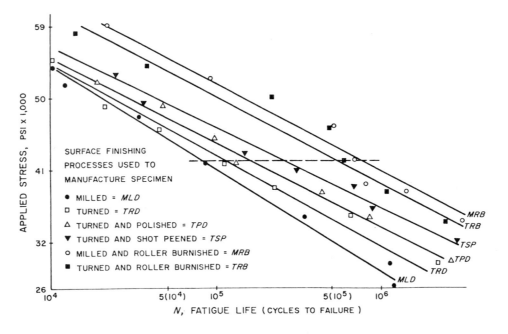

Review Questions

1. Name three manufacturing processes that inherently result in the need for surface-cleaning and smoothing operations.
2. Why should a product designer be concerned about the finishing operations that may be required on the product?
3. What are three basic cleaning methods?
4. What are the primary limiting factors in the use of abrasive cleaning?
5. The barrel should not be filled too full when tumbling is used for cleaning. Why?
6. What materials are used as compounds in mass finishing?
7. How are delicate parts cleaned by tumbling?
8. When wire brushing is used for surface cleaning, what is the effect of that process on the surface?
9. On what type of surfaces is belt sanding effective?
10. What is the requisite condition for barrel finishing to be effective?
11. What is the basic differences in the action obtained in rolling and vibratory finishing barrels?
12. Basically, what type of process is buffing? That is, what other processes are closely related to buffing?
13. How is the stiffness of buffing wheels controlled?
14. Why is part design so important if buffing or belt sanding must be used to finish the part?
15. What type of surface is produced by barrel burnishing?
16. Explain how electropolishing produces a smooth surface.
17. What is cavitation?
18. Why is more than one of the three basic chemical cleaning processes often used on a given product?
19. Why may vapor degreasing not work well on parts made of thin aluminum?
20. What is the absolute condition that must be met for vapor degreasing to clean satisfactorily?
21. What is the purpose of pickling?
22. Under what conditions is ultrasonic cleaning usually employed?
23. Why have synthetic resin paints largely replaced oil-base paints?
24. Why are prime coats used in painting?
25. What four methods commonly are used for applying paints in manufacturing?
26. What is the reason for using the electrostatic principle in spray painting?
27. Why is rotating-disc atomization replacing air atomization in spray painting?
28. Why is dip painting limited in its use?
29. What are the principle advantages of electrocoating?
30. Why is infrared drying seldom used for drying paints on wood products?
31. Why are paints that are excellent for finish coats usually not satisfactory for primers?
32. Why should the thickness of galvanizing be carefully controlled?
33. What are rimmed steels, and why should they not be galvanized?
34. What are the differences in the purposes of Parkerizing and Bonderizing?
35. Why is it difficult simultaneously to put parts of widely differing shape through an automatic electroplating system?
36. What are two reasons why electroless plating is preferred over electroplating?
37. Why is part design especially important for parts that must be electroplated?
38. Why is vacuum coating usually preferred over electroplating when costly metals are the plating material?
39. What is the major advantage of sputtering over vacuum evaporative coating?
40. At Wally's Wood Furniture Finishing Factory, Harry is suggesting we replace our spray painters with an electrostatic painting system. He read about these processes in the June 1986 issue of *Manufacturing Engineering*. What is wrong with this idea?
41. In mass finishing, compounds can be acidic or neutral or alkaline. In terms of pH, what does this mean?
42. What is the difference between polishing, coloring, and burnishing?
43. What is a spangle pattern in galvanizing?
44. Why does a burr always form at the intersection of two holes?
45. What would be the effect of switching from ceramic media to plastic or resin-bonded media in a mass-finishing process, assuming that the same shape and size of media was being used?
46. What is the effect of surface finishing on fatigue life?
47. Why do you think that electrodischarge machining results in such poor fatigue life?

Electronics Fabrication

Introduction

The evolution of electronic technology over the past four decades has been so rapid that describing electronic manufacturing processes is like trying to paint a landscape while riding on a train. The whirlwind computer used to generate digital data for the first NC machine in 1952 was based on vacuum tube technology. In the 1960s, transistors overtook the bulky, slow vacuum tube assemblies. See Figure 41-1. In the 1960s, the first integrated circuits (ICs) were fabricated on silicon. Why silicon? Because silicon dioxide, a glassy material, is an excellent, easily grown insulation material which acts as a mask (a barrier) for the selective introduction of dopants. The oxidation process is readily (temperature-) controlled and inexpensive. In the 1970s, the small ICs gave way to larger ICs and these in turn are now being replaced by large-scale (LSI) circuits and very large-scale integrated circuits (VLSIs). See Figure 41-2. The tiny

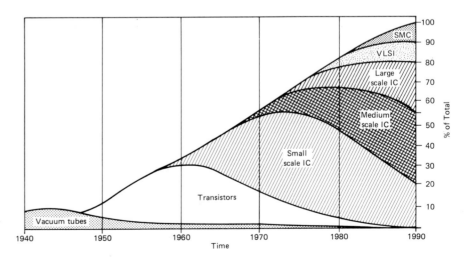

FIGURE 41-1 Evolution of electronics from vacuum tubes to very large integrated circuits.

992

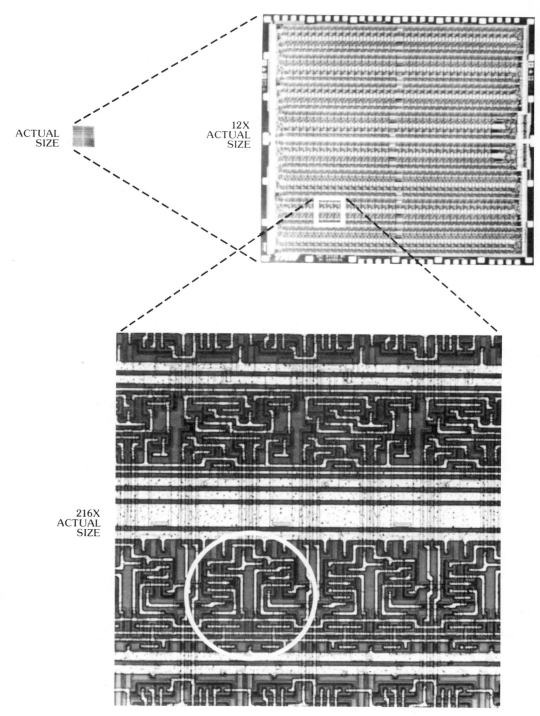

ACTUAL SIZE

12X ACTUAL SIZE

216X ACTUAL SIZE

FIGURE 41-2 VLSI chip containing 22,000 devices. Circled area in lower photo contains about 70 resistors and transistors. (*TRW/QUEST,* Spring 1978.)

microelectronic circuits are called *chips*. They are mounted on chip carriers, also known as packages. The most popular chip carrier, the DIP, has two rows of evenly spaced pins and provides the means of connecting the tiny solid-state electronic circuit to the circuit board. The first printed circuit cards were also introduced in the late 1950s. The pins on the packages were inserted through holes in boards. The pins are soldered to the electrical circuit that has been ''printed'' on the board.

The introduction of the IBM System 360 computers in 1964 typified the state of the art. Solid-state ICs were attached to pinned ceramic substrates, which were soldered into plated holes in multilayer cards. It quickly became evident that the smaller and more integrated the circuit, the faster it could perform electronic switching processes. Larger integrated circuit chips were produced on silicon wafers. The wafers were cut up (diced) to form chips, and the chips were bonded to carriers (DIPS) or substrates. These modules were sealed and soldered to cards. The cards were plugged into boards, and several boards were attached to the machine in a frame, or gate. Power and signal cables connected the boards. Figure 41-3 shows the levels in the hardware. Smaller components are assembled into successively larger components. In 1979 the IBM 4241 used large-scale integration chips, each consisting of a 704 logic-gate array.

By 1981, IBM had developed the 3081 processor, which uses up to 118 LSI chips with a larger (90-millimeter) multilayer ceramic module. Six or nine modules are plugged into a 24 by 28 inch board, reducing chip-to-chip, module-to-module, and board-to-board delays. The card, a packaging device that had been in use since the development of solid logic technology, was eliminated. However, most electronic fabricators, including IBM, are using the approach outlined in Figure 41-3.

Hierarchy in Hardware

The VLSI chip is a prime example of the real reasoning behind microelectronics. Making electronic components many times smaller makes them many times cheaper to produce than individual components. Combining the different electrical elements on the same chip is *integration*. *The higher the level of circuit integration, the less expensive it is to produce in terms of cost per functional element because the manufacturing cost per IC chip is about the same regardless of how many components are packed onto the chip.* As more circuits are integrated into the chip, fewer connections need be made on the circuit boards and panels. Transistor technology was more reliable than vacuum tube technology. The IC, in turn, was many times more reliable than the individual transistors and very large ICs are more reliable than small ICs. Also, as a general rule, the smaller the device, the less power it consumes.

Packages are simply ''chip carriers.'' Figure 41-4 shows two examples of IC packaging currently being used, the DIP and the flat pack. Figure 41-5 shows a third example of a chip carrier. The chips are mounted directly on a ceramic substrate. The flat pack configuration has all the leads on one plane and is used where the chip is permanently soldered into a multilayer circuit board. The DIP, with its two rows of evenly spaced pins, has been able to accommodate rapid

Chip in wafer, untested

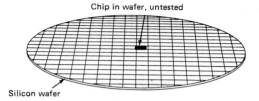

Silicon wafer

Testing and yield per good chip

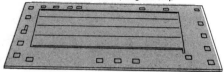

Package, packaging and testing

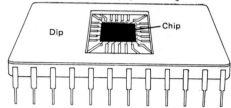

Dip   Chip

Printed-circuit board

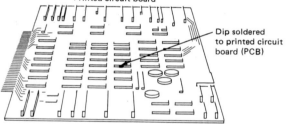

Dip soldered
to printed circuit
board (PCB)

Back panel and wiring
cabinet and power supply

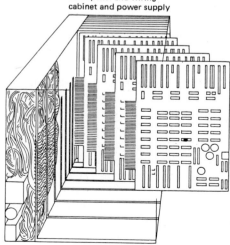

FIGURE 41-3 The ascending hierarchy of hardware begins with an integrated circuit, a chip. (Source: *Scientific American,* September 1977.)

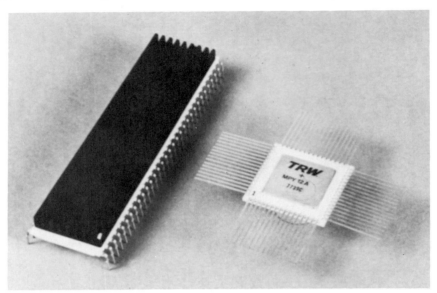

FIGURE 41-4   Two common packaging configurations: dual in-line package on left and standard flat pack on right. (*TRW/QUEST*, Spring 1978.)

FIGURE 41-5   Chips are attached directly to a ceramic substrate, the pins of which are inserted into the circuit board and soldered. (*Mechanical Engineering*, October 1986.)

changes in the integrated circuit technology. While the integrated transistor has shrunk in size, semiconductor devices have grown larger with increasing levels of integration. The number of necessary input/output pins has steadily increased, and a range of DIPs of different lengths and widths has been developed to handle these changes. DIPs now routinely have up to 64 pins. The higher the pin count on the chip carrier, the greater the necessary circuit board area and number of holes, as well as the complexity of routing between them. Thus, circuit boards are becoming more expensive and the DIP may now be approaching the limit of its usefulness. Figure 41-6 shows an example of a very high speed IC which is basically a computer on a chip. The chip is an example of a fault-tolerant design, which means that it has redundancy in its circuitry which improves its reliability. The wires connecting the IC to the package can easily be seen. Figure 41-7 shows three examples of a new type of chip carrier. Leadless chip carriers change the assembly process of the boards because a leadless chip carrier can be surface-mounted on the board, instead of through-hole mounted, and takes up much less board area. Surface mounting also eliminates the drilling of many holes and thus reduces the distance between leads on the package from about 0.100 in. to 0.050 in. or less.

Printed circuit boards are used to carry power and signals to and from the electronic devices. The fabrication of the boards involves additive plating or selective etching to form the circuit on the board.

Circuit boards replace loose wires and connecting cables inside electric equipment. The wire circuit is plated (printed) on the backside of a thin epoxy or fiberglass board. The packages are placed on the top side and holes are drilled in the board for the pins on the packages. The connections between the pins and the circuit are made by soldering. A single board may need 40,000 holes

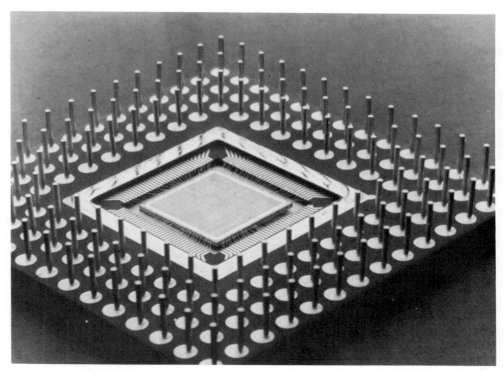

FIGURE 41-6 Example of CMOS VHSIC chip mounted in TRW pin-grid array package for fault tolerance testing. (Source: *TRW/QUEST,* Winter 1986,87.)

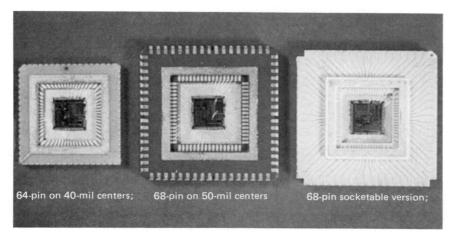

64-pin on 40-mil centers;   68-pin on 50-mil centers   68-pin socketable version;

FIGURE 41-7 Leadless ceramic chip carriers can eliminate the hole-drilling process in circuit board fabrication.

Input board

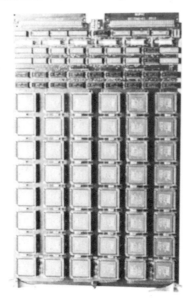

Array board

Output board

Rack-mountable enclosure

FIGURE 41-8 Circuit boards are mounted in racks. (Source: *TRW/ QUEST,* Winter 1986-87.)

drilled in it for layer-to-layer connections as well as for attaching devices. Some examples of boards and cabinets are shown in Figure 41-8.

Fundamentals of Transistors. Today's integrated circuits are an outgrowth of the basic MOS (metal oxide semiconductor) transistor. Almost without exception, ICs consist of transistors and resistors. Aluminum serves as an electrical conductor. Silicon oxide acts as an electric insulator on the surface of the device and under the gate, as shown in Figure 41-9. The semiconductor can be

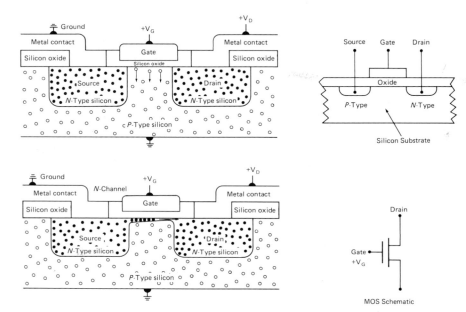

FIGURE 41-9   Metal oxide semiconductor (MOS) controls the flow of electrons between the source and the drain with a positive voltage.

either N-type silicon or P-type silicon. N-type silicon has excess free electrons (dots) and P-type silicon has excess "holes," or vacancies in its electron shells (circles). The designations of "N" and "P" identify the types of impurities (atoms) that are added to the pure silicon. These materials are called dopants. That is, *doping* is the process of introducing non-silicon atoms into the silicon.

Here is how a transistor works. MOS transistors are normally off when the gate-to-source voltage (V_{GS}) is zero. The operation of this gate can be viewed as a capacitor. One plate of this capacitor is the material used to form the gate of the transistor. The second plate is the silicon substrate itself. The electric field that exists between these two plates is used to control the flow of signal carriers from the source to the drain. The source and the drain are strongly "doped" N-type silicon. When a positive voltage is gradually applied to the gate, holes are driven away from the interface of the oxide and silicon under the gate and negative charge is attracted. With a sufficiently high gate voltage, an inversion occurs that changes a very thin region of the silicon under the gate from positive P-type (with holes) to negative N-type (with electrons). The inverted region is called an N channel and provides a conduction path for electrons. If a positive voltage is applied to the drain, current (electrons) will flow from the source to the drain through the channel. The gate voltage controls the amount of the current: the higher the voltage, the higher the current. Keeping the gate voltage constant (above the threshold for creating the channel) and increasing the drain voltage also increases the current from source to drain. As the drain voltage increases, the channel eventually becomes "pinched off." The current is then said to be saturated and remains at a constant value as the drain voltage is further increased.

Historically, MOS transistors were first developed as PMOS devices (the "P" refers to the class of impurities or dopants). Later another type of device was developed using the "N" class of dopants, known as NMOS. Still later, the N- and P-type devices were combined into a single structure known as CMOS for Complementary MOS. CMOS devices take maximum advantage of the low-power aspects of the technology.

Bipolar transistors used in VLSIs may be either a PNP or NPN structure, as shown in Figure 41-10. Both types are frequently used on the same chip. If a junction of N and P materials is formed, and an electrical potential is applied across the junction such that the P section is positive and the N section negative, electrons will move from the N material into the P material and holes will move from the P to the N material. If the potential is reversed, very little current flows. This is the principle of operation for semiconductor diodes used as detectors, logic gates, and switches. A transistor consists of two such junctions, forming either the PNP or NPN configuration. When the three elements are properly biased, the transistor functions either as an analogue amplifier or a

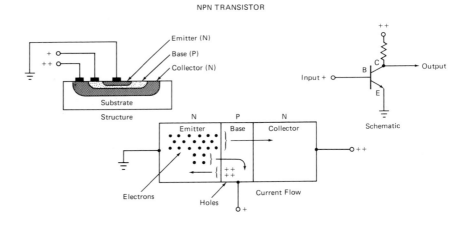

FIGURE 41-10 Structure and basic principle of bipolar transistors. (Source: *TRW/QUEST,* Summer 1982.)

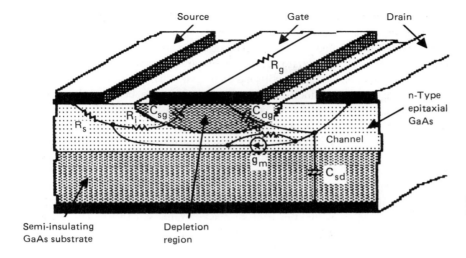

Source   Gate   Drain

R_g

n-Type
epitaxial
GaAs

R_i C_{sg} C_{dg}

R_s

Channel

g_m C_{sd}

Semi-insulating   Depletion
GaAs substrate   region

FIGURE 41-11   Field-effect transistor showing cross section with superimposed equivalent circuit. (Source: *TRW/QUEST,* Winter 1983,84.)

digital switch. The center section, or base, serves as the control element for current flow between the other two elements (emitter and collector).

Figure 41-11 shows a field-effect transistor (FET), which operates like a bipolar device in circuits but uses a different mode of operation. In general, the FET is easier to make than the bipolar transistor. The gate is actually insulated from the source and drain elements so that no current can flow in the gate circuit. In this respect, the device is similar to the vacuum tube, in which an electrostatic field around an open-mesh grid controls the flow of electrons between the cathode and the plate.

FETs are preferred over bipolar transistors for low-power and lower-speed applications.

Integrated Circuit Fabrication Process.   The fabrication of integrated circuits requires a method for accurately forming patterns on the silicon wafer. The microelectronic circuit is built up layer by layer, each layer receiving a pattern from a mask prescribed in the circuit design. The process of photoengraving known as *photolithography,* or simply masking, is employed for this purpose.

The basic material for the circuits is raw silicon, chemically purified to a purity of 99.9999999%. To grow silicon rods, a charge of purified silicon is placed in a crucible and brought up to the melting point of silicon, 1420°C. It is necessary to maintain an atmosphere of purified inert gas over the silicon while it is melted, both to prevent oxidation and to keep out unwanted impurities. Desired impurities, known as dopants, are added to the silicon at this point to produce a specific type of conductivity, characterized by either P-type charge carriers or N-type ones.

A large-diameter single crystal is grown from the melt by inserting a perfect single crystal ''seed'' and slowly turning and withdrawing it. Single crystals from 2 to 6 in. in diameter and several feet long can be pulled from the melt.

The uneven surface of a crystal, as grown, is ground to produce a cylinder of standard diameter.

The crystal is mounted in a fixture and cut into wafers with a thin, high-speed diamond saw. In the finishing step the wafers are first smoothed on both sides by grinding and then are highly polished on one side. The final wafer is typically about half a millimeter thick. The final steps must also be carried out in an absolutely clean environment. There can be no defects, polishing damage, scratches, or even chemical impurities on the finished surface.

For the purposes of explaining the basic processes, the fabrication steps to manufacture a dual-gate MOS transistor will be outlined. Most IC fabricators purchase finished silicon wafers. As a first step, to make an N-channel transistor, a layer of silicon oxide is grown in an oxidation furnace on P-type silicon as shown in Figure 41-12(a). A layer of photoresist material is applied and exposed to ultraviolet radiation through a glass mask, Figure 41-12(b). The areas of photoresist that have been exposed to radiation harden when developed. Next, the unhardened photoresist is removed, thereby exposing the SiO_2. Unprotected silicon oxide is etched away by acid, Figure 41-12(c), and the remaining photoresist is removed. Phosphorus atoms are diffused into the unprotected silicon to create the source and drain areas, Figure 14-12(d). After the photoresist has been removed, Figure 41-12(e), a layer of silicon oxide, doped with phosphorus, is deposited, using chemical vapor deposition (CVD), followed by a layer of undoped silicon oxide and a layer of photoresist. The photoresist is developed to form channels for the gates, Figure 41-12(f). An acid etch removes the oxide in the channels and photoresist is removed, Figure 41-12(g). Next, the wafer undergoes thermal oxidation to grow a thin insulating oxide layer in the channels. At the same time, phosphorus atoms from the P-doped silicon oxide layer diffuse into the silicon substrate and extend the source and drain, Figure 41-12(h). Photoresist is added and developed so that the source and drain regions are open and an acid etch can remove the surface oxide, Figure 41-12(i). The photoresist is removed and a layer of aluminum is deposited by vacuum evaporation to provide electrical contacts to the circuit. Photoresist is added and developed to protect the contacts and channels, Figure 41-12(j). The unprotected aluminum is etched away, the photoresist is removed, and the device is essentially complete, Figure 41-12(k). The broken lines in the top view of the entire dual-gate MOS transistor device, Figure 41-12(l), indicate the area shown by the cross sections in the fabrication steps.

In the actual process, thousands of circuits are simultaneously manufactured on a single chip. From 100 to 1000 chips, each with an IC, are contained on each wafer, and as many as 24 wafers can be processed simultaneously. The wafers are placed in material-handling devices called *boats* for insertion in furnaces and vapor-deposition processes. It is this capability for mass processing that makes these complex and versatile devices so inexpensive. Another important factor is that virtually every process can be automated at the A(2) level. The only handling required is in transferring the wafers between processing stations, which means fewer rejects and lower production costs.

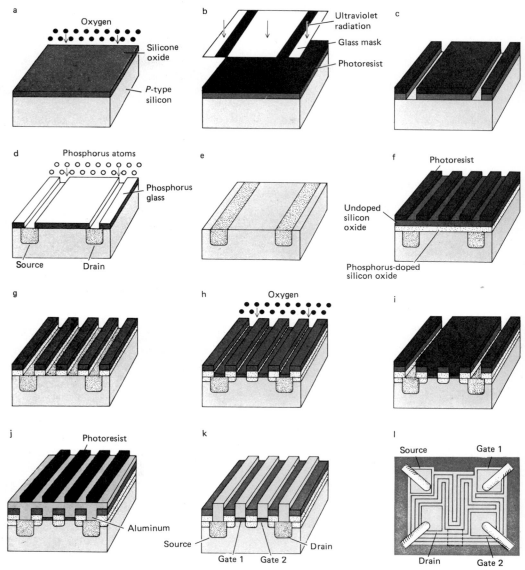

FIGURE 41-12 Fabrication steps for a discrete dual-gate MOS transistor (Wm. C. Hittinger, MOS Technology, *Scientific American.*)

Lithography. The process of photolithography or optical photolithography is explained in more detail in Figure 41-13. This process of pattern transfer and pattern definition can be repeated many times during IC fabrication. The figure shows the basic steps, using as an example the etching of a pattern in a silicon dioxide layer. Each masking step requires that the wafer be coated with a photosensitive emulsion (photoresist). The light-sensitive material is exposed to ultraviolet radiation through the photomask.

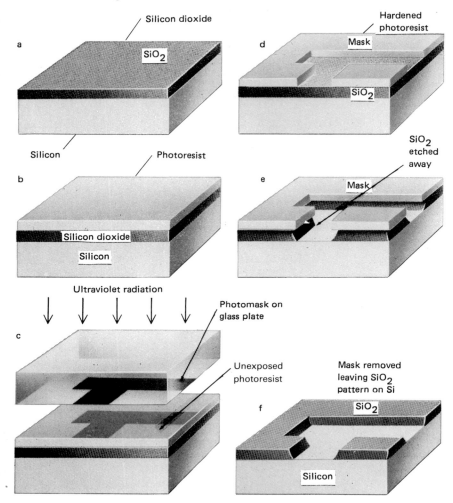

FIGURE 41-13   Steps in photolithography used to transfer a pattern from a glass plate (photomask) to the silicon wafer. (Wm. G. Oldham, *Scientific American*, Sept. 1977.)

Figure 41-14 shows one such photomask after its photographic reduction onto a photographic glass plate. A set of glass plates are fabricated, each holding the pattern for a single layer of the circuit. Prior to reduction, the highly magnified layouts of circuits are carefully inspected to ensure that each circuit element is correctly placed and is the correct size. The primary pattern, called the reticle, is typically ten times the final size of the circuit. The perfect reticle pattern is then reproduced side by side, photographically reduced, and reproduced hundreds of times in a "step and repeat" process to yield a set of final-size master masks. Master masks are used to make working masks. The working plate may be either a fixed image on an ordinary photographic emulsion or a more durable glass substrate with an etched chromium film forming the pattern. The working photomask serves to block out light. Exposure to ultraviolet light renders the photoresist insoluble (that is, hardened) when immersed in a devel-

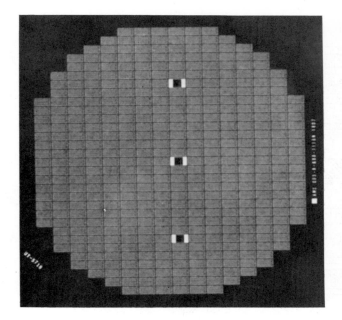

oper solution. Hence a pattern of the photoresist is left whenever the photoresist is exposed; that is, the mask is not opaque. The wafer is next immersed in a solution of hydrofluoric acid, which selectively attacks the exposed (unmasked) silicon dioxide, leaving the photoresist pattern and the silicon substrate unaffected. In the final step, the photoresist pattern is removed by means of another chemical treatment. Each layer of the microcircuit has a different set of photomasks.

As circuit feature sizes approach 1 μm and smaller, microscopic particles of dirt from the air, from people, and from wafer-transporting machines can have a disastrous effect on integrated circuit yield. Particles that land on a mask are especially troublesome because they are repeated every time the mask image is projected on a wafer. To minimize that problem, many companies that use projection lithography have adopted the use of *pellicles*.

A pellicle consists of a transparent thin membrane that is stretched over a supporting frame on the top of the reticle (see Figure 41-15). The membrane

FIGURE 41-14 (Left) Photomask for 16,384 bit IC. Glass plate with pattern for single layer IC etch in chromium film. (Right) Enlargement of plate (10×). (Source: Wm. G. Oldham, *Scientific American,* September 1977.)

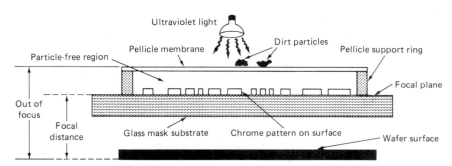

FIGURE 41-15 Pellicles, transparent thin membranes, are used to keep dust and dirt from ruining the lithography process. (*Electronic Design,* June 14, 1984.)

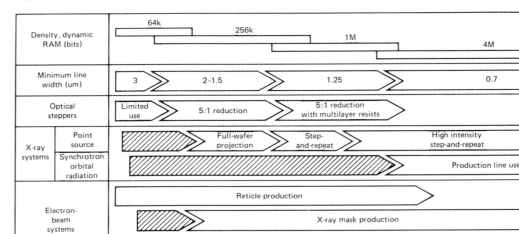

Time Bar

FIGURE 41-16 As circuit density increases and line widths decrease, the optimum lithographic technique for each generation of circuits will change. (Source: Mitsubishi Electric Corp.)

and the frame form a dust-free chamber above the patterned mask surface. Any particle that falls on the surface of the pellicle membrane is outside the focal plane of the pattern and is therefore not imaged on the wafer surface.

In other attempts to hold down particle dirt, equipment manufacturers are examining new wafer material-handling mechanisms, new physical arrangements of equipment to provide cleaner air flow, robotic additions to the equipment to reduce the human contribution, and even new materials that generate fewer dust particles.

Advances in lenses have carried optical lithography systems to the submicrometer region. The theoretical limits of optics suggest that the minimum feature size that can be projected is about 0.5 μm. For working manufacturing systems, however, a limit of about 0.75 μm is more realistic. To go below this, electron-beam, X-ray, and focused ion-beam systems are needed, as shown in Figure 41-16. E beams can be controlled very accurately with spot sizes of 0.1 μm and smaller, and they are capable of writing patterns directly in a resist. X-rays have a much shorter wave length than ultraviolet light and can pass through most types of dirt particles that might otherwise appear on masks. Focused ion beams can be very finely controlled and used to fabricate submicrometer structures directly, without masks or resists. Synchrotron orbital radiation and focused ion beams are the "farthest out" technologies, with the first production uses not expected until the early 1990s.

Computer-Aided Design in Microelectronics.

Figure 41-17 shows the manufacture of a LSIC with the design carried out with the aid of a computer. Before 1970, electronic circuits were analyzed and designed almost exclusively by hand. Rapid growth in the complexity of integrated circuits has made computer aids essential to the design process today.

Most of the computer aids in use today do not do the actual *design*. Instead, they can perform rapid analyses (system simulation, logic checking) of a given design, varying parameters as specified by the engineer. Computer aids can also cope with masses of data that would overwhelm the unaided engineer. Computer-based *layout design systems* are extremely useful in preparing and modifying the geometric patterns required for integrated circuit masks, helping to determine the most space-conserving layout of the circuit elements.

The description of all the types of computer-aided design (CAD) tools is beyond the scope of this book. However, at no time should CAD eliminate the need for manual analysis and design because it is only through the latter type of experience that designers develop the imagination and insight needed for skill in design. As suggested in Figure 41-17, the photomask patterns can be gen-

FIGURE 41-17 Sequence of steps in the computer-aided design (CAD) manufacture of integrated circuits. (Source: Wm. G. Oldham, *Scientific American,* September 1977.)

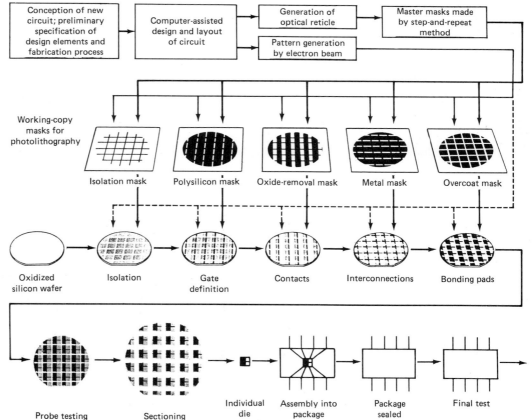

erated by electron-beam machining whereby the pattern for the mask is "written" directly on the working mask from information stored in the computer.

When first introduced, direct-writing electron-beam lithography was expected rapidly to overtake optical systems for exposing the photoresist material on wafers, since it eliminates the masks and offers more flexibility and higher resolution. Up to now, however, optical systems have prevailed because of major improvements in lens quality and electronic and mechanical enhancements in the step-and-repeat systems. Alignment tolerances have been reduced to 0.35 μm, and throughput has been boosted to more than 40 five-inch wafers an hour, while patterns with sub-2 μm feature sizes are being created.

Nevertheless, E-beam systems have made their mark in two areas: mask making, where they are used extensively to create very fine pattern reticles, and applications that require limited production and quick turnaround, such as gate arrays.

After the ICs have been fabricated, the wafer is probe-tested to determine the good circuits. The defective circuits are marked with an ink spot. Individual "dice" bearing good circuits are then selected from the sectioned wafer and assembled into packages. A final test is performed to ensure that the packaged circuit works properly.

Fabrication of Printed Circuit Boards (PCB).   As electronics technology developed—from vacuum tubes to discrete transistors to integrated circuits to the microcomputer-on-a-chip—the printed circuit board emerged as the interconnecting link between the devices and the functioning component.

A printed circuit board is produced from a copper-clad laminate by means of a series of mechanical and chemical operations outlined in Figure 41-18. The copper-clad base laminate is prepared by combining a plastic resin, reinforcing material, and an electrodeposited copper foil under precise heat and pressure conditions. The laminate is next drilled, usually on a NC drilling machine, and additional copper is added by electroless and electroplating processes. Photoresist material is added, covered with the circuit image mask, and exposed. The mask is removed, the resist developed, and the unwanted copper foil is selectively removed by chemical etching processes to produce the desired circuit pattern. The electronic components are then attached using wave-soldering techniques to produce the printed wiring board. Figure 41-19 shows how conventional discrete parts with leads and IC packages are inserted through the PCB for attachment. The devices are wave-soldered by passing the boards over a molten bath of solder which has a standing wave of molten metal that contacts the lead tips. Circuit patterns may be either single-sided or double-sided, depending on the type of laminate used and the end application. More complex double-sided circuit boards are most common in high-performance applications such as computers where high packaging densities are required.

Manufacture of the Laminate.   Electrical, thermal, mechanical, and functional properties of the printed circuit board are determined by the copper-clad laminate from which the printed circuit board is produced. The copper-clad

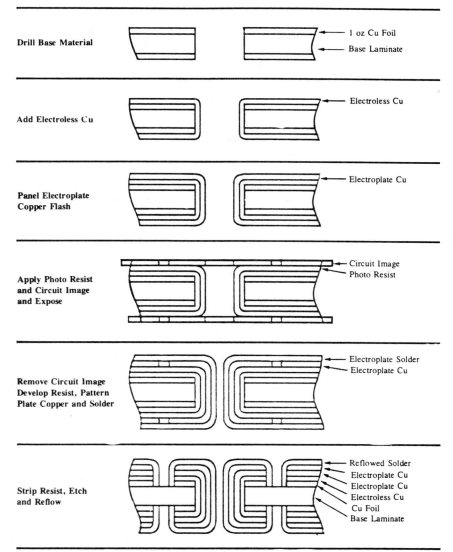

Drill Base Material — 1 oz Cu Foil / Base Laminate

Add Electroless Cu — Electroless Cu

Panel Electroplate Copper Flash — Electroplate Cu

Apply Photo Resist and Circuit Image and Expose — Circuit Image / Photo Resist

Remove Circuit Image Develop Resist, Pattern Plate Copper and Solder — Electroplate Solder / Electroplate Cu

Strip Resist, Etch and Reflow — Reflowed Solder / Electroplate Cu / Electroplate Cu / Electroless Cu / Cu Foil / Base Laminate

FIGURE 41-18 Basic steps in the making of a plated-through-hole double-sided printed circuit board.

laminate is composed of resin, reinforcing agent, and copper foil. To a large degree, it is the properties of the laminate that determine the type of printed circuit board that can be produced and in what type of electronic equipment it can reliably be installed. Properties like flexural strength, water absorption, arc resistance, and dielectric strength are functions of the laminate and its construction.

The most often used materials for the resin are epoxy and phenolic. The reinforcing agent (the substrate) provides strength. Glass fiber and cellulose paper are commonly used. Figure 41-20 shows the steps in the process. The

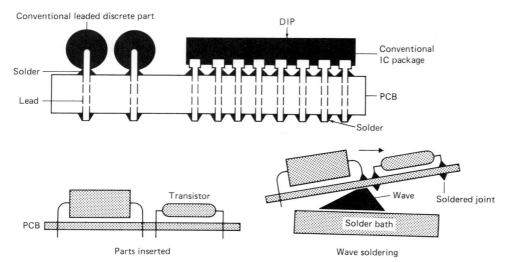

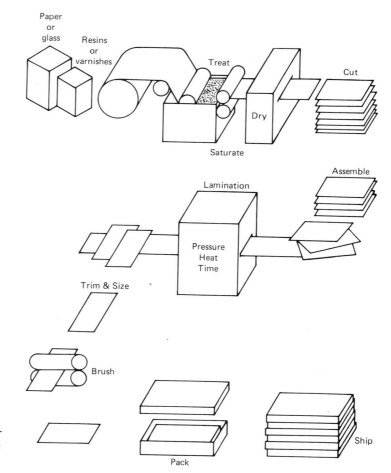

FIGURE 41-19 Wave-soldering parts and DIPs to the conventional through-hole PCB.

FIGURE 41-20 Process for the manufacture of the base laminate for printed circuit boards. (*GE Materials Technology,* Spring 1980.)

process appears simple. It becomes complicated, however, when one tries to achieve the properties needed for a specific use. A laminate is a compromise, with an improvement in one property often coming at the expense of another. See Table 41.1.

A roll of substrate is saturated in a resin solution, an operation referred to as *treating*. The treated material is dried in an oven to eliminate the solvent and partially polymerize or "cure" the resin. At this point the roll of treated material is not tacky and can be handled.

The semicured, resin-impregnated substrate is then cut into sheets, stacked between stainless steel press plates with copper foil placed on either side of the sheet. The assembly is placed in a laminating press. Heat and pressure are applied so that the resin will flow, bonding the individual plies together. The press cycle is timed to cure the resin fully and make the plies inseparable.

Although copper foils utilized in laminate manufacture range in thickness from 9 μm to nearly 200 μm, the most common thickness is 35 μm or about 0.0014 in. Relatively thin foils permit finer conductor widths and closer spacings of printed circuits. As circuit miniaturization continues to be more prevalent, so will the use of thinner copper foils.

Component mounting holes—an important part of the board—are produced by either punching or drilling. Drilling is more expensive, normally requiring NC drilling machines. Because of the high strength and abrasiveness of the woven glass fabric laminates, holes must be drilled rather than punched. However, even carbide-tipped drill bits or TiN-coated drills wear out quickly, and the quality of the holes is frequently marginal. The variables affecting drill life and hole quality include (1) adhesion between resin and glass, (2) glass content, and (3) feed rate.

TABLE 41.1 Laminate Properties

Paper/Phenolic
- Low cost
- Adequate electrical properties
- Low strength
- Moisture-sensitive

Woven glass/Epoxy
- High cost
- Excellent electrical properties
- High strength
- Moisture-insensitive
- High thermal stability
- High dimensional stability

Packaging. Historically, semiconductor manufacturers have tried to build packages that preserve or increase the density of components on circuit boards, that lower material costs, and that improve performance at higher operating frequencies. Packaging for today's ICs and power devices involves a great deal more than just a knowledge of mounting technology and chip carriers. It requires an understanding of connector placement equipment, soldering processes, and PCBs, as well as the ability to integrate them all. Often, the most suitable package for a given component not only must fulfill the requirements of the increasingly high pin counts but must also be specifically made for automatic placement, must be able to withstand the rigors of mass soldering, and must have thermal expansion characteristics that match those of the PCB.

The through-hole mounting technology is slowly evolving toward smaller leadless devices. For example, leadless plastic chip carriers with 40-, 25-, and even 20-mil (0.020 in.) center-to-center pin spacings are beginning to appear. Additionally, the four-sided flat IC, or quad pack, is being used for signal-processing chips, thereby ensuring easy handling and maximum density. Single in-line packages have emerged for discrete power chips. New connection techniques, which include the use of elastomerics, have been developed that prevent

TABLE 41.2   **How ICs and Discrete Components Affect Packaging**

Parameters	Year			Packaging concern
	1983	1986	1991	
Gate delay, bipolar	330	75	50	Small size
				Routing of package pins to chip leads
Power dissipation	10	25	30	Thermal management (especially
Bipolar				important for small packages)
MOS	1.5	1.5–2.0	2.0	
Chip size (mils)	270	500	600	Assembly
				Package-to-chip stresses
Number of gates	<1k	10k–20k	40k	High pin counts (I/O ≈ 4.5 [gates]$^{0.4}$)
(gate arrays)				

Source: Electronic Designs, June 14, 1984.

damage to a package during installation and that reduce electrical and mechanical stresses in normal operations.

Central to the development of all packaging systems are the packages themselves and the chips they contain. Whether meant to complement either the physical or the electrical characteristics of ICs and discrete parts, packaging goals are simply defined (see Table 41.2). Plastic chip carriers, from the standpoint of power handling and temperature cycling, must become more rugged and must offer a higher degree of hermetic isolation, while retaining their cost advantage over competing designs. High-quality ceramic chip carriers must have better power-dissipating structures and simultaneously lower costs. Therefore, they become more economically attractive while maintaining their electrical and thermal superiority to their plastic counterparts.

Circuitry size and weight reduction can be achieved in electronic devices by using *surface mounting technology (SMT)* packaging concepts. Surface-mounted components (SMC) are much smaller than their conventional counterparts. The SMT integrated circuit is packaged in a chip carrier that has either physically short leads or leadless conductors to maintain the carrier's low profile. See Figure 41-21. In SMCs, the leads or conductors run along all four sides of the device to make accessible many more interface signals. Surface-mounted components are attached to matching solder pads on the PCB surface, eliminating the conventional method of inserting leads through holes in the PCB for physical and electrical part connection. Hot-vapor soldering replaces wave soldering. The advantage of SMT is that less surface area is needed and both sides of the board are available for component attachment, resulting in significant area reductions.

SMC technology also lends itself to lower-cost automated manufacturing assembly techniques because the size and shape of SMC parts (basically squares and rectangles) are uniform. In fact, SMT was originally developed in Japan as a way to lower manufacturing costs and improve quality control through automated assembly, and not as a miniaturization scheme.

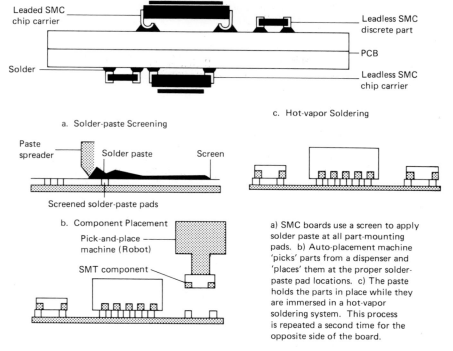

a. Solder-paste Screening

c. Hot-vapor Soldering

b. Component Placement

a) SMC boards use a screen to apply solder paste at all part-mounting pads. b) Auto-placement machine 'picks' parts from a dispenser and 'places' them at the proper solder-paste pad locations. c) The paste holds the parts in place while they are immersed in a hot-vapor soldering system. This process is repeated a second time for the opposite side of the board.

FIGURE 41-21 Surface mount technology is gradually replacing through-hole PCB technology.

Smaller spacing between leads results in smaller components. The 64-pin DIP is three inches long and nearly an inch wide. Automatic insertion of a package with a large number of pins is difficult and the parts must therefore be assembled by hand. A surface-mounted chip carrier with 0.050 in. lead centers will require only about 1 square inch of board area.

Reduced board area and automation are not the only benefits of surface-mounted components. Small packages also mean shorter electronic connections between the board and the integrated-circuit chip inside the package. Shorter signal tracks have less capacitance and less inductance, reducing the electrical load carried by the on-chip driver circuits. Small loads reduce the turn-on time, permitting faster communication between chips. Smaller loads also require less power for signal operation.

Surface-mounted IC packages have taken many forms since they were introduced in the late 1970s. Ceramic or plastic, square or rectangular, and leadless or leaded types are available, with leads of several bend shapes on either two or four sides.

Obviously many factors affect the choice of a particular package. For mass connection and when temperature management is a concern, leaded packages are chosen. Pin-grid arrays are used when convenience and ease of connection are the main considerations. Leadless carriers are most attractive when cost, space, and operating frequency are the controlling factors. Figure 41-22 shows how elastomeric connectors are used to unite high-density leadless chips to PCBs. Until now, elastomerics have been used chiefly to link boards to one

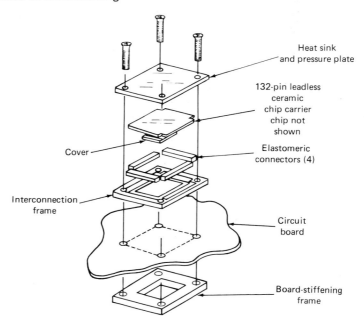

Heat sink and pressure plate

132-pin leadless ceramic chip carrier chip not shown

Cover

Elastomeric connectors (4)

Interconnection frame

Circuit board

Board-stiffening frame

FIGURE 41-22 Elastomeric connector with 33 connectors (.003 in. wide on .006 in. centers) for uniting chip carrier to PCB. (Textronix).

another in a system. Now they are being employed—primarily with leadless chip carriers—to connect carriers to the boards themselves, in turn improving all of the aspects of electrical and mechanical performance.

Elastomeric connectors are already being used in high-density systems and may turn out to be more acceptable than pin-type sockets.

Integrated Circuit Economics. The larger the IC, the greater the chance for a defect to appear and render the circuit inoperative. The yield (the fraction of good circuits per wafer) decreases with the size of the chip because larger circuits are more likely to incorporate a defect.

At first, it might appear most economical to build very simple, and therefore very small, circuits on the grounds that more of them would likely be good ones. It is true that small circuits are inexpensive; simple logic circuits are available for as little as 10 cents each. However, the cost of testing, packaging, and assembling the completed circuits into an electronic system must also be taken into account. Once the circuits are separated by breaking the wafer into individual chips, each chip or die must be handled individually. From that point on, the cost of any process such as packaging or testing is not shared by hundreds or thousands of circuits. Thus packaging and testing costs often dominate the other production costs in the fabrication of integrated circuits.

The typical IC manufacturing facility might employ about 100 direct laborers and be able to manufacture several thousand wafers per week. Assuming that there are 50 working chips per wafer, such a plant can produce five million ICs per year.

A wafer-fabrication facility must be extremely clean and orderly. Because of the smallness of the structures being manufactured, even the tiniest dust particles cannot be tolerated. A single dust particle can cause a defect that will result in the malfunction of a circuit. Special clothing is worn to protect the manufacturing environment from dust carried by humans. The air is continuously filtered and recirculated to keep the dust level at a minimum. For the purpose of comparison the dust level in a modern hospital is on the order of 10,000 particles per cubic foot while that in a wafer fabrication clean room is less than 100 particles per cubic foot.

Summary. With wafer diameters expanding to 6 inches and minimum chip features shrinking to 3 μm and below, semiconductor fabrication processes have improved in these three areas: lithography, where improved lens systems are keeping optical methods in the lead; etching, where dry plasma and reactive-ion techniques are producing the vertical profiles essential to fine lines; and deposition processes, with much improved sputtering and chemical vapor deposition.

Improved packaging techniques have resulted in increased board density (more components per board), calling for an integrated approach—one that guarantees the device's electrical and mechanical integrity by meshing its connector, soldering, and component placement needs.

Review Questions

1. What do the letters IC, LSI, and VLSI stand for?
2. The term *packaging* has a very different meaning in microelectronics. What are packages, what is packaging, and why is packaging needed in microelectronics?
3. How is a DIP typically mounted on a PCB?
4. If a single PCB has 40,000 holes drilled in it, which process would you pick to do the job, multiple spindle drilling or NC drilling?
5. What is the difference between N-type silicon and P-type silicon?
6. What is a CMOS device?
7. The process of photolithography was described in Chapter 32 under a different name. What was this process called?
8. Why must the silicon wafer be a single crystal with no plastic deformation in the surface?
9. In the integrated circuits, what is the function of silicon oxide?
10. Why is the photoresist material light-sensitive?
11. What is the reticle?
12. Why are pellicles used in projection lithography?
13. What advantages do surface-mounted devices offer over through-hole-mounted devices?
14. In Table 41.1 what are the adequate electrical properties referred to for the laminate?
15. Many manufacturers of ICs use robots to put the devices on the boards. This is called populating a PCB with devices. What do you suspect is the major problem with using a robot for device insertion in the PCB? Robots are discussed in Chapter 42.
16. Why must an IC fabrication facility be very clean and dust-free?
17. What is another name for an emitter?
18. In Figure 41-11, N-type GaAs was termed epitaxial. What is GaAs and what does the term *epitaxial* mean? That is, what is epitaxial growth? (Hint: This was not discussed in the text.)
19. What is the significance of being able to reduce the line width in ICs?
20. What techniques are being used to accomplish reduced line widths?

Problems

1. In Figure 41-8, the output PCB is going to be drilled on a NC drilling machine. Assume that it takes 3 seconds to drill one hole for one DIP pin. How long do you estimate that it will take to drill all the holes in one output PCB?
2. Estimate the cost of making one IC chip at the typical IC manufacturing facility having about 100 direct-labor employees. Assume that 10% of the employees make chips.

Manufacturing Systems and Automation

Introduction

In recent years, machine tools have become more powerful and versatile. Cutting tools have become more effective, capable of cutting at faster speeds with reduced cutting forces. Presses have become larger and tooling more sophisticated. However, the ability of a company to reduce costs by taking time out of the machining cycle (reducing machining time or tool change time) has done little to reduce overall manufacturing costs or make marked improvements in productivity. On an automatic screw machine producing a million-piece run, saving a couple of seconds per piece can be the difference between a healthy profit and a heavy loss. However, on a 25-piece lot, even a 10% reduction in machining time will have very little impact on cost for a part that takes two months to get through the plant. Table 42.1 compares time utilization in numerically controlled machining centers versus machining on conventional machine tools and machining in transfer machines. Observe that for machine tools used in lot production, only about one fourth of the time at best is spent adding value (making chips). NC does reduce setup time, but considerable improvement can be made in the noncutting areas. Efficient utilization of the workers, materials, and machines is critical to a successful organization.

One of the reasons for this situation is the complexity of the manufacturing system, which requires elaborate schedules and long lead times in the planning. Superimposed on the inherent complexity are delays and disturbances caused by engineering design changes, vendor failures, material shortages, emergency orders, machine breakdowns, quality problems, and so on. The traditional way to cope with these factors has been to create a buffer at each machine to ensure that each machine has work available but this in turn creates a large in-process inventory. Another strategy has been to use daily lists, colored tags, and people to expedite orders through the shop. Needless to say, this solution itself acts as a disturbance to the regularly scheduled work in progress. What is needed is

1016

TABLE 42.1 Time Utilization of Conventional Equipment versus NC Machining Centers and Transfer Machines as Percentage of Overall Machining Cycle

Activity	Machining Center (Small-Medium Lots)	Conventional Machine (Small Lots)	Transfer Line (Large Lots)
Metal cutting	23% of time	20% of time	50% of time
Positioning or tool changing or transfer	27% of time	10% of time	12% of time
Gaging	8% of time	15% of time	1% of time
Loading/unloading	10% of time	15% of time	—
Setup[a]	5% of time	20% of time	—
Waiting or idle[b]	14% of time	13% of time	22% of time
Repair and technical problems	13% of time	7% of time	15% of time

[a]Actual setup of a transfer line may take 6 to 24 months.
[b]Due to stockouts, personnel allowance, slow machine cycles, etc.
Source: After C. E. Carter, "Toward Flexible Automation," *Manufacturing Engineering,* August 1982, p. 75.

effective shop floor management and organization to make sure that machines and people are utilized to the greatest possible extent. That is, more is to be gained through better overall management and planning of the manufacturing system with machines running at moderate speeds and feeds than operating at extremely fast but sporadic machining rates.

These final three chapters explain how manufacturing systems can be designed and integrated for low cost, superior quality, and on-time delivery. We will begin with a discussion of the manufacturing systems, show how they are tied to production systems, and conclude with a detailed discussion of integrated manufacturing production systems.

Trends in Manufacturing Systems

Significant changes are taking place in the design of manufacturing systems, fueled by the following trends:

1. The implementation of cellular manufacturing systems as the first step in just-in-time manufacturing will increase.
2. Proliferation of the number and variety of products will continue, resulting in a decrease in quantities (lot size) as variety increases.
3. Requirements for closer tolerances (more precision, better quality) will increase.
4. Increased variety in *materials,* including composite materials with widely diverse properties, will cause further proliferation of the manufacturing processes.
5. The cost of materials, including material handling, and energy will continue to be a major part of the total product cost, while direct labor will be 5% to 10% of the total.

6. Product reliability will increase in response to the excessive number of product liability suits.
7. Customer demand for superior quality with reduced cost (relatively speaking) and on-time delivery of products will continue.

Manufacturing Systems Optimization

In general, a manufacturing system should be an integrated whole, composed of integrated subsystems, each of which interacts with the whole system. The system will have a number of objectives and its operation must optimize the whole.

Optimizing pieces of the system (i.e., the processes or the subsystems) does not optimize the whole system.

System operation requires information gathering and communication with decision-making processes that are integrated into the manufacturing system.

Each company will have many differences resulting from differences in subsystem combinations, people, product design, and materials. From company to company, the differing interactions of the social, political, and business environments make each company a unique manufacturing production system with its own set of problems. Clearly, there is a danger in grouping all the companies on a functional or departmental basis. In this chapter, we will try to extract some general principles. Let us begin with a historical perspective.

How the Job Shop Design Evolved

In the first industrial revolution, basic machine tools were invented and developed. With them came the first levels of mechanization and automation. Factories developed along with the manufacturing processes. Factories were needed to focus the resources (materials, workers, and processes) at the site where power was available. These early plants were arranged according to the kinds of machines, that is, functionally, because the machines were an extension of some human capability or human attribute. A machinist developed different skills from those of a leatherworker, an ironworker, or a foundry worker. The processes were divided according to the kinds of skills needed to operate the processes.

The early factories were placed beside rivers where water was used to turn waterwheels which in turn drove overhead shafts that ran the length of the factory. A belt from the main shaft powered each machine. The grouping of similar machines which needed to run at about the same speed was logical and expedient. Steam engines and, later, electric motors replaced other types of machine power and greatly increased manufacturing system flexibility. However, the functional arrangement persisted and became known as the job shop.

As product complexity increased and the factory grew larger, separate departments evolved for product design, accounting (bookkeeping), and sales. Later in the scientific management era of Taylor and Gilbreth,[1] departments for production planning, work scheduling, and methods improvement were added

---

[1]F. W. Taylor and Frank and Lillian Gilbreth are generally recognized as the founders of the industrial engineering profession.

to what became the *production system*. The production system services the manufacturing system; refer to Figure 1-4 and related discussion in Chapter 1. Because the production system was composed of functional areas to serve the functionally designed job shop, it naturally evolved with a functional structure.

Systems Defined.   The best definition of the concept *system* comes from Moshe F. Rubinstein's *Patterns of Problem Solving*, Prentice-Hall, Inc., 1975, and is quite appropriate for manufacturing systems.

The word *system* is used to define abstractly a relatively complex assembly (or arrangement) of physical elements characterized by measurable parameters. In order to model the system:

1. The system's boundaries or constraints must be defined.
2. The system's behavior in response to excitations or disturbances from the environment must be predictable through its parameters.

Models are used to describe how the system works. Mathematical models for use in a control computer generally require a ''theory'' or equations that describe the system's boundaries and behavior through its input parameters. In short, if no theory exists, the model is not viable and the system is not controllable.

Manufacturing systems modeling and analysis is desirable but difficult because:

1. Objectives are difficult to define, particularly in systems that impact on social and political issues. Goals may conflict.
2. The data or information may be difficult to secure, inaccurate, conflicting, missing, or even too abundant to digest.
3. Relationships may be awkward to express in analytical terms, and interactions may be nonlinear; thus, well-behaved functions embodied in many analysis tools do not apply.
4. System size may inhibit analysis due to implied time expenditures.
5. Systems are always dynamic and change during analysis. The environment can change the system or vice versa.
6. All systems analysis will be subject to errors of omission and commission. Some of these will be related to breakdowns or delays in feedback elements because manufacturing systems include people in information loops.

Because of these difficulties, simulation is a widely used technique for manufacturing systems modeling and analysis as well as for systems design. The integrated manufacturing production system (IMPS) approach described in Chapter 44 overcomes many of these problems.

Manufacturing Versus Production Systems

In general terms, a manufacturing system inputs materials, information, workers, and energy to a complex set of elements known as machines or machine tools which can be characterized. The materials are processed and gain value. Manufacturing system outputs may be either consumer goods or inputs to some other process (producer goods). Manufacturing systems are very interactive and

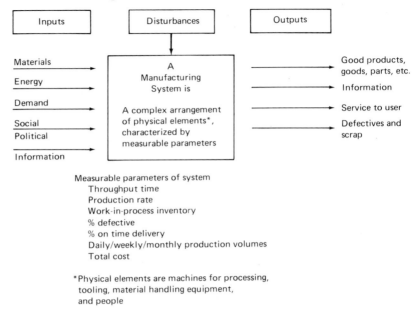

FIGURE 42-1   Generalized view of a manufacturing system with its input/output conditions.

dynamic. They involve many people who carry the process know-how. Change seems to be the only constant in this system.

Figure 42-1 gives a general picture of the manufacturing system. Observe that many of the inputs cannot be fully controlled (by management), and the effect of the disturbances must be counteracted by manipulating the controllable inputs or the system itself. Controlling material availability or predicting demand fluctuations may be difficult. The national economic climate can cause shifts in the business environment which can seriously change any of these inputs. In other words, in manufacturing systems, not all inputs are fully controllable. The manufacturing systems themselves differ in structure or physical arrangement, as we shall see in the next section. However, all the manufacturing systems are serviced by a production system. Figure 42-2 shows this schematically. Because the oldest and most common manufacturing system is functionally organized, most production systems are functionally organized. Walls usually separate the people in these functional areas from all other areas. Breakdowns in communication links are common. Long lags in feedback loops result in manufacturing system problems.

Classification of Manufacturing Systems (MS)

Table 42.2 shows manufacturing industries that differ by the products they make or assemble. Table 42.3 provides a partial list of service industries. Examination of all these industries reveals four kinds of classical manufacturing systems (or hybrid combinations thereof) and one new manufacturing system that covers all these industries. Table 42.4 lists examples of five types of manufacturing sys-

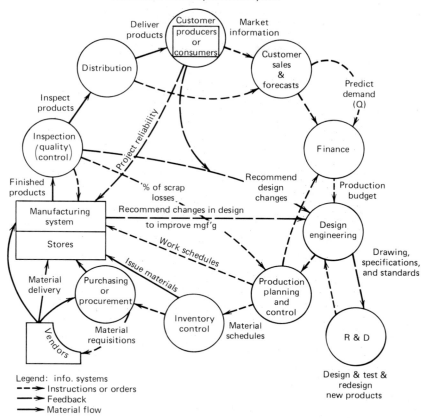

FIGURE 42-2 The manufacturing system is serviced by a production system, which is typically functionally designed.

tems. More examples can be added to Table 42.2 and Table 42.3; neither list is meant to be complete. The classical systems here are the *job shop,* the *flow shop,* the *project shop,* and the *continuous process.* The linked-*cell* is a new kind of system. The *assembly line* and a system that has *batch flow* might also be added as another MS.

Figure 42-3 shows schematics of the four classical systems. *Cellular manufacturing systems* will be discussed later.

Job Shops.
The job shop's distinguishing feature is the production of a wide variety of products which results in small manufacturing lot sizes, often one of a kind. Job shop manufacturing is commonly done to specific customer order, but, in truth, many job shops produce to fill finished goods inventories. Because the plant must perform a wide variety of manufacturing processes, general-purpose production equipment is required. Workers must have relatively high skill levels to perform a range of different work assignments. Job shop products include space vehicles, aircraft, machine tools, special tools, and equipment.

TABLE 42.2 **Partial List of Producer, Consumer Goods and Industries**

Aerospace/Airplanes
Appliances
Automotive (cars, trucks, vans, wagons, etc.)
Beverages
Building supplies (hardware)
Cement/Asphalt
Ceramics
Chemicals and allied industries
Clothing (garments)
Construction and construction materials
Computers
Drugs, soaps, cosmetics
Engineering
Electrical/Microelectronics
Equipment and machinery (agricultural, construction, and electrical products, electronics, household products, industrial machine tools, office equipment, computers, power generation)
Foods (canned, dairy, meats, etc.)
Footwear
Furniture
Glass
Hospital supplies
Leather and fur goods
Machines
Marine engineering
Metals (steel, aluminum, etc.)
Natural resources (oil, coal, forest, pulp and paper)
Publishing and printing (books, records)
Restaurants
Retail (food, department store, etc.)
Ship building
Textiles
Tire and rubber
Tobacco
Transportation vehicles (railroad, airline, truck, bus)
Utilities (electric power, natural gas, telephone)
Vehicles (bikes, cycles, ATVs, snowmobiles)

TABLE 42.3 **Types of Service Industries**

Advertising/Marketing
Education
Entertainment (radio, TV, movies, plays)
Equipment/Furniture rental
Financial (banks, investment companies, loan companies)
Health care
Insurance
Transportation/Car rental
Travel (hotel, motel)

TABLE 42.4 Types and Examples of Manufacturing Systems

Types of Manufacturing Systems	Examples	
	Service[a]	Products[b]
Job shop	Auto repair Hospital Restaurant University	Machine shop Metal fabrication Custom jewelry FMS
Flow shop or flow line	X-ray Cafeteria College registration Car wash	TV factory Auto assembly line
Cellular shop[c]	Fast food restaurant (Kentucky Fried Chicken, Wendy's Hamburgers) Law offices Midas Muffler	Family of "turned" parts Composite part families Line families Design families Manned cells Robotic cells
Project shop	Movie Broadway play TV show	Locomotive assembly Bridge construction House construction
Continuous processes	Telephone company Power company	Oil refinery Chemical plant

[a]Customer receives a service or perishable product.
[b]Products can be for customers or for other companies.
[c]New design of manufacturing system.

FIGURE 42-3 Four classical designs of manufacturing systems.

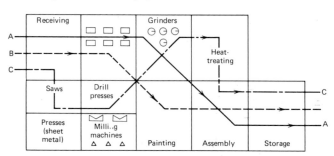

(a) Job shop — functional or process layout

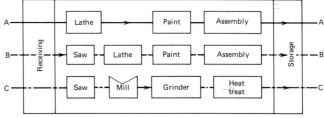

(b) Flow shop — line or product layout

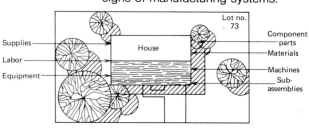

(c) Project shop — fixed position layout

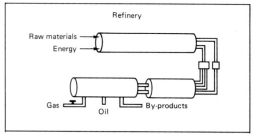

Continuous-process layout

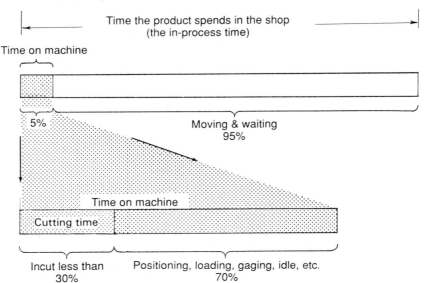

FIGURE 42-4 In the job shop, the parts spend most of the time waiting or moving and only 5% on the machine. (Source: C. F. Carter, Cincinnati Milacron.)

Figure 42-3 shows the functionally arranged job shop. Production machines are grouped according to the general type of manufacturing process. The lathes are in one department, drill presses in another, plastic molding in still another, and so forth. The advantage of this layout is its ability to make a wide variety of products. Each different part requiring its own unique sequence of operations can be routed through the respective departments in the proper order. *Route sheets* are used to control the movement of the material. Forklifts and handcarts are used to move materials from one machine to the next. As the company grows, the job shop evolves into a production job shop (PJS).

The production job shop (PJS) becomes extremely difficult to manage as it grows, resulting in long product throughput times and very large in-process inventory levels. In the job shop, parts spend 95% of the time waiting (delay) or being transported and only 5% of the time on the machine. See Figure 42-4. Of the time during which parts are on the machine, only 30% involves machining.

The PJS manufacturing system builds large volumes of products but still builds in lots or batches, usually medium-sized lots of 50 to 200 units. The lots may be produced only once, or they may be produced at regular intervals. The purpose of batch production is often to satisfy continuous customer demand for an item. This system usually operates in the following manner. Because the production rate can exceed the customer demand rate, the shop builds an inventory of the item, then changes to other products to fill other orders. This involves tearing down the setups on many machines and resetting them for new products. When the stock of the first item becomes depleted, production is repeated to build the inventory again.

General-purpose manufacturing equipment is designed for higher production rates. For example, automatic lathes capable of holding many cutting tools are used rather than engine lathes. The machine tools are often equipped with specially designed workholding devices, jigs, and fixtures, which increase process output rate, precision, accuracy, and repeatability.

Industrial equipment, furniture, textbooks, and components for many assembled consumer products (household appliances, lawn mowers, and so on) are made in production job shops. Such systems are called machine shops, foundries, plastic molding factories, and pressworking shops.

As much as 75% of all piece part manufacturing is estimated to be in lot sizes of 50 pieces or fewer. Hence, production job shop production constitutes an important portion of total manufacturing activity.

Flow Lines. The flow line has a product-oriented layout. When the volume gets very large, especially in an assembly line, this is called *mass production*. This kind of system can have (very) high production rates. Specialized equipment, dedicated to the manufacture of a particular product, is used. The entire plant may often be designed exclusively to produce the particular product, using special-purpose rather than general-purpose equipment. The investment in the specialized machines and specialized tooling is high. Many production skills are transferred from the operator to the machines so that the manual labor skill level in a flow shop tends to be lower than in a production job shop. Items are made to "flow" through a sequence of operations by material-handling devices (conveyors, moving belts, and transfer devices). The items move through the operations one at a time. The time the item spends at each station or location is fixed and equal (balanced).

Most factories are mixtures of the job shop with flow lines. Obviously, the demand for products can precipitate a shift from batch to high-volume production, and much of the production from these plants is consumed by that steady demand. Subassembly lines and final assembly lines are further extensions of the flow line, but the latter are usually much more labor-intensive.

In the flow line manufacturing system, the processing and assembly facilities are arranged in accordance with the product's sequence of operations. Work stations or machines are arranged in line with only one work station of a type, except where duplicates are needed for balancing the time products take at each station. The line is organized by the processing sequence needed to make a single product or a regular mix of products. A hybrid form of the flow line produces batches of products moving through clusters of work stations or processes organized by product flow. In most cases, the setup times to change from one product to another are long and often complicated.

Transfer Lines. Various approaches and techniques have been used to develop machine tools that would be highly effective in large-scale manufacturing. Their effectiveness was closely related to the degree to which the design of the products was standardized and the time over which no changes in the design were permitted. If a part or product is highly standardized and will be manufactured in large quantities, a machine that will produce the parts with a minimum of skilled labor can be developed. A completely tooled automatic screw machine is a good example for small parts. Figure 42-5 shows an example of an *automated transfer machine* for the production of V-8 engine blocks at the rate of 100 per hour. There are five independent sections that perform 265 drilling, 6 milling, 21 boring, 56 reaming, 101 counterboring, 106 tapping, and 133 inspection operations. These are performed at 104 stations, including 1 loading, 53 machining, 36 visual inspection, and 1 unloading. Provision is made for banking parts between each section. However, such specialized machines are expensive to design and build and may not be capable of making any other products. Consequently, to be economical, such machines must be operated for considerable periods of time to spread the cost of the initial investment over many units. These machines and systems, although highly efficient, can be utilized only to make products in very large volume, and desired changes of design in the products must be avoided or delayed because it would be too costly to scrap the machines.

However, as we have already noted, products manufactured to meet the demands of the free-economy, mass-consumption markets need to have changes in design for improved product performance as well as style changes. Therefore, hard automation systems need to be as flexible as possible while retaining the ability to mass-produce. This has led to the following developments.

First, the machines are constructed from basic building blocks or modular units that accomplish a function rather than produce a specific part. The production machine modules are combined to produce the desired system for making the product. These units are called self-contained power-head production machines. Such machines are then connected by automatic transfer devices to handle the material (that is, the thing being produced), moving it automatically from one process to the next.

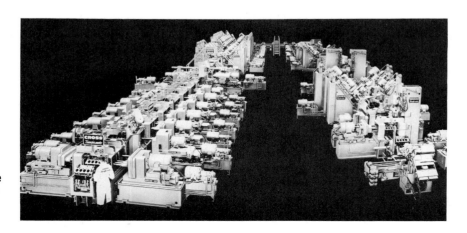

FIGURE 42-5   The transfer machine is an example of a flow line manufacturing system. (Courtesy The Cross Company.)

Second, the incorporation of programmable logic controllers (PLCs) and feed-back control devices have made these machines more flexible. Modern PLCs have the functional sophistication to perform virtually any control task. These devices are rugged, reliable, easy to program, and economically competitive with any alternate control device, and they have replaced conventional hard-wired relay panels in many applications. Relay panels are hard to reprogram, whereas PLCs are very flexible. Relays have the advantage of being well under-stood by maintenance people and are invulnerable to electronic noise, but con-struction time is long and tedious. PLCs allow for mathematical algorithms to be included in the closed-loop control system and are being widely used for single-axis, point-to-point control as typically required in straight-line machin-ing, robot handling, and robot-assembly applications. They do not at this time challenge computer numerical controls (CNC) used on multiaxis contouring machines. However, PLCs are used for monitoring temperature, pressure, and voltage on such machines. PLCs are used on transfer lines to handle complex material movement problems, gaging automatic tool setting, on-line tool wear compensation, and automatic inspection, giving these systems flexibility that they never had before.

Project Shop.   In the typical project manufacturing system, a product must remain in a fixed position or location during manufacturing because of its size and/or weight. The materials, machines, and people used in fabrication are brought to the site. Locomotive manufacturing, large aircraft assembly, and shipbuilding use fixed-position *layout*.

Fixed-position *fabrication* is also used in construction jobs (buildings, bridges, and dams). As with the fixed-position layout, the product is large, and the construction equipment and manpower must be moved to it. When the job is completed, the equipment is removed from the construction site.

The project shop invariably has a job shop/flow shop manufacturing system making all the components for the large complex project and thus has a func-tionalized production system.

Continuous Processes.   In the continuous process, the project physically flows. Oil refineries, chemical-processing plants, and food-processing operations are examples. This system is sometimes called ''flow production'' when the manufacture of either complex single parts (such as a canning operation or automotive engine blocks) or assembled products (such as television sets) is described. However, these are not continuous processes, but simply high-volume flow lines. In continuous processes, the products really do flow because they are liquids, gasses, or powders.

Continuous processes are the most efficient but least flexible kinds of man-ufacturing systems. They usually have the leanest, simplest production systems because these manufacturing systems are the easiest to control, having the least work-in-process.

The Cellular Manufacturing System. A cellular manufacturing system (CMS), composed of linked-cells, is the newest kind of manufacturing system. In cells, processes are grouped according to the sequence and operations needed to make a product. This arrangement is much like that of the flow shop but is designed for flexibility. The cell is often laid in a U-shape so that the workers can move from machine to machine, loading and unloading parts. Figure 42-6 shows an example of a manned cell. The machines in the cell are all of the A(2) or higher level of automation so that they can complete the desired processing untended, turning themselves off when done with a machining cycle. The cell usually includes all the processing needed for a complete part or subassembly.

Cells are typically manned, but unmanned cells are beginning to emerge with a robot replacing the worker. A robotic cell design is shown in Figure 42-7 with one robot and four CNC machines. For the cell to operate autonomously, it must have adaptive control capability, that is, A(5) level of automation.

To form a cellular manufacturing system, the first step is to restructure portions of the job shop, converting it in stages into manned cells. See Figure 42-8.

FIGURE 42-6 Example of a manned manufacturing cell with six machines and one multifunctional worker. The lathe operation is duplicated.

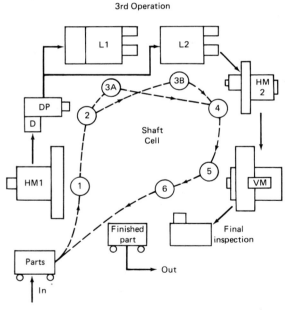

Work Sequence	Name of Operation	Time		
		Manual	Walking	Machine
①	Mill ends on work on HM (1)	12"	5"	30"
②	Drill hole on DP	15"	5"	20"
③A or ③B	Turn – bore on L1 or L2	'13"	5" 8"	180"
④	Mill flats on HM2	12"	8" 5"	20"
⑤	Mill steps on VM	13"	7"	30"
⑥	Final inspect	10"	5"	
		75"	35"	280"

Cycle Time = 75 sec + 35 sec = 110 seconds

Longest Machining Time = 180 seconds

Total Machining Time = 280 seconds

All machines in the cell are capable of running untended while the operator(s) are doing manual operations (unload, load, inspection, deburr) or walking from machine to machine. The time to change tools and workholders (perform setup) is not shown.

Key

DP = Drill press
L = Lathes
HM = Horizontal milling machine
VM = Vertical milling machine

— -→ Paths of workers moving within cell

——→ Material movement paths

① Operation sequence

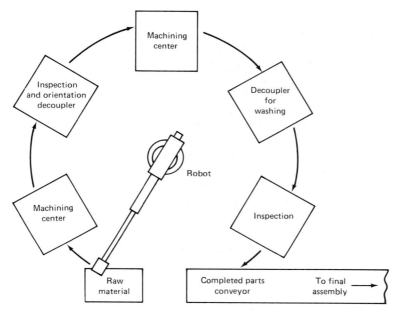

FIGURE 42-7 Example of robotic cell. Decoupler checks part, reorients part for next machine, and controls inventory in the cell.

Cells are designed to manufacture specific groups or families of parts. As will be described in Chapter 44, cells are linked either *directly* to each other or to subassembly points, or *indirectly* by the pull system of inventory control, called Kanban. The flow shop manufacturing within the plant is also redesigned to make these systems operate like cells. To do this, the long setup times typical in flow lines must be vigorously attacked and reduced so that the flow lines can be changed quickly from making one product to making another. The need to line-balance the flow line must be eliminated. This makes them flexible and compatible with the cells designed to make piece parts and with the subassembly lines and final assembly lines.

Operators move around the cells, servicing different processes. Figure 42-9 shows a more elaborate layout of a five-machine cell with one worker.

How Cells Are Different

Cells have many features that make them unique and different from other manufacturing systems. Parts move from machine to machine *one at a time* within the cell. For material processing, the *machines are typically A(2) or higher—* capable of completing a machining cycle initiated by a worker. The *U* shape puts the start and finish points of the cell next to each other. Every time the operator completes a walking trip around the cell, a part is completed. This time defines the cycle time (CT). The machining time (MT) for each machine needs only to be less than the time it takes for the operator to complete the walking trip around the cell. As shown in Figure 42-9, *machining times are overlapping and need not be equal (balanced)* so long as no MT is greater than the CT. The

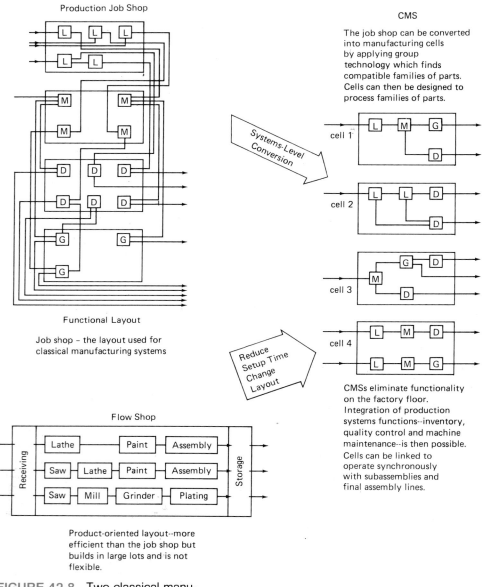

Production Job Shop

Functional Layout

Job shop – the layout used for
classical manufacturing systems

Systems-Level Conversion

CMS

The job shop can be converted
into manufacturing cells
by applying group
technology which finds
compatible families of parts.
Cells can then be designed to
process families of parts.

cell 1
cell 2
cell 3
cell 4

CMSs eliminate functionality
on the factory floor.
Integration of production
systems functions--inventory,
quality control and machine
maintenance--is then possible.
Cells can be linked to
operate synchronously
with subassemblies and
final assembly lines.

Reduce
Setup Time
Change
Layout

Flow Shop

Receiving	Lathe		Paint	Assembly	Storage
	Saw	Lathe	Paint	Assembly	
	Saw	Mill	Grinder	Plating	

Product-oriented layout--more
efficient than the job shop but
builds in large lots and is not
flexible.

FIGURE 42-8 Two classical manufacturing systems in common use today—the job shop and the flow
shop—require a systems level conversion to reconfigure them into
CMS.

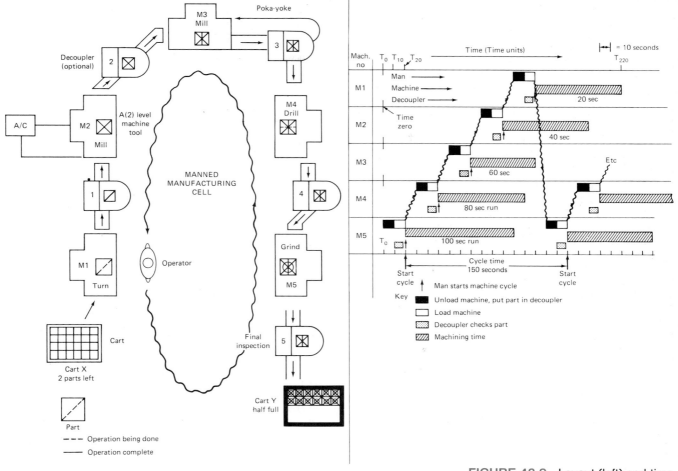

FIGURE 42-9 Layout (left) and time bar diagram of a five-machine, U-shaped cell with decouplers, operator upstreaming, and variable machine times (MT).

cycle time is 150 seconds. The total machine time was $60 + 80 + 100 + 80 + 80 = 400$ seconds. A NC machining center capable of performing the five machining operations could easily replace the cell. However, the cycle time for a part would jump from 150 seconds to over 400 seconds because combining the processes into one machine prevents *overlapping of the machining times*. The CT for the cell can be readily altered by adding (a portion of) an additional worker to the cell. See Chapter 44. The standard operations routine sheet, shown in Figure 42-10, is used to plan the manufacture of one of the parts in the family within the cell. The plan shows the relationships between the manual operations performed by the worker, the machining operations performed by the machine, and the time spent by the worker walking from machine to machine. The manual operations include loading and unloading the machine, checking quality, deburring and taking chips out of fixtures. The cycle time is determined by the system's need for the parts made by the cell.

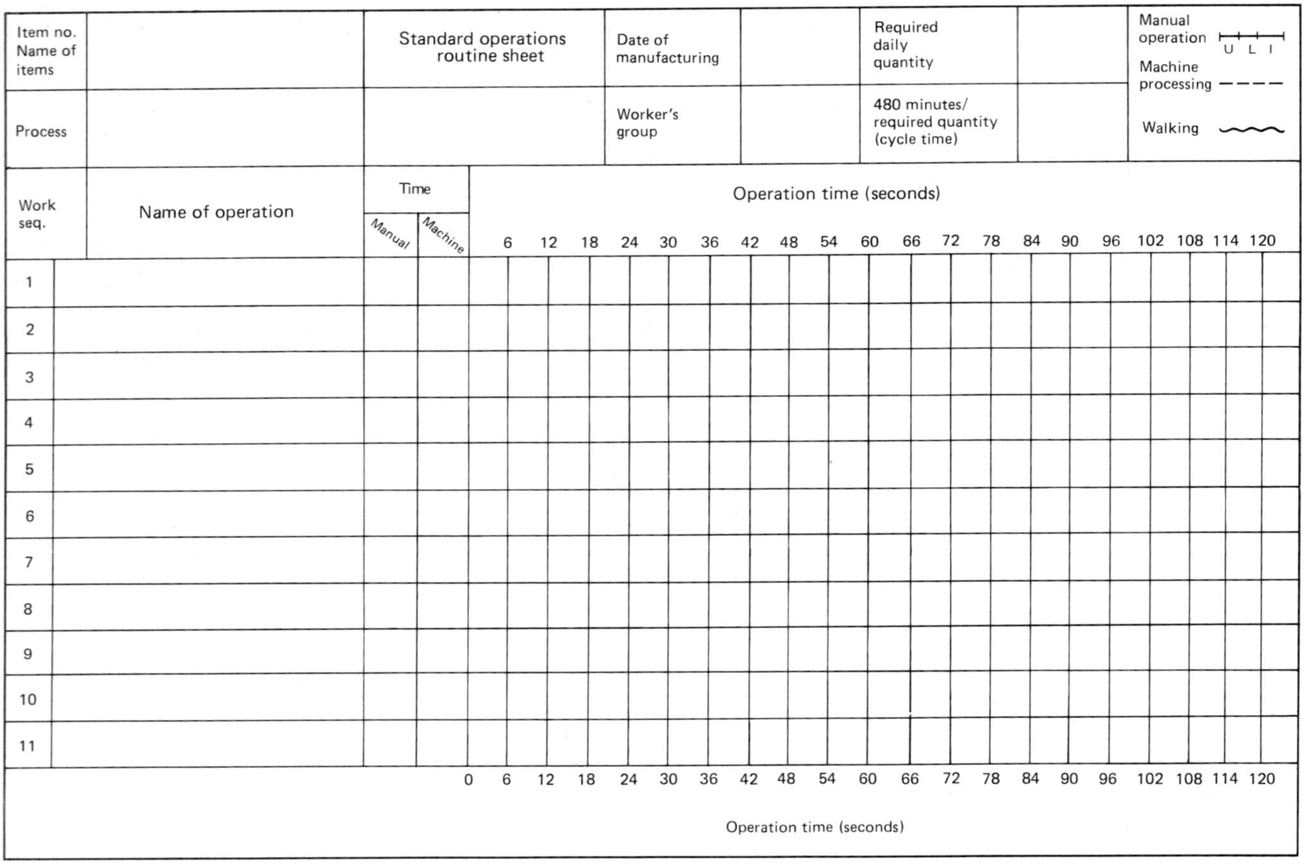

Item no. Name of items		Standard operations routine sheet	Date of manufacturing		Required daily quantity		Manual operation U L I Machine processing
Process			Worker's group		480 minutes/ required quantity (cycle time)		Walking

Work seq.	Name of operation	Time Manual	Time Machine	Operation time (seconds) 6 12 18 24 30 36 42 48 54 60 66 72 78 84 90 96 102 108 114 120
1				
2				
3				
4				
5				
6				
7				
8				
9				
10				
11				

0 6 12 18 24 30 36 42 48 54 60 66 72 78 84 90 96 102 108 114 120

Operation time (seconds)

FIGURE 42-10   Example of standard operations routine sheet.

Workers in the cells are multifunctional; each worker can operate more than one kind of process (or multiple versions of the same process) and also performs inspection and machine maintenance duties. Cells eliminate the job shop concept of one person/one machine and thereby greatly increase worker productivity and utilization.

The restriction of the cell to a parts family makes reduction of setup in the cell possible. This is often called flexible fixturing. The general approach to setup reduction will be discussed in Chapter 44.

In some cells *decouplers* are placed between the processes, operations or machines to provide flexibility, quality control, and process delay for the manufacturing cell. For example, a process delay decoupler would delay the part movement to allow the part to cool down, heat up, cure, or whatever is necessary for a period of time greater than the cycle time for the cell. Decouplers and flexible fixtures are vital parts of unmanned cells.

Table 42.5 provides a summary of the chief characteristics of the five basic manufacturing systems. Figure 42-11 summarizes the discussion on manufac-

TABLE 42.5 Characteristics of Basic Manufacturing Systems

Characteristics	Job Shop	Flow Shop	Project Shop	Continuous Process	Linked Cells
Types of machines	Flexible, general purpose some specialization	Single purpose, single function	General purpose; mobile, manual	Specialized, high technology	Simple, customized single-cycle automation
Design of processes	Functional or process	Product flow layout	Project or fixed-position layout	Product	U-shaped, sequenced to flow for family of parts, overlapping
Setup time	Long, variable, frequent	Long and complex	Variable, every job a new setup	Rare and expensive	Short, frequent, one touch
Workers	Single-functioned; highly skilled; one man—one machine	One function, lower-skilled, one man—one machine	Specialized, highly skilled	Skill level varies	Multifunctional, respected
Inventories (WIP)	Large inventory to provide for large variety	Large to provide buffer storage	Variable, usually large	Very small	Small
Lot sizes	Small to medium	Large lot	Small lot	Very large	Small
Manufacturing lead time	Long, variable	Short, constant	Long, variable lead time	Very fast, constant	Short, constant

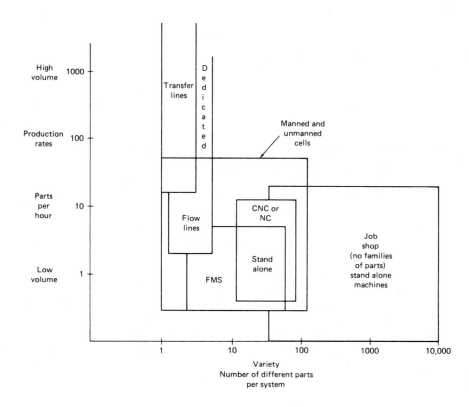

FIGURE 42-11 Part variety versus production rates for various manufacturing systems.

turing systems by comparing different systems based on their production rate and product flexibility—that is, the number of different parts the system can handle. The project shop and continuous processes are not shown. The widely publicized Flexible Manufacturing System (FMS) can be classified as a job shop because random order of part movement is permitted and therefore scheduling of parts and machines within the FMS is necessary, just as it is in the job shop. This feature can make FMS designs difficult to link to the rest of manufacturing systems.

Additional material on manufacturing cells and CMS can be found in Chapter 44.

Let us summarize the characteristics of manufacturing cells. The cell makes parts one at a time in a flexible design. Cell capacity (the cycle time) can be quickly altered to respond to changes in customer demand. The cycle time does not depend upon the machining time.

Families of parts with similar designs, flexible workholding devices, and tool changers in programmable machines allow rapid changeover from one component to another. Rapid changeover means that quick or one-touch setup is employed, often like flipping a light switch. Significant inventory reductions between the cells is possible, and the inventory level can be directly controlled. Quality is controlled within the cell and the equipment within the cell is routinely maintained by the workers.

For robotic (unmanned) cells, the robot typically loads and unloads parts for one to five CNC machine tools, but this number may be increased as the robot becomes mobile. A machining center represents a cell of one machine but is not as flexible as a cell composed of multiple simple machines. Cellular layouts facilitate the integration of critical production functions while maintaining flexibility in producing superior-quality products.

The product designer can easily see how parts are made in the cell since all the processes are together. Because quality-control techniques are also integrated into the cells, the designer knows exactly the cell's process capability. The designer can easily configure the future designs to be made in the cell. This is truly designing for manufacturing.

Automation

The term *automation* has many different definitions. Apparently it was first used in the early 1950s to mean automatic handling of materials, particularly equipment used to unload and load stamping equipment. It has now become a general term referring to services performed, products manufactured and inspected, information handling, materials handling, and assembly, all done automatically as an automatic operation.

In Chapter 1, Amber and Amber's Yardstick for Automation was presented. This classification is based on the concept that all work requires energy and information, and certain functions must be provided by man or machine. Whenever a machine replaces a human function or attribute, it is considered to have taken an "order" of automaticity. In today's factory, levels A(2), A(3), A(4), and A(5) are found, with occasional A(6) levels. This portion of the yardstick

ORDER OF AUTOMATICTIY	HUMAN ATTRIBUTE MECHANIZED	DISCUSSION	EXAMPLES
A (3) Automatic repeat cycle or Open loop control	DILIGENCE Carries out routine instructions without aid of man Open end or nonfeedback	All automatic machines Loads, processes, unloads, repeats— System assumed to be doing okay. Probability of malfunctions negligible Obeys fixed internal commands or external program	• Record player with changer • Automatic screw mach. • Bottling machines • Clock works • Donut maker • Spot welder • Engine production lines • Casting lines • Newspaper printing machines • Transfer machines
A (4) Self-measuring and adjusting feedback or closed loop systems	JUDGEMENT Measures and compares result (output) to desired size or position (input) and adjusts to minimize any error	Self-adjusting devices Feedback from product position, size, velocity, etc. 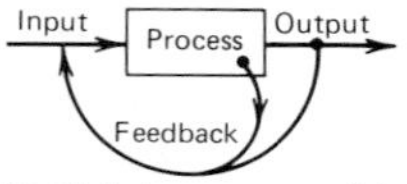 Input → Process → Output, Feedback Multiple loops are possible	• Product control • Can filling • N/C machine tool with position control self-adjusting grinders • Windmills • Thermostats • Waterclock • Fly ball governor on steam engine
A (5) Adaptive or computer control: Automatic Cognition	EVALUATION Evaluates multiple factors on process performance, evaluates and reconciles them Use mathematical algorithms	Process performance must be expressed as equation 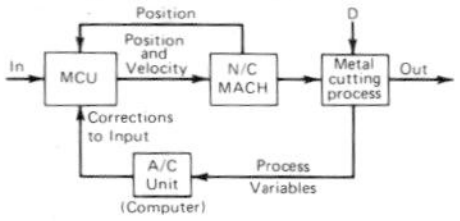 [diagram: In → MCU → Position and Velocity → N/C MACH → Metal cutting process → Out; Corrections to Input; A/C Unit (Computer); Process Variables]	N/C Machine with A/C capability Maintaining pH level Turbine fuel control
A (6) Expert systems or limited self-programming	LEARNING BY EXPERIENCE	Subroutines are a form of limited self-programming Trial and error sequencing Develops history of usage	Phone circuits Elevator dispatching

FIGURE 42-12   Yardstick for Automation: Levels A(3) to A(6).

is given in greater detail in Figure 42-12. *Mechanization* refers to the first and second orders of automaticity, which includes semiautomatic machines. Virtually all of the machine tools described in the previous chapters are A(2) machines. The machines used in manufacturing cells are basically A(2) machines. Automation as we know it today begins with the A(3) level. In recent years, this level has taken on two forms: *hard* (or *fixed-position*) *automation* and *soft* (or *flexible* or *programmable*) *automation*. Instructions to the machine, telling it what to do, how to do it, and when to do it, are called the *program*. In hard automation transfer lines, the programming consists of cams, stops, slides, and hard-wired electronic circuits using relay logic. An example of this level of automation is the automatic screw machine. If the machine is programmed with a tape, a programmable controller (PLC), a hand-held control box, microprocessor, or computers, control instructions are easily changed, with the software making the system or device much more adaptable.

Self-adjusting and measuring machines are of the A(4) level replacing human judgment and allowing these machines to be self-correcting. This is commonly known as *feedback control* or *closed loop,* meaning that information about the performance is fed back (or looped back) into the process. Examples of A(4) machines were given in Chapter 29, Machining Centers. A simple A(4) level is one in which some parameter of the process is measured using a detection device (sensor). This information is fed back to a comparator, which makes comparisons with the desired level of operation. If the output and the input are not equal, an error signal is created and the process adjusts to reduce the error. An automatic grinder that checks the diameter of a part and automatically repositions the grinding wheel to compensate for wheel wear would be an example of an A(4) machine.

The A(5) level, typified by *adaptive control* (AC), replaces human evaluation. Refer to Figure 29-6. Basically, A(5) machines are capable of adapting the process itself so as to optimize it in some way. This level of automation requires that the system have a computer. Programmed into the computer are models (mathematical equations) that describe how this process or system behaves, how this behavior is bounded, and what aspect of the process or system is to be optimized. *This modeling obviously requires that the process be sufficiently well understood theoretically so that equations (models) can be written that describe how the real process works.* This level of automation has been achieved for continuous processes (oil refineries, for example) where the theory (of heat transfer and fluid dynamics) of the process is well understood and parameters are easier to measure. Unfortunately, the theory of metal forming and metal removal is less well understood and parameters are usually difficult to measure. These processes have resisted adequate theoretical modeling, and as a consequence there are very few A(5) machines on the shop floor. The basic elements of the adaptive control loop are

1. *Identification:* Measurements from the process itself, its output, or process inputs using sensors.
2. *Decision analysis:* Optimization of the process in the computer.
3. *Modification:* Signal to the controller to alter the inputs.

To raise the automatic grinder to the A(5) level, assume that it is a cylindrical center-type grinder on which part deflection as well as part size are measured. In this process, the cutting forces tend to deflect the part more as the grinding wheel gets further from the centers. The adaptive control software program would have equations that relate deflection to grinding forces and infeed rates. The infeed would be decreased to reduce the force and minimize deflection. Notice that the overall system would still have to compensate for grit dulling and grit attrition, which will also alter the grinding forces. The A(5) level reflects deductive reasoning whereby particular outcomes are deduced from general principles.

The A(6) level reflects the beginnings of artificial intelligence (AI) in which the control software is infected with elements (subroutines) that permit some ''thinking'' on the part of the software. Few if any such systems exist on the factory floor.

The A(6) level tries to *relate cause to effect*. Suppose the effect was tool failure. At the A(5) level, the system would detect the increase in deflection due to increased forces (due to the tool's dulling) and reduce the feed to reduce the force. However, it might simultaneously try to increase the speed to maintain the MRR constant, which would increase the rate of tool wear. This was not the desired result. At the A(6) level, multiple factors are evaluated so that the system can recognize the need to change the tool rather than reduce the cutting feed. That is, the system learned by experience that feed reduction/speed increase was the wrong decision, that it was tool wear that caused force to increase. Such systems are often called *expert systems* when the software contains the collective experience of human experts or prior replications of the same process. The A(7) level reflects the next level of AI whereby inductive reasoning is used. The system software can determine a general principle (the theory) based on the particular facts (the data base) collected. NC and CNC machines are discussed in Chapter 29. Our next discussion will concern transfer lines and FMS.

Transfer Machines.   A *transfer machine* is an automated flow line. Workpieces are automatically transferred from station to station, from one machine to another. Operations are performed sequentially. Ideally, work stations perform the operation(s) simultaneously on separate workpieces with the number of parts equal to the number of stations. Each time the machine cycles, a part is completed. Figure 42-6 shows an example of a transfer line.

Transferring usually is accomplished by one or more of four methods. Frequently, the work is pulled along supporting rails by means of an endless chain that moves intermittently as required. In another method the work is pushed along continuous rails by air or hydraulic pistons. A third method, restricted to lighter workpieces, is to move them by an overhead chain conveyor, which may lift and deposit the work at the machining stations.

A fourth method often is employed when a relatively small number of operations, usually three to ten, are to be performed. The machining heads are arranged radially around a rotary indexing table, which contains fixtures in which the workpieces are mounted (see Figure 42-13). The table movement may be continuous or intermittent. Face-milling operations sometimes are performed by moving the workpieces past one or more vertical-axis heads. Such circular configurations have the advantage of being compact and of permitting the workpieces to be loaded and unloaded at a single station without having to interrupt the machining.

Means must be provided for positioning the workpieces correctly as they are transferred to the various stations. One method is to attach the work to carrier pallets or fixtures that contain locating holes or points that mate with retracting pins or fingers at each work station. Excellent station-to-station precision is obtained as the fixtures thus are located and then clamped in the proper positions. However, carrier fixtures are costly. When it is possible they are eliminated and the workpiece transported between the machines on rails, which locate the parts by self-contained holes or surfaces. This procedure eliminates the labor required for fastening the workpieces to the carrier pallets, as well as the pallets.

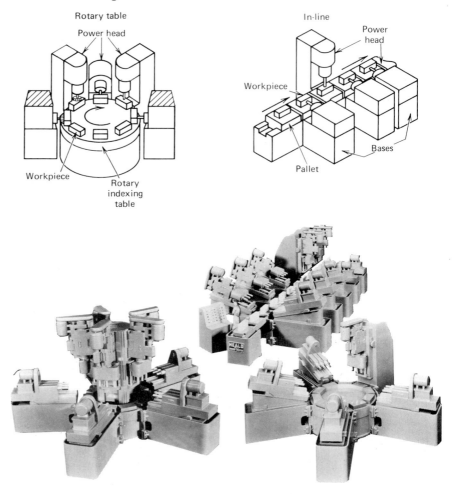

FIGURE 42-13 Examples of rotary and in-line transfer units that can be constructed from components. (Courtesy Heald Machine Company.)

In the design of large transfer machines, the matter of the geometric arrangement of the various production units must be considered. Whether or not transfer fixtures or pallets must be used is an important factor. These fixtures and pallets usually are quite heavy. Consequently, when they are used, a closed rectangular arrangement is often employed so that the fixtures are automatically returned to the loading point. If no fixtures or pallets are required, straight-line configurations can be employed. Whether pallets or fixtures must be used is dependent primarily on the degree of precision needed as well as on the size, rigidity, and design of the workpieces. If no transfer pallet or fixture is to be used, locating bosses or points must be designed or machined into the workpiece.

The matter of tool wear and replacement is of great importance when a large number of operations are incorporated in a single production unit. Tools must be replaced before they become worn and produce defective parts. Transfer machines often have more than 100 cutting tools. If the entire complex machine had to be shut down each time a single tool became dull and had to be replaced, overall productivity would be very low. This is avoided by designing the tooling so that certain groups have similar lives, monitoring tool thrust, and torque and by shutting down the machine before the tooling has deteriorated. All the tools in the affected group can be changed so that repeated shutdowns are not necessary. If the transfer machine is equipped with AC drives, diagnostic feedback information of individual processes is easily accomplished. See Figure 42-14. Programmable drives like these replace feedboxes, limit switches and hydraulic cylinders, and eliminate changing belts, pulleys, or gears to change feed rates and depths of cut and make the system much more flexible.

Methods have been developed for accurately presetting tools and for changing them rapidly. Tools can be changed in a few minutes, thereby reducing machine downtime. Increasingly, tools are preset in standard quick-change holders with excellent accuracy, often to within 0.0002 in. (0.005 mm.).

In the use of multiple-station machines, the entire machine may have to be shut down when one or two stations become inoperative. This is usually avoided by arranging the individual units in groups, or sections with 10 to 12 stations per section, and providing for a small amount of buffer storage (*banking* of workpieces) between the sections. This permits production to continue on all remaining sections for a short time while one section is shut down for tool changing or repair.

The most significant problem deals with designing the line itself so that it operates efficiently as a whole. Processes are grouped together at stations. Transfer machines or, for that matter, any system wherein a number of processes are connected sequentially, will require that the line be *balanced*. Line balancing

(1) a machine-information center

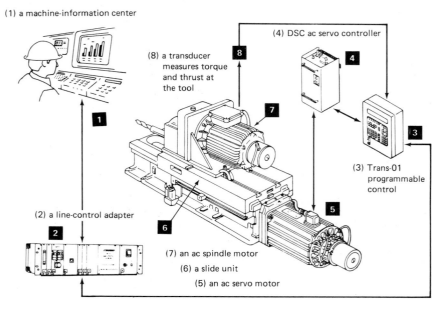

(4) DSC ac servo controller

(8) a transducer measures torque and thrust at the tool

(3) Trans-01 programmable control

(2) a line-control adapter

(7) an ac spindle motor

(6) a slide unit

(5) an ac servo motor

FIGURE 42-14   Distributed control systems monitor torque and thrust at the tool; tool breakage is therefore automatically prevented by reducing feed rate. (*American Machinist,* July 1985.)

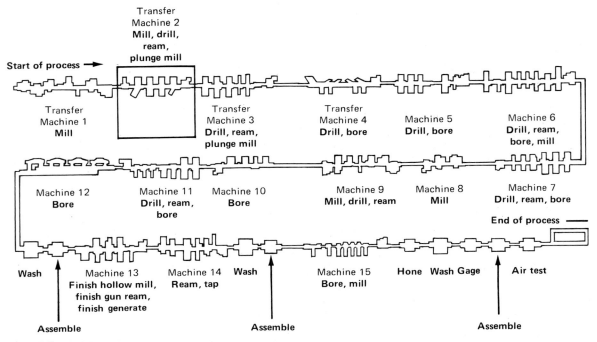

FIGURE 42-15 This megasystem at Ford is one of the largest flow lines in the world. It produces cylinder heads. (*American Machinist*, July 1985.)

means that the process time at each station must be the same, with the total nonproductive time for all other stations minimized. Theoretically, there will be one station that will have the longest time and this station will control the cycle time for all the stations. Computer algorithms have been developed to deal with line balancing. When the line is the size of the one shown in Figure 42-15, this becomes a rather complex problem. This line produces cylinder heads with 15 transfer machines, two assembly machines, part washers, gages, and inspection equipment.

Transfer machine 1, which has 30 stations, performs rough-milling. Machines 2 and 3, with 23 and 25 stations, respectively, perform a variety of drill, plunge-mill, and spot-face operations. Valve throats are finished by 23-station Machine 4, which also drills oil-feed holes. Transfer machine 6, with 31 stations, does rough- and finish-boring, drilling, reaming, and milling. This system produces engine blocks and cylinder heads at a rate of over 100,000 units per year—100 cylinder heads per hour.

In Figure 42-16, the design of the horizontal NC transfer machine has CNC machines replacing slide units. In addition to CNC units with tool changers, this system features multiaxis fixture positioning, palletized automation, in-process gaging, size control, fault diagnostics, and excellent process capability.

Transfer machines are either A(3), A(4), or A(5) level machines, depending upon whether they have the built-in capacity for sensing when corrective action is required and how such corrections are made. Sensing and feedback control

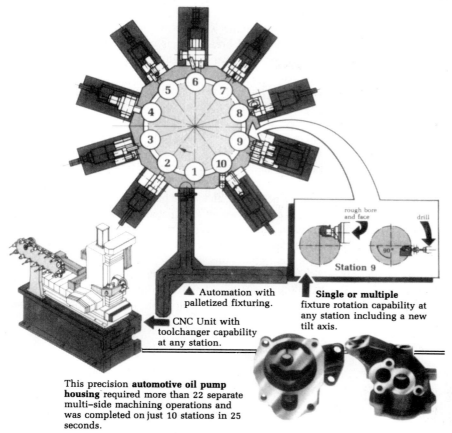

Station 9

▲ **Automation with palletized fixturing.**

■ **Single or multiple** fixture rotation capability at any station including a new tilt axis.

■ CNC Unit with toolchanger capability at any station.

This precision **automotive oil pump housing** required more than 22 separate multi–side machining operations and was completed on just 10 stations in 25 seconds.

FIGURE 42-16 Horizontal rotary transfer machine with CNC units in place of conventional slide units. (Courtesy Jestadt.)

systems are essential requirements for the fourth level and all higher levels of automation.

Many machines have *feed-forward* devices built into them. This means that the system takes information from the input side of the process rather than the output side of the process and uses that information to alter the process. For example, the temperature of the billet as it enters a hot-rolling or hot-extrusion process can be sensed and used as feed-forward information to alter the process parameters. This would be an A(5) or adaptive control example. Sensors can be located in three positions—ahead of the process, in the process, or on the output side of the process. The feed-forward concept can also be applied at the A(4) level. Suppose that you have a transfer line that processes two types of flywheel housings that are similar but require different machining operations. When the housings are fed into the machine, in mixed order, a sensing device contacts a distinguishing boss that is on one type but absent from the other. The sensing and feed-forward system then sets the proper tooling for that housing and omits operations that are required only for the other type.

It is common practice to equip transfer machines with automated inspection stations or probing heads that determine whether the operation was performed correctly and detect whether any tool breakage has occurred that might cause damage in subsequent operations. For example, after drilling, a hole is checked to make certain it is clear prior to tapping.

Automation and transfer principles are also used very successfully for assembly operations. In addition to saving labor, automatic testing and inspection can be incorporated into such machines at as many points as desired. Such in-process inspection should be used to prevent defects from being made rather than finding defects after they are made. This assures superior quality. When defective assemblies are simply discovered and removed for rework or scrapped, the cause of the problem is not necessarily corrected.

The range of products now being completely or partially assembled automatically is very great. Figure 42-17 shows an automated assembly machine with a circular transfer table and a linear transfer line serviced by several robots, each performing a different task. Both tactile and visual sensors are needed.

In many cases, some manual operations are combined with some automatic operations. For example, one transfer machine for assembling steering knuckle, front wheel hub, and disk-brake assemblies has 16 automatic and 5 manual work stations. As with manufacturing operations, automatic assembly often can be greatly improved through proper part design.

FIGURE 42-17 Robotic assembly system using tactile and visual sensors. (Source: *Computers in America,* January 1984).

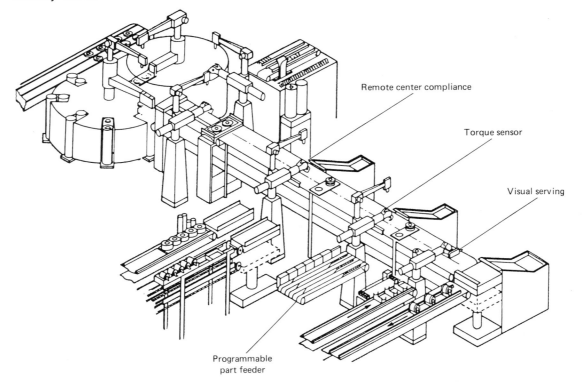

Remote center compliance

Torque sensor

Visual serving

Programmable
part feeder

The Flexible Manufacturing System (FMS).   The most publicized type of modern manufacturing systems is known as the *flexible manufacturing system* or *FMS*. The development of FMSs began in the United States in the 1960s. The idea was to combine the high reliability and productivity of the transfer line with the programmable flexibility of the NC machine in order to be able to produce a variety of parts. The first systems, called variable mission or flexible manufacturing systems, were built by Cincinati Milacron and Sundstrand. In the late 1960s Sundstrand installed a system for machining aircraft speed drive housings that is still in use today. Overall, however, very few of these systems were sold until the late 1970s and early 1980s, when a worldwide FMS movement began. Table 42.6 shows the FMS installations and FMS exports and

TABLE 42.6   Number of Flexible Manufacturing Systems by Country and FMS Imports/Exports

Country	Number of Systems	Comments
United States	70	First FMS installation in 1963. FMS mainly in large companies. Cost is typically $1 million per machine
Japan	70	7–150 different parts per FMS; most systems home-built; perhaps as many as 100 FMSs in Japan
Italy	25	Extensive FMS installations in automotive industry
West Germany	30	30% of FMS cost in material handling; 50–250 different parts per FMS
United Kingdom	30	Government sponsoring—20% to 50% of systems cost; 5 to 120 parts/system. First installation in 1967
Sweden	8	Most installations in automotive industry
Others in Eastern Europe	6	Czechoslovakia has about 6 FMS in operation and 20 more under development
Others in Western Europe	26	The Netherlands and Switzerland have FMS installations
France	25	Most installations in automotive, aerospace, and machine tool industries
U.S.S.R.	60	Pioneered GT: Have developed standardized FMS for any turned part
East Germany	20	First system installed in 1971; 14 to 200 different parts/FMS.
Korea	1	Just a matter of time before these
Taiwan	1	countries increase the FMS involvement
China	0	
Total	372	

Sources: V. Raju, *Manufacturing Engineering*, April 1986; U.S. Department of Commerce; FMS, IFS Publications, Ltd., United Kingdom.

imports for 14 countries. International trade in FMS is not significant. Despite all that has been written, there are fewer than 400 systems in the world (less than .1% of the machine tool population). Most remarkable is the fact that there are about 60 suppliers of FMS for about 150 to 200 users (companies). There is also some evidence that the market for these large expensive systems became saturated in 1984.

The FMS differs from the cell in that the product can take random paths through the machines. This system is fundamentally an automated, conveyorized, computerized job shop. The system is complex to schedule. Because the machining time for different parts varies greatly, the FMS is difficult to link to an integrated system and often remains an island of expensive automation.

In about 80% of installations, the collection of components being machined consists of nonrotational (prismatic) parts like crankcases and transmission housings. The number of machine tools in a FMS varies from 2 to 10, as shown in Figure 42-18, with 3 or 4 being typical. Annual production volumes for the systems are usually in the range of 3,000 to 10,000 parts with the number of different parts ranging from 2 to 20, with 8 being typical. The lot sizes are typically 20 to 100 parts and the typical part has a machining time of about 30 minutes with a range of 6 to 90 minutes per part. Each part typically needs 2 or 3 chuckings and needs about 40 operations. Invariably, NC machining centers dominate older FMSs. In recent years, CNC machine tools have been favored, leading to a considerable number of systems' being operated under direct numerical control (DNC). The machining centers always have tool changers. In order to overcome the limitation of a single spindle, some systems are being built with head changers.

Other common features of FMSs, as shown in Figure 42-19, are pallet changers, underfloor conveyor systems for the collection of chips (not shown), and a conveyor system that delivers parts to the machine. This is also an expensive part of the system, as the conveyor systems are either powered rollers, me-

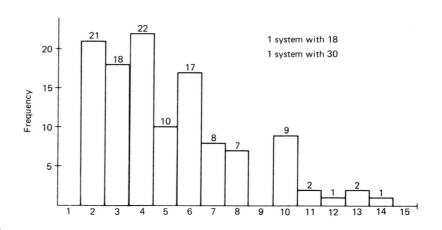

FIGURE 42-18 Histogram of the number of machine tools per FMS, reflecting systems installed worldwide. (Data from U.S. Department of Commerce.)

No. of machine tools in the FMS

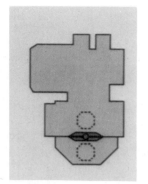

Machining center with
pallet changer and tool
changer

Machining
center
with automatic
material
handling
system

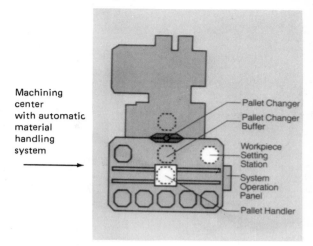

Pallet Changer

Pallet Changer
Buffer

Workpiece
Setting
Station

System
Operation
Panel

Pallet Handler

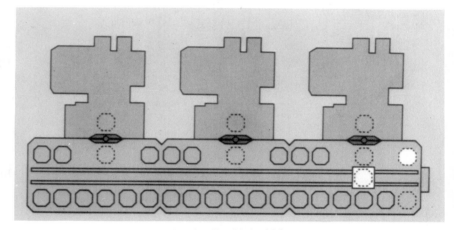

FMS with rail guided vehicles

FIGURE 42-19 The FMS often
uses a pallet changer with the conveyor system to deliver parts to the
machine tool.

chanical pallet transfer conveyors, or, more recently, wire-guided carts (also
called AGVs—automated guided vehicles) operating on underground towlines
or buried guidance cables. See Figure 42-20 and Figure 42-21. The carts are
more flexible than the conveyors. The AGVs also serve to connect the islands
of automation, operating between FMSs, replacing human guided vehicles (forklift trucks).

Pallets are a significant cost item for the FMS because the part must be
accurately located in the pallet and the pallet accurately located in the machine.
Since many pallets are required for each different component, pallets typically
represent anywhere from 15% to 20% of the total system cost. FMSs cost about
$1 million per machine tool. Thus the seven-machine FMS shown in Figure
40-20 cost $6 million for hardware and software with the transporter costing
over $1 million.

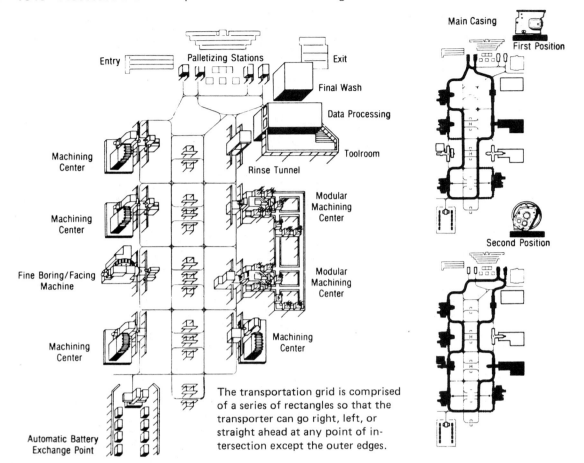

FIGURE 42-20   Renault's seven-machine FMS, with AGV transporter, routes different parts through different machines. (*Modern Machine Shop,* April 1984.)

The CNC machines receive programs as needed from a host minicomputer which acts as a supervisory computer for the system, tracking the status of any particular machine in the system. In recent years, in-process inspection and automatic tool position correction for tool wear have been added features along with diagnostic routines to computer-monitor the condition of the machines. However, a common problem with these systems is the monitoring of the tool condition and performance. Most installations also incur problems in the performance and reliability of the software and the control systems. It takes just as long to debug software as it does to debut hardware and delays of two to six months in startup are not uncommon.

Figure 42-22 shows the levels of computer control in the FMS. While human labor is usually incorporated to load workpieces, unload finished parts, change worn tools, and perform equipment maintenance and repair, CNC and DNC functions can be incorporated into a single FMS. The system can usually monitor piece part counts, tool changes, and machine utilization, with the computer also

Types of AGVs

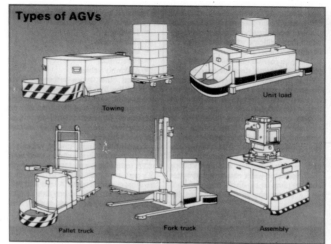

Towing

Unit load

Pallet truck

Fork truck

Assembly

AGV components

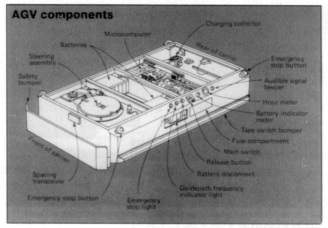

Steering principle

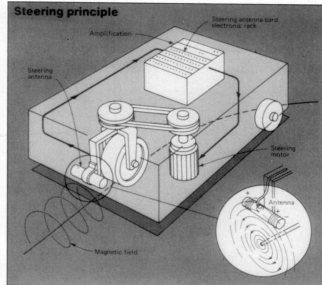

FIGURE 42-21 Automated guided vehicles can be used as a materials handling system for automated assembly as well as in FMSs, delivering workpieces and tooling to the machines. The means of guidance is a buried-wire guidepath tracked by an antenna on the vehicle. The need for guidepath flexibility has inspired new guidance systems (infrared and inertial), further adding to the cost of the FMS. (American Machinist, March 1986.)

providing supervisory control of the production. The workpieces are launched randomly into the system, which identifies each part in the family and routes it to the proper machines. The systems generally display reduced manufacturing lead time, low in-process inventory, and high machine tool utilization, with reduced indirect and direct labor. The materials-handling system must be able to route any part to any machine in any order and provide each machine with a small queue of ''banked parts'' waiting to be processed so as to maximize machine utilization. Convenient access for loading and unloading parts, compatibility with the control system, and accessibility to the machine tools are other necessary design features for the materials-handling system. The computer control for an FMS system has three levels. The master control monitors the entire system for tool failures or machine breakdowns, schedules the work, and routes the parts to the appropriate machine. The DNC computer distributes

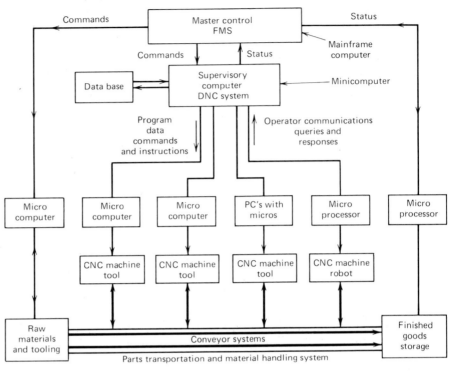

FIGURE 42-22 Three levels of computer control in a DNC system.

programs to the CNC machines and supervises their operations, selecting the required programs and transmitting them at the appropriate time. It also keeps track of the completion of the cutting programs and sends this information on to the master computer. The bottom level of computer control is at the machines themselves.

It is difficult to design an FMS because it is in fact a very complex assembly of elements that must work together. Designing one to be flexible is difficult. Many companies have found that between the time they ordered their system and had it installed and operational, design changes had eliminated a number of parts from the FMS. That is, the system was not as flexible as they thought. Figure 42-23 shows some typical FMS designs.

The use of GT (Group Technology) to identify the initial family of parts around which the FMS is designed greatly improves the FMS design. One might say that FMSs were developed before their time, because they are being much more readily accepted since GT has been used (at least conceptually) to identify families of parts.

An FMS generally needs about three or four workers per shift to load and unload parts, change tools, and perform general maintenance; they are not, therefore, really unmanned. They are rarely left untended, as in third-shift operations. Other than the personnel doing the loading and unloading, the workers in the FMS are usually highly skilled and trained in NC and CNC. Most in-

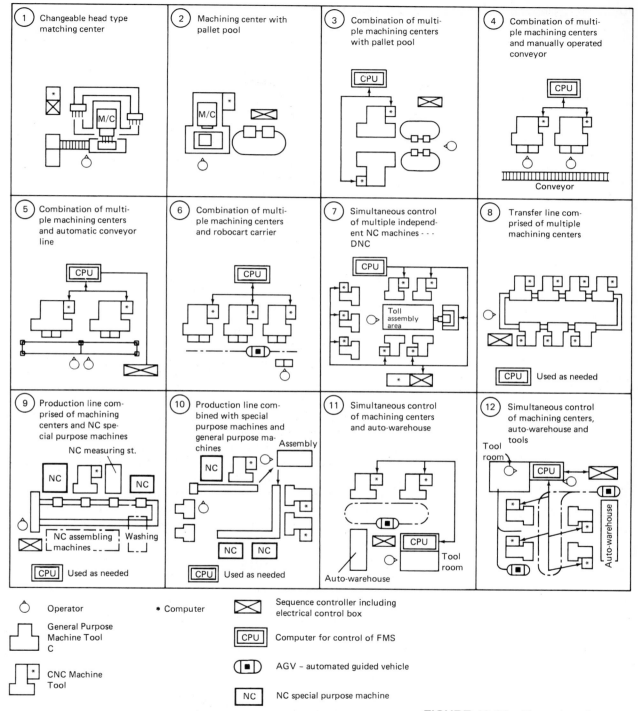

FIGURE 42-23 Examples of machining centers and FMS designs.

1049

TABLE 42.7   What Users Expect from a Flexible Manufacturing System

Lower Unit Cost Due to:
  Labor saving
  Reduced setup time
  Ability to accommodate model changes and design modification of parts
  Reduction of floor space
  Improvement in machine utilization (30% typical)
  Faster throughput times (increased output)

Product delivery
  Continuous check of production conditions
  Reduction of work-in-process (40% typical)
  Reduction of time to transfer products between processes
  Ability to diversify production parts
  Faster supply of spare or replacement parts
  Early detection of defects and system malfunctions

Improvement in quality
  Consistent process capability (machine accuracy and precision)
  Standardization
  Inprocess inspection of products
  First part right

stallations run fairly reliably (once they are debugged) over three shifts, with up-time ranging from 70% to 80%. While the typical FMS installation may not be as flexible (in terms of the different parts it can make) as was thought to be when designed, it probably will yield the kinds of results shown in Table 42.7.

Table 42.8 gives the primary characteristics of FMSs. FMSs are, in fact, classic examples of supermachines. Such large expensive systems must be examined with careful and complete planning. It is important to remember that even though they are often marketed and sold as a ''turnkey'' installation (the buyer pays a lump sum and receives a system that he can turn on and run), this is only rarely possible with a system that has so many elements that must work together reliably. Taken in the context of integrated manufacturing systems,

TABLE 42.8   Characteristics of Flexible Manufacturing Systems

Multiple machine tools—NC or CNC
Automated material handling system
Computer control for system—DNC
Multiple parts—Medium-sized lots (200–10,000) with family of parts
Random sequencing of parts to machines (optional)
Automatic tool changing
In-process inspection
Parts washing (optional)
Automated storage and retrieval (optional)

large FMSs restrict the flow of parts. In the long run, smaller manned or unmanned cells may well be the better solution in terms of system flexibility.

A better name for these systems is *variable mission* or *random path manufacturing systems*.

Robotics

Robots are steel-collared workers. As defined by the Robot Institute of America, "A *robot* is a reprogrammable, multifunctional manipulator designed to handle material, parts, tools or specialized devices through variable programmed motions for the performance of a variety of tasks." The word *robot* was coined in 1921 by Karel Capek in his play *R. U. R.* ("Rossum's Universal Robots"). The term is derived from the Czech word for "worker." Another famous author, Isaac Asimov, depicted robots in many of his stories and gave three laws which hold quite well for industrial robotic applications. Asimov's Three Laws of Robotics were

1. A robot may not injure a human being or, through inaction, allow a human being to be harmed. (Safety first.)
2. A robot must obey orders given by human beings except when that conflicts with the First Law. (A robot must be programmable.)
3. A robot must protect its own existence unless that conflicts with the First or Second Law. (Reliability.)

In considering the use of a robot, the following points should be considered. Anything that makes a job or task easy for the robot to do makes the job easy for a human to do as well. The robot is a severely handicapped worker and is several orders of magnitude less flexible than a human. Robots cannot think or solve problems on the plant floor. The big advantage of the robot is that it will do a job in an exact cycle time whereas a human often cannot. This is an important feature in cells.

For our purposes, if a machine is programmable, capable of automatic repeat cycles, and can perform manipulations in an industrial environment, it is an industrial robot.

All robots have the following basic components:

1. *Manipulators:* the mechanical unit, often called the "arm," that does the actual work of the robot. It is composed of mechanical linkages and joints with actuators to drive the mechanism directly or indirectly through gears, chains, or ball screws.
2. *Feedback devices:* transducers that sense the positions of various linkages and joints and transmit this information to the controllers in either digital or analogue form [A(4) level robots].
3. *End effector:* The "hand" or "gripper" portion of the robot which attaches the end of the arm and performs the operations of the robot.
4. *Controller:* the brains of the system that direct the movements of the manipulator. In higher-level robots, computers are used for controllers. The functions of the controller are to initiate and terminate motion, store

data for position and motion sequence, and interface with the "outside world," meaning other machines and human beings.

5. *Power supply:* electric, pneumatic, and hydraulic power supplies used to provide and regulate the energy needed for their manipulator's actuators.

Manufacturers of industrial robots have proliferated in recent years. Most commercially available robots have one of four mechanical configurations, three of which are shown in Figure 42-24. Cylindrical coordinate robots have a work envelope (shaded region) that is a portion of a sphere. Jointed-spherical coordinate robots have a jointed arm and a work envelope that approximates a portion of a sphere. Rectangular coordinate robots (not shown) with a rectangular work envelope have been developed for high precision in assembly applications. Figure 42-25 shows the six axes of motion typical for a jointed arm robot. In addition, robots may have two or three additional minor axes of motion at the end of the arm (commonly called the wrist). These three movements are "pitch" (vertical movement), "yaw" (horizontal motion), and "roll" (wrist rotation).

The hand (or gripper or end effector), which is usually custom-made by the user, attaches to the wrist.

Industrial robots used in industry today fill three main functions: material handling, assembly, and materials processing. They have, for the most part, very primitive motor and intelligence capabilities, with most robots being A(3) level machines. The sensory-interactive control, decision-making, strength-to-size, and artificial intelligence capabilities of robots are far inferior to those of human beings at this time. With regard to shortcomings in performance, the robot's major stumbling blocks are its accuracy and repeatability (that is, process capability or dexterity).

Nevertheless, robots are having a strong impact in the industrial environment, often doing those jobs which are hazardous, extremely tedious, and unpleasant. Robots perform well doing paint spraying, loading and unloading small forgings or die-casting machines, spot welding, and so forth.

The A(3) level robot is usually called a "pick-and-place" machine and is capable of performing only the simplest repeat-cycle movements, on a point-to-point basis, being controlled by an electronic or pneumatic control with manipulatory movement controlled by end stops.

All A(3) robots are usually small robots with relatively high speed movements, good repeatability (0.010 in.), and low cost. They are simple to program, operate, and maintain but have limited flexibility in terms of program capacity and positioning capability.

In order to raise the robot to an A(4) machine, sensory devices must be installed in the joints of the arm(s) to provide positional feedback and error signals to the servomechanisms, just as was the case for the NC machines. The addition of an electronic memory and digital control circuitry allows this level of robot to be programmed by a human being guiding the robot through the desired operations and movements using a "hand controller." See Figure 42-26. The hand-held control box has rate-control buttons for each axis of motion of

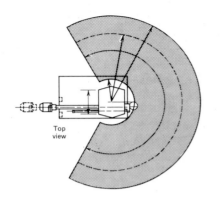

Top view

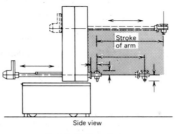

Stroke of arm

Side view

A. Cylindrical coordinate

A

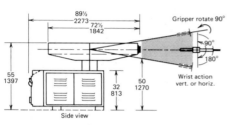

89½
2273
72½
1842

Gripper rotate 90°

90°

180°

Wrist action vert. or horiz.

55
1397

32
813

50
1270

Side view

B. Spherical coordinate

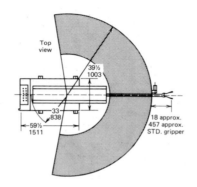

Top view

39½
1003

33
838

59½
1511

18 approx.
457 approx.
STD. gripper

B

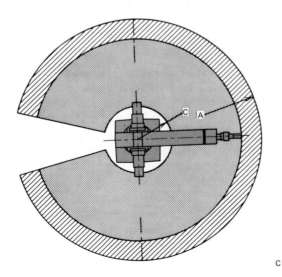

C A

C

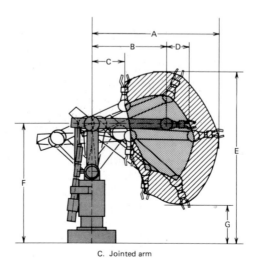

A

B

D

C

E

F

G

C. Jointed arm

FIGURE 42-24 Work envelopes (shaded regions) for three typical industrial-robot designs.

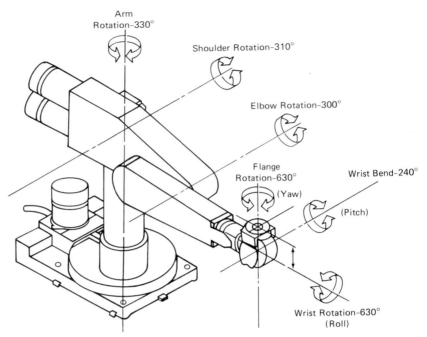

Arm
Rotation-330°

Shoulder Rotation-310°

Elbow Rotation-300°

Flange
Rotation-630°

(Yaw)

Wrist Bend-240°

(Pitch)

Wrist Rotation-630°
(Roll)

FIGURE 42-25 The six axes of motion of a robot include roll, pitch, and yaw movement of the wrist.

the robot arm. When the arm is in the desired position, the "record" or "program" button is pushed to enter that position or operation into the memory. This is similar to point-to-point NC machines as the path of the robot arm movement is defined by selected end points when the program is played back. The electronic memory can usually store multiple programs and randomly access the required one, depending on the job to be done. This allows for a product mix to be handled without stopping to reprogram the machine. The addition of a computer, usually a minicomputer, makes it possible to program the robot to move its "hand" or "gripper" in straight lines or other geometric paths between given points, but the robot is still essentially a point-to-point machine.

There are three ways of controlling point-to-point motion independently: (1) sequential joint control, (2) uncoordinate joint control, and (3) terminally coordinate joint control.

Point-to-point servocontrolled robots have the following common characteristics: high load capacity, large working range, and relatively easy programming; but the path followed by the manipulator during operation may not be the path followed during teaching.

In order to make the A(4) robot continuous-path, position and velocity data must be sampled on a time base rather than as discretely determined points in space. Due to the high rate of sampling, many spatial positions must be stored in the computer memory, thus requiring a mass storage system.

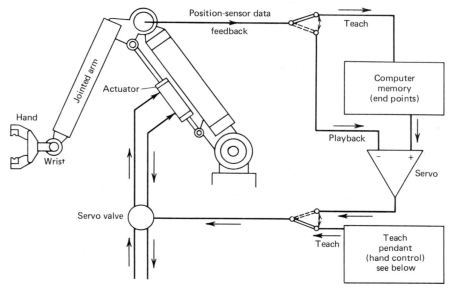

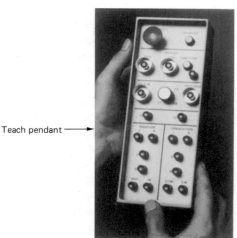

Teach pendant ──▶

FIGURE 42-26   The manual programming of an industrial robot is accomplished by using a hand-held "teach pendant." The robot is guided through a sequence of operations. The successive positions of all the robotic joints are stored in an electronic memory. This establishes the end points of the motions of the arm. By switching from "teach" to "playback," the stored end points are replayed. (Courtesy Cincinnati Milacron, Inc.)

Continuous path control techniques can be divided into three basic categories based on how much information about the path is used in the motor control calculations. These are illustrated in Figure 42-27.

The conventional or servocontrol approach uses no information about where the path goes in the future. The drive signals to the robot's motors are based on the past and present path tracking error. This is the control design used in most of today's industrial robots and process control systems.

The second approach is called preview control, also known as ''feed-forward''

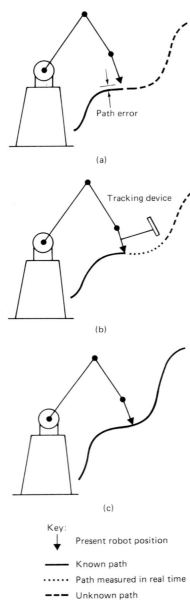

Key:

↓ Present robot position

—— Known path

······ Path measured in real time

--- Unknown path

FIGURE 42-27 Continuous-path robots can use (a) basic servocontrol; (b) tracing or preview control; or (c) trajectory control to position the end effector.

control, since it uses a tracking device to determine how the path is changing in addition to the past and present tracking error used by the servocontroller.

The third type of path control is the "path planning" or "trajectory calculation" approach. Here the controller has available a complete description of the path the manipulator should follow from one point to another. Using a mathematical-physical model of the arm and its load, it precomputes an acceleration profile for every joint, predicting the nominal motor signals that should cause the arm to follow the desired path. This approach has been used in some advanced research robots to achieve highly accurate coordinate movements at high speed.

Continuous-path machines tend to be smaller than point-to-point machines, with ligher load capacity, greater precision (± 0.0020 to ± 0.020 in.), and somewhat higher end-of-arm speed.

Many of the industrial robots in use at this time are A(4) point-to-point machines. Most robots operate in systems wherein the items to be handled or processed are placed in precise locations with respect to the robots. Even robots with computer control which can follow a moving auto conveyor line while performing spot-welding operations have point-to-point feedback information, but this is satisfactory for most industrial applications.

To expand the capability of this handicapped worker made of steel, sensors are used to obtain information regarding position and component status. Tactile sensors provide information about force distributions in the joints and in the hand of the robot during manipulations. This information is then used to control movement rates. Visual sensors collect data on spatial dimensions by means of image recording and analysis. Visual sensors are used to identify workpieces; determine their position and orientation; check position, orientation, geometry, or speed of parts; determine the correct welding path or point; and so forth.

In order to provide a robot with tactile or visual capability, powerful computers and sophisticated software are required, but this appears to be the most logical manner to raise the robot to the A(5) level of automation at which it can adapt to variations in its environment. Vision systems can locate parts moving past a robot on a conveyor, identify those parts that should be removed from the conveyor, and communicate this information to the robot. The robot tracks the moving part, orients its gripper, picks up the part, and moves it to the desired work station.

Table 42.9 represents a chronological listing of desirable attributes for robots. Table 42.10 provides some additional information on the current state of robotic vision and tactile sensing.

The current generation of robots is finding applications in the areas described below.

Die casting. In single- or multishift operations, custom or captive shops, robots unload machines, quench parts, operate trim presses, load inserts, ladle metal, and perform die lubrication. Die life is increased because die-casting machines can be operated without breaks or shutdowns. Die temperature remains stable and better controlled with uniform cycle times.

TABLE 42.9 Robot Attributes[a]

Work space command with six infinitely controllable articulations between the robot base and the hand
Fast, ''hands-on'' programming
Local and remote program memory
Random program selection by external stimuli
Process capability (accuracy and repeatability) to meet application needs
Weight handling capability (50 to 200 lbs.)
Intermixed point-to-point and path-following control
Synchronization with moving targets
Compatible computer interface for off-line programming
High reliability (at least 400 mean time hours between failures)
Vision capability
Tactile sensing
Multiple appendage hand-to-hand coordination
Computer-directed appendage trajectories
Mobility
Minimized spatial intrusion for multiple robots
Energy-conserving musculature
General purpose hands
Man–robot voice communication
Inherent safety (Asimov's laws of robotics)

[a]J. F. Engelberger, *The Industrial Robot,* September, 1979, with authors modifications.

TABLE 42.10 Sensors Used on Robots with Some Typical Applications

Sensor Type	Design	Application
Visual	Video pickup tubes (TV camera)	Position detection, part inspection
	Semiconductor sensors (lasers)	Parts detection, identification, sorting
	Fiber optics	Consistency testing (e.g., in manipulation, welding, and assembly)
		Guidance and control
Tactile	Feelers	Position detection
	Pin matrix	Tool monitoring, (e.g., in casting, cleaning, grinding, manipulation, and assembly)
	Load cells (piezo, capacitance)	
	Conductive elastomers	
	Silicon	Force sensing, part identification
		Object recognition, pressure
Electrical (inductive, capacitative)	Shunt (current determination)	Position detection
	Capacitor	Status determination (e.g., in manipulation and welding)
	Coil	

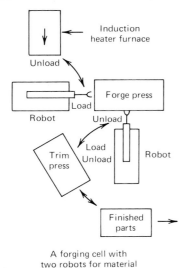

A forging cell with two robots for material handling.

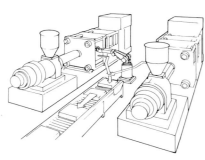

Unloading injection molders

FIGURE 42-28 Application of robots to materials handling, loading, and unloading hot parts from forge and unloading two plastic-injection molding machines. (Mert Corwin, "A Computer Controlled Robot for Automotive Manufacturing," Wolfsburg, West Germany, September 12–15, 1977; courtesy Cincinnati Milacron, Inc.)

Press transfer. Robots in sheet metal press transfer lines guarantee consistent throughput shift after shift. Large and unwieldy parts can be handled at piece rates as high as 400 per hour with no change in cycle time due to fatigue. Robots are adaptable for long-run or short-run operations. Programming for new part sizes can be accomplished in minutes.

Materials handling. Strength, dexterity, and a versatile memory allow robots to pack goods in complex palletized arrays or to transfer workpieces to (and from) moving or indexing conveyors from machines. Savings are dramatic in these labor-intensive operations. See Figure 42-28. Operating costs are reduced when robots feed forge presses and upsetters. They work continuously without fatigue or the need for relief in the hot, hostile environments commonly found in forging. Robots can easily manipulate the hot parts in the presses. See Figure 42-28.

Investment casting. Scrap rates as high as 85% have been reduced to less than 5% when molds are produced by robots. The smooth, controlled motions of the robot provide consistent mold quality impossible to achieve manually.

Material processing. Product quality is improved and sustained with point-to-point, continuous-path robots in jobs such as routing, flame cutting, mold drying, polishing, and grinding. Once programmed, the robot will process each part with the same high quality. Figure 42-29 shows a robot cell with vision system, for drilling, tapping, buffing, filing, and deburring a family of plastic workpieces. The deburring program and the robot's hand are changed automatically when the robot retrieves a different workpiece. The input conveyor holds about a three-hour supply of parts for the cell. The camera equipment is placed about eight feet over the belt. When the controller receives a request for certain parts from assembly, the vision system identifies the raw material on the input side and the robot retrieves it and starts it through the cell.

Welding. Robots spot-weld cars and trucks for almost every major manufacturer in the world, with uniformity of spot location and weld integrity. In arc

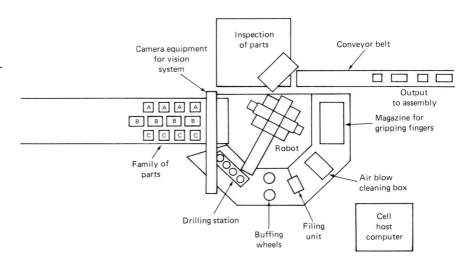

FIGURE 42-29 Robotic CMS for the drilling, tapping, and deburring family of parts.

welding, robots increase arc time, free operators from hazardous environments, reduce the cost of worker protection, and improve consistency of weld quality.

Machine tool loading. The most efficient use of robots may well be achieved by integrating them into cellular arrangements of machine tools being used to process families of component parts. The robot can provide part loading and unloading of two to five CNC machines grouped properly in a machining cell. These unmanned cells facilitate maximum automation and productivity while maintaining programmable flexibility in producing small to medium-size production lots of parts from compatible parts families. The robots can also change tools in the machines and, in the future, even the workholding devices, thereby adding more flexibility to the cell. These cellular arrangements (or layouts) help to achieve maximum machine tool utilization by greatly increasing the percentage of time the machines spend cutting, which in turn increases the output for the same investment. This is the name of the game in productivity. Examples of this concept are shown in Chapter 29 and Chapter 44.

Economics of Programmable Robots.
Debugging hard automation systems can be costly and time-consuming, taking an average of 12 months to bring a hard automation system on stream. If products change regularly, it is quite possible that the system will be obsolete before it becomes operational. Programmable automation generally takes less debugging and is less subject to obsolescence. As the robots become smarter, they will be able to replace more manual activities. As they become cheaper, it becomes more economical for them to replace hard automation types of functions as well as human labor.

Figure 42-30 shows a comparison of the cost of the steel-collar worker and that of the blue-collar worker in the U.S. automobile industry, beginning in 1961, when the first Unimate robot was used in an automobile plant, up to 1984. In 1961, labor cost about $3.80 per hour, including all fringe benefits. This figure rose to about $14 per hour, including fringes. Throughout the same period, the cost of robot labor runs about $4.80 per hour. This figure includes capitalized cost for an eight-year life, maintenance, repair, installation, cost of power, and so forth. This figure is conservative, since it is based on 32,000 hours of operating life, and many robots have exceeded 80,000 hours of in-plant service with

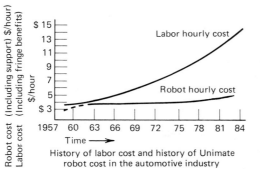

History of labor cost and history of Unimate robot cost in the automotive industry

Unimate price	$40,000
Useful life — 8 yrs @ 2 shifts	
Cost of money — 11%	
Installation cost — two @	$12,000
Maintenance cost	1.05/hr.
Power cost	0.35/hr.
Overhaul — two	0.40/hr.
Depreciation	1.25/hr.
Installation	0.80/hr.
Money cost	0.65/hr.

FIGURE 42-30 Cost per hour for a robot versus human labor in the automotive industry. (Data and plot from J. F. Engelberger, *The Industrial Robot,* September 1979, p. 115).

up-times exceeding 95%. Of course, economics is not the only reason to implement a robot. The loss of human capability and manufacturing flexibility in the system when a robot replaces a human must be carefully evaluated.

The robot and computers are seen as critical elements of advanced manufacturing technologies available for the next decade. Technology abounds: computer-aided design, computer-aided manufacturing (NC, CNC, AC, DNC), computer-aided testing and inspection (CATI), automatic assembly and warehousing, robots, and much more. But anyone can buy computers, robots, and other pieces of automation hardware and software. The secret to manufacturing success lies in the redesign of the manufacturing system so that it can achieve superior quality at low cost, with on-time delivery, and still be flexible. This requires a visionary management team. No better example of this can be found than at the Toyota Motor Company. Led by their vice-president for manufacturing, Mr. Taiichi Ohno, who conceived, developed, and promoted Toyota's unique manufacturing production system, this company has emerged as the world leader in car production. The Toyota system is unique and as revolutionary today as were the Taylor system of scientific management or the Ford system for mass assembly in their day. It is significant that virtually every manufacturing system or technology cited in this chapter is practiced at Toyota. This new system will be discussed in Chapter 44. Students of manufacturing engineering are well advised to be knowledgeable of this unique system.

Review Questions

1. Why does the elimination or reduction of setup time greatly improve the productivity of short-run or small-lot operations?
2. In Table 42.1, the transfer line gave 0% of time to loading/unloading and setup. Why?
3. What are four classical forms for manufacturing systems?
4. Can you give a service example of each of the four classical forms for manufacturing systems?
5. What is the new form of manufacturing system that has emerged in the last decade, and how does this new form differ from the previous classical forms?
6. What is meant by the statement, "The manufacturing system is usually a mixed system"?
7. What are the trends that are driving companies toward smaller lot production?
8. What is meant by the statement, "Processes proliferate"?
9. What is group technology, and how does it convert the functional job shop into a cellular flexible shop? Do you think that 100% of the job shop processes will be converted?
10. What is meant by the statement, "Optimizing subsystems does not optimize the whole"?
11. What have been the benefits generally experienced by companies that have undergone conversions of their systems through GT?
12. How does the "do-nothing" alternative act as a constraint on the implementation of a GT program?
13. Name examples of A(3) and A(4) systems that exist in the home. What about an A(5) or A(6) system?
14. Why is it so difficult to model manufacturing systems?
15. What is a route sheet, and in which manufacturing system is it used and for what purpose?
16. What is meant by the term *line balancing*?
17. Flexible manufacturing systems use NC machines with either robots or conveyors. What differentiates this system from a transfer line?
18. Why do you think that transfer lines are using more programmable machines and PLCs?
19. In manned machining cells, why are the machines at least single-cycle automatics? What must be done if they are not?
20. How is it possible that the total machining time to make a part exceeds the cycle time for the part in the manufacturing

cell? Recall from Table 42.1 that metal cutting represented about 23% of the cycle time in a machining center. Thus, a part with a four-minute cycle time would typically have about one minute of machining time. In the cell, a part with four minutes of machining time might have a one-minute cycle time!

21. Define *flexibility* as a design criterion for the CMS.

22. Cells permit the product design function to be readily integrated—i.e., the designer knows beforehand the process capability of the set of processes which will make the part. How/why does he know this for cells but not for functional job shops?

23. The A(6) level tries to relate causes to effect using AI. What is AI?

24. What is a feed-forward device?

25. What are the main areas in which robots are utilized?

26. What are the basic components of all robots?

27. What are the four common work envelopes for industrial robots?

28. How is positional feedback obtained in robots?

29. What are the primary differences between an instructional robot and an industrial robot? See Chapter 44.

30. What is simulation, and how is it used?

31. Compare a rotary transfer machine with an unmanned robotic cell.

32. Compare a human worker in a manned cell to the robotic worker. What are the advantages of one over the other?

33. In the analogy of manufacturing to football given in Chapter 1, what is the MPS equivalent to the kicking tee used by the placekicker on the football team?

34. What is tactile sensing? Name some examples of tactile sensing in the home.

35. The vast majority of robots are used in automobile fabrication for welding. What kind of welding is routinely done robotically?

Production Systems

Introduction

In the previous chapter, the basics of production systems and manufacturing systems were reviewed. The manufacturing processes make or assemble products while the systems integrate the processes (with their operations) into selected sequences to produce entire products. We can detail and name the basic processes (casting, forming, machining, and so on) and the operations (transportation, inspection, storage). We can categorize the manufacturing systems (e.g., job shop, flow shop, linked cells), but we cannot yet clearly define the classical production systems, even though many such systems exist. Classical production systems are mostly collections of subsystems devised to service the manufacturing systems. Their objectives are to aid the manufacturing systems in producing products to meet delivery schedules at costs that maximize profits.

In Chapter 1, the analogy of the university football program was presented. The athletic department was equated to the production system. Recall that the athletic department does many things for the football team (and other teams as well) to help it get ready to play (or produce). Thus, the production system (PS) serves the manufacturing system(s) (MS) and the individual processes but does not actually make products. The PS may design the product (even choose the color scheme), purchase materials and supplies, plan its implementation, sell it, forecast its demand, maintain inventory, hire and fire workers, pay employees, and so on. This chapter briefly examines the functions typically performed by the production system.

Typical Functional Elements

Production systems or production operation systems serve and support the manufacturing processes and manufacturing systems by providing and transmitting information, energy, knowledge, skills, and services to the plant areas, to the company's vendors, and to its customers. Traditionally, the PS included the following areas (see Figure 43-1).

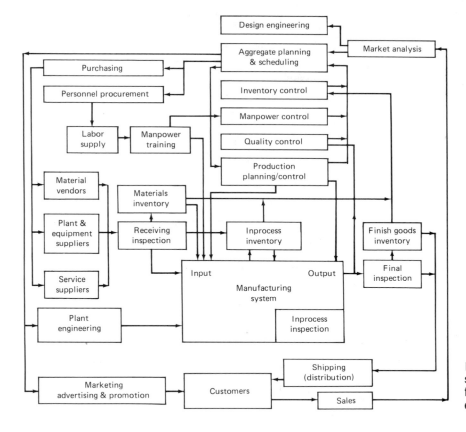

FIGURE 43-1 Classical production system showing inclusion of manufacturing system and major functional elements.

- Marketing and sales.
- Finance (not shown).
- Accounting.
- Personnel procurement.
- Research and development (not shown).
- Design engineering (product design).
- Purchasing.
- Aggregate planning and scheduling.
- Production scheduling.
- Quality control.
- Production planning and control.
- Inventory control.
- Plant engineering (maintenance).

In most companies, the workers in these areas are called *staff* or indirect labor to distinguish them from the *line personnel* of the manufacturing system. Although production systems have no standard design, they are usually arranged functionally. To connect the functional areas, informal lines of communication

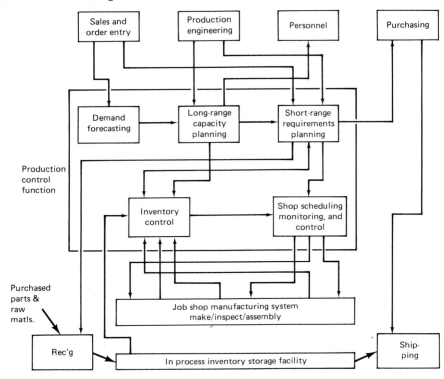

FIGURE 43-2 Flow of information in a PJS MP system, with emphasis on the production control.

(information flows) are developed. Figure 43-2 shows the necessary communication links for a production job shop just in the areas of production control. This network for communication can become complex. Most managerial workers are in the PS, except for foremen, line supervisors, and manufacturing managers. Companies with a poorly defined collection of subsystems and functions may not serve the manufacturing system well. Often the production worker views the services of quality and inventory control with distrust. The production worker may not understand how control charts monitor his process or how the MRP system controls the work-in-process. An adversarial relationship develops between the people in MS and people in the PS. Computerizing this function merely complicates the problem. This is perhaps the most important difference between their integrated manufacturing production system (IMPS) discussed in Chapter 44 and the classical production system. That is, in IMPS, the key functions of the production system are infused into the manufacturing system. They not only serve the manufacturing system but become an integral operational part of the making of the product.

Marketing.   The chief activities of *marketing* are forecasting sales, advertising, and estimating future demand for existing products. Selling the product is the primary interest of marketing. Promotional work, a highly specialized activity, involves advertising and customer relations. Customer service is a critical function for any manufacturing company.

Thus, marketing provides information and services concerning

1. Sales forecast of future demand.
2. Sales order data.
3. Customer quality requirements.
4. Customer reliability requirements.
5. New products or modifications for existing products.
6. Customer feedback on products.
7. Customer service (repair or replace defective products).

There is no piece of information that is more vital to a company, nor harder to come by accurately, than future demand. This information is required in order to plan effectively how much should be produced and to schedule that production when changes in demand are predicted. The faster the MS can respond to changes in product demand the better because the quick response reduces the need to develop accurate long-range forecasts. Short-range forecasts are more accurate than long-range forecasts.

Sales order information is central to *production planning and control*. Products are either made to stock (finished goods inventory) or made to fill customer orders. Therefore the orders determine how much and what kinds of products or services must be produced.

Marketing develops information on new products or new uses for old products. This information usually goes to *research and development* or to product design engineers.

Marketing also gathers customer feedback on existing products. The marketing department, which is in direct customer contact, gathers complaints about product performance and communicates them to design and/or manufacturing. Often the long-time users are the ones who identify product characteristics that create problems in its use. Clearly, customers want superior-quality products which give them reliable service. In this sense, quality and reliability are related, but they are functionally different. Quality is concerned with the prevention of defects and the conformance to specifications at the time the product is made or sold. Reliability is concerned with the performance of the product over time, while "in service" with the customer. In general, a superior-quality product is more reliable if the design is good. Failures of products that are well made but perform badly because of faulty design occur infrequently but are usually spectacular and newsworthy.

Finance.   *Financial* functions involve the management of the company's assets. For the production system, finance provides information and services concerning the following:

1. Internal capital financing.
2. Budgeting.
3. Investment analysis.

Internal financing includes the review of budgets for operating sections, evaluation of proposed capital investments for production facilities, and preparation of financial statements such as balance sheets or profit-and-loss (or income-and-expense) statements.

Periodically, the manufacturing manager, as well as other managers, must submit budgets of expected financial requirements and expenditures to the finance department. The decisions made during budget preparation and the discussions on budget adjustments have a significant impact on the manufacturing system's operation. One of the strongest criticisms of the American system is that *the decision makers know little about manufacturing processes or systems* and therefore make poor investment decisions. Few MBA programs have a course in manufacturing processes. Very few undergraduate business students take courses in manufacturing processes. Managers do not really know what business they are in when they do not understand what they do. However, American managers usually do have problem-solving and decision-making skills for handling investment alternatives which require knowledge of such concepts as rate of return, depreciation, sinking funds, payback periods, and compound interest. Managers must have the financial expertise to understand the very complex and constantly changing tax structure, tax regulations, and tax court decisions that affect the company's capital investment decisions.

Accounting. The accounting department maintains the company's financial records. Money is used to keep score, so to speak. Accounting also provides data needed for decision making. For the PS, accounting provides information and services on the following:

1. Cost accounting.
2. Special reports.
3. Data processing.

The cost-accounting information indicates the level of efficiency of various departments and also the cost of the products being manufactured. The unit-cost data (cost of materials, direct labor, and overhead) help the company to establish prices. Most American companies view this classical equation as follows:

$$\text{Unit cost} + \text{Profit} = \text{Sales price}$$

If unit cost goes up, then, to maintain profit, the sales price goes up. How many times have you heard this statement, ''You'd better buy it now; next year it will cost you more''? Such a statement is an open invitation to those who view the equation this way:

$$\text{Sales price} - \text{cost} = \text{Profit}$$

The marketplace and the customer dictate the sales price. The only way to maintain or improve profit is to reduce the cost (unit cost)! The purchasing (procurement) department uses manufacturing cost data in analyzing whether a product should be manufactured by the company (in-house) or purchased from a vendor (the classical make-or-buy decision).

Accounting also produces special reports that monitor the status of the scrap and rework levels, raw-materials inventories, work-in-process inventories, finished-goods inventories, direct-labor hours and overtime, and so on. These reports provide quantitative measures of performance which can be compared with the original plans (estimates).

In large companies, the accounting department often controls the data-processing equipment. In companies that use computers for problem solving instead of for record keeping, data processing is a separate function.

Personnel Functions.   The *personnel* department represents manpower, one of the essential inputs to the manufacturing system, and provides information and services concerning

1. Recruitment.
2. Training.
3. Labor relations.
4. Safety.

Although the personnel department may not hire people directly, it assists the company managers by recruiting, screening, and testing potential employees. It also handles the details of terminations and department transfers.

The personnel department can assist in training workers. Industrial accidents are both costly and disruptive to the work force and the production schedules. By working closely with the personnel department, management develops and institutes programs that can minimize safety problems.

If the company has a union, the personnel department will handle labor relations—collective bargaining, and problems with the shop stewards and union officials.

Research and Development (R&D).   *Research and development* involves invention or discovery and innovation, and their development in terms of achievable ends, such as new materials, products, processes, tools, and techniques. Many industries show the impact of R&D on their manufacturing systems. For example, for many years, the wood-products industry's manufacturing system produced only lumber products. In recent years R&D efforts have produced new products and processes for making plywood, particleboard (from wood chips), gardening mulch (from bark), laminated beams and panels, and chemicals (from wood).

The manufacturing manager must rely on R&D for ideas on the manufacture of new products and processes, and on the implementation of new production technology. Often, R&D also provides ideas for product improvement and may answer questions on economical uses of by-products and waste products from manufacturing operations.

Generally, *engineering functions* are staff functions in the production system, providing information and services on the following:

Engineering Functions

1. Product design engineering or design.
2. Manufacturing engineering.
3. Industrial engineering.
4. Quality engineering.
5. Plant engineering.

Product Design Engineering.   In discussing design, we must review design in relation to manufacturing. Through recognition of the systems approach, the design stage can save many dollars and much time later in manufacturing, inspection, assembly, packaging, and even distribution and marketing. Let us examine some of the simpler aspects of designing for manufacture, sometimes termed ''producibility'' in design.

Traditionally, a design will often develop in three phases, as was outlined in Chapter 9. In the *conceptual* or *idea* phase, the designer conceives of an idea for a device that will accomplish some function. This stage establishes the functional requirements that must be met by the device.

In the second, *functional-design* stage, a device is designed that will achieve the functional requirements established in the conceptual stage. Often more than one functional design will be made, suggesting alternative ways in which the functions can be met. At this stage, the designer is usually more concerned with materials than with processes and may ignore the fact that the designed configuration cannot be produced economically utilizing the material being considered.

The third phase of design is called *production design*. Although attention should also be given to the appearance of the product at this stage, particularly if sales appeal is important, the major emphasis is on providing a design that can be manufactured and assembled economically. The *design engineer* must, of course, know that certain manufacturing processes and operations exist that can manufacture the desired product. However, merely knowing that feasible processes exist is not sufficient. He also must know their limitations, relative costs, and process capabilities (accuracy, tolerance requirements, and so forth), in order to design for producibility. If maximum economy is to be achieved, the designer should be aware of the intimate relationship between design details and production operations.

It is extremely important that the relationship between manufacturing, including assembly, and design be given careful consideration throughout the design phase. Changes can be made for pennies in the design room that would cost hundred or thousands of dollars to effect later in the factory. This type of consideration should be an integral and routine part of planning for manufacturing. Having the design engineers make a working prototype of each and every new model before any production drawings are made is one approach. If the model performs in accordance with the conceptual requirements, a second model is made, using, insofar as possible, the same manufacturing and assembly methods that will be used later for production. Any changes that will permit easier and more economical production are incorporated into this second model. If the second model meets the functional requirements of the engineering design group, it is then sent to the drafting room, and production drawings are made from it. This practice eliminates details from products that are costly to produce and the need for a lot of design changes after the part has gone into production.

The designer plays a key role in determining what processes and equipment must be used to manufacture the product, although often indirectly. Clearly, one of the ways by which he can indirectly determine the process is through the selection of material. For example, the chassis for record players (stereos) are usually made from zinc and are die cast. Suppose, however, that the designer specifies a composite chassis to be made from fiber-reinforced plastic; then a form-molding process is needed instead of a die-casting machine. He specifies a particular joining process when he calls for a welded joint. These kinds of direct relationships are pretty obvious. However, other equipment and processes may be specified just as certainly in not so obvious ways.

One of the most common ways in which equipment and processes may be specified indirectly is through dimensional tolerances placed on a drawing. If a tolerance of 0.0002 in. is shown, a grinding operation may be specified just as definitely as if the word *grind* were placed on the drawing. Designers often fail to realize this fact and specify unnecessarily close tolerances; expensive and unnecessary operations result. On the other hand, indefinite dimensions and tolerances can lead to important requirements' being ignored by the factory. Designers should realize that the dimensions and tolerances they place on a drawing may have implications and results far beyond what they anticipated.

Design details are directly related to the processing that will be used, making the processing easy, difficult, or impossible and affecting the cost and/or quality. Some simple examples help to amplify the discussion.

Design Details Related to Casting.   Some design details that relate to casting are shown in Figure 43-3. These emphasize the basic fact that castings shrink

FIGURE 43-3 Good and bad details in casting design.

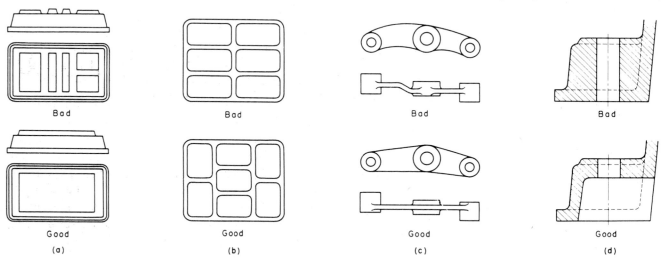

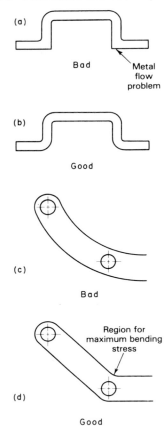

(a)

Bad   Metal
flow
problem

(b)

Good

(c)

Bad

(d)

Region for
maximum bending
stress

Good

FIGURE 43-4 Examples of good and bad design details for forgings.

during solidification and cooling, and that difficulties may be experienced unless uniform sections are employed and adequate provision made to avoid excessive contraction stresses. Also, by remembering this principle, one often may reduce the weight and cost of castings, as illustrated. Figure 43-3c emphasizes that mold parting lines should be in a single plane, if practical. This also applies to forgings.

Design Details Related to Forgings. When designing forgings, two primary factors should be kept in mind: (1) metal flow and (2) minimizing the number of operations and dies. These two factors often are closely related. As illustrated in Figure 43-4a, the shape that would be very satisfactory for casting may be quite poor for forging, requiring more than one operation. This difficulty can be eliminated by the change shown in Figure 43-4b. Figure 43-4c and d illustrate how a shape that is easy to design may be costly to produce because of added difficulties in machining the forging die. Contouring is needed instead of straight-line machining.

Design Details Related to Machining. The ways in which design details can affect machining are almost unlimited. Figure 43-5 illustrates just a few. It is far better for the designer to visualize how the workpiece will be machined and make minor modifications that will permit easy and economical machining than to force the manufacturing department to find some way to machine a needlessly bothersome detail, usually at excessive cost. Too often manufacturing will take for granted that the part has to be made as designed rather than take the trouble to contact the designer to ascertain whether some modification can be made that will facilitate machining.

Grinding may be made difficult, or virtually impossible, through poor design. See Figure 43-6. With no provision for entry of the grinding wheel or its supporting arbor (top views), the required surfaces can not be ground by ordinary procedures. A change in the design corrects the problem but may add another process. Another common deficiency in the design of parts that require the grinding of external cylindrical surfaces is the failure to provide any means to grip or hold the workpiece during grinding. This happens most frequently when only one or a few pieces are involved.

Through this kind of thinking product design engineering prepares the product design for the customer. If the product is proprietary, the manufacturer is responsible for developing and designing the product.

The product design is documented with component drawings, specifications, and bill of materials (see Figure 43-7) that defines how many of each component go into the product. Initial designs may be based on information from R&D. Prototypes are often built for testing and demonstrating. Manufacturing engineering should be consulted on matters of producibility. In a further step sometimes referred to as ''value engineering,'' engineers look for design changes that could reduce production costs and maintain quality or function. Manufacturing cost estimates are prepared at this point to help determine the market situation for the product.

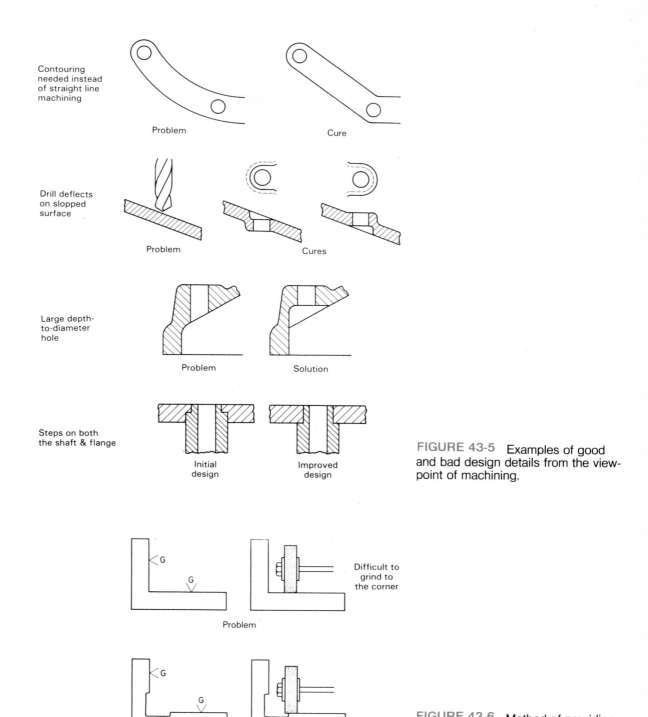

Contouring
needed instead
of straight line
machining

Problem   Cure

Drill deflects
on slopped
surface

Problem   Cures

Large depth-
to-diameter
hole

Problem   Solution

Steps on both
the shaft & flange

Initial
design   Improved
design

FIGURE 43-5 Examples of good
and bad design details from the viewpoint of machining.

G   G

Difficult to
grind to
the corner

Problem

G   G

Solution

FIGURE 43-6 Method of providing
proper clearance for grinding-wheel
mounting and for overtravel.

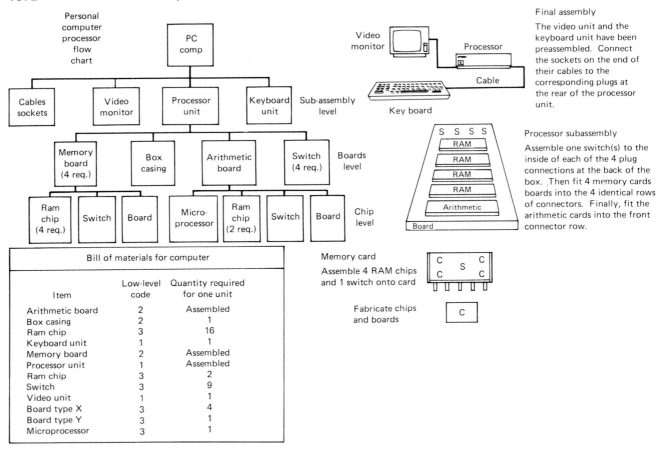

Personal computer processor flow chart

Final assembly

The video unit and the keyboard unit have been preassembled. Connect the sockets on the end of their cables to the corresponding plugs at the rear of the processor unit.

Processor subassembly

Assemble one switch(s) to the inside of each of the 4 plug connections at the back of the box. Then fit 4 memory cards boards into the 4 identical rows of connectors. Finally, fit the arithmetic cards into the front connector row.

Memory card

Assemble 4 RAM chips and 1 switch onto card

Fabricate chips and boards

Bill of materials for computer		
Item	Low-level code	Quantity required for one unit
Arithmetic board	2	Assembled
Box casing	2	1
Ram chip	3	16
Keyboard unit	1	1
Memory board	2	Assembled
Processor unit	1	Assembled
Ram chip	3	2
Switch	3	9
Video unit	1	1
Board type X	3	4
Board type Y	3	1
Microprocessor	3	1

FIGURE 43-7 The process flow chart and the bill of materials (BOM) for a personal computer. See also Figure 41-3.

Upon completion of the design and fabrication of the prototype, company management reviews the design and decides whether to manufacture the item. Engineering management must review and approve the product's design. Many companies call this an "engineering release."

Corporate management must review and approve the product's general suitability. This second decision represents an authorization to produce the item. The design process discussed in Chapter 9 covered material selection factors, manufacturing considerations, materials substitutions, and product liability.

Manufacturing Engineering. Manufacturing engineers address the planning and management of all manufacturing processes and systems. Using the specifications, *manufacturing engineers* plan the manufacture of the product, determining which machines, workers, tools, and many other manufacturing-system components should be used to meet quality and functional requirements. In the classical sense, manufacturing engineering includes responsibility for working with and giving advice to product designers on product producibility. Once production has started, engineering design changes are expensive. Man-

DARVIC INDUSTRIES

ROUTING SHEET

NAME OF PART _____ Punch _____   PART NO. _____ 2 _____

QUANTITY _____ 1,000 _____   MATERIAL _____ SAE 1040 _____

OPERATION NUMBER	DESCRIPTION OF OPERATION	EQUIPMENT OR MACHINES	TOOLING
1	Turn $\frac{5}{32}$, 0.125, and 0.249 diameters	J & L turret lathe	#642 box tool
2	Cut off to $1\frac{3}{32}$ length	"	#6 cutoff in cross turret
3	Mill $\frac{3}{16}$ radius	#1 Milwaukee	Special jaws in vise $\frac{3}{16}$ form cutter x 4" D
4	Heat treat. 1,700° F for 30 minutes, oil quench	Atmosphere furnace	
5	Degrease	Vapor degreaser	
6	Check hardness	Rockwell tester	

FIGURE 43-8 Routing sheet for making the punch shown in Figure 43-12.

ufacturing engineering may also design individual processes for machine tool design (or modification), for tooling design and specifications (workholding devices and cutting tools and dies) and for production process and operation planning (called process planning) and may solve processing problems on the plant floor as well.

Process planning consists of determining the *sequence* of individual manufacturing processes and operations needed to produce the parts in the job shop manufacturing system. The *route sheet* and the *operations sheet* are the documents that specify the process sequence through the job job. The route sheet lists manufacturing operations and associated machine tools for each workpart. See Figure 43-8. It travels with the parts which move in batches between the processes. When a cart of parts has to be moved from one point in the job shop to another, the route sheet provides routing (travel) information, telling the material handler which machine in which department the parts must go to next. The operations sheet describes what machining or assembly operations are done to the parts at particular machines. See Figure 43-9. Over time, many different people plan the same part; therefore there can be many different process sequences and many different routes for the same or similar parts.

Basic Requirements for Process Planning.   The first step in planning is to determine the basic job requirements that must be satisfied. These usually are determined by analysis of the drawings and the job orders. They involve consideration and determination of the following:

1. Size and shape of the geometric components of the workpiece.
2. Tolerances.
3. Material from which the part is to be made.
4. Properties of material being machined.
5. Number of pieces to be produced. See the section below on quantity versus process and case studies on economic analysis.

FIGURE 43-9   Operation sheet for the threaded shaft shown in Figure 43-10.

DARVIC INDUSTRIES

OPERATION SHEET

PART NAME: Threaded Shaft 1340 Cold Rolled Steel   Part No. 7358-267-10

OPER. NO.	NAME OF OPERATION	MACH. TOOL	CUTTING TOOL	CUTTING SPEED		FEED ipr	DEPTH OF CUT Inches	REMARKS
				ft/min	rpm			
10	Face end of bar	Engine Lathe		120	458	Hand		Use 3-jaw Universal chuck
20	Center Drill End	"	Combination center drill		750	Hand		
30	Cut off to $3\frac{9}{16}$ length	"	Parting tool	120	458	Hand		To prevent chattering, keep overhang of work and tool at a minimum and feed steadily. Use lubricant.
40	Face to length	"	RH facing tool (small radius point)	120	458	Hand	(R)$\frac{1}{8}$ max. (F).005	Before replacing part in 3-jaw chuck, scribe a line marking the $3\frac{1}{2}$ inch length.
50	Center Drill End	"	Combination center drill		750	Hand		
60	Place between centers, turn $.501\atop.499$ diameter, and face shoulder	"	RH turning tool (small radius point)	120 160	(R)458 (F)611	(R).0089 (F).0029	(R).081(3) (F).007	
70	Remove and replace end for end and turn $.877\atop.873$ diameter	"	RH tools (R)(small radius point) (F) Round nose tool	120 160	(R)458 (F)611	(R).0089 (F).0029	(R).057 (F).005	
80	Produce 45°-chamfer	"	RH round nose tool	120	458	Hand	(R)$\frac{1}{8}$ max. (F).005	
90	Cut $\frac{7}{8}$-14 NF-2 thread	"	Threading tool	60	208		(R).004 (F).001	(1) Swivel compound rest to 30 degrees. (2) Set tool with thread gage. (3) When tool touches outside diameter of work set cross slide to zero. (4) Depth of cut for roughing = .004. (5) Engage thread dial indicator on any line. (6) Depth of cut for finishing = .001 Use compound rest.
	Remove burrs and sharp edges	"	Hand file					

Date _____   _____

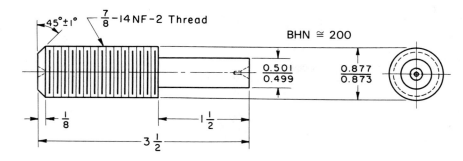

45°±1° $\frac{7}{8}$-14NF-2 Thread

BHN ≅ 200

$\frac{0.501}{0.499}$

$\frac{0.877}{0.873}$

$\frac{1}{8}$

$1\frac{1}{2}$

$3\frac{1}{2}$

Matl. AISI 1340 Medium carbon steel

Tolerance BHN ≅ 200 specified = $\pm\frac{1}{64}$

FIGURE 43-10   Threaded shaft, to be manufactured in quantities of 25 units.

Such an analysis for the threaded shaft shown in Figure 43-10 would be as follows:

1. **a.** Two concentric and adjacent cylinders, having diameters of 0.877/0.873 and 0.501/0.499, respectively, and lengths of 2 in. and 1½ in.
   b. Three parallel, plain surfaces forming the ends of the cylinders.
   c. A 45 degree ⅛-in. bevel on the outer end of the ⅞ in. cylinder.
   d. A ⅞-in. NF-2 thread cut on the entire length of the ⅞-in. cylinder.
2. The tightest tolerance is 0.002 in., and the angular tolerance on the bevel is 1 degree.
3. The material is AISI 1340 cold-rolled steel, Bhn 200.
4. The job order calls for 25 parts.

A number of conclusions regarding the processing can be drawn from this analysis. First, because concentric, external, cylindrical surfaces are involved, turning operations are required and the piece should be made on some type of lathe. Second, because 25 pieces are to be made, the use of a turret lathe would not be justified. As the company's NC lathe is in use, an engine lathe will be used. Third, because the maximum required diameter is approximately ⅞ in., 1-in.-diameter cold-rolled stock will be satisfactory; it will provide about 1/16 in. for rough and finish turning of the large diameter. From this information, the operations sheet(s) for a particular engine lathe is prepared.

Operation Sheets.   The *operation sheet* shown in Figure 43-9 lists, in sequence, the operations required for machining the threaded shaft shown in Figure 43-10. Typically, a single operation sheet lists the operations that are done in sequence on a single machine. However, the sheets may cover all the operations for a given part for a group of machines in a manned cell. See Chapter 44.

Operation sheets vary greatly as to details. The simpler types often list only the required operations and the machines to be used. Speeds and feeds may be left to the discretion of the operator, particularly when skilled workers and small

Turning, Single Point and Box Tools　　　　　　　　　　　　　　　MEDIUM-CARBON ALLOY STEELS, WROUGHT

MATERIAL	HARD-NESS Bhn	CONDITION	DEPTH OF CUT* in / mm	HIGH SPEED STEEL TOOL SPEED fpm/m/min	FEED ipr/mm/r	TOOL MATERIAL AISI/ISO	CARBIDE TOOL — UNCOATED SPEED BRAZED fpm/m/min	INDEX-ABLE fpm/m/min	FEED ipr/mm/r	TOOL MATERIAL GRADE C/ISO	CARBIDE TOOL — COATED SPEED fpm/m/min	FEED ipr/mm/r	TOOL MATERIAL GRADE C/ISO
5. ALLOY STEELS, WROUGHT (cont.)	175 to 225	Hot Rolled, Annealed or Cold Drawn	.040	135	.007	M2, M3	375	500	.007	C-7	650	.007	CC-7
			.150	105	.015	M2, M3	300	400	.020	C-6	525	.015	CC-6
Medium Carbon			.300	80	.020	M2, M3	240	315	.030	C-6	400	.020	CC-6
1340 4340 81B45			.625	65	.030	M2, M3	190	250	.040	—	—		
1345 50B40 8640			1	41	.18	S4, S5	115	150	.18	P10	200	.18	CP10
			4	32	.40	S4, S5	90	120	.50	P20	160	.40	CP20
			8	24	.50	S4, S5	73	95	.75	P30	120	.50	CP30
			16	20	.75	S4, S5	58	76	1.0	—	—		
4042 50B44 8642	225 to 275	Annealed, Normalized, Cold Drawn or Quenched and Tempered	.040	115	.007	M2, M3	350	465	.007	C-7	600	.007	CC-7
4047 5046 8645			.150	90	.015	M2, M3	280	365	.020	C-6	475	.015	CC-6
4140 50B46 86B45			.300	70	.020	M2, M3	220	285	.030	C-6	375	.020	CC-6
4142 5140 8740			.625	55	.030	M2, M3	170	225	.040	—	—		
4145 5145 8742			1	35	.18	S4, S5	105	140	.18	P10	185	.18	CP10
4147 5147			4	27	.40	S4, S5	85	110	.50	P20	145	.40	CP20
			8	21	.50	S4, S5	67	87	.75	P30	115	.50	CP30
			16	17	.75	S4, S5	52	69	1.0	—	—		
			.040	90	.007	T15, M42†	330	440	.007	C-7	575	.007	CC-7
			.150	70	.015	T15, M42†	260	340	.015	C-6	450	.015	CC-6
			.300	55	.020	T15, M42†	200	270	.020	C-6	350	.020	CC-6
			.625	—									
			27		.18	S9, S11†	100	135	.18	P10	175	.18	CP10
			21		.40	S9, S11†	79					.40	CP20
					.50								CP30

FIGURE 43-11 Portion of table from MDC *Machining Data Handbook,* 3rd ed., which gives recommended speeds and feeds for a given material; broken down by operation, material hardness, and condition according to what tool material is being used at a given depth of cut. Such data are considered to be reliable but their accuracy is not guaranteed. (Courtesy Metcut Research Associates, Inc.)

quantities are involved. However, it is common practice for complete details to be given regarding tools, speeds, and often the time allowed for completing each operation. Such data are necessary if the work is to be done on NC machines, and experience has shown that these preplanning steps are advantageous when ordinary machine tools are used.

The selection of speeds and feeds required to manufacture the part may require referring to handbooks in which tables of the type shown in Figure 43-11 are given. For the threaded shaft, this table will give suggested values for the turning and facing operations for either high-speed steel or carbide tools. Note that the tables are segregated by workpiece material (medium-carbon alloy steels, wrought or cold-worked) and then by process; in this case, turning. For these materials, additional tables for drilling and threading would have to be referenced. Notice that the depth of cut dictates the speed and feed selection and that rough and finish cuts are used. That is, operation 60 requires three roughing cuts and one finishing cut. While the operation sheet does not specifically say so, high-speed steel tools are being used throughout. If the job was being done on an NC lathe, it is likely that all the cutting tools would be carbides. Notice also that the job as described for the engine lathe required two setups—a three-jaw chuck to get the part to length and produce the centers and a between-centers setup to complete the part between each setup. It was also necessary to stop the lathe to invert the part. Do you think that this part could be manufactured more efficiently in an NC lathe?

Data of the type shown in Figure 43-11 are currently being computerized to be compatible with computer-aided manufacturing and computer-aided process planning systems. Table 43.1 shows the most frequently reported system inputs and output for this kind of data.

As noted in Chapter 42, the cutting time, as computed from the machining parameters, represents only 20% to 30% of the total time needed to complete

TABLE 43.1 **Typical Inputs and Outputs
for Computerized Machinability Data Systems
in Use Today**

Inputs	Outputs
Machine tool code	Recommended speed/feed
Cutting tool code	Production-time calculations
Material cost	Number and size of cuts
Operation code	Optimum speed/feed
Depth of cut	Cutting tool/grade
Material hardness	Machine tool

Source: "Machining Briefs," Metcut Research Associates, Inc., March/April 1982.

the part. To refer back to Figure 43-9, the operator will need to pick up the part after it is cut off at operation 30, stop the lathe, open up the chuck, take out the piece of metal in the chuck, scribe a line on the part marking the desired 3½-in. length, place the part back in the chuck, change the cutoff tool to a facing tool, adjust the facing tool to the right height, and move the tool to the proper position for the desired facing cut in operation 40. All these operations take time and someone must estimate how much time is required for such noncutting operations if an accurate estimate of total time to make the part is to be obtained.

When an operation is machine-controlled, as in making a lathe cut of a certain length with power feed, the required time can be determined by simple mathematics. For example, if a cut is 10 in. in length and the turning speed is 200 rpm, with a feed of 0.005 in. per revolution, the cutting time required will be 10 minutes. Procedures are available for determining the time required for people-controlled machining elements, such as moving the carriage of a lathe by hand back from the end of one cut to the starting point of a following cut. Such determinations of time estimates are generally considered the job of the industrial engineer, and space does not permit us to cover this material in this text; but the techniques are well established. Actual time studies, accumulated data from past operations, or some type of motion-time data, such as MTM (methods–time–measurement), can be employed for estimating such times. Each can provide accurate results that can be used for establishing standard times for use in planning. Various handbooks and books on machine-shop estimating contain tables of average times for a wide variety of elemental operations for use in estimating and setting standards. However, such data should be used with great caution, even for planning purposes, and they should never be used as a basis of wage payment. The conditions under which they were obtained may have been very different from those for which a standard is being set.

Route Sheets. *Route sheets* are very useful for general planning in the job shop even though they do not provide detailed information for the operation. The route sheet lists the operations that must be performed in order to produce

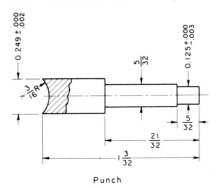

Punch

Matl. – 0.250 dia. AISI 1040

H.T. to 50 R.C. on 0.249 dia.

FIGURE 43-12 Part drawing of a punch. See Figure 43-8 for the route sheet.

the part, in their sequential order, and the machines or work stations and the tooling that will be required for each operation. For example, Figure 43-12 shows the drawing of a small, round "punch," and Figure 43-8 is a routing sheet for making this part. Once the routing of a part has been determined, the planning of each operation in the processing can then be done and the operations sheet prepared. The route sheet is also needed by *production planning,* another functional group in the PS that schedules the products (parts) through the machines and the personnel to run the machines. Notice that the route sheet does not provide information on when the parts are to be made.

Quantity Versus Process and Material Alternatives. Most processes are not equally suitable and economical for producing a range of quantities for a given product. Consequently, the quantity to be produced should be considered, and the design should be adjusted to the process that actually is to be used before it is "finalized." As an example, consider the part shown in Figure 43-13. Assume that, *functionally,* brass, bronze, a heat-treated aluminum alloy, or ductile iron would be suitable materials. What material and process would be most economical if 10, 100, or 1000 parts were to be made?

If only ten parts were to be made, contour sawing, followed by drilling the ¾-in. hole, would be very economical. The irregular surface would be difficult to produce by other machining processes. Casting would require the making of a pattern, which would be about as costly to produce as the part itself. It is unlikely that a suitable piece of ductile iron or heat-treated aluminum alloy would be readily available. Brass would be considerably cheaper than bronze. Therefore, brass, contour sawing, and drilling most likely would be the best combination. For ten parts, the excess cost of brass over ductile iron would not be great, and this combination would require no special consideration on the part of the designer.

For a quantity of 100 parts, an effort to minimize machining costs may be worthwhile. Casting might be the most economical process followed by machining. Ductile iron would be cheaper than any of the other permissible materials. Although the design requirements for casting this simple shape would be minimal, the designer would want to consider them, particularly as to whether the hole should be cored.

For 1000 parts, entirely different solutions become feasible. The use of an aluminum extrusion, with the individual 2-in. units being sawed-off, might be the most economical solution assuming that the time required to obtain a special extrusion die was not a factor. The hole in the part could be extruded, so this process is eliminated, but what are the trade-offs? What material would you recommend for the part now? How many feet of extrusion would be required, including sawing allowance? How much should be spent on the die so that the per-piece cost would not be great. Would the cost of the die be offset by the savings in machining costs? If extrusion is used, the designer should make sure that any tolerances specified are well within commercial extrusion tolerances.

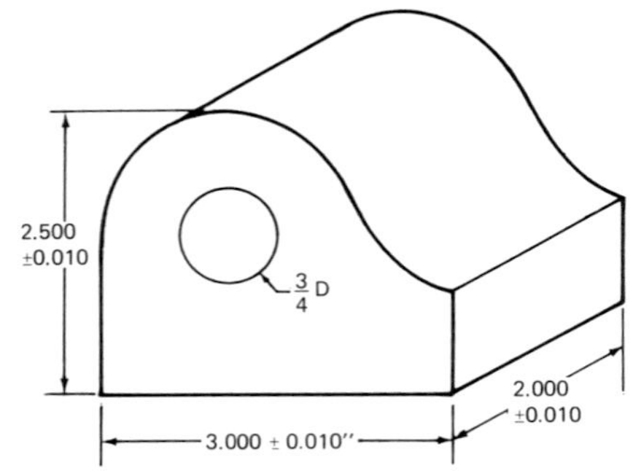

This simple example clearly illustrates how quantity can affect both material and process selection, and the selection of the process may require special

2.500
±0.010

3.000 ± 0.010″

2.000
±0.010

¾ D

FIGURE 43-13 Part to be analyzed for production.

considerations and design revisions on the part of the designer. Obviously, if the dimensional tolerances were changed, entirely different solutions might result. When more complex products are involved, these relations become more complicated, but they also are usually more important and require detailed consideration by the designer.

Other duties of the manufacturing engineer may include the responsibility for the design of tools, jigs, and fixtures to produce the product. Just as the engineer who is concerned with manufacturing must understand the functioning, capabilities, and proper utilization of machine tools but almost never designs them, similarly he or she should have a thorough understanding of the basic principles of jigs and fixtures so as to utilize them effectively. However, in most cases, the design of the tooling is left to the tool design specialists. In most companies, the manufacturing engineer makes recommendations on new machine tools and other equipment.

After the product is in production, manufacturing problems invariably arise, and because of the functional design and operation of the existing system (the job shop), the manufacturing engineer has the responsibility for solving them. A typical scenario might entail poor-quality materials being received from a vendor and accepted (in error). Use of these poor-quality materials causes fixtures to work improperly, producing defective parts and resulting in components that cannot be assembled. Assembly workers, on incentive pay, will sacrifice quality for the sake of the piece rate, and the company ships defective products. Finding the cause of such problems is part of the manufacturing engineer's job and is sometimes called ''troubleshooting'' or ''fire fighting.'' The actual causes of problems may not be eliminated because of the pressure to keep on schedule. Since the causes are not eliminated, the defects keep coming back (in greater numbers) when material shortages occur in the system. The manufacturing engineer can find himself responsible for a system he cannot possibly control because of its functional design.

Industrial Engineering.   *Industrial engineers* are usually responsible for determining the level of workers, machines, and materials needed on the plant floor to turn the ideas developed in R&D, marketing, and procurement into real products. Industrial engineers look for the ''better way'' to produce products and services under uncertain conditions and constraints such as the nature of the plant, materials, machines on hand, manpower, and available capital. The industrial engineering department is responsible for many elements of the MS and PS, including some which overlap with those of the manufacturing engineer. These include

1. Production methods analysis (motion study).
2. Work measurement (time study and time standards).
3. Plant design (layout) and material handling.
4. Quality engineering (may be a separate functional group).
5. Plant maintenance information.

Part of the industrial engineering function is to determine, standardize, and analyze the methods used to produce particular products and services. Infor-

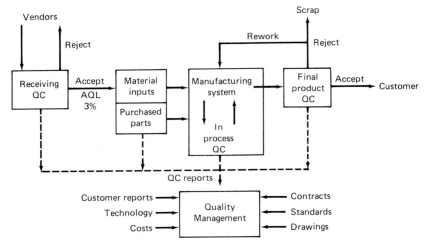

FIGURE 43-14 The classical quality control system.

mation from motion-study techniques, videotapes and movies, SMED (see Chapter 44), and other techniques are used to determine how a product or service can best be produced and to develop efficient work methods. In other words, the sequence of activities as well as the machines, tools, and materials to be used must all be specified in the job shop environment.

After standardization of the job's content, information is obtained on how much time is required to do the job. This information is based on the time required for an average person to produce a given product or service, using average effort under normal working conditions. Time standards and studies are used to develop standard times for the standard methods. The method for a particular job can be analyzed using an *operations analysis sheet*.

Designing the plant layout and the associated materials-handling equipment falls to industrial engineering. Layouts that reduce manufacturing costs with minimized material handling and inventory are fundamental to integrated manufacturing systems.

Plant Engineering. Plant engineering, another functional engineering group within the classical production system, is responsible for in-plant construction and *maintenance,* machine tool and equipment repair, heating and air-conditioning system maintenance, and any other mechanical, hydraulic, or electric problems not necessarily related to the manufacturing system.

Quality Engineering. Quality engineering is responsible for assuring that quality of product and its components meets standards specified by the designer before, during, and after manufacturing. See Figure 43-14. In-process quality control inspections are performed throughout the manufacturing system. In the classical system, materials and parts purchased from outside suppliers are inspected when they are received. The acceptable quality level (AQL) is traditionally around 3%, which means the company is willing to accept 3% defectives

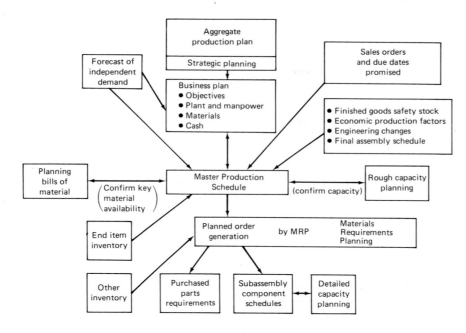

FIGURE 43-15 Master production schedule indicates what products to produce and when the products are needed.

from their vendors. Many companies feel that it costs too much to reduce defectives below this level. Parts fabricated inside the company are inspected many times during processing. Final inspection and testing of the finished product is performed to determine overall functional performance and appearance quality.

Procurement/Purchasing. The *procurement/purchasing* functions in a company primarily involve the acquisition of specified materials, equipment, services, and supplies of the proper quality, in the correct quantities, at the best prices, at the correct time. Many departments are involved in procurement/purchasing; manufacturing, marketing, finance, accounting, research and development, and engineering. For production systems, procurement provides information on vendors, prices, new products, and materials and determines the delivery schedule for purchased items.

Production Planning and Control. *Production planning* translates sales into forecasts by part number. The authority to manufacture the product is translated into a *master production schedule,* a key planning document specifying the products to be manufactured, the quantity to be produced, and the delivery date to the customer. See Figure 43-15. The master schedule is converted into purchase orders for raw materials, orders for components from outside vendors, and production schedules for parts made in the manufacturing system. *Production control* develops the timing and coordination to ensure that delivery of the final product meets customer demand. Because of the complexity of the job shop MS, production is not controlled very well; therefore, many other control functions are needed.

The scheduling periods used in the master schedule are usually months. The master schedule must take into account the production capacity of the plant (how much can be built in a given period of time). The capacity of a job shop is tremendously variable and flexible and is not well controlled. Because of this characteristic, larger quantities of products are often requested in violation of the master schedule.

Based on the master schedule, individual components and subassemblies that make up each product are planned. Raw materials are ordered to make the various components. Purchased parts are ordered from vendors. Planning is a must if the components and assemblies are to arrive when needed, not months before or, even worse, days or weeks late. The frequently used *material requirements planning* (mrp) technique is discussed later.

The next task is *production scheduling* in which start dates and due dates are assigned for the various components to be processed through the factory. Many factors make the scheduling job complex. The number of individual parts can be in the thousands. Each part seems to have its own individual process route through the plant. Parts are often routed through dozens of separate machines in many different departments. The number of machines in the shop is limited, and the machines are different, perform different operations, and have different features, capacities, and capabilities. In effect, the orders compete with each other for machines. In addition to these factors, parts become defective during the processing, cycle times vary, machines break down, operators expand job times to fit the time available. All these factors destroy the validity of the planning schedule and require (in the classical system) huge amounts of resources (lots of people and paperwork) to manage all the exceptions. Chaos reigns supreme in the large job shop.

Dispatching is the production planning, and control function requiring voluminous paperwork whereby individual orders such as order tickets, route sheets, part drawings, and job instructions are sent promptly to the machine operators or foremen.

Expediters find lost or late materials by tracking the progress of the order against the production schedule. To speed up late orders, the expediter may rearrange the order-processing sequence for a certain machine, coax the foreman to tear down one setup so that another order can be run, or hand-carry parts from one department to the next just to keep production going. Obviously, the master schedule is disrupted. The size of the production control department and the number of expediters in a company is an informal measure of the level of chaos and inefficiency in the manufacturing system.

The product, when finished, is either shipped directly to the customer or stocked in inventory. *Inventory control* is used to ensure that enough products of each type are available to satisfy customer demand. However, competing with this objective is the company's desire to minimize its financial investment in inventory. Inventory control interfaces with marketing and production control since coordination must exist between the various products' sales, production, and inventory levels. Although none of these three functions can operate effectively without information about what the others are doing, this information is often missing or out of date. Inventory control is often done by the production planning and control department. Because of its importance to all existing systems, inventory control is discussed in more detail in the next section.

Inventory Control.   The life blood of a manufacturing system is its inventory. Inventory represents a major portion of a manufacturing facility's assets. For most companies, however, excess inventory represents idle investment dollars and wasted storage space. Even though the cost of carrying large inventories is substantial, many reasons are given for having them:

1. Fluctuation in demand and/or supply.
2. Protection against process breakdowns.
3. Replacement parts for lost batches or defective lots.
4. Overproduction in anticipation of future demand.
5. Protection from defective parts.
6. Goods in transport.
7. Just in case they are needed.
8. Quantity purchasing.

Inventory control governs finished goods, raw materials, purchased components, and work-in-process within the factory. The idea is to achieve a balance between too little inventory (with possible stock-outs of raw materials) and too much inventory (with investments and storage space tied up).

Classical Inventory Control Requirements.   The requirements for a total inventory control system are that it

1. Analyze and plan inventory requirements.
2. Purchase raw materials and component parts in the amounts needed, according to scheduled usage.
3. Receive and record the receipt of purchased materials.
4. Provide adequate facilities to store raw material, work-in-process, and finished goods inventory.
5. Maintain accurate records of inventories on hand and on order.
6. Install realistic controls for materials in stores and for the issuance of materials, parts, and supplies.

A good inventory control system provides the correct quantity and quality of material at the correct time. This system also maintains accurate rercords/control of these materials.

Inventory Models   In 1915 Ford Harris and R. H. Wilson derived, independently, the "simple lot size formula." This model states (see Figure 43-16):

Total annual cost = purchase cost + order cost + holding cost
$$TC = RP + RC/Q + QiP/2$$

where   (43-1)

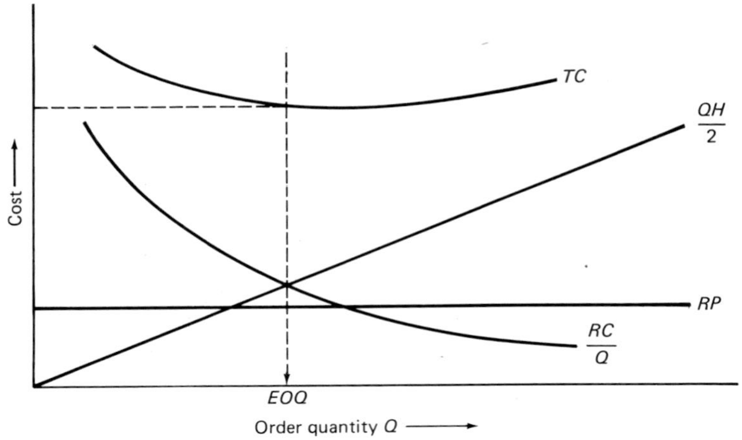

FIGURE 43-16   The economic order quantity (EOQ) model minimizes the total (annual) cost of inventory.

R = annual demand in units,
P = unit cost of an item,
C = ordering cost per order,
$H = iP$ = holding cost per unit per year,
Q = lot size or order quantity in units,
i = annual holding cost as a fraction of unit cost.

To find the Q which minimizes the total cost,

$$d(TC)/dQ = 0 - RC/Q^2 + iP/2 = 0$$
$$RC/Q^2 = iP/2$$

$$Q = \sqrt{\frac{2RC}{iP}}$$

(43-2)

This model assumes that the ordering cost is fixed when in fact it is not. This model has also been extended to cover economic production quantity by letting C equal setup cost. (A better solution is to eliminate setup and its cost. The best lot size is then the smallest lot size that permits a smoothly running process.) Many other models have been developed for the inventory process, and many models and systems have since evolved for stock or inventory replenishment including the reorder point (ROP) model. Figure 43-17 illustrates the relationship between lead time (L); order size (Q), safety stock (SS), expected demand rate (D), and the reorder point (ROP). The reorder point is based on how long it takes to obtain parts (lead time) and how many parts will be used up during this (lead) time.

One of the problems with the reorder point model is the dependence of the model on usage. The model shows that parts are used linearly. In fact, in batch processing, utilization is usually uneven, with large spikes or peaks. That is, 200 of an item may be used in week 2, zero in weeks 3 through 9, and 200 again in week 10. The order point model would cause the system to run out of parts in week 2, precipitating a rush order for parts which would remain on the shelf unnecessarily until week 10.

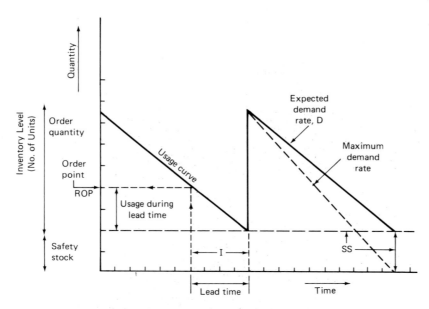

To Find ROP
1. Establish lead time.
2. Back lead time off from order receipt time.
3. Go to usage curve.
4. Find order point on vertical scale.

FIGURE 43-17 Classical inventory model for ROP (reorder point) methods.

The model also breaks down if the average utilization changes. To use the model effectively, demand must be constantly recalculated.

The utilization of certain parts is dependent upon the demand for other parts. For example, in the building of cars, the demand for tires depends upon how many cars are built. While demand for cars must still be estimated, the demand for tires is known, based on the demand for cars.

These two types of demand are categorized as *independent* and *dependent*. The first must be "guessed" (forecasted). The second can be computed. The basis for these computations is a record of the relationship between the independent demand item (cars) and its components (tires, horn, windows). This record is the "bill of materials." See Figure 43-7.

Manufacturing Resource Planning (MRP). In order to control inventory within the job shop, a computerized system called *material requirements planning* (mrp) was developed. See Figure 43-18. Given a schedule showing the expected demand of independent demand items (a master production schedule) and given the relationship between independent and dependent demand items (bills of materials), mrp will calculate the quantities of dependent demand items needed, and when they will be needed.

The potential value of this concept is significant. Suppose a company has 100 finished goods items, 400 assemblies and subassemblies, and 1,000 raw material

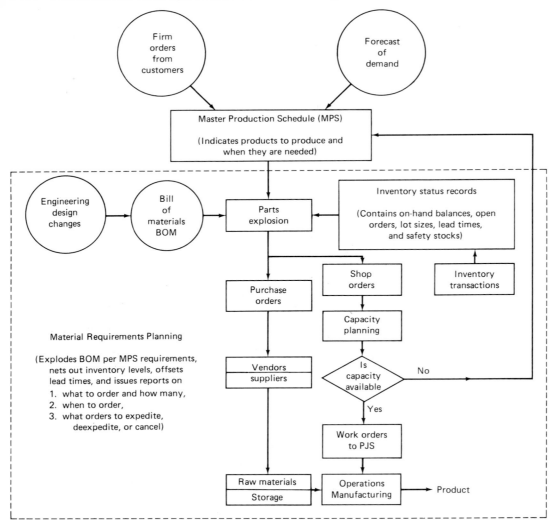

FIGURE 43-18   mrp is a computerized inventory system for the job shop.

items. Using statistical stock replenishment, it will need to forecast the average demand for 1,500 items. Many of these items will have "lumpy" demand which will cause the stockout/expedite/overstock cycle discussed earlier. With mrp, a master production schedule (MPS) for 100 items must be maintained; the other 1,400 items will have their exact demand computed for every period.

Almost at the same time that mrp was developed, practitioners began to expand the concept's scope. Just as bills of materials could establish the usage of dependent materials, other records could be developed to tell the dependent requirements of labor hours, machine hours, capital, shipping containers—in fact any of the resources required to support the job shop MPS. Material requirements planning has become MRP, or manufacturing resource planning. This system is responsible for final scheduling of production, dispatching, and releasing purchase orders. It will also maintain stock status, monitor output and scrap levels, and compare performance against the plan.

MRP and mrp are good *planning* techniques. There is no inherent capability to *control* and replan. Subsequent sophistication of MRP by adding feedback of actual results has led to "closed-loop manufacturing resource planning" or MRP II. Shop floor control and vendor control systems are added to the existing software so that revisions of dates and quantities will be taken into account in the next *planning cycle*.

The job shop is a complex dynamic manufacturing system. Inventory is its life blood. MRP is an attempt to control inventory using computers. Unfortunately, the MRP systems that have evolved are so complex that very few people in the companies that use them really understand them. Would you trust a system you did not understand? What is the value of an inventory control system that the majority of the users do not comprehend?

What makes a good inventory control system? First, everyone who uses it must understand how it works. It must have accurate information and make accurate predictions or forecasts. The users of the system must act on the information that the system produces. In Chapter 44 a simple, remarkable system will be described that is designed to "eat its own paperwork" and operate as a paperless system. But a word of caution: This inventory control system is not useable in the job shop. It is for a new MS called "linked-cell" MS, which can replace the job shop.

Computer-Integrated Manufacturing (CIM).

A number of definitions have been developed for *computer-integrated manufacturing (CIM)*. However, a CIM system is commonly thought of as an integrated system that encompasses all the activities in the production system from the planning and design of a product through the manufacturing system, including control. CIM is an attempt to combine existing computer technologies in order to manage and control the entire business. CIM is the approach that many companies are using to get to the automated factory of the future.

As with traditional manufacturing approaches, the purpose of CIM is to transform product designs and materials into saleable goods at a minimum cost in the shortest possible period of time. CIM begins with the design of a product and ends with the manufacture of that product. With CIM, the customary split between the design and manufacturing functions is (supposed to be) eliminated.

Elements of CIM.

CIM differs from the traditional job shop manufacturing system in the role the computer plays in the manufacturing process. Computer-integrated manufacturing systems are basically a network of computer systems tied together by a single integrated data base. Using the information in the data base, a CIM system can direct manufacturing activities, record results, and maintain accurate data. CIM is the computerization of design, manufacturing, distribution, and financial functions into one coherent system.

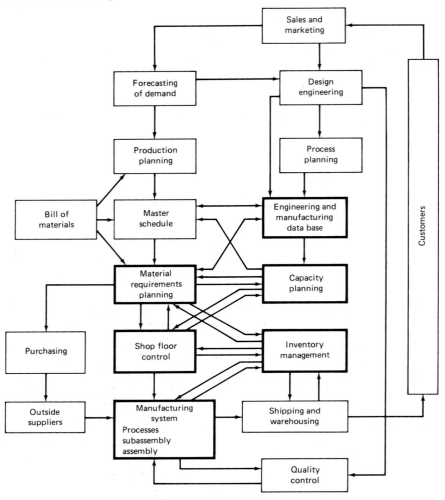

FIGURE 43-19 Cycles of activities in a computer-integrated manufacturing system.

Figure 43-19 presents a block diagram illustrating the functions and their relationship in CIM. These functions are identical to those found in a traditional production (planning and control) system for a job shop MS. With the introduction of computers, changes have occurred in the organization and execution of production planning and control through the implementation of such systems as mrp, capacity planning, inventory management, shop floor control, and cost planning and control.

Engineering and manufacturing data bases contain all the information needed to fabricate the components and assemble the products. As shown in Figure 43-19, the design engineering and process planning functions provide the inputs for the engineering and manufacturing data base. This data base includes all the data on the product generated during design such as geometry data, parts lists,

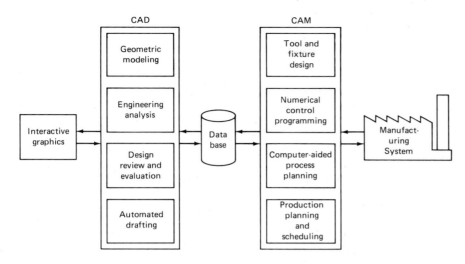

FIGURE 43-20 Data base relationship to design and manufacturing.

and material specifications. The bill of materials is shown separately but it is a key part of the data base. Figure 43-20 shows how the CAD/CAM data base is related to the design and manufacturing activities.

Capacity planning is concerned with determining what labor and equipment capacity is required to meet the current master production schedule as well as the long-term future production needs of the firm. Capacity planning is typically performed in terms of labor and/or machine hours available. The master schedule is transformed into material and component requirements using mrp. These requirements are then compared with available plant capacity over the planning horizon. If the schedule is incompatible with capacity, adjustments must be made either in the master schedule or in plant capacity. The possibility of adjustments in the master schedule is indicated by the arrow in Figure 43-19 leading from capacity planning to the master schedule.

The term "shop floor control" in Figure 43-19 refers to a system for monitoring the status of manufacturing activities on the plant floor and reporting the status to management so that effective control can be exercised.

The cost planning and control system consists of the data base to determine expected costs to manufacture each of the products of the firm. It also consists of the cost collection and analysis software to determine what the actual costs of manufacturing are and how these actual costs compare with the expected costs.

Computer-Assisted Design (CAD).   A major element of a CIM system is a computer-assisted design (CAD) system. CAD involves any type of design activity that makes use of the computer to develop, analyze, or modify an engineering design. The design-related tasks performed by a CAD system are

- Geometric modeling.
- Engineering analysis.

- Design review and evaluation.
- Automated drafting.

Geometric modeling corresponds to the synthesis phase of the design process. To use geometric modeling the designer constructs the graphical image of the object on the CRT. The object can be represented using several different methods. The basic method uses wire frames to represent the object. The wire frame geometric modeling is classified as

- Two-dimensional representation used for flat objects.
- Three-dimensional modeling of more complex geometries.

Other enhancements to wire-frame geometric modeling are:

- Color graphics.
- Dashed lines to portray rear edges which would be invisible from the front.
- Removal of hidden lines.
- Surface representation that makes the object appear solid to the viewer.

The most advanced method of geometric modeling is solid modeling in three dimensions. This method uses solid geometry shapes called *primitives* to construct the object. See Figure 43-21.

Engineering analysis is required in the formulation of any engineering design project. The analysis may involve stress calculations, finite element analysis for heat transfer computations, or the use of differential equations to describe the dynamic behavior of the system. Generally, commercially available general-purpose programs are used to perform these analyses.

Design Review and Evaluation techniques check the accuracy of the CAD design. Semiautomatic dimensioning and tolerancing routines which assign size specification to surfaces help to reduce the possibility of dimensioning errors. The designer can zoom in on the part design details for close scrutiny. Many

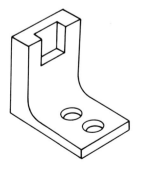
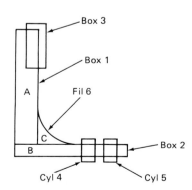

FIGURE 43-21 The CAD (computer-assisted design) part design is constructed by summing basic solid geometry shapes.

Part = Sum of Vols A, B, C
    Volume A = Box 1 − Box 3
    Volume B = Box 2 − Cyl 4 − Cyl 5.
    Volume C = Fil 6

systems have a layering feature which involves overlaying the geometric image of the final shape of a machined part on top of the image of a rough part. Other features are interference checking and animation that enhances the designer's visualization of the operation of the mechanism and helps to ensure against interferences.

Automated drafting involves the creating of hard-copy engineering drawings from the CAD data base. Typical features of CAD systems include automatic dimensioning; generation of crosshatched areas; scaling of drawings; ability to develop sectional views; enlarged views of particular part details; and the ability to rotate the part and perform transformations such as oblique, isometric, and perspective views.

Parts Classification and Coding for Group Technology.   A group technology coding and classification system can also be developed using the CAD data base. Parts coding involves assigning letters or numbers to parts to define their geometry or manufacturing process sequence. The codes are to classify or group similar parts into families. Designers can use the classification and coding system to retrieve existing part designs rather than redesigning new parts.

Computer-Aided Manufacturing.   Another major element of CIM is computer-aided manufacturing (CAM). An important reason for using a CAD system is that it provides a data base for manufacturing the product. However, not all CAD data bases are compatible with manufacturing software. The tasks performed by a CAM system are

- Numerical control or CNC programming.
- Computer-aided process planning (CAPP).
- Production planning and scheduling.
- Tool and fixture design.

Numerical control, discussed in Chapter 29, can use special computer languages. APT and COMPACT II are the two most common language-based computer-assisted programming systems used in industry today. These systems take the CAD data and adapt it to the particular machine control unit/machine tool combination used to make the part.

Computer Aided Process Planning (CAPP).   CAPP uses computer software to determine how a part is to be made. If GT is used, then parts are grouped into part families according to how they are to be manufactured. For each part family, a standard process plan is established. That is, each part in the family is a variation of the same theme. The standard process plan is stored in computer files and then retrieved for new parts that belong to that family.

Figure 43-22 explains the CAPP process. The user initiates the procedure by entering the part code. The CAPP program then searches the part family file to determine whether a match exists. If the file contains an identical code number, the standard machine routing and operation sequence are retrieved for display

```
                              ┌──────────────────┐
                              │   Raw material   │
                              │    selection     │
                              └──────────────────┘
                                        │
                                        ▼
┌────────────┐          ┌──────────────────────────┐          ┌────────────┐
│  Standard  │          │ Create operation sequence │          │            │
│ operation  │ ───────▶ │ ─ ─ ─▶   variant          │          │Design logic│
│ sequence   │          │        generative  ◀─ ─ ─ │          │            │
└────────────┘          └──────────────────────────┘          └────────────┘
                                        │
                                        ▼
                              ┌──────────────────┐
                              │ Create operation │
                              │     details      │
                              └──────────────────┘
           ┌──────────┬──────────┼──────────┬──────────┐
           ▼          ▼          ▼          ▼          ▼
      ┌─────────┐┌─────────┐┌─────────┐┌─────────┐┌─────────┐
      │Tool plan││  Tool   ││   Cut   ││ Feeds & ││ Create  │
      │         ││ layout  ││sequence ││ speeds  ││detailed │
      └─────────┘└─────────┘└─────────┘└─────────┘│operation│
                      ▲           ▲        ▲      │instruct.│
                 ┌─────────┐  ┌─────────────┐    └─────────┘
                 │Interactive│ │Machinability│
                 │ graphics  │ │ data base   │
                 └─────────┘  └─────────────┘
```

FIGURE 43-22 CAPP (computer-aided process planning) can be variant or generative.

to the user. The standard operation sequence is examined by the user to permit any necessary editing of the plan to make it compatible with the new part design. This is variant CAPP. After editing, the process plan formatter prepares the paper documents (the route sheet and operation sheets for the job shop).

If an exact match cannot be found, a new plan will have to be *generated* based on design logic and the examination of geometry and tolerance information and then comparing these requirements to the machine capability. The software determines what processes can produce what surfaces and establishes the machining constraints. Once the process plan for a new part code number has been entered and verified, it becomes the standard process for future parts of the same classification. The process plan formatter may include software to compute machining conditions, layouts, and other detailed operation information for the job shop.

Manufacturing Automation Protocol (MAP).   The CIM approach results in a vast collection of computers throughout the company. Because these computers are purchased at different times, they either cannot communicate with each other or require extensive (costly) custom interfaces. To deal with this problem, General Motors initiated MAP, an acronym for Manufacturing Automation Protocol, which is a communication network specification defining a local area computer network—a network that offers reliable high-speed communication channels that are optimized for connecting different items of information-processing equipment within a limited geographic area. The major thrust of MAP is to provide a LAN (local area network) that is based on open rather than proprietary standards so that equipment from multiple vendors can be connected through the same network. To do this, each vendor's equipment must

ISO Protocol Layer	MAP 2.1 Protocol Selection
User Program	MMFS/EIA 1393A
7. Application	ISO Case Kernel, FTAM
6. Presentation	Not Used Yet (Null)
5. Session	ISO Session Kernel
4. Transport	ISO Transport Class 4
3. Network	ISO CLNS
2. Datalink	IEEE 802 2 Link Level Control Class 1
1. Physical	IEEE 802.4 Token Access On Broadband Media

FIGURE 43-23 International Standards Organization (ISO) seven-layer model for open-system interconnects. Current MAP protocols selected for MAP version 2.1 shown.

meet the universal MAP specification—a specification that most users feel is long overdue.

The network architecture created by most vendors in recent years has conformed to the International Standards Organization (ISO) model for open systems interconnection (OSI). See Figure 43-23. This reference model formalizes a developing consensus among communication network designers, provides the framework for layered networks, and introduces a uniform terminology for naming the various utilities involved. MAP uses the same ISO seven-layer reference model and further defines the protocol standards to be followed at each layer.

The block diagram in Figure 43-24 shows how computer devices are connected on a MAP network for a manufacturing cell or system. In general, layer 1 connects to the coaxial cable, which carries the signals to and from other computer devices. Layer 7 communicates with the computer device's application program. The arrangement of hardware and software required for ISO seven-layer communications at each computer device is a MAP interface.

When communications occur, each ISO layer (except any layer that is null) must pass the signals to the next so that the correct data are transferred from the coaxial cable to the computer—or from the computer to the coaxial cable, depending on whether the device is sending or receiving.

To aid understanding of the function performed at each layer, analogies have been made to sending a letter through the United States postal system. Figure 43-25 defines the primary function of each ISO layer and a corresponding function of the postal system. The example relates to a message placed inside an envelope, the envelope addressed to a receiver and mailed.

In the next chapter, a different approach to the factory of the future will be presented wherein many of the functions of the production system are integrated into the manufacturing system. This requires that the job shop MS be replaced with a linked-cell MS. The functions of production control, inventory control, quality control, and machine tool maintenance are the first to be integrated.

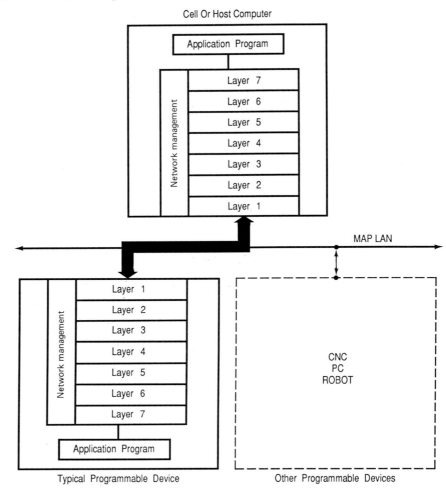

FIGURE 43-24 Typical block diagram of a manufacturing cell or system connected by a MAP local area network. The data in and out of each programmable device must travel through each ISO layer (except null layers).

Review Questions

1. How does a production system differ from a manufacturing system?
2. How is forecasting related to the manufacturing system and the production system as a whole?
3. Why is it so important for the designer to have intimate knowledge of the available manufacturing processes at the various stages of the design activity?
4. Explain how function dictates design with respect to the design of footwear. Use examples of different kinds of footwear (shoes, sandals, high heels, boots, etc.) to emphasize your

points. For example, cowboy boots have pointed toes so that they slip into the stirrups easily and high heels to keep the foot in the stirrup.
5. Most companies, when computing or estimating costs for a job, will add in an overhead cost, often tying that cost to some direct cost, like direct labor. What is included in this overhead cost?
6. How is this overhead cost, which is often given as 100% to 200% of direct labor cost, impacted by better planning and scheduling?

ISO LAYER	ISO FUNCTION	U.S. POST OFFICE EQUIVALENT FUNCTION
Layer 7 Application	Provides all services directly comprehensible to application programs.	Message placed inside envelope, envelope addressed to receiver and sent on its way by mailing.
Layer 6 Presentation	Restructures Data to/from standardized format used within the network.	Format and language of the letter, includes translation into proper language if required.
Layer 5 Session	Names/address translation access security, and synchronize/manage data.	Name, address and zip code of both the receiver and the sender. The to and from.
Layer 4 Transport	Provides transparent, reliable data transfer from end node to end node.	Certified or registered mail providing verification to the sender that the letter arrived at the correct destination.
Layer 3 Network	Performs message routing for data transfer between nodes not in the same LAN.	The segment of the postal distribution system which transfers a letter outside the local postal system to a system in another city or country.
Layer 2 Datalink	Provides the means to establish, maintain and release logical datalinks between systems, transfer data frames between nodes in the same LAN, and detect and correct errors.	The segment of the postal distribution system which transfers a letter from the sender to a destination within the same postal system. (Either a receiver in the same system or to a distribution center for forwarding to a receiver in the system or to a distribution center for forwarding to a receiver in another system.)
Layer 1 Physical	Encodes and physically transfers messages between adjacent nodes.	Conveyance: postman, auto, truck, airplane, etc.

FIGURE 43-25   A comparison of each ISO layer to the U.S. Postal System equivalent function.

7. How does the design of the product influence the design of the manufacturing system, including assembly and the production system?

8. Discuss this statement: "Software can be as costly to design and develop as hardware and will require long production runs to recover, even though these costs may be hidden in the overhead costs."

9. What is meant by this statement? "All processing schemes have a distinct technology and history and call for a distinct set of optimally suited materials."

10. If a designer called for a part to be made from cast steel, what processes has he eliminated from the possible list for casting this part?

11. Give an example of how the available cooking processes might limit what the chef can put on the hotel menu.

12. Give an example in design (detail) that might lead to a part failure or defect.

13. In Figure 43-5, how would you have designed the shaft and flange pair if you were going to join these parts by friction welding?

14. Under what assembly conditions (again referring to Figure 43-5) might the initial design of the flange and shaft have been better than the improved design?

15. Can the part shown in Figure 43-10 be made more efficiently in an NC lathe? Why or why not?

16. What prevents the part shown in Figure 43-10 from being made by an automatic lathe?

17. Outline the manufacturing processes that you think might be needed to fabricate a razor blade and estimate what you think the blade itself actually costs. Razor blades have a final edge-

cutting radius of 1000 angstroms or better. Did you determine the basic job requirements?

18. Figure 43-12 shows a drawing of a punch. What are the critical dimensions? Why are they critical? Why is overall length not critical?

19. Suppose that the part in Figure 43-10 were to be made in quantities of 250 rather than 25. What processes and machine(s) would you have selected? Suppose that the quantity were 25,000; what then?

20. Can the part in Figure 43-13 be made by extrusion and cutoff, considering the tolerances?

21. Find an example of a product or item the design of which has changed so that the item can be automatically assembled rather than manually assembled. (This is called design for assembly (DFA).)

22. Manufacturing engineers create process plans. For the part shown in Figure 43-21, provide a process plan to make 1000 of these brackets. The brackets go into a metal bookcase and support the shelves.

23. Explain the difference between a route sheet and an operations sheet.

24. The IEs at the plant should assist the workers who are trying to reduce setup. What techniques are available to them for this effort?

25. What kinds of inspection systems have an AQL?

26. What is the difference between an expediter and a dispatcher in a job shop?

27. What is the difference between inventory control and production control?

28. What is the difference between dependent and independent demand?

29. How does the mrp use the BOM and the MPS to determine how many of an item should be ordered?

30. Define CIM in your own words.

31. A standard piece of CAD analysis is called FEA. What does a finite element analysis do?

32. What is the difference between the two types of CAPP?

33. What is MAP?

34. What is LAN?

Integrated Manufacturing Production Systems

Many people have been describing the factory of the future in current literature. There has been so much written about computers, CIM (computer-integrated manufacturing), FMS, and robotics in the past ten years that it has become impossible to keep up with all the acronyms or to read all the journal articles. Invariably, these articles describe a strategy in which one is automating, computerizing, or robotizing the existing manufacturing system—that is, the job shop. The implementation of integrated manufacturing production systems (IMPSs) requires a very different strategy from that required for CIM. IMPS requires a systems-level change for the factory—a change that will impact every segment of the company, from accounting to shipping—that begins with the manufacturing system. In this chapter, the IMPS implementation strategy is boiled down into ten steps. However, before embarking on this approach to the "factory with a future," we must discuss some preliminary, preparatory steps.

Integration of the production system functions into the manufacturing system requires commitment from top-level management and communication with everyone, particularly manufacturing. Total employee and union participation is absolutely necessary, but it is not usually the union leadership or the production workers who raise barriers to IMPS. It is those in middle management who have the most to lose in this systems-level change. The preliminary steps are as follows:

- All levels in the plant, from the production workers to the president must be educated in IMPSs philosophy and concepts;
- Top management must be totally committed to this venture and everyone involved must be motivated;
- Everyone in the plant must understand that cost, not price, determines profit; Customers determine price and want low cost, superior quality, and on-time delivery.

Introduction

1097

- Everyone must be committed to the elimination of waste as the way to reduce cost.

Step I. Form Cells—Build the Foundation

In Chapter 42, the concept of cellular manufacturing was discussed. Cells replace the production job shop. The first task is to restructure and reorganize the basic manufacturing system into cells which fabricate families of parts. Next, systematically create a "linked-cell" system that is networked for one-piece movement of parts within cells and for small-lot movement between cells. Creating cells is the first step in designing a manufacturing system in which inventory control and quality control are integral parts.

Step II. RETAD—Rapid Exchange of Tooling and Dies

Everyone on the plant floor must be taught how to reduce setup time using SMED (single-minute exchange of die). A setup reduction team assists production workers and foremen and demonstrates a project on the plant's "worst" setup problem. Reducing setup time is critical to reducing lot size.

Step III. Integrate Quality Control

A multiprocess worker can run more than one kind of process. A multifunctional worker can do more than operate machines. He is also an inspector who understands process capability, quality control, and process improvement. In IMPS, every worker has the responsibility and the authority to make the product right the first time and every time and the authority to stop the process when something is wrong. This integration of quality control into the manufacturing system markedly reduces defects while eliminating inspectors. Cells provide the natural environment for integration of quality control.

Step IV. Integrate Preventive Maintenance/ Machine Reliability

Making machines operate reliably begins with the installation of an integrated preventive maintenance program, giving workers the training and tools to maintain equipment properly. The excess processing capacity obtained by reducing setup time allows operators to reduce the equipment speeds or feeds and to run processes at less than full capacity. Reducing pressure on workers and processes to produce a given quantity fosters in workers a drive to produce perfect quality.

Housekeeping rules: 1. A place for everything and everything in its place. Everything should be put away so it is ready to use the next time. 2. Each worker is responsible for cleanliness of workplace and equipment.

Step V. Level and Balance Final Assembly

Level and balance the entire manufacturing production system by producing a mix of final assembly products in small lots. This will reduce the lumpiness of the demand for the component parts and subassemblies. Standardize the cycle time. Use a simplified and synchronized system to produce the proper number of components everyday, as needed. Begin at mixed-model final assembly and work backward through subassembly and cells. Each process, cell, and subassembly has essentially the same cycle time as the final assembly.

Integration of production control is materialized by linking the cells, subassemblies, and final assembly elements, utilizing Kanban. The structure of the manufacturing system now defines paths that parts can take through the plant. Begin by connecting the elements with Kanban links. The need for route sheets is eliminated. The parts, that is, the in-process inventory, flow within the structure. All the linked-cells, processes, subassemblies, and final assemblies start and stop together. They are synchronized. This is the integration of production control into the MS.

Step VI. Link Cells

The people on the plant floor directly control the inventory levels in their areas through the control of the Kanban. This is the integration of the inventory control system to reduce systematically the work-in-process (WIP). The reduction of WIP exposes problems that must be solved before more inventory can be further reduced. The minimum level of WIP (which is actually the inventory *between* the cells and subassemblies) is determined by the percent defective, the reliability of the equipment, and the setup time.

Step VII. Reduce WIP/ Expose Problems

Educate and encourage suppliers (vendors) to develop their own IMPS for superior quality, low cost, and rapid on-time delivery. They must be able to deliver material to the customer when it is needed and where it is needed, without incoming inspection. The linked-cell network ultimately should include every vendor; every vendor, therefore, should become a remote cell.

After these eight steps have been completed, the manufacturing system has been redesigned and infused (integrated) with the critical production functions of quality control, inventory control, production control, and machine tool maintenance. Automation of the integrated system is next.

Step VIII. Extend IMPS to Include Vendors

Converting manned cells to unmanned cells is an evolutionary process, initiated by the need to solve problems in quality, reliability, or capacity (eliminate a bottleneck). It begins with mechanization of operations such as load, unload, inspect, and clamp and moves toward automatization of human thinking and automatic detection and correction of problems and defects. This is called *autonomation* or computer-aided testing and inspection (CATI).

Step IX. Automate and Robotize to Solve Problems

Total computerization of the redesigned linked-cell manufacturing system converts it to a computerized IMPS or CIMPS. The computer communications for CAD and DNC arrangements can use MAP/TOP strategies as can the robotic cells themselves. The secret here is not to computerize the existing job shop (functionalized) manufacturing system. The good news is that computerizing the integrated system will be easier because the IMPS is simpler. The ultimate integration of the manufacturing system with the rest of the production system results in computerized integrated manufacturing production systems (CIMPS).

Step X. Computerize

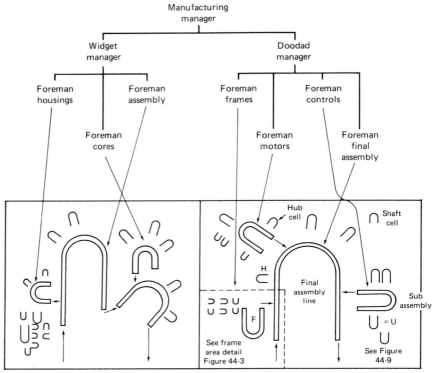

FIGURE 44-1   Plant layout for integrated manufacturing with cells linked by Kanban or by point of use.

Forming the Linked-Cell Factory

In Chapter 42, the operation of a manned cell was described. This cell, designed for the manufacture of a family of shafts, has the unique feature of decoupling the machine time (MT) from the cycle time (CT) while building products one at a time. This relaxes the line-balancing problem common to flow lines and transfer lines while greatly enhancing the flexibility. The cycle time is controlled by the time it takes the worker or workers to complete a walking loop through the cell or cells. Therefore, the cycle time can be altered by adding or deleting workers. The details of this system are shown in Figures 44-1 thru 44-5 and Figure 42-6.

Beginning with Figure 44-1, the linked-cell plant is composed of cells to fabricate components, subassembly lines, and final assembly lines. Figure 44-2 shows the details of a hub cell which produces hubs described by means of a standard operations routine sheet. (See also Figure 42-10.) The design of the cell is shown at the lower left. Hubs are made by a four-station cold header, followed by three machining operations. The cell combines forming and machining. The third operation (finish milling) has a long MT—70 seconds—compared to the needed cycle time of 55 seconds for the cell; therefore this operation, the finish milling, is duplicated in the cell. The cold header can produce a part every 5 seconds. Its capacity production is much greater than

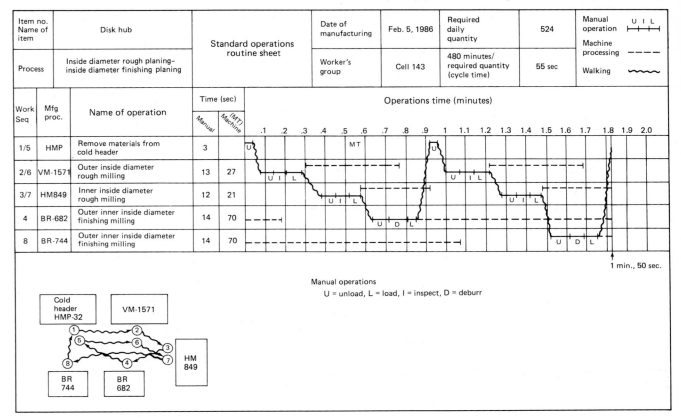

Item no. Name of item	Disk hub		Standard operations routine sheet		Date of manufacturing	Feb. 5, 1986	Required daily quantity	524	Manual operation	U I L
Process	Inside diameter rough planing–inside diameter finishing planing				Worker's group	Cell 143	480 minutes/required quantity (cycle time)	55 sec	Machine processing / Walking	— — — / ∿

Work Seq	Mfg proc.	Name of operation	Time (sec) Manual	(MT) Machine	Operations time (minutes)
1/5	HMP	Remove materials from cold header	3		
2/6	VM-1571	Outer inside diameter rough milling	13	27	
3/7	HM849	Inner inside diameter rough milling	12	21	
4	BR-682	Outer inner inside diameter finishing milling	14	70	
8	BR-744	Outer inner inside diameter finishing milling	14	70	

1 min., 50 sec.

Manual operations
U = unload, L = load, I = inspect, D = deburr

FIGURE 44-2 Example of a standard operations routine sheet for a manned cell. The hub manufacturing machine is a four-station cold header which makes parts as needed by the first milling machine (VM-1571).

needed by the cell but it is *automated* to produce parts exactly at the rate the cell needs the parts and no faster. Overproduction will result in the need to store parts, transport parts to storage, retrieve the parts when needed, keep track of the parts (paperwork), and so on. All this requires people and costs money but adds no value. The cold header, like all the machines in the cell, is underutilized. The objective is to minimize the resources within the system that do not depreciate—direct material and labor.

As discussed in Chapter 42, there is no need to balance the MTs for the machines. It is necessary only that no MT be greater than the required CT. Since the MT for finish milling is greater than the cycle time (70 > 55), the finish-milling process is duplicated and the worker alternates machines on his trips through the cell. The machines must do identical work. The cycle time is determined by the system's requirements since this cell, like all the others in the plant, is geared to produce parts as needed, when needed, by the subsequent process.

The machining speeds and feeds can be relaxed to extend the tool life of the cutting tools and reduce the wear and tear on the machines so long as the MT

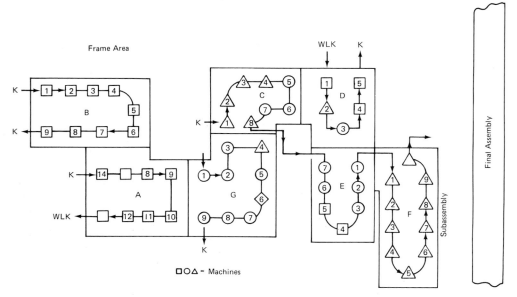

FIGURE 44-3 Seven manned cells showing the movement of parts. Cells C and E directly linked.

for a particular machine does not exceed CT for the cell. This increases the reliability of the process, reducing the probability of a breakdown. There is no mystery as to which process within the cell is the bottleneck, the machine with the longest MT. There is no need for any computer analysis to find the bottlenecks. Everyone in the manufacturing system can see and understand how the cell functions and therefore which process is the most likely to delay the cell's CT.

The cycle time is determined from the demand rate for the parts according to the following calculations:

$$\frac{\text{Daily demand}}{\text{for parts}} = \frac{\text{Monthly demand (forecast plus customer orders)}}{\text{Number of days in month}} \qquad (44.1)$$

$$\text{CT} = \frac{1}{\text{PR}} \text{ where PR} = \frac{\text{Daily demand (parts)}}{\text{Hours in day (hrs)}} \qquad (44.2)$$

You may think that this incredibly simple approach cannot possibly be the way in which IMPS companies calculate cycle time, but life is simpler when the PJS has been eliminated and a linked-cell system has been installed.

Figure 44-3 shows the frame area of the plant in more detail. This area contains six component cells and one subassembly cell which directly feeds the mixed-model final assembly line. These seven manned cells are "linked" by Kanban or linked directly to another nearby cell just as the subassembly line is directly linked to a "point of use" in the final assembly line. Cell E is directly linked to cell F. Cell D in this area is feeding parts to cell H, as shown in Figure 44-4, using Kanban. Cell H withdraws parts from cell D as needed. In the linked-cell system, the WIP between the cells is controlled by the Kanban. The

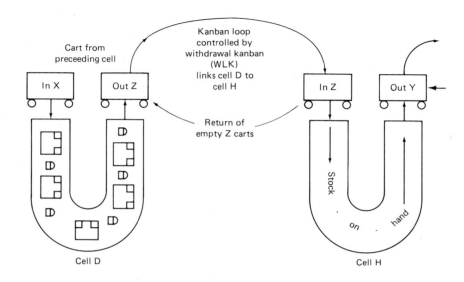

FIGURE 44-4 "Kanban-linked cells" control the WIP between the cells. The inventory in the cell is the stock on hand or SOH.

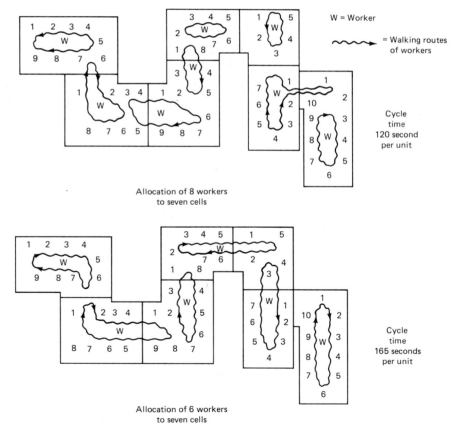

Allocation of 8 workers to seven cells

Cycle time 120 second per unit

W = Worker

$\sim\!\!\sim\!\!\sim\!\!\rightarrow$ = Walking routes of workers

Allocation of 6 workers to seven cells

Cycle time 165 seconds per unit

FIGURE 44-5 Number of workers reduced as demand decreased and cycle time increased.

inventory within the cells is called the *stock-on-hand*. When the cell can make a very high percentage of perfect parts, have no machine breakdowns and virtually no setups, the Kanban link can be replaced with a direct link. See the discussion on Step VIII.

Figure 44-5 shows the frame area with two different allocations of workers. In the upper diagram, eight workers are tending the machines in seven cells. The cycle time was 120 seconds per unit. The next month a slower cycle time was needed because the demand decreased. Six workers were allocated to the seven cells; the result is a decrease in the production rate and an increase in the cycle time to 165 seconds. Notice that all the cells have the same cycle time in this area. For a cycle time of 120 seconds, each worker is tending 7 machines. For a cycle time of 165 seconds, each worker is tending 9 or 10 machines. The workers in these cells spend from 10 to 15 seconds at a machine and 5 seconds walking to the next machine. The frame foreman allocates the number of workers in the frame area. He tries to allocate the minimum number of workers to the area.

Workers can be added if a problem arises or if some cell continually falls behind. This is not a new concept. Fast-food restaurants like Wendy's add workers to their cells when demand increases and remove workers when business is slow. Wendy's is an example of CMS which produces hamburgers. The workers in the IMPS factory are multifunctional, just like the workers at Wendy's. The machines at Wendy's are A(2) level of automation. The fried potato and drink machines are loaded and started by the worker, but they shut themselves off automatically when the cycle is complete.

The Design of Cellular Manufacturing Systems.   The conversion of the functional system into a flexible, linked-cell system (LCS) is a design task. Table 44.1 outlines the method by which manufacturing cells can be formed. Most companies "design" their first cell by one of the trial-and-error techniques for expediency in gaining experience in cells. *Digital simulation* is gaining wider usage in designing and analyzing manufacturing systems with the advent of newer, more versatile languages. Another technique being extensively researched is called *physical simulation*. This approach uses small robots and scaled-down versions of machine tools (minimachines) to emulate real world systems. The machines employ essentially the same minicomputers and software as the full-scale systems. See Figure 44-6. In this way, the development of the software needed to integrate the machines and design of the cell can be done prior to the installation of the full-scale system on the shop floor.

Minimachine tool laboratories are ideal for providing hands-on instruction for students in manufacturing systems. Unmanned cells and FMSs can be simulated in the laboratory at quite reasonable cost. Generally speaking, the industrial robots and full-size machines are expensive and may be considered dangerous for student usage. *Instructional robots* make it possible for education systems and small businesses to gain hands-on experience with this technology. These robots cost but a fraction of the industrial versions but use essentially the same electronic controls with stepping motors or low-pressure hydraulics. The micro-

TABLE 44.1   How to Form Cellular Manufacturing Systems

Integrated manufacturing production systems (IMPS) are based on a linked cellular manufacturing system design. Knowing how to design the cells to be flexible is the key to successful manufacturing.

I. Make tacit judgments based on axiomatic design principles.
- A. Minimize function requirements (flexibility is chief criterion).
- B. Simplify the design of system.
- C. Minimize the design information in the system.
- D. Decouple those elements that are functionally coupled.

II. Use group technology methodology.
- A. Production flow analysis finds families and defines cells.
- B. Coding/Classification is more complete and expensive.
- C. Other GT methods, including "eyeball" or tacit judgments.
   1. Find the key machine, often a machining center, and declare all parts going to this machine a family. Move machines needed to complete all parts in family around key machine.
   2. Build cell around a common set of components such as gears, splines, spindles, rotors, hubs, shafts, etc.
   3. Build cell around common set of processes, e.g., drill, bore, ream, keyseat, chamfer holes.
   4. Build cell around set of parts that eliminates the longest (most time-consuming) element in setups between parts being made in the cell.

III. Simulation
- A. Digital simulation of the system.
- B. Physical simulation of the system.
- C. Object-oriented and graphical simulation.

IV. Pick a product or products.
   Design a linked-cell manufacturing system beginning with the final assembly line (convert FA to mixed model) and move backward through subassembly to component parts and suppliers.

Cells are linked together using Kanban or directly.

---

processors in these machines use a small personal computer for program control. The trade-off here is one of scaling, as these robots will have much lower speed of response, lower weight-carrying capacity, poorer process capability, and poor reliability than their industrial counterparts but may cost only one tenth of the industrial machine.

Group Technology.   *Group technology* offers a systems solution to the re-organization of the functional system, restructuring the job shop into cellular manufacturing systems. These conversions represent systems-level changes, which will create the potential for tremendous savings; but, because of the magnitude of the changes, careful planning and full cooperation from everyone involved are absolutely required.

The application of this concept to a manufacturing facility results in the grouping of units or components into families wherein the components have similar

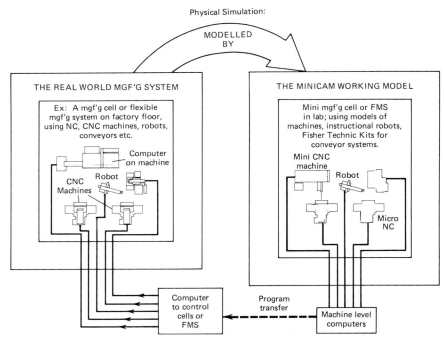

FIGURE 44-6   Schematic defining physical simulation, a new technique to study manufacturing systems, using scaled-down but fully functional models.

design or manufacturing sequences. Machines are then collected into groups or cells (machine cells) to process the family. See Figure 44-7. By grouping similar components into families of parts, a group or set of processes can be collected together to make a family. This is a cell. Refer back to Figure 44-2. The order of machines in the cell now defines the manufacturing sequence.

Changing the manufacturing system will necessitate restructuring the *production system* and all its functions. In shifting from one type of system to another, the change will affect product design, tool design and engineering, production planning (scheduling) and control, inventories and their control, purchasing, quality control and inspection, and of course the production worker, the foreman, the supervisors, the middle managers, and so on, right up to top management. Such a conversion cannot take place overnight and must be viewed as a *long-term transformation* from one type of *manufacturing production system* to another.

The entire shop will not be able to convert into families immediately. Therefore, the total collection of manufacturing systems will be a mix, evolving toward a perfect linked-cell system over time. This will create scheduling problems, as in-process times for components made by cells will be vastly different from those made under traditional job shop conditions. However, as the volume in the functional area is decreased, the total system will become more efficient.

Finding families of parts is one of the first steps in converting the functional system to a cell. Refer again to Table 44-1.

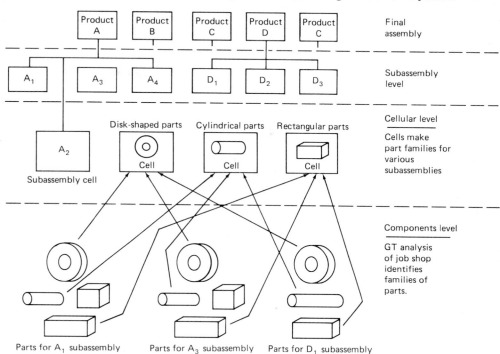

FIGURE 44-7 Short course on cell formation by group technology.

Judgment methods using axiomatic design principles are, of course, the easiest and least expensive, but also the least comprehensive. "Eyeball" techniques clearly work for restaurants, but not in large job shops where the number of components may approach 10,000 and the number of machines 300 to 500.

Product flow analysis (PFA) uses the information available on route sheets or cards (see Figure 43-8 for an example). The idea is to sort through all the components and group them by a matrix analysis, using product-routing information. See Figure 44-8. This method is more analytical than tacit judgment, but not as comprehensive as coding/classification. PFA is a valuable tool in the systems reorganization problem. For example, it can be used as an "up-front" analysis, a sort of "before the fact" analysis which will yield some cost/benefit information. Decision makers would have some information on what the company could expect in terms of the percentage of their product that could be made by cellular methods, what would be a good "first cell" to undertake, what coding/classification system would work best for them, how much money they might have to invest in new equipment, and so forth.

In short, PFA can greatly reduce the uncertainty in making the decision on reorganization. As part of this technique, an analysis of the flow of material in the entire factory is performed, laying the groundwork for the new linked-cell layout of the entire plant.

Job Number	Machine Code Letter									
	A	B	C	D	E	F	G	H	I	J
1								X		
2		X	X							
3				X						
4						X	X			
5	X	X	X							
6									X	X
7	X		X							
8						X		X		
9									X	X
10				X	X					
11	X	X	X					X		
12						X	X			
13								X		
14				X	X					
15									X	X
16		X				X	X	X		
17										X
18	X	X								
19						X	X	X		
20					X					

Job Number	Machine Code Letter									
	A	B	C	D	E	F	G	H	I	J
7	X		X							
11	X	X	X				X (exception)			
2		X	X							
5	X	X	X							
18	X	X								
14				X	X					
3				X						
10				X	X					
20					X					
12						X	X			
4							X	X		
19						X	X	X		
16						X	X	X		
8			X (exception)			X		X		
1								X		
9									X	X
13									X	
6									X	X
15									X	X
17										X

Cell will have 3 machines F, G, H, for manufacture of 6 jobs.

FIGURE 44-8 Schematic to explain production flow analysis, a technique to find families of parts based on the production routing.

Many companies converting to a cellular system have used a *coding/classification* method. There are design codes, manufacturing codes, and codes that cover both design and manufacturing.

Classification sorts items into classes or families based on their similarities. It uses a code to accomplish this goal. Coding is the assignment of symbols (letters or numbers or both) to specific component elements based on differences in shape, function, material, size, processes, and so on.

No attempt to review coding/classification (C/C) methods will be made here. C/C systems exist in bountiful number in published literature and from consulting firms.

Whatever C/C system is selected, it should be tailored to the particular company and should be as simple as possible so that everyone understands it. It is not necessary that old part numbers be discarded, but every component will have to be coded prior to the next step in the program, finding families of parts.

This coding procedure will be costly and time-consuming, but most companies that choose this conversion understand the necessity of performing this analysis.

The families of parts will not be all the same with regard to their material flow and therefore will require different designs (layouts). In some families, every part will go to every machine in exactly the same sequence, and no machine will be skipped and no back flow will be allowed. This is, of course, the purest form of a cellular system. Other families may require that some components skip some machines and that some machines be duplicated. However, back flow is still not allowed.

The formation of families of parts leads to the design of cells, but cell design is by no means automatic. It is the critical step in the reorganization and must be carefully planned. Many companies begin with a pilot cell so that everyone can see how cells function. The company should proceed with developing manned cells, not waiting until all the parts have been coded. Simply select a product or group of products that seems most logical. Only in this way will everyone learn how cells operate and how to reduce setup time on each machine. Machines will not be utilized 100%. Machine utilization rate usually improves but may not be what it was in the functional system. *The objective in manned cellular manufacturing is fully to utilize the people,* enlarging and enriching their jobs by allowing them to become multifunctional. In fact, one of the inherent results of the conversion to a cellular system is that the worker becomes multifunctional. That is, he learns to operate many machines and/or perform many duties or tasks. In unmanned cells, utilization of the equipment is more important because the most flexible element in the cell has been removed, that is, the worker, and has been ''replaced'' by a robot.

The manned cellular system provides the worker with a natural environment for job enlargement. Greater job involvement enhances job enrichment possibilities and clearly provides an ideal arrangement for improving quality. Part quality can be checked between each step in the process.

The similarity in shape and processes needed in the family of parts allows setup time to be reduced or even eliminated. Being able to do the setup in less than 10 minutes is called *single-minute exchange of dies* (SMED).

In *robotic cells,* the microcomputers of the CNC machine tools and a robot are networked together with a cell host computer. It is difficult if not impossible to conceive of this kind of arrangement without resorting to some method that collects the work into compatible families. All the machines in the cell are programmable, and therefore this kind of automation is very flexible. See Figure 44-9 for an example of a robotic cell.

While it is an easy task to draw the boxes and connecting lines in Figure 44-9, it is quite a different matter to realize all of the mechanical, electrical, and computer-engineering interfaces required to arrive at a fully unmanned, flexible, autonomous cell. In the real manufacturing environment, things are not as prescribed. The incoming work materials vary in geometry and specifications. The end effectors of the robot may not be able to accommodate such changes or take note of random variations in the presentation of the components. Parts

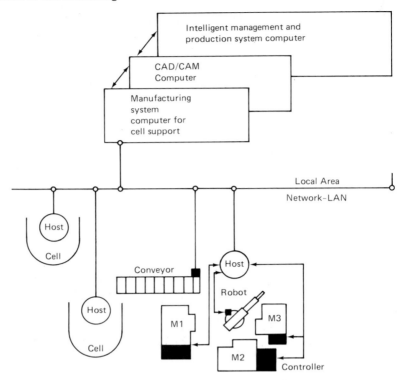

FIGURE 44-9 Manned cells evolve into unmanned cells. The cell host computer is linked directly to higher-level computers in the production system.

of the system will break, cutting tools will wear, and the quality control requirements will undoubtedly call for measuring more than one or two diameters.

A typical cell might have three different parts, with each part having six different sizes. To accommodate this family of 18 different parts may require that the process capability of robots be improved, that flexible grippers and workholders be designed, that controllers for existing machines be modified to allow supervision by a cell host computer and that reliable cell-control software be developed.

Elimination of Setup

The IMPS approach to productivity demands that small lots be run in manufacturing. This is impossible to do if machine setups take hours to accomplish. The economic order quantity (EOQ) formula has been used in the United States to determine what quantity should run to cost-justify a long and costly setup time. The EOQ was a faulty suboptimal approach. Instead of accepting setup times as a large fixed number, we should have reduced setup times. This results in reduced lot sizes.

Successful setup reduction is easily achieved when approached from a methods engineering perspective. Much of the initial work in this area has been done by Shigeo Shingo and Taiichi Ohno. Large reductions can be achieved by ap-

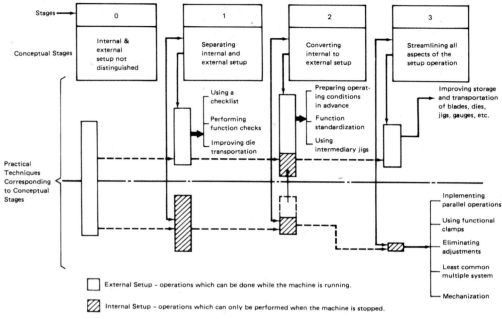

Conceptual stages and practical techniques of the SMED system
for the rapid exchange of tooling and dies, after Shingo

FIGURE 44-10 Conceptual stages of the SMED system for the rapid exchange of tooling and dies.

plying time and motion studies and Shingo's *SMED* rules for rapid exchange of dies. See Figure 44-10. Setup time reduction occurs in four stages. The initial stage is to determine what currently is being done in the setup operation. The key first step is to separate all setup activities into two categories, *internal* and *external*. Internal elements can be done only when the machine is not running. External elements can be done while the machine is running. This elementary division will usually shorten the lead time considerably. Stages 2 and 3 focus on reducing the internal time. The key here is for workers and setup people to learn how to reduce setup times and apply the simple principles and techniques. If a company must wait for the setup reduction team to examine every process, an IMPS will never be achieved.

In the last stages of SMED, it may be necessary to invest capital to drive the setup times below one minute. Automatic positioning of dies, bolster plates on rollers, intermediate jigs, and duplicate work holders represent the typical kinds of hardware needed. The result is that two-hour setup times can be reduced to under five minutes in relatively short order.

The initial objective here is to reduce the setup time until it is equal or less than the cycle time (1 to 2 minutes). This will permit a significant initial reduction in lot size. After this, setup time reductions will result in further reductions in the lot size. The next goal is to get the setup time down to less than the time needed to load, unload, inspect, deburr, and so forth at a machine. As shown in Table 44.2, when numerous processes are involved in cellular man-

TABLE 44.2   Example of Change from Part A to B for Four-Process Cell

Process	No. 1	No. 2	No. 3	No. 4
	A	A	A	A
	A	A	A	A
Setup change no. 1	Setup change	A	A	A
Setup change no. 2	B	Setup change	A	A
Setup change no. 3	B	B	Setup change	A
Setup change no. 4	B	B	B	Setup change
	B	B	B	B
	B	B	B	B

ufacturing of a family of parts, sequential setup changes are utilized. At the outset, the setup times may be long compared to the machining time (the run times). The setups are done sequentially. After setup, defect-free products should be made right from the start. The first part will be good. Ultimately, the ideal condition would be to eliminate setup between parts.

Figure 44-11 shows the drilling machine and the vertical milling machine from the cell described back in Chapter 42 in Figure 42-6. This cell was designed for a family of four parts. In order to reduce the setup time to a few seconds, the machines were modified to eliminate setup as far as possible. The total cost to modify the two machines was around $300. The table on the vertical mill was equipped with a digital readout device so that the four starting locations of the four machining operations was always the same. The fixtures are never removed from the oversize table. The first part out of these fixtures was always a good part.

Other setup reduction techniques include using group jigs (one jig to accommodate different parts, using adapters), training operators in rapid setup techniques, practicing rapid setups, and the intermediate jig concept.

The *intermediate jig* is like the cassette for a VCR. The cassette is quickly loaded and locked in place. The tape inside every cassette can be different. To the machine (the tape player), every different fixture (tape) that is placed in the intermediate jig (cassette) looks the same. To the cutting tool (the tape head), every workholder looks different. Designing workholders so that they appear the same to the machine tool usually requires one to construct an intermediate jig or fixture plate to which the fixture itself is attached.

In summary, the savings in setup times are used to decrease the lot size and increase the frequency at which the lot is produced. The smaller the lot, the shorter the throughput time and the lead time. This makes the use of IMPS principles feasible and makes the Kanban inventory control system practical.

Integrated Quality Control

When management and production workers trust each other, it is possible to implement an integrated quality control program of the kind found in IMPSs.

Japan was started on the road to superior quality with the visits of W. Edwards Deming to Japan in the late 1940s and early 1950s. The Japanese were desperate

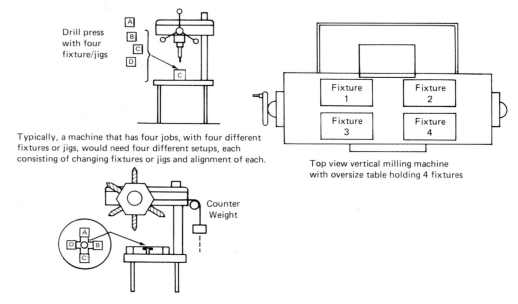

Drill press with four fixture/jigs

Typically, a machine that has four jobs, with four different fixtures or jigs, would need four different setups, each consisting of changing fixtures or jigs and alignment of each.

Top view vertical milling machine with oversize table holding 4 fixtures

Counter Weight

With redesign, the four fixtures were mounted on a turntable and are permanently aligned to the spindle when locked in position. Turret replaces single spindle. Automatic feed replaces hand wheel.

FIGURE 44-11 Machines in cell process families of parts. Reducing the variety of parts coming to the machine permits one to modify the machine so that setup times can be reduced or eliminated.

to learn about quality, and statistical quality control techniques were readily accepted. They believed that everyone in America used statistical process control techniques. However, they did something we did not do. They taught the techniques and concepts to everyone, including top management and the production workers, who even had a quality journal on the subject. At Toyota, under the leadership of Taiichi Ohno, a new idea took hold, quite different in concept from our inspection philosophy. *Inspect to prevent the defect* from occurring rather than to *find* the defect after it has occurred. Ultimately, the concept of *autonomation* evolved.

Autonomation means the autonomous control of quality and quantity. Stop everything immediately when something goes wrong. Control the quality at the source instead of using inspectors to find the problem that someone else may have created. The workers in the IMPS factory inspect each other's work. Ohno, former Vice President of Manufacturing for Toyota, was convinced that Toyota had to raise its quality to superior levels in order to penetrate the world automotive market. He wanted every individual to be personally responsible for the quality of the piece, part, or product that was produced.

The best approach is to make every worker an inspector and to give each person only one part to work on at a time, so that under no circumstance can a worker bury problems by working on alternative parts. Cells produce parts one at a time, just like assembly lines. Push buttons are installed on the assembly lines to stop the lines if anything goes wrong. If workers find defective parts, if they cannot keep up with production, if production is going too fast compared to the quantity needed for the day, or if a safety hazard is found, they are obligated to stop the lines. The problem is fixed immediately. Meanwhile, the other workers maintain their equipment, change tools, sweep the floor, or practice setups; but the line does not move until the problem is solved.

In IMPSs, the number of inspectors on the plant floor is markedly reduced. Products that fail to conform to specification are immediately uncovered because they are used immediately.

Autonomation also encompasses inspection by machines, automatically, again to prevent the defect from occurring rather than to inspect to find the defect after the part is made. Inspection by a machine instead of by a person is faster, easier, and more repeatable. This is called *in-process control inspection*.

Inspection becomes part of the production process and does not involve a separate location or person to perform it. Parts are 100% inspected by devices which either stop the process if a defect is found or correct the process before the defect can occur. The machine automatically shuts off when a problem arises. This prevents mass production of defective parts. The machine may also shut off automatically when the necessary parts have been made. This is part of inventory control and will be discussed later.

For manual work, another system for preventing defective work is called *Andon*. Andon is actually an electric light board which hangs high above the conveyor assembly lines so that everyone can see it. When a worker on the line needs help, he can turn on a yellow light. Other nearby (multifunctional) workers who have finished their jobs within the allotted cycle time move to assist workers having problems. If the problem cannot be solved within the cycle time, a red light comes on and the line automatically stops until the problem is solved. In most cases, the red lights go off within 10 seconds and the next cycle begins— a green light comes on—with all the processes beginning together. The name for this system is *Yo-i-don,* which literally means ''ready–set–go.'' Such systems are built on teamwork and a cooperative spirit among the workers, fostered by a management philosophy based on harmony and trust.

Contrast that with many operations in the United States. How long does it take us to find a problem, to convince somebody that it is a problem, to get the problem solved, and to get the fix implemented? How many defective parts are produced in the meantime? Line shutdowns in IMPSs are encouraged to protect quality. Management must have confidence in the individual worker.

Another interesting technique, with which many American companies are already familiar, is Quality Circles. The Japanese call them ''small group improvement activities'' (SGIA). A Quality Circle is a group of employees who meet on a scheduled basis (daily, weekly) to discuss production and quality problems, to try to devise a solution to those problems, and to propose the solutions to their management. The group may be led by a foreman or by a production worker. It usually includes people from a given discipline or a given production area. Quality Circles should be a natural entity within the manufacturing production system and not artificially created. They are basically a technique to identify causes of problems and to generate ideas and suggestions to solve problems on the local level. They are the "last brick" on the IQC wall and are not the "savior" that many American companies believe they are.

Integrate Preventive Maintenance

The operators are trained to perform routine machine tool maintenance. Just adding lubricants (oiling the machine), checking for wear and tear, replacing damaged nuts and bolts, routinely changing and tightening belts and bolts, and listening for telltale whines and noises which signify impending failures can do wonders for machine tool reliability. The maintenance department must instruct the workers on how to do these things and help them prepare the routine check lists for machine maintenance. The workers are also responsible for keeping their areas of the plant clean and neat. Thus, another function that is integrated into the MS is maintenance and housekeeping.

Naturally, the machines still need attention from the experts in the maintenance department, just as the airplane is taken out of service periodically for engine overhaul and maintenance. One alternative here is to switch to two eight-hour shifts separated by two four-hour time blocks for machine maintenance, tooling changes, restocking, long setups, overtime, earlytime, and so forth. The main advantage that equipment has over people is that it can decrease variability, but it must be reliable and dependable. Smaller machines are simpler and easier to maintain and therefore are more reliable. Small machines in multiple copies add to the flexibility of the system as well. The linked-cell system permits certain machines in the cells to be slowed down and therefore, like the long-distance runner, to run farther and easier without breakdown. Many observers of IMPSs come away with the feeling that the machines are "babied." In reality, they are being run at the pace needed to meet the demand.

IMPS companies try to make or modify equipment when possible in multiple copies because two teams and two sets of equipment making the same products or families of parts are more flexible in the event of a machine failure and in terms of capacity. Thus in IMPSs, *capacity is replicated in proven increments as demand grows.* Because the increment has an optimal design, this is an economic choice as well as having the security of dealing with a proven technology. Modifying existing equipment shortens the lead time. Manufacturing in multiple versions of small capacity machines retains the expertise and permits the company to keep improving and mistake-proofing the process. In contrast to this approach is the typical job shop, where a new supermachine would be purchased and installed when product demand increases. That is, we opt for a new, untried manufacturing system which takes months, even years, to debug and make reliable.

Level and Balance the MS

IMPS is the amalgamated experience of many companies which have Americanized and implemented some version of the Toyota production system. Machine layout follows the flow of processes wherein products having common or similar processes are grouped together. Quick conveyance between the processes

TABLE 44.3   Example of Product Mix on Final Assembly Line Which Determines the Cycle Time by Model

Car Mix for Line		Cycle Time by Model (Min)	Production Min by Model	Sequence (24 Cars)
Q	Model			
50	Two-door coupe	9.6	100	TDC, TDF, TDF, FDS, FDW,
100	Two-door fastback	4.8	200	TDC, TDF, TDF, FDW, FDW,
25	Four-door sedan	18.2	50	TDC, TDF, TDF, FDS, FDW,
65	Four-door wagon	7.7	130	TDC, TDF, TDF, FDW, ...
240	Cars/8 hours	480 min/240 = 2 min per car		

is provided, along with the means to reduce setup time. The basic premise of the system is to produce the kind of units needed in the quantities needed at the time needed. The system is called *just-in-time* manufacturing (JIT) and depends upon *smoothing of the manufacturing system.* In order to eliminate variation or fluctuation in quantities in feeder processes, it is necessary to eliminate fluctuation in final assembly. Small lot sizes, made possible by setup reduction between the cells, single-unit conveyance within the cells, and standardized cycle times are the keys to accomplishing a smoothed manufacturing system. Every part, sequence of assembly operations, or subassembly has the same number of specified minutes. The minimum number of workers needed to produce one unit of output in the cycle time is used.

Cycle time is determined as follows. Suppose that the forecast is for 240 cars per day and 480 production minutes are available (60 minutes × 8 hour/day). Thus cycle time = 2 minutes. Every 2 minutes a car rolls off the line. Suppose that the mix is as given in Table 44.3.

The subprocesses that feed the two-door fastback are controlled by the cycle time for this model. Every 4.8 minutes the rear deck line will produce a rear hatch for the fastback version. Every 4.8 minutes two doors are made. Every car, regardless of model type, has an engine. Each engine needs four pistons; therefore every 2.0 minutes, four pistons are produced. Parts and assemblies are produced in their minimum lot sizes and delivered to the next process, under the control of Kanban.

Two disk hubs, made in cell 143, are used in every car; the CT, then, for the cell is about one minute, actually 55 seconds. The extra five seconds per cycle provide for some spare capacity for making spare hubs for replacement parts. The cycle for a cell making parts exclusively for the wagon would be 7.7 minutes. The objective here is to make the same amount of product every day. Leveling the daily schedule takes the lumps out of the feeder process.

Link the Cells

The cells are linked to each other by the pull system of inventory control called *Kanban.* Kanban is a visual control system that is only good for IMPSs with linked cells and its namesakes (see Table 44.4) and is not good for the job shop. Figure 44-12 shows how the elements are linked together by Kanban, thus

TABLE 44.4 Other Names for Integrated Manufacturing Production Systems

- Just-in-time/Total quality control
- ZIPS (Zero inventory production system)—Omark Industries
- MAN (Material as needed)—Harley-Davidson
- MIPS (Minimum inventory production system)—Westinghouse
- Ohno System (after Toyota's Taiichi Ohno, mastermind of the system)—many companies in Japan
- Toyota production system (the "model" in reality)
- Stockless production—HP
- Kanban—Many companies in the United States and Japan

providing control over the route that the parts must take (while doing away with the route sheet), control over the amount of material flowing between any two points, and information about when the parts will be needed. In order to accomplish this, there are two kinds of Kanban: *Withdrawal* (or conveyance) *Kanban* (WLK) and *Production-Ordering Kanban* (POK). One can think of the Kanban as a loop connecting the output side of one cell with the input side of the next cell. The loop is filled with carts or containers which hold parts in specific numbers—every cart has the same number of parts. Each cart has one WLK and one POK. Here is how Kanban works.

Referring to Figure 44-13, the assembly line is being fed at two points by the part-machining cell. In Kanban, the cell makes parts for the stations only when requested by means of a Production-Ordering Kanban. That is, production is pulled, as needed, from the manufacturing cell. Suppose assembly station A uses up a container of parts. The empty container is moved back to the manu-

FIGURE 44-12 The processes are linked together by WLK Kanban.

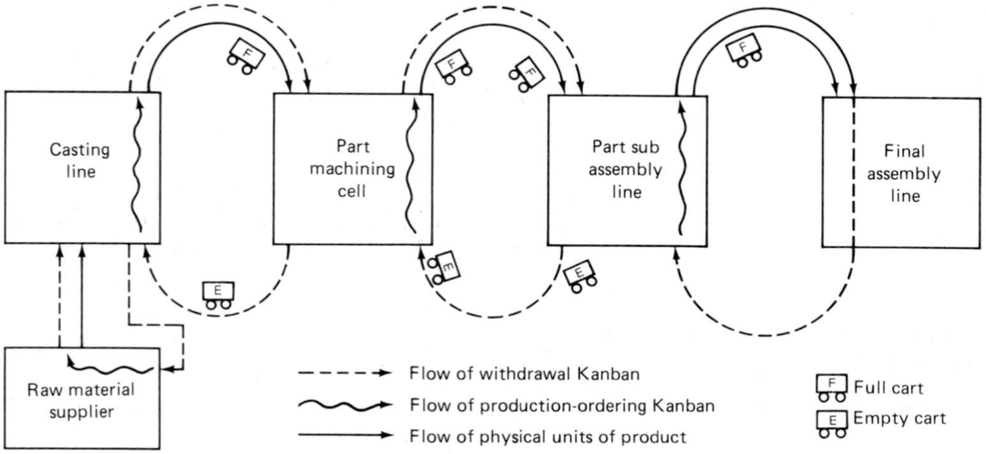

- - - → Flow of withdrawal Kanban
~~~ → Flow of production-ordering Kanban
——— → Flow of physical units of product

Full cart
Empty cart

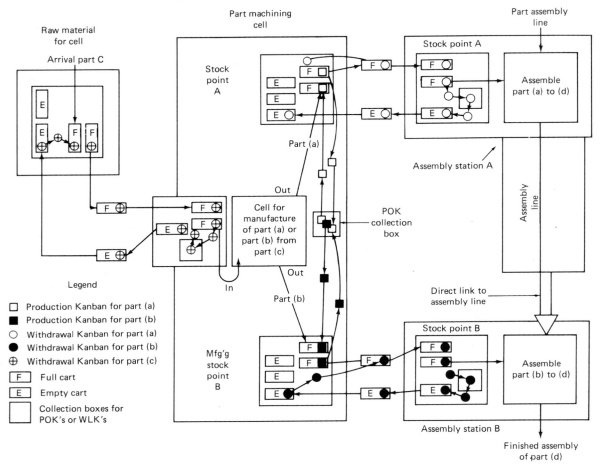

Part machining
cell

Part assembly
line

Raw material
for cell

Arrival part C

Stock
point
A

Stock point A

Assemble
part (a) to (d)

Part (a)

Assembly station A

Out

Cell for
manufacture
of part (a) or
part (b) from
part (c)

POK
collection
box

Assembly
line

In

Out

Part (b)

Direct link to
assembly line

Legend

□   Production Kanban for part (a)
■   Production Kanban for part (b)
○   Withdrawal Kanban for part (a)
●   Withdrawal Kanban for part (b)
⊕   Withdrawal Kanban for part (c)

F   Full cart
E   Empty cart
☐   Collection boxes for
     POK's or WLK's

Mfg'g
stock
point
B

Stock point B

Assemble
part (b) to (d)

Assembly station B

Finished assembly
of part (d)

FIGURE 44-13   Two-card Kanban
system—WLKs and POKs.

facturing cell by means of a WLK. The WLK is detached from the empty cart
and attached to a full cart. The POK that was attached to the full cart is placed
in the POK collection box. Thus, an order is placed to replace the cart of parts
that is being removed. The sequence in which the POKs are placed in the
collection box dictates the order in which the lots will be made. The full cart
with the WLK attached is taken to the subassembly area. When subassembly
begins to use this cart of parts, the WLK is detached and placed in a collection
box. The WLK can be used with an empty cart to withdraw more parts from
the machining cell. The number of POKs equals the number of WLKs equals
the number of carts. The carts hold a set number of parts. The exact amount of
inventory at any place in the system is known.

$$\text{Maximum inventory} = \text{Number of carts} \times \text{Number of parts in cart}$$

The same kind of loop is connecting the final assembly cell B to the part-
machining cell and all the other cells in linked-cell MS.

Work-in-process inventory has been analogized to the water in a river as shown in Figure 44-14. A high river level is equivalent to a high level of inventory in the system. The high river level covers the rocks in the riverbed. Rocks are equivalent to problems. Lower the level of the river (inventory) and the rocks (problems) are exposed. This analogy is quite accurate. The problems receive immediate attention when exposed. When all the rocks are removed, the river can run very smoothly with very little water. However, if there is no water, then the river has dried up. The notion of zero inventory is misleading. While zero defects is a proper objective, zero inventory is not possible. The idea is to minimize the necessary WIP between the cells. (Within the cell, parts are already handled one at a time, just as they are in assembly lines.)

The level of WIP between the stand-alone process, cells, subassembly, and assembly actually is controlled by the foremen in the various departments. The

Reduce Work-In-Process

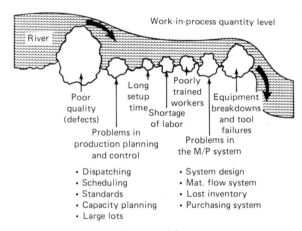

Large work-in-process.

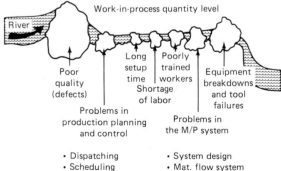

Lower level of work-in-process.

FIGURE 44-14 Lowering the work-in-process inventory causes problems in the system to be uncovered, just as lowering the level of a river causes the rocks to be uncovered.

control is integrated and performed at the point of use. Using the part A loop from Figure 44-13 as an example, observe that there are 10 carts in the loop. Each cart hold 20 parts. The maximum inventory in this area is therefore 200 parts. The foreman goes to the stock area and picks up the Kanban cards (one WLK, one POK), which puts one full cart of parts out of commission. The (maximum) inventory level is now 9×20 or 180 parts. The foreman waits until a problem appears. When it appears, he immediately restores the Kanban which restores the inventory to its previous level. The cause of the problem may or may not be identified by the restoration of the inventory but the condition is relaxed until a solution can be enacted. Once the problem is solved, the foreman repeats this procedure. If no other problems occur, he then tries to drop the inventory to $8 \times 20 = 160$ parts. This procedure is repeated daily all over the plant. After a few months, the foreman in the frame area may be down to 5 carts of 20 parts. Over the weekend the system will be restored to 10 carts between the two points, but this time each cart will hold only 10 parts. If everything works smoothly, with the reduced WIP lot size, the foreman will then remove a cart to see what happens. More than likely, he will find that some setup times should be reduced. In this way, the inventory in the linked-cell system is continually reduced, exposing problems. The problems are solved one by one.

The minimum level of inventory that can be achieved reflects the quality level, the probability of a machine breakdown, the length of the setups, the variability in the manual operations, the number of workers in the cell, parts shortages, and so on. It appears that the minimum number of carts is three, and, of course, the minimum lot size is one.

The significant point here is that inventory becomes a controllable independent variable rather than an uncontrollable variable dependent on cravings of the users of the manufacturing system for more inventory.

Vendor Relationships

The traditional purchasing department permits its vendors to make weekly/monthly/semiannual deliveries with long lead times—weeks/months are not uncommon. A large safety stock is kept just in case something goes wrong. Quantity variances are large and late and early deliveries are the norm. This situation leads to expediting.

As a hedge against vendor problems, multiple sources are developed. This may happen because one vendor cannot handle all of the company's work. The purchasing department may claim that pitting one vendor against another gives the company a competitive advantage.

IMPSs use *just-in-time purchasing*. JIT purchasing is a program of continual long-term improvement. The buyer and the vendor work together to reduce lead times, lot sizes, and inventory levels. Both companies become more competitive in the world marketplace.

In this environment, longer-term (18–24 month) flexible contracts are drawn up with 3 or 4 weeks' lead time at the outset. The buyer provides updated forecasts every month which are good for 12 months, commits to long-term quantities, and perhaps even promises to buyout any excess materials. Exact delivery is specified by mid-month for the next month. Frequent communication between the buyer and the vendor is typical. Kanban controls the material movement between the vendor and the buyer. The vendor is a remote cell. Long-range forecasting for six months to one year is utilized. As soon as the buyer sees a change, the vendor is informed; this knowledge gives the vendor better visibility instead of a limited lead-time view. The vendor thus has *build-schedule stability*—no "jerking" up and down of the build-schedule.

The buyer moves toward fewer vendors, often going to local, sole sourcing. Frequent visits are made to the vendor by the buyer, who may supply engineering aid (quality, automation, setup reduction, packaging, and the like) to help the vendor become more knowledgeable on how to deliver, on time, the right quantity of parts that require no incoming inspection. The buyer and the seller must be willing to work together to solve problems.

The advantages of single sourcing are that resources can be focused on selecting, developing, and monitoring one source instead of many. When tooling dollars are concentrated in one source, there is a savings in tooling dollars. The higher volume should lead to lower costs. The vendor is more inclined to do special things for the buyer. The buyer and the vendor learn to trust each other. The quality is more consistent and easier to control and monitor.

Forecasting Demand in IMPS.   With regard to forecasting, many people do not know that Toyota has a very strong market research program (a long-range forecasting method) to predict the long-term demand for their cars. These forecasts are highly reputed for their accuracy. Toyota performs a long-range forecast and market survey covering over 60,000 people twice a year and investigates other various trends every two months. As a result of this forecast, Toyota constructs a monthly production plan that is fixed and breaks up that plan into daily manufacturing orders. The decision on the number of cars to be produced in the month of March is finalized in early February. Precise daily schedules are planned and production is leveled (make the same amount of every product every day). This daily schedule is communicated only to final assembly, and *information* regarding the specific demand for subassembly and component parts is communicated back from final assembly through their pull system of inventory control, Kanban. In this way, the forecasted plan is converted to a production plan only for orders received and accepted.

Toyota can produce a special-order car in two days or less but the production period for the processing of the raw material to completion will exceed this time. The body, frame, and various other parts are already processed according to a fixed production plan, while painting, certain subassembly, and final assembly can take place over one or two days.

Benefits of Conversion.   The conversion to cellular manufacturing and IMPSs results in significant cost savings over a two- to three-year period. Specifically, manufacturing companies report significant reductions in raw materials and in-process inventories, setup costs, throughput times, direct labor, indirect labor, staff, overdue orders, tooling costs, quality costs, and the cost of bringing new designs on line.

However, this reorganization has a greater and immeasurable benefit. It prepares the way for automation (Steps IX and X). The progression from the functional shop to the factory with linked-cells and ultimately to robotic cells with computer control for the entire system must be accomplished in logical, economically justified steps, each building from the previous stage.

Constraints to Conversion. Aside from the failure to recognize cells as a new form of manufacturing system, a major effort on the part of a business is required to undertake a conversion to IMPSs. The constraints on implementation of an IMPS are as follows:

1. *Systems changes are inherently difficult to implement.* Changing the *entire* manufacturing production system is a huge job.
2. *Companies spend freely for product innovation but not for process innovation.* It is easier to justify new hardware for the old manufacturing system than to rearrange the old hardware into a new manufacturing system (linked-cells). However, anyone with capital can buy the newest equipment, often creating another island of automation.
3. *Fear of the unknown.* Decision making is choosing among the alternatives in the face of uncertainty. The greater the uncertainty, the more likely that the "do-nothing" alternatives will be selected. While converting to linked-cells will free up additional capacity (setup time saved) and capital (funds not tied up in inventory), such conversions will require expenditure of funds for equipment modifications, employee training (in quality, maintenance, and setup reduction), and so forth. *The long-term payback equals a high-risk situation in the minds of the decision makers.*
4. *Faulty criteria for decision making.* Decisions should be based on the ability of the company to compete (quality, reliability, delivery time, flexibility for product change or volume change) rather than on output or cost alone.
5. The conversion to IMPS represents a *real threat to middle managers.* The functional areas for which they have been responsible are being shifted and integrated into the manufacturing system. Also, the short-term life of the financially oriented middle managers in conflict with the long-term nature of the program results in resistance to change, in addition to the erosion of their functional empires.
6. *Lack of blue-collar involvement in the decision-making process of the company.* Getting the production workers involved in the decision-making process is in itself a significant change. The managers of the manufacturing system have had problems adjusting to this situation.

Clearly, education is needed to overcome these constraints. The attitudes of management and workers must change. Changing to cellular MSs and IMPSs requires an evolutionary, dynamic philosophy, but such conversions offer great potential for markedly improving quality and productivity.

Summary.   The factory with a future will require much higher levels of knowledge and more effective modes of information transfer about the quality and quantity of goods being manufactured.

The knowledge base of the factory worker must be increased to improve the productivity of the workers and, ultimately, the productivity of the company. Knowledge has a market value. Japan bought the technical knowledge it needed to build cars, electronics, machine tools, motorcycles, and even electron microscopes. The early Japanese transmission electron microscopes were virtually duplicates of the precision-made German instruments. A common thread that ties all the Japanese product areas together is high (sophisticated) technology. For example, the precision machining of the magnetic lens and the fabrication of high-voltage electronics were the key to the construction of quality electron microscopes.

Information clearly has value because people are willing to pay for it. Information also has cost because it costs something to produce the information and/or knowledge.

The factory with a future will need superior information systems and people who can analyze, program, and otherwise deal with the information on the factory floor. The unique thing about knowledge (and information) is that, unlike energy, it does not follow the laws of conservation. *Knowledge is synergistic, breeding on itself.* Thus, as the factory worker becomes better educated and more knowledgeable about how the entire manufacturing production system works, the system will rapidly improve.

Through the use of IMPS methodology, manufacturing systems become simpler and therefore easier to automate. By the next century, we will see significantly fewer workers on the plant floor. These workers will be far better educated and more productive. They will be involved in solving daily production problems, in working to improve the entire system, making decisions about how to improve their jobs, the processes, and the manufacturing system.

Many American industries are now undertaking massive educational programs to teach their employees about quality control, machine maintenance, setup reduction, and manufacturing systems. These educational steps are the key to improving productivity and are the cornerstone to IMPSs. This system is revolutionary and not just a passing fad. It represents the model for the third industrial revolution.

The attitude toward the production worker is critical. People must not be viewed as human machines. They must be respected. A job that a machine can do better than a human should not be done by humans. It is below their dignity. In the United States, we believe in the value of human worth, but our workers have had to surrender these rights when they go on the factory floor. Workers are important as people. *People are America's greatest untapped resource.* Our production people can do much more than we are now giving them the opportunity to do. Management must give workers an opportunity to do more. The management system must provide all the workers with opportunities to display maximum abilities and to make contributions to improve their job and the entire system. This must be practiced, not just preached. More employee training and education, at all levels, is absolutely necessary. The company must be prepared for large training costs in addition to large capital equipment costs. Workers need to be trained in quality control techniques and in setup reduction. All workers must be involved in the changes because there is no other way that the gains needed in quality and productivity can be achieved. Implementation is easy when the suggestions for changes needed to enhance quality or lower cost come from the workers. The step-by-step methodology that has been outlined will result in a company's becoming a factory with a future.

Review Questions

1. Why is setup reduction one of the first steps needed to convert current systems into IMPSs?
2. What is the difference between the Ford system and the Ohno system in terms of lot size?
3. In his book *Megatrends,* Naisbitt[1] argues that we are moving from a condition of forced technology toward high tech/high touch. What are high-touch aspects of cellular manufacturing?
4. What does the acronym RETAD stand for?
5. In IMPS, *integration* has a different meaning from the one you have previously heard or used. In IMPSs, what is being integrated into what?
6. How is production control, a classical function of the production planning and control department, integrated into the manufacturing system?
7. What do we mean when we say a worker can operate multiple processes? Is this the same thing as being multifunctional? Are we all really multifunctional?
8. What is involved in the integration of quality control into the manufacturing system?
9. How does the strategy of less-than-full capacity loading or scheduling of the manufacturing system assist in improving quality?
10. Suppose you are working in a cell within a linked-cell JIT manufacturing system. How do you find out what you are to make and when you are to make it? (What to make and when to make it are the primary things you need to know).
11. The effect of reducing the lot size is to take the lumps out of the manufacturing system's flow, making the flow smoother and more laminar. What other technique is used to level and balance the flow?
12. Perhaps the most powerful aspect of the linked-cell JIT system is its ability truly to control the level of the inventory. No system can operate without inventory. It is the life blood

of the system, but how is the inventory level controlled and, in fact, continuously reduced in IMPSs?
13. What is the effect on the inventory in this or any manufacturing system of poor quality? Of unreliable machines? Of long setups?
14. What determines the minimum level of the inventory in a linked-cell JIT manufacturing system?
15. What is meant in stating that a vendor is just a remotely linked cell?
16. What is *autonomation?* (Note that this is not *automation.*)
17. Many factories have undertaken massive programs to computerize and automate their job shops. In fact we have many examples wherein we have used the computer to improve a production system function. CAD is an excellent example of this. How is the CIM technique different from CIMPS?
18. How is cycle time determined for the cell?
19. How is the cycle time for the cell related to the number of workers in the cell?
20. Discuss the axiomatic design principles as they are used to design cells.
21. What is PFA?
22. What is an example of a code that you use every day which classifies you (that is, tells the reader of the code something about you)?
23. What do we mean when we say that the worker is decoupled from the machine?
24. Why is a robotic cell not as flexible as a manned cell?
25. Is a robotic cell the same thing as a FMS? Explain why or why not. (Hint: Review material in Chapter 42.)
26. What is the difference between internal and external setup?
27. Explain the intermediate jig concept.
28. What is the key to implementing a successful setup reduction program?

[1]Naisbitt, John, *Megatrends,* Warner Books, Inc., 1982.

29. How is the IQC approach different from classical quality control?
30. Why are Taguchi methods referred to as off-line QC methods?
31. What are some other names for IMPSs?
32. How is work-in-process (WIP) or in-process inventory reduced?
33. Who is Taiichi Ohno?
34. How is JIT purchasing different from other purchasing strategies? That is, what is the goal of JIT purchasing?
35. What are some of the constraints on the implementation of IMPSs?

Problem

1. Figure 44-2 shows the standard operations routine sheet for the machining of disc hubs. The machines for the cells are shown. The machine labeled "Hub cold header" cold-forms the hubs which then require three machining processes, two roughing and one finishing.
   a. What is the cycle time for these parts?
   b. What is the total machining time, exclusive of all delays, for one part?
   c. Is the cell capable of meeting the required daily quantity?
   d. Why does the worker visit BR744 every other trip or cycle through the cell?
   e. What percentage of the cycle time is spent walking?

Case Studies

This study is designed to get you to question why parts are made from a particular material and how they could be fabricated to their final shape. For one or more of the products listed below, write a brief evaluation that addresses the following questions.

Case Study 1
Material Selection

1. What is the normal use (or uses) of this product or component? What are its normal operating conditions? What would be some of the extreme operating conditions in terms of temperatures, loadings, impacts, corrosive media, etc.?
2. What are the major properties or characteristcs that the material must possess in order to function in this application?
3. What material (or materials) would you suggest and why?
4. How might you propose to fabricate this product?
5. Would the product require heat treatment? For what purpose? What kind of treatment?
6. Would this product require any surface treatment or coating? For what purpose? What would you recommend?

Products:

A. The head of a carpenter's claw hammer.
B. The lid for a top-loading washing machine.
C. Residential interior door knob.
D. A paper clip.
E. A thumb tack.
F. A pair of scissors.
G. A moderate-to-high-quality household cookpot.
H. Case for a jeweler-quality wrist watch.
I. Jet engine turbine blade to operate in the exhaust region of the engine.
J. Standard open-end wrench.
K. A socket-wrench socket to install and remove spark plugs.
L. The frame of a 10-speed bicycle.
M. Interior panels of a microwave oven.
N. Handle segments of a retractable blade utility knife with internal storage for additional blades.
O. The outer skin of an automobile muffler.
P. The interior crank handle for an automobile window.
Q. The basket section of a grocery-store shopping cart.
R. The body of a child's toy wagon.
S. Decorative handle for a kitchen cabinet.
T. Automobile engine block.
U. Front bumper for an automobile.
V. Nails for the installation of aluminum siding on homes.
W. Household dinnerware (knife, fork, and spoon).
X. The blades on a high-quality cutlery set.
Y. The basin of a kitchen sink.
Z. The base plate (with heating element) for an electric steam iron.

Case Study 2
The Aluminum
Connecting Rods

Winning Racing, Inc., is a manufacturer of high-performance automotive components, specially designed for racing applications. One highly successful product line is a series of designed connecting rods, made of forged alloy steel.

Noting the successful use of aluminum alloys in certain racing applications, Team Rabbit has requested Winning Racing to produce a special set of lightweight aluminum connecting rods, using their highly successful existing design. The rods for three engines were made on a special run, using the existing dies, and were put into the engines for testing. During dynamometer testing, however, the engines failed in under 30 minutes, and the failures were attributed to the connecting rods, although none of the rods broke. What had been overlooked that caused the trouble?

Case Study 3
A Flying Chip from
a Sledgehammer

Industrial sledgehammers are used throughout JCL Industries, most having a 15-pound head of AISI 1060 steel. To reduce tool replacement costs, the company machine shop periodically gathers hammers with heavily deformed (mushroomed) heads and grinds off the deformed segment.

A reground hammer was placed back in use. When it struck a metal plate, a chip flew off a corner of the hammer head and lodged in the eye of a worker. A lawsuit resulted.

Investigation showed that the head of a new hammer should have a bulk hardness between R_c 44 and 55. The chip, however, had a hardness of R_c 65 on the fracture surface. Inspection of other hammers showed numerous chipped regions, all on redressed hammers.

What do you suspect to be the problem? How would you alter the procedures or policies of JCL Industries to eliminate a possible recurrence, yet minimize expense?

Case Study 4
Improper Utilization
of Phase Diagrams

Harry Simon, a production engineer with Missouri Machine Co., needed to reduce the thickness of a standard strip of aluminum—3% copper alloy for use in a shimming application. Noting that the available rolling equipment had limited capacity, he decided to hot-roll the material in an effort to reduce the required forces. In selecting the heating temperature, he consulted the phase diagram in Figure CS-4 and selected 575°C (single-phase α region, 25°C below where liquid would form). While being rolled at this temperature, however, the strip fragmented, breaking into pieces, rather than deforming uniformly. What had Harry overlooked?

Case Study 5
Heat-Treated Axle Shafts

The high strength and fatigue requirements of automobile axle shafts generally require a heat-treated steel with a tempered martensite structure and a surface hardness of approximately R_c 50. The shafts are approximately 35 mm (1⅜ in.) in diameter and 1.06 meters (3½ feet) in length, and distortion or warpage must

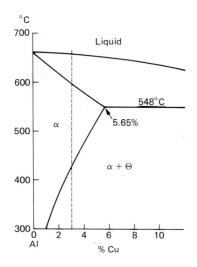

FIGURE CS-4   Aluminum-rich section of the aluminum–copper equilibrium phase diagram.

be minimized. In the 1960s, these requirements were met by oil-quenching bars of medium-carbon alloy steels, such as 4140, 8640, and 5140. Now they are frequently manufactured from the less expensive plain-carbon steels, such as 1038 and 1040. How would you propose to heat-treat these lower-hardenability metals to the desired hardness and structure and still limit distortion?

A further cost saving can be obtained if a small segment of the axle shaft surface can serve as the inner race of a bearing. This, however, requires a surface hardness of R_c 60 in this region. How would this additional requirement modify your material selection? Assuming that plain-carbon steel is still used, how might this region of high surface hardness be obtained?

Figure CS-6 shows one end of a wire cable that broke in service, permitting the large boom of a crane to fall. Enlarged views of two typical wire ends are shown in the insets.

The company that manufactured the cable claimed that their product was of high quality and had simply been loaded beyond its rated capacity. An employee of the construction firm who was operating the crane was injured by the falling boom. He maintained that no overload had been applied and filed suit against the cable manufacturer, claiming that the cable was defective at that location.

Who do you feel was correct in this case? What evidence supports your position? From your knowledge of wire and cable manufacture, why is it unlikely that a wire cable would be defective at a specific location when new? What things could have happened to the cable while it was in use that could have contributed to the failure? Is it possible that the cable could have failed while under a load that was within its rated capacity? If so, who would have been at fault and responsible for the failure?

Case Study 6
The Broken Wire Cable

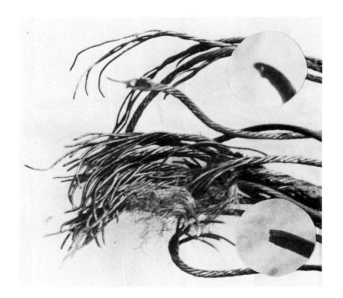

FIGURE CS-6 Wires at one end of a broken wire cable. Insets show enlarged views of two individual wire ends, typical of all wires in the cable.

Case Study 7
The Short-Lived Gear

A 10-in. (250 mm) diameter gear has been fabricated from AISI 1080 steel. The gear blank has been hot-forged, air-cooled, and then full annealed in preparation for machining. Following finish machining, the gear teeth are flame-hardened and quenched to produce a surface hardness of Rockwell C 55. After a brief period of use, the teeth of the gear begin to deform and the gear fails to mesh properly.

1. What do you feel is the probable cause of the failure? How would you attempt to prove your hypothesis?
2. If you feel that the gear had been improperly manufactured, what change or changes in the manufacturing process would you recommend? How would your product be different from the one described above?

Case Study 8
The Broken Marine Engine Bearings

The bearings on a small shipboard marine engine have been manufactured from AISI 52100 (bearing quality) steel that has been austenitized, quenched, and tempered to establish the desired final properties. Performance under normal operating conditions has been adequate for several months. After exposure to a period of subzero temperatures (a New England winter), the engine fails during its first operation. Tear-down reveals rather brittle cracks in the bearings. Micrometer measurements further reveal that the dimensions of the bearings are somewhat larger than manufacturer's specifications.

1. What do you suspect to be the cause of the failure? How would you go about proving your supposition? Who was at fault, the manufacturer of the bearings or the customer?
2. Assuming your conclusion in question 1 is correct, how would you attempt to prevent future occurrences of this type?

The Terrific Tool Company is considering an expansion of its line of conventional hand tools to provide a set of safety tools, capable of being used in areas of gas leaks and the like where the potential of explosion or fire exists. Conventional irons and steels are pyrophoric (small slivers or fragments can burn in air—that is, form sparks if dropped or impacted on a hard surface).

You are asked to specify the materials and processes to manufacture a nonsparking pipe wrench. The new product must be of the same shape and size as a conventional wrench and retain all of the characteristic properties (strength in the handle, hardness in the teeth, fracture resistance, corrosion resistance, and so on). In addition, the new safety wrench must be nonsparking (or nonpyrophoric).

You find yourself restricted to nonferrous metals, such as aluminum- and copper-based alloys. Both casting and forging appear to be viable manufacturing options. You find that copper–2% beryllium can be age-hardened to provide the strength and hardness properties necessary for the jaws of the wrench, but you find also that the cost of this material is quite high. You feel that the wrench must be made in the most economical manner possible if the new line of safety tools is to be accepted by potential customers.

Suggest some alternative manufacturing systems that could be used to produce such a wrench. What might be the advantages and disadvantages of each? Which of them would you recommend to your supervisor?

Case Study 9
Nonsparking Wrench

Lavatory wash basins (that is, bathroom sinks) have been successfully made from a variety of engineering materials, including cast iron, steel, stainless steel, ceramics, and polymers (such as melamine). Your company, Diversified Household Products, Inc., is considering a possible entrance into this market and has assigned you the tasks of (1) assessing the competition and (2) recommending the "best" approach toward producing this product.

1. For each of the above materials (or families of materials), describe the possible means of fabricating lavatory wash basins. Consider sheet metal forming, casting, molding, joining, and other types of fabrication processes.
2. For each of the materials (or families of materials), select what you feel is the best approach and list some of the pros and cons of the material, the process, the cost, and the performance of the product.
3. The basins generally require a surface that is nonporous and stain resistant, scratch resistant, corrosion resistant, and attractive (and possibly available in a variety of colors). One approach to providing these qualities is through a coating of porcelain enamel. For which of the materials above would such a coating be required to produce the desired properties? Which would require no additional surface treatment? For which might the desired surface properties be compromised?
4. If your company were to consider producing lavatory wash basins on a competitive basis, which of the alternative manufacturing systems (material and manufacturing process) would you recommend? What are its most attractive features?

Case Study 10
Fabrication of Lavatory Wash Basins

FIGURE CS-11   Diesel engine fuel
metering lever.

Case Study 11
Diesel Engine Fuel Metering Lever

The component pictured in Figure CS-11 is a fuel-metering device lever for a large diesel engine. The long dimension of the part is approximately 2½ in. and the width is ⅞ in. The large hole through the center is 0.368 in. in diameter and the two small holes are to be tapped to accept 10–32 threaded bolts. A realistic production run would be for approximately 20,000 pieces.

1. Describe in a general way, three possible means of manufacturing the desired shape. For each, discuss briefly the pros and cons of the approach.
2. If the product were to require mechanical properties of a 55,000 psi tensile strength, a surface hardness of Rockwell B 55, and an elongation of 1%, what types of materials would seem appropriate for this part?
3. Briefly discuss the suitability of the materials proposed in Question 2 to the various processes proposed in Question 1. Are there any significant limitations or incompatibilities?
4. What do you feel would be the best material-and-process solution among those proposed above? For this solution, in what form would you purchase the starting material? Would heat treatment be required to achieve the necessary final properties?

Case Study 12
Shaft and Circular Cam

The Cab-Con Corporation makes the part shown in Figure CS-12 in large quantities. (Only the essential dimensions are shown.) Currently, it is machining the part from AISI 1120 bar stock that is 76.2 mm (3 in.) in diameter and then carburizing the outer periphery of the disk cam, which must have a hardness of at least $55R_c$, although the depth of the hardness need not exceed 0.15 mm (0.006 in.). Obviously, the machining cost and the wastage of material are excessive.

The chief engineer has assigned you to devise a more economical procedure for producing this part. How would you propose to produce the part? (Assume that the volume is sufficient to justify any required new equipment, or that some operations can be subcontracted, if that would be more economical.)

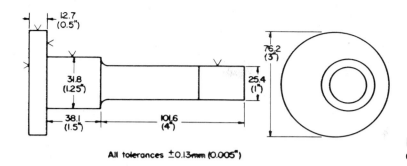

12.7
(0.5")

76.2
(3")

31.8
(1.25")

25.4
(1")

38.1
(1.5")

101.6
(4")

All tolerances ±0.13 mm (0.005")

FIGURE CS-12 Shaft with eccentric circular cam.

Foreign competition has become a serious threat to the Lunker Fishhook Company, where you are employed as a manufacturing engineer. Your company has produced fishhooks for over 35 years and manufactures a standard-type hook in a variety of sizes. The marketing staff confirms a well-known fact that fishermen are highly superstitious individuals. Your competitors are offering their hooks in a variety of colors and it is confirmed that the customers have definite but varied preferences.

You are being asked to devise a series of surface treatments that would allow the company to offer their product in a variety of colors and finishes. The hooks are fabricated from 1070 carbon steel wire, formed to precision shape (eye, barbs, bends, and point), and heat-treated by a quench and temper to a final tensile strength of 240 ksi (1650 MPa). The final product must be strong enough to resist bending and yet not be brittle, where it might break. The proposed surface treatments must provide both corrosion resistance and appearance without fouling the point or the barbs or altering the mechanical properties of the hook. If applied before heat treatment, the coating must endure that process and maintain its appearance. If applied after heat treatment, it cannot weaken or embrittle the hook.

Of the available surface modification processes, which ones might be attractive for this application? What specific types of coatings or treatments could be used to produce the desired variety of colors and finishes?

Case Study 13
Surface Treatments for Fishhooks

The propeller of a moderately large pleasure boat has been cast from a nickel-aluminum-bronze alloy (82% Cu, 9% Al, 4% Ni, 4% Fe, 1% Mn). It is 13 in. (330 mm) in diameter with three 10-pitch blades and has been designed for both fresh- and saltwater usage.

1. One of the blades has struck a rock and is badly bent. A replacement propeller is quite expensive and cannot be obtained for several weeks. The owner, therefore, wishes to have the damaged propeller repaired. Can it simply be hammered back into shape? Would you recommend any additional processing, either before or after the repair? If you would attempt a repair, how would you proceed? If not, why?
2. A propeller identical to the one above has also been damaged by an impact, but this

Case Study 14
Repairs to a Damaged Propeller

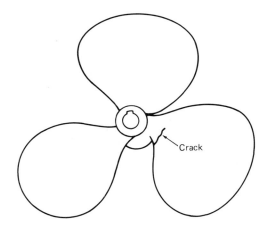

FIGURE CS-14   Ship propeller showing location of a crack at the base of one of the blades.

time the damage is in the form of a crack at the base of one of the blades, as shown in Figure CS-14. Since the crack does not penetrate the hub, a repair is suggested, using a welding or brazing technique. Would you recommend such a repair? If so, how would you suggest that the repair be made? Outline the procedure that you would recommend and the justification for your recommendations. Would there be any sacrifice in quality or performance with the repaired propeller? If you would not perform a repair, why not?

Case Study 15
Repair to a 10-Speed Bicycle

The frame of a high-quality, 10-speed bicycle has been made of cold-drawn alloy steel tubing to take advantage of the additional strength that is provided by the cold working of the metal. As the result of abuse, the frame fractured and a repair was made by conventional arc welding. The repair seemed to be of good quality, but shortly thereafter the frame again broke. This time the break was adjacent to the repair weld and the characteristics of the break were different. While the first fracture was somewhat brittle in nature, the second appears to be somewhat ductile, with considerable evidence of metal flow prior to fracture.

1. What do you think was the probable cause of the second fracture? Was the weld in any way defective? Was the second failure the result of the welding repair? Explain.
2. Was there a better means of repairing the original fracture? What would your recommendation have been?

Case Study 16
The 6-Second Pinion Gear

In 1970, the manufacturers of a new type of gear-making machine invite you, an independent consultant, to attend a meeting at company headquarters to examine a problem they are having with their new gear-cutting process. They claim that their process generates finished helical pinion gears (the kind used in automatic transmissions in cars) ten times faster than conventional processes (see Chapter 31). In their system, a rotating pinion blank feeds between two

Case Studies **1137**

3. The linear motion of the cutting inserts, along with the rotation of the gear, generates a helical profile.

1. Two cutter heads, each studded with 90 carbide inserts, turn in opposite directions to cut teeth in a rotating blank as it passes through, and again as it returns, generating a finished pinion.

2. Tools shown together here are actually on opposite sides of the gear. The effective cutting pattern for the roughing cut appears on the left; and the pattern for the finish cut, on the return stroke of the spindle slide, is on the right.

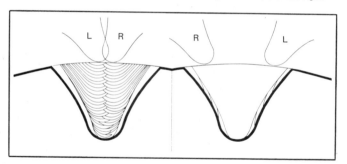

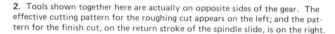

4. Machine uses cutter heads of large diameter with a series of small radially clamped carbide inserts.

FIGURE CS-16 Exhibits for the 6 second pinion.

cutter heads that turn in opposite directions, as shown in Figure CS-16(1). Each of the two cutter heads contains 90 radially mounted carbide insert tools. As shown in Figure CS-16(2), the cutting tools make a series of roughing and finishing cuts while the pinion blank rotates. As shown in Figure CS-16(3), this combination of the linear motion of the carbide inserts, along with the rotation of the gear blank, generates the helical profile. This machine was designed to produce 600 gears per hour. The carbide insert teeth are clamped in the cutter bodies, as shown in Figure CS-16(4). Pockets ground into the steel cutter body provide accurate insert location for the carbide tools.

At the meeting, the company describes the nature of their problem. They have purchased carbide tools from two vendors. These tools have very long tool life at the speeds and feeds that are to be used to machine the gears. At the meeting are representatives from the two carbide tool manufacturing companies who supplied the tools. Both provide data confirming that the tools should last 40 hours or more without appreciable flank or crater wear. They present results of turning tests, cutting the same steel to be used in the pinions with the same structure and properties (hardness). These results confirm the expected good performance. In addition, the geometry of the inserts provided by both companies is identical.

The gear machine company has conducted tests on the machine using HSS (high-speed-steel) tools and aluminum gear blanks. The machine ran continuously for over 100 hours, producing over 600 gears per hour with no problems. (They now have a room full of aluminum pinion gears.) When they switched to steel pinion gears and carbide tools, however, they experienced rapid tool failure and could not machine more than six pinions before they had to stop the machine, even though the same cutting fluids were used on the machine as had been employed during the favorable tool life tests. The machine operator observed that "the thing spit out carbide teeth like a machine gun."

Also present at the meeting was another consultant who had performed a complete vibration analysis of the machine. His results, however, could neither confirm nor disprove that chatter was the cause of the rapid tool failure.

You request that the following test be run. The machine is to be setup to cut the gears at a rate of *one per minute* instead of ten per minute. This requires some gear changes and adjustments to the machine. Again, the machine performed perfectly for the HSS tools and aluminum gears. When the carbide cutters were inserted in place of the HSS tools, the carbides again experienced rapid tool failure. Everyone now knew the cause of the problem. Explain the real reason why this machine failed to meet the performance specifications and the cause of the rapid tool failure.

Case Study 17
Estimating the Machining Time for Turning

As the plant manufacturing engineer, you have been called into the production department to provide an expert opinion on a machining problem. Unfortunately, the only tool or instrument you have available at the time is a one-inch micrometer.

The production manager would like to know the minimum time required to machine a large forging. The 8-foot-long forging is to be turned down from an original diameter of 10 in. to a final diameter of 6 in. The forging has a BHN of 300 to 400. The turning is to be performed on a heavy-duty lathe, which is equipped with a 50 HP motor and a continuously variable speed drive on the spindle. The work will be held between centers, and the overall efficiency of the lathe has been determined to be 75%.

The forging (or log) is made from medium-carbon, 4345 alloy steel. The steel manufacturer, some basic experimentation, and established knowledge of the product and its manufacture have provided the following information:

a. A tool-life equation developed for the most suitable type of tool material at a feed of 0.020 ipr and a rake angle of $\alpha = 10$ degrees. The equation VT^n

= C generally fits the data, with V = cutting speed and T = the time in minutes to tool failure. Two test cuts were run, one at V = 60 sfpm where T = 100 minutes and another at V = 85 sfpm, where T = 10 minutes.

b. According to the vendor, the dynamic shear strength of the material is on the order of 125,000 psi.

c. For the standard feed of 0.020 ipr, the chip thickness ratio varies almost linearly between the speeds of 20 and 80 fpm, the values being 0.4 at the speed of 20 fpm and 0.6 at 80 fpm. The chip thickness values were determined by micrometer measurement in order to determine the value of r_c.

d. The machined forging (log) will be used as a roller in a newspaper press and must be precisely machined. If the log deflects during the cutting more than 0.005 in., the roll will end up barrel-shaped after final grinding and polishing.

Estimate the minimum time required to machine this forging, assuming that one finishing pass will be needed when the log has been reduced to 6 in. in diameter. The deflection due to cutting forces must be kept below 0.005 in. at the mid-log location.

You can assume that $F_c \times 0.5 = F_f$ and $F_f \times 0.5 = F_R$ and that F_R causes the deflection.

Figure CS-18 shows the design of one of four legs on a casting made by the Hardhat Company. These legs are used to attach the device to the floor. The section drawing to the right shows the typical loading to which the leg is subjected. The company is currently drilling the bolt hole and then counterboring the land. Manufacturing has experienced some difficulty in machining these legs. They report a lot of drill breakage. Quality control reports that distances between the four holes are frequently too large. Sales has recently reported that a substantial number of in-service failures have occurred with these legs. You

Case Study 18
Bolt-Down Leg
on a Casting

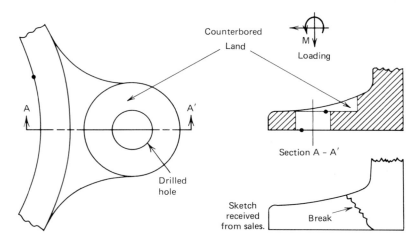

FIGURE CS-18 Schematic design of bolt-down leg

have obtained a sketch from one of the salesmen showing where the legs typically fail. The leg is manufactured from gray cast iron.

1. What machining difficulties would you expect this leg to have?
2. What do you think has caused the failures?
3. What do you recommend to solve this problem in the future in terms of materials, design, and manufacture?
4. What do you recommend to be done with the units in the field to stop the failures?

Case Study 19
Overhead Crane
Installation

In order to install an overhead crane (Figure CS-19) in one bay of an assembly plant, brackets for the rails of the crane are to be mounted on eight columns, four on each side of the bay area facing each other. The rails for the crane will span four columns. Each bracket on each column will need six holes in a circular pattern. The holes must be accurately spaced within 5 minutes of the arc of each other. The axis of the holes must be parallel and normal to the face of the columns. The center of the bolt-hole circle must be at a height of at least 20 feet from the floor, but the centers of all the eight bolt-hole circles must be on the same parallel plane so that the rails for the crane are level and parallel with each other. Four of the columns along the wall have their faces flush with the wall surface so that mechanical clamping or attachments cannot be used. The building code will permit no welding of anything to these columns.

1. How would you proceed to get the bolt holes located in the right position on the beams?
2. How would you get the hole patterns located properly with respect to each other on all eight beams?
3. List the equipment you will need.
4. Make a sketch of any special tool you recommend.
   Hints: Check Chapter 28 for drill jig designs and ask your favorite civil engineer for suggestions.

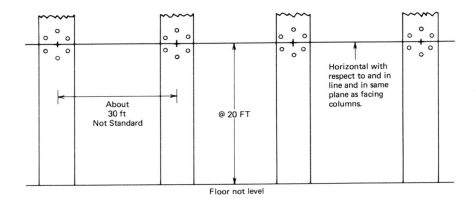

FIGURE CS-19   Hole pattern locations on four of the eight columns. The other four columns face these four.

The Quality Machine Works, which does job shop machining, has received an order to make 40 duplicate pieces, made of AISI 4140 steel, which will require one hour per piece of actual cutting time if an ordinary HSS milling cutter is used. John Young, a new machinist, says the cutting time can be reduced to not over 25 minutes per piece if the company will purchase a suitable tungsten carbide milling cutter. Hans Oldman, the foreman for the milling area, says he does not believe that John's estimate is realistic, and he is not going to spend $450 of the company's money on a carbide cutter that probably would not be used again. The machine-hour rate, including labor, is $30 per hour. John and Hans have come to you, the supervisor of the shop, for a decision on whether or not to buy the cutter, which is readily available from a local supplier.

What are the things you should consider in this situation? Who do you think is right, John or Hans?

Cast Study 20
HSS Versus Tungsten Carbide

You have received the part drawing for a typical lathe part that will require turning, facing, grooving, boring, and threading as it is machined from a casting. Unfortunately, you do not yet know what the quantity will be. To be prepared, you have developed some cost data for the manufacture of the part by four different lathe processes (see Table CS-21). Complete the table by determining the run cost per batch, the cost per unit at the various quantities, and the total cost per batch.

Answer the following questions regarding this situation:

1. Of the four costs listed for each process, which costs are fixed and which are variable?
2. For which of these costs would you have to estimate the machining time per piece and the cycle time per piece, including the time to change parts and setups?

Case Study 21
Break-Even Point Analysis of a Lathe Part

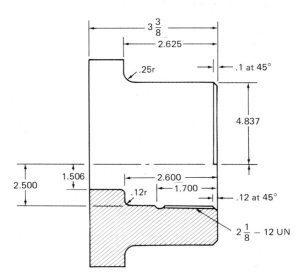

FIGURE CS-21   Drawing of part to be produced on a lathe.

TABLE CS-21

	Make Quantity				
	10,000 Units	1,000 Units	100 Units	10 Units	1 Unit
Cost to Produce on Six-Spindle Automatic					
Total Cost of Batch	—	—	—	—	—
Engineering 2.5 hrs at $40/hr	50.00	50.00	50.00	50.00	50.00
Tooling (Cutting tools and workholder)	600.00	600.00	600.00	600.00	600.00
Setup 8 hrs at $15.00/hr	120.00	120.00	120.00	120.00	120.00
Run cost per batch: 50¢ per piece	—	—	—	—	—
Cost each	—	—	—	—	—
Cost to produce on turret lathe					
Total cost of batch	—	—	—	—	—
Engineering 2 hrs at $20/hr	40.00	40.00	40.00	40.00	40.00
Tooling	150.00	150.00	150.00	150.00	150.00
Setup 4 hrs at $20.00/hr	48.00	48.00	48.00	48.00	48.00
Run cost per batch: $8.00 per piece	—	—	—	—	—
Cost each	—	—	—	—	—
Cost to produce on engine lathe					
Total cost of batch	—	—	—	—	—
Engineering 1 hr at $20/hr	20.00	20.00	20.00	20.00	20.00
Tooling	—	—	—	—	—
Setup 2 hrs at $12.00/hr	24.00	24.00	24.00	24.00	24.00
Run cost per batch: $12 per piece	—	—	—	—	—
Cost each	—	—	—	—	—
Cost to produce on NC lathe					
Total cost of batch	—	—	—	—	—
Engineering and programming	150.00	150.00	150.00	150.00	150.00
Tooling	100.00	100.00	100.00	100.00	100.00
Setup 1 hr at $20.00/hr	20.00	20.00	20.00	20.00	20.00
Run cost per batch: $2 per piece	—	—	—	—	—
Cost each	—	—	—	—	—

3. How would you go about estimating this time, and what time elements might be included in the cycle time in addition to the machining time?
4. How would you use this estimate of time in the cost table? Show the calculation.
5. Make a plot of cost in dollars versus quantity, with all four methods on one plot.
6. Make a plot of cost per unit versus quantity, again with all four methods on one plot. Find the break-even quantities. *Hint:* Did you plot the data on log paper?
7. Discuss these plots and the break-even quantities that you found in part 5 versus part 6.
8. When would you use the turret lathe? When would you use the turret lathe if you had no NC lathe?

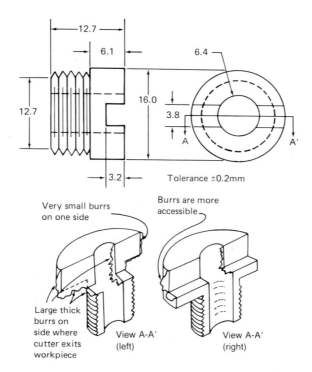

FIGURE CS-22   The Yo-Gi Collar

Case Study 22
Burrs on Yo-Gi's Collar

The Yo-Gi Company requires 25,000 collars as shown in Figure CS-22. They are to be made from 18-8 (AISI 304) stainless steel by machining on a screw machine using transfer and slotting attachments. The slotting process is creating large, thick burrs on the surface where the cutter exits the workpiece, as shown in sections A–A′ (left). The part designer has recommended a redesigned part to eliminate the undercut and make the burrs more accessible; see view A–A′ (right). The burrs on the internal surfaces will still remain, however, and are still inaccessible.

1. What are the methods that are commonly used in industry to remove burrs?
2. Outline the sequence of steps needed to produce this part on a screw machine from ⅝-in. stainless steel bar that would serve to eliminate the burrs. It is helpful to show the part progression.
3. Determine two other practical ways of making these collars. Estimate the cost of these alternative methods versus the screw machine method.

Case Study 23
The Vented Cap Screws

The machine shop at the Hy-Fly Space Laboratories received an order from their engineering department for 10 of the vented cap screws shown in Figure CS-23. The machine shop foreman returned the order, stating that there was no practical way to make these other than by hand. The design engineer insisted that the

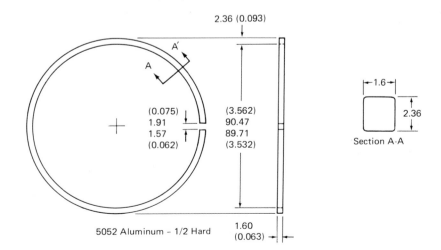

FIGURE CS-23   Original design of the vented cap screw.

vent slot, as designed, was essential to assure no pressure buildup around the threads of the screw body in the intended application, and that he knew they could be made because he had seen such cap screws.

To accommodate the shop, however, the design engineer redesigned the part, now specifying that a vent hole be drilled parallel to the axis of the cap at location D. The small hole, with cross-sectional area equal to the previous 0.062 in. × 0.031 in. slot, would be drilled ½ in. deep through the cap to connect to the longitudinal slot. This would eliminate the right-angle slot geometry.

1. What do you think of this solution? *Hint:* How big is the hole size compared to its depth?
2. Can you suggest an alternative redesign that would make a slot practical to manufacture? Show a sketch of your redesigned cap screw.

Case Study 24
Aluminum Retainer Rings

The Owlco Corporation has to make 5000 retainer rings, as shown in Figure CS-24. It is essential that the surfaces be smooth with no sharp corners on the circumferential edges. Determine the most economical method for manufacturing these rings.

FIGURE CS-24   Aluminum snap-in retainer ring.

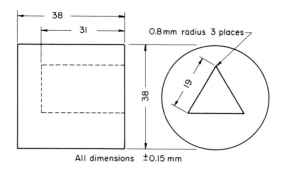

All dimensions ±0.15 mm

FIGURE CS-25 Component containing dead-end triangular hole.

Case Study 25
The Component with the Triangular Hole

The Johnstone Company estimates that its annual requirements for the socket component shown in Figure CS-25 will be at least 50,000 units and that this volume will continue for at least 5 years. Consequently, it wants to consider all practicable methods for making the component and has assigned you the task of determining these methods and of recommending which should be explored in detail to determine the most effective and economical process. The specifications call for a lightweight metal, such as an aluminum alloy, that must have a tensile strength of at least 138 Mpa and an elongation of at least 3%.

Determine at least five practicable methods for producing the part. Suggest which two appear most likely to be economical and thus should be investigated fully. (Give the reasons for your selections.)

Case Study 26
Portable Failure Analysis Kit

You are a member of a corporate failure analysis group, working out of the home office of Big-One Petroleum Corporation, a large petroleum company with production sites throughout North America. These sites generally contain such equipment as pumps, valves, pipelines, and storage tanks, which occasionally experience failure under both normal and abnormal production conditions. Much of this equipment is large and cumbersome; therefore when a failure occurs, you must frequently travel to the site of the failure to begin your investigation. Your primary objective is to collect information and acquire specimens or samples that can be taken back to your base laboratory for further investigation and study.

Your present assignment is to design and equip a portable failure analysis kit for on-site investigation. This would take the form of a suitcase that could be stored in your laboratory. When a problem occurs, you could simply pick up the kit and be on the next flight to that location.

1. Your container is restricted to the size of a typical suitcase and must be carried by a single individual (suggested weight of less than 50 pounds). Moreover, you are expected to purchase and equip this case for less than $600. With these constraints, develop a list of the items that you would include in your case. For each item, discuss briefly a description of the item and its intended use or uses; justify the cost (if over $25), the weight (if over 5 pounds), and the size (if it occupies over 20% of the suitcase).

2. If additional funds were subsequently provided to upgrade your suitcase, but the size and weight restrictions were maintained, what additional or upgraded items would you want to include? Discuss each in the manner used in question 1.

Case Study 27
Fire Extinguisher Pressure Gage

As a materials engineer for the Fyre-Pro Extinguisher Company, you have recently been made aware of several potentially hazardous failures that have occurred in your company's products. Bourdon tube pressure gages (Figure CS-27) are used to monitor the internal pressure of your sodium bicarbonate (dry chemical) fire extinguishers. In several extinguishers, a longitudinal crack has formed along the axis of the bourdon tube, a curved tube of elliptical cross section that has been fabricated from phosphor bronze tubing. These cracks are particularly disturbing for several reasons. First, they allow the fire extinguisher to lose pressure and become inoperable. More significantly, however, the cracks allow the tube to deflect elastically in such a manner that the gage still indicates high internal pressure. Thus, while the extinguisher has actually lost all internal pressure and is useless, the reading on the gage would cause the owner to believe that he still had an operable fire-fighting device.

Your task is to determine the cause of these failures and suggest appropriate corrective measures.

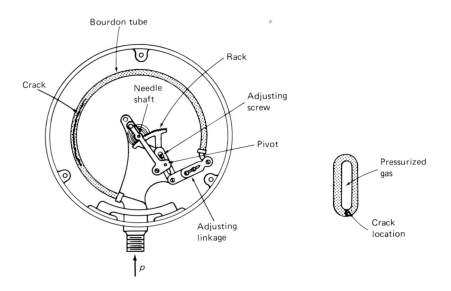

FIGURE CS-27 Schematic of a bourdon-tube pressure gage indicating location of the crack and cross section of the bourdon-tube.

(a) Schematic of a bourdon-tube pressure gage.

(b) Cross section of bourdon tube

1. What additional information would you like to have regarding the failed components, their fabrication history, and their service history? Why?
2. What might be some of the possible causes of these failures? What type of evidence would support each possibility? What types of additional tests or investigations might you propose?
3. Could these failures have occurred in "normal" use, or is it likely that some form of negligence, abuse, or misuse was involved?
4. What possible "corrective" or "preventative" measures might you suggest to prevent a recurrence?

The impeller of a large industrial disposal unit is made of a circular segment of hot-rolled steel plate. Pieces of tungsten carbide are brazed into recesses in the plate with a copper–base brazing alloy as shown in Figure CS-28. In service, the impeller rotates at approximately 700 rpm, shredding the refuse (paper, bottles, chemicals, etc.) from a pharmaceutical manufacturer, to a pulp, eliminating all identification and permitting discharge into a refuse dump. After 5 weeks of service, the impeller must be removed. Several pieces of tungsten carbide have broken free and are chewing up the unit. Large craters have formed in the steel plate adjacent to the carbide inserts.

What do you suspect was the cause of the problem? How would you alter the design, materials, or fabrication to prevent its recurrence?

Case Study 28
The Industrial Disposal Impeller

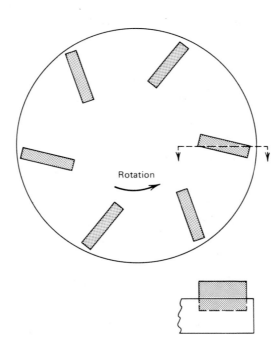

Rotation

FIGURE CS-28   Schematic of impeller design.

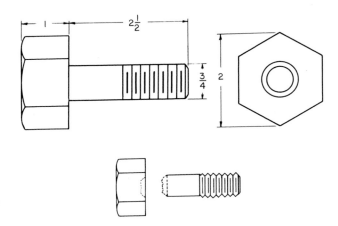

FIGURE CS-29 Bolt design and location and nature of fracture.

Case Study 29
The Bronze Bolt Mystery

You are employed by the Mountainous Irrigation Company, which had 10,000 special "oversized-head" bolts made from phosphor bronze (10% tin) for use in its various pumping plants. See Figure CS-29 for the bolt design. Within a few weeks after a number of these bolts were installed, the heads broke off several of them, the fracture being along the dashed line shown in the figure. The broken bolts were replaced, with special precautions being taken to assure that they were not overloaded. Again, after only a few days, several of these replacement bolts broke in the same manner. You have been assigned the task of determining the cause of the failures. Upon examining several unused bolts, you discover that many of them have fine cracks at the intersection of the head and body. Upon checking with the manufacturer, you learn that the bolts were made by an ordinary upset-forging bolt-manufacturing process.

1. How could you determine the cause of the cracks? What procedures would you recommend?
2. What do you suspect is the cause?
3. How could the difficulty have been avoided?

Case Study 30
The Undergroud Steam Line

An underground steam line for a military base in Alaska, approximately 1.6 km (1 mile) in length, utilized the units depicted in Figure CS-30, each unit being 9.14 meters (30 feet) in length. As shown, the steam line for each unit was enclosed in a cylindrical conduit made of sheet steel, and weighed approximately 356 kg (785 pounds). The conduit was to serve as the return drain line and to provide insulation. Each length of steam line was supported within the conduit by means of three U-shaped legs, made by cold-bending hot-rolled steel bar stock. See section A–A′ for cross section. These legs were welded to the pipe about 2 meters from each end.

The units were fabricated in California and were transported to Alaska by a sequence of truck, barge, and railroad. Because of an early winter, only one-

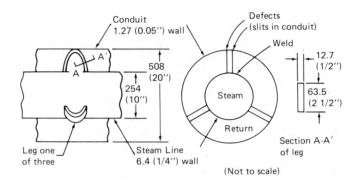

FIGURE CS-30 Schematic diagram showing method of supporting steam line in drainage conduit for an underground steam line.

half of the line was installed the first summer. Before work was resumed the following spring, someone decided to test the line and found that there were numerous holes in the conduit. The holes (slits in casing) were located where the edges of the U-shaped supporting legs contacted the outside conduit (casing) on the tops of the casing. As the result of these holes, the project was delayed and a cost litigation ensued.

1. Why do you think these holes occurred?
2. Was 'it a design error? A fabrication error? A service error? A service environment error? What error(s) caused the failures? When did the defects actually develop?
3. What action would you recommend to avoid this problem in the future?
4. What should be done about the part of the line already installed?

Mary M. is suing the SnoCat Snowmobile Company and ACME Components for $750,000 over her husband's death. He was killed while racing his snowmobile through the woods in the upper peninsula of Michigan. Her lawyer claims that he was killed because a tierod broke, causing him to lose control and crash into a tree, breaking his neck. While it was impossible to determine whether the tierod broke before the crash or as a result of the crash, the following evidence has been put forth. The tierod was originally designed and made entirely out of low carbon steel (heat-treated by case hardening) in three pieces as shown in Figure CS-31. These tierods were subcontracted by SnoCat to ACME Components. ACME Components changed the material of the tierod bolts from steel to a heat-treated aluminum having the same UTS as the steel. They did this because aluminum rods were easier to thread-roll rather than form by a

Case Study 31
The Case of the Snowmobile Accident

FIGURE CS-31   Assembled tie rod for snowmobile.

thread-cutting operation. It was further found that threads on one of the tierod bolts were not as completely formed as they should have been. The sleeve of the tierod in question was split open and one of the tierod bolts was bent.

Mary's lawyer claimed that the tierod was not assembled properly. He claimed that one rod was screwed into the sleeve too far and the other not far enough, thereby giving it insufficient thread engagement. ACME testified that these tierods are hand-assembled and only checked for overall length and that such a misassembly was possible. Mary's lawyer stated that the failure was due to a combination of material change, manufacturing error, and bad assembly, all combining to result in a failure of the tierod.

A design engineer for SnoCat testified that the tierods were ''way overdesigned'' and would not fail even with slightly small threads or misassembly. SnoCat's lawyer then claimed that the accident was caused by driver failure and that the tierod broke upon impact of the snowmobile with the tree. One of the men racing with Mary's husband claimed that her huband's snowmobile had veered sharply just before he crashed but under cross examination, he admitted that they had all been drinking that night because it was so cold (he guessed 20 below). Since this accident had taken place over five years ago, he could not remember how much they had had to drink.

You are a member of the jury and have now been sequestered to decide if SnoCat is guilty of negligence resulting in death. The rest of the jury, knowing you are an engineer, has asked for your opinion. What do you think? Who is really to blame for this accident? What actually caused the accident?

Appendix
Selected References
for Additional Study

Handbooks and General References

1. *Metals Handbook,* 9th ed., ASM, Metals Park, Ohio.
    Vol. 1, Properties and Selection: Irons and Steels (1978)
    Vol. 2, Properties and Selection: Nonferrous Alloys and Pure Metals (1979)
    Vol. 3, Properties and Selection: Tool Materials and Special-Purpose Metals (1980)
    Vol. 4, Heat Treating, (1981)
    Vol. 5, Surface Cleaning, Finishing, and Coating, (1982)
    Vol. 6, Welding, Brazing, and Soldering, (1983)
    Vol. 7, Powder Metallurgy, (1984)
    Vol. 8, Mechanical Testing, (1985)
    Vol. 9, Metallography and Microstructures, (1985)
    Vol. 10, Materials Characteristics, (1986)
    Vol. 11, Failure Analysis and Prevention, (1986)
    Vol. 12, Fractography, (1987)
    Vol. 13, Corrosion, (1987)
2. *Metals Handbook,* Desk Edition, ASM, Metals Park, Ohio (1984).
3. *Metals Handbook,* 8th ed., ASM, Metals Park, Ohio.
    Vol. 1, Properties and Selection of Materials, (1961)
    Vol. 2, Heat Treating, Cleaning and Finishing, (1964)
    Vol. 3, Machining, (1967)
    Vol. 4, Forming, (1969)
    Vol. 5, Forging and Casting, (1970)
    Vol. 6, Welding and Brazing, (1971)
    Vol. 7, Atlas of Microstructures, (1972)
    Vol. 8, Metallography, Structures and Phase Diagrams, (1973)
    Vol. 9, Fractography and Atlas of Fractographs, (1974)

Vol. 10, Failure Analysis and Prevention, (1975)

Vol. 11, Nondestructive Inspection and Quality Control, (1976)

4. *ASM Metals Reference Book* 2nd ed., ASM, Metals Park, Ohio (1983).

5. *Tool and Manufacturing Engineers Handbook,* 4th ed, Society of Manufacturing Engineers, Dearborn, Michigan.

Vol. 1, Machining, (1983).

Vol. 2, Forming, (1984)

Vol. 3, Materials, Finishing and Coating, (1985)

Vol. 4, Quality Control and Assembly, (1987)

Vol. 5, Manufacturing Management, (1987)

6. *SAE Handbook, Part 1—Materials* (issued annually), Society of Automotive Engineers, Warrendale, PA.

7. "Materials Selector" Issue of *Materials Engineering,* (issued annually), Penton/IPC Publications.

8. "Materials and Processing Databook" Issue of *Metals Progress* (annually through 1985), ASM, Metals Park, Ohio.

9. "Guide to Engineered Materials", (annually since 1986), ASM, Metals Park, Ohio.

10. *Smithells Metals Reference Book,* 6th ed., E. A. Brandes, Butterworths, (1983).

11. *Materials Handbook,* 12th ed., George S. Brady and Henry R. Clauser, McGraw-Hill (1985).

12. *Woldeman's Engineering Alloys,* 6th ed., R. C. Gibbons, ASM, Metals Park, Ohio (1979).

13. *Materials and Processes—Part A: Materials; Part B: Processes,* 3rd ed., James F. Young and Robert S. Shane (eds.), Dekker (1985).

14. *Production Handbook,* 4th ed., John A. White (ed.), John Wiley and Sons, (1987).

15. *Production Processes: The Productivity Handbook,* 5th ed., Roger W. Bolz (ed.), Industrial Press (1981).

16. *ASME Handbook,* McGraw-Hill, New York.

Vol. 1, Metals Engineering: Design, (1965).

Vol. 2, Metals Properties

Vol. 3, Engineering Tables

Vol. 4, Metals Engineering: Processes

17. *Handbook of Printed Circuit Manufacturing,* R. H. Clark, Van Nostrand Reinhold, (1985).

18. *ASTM Standards* (Multiple Volumes), published annually, ASTM, Philadelphia, PA.

19. *Source Book on Materials Selection,* Vols. 1 & 2, ASM, Metals Park, Ohio (1977).

20. *Source Book on Industrial Alloy and Engineering Data,* ASM, Metals Park, Ohio (1978).

21. *Proceedings of the North American Metalworking Research Conferences*

Vol. 2—Univ. of Wisconsin

Vol. 3—Carnegie Press

Vols. 4 thru 15–Society of Manufacturing Engineers

Basic and General Textbooks

22. *The Science and Engineering of Materials,* Donald R. Askeland, Brooks Cole (1984).

23. *Structures and Properties of Engineering Alloys,* W. F. Smith, McGraw-Hill, New York (1981).

24. *Introduction to Materials Science for Engineers,* James F. Shackelford, Macmillan (1985).
25. *Engineering Materials and Their Applications,* R. A. Flinn and P. K. Trojan, Houghton Mifflin, Boston (1975).
26. *Manufacturing Processes for Engineering Materials,* Serope Kalpakjian, Addison-Wesley (1984).
27. *Engineering Materials—Properties and Selection,* 2nd ed., Kenneth Budinski, Reston, Reston, VA. (1983).
28. *Manufacturing Engineering,* Kenneth C. Ludema, R. M. Caddell, and A. G. Atkins, Prentice-Hall (1987).
29. *Manufacturing Engineering Processes,* Leo Alting, Marcel Dekker, New York (1982).
30. *Manufacturing Processes,* 8th ed., B. H. Amsted, Phillip F. Ostwald and Myron L. Begeman, Wiley, New York (1987).
31. *Manufacturing Processes and Materials for Engineers,* 3rd ed., L. E. Doyle et. al., Prentice Hall, NJ (1984).
32. *Processes and Materials of Manufacture,* 2nd ed., R. A. Lindberg, Allyn and Bacon, Boston, Mass. (1983).
33. *Introduction to Manufacturing Processes,* 2nd ed., John Schey, McGraw-Hill (1987).
34. *Mechanical Metallurgy,* 3rd ed., G. E. Dieter, McGraw-Hill, New York (1986).
35. *Principles and Applications of Tribology,* Desmond F. Moore, Pergamon Press.

Ferrous Metals

36. *Engineering Properties of Steel,* ASM, Metals Park, Ohio (1982).
37. *The Making, Shaping and Treating of Steel,* 10th ed., AISI (1964).
38. *Steel Selection: A Guide for Improving Performance and Profits,* R. F. Kern, and M. E. Suess, Wiley-Interscience (1979).
39. *Handbook of Stainless Steels,* Donald Peckner and I. M. Bernstein (eds.), McGraw-Hill (1977).
40. *Stainless Steel,* R. A. Lula, ASM, Metals Park, Ohio (1985).
41. *Tool Steels,* 4th ed., G. A. Roberts and R. A. Cary, ASM, Metals Park, Ohio (1980).
42. *Metallurgy and Heat Treatment of Tool Steels,* Robert Wilson, McGraw-Hill (1975).
43. *Cast Iron: Physical and Engineering Properties,* H. T. Angus, Butterworths (1976).

Nonferrous Metals

44. *Aluminum Standards and Data,* The Aluminum Association, Washington, D.C. (1984).
45. *Source Book on Selection and Fabrication of Aluminum Alloys,* ASM, Metals Park, Ohio, (1978).
46. *Aluminum Alloys: Structure and Properties,* L. F. Moldolfo, Butterworths, (1975).
47. *Aluminum,* ASM, Metals Park, Ohio.
    Vol. 1: Properties, Physical Metallurgy and Phase Diagrams (1967).
    Vol. 2: Design and Application
    Vol. 3: Fabrication and Finishing

48. *Aluminum: Properties and Physical Metallurgy,* John E. Hatch (ed.), ASM, Metals Park, Ohio (1984).

49. *Standards Handbook: Copper, Brass and Bronze* (7 volumes), Copper Development Association.

50. *Source Book on Copper and Copper Alloys,* ASM, Metals Park, Ohio (1979).

51. *Understanding Copper Alloys: The Manufacture and Use of Copper and Copper Alloy Sheet and Strip,* Olin Brass (1977).

52. *Titanium Alloys Handbook,* Metals and Ceramics Information Center, Battelle Columbus Laboratories, Columbus, Ohio (1972).

53. *Titanium and Titanium Alloys Source Book,* ASM, Metals Park, Ohio (1982).

54. *Engineering Properties of Zinc Alloys,* International Lead Zinc Research Organization, New York (1980).

Plastics and Composites

55. *Composite Materials Handbook,* M. M. Schwartz, McGraw-Hill, New York (1984).

56. *Handbook of Composites,* George Lubin (ed.), Van Nostrand Reinhold, New York (1982).

57. *Engineers' Guide to Composite Materials,* John W. Weeton (ed.)., ASM, Metals Park, Ohio (1986).

58. *Fabrication of Composite Materials—Source Book,* M. M. Schwartz (ed.), ASM, Metals Park, Ohio (1985).

59. *Composites, Engineered Materials Handbook,* Vol. 1, ASM International, (1987).

60. *Polymer Processing: Analysis and Innovation,* ASME PED–Vol. 5, (1982).

61. *Plastics Process Engineering,* James L. Throne, Marcel Dekker, New York (1979).

62. *Plastic Materials and Processes,* Seymour Schwartz and Sidney H. Goldman, Van Nostrand Reinhold, New York (1982).

63. *Handbook of Thermoset Plastics,* Sidney H. Goodman, Noyes Publications (1986).

64. *Plastics Product Design Engineering Handbook,* 2nd ed., Sidney Levy and J. H. DuBois, Chapman & Hall (1984).

Elevated Temperature Applications

65. *Engineer's Guide to High Temperature Materials,* F. J. Clauss, Addison Wesley, Reading, Mass. (1969).

66. *Source Book on Materials for Elevated Temperature Applications,* ASM, Metals Park, Ohio (1979).

Design

67. *Engineering Design: A Materials and Processing Approach,* George Dieter, McGraw-Hill (1983).

68. *Handbook of Product Design for Manufacturing—A Practical Guide for Low-Cost Production,* James G. Bralla (ed.), McGraw-Hill, New York (1986).

69. *Designing for Economical Production,* H. E. Trucks, Society of Manufacturing Engineers, Dearborn, Mich. (1976).

70. *Product Design and Process Engineering,* B. W. Niebel and A. B. Draper, McGraw-Hill, New York (1974).
71. *Mechanical Engineering Design,* J. E. Shiglie, McGraw-Hill, New York (1963).
72. *Fundamentals of Tool Design,* 2nd ed., Edward G. Hoffman (ed.), Society of Manufacturing Engineers, (1984).
73. *Tool Design,* Donaldson, LeGain and Goold, McGraw-Hill.
74. *Die Design Handbook,* 2nd ed., F. W. Wilson, et. al. (eds), Society of Manufacturing Engineers, Dearborn, Mich.
75. *Atlas of Stress-Strain Curves,* H. E. Boyer (ed.), ASM, Metals Park, Ohio (1986).
76. *Atlas of Fatigue Curves,* Howard E. Boyer (ed.), ASM, Metals Park, Ohio (1985).
77. *Application of Fracture Mechanics for Selection of Metallic Structural Materials,* J. E. Campbell, et. al., ASM, Metals Park, Ohio (1982).

Casting

78. *Cast Metals Handbook,* 4th ed., American Foundryman's Society (1957).
79. *Iron Castings Handbook,* Charles F. Walton, Iron Castings Society (1981).
80. *Gray and Ductile Iron Casting Handbook,* Gray and Ductile Iron Founder's Society, Cleveland, Ohio (1971).
81. *Steel Castings Handbook,* 5th ed., Steel Founder's Society of America, Ohio (1980).
82. *Foundry Technology Sourcebook,* ASM and AFS, Metals Park, Ohio (1982).
83. *Investment Casting Handbook—1980,* Investment Casting Institute, Chicago, IL (1979).
84. *Principles of Metal Casting,* 2nd ed., Heine, Loper and Rosenthal, McGraw-Hill.

Welding and Joining

85. *Welding Handbook,* 7th ed., American Welding Society, New York (1976)—3 volumes.
86. *Welding Handbook,* 6th ed., American Welding Society, New York (1968)—5 volumes.
87. *Welding Encyclopedia,* L. B. MacKenzie, 16th ed.—revised and re-edited by T. B. Jefferson, Monticello Books, Morton Grove, IL (1968).
88. *Metals Joining Manual,* M. M. Schwartz, McGraw-Hill (1979).
89. *Modern Welding Technology,* H. B. Cary, Prentice-Hall, NJ (1979).
90. *Source Book on Electron Beam and Laser Welding,* ASM, Metals Park, Ohio (1980).
91. *Source Book on Brazing and Brazing Technology,* ASM, Metals Park, Ohio (1980).
92. *Solders and Soldering,* 2nd ed., H. H. Manko, McGraw-Hill (1979).
93. *Handbook of Adhesives,* 2nd ed., Irving Skiest (ed.), Van Nostrand Reinhold (1977).
94. *Adhesives Technology Handbook,* Arthur H. Landrock, Noyes Publications, (1985).
95. *Adhesives in Modern Manufacturing,* Society of Manufacturing Engineers, Dearborn, Mich. (1970).
96. *Standard Handbook of Fastening and Joining,* Robert E. Parmley (ed.), McGraw-Hill, New York (1977).

Fabrication

97. *Handbook of Metal Forming*, Kurt Lange (ed.), McGraw-Hill, New York (1985).

98. *Handbook of Metalforming Processes*, B. Avitzur, Wiley-Interscience, New York (1983).

99. *Metal Forming: Fundamentals and Applications*, Altan, Oh and Gegel, ASM, Metals Park, Ohio (1983).

100. *Principles of Industrial Metalworking Processes*, G. W. Rowe, Arnold, London (1977).

101. *Mechanics of Sheet Metal Forming: Material Behavior and Deformation Analysis*, D. Koistionen, Plenum (1977).

102. *Techniques of Pressworking Sheet Metal*, 2nd ed., D. F. Eary and E. A. Reed, Prentice Hall, NJ (1974).

103. *Forging Industry Handbook*, 3rd ed., T. G. Bryer (ed.), Forging Industry Association, Cleveland, Ohio and ASM (1985).

104. *Forging Equipment, Materials and Practice (MCIC-HB-03)*, Metals and Ceramics Information Center, Battelle Columbus Labs.

105. *Forging Materials and Practices*, A. M. Sabroff, et. al., Reinhold, N.Y. (1968).

106. *Open Die Forging Manual*, 3rd ed., Forging Industry Association, Cleveland, Ohio (1982).

107. *Extrusion*, K. Laue and H. Stenger, ASM, Metals Park, Ohio (1981).

108. *Fine-Blanking: Practical Handbook*, Feintool, A. G. Lyss, Switzerland (1972).

109. *Plasticity for Engineers*, C. R. Calladine, Ellis Horwood (1985).

110. *Applications of Lasers in Materials Processing*, ASM, Metals Park, Ohio (1979).

Powder Metallurgy

111. *Powder Metallurgy, Principles and Applications*, F. V. Lenel, Metal Powder Industries Federation, Princeton, NJ (1980).

112. *Introduction to Powder Metallurgy*, J. S. Hirschorn, American Powder Met. Inst., New York, (1969).

113. *Powder Metallurgy Processing: New Techniques and Analyses*, H. A. Kuhn and A. Lawley, Academic Press (1978).

114. *Source Book on Powder Metallurgy*, ASM, Metals Park, Ohio (1979).

115. *Handbook of Powder Metallurgy*, H. H. Hausner, Chemical Publishing Co. (1973).

116. *Powder Metallurgy Equipment Manual—3*, Samuel Bradbury (ed.), MPIF, (1986).

Machining

117. *Machining Data Handbook*, 3rd ed., Machinability Data Center, Cincinnati, Ohio (1980)—2 volume set.

118. *The Machining of Metals*, E. J. A. Armarego and R. H. Brown, Prentice-Hall, Engelwood Cliffs, NJ (1969).

119. *High Speed Machining*, ASME PED Vol. 12 (1984).

120. *Machining of Plastics*, Akira Kobayashi, Krieger Pub. Co. (1981).

121. *Design of Cutting Tools*, A. Bhattacharyya and Inyong Ham, Society of Manufacturing Engineers (1969).

122. *New Developments in Grinding*, M. C. Shaw (ed.), Carnegie Press (1972).

123. *On the Art of Cutting Metals—75 Years Later*, ASME, PED Vol. 7 (1982).

124. *Fundamentals of Metal Machining and Machine Tools,* G. Boothroyd, McGraw-Hill (1975).
125. *Gear Handbook,* Dudley, McGraw-Hill.
126. *Metal Cutting,* 2nd ed., E. M. Trent, Butterworths (1984).
127. *Jigs & Fixtures,* William E. Boyes (ed.), Society of Manufacturing Engineers.

Heat Treatment

128. *Heat Treater's Guide, Standard Practices and Procedures for Steel,* Paul M. Unterweiser, ASM, Metals Park, Ohio (1982).
129. *Heat Treatment, Structure and Properties of Nonferrous Alloys,* Charles R. Brooks, ASM, Metals Park, Ohio (1982).
130. *Steel and Its Treatment,* Bofors Handbook, K. E. Thelning, Butterworths (1975).
131. *Principles of Heat Treatment of Steel,* G. Krauss, ASM, Metals Park, Ohio (1980).
132. *Principles of Heat Treatment,* M. A. Grossman and E. C. Bain, ASM (1964).
133. *Practical Heat Treating,* Howard E. Boyer, ASM, Metals Park, Ohio (1984).
134. *The Heat Treating Source Book,* ASM, Metals Park, Ohio (1986).
135. *Source Book on Heat Treating,* ASM, Metals Park, Ohio (1975).
   Vol. 1: Materials and Processes
   Vol. 2: Production and Engineering Practices
136. *Atlas of Isothermal and Cooling Transformation Diagrams,* ASM, Metals Park, Ohio (1977).
137. *Atlas of Continuous Cooling Transformation Diagrams for Engineering Steels,* ASM, Metals Park, Ohio (1980).

Surfaces and Finishes

138. *Metal Finishing: Guidebook Directory,* Metals and Plastics Publications, Inc., Westwood, NJ (published annually).
139. *Surface Preparation and Finishes for Metals,* J. A. Murphy (ed.), Society for Manufacturing Engineers, Dearborn, Mich.
140. *Surface Finishing Systems,* George Rudski, ASM, Metals Park, Ohio (1983).
141. *Carburizing and Carbonitriding,* ASM, Metals Park, Ohio (1977).
142. *Source Book on Nitriding,* ASM, Metals Park, Ohio (1977).
143. *Electroplating Engineering Handbook,* 4th ed., Lawrence Durney (ed.), Van Nostrand-Reinhold (1984).

Corrosion

144. *An Introduction to Metallic Corrosion,* 3rd ed., Ulick R. Evans, Edward Arnold Ltd. and ASM (1981).
145. *Corrosion and Its Control—An Introduction to the Subject,* Atkinson and Van Droffelaar, National Association of Corrosion Engineers, Houston, TX (1982).
146. *Corrosion Engineering,* 3rd ed., M. G. Fontana, McGraw-Hill, New York (1986).
147. *Corrosion and Corrosion Control,* 3rd ed., H. H. Uhlig and R. Winston Review, Wiley-Interscience, New York (1985).
148. *Corrosion (2nd Ed)—Vol. 1: Corrosion of Metals and Alloys and Vol. 2: Corrosion Control,* L. L. Shrier (ed.).
149. *Design and Corrosion Control,* V. R. Pludek, Wiley (1977).

150. *Corrosion Resistance Tables,* 2nd ed., Philip A. Schweitzer, Dekker (1986).

151. *Corrosion Prevention by Protective Coatings,* Charles G. Munger, National Association of Corrosion Engineers (1984).

Testing, Inspection, and Quality Control

152. *The Testing of Engineering Materials,* 4th ed., H. E. Davis, G. E. Troxell, and G. F. W. Hauck, McGraw-Hill, New York (1982).

153. *Nondestructive Testing Handbook,* 2nd ed., (Multiple Volumes), American Society for Nondestructive Testing (1982–   ).

154. *Metallography: Principles and Practice,* George VanderVoort, McGraw-Hill (1984).

155. *Handbook of Industrial Metrology,* Prentice-Hall.

156. *Handbook of Dimensional Measurement,* 2nd ed., F. T. Farago, Industrial Press (1982).

157. *ISO System of Limits and Fits, General Tolerances and Deviations,* American National Standards Institute.

158. *Introduction to Quality Engineering,* Genichi Taguchi, Kraus Int. Pub. (1986).

159. *Introduction to Statistical Quality Control,* D. C. Montgomery, John Wiley (1985).

160. *Statistical Quality Control,* 6th ed., Eugene I. Grant and Richard S. Leavenworth, McGraw-Hill (1987).

161. *Guide to Quality Control,* Kaoru Ishikawa, Asian Prod. Organization (1984).

162. *Quality Control Handbook,* Juran, McGraw-Hill (1979).

163. *Quality, Productivity, and Competitive Position,* W. Edwards Deming, MIT Press (1982).

164. *Quality Control Source Book,* ASM, Metals Park, Ohio (1982).

165. *Quality Planning and Analysis,* 2nd ed., Juran and Gryan, McGraw-Hill (1980).

Failure Analysis and Product Liability

166. *Understanding How Components Fail,* Donald J. Wulpi, ASM, Metals Park, Ohio (1985).

167. *Case Histories in Failure Analysis,* ASM, Metals Park, Ohio (1979).

168. *Failure Analysis: The British Engine Technical Reports,* F. R. Hutchings and P. M. Unterweiser, ASM, Metals Park, Ohio (1981).

169. *Failure Analysis: Case Histories and Methodology.* Naumann, ASM, Metals Park, Ohio (1983).

170. *Why Metals Fail,* R. D. Barer and B. F. Peters, Gordon and Breach, New York (1970).

171. *Analysis of Metallurgical Failures,* V. J. Conangelo and F. A. Heiser, Wiley-Interscience (1974).

172. *Source Book on Failure Analysis,* ASM, Metals Park, Ohio (1975).

173. *Failure of Materials in Mechanical Design,* J. A. Collins, Wiley-Interscience (1981).

174. *Engineering Aspects of Product Liability,* V. J. Conangelo and P. A. Thornton, ASM, Metals Park, Ohio (1981).

175. *Product Liability and the Reasonably Safe Product,* A. S. Weinstein et. al., Wiley-Interscience, New York (1978).

Robotics

176. *Robotics for Engineers,* Yoram Koren, McGraw-Hill (1985).

177. *Industrial Robots; Computer Interfacing and Control,* Wesley E. Snyder, Prentice-Hall (1985).

178. *Industrial Robotics,* Mikell P. Groover, et. al., McGraw-Hill (1986).

179. *Industrial Robots and Robotics,* Edward Kafrissen and Mark Stephens, Reston (1984).

180. *Handbook of Industrial Robotics,* Nof (ed.), John Wiley (1985).

181. *Assembly with Robots,* Tony Owen, Prentice-Hall (1985).

182. *Robotics: Control Sensing, Vision, and Intelligence,* K. S. Fu, R. C. Gonzalez and C. S. G. Lee, McGraw-Hill (1987).

183. *Industrial Robots,* Vols. 1 and 2, Society of Manufacturing Engineers (1981).

Integrated Manufacturing Production Systems

184. *Anatomy of Automation,* George H. Amber and Paul S. Amber, Prentice-Hall (1962).

185. *Automation in Manufacturing Systems, Processes and Computer Aids,* ASME PED Vol. 4 (1981).

186. *Automation, Production Systems and Computer Integrated Manufacturing,* 2nd ed., Mikell P. Groover, Prentice-Hall (1987).

187. *Computer Applications in Manufacturing Systems,* ASME PED Vol. 2 (1980).

188. *Numerical Control and Computer Aided Manufacturing,* R. S. Pressman and J. E. Williams, Wiley (1977).

189. *Principles of Numerical Control.* James L. Childs, Industrial Press (1982).

190. *Group Technology,* E. A. Arn, Springer-Verlag (1975).

191. *Group Technology,* Gallagher and Knight, Butterworths (London) (1973).

192. *The Introduction of Group Technology,* John L. Burbridge, Halsted Press Book, John Wiley and Sons (1975).

193. *Manufacturing Cost Estimating Guide,* P. R. Ostwald, American Machinist (1982).

194. *Manufacturing Systems Engineering,* K. Hitomi, Taylor & Francis Ltd. (London) (1979).

195. *Production Systems: Planning, Analysis and Control,* 2nd ed., James L. Riggs, John Wiley-Interscience (1976).

196. *The Role of Computers in Manufacturing Processes,* Gideon Halevi, John Wiley-Interscience (1980).

197. *Toward the Factory of the Future,* ASME, PED Vol. 1 (1980).

198. *Flexible Manufacturing Systems Handbook,* Noyes Publishing (1984).

199. "Cellular Manufacturing Systems—An Overview", J T. Black, *IIE Journal,* Nov. 1983, p. 36.

200. *Group Technology in the Engineering Industry,* J. L. Burbidge, Mechanical Engineering Publications LTD., London (1979).

201. *Group Technology—Applications to Production Management,* I. Ham, K. Hitomi and T. Yoshida, Kluwer-Nijhoff Publishing Co., Boston (1985).

202. *Zero Inventories,* R. W. Hall and APICS, Dow Jones-Irwin, Homewood, IL (1983).

203. *Kanban and Just-In-Time at Toyota,* Japan Management Association, Trans. D. J. Lu, Productivity Press, Cambridge, MA. (1986).

204. *Toyota Production System,* Y. Monden, IIE Press, Norcross, GA. (1983).

205. *The Design and Operation of FMS—Flexible Manufacturing Systems*, P. G. Ranky, IFS Publications LTD (UK) and North-Holland Publishing Company, New York (1983).

206. *Japanese Manufacturing Techniques*, R. J. Schonberger, Free Press, New York (1982).

207. *World Class Manufacturing*, R. J. Schonberger, Free Press (1986).

208. *A Revolution in Manufacturing: The S.M.E.D. System*, S. Shingo, Productivity Press, Cambridge, MA (1985).

209. *Zero Quality Control: Source Inspection and the Poka-Yoke System*, S. Shingo, Productivity Press, Cambridge, MA (1986).

210. *Study of 'Toyota' Production System From Industrial Engineering Viewpoint*, Japan Management Association, Tokyo (Also available from Productivity Press, Cambridge, MA (1981).

211. *The New Manufacturing Challenge*, K. Suzaki, Free Press (1987).

212. *Just-In-Time Manufacturing*, C. A. Voss, IFS Publications LTD. (UK), Springer-Verlag (1987).

Index